STARS AND STELLAR SYSTEMS

Compendium of Astronomy and Astrophysics

(IN NINE VOLUMES)

GERARD P. KUIPER, *General Editor*
BARBARA M. MIDDLEHURST, *Associate General Editor*

I
TELESCOPES

II
ASTRONOMICAL TECHNIQUES

III
BASIC ASTRONOMICAL DATA

IV
CLUSTERS AND BINARIES

V
GALACTIC STRUCTURE

VI
STELLAR ATMOSPHERES

VII
NEBULAE AND INTERSTELLAR MATTER

VIII
STELLAR STRUCTURE

IX
GALAXIES AND THE UNIVERSE

CONTRIBUTORS

L. H. ALLER
G. L. BERGE
R. C. BLESS
S. J. CZYZAK
LEVERETT DAVIS
HERBERT FRIEDMAN
J. MAYO GREENBERG
G. HARO
H. L. JOHNSON
HUGH JOHNSON
F. J. KERR
WILLIAM LILLER
B. T. LYNDS
R. MINKOWSKI
GUIDO MÜNCH
E. N. PARKER
LYMAN SPITZER, JR.
E. K. L. UPTON

NEBULAE AND INTERSTELLAR MATTER

Edited by

BARBARA M. MIDDLEHURST

and

LAWRENCE H. ALLER

THE UNIVERSITY OF CHICAGO PRESS

CHICAGO & LONDON

This publication has been supported in part by the

NATIONAL SCIENCE FOUNDATION

Library of Congress Catalog Card Number: 66-13879

The University of Chicago Press, Chicago & London
The University of Toronto Press, Toronto 5, Canada

Preface to the Series

THE SERIES "Stars and Stellar Systems, Compendium of Astronomy," comprising nine volumes, was organized in consultation with senior astronomers in the United States and abroad early in 1955. It was intended as an extension of the four-volume "Solar System" series to cover astrophysics and stellar astronomy. In contrast to the "Solar System" series, separate editors have been appointed for each volume. The volume editors, together with the general editors, form the editorial board that is responsible for the over-all planning of the series.

The aim of the series is to present stellar astronomy and astrophysics as basically empirical sciences, co-ordinated and illuminated by the application of theory. To this end the series opens with a description of representative telescopes, both optical and radio (Vol. **1**), and of accessories, techniques, and methods of reduction (Vol. **2**). The chief classes of observational data are described in Volume **3**, with additional material being referred to in succeeding volumes, as the topics may require. The systematic treatment of astronomical problems starts with Volume **4**, as is apparent from the volume titles. Theoretical chapters are added where needed, on dynamical problems in Volumes **4**, **5**, and **9**, and on astrophysical problems in Volumes **6**, **7**, and **8**. In order that the chapters may retain a greater degree of permanence, the more speculative parts of astronomy have been de-emphasized. The level of the chapters will make them suitable for graduate students as well as for professional astronomers and also for the increasing number of scientists in other fields requiring astronomical information.

The undersigned wish to thank both the authors and the volume editors for their readiness to collaborate on this series, which it is hoped will stimulate the further growth of astronomy.

The editors wish to acknowledge the support by the National Science Foundation both in defraying part of the costs of the editorial offices and in providing a publication subsidy.

GERARD P. KUIPER
BARBARA M. MIDDLEHURST

Preface to Volume 7

THE interstellar medium is now believed to contain at least 20 per cent of the mass of the Galaxy. Since stars must be formed out of it and, in turn, eject and return matter to it from their surfaces, interplay of material must be taken into consideration in any theoretical studies. Progress in understanding the gas clouds, proto-stars, nebulae, and other constituents less well-defined than stars has been rapid in recent years. On the one hand, the importance of the role of interstellar material in relation to the formation of stars is recognized and treated here. On the other, the very high energies reached through the acceleration of charged particles in magnetic fields of gas clouds are of considerable interest to both astronomers and physicists and form the subject of several chapters. The significant range of wavelengths available for study of interstellar matter has recently been greatly extended at both ends of the spectrum by advancement in techniques of X-ray rocket experiments and further improvements in radio methods. Several chapters deal with these topics, and we are particularly grateful to the authors, here and elsewhere, who have so generously revised their chapters to bring them up to date as far as possible in this very active field.

The first (keynote) chapter by Spitzer treats the dynamics of interstellar clouds and some considerations underlying star formation. Chapter 16 on the early stages of stellar evolution by Upton was intended to be included in Volume 8 (*Stellar Structure*) of this series but was not completed in time. It is also relevant here, and ends our present volume. This chapter is concerned with the stages in the life of a star at which gravitational attraction has become dominant and at which the condensation and contraction of a hitherto amorphous blob into a well-defined star is assured.

Diffuse and dark nebulae are treated from the observational point of view in chapters 2 and 3, by Hugh Johnson and Beverly Lynds, respectively. Flare stars, discussed by Haro in chapter 4, are associated with both types of objects, which seem to modify their characteristics; flare stars, as also T Tauri stars, appear to be in relatively early stages of stellar evolution.

The solid components of the interstellar medium, in the form of grains or "dust" particles, scatter and extinguish starlight. Stellar extinction was treated by Sharpless in Volume 3 (*Basic Astronomical Data*) of the series, and the subject has been brought up to date by Harold Johnson in chapter 5, which contains much recent data on work in the infrared, which is of particular relevance to the determination of the ratio, *R*, of total to selective absorption in space, and hence to the dimensions of the Galaxy. Controversy on this subject remains, but the editors and the author are convinced that the highly important experimental results presented here are extremely relevant and that though the final values of *R* may differ from Johnson's, they will not do so by very large amounts. Greenberg gives the basic physical theory and discusses the probable nature of particles that are believed to constitute the interstellar grains in chapter 6.

In chapter 7, G. Münch discusses interstellar absorption lines at visual wavelengths. These not only provide important tools for galactic structure research but also supply essential information on the physical state of the medium.

Bright nebulae can be either of the reflection or emission type. Any correct interpretation of the emission-line spectrum of the interstellar gas requires accurate values of the necessary atomic parameters. Recombination lines of hydrogen and helium can be treated without particular difficulty, but the forbidden lines of abundant ions such as O^{++}, N^{+}, or S^{+} yield information on the electron temperature, density, or chemical composition of the interstellar medium or a gaseous planetary nebula only when both transition probabilities and target areas for collisional excitation of the metastable levels are known. These subjects are treated in chapter 8 by Czyzak. The theory of physical processes has been developed particularly for planetary nebulae where the conditions are somewhat simpler than in typical diffuse irregular nebulae. Various aspects of the problems of planetary nebulae are discussed by Aller and Liller in chapter 9.

Radio astronomy supplies increasingly important data on the physical state of the interstellar medium by reaching regions (such as those containing mainly neutral hydrogen) which may be inaccessible at visible wavelengths. In chapter 10, Kerr reviews the important information yielded by study of the 21-cm hydrogen line and the more elusive features recently observed in both absorption and emission due to OH. Some of these features at present appear to defy interpretation in terms of any simple models or processes of thermal excitation and emission.

The important nonthermal emission processes occurring in interstellar gas clouds and what is known from observation about the character of these sources are discussed by Minkowski in chapter 11 on nonthermal galactic radio sources, by Bless in chapter 12 on the theory of synchrotron radiation, particularly in old supernovae, and by Friedman in chapter 13 on X-ray sources. In chapter 14 Parker discusses the origin of galactic cosmic rays, illustrating the close connection between a traditional field of physics and modern astrophysics. All nonthermal mechanisms invoke "local" or galactic-sized magnetic fields, the evidence for whose existence is discussed by Davis and Berge in chapter 15.

The active state of the field covered by the present volume, particularly with regard to both theoretical and observational phases of what might be called high-energy astrophysics has made the preparation of this volume difficult; we are particularly grateful to the authors who have struggled valiantly to keep their chapters up to date and to supply new and important material down to the firing of the finish gun. In any enterprise such as this, where a large number of authors participate and all of them have their regular tasks, it is inevitable that there should have been delays and frustrations in the preparation and submission of manuscripts. But delays and frustrations have not been the monopoly of the authors; the editors have had their problems also.

The usual difficulties with regard to variations in notation have been encountered. The general policy followed here has been to tamper with the author's notation as little as possible and to trust in the reader's intelligence and the self-consistency within the chapters to carry him through. Where the notation is consistent with the literature, no attempt has been made to revise it, since such revisions introduce their own problems and allow the possibility of introducing errors. The attention of the reader is drawn to the difference in the meaning of the term *spectral index* as used in the chapter by Bless and in other chapters. Some confusion may arise because many nebulae are referred to under a variety of names; thus, the Crab Nebula also appears as M 1 and NGC 1952. To help readers who are not familiar with the designations, a list of alternatives for a number of these nebulae is given on page xi.

The great progress achieved in the study of the interstellar medium during recent years has been made possible by the application of a number of different complementary observational techniques and advances in theories of cosmic rays, atomic structure and interaction with radiation, plasma physics, and hydrodynamics. Particularly striking has been the interplay between radio and optical astronomy. The 21-cm radio frequency and Hα (optical) radiation supply complementary information on the cold neutral hydrogen and the hot ionized hydrogen in adjoining regions in the medium. Both velocities and emissivities can be measured; hence it is possible to construct models for gaseous nebulae and regions of the interstellar cloud complexes. Even more important, probably, has been the recognition of nonthermal sources as abundant features of the observable Universe—and as a corollary—the presence of ubiquitous magnetic fields. The most spectacular progress has been made where it is possible to secure optical identifications of radio-frequency sources, for example, in the Crab Nebula. Even more startling progress can be anticipated when observations can be secured from above the Earth's atmosphere. To date, very few X-ray sources have been identified with optical objects, but as knowledge of their positions is improved, identification efforts will prove more and more fruitful. The ultraviolet region of the spectrum has been studied in only a fragmentary way. At the present time, the clues turned up by radio telescopes have put a very heavy burden on the few optical telescopes adequate for pursuing them. The situation is certain to grow worse, especially when observations from space vehicles become available.

Probably the most striking feature of the interstellar medium observable from our part of the Universe is the great 30 Doradus complex in the Large Magellanic Cloud. In size and mass it strikingly dwarfs the Orion complex and doubtless contains many clues to the nature of the interstellar clouds. There are a number of other extended diffuse nebulae in the Magellanic Clouds, and it is in these systems that the process of star formation can be more easily followed, probably, than is possible in local areas in the Galaxy.

Yet, although fine radio observations are possible, particularly with the great Parkes dish, the existing optical telescopic equipment in the southern hemisphere is pitifully inadequate. A few steps are being taken to remedy this situation, notably in Chile, South Africa, and Australia, but the instruments so far provided fall grievously short of what is needed, and the editors share the conviction of many astronomers that much more must be done before this advance can be extended to meet requirements in a satisfactory way. It has been stated that some 90 per cent of all scientists who have ever lived are now active, and we might expect that the available facilities should be increased in a reasonably commensurate way. Granting that the figure may be high for astronomy, which is the oldest science other than medicine, the increase in necessary capital equipment falls far short of the needs generated by the increase in the number of astronomers.

Optical astronomy is still largely dependent on the legacy of George Ellery Hale, who founded the Mount Wilson Observatory two-thirds of a century ago. The largest optical telescopes in operation in the U.S.A. are provided by private funds, or in the instance of the Lick 120-inch, provided by the State of California. In the mid-sixties, astronomy occupies a unique position among all sciences, in that in America its most essential tools were provided by nonfederal funds. A parallel situation exists in many other countries.

Although a large number of small telescopes have been provided, and a few instruments of intermediate size, for example, two 61-inch reflectors (one for work in the infrared) and an 84-inch reflector in Tucson, have been funded by government agencies, the large telescopes which are indispensible for progress in astronomy have yet to appear. As a consequence of inadequate observing facilities, an increasing proportion of young people turn to theory, even though they may have no particular talents in this field; we witness an ever-growing number of theories chasing an inadequate number of observations. In the present volume, we have tried to present a balanced picture and to give appropriate weight to both theory and observation.

The editors wish to record their regrets concerning the death in November, 1965, of Dr. Dean McLaughlin, who was a member of the editorial group concerned with the present volume in its early stages. He was a valued friend and colleague of both editors, and the astronomical community has lost much by his passing.

Finally, we wish to extend thanks to the Press for their continuing co-operation, to Mrs. Betty Fink for editorial assistance at Tucson over a long period, to the authors, and to many others who have assisted in various ways to make this as comprehensive and accurate a volume as possible.

BARBARA M. MIDDLEHURST
LAWRENCE H. ALLER

Some Well-known Nebulae

The list of nebulae that follows will assist identification where two or more alternative designations exist. The use of one or other of these in the text could be confusing for some readers, and for these a summary of the alternatives may be welcome. There was no intention of providing a full index here. Additional information can be obtained from many handbooks and atlases, notably those mentioned in the chapters and the following:

Bečvář, A. 1959, *Atlas Coeli II* (Prague: Czechoslovakian Academy of Sciences).
Norton, A. P., and J. G. Inglis, 1943, *Norton's Star Atlas* (London: Gall & Inglis).
Perek, L., and L. Kohoutek, 1967, *Catalogue of Planetary Nebulae* (Prague: Czechoslovakian Academy of Sciences).

SHORT LIST OF NEBULAE: IDENTIFICATIONS

Name of Nebula	Type*	NGC or IC Number	Messier Number	Other Designations [Exciting Star]	Notes
η Carinae........	E	3372	....	Key-hole Nebula	....
Crab............	E, F	1952	1	3C144, Tau X-1	....
Cygnus Loop.....	E, F	6960, 6992, 6995, 6979	....	Veil Nebulae	....
30 Doradus......	E	2070	....	30 Doradus Loop	1
Dumbbell........	P	6853	27	....	....
Electra..........	R	....	....	[HD 23302]	....
Hind's...........	E(var)	1554–5	....	[T Tau]	....
Horsehead.......	D with E rim	IC 434	....	B 33	2
Lagoon..........	E	6523	8	HS 126	3
Maia............	R	....	....	[HD 23408]	....
Merope..........	R	IC 349	....	[HD 23480]	....
North America...	E	7000, IC 5067, 5068	....	....	....
Omega..........	E	6618	17	Horseshoe, Swan	....
Orion............	E	1976	42	HS 66	3
Owl.............	P	3587	97	....	....
Pelican..........	E	IC 5067–5070	....	....	....
Ring............	P	6720	57	....	....
Rosette..........	E	2237–9	....	....	....
Saturn...........	P	7009	....	....	....
Trifid............	E	6514	20	....	....
....	P	7662	....	Ring Nebula in Andromeda	....
....	E	6611	16	HS 142	3

* D = dark; E = emission; P = planetary; R = reflection.
1. In Large Magellanic Cloud.
2. B: Barnard.
3. HS: Haze-Shajn.

Table of Contents

Table: Some Well-known Nebulae xi

1. Dynamics of Interstellar Matter and the Formation of Stars 1
Lyman Spitzer, Jr.

1. Introduction 1

2. Observed Properties of the Interstellar Medium 4
2.1. Spatial Distribution of Gas and Dust 4
2.2. Composition and Physical State of the Gas 7
2.3. Magnetic Field in the Galaxy 10
2.4. Velocities of Interstellar Clouds 11
2.5. Internal Velocities within Interstellar Clouds 13

3. Dynamics of Interstellar Matter 16
3.1. Energy Dissipation 17
3.2. Sources of Kinetic Energy 19
3.3. Expansion of Gas around a Young O Star 22
3.4. Interaction of an Isolated Cloud with a Young O Star 28
3.5. Supernova Shells 33

4. Formation of Clouds, Protostars, and Stars 35
4.1. Formation and Equilibrium of Clouds 36
4.2. Gravitational Instability in a Uniform Medium 40
4.3. Gravitational Instability in Isolated Systems 42
4.4. Effect of Magnetic Fields on Gravitational Contraction 48
4.5. Angular Momentum and Turbulence 53
4.6. Origin of Associations 56
4.7. Summary 57

References 58

2. Diffuse Nebulae 65
Hugh M. Johnson

1. Introduction 65

2. Reflection Nebulae 73
2.1. Introduction 73
2.2. Interpretation in Terms of Scattered Light 79
2.2.1. The Pleiades 79
2.2.2. NGC 7023 81

3. Emission Nebulae 82
3.1. Introduction 82

3.2. The Bright-Line Spectrum of NGC 1976 . . . 91
3.3. The Bright-Line Spectra of Other Diffuse Nebulae . . . 98
3.4. The Continuous Spectrum . . . 100
3.5. Forms and Structural Features . . . 102
3.6. Radial Velocities . . . 106
3.7. Emission Nebulae in External Galaxies . . . 109
3.8. Conclusions and Suggestions for Further Observations . . . 112

References . . . 113

3. DARK NEBULAE . . . 119
B T. Lynds

1. Dark Markings; Early History . . . 119

2. Catalogues of Dark Nebulae . . . 120

3. Distribution of Dark Nebulae . . . 124
3.1. Distances . . . 124
3.2. Galactic Distribution . . . 130

4. Physical Properties . . . 133
4.1. Size and Shape . . . 133
4.2. Globules . . . 134
4.3. Masses and Densities of Dark Nebulae . . . 135

References . . . 138

4. FLARE STARS . . . 141
G. Haro

1. Introduction . . . 141

2. Flare Stars in Stellar Aggregates and in the Solar Vicinity . . . 142
2.1. The Orion Aggregate . . . 142
2.2. NGC 2264 . . . 149
2.3. The Taurus Dark Clouds . . . 150
2.4. The Pleiades Region . . . 150
2.5. The Coma Berenices Cluster . . . 155
2.6. Praesepe . . . 156
2.7. The Hyades . . . 156
2.8. The Solar Vicinity . . . 157

3. Further Discussion and Summary . . . 161

References . . . 164

5. INTERSTELLAR EXTINCTION . . . 167
H. L. Johnson

1. Introduction . . . 167

2. Previous Work and Results . . . 168

3. The Cluster-Diameter Method 170

4. The Variable-Extinction Method 174

5. The Color-Difference Method 185

6. Discussion 212

7. Consequences of the Variation in Interstellar Extinction 213
   7.1. Range of Values of R 213
   7 2. Distances of Clusters and the Galactic Center, and the Oort Constant A . 215

8. Conclusion 218

References 219

6. INTERSTELLAR GRAINS 221
J. Mayo Greenberg

1. Introduction 221

2. Observational Evidence for Grains 222
   2.1. Extinction 223
   2.2. Polarization 229
   2.3. Reflection Nebulae; Scattering and Extinction in Other Galaxies. . . 235

3. General Formulation of the Theory of Extinction and Polarization by Grains 235

4. Physical and Chemical Properties of the Interstellar Medium 238

5. Physics and Chemistry of Grains 238
   5.1. Grain Characteristics 238
      5.1.1. Nucleation 239
      5.1.2. Growth 240
      5.1.3. Destruction 240
   5.2. Dirty Ice 243
   5.3. Metallic Grains 259
   5.4. Graphite 261
   5.5. Free Radicals 265

6. Orientation Mechanism 267

7. Grain Optics 274
   7.1. Pure and Impure Dielectric Grains 276
   7.2. Metallic Particles 293
   7.3. Particles with Non-Isotropic Optical Characteristics 302
   7.4. Free Radicals 303
   7.5. Core-Mantle Particles 304

8. Interpretations of Interstellar Extinction and Polarization 308
   8.1. Dielectric Grains 308
   8.2. Metallic Grains 332
   8.3. Graphite Flakes 337
   8.4. Free Radicals 340
   8.5. Core-Mantle Particles 343

9. Reflection Nebulae . . . 346

10. Diffuse Interstellar Lines . . . 355

11. Summary and Conclusions . . . 358

References . . . 361

7. INTERSTELLAR ABSORPTION LINES . . . 365
Guido Münch

1. Introduction . . . 365

2. The Data of Observations . . . 367
   2.1. Identification . . . 367
   2.2. Catalogues . . . 369
   2.3. Structure of the Interstellar Absorption Lines . . . 370

3. Methods of Analysis . . . 373
   3.1. Curves of Growth for Single Clouds . . . 373
   3.2. The Doublet-Ratio Method . . . 374
   3.3. Curves of Growth for More than One Cloud . . . 377
   3.4. Statistical Curve of Growth . . . 378
   3.5. Ionization Equilibrium in Interstellar Space . . . 380
   3.6. Excitation in Interstellar Space . . . 385

4. Interpretation of Observations . . . 388
   4.1. Doublet-Ratio Method . . . 388
   4.2. Intensity-Distance Relations . . . 391
   4.3. The Mean Space Densities of Ca II and Na I . . . 393
   4.4. Correlation between Interstellar Line Strengths and Reddening . . . 395
   4.5. The Intercloud Medium . . . 395
   4.6. The Chemical Composition of the Interstellar Gas . . . 397
   4.7. The High-Velocity Clouds . . . 399

References . . . 401

8. ATOMIC PROCESSES WITH SPECIAL APPLICATION TO GASEOUS NEBULAE . . . 403
S. J. Czyzak

1. Introduction . . . 403

2. Atomic Structure Calculations . . . 406
   2.1. Wave Functions for Hydrogen . . . 406
   2.2. Wave Functions for Atoms and Ions Other than Hydrogen . . . 407

3. Theory of Transition Probabilities . . . 426

4. Calculation of Collision Cross-Sections Due to Electron Impact . . . 454

5. Summary . . . 479

References . . . 480

9. PLANETARY NEBULAE . . . 483
L. H. Aller and William Liller

1. Introduction . . . 483
1.1. Survey of the Data . . . 483
1.2. Catalogues . . . 484

2. Distances of Planetary Nebulae . . . 486
2.1. Trigonometric Parallaxes . . . 486
2.2. Proper Motions . . . 486
2.3. Parallaxes of Individual Objects . . . 487
2.4. Statistical Methods for Distance Determinations . . . 487

3. Methods of Observation of Planetary Nebulae . . . 491
3.1. Visual and Direct Photographic Observations . . . 492
3.2. Spectroscopic Methods . . . 492
3.3. Polarization Measurements . . . 493
3.4. Photoelectric Methods . . . 493
3.5. Summary . . . 494

4. Origin of Spectra of Planetary Nebulae . . . 496
4.1. Primary Mechanisms . . . 496
4.2. Collisional Excitation . . . 496
4.3. Bowen's Fluorescent Mechanism . . . 496
4.4. The Continuum . . . 497
4.5. Differences between the Spectra of Different Objects . . . 497
4.6. The Ultraviolet and Infrared Spectra of Planetary Nebulae . . . 499

5. The Primary Mechanism: The Recombination Lines of Hydrogen and Helium 499
5.1. The Hydrogen Lines in Optically Thin Nebulae . . . 499
5.2. The Hydrogen Lines in Optically Thick Nebulae . . . 509
5.2.1. Lyα Scattering . . . 510
5.2.2. The Role of the Metastable 2s-Level . . . 512
5.2.3. Effect of Optical Thickness . . . 512
5.3. Collisional Excitation . . . 513
5.4. The Helium Spectrum in Planetary Nebulae . . . 514

6. The Thermal Balance of a Gaseous Nebula . . . 518
6.1. General Considerations . . . 518
6.2. Statistical Equilibrium in the Continuum . . . 519
6.3. Radiative Equilibrium . . . 520
6.4. Collisional Excitation of Forbidden Lines . . . 520
6.5. Collisional Excitation of Discrete Levels of Hydrogen . . . 521

7. The Forbidden Lines; Electron Temperatures and Densities . . . 523
7.1. Theoretical Considerations . . . 523
7.2. Electron Temperatures and Densities from Observed Intensity Ratios . 528
7.3. The 3729 A/3726 A Ratio of [O II] . . . 529
7.4. Comparison of Ionic Densities . . . 530

8. Continuous Spectra of the Planetary Nebulae . . . 531

9. Radio-Frequency Emission . . . 536
9.1. Theoretical Considerations . . . 536
9.2. Comparison with Observations . . . 538
9.3. Nonthermal Radio-Frequency Radiation . . . 541

10. Chemical Composition of Planetary Nebulae . . . 541
10.1. Interpretation of Data; Special Difficulties . . . 541
10.2. Filamentary Structure . . . 544
10.3. Abundance Ratios . . . 544

11. Central Stars of Planetary Nebulae . . . 546
11.1. Types of Spectra . . . 546
11.2. Temperatures . . . 547
11.2.1. Ionization Temperature . . . 547
11.2.2. Indirect Temperature Determination . . . 549
11.3. Model Atmospheres . . . 555
11.4. Evolution of the Central Star . . . 556

12. Structure and Magnetic Fields . . . 559

13. Internal Motions . . . 565

References . . . 569

10. RADIO-LINE EMISSION AND ABSORPTION BY THE INTERSTELLAR GAS . . . 575
F. J. Kerr

1. Introduction . . . 575

2. Short Review of Observations . . . 576

3. Excitation of the Hydrogen Line and OH Lines . . . 577
3.1. The Hydrogen Line . . . 577
3.2. Transition Probability . . . 578
3.3. Spin Temperature . . . 579
3.4. OH Lines . . . 580

4. The Formation of Emission and Absorption Lines . . . 582
4.1. Emission Lines . . . 582
4.2. Number of Emitting Atoms . . . 583
4.3. Line Broadening . . . 584
4.4. Absorption Lines . . . 585

5. Observational Techniques . . . 586
5.1. Receivers . . . 586
5.2. Antennas . . . 588
5.3. Over-All System . . . 588

6. Observational Results, H and OH . . . 589
6.1. Hydrogen Emission . . . 589
6.2. Hydrogen Absorption . . . 593
6.3. Zeeman Splitting and the Galactic Magnetic Field . . . 596
6.4. Neutral Hydrogen in Other Galaxies . . . 597
6.5. OH Observations . . . 599

7. Physical Properties of the Gas . . . 602
7.1. Temperature . . . 602
7.2. Number of Atoms; Mean Density . . . 606
7.3. Motions . . . 607
7.4. Cloud Structure . . . 608

8. Correlation with Other Population I Constituents . . . 611
8.1. Star Groupings . . . 611
8.2. Dust . . . 614
8.3. Interstellar Calcium . . . 615
8.4. Ionized Hydrogen . . . 617

9. Other Line Possibilities . . . 617

References . . . 619

11. NONTHERMAL GALACTIC RADIO SOURCES . . . 623
R. Minkowski

1. Introduction . . . 623

2. Thermal Emission and Absorption . . . 624

3. Nonthermal Emission . . . 627

4. Supernovae . . . 628

5. The Remnants of Observed Supernovae . . . 629
5.1. Tycho's Nova . . . 630
5.2. Kepler's Nova . . . 633
5.3. The Supernova of +1006 . . . 636
5.4. The Crab Nebula . . . 637
5.4.1. The Supernova of +1054 . . . 637
5.4.2. General Remarks on the Crab Nebula . . . 638
5.4.3. The Filamentary System . . . 639
5.4.4. The Diffuse Mass . . . 642
6. Cassiopeia A . . . 652
7. Supernovae of Type II . . . 656

8. Old Supernova Remnants . . . 656
8.1. The Cygnus Loop . . . 657
8.2. IC 443 . . . 659
8.3. Miscellaneous Objects . . . 659
8.4. Supernova Remnants in the Large Magellanic Cloud . . . 661

References . . . 663

12. THE THEORY OF SYNCHROTRON RADIATION . . . 667
R. C. Bless

1. Introduction . . . 667

2. The Theory of Synchrotron Radiation . . . 668
2.1. The Total Power Radiated . . . 669

2.2. The Angular Distribution of the Radiation 670
2.3. The Spectral Distribution of the Radiation 671
2.4. The Polarization 674
2.5. Quantum Effects 675

3. Astronomical Application 675
3.1. Radiation from a Single Electron 675
3.2. Synchrotron Radiation from an Assemblage of Electrons 676
3.3. Determination of the Energy in the Particles and in the Magnetic Field . 680

Appendix I. Electron Energy Losses by Other Processes 681

Appendix II. Application to the Crab Nebula 682

References 684

13. Discrete X-Ray Sources 685
Herbert Friedman

1. Introduction 685

2. Early Observations 686

3. X-Ray Emission Mechanisms 688

4. Occultation of the Crab Nebula 688

5. Cosmic X-Ray Surveys 689

6. Distribution of X-Ray Sources 693

7. The Crab Nebula 697

8. Sco XR-1 700

9. Extragalactic Sources 704

10. Conclusion 705

References 706

14. Dynamical Properties of Cosmic Rays 707
E. N. Parker

1. Introduction 707

2. Local Properties of the Cosmic-Ray Gas 710

3. Galactic Environment of the Cosmic-Ray Gas 715
3.1. The Galactic Magnetic Field 716
3.2. The Interstellar Gas 717
3.3. Theoretical Considerations 717
3.4. Nonthermal Radio Emission 719

4. The Extension of Cosmic Rays through Space and Time 720
4.1. Observational Evidence 720
4.2. Theoretical Considerations 723

4.2.1. Galactic Theory 724
4.2.2. Local Group Theory 724
4.2.3. Universal Theory 725
4.2.4. Transience Theories 726

5. Cosmic-Ray Life, Production Rate, and Isotropy 727
5.1. Production Rate 727
5.2. Isotropy and the Galactic Halo 728

6. Cosmic-Ray Origin 730
6.1. General Considerations 730
6.2. Theoretical Cosmic-Ray Acceleration Mechanisms 733
6.2.1. The Fermi Mechanism 733
6.2.2. Hydrodynamic Mechanism 735
6.2.3. Wave Acceleration 735

7. The Dynamical Properties of Cosmic-Ray Gas 737
7.1. General Remarks 737
7.2. Hydromagnetic Waves 738
7.3. Flow out of Strong Fields 740
7.4. Inflation of a Galactic Halo 741
7.4.1. Non-Existence of Static Equilibrium 741
7.4.2. Application to Galaxy 742
7.4.3. Discussion 745

Appendix I. Equations of Motion of Collisionless Gas 747

Appendix II. Inflation of a Simple Two-Dimensional Field 749

References 750

15. EVIDENCE FOR GALACTIC MAGNETIC FIELDS 755
Leverett Davis, Jr., and G. L. Berge

1. Introduction 755

2. Direct Connections between Magnetic Fields and Radiation 755
2.1. Synchrotron Radiation within the Galaxy 756
2.2. Faraday Rotation 758
2.3. Zeeman Effect 762

3. Polarization of Starlight 765

4. Indirect Arguments 768

References 769

16. PRIMORDIAL STELLAR EVOLUTION 771
E. K. L. Upton

1. General Considerations 771

2. Contraction with Radiative Transport 775

3. Conditions for Homologous Contraction . . . 777
4. Results for Homologous Contraction . . . 780
5. Nonhomologous Contraction and Thermal Relaxation . . . 788
6. Hayashi's Theorem and Convective Transport . . . 797
6.1. The Radius-Entropy Relation . . . 798
6.2. The Radius-Luminosity Relation . . . 799
6.3. The Completely Convective Model as a Limit . . . 799
6.4. The Effect of Superadiabatic Convection . . . 805
7. Results for Convective Transport . . . 808
8. Pre–Main-Sequence Nuclear Reactions . . . 815
9. Thermodynamic Collapse . . . 819
References . . . 823
Index of Subjects and Definitions . . . 825

CHAPTER 1

*Dynamics of Interstellar Matter and the Formation of Stars**

LYMAN SPITZER, JR.
Princeton University Observatory

1. INTRODUCTION

SPECULATION on the origin of stars is as old as astronomy. A detailed theory of star formation must rest on an understanding of the motions of interstellar gas, one of the youngest fields of astrophysics as well as one of the most intricate. This chapter summarizes our present, very tentative knowledge on interstellar gas dynamics together with what is known about the mechanism of star creation.

The most definite single element in this picture is that new stars have in fact been produced in abundance during the last 10^8 years and are presumably being formed at the present time. This introductory section reviews the evidence on star formation, which is based on the study of the stars themselves and not on any properties of the interstellar medium. Subsequent sections discuss first the observational evidence on the structure of the interstellar medium and on the observed motions of the gas, next the rudimentary theory of these motions, and finally the possible mechanisms by which gas is compressed into clouds and clouds into protostars and stars.

Proof of recent star formation is based on two methods for determining the age of a star. The first method utilizes nuclear physics, which indicates how much energy a star can generate from thermonuclear reactions and thereby sets a limit to the time a star of known mass can radiate at a given rate. The second method, less conclusive than the first, is based on the theory of stellar motions, which indicates that a system of stars cannot have endured for more than a certain length of time.

The first method has been summarized by Schwarzschild (1958). The theory of internal constitution has been used to determine the length of time a star remains on the main sequence, that is, before conversion of hydrogen into helium in the core takes the star off the main sequence. The resultant times are given in the fourth column of Table 1, which is essentially Schwarzschild's Table 30.4. Evidently all stars of type F1 or earlier

* This chapter, originally written in 1959, was substantially revised in 1962 while the author was a Research Associate at the Mount Wilson and Palomar Observatories.

have been formed within the past 3×10^9 years and are appreciably younger than the Galaxy as a whole. From the observed numbers of these early main-sequence stars in the solar neighborhood, it is possible to compute the rate at which these stars must be forming on the assumption that this birth rate has been constant. Such computations were first carried out by Salpeter (1955). The computed values of the birth-rate function ψ, in stars per year per cubic parsec per mass interval, are given in the fifth column of Table 1. For the B stars, the rate of formation per unit mass interval is considerably less than for later-type stars; the relative number of such early-type stars observed is even smaller because of the short life of such stars.

Observations of stellar motions generally confirm the belief that the early main-sequence stars are relatively young. As pointed out originally by Ambartsumian (1947, 1949) and confirmed more recently by Blaauw (1956*a*, 1958), most O stars are in loose stellar associations which are generally believed to be expanding (Blaauw 1952*a;* Blaauw and Morgan 1953). The measured expansion velocities, ranging up to about 10 km/sec,

TABLE 1

LIFETIMES AND BIRTH RATES FOR MAIN-SEQUENCE STARS

Spectral Type	M_{vis}	Mass ($\mathfrak{M}_\odot$)	τ* (years)	ψ† [year^{-1}pc^{-3}($\mathfrak{M}/\mathfrak{M}_\odot$)$^{-1}$]
B2.......	−2	18	4.0×10^{7}	0.04×10^{-13}
B5.......	−1	7.2	1.0×10^{8}	0.19×10^{-13}
B8.......	0	4.0	2.4×10^{8}	0.8×10^{-13}
A0.......	+1	2.5	4.7×10^{8}	3.5×10^{-13}
A5.......	+2	1.7	1.5×10^{9}	5.4×10^{-13}
F1.......	+3	1.4	3.6×10^{9}	8.0×10^{-13}

* Time interval that a star remains on the main sequence.
† Rate at which new stars are formed.

give ages of about 10^6 to 10^7 years for these systems, and it is difficult to see how the stars can be much older. The theory of stellar constitution shows that the lifetime of an O star does not exceed a few million years while it is on the main sequence. In general, this interval agrees with the kinematic ages of the associations. Evidently these main-sequence stars are younger than the Galaxy, and their formation must somehow be explained.

Studies of these young stars reveal a number of important clues about the process of star formation. Perhaps the most important result of this sort is the distinction between "extreme Population I," characterized by young early-type stars and interstellar gas and dust, and "halo Population II," characterized by the absence of main-sequence stars of type F1 and earlier and by the relative scarcity of gas and dust. While the original classification by Baade (1944) has now been broadened to provide a more detailed breakdown (conclusions of Conference on Stellar Populations; see Oort *et al.* 1958), the strong correlation in space between young stars and interstellar clouds, emphasized by Baade, is undisputed. This correlation is most evident in the nearest associations, such as in Orion and Taurus, where many faint stars, including many of T Tauri type, can be seen imbedded in a dense interstellar cloud (Herbig 1958). These stars are substantially above the main sequence, consistent with the assumption that they are still contracting gravitationally and that they have been formed relatively recently. The data strongly suggest that new stars form continuously from interstellar clouds. On the other hand, Ambartsumian (1955, 1958*a*) has suggested that new stars are created in some other way, with a surrounding nebula forming in the same process, and that possibly entirely new and unsuspected properties of matter may be involved. This imaginative point of view, which Ambartsumian (1958*b*) suggests may also be relevant to the formation of galaxies, has not as yet led to any detailed analyses and will not be considered further here.

If it is assumed that new stars form from the interstellar gas, one is led to inquire how the star formation rate ψ depends on the gas density. This problem has been investigated by Schmidt (1959) and Salpeter (1959). Since Type I stars are more concentrated toward the galactic plane than is the interstellar hydrogen, the conclusion has been drawn that ψ varies about as n^2, where n is the number of H atoms/cm^3. Other less direct arguments also indicate that ψ varies at least as rapidly as n.

Another important element in the picture is whether stars are born individually or in groups. It has been shown by Roberts (1957) that the number of B stars formed in clusters and associations accounts approximately for all stars of this type formed in the solar neighborhood. This same conclusion was extended by Ebert, von Hoerner, and Temesváry (1960) to all early main-sequence stars. While formation of single stars cannot be excluded observationally, the evidence is consistent with the assumption that all Type I stars have formed in groups—with a mass of about $10^3\ \mathfrak{M}_\odot$ in each group. Fessenkov and Razhkovsky (1952) maintain that in many cases groups of new stars are formed along a line, thereby producing a star chain; this suggestion has not yet been generally accepted.

Another observation that should be explained is the high incidence of binaries among young stars. The relatively rapid rotation among young stars may also be attributed to circumstances associated with the process of birth.

A fact that may require special attention is the occurrence of a few Type I stars with space motions as high as 100 km/sec. Blaauw and Morgan (1954) have pointed out that AE Aur and μ Col are moving away from the Orion Nebula in opposite directions with nearly the same speed, 127 km/sec. Several other stars which have comparable velocities but whose spectra are apparently those of normal, early Population I stars, have been discussed by Greenstein, MacRae, and Fleischer (1956) and by Bonsack and Greenstein (1956). In fact, Blaauw has shown (1956*a*, *b*) that one out of every four stars of type B0 or earlier in the solar neighborhood is a high-velocity object (space velocity exceeding 30 km/sec). The remaining O and B0 stars and virtually all stars of types B1 to B5 have relatively low random velocities of about 10 km/sec (Vyssotsky 1957). The small increase in space velocity from type B to F can probably be explained as an age effect resulting from interactions between stars and cloud complexes (Spitzer and Schwarzschild 1953) or possibly between stars and large-scale inhomogeneities in the galactic potential field that are associated with dynamical instabilities (Toomre 1963).

Stars of Population II have not been treated in this section. These stars were presumably formed during the early life of our Galaxy, when the conditions were quite different from those prevailing now. The density, temperature, and velocity of the interstellar medium at this early date are extremely uncertain. It has even been suggested

(Dirac 1938; Dicke 1962) that the ratio of gravitational to electric forces was much greater at this time. The detailed mechanism of star formation at this early stage is therefore rather speculative and will not be discussed here. While it seems apparent that the formation of Population II stars must have occurred rather rapidly in the early years of the Galaxy, there seems little hope of analyzing this process in detail until the formation of Population I stars, about which we have substantially more evidence, is better understood.

2. OBSERVED PROPERTIES OF THE INTERSTELLAR MEDIUM

As a necessary background for discussions of interstellar dynamics and star formation, we will here briefly review current knowledge on the density, composition, physical state, and velocity of the interstellar medium. Such information is obtained from emission and absorption by neutral H at 21 cm; from obscuration and reddening of light by dust grains; from interstellar absorption lines of Na, Ca^+, and other atoms; and from emission lines observed in the spectra of diffuse nebulae. These topics are discussed more fully elsewhere (e.g., chap. by Münch, this vol.), and only a brief summary will be given here. The discussion will be devoted almost entirely to properties of the medium in the solar neighborhood; that is, in the Orion spiral arm. Information on gas and dust elsewhere in and around the Galaxy is very tentative.

2.1. Spatial Distribution of Gas and Dust

Basic information on the density of interstellar matter has been obtained from the intensity of the hydrogen emission line at 21 cm. The data obtained by van de Hulst, Muller, and Oort (1954) indicate a mean density of 0.7 neutral H atom/cm^3 in the solar neighborhood. A more detailed study by Schmidt (1957) shows widespread spatial variations of the hydrogen density but yields about the same average value. Since the 21-cm line may be more nearly saturated than has been assumed and since the ionized H atoms will increase the mean density slightly (Westerhout 1958), we shall take the total mean interstellar gas density as equivalent to 1 H atom/cm^3 near the galactic plane. This value, corresponding to 0.025 $\mathfrak{M}_\odot/pc^3$, is substantially less than that implied by the gravitational measurement of 0.15 $\mathfrak{M}_\odot/pc^3$ for stars and gas found from the motions and spatial distributions of stars perpendicular to the galactic plane (Oort 1932, 1960). Since the density of known stars is about half this measured total value, an interstellar gas density as great as 0.075 $\mathfrak{M}_\odot/pc^3$, or 3 H atoms/cm^3, would not be in conflict with this gravitational limit and the densities of known stars.

The grains responsible for the extinction of starlight have a total density of about 1.4×10^{-26} gm/sec^3, about 1 per cent of the hydrogen density, according to van de Hulst (1949). This value is based on the assumption that the optical properties of small particles can be obtained from classical theory using the macroscopic optical constants. In this picture, the radii of the particles responsible for absorption and scattering of starlight average about 3×10^{-5} cm and, at the required density, an appreciable fraction, possibly most, of the atoms heavier than helium are in the grains. Platt (1956) has suggested a quite different picture, where the radius of the particles is about 10^{-7} cm and the total mass of the dust is less by two orders of magnitude than van de Hulst's value. It is difficult to see how such small grains could avoid growing to larger sizes as long as most of the heavier elements are still present in the gas. Moreover, observational evidence on the composition of the gas, discussed below, suggests that an appreciable fraction of beryllium and calcium atoms may be locked up in the grains. At the present time, however, this lower mean density for the dust cannot be excluded.

While it is sometimes useful to discuss a mean density, the actual space distribution of interstellar material is highly irregular, as may readily be seen from photographs of the Milky Way (see Morgan, Strömgren, and Johnson 1955; Rodgers *et al.* 1960). Irregularities of all discernible scales, ranging from globules a few thousand astronomical units across to complexes of clouds extending over several hundred parsecs, are evident. A number of models have been proposed to describe this partially chaotic, partially organized, situation. Such models, while clearly not exactly corresponding to reality, can be helpful in providing a theoretical fit for some of the data.

The simplest and most widely used model is the discrete-cloud model, originally developed by Ambartsumian and Gordeladse (1938) and by Ambartsumian (1940). In this picture, spherical clouds of the same radius R are dispersed through space. Many of the observed statistical properties of obscuration by dust clouds are reproduced by this model, with parameters about equal to those in Table 2. The value for the mass of

TABLE 2

PARAMETERS OF "STANDARD" CLOUD

Radius (R)	7 pc
Number per cubic kiloparsec, n	5×10^4
Number in line of sight per kiloparsec, $\pi R^2 n$	8
Fraction of volume occupied, $4\pi R^3 n/3$	0.07
Photographic extinction in single cloud	0.2 mag
Mass	400 $\mathfrak{M}_\odot$
Density of neutral H	10 cm^{-3}

a "standard cloud" is computed on the assumption that the ratio of dust to hydrogen by mass has the average value of 1 per cent. As shown below, the number of such standard clouds in a line of sight, per kiloparsec, is consistent with the number of separate absorption components observed in the interstellar calcium lines.

As expected from the simplicity of this model, further complications are needed if close agreement with observation is to be obtained in detailed investigations. The statistics of stellar color excesses have been investigated by Schatzman (1950) and Münch (1952), who found that to obtain reasonable agreement with the observations, a number of large clouds must be assumed with a photographic extinction through each such cloud of about 1 mag, and about 0.7 such clouds in the line of sight per kiloparsec. If the density within these large clouds were twice that in a typical smaller cloud, the radius of a large cloud would be about 15 pc and its mass about 7×10^3 $\mathfrak{M}_\odot$. Similar large clouds have been found by Clark, Radhakrishnan, and Wilson (1962) in their analysis of 21-cm absorption spectra of twelve radio sources. The densities, radii, and masses which they found were 20 H atoms/cm^3, 6.5 pc, and 10^3 $\mathfrak{M}_\odot$; the number of such clouds in the line of sight per kiloparsec was 0.8. While these absorbing clouds of hydrogen are about half as big in diameter as those found from the optical color excesses, their general properties are similar.

Spitzer (1948*a*) has shown that the interstellar sodium absorption lines are generally consistent with the standard-cloud model, with one such cloud producing a somewhat saturated pair of D lines, but that in addition, less opaque clouds, either smaller or more rarefied, must be assumed present in order to fit the observations. Strömgren (1948) has used interstellar absorption lines data in analysis of physical conditions in a standard cloud; Table 2 incorporates his results. In addition, he found it necessary to assume the presence of a single dense cloud between the Sun and χ^2 Ori. If the hydrogen is assumed to be neutral, which seems most likely, n_H is about 60/cm³; with an assumed diameter of 15 pc, the total mass is about $2 \times 10^3\ \mathfrak{M}_\odot$. It is clear that for increasingly detailed comparison with observations, any model must be made more and more complicated.

A different point of view has been taken by Bok (1948), who divided the dark clouds that could readily be seen on photographs into different categories. His list, with some modifications, is given in Table 3. The masses are greater by about a factor of 50 than the values given by Bok. In accordance with the results by van de Hulst (1949), the photographic extinction produced by 1 gm of dust/cm² has been taken to be 4.5×10^4 mag, about twice the value adopted by Bok. The ratio of dust to gas by mass has again been set equal to 1 per cent. The values of $\mathfrak{M}/\pi R^2$ in the last column are proportional to

TABLE 3

VISIBLE DARK NEBULAE

Type	Mass $\mathfrak{M}/\mathfrak{M}_\odot$	Radius R (pc)	n (H/cm³)	$\mathfrak{M}/\pi R^2$ (gm/cm²)
Small globule..............	>0.1	0.03	$>4\times10^4$	$>10^{-2}$
Large globule..............	3:	0.25	1.6×10^3:	3×10^{-3}:
Intermediate cloud.........	8×10^2	4	100	3×10^{-3}
Large cloud................	1.8×10^4	20	20	3×10^{-3}

the mean total extinction through each cloud, which is about 1.4 mag for the three larger nebulae. Since the discrete clouds obtained on the Ambartsumian model have too small an extinction to be seen on a photograph, they do not appear in Table 3.

The large cloud listed in Table 3 has properties similar to the large clouds which, as we have already seen, are required both by the distribution of observed color excesses and by the 21-cm absorption data. The dense cloud which Strömgren found to produce interstellar absorption lines in χ^2 Ori is rather similar to, but somewhat denser than, these large clouds. The extended clouds listed by Greenstein (1937), with radii of about 50 pc, are also somewhat similar to these idealized large clouds but considerably bigger. The diffuse nebula in Monoceros, observed by Minkowski (1949), is similar to the "large cloud" in Table 3, except that it is predominantly ionized; Minkowski found a diameter of 17 pc, a proton density of 32/cm³, and a mass of about $10^4\ \mathfrak{M}_\odot$—values roughly similar to those listed in Table 3. The Orion Nebula, on the other hand, corresponds more closely to the emission counterpart of the large globule shown in the table. According to Greenstein (1946), Osterbrock and Flather (1959), and Dokuchaeva (1959), the mean density within a radius of 0.5 pc from the exciting Trapezium stars (4′ of arc at a distance of 400 pc) is about 10^3 protons/cm³, which yields a mass of some 10 to 15 $\mathfrak{M}_\odot$. The fainter, less dense extensions of the nebula out to a distance of 1 to 2 pc may have a greater total mass, perhaps as great as 100 $\mathfrak{M}_\odot$.

While the four types of clouds listed in Table 3, taken together with the hypothetical standard cloud in Table 2, are useful in describing the observations, they certainly do not exhaust the variety of interstellar structures. Clouds tend to be gathered together in groups, small clouds occurring as concentrations within large clouds and large clouds occurring together in vast complexes that together make up a spiral arm. As van de Hulst (1955) has pointed out, structural details of all sizes can be discerned from the diameter of a spiral arm in the galactic plane (500–1000 pc) down to emission filaments as sharp as the resolving power of the plate and less than 1000 a.u. across. These relatively sharp filaments are certainly of interest in interstellar gas dynamics but are probably not important in star formation.

Of particular dynamical interest are the so-called elephant-trunk structures which appear at the edges of bright nebulae surrounding one or more early-type stars. These are generally somewhat elongated, dark nebulae pointing toward the exciting star and marked by bright edges showing an emission spectrum. These clouds, whose structure has been studied by Pottasch (1956) and Osterbrock (1957*b*), are discussed in some detail in Section 3.

In addition to the gas and dust in spiral arms, a medium of lower density must also extend between the arms and to considerable distances from the galactic plane. Direct observational evidence on this low-density medium is very scanty, although indirect evidence has been put forward both on the galactic halo, or corona (Pikelner, 1953; Spitzer, 1956; Pikelner and Shklovsky, 1959; Münch and Zirin, 1961; Grzedzielski and Stepien, 1963), and on the intergalactic medium (Kahn and Woltjer, 1959). Gas outside spiral arms may play an important role in the evolution of the galaxy, but it is not likely to condense directly into new stars. Hence we shall not discuss this interesting but somewhat speculative topic.

2.2. Composition and Physical State of the Gas

We pass now to a discussion of the chemical composition, ionization, and kinetic temperature of the gas. The ratio of dust to neutral hydrogen has been shown by Lilley (1955) to be roughly constant between one large region and another despite considerable changes in the total obscuration and 21-cm emission. On the other hand, Bok, Lawrence, and Menon (1955), Helfer and Tatel (1959), and others have shown that this proportionality between extinction and 21-cm emission fails when smaller regions are examined. Lambrecht and Schmidt (1958) have concluded that the ratio of dust to neutral hydrogen varies about as the square root of the hydrogen density. In relatively dense dark clouds either the hydrogen becomes molecular, which is not unexpected, or the dust is concentrated without a corresponding concentration of the gas, which seems less likely.

Relatively precise information on the composition of the interstellar gas is limited to the Orion Nebula, where the work by Aller and Liller (1959) shows that the abundance of the lighter elements relative to hydrogen is about the same as in the B stars. A similar conclusion for the abundance of hydrogen relative to helium in the Orion Nebula was obtained by Mathis (1957). Estimates of chemical composition from the strengths of interstellar absorption lines are somewhat less certain. The results obtained by Strömgren (1948) and Seaton (1951) are consistent with the assumption that the composition of the gas is about the same as in Population I stars generally, except that calcium seems underabundant relative to sodium by about an order of magnitude. This relative scarcity of observable calcium can be interpreted, perhaps, as the result of calcium atoms being predominantly locked up in the grains. For the high-velocity components (Routly and Spitzer 1952), the ratio is more nearly normal, which might be ascribed to the evaporation of calcium atoms by radiation from the O stars which, as we shall see below, has presumably been responsible for accelerating the cloud. This point of view receives some support from the underabundance of beryllium (Spitzer and Field 1955) by at least one order of magnitude; beryllium and calcium are chemically somewhat similar, and if calcium is predominantly attached to solid particles, beryllium should be also.

The ionization of the gas is dominated by a phenomenon first pointed out by Strömgren (1939). In a hydrogen gas of uniform density surrounding a star of early spectral type, the transition between 90 per cent and 10 per cent ionization of the hydrogen is relatively sharp, occurring in a transition layer about 2 pc thick, if the hydrogen density is about 1 atom/cm^3. For comparison, the radius of the sphere of ionized hydrogen for this same assumed density is about 90 pc for an O7 star. The physical reason for this sharp transition is that the optical thickness of a thin layer of neutral gas for radiation in the Lyman continuum is high, and once the density of neutrals rises above a certain level, the intensity of the ionizing radiation is rapidly depleted. Within this sphere of ionized hydrogen, called an H II region, or a *Strömgren sphere*, most elements other than helium will be ionized at least once. Outside this sphere, in the H I region, not only hydrogen but also oxygen and nitrogen will be neutral, carbon and silicon being the most abundant ionized elements. Since the interstellar density is rarely completely uniform one would not expect the H II region to show the ideal spherical symmetry of the theory, but with this modification the observed emission regions around O stars seem generally consistent with the predictions of the Strömgren theory. Revised numerical values for the radii of H II regions have been given by Pottasch (1960).

An important distinction between H I and H II regions is the difference in equilibrium kinetic temperatures, a conclusion first reached by Spitzer and Savedoff (1950). In both regions, kinetic energy is gained by photoelectric ionization of neutrals following electron capture by an ion; the ejected photoelectron generally leaves with a few volts of kinetic energy. In H II regions, the abundant element, hydrogen, is responsible for the energy gain, while oxygen and neon are responsible for most of the energy radiated. The gains and losses are equal at a temperature in the neighborhood of 10,000° K. According to recent detailed computations by Burbidge, Gould, and Pottasch (1963), the equilibrium temperature averages more nearly 6000° K, with the precise value depending on the color temperature of the exciting star as well as on the abundances of oxygen and neon relative to hydrogen.

In H I regions, the hydrogen remains neutral and the energy gain is due to the much less abundant atoms, carbon and silicon. As a result, the rate of energy gained by the gas per free electron is reduced by several orders of magnitude in H I regions as compared to H II regions. Since the ions responsible for radiation by electron excitation are about as abundant in the two regions, the equilibrium temperature in a region of neutral hydrogen is much less than in an ionized region. According to cross-sections obtained by Seaton, the equilibrium temperature (Spitzer 1954; Seaton 1955) is at most about 20° K, if C^+ is present with its anticipated abundance; if the amount of C^+ is negligible, an equilibrium temperature between 50° and 100° K is likely as a result of other cooling mechanisms, in H I regions, particularly collisions of H atoms with grains or with molecules of H_2 which may be a major constituent of the gas in these cool regions (McCrea and McNally 1960; Gould, Gold and Salpeter 1963). As pointed out by Hayakawa (1960), these equilibrium temperatures would be somewhat increased, especially for the less dense H I regions, if the intensity of cosmic rays below 10^{10} ev were very much greater than the intensity observed at higher energies; however, extrapolation of the cosmic-ray spectrum to lower energies is very uncertain. In any case, the equilibrium temperature of the dense H I clouds must apparently be some two orders of magnitude less than that of H II regions. An analysis by Ebert (1955*a*) indicates that the temperature in an H I cloud is relatively insensitive to density up to 10^6 H atoms/cm^3, if the cloud remains transparent.

Kahn (1955) has pointed out a substantial modification of these results for H I regions. As a result of the velocities of the interstellar clouds, collisions between clouds should occur and produce appreciable heating. Following each such collision the rate of cooling will be rapid, but as the temperature falls the cooling rate drops. Computations by Kahn (1955) and Seaton (1955) indicate that the harmonic mean temperature of the clouds in the presence of intermittent collisional heating should be about 100° K when the theoretical equilibrium value is much less. Detailed computations of the temperature in an H I cloud, as a function of time after heating to several thousand degrees, have been made by Takayanagi and Nishimura (1960).

The predictions of these theoretical studies are apparently in rough accord with observational evidence on interstellar temperatures. According to van de Hulst, Muller, and Oort (1954), the saturation intensities of the 21-cm line indicate a temperature of 125° K; as pointed out by Kahn (1955), this is a harmonic mean. Some dense clouds apparently have lower temperatures; Davies (1956) has observed regions in Cygnus and Auriga where the diminished 21-cm intensity indicates a gas temperature less than 65° K. According to Clark *et al.* (1962), measures of the 21-cm line absorption indicate a temperature of about 60° K for the absorbing H I clouds. The temperatures of H II regions are less well explored. Burbidge *et al.* (1963) have shown that for two diffuse galactic nebulae the observed intensity ratios of O II and O III lines relative to Hα indicate a temperature of 6000° K, and this is in agreement with theoretical expectations. Near the galactic center, ratios of N II to Hα indicate temperatures as great as 20,000° K, which these authors attribute to heating by energetic particles. In the Orion Nebula, the observations suggest a somewhat higher temperature than the computed equilibrium values. An electron temperature of about 10,000° K has been deduced by Aller (1946), Greenstein (1946), and by the Andrillats (1959) from an analysis of optical data on the hydrogen spectrum from this nebula. Aller and Liller (1959) obtained 9000° K from measured ratios of line intensities in the O III spectrum from Orion. Pariiskii (1962) has obtained a value of 12,000° K from measured radio fluxes at 9.4 and 75 cm with an uncertainty of about 1000° K resulting from uncertainty in the spatial structure of the Orion radio source. Radio measures of other H II regions give results generally consistent with

10,000° K (Mills, Little, and Sheridan 1956; Haddock 1957; Wade 1958*b*) but are not sufficiently detailed to give precise values.

While detailed agreement between theory and observation has not yet been established, the large difference in temperature between H I and H II regions predicted by theory can be taken as fully confirmed observationally. As we shall see below, this difference has major dynamical consequences in the interstellar gas.

2.3. Magnetic Field in the Galaxy

The interstellar polarization observed by Hiltner (1949, 1951) and Hall (1949; see also Hall and Mikesell 1950) almost certainly requires a magnetic field of some type for its explanation, but both the value of the field and its topography are quite uncertain. According to the paramagnetic relaxation theory of Davis and Greenstein (1951), a field strength B of 10^{-5} gauss is needed to explain the observations. Ferromagnetic relaxation, analyzed by Henry (1958), can also align grains but, according to Cugnon (1963), about the same field is required to explain polarization with this mechanism as with ferromagnetic relaxation. In both theories the plane of polarization tends to be perpendicular to the magnetic field with the plane of vibration parallel to the field.

The observed angles of polarization suggest that the lines of magnetic force in the solar neighborhood are preferentially oriented along a spiral arm. This result may be explained in two ways. One may assume, following Chandrasekhar and Fermi (1953*a*), that the mean magnetic field is a vector oriented parallel to the spiral arm. That deviations of the magnetic vector are apparently small, despite the mean space velocity of 14 km/sec observed for the interstellar clouds as described in the next section, indicates that the field strength must be about 2×10^{-5} gauss, according to the arguments of Chandrasekhar and Fermi (1953*a*). Alternatively, one may adopt the magnetic topography suggested by Spitzer and Tukey (1951) and assume an ellipsoidal distribution of magnetic vectors at different points with the mean field vanishing over a large region but with the mean-square component of $\mathbf{B}$ along the spiral arm exceeding that in other directions, a result which would arise naturally from the shearing effect of differential galactic rotation (Spitzer 1954). Such a chaotic arrangement of lines of force is possible only if the mean field B does not much exceed 10^{-6} gauss. Arguments for this relatively low field have been summarized by Spitzer (1962*b*), who points out, in particular, that the initial stages of star formation become difficult to understand if the magnetic field throughout a spiral arm is as great as 2×10^{-5} gauss. Woltjer (1962, 1965) finds that a field less than 2×10^{-5} gauss can apparently not be reconciled with the generation of galactic radio noise by synchrotron radiation, in view of the small number of electrons observed in the primary cosmic radiation (Earl, 1961). However, some recent observational evidence would seem to favor a lower value of B. An attempt by Davies, Slater, Shuter, and Wild (1960) to measure the circular polarization of the 21-cm line gives an upper limit of 5×10^{-6} gauss for the mean magnetic field intensity within absorbing H I clouds. More recent measures by Davies, Verschuur, and Wild (1962) confirm this upper limit for most clouds; while these authors find a field of 2.5×10^{-5} gauss in one of the 21-cm components in the spectrum of the Taurus A radio source, Morris, Clark, and Wilson (1963) apparently find a much smaller field in this nebula. The apparent absence of Faraday rotation noted by Westerhout, Seeger, Brouw, and Tinbergen (1962) in the galactic radiation at 75 cm is consistent with a relatively low average value for B. In a dense cloud, which has contracted gravitationally, field strengths substantially above the average values are to be expected.

Additional evidence that the magnetic lines of force have a systematic orientation of some sort is provided by the work of Shajn (1955, 1956). His comparison between the direction of the magnetic field, determined from the plane of polarization for light from adjacent stars, and the direction of filamentary structure in various galactic nebulae shows significant agreement in a number of cases. Shajn proposed that all filaments are oriented along lines of magnetic force and that the magnetic field is largely responsible for the appearance of filamentary structure. While this point of view has not been definitely established, it seems likely to be correct.

2.4. Velocities of Interstellar Clouds

It has been known since the days of Plaskett and Pearce (1933) that the interstellar medium shares in the galactic rotation. The velocities of the K and H lines in spectra of early-type stars show one-half the rotational shift of the stars themselves, a result consistent with the assumptions that interstellar calcium is uniformly distributed, on the average, between the Sun and the stars and that the medium rotates about the galactic center at the same rate as the early-type stars. Adams (1949) has shown that even in the relatively near stars, within 600 pc, the K and H lines yield rotational constants substantially equal to the values obtained from stellar velocities.

The 21-cm line has been used to extend this same conclusion throughout most of the Galaxy. As shown by van de Hulst *et al.* (1954), the observations are generally consistent with the assumption that most of the hydrogen gas is moving around the galactic center with the circular velocity. Some definite examples of motion outward from the galactic center have also been found. Van Woerden, Rougoor, and Oort (1957) have observed an extended arm or filament moving away from the galactic center, toward the Sun, at a velocity of some 50 km/sec. A detailed discussion by Oort and Rougoor (1960) of the hydrogen gas near the galactic center indicates that the distance of this filament from the center is probably about 3000 pc.

Within a spiral arm, different regions of gas may have different velocities with respect to each other. We shall denote a moving mass of gas by the word "cloud." Evidence for the existence of separate moving clouds was first obtained directly by Beals (1936) and indirectly by Merrill and Wilson (1937). The most complete information on velocities of interstellar clouds was obtained by Adams (1949), who measured the velocities of various components of interstellar K and H absorption lines in the spectra of 300 B and O stars in the solar neighborhood. The resolving power of the spectrograph corresponded to a velocity separation of about 9 km/sec. Figure 1 shows the number of components observed in each 5 km/sec interval from -50 to $+50$ km/sec; the velocities have been reduced to the local standard of rest by correction for solar motion. In addition to the components shown in the figure, four components were found with velocities between 50 and 100 km/sec, and five between -50 and -80 km/sec. These data have since been supplemented by Münch (1957), who observed interstellar K, H, D_1, and D_2 in 112 stars in the direction of the Perseus spiral arm.

These results have been analyzed by a number of authors. Whipple (1948), considering less extensive preliminary data by Adams (1943), found that the data indicated six to nine clouds in the line of sight per kiloparsec on the average. The velocity distribution was found to be non-Maxwellian, the stronger components showing lower space velocities relative to the local standard of rest than the weaker components. Whipple attributed this effect to overlapping of the components produced in different clouds, and partly to physically smaller velocities for the larger clouds. This second effect would be expected dynamically, he pointed out, if small clouds amalgamated to form bigger ones.

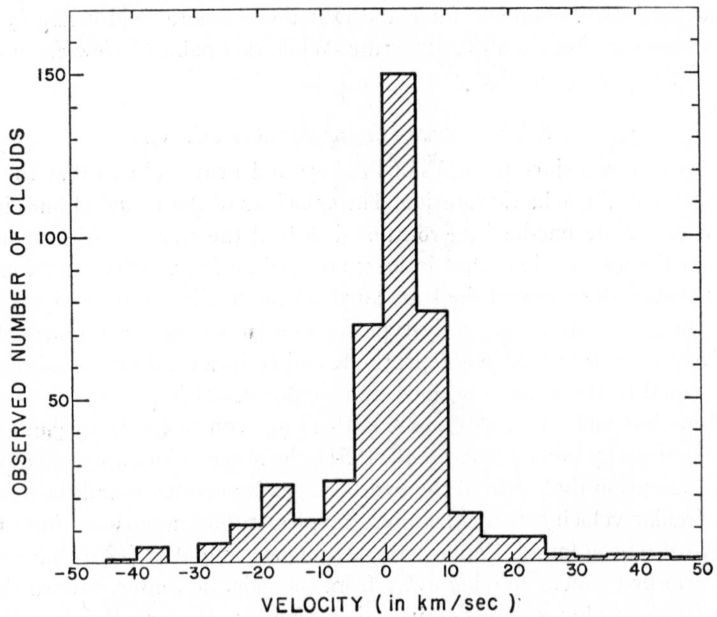

FIG. 1.—Velocity distribution of interstellar clouds. The histogram represents the observed number of absorption components observed in the interstellar Ca^+ lines by Adams (1949). Each vertical bar shows the number of such components within a velocity interval of 5 km/sec. The radial velocities are corrected for solar motion.

The effect of overlapping components was taken into account in detail by Blaauw (1952*b*), who fitted the data to a simple theoretical model of identical clouds. With $P(v)dv$ representing the fraction of clouds with a radial velocity between v and $v + dv$, he found that the data could best be fitted by the non-Maxwellian function

$$P(v) = \frac{1}{2\eta} e^{-|v|/\eta}, \tag{1}$$

with a value of η equal to 5 km/sec for the close stars (about 250 pc) and 8 km/sec for the more distant ones (about 900 pc) yielding dispersions of 7 and 11 km/sec, respectively. The number of clouds per kiloparsec he found to be from eight to ten.

A different point of view has been put forward by Schlüter, Schmidt, and Stumpff (1953). In a detailed statistical analysis these authors conclude that three quite different types of clouds may be distinguished: (1) slow clouds, producing a strong, relatively undisplaced component in almost all stars; (2) fast clouds, producing strongly displaced components; and (3) circumstellar clouds, predominantly around stars of early spectral type and producing components with negative radial velocities. The predominance of a strong undisplaced component of K and H in virtually all stars had been emphasized earlier by Donn (1951).

It seems evident that Blaauw's model, while providing a convenient first approximation, is unrealistic in its neglect of physical differences between clouds of different velocities. As we shall see below, the 21-cm absorption data provide strong support for such differences. Nevertheless, his exponential velocity distribution (eq. [1]) has been useful in several analyses of cloud velocities, observed both optically and at 21 cm.

Münch has analyzed equivalent widths of Na and Ca^+ lines to obtain values of the parameter η in equation (1). For stars in the solar neighborhood, he found values of 3.3 km/sec for Na and 5.0 km/sec for Ca^+, corresponding to dispersions of 4.6 and 7 km/sec, respectively. This apparent difference of kinematic behavior between Na and Ca^+ had been pointed out earlier by Wilson (1939). As we have seen in the previous section, this difference apparently arises from the higher relative abundance of Ca^+ in the faster clouds.

The velocity dispersion of the separate clouds may be found from the 21-cm results also. Van de Hulst *et al.* (1954), on the basis of equation (1) for $P(v)$, found a value of 8.5 km/sec for η yielding a dispersion of 12 km/sec. On the other hand, Westerhout (1957) observed the much smaller dispersion of 6 km/sec, in good agreement with the result by Münch. In view of the complexities of the motions involved, no simple number can characterize the situation, but it would appear that a root-mean-square radial velocity of 8 km/sec is a reasonable value of the dispersion both for Ca^+ ions and neutral H atoms. The root-mean-square space velocity is then 14 km/sec, if a roughly isotropic distribution is assumed. This value may be too great for the relatively denser cooler clouds.

It is evident from Figure 1 that some of the components of K and H observed by Adams have radial velocities much too great to be consistent either with a Maxwellian velocity distribution or even with Blaauw's exponential distribution. It was pointed out by Blaauw (1952*b*) and independently by Searle (1952) and Schlüter, Schmidt, and Stumpff (1953) that negative velocities preponderate among these highly displaced components. Of 64 stars with such high-velocity components (velocities greater than 15 km/sec), in 41 the residual velocities of these components are negative, in 18 they are positive, while in 5 both positive and negative components appear. This same asymmetry is present if we take the difference: cloud velocity minus star velocity. This evidence indicates strongly that these high-speed clouds are accelerated in the neighborhood of the stars in whose spectra they are viewed. In this sense, these fast clouds may be regarded as circumstellar, in accordance with the suggestion by Schlüter, Schmidt, and Stumpff (1953). A possible theoretical explanation for the acceleration of clouds away from hot, young stars is discussed in Sections 3.3 and 3.4 below.

2.5. Internal Velocities within Interstellar Clouds

The width of an absorption line produced by a single cloud should provide a good determination of the velocity dispersion within the cloud. Strongly displaced components of K and H with a residual velocity exceeding 30 km/sec may safely be attributed to single clouds. The mean dispersion found for three such components by Spitzer and Skumanich (1952) was 3.5 km/sec. From the ratios of equivalent widths of K and H in two such stars, Routly and Spitzer (1952) obtained 2.4 km/sec. While all photometric measures of these weak components have been quite uncertain, a dispersion of 3 km/sec in the line-of-sight velocity within a single cloud seems indicated.

Quantitatively more precise estimates are provided by the width of the 21-cm line in absorption. This technique has been applied to the spectrum of Cas A by Hagen, Lilley, and McClain (1955). The Cas A source is close to the galactic plane in the direction of the Perseus Arm. Three strong components appear, one produced by the Orion Arm in which the Sun is located, the other two by the Perseus Arm. The velocity dispersion for the strong local component is about 1.6 km/sec with about 2.1 km/sec for the average of the two Perseus components. Weaker components and components in several other sources, observed by Muller (1959), show about the same dispersion; the mean internal velocity dispersion in the 20 absorption components measured by Muller in four radio sources is 1.9 km/sec. More recently, Clark, Radhakrishnan, and Wilson (1962) have observed 21-cm absorption in 12 radio sources and have found a wide range of velocity dispersion from about 1 to 5 km/sec with a median at 2.0. This evidence leads one to conclude that for the hydrogen atoms in these particular clouds the internal dispersion of velocities in the line of sight averages about 2 km/sec.

A more detailed consideration of these absorption data poses a number of problems. The line of sight to the Cas A source, for example, passes through about 1 kpc of the spiral arm in the solar neighborhood and should pass through about 10 clouds, according to the data on K and H. The spread of radial velocities of such clouds resulting from galactic rotation alone should be about 10 km/sec. Neutral hydrogen atoms with about this spread of random velocities are known to be present from the 21-cm emission data. Since the mean density of such atoms in this local arm is about 0.7 H atoms/cm^3, according to van de Hulst, Muller, and Oort (1954), the total number of such atoms in a 1-cm^2 column from the Sun to Cas A should be about 2×10^{21}. Yet the absorption data indicate that the number of such H atoms with this wide velocity spread does not exceed 0.5×10^{21}.

The most likely explanation of this discrepancy is that the kinetic temperature in these more rapidly moving, relatively transparent clouds is substantially greater than the 125° K assumed, perhaps about 500° K in contrast to the 60° K found by Clark *et al.* (1962) in the more quiescent, more opaque clouds revealed by the absorption. Since the absorption coefficient for 21-cm radiation varies as $1/T$, this assumption seems to explain the observations. The observed velocity dispersions within single clouds do not contradict this point of view, since a velocity dispersion of 2 km/sec for H atoms corresponds to a kinetic temperature of about 500° K; the temperature can be less than this value in the more opaque clouds, if differential mass motions are also present.

Finally we discuss the internal motions in those few systems where more detailed measures are available. The 21-cm data are limited to a few large regions around stellar associations. The optical data are limited to the Orion Nebula and some supernova remnants; apart from planetary nebulae, which are not truly interstellar, these are the only nebulae of sufficiently high surface brightness to permit spectrograms of adequate resolution.

The 21-cm measures give important information on the kinematic properties of gas around stellar associations. The most complete picture has been obtained for the Orion complex, inclosing the Orion Nebula and the Orion association, studied by Menon (1958). The observations are consistent with a central region approximately 80 pc in diameter which is not expanding and is surrounded by a shell about 30 pc thick that is expanding at 10 km/sec. The Orion Nebula and Trapezium are about 20 pc to one side of the center of the system. The distribution of mass is asymmetrical, and the ring is rather elliptical and has its major axis parallel to the galactic equator. The density of neutral hydrogen averages about 3 atoms/cm^3 within this region—about three times the mean value in the local spiral arm. The total mass in the expanding shell is about 6×10^4 $\mathfrak{M}_\odot$. Evidently this shell has somewhat more mass than the large cloud in Table 3 but about three times the diameter and a correspondingly reduced density.

Somewhat similar results were obtained by Wade (1958*a*) for the neutral gas surrounding the H II region about the O8 star λ^1 Orionis. In this case the observations are best explained by a shell of neutral gas with inner and outer diameters of about 15 and 35 pc expanding at 8 km/sec. Inside the shell, the gas is ionized. The total mass of neutral gas is about 5×10^4 $\mathfrak{M}_\odot$.

Extensive optical data on internal motions of the Orion Nebula, which is located within the H I region discussed by Menon, have been reported by Wilson, Münch, Flather, and Coffeen (1959), using spectrograms obtained with 31 parallel slits. The widths of the observed emission lines are variable across the nebula. Measures in those regions where the lines are sharpest are consistent with a kinetic temperature of 10,000° K, when the root-mean-square turbulent velocity (in one dimension) is assumed to be about 6 km/sec for O III and about 8 km/sec for H. Spatial variations of the radial velocity are inconsistent with the Kolmogoroff distribution for incompressible turbulence. The data indicate splitting of the lines in certain regions with a separation amounting to as much as 25 km/sec, attributed to shock waves. The gas in the nebula appears to be expanding into the surrounding H I region at 10 km/sec.

The kinematics of two supernova remnants, the Crab Nebula and the Cygnus Loop, have been discussed by Münch (1958) and Minkowski (1958), respectively. In the Crab Nebula, no simple velocity pattern of the type that might result from uniform expansion, for example, was noted. The velocities extended from +2200 to −1150 km/sec, a range of about 3000 km/sec. Along each filament, however, the range of velocities was much less—about 300 km/sec. The internal velocity distribution across a filament was not more than 30 km/sec. Münch points out that the major axis of the elliptical envelope of the filaments is parallel to the galactic equator to within 2° and suggests that this effect may be due to the influence of a magnetic field parallel to the galactic plane. The detailed analysis by Woltjer (1958) indicates that the composition of the filaments is probably not very abnormal; the electron density is as great as 10^3/cm^3 in the brighter filaments, and the temperature is of order 10,000° K, a typical value for planetary nebulae. The total mass in the filaments is about 0.1 $\mathfrak{M}_\odot$, about the same value found by Osterbrock (1957*a*). However, the gas between the filaments might have a mass greater by an order of magnitude.

Minkowski's radial velocities in the Cygnus Loop, on the other hand, show a variation with distance from the center consistent with a simple radial expansion. These data would seem to verify an earlier suggestion by Oort (1946) that this object is an old supernova shell slowed down by interaction with the continuous medium. Measures indicate that this is an incomplete thick shell with inner and outer diameters about 20 and 40 pc with an expansion velocity varying from about 50 km/sec at the inner boundary to 120 km/sec at the outer boundary. As in the Crab Nebula, the shell is mostly composed of filaments with lengths and diameters of the order of 1 pc and 0.01 pc, respectively, with an electron density of $10^3/cm^3$ and a total mass per filament of about $10^{-2}\,\mathfrak{M}_{\odot}$. The mass of the visible Cygnus Loop, which may be only a fraction of the expanding shell, probably amounts to about one solar mass.

3. DYNAMICS OF INTERSTELLAR MATTER

Any detailed explanation of the motions described in the preceding section is out of the question at the present time. Even if our information on cloud motions were more definite and even if the conditions in interstellar space, in particular the magnetic field topography and strength, were better known, lack of basic knowledge on such matters as turbulence of a compressible gas, plasma physics, and hydromagnetics would make a precise and realistic theory impossible. Discussions of many interesting topics in interstellar dynamics, including such items as the role of turbulence, equipartition of kinetic and magnetic energy, and other interrelations between velocities and magnetic fields, are reproduced in the *Proceedings of the Third Symposium on Cosmical Gas Dynamics* (I.A.U. Symp. No. 8), particularly in the summarizing paper by Parker (1958). Detailed analyses of the various magnetofluid dynamical problems relevant to interstellar gas dynamics are given in the book by Kaplan (1958*a*).

This section concentrates primarily on the mechanisms responsible for accelerating interstellar clouds. First, the energetics of the interstellar motions will be discussed with a treatment of possible energy sources following an analysis of energy dissipation. Second, the spherical expansion of gas around a newly born hot star will be considered, since it appears that this is probably the main driving mechanism for the interstellar velocities. Another subsection discusses the interaction between a single cloud and an O star; this analysis accounts for the observed properties of the bright rims observed near early-type stars and may account for the acceleration of a few clouds to velocities as high as the 100 km/sec observed by Adams. A final subsection treats the expansion of supernova shells and the effects likely to be produced by the magnetic fields observed in such shells; there is a possibility that expanding shells around Type II supernovae provide the dominant mechanism for maintaining the interstellar cloud velocities.

Table 4 summarizes physical conditions of dynamical interest in the interstellar medium. The H I and H II clouds are directly observed. The rarefied H II medium is hypothetical; this gas between the clouds is not directly observed but may be described by the values in the last column. The sound velocity in a cloud is taken to be that for an isothermal disturbance, since the cooling time is less than 10^6 years (Spitzer and Savedoff 1950). The kinematic viscosity in H II regions is computed (Piddington 1957*a*) on the assumption that helium is neutral; if the helium is ionized, the kinematic viscosity will be less by several orders of magnitude. The particle abundance of helium relative to hydrogen is taken to be 0.1 throughout; the electron density in an H I region is assumed to be 1/2000 times the particle density of hydrogen atoms.

The pressures in the H I clouds and in the rarefied gas between the clouds are assumed equal in Table 4. Clearly there will be a tendency toward pressure equilibrium, since any inequalities of pressure tend to propagate away at the sound velocity. On the other hand, Kahn (1960) has suggested that the tendency of an H I cloud to expand is offset by collisions with other clouds and that pressure equilibrium between the clouds and the surrounding medium is seldom reached. The evidence is insufficient at the present time to decide between these two possibilities.

3.1. Energy Dissipation

The kinetic energy of the interstellar clouds may be computed from the data in the preceding sections. A mean density of one H atom/cm^3 and a root-mean-square space velocity of 14 km/sec yields a kinetic energy density of 1.7×10^{-12} erg/cm^3. It is of

TABLE 4

Physical Properties of the Interstellar Material

Physical Properties	Standard H I Cloud	H II Cloud	Rarefied H II Gas
H atoms (or ions)/cm^3	10	10	0.05
Temperature (° K)	100	10,000	10,000
Mean free paths (a.u.)			
Electrons	4.6×10^{-3}	2.3×10^{-3}	0.45
Protons	......	5.6×10^{-3}	1.1
Neutral H	20	......	......
Kinematic viscosity (cm^2/sec)	2.4×10^{18}	7.5×10^{18}	1.5×10^{21}
Resistivity (e.m.u.)	1.3×10^{11}	1.4×10^{8}	1.4×10^{8}
Sound velocity (km/sec)	0.9	11	11
Radius of gyration (km) at 10^{-5} gauss			
Electrons	0.31	3.1	3.1
Protons	......	1.4×10^{2}	1.4×10^{2}

interest that this value is roughly equal both to the energy density of known cosmic radiation in our neighborhood, amounting to 10^{-12} erg/cm^3 according to Peters (1959), and to the energy density of starlight at the earth (3×10^{-13} erg/cm^3). This value for the kinetic energy density of interstellar clouds is not firmly established. If the 21-cm emission and the Ca^+ absorption are selectively produced by the more rapidly moving clouds with a relatively high kinetic temperature and a relatively low density, this value might be reduced somewhat. The magnetic energy density in a spiral arm would appear to be either substantially greater or considerably less than these values, amounting to 1.7×10^{-11} and 4×10^{-14} erg/cm^3 for a field density of 2×10^{-5} and 10^{-6} gauss, respectively.

Since the assumed velocity of the clouds is about ten times the internal sound velocity, collisions between clouds will tend to be inelastic and will dissipate energy. The rate of this dissipation has been computed by Kahn (1955) on the assumption that spherical H I clouds are moving in a vacuum. Thus whenever an element of gas collides with an H I cloud, its kinetic energy will be converted into heat. If we assume that the 21-cm line is produced by the same standard clouds deduced from color excess and interstellar line studies, we see (Table 2) that a straight line intersects 8 clouds/kpc. A velocity of 14 km/sec (the r.m.s. value in 3 dimensions) gives a collision every 9×10^6 years for each element of a cloud, if all other clouds were stationary. Consideration of the motions of other clouds increases the collision rate by $\sqrt{2}$ but indicates that only half the total kinetic energy is dissipated, on the average, in each collision. The final energy dissipation rate is 4×10^{-27} erg cm^{-3} sec^{-1}, which is about the value found by Kahn with roughly the same parameters as those used here.

The accuracy of this computation depends on the value of the interstellar magnetic field. If this field is relatively weak, the collisions will be inelastic, as assumed; the density of the intercloud medium is too low and the cloud velocity too high to permit any cushioning of the impact by the gas between the clouds.

An entirely different situation arises if the magnetic field is so strong that the lines of force remain predominantly parallel to the spiral arm despite the random cloud velocities. As we have seen, the magnetic energy density $B^2/8\pi$ must then be large compared to the kinetic energy density of the clouds and B must be about 2×10^{-5} gauss. To avert a collapse of the clouds, the field strength inside and outside the clouds should be about equal. The velocity of Alfvén (transverse) and magnetosonic (compressional) waves would be 14 km/sec inside a cloud and about 200 km/sec outside.

In this situation, transverse cloud velocities are completely subsonic. As shown by Pikelner (1957), the rate of dissipation of energy in collisions between clouds is reduced by an order of magnitude if the clouds are moving transverse to the magnetic field and if the magnetic energy density is increased to a value about equal to the kinetic energy density in each cloud. Thus collisions will dissipate primarily the energy of motion along the lines of force. The total rate of dissipation is then reduced by a factor of $3^{3/2}$, becoming 8×10^{-28} erg cm^{-3} sec^{-1}. Transverse velocities would be damped indirectly as a result of coupling to the parallel velocities; this coupling results from the finite amplitude of the oscillations, which are essentially Alfvén waves.

These transverse hydromagnetic waves are also subject to a direct dissipation of energy resulting from the presence of neutral H, as pointed out by Piddington (1955, 1957*a*) and Cowling (1956). Piddington has shown that within an H I cloud a wave with a length of 1 parsec will be dissipated in 10^6 years. For transverse motion of the cloud as a whole, a wavelength of 30 to 100 pc must be considered, and for these longer waves this additional dissipation is unimportant. We conclude that if a strong magnetic field is present, dissipation occurs primarily from cloud motions parallel to the lines of force with a resulting marked diminution of the total dissipation rate.

All the kinetic energy dissipated in cloud collisions presumably appears as radiation. Thus a determination of the rate of radiation yields, in principle, the rate of collisional dissipation. As pointed out by Seaton (1958), theoretical estimates of the energy radiated by C^+ ions in a standard cloud at an interstellar temperature of 125° K give a mean rate of energy radiation equal to 6×10^{-28} erg cm^{-3} sec^{-1}, far greater than can be accounted for by direct absorption of starlight in H I regions. Since an H I cloud is presumably heated by a collision to several thousand degrees and then cools off by radiation, the correct total rate of radiation will presumably be substantially greater than 6×10^{-28} erg cm^{-3} sec^{-1}. Hence the rate of radiation would appear to be somewhat greater than the value of 8×10^{-28} erg cm^{-3} sec^{-1} computed for the case of a strong magnetic field, a result which would seem to imply that the actual magnetic field must be relatively weak. This reasoning is by no means conclusive, especially since the composition of the interstellar gas is so uncertain.

3.2. Sources of Kinetic Energy

To replace the energy which is constantly being radiated by heated H I clouds, a large store of energy must be available. Four such sources have been proposed in the Galaxy: (*a*) the mass energy of relativistic particles, stored mostly in the galactic corona, (*b*) the kinetic energy of galactic rotation, (*c*) kinetic energy produced during a supernova explosion, and (*d*) radiant energy streaming out from the stars. As we shall see, the last two mechanisms seem much the most important.

We consider first the energy from cosmic rays. As we have already noted, the energy density of this radiation is about equal to that of the gas kinetic energy. The rate of energy dissipation in the galactic plane, according to the data of Morrison, Olbert, and Rossi (1954), is about 1.3×10^{-27} erg cm^{-3} sec^{-1} for a proton density of 1/cm^3; between 10 and 20 per cent of this energy is available for heating the gas, according to Ginzburg (1958). Thus the power available from the known dissipation of cosmic rays is apparently somewhat less than the power dissipated by colliding clouds. Particles of energies below 10^9 ev, which are not detected on the Earth, could increase the rate of dissipation and might heat the interstellar gas significantly, but there is no reason to assume large numbers of such particles. Biermann and Davis (1958) have suggested that the present cosmic radiation was all produced at the beginning of the Galaxy and has been stored since then in an extended corona some 10^5 pc in diameter that is confined by an extended magnetic field. Such a store of cosmic rays is still conjectural, but if it existed and if, in addition, there were a mechanism for transforming particle kinetic energy into macroscopic kinetic energy of the clouds, cosmic rays might conceivably provide the energy needed for cloud motions. The mechanism originally proposed by Fermi (1949) for converting cloud energy into particle energy is well known, but no mechanism has yet been proposed for the converse process. It is not unlikely that the presence of cosmic rays, tied to lines of magnetic force, may have important hydromagnetic effects throughout interstellar gas dynamics, especially if, as seems likely, the energy density of these relativistic particles exceeds that of the magnetic field within spiral arms. This energy supply, however, would appear to be marginal for the maintenance of the cloud velocities.

The second source, the kinetic energy of galactic rotation, is scarcely adequate, as can be seen by a very simple argument. The velocity of rotation in the solar neighborhood is about 250 km/sec, giving an energy density of 5×10^{-10} erg/cm^3 for a density of 1 H atom/cm^3. A dissipation rate of 4×10^{-27} erg cm^{-3} sec^{-1} would exhaust this supply in about 4×10^9 years. It has been suggested that galactic rotation could drive a hierarchy of turbulent eddies of decreasing size—the largest eddy being the Galaxy itself. Since neither the Ca^+ absorption data nor the 21-cm emission data indicate such a hierarchy of eddies, this mechanism seems unlikely. In any case the total stored energy appears inadequate, and we shall not consider this possibility.

Supernovae have not been widely considered as a source of energy for maintaining the motion of the interstellar gas. Of the two generally recognized types of supernovae, those of Type II eject material at the higher velocity—about 7000 km/sec as compared to 1000 km/sec for those of Type I. While there is little definite information about the frequency of supernovae or about the mass of ejected material, a frequency of 1 per galaxy per 200 years for each of the two types and an ejected mass of 1 $\mathfrak{M}_\odot$ would be not inconsistent with present information (Minkowski, informal communication, 1963). Hence we may neglect supernovae of Type I. If the supernovae of Type II are assumed to be distributed uniformly over a cylinder 300 pc thick and 10,000 pc in radius, the energy output in these shells is 2.8×10^{-26} erg cm^{-3} sec^{-1}, about seven times the rate of dissipation in the interstellar clouds. This value should be increased somewhat to allow for the concentration of Type II supernovae in spiral arms; on the other hand, the concentration in the solar neighborhood is probably less than that closer to the galactic center, and we shall assume here that these two correction factors compensate each other. [See also chapter by Parker, this volume.—*Eds.*]

TABLE 5

RADIANT ENERGY AVAILABLE FOR HEATING INTERSTELLAR GAS

Spectral Type	Ultraviolet Temp., T_u(° K)	Black Body Flux Shortward of ν_0 (ergs/sec)	No. of Stars within 10^3 pc	Power Available for Heating Gas (ergs/sec)
O5.......	56000	100×10^{37}	1.5	54×10^{37}
6.......	44000	42	6	70
7.......	36000	17	4.5	17
8.......	30000	6.4	9	11
9.......	25000	1.48	26	6.1
B0.......	21000	0.16	30	0.64
1.......	18000	0.020	44	0.10
				158.8

The effect of ordinary novae is apparently considerably smaller. According to estimates by Thackeray (1948) and Biermann (1955), the total mass ejected from novae per year in the Galaxy is an order of magnitude less than from supernovae and the ejection velocity is substantially less than 7000 km/sec.

The fourth energy source is provided by the radiant energy of the stars. Spitzer (1951) pointed out that the heating of H II regions by ultraviolet radiation from early-type stars could have important dynamical consequences. Oort (1954) has proposed a specific mechanism whereby the heating produced by a newly born O star would produce outward expansion of all the surrounding gas. A similar, less detailed suggestion was made by Schlüter (1955) in collaboration with Biermann. Here we consider the energy available for this process.

Specifically, we consider the rate at which energy is available for this process from the O and B stars within 1000 pc of the Sun. Table 5 gives the information needed to compute this rate. For the different spectral types given in the first column, the second column gives the ultraviolet temperature T_u, defined as the temperature of a black body whose luminosity below the Lyman limit just equals that of the star. For the stars of type O, these values of T_u have been taken from a recent discussion by Pottasch (1965), who uses the well-known Zanstra method for stars imbedded in diffuse nebulae; his values, which have been averaged and smoothed, may be lower limits if an appreciable amount of ultraviolet flux escapes from the nebulae. The reasonable agreement which he finds between the values of T_u obtained by the Zanstra method and those obtained from the ratio of the O II and O III lines indicates that these values are probably not far from the truth. The value of T_u for a B1 star was taken from a theoretical model by Pecker (1950), on the assumption that T_u equals the surface temperature because of the high opacity below the Lyman limit. The third column gives values of the ultraviolet flux corresponding to T_u. The radii of the stars were taken to vary from 5 $R_\odot$ for B1 stars to 7 $R_\odot$ for stars of type O5. These are roughly the same as the values assumed in corresponding calculations by Oort and Spitzer (1955) and by Pottasch (1958). Consideration of the supergiant B stars, with larger radii, would increase slightly the computed power available.

The fourth column gives the observed number of stars of each type within 1000 pc from the Sun. These numbers are taken from a list by Morgan (subsequently included in the catalogue by Morgan, Code, and Whitford 1955) and multiplied by 1.5 to allow for the unobserved section of the sky. The last column gives the total power available for heating the gas, obtained by multiplying columns three and four together with a factor $kT_0/h\nu_0$. This factor allows approximately for the fact that when a photon of energy $h\nu$ is absorbed, only the excess energy $h(\nu - \nu_0)$ is available for heating the gas, where ν_0 is the frequency at the Lyman limit; for ionizing radiation from a black body at a temperature T_0, this mean excess energy does not differ greatly from kT_0.

The total power available for heating the gas within 10^3 pc, according to the sum of the numbers in the last column in Table 5, is 1.59×10^{39} ergs/sec; about three-fourths of this total comes from O5 and O6 stars. On the assumption that these early-type stars are mostly within 150 pc of the galactic plane, we may divide the total available power by 2.8×10^{64} cm^3, the volume of a cylinder 1000 pc in radius and 300 pc thick, and obtain 5.7×10^{-26} erg cm^{-3} sec^{-1} as the energy available for heating the interstellar gas. Evidently this value exceeds by more than an order of magnitude the computed rate of dissipation in cloud collisions for a weak magnetic field. Since the efficiency with which the interstellar gas converts thermal energy into kinetic energy may be as great as 10 per cent, it appears that the ultraviolet energy from the early-type stars may be about sufficient to account for the velocities of the interstellar clouds. This tentative conclusion is somewhat uncertain, since the estimates of the energy dissipated in cloud collisions and of the ultraviolet luminosity of the O stars could conceivably be in error by an order of magnitude.

It is possible for the early-type stars to accelerate clouds by other means in addition to heating the gas. For example, radiation pressure on the grains can lead to outward acceleration of a cloud, an effect recently analyzed by Harwit (1962). While the effect might be important in special situations, it cannot be important for interstellar clouds generally; the efficiency of conversion of radiant energy into mechanical energy by this process is equal to the fractional Doppler shift, or v/c, and is about equal to 10^{-4}. With so low an efficiency, the total energy available is inadequate to maintain the observed cloud velocities.

The more important energy input rates found in this section and the energy dissipation rates found in the previous section are summarized in Table 6. Either the kinetic energy in supernova shells or the ultraviolet radiant energy from early-type stars would be more than adequate to maintain the velocities of the interstellar clouds, although the former energy source is quantitatively uncertain by at least an order of magnitude.

In the next two sections we discuss the dynamical phenomena associated with early-type stars, especially those of type O; dynamical effects associated with supernova shells are treated in a subsequent section.

3.3. Expansion of Gas around a Young O Star

We now consider the detailed mechanism by which an O star can accelerate the interstellar material in its immediate vicinity. The basic concept, due originally to Oort (1954), is that an O star in a relatively dense nebula begins to radiate rather abruptly in a time not exceeding 10^5 years. To idealize the problem at first, we shall ignore inhomogeneities in the initial density distribution, assuming complete spherical symmetry, and shall also neglect the presence of a magnetic field. These neglected but important effects, which are difficult to analyze, are considered briefly at the end of this section.

TABLE 6

Energy Dissipation Rates in H I Clouds and Energy Sources Available in ergs cm^{-3} sec^{-1}

Energy dissipated by H I clouds:	
Computed from collision rate, no magnetic field	40×10^{-28}
Computed from collision rate, in strong magnetic field	8×10^{-28}
Computed from radiation by C^+ at 125° K	6×10^{-28}
Energy dissipated by cosmic rays	13×10^{-28}
Energy available for accelerating interstellar gas:	
Kinetic energy of Type II supernova shells	280×10^{-28}
Excess ultraviolet energy from early-type stars	570×10^{-28}

With these idealizations, the dynamical problem is, in principle, straightforward. While no complete analytical solution is available, the appropriate equations of motion, of ionization and temperature equilibrium, and of continuity could be solved numerically for any assumed set of initial conditions. Because of the lack of such solutions, we shall survey the information that has been obtained from more approximate considerations.

Two qualitatively important processes will occur when an O star begins to radiate within a cloud of originally cool gas. In the first place, the gas will be progressively ionized. Since the mean free path of an ultraviolet photon is very short when hydrogen is predominantly neutral, there will be a relatively sharp ionization front, exactly as in the equilibrium situation considered by Strömgren. This ionization front will move rapidly outward until its distance from the star approaches the equilibrium value at which the energy radiated in electron captures between the star and the front just equals the available ultraviolet radiation of the star. In the second place, the heated gas will obviously expand since the pressure in the surrounding cool gas is only about 1 per cent of the pressure in the ionized region. As in explosions on the Earth, the heated gas generates a shock wave that moves outward into the cool gas, compressing it and heating it. These two effects react on each other. The outward motion of the gas decreases the density in the ionized region thereby increasing the mass of gas that can be present within the H II region in radiative equilibrium. Conversely, the outward motion of the ionization front and the progressive fall of density and pressure in the ionized region modify the strength of the outward moving shock.

While processes between the ionization and shock fronts and also within the ionized zone are important, the physical situation is dominated by these two narrow regions. The structure of such fronts, including equipartition between ions and electrons, emission of radiation, temperature and density profiles, etc., has been considered in detail by Kaplan (1958*a*, Sec. 10). The structure of the ionization front has also been considered by Axford (1961). Since it has not been possible to make observations on these detailed features, we shall treat these fronts as discontinuities. We now derive some of the properties of these two fronts in the relatively simple one-dimensional case.

In both fronts, the physical situation differs greatly from that in terrestrial shocks and detonation fronts, in that the temperature is determined predominantly by radiative effects. In terrestrial shocks, radiation processes can generally be neglected; either the rate of radiation is low or the mean free path of a photon is very short. In relatively dense interstellar clouds, on the other hand, the times required to reach equilibrium temperatures are short compared to the times required for shock waves to travel appreciable distances. According to the computations by Spitzer and Savedoff (1950), in an H II region the time required to approach an equilibrium temperature of some 10^4 ° K is about $2 \times 10^4/n_e$ years, where n_e is the electron density per cubic centimeter. Since shock velocities of interest are generally in the neighborhood of 10 to 20 km/sec, the interstellar gas will travel less than 0.4 pc before it is essentially at an equilibrium temperature. While this distance is large compared to the mean free path, it is much smaller than the dimensions of an H II region. In an H I region, the equilibrium time is somewhat longer for the same gas density, but in a shock the density will generally be increased to a high value. For a hydrogen density of 10^2 atoms/cm^3, for example, the time to approach an equilibrium temperature of 50° K is likely to be about 4×10^4 years. This result is not entirely conclusive, since with extreme assumptions on interstellar abundances the equilibrium time could be increased by at least an order of magnitude. Even for clouds of pure hydrogen, however, Zimmerman (1959) has shown that the rate of cooling is very rapid if the temperature is increased above 2×10^4 ° K to as much as 10^5 ° K; a shock wave passing into neutral hydrogen at 100 km/sec would produce such an increase. While under some interstellar conditions, deviations from radiative equilibrium may be important in shock and ionization fronts, we shall neglect such deviations in a first approximation.

With this restriction, the conditions to be satisfied at both a shock front and at an ionization front become very similar. We let n_H, ρ, and p denote the number of hydrogen atoms (including both neutral and ionized atoms) per cubic centimeter, the gas density, and the gas pressure, respectively. Subscripts 1 and 2 denote quantities ahead of and behind the front, respectively. If U_1 and U_2 denote velocities relative to the front, conservation of matter and momentum yields

$$\rho_1 U_1 = \rho_2 U_2 = mJ\,, \tag{2}$$

$$p_1 + \rho_1 U_1^2 = p_2 + \rho_2 U_2^2\,. \tag{3}$$

The mass m is defined as ρ/n_H. Evidently J is the number of hydrogen atoms crossing the front per cm^2 per sec. Instead of the conservation of energy, we have simply the condition that the temperatures T_1 and T_2 are known on each side of the front. In a shock through a dense cloud, T_1 and T_2 are about equal, while in an ionization front T increases from a value between 50° and 150° K ahead of the front to about 10,000° K behind the front. Solutions of these equations have been discussed by Kaplan (1954), by Kahn (1954), and by Savedoff and Greene (1955) for a shock, and by Goldsworthy (1958, 1961) for an ionization front.

If we introduce the isothermal sound velocity C defined by

$$C^2 = p/\rho\,, \tag{4}$$

equations (2) and (3) may be combined to yield for ρ_2/ρ_1, the compression across the discontinuity,

$$\frac{\rho_2}{\rho_1} = \frac{(C_1^2 + U_1^2) \pm [(C_1^2 + U_1^2)^2 - 4U_1^2C_2^2]^{1/2}}{2C_2^2}\,. \tag{5}$$

For an isothermal shock, C_1 equals C_2 and we have, disregarding the trivial solution $\rho_2 = \rho_1$,

$$\frac{\rho_2}{\rho_1} = \frac{U_1^2}{C_1^2}\,. \tag{6}$$

The compression is determined by equation (6) for any shock velocity. In contrast with the more familiar shocks in which no radiation of energy occurs, very large compressions are possible.

At an ionization front, C_2 considerably exceeds C_1 and the condition that equation (5) gives a real value places a restriction on U_1. It is readily shown that U_1 must satisfy either

$$U_1 \geq U_R \equiv C_2 + (C_2^2 - C_1^2)^{1/2} \approx 2C_2 \tag{7}$$

or

$$U_1 \leq U_D \equiv C_2 - (C_2^2 - C_1^2)^{1/2} \approx \frac{C_1^2}{2C_2}\,. \tag{8}$$

There is another restriction on U_1 at an ionization front. The rate J at which atoms cross the front must equal the available flux of ionizing photons in the far ultraviolet. Let us consider, for example, a situation in which an O star starts to shine in a dense H I cloud. If the radius r_i of the ionized region is much less than its value in radiative equilibrium, then absorption of the ionizing radiation before this radiation reaches the ionization front is negligible and $4\pi r_i^2 J$ is simply the total number of ionizing photons emitted by the star per second.

Since J is determined independently, U_1 depends through equation (2) on the density ρ_1. The nature of the solutions for different values of ρ_1 has been investigated by Kahn (1954); the energy equation adopted by Kahn is not realistic for the interstellar conditions of interest, and his results are modified here to apply to the conditions where C_1 and C_2 are determined by radiative equilibrium.

First we consider the situation in which U_1 exceeds U_R; this case is called R-type by Kahn. The R refers to the "rarefied" density for which this case applies. Since U_1 exceeds C_2 and C_2 exceeds C_1, the motion is evidently supersonic relative to the gas ahead of the ionization front. No disturbances can precede the front. The negative sign must be used in equation (5) to yield solutions of physical interest for interstellar space. In this situation, ρ_2/ρ_1 equals $1 + C_2^2/U_1^2$; the velocity change, $U_1 - U_2$, is C_2^2/U_1. The ionization front moves rapidly outward with a velocity that may vastly exceed C_2 and with relatively little change in density. As an example, we consider an O7 star beginning abruptly to radiate in a cloud of density 10 H atoms/cm^3. When the ionization front is at a distance of 5 pc (half its equilibrium value), the velocity of the front is 1100 km/sec. The relative increase in density will be only about 10^{-4}, and the gas will be accelerated outward by only 0.1 km/sec upon passing through the front.

When U_1 is less than U_R, the motion of the ionization front relative to the gas ahead is subsonic. Disturbances can now precede the front. If U_1 computed from equation (2) is between U_R and U_D, a shock wave will go ahead, increasing ρ_1 and decreasing U_1 so that inequality (8) can be satisfied. Similarly, if U_1 is much less than U_D, an expansion wave can precede the front and increase U_1. This situation is similar to that of a detonation wave, which has been extensively studied (see Courant and Friedrichs 1948). The fronts which appear physically are generally those of the so-called Chapman-Jouguet type and correspond to U_1 equal to U_D, a situation called D-critical by Kahn. Following Kahn, we shall assume that unless U_1 exceeds U_R ahead of the front, the flow of gas in front of a steady one-dimensional ionization front will adjust itself so that U_1 equals U_D at the front. For conditions in interstellar space, which are neither one-dimensional nor steady, it is not obvious that this assumption of D-critical conditions is always valid (see Axford 1961).

A D-critical ionization front has remarkably simple properties. The velocity U_2 is exactly equal to the velocity C_2, the isothermal sound speed behind the shock. If C_2 much exceeds C_1, ρ_2/ρ_1 is about equal to $C_1^2/2C_2^2$, or $T_1/4T_2$ (the ionization of hydrogen changes the mean particle weight by a factor of ~ 2), and p_2/p_1 is one-half. When an ionization front has reached a quasi-equilibrium situation, with the radius of the H II region given by the Strömgren theory but gradually increasing as the gas expands and the density drops, the front will tend, perhaps, to be D-critical. The density ratio will then be about 1/400, and the ionization front will move through the compressed gas ahead at the relatively leisurely pace of 0.03 km/sec.

The compression of the gas ahead of the front is largely accomplished by an isothermal shock front, as pointed out by Kahn (1954) and by Savedoff and Greene (1955). The shock velocity is about equal to that of the ionization front, since the velocity of the gas between the two fronts, relative to each front, is small. If we assume that the gas is nearly motionless behind the ionization front, then the outward velocity of both fronts is equal to C_2 or 11 km/sec. From equation (6) we see that ρ_2/ρ_1 in the shock front is about equal to 200. If the initial density of the neutral cloud in which the O star forms is several hundred atoms per cm^3, as might be anticipated in a large cloud contracting under its own gravitational force, the compressed region between the two will be raised to a density of about 10^5 atoms/cm^3.

One might attempt to construct a solution for the flow around a newly formed O star by combining a D-critical ionization front with an isothermal shock front, assuming uniform velocities or uniform expansion outside the two fronts. Solutions of this type have been investigated by Kaplan (1957, 1958a) who used the simplifying assumption that the flow is one-dimensional. From the rather restrictive conditions which his solutions must satisfy, he concluded that the actual flow is likely to be turbulent and that the observed clouds may be formed in this way. In the three-dimensional situation, no simple solutions of this type exist for gradually decreasing density behind the ionization front; dynamical effects outside the two fronts must presumably be considered in detail.

Such dynamical effects have been analyzed by Savedoff and Greene (1955), who replaced the ionization front by a rigid expanding sphere and made a number of other approximations. Their results showed that the thickness of the compressed cold region (between the two fronts) was much smaller than the radius of the ionized region and that the mass in this compressed region increased steadily with time.

The change of V_s, the outward velocity of the isothermal shock as the gas expands, may be computed very simply. According to the computations by Strömgren, the radius r_i of the ionized region and the proton particle density n_i within this region are related by the simple formula

$$n_i = \left(\frac{s_0}{r_i}\right)^{3/2} ; \quad (9)$$

s_0 is defined as the value of r_i for n_i equal to unity. We may assume that across the ionization front the pressure will not change by a large factor; for the *D*-critical case we have already seen that the pressure decreases by one-half. Since the temperature in each region is constant with time, the pressure in the cold compressed region must vary about as n_i, and hence the ratio of densities across the shock front will also vary as n_i. From equations (6) and (9) it follows that

$$V_s \propto n_i^{1/2} \propto \frac{1}{r_i^{3/4}} . \quad (10)$$

Thus the outer parts of the cloud acquire less velocity from the shock front than do the inner parts.

The 21-cm observations on the expansion of the Orion cloud have been compared by Savedoff (1956) with these theoretical expectations. The present expansion velocity of 10 km/sec and radius of 60 pc is consistent with an initial radius and particle density of 38 pc and 4.6 cm^{-3} some 2×10^6 years ago, if one assumes that the initial outward velocity of the shock front was 14 km/sec. The expansion is attributed to an O7.5 star which has since burned out. Savedoff suggests that most of the observed O and B stars in the Orion region are second-generation objects. As we shall see subsequently, stars would be unlikely to form in the compressed shell if it were nearly homogeneous. However, contracting protostars are likely to be distributed about within the nebula, and these might well start to shine after the first O-type star had burned out.

Dynamical effects behind the expanding ionization front have been considered by Goldsworthy (1961), using a similarity solution, and by Vandervoort (1963*a*, *b*), using a more direct integration of the equations. Both these analyses indicate that a density minimum should develop very soon near the exciting star. Kahn and Menon (1961) point out that radio measurements indicate a density decrease at the center of the Rosette Nebula, thereby agreeing with theoretical predictions for an age of about 5×10^4 years. At the center of the Orion Nebula, on the other hand, the electron density rises steeply to within less than 0.5 pc from the center, an observational result obtained also by Osterbrock and Flather (1959) from optical data and by Pariiskii (1962) from radio observations. If a simple dynamical model is assumed, the age of the Orion Nebula, as measured from the time the Trapezium stars attained their full luminosity, is of the order of 10^4 years at most.

About the later stages of expansion around an O star there is relatively little information. Complicated phenomena arise when the shock front reaches the outer boundary of the neutral cloud; an expansion wave presumably starts inward, producing a new shock on reaching the ionization front. In these later stages, departures from spherical symmetry probably become important. Evidently the transition from a uniformly expanding spherical shell to the relatively disordered interstellar motion requires a fragmentation into smaller clouds.

Spitzer (1954), Frieman (1954), and Klein (1954) have proposed that this fragmentation results from the familiar Rayleigh-Taylor instability, which arises when a light fluid is accelerating a denser one. Detailed studies by Osterbrock (1957*b*) and Pottasch (1958) show many discrepancies, however, between theoretical predictions based on Rayleigh-Taylor instability and the observed bright rims, elephant trunks, and other irregularities at the boundaries of H II regions. Most of the acceleration of the gas apparently is produced in the shock front. Vandervoort (1962) has shown that a *D*-type ionization front can be unstable, and this effect, possibly reinforced by large initial density fluctuations within the neutral cloud and perhaps also by a thermal instability of the type discussed in the next section, may well be responsible for the fragmentation of the expanding shell into separate clouds.

The interstellar magnetic field, which has been neglected in the preceding discussion, will certainly have a profound influence on the expansion of gas around a young O star. If the magnetic field is assumed uniform initially, the dynamical problem is, in principle, straightforward although complex. Since this problem has not been considered in detail, it is not certain what modifications in the expansion will result from the presence of the magnetic field. However, the general nature of the phenomena to be expected under these conditions can be examined on the basis of the following simple arguments.

The relative importance of gas kinetic forces to magnetic forces is indicated by the ratio of the gas pressure nkT to the magnetic energy density $B^2/8\pi$. We denote this ratio by β. If β is large, magnetic forces have only a secondary influence on the dynamics; if β is small, the magnetic forces will dominate. If an O star is formed in a cloud with a density of 30 H atoms/cm^3, the initial gas pressure in the H II region is 8.3×10^{-11} dyne/cm^2. If the initial magnetic field strength is 10^{-5} gauss, the magnetic energy density is 4.0×10^{-12} erg/cm^3 and β is 21. For a field strength of 10^{-6} gauss, β is 2100. In either case, the excess gas pressure, which drives the outward expansion, is much larger than the surrounding magnetic pressure and one may expect the H II region to expand at about the normal sound speed. On this basis one may argue that the magnetic field will not greatly affect the outward initial acceleration of the gas.

The detailed nature of the expansion, however, will certainly be completely modified by the magnetic field. We have seen that a shock wave precedes the ionization front so that the gas pressure behind the shock is increased to about the same value as in the H II region. In the absence of a magnetic field, the density increases across this shock by a factor of 200. If we now assume that this shock is moving across a magnetic field, the field strength will increase by the same factor as the density, since the conducting material is frozen to the lines of force. If B increased by a factor 200, the magnetic energy density would increase by a factor 4×10^4 and would much exceed the driving pressure in the H II region. Evidently the amount of compression between the shock front and the ionization front will be entirely determined by the initial magnetic field strength, and the actual compression may be much smaller than the factor 200 found above. It is not even certain that under these conditions a shock front must necessarily precede the ionization front.

Thus, as the expansion proceeds, the direction and magnitude of the outward velocities will be modified by the field. While the expansion along the field may not be much affected, expansion across the field will produce disturbances along the lines of force; there will be some tendency for the lines of force to remain straight, thereby producing cylindrical expansion. The expanding H II region may tend to form a prolate spheroid with its major axis along the magnetic field. Analysis of the apparent ellipticity of expanding H I regions, for example, the Orion ring studied by Menon (1958), might yield a determination of the initial magnetic field strength.

The detailed development of an expanding shell around an O star is by no means understood. It is clear, however, that this process can accelerate very large amounts of gas to a velocity of about 10 km/sec. Acceleration seems likely to persist at least until the ion density n_i falls to about 1 H atom/cm^3 when the gas pressure in the H II region is about equal to the surrounding magnetic pressure at a field strength of roughly 10^{-5} gauss. The total mass within such an H II region is then of the order of $10^5\ \mathfrak{M}_\odot$ for an early O-type star. The amount of outward-moving neutral gas would be somewhat less, in agreement with the values of 6×10^4 and $5 \times 10^4\ \mathfrak{M}_\odot$ observed for the expanding shells surrounding the Orion Nebula and λ^1 Orionis, respectively.

3.4. Interaction of an Isolated Cloud with a Young O Star

We have seen that radial expansion of hydrogen gas around a young hot star can accelerate a large mass of gas to velocities comparable with the mean value observed in interstellar clouds. However, this mechanism does not seem capable of accounting for the substantially higher velocities, in excess of some 20 km/sec, which are shown in Figure 1. Here we discuss a mechanism, proposed by Oort and Spitzer (1955), by which a single cloud in the neighborhood of a newly formed O star can reach very high velocities. Ionized gas streams off the side of the cloud facing the O star, and the reaction accelerates the cloud outward as in a rocket. First we discuss the nature of the ionization and the shock fronts in such an isolated cloud, following the analysis by Pottasch (1958) and treating subsequently the dynamical effects on the cloud.

Let us consider a cloud near enough to an O star that absorption of ionizing photons by the hydrogen within the H II region may be ignored. As in the preceding analysis, the magnetic field will be neglected. Conditions at the ionization front, on the side of the cloud facing the O star, may be assumed D-critical. The velocity of the ionized gas relative to the ionization front will be C_i, the isothermal sound velocity in the ionized hydrogen; since the ionization front must be moving rather slowly into the cloud, C_i will also be the velocity of the gas relative to the cloud's center of gravity.

The number of atoms per cm^2 per sec streaming toward the star, away from the cloud, is equal to the flux J of photons reaching the ionization front. This stream of atoms forms an insulating layer that absorbs the ultraviolet light and reduces J. We let J_0 be the ultraviolet flux in the absence of the insulating layer and f be the ratio of J_0 to J. The difference between J_0 and J is caused by electron-ion recombination followed by subsequent absorption and photoelectron ionization. The rate of such recombinations per cm³ per sec is αn_i^2, where α is the recombination coefficient. If we assume spherical divergence of the atoms streaming away from the cloud, the proton density n_i is given by

$$n_i = \frac{J}{C_i}\left(\frac{R_c}{R}\right)^2, \tag{11}$$

where R_c is the radius of curvature of the ionization front and R is the distance of the gas from this center of curvature. We then obtain

$$J_0 - J = \int_{R_c}^{\infty} \alpha n_i^2 dR = \frac{\alpha J^2 R_c}{3C_i^2}. \tag{12}$$

Equation (12) yields

$$f^2 - f = \frac{\alpha J_0 R_c}{3C_i^2}. \tag{13}$$

TABLE 7

COMPUTED CHARACTERISTICS OF BRIGHT RIMS NEAR AN O STAR

Distance r (pc)	$J_0/J = f$	n_i(cm⁻³)	n(cm⁻³)
1.	14.2	2,700	5.1×10^6
2.	7.9	1,320	2.1×10^6
4.	4.7	620	8.0×10^5
8.	2.7	320	3.2×10^5
12.	2.2	170	1.5×10^5
20.	1.8	81	5.7×10^4

From equations (11) and (12) we see that $n_i(R_c)$, the density just behind the ionization front, is given by

$$n_i^2(R_c) = \frac{3J_0}{\alpha R_c}\left(1 - \frac{1}{f}\right). \tag{14}$$

These equations are applied by Pottasch (1958) to sharply curved bright rims in diffuse nebulae. The calculations show large variations in f and n_i with varying distance from an O star. Variations with spectral type, from O5 to O9.5, are somewhat less marked, and if an average value of J_0 is taken for O stars, the results in Table 7 are obtained. Variations of C_i with f and with distance r from the exciting star were taken into account in the table but do not greatly affect the results. The values in the last column give the density of neutral hydrogen ahead of the ionization front. These densities are very great. Evidently the material is much compressed by the shock front in order to satisfy equation (2) and yield a D-critical ionization front.

It is not yet possible to check directly the computed densities in Table 7. The "observed densities" given earlier by Pottasch (1956) are determined essentially from equation (14) with f assumed large and with use of the observed rim thickness instead of R_c.

Thus any comparison between these values and the theory is primarily a consistency check on the calculations. As Pottasch (1963) points out, however, the computed values of the neutral density n are consistent with the lower limit found by Bok for a small globule (see Table 3); these globules may be closely related to the bright rims.

It is possible to compare the observed values of rim thickness with theoretical predictions based on observed values of R_c, the radius of curvature of the rim. It is evident from equation (11) that when f is nearly unity and J is about equal to J_0, n_i varies as $(R_c/R)^2$. If the line of sight passes at a distance R_A from the cloud center, the apparent brightness, obtained by integrating n_i^2 along the line of sight, varies as $(R_c/R_A)^3$. Pottasch finds the best fit between his observed thicknesses and theoretical predictions if the thickness is assumed to be measured at a point at which the brightness is 0.7 times that at the front,

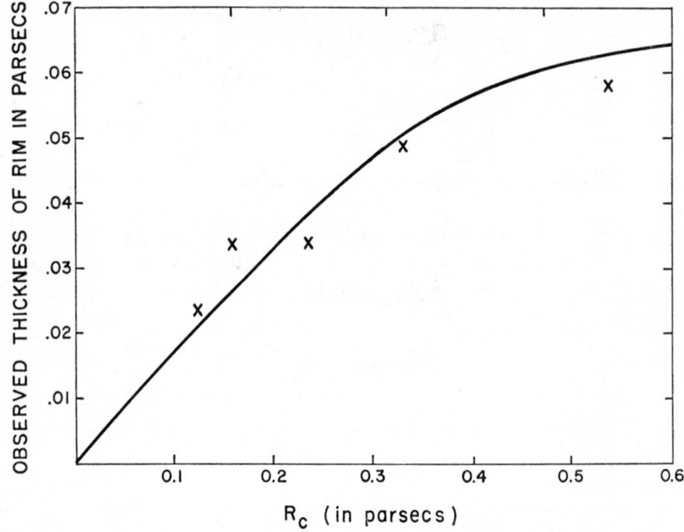

FIG. 2.—Thickness of bright rims in diffuse nebulae. The average thicknesses observed by Pottasch for bright rims near early-type stars are plotted against R_c, the radius of curvature of the rim. The theoretical curve is based on the dynamical theory taking into account the formation of an insulating layer between the cloud and the exciting star.

where R_A equals R_c. On this basis, the ratio of the rim thickness to R_c should be $0.7^{-1/3} - 1$, or 0.126. When f greatly exceeds unity, the rim becomes thicker; Pottasch computes a rim thickness of about $0.2\,R_c$ when f is 10. If J_0 is again set equal to its average value for O stars, f may be computed as a function of R_c from equation (13) and a theoretical relationship may be found between R_c and the rim thickness. This relation is plotted in Figure 2 with the average rim thicknesses observed for rims of different curvature. Since the observed mean value of r, the distance to the exciting star, increases from 2.6 pc for the most sharply curved rims ($R_c = 0.13$ pc) to 8.9 pc ($R_c = 0.55$ pc), this correlation is taken into account in the theoretical curve. The agreement between theory and observation shown in Figure 2 is relatively good. It would be of interest to obtain more quantitative measurements of surface brightness and brightness profiles for a few rims to supplement and calibrate the existing rough thickness measurements.

An important confirmation of this theory has been obtained by Courtès, Cruvellier, and Pottasch (1962), who observed the velocity of expansion of the hydrogen gas away from an elephant-trunk structure in M16. The Hα emission line seen against the background of this dark cloud was observed shifted, giving an approach velocity of 13 km/sec relative to the line at neighboring points, in agreement with the isothermal sound speed shown in Table 4.

While the theory of these bright rims could be improved in a number of ways, the neglect of a magnetic field is likely to be the most serious approximation. We have already seen that such a field is likely to reduce very substantially the maximum compression of the cold gas between the shock front and the ionization front. Thus the values of n in the last column of Table 7 are not likely to be correct. One may hope that the theoretical relationships between n_i and r shown in Table 7 and between the rim thickness and R_c shown in Figure 2 remain valid provided that β, the ratio of gas pressure to magnetic energy density, is large within the H II region. When n_i is $10^3/\text{cm}^3$, and the predicted value for r about equal to 2 pc, β will exceed unity if B is less than 2×10^{-4} gauss. Since B and n vary together, this would require a field smaller than 2×10^{-6} gauss in the initial uncompressed cloud with a density of 10 atoms/cm^3. The relatively close agreement between theory and observations shown in Figure 2 suggests that the initial magnetic field in a cloud giving birth to an O star is at most of the order of 10^{-6} gauss, although more precise observations as well as a more detailed theoretical analysis would be required to make this conclusion definite.

We consider next the dynamical consequences of the interaction between a single cloud and an O star, in particular the acceleration of the cloud. This is a much simpler problem than the expansion of a spherical shell, since such a cloud can be analyzed, at least in a first approximation, as an isolated system and radial accelerations produced by the surrounding medium can be ignored. Then the outward acceleration of the cloud is entirely determined by the reaction of the ionized gases which shoot out toward the O star, exactly as with an ordinary rocket; for an isolated cloud well within an H II sphere, the effect of radiation pressure on the dust grains is generally quite unimportant. If $\mathfrak{M}$ is the mass of the cloud and V is the velocity of the ionized gases relative to the center of gravity of the cloud, then the outward acceleration of the cloud dv/dt is given by

$$\mathfrak{M} \frac{dv}{dt} = -V \frac{d\mathfrak{M}}{dt}. \tag{15}$$

If $\mathfrak{M}_0$ and v_0 are the initial mass and velocity, we have the usual rocket equation

$$\mathfrak{M} = \mathfrak{M}_0 e^{-(v-v_0)/V}. \tag{16}$$

For sufficiently small values of $\mathfrak{M}/\mathfrak{M}_0$, v can be made arbitrarily large. In this way, as proposed by Oort and Spitzer (1955), individual clouds can be accelerated to very high velocities.

To compute the rate at which acceleration takes place, we may determine $d\mathfrak{M}/dt$ from the previous discussion. If the projected cloud area is πR_c^2,

$$\frac{d\mathfrak{M}}{dt} = \frac{\pi R_c^2 J_0 m}{f}, \tag{17}$$

where m is the mass of a hydrogen atom; heavier atoms are ignored here. Since J_0 varies as $1/r^2$, where r is the distance from the star, we introduce the constant

$$g = mJ_0 r^2 . \tag{18}$$

If now we substitute equations (16) and (17) in (15) and use equation (13) to eliminate f, neglecting $1/f$ as compared to unity, we obtain on integration

$$(v + V)\, e^{-v/V} = V - \frac{\pi V R_c^{3/2}}{\mathfrak{M}_0} \left(\frac{3mg}{a}\right)^{1/2} \ln\left(\frac{r}{r_0}\right), \tag{19}$$

where r/r_0 is the ratio of final and initial distances from the central star. In this result we have set v_0 equal to zero and equated C_i to V. Equation (19) may be used to determine the value of the initial mass which will just evaporate for very large V, that is, for which the right-hand side vanishes. Inserting numerical values ($a = 5 \times 10^{-13}$, $r/r_0 = 10$, $g = 2 \times 10^{24}$ for the mean radiation field used by Pottasch, corresponding to a star of type O6–O7), we find that if R_c is 5 pc, $\mathfrak{M}_0$ is about 1000 $\mathfrak{M}_\odot$. On the other hand, if we let R_c be 0.2 pc, about the mean value for the bright rims investigated by Pottasch, then $\mathfrak{M}_0$ falls to about 10 $\mathfrak{M}_\odot$. Masses slightly greater than these critical masses can be accelerated to very high velocities.

This simple theoretical model admittedly has a number of shortcomings. If a dense cloud is literally in a vacuum, it will expand in all directions at the speed of sound and gradually dissipate. If, as seems likely, there is a tendency toward pressure equilibrium in the interstellar material, a dense cold cloud will be surrounded by a less dense medium at a high temperature. Radial expansion of the cloud will then not occur. Because of its low density, the material surrounding the cloud will not much affect the accelerations discussed above.

An attempt to analyze in more detail this rocket-like acceleration of a cloud has been made by Kahn (1954), who developed, in this connection, the analyses of ionization and shock fronts already described. His results are not very realistic, however, for two reasons which he points out. First, his one-dimensional treatment exaggerates the importance of the insulating layer, which in his analysis grows steadily with time. Second, he follows the dynamics only until the shock front reaches the far boundary of the cloud and neglects reflections back and forth; in fact, acceleration to a very high velocity must be produced by many shocks traveling outwards and expansion waves traveling back.

There seems no strong reason to doubt that the rocket effect can account for the occasional clouds of relatively high velocity, from 20 to 100 km/sec, which have been observed by Adams. As Oort and Spitzer (1955) point out, the number of such clouds is consistent with the rate of formation of O stars and the observed distribution over velocities is consistent with equation (19). From this equation, it is readily shown that the number of clouds accelerated to a velocity between v and $v + dv$ varies asymptotically as exp $(-v/V)$, if the distribution of initial masses and other quantities is reasonably uniform. If the mean free path for slowing down is independent of velocity, then this same law should apply to the number of high-velocity clouds in the line of sight. The number of high-velocity components observed by Adams follows this relationship roughly when V is set equal to C_i, or 11 km/sec; a value of V as high as 20 km/sec is required to fit the components of very high velocity (more than 50 km/sec).

On the other hand, it is also possible that shock waves can account for many of the high-velocity components observed. It is well known that as a shock wave travels through a medium of decreasing density the strength and velocity of the shock tends to increase. If, as suggested by Kaplan (1958*b*) and Kahn (1960), the displaced components of K and H are produced by shocks rather than by discrete clouds, one might expect that some of the gas, in regions of relatively low density, would be accelerated up to very substantial velocities. No theoretical analysis of this effect has appeared as yet.

3.5. Supernova Shells

When a supernova of Type II explodes, a shell of material is expelled at a very high velocity between 5,000 and 10,000 km/sec. Within a thousand years, the shell diameter reaches about 10 pc if no deceleration has occurred, and the expanding gas should form a striking feature of the interstellar medium. The Cygnus Loop is believed to be the remnant of such a shell, while the Crab Nebula is believed to have been formed by the relatively slower shell ejected from a Type I supernova. While the genesis of an expanding shell is outside the scope of the present survey, we shall discuss the dynamics of the shell when it has reached interstellar dimensions.

There are a number of differences between the expansion of gas around a supernova and the corresponding phenomenon around a newly formed O star. In the former case, all the kinetic energy is put in initially at a very high value of energy per gram. In the latter, the energy is fed in steadily at a much lower concentration. We have seen in Section 3.2 that the total mass of a Type II supernova shell may be as great as 1 $\mathfrak{M}_\odot$. The mass of material accelerated by an O star, on the other hand, is probably between 10^4 and 10^5 $\mathfrak{M}_\odot$. Since the velocities differ by a factor of almost a thousand in the two cases, a Type I supernova will produce more kinetic energy than will a single O star. However, the total radial momentum produced around an O star exceeds the initial momentum of a supernova shell by at least one order of magnitude.

The simplest model for analyzing the motion of a shell is the "snowplow" model. In this model, which has been applied to supernovae by Oort (1946), Öpik (1953), and Minkowski (1958), the outgoing shell of material, of initial mass $\mathfrak{M}_0$, sweeps up the interstellar gas, of density ρ, and the outward momentum $\mathfrak{M}v$ remains constant and equal to $\mathfrak{M}_0 v_0$. The increase of mass $\Delta\mathfrak{M}$ is given by

$$\Delta\mathfrak{M} = \tfrac{4}{3}\pi r^3 \rho \,. \tag{20}$$

If equation (20) is substituted into the momentum conservation condition, and an integration performed, one finds that the time t to travel the distance r is given by

$$t = \left(1 + \frac{\Delta\mathfrak{M}}{4\,\mathfrak{M}_0}\right)\frac{r}{v_0}. \tag{21}$$

From equations (20) and (21) we see that a shell of 1 $\mathfrak{M}_\odot$ moving at a velocity of 7000 km/sec through interstellar material with a density n of 1 H atom/cm^3 will double its mass, and halve its velocity when it has moved 2.1 pc, taking 370 years in the process. After 3.6×10^4 years, the shell will have moved outward 10 pc, increasing its mass to 100 $\mathfrak{M}_\odot$, and reducing its velocity to 70 km/sec. To explain on the basis of these equations the observed radius and velocity of the Cygnus Loop (20 pc and 100 km/sec), Minkowski (1958) points out that we must take a somewhat lower value for n; a value of 0.1 atom/cm^3 with the same $\mathfrak{M}_0$ and v_0 assumed above will fit the observed r and v, leading to a value of 70 $\mathfrak{M}_\odot$ for the present mass. This result is not greatly in disagreement with the spectroscopic mass estimate of about 1 $\mathfrak{M}_\odot$ for the Cygnus Loop, which is only a small section of the active shell.

This model, if correct, has the immediate consequence that supernova shells cannot account for the observed velocities of most of the interstellar clouds. If the radial momentum of the expanding shell remains constant, as the velocity decreases by two orders of magnitude, the kinetic energy will also decrease by two orders of magnitude and will then be entirely insufficient to offset the energy dissipated by collisions between clouds. To discuss whether or not the kinetic energy actually decreases in this way, we must turn briefly to the actual dynamical and thermal processes.

When a shell is expanding rapidly into quiescent, less dense gas, one would expect a shock to precede the shell in the same way that a shock precedes the ionization front around an O star. In fact, Kaplan (1956, 1958*a*) finds that the outward-moving shell is compressed to about 1 per cent of the radius. Moreover, both the shell and the material through which the shock passes should be heated to very high temperatures, amounting to about 10^9 ° K for a velocity of 7000 km/sec. As pointed out by Kahn (1960), the resulting high pressure would be expected to maintain the shock and to increase the total radial momentum above its initial value. In the complete absence of radiation by the shell, one would expect a large part of the kinetic energy of the initial cloud to be transformed first into thermal energy and then back into kinetic energy in a much greater mass of material. In this case the total energy output from supernovae might well suffice for maintaining the interstellar cloud motions. If this thermal energy is quickly radiated, however, the snowplow model is realistic and supernova shells are not of major importance in maintaining the velocities of the interstellar clouds. At the present time the evidence is insufficient to decide between these two possibilities.

Shklovsky (1962) has developed a theory of adiabatic expansion of the gas which has been heated to 10^8 ° K by the shock from a supernova; he proposes that the expanding bubble of hot gas produced in this way may rise away from the galactic plane because of its buoyancy. However, the radiation from atomic excitations, which may substantially exceed the radiation from free-free transitions, is not considered, and it is not clear whether the gas will stay hot as long as Shklovsky assumes. A more detailed theory of the dynamical and radiative phenomena produced by a supernova shell would be very desirable.

Any realistic theory of supernova shells must probably take into account filamentary structure and magnetic fields. These effects are probably equally important all through the study of interstellar gas dynamics, but their importance is particularly obvious in supernova shells, where the presence of magnetic fields is well established from the observed synchrotron radiation (see chaps. by Bless and Minkowski, this vol.). Intricate filaments are conspicuous both in the outer parts of the Crab and in the Cygnus Loop. There is at present no understanding of the dynamics of these phenomena. Woltjer has suggested that the filaments in the outer part of the Crab Nebula are lines of current flow and that the material is held in by the "pinch effect." (For a simple description, see Spitzer 1962*a* Sec. 4.3, and Woltjer 1965.) However, a pinched discharge is known to be violently unstable, both in theory and in observed laboratory discharges.

It is not clear whether the appearance of the filaments can be reconciled with this instability. Another possibility is that the filaments represent lines of magnetic force on which the temperature is lower and the density higher than on adjacent lines. As we have already noted in Section 2.3, Shajn (1955, 1956) has shown, by comparing polarization data with photographs of galactic nebulae, that this may well be the correct explanation for filaments in the general galactic field as well as in the nebulae.

In the magnetic field in the Crab Nebula, Woltjer has shown that within the amorphous region, where the density is very low, this field must be nearly force free; otherwise very large accelerations would occur in this inner region. Outside the amorphous region, the magnetic stresses will be transmitted to the surrounding material. Piddington (1957*b*) has pointed out that if the mass of the shell were substantially less than 1 $\mathfrak{M}_\odot$, the radial acceleration produced by these magnetic fields would be greater than is consistent with the observations; a mass of 1 $\mathfrak{M}_\odot$ for the shell is consistent with the spectroscopic evidence discussed in the preceding section.

How the magnetic field within the nebula can be produced is still a mystery. One possibility appears to be a dynamical one; i.e., the motions of the ionized gas can amplify the magnetic field, as discussed by Batchelor (1950), Schlüter and Biermann (1950), and others (see Cowling 1957). While it is obvious that a magnetic field can be amplified by motions in a conducting fluid, all examples of such amplification result in a reduction in the scale of the field—the distance over which the field changes sign, for example. The assumption is usually made that the polarization is produced by a large-scale field, although it is not clear whether a small-scale field in the Crab Nebula might be consistent with the observations.

Another possibility has been proposed by Piddington (1957*b*). He suggests that the magnetic amplification is produced by differential rotation of the nebula and the central star, the energy being provided by streams of relativistic particles shot out along the force lines from the star. If the original stellar field is that of a dipole, the scale of the amplified field will again be very small in the radial direction.

Yet another possibility has been suggested by Van Wijk (private communication). One might suppose that flux tubes in the galactic medium penetrate the expanding nebula, as a result of some instability, and are then compressed by the high pressure in the surrounding material. While this mechanism can readily produce a magnetic field of 10^{-3} gauss, it is not clear how the observations on the optical and radio emission from the inner region of the nebula can be made consistent with such a model. The detailed explanation of the magnetic field in the Crab Nebula is a challenging problem for the theorist.

If the galactic magnetic field has a large-scale structure with the mean magnetic field directed along the spiral arms, the origin of the galactic magnetic field is an equally difficult problem. In this case, however, one may at least assume that the field was primitive, and has been present since the beginning of the Galaxy and possibly of the Universe.

4. FORMATION OF CLOUDS, PROTOSTARS, AND STARS

The preceding discussion has summarized the more important aspects of our present knowledge of the structure and motions of the interstellar medium. In this concluding section we explore the processes that may be involved in the formation of clouds from rarefied gas and in the condensation of these clouds to form new stars. It seems reasonably likely, on the basis of the evidence presented in the introductory section, that such processes occur, and it is therefore of interest to investigate the physical mechanisms that may be important.

In the ensuing discussion, two important topics have been omitted. The first is the replenishment of the interstellar gas by ejection from stars and by condensation from the galactic halo, or corona. The summaries by Schwarzschild (1958) and Weymann (1963) indicate that ejection from stars has doubtless played an important part in the evolution of the interstellar medium. So little is known about this topic, however, that any detailed physical discussion would be premature. The same argument applies with even greater force to the galactic halo, whose properties are still highly uncertain.

The second omitted topic is the accretion of matter by existing stars. The ingenious theory by Hoyle and Lyttleton (1939) proposed that most early-type stars were formed not by condensation of clouds into new stars, but by accretion of dense interstellar matter by slow-moving old stars. Schatzman (1955) and Mestel (1954) have shown that a star of high surface temperature ionizes and heats the neighboring gas so effectively that the accretion rate is sharply reduced and becomes virtually negligible. It appears unlikely that this process can account for any appreciable number of early-type stars.

4.1. Formation and Equilibrium of Clouds

A challenging theoretical problem is why the interstellar material is distributed in irregular cloud formations with apparently such widely varying densities. One possibility is that all such irregularities are highly transient, resulting from the propagation of compressional waves through the gas. According to this picture, discussed by Kaplan (1958*a*, *b*) and Kahn (1960), the observed high-density regions have been produced by isothermal shocks and will soon dissipate. As we have already seen, this mechanism may be responsible for some of the high-velocity components of K and H observed by Adams. Shocks may also be responsible for the existence of the globules discussed in Section 2. It seems difficult to believe that shocks alone can account for most of the irregular features observed in photographs of the Milky Way or for the bulk of the low-velocity H I clouds that are believed to produce most of the observed interstellar absorption lines. It seems likely that any interstellar shocks are traveling through a medium in which both the density and temperature are non-uniform.

We therefore consider first how inhomogeneities can be expected to arise in the interstellar material. To idealize the problem, we shall consider that the gas initially has a uniform temperature and density in the galactic plane. Let us investigate the various mechanisms that can produce inhomogeneities.

In Section 2 we saw that hot H II regions surround early-type stars, with cooler H I clouds in other regions. Thus the first and most obvious factor producing inequalities of temperature between different regions is the distribution of early-type stars. In the absence of motions, equilibrium would be reached with a hot rarefied gas surrounding such stars and with a cooler, denser gas of neutral hydrogen elsewhere. When motions of stars and gas are considered, the hot and cool regions become less regular, although it would not appear that this process can by itself produce density inhomogeneities outside the spherical H II regions which surround O and B stars. To explain why the material outside Strömgren spheres is characterized by density inhomogeneities, we must seek another explanation.

One possibly important mechanism may be associated with thermal instabilities of different types. Under some conditions, a uniform gas at the same temperature throughout is not an equilibrium configuration; some regions will expand and heat up, while others will contract and cool off. Such processes have been considered by Parker (1953*b*) and have been analyzed in detail for diffuse nebulae and for interstellar matter by Zanstra (1955*a*, *b*) and by Field (1962). To understand this mechanism, we must consider the processes by which kinetic energy is gained or lost by the electrons. We may let G_e be the energy gain per free electron per sec, owing primarily to electron capture by an ion followed by photoelectric ejection; L_e denotes the corresponding energy loss, owing primarily to electron excitation of an atom or ion followed by emission of a photon. Then conservation of energy yields the equation

$$\frac{3}{2}nk\frac{dT}{dt}-kT\frac{dn}{dt}=n_e(G_e-L_e)\,, \tag{22}$$

where n is the total number of particles per cubic centimeter. G_e and L_e are both functions of T as well as of the particle densities of other constituents. Changes of potential energy are ignored in equation (22), a valid assumption if the dominant constituents (hydrogen and helium) remain either completely ionized or completely neutral. Let us assume that the right-hand side of this equation vanishes for T equal to T_0, that is, that T_0 is an equilibrium temperature. Let us also assume that the pressure p is constant during any change. Then equation (22) yields, for T in the vicinity of T_0,

$$\frac{5}{2}\frac{p}{T}\frac{dT}{dt}=n_eA\,(T-T_0)\,, \tag{23}$$

when we have employed a Taylor expansion of $G_e - L_e$ and retained only the first-order term. If A is negative, T approaches T_0 exponentially and the equilibrium is stable. This corresponds to the physical situation where the losses exceed the gains if the temperature is raised. On the other hand, if A is positive, the losses drop off faster than the gains with increasing temperature. This situation is evidently unstable.

If the increase of ionization level with increasing T (and decreasing n) is ignored, A tends to be negative, since the excitation of atoms from the ground level tends to increase sharply with increasing T. Zanstra (1955*a*, *b*) has pointed out that thermal instability can appear if the atoms responsible for the cooling become ionized with increasing T, provided that the more highly ionized atoms are less readily excited. However, detailed computations by Daub (1963) show that, when all the ions present in planetary nebulae are taken into account, no instability appears.

The analysis by Field (1962) indicates that in a rarefied interstellar gas, thermal instability can result when cosmic rays are the dominant heating source, which may be anticipated in H I clouds at a density of about 0.1 cm^{-3} or less. In this situation, the energy gain G_e per free electron per second is independent of density and temperature (n_e/n_H is assumed fixed), while L_e, resulting from collisions between electrons and other constituents of the gas, is proportional to the density. For perturbations at constant pressure, the decrease of density with increasing temperature will reduce L_e and will produce instability, unless the variation of L_e with temperature is sufficiently steep to offset this decrease. The detailed computations reported by Field indicate that, for actual conditions in H I regions, an instability will appear if the equilibrium temperature exceeds 100° K, corresponding to a particle density less than 0.1 H atom/cm^3. As a result of this instability, condensations in which the temperature is less than 100° K should form, separated by a rarefied gas in which the temperature exceeds 10^4 ° K. This theoretical model seems to be similar to actual interstellar clouds, although the gas density at which the instability appears is substantially less than the mean density of the gas.

Another type of instability is associated with the rate of cooling. If a uniform gas has been heated by some transient process, such as a shock wave, for example, and then cooled by radiation, any initial density perturbation will tend to grow. The cooler regions will be denser, if the pressure is assumed to remain constant, and will therefore cool more rapidly and become relatively denser than the surrounding gas.

A related type of instability has been considered by Schatzman (1958), who considers the reduction in pressure by a factor of 2 as hydrogen recombines. To obtain an instability, one must consider the variation of the ionizing radiation as the material recombines and its optical thickness increases. Schatzman finds that instabilities appear in a mass of gas whenever the initial optical thickness of the gas in the Lyman continuum is large. This condition will tend to be fulfilled near the boundary of an H II region. In Schatzman's analysis, the temperature remains constant and the density increases by a factor of 2 only as the hydrogen recombines. Actually, as the gas becomes neutral, the temperature will drop and much greater increases in density may be expected. The time scale of this process in relation to the rate of expansion of the H II region has not been studied, but it is possible that this instability may account for some of the irregularities observed near the rims of bright H II regions.

A somewhat different group of mechanisms that may be involved in the formation of clouds is that concerned with dust grains. These grains can have an important effect in depressing the temperature, either as a result of direct collisons between H atoms and grains, or indirectly because of their catalytic action in forming H_2 molecules at the solid surface. The rate at which grains grow varies with the density, yielding another possible thermal instability. Within a denser cloud, the grains will grow more rapidly and the temperature will fall further. This process is presumably limited by the relatively small amount of heavy elements present in the gas. Within a dense cloud, most of these heavy atoms may be locked up within the grains.

Another source of instability that would tend to push the dust into separate clouds in the absence of other effects is the force between dust grains arising from the galactic radiation field. Because of the shadow of each grain on the other, the grains are pushed together by radiation pressure and diffuse through the gas. This mutual attraction between grains, analyzed by Spitzer (1941, 1948*b*) and by Harwit (1962), is about a hundred times greater than the mutual gravitational attraction and would lead to a significant concentration of grains relative to gas in about 4×10^7 years in a hydrogen gas of density 1 atom/cm^3. Inside interstellar clouds, the gas density is generally greater by an order of magnitude and the effect is probably negligible (Savedoff 1955). Thus concentration of grains is unlikely to lead directly to star formation (Whipple 1946; Spitzer 1948*b*) within the observed interstellar clouds. This effect should, however, sweep most dust grains out of any rarefied intercloud medium and into the clouds, accentuating the inhomogeneity of the interstellar material. Near a group of O stars, the radiation intensity is much enhanced and the increased radiation force between grains may lead to the collapse of additional protostars (Harwit 1962).

Evidently a considerable variety of processes may be expected to produce inhomogeneities in the interstellar gas. Once these arise, their equilibrium state and further development are naturally of interest. We explore these subjects here.

First we consider the equilibrium of the various clouds discussed in Section 2. The standard cloud of Table 2 will be in equilibrium with the surrounding gas, if the intercloud medium is ionized at a temperature of 10,000° K and has a density of about 0.05 H atom/cm^3; these values have been adopted in Table 4. Most of the clouds in Table 3, however, require somewhat lower temperatures for pressure equilibrium. For a large cloud, a temperature of 50° K is required if n is set equal to mp/kT_1, where p is the pressure in the H II regions. Actually, n may well exceed mp/kT by a factor of about 2 because of self-gravitational effects (see eq. [47]) that increase T for equilibrium to about 100° K. In an intermediate cloud, a temperature between 20° and 40° K would be required, provided we again assume gravitational self-attraction; the exact value depends on whether the hydrogen is molecular or atomic. Temperatures in this range, comparable to the theoretical equilibrium temperature in an H I region, seem not unreasonable, since a large dense cloud, probably moving rather slowly, might be expected to reach a temperature close to the equilibrium value.

The equilibrium state of the globules poses a difficult problem. Pressure equilibrium requires either that the temperatures of the globules be extremely low or that the ratio of gas to dust in these particular objects be much less than assumed. Possibly the globules are regions which have been compressed temporarily by an expanding H II region or by a passing shock wave (Gershberg 1962). Alternatively, they could be systems of much greater mass than the tentative values in Table 3 and could then be in gravitational equilibrium at substantial internal temperatures. Further data are required.

Next we pass to the problem of cloud evolution. The development of a cloud is likely to be dominated by collisions with other clouds. As we have seen in Section 3.1, the time interval between collisions is about 10^7 years. It is difficult to predict what the results of a collision will be, largely because of a lack of definite information on the strength and topography of the magnetic field. It seems likely, however, that head-on collisions will be inelastic, at least along the lines of magnetic force. While grazing impacts might produce some fragmentation of clouds, one may surmise that, on the whole, large clouds tend to grow by accretion of small ones. Clouds may also grow by accretion of intercloud material. This rarefied gas between the clouds is assumed to be hot simply because it cools so slowly. The interface between compressed cool gas and the hot rarefied gas is not in equilibrium but will move slowly outward, at least in the direction parallel to the magnetic field.

Collisions between clouds will also affect the internal state of each cloud by producing motion as well as heat. Parker (1953*a*) suggests that the rotational velocity will steadily increase, because of the tendency toward equipartition of rotational and translational energy in successive collisions. The magnetic field will probably counterbalance any tendency of this sort, and it is not clear, in any case, that clouds have enough inner cohesion to use arguments of equipartition.

These considerations do not lead to any very definite results except that large clouds probably grow from small ones. Much more information is needed before any firm statements can be made about the evolution of interstellar clouds.

4.2. Gravitational Instability in a Uniform Medium

Gravitational contraction of a mass of gas is obviously an important step in the formation of new stars. The detailed processes occurring during the collapse of a protostar under gravitational forces are very complicated. One relatively simple question that can be answered theoretically, at least in principle, is under what situation a gas in equilibrium becomes unstable and starts to collapse catastrophically.

To analyze this situation, one considers infinitesimal perturbations about an initial equilibrium state in which the pressure p_0, density ρ_0, velocity $\boldsymbol{v}_0$, and gravitational potential Φ_0 are independent of time. Small perturbations, p_1, ρ_1, $\boldsymbol{v}_1$, and Φ_1, are then considered; the equations for these infinitesimal quantities can thus be linearized and, under certain conditions, solved. If the equilibrium quantities p_0, ρ_0, $\boldsymbol{v}_0$, and Φ_0 are functions of position, however, the solution of the differential equations for p_1, ρ_1, $\boldsymbol{v}_1$, and Φ_1 becomes very complex.

A drastic simplification of the problem which gives approximate results for a wide variety of situations was introduced by Jeans (1928). In this approach, the equilibrium quantities, p_0, ρ_0, Φ_0, are assumed constant throughout space and $\boldsymbol{v}_0$ is set equal to zero. The resultant equations for the perturbed quantities can then be solved directly to yield conditions for instability in an idealized, uniform, infinite medium. The one defect of this approach is that the assumed constant values for ρ_0 and Φ_0 are mutually inconsistent, since $\nabla^2\Phi_0$ should equal $4\pi G\rho_0$; physically, an infinite uniform medium cannot be in equilibrium, even momentarily, under gravitational forces. Evidently the results obtained for the perturbed quantities by this method are formally incorrect, since the unperturbed medium does not satisfy the conditions of the problem. The formulae obtained in this way for the onset of instability are, however, frequently rather similar to the correct criteria for finite systems and have a certain usefulness if viewed with caution. We review in this section the numerous analyses that have been presented along these lines.

On the assumption that $\boldsymbol{\nabla}\Phi_0$ and $\boldsymbol{v}_0$ both vanish, the familiar linearized equations become

$$\rho_0 \frac{d\boldsymbol{v}_1}{dt} = -\boldsymbol{\nabla} p_1 - \rho_0 \boldsymbol{\nabla}\Phi_1 , \tag{24}$$

$$\frac{\partial \rho_1}{\partial t} + \rho_0 \boldsymbol{\nabla} \cdot \boldsymbol{v}_1 = 0 , \tag{25}$$

$$\nabla^2\Phi_1 = 4\pi G\rho_1 , \tag{26}$$

where Φ_1 is the perturbation in the gravitational potential. To solve these equations we must assume some relationship between p_1 and ρ_1. Since relatively slow changes in the interstellar gas tend to be isothermal rather than adiabatic, we may write

$$p_1 = \rho_1 C^2 = \rho_1 \left(\frac{kT}{m}\right), \tag{27}$$

where C is the isothermal sound velocity, which we take to be constant. If we consider a wave front in the yz-plane and let

$$\rho_1 = Ae^{i(\kappa x + \omega t)}, \tag{28}$$

with similar functional equations for v_1, p_1, and Φ_1, we obtain by simple algebra,

$$\omega^2 = \kappa^2 C^2 - 4\pi G\rho_0 . \tag{29}$$

Evidently if the wave number κ is less than a critical value, ω is imaginary and the perturbation grows exponentially. The minimum thickness l of a slab of matter which will contract gravitationally may be taken as half the wavelength for the critical wave number, and we find

$$l^2 = \frac{\pi kT}{4mG\rho_0}. \tag{30}$$

This analysis has been extended to include a variety of other effects. Chandrasekhar (1951) has found that, if isotropic turbulence is present with a root-mean-square velocity U, then $U^2/3$ must be added to C^2 in equation (29). Chandrasekhar and Fermi (1953*b*) have discussed the influence of an initially uniform magnetic field **B.** If the magnetic field is parallel to the wave front, they find

$$\omega^2 = \kappa^2(C^2 + V_A^2) - 4\pi G\rho_0 , \tag{31}$$

where V_A is the Alfvén hydromagnetic velocity, given by

$$V_A^2 = \frac{B^2}{4\pi\rho_0}. \tag{32}$$

In this case the perturbation compresses the field and, if the magnetic pressure $B^2/8\pi$ much exceeds the material pressure nkT, the minimum size of the gravitationally unstable slab will be greatly increased. If **B** is at some angle to the wave front, two modes are present and the instability of one of these will be determined by the usual Jeans criterion in the absence of a magnetic field. When **B** is very strong, the particle motions in this mode are nearly parallel to **B** and the magnetic field does not affect the stability.

In a medium rotating with an angular velocity $\mathbf{\Omega}$, the centrifugal force must affect the initial equilibrium state. Since the Jeans method ignores all gravitational forces in the initial unperturbed state, one may also ignore centrifugal force both in the initial and in the perturbed medium and consider only the Coriolis force $2\mathbf{\Omega} \times v_1$. With this idealization, the problem becomes tractable. Analysis of this situation by Chandrasekhar (1955) for $\mathbf{\Omega}$ constant in space gives results very similar to the case of the uniform magnetic field. When the wave front and the axis of rotation are at some angle, the Jeans criterion is unchanged; for rapid rotation, the velocities of the unstable perturbation are parallel to the axis of rotation. If the wave front is parallel to the axis of rotation, then the velocities must necessarily be transverse to the rotation vector $\mathbf{\Omega}$, and the dispersion relation becomes

$$\omega^2 = \kappa^2 C^2 - 4\pi G\rho_0 + 4\Omega^2 , \tag{33}$$

where Ω is the angular velocity in the unperturbed system. For motions transverse to the rotational axis, gravitational instability can begin only if

$$\Omega^2 < \pi G\rho_0 . \tag{34}$$

Bel and Schatzman (1958) have considered the situation where Ω depends on r, the distance from the axis in the unperturbed situation. From their equation (12) it is readily seen that in our equation (33) the $4\Omega^2$ term must be multiplied by $1 + rd\Omega/2\Omega dr$, or $-B/\Omega$, where B is Oort's galactic-rotation constant; in the solar neighborhood this correction factor is about one-fourth its value for an inverse-square force. Hence Ω can have twice the value given by equation (34) without destroying the instability.

One might suppose that, if both a Coriolis force and a uniform magnetic field were present with some angle between **B** and $\mathbf{\Omega}$, the velocity in the disturbance could not be parallel to both **B** and $\mathbf{\Omega}$ and the motion would tend to be stabilized. The analysis by Chandrasekhar (1954) shows that this expectation is not fulfilled and the Jeans criterion, equation (30), is still applicable, provided only that **B** is not parallel to the wave front. In the unstable mode, the particle velocity is parallel to **B**, if **B** is large, but need not be parallel to $\mathbf{\Omega}$. Evidently rotation has no effect on gravitational instability, if a magnetic field is present along which the fluid velocity may be directed. Physically, a slight bending of the lines of force produces magnetic stresses which cancel out the Coriolis force, the angular momentum of each fluid element is no longer constant, and gravitational condensation is not impeded by the rotation. If the wave front is perpendicular both to **B** and to $\mathbf{\Omega}$, then $\kappa^2 V_A^2$ must be added on the right-hand side of dispersion relation (33), a result established by Anand and Kushwaha (1962).

The effects of viscosity and finite conductivity have been considered by Pacholczyk and Stodolkiewicz (1959, 1960). Their analysis shows that these dissipative phenomena do not significantly modify the criterion for instability. Similar conclusions have been reached by Kumar (1960, 1961) and Stephenson (1961).

4.3. Gravitational Instability in Isolated Systems

In any real situation, masses of gas are more or less isolated, and we turn now to a consideration of gravitational instabilities in such systems. Effects produced by a magnetic field and by turbulent motions are postponed to following subsections as are most of the problems associated with angular momentum.

Results on the gravitational instability of isolated systems are now available for two idealized situations. Instability of a rotating mass of gas held in equilibrium in the z-direction (parallel to the axis of rotation) by its own self-gravitational force has been considered by Fricke (1954) and by Safranov (1960); this situation is relevant to condensation of the gas in a rotating galaxy. Gravitational instability of an isothermal gaseous sphere surrounded by a tenuous gas at a constant pressure has been considered by Ebert (1955*b*, 1957), independently by Bonnor (1956), and most recently by McCrea (1957). We summarize their results briefly.

In a rotating isothermal gas, the unperturbed material is assumed to be in hydrostatic equilibrium in the z-direction, parallel to the axis of rotation. If the gravitational force is assumed to be produced by the gas in the neighborhood, then from the equation of hydrostatic equilibrium we may readily derive (Spitzer 1942; Ledoux 1951)

$$\frac{\rho}{\rho(0)} = \mathrm{sech}^2\left(\frac{z}{H}\right), \tag{35}$$

where $\rho(0)$ is the density in the galactic plane at z equal to 0 and H is given by

$$H^2 = \frac{kT}{2\pi G m \rho(0)}. \tag{36}$$

Comparison of equations (30) and (36) indicates that H/l is $2^{1/2}/\pi$ or 0.45. Thus the thickness of the medium is somewhat less than that of the thinnest slab that would be gravitationally unstable on the Jeans analysis, and the gas is stable against condensation into slabs of shorter wavelength in the z-direction. If stars, or the galactic nucleus, contribute to the gravitational potential in the z-direction, H is further reduced and this conclusion is reinforced.

Fricke (1954), taking into account the density distribution in equation (35) and the presence of Coriolis forces and following work by Ledoux (1951), has analyzed the instability of the gas for perturbations in the xy-plane. All quantities, including T, ρ_0, and the angular velocity Ω, were assumed constant with r. Instability was found if the density $\rho(0)$ in the plane of symmetry exceeded by a factor 1.4 the value obtained from condition (34) for ρ_0. This same result, with a numerical factor of 1.5, was obtained by Safranov (1960) with more general assumptions about $\Omega(r)$. Safranov also investigated the dependence of critical density on the radial wavelength λ of the perturbation; on the assumption that the rotation of the system is determined primarily by the central mass, he found that the critical density would be least for λ an order of magnitude greater than the scale height H.

These results may be applied to the gas in the galactic plane. Since Ω near the Sun is about 10^{-15} sec^{-1}, which corresponds to a rotational period of 2×10^8 years, while H is several hundred parsecs, gravitational instability may be expected if $\rho(0)$ exceeds 10^{-23} gm/cm^3 over a radial extension of at least a thousand parsecs. Such high densities over such extended regions are apparently not present in the solar neighborhood but may have played an important role in condensation at other times or other places. In our region of the Galaxy, where smaller, somewhat separated clouds are observed, it would appear that instabilities of relatively small isolated clouds are more likely to play an important part in the formation of the young stars around us.

We now turn to a discussion of gravitational instability of an isolated cloud within the Galaxy. The physical picture is that of a cool neutral cloud surrounded by a hot H II region. We idealize the problem somewhat and consider a quiescent isothermal sphere of gas imbedded in a rarefied medium at constant pressure. As a first approximation, we neglect both magnetic fields and rotation, considering these effects in later subsections. The situation in which this sphere will be unstable has been analyzed by Ebert (1955*b*) and independently by Bonnor (1956). Both authors have pointed out that for a given mass $\mathfrak{M}$ at a temperature T there is a maximum surface pressure p_m which an isothermal sphere can maintain. For a surface pressure somewhat less than p_m, two configurations are possible—an extended one with relatively uniform density, which is stable, and a small configuration much more centrally condensed, which is unstable.

In view of the importance of the result, we shall give the essentials of the derivation. For an isothermal sphere of perfect gas in hydrostatic equilibrium, Poisson's equation may be written (Emden 1907)

$$\frac{1}{\xi^2}\frac{d}{d\xi}\left(\xi^2\frac{du}{d\xi}\right)=e^{-u}, \tag{37}$$

where ξ, the dimensionless radius, is given by

$$\xi=\left(\frac{4\pi G m\rho_0}{kT}\right)^{1/2} r; \tag{38}$$

ρ_0 is the central density, and u may be defined by the relationship

$$\rho = \rho_0 e^{-u} . \tag{39}$$

The quantity $kTu(\xi)$ is the change of gravitational potential from the center to the distance ξ. The pressure is assumed to be related to ρ by the usual gas law; m denotes the mean mass per particle.

From equation (37) it is readily shown that $\mathfrak{M}(r)$, the mass interior to r, is given by

$$\mathfrak{M}(r) = 4\pi \int \rho r^2 d r = \frac{1}{(4\pi \rho_0)^{1/2}} \left(\frac{kT}{mG}\right)^{3/2} \xi^2 \frac{du}{d\xi} . \tag{40}$$

Let us now consider a sequence of models in which the total mass $\mathfrak{M}(R)$ is constant but ξ_1, the value of ξ at the boundary, is varied. For each ξ_1, the value of ρ_0 will be determined from equation (40), thus determining R, the cloud radius, through equation (38) and also the pressure p. We have, eliminating ρ_0 by use of equation (40),

$$R = \frac{mG\,\mathfrak{M}(R)}{kT} \times \frac{1}{\xi_1 du/d\xi} , \tag{41}$$

$$p(R) = \rho \frac{kT}{m} = \frac{k^4 T^4}{4\pi [\,\mathfrak{M}(R)]^2 m^4 G^3} \times \xi_1^4 \left(\frac{du}{d\xi}\right)^2 e^{-u_1} , \tag{42}$$

$$\frac{\rho_0}{\bar{\rho}} = \frac{4\pi \rho_0 R^3}{3\,\mathfrak{M}(R)} = \frac{\xi_1}{3 du/d\xi} . \tag{43}$$

The three dimensionless functions of ξ which occur in these three equations are given in Table 8, which is taken from the tabulation by Chandrasekhar and Wares (1949). Since $du/d\xi$ is about equal to $\xi/3$ for small ξ, R varies inversely as ξ_1^2. Thus for constant $\mathfrak{M}(R)$ the configurations with small ξ_1 are very extended ones with low pressures and densities and very little density concentration. With increasing ξ_1, the radius contracts, the pressure and density increase, the self-gravitational force becomes relatively larger, and the sphere becomes more centrally condensed. It will be noted from the table that $p(R)$ has a maximum value at $\xi_1 = 6.5$; $\rho_0/\bar{\rho} \approx 6$ for this critical case. Further increases in ξ_1 increase the self-gravitational attraction more than the density, and the pressure falls as the central condensation continues to increase. It is readily shown that all the configurations for ξ greater than ξ_1 are unstable. These general conclusions also follow qualitatively from the virial theorem, as pointed out by McCrea (1957). The virial theorem has also been used by Parker (1953*a*) and by Huang (1954); the latter showed that a non-rotating cloud would be stable against pulsation if the density were uniform; as we have seen, $\rho_0/\bar{\rho} \approx 6$ for the configuration which is barely unstable.

From equation (42) and the values given in Table 8, we obtain for p_m, the maximum pressure,

$$p_m = 1.40 \left(\frac{kT}{m}\right)^4 \frac{1}{G^3\,\mathfrak{M}^2} , \tag{44}$$

while the radius R_m corresponding to this critical condition is

$$R_m = 0.41 \frac{mG\,\mathfrak{M}}{kT} . \tag{45}$$

Evidently R_m is the minimum radius at which a cloud of mass $\mathfrak{M}$ surrounded by a medium at some uniform pressure can exist in gravitational equilibrium. If $\mathfrak{M}$, the total mass, is expressed in terms of the mean density ρ, we obtain

$$R_m^2 = 0.57 \frac{kT}{mG\bar{\rho}}, \tag{46}$$

a result very similar to the Jeans criterion, equation (30). From equations (44) and (45) it follows that

$$\bar{\rho} = 2.44 \frac{m p_m}{kT}. \tag{47}$$

Evidently mp_m/kT is the mean density consistent with pressure equilibrium without gravitational forces.

TABLE 8

PROPERTIES OF A BOUNDED ISOTHERMAL GAS SPHERE OF CONSTANT MASS

ξ	Radius $(\xi\, du/d\xi)^{-1}$	Pressure $\xi^4(du/d\xi)^2e^{-u}$	$\rho_0/\bar{\rho}$ $(3\, du/\xi d\xi)^{-1}$
0.00......	∞	.000	1.000
0.50......	12.30	.0016	1.025
1.00......	3.301	.0783	1.100
1.50......	1.636	.599	1.227
2.00......	1.055	2.052	1.407
2.50......	0.788	4.493	1.642
3.00......	0.645	7.479	1.934
3.50......	0.560	10.44	2.285
4.00......	0.506	12.97	2.699
4.50......	0.471	14.91	3.176
5.00......	0.446	16.24	3.720
5.50......	0.430	17.06	4.337
6.00......	0.418	17.46	5.018
6.50......	0.410	17.56	5.775
7.00......	0.405	17.44	6.608
7.50......	0.401	17.18	7.518
8.00......	0.399	16.82	8.506
8.50......	0.398	16.40	9.574
9.00......	0.397	15.95	10.725
9.50......	0.398	15.49	11.958
10.00......	0.398	15.02	13.277
11.00......	0.401	14.13	16.173
12.00......	0.405	13.31	19.423

The instability of a centrally condensed isothermal sphere may also be derived from analysis of radial motions in an isolated isothermal sphere extending to infinity. Ebert (1957) has shown that such a configuration is unstable against radial pulsations in which a sphere of dimensionless radius ξ about equal to 4.1 contracts (or expands), while more distant shells move in the opposite direction. This instability of an isothermal sphere is a special case of the well-known radial instability of a star in which the pressure varies as ρ^γ, and γ is less than 4/3.

Instability of an isolated mass cannot occur for an isothermal sheet or cylinder, if the perturbations are limited to changes in the sheet thickness h or in the cylinder radius R. This result may be seen immediately by comparing $\nabla p/\rho$ with the gravitational acceleration g. In a sheet, g at the boundary is independent of the thickness h while $\nabla p/\rho$ varies as $1/h$; hence gravitational forces become less important as the sheet is compressed, a result qualitatively different from that found in the Jeans case. Similarly, in an isothermal cylinder with a radius R and a fixed mass per unit length, g and $\nabla p/\rho$ both vary as $1/R$. If gravity is unimportant for one value of R, it will be unimportant for any R. We have already seen that a sheet may be unstable for perturbations in the plane of the sheet. For a spherical cloud only two mechanisms are apparently available for producing gravitational instability: (*a*) three-dimensional radial compression with T remaining constant, and (*b*) a decrease of T on compression, which will always tend to make the internal pressure relatively less important than gravity.

The assumption that the temperature T is rigorously constant in a contracting sphere is, of course, only an approximation. Parker (1953*b*, 1958) has pointed out that under some conditions the H I temperatures, as computed by Spitzer and Savedoff (1950), show a decrease with increasing density corresponding to a value of γ as low as 0.4 in the pressure-density relationship; he has computed the criteria for instability under these conditions. On the other hand, Ebert's (1955*a*) calculation shows that for densities in excess of 100 H atoms/cm^3, γ is generally close to unity. All these calculations assume that the material is transparent to the radiation which ionizes the atoms and heats the grains. A discussion of the instabilities to be expected when a cloud is subject to pure surface heating has been given by Hatanaka, Unno, and Takebe (1961). Since they assume that all the heating is caused by cosmic-ray particles of energies below 10 Mev (Hayakawa 1960), their detailed assumptions may not be realistic, but their analysis indicates the reduction of critical mass which results if the cloud is nearly opaque to the heating source.

How interstellar clouds approach the critical conditions is an intricate question. One may suppose that a cloud of mass somewhat less than the maximum value (i.e., with a surface pressure somewhat less than p_m) is subject to a transitory increase in pressure which compresses the cloud and increases its central condensation, so that when the pressure decreases to its earlier value the cloud is unstable. Alternatively, either the mass of the cloud or the external pressure may rise and remain above the values permitted by equation (44) and the cloud may collapse; this process cannot be termed an instability, since no equilibrium solution of any sort exists. In either case, whether collapse takes place because the equilibrium is unstable or because no equilibrium exists, the condition for gravitational contraction is that the external pressure must be near the value of p_m given in equation (44).

We now consider how close the observed clouds are to this critical condition for collapse. To find a criterion that is independent of T, we may eliminate T from equations (44) and (45), obtaining

$$\frac{G\mathfrak{M}^2}{R_m^4} = 25p_m . \tag{48}$$

If for the pressure in the surrounding H II region we assume the value 1.4×10^{-13} dyne/cm^2, corresponding to a proton density of 0.05 per cm^3, we obtain

$$\frac{\mathfrak{M}}{\pi R_m^2} = \frac{5}{\pi}\left(\frac{p_m}{G}\right)^{1/2} = 2.3 \times 10^{-3}\,\text{gm/cm}^2 , \tag{49}$$

corresponding to a mean extinction of about 1 mag, if the ratio of grains to hydrogen in the dense clouds is about the same as in interstellar matter generally. The extinction through the cloud center, where the density is relatively high, exceeds the mean extinction by a factor of 3.6, amounting to 4 magnitudes. Since the pressure in the region surrounding the H I clouds is quite uncertain, these computed extinctions through a cloud which is on the verge of gravitational collapse may be incorrect by more than a factor of 2.

It is rather remarkable that, according to the criterion (49), the clouds listed in Table 3 are not far from the critical condition where collapse becomes possible. This result is probably only a coincidence, since it implies that these clouds are all in pressure equilibrium and have a temperature inversely proportional to the central density. We have already seen that pressure equilibrium is improbable for the globules and not really certain for the larger clouds. Moreover, if the temperatures are in fact quite different in clouds of different density, the assumption of isothermal contraction, on which all the previous analysis has been based, becomes very questionable. Observational selection may well provide at least a partial explanation for the general agreement between most of the $\mathfrak{M}/\pi R^2$ values in Table 3, corresponding to an extinction of about 1.4 mag.

Even for the large cloud, observational selection influences our interpretation of the data. On the one hand, one may suppose that the maximum pressure in a cloud is correctly given by equation (44), but that the constant in this equation should be somewhat increased because of internal turbulent motions and magnetic fields. The critical value of $\mathfrak{M}/\pi R_m^2$ would also be increased. In this picture, it is only occasionally, as a result of exceptional circumstances, that one of the large clouds reaches the critical condition and promptly collapses. On the other hand, one may suppose that the large cloud represented in Table 3 is in fact on the verge of collapse and represents the most conspicuous one of its type. If this were true, most clouds of this type would be characterized by smaller values of $\mathfrak{M}/\pi R^2$. The true situation probably embodies elements from each.

We close this section with a brief discussion of what happens when the critical condition is passed and the cloud begins to contract radially. The gas pressure is soon unimportant compared to gravity and the material undergoes free fall. In the absence of perturbations, including rotation or magnetic fields, and with spherical symmetry and isothermal conditions assumed, the motions remain radial, and as shown by Ebert (1955*b*) the gas initially at some radius r reaches the center in a time equal to $(3\pi/32G\rho)^{1/2}$, where ρ is the initial density interior to r; for ρ equal to 3×10^{-23} gm/cm^3, this time interval is about 10^7 years.

Actually, as first pointed out by Hoyle (1953), as the gas density increases, somewhat smaller masses become unstable. These smaller condensations each collapse in turn into somewhat smaller aggregations, and so on; according to equation (30), the critical distance l decreases as $1/\rho^{1/2}$, provided that T remains constant. The mass ρl^3 also decreases as $1/\rho^{1/2}$. This process of "fragmentation" has been followed analytically by Hunter (1962), who discusses the growth of infinitesimal perturbations in a collapsing sphere of cold gas with nearly uniform density. He shows that the density perturbations increase more rapidly than does the unperturbed density. This same conclusion has been established by Savedoff and Vila (1962) for a uniform sphere contracting adiabatically, provided that the ratio of specific heats is less than 5/3. As the cloud contracts, there will be some tendency for the fragments to collide again and coalesce, especially if, as pointed out by Layzer (1963), the angular momentum acquired naturally by the fragments during their formation prevents their further contraction. On the other hand, some mechanism for disposing of angular momentum is required for star formation in general (see Sec. 4.5), and if this process is sufficiently rapid, collisions between the contracting fragments are probably unimportant. Evidently it is not yet certain that this complex fragmentation process can occur at all, but because of the lack of any alternative theory we shall assume that this is the process by which a cloud of large mass may condense into many stars.

As shown by Hoyle, fragmentation should cease if the gas becomes sufficiently opaque to radiation, so that further contraction is adiabatic rather than isothermal. More specifically, a free-fall collapse ends and Helmholtz contraction begins when the increase of opacity decreases the rate of radiation loss below the rate of compressional heating, so that the protostar can heat up and reach hydrostatic equilibrium. Following preliminary analyses of this problem by Rouskol (1955) and by Mestel and Spitzer (1956), a detailed study has been carried out by Gaustad (1963). His results show that radiative opacity is inadequate to terminate collapse for protostars more massive than 0.1 $\mathfrak{M}_{\odot}$. On the basis of this analysis one would expect fragmentation to continue until the masses of the fragments were of this order.

For a gas cloud of such low mass, one might expect that a Helmholtz contraction phase would begin after fragmentation had ceased and that the protostar would gradually shrink down to stellar dimensions. The duration of such a contraction phase is limited, however, by an instability, as pointed out by Cameron (1962), which results from dissociation and ionization of hydrogen at temperatures higher than about 2000° K. This process lowers γ, the ratio of specific heats, below 4/3, and the system becomes unstable against collapse. If the abundance of helium is appreciable, ionization of helium will maintain instability until the second ionization of helium is complete at a temperature of several hundred thousand degrees. Further fragmentation during this second collapse is possible, although if the protostar has spent some time in a Helmholtz contraction phase the high radial symmetry acquired would tend to slow down the rate of growth of subcondensations. The final process of star formation, therefore, is a rather violent collapse to stellar dimensions, with the large kinetic energy acquired leading to a substantial overshoot of temperature. The dynamical equations governing this contraction of a spherical mass have been integrated in special cases by McVittie (1956), who made simplifying non-physical assumptions about the radiation flow.

While these discussions give a preliminary orientation to the problem of contraction, they are all somewhat academic, as they neglect magnetic fields, rotation, and turbulence. These effects, which are considered briefly below, may be expected to alter very greatly the various stages of star formation.

4.4. Effect of Magnetic Fields on Gravitational Contraction

The galactic magnetic field will modify the conditions for gravitational contraction and the subsequent fragmentation of the material into progressively smaller masses. We consider these effects briefly here, postponing to the final subsection the influence of the magnetic field on the disposal of angular momentum.

We have seen that in the Jeans analysis a magnetic field has no effect on gravitational instability, since the condensations may occur along the lines of magnetic force. In an isolated system this conclusion must be changed. If the magnetic field is very strong, any contraction must be one-dimensional and parallel to the lines of force. In the preceding subsection, where magnetic fields were ignored, we have seen that one-dimensional contraction can lead to gravitational instability of a rotating galaxy whose extension in the galactic plane is much greater than the wavelength of the instability. Such one-dimensional contraction might well be associated with the formation of such large-scale features as spiral arms but does not seem likely to form an isolated cloud within the Galaxy. If an H I cloud is given such a low temperature that gravitational forces are important, the cloud will contract along the lines of force. Such motion is not unstable, however, if the temperature remains constant. The increase in density resulting from the contraction will make possible a new equilibrium, with the cloud flattened in the direction of the magnetic field. It does not appear that contraction of this sort can give rise to very high densities, and somewhat higher densities will not directly facilitate contractions transverse to the magnetic field.

We shall assume, then, that if a cloud is to condense into new stars it must condense radially, that is, across the lines of magnetic force as well as along them. In an idealized infinite medium the minimum thickness l of a slab that will be gravitationally unstable transverse to a magnetic field is equal to π/κ, where κ is given by equation (31) with ω set equal to zero. We have

$$l^2 = \frac{\pi}{4G\rho_0}(C^2 + V_A^2) = \frac{\pi}{4G\rho_0}\left(\frac{kT}{m} + \frac{B^2}{4\pi\rho_0}\right). \tag{50}$$

No detailed analytical results have been obtained on the gravitational instability of an isolated cloud with a magnetic field. In general, however, the sound speed C and the Alfvén speed V_A will occur quadratically in the same combination as in equation (50), and it will therefore suffice for our present purposes to deal with the ratio V_A/C. If T is 10^2 ° K, and ρ_0 corresponds to 10 H atoms/cm³, we have

$$\frac{V_A}{C} = 7.6 \times 10^5 B. \tag{51}$$

Evidently if B were much less than 10^{-6} gauss, V_A/C would be small and magnetic forces would not affect the onset of gravitational instability. Even if B is as great as 2×10^{-6} gauss, l in equation (50) is increased by only about 1.6, and one may suppose that the criterion found above for the instability of an H I cloud will not be much affected.

On the other hand, if we choose for B the value of 2×10^{-5} gauss, as suggested by the work of Chandrasekhar and Fermi (1953*b*), V_A exceeds C by a full order of magnitude and gas pressure becomes entirely negligible in comparison to magnetic forces. The detailed criteria for gravitational contraction of an isolated cloud have not been worked out, but it is clear that a configuration will collapse only if the magnetic pressure $B^2/8\pi$ is less than the pressure needed to sustain the material against gravitational forces. Thus we find, as a criterion for collapse,

$$\frac{B^2}{8\pi} < \gamma \frac{G\mathfrak{M}^2}{R^4}, \tag{52}$$

where γ is a numerical constant depending on the distribution of density and magnetic field strength. For example, if both pressures are averaged over the volume of the polytrope $n = 2$, γ equals $1/4\pi$; alternatively, if we replaced p_m by $B^2/8\pi$ in equation (48) we obtain equation (52) with γ equal to 0.04. The former, higher value is relevant if we are interested in the value of the compressed field inside the cloud. Since we are concerned here with the value of the external magnetic field outside the contracting sphere, we shall adopt the latter, smaller value for γ and obtain

$$\frac{\mathfrak{M}}{\pi R^2} > \frac{5B}{\pi(8\pi G)^{1/2}} = 1.2 \times 10^3 \, B \text{ gm/cm}^2 . \tag{53}$$

A related stability analysis has been done by Chandrasekhar and Fermi (1953*b*) for the gravitational instability of an infinite cylinder of radius R with an axial magnetic field B. They consider the particular longitudinal and lateral oscillations which transform the cylinder into a train of sausages without compressing the gas. This analysis shows that instability always sets in for wavelengths somewhat longer than a certain minimum value but that this wavelength grows exponentially as the magnetic field decreases much below a value comparable with that found in equation (53), where we take $\mathfrak{M}$ to be the mass of gas in a length R of the cylinder.

If we apply equation (53) to actual clouds, on the assumption that these are on the verge of collapse, we find that $\mathfrak{M}/\pi R^2$ is rather high if a high magnetic field is assumed. With B set equal to 2×10^{-5} gauss, $\mathfrak{M}/\pi R^2$ in an unstable cloud must exceed 2.4×10^{-2} gm/cm^2, some ten times the value found in equation (49), corresponding to an extinction of 11 magnitudes if the ratio of gas to dust has its usual value. The extinction through the center would exceed 40 mag if the relative central concentration of a magnetofluid cloud, just before collapsing, were about the same as for a cloud with B equal to zero. With a mean density of 20 atoms/cm^3 the radius would be at least 170 pc, and the mass $10^7 \, \mathfrak{M}_\odot$. Even if a mean density of 10^2 atoms/cm^3 were assumed, the radius would be at least 34 pc and the mass $4 \times 10^5 \, \mathfrak{M}_\odot$.

Such very large, heavily obscuring clouds should be very noticeable, if present. According to Blaauw (1962), extinctions of 5 to 8 magnitudes are present in regions extending over some 5 pc. As emphasized by Cameron (1962), such clouds, which have a density in the neighborhood of 10^3 atoms/cm^3, could contract gravitationally across a field of 2×10^{-5} gauss, but it is difficult to understand how these clouds could ever have formed at all if the magnetic field had been so great before compression began. In view of the many complexities in this subject, one cannot draw firm conclusions, but the absence of the extended highly absorbing regions which would be required for instability of H I regions at their typical density, if B is assumed to be 2×10^{-5} gauss, may be regarded as evidence in favor of a substantially weaker magnetic field (Spitzer 1962*b*). If B is 10^{-6} gauss, the magnetic field has no large effect on the instability of the interstellar clouds, since with this field, $B^2/8\pi$ does not much exceed the interstellar gas pressure and may be somewhat less.

We consider next the processes occurring during the collapse of a magnetofluid protocluster. In the absence of a magnetic field we have seen that fragmentation occurs until the gas has split up into elements of about $1 \, \mathfrak{M}_\odot$. This process can be greatly modified by the presence of a magnetic field. If in inequality (53) we express R in terms of the mass $\mathfrak{M}$ and the mean density $\bar{\rho}$ for a spherical condensation, we obtain

$$\mathfrak{M} > \mathfrak{M}_c \equiv 0.057 \frac{B^3}{G^{3/2}\bar{\rho}^2} . \tag{54}$$

As the cloud collapses, B varies as $1/R^2$, where R is the cloud dimension perpendicular to **B**. If the collapse is entirely radial, ρ varies as $1/R^3$, B^3/ρ^2 is constant, and $\mathfrak{M}_c$ does not change. Hence, for a spherical collapse, fragmentation into masses less than $\mathfrak{M}_c$ is not possible.

There are two ways in which a protocluster may manage to condense into stars of relatively small mass. The first, recently pointed out by Mestel (1962), relies on a non-spherical collapse to decrease B^3/ρ^2. The second, suggested by Mestel and Spitzer (1956), relies on the separation of magnetic field and gas when the fractional ionization becomes sufficiently low.

The physical process underlying the first mechanism is easily explained, although the details have not as yet been accurately analyzed. If a protocluster with $C \gg V_A$ becomes unstable, the magnetic field will be a small perturbation initially and fragmentation will proceed with spherical symmetry, at least on the average. The ratio of B^3 to ρ^2 will remain constant until the mass of the fragment approaches $\mathfrak{M}_c$ computed from equation (54). At this stage magnetic forces will slow down the contraction across the field and as a result B^3/ρ^2 will decrease. In the next fragmentation the separate masses will probably be somewhat lower than would correspond to the initial values of $\mathfrak{M}_c$. It is not clear whether the flattened disks which start to contract will fragment into smaller disks or into spheres, nor is it clear how far fragmentation can continue beyond this stage.

The second mechanism, which has been analyzed in considerable detail, is based on the fact that the magnetic lines of force have no direct effect on the neutral particles but influence only the electrons and positive ions. Under conditions of low relative ionization, the ions and electrons gyrating around a particular line of force can drift at appreciable rates through the neutral gas. Thus the neutral gas can contract inward, leaving the magnetic field and the ionized gas behind.

A simple analysis of this process will be given here (Spitzer, 1958*b*). Let us suppose that a cloud is contracting isothermally and has already decreased its radius by a factor of 10 and increased its density and its internal magnetic field by factors of 10^3 and 10^2, respectively. The compression of the field will produce a force F/cm^3 on the ions and electrons, given approximately by

$$F = \frac{B^2}{8\pi R}, \tag{55}$$

where R is the cloud radius. If we neglect inertial forces on the ions, then F is balanced by the frictional force produced by collisions between ions and neutral atoms. Let v_D be the relative drift velocity of ions with respect to the atoms and n_i and n_H the particle densities of ions and hydrogen atoms. If σ is the collision cross-section and v the relative velocity, essentially the thermal speed of the hydrogen atoms, then we have

$$F = n_i n_{\mathrm{H}} \langle \sigma v \rangle m_{\mathrm{H}} v_D , \tag{56}$$

where $\langle\sigma v\rangle$ denotes an average over all relative speeds; we have assumed that the momentum transferred per collision of a hydrogen atom with a heavy positive ion is $m_{\rm H}v_D$. Elimination of F from these two equations yields

$$v_D = \frac{B^2}{8\pi R n_i n_{\rm H}\langle\sigma v\rangle m_{\rm H}}. \tag{57}$$

To evaluate v_D we shall assume that the internal magnetic pressure $B^2/8\pi$ just balances the gravitational self-attraction of the cloud. If we consider a cylindrical configuration of radius R with the magnetic field lines parallel to the axis, then the assumed balance of gravitational and magnetic forces on the gas as a whole yields the condition

$$\frac{B^2}{8\pi R} = \frac{2\rho G\mathfrak{M}}{R}, \tag{58}$$

where $\mathfrak{M}$ is the total mass of gas per unit length of the cylinder, equal to $\pi R^2\rho$, and where the density ρ equals $n_{\rm H}m_{\rm H}$; variations of density with distance from the axis are ignored here. Substituting equation (58) into equation (57) yields for R/v_D, which we define as the time constant, t_c, for ambipolar diffusion (Spitzer 1962*b*)

$$t_c = \frac{\langle\sigma v\rangle}{2\pi G m_{\rm H}}\frac{n_i}{n_{\rm H}}. \tag{59}$$

Osterbrock (1961) has shown that for encounters between protons and hydrogen atoms at relatively low energies, corresponding to temperatures of about 100° K, σ has the relatively high value of 10^{-14} cm^2. In the normal H I cloud, $n_i/n_{\rm H}$ is about 5×10^{-4}, yielding a value of 2×10^{10} years for t_c. In a contracting cloud, however, n_i may be much reduced. The extinction through the cloud will amount to about 10^2 mag, for the conditions under consideration, and there will be essentially no ultraviolet light to produce ionization. The ion density will be determined primarily by the rate of recombination. Radiative recombination is relatively slow, producing a linear increase of $1/n_i$ with time amounting to 1 cm^3 every 3000 years. If the ions are molecular, dissociative recombination may occur, with a rate several orders of magnitude greater. When the ion density has fallen appreciably, recombination presumably occurs primarily at the surface of the grains, for which the time constant is about 5×10^4 years at a hydrogen density of 2×10^4 cm^{-3}, if the ions are assumed to be C^+, and T is again set equal to 100° K. Unless T is less than 10° K, n_i will drop to a very low value in 10^6 years. If $n_i/n_{\rm H}$ is less than 5×10^{-9}, a reduction by 10^{-5} from its value in the absence of recombination, magnetic lines of force could move out of the cloud in 2×10^5 years, a time interval comparable with the time that would be required for the cloud to fall freely to the center from the compressed state considered here, if the magnetic field were zero.

So large a reduction in n_i is difficult to achieve if the intensity of cosmic rays is appreciable (Cameron 1962). The intensity of cosmic rays measured at the Earth would maintain an ion density of about 6×10^{-4} per cm^3, if recombination occurs at the surface of the grains; if the ratio of grains to gas is constant, this ion density is independent of the gas density. For the hydrogen density considered above, $n_i/n_{\rm H}$ is 3×10^{-8} for this ion density, and t_c is about 10^6 years. A number of influences may either increase or decrease t_c. On the one hand, the cosmic-ray particles with energies below a few Bev, which were ignored in the above estimate, may increase the ionization rate by an order of magnitude

(Cameron 1962) or more. On the other hand, the ions produced (probably mostly H_3^+, which is formed rapidly in collisions between H_2^+ and H_2) may disappear by dissociative recombination at a rate sufficiently fast to reduce n_i below 10^{-4} per cm^3; the rate coefficient for this process is uncertain. Moreover, if the cloud contracts by another factor ten in radius, increasing the particle density to 2×10^7 cm^{-3}, the amount of material down to the center of the cloud will be about 50 gm/cm^2 and the cosmic rays will begin to be absorbed. According to Cameron (1962), the rate of ionization produced by K^{40} radioactivity at the present time is about three orders of magnitude less than that produced by the cosmic rays measured at the Earth. In the absence of cosmic rays, the rate of separation of ions and neutrals would become relatively rapid, provided that some charged grains do not themselves carry enough current to maintain the magnetic field throughout the cloud (Spitzer 1963).

The separation of magnetic field and ions from the neutrals in a cloud is about as uncertain, at the present time, as the process of continued fragmentation in the presence of an appreciable magnetic field. We are fairly sure from the observational evidence that one of these two processes, or perhaps some other unsuspected mechanism, must somehow make possible the condensation of stars from a cloud of gas in which a magnetic field is initially present. The later stages of evolution of a protocluster, when fragmentation ceases and the fragments contract into stars, are even more uncertain than the earlier phases. Among the many unsolved problems is the role of angular momentum, which is considered in the following subsection.

4.5. Angular Momentum and Turbulence

In the earliest discussions of star formation it was realized that angular momentum posed serious problems in the theory of condensation. In the absence of external torques, the rotational velocity of a mass of gas varies inversely as the radius R. Stars of early spectral type are observed to rotate at speeds of about 100 km/sec. Since the present mean density of these stars exceeds the density of interstellar clouds by a factor of at least 10^{21}, their radii are less by a factor of 10^7 than they were before contraction began. There are two possibilities. Either the initial angular momentum per unit mass was very low, corresponding to the extremely low velocity of 1 cm/sec, or else the angular momentum was decreased by some process. Since the turbulent velocities in the interstellar medium are probably at least 0.1 km/sec in a cloud with a mass of several $\mathfrak{M}_\odot$, it seems unlikely that much material can be sufficiently quiescent to condense into stars without loss of angular momentum; in particular, the rate of star formation derived in the first section seems quite inconsistent with this picture. Hence we are driven to the conclusion that the angular momentum per unit mass is decreased by some four orders of magnitude in the course of contraction.

In the absence of a magnetic field, disposal of so much angular momentum is difficult. The most powerful mechanism in this respect is that proposed by von Weizsäcker (1947), who has applied modern concepts of turbulence to the problem. The general picture is that a rotating protostar forms a concentration in a gaseous medium that is in turbulent motion. As a result of turbulent viscosity, the protostar is slowed down and contracts gravitationally, while the angular momentum is communicated to the surrounding medium, which moves outward. The medium is assumed to be replenished continually from the outer layers of the protostar. The energy required to drive the turbulent motions is assumed to come from the contraction. Only a small fraction of the initial mass is assumed to contract to a star; most of the mass escapes to infinity, carrying practically all the angular momentum. The gain in energy of the escaping mass is less, of course, than the energy released from the contraction.

Von Weizsäcker has given a rough estimate of the time constant for this process in terms of the *mixing length* l, defined as the distance traveled by a turbulent eddy before it becomes slowed down by the surrounding fluid and loses its identity. If R_0 is the cloud radius, then the number of steps required to go a net distance R_0 is $(R_0/l)^2$, where we assume that l is somewhat less than R_0. Since the time for a single step is l/v, where v is the root-mean-square turbulent velocity, the time τ for a fluid element to reach the surface is roughly equal to $(R_0/l)^2$ times l/v, or

$$\tau = \frac{kR_0^2}{lv}, \tag{60}$$

where k is a constant of order of magnitude unity. A detailed analysis by Lüst (1952) for a cloud rotating about a central mass gives essentially equation (60), apart from numerical factors of order unity. The time required to reduce the angular momentum by a factor of 2 should about equal the value of τ found from this equation. If R_0 is 1 pc, v is 1 km/sec and R_0/l is 5 (the value assumed by von Weizsäcker), then τ becomes 5×10^6 years. Evidently to reduce the angular momentum by orders of magnitude would require more than 10^7 years, although a decrease of τ in the process of contraction might shorten the interval somewhat. Angular momentum may also be carried off by particle ejection from the surface, a process considered by Schatzman (1954*a*). It is doubtful whether this stellar-type process can begin sufficiently early in the life of a protostar to facilitate contraction to stellar size by first disposing of the excess angular momentum.

The presence of turbulence would also affect the rate of condensation. We have already noted that if the turbulent velocity exceeds the thermal velocity a greater mass is needed within a fixed radius to yield gravitational instability. Schatzman (1954*b*) has suggested that fragmentation into smaller masses may be hindered by turbulence. A relatively simple argument shows that this will not occur if the turbulent velocity remains constant during the contraction. In equation (46), a criterion for gravitational instability, kT/m may be replaced by C^2 (the square of the sound velocity) to which must be added the mean-square turbulent velocity within the region in question. This latter quantity will not exceed its value for the entire cloud and will generally be less. Thus as ρ increases, R_m decreases at least as rapidly as $\rho^{-1/2}$, which is a sufficient condition for fragmentation to occur. It may be expected that if the temperature remains constant, the turbulent velocity will not increase much as the contraction proceeds, since the dissipation increases very rapidly when the velocity dispersion rises much above the sound velocity.

McCrea (1961) has suggested that the condensation of individual stars will be facilitated if the turbulence is strongly supersonic. McCrea pictures the cloud as composed of individual elements, which he calls "floccules," that move nearly independently and form protostars by mutual collisions. He points out that the protostars so formed may have much less angular momentum than if they had condensed from an initially rotating uniform gas cloud. Disposing of the residual angular momentum of each such protostar is apparently still a problem.

Since it now seems likely that an appreciable magnetic field is present in interstellar space, our concepts of turbulence in an interstellar cloud must be appreciably modified. In particular, the disposal of the angular momentum may be a direct consequence of the magnetic field. In principle, even a very weak magnetic field could ultimately stop the rotation of some of the gas within a cloud if the conductivity were assumed infinite. If a protostar is joined to distant matter by magnetic lines of force, continued rotation will continue to twist the lines and increase the magnetic stresses until the rotation is completely stopped.

The important problem, of course, is whether this retardation occurs rapidly enough to be important in a contracting protostar. Several analyses of magnetic retardation have been carried out. Lüst and Schlüter (1955) have treated the case of a rotating star whose angular momentum is transferred outward by means of a nearly force-free magnetic field. Significant retardation is found in a time of the order of 10^7 years.

The analysis by Ebert *et al.* (1960), more relevant to the problem of star formation, considers the transfer of angular momentum away from a rotating cloud of radius R; a uniform magnetic field parallel to the axis of rotation is assumed. The twisting of the lines of force travels outward at the Alfvén speed V_A accelerating the interstellar material within a radius R up to the same angular velocity as the cloud. Since the amount of material so accelerated per second is $\pi\rho R^2 V_A$, if ρ is the interstellar density, the time constant for retardation of the cloud rotation is given by

$$\tau = \frac{c\,\mathfrak{M}}{\pi\rho R^2 V_A}, \tag{61}$$

where $\mathfrak{M}$ is the mass of the cloud and c is a numerical constant which Ebert finds to be 0.4. The rotational velocity will decrease with time t as $\exp(-t/\tau)$ in this simple picture.

When equation (61) is applied to clouds in various stages of contraction and fragmentation, Ebert concludes that within 10^7 years a reduction in angular momentum by an order of magnitude may be expected if the initial magnetic field is 10^{-5} gauss. Mestel (1962) points out that this time is comparable to the time of free fall for an initial density of 20 H atoms/cm^3 and concludes that a collapsing cloud probably does not have enough time to dispose of much angular momentum in this particular way. As Mestel emphasizes, however, if the magnetic field has a component perpendicular to the axis of rotation, the cloud may be supported entirely by magnetic and centrifugal forces and will have a much longer time available for loss of its angular momentum by magnetic stresses. This longer time of suspended contraction might also facilitate the drift of the ions and the magnetic field out of the cloud. The rate of contraction in such a case might well be determined by the rate at which either angular momentum or magnetic flux is lost. While detailed proof is lacking, we may reasonably adopt the working hypothesis that a magnetic field can effectively reduce the angular momentum and permit the contraction of the cloud into stars.

4.6. Origin of Associations

A detailed dynamical theory of star formation including effects of turbulence, magnetic fields, and radiation would presumably also explain the mass and velocity distributions of newly formed stars, the observed frequency of binary stars, and, in particular, the tendency for expanding associations. There is at present no detailed theory of formation of stellar associations with outward stellar velocities, but several suggestions have been put forward.

Zwicky (1953) and McCrea (1955) have suggested that the stars were originally bound in a stable configuration by a much greater mass of gas that was subsequently swept away. One objection to this point of view is the relatively high value of about 10 km/sec observed for the expansion velocities in the stellar associations. As Oort (1954) has pointed out, these velocities must be about equal to the random velocities of the stars before the gas escaped. From the virial theorem it follows that these velocities correspond to a rather compact mass—$10^3\ \mathfrak{M}_\odot$ within a radius of 0.04 pc, for example. It is rather surprising that an extended cloud should contract to so small a radius in view of the fragmentation into protostars that presumably begins much earlier. It is conceivable, however, that friction between the protostars and the residual gas might lead to such a marked general contraction of the protocluster.

A second objection is that the gas must be taken away rather rapidly if the velocities of the stars are to remain unchanged. If the gas moves away slowly, the stellar orbits will change adiabatically, the kinetic energy decreasing, and bound orbits remaining bound. The stars must be assumed to move at 10 km/sec. To get rid of the gas fast enough requires expansion velocities of this order close to the cluster. Neither of these objections seems insuperable, but on the whole this picture is somewhat unlikely.

Schatzman (1954*b*) has proposed that the stars escaping from associations are those which have been accelerated to escape velocities by mutual encounters during the contraction phase. His analysis is of necessity somewhat idealized, but a more complete investigation, taking into account magnetic fields, turbulence, and radiative effects, might indicate that this effect can be important. More generally, however, this picture suffers from the same objection cited above: to obtain the observed high velocities of expansion, the contracting protocluster must be unexpectedly compact. Moreover, it is not clear observationally that the massive, tightly bound clusters required by this theory at the center of each association are consistent with observations.

The most promising explanation of observed associations would appear to be Oort's (1954). He suggests that one or more early-type stars begin to shine before the contraction and fragmentation of the rest of the cloud has progressed very far. The ionization and heating of the surrounding gas will then accelerate material outwards, as proposed by Oort and Spitzer (1955). In this way masses of gas may be given substantial outward velocities before they condense into stars. In accordance with the discussion in the preceding section, an outward velocity of 10 km/sec appears a reasonable expectation on the basis of these assumptions. A related suggestion has been put forward by Öpik (1953), who suggests that associations are produced by star formation in the expanding shell of gas produced by a supernova explosion; star formation would presumably not occur until the shell had decelerated to a velocity of about 10 km/sec (see Sec. 3.5).

As a variant of this theory, Oort and others have suggested that the increased pressure of the heated gas could trigger a series of star-forming condensations. We have already seen that one-dimensional compression of an isolated cloud of matter cannot lead to gravitational instability of the cloud as a whole. More specifically, we may consider the expanding shell around the Orion Nebula, analyzed by Savedoff (1956). From his estimate of 5 atoms/cm^3 for the initial density and a compression ratio of 200 across the shock, we obtain a density of 10^3/cm^3 in the compressed region between the shock and the ionization front. From equation (30) we find that for T equal to 100° K, the minimum thickness of the slab that is gravitationally unstable is 2.5 pc, an order of magnitude greater than the thickness of the compressed region. Hence the shell is gravitationally stable, a conclusion emphasized by Dibai (1958).

There are a number of ways in which an expanding cloud of dense gas can affect star formation in the surrounding region. Oort (1954) has pointed out that when a cloud is strongly compressed the angular momentum may be decreased, facilitating star formation. Dibai (1958) suggests that if a globule with a density much higher than in the surrounding gas is present in the cloud, the passing shock wave will compress the globule and facilitate its gravitational contraction. Star formation might also be triggered by the effect of radiation pressure tending to force grains together, as suggested by Harwit (1962) for clouds of gas and dust in an O star association, or by Öpik (1953) for an expanding shell of gas and dust around a supernova. Uncertainty in the conditions under which these processes might occur makes it difficult to evaluate all these ideas.

It is tempting to infer that the high-velocity O and B stars with velocities ranging up to about 100 km/sec discussed in Section 1 may have condensed from a few clouds that have been accelerated to such high velocities. The high fraction of such stars, about a quarter of the early-type stars, is unfavorable to this hypothesis. Alternatively, as suggested by Blaauw (1961), a high-velocity star may result when one component of a close binary becomes a supernova, releasing the other component at its high orbital velocity. It seems somewhat unlikely, however, that the remarkable star pair, AE Aur and μ Col, can be explained in this way. The fact that the momenta of these two stars appear to be rather precisely equal and opposite suggests some sort of stellar fission.

4.7. Summary

We close this section with a tentative description of how interstellar material may condense into new stars. Despite substantial progress in recent years, it is evident that our understanding of star formation is still very fragmentary. The following outline of the process of star formation is, therefore, quite speculative and is given only to provide a coherent framework to hold together and summarize the various concepts presented above.

We may suppose that gas is initially distributed in the galactic plane, in part expelled from stars, in part condensing into the plane from the corona or from intergalactic matter. There is no particular reason why this gas should be uniform in its distribution or physical state, and in fact there are numerous reasons to expect inhomogeneities to be present. Clouds will form and will grow by collisions with other clouds and by accretion of gas from the intercloud medium. The temperature of such H I regions will tend to be low, and the formation of H_2 molecules in the denser clouds, facilitated by the higher relative number of grains, will reinforce this tendency.

When a large cloud or group of clouds surpasses a certain limiting mass, it starts to collapse gravitationally under the pressure of the surrounding material. If we assume pressure equilibrium with p_m equal to 1.4×10^{-13} dynes/cm², then for a typical large H I cloud with a mean density of 20 H atoms/cm³ (corresponding to a temperature of 125° K), collapse begins if the mass exceeds $10^4\,\mathfrak{M}_\odot$. Thus the normal process of condensation produces a group of stars, either a cluster or association.

When a cloud begins to collapse under gravitation, perhaps because of a collision with one or more small clouds, or because of compression by a passing shock wave, the initial rate of gravitational contraction will be slow. As the density increases, however, the contraction becomes more rapid and condensations of smaller scale start to become unstable. Fragmentation into smaller masses will then proceed rapidly until the critical mass $\mathfrak{M}_c$ set by the magnetic field is reached. We shall assume arbitrarily that this magnetic field is 10^{-6} gauss initially; if the field strength before gravitational collapse significantly exceeds this value, the general picture of star formation presented here will require very substantial revision. With this assumed initial value for B, the critical mass is 1.5×10^3 $\mathfrak{M}_\odot$ for an assumed initial density of 20 H atoms/cm³. Hence for these assumed parameters the critical mass is reached when the cloud fragments into about seven separate condensations. The time required to reach this stage, once gravitational instability has been reached, should somewhat exceed 10^7 years, the time for an isolated sphere of the assumed initial density to collapse under free fall.

Beyond this stage the condensation becomes difficult to follow, since a number of processes are under way simultaneously. Because of the magnetic forces, the fragments may tend to become flattened disks. The centrifugal force may also become important at about this same time, and the combination of magnetic and centrifugal forces may hold up the collapse. During such an interim period of suspended contraction, the magnetic field, and with it the charged particles, may slowly diffuse out of the cloud while at the same time angular momentum is carried out of the cloud by magnetic stresses. Alternatively, fragmentation of flattened disks may continue without serious interruption until masses of stellar size are formed. The collapse of each fragment may even continue catastrophically until the individual protostars have radii of about an astronomical unit and the objects are essentially stellar.

There are so many uncertainties in this picture that at present we do not really have a theory of star formation. In a general way, however, we can at least visualize some of the important physical processes which act in the formation of stars from interstellar clouds.

REFERENCES

Adams, W. S. 1943, *Ap. J.*, **97**, 105.
———. 1949, *ibid.*, **109**, 354.
Aller, L. H. 1946, *Pub. A.S.P.*, **58**, 165.
Aller, L. H., and Liller, W. 1959, *Ap. J.*, **130**, 45.
Ambartsumian, V. A. 1940, *Bull. Abastumani*, **4**, 17.
———. 1947, *Stellar Evolution and Astrophysics* (U.S.S.R.: Erevan).
———. 1949, *A.J. (U.S.S.R.)*, **26**, 3.
———. 1955, *Observatory*, **75**, 72.
———. 1958*a*, *Rev. Mod. Phys.*, **30**, 944 (*Proc. I.A.U. Symp. No. 8*).
———. 1958*b*, *Acad. Sci. Armen. S.S.R.* (Phys.-Math. Series), **11**, No. 5, 9.

Ambartsumian, V. A., and Gordeladse, S. G. 1938, *Bull. Abastumani*, **2**, 37.
Anand, S. P. S., and Kushwaha, R. S. 1962, *Ann. d'ap.*, **25**, 118.
Andrillat, Y., and Andrillat, H. 1959, *Ann. d'ap.*, **22**, 104.
Axford, W. I. 1961, *Phil. Trans. R. Soc. London, A*, **253**, 301.
Baade, W. 1944, *Ap. J.*, **100**, 137.
Batchelor, G. K. 1950, *Proc. R. Soc. London, A*, **201**, 405.
Beals, C. S. 1936, *M.N.*, **96**, 661.
Bel, N., and Schatzman, E. 1958, *Rev. Mod. Phys.*, **30**, 1015 (*Proc. I.A.U. Symp. No. 8*).
Biermann, L. 1955, *Gas Dynamics of Cosmic Clouds* (*Proc. I.A.U. Symp. No. 2*) (Amsterdam: North-Holland Publishing Co.), p. 212.
Biermann, L., and Davis, L. 1958, *Zs. f. Naturforsch.*, **13a**, 909.
Biermann, L., and Schlüter, A. 1954, *Zs. f. Naturforsch.*, **9a**, 463.
Blaauw, A. 1952*a*, *B.A.N.*, **11**, No. 433, 405.
———. 1952*b*, *ibid.*, **11**, No. 436, 459.
———. 1956*a*, *Ap. J.*, **123**, 408.
———. 1956*b*, *Pub. A.S.P.*, **68**, 495.
———. 1958, *Ric. Astr. Vatican Obs.*, **5**, 105; *Pontif. Acad. Scient. Scripta Varia*, **16**, 105.
———. 1961, *B.A.N.*, **15**, No. 505, 265.
———. 1962, *Interstellar Matter in Galaxies*, ed. L. Woltjer (New York: W. A. Benjamin, Inc.), p. 107.
Blaauw, A., and Morgan, W. W. 1953, *Ap. J.*, **117**, 256.
———. 1954, *ibid.*, **119**, 625.
Bok, B. J. 1948, Centennial Symposia. *Harvard Obs. Monograph* No. 7, p. 53.
Bok, B. J., Lawrence, R. S., and Menon, T. K. 1955, *Pub. A.S.P.*, **67**, 108.
Bonnor, W. B. 1956, *M.N.*, **116**, 351.
Bonsack, W. K., and Greenstein, J. L. 1956, *Pub. A.S.P.*, **68**, 249.
Burbidge, G. R. 1958, *Pub. A.S.P.*, **70**, 83.
———. 1960, *Die Entstehung von Sternen durch Kondensation diffuser Materie* (Heidelberg: J. Springer).
Burbidge, G. R., Gould, R. J., and Pottasch, S. R. 1963, *Ap. J.*, **138**, 945.
Burgers, J. M. 1946, *Proc. Akad. Sci. Amsterdam*, **49**, 589, 600.
Cameron, A. G. W. 1962, *Icarus*, **1**, 13.
Chandrasekhar, S. 1951, *Proc. R. Soc. London, A*, **210**, 26.
———. 1954, *Ap. J.*, **119**, 7.
———. 1955, *Vistas in Astronomy*, ed. A. Beer (London: Pergamon Press), **1**, 344.
Chandrasekhar, S., and Fermi, E. 1953*a*, *Ap. J.*, **118**, 113.
———. 1953*b*, *ibid.*, **118**, 116.
Chandrasekhar, S., and Wares, G. W. 1949, *Ap. J.*, **109**, 551.
Clark, B. G., Radhakrishnan, V., and Wilson, R. W. 1962, *Ap. J.*, **135**, 151.
Courant, R., and Friedrichs, K. O. 1948, *Supersonic Flow and Shock Waves* (New York: Interscience Publishers, Inc.).
Courtès, G., Cruvellier, P., and Pottasch, S. R. 1962, *Ann. d'ap.*, **25**, 214.
Cowling, T. G. 1956, *M.N.*, **116**, 114.
———. 1957, *Magnetohydrodynamics* (New York: Interscience Publishers).
Cugnon, P. 1963, *Bull. Soc. R. Sci. Liège*, **32**, 228.
Daub, C. T. 1963, *Ap. J.*, **137**, 184.
Davies, R. D. 1956, *M.N.*, **116**, 443.
Davies, R. D., Slater, C. H., Shuter, W. L. H., and Wild, P. A. T. 1960, *Nature*, **187**, 1088.
Davies, R. D., Verschuur, G. L., and Wild, P. A. T. 1962, *Nature*, **196**, 563.
Davis, L., and Greenstein, J. L. 1951, *Ap. J.*, **114**, 206.
Dibai, E. A. 1958, *A.J.* (*U.S.S.R.*), **35**, 469 (*Soviet Astron.—A.J.*, **2**, 429).
Dicke, R. H. 1962, *Science*, **138**, 653.
Dirac, P. A. M. 1938, *Proc. R. Soc. London, A*, **165**, 199.
Dokuchaeva, O. D. 1959, *A.J.* (*U.S.S.R.*), **36**, 461 (*Soviet Astron.—A.J.*, **3**, 451).
Donn, B. 1951, *A.J.*, **56**, 124.
Earl, J. A. 1961, *Phys. Rev. Letters*, **6**, 125.
Ebert, R. 1955*a*, *Zs. f. Ap.*, **36**, 222.
———. 1955*b*, *ibid.*, **37**, 217.
———. 1957, *ibid.*, **42**, 263.
Ebert, R., Hoerner, S. von, and Temesváry, S. 1960, *Die Entstehung von Sternen durch Kondensation diffuser Materie* (Heidelberg: J. Springer).
Emden, R. 1907, *Gaskugeln* (Leipzig and Berlin: B. G. Teubner).
Fermi, E. 1949, *Phys. Rev.*, **75**, 1169.
Fessenkov, V. G., and Razhkovsky, D. A. 1952, *A.J.* (*U.S.S.R.*), **29**, 381.
Field, G. B. 1962, *Interstellar Matter in Galaxies*, ed. L. Woltjer (New York: W. A. Benjamin, Inc.), 183.
Fricke, W. 1954, *Ap. J.*, **120**, 356.
Frieman, E. A. 1954, *Ap. J.*, **120**, 18.
Gaustad, J. E. 1963, *Ap. J.*, **138**, 1050.
Gershberg, R. E. 1962, *A.J.* (*U.S.S.R.*), **38**, 819 (*Soviet Astron.—A.J.*, **5**, 626).

Ginzburg, V. L. 1958, *Progress in Elementary Particle and Cosmic Ray Physics* (Amsterdam: North-Holland Publishing Co.), **4**, chap. 5, 339.
Goldsworthy, F. A. 1958, *Rev. Mod. Phys.*, **30**, 1062 (*Proc. I.A.U. Symp. No. 8*).
———. 1961, *Phil. Trans. R. Soc. London, A*, **253**, 277.
Gould, R. J., Gold. T., and Salpeter, E. E. 1963, *Ap. J.*, **138**, 408.
Greenstein, J. L. 1937, *Harvard Ann.*, **105**, 359.
———. 1946, *Ap. J.*, **104**, 414.
Greenstein, J. L., MacRae, D. A., and Fleischer, R. 1956, *Pub. A.S.P.*, **68**, 242.
Grzedzielski, S., and Stepien, K. 1963, *Acta Astr.*, **13**, 143.
Haddock, F. T. 1957, *Radio Astronomy* (*Proc. I.A.U. Symp. No. 4*) (Cambridge: Cambridge University Press), p. 192.
Hagen, J. P., Lilley, A. E., and McClain, E. F. 1955, *Ap. J.*, **122**, 361.
Hall, J. S. 1949, *Science*, **109**, 166.
Hall, J. S., and Mikesell, A. H. 1950, *Pub. U.S. Naval Obs.*, Vol. **17**, Part I.
Harwit, M. 1962, *Ap. J.*, **136**, 832.
Hatanaka, T., Unno, W., and Takebe, H. 1961, *Pub. Astr. Soc. Japan*, **13**, 173.
Hayakawa, S. 1960, *Pub. Astr. Soc. Japan*, **12**, 110.
Helfer, H. L. and Tatel, H. E. 1959, *Ap. J.*, **129**, 565.
Henry, J. 1958, *Ap. J.*, **128**, 497.
Herbig, G. H. 1958, *Ric. Astr. Vatican Obs.*, **5**, 127; *Pontif. Acad. Scient. Scripta Varia*, **16**, 127.
Hiltner, W. A. 1949, *Science*, **109**, 165.
———. 1951, *Ap. J.*, **114**, 241.
Hoyle, F. 1953, *Ap. J.*, **118**, 513.
Hoyle, F., and Lyttleton, R. A. 1939, *Proc. Cambridge Phil. Soc.*, **35**, 405, 592.
Huang, S. 1954, *A.J.*, **59**, 137.
Hulst, H. C. van de. 1949, *Rech. Astr. Obs. Utrecht*, Vol. **11**, Part 2.
———. 1955, *Gas Dynamics of Cosmic Clouds* (*Proc. I.A.U. Symp. No. 2*) (Amsterdam: North-Holland Publishing Co.), p. 111.
———. 1958, *Rev. Mod. Phys.*, **30**, 913 (*Proc. I.A.U. Symp, No. 8*).
Hulst, H. C. van de, Muller, C. A., and Oort, J. H. 1954, *B.A.N.*, **12**, No. 452, 117.
Hunter, C. 1962, *Ap. J.*, **136**, 594.
Jeans, J. H. 1928, *Astronomy and Cosmogony* (Cambridge: Cambridge University Press), p. 345.
Kahn, F. D. 1954, *B.A.N.*, **12**, No. 456, 187.
———. 1955, *Gas Dynamics of Cosmic Clouds* (*Proc. I.A.U. Symp. No. 2*) (Amsterdam: North-Holland Publishing Co.), p. 60.
———. 1958, *Rev. Mod. Phys.*, **30**, 1058 (*Proc. I.A.U. Symp. No. 8*).
———. 1960, *Die Entstehung von Sternen durch Kondensation diffuser Materie* (Heidelberg: J. Springer).
Kahn, F. D., and Menon, T. K. 1961, *Proc. Nat. Acad. Sci.*, **47**, 1712.
Kahn, F. D., and Woltjer, L. 1959, *Ap. J.*, **130**, 705.
Kaplan, S. A. 1954, *A.J.* (*U.S.S.R.*), **31**, 31.
———. 1956, *ibid.*, **33**, 646.
———. 1957, *ibid.*, **34**, 183 (*Soviet Astron.-A.J.*, **1**, 183).
———. 1958*a*, *Interstellar Gas Dynamics* (Moscow: Phys.-Math. State Publishing House; translated by Technical Information Center, Wright-Patterson Air Force Base, Ohio).
———. 1958*b*, *Rev. Mod. Phys.*, **30**, 943 (*Proc. I.A.U. Symp. No. 8*).
Klein, R. 1954, *Ann. d'ap.*, **17**, 1.
Kumar, S. S. 1960, *Pub. Astr. Soc. Japan*, **12**, 552.
———. 1961, *ibid.*, **13**, 121.
Lambrecht, H., and Schmidt, K. H. 1958, *Wiss. Zs. Univ. Jena*, **8**, 1.
Layzer, D. 1963, *Ap. J.*, **137**, 351.
Ledoux, P. 1951, *Ann. d'ap.*, **14**, 438.
Lilley, A. E. 1955, *Ap. J.*, **121**, 559.
Lüst, R. 1952, *Zs. f. Naturforsch.*, **7a**, 87.
Lüst, R., and Schlüter, A. 1955, *Zs. f. Ap.*, **38**, 190.
Mathis, J. S. 1957, *Ap. J.*, **125**, 328.
McCrea, W. H. 1955, *Observatory*, **75**, 206.
———. 1957, *M.N.*, **117**, 562.
———. 1961, *Proc. R. Soc. London, A*, **260**, 152.
McCrea, W. H., and McNally, D. 1960, *M.N.*, **121**, 238.
McVittie, G. C. 1956, *A.J.*, **61**, 451.
Menon, T. K. 1958, *Ap. J.*, **127**, 28.
Merrill, P. W., and Wilson, O. C. 1937, *Ap. J.*, **86**, 44.
Mestel, L. 1954, *M.N.*, **114**, 437.
———. 1962, Lectures at Institute for Advanced Study, Princeton University.
Mestel, L., and Spitzer, L. 1956, *M.N.*, **116**, 503.
Mills, B. Y., Little, A. G., and Sheridan, K. Y. 1956, *Australian J. Phys.*, **9**, 218.

Minkowski, R. 1949, *Pub. A.S.P.*, **61**, 151.
———. 1958, *Rev. Mod. Phys.*, **30**, 1048 (*Proc. I.A.U. Symp. No. 8*).
Morgan, W. W., Code, A. D., and Whitford, A. E. 1955, *Ap. J. Suppl.*, **2**, No. 14, 40.
Morgan, W. W., Strömgren, B., and Johnson, J. M. 1955, *Ap. J.*, **121**, 611.
Morris, D., Clark, B. G., and Wilson, R. W. 1963, *Ap. J.*, **138**, 889.
Morrison, P., Olbert, S., and Rossi, B. 1954, *Phys. Rev.*, **94**, 440.
Münch, G. 1952, *Ap. J.*, **116**, 575.
———. 1957, *ibid.*, **125**, 42.
———. 1958, *Rev. Mod. Phys.*, **30**, 1042 (*Proc. I.A.U. Symp. No. 8*).
Münch, G., and Zirin, H. 1961, *Ap. J.*, **133**, 11.
Muller, C. A. 1959, *Paris Symposium on Radio Astronomy* (*Proc. I.A.U. Symp. No. 9*) (Stanford: Stanford University Press), p. 360.
Öpik, E. J. 1953, *Irish A.J.*, **2**, 219.
Oort, J. H. 1932, *B.A.N.*, **6**, No. 238, 249.
———. 1946, *M.N.*, **106**, 159.
———. 1954, *B.A.N.*, **12**, No. 455, 177.
———. 1960, *ibid.*, **15**, No. 494, 45.
Oort, J. H., *et al.* 1958, *Ric. Astr. Vatican Obs.*, **5**, 531; *Pontif. Acad. Scient. Scripta Varia*, **16**, 531.
Oort, J. H., and Rougoor, G. W. 1960, *Proc. Nat. Acad. Sci.*, **46**, 1.
Oort, J. H., and Spitzer, L. 1955, *Ap. J.*, **121**, 6.
Osterbrock, D. E. 1957*a*, *Pub. A.S.P.*, **69**, 227.
———. 1957*b*, *Ap. J.*, **125**, 622.
———. 1961, *ibid.*, **134**, 270.
Osterbrock, D. E., and Flather, E. M. 1959, *Ap. J.*, **129**, 26.
Pacholczyk, A. G., and Stodolkiewicz, J. S. 1959, *Bull. Acad. Polon. Sci.*, **7**, 429, 681.
———. 1960, *Acta Astr.*, **10**, 1.
Pariiskii, Y. N. 1962, *A.J.* (*U.S.S.R.*), **38**, 798 (*Soviet Astron.—A.J.*, **5**, 611).
Parker, E. N. 1953*a*, *Ap. J.*, **117**, 169.
———. 1953*b*, *ibid.*, **117**, 431.
———. 1958, *Rev. Mod. Phys.*, **30**, 955 (*Proc. I.A.U. Symp. No. 8*).
Pecker, J. C. 1950, *Ann. d'ap.*, **13**, 433.
Pecker, J. C., and Schatzman, E. 1958, *Rev. Mod. Phys.*, **30**, 1101 (*Proc. I.A.U. Symp. No. 8*).
Peters, B. 1959, *J. Geophys. Res.*, **64**, 155.
Piddington, J. H. 1955, *M.N.*, **114**, 638, 651.
———. 1957*a*, *Australian J. Phys.*, **10**, 515.
———. 1957*b*, *ibid.*, **10**, 530.
Pikelner, S. B. 1953, *Dokl. Akad. Nauk* (*S.S.S.R.*), **88**, 229.
———. 1957, *A.J.* (*U.S.S.R.*), **34**, 314 (*Soviet Astron.—A.J.*, **1**, 310).
Pikelner, S. B., and Shklovsky, I. S. 1959, *Ann. d'ap.*, **22**, 913.
Plaskett, J. S., and Pearce, J. A. 1933, *Pub. Dom. Ap. Obs. Victoria*, **5**, 167.
Platt, J. R. 1956, *Ap. J.*, **123**, 486.
Pottasch, S. R. 1956, *B.A.N.*, **13**, No. 471, 77.
———. 1958, *ibid.*, **14**, No. 482, 29.
———. 1960, *Ap. J.*, **132**, 269.
———. 1965, *Vistas in Astronomy*, ed. A. Beer (London: Pergamon Press), **6**, 149.
Roberts, M. S. 1957, *Pub. A.S.P.*, **69**, 59.
Rodgers, A. W., Campbell, C. T., Whiteoak, J. B., Bailey, H. H., and Hunt, V. O. 1960, *An Atlas of Hα Emission in the Southern Milky Way* (Mt. Stromlo Obs. Pub.).
Rouskol, E. L. 1955, *Les Particules Solides dans les Astres*, Sixth International Astrophysical Colloquium, Liége, p. 650.
Routly, P. McR., and Spitzer, L. 1952, *Ap. J.*, **115**, 227.
Safranov, V. 1960, *Dokl. Akad. Nauk* (*S.S.S.R*), **130**, 53; *Ann. d'ap.*, **23**, 979.
Salpeter, E. E. 1955, *Ap. J.*, **121**, 161.
———. 1959, *ibid.*, **129**, 608.
Savedoff, M. P. 1955, *Gas Dynamics of Cosmic Clouds* (*Proc. I.A.U. Symp. No. 2*) (Amsterdam: North-Holland Publishing Co.), p. 218.
———. 1956, *Ap. J.*, **124**, 533.
Savedoff, M. P., and Greene, J. 1955, *Ap. J.*, **122**, 477.
Savedoff, M. P., and Vila, S. 1962, *Ap. J.*, **136**, 609.
Schatzman, E. 1950, *Ann. d'ap.*, **13**, 367.
———. 1954*a*, *ibid.*, **17**, 300.
———. 1954*b*, *ibid.*, p. 382.
———. 1955, *Gas Dynamics of Cosmic Clouds* (*Proc. I.A.U. Symp. No. 2*) (Amsterdam: North-Holland Publishing Co.), p. 193.
———. 1958, *Rev. Mod. Phys.*, **30**, 1012 (*Proc. I.A.U. Symp. No. 8*).
Schlüter, A. 1955, *Gas Dynamics of Cosmic Clouds* (*Proc. I.A.U. Symp. No. 2*) (Amsterdam: North-Holland Publishing Co.), p. 144.

Schlüter, A., and Biermann, L. 1950, *Zs. f. Naturforsch.*, **5a**, 237.
Schlüter, A., Schmidt, H., and Stumpff, P. 1953, *Zs. f. Ap.*, **33**, 194.
Schmidt, M. 1957, *B.A.N.*, **13**, No. 475, 247.
———. 1959, *Ap. J.*, **129**, 243.
Schwarzschild, M. 1958, *Structure and Evolution of the Stars* (Princeton: Princeton University Press).
Searle, L. 1952, *Ap. J.*, **116**, 650.
Seaton, M. J. 1951, *M.N.*, **111**, 368.
———. 1955, *Ann. d'ap.*, **18**, 188.
———. 1958, *Rev. Mod. Phys.*, **30**, 1100 (*Proc. I.A.U. Symp. No. 8*).
Shajn, G. A. 1955, *A.J.* (*U.S.S.R.*), **32**, 110, 381.
———. 1956, *ibid.*, **33**, 305.
Shklovsky, I. S. 1962, *A. J.* (*U.S.S.R.*), **39**, 209 (*Soviet Astron.—A.J.*, **6**, 162).
Spitzer, L. 1941, *Ap. J.*, **94**, 232.
———. 1942, *ibid.*, **95**, 329.
———. 1948*a*, *ibid.*, **108**, 276.
———. 1948*b*, Centennial Symposia, *Harvard Obs. Monograph* No. 7, p. 87.
———. 1951, *Problems of Cosmical Aerodynamics* (Proceedings of the First Symposium on the Motion of Gaseous Masses of Cosmical Dimensions, Paris, 1949; Central Air Documents Office, Dayton, Ohio), p. 31.
———. 1954, *Ap. J.*, **120**, 1.
———. 1956, *ibid.*, **124**, 20.
———. 1958*a*, *Ric. Astr. Vatican Obs.*, **5**, 445; *Pontif. Acad. Scient. Scripta Varia*, **16**, 445.
———. 1958*b*, *Electromagnetic Phenomena in Cosmical Physics* (*Proc. I.A.U. Symp. No. 6*) (Cambridge: Cambridge University Press), p. 169.
———. 1962*a*, *Physics of Fully Ionized Gases* (2d rev. ed.; New York: Interscience Publishers, Inc.).
———. 1962*b*, *Interstellar Matter in Galaxies*, ed. L. Woltjer (New York: W. A. Benjamin, Inc.), p. 98.
———. 1963, *Origin of the Solar System*, ed. R. Jastrow and A. G. W. Cameron (New York: Academic Press, Inc.), p. 39.
Spitzer, L., and Field, G. B. 1955, *Ap. J.*, **121**, 300.
Spitzer, L., and Savedoff, M. P. 1950, *Ap. J.*, **111**, 593.
Spitzer, L., and Schwarzschild, M. 1953, *Ap. J.*, **118**, 106.
Spitzer, L., and Skumanich, A. 1952, *Ap. J.*, **116**, 452.
Spitzer, L., and Tukey, J. W. 1951, *Ap. J.*, **114**, 187.
Stephenson, G. 1961, *M.N.*, **122**, 455.
Strömgren, B. 1939, *Ap. J.*, **89**, 526.
———. 1948, *ibid.*, **108**, 242.
Takayanagi, K., and Nishimura, S. 1960, *Pub. Astr. Soc. Japan*, **12**, 77.
Thackeray, A. D. 1948, *Observatory*, **68**, 22.
Toomre, A. 1963, *Ap. J.*, **138**, 385.
Underhill, A. B. 1950, *Pub. Dom. Ap. Obs. Victoria*, **8**, 357.
Vandervoort, P. O. 1962, *Ap. J.*, **135**, 212.
———. 1963*a*, *ibid.*, **137**, 381.
———. 1963*b*, *ibid.*, **138**, 426.
Vyssotsky, A. N. 1957, *Pub. A.S.P.*, **69**, 109.
Wade, C. M. 1958*a*, *Rev. Mod. Phys.*, **30**, 946 (*Proc. I.A.U. Symp. No. 8*).
———. 1958*b*, *Australian J. Phys.*, **11**, 388.
Weinreb, S. 1962, *Ap. J.*, **136**, 1149.
Weizsäcker, C. F. von. 1947, *Zs. f. Ap.*, **24**, 181.
Westerhout, G. 1957, *B.A.N.*, **13**, No. 475, 201.
———. 1958, *ibid.*, **14**, No. 488, 215.
Westerhout, G., Seeger, C. L., Brouw, W. N., and Tinbergen, J. 1962, *B.A.N.*, **16**, No. 518, 187.
Weymann, R. 1963, *Ann. Reviews of Astronomy and Astrophysics*, **1**, 97.
Whipple, F. L. 1946, *Ap. J.*, **104**, 1.
———. 1948, Centennial Symposia, *Harvard Obs. Monograph* No. 8, p. 109.
Wilson, O. C. 1939, *Ap. J.*, **90**, 244.
Wilson, O. C., Münch, G., Flather, E. M., and Coffeen, M. F. 1959, *Ap. J. Suppl.*, **4**, No. 40, 199.
Woerden, H. Van, Rougoor, W., and Oort, J. H. 1957, *C.R.* (*Paris*), **244**, 1691.
Woltjer, L. 1958, *B.A.N.*, **14**, No. 483, 39.
———. 1962, *Interstellar Matter in Galaxies*, ed. L. Woltjer (New York: W. A. Benjamin, Inc.), p. 88.
———. 1965, In *Galactic Structure*, ed. A. Blaauw and M. Schmidt, Vol. **5**, of *Stars and Stellar Systems* (Chicago: University of Chicago Press), chap. 23.
Zanstra, H. 1955*a*, *Gas Dynamics of Cosmic Clouds* (*Proc. I.A.U. Symp. No. 2*) (Amsterdam: North-Holland Publishing Co.), p. 70.
———. 1955*b*, *Vistas in Astronomy*, ed. A. Beer (London: Pergamon Press), **1**, 256.
Zimmerman, H. 1959, *A.N.*, **285**, 129.
Zwicky, F. 1953, *Pub. A.S.P.*, **65**, 205.

GENERAL REFERENCES

Detailed analyses and full discussions of basic observational and theoretical problems of interstellar gas dynamics appear in the proceedings of the following three symposia jointly organized by the International Astronomical Union and the International Union of Theoretical and Applied Mechanics:

Symposium on the Motion of Gaseous Masses of Cosmical Dimensions (Paris, Aug. 16–19, 1949), *I.A.U. Symposium No. 1,* edited by J. M. BURGERS and H. C. VAN DE HULST. Dayton, Ohio: Central Air Documents Office, Army-Navy-Air Force, 1951.

Gas Dynamics of Cosmic Clouds (Cambridge, England, July 6–11, 1953), *I.A.U. Symposium No. 2,* edited by J. M. BURGERS and H. C. VAN DE HULST. Amsterdam: North-Holland Publishing Co., 1955.

Proceedings of the Third Symposium on Cosmical Gas Dynamics (Cambridge, Mass., June 24–29, 1957), *I.A.U. Symposium No. 8,* Smithsonian Symposium Publ. No. 1, edited by J. M. BURGERS and R. N. THOMAS, *Rev. Mod. Phys.,* **30,** 905–1108, 1958.

In addition, a detailed analytical discussion of many of the fluid-dynamic problems important in interstellar space has been given in the following monograph:

Interstellar Gas Dynamics by S. A. KAPLAN (State Publishing House for Physico-Mathematical Literature, Moscow); translated by Technical Information Center, Wright-Patterson Air Force Base. Ohio, 1958.

Surveys of the star-formation problem with extensive references have appeared in three essays: one by G. R. Burbidge; one by F. D. Kahn; and one by R. Ebert, S. von Hoerner, and S. Temesváry. These essays, which were awarded prizes by the *Gesellschaft deutscher Naturforscher und Ärzte,* have been published by J. Springer (Heidelberg, 1960) under the title *Die Entstehung von Sternen durch Kondensation diffuser Materie.*

Many problems of interstellar gas dynamics have been considered at a conference on The Distribution and Motions of Interstellar Matter in Galaxies, held at the Institute for Advanced Study, Princeton, N.J., on April 10–20, 1961. The papers presented at this conference have been published by W. A. Benjamin (New York, 1962) under the title *Interstellar Matter in Galaxies,* edited by L. WOLTJER.

CHAPTER 2

Diffuse Nebulae

HUGH M. JOHNSON
Lockheed Palo Alto Research Laboratory

1. INTRODUCTION

STARS appear to account for most of the matter in the universe. They are organized into a hierarchy of systems from clusters of galaxies to double and single stars. At the galactic level of organization there is a minor but important residue of interstellar matter which is organized fairly coherently with the Stellar System. This material tends to share the equatorial concentration of regular galaxies, the spiral arm concentration of spiral galaxies, the young star concentrations within the arms, and the concentration of stars in clusters. In part these associations may result from effects of gravitational and magnetic fields which bind the system together; and in part the stellar and interstellar matter are directly and closely interchanged.

Most interstellar matter is in the form of gas or of solid grains, often called "dust." The gas and dust can be detected because they absorb and re-emit or scatter some of the energy radiated by stars. When these processes occur in a volume that subtends a finite solid angle, an ordinary diffuse nebula is optically defined. An undetermined percentage of nebulae are extraordinary objects: supernova remnants that may emit independently of current stellar radiation. Earlier generations of astronomers also used the generic term "nebulae" for other objects composed of unresolved stars, but this is no longer good usage.

Nebulae have been successively believed to be (*a*) ultimately resolvable as stars (seventeenth and eighteenth centuries), (*b*) a luminous fluid (W. Herschel, early nineteenth century), (*c*) ultimately resolvable as stars (Lord Rosse, middle nineteenth century), (*d*) gaseous (Huggins, late nineteenth century), (*e*) gas and dust (Slipher, Hertzsprung, early twentieth century), (*f*) distinct from an extragalactic species (Curtis 1919), and finally (*g*), in some cases, supernova remnants (Mayall and Oort 1942).

Hubble stated in 1922 that the "more detailed classification [of galactic nebulae] will depend upon form, spectrum, and relation to stars. . . ." He reported that "published results list emission spectra for thirteen diffuse nebulae as against continuous or absorption spectra for eight." Hubble's (1922*a*) classification scheme is shown on page 66.

	Galactic Nebulae			Non-Galactic Nebulae
	Planetary	Diffuse		
		Luminous	Dark	
Example	NGC 7662	NGC 1976	Barnard 86	Details omitted here

When a luminous nebula is observed with a slit spectrograph, the spectrum is generally a continuum on which bright or dark lines are superimposed. The bright lines form a characteristic "nebular spectrum," known since Huggins (1864) but in 1922 still mostly unidentified, except for hydrogen and helium. In the bright-line portion, the continuum may be very weak or seem to be absent. When the continuum is present, dark lines may be seen. The dark-line spectrum was shown by Slipher (1912) to be a replica of the spectrum of the star that presumably illuminates the nebula. When a wide-slit or slitless spectrograph is used, absorptions are no longer visible and monochromatic emissions look like a continuum when the nebular image is large compared to the length of the spectrum. For small images, the slitless spectrograph distinguishes between (monochromatic) "emission" nebulae and those with "continuous" spectra. The advantage of the slitless spectrograph in a survey is that the continuous type is fully exposed and there is no confusion with underexposure of the emission type.

The source of nebular energy was not traced directly to imbedded stars until the light of Merope, Maia, ρ Ophiuchi, $+0°1177$ (M78), and HD 200775 (NGC 7023) was shown to be reflected by nearby nebulae; the presence of grains in some nebulae and the direct agency of starlight were thus demonstrated. Hubble (1922*a*, *b*) proved the association of stars and nebulae in two ways. First, he showed that, in a list of 33 diffuse nebulae with continuous spectra, the spectral type of the associated star was always B1 or later, whereas in a list of 29 emission nebulae, the star was always of spectral type B0 or earlier. Second, there was a relation between angular extent a of a nebula from the associated star and the photographic magnitude m of the star which had the form

$$m + 4.90 \log a = 11.02 \pm 0.10 \tag{1}$$

where a is measured in minutes of arc on a Seed 30 (blue-sensitive) plate from a 1-hour exposure with an $f/5$ reflector. This agrees satisfactorily with an inverse-square illumination of randomly distributed clouds that scatter light perfectly. Relation (1) held for about 80 nebulae including continuous and emission types, which is puzzling. Failures to conform to relation (1) will be mentioned later.

The investigations outlined in the preceding paragraphs have far-reaching implications. First, the two types of nebular spectra suggest the presence of two distinct components of matter other than stars: gas and particles of unspecified size.

Second, the dependence of the spectral type of nebulae on the spectral type (or temperature) of the involved stars suggests that gas and dust are found together in interstellar space; the component which dominates the spectrum depends on the quality of the starlight. Later dynamical arguments support the well-mixed gas and dust concept

(Spitzer 1941). The alternative hypothesis that low-temperature stars are surrounded only by solid particles and high-temperature stars only by gas seems less tenable. The weakness of the continuous component in the predominantly emission-type spectra may be attributed primarily to underexposure.

Third, the dependence of the maximum radius of a nebula on the brightness of the star suggests that nebulous matter could be found to indefinite distances from stars, and that there is a true interstellar medium. This third implication was not generally accepted until Trumpler (1930) demonstrated a photographic extinction of starlight of about 0.79 mag/kpc over a large range of distances and directions in or near the galactic plane.

It became evident that the material which scattered starlight in the line of sight to a star was identical in type to that which appeared luminous where it happened to be near a bright star. Probably a similar behavior was exhibited by the gas.

The gaseous counterpart to Trumpler's interstellar matter was easily recognizable in the nonstellar absorption lines discovered by Hartmann (1904) (see chapter by Münch).

Fourth, the absence of nebulae around bright objects far from the galactic plane, such as globular clusters, showed that the interstellar medium is strongly concentrated toward the galactic plane. Diffuse nebulae at high galactic latitudes are rare, but when they are present they are found nearby.

Fifth, the results from 1912 to 1922 implied that dark nebulae differ from bright continuous nebulae only in the amount of illumination. The difference may be partly a matter of contrast, but dark nebulae must also be unusually dense dust clouds to produce the silhouette effect against star fields or bright nebulae. Because they are so faint relatively little is known from direct observation about the physical conditions in dark nebulae (but see chapter by Mrs. Lynds).

One might expect a nebula to appear wherever a star of high absolute brightness shines near the galactic plane. The absence of bright nebulae around h and χ Persei, or around Betelgeuse, for example, shows this view to be incorrect. Lacunae as well as concentrations evidently characterize the structure of the interstellar medium. Next, one may ask whether bright stars at least tend to associate with concentrations of interstellar matter.

If stars form most easily in the densest parts (Schmidt 1959) and if bright stars are short-lived (Sandage 1958) and have motions slow relative to the interstellar clouds (Plaskett and Pearce 1930), an association of bright stars with the concentrations of interstellar matter would be expected. If this is true, statistics such as the numbers of clouds per cubic parsec, collision rates, and so forth, based on a theory of random encounters between clouds and stars (Oort and van de Hulst 1946), should be viewed with caution.

Important theoretical developments followed Hubble's papers (1922). Menzel (1926) and Zanstra (1926) outlined the mechanism of photoionization in gaseous nebulae; Bowen (1927) identified the chief emissions in the nebular spectrum, aside from hydrogen and helium, as due to forbidden transitions from metastable levels only a few electron volts above the ground level in abundant ions, free electrons from photoionization of hydrogen providing the excitation energy. In 1937, Menzel and his associates initiated an intensive series of studies of physical processes in gaseous nebulae (Aller 1956, Menzel

1962), following up Bowen's discoveries. Although the theory and associated observations (see chapter 9) were primarily for planetary nebulae, many of the results are applicable to diffuse gaseous nebulae.

About 1936, a series of spectrographic and photometric observations of diffuse nebulae was made at Yerkes Observatory. In final form the Yerkes nebular spectrograph (essentially only a fast camera, but here a novel device without a collimator) scanned an area of $1\overset{\circ}{.}5 \times 6'$ of the sky (Struve, Van Biesbroeck, and Elvey 1938). Hydrogen-line emission was discovered (Struve 1941) extending over regions of several degrees in the Milky Way although bright nebulae previously known did not do so, except for the Great Loop in Orion discovered by Pickering (1895). Availability of fast ($f/1$) Schmidt cameras and fast red-sensitive emulsions after 1936 enabled observers to obtain full exposures in reasonable times on Hα, the brightest diffuse nebular emission. Thus the highest contrast possible between the background night-sky continuum and the nebula could be obtained.

Since the extended Hα-emission regions were much fainter and larger than the catalogued nebulae, they seemed to represent a more tenuous medium between the nebulae. Struve (1941), using absolute surface brightness in Hα, computed interstellar densities of about 2 atoms cm^{-3} and confirmed an early theoretical estimate of the density (Eddington 1926). Strömgren (1939), developing the theory of H II spheres, gave the ion density N_{H} and the radius of ionization s_0 as a function of the properties of the exciting star; that s_0 is defined rather sharply for a hydrogen ionization region is a special feature of this theory. A clear-cut distinction can be made between neutral (H I) and ionized (H II) hydrogen regions. The theory also explains the sharp transition between emission and reflection nebulae accompanying the transition from stellar spectral class B0 to B1. With advancing spectral class, s_0 shrinks rapidly.

Several surveys have been made with wide-field cameras equipped with narrow-band Hα filters. Details are summarized in Table 1. Of these cameras, only the 48-inch Schmidt compares in resolution with the large reflectors that had provided the earlier narrow-field blue-plate studies of many bright nebulae. None of the narrow-band cameras has achieved the spectral purity (a few angstroms) of the spectrographic surveys, but their immense advantage over spectrographs is in scanning a circular field of several degrees, instead of only a narrow strip. The photoelectric technique is not so good for extended and amorphous objects. The choice is between a laborious point-by-point mapping with large gaps (Strömgren 1951) and a large-scale integration of the light (Osterbrock and Stockhausen 1960, 1961). The combination of photoelectric calibration and photographic interpolation, often used for star clusters, has been somewhat neglected (see, however, Faulkner [1964] and Dickel [1965]). Image-tube techniques have now been applied to galactic nebulae. Combination of several techniques would improve the usefulness of the optical data on the structure of nebulae or H II regions in one or more wavelengths.

Apart from the fundamental recognition of bright nebulae as reflection and emission types, spectroscopy of the diffuse nebulae has yielded little information. Planetary nebulae with much higher surface brightnesses have provided more data. Only the Orion Nebula rivals the brightest planetary nebulae. Gaseous nebulae observed during the preparation of the *Henry Draper Catalogue* were arranged by Miss Cannon (1916) into six classes, Pa through Pf, which approximately correspond to Aller's (1956) excitation classes 3 through 10. Of 117 objects in her Table 1—*Spectra of Gaseous Nebulae, Class P*—only three are galactic diffuse nebulae, and they are all classed as Pb (low-excitation). Thus by 1916 a systematic difference between the spectra of diffuse and planetary nebulae, accompanying the difference in structure, had been recorded, but little more recent quantitative information can be found. For example, helium abundance relative to hydrogen has been published only for the nebulae listed in Table 2. Abundances of the elements in diffuse nebulae have been discussed for NGC 1976 (Aller and Liller 1959), in some possible supernova remnants (Parker 1962), in diffuse nebulae in the Magellanic Clouds (Dickel, Aller, and Faulkner 1964), 30 Doradus (Mathis 1965*b*, Faulkner and Aller 1965), and the η Carinae Nebula (Faulkner and Aller 1965). Electron temperatures and densities have been estimated for very few diffuse nebulae (cf. Sec. 3.3), including the Cygnus Loop, IC 443, and the Crab Nebula (Osterbrock 1957*a*). The Cygnus Loop and IC 443 are probably supernovae remnants and not diffuse nebulae in the usual sense,

TABLE 1

EXTENDED SURVEYS BY DIRECT PHOTOGRAPHY IN Hα

Camera	Aperture (inches)	Focal Length (inches)	Maximum Exposure (min)	Filter	Reference
Schmidt..........	48 (circ)	120	60	Plexiglas	Minkowski (1946, 1947, 1948)
Saphis Boyer O.G..	2 (sq)	3.0	240	RG2+Baird*	Courtès (1951)
Greenstein-Henyey wide-angle......	0.17 (circ)	0.34	120	Corning 2403	Sharpless and Osterbrock (1952)
Maksutov.........	20 (circ)	48	....	KC15	Fessenkov and Rozhkovsky (1953)†
Zeiss Sonar O.G....	2 (sq)	4.5	360	Corning 2462 +Baird*	Bok, Bester, and Wade (1955)
Maksutov (*a*)......	17.8 (circ)	25.4	70	KC5	Shain and Haze (1952*d*)‡
(*b*)......	17.8 (circ)	25.4	180	KC5+Baird*	Haze and Shain (1955)
Schmidt..........	2.4 (circ)	4.0	60	VR2 or Wratten 29	Gum (1955)
Flat-field Schmidt (*a*).............	6.4 (sq)	7.7	40	Chance	Morgan, Strömgren, and Johnson (1955)
(*b*).............	5.7 (circ)	7.7	360	Chance+ Baird*	Johnson (1955, 1956)
Schmidt..........	48 (circ)	120	60	Plexiglas	Sharpless (1959)
Flat-field Schmidt..	8 (circ)	8	20	OR 1	Johnson (1960*b*)§
Flat-field Schmidt..	8 (circ)	8	20	OR 1	Rodgers, Campbell, and Whiteoak (1960), Rodgers, Campbell, Whiteoak, Bailey, and Hunt (1960)‖

All observers used 103aE emulsion except Sharpless and Osterbrock, who used 103aF. In all surveys, except the 48-inch blue comparison, a yellow comparison passband was used.

Catalogues of the emission nebulae in the Magellanic Clouds have been published by K. G. Henize, 1956, *Ap. J. Suppl.*, 2, 315. A chart showing location of emission nebulosities in Messier 33 is given in Figure 1, p. 232, in Aller (1954).

* Interference filter.

† *Atlas of Gaseous-Dusty Nebulae.*

‡ *Atlas of Diffuse Gaseous Nebulae.*

§ Atlas of absolute isophotes in three passbands.

‖ *Atlas of H-Alpha Emission in the Southern Milky Way.*

TABLE 2

ABUNDANCE OF ELEMENTS (RELATIVE TO 10^4 ATOMS H)

Object	$T_e/10^4$ (° K)	He/10^4	N	O	Ne	S	Cl	Ar	Ref.
Sun	...	...	0.96	9.1	...	0.20	...	...	1
Early-type stars	...	0.16	1.5	5.9	5.2	0.30	0.016	0.081	1
Planetary nebulae	...	0.19	2.4	5.9	1.1	0.53	0.020	0.076	1, 2
IC 1569	...	0.1–0.2	...	...	...	...	...	...	3
NGC 346 (SMC)	$\geq$1.36	0.11	...	...	...	...	...	...	4, 12
	1.5	0.076–0.07	...	3	...	...	...	...	5
NGC 604 (M33)	...	0.102	...	...	...	...	...	...	6
NGC 1952 (Crab)	1.7	0.62	3.4	3.1	1.2	0.84	...	...	7
	1.0	0.45	6.1	11	6.2	3.5	...	...	7
	0.8	0.42	7.0	27	20	8.7	...	...	7
NGC 1976 (Orion)	1.0	0.117	...	...	...	...	...	...	6
	0.9	...	0.48	3.62	0.58	0.96	0.006	0.0098	8
NGC 2070 (LMC)	1.4	0.13	...	...	...	...	...	...	9
	1.0	0.14	...	2.1	...	...	...	...	6,* 13
	1.08	0.082	...	2.5	...	...	...	...	12
NGC 3372 (η Car)	1.0	0.122	...	1.8	...	...	...	...	12
NGC 4214	...	0.17	...	...	...	...	...	...	3
NGC 4278†	2	...	0.085	0.247	0.040	...	...	...	10
	1	...	0.34	2.03	0.33	...	...	...	10
LMC (average)	1.0	(0.08)	...	3.0	1.4	...	...	...	5
NGC 6514 (M20)	1.0	0.094	...	...	...	...	...	...	6
NGC 6523 (M8)	1.0	0.11	...	...	...	...	...	...	6
NGC 6822	...	~0.10	...	...	...	...	...	...	11*

* Also announces that He/H is constant in M31 for radius from 25′ to 89′.

† *E*-galaxy core.

(1) Aller (1961); (2) Aller (1964); (3) Mathis (1965*a*); (4) Aller and Faulkner (1962); (5) Dickel, Aller, and Faulkner (1964); (6) Mathis (1962); (7) Woltjer (1958); (8) Aller and Liller (1959); (9) Johnson (1959*c*); (10) Osterbrock (1960); (11) Schmidt (1962); (12) Faulkner and Aller (1965); (13) Mathis (1965*b*).

but it is then surprising that IC 443 should associate with emission-line stars (Dolidze 1960). The Crab Nebula is certainly a supernova remnant and is discussed from other viewpoints in the chapters by Friedman, Bless, Parker, and Minkowski.

The central stars of planetary nebulae, which are fast-moving stars far-flung throughout the Galaxy, create their own envelopes, as do novae, and energize them. Therefore a definite percentage of distant planetaries are seen because the line of sight often extends out of the equatorial sheet of dust. By contrast, the diffuse nebulae are associated with slow-moving stars that are frequently immersed in the thick of the dust. Hence we can distinguish diffuse nebulae over a radius hardly exceeding 2 kpc or, say, 3 per cent of the area of the galactic disk. This is too small to test adequately the predictions of the rates of change of chemical composition, density, and so forth with galactic radius in the disk (Schmidt 1959). It is also slightly too small to give definitive answers about "spiral" structure in the solar neighborhood because, although spectroscopic parallaxes of hot stars are used in optical models of the Galaxy, the excited H II regions indicate the presence of hot stars just as they trace spiral arms in M31 (Baade 1951).

The discovery of the 21-cm radiation of galactic neutral hydrogen in 1951 provided a means for piercing the galactic dust. The greater part of the galactic disk was now available for studies of the kinematics, distribution, and density of interstellar H I regions. H I had previously been unobservable because, although it constitutes about 97 per cent of all hydrogen gas (Westerhout 1958*a*, *b*), it is almost never excited to the second or higher energy levels from which a Balmer absorption line may be produced. The main disadvantage with observations of the 21-cm line is that the absence of hydrogen atoms, presence of H^+, or H_2 may be indistinguishable. The absence of hydrogen *atoms* does not necessarily imply the absence of interstellar gas. Optical studies of distant diffuse nebulae are still desirable.

High-speed particles in interstellar space interacting with magnetic fields give rise to most of the continuous radiation at radio frequencies in our Galaxy and in extragalactic systems. According to Shklovsky (1960), the concentration of relativistic electrons is a few times 10^{-11} cm^{-3}, about the same as the concentration of heavy nuclei in primary cosmic rays observed at the Earth. The intensity of their synchrotron radiation decreases toward shorter wavelengths, although the intensity of the thermal emission of an ionized diffuse nebula does not. In the centimeter range, H II regions are therefore usually seen in emission, and, in the meter range, in absorption. At the intermediate wavelength, in which the electron temperature of an H II region equals the brightness temperature of the background, the ionized region is undetectable. In combining radio and optical studies of nebulae, two difficulties are noteworthy: radio resolving power is usually inferior to optical, whereas optical penetrating power is much inferior to radio. Nevertheless the methods successfully complement one another.

Rocket and satellite researchers are now exploring new wavelength regions. A new class of nebulosities having a high surface brightness in the passband 1225 to 1350 A has already been discovered (Boggess 1961). The distribution of these nebulosities is reproduced in Figure 1. Some nearly coincide with known H II regions, but the explanation of their radiation has not yet been given. Indeed, Byram, Chubb, and Friedman (1964) argue that the results of the 1957 rocket data shown in Figure 1 are suspect; therefore, reliable rocket data for nebulae may yet be in the future. Figures 2 through 5 continue with sky maps for four wavelength regions in which diffuse nebulae have been detected. The characteristics of these regions are summarized in Table 3.

Table 4 gives a résumé of dates important for the study of galactic diffuse nebulae.

TABLE 3

DIFFUSE EMISSION NEBULAE IN FOUR SPECTRAL REGIONS

Figure	Wavelength Region of Survey	Method	Resolution	Characteristics of Region	Reference to Figure
1	1225–1350 A	Rocket	3°	Origin and spectral features unidentified	Boggess (1961)
2	3100–11000 A (6563 A)	Direct photography	1′	All atomic processes (solid grains present)	Johnson (1960*b*)
3	1 cm–3.5 m (22 cm)	Radio paraboloid	0°6	Thermal (free-free) emission on weak nonthermal background	Westerhout (1958*a*)
4	3.5 m–15 m (3.5 m)	Radio interferometer	0°8	Weak thermal absorption of strong nonthermal background	Hill, Slee, and Mills (1958)
5	(15 m)	Radio interferometer	1°4	Strong thermal absorption of strong nonthermal background	Shain, Komesaroff, and Higgins (1951)

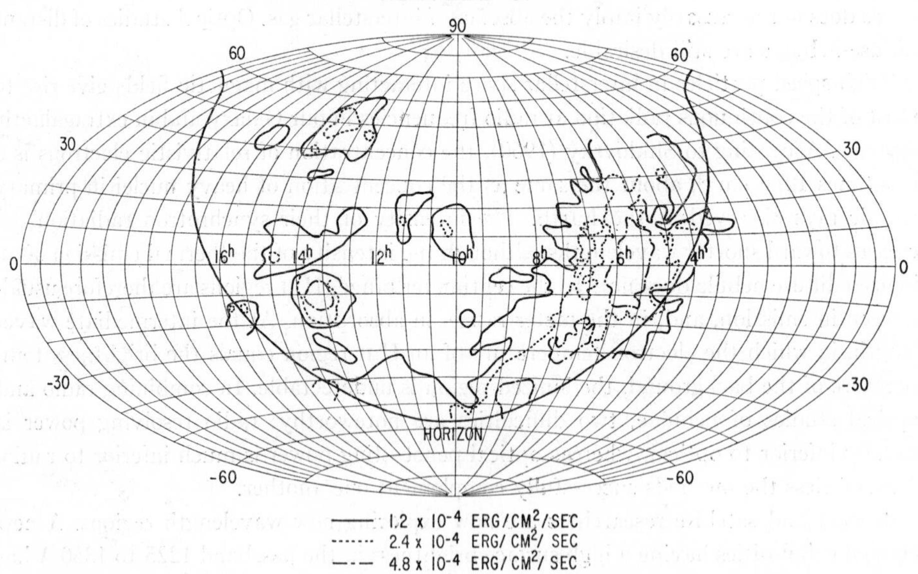

FIG. 1.—Map of the sky at 1225–1350 A in equatorial coordinates. Contours: 1.2 (———), 2.4 (------), and 4.8 × 10^{-4} (—•—•—•) erg cm^{-2} sec^{-1} per hemisphere. The heavy smooth curve indicates the horizontal (the great circle 90° from the zenith); the horizon is depressed about 15° at the altitude of the rocket (Boggess 1961. By permission of McGraw-Hill Book Co., Inc.).

2. REFLECTION NEBULAE

2.1. Introduction

Quantitative surveys for continuous nebulae are summarized almost completely in Table 5. The data refer to colorimetry or polarimetry of the scattered light by photographic (*pg*), photoelectric (*pe*), or spectrophotometric (*sp*) means. Nebulae such as IC 405 or NGC 1976 (the Orion Nebula) are combined reflection and emission types. The components of IC 405 are separate (Herbig 1958 and Plates 2 and 3); those of NGC 1976 mix inseparably (cf. Sec. 3.4). Monochromatic and atomic continuous emission may affect the scattering measures; such components are discussed in chapter 9 (Aller and Liller) and briefly for the Orion Nebula in Section 3.4 below. For the basic data, Cederblad's

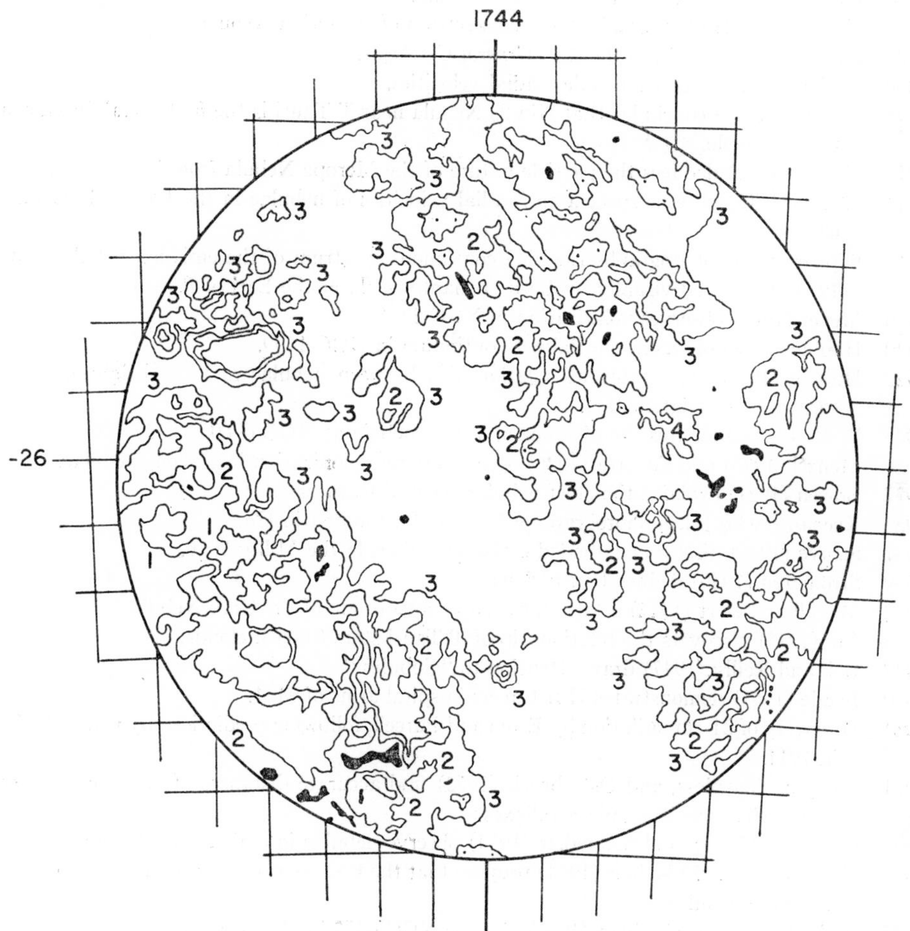

Fig. 2.—Sky map centered on $\alpha = 17^h44^m$, $\delta = -26°$ (1950), and Hα light (passband 326 A). Isophotes are at (1) 41, (2) 30, (3) 26, and (4) 11×10^{-5} erg cm^{-2} sec^{-1} sterad^{-1}. From a plate taken with an 8-inch $f/1$ Schmidt camera (Johnson 1960*b*).

(1946) catalogue of bright nebulae is supplemented by Rozhkovsky's (1957*b*) list of reflection nebulae. Dark nebulae have been catalogued and charted by Barnard (1927), Ruskol (1950), Khavtassi (1960), and Mrs. Lynds (1962).

No convenient and logical classification of reflection nebulae has so far been published. Factors such as the degree of mixing of dust with gas, associations with dark nebulae, and types of associated stars are not easily related. A classification sequence,

TABLE 4

SOME HISTORICAL EPOCHS FOR GALACTIC DIFFUSE NEBULAE

1610	Peiresc (Bigourdan 1916) discovers the Orion Nebula.
1864	Huggins (1864) discovers the bright-line spectrum of the Orion Nebula (NGC 1976).
1880	Draper (1880) photographs the Orion Nebula.
1882	Huggins (1882) photographs the line spectrum of the Orion Nebula.
1888	Dreyer (1953) publishes the *New General Catalogue.*
1890	Keeler (1894) measures nebular radial velocities.
1895	Barnard (1895) concludes that Hind's Nebula near T Tauri is the first certainly known to be variable.
1912	Slipher (1912) deduces the particle nature of the Merope Nebula from its spectrum.
1914	Wilson (1915) obtains spectra and radial velocities of nebulae in the Large Magellanic Cloud.
1919	Slipher (1919) photographs the blue continuous spectrum of the emission nebula NGC 1976, discovered visually by Lord Oxmantown (later the Earl of Rosse) in 1868.
1920	Meyer (1920) observes polarization in NGC 2261.
1920	Hubble (1920) observes the Balmer continuum in NGC 1499.
1922	Hubble (1922*a*, *b*) establishes relationships between nebulae and the brightness and spectra of associated stars.
1923	Wolf (1923) measures the distance of a dark nebula at NGC 6960.
1926	Menzel (1926) and Zanstra (1926) investigate photoionization in gaseous nebulae.
1927	Bowen (1927) explains the chief nebular spectral features.
1936	Keenan (1936) takes photograph of NGC 6618, the first in Hα.
1937	Struve (1941) discovers extended emission regions in the Milky Way.
1939	Strömgren (1939) defines H II regions.
1942	Mayall and Oort (1942) establish the supernova origin of the Crab Nebula.
1945	Joy (1945) discovers the relationship of T Tauri stars to dark clouds.
1947	Bok and Reilly (1947) draw attention to "globules."
1950	Baade (1951) demonstrates H II tracers in spiral arms of M31.
1951	Discovery of 21-cm radiation by Ewen and Purcell, following predictions by van de Hulst in 1944.
1951	Morgan, Sharpless, and Osterbrock (1952) find accurate distances of emission nebulae from stellar spectroscopic parallaxes.
1953	Haddock, Mayer, and Sloanaker (1954) observe nebulae in emission at 9.4 cm.
1954	Shain, Haze, and Pikelner (1954) propose that the two-quantum continuum is dominant in diffuse nebulae.
1955	Mills, Little, and Sheridan (1956) observe NGC 6357 in absorption at 3.5 m.
1956	Mathis (1957) measures the ratio of He to H in the Orion Nebula.
1958	Kupperian *et al.* (Boggess 1961) observe ultraviolet nebulae from a rocket.
1965	Weaver, Williams, Dieter, and Lum discover OH in emission in H II regions.

TABLE 5—AN INDEX TO QUANTITATIVE COLORIMETRY AND POLARIMETRY OF CONTINUOUS NEBULAE

Nebula	Ref. (Color)	Ref. (Polarization)
BD+31°597	29 *pg*	29 *pg*
NGC 1333	6 *pg*, 29 *pg*, 32 *pe*	29 *pg*, 32 *pe*
IC 348	6 *pg*, 29 *pg*, 32 *pe*	29 *pg*, 32 *pe*
Electra	32 *pe*	3 *pg*, 32 *pe*
Maia	2 *pe–sp*, 32 *pe*	3 *pg*, 32 *pe*
Merope	2 *pe*, 6 *pg*, 7 *sp–pg*, 12 *pg*, 32 *pe*	3 *pg*, 32 *pe*
IC 359 (?)	6 *pg*	....
B 10	6 *pg*, 32 *pe*	32 *pe*
B 214	6 *pg*	....
B 14	6 *pg*	....
NGC 1786	6 *pg*	....
IC 405	32 *pe*	30 *pe*, 32 *pe*
IC 410	....	30 *pe*
Ced 44	32 *pe*	32 *pe*
NGC 1976	7 *sp*, 9 *sp*, 10 *sp*, 11 *sp*, 14 *sp*, 22 *pg*, 26 *pg*, 31 *pg*	1 *pg*, 16 *pe*, 18 *pe*, 20 *sp*, 22 *pg*
NGC 1982	....	23 *pg*
NGC 1999	11 *sp*	....
IC 431	38 *pg*	....
IC 432	....	21 *pg*
NGC 2023	6 *pg*	....
IC 435	6 *pg*, 38 *pg*	....
FU Orionis	....	8 *pg*
NGC 2068	6 *pg*	32 *pe*
NGC 2245	43 *pg*	....
NGC 2071	6 *pg*	....
NGC 2247	43 *pg*	....
NGC 2261	11 *sp*, 32 *pe*, 37 *pg*, 42 *pe–pg*, 44 *pe*, 45 *pe*	1 *pg*, 28 *pg*, 32 *pe*, 36 *pg*, 44 *pe*, 45 *pe*
BD–12°1771	6 *pg*	....
NGC 3034	....	27 *pe*
η Carinae	....	25 *pg*, 35 *sp*, 40 *pe*
ν Scorpii	5 *pg*	....
BD–19°4357	6 *pg*, 29 *pg*	29 *pg*
IC 4601*a*	6 *pg*, 29 *pg*	29 *pg*
IC 4601*b*	6 *pg*, 29 *pg*	29 *pg*
σ Scorpii	5 *pg*	....
CD–24°12684	5 *pg*	....
ρ Ophiuchi	5 *pg*	5 *pg*
Antares	5 *pg*, 32 *pe*	....
22 Scorpii	5 *pg*	....
NGC 6514	....	19 *pe*, 23 *pg*, 24 *pg*
NGC 6523	34 *pg*	19 *pe*, 23 *pg*
NGC 6589	29 *pg–pe*	29 *pg*
NGC 6590	29 *pg–pe*	....
IC 1284	29 *pg–pe*	29 *pg*
NGC 6618	34 *pg*	27 *pe*, 39 *pg*
IC 1287	29 *pg–pe*, 32 *pe*, 38 *pg*	29 *pg*, 32 *pe*
NGC 6729	11 *sp*	13 *pg*
Ced 167	38 *pg*	....
γ Cygni	3 *pg*, 5 *pg*	3 *pg*
NGC 6914*a*	6 *pg*	....
NGC 7023	4 *pg*, 6 *pg*, 11 *sp*, 29 *pg–pe*, 32 *pe*	3 *pg*, 15 *pg*, 17 *pg*, 29 *pg*, 32 *pe*, 33 *pe*
BD+67°1332	38 *pg*	....
NGC 7129	6 *pg*	....
IC 5146	6 *pg*	....

Chronological References: (1) Meyer (1920); (2) Struve, Elvey, and Keenan (1933); (3) Henyey (1936); (4) Keenan (1936); (5) Struve, Elvey, and Roach (1936); (6) Collins (1937); (7) Greenstein and Henyey (1939*a*); (8) Walter (1941); (9) Barbier (1944); (10) Greenstein (1946); (11) Greenstein (1948); (12) Schalén (1948); (13) Whitney and Weston (1948); (14) Barbier (1949); (15) Gliese and Walter (1951); (16) Hall (1951); (17) Weston (1952); (18) Dombrovsky (1955*a*); (19) Dombrovsky (1955*b*); (20) Gurzadian (1956*b*); (21) Khachikian (1956); (22) Rozhkovsky (1956*a*); (23) Rozhkovsky (1956*b*); (24) Rozhkovsky and Matiagin (1956); (25) Thackeray (1956); (26) Wurm and Rosino (1956); (27) Dombrovsky (1958*a*); (28) Khachikian (1958); (29) Martel (1958); (30) Dombrovsky (1958*b*); (31) Andrillat and Andrillat (1959); (32) Johnson (1960*a*); (33) Gehrels (1960); (34) Gershberg *et al.* (1961); (35) Thackeray (1961); (36) Khachikian and Kallogian (1962); (37) Parsamian (1962); (38) Rozhkovsky (1962); (39) Rozhkovsky, Glushkov, and Jakusheva (1962); (40) Wesselink (1962); (41) Gehrels and Teska (1963); (42) Martel (1963); (43) Parsamian (1963); (44) Hall (1964); (45) Hall (1965).

correlated with an age sequence of star clusters, may be developed from the observation that dark nebulae often appear as lanes and sometimes terminate bulbously:

1. An example of a prestellar bulb is in Morgan, Strömgren, and Johnson (1955, bottom of Fig. 10).

2. IC 5146 is an extremely young cluster and nebula (Walker 1959) inside the bulb at the end of a strong dark lane, B 168*,[1] 1°.7 × 9′. The central star, BD+46°3474, of spectral type B1 V, produces a partly emission-type and partly reflection-type spectrum in the nebula.

3. NGC 7023 is a reflection nebula inside a diffuse bulb at the end of a weak lane of obscuration (see the Palomar Sky Survey). The illuminating star, HD 200775, is of

[1] Barnard catalogue number.

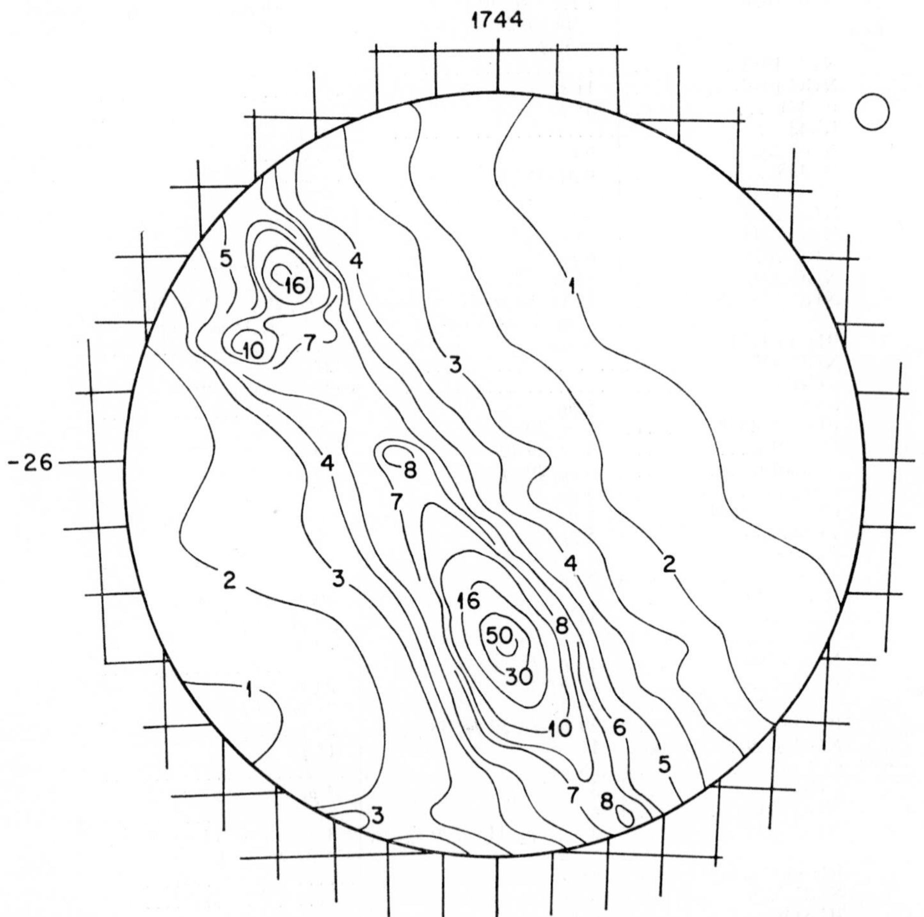

Fig. 3.—Sky map (same area as in Fig. 2) showing contours at 22 cm. The numbers in the contours are multiples of 3.25° K in T_b. Sagittarius A, NGC 6523, and NGC 6514 (peaks near bottom-center, lower top left, and top left, respectively) are seen in the galactic ridge (Westerhout 1958*a*). The circle, top right, shows the antenna beamwidth.

spectral type dB3ne, a shell spectrum according to Greenstein and Aller (1947) and Weston (1949).

4. Finally, the Pleiades stars and nebulae are surrounded by broken dark nebulae, well shown on the Palomar Sky Survey charts. It can hardly be doubted that the stars in dark bulbs have formed *in situ* but have not used all the interstellar material. The above may be an age sequence up to 2×10^7 years for the Pleiades, but we can also find extremely young clusters, for example, h and χ Persei, and NGC 2362, which lie outside any sequence involving optical nebulae.

Next to the Orion Nebula itself, the Pleiades nebulae and NGC 7023 have been studied the most extensively. They differ from one another in several respects and will be discussed as examples.

The Pleiades lie in the extensive Taurus nebulosities that are part of Gould's Belt, or

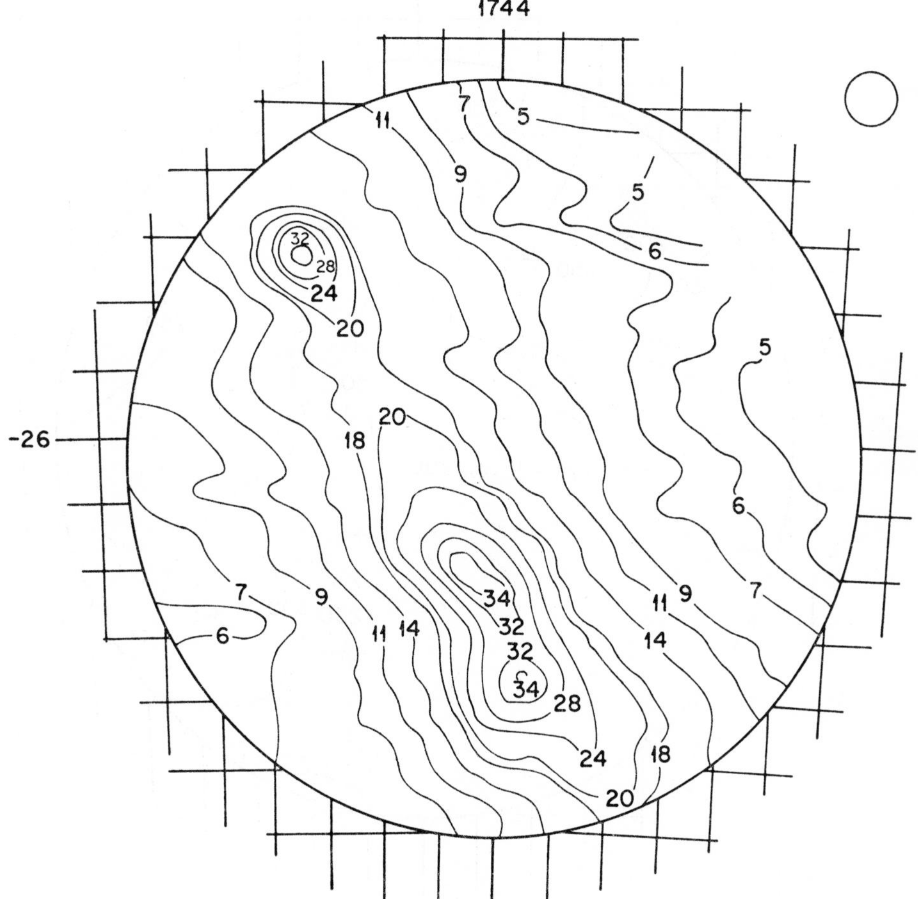

FIG. 4.—Sky map (same area as in Fig. 2) showing contours at 3.5 m. Brightness temperatures are entered into the diagram in units of 1000°K. NGC 6523, a thermal source, is weak, but there is a strong nonthermal source in the direction of NGC 6514. The circle, top right, shows the antenna beamwidth.

the Local System (Lund $l^{\mathrm{I}} = 135°$, $b^{\mathrm{I}} = -22°$), and NGC 7023 is near a node of the Belt at $l^{\mathrm{I}} = 71°$, $b^{\mathrm{I}} = +14°$. The Pleiades contain several bright star-and-nebula centers as well as hundreds of fainter stars; NGC 7023 is illuminated by a single star, although Collinder (1931) has classified the object as a dubious nebulous cluster of 15 stars and Weston (1953) has found that the cluster contains Hαe stars. In general, reflection nebulae tend to be amorphous and emission nebulae filamentary, but the extensive filamentary structure of the Pleiades makes it an exception to this rule. NGC 7023 is mostly amorphous. The distance of the Pleiades is exactly 126 pc, but widely different estimates have been made for the distance of NGC 7023. The apparent visual magnitude of HD 200775 is 5.65 after correction for extinction (Mendoza 1958). If the absolute visual magnitude is -3, from an interpolation of B1e and B2e magnitudes in Scorpio-Centaurus

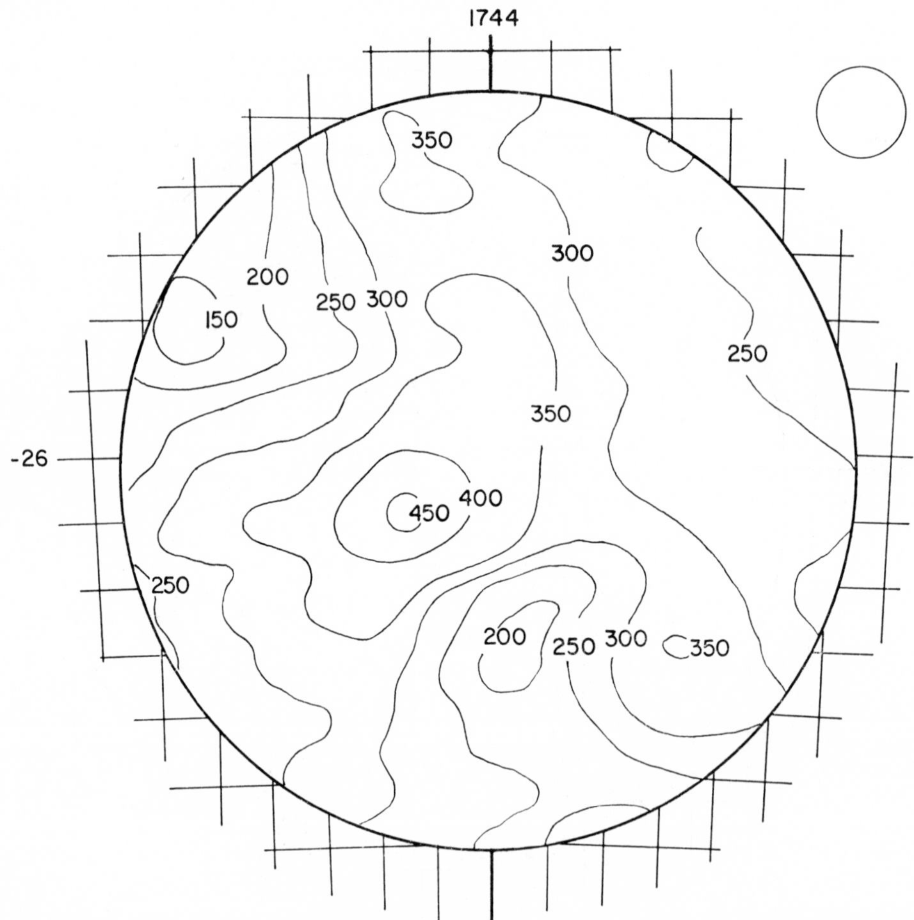

FIG. 5.—Sky map (same area as in Fig. 2) at 15 m, after C.A. Shain of Radiophysics Laboratory, C.S.I.R.O., Sydney. Brightness temperatures are given in units of 1000°K. Thermal features are not present. The circle, top right, shows the antenna beamwidth.

and B6e magnitudes in the Pleiades, the distance of NGC 7023 is 540 pc, and the z-distance is +130 pc—rather high since it is not in Gould's Belt. From the blue charts of the Palomar Sky Survey, the angular and corresponding linear dimensions of the nebulae are as follows: total Pleiades nebulosity 90′ (3.3 pc); around individual stars 30′ (1.1 pc); individual filaments 15′ (0.5 pc) to 10″ (0.005 pc). The NGC 7023 nebulosity is 23′ (3.6 pc) and is inside the dust bulb of 34′ (5.3 pc). The red-chart images of both nebulae are less exposed than the blue-chart images, but, apart from the effect of exposure differences, the structures appear to be alike in the two wavelengths.

The most striking feature of the Pleiades nebulosity is the system of ray-like filaments permeating the quasiparallel series. The strongest rays lie in an east-west direction without systematic relationships to the principal stars of the cluster. Groups of finer, somewhat curved rays are associated, especially with Maia, in position angles $PA = 50°$ and $PA = 0°$ (north) and, with Merope, from $PA = 150°$ to 200° curving southward. The filaments around Alcyone are shorter, thicker, and in many directions. Very regular but very weak east-west rays pass near the star Atlas. Electra lies immediately north of the strongest of the east-west rays, otherwise being centered in a more diffuse and restricted area. Taygeta and Pleione are essentially free of nebulosity. Barnard's (1927) *Atlas* photograph and excellent chart of the region are nearly on the same scale as the Palomar Sky Survey prints.

2.2. Interpretation in Terms of Scattered Light

The evidence for scattering in the Pleiades nebulae and in NGC 7023 will be reviewed here; Greenberg (chapter 6) discusses the scattering properties of solid particles more fully. Both colorimetry and polarimetry, especially color measured relative to the color of the illuminating star, and direction of the magnetic vector of the light, contribute useful evidence.

2.2.1. *The Pleiades.*—Following Slipher's (1912) discovery that the Merope Nebula spectrum imitates that of Merope itself, Hertzsprung (1913) observed that the surface brightness of the nebula was 3 or 4 mag fainter than that of a white, diffusely reflecting surface normally illuminated by Merope would be. He estimated that particles of radius 2 mm and albedo 0.10 would satisfy both the spectroscopic and photometric observations. Struve, Elvey, and Roach (1936) observed the Maia and Merope nebulae in two wavelengths and found the nebulae bluer than the stars by $-0\overset{m}{.}2$ to $-0\overset{m}{.}3$. Henyey (1936) found no polarization of the nebulae but found a trace of polarization in the strong ray near Electra. Collins (1936) measured two points about 10′ south of Merope, finding a photovisual surface brightness of $21\overset{m}{.}51/\text{sec}^2$ of arc and an international CI of $-0\overset{m}{.}62$, which is $-0\overset{m}{.}49$ bluer than Merope. Greenstein and Henyey (1939*a*) corrected the preceding color difference to $-0\overset{m}{.}27$ by a revision of the sky background and obtained $-0\overset{m}{.}35 \pm 0\overset{m}{.}18$ for the same quantity 10′ from Merope. Their value of photovisual surface brightness is $22\overset{m}{.}0/\text{sec}^2$ of arc.

Schalén (1945, 1953) has published a theory for scattering in nebulae, predicting the color difference $C_{\text{neb}} - C_{\text{star}}$ as a function of the distance r from the star, the distribution of particle size d, and the geometry of the star-nebula system. The situations of "star in front," "star behind," and "star inside a plane-parallel nebula" are considered. Unfortunately only metallic particles in the distributions $\Phi(d) = (2\pi)^{-1/2}\sigma^{-1} \exp -(d - d_0)^2/2\sigma^2$

and $\Phi(d) = Ad^{-3}$ in the range 20 m$\mu \leq d \leq$ 500 mμ were investigated by Schalén. An earlier theory by Henyey (1937) is restricted to $d \gg \lambda$ with the star distant from the nebula. Schalén provides curves for $C_{\text{neb}}(r) - C_{\text{star}}(r)$ in the range 4400 to 6300 A. The color difference increases with r in different ways, according to the geometry, and from different zero points, according to $\Phi(d)$. Schalén (1948) photographed the Merope Nebula to compare its color with the theory. The color index increases about 0.3 mag from $r = 3'$ to $r = 19'$, but the zero point of the color difference seems to be undetermined. Schalén concluded that good agreement is obtained with $\Phi(d) = Ad^{-3}$, the star situated behind the nebula, and with the ratio of the distance from the nebula to the thickness of the nebula about 1/10. Schalén's estimate of the grain density in this model was $\rho = 6 \times 10^{-23}$ gm/cm^3; he noted that this is much higher than is generally assumed for dark clouds.

Fessenkov (1955) observed the surface brightness of the Merope Nebula as a function of r and decided that the nebula can be treated as a flat surface perpendicular to the line of sight with Merope at the edge of the material. In the same way, IC 431, IC 432, and IC 435 were found to be spherical in shape (Fessenkov 1955, Minin 1961).

TABLE 6

MEAN PHOTOELECTRIC MAGNITUDES AND COLORS IN THE PLEIADES AND NGC 7023

Nebula at $r=3'$	V/sec^2 of arc	$B-V$	$U-B$	$(B-V)_{\text{neb}}-(B-V)_{\text{star}}$	$(U-B)_{\text{neb}}-(U-B)_{\text{star}}$
Maia............	21^{m}4	−0^{m}52	−0^{m}65	−0^{m}40	−0^{m}25
Electra..........	21.2	−0.30	−0.67	−0.19	−0.26
Merope..........	21.0	−0.17	−0.37	−0.11	+0.06
NGC 7023........	23.2	+0.14	−1.10	−0.31	−0.62

Johnson (1960*a*) obtained the colors of the Maia, Electra, and Merope nebulae photoelectrically at $r = 3'$ (Table 6). He also found that the polarization of the nebular light is not strictly radial; that is, the plane containing the line of sight and the star does not contain the magnetic vector. The polarization of light scattered by spherical particles, or nonspherical particles randomly oriented, should be radial in this sense; therefore, it may be assumed that in the Pleiades nebulae, as in some other nebulae (Martel 1958), nonspherical particles are not randomly oriented. Polarization of starlight by interstellar grains in the line of sight suggests a similar result. The particle-aligning mechanism of interstellar magnetic fields may apply to local areas in diffuse nebulae as well.

The filamentary structure in the Pleiades nebulae suggests that the motion of the gas is influenced by magnetic fields (but cf. Gershberg 1960). Such fields may bend and tangle in distances of the order of 1 pc. Gas, in turn, interacts with and controls the motion of the interstellar grains (Spitzer 1941). However, ionized hydrogen has not been detected in the Pleiades (Greenstein and Henyey 1939*a*, Johnson 1953); so arguments for the presence of gas depend on the grains. The neutral hydrogen, equivalent to 470 solar masses, observed in the Pleiades by Drake (1958), is believed to be in the form of protostars, whose motions would be controlled by the gravitational field of the cluster.

Hall (1955) has found that the magnetic vectors of the light of four Pleiades and six background stars are nearly perpendicular to the Merope filaments near them. According to the Davis and Greenstein (1951) theory, the long axes of the scattering particles will then be oriented at right angles to the filaments. Some cluster members are unpolarized and must be in front of the nebula. According to van den Bergh (1956), the mean electric vector for 217 Pleiades stars is in $PA = 122°$, and the circle parallel to the galactic equator is at $PA = 134°$. Van den Bergh found deviations of the vector from the mean to be correlated with the filamentary structure of the Pleiades nebulae.

From the data of Table 6 it may be seen that the Pleiades nebulae provide up to half of the light above the level of sky light in a diaphragm of 10″ diameter that includes a star of visual magnitude 16. The rather high polarization of the light of the nebulae can thus affect the measurement of stellar polarization up to several magnitudes brighter than 16.

To conclude this discussion of the Pleiades, let us note Adams' (1949) observation of the unique strength of interstellar CH II 3957–4232 A relative to interstellar H and K in the Pleiades. If the peculiarity depends on the radiation of the stars, the gas is near the cluster but its radial velocity exceeds the radial velocity of the cluster stars by about 10 km/sec.

2.2.2. *NGC 7023.*—Pease (1915) discovered that the continuous spectrum of NGC 7023 was crossed by absorption lines also present in HD 200775. Slipher (1918) and Greenstein and Aller (1947) found Hβ and Hγ in emission; therefore, in reflecting the spectrum of the shell star, the nebula may spuriously appear to be an emission type. However, Shain, Haze, and Pikelner (1954) classify NGC 7023 as a mixed continuous and emission type, and Mme. Martel (1958) reports that her Hα plates show an H II region around HD 200775.

The photographic intensity of NGC 7023 was first measured by Keenan (1936) as a function of the distance r from the star. The mean intensity I decreases from $19^{m}.5/\text{sec}^2$ of arc at $r = 0'.5$ to $21^{m}.7/\text{sec}^2$ of arc at $r = 3'.2$. Keenan also observed at 6300 A, and his color differences as a function of r have been reported and discussed by Schalén (1945) in terms of his theory. The difference $C_{\text{neb}} - C_{\text{star}}$ increases from $-0^{m}.03$ at $r = 0'.5$ to $+0^{m}.47$ at $r = 1'.0$, then drops to $+0^{m}.28$ at $r = 1'.7$. This behavior is characteristic of the scattering of the light of a star behind a nebula; the large color excess of the star also indicates that it is behind the nebula. The characteristic radius of the maximum color difference is a function of the absorption coefficient, which in turn contains the number of particles per cubic centimeter. From these relations Schalén estimates 6×10^{-7} metallic particles cm^{-3}, with diameters d such that $50\ \text{m}\mu \leq d \leq 100\ \text{m}\mu$, in NGC 7023.

Unfortunately the photometric data for NGC 7023 are not in agreement. Collins (1937), using the photographic and photovisual passbands, observed $I_{pv} = 21^{m}.78/\text{sec}^2$ of arc and $C_{\text{neb}} - C_{\text{star}} = -0^{m}.3$ at $1'.5 < r < 2.3$. Greenstein (1948) also found spectrophotometrically that the nebula is bluer than the star. Mme. Martel (1958) observed $I_{pv} = 20^{m}.58/\text{sec}^2$ of arc at $r = 0'.8$ south, falling to $22^{m}.79/\text{sec}^2$ of arc at $r = 1'.6$ east; although the international color difference depends strongly on direction from the star in the wide range $-0^{m}.56 < C_{\text{neb}} - C_{\text{star}} < -0^{m}.02$ at $r = 0'.6$, but becomes $-0^{m}.47$ at $r = 1'.6$ east and west of HD 200775. Mme. Martel made photoelectric scans over areas 34″ in diameter and gave her data as international magnitudes and colors. Johnson's (1960*a*) U, B, V data are in Table 6. The nebula appears to be bluer than the star at all radii, contrary to the data analyzed by Schalén.

Henyey (1936) measured a mean polarization of 12 per cent in NGC 7023. Gliese and Walter (1951), Weston (1952), and Mme. Martel (1958) measured polarization in many areas throughout NGC 7023 to $r = 4'.7$. In areas of 0.014 to 0.28 min^2 of arc, polarization ranges from -5 to $+56$ per cent, averaging 14 per cent according to Mme. Martel and 16 per cent according to Gliese and Walter—about the same as found in other nebulae. The polarization seems to be approximately radial over much of the nebula in the sense defined in the Pleiades discussion, but Mme. Martel finds that in a third of the areas examined carefully in NGC 7023 the plane of polarization is significantly nonradial. In all areas with filamentary structure, the plane of polarization coincides with the direction of the rays. As expected (Elvius 1960), this is at right angles to the plane of polarization of the light of stars seen through filaments (Hall 1955). Gliese and Walter note that the degree of polarization is independent of radius in NGC 7023, and Mme. Martel finds that polarization is unaffected by the presence of an H II region. The light of HD 200775 is 0.7 per cent polarized (Hall and Mikesell 1950); so extraneous scattering evidently plays little part in the results quoted above.

Theoretical and observational approaches to the problem of light scattered by grains near bright stars seem to be deficient. A theory for small dielectric particles which includes polarization effects is needed. Such a theory, if available today, would find observational data still too sparse for a good test. Rozhkovsky (1960) has begun to supply the needed data. Many parameters will appear in the theory, and even with very accurate and complete data it will not be easy to derive unambiguous models. Van Houten (1961) gives a clear discussion of a model of NGC 7023 that embodies some of the observations.

3. EMISSION NEBULAE

3.1. Introduction

Sharpless (1959) avoids the terms "diffuse emission nebula" or "extended emission region" and replaces them with "H II region," defined to include the hot stars that ionize the gas. This is physically sensible and observationally useful, for it organizes separate bits and pieces of nebulae into coherent systems. With irregular densities and absorption, a region ionized by a single star or by an associated group of stars might be catalogued as a large number of "nebulae." The North America Nebula, Pelican Nebula, and several objects in Cygnus (Haze and Shain 1955) exemplify parts of one H II region (cf. Morgan, Strömgren, and Johnson 1955 for a reassembly). Sharpless' usage may be confusing with reference to supernovae remnants, and it is by no means certain that the ionized gas of supernovae remnants is always recognizable as such. For example, YM 29 (Johnson 1955) is a symmetrical emission nebula, without a central star and well isolated ($b^{\text{II}} = +14°$) from external photoionization; it has not been detected at 750 Mc/s with the 300-ft telescope of the National Radio Astronomy Observatory. Minkowski (1958) suggested that this object is a supernova remnant. On the other hand, the Great Loop in Orion has also been suggested as the product of supernovae explosions (Blaauw 1961).

Most individual optical nebulae can be clearly identified by x- and y-coordinates in millimeters from a corner of a Palomar Sky Survey chart. Although several surveys have recorded Hα more strongly than can be done in 1 hour with the $f/2.5$ Palomar Schmidt, its great aperture and resolving power in general compensate for relative underexposure because of the fine structure of many nebulae.

Parts of an H II region may be intrinsically faint or obscured by dust. The exciting star may be difficult to identify because spectral types are not available for all stars in the vicinity of the gas. When spectral types are known, it may still be hard to assign the same modulus to all of them in the H II region (Kopylov 1958). Hoffleit (1956) finds, for example, that in the direction of the η Carinae Nebula the evidence is rather conclusive that the hot stars are spread over a wide distance.

The full analysis of an H II region ideally combines observations by several instruments and of several kinds. A direct photograph in Hα light can be supplemented by direct photographs in other monochromatic emissions (Wurm and Rosino 1959). The angular size and structure of the various ionic and stellar parts can then be compared. A spectrogram of the gas gives the degree of excitation, the electron temperature, the electron density, some ionic abundances, and the radial velocity. Spectrograms of the exciting stars give spectral types, absolute magnitudes, and radial velocities. Photometry of the nebula gives the absolute intensity of one or more monochromatic emissions and the continuum. Such photometry may be done from a spectrogram, direct photograph, or photoelectric equipment. Photometry of the exciting stars gives the apparent magnitudes, interstellar extinction, and true distance moduli when combined with the stellar spectral data. From these sources we may deduce the linear dimensions, distribution, densities, mass, and motions of the gas, as well as the character of the ultraviolet radiation of the involved stars. Free-free transitions in the H II region observed at microwavelengths give independent estimates of angular size and brightness of the mass of gas, free of interstellar optical extinction. But the radio data ultimately depend on the optical data for the stellar component through the derivation of densities, masses, and so forth. If concentrations of H I and H II are mingled, then 21-cm radio observations provide information about the motions and masses in both.

The fundamental assumption that radiofrequency emission from H II regions depends on a free-free thermal emission coefficient has been tested by Boggess (1954) and by Haddock (1957) on NGC 6514, NGC 6523, NGC 6611, and NGC 6618, and by Osterbrock and Stockhausen (1960, 1961) on NGC 2175 and NGC 281. The test compares predicted and observed ratios of centimeter and Hα and Hβ intensities, the latter corrected for interstellar extinction. Unfortunately, the data do not allow a very critical test. Haddock has also shown that the form of the radio spectrum of NGC 1976 is purely thermal. Table 7 has been prepared from both optical and radio data available for the H II regions listed in Westerhout's (1958*a*) catalogue. This was compiled from a survey made with a 25-m radio telescope at 1390 Mc/s in the area $320° < l^{\mathrm{I}} < 56°$ by $-8° < b^{\mathrm{I}} < +6°$. The beamwidth is 0°.57, and the limit of detection, 20° K, is expressed in other terms later. Each column of data in Table 7 is explicitly attributed to Westerhout or to the optical sources of Sharpless (1959) or Gershberg and Metik (1960), who used the material of Haze and Shain (1955). Table 1 lists the optical equipment.

Column 1 of Table 7 gives the number of the object in the final catalogue of Haze and Shain (1955). Column 2 gives the corresponding name of the nebular part of the H II region, usually from the NGC or IC catalogues. Columns 3 and 4 give Westerhout's coordinates. Columns 5 and 6 give the distance r according to Westerhout or Gershberg

TABLE 7

PHYSICAL DATA FOR THERMAL SOURCES IN THE 1390 MC/S SURVEY*

(1) HS No.	(2) H II	(3)† α (1950)	(4)† δ	(5)† r (pc)	(6)‡ r (pc)	(7)‖ φ	(8)‡ Diam.	(9)† $10^{-3}E\left(\frac{cm^{-6}}{pc}\right)$	(10)† $N_e(cm^{-3})$	(11)‖ $N_e(cm^{-3})$	(12)† 𝔐/𝔐⊙	(13)‖ 𝔐/𝔐⊙	(14)‡ Exc. Stars	(15‡) Sp. Types
15	IC 1795	2 22ᵐ7	+61°51′	2000	1500	......	150′	72	85	23–55	1850	92	8	B
16	IC 1805	2 29.5	+61 13	2000	1500	......	150	2	5	30	15000	380	8	B
25	IC 1848	2 47.0	+60 10	2000	1700	......	120	4	10	23	10800	800	9	Oe5–B
55	IC 410	5 19.4	+33 20	2000	3000	......	55	24	45	21	1540	3100	5	B0–B8
66	NGC 1976	5 32.8	−05 27	500	400	......	60	660	700	740	43	760	3	Oe5–B1
80	NGC 2024	5 38.4	−01 54	500	400	......	120	30	100	15–75	55	0.6	1	B0
87	NGC 2175	6 06.0	+20 30	1700	2000	......	40	11	27	11	1700	350	1	Oe5
97	NGC 2237	6 29.3	+04 57	1400	1660	......	100	6	13	31–50	8600	9500	9	B0–B2
120	NGC 6357	17 22.2	−34 17	1000	1100	......	90	35	50	58–220	3200	0.8+6	2	Oa–B
121	NGC 6383	17 30.8	−32 42	800	......	......	120	2	10	......	4000	......	9	Oe5–B5
......	No. 25†	17 46.8	−31 24	1000	2300	......	30	8	35	37	150	280	3	B0–B5
124?	......	17 58.2	−23 22	......	1600	......	40	......	......	13	......	250	......	......
126	NGC 6523	18 01.0	−24 22	1200	1600	......	90	15	32	30	2750	3000	14	Oe5–B5
129?	......	18 02.6	−21 37	......	1500	29′	90	20	......	13	......	220	1	B1
130	NGC 6559?	18 06.4	−23 54	1200	1500	......	40	6	30	18	240	310	3	B0–B5
137	IC 4701	18 14.4	−16 24	1700	2900	......	60	7	16	10	7900	2500	2	Oe5–B3
140	IC 1284	18 14.8	−19 42	1800	......	......	20	20	59	......	275	......	3	B2–B5
141	......	18 15.0	−11 55	1400	1800	34	140	30	46	38–70	2500	840	7	Oa–B
142	NGC 6611	18 16.3	−13 45	2000?	1700	25	90	60	66?	70	3600?	300	10	B0–B8
145	NGC 6618	18 17.8	−16 09	1700	1900	......	60	200	125	95	5600	1100	6	B0–B5
147?	......	18 23.4	−12 40	......	1900	38	15	10?	......	50	......	4.5	......	......
152	......	18 29.6	−02 12	......	......	......	25	20	......	......	......	......	......	......
155?	......	18 33.6	−07 30	......	......	......	20	20	......	......	......	......	......	......

* Data for the η Car Nebula (NGC 3372) is given by Hindman and Wade (1959).
† Westerhout (1958*a*).
‡ Sharpless (1959).
‖ Gershberg and Metik (1960).

TABLE 7—*Continued*

(1) HS No.	(2) H II	(3)† α (1950)	(4)† δ	(5)† r (pc)	(6)‖ r (pc)	(7)† φ	(8)‡ Diam.	(9)† $10^{-3}E\left(\frac{cm^{-6}}{pc}\right)$	(10)† $N_e(cm^{-3})$	(11)‖ $N_e(cm^{-3})$	(12)† $\mathfrak{M}/\mathfrak{M}_\odot$	(13)‖ $\mathfrak{M}/\mathfrak{M}_\odot$	(14)‡ Exc. Stars	(15‡) Sp. Types
158?	...	18h45m4	−02°02′	...	...	18′	8′	125?	...	...	...	...	...	...
...	S 76?‡	18 53.7	+07 47	...	...	...	7	?	...	...	...	...	...	...
...	S 79‡	19 20.8	+14 08	...	...	38	40	67?	...	...	...	...	...	...
166	NGC 6823	19 40.1	+23 05	1800	2500	...	40	7	25	24	700	54	3	B
176	...	20 00.1	+33 23	...	1900	42	1080	7	...	7	...	240	...	...
177	...	20 05.3	+33 59	...	1900	...	1080	4	...	18	...	200	...	...
185	NGC 6888	20 05.8	+37 59	1500	1600	...	18	16	55	71	250	40	...	...
191	...	20 15.0	+43 29	1500?	1500	...	...	12	40?	55	360?	90	...	...
193	IC 1318*a*	20 15.8	+41 47	1500?	1500	...	1080	12	35?	27	770?	300	...	...
...	No. 63†	20 17.4	+45 23	1500?	...	84	...	2	7?	...	7000?	...	...	...
197	...	20 17.7	+36 53	1500?	1600	...	...	22	68?	28	180?	48	...	...
196	...	20 19.0	+39 11	900?	1500	24′	...	11	42?	21	200?	300	...	...
...	Cyg X	20 20.8	+40 13	6000?	...	56	...	25	16?	...	3×10^5?	...	...	...
201	IC 1318*b*	20 25.6	+39 53	900?	...	...	1080	8	24?	31	1300?	2100	...	...
204?	...	20 29.9	+38 56	...	...	42	...	15	...	8	...	160	...	...
207	...	20 30.4	+43 52	1500?	1300	30	...	18	39?	43	1500?	66	...	...
215	...	20 33.4	+46 52	...	...	46	...	6	...	...	...	...	...	...
...	No. 72†	20 33.8	+42 28	...	...	30	...	16	...	...	...	...	...	...
...	No. 73†	20 34.1	+41 10	1500?	...	56	...	10	21?	...	6000?	...	...	...
218?	...	20 38.6	+41 57	1500?	...	42	...	20	35?	...	3900?	...	...	...
220	...	20 41.7	+39 04	...	...	...	...	11	...	...	...	...	...	...
...	No. 77†	20 44.2	+40 52	900	...	...	...	5	26	...	200	...	...	...
224	NGC 6990	20 46.1	+30 28	400?	...	84	210	2	14?	...	240?	...	...	...
...	No. 79	20 47.8	+41 40	900	...	...	...	21	84	...	50	...	...	...
223	IC 5068	...	...	...	720	...	...	...	...	32	...	76	...	...
229	IC 5070	...	...	...	720	...	...	...	...	19	...	150	...	...
234	NGC 7000	20 53.4	+43 52	900	1000	(150)	240	3	9	7.5	11000	3000	1	Oe5

and Metik, who in turn have quoted from many sources of spectroscopic parallax; the discrepancies and blanks indicate the uncertainty in the distances. In column 7, ϕ is the geometric mean of the widths of the cross-sections and/or contours at half intensity observed at 1390 Mc/s after subtraction of the background. Column 8 gives the maximum diameter in Hα according to Sharpless.

Column 9 gives a radio value of 10^{-3} E averaged over the nebula, where

$$E = \int_a^b N_e^2(s)\,ds \tag{2}$$

is the *emission measure* (Strömgren 1948) for a depth of $(b - a)$ in parsecs through the H II region. The derivation of E proceeds from the classical absorption coefficient $\kappa(\nu)$ for free-free transitions in an ionized gas with electron temperature T_e and $N_e = N_i$, or equal volume densities of electrons and ions. Thus (see chapter 9 [Aller and Liller], Section 9),

$$\kappa(\nu) = \frac{4e^6 N_e^2 L}{3\sqrt{2\pi}(M_e K T_e)^{3/2} c\nu}, \tag{3}$$

where the factor L changes only slightly with N_e and ν, so that approximately

$$\kappa(\nu) = 0.12\, N_e^2 T_e^{-3/2} \nu^{-2}. \tag{4}$$

The brightness temperature of the part of the H II region where the optical depth is τ is

$$T_b = T_e(1 - e^{-\tau}), \tag{5}$$

and

$$\tau = \int_a^b \kappa(\nu)\,ds = 0.12 T_e^{-3/2} \nu^{-2} \int_a^b N_e^2\,ds \tag{6}$$

where $(b - a)$ in centimeters is the length of the column of gas in the line of sight. Setting $\nu = 1390$ Mc/s, $T_e = 10{,}000^\circ$ K, and converting the measure of length from centimeters to parsecs, we have

$$\tau = 1.86 \times 10^{-7} E. \tag{7}$$

For $E < 10^6$, $T_b = \tau T_e$ approximately, while the flux density of the total radiation from a discrete source subtending a solid angle of Ω radians, is

$$S = 2\kappa\nu^2 c^{-2} \int T_b d\Omega, \tag{8}$$

when we integrate over the whole source. Substituting ϕ^2 for the solid angle in equation (8), we have

$$10^{-3} E = 38 \times 10^{23} S\phi^{-2}, \tag{9}$$

to be found in column 9 of Table 7. Because the emission in a Balmer line of the recombination spectrum of a highly ionized gas that is mostly hydrogen is proportional to $N_i N_e T_e^{-3/2} = N_e^2 T_e^{-3/2}$, definition (2) means that E also measures the brightness of an H II region in Hα, provided that interstellar extinction does not affect the optical result. In some instances—for example, Cyg X—the H II region is entirely hidden by dark clouds of dust. In determining whether S pertains to a thermal (free-free transitions) or a nonthermal (synchrotron) source, Westerhout has relied on positive identification of a radio source with an optical emission nebula and on the character of its radio spectrum, or on the radio spectrum alone for Cyg X. Westerlund (1960*a*) has revealed an interesting H II complex in Carina, about 3° from NGC 3372, with nearly the same flux density as NGC 3372 at 1400 Mc/s and also at 3.5 m. The field is shown in Plate 1.

The electron density N_e is a function of E and, therefore, of S and ϕ through definition (2) and equation (9), provided that the function $N_e(s)$ is known or approximated by a model of the H II region. Wade (1958) proposes to redefine E by using N_{eq} in place of N_e in equation (1), where N_{eq} is the "equivalent" density of an H II region that is needed to produce the observed radio emission. In the special case in which T_e is isothermal, N_{eq} equals the root-mean-square density. Westerhout assumes constant N_e in an H II region. He also assumes the shape of the H II region to be cylindrical with a circular cross-section of diameter ϕ degrees and a length $r\phi/57.3$ that is equal to the linear diameter. Then

$$N_e = 4.8 \times 10^{13} S^{1/2} T_e^{1/4} \phi^{-3/2} r^{-1/2} \tag{10}$$

and the total mass in solar units is

$$\mathfrak{M} = 4.8 \times 10^{6} S^{1/2} T_e^{1/4} \phi^{3/2} r^{5/2} \,. \tag{11}$$

Table 7 (cols. 10 and 12) gives the results of Westerhout's application of these formulae with $T_e = 10{,}000°$ K.

Gershberg and Metik have reduced the plates of Haze and Shain to a photometric scale, standardized in absolute units from other observations of the following three diffuse nebulae: NGC 1976 (Razmadze 1958; Aller and Liller 1959), NGC 6523 (Pronik 1960*a*), and NGC 7000 (Gershberg and Pronik 1959*a*). The observations lead to $I_{obs}(H\alpha)$, the observed intensity of Hα in erg cm^{-2} sec^{-1} sterad^{-1}, and they are approximately corrected to true $I(H\alpha)$ by an application of the observed color excesses of hot stars in the respective H II regions. Gershberg and Metik do not give the value of E for each of the objects in their list, but they derive formulae for the mean electron density and the total mass of H II regions observed in Hα. They start from Burgess' (1958) theoretical value for the power emitted by a cubic centimeter of ionized hydrogen in Hα at $T_e =$ 10,000° K,

$$\epsilon(H\alpha) = 2.6 \times 10^{-26} N_e^2 \tag{12}$$

in erg cm^{-3} sec^{-1} sterad^{-1}, according to "case B" of Baker and Menzel (1938), where, in a nebula opaque to Lyman-line radiation, one takes account of the nebular radiation field, but not the stellar, by assuming that absorptions from energy level 1 to level r are exactly balanced by the inverse spontaneous transitions. See chapter 9, Section 5 (Aller and Liller).

It follows that

$$N_e = 0.63 \times 10^{13} I(H\alpha)^{1/2} (b - a)^{-1/2} \tag{13}$$

where $(b - a)$ is the length of the line of sight through the H II region in centimeters. The relation

$$\frac{I(H\alpha)}{I_{obs}(H\alpha)} = 10^{0.4\,\Delta m} = 10^{0.4 A_v 5500/6563} = 10^{0.34 A_v} \tag{14}$$

gives $I(H\alpha)$ from $I_{obs}(H\alpha)$ once the visual extinction A_v is estimated from photoelectric observations of stars in the H II region. Finally, converting the size l (mm) of the image of the H II region on the plate (scale 232″/mm) to the size of the object through the distance r, Gershberg and Metik have the formulae

$$N_e = 1.1 \times 10^5 I_{\rm obs}(\mathrm{H}\alpha)^{1/2}(rl)^{-1/2} \times 10^{0.17A_v} \quad (15)$$

and

$$\mathfrak{M} = 3.7 \times 10^{-6} I_{\rm obs}(\mathrm{H}\alpha)^{1/2} r^{5/2}(vl)^{-1/2} \times 10^{0.17A_v} \quad (16)$$

where v is the volume corresponding to the size l and $\mathfrak{M}$ are in solar units. Table 7 gives the results of Gershberg and Metik's application of these formulae in columns 11 and 13, respectively.

For $\phi > 0\overset{\circ}{.}5$, the limit of detection in Westerhout's survey is about $E = 2000$, and for $\phi < 0\overset{\circ}{.}5$ it is $E\phi^2 = 850$. The limit in several optical surveys has been near $E = 400$, with resolving power of about 1′.

TABLE 8

GALACTIC H II REGIONS FOR WHICH N_e HAS BEEN DERIVED FROM SURFACE INTENSITY

Region	N_e (cm^{-3})
NGC 281	>15 (Johnson 1953) *H;* 28–45* (Pottasch 1960*b*) 2*q*
NGC 1499	>14–31 (Johnson 1953) *H;* 42 (Pottasch 1960*b*) 2*q*
IC 434	36 (Pottasch 1960*b*) 2*q;* 10 (Rishbeth 1958) *rf*
NGC 2237	14 (Minkowski 1949, 1955*a*) *H;* 17 (Wade 1958) *rf*
NGC 3372	120 (Bok 1932) *H;* 71 (Wade 1959) *rf*
NGC 6514	87 (Boggess 1954) *H;* 214–450* (Pottasch 1960*b*) 2*q*
NGC 6523	92–180* (Boggess 1954) *H;* 65–160* (Pottasch 1960*b*) 2*q*
	30 (Pronik 1960*a*) *H;* 49 [70]–110* [160] (Shain and Haze 1952*b*) *H* (cf. Boggess 1954)
NGC 6611	61 (Boggess 1954) *H;* 120–227* (Pottasch 1960*b*) 2*q*
NGC 6618	180–360* (Boggess 1954) *H;* 100* (Razmadze 1958) *H;* 78* [280] (Shain and Haze 1952*b*) *H* (cf. Boggess 1954)
NGC 7000	11–14 (Pikelner 1954*a*) *H;* 23–36* (Pottasch 1960*b*) 2*q*

* Maximum value. N_e derived from: Balmer line (*H*); two-quantum continuum (2*q*) or radiofrequency intensity (*rf*).

The radio and optical results do not always agree within the limits of error assigned by the respective authors: a factor ≤ 2 in $\mathfrak{M}$ is estimated in both surveys.

Table 7 concludes in columns 14 and 15, respectively, with the number of the exciting stars in the H II region and their range of spectral types from the *Henry Draper Catalogue* or its extension where these are known. Some stars which contribute strongly to the excitation of an H II region may not be visible (for example, there are almost certainly such stars in NGC 7000) or their spectra not observed. The exciting stars of an H II region are merely the hottest ones in it. They may contribute little to the total mass of the stars in the system, and the ionized gas may be a minor part of the total mass within an H II region, as is true for NGC 1976 (Johnson 1961).

Table 8 gives additional values of N_e but excludes NGC 1976 and nebulae for which only one value has been published (see Chamberlain 1953, Johnson 1953, Böhm 1956, Drake 1959, Gurzadian 1959, and Pike and Drake 1964). Higgs *et al.* (1964), Pike and Drake (1964), and Yang and West (1964) gave detailed radio maps of Cygnus X; Pike and Drake derive electron densities and masses of 20 nuclei of it. The surface brightness on which Table 8 is based may have been measured in a Balmer line (*H*), in the two-quantum continuum (2*q*), or in a radio-frequency (*rf*). Discrepancies of as much as a factor of 3 in N_e are common in Table 8. Most data are not expressed in such a way that one can tell whether nebular inhomogeneities explain the disagreements. Refined correction for extinction alone can alter N_e by a factor of 3: witness Boggess' (1954) reduction of the Shain and Haze (1952*b*) data for NGC 6523 and NGC 6618.

Several statistical conclusions have been drawn from the surveys on which Table 6 is based.

Figure 6 shows the number of H II regions, from 159 in Gershberg and Metik's catalogue, as a function of the logarithm of ionized-gas mass in solar units. The most abundant H II regions have a mass of about 100 $\mathfrak{M}_\odot$, and none was found more massive than

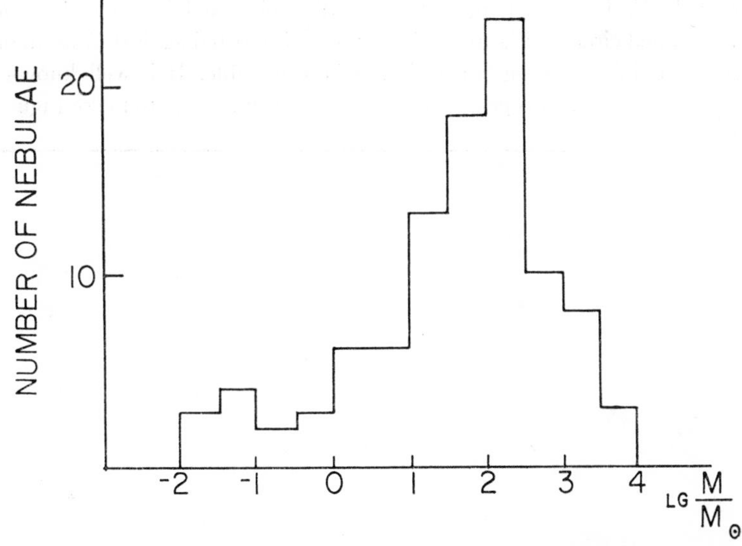

FIG. 6.—The frequency function for ionized-gas nebular masses (Gershberg and Metik 1960)

10,000 $\mathfrak{M}_\odot$. The lower limit is uncertain because the limit of detectability may occur before the lower limit is reached. The assumption of uniform electron density leads to systematic errors in the derived values of N_e and $\mathfrak{M}$. If the H II region is composed of gas clouds that fill only the fraction α of its apparent volume, it will resemble a uniform H II region in both size and mean surface brightness if the cloud density is $\alpha^{-1/2}N_e$ and the total mass is $\alpha^{1/2}\mathfrak{M}$. These factors apply equally to the optical and radio data.

Strömgren (1939) has predicted the radius s_0 of an H II zone as a function of the effective temperature and radius of the exciting star,

$$N_e^{2/3}s_0 = U\,. \tag{17}$$

The theory is discussed in Aller (1954, 1956) and in chapter 9 of this volume, in Gershberg and Pronik (1959*b*) and in Seaton (1960). For *n* stars, the radius becomes

$$S_0 = \left(\sum_{i=1}^{n} s_{0i}^3\right)^{1/3}. \tag{18}$$

By including the Lyman-continuum radiation field scattered by the gas cloud and the exact dependence of optical depth in the Lyman continuum on frequency below the series limit, Pottasch (1960*c*) showed that Strömgren's approximation of the function U might be increased by a factor of about 1.6. Münch (1960*a*) has published a short criticism of the work of Pottasch, but he has left Pottasch's result practically unchanged. Gould, Gold, and Salpeter (1963) treat the problem anew, but they change Strömgren's numerical results very little. A further improvement of the original Strömgren tabulation of U might be made by using modern values (Arp 1958, Underhill 1951) of absolute magnitudes, bolometric corrections, and effective temperatures, and by setting $T_e = 10^4$ °K rather than equal to the stellar effective temperature, but there is now much uncertainty about the ultraviolet flux of hot stars (Stecher and Milligan 1962).

Westerhout's N_e is plotted against the radius observed by Sharpless, $R = r \sin(\text{diam}/2)$, as a filled circle in Figure 7. The same N_e is plotted against S_0 as an open circle in the figure and joined to the filled circle where possible. It is well known that the *Henry Draper* early spectral types, all taken from Sharpless' list to make Figure 7 homo-

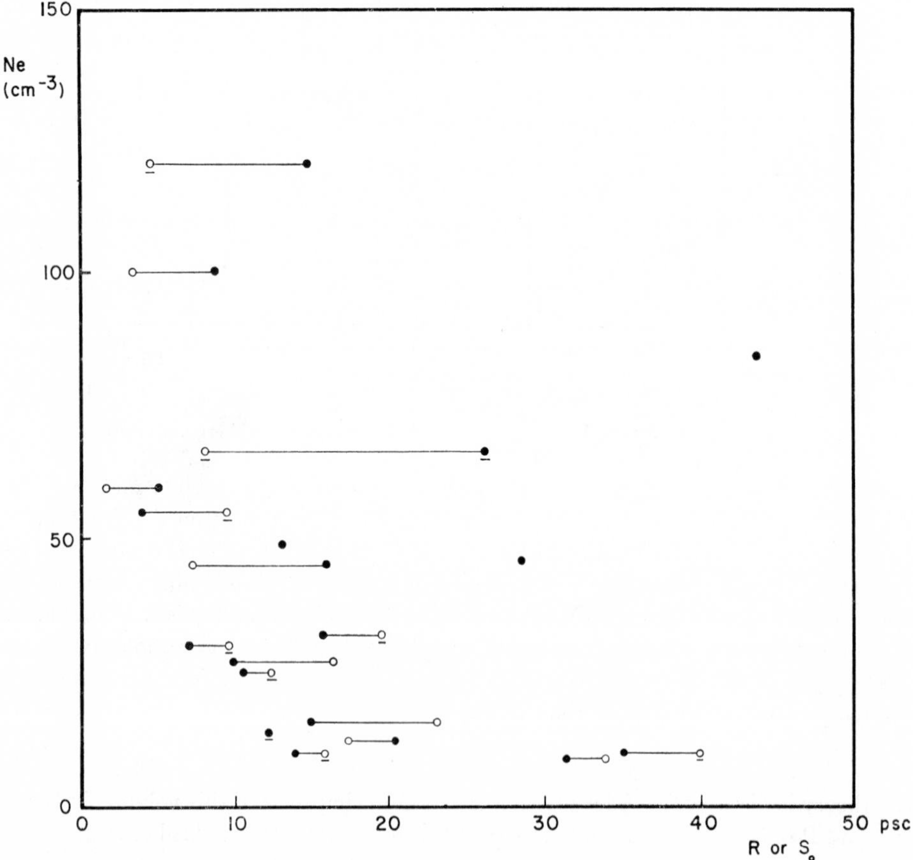

FIG. 7.—Electron density in H II regions plotted as a function of observed radius R (filled circles) or predicted radius s_0 (open circles). The most uncertain values are underlined. (Diagram prepared by G. E. McCluskey.)

geneous, tend to be too cool. For example, the nine exciting stars in NGC 2237 are classified B0–B2 in the *Henry Draper Catalogue*, but as O5–O9 V by Morgan, Whitford, and Code (1953). A general consequence of this should be to make $R > S_0$, as observed at large densities, but there is a tendency toward $R < S_0$ at small densities. Of all the sources of error that can enter here, it should be specially noted that deviations below 912 A from blackbody radiation at the adopted effective temperature of the star will be important when the model stellar atmospheres used are not reliable. Boggess (1961) has observed less radiation in the near ultraviolet in early B stars than current model atmospheres predict. Pronik (1960*b*) finds good agreement between the quantities R and S_0 above from the predictions of model stellar atmospheres, but he has checked only four H II regions: NGC 1976, NGC 2237, NGC 6514, and NGC 6523.

The question whether so-called H II spheres are generally ionization bounded or sometimes density bounded, also must be answered first. In the latter case, $R < S_0$. We cannot confidently maintain that the low N_e's in Figure 7, for example, are density bounded, because there are other possibilities—such as mere failure to detect the outermost and faintest parts of H II zones.

The lower left corner of Figure 7 is empty because it is outside the area of instrumental sensitivity. The emptiness of the upper right corner of Figure 7, together with the condition $R > S_0$, suggests that the census of hot stars in dense H II regions is incomplete, or that the known stars are overluminous in the far ultraviolet.

3.2. The Bright-Line Spectrum of NGC 1976

NGC 1976 is the most intense H II region in the Galaxy as viewed from the Sun. Its spectrum has been studied in much more detail than is possible for weaker H II regions. Ordinary diffuse nebulae and extended-emission regions are 1000 and 20,000 times fainter than the center of the Orion Nebula (Strömgren 1951); but if NGC 1976 is regarded as representative, except for its greater density, it can be used as a guide in many problems concerning interstellar matter (cf. "Joint Discussion B on the Orion Nebula," *Trans. I.A.U.*, Vol. XIIIB). Table 9 combines the Orion spectral-line data published by Aller and Liller (1959) in the wavelength region from 3726 to 10938 A with those of Kaler, Aller, and Bowen (1965) in the region between 3187 and 5016 A. The latter data are calibrated with the former. There are other recent but less comprehensive lists of wavelengths, identifications, and intensity estimates (Andrillat and Andrillat 1959; Chopinet and Fehrenbach 1961; Flather and Osterbrock 1960). The spectra were taken in the brightest central part of the Orion Nebula, within about 2′ of arc from the Trapezium.

If the theory of the recombination spectrum of hydrogen is correct, one could compare the data with the theoretical Balmer decrement or with the theoretical ratios of the intensities of Paschen and Balmer lines that originate from a given level and in this way derive reddening corrections for all emissions. An independent estimate of reddening is required in order to use the data as a check on the theory. Mathis (1957) derived the reddening of the Trapezium from the six-color photometry of Stebbins and Whitford (1945) and has applied one-half its reddening to nebular emissions near it. He argues that the Trapezium is probably behind most of the visible nebular material, because the observed radial velocities of the gas are not symmetrical around those of the Trapezium.

TABLE 9

Emission Lines in the Orion Nebula

Measured Wavelength (A)	I (obs)	I (0.5 cor)	I (1.0 cor)	Identification		
				Wavelength (A)	Element	Transition
3187.74......	6.3	8.2	10.9	3187.74	He I	$2^3S–4^3P$
3286.21......	......	......	......	3286.2	[Fe III]	$a^5D_3–a^3D_3$
3296.69......	0.25	0.3	0.4	3296.79	He I	$2^1S–8^1P$
3323.98......	......	......	......	3324.01	S III	$3d^3P_1–4p^3P_2$
3324.96......	0.18	0.2	0.3	3324.87	S III	$3d^3P_2–4p^3P_2$
3334.87......	0.17	0.2	0.3	3334.87	Ne II	$3s^4P_{5/2}–3p^4D_{7/2}$
......	......	......	......	3334.90	[Fe III]	$a^5D_2–a^3D_2$
3342.92......	0.17	0.2	0.3	3342.5	[Ne III]	$^1D_2–^1S_0$
......	......	......	......	3342.9	[Cl III]	$^4S–^2P_{3/2}$
3353.34......	0.08	1.0	1.3	3353.33	[Cl III]	$^4S–^2P_{1/2}$
3354.60......	0.35	0.4	0.6	3354.55	He I	$2^1S–7^1P$
3356.69......	......	......	......	3356.6	[Fe III]	$a^5D_1–a^3D_2$
3360.47......	......	......	......	3360.63	Ne II	$3s^4P_{1/2}–3p^4D_{3/2}$
3370.39......	......	......	......	3370.38	S III	$3d^3P_2–4p^3P_1$
3387.01......	0.27	0.3	0.4	3387.13	S III	$3d^3P_1–4p^3P_0$
3447.56......	0.61	0.8	0.9	3447.59	He I	$2^1S–6^1P$
3453.26......	......	......	......	3453.21	He I	$2^3P–20^3D$
3465.89......	......	......	......	3465.89	He I	$2^3P–17^3D$
3471.86......	0.13	0.2	0.2	3471.80	He I	$2^3P–16^3D$
3478.90......	0.13	0.2	0.2	3478.97	He I	$2^3P–15^3D$
3487.73......	0.18	0.2	0.3	3487.72	He I	$2^3P–14^3D$
3497.33......	0.06	0.07	0.09	3497.34	S III	unclassified
3498.62......	0.22	0.3	0.3	3498.64	He I	$2^3P–13^3D$
3512.60......	0.27	0.3	0.4	3512.51	He I	$2^3P–12^3D$
3530.47......	0.34	0.4	0.5	3530.49	He I	$2^3P–11^3D$
3554.49......	0.41	0.5	0.6	3554.39	He I	$2^3P–10^3D$
3587.33......	0.59	0.7	0.9	3587.25	He I	$2^3P–9^3D$
3613.65......	0.89	1.1	1.3	3613.64	He I	$2^1S–5^1P$
3634.27......	0.84	1.0	1.3	3634.24	He I	$2^3P–8^3D$
3652.05......	0.13	0.2	0.2	3651.97	He I	$2^3P–8^3S$
3654.69......	0.19	0.2	0.3	......	......	......
3656.05......	......	......	......	3656.14	H	$2^2P–38^2D$
3656.68......	0.15	0.2	0.2	3656.67	H	$2^2P–37^2D$
3657.32......	0.12	0.1	0.2	3657.27	H	$2^2P–36^2D$
3657.97......	0.37	0.4	0.5	3657.93	H	$2^2P–35^2D$
3658.68......	0.42	0.5	0.6	3658.64	H	$2^2P–34^2D$
3659.45......	0.41	0.5	0.6	3659.42	H	$2^2P–33^2D$
3660.30......	0.44	0.5	0.6	3660.28	H	$2^2P–32^2D$
3661.24......	0.49	0.6	0.7	3661.22	H	$2^2P–31^2D$
3662.22......	0.59	0.7	0.9	3662.26	H	$2^2P–30^2D$
3663.40......	0.63	0.8	0.9	3663.41	H	$2^2P–29^2D$
3664.70......	0.70	0.8	1.0	3664.68	H	$2^2P–28^2D$
3666.10......	0.70	0.8	1.0	3666.10	H	$2^2P–27^2D$
3667.70......	0.79	1.0	1.2	3667.68	H	$2^2P–26^2D$
3669.49......	0.89	1.1	1.3	3669.47	H	$2^2P–25^2D$
3671.48......	0.96	1.2	1.4	3671.48	H	$2^2P–24^2D$
3673.75......	1.06	1.3	1.6	3673.76	H	$2^2P–23^2D$
3676.37......	1.18	1.4	1.7	3676.37	H	$2^2P–22^2D$
3679.38......	1.30	1.6	1.9	3679.36	H	$2^2P–21^2D$
3682.80......	1.49	1.8	2.2	3682.81	H	$2^2P–20^2D$
3686.83......	1.69	2.0	2.5	3686.83	H	$2^2P–19^2D$
3691.56......	1.90	2.3	2.8	3691.56	H	$2^2P–18^2D$
3694.15......	0.10	0.12	0.15	3694.22	Ne II	$3s^4P_{5/2}–3p^4P_{5/2}$
3697.16......	2.11	2.5	3.1	3697.15	H	$2^2P–17^2D$
3703.89......	2.34	2.8	3.4	3703.86	H	$2^2P–16^2D$

TABLE 9—*Continued*

Measured Wavelength (A)	*I* (obs)	*I* (0.5 cor)	*I* (1.0 cor)	Identification: Wavelength (A)	Element	Transition
3705.00	1.12	1.3	1.6	3705.00	He I	2^3P-7^3D
3709.46	0.10	0.12	0.14	3709.64	Ne II	$3s^4P_{3/2}-3p^4P_{1/2}$
	...	...	...	3709.37	S III	$3d^3P_1-4p^3D_2$
3711.99	2.85	3.4	4.1	3711.97	H	2^2P-15^2D
3717.84	0.20	0.24	0.29	3717.78	S III	$4s^3P_2-4p^3S$
3721.92	4.1	4.9	5.9	3721.94	H	2^2P-14^2D
	...	...	...	3721.69	[S III]	$^3P_1-^1S$
3726.08 }	127	152	183	3726.05	[O II]	$^4S-^2D_{3/2}$
3728.76 }				3728.80	[O II]	$^4S-^2D_{5/2}$
3732.88	0.13	0.2	0.2	3732.86	He I	2^3P-7^3S
3734.30	3.5	4.2	5.0	3734.37	H	2^2P-13^2D
3747.98	...	...	...	3747.90	S III	$3d^3P_0-4p^3D_1$
3750.11	4.4	5.2	6.3	3750.15	H	2^2P-12^2D
3756.21	0.09	0.10	0.13	3756.10	He I	2^1P-14^1D
3758.49	0.06	0.07	0.09	...	...	...
3770.61	5.4	6.4	7.7	3770.63	H	2^2P-11^2D
3777.05	0.06	0.07	0.09	3777.16	Ne II	$3s^4P_{1/2}-3p^4P_{3/2}$
3784.93	0.09	0.10	0.13	3784.89	He I	2^1P-12^1D
3797.88	7.8	9.1	11	3797.90	H	2^2P-10^2D
3805.73	0.10	0.12	0.15	3805.77	He I	2^1P-11^1D
3806.46	0.08	0.09	0.11	...	...	...
3819.64	1.88	2.2	2.7	3819.61	He I	2^3P-6^3D
3829.80	0.05	0.06	0.07	3829.77	Ne II	$3p^2P_{3/2}-3d^2D_{5/2}$
3831.79	0.08	0.09	0.11	3831.85	S III	$4s^3P_0-4p^3P_1$
3833.58	0.13	0.15	0.18	3833.57	He I	2^1P-10^1D
3835.36	10.9	12.8	15.1	3835.39	H	2^2P-9^2D
3837.79	0.10	0.12	0.14	3837.80	S III	$4s^3P_1-4p^3P_1$
3838.35	0.13	0.15	0.18	3838.39	N II	$3p^3P_2-4s^3P_2$
	...	...	...	3838.32	S III	$4s^3P_2-4p^3P_2$
3853.88	0.02	0.02	0.03	3853.66	Si II	$2p^2\ ^2D_{3/2}-4s^2P_{3/2}$
3856.14	0.34	0.40	0.48	3856.02	Si II	$2p^2\ ^2D_{5/2}-4s^2P_{3/2}$
3862.70	0.19	0.22	0.24	3862.59	Si II	$2p^2\ ^2D_{3/2}-4s^2P_{1/2}$
3867.73	...	...	...	3867.48	He I	2^3P-6^3S
3868.76	19.7	23	27	3868.76	[Ne III]	$^3P_2-^1D$
3871.76	0.16	0.19	0.20	3871.82	He I	2^1P-9^1D
3882.49	...	...	...	3882.20	O II	$3p^4D_{7/2}-3d^4D_{7/2}$
3888.92	18.1	21	25	3889.05	H	2^2P-8^2D
				3888.65	He I	2^3S-3^3P
3918.94	0.13	0.15	0.18	3918.98	C II	$3^2P_{1/2}-4^2S$
3920.66	0.26	0.30	0.35	3920.68	C II	$3^2P_{3/2}-4^2S$
3926.53	0.22	0.25	0.30	3926.53	He I	2^1P-8^1D
3928.53	0.06	0.07	0.08	3928.62	S III	$3d^3D_3-4p^3P_2$
3933.76	...	...	...	3933.66	Ca II	$4^2S-4^2P_{3/2}{}^{abs}$
3964.80	1.35	1.6	1.8	3964.73	He I	2^1S-4^1P
3967.47 }	24.4	28	32	3967.47	[Ne III]	$^3P_1-^1D_2$
3969.97 }				3970.07	H	2^2P-7^2D
3982.61	0.02	0.02	0.03	...	...	...
3983.72	0.075	0.09	0.10	3983.77	S III	$3d^3D_2-4p^3P_1$
3985.87	0.075	0.09	0.10	3985.97	S III	$3d^3D_1-4p^3P_0$
3993.18	0.065	0.07	0.09	...	...	...
4009.27	0.28	0.32	0.37	4009.27	He I	2^1P-7^1D
4026.19	2.7	3.1	3.5	4026.19	He I	2^3P-5^3D
4068.69	1.49	1.7	1.9	4068.60	[S II]	$^4S_{3/2}-^2P_{3/2}$
4069.76	0.11	0.12	0.14	4069.90	O II	$3p^4D_{3/2}-3d^4F_{5/2}$
	...	...	...	4069.64	O II	$3p^4D_{1/2}-3d^4F_{3/2}$
4072.11	0.12	0.14	0.15	4072.16	O II	$3p^4D_{5/2}-3d^4F_{7/2}$
4075.82	...	...	...	4075.87	O II	$3d^4D_{7/2}-3d^4F_{9/2}$

TABLE 9—*Continued*

Measured Wavelength (A)	I (obs)	I (0.5 cor)	I (1.0 cor)	Identification		
				Wavelength (A)	Element	Transition
4076.44......	0.69	0.78	0.88	4076.35	[S II]	$^4S_{3/2}-{}^2P_{1/2}$
4077.75......	0.03	0.03	0.04	......	......	......
4087.12......	0.015	0.02	0.02	4087.16	O II	$3d^4F_{3/2}-4f^4G_{5/2}$
4089.21......	0.03	0.03	0.04	4089.30	O II	$3d^4F_{9/2}-4f^4G_{11/2}$
4097.27......	0.06	0.07	0.08	4097.31	N III	$3^2S_{1/2}-3^2P_{3/2}$
	......	......	......	4097.26	O II	{$3p\ P_{1/2}-3d^4D_{3/2}$ {$3d^4F_{7/2}-4f^4G_{9/2}$
4099.48......	0.05	0.06	0.06	......	......	......
4101.64......	25	28	32	4101.74	H	2^2P-6^2D
4104.96......	......	......	......	4104.74	O II	$3p^4P_{3/2}-3d^4D_{5/2}$
	......	......	......	4105.00	O II	$3p^4P_{3/2}-3d^2D_{3/2}$
4110.69......	0.05	0.06	0.06	4110.80	O II	$3p^4P_{3/2}-3d^4D_{1/2}$
4119.24......	0.06	0.07	0.08	4119.22	O II	$3p^4P_{5/2}-3d^4D_{7/2}$
4120.82......	0.30	0.34	0.38	4120.81	He I	2^3P-5^3S
4131.72......	0.05	0.06	0.06	......	......	......
4132.75......	0.06	0.07	0.07	4132.81	O II	$3p^4P_{1/2}-3d^4P_{3/2}$
4143.69......	0.45	0.50	0.56	4143.76	He I	2^1P-6^1D
4153.27......	0.08	0.09	0.10	4153.30	O II	$3p^4P_{3/2}-3d^4P_{5/2}$
4156.34......	0.13	0.14	0.16	4156.54	O II	$3p^4P_{5/2}-3d^4P_{3/2}$
4169.09......	0.07	0.08	0.09	4168.97	He I	2^1P-6^1S
4185.46......	0.015	0.02	0.02	4185.46	O II	$3p^2F_{5/2}-3d^2G_{7/2}$
4189.71......	0.03	0.03	0.04	4189.79	O II	$3p^2F_{7/2}-3d^2G_{9/2}$
4244.09......	0.05	0.05	0.06	4243.98	[Fe II]	$a^4F_{9/2}-a^4G_{11/2}$
4253.73......	0.06	0.07	0.08	4253.74	O II	$3d^2G_{9/2}-4f^2H_{11/2}$
	......	......	......	4253.98	O II	$3d^2G_{7/2}-4f^2H_{9/2}$
	......	......	......	4253.59	S III	$4s^3P_2-4p^3D_3$
4267.13......	0.36	0.39	0.43	4267.15	C II	3^3D-4^2F
4276.95......	0.03	0.03	0.04	4276.83	[Fe II]	$a^4F_{7/2}-a^4G_{9/2}$
4284.96......	0.03	0.03	0.04	4284.99	S III	$4s^3P_1-4p^3D_2$
4287.53......	0.11	0.12	0.13	4287.40	[Fe II]	$a^6D_{9/2}-a^6S_{5/2}$
4303.77......	0.03	0.03	0.04	4303.82	O II	$3d^4P_{5/2}-4f^4D_{7/2}$
4305.78......	......	......	......	4305.90	[Fe II]	$a^4F_{5/2}-a^4G_{5/2}$
4317.05......	0.06	0.07	0.07	4317.14	O II	$3s^4P_{1/2}-3p^4P_{3/2}$
4319.56......	0.03	0.03	0.04	4319.63	O II	$3s^4P_{3/2}-3p^4P_{5/2}$
	......	......	......	4319.62	[Fe II]	$a^4F_{5/2}-a^4G_{7/2}$
4326.13......	0.06	0.07	0.07	......	......	......
4332.57......	0.03	0.03	0.04	4332.71	S III	$4s^3P_0-4p^3D_1$
4340.27......	41	44	48	4340.47	H	2^2P-5^2D
4342.68......	0.04	0.04	0.05	......	......	......
4345.49......	0.11	0.12	0.13	4345.56	O II	$3s^4P_{3/2}-3p^4P_{1/2}$
4346.75......	0.03	0.03	0.04	4346.85	[Fe II]	$a^4F_{7/2}-a^4G_{11/2}$
4347.50......	0.03	0.03	0.04	4347.43	O II	$3s^2D_{3/2}-3p^2D_{3/2}$
4349.39......	0.08	0.09	0.09	4349.43	O II	$3s^4P_{5/2}-3p^4P_{5/2}$
4351.32......	0.015	0.02	0.02	4351.27	O II	$3s^2D_{5/2}-3p^2D_{5/2}$
4359.45......	0.09	0.10	0.10	4359.34	[Fe II]	$a^6D_{7/2}-a^6S_{5/2}$
4363.18......	1.55	1.7	1.8	4363.21	[O III]	$^1D-{}^1S$
4366.95......	0.06	0.06	0.07	4366.90	O II	$3s^4P_{5/2}-3p^4P_{3/2}$
4368.45......	0.09	0.10	0.10	......	......	......
4387.93......	0.6:	0.6	0.7	4387.93	He I	2^1P-5^1D
4409.12......	0.03	0.03	0.03	......	......	......
4413.91......	0.08	0.09	0.09	4413.78	[Fe II]	$a^6D_{5/2}-a^6S_{5/2}$
4414.98......	0.06	0.06	0.07	4414.91	O II	$3s^2P_{3/2}-3p^2D_{5/2}$
4416.29......	0.06	0.06	0.07	4416.27	[Fe II]	$a^6D_{9/2}-b^4F_{9/2}$
4416.91......	0.05	0.05	0.06	4416.98	O II	$3s^2P_{1/2}-3p^2D_{3/2}$
4437.66......	0.10	0.11	0.11	4437.55	He I	2^1P-5^1S
4446.87......	......	......	......	4447.03	N II	$3p^1P-3d^1D$

TABLE 9—*Continued*

Measured Wavelength (A)	I (obs)	I (0.5 cor)	I (1.0 cor)	Identification		
				Wavelength (A)	Element	Transition
4452.30......	0.06	0.06	0.07	4452.38	O II	$3s^2P_{3/2}-3p^2D_{3/2}$
	....	....	....	4452.11	[Fe II]	$a^6D_{3/2}-a^6S_{5/2}$
4458.12......	0.05	0.05	0.06	4457.95	[Fe II]	$a^6D_{7/2}-b^4F_{7/2}$
4471.43......	4.6	4.9	5.1	4471.48	He I	2^3P-4^3D
4590.89......	0.04	0.04	0.04	4590.97	O II	$3s^2D_{5/2}-3p^2F_{7/2}$
4607.06......	0.06	0.06	0.06	4607.15	N II	$3s^3P_0-3p^3P_1$
	....	....	....	4607.0	[Fe III]	$a^5D_4-a^3F_3$
4621.41......	0.03	0.03	0.03	4621.39	N II	$3s^3P_1-3p^3P_0$
	....	....	....	4621.5	[C I]	$^3P_1-^1S$
4630.60......	0.08	0.08	0.09	4630.54	N II	$3s^3P_2-3p^3P_2$
4634.01......	0.04	0.04	0.04	4634.16	N III	$3^2P_{1/2}-3^2D_{3/2}$
4638.81......	0.08	0.08	0.09	4638.85	O II	$3s^4P_{1/2}-3d^4D_{3/2}$
4640.73......	0.08	0.08	0.09	4640.64	N III	$3^2P_{3/2}-3^2D_{5/2}$
4641.86......	0.15	0.15	0.16	4641.81	O II	$3s^4P_{3/2}-3p^4D_{5/2}$
4649.15......	0.25	0.26	0.27	4649.14	O II	$3s^4P_{5/2}-3p^4D_{7/2}$
4650.85......	0.08	0.08	0.09	4650.84	O II	$3s^4P_{1/2}-3p^4D_{1/2}$
4658.18......	1.2	1.2	1.3	4658.1	[Fe III]	$a^5D_4-a^3F_4$
4661.67......	0.08	0.08	0.09	4661.64	O II	$3s^4P_{3/2}-3p^4D_{3/2}$
4676.11......	0.06	0.06	0.06	4676.23	O II	$3s^4P_{5/2}-3p^4D_{5/2}$
4677.99......	....	....	....	....	....	....
4699.12......	0.04	0.04	0.04	4699.21	O II	$3p^2D_{3/2}-3d^2F_{5/2}$
4701.69......	0.24	0.25	0.25	4701.5	[Fe III]	$a^5D_3-a^3F_3$
4705.54......	0.04	0.04	0.04	4705.36	O II	$3p^2D_{5/2}-3d^2F_{7/2}$
4711.48......	0.17 } 1.1	0.2	0.2	4711.34	[Ar IV]	$^4S_{3/2}-^2D_{5/2}$
4713.19......	1.05	1.0	1.0	4713.14	He I	2^3P-4^3S
4733.93......	0.09	0.09	0.09	4733.9	[Fe III]	$a^5D_2-a^3F_2$
4740.28......	0.17	0.17	0.18	4740.20	[Ar IV]	$^4S_{3/2}-^2D_{3/2}$
4754.87......	0.11	0.11	0.11	4754.7	[Fe III]	$a^5D_3-a^3F_4$
4769.61......	0.07	0.07	0.07	4769.4	[Fe III]	$a^5D_2-a^3F_3$
4774.42......	....	....	....	4774.74	[Fe II]	$a^4F_{9/2}-b^4F_{7/2}$
4777.88......	0.05	0.05	0.05	4777.7	[Fe III]	$a^5D_1-a^3F_2$
4814.69......	0.10	0.10	0.10	4814.55	[Fe II]	$a^4F_{9/2}-b^4F_{9/2}$
4861.24......	100	100	100	4861.33	H	2^2P-4^2D
4864.13......	0.08:	0.08	0.08	....	....	....
4881.14......	0.31	0.31	0.31	4881.0	[Fe III]	$a^5D_4-a^3H_4$
4921.97......	1.5	1.5	1.5	4921.93	He I	2^1P-4^1D
4931.05......	0.09	0.09	0.09	4931.0	[O III]	$^3P_0-^1D$
	....	....	....	4930.5	[Fe III]	$a^5D_1-a^3P_0$
4958.90......	113	112	111	4958.92	[O III]	$^3P_1-^1D$
(5007)........	342	335	332	5006.85	[O III]	$^3P_2-^1D$
(5016)........	2.24	2.2	2.1	5015.68	He I	2^1S-3^1P
5047.5.......	0.3	0.3	0.3	5046.1	He I	2^1P-4^1S
5056.4.......	0.3	0.3	0.3	5056.1	Si II	$4p^2P_{3/2}-4d^2D$
5157.0.......	0.3	0.3	0.3	....	....	....
5198.6.......	0.7	0.7	0.7	5198.5	[N I]	$2p^3\,^4S-2p^3\,^2D$
5269.4.......	1.6	1.5	1.5	5270.3	[Fe III]	$a^5D_3-a^3P_2$
5517.4.......	6.5	6.1	5.7	5517.7	[Cl III]	$2p^3\,^4S-2p^3\,^2D_{5/2}$
5536.9.......	16	15	14	5537.6	[Cl III]	$2p^3\,^4S-2p^3\,^2D_{3/2}$
5754.9.......	16	15	13	5754.6	[N II]	$2p^2\,^1D-2p^2\,^1S$
5875.5.......	31	28	25	5875.6	He I	2^3P-3^3D
5957.5.......	0	0	0	5957.6	Si II	$4p^2P_{1/2}-5s^2S$
5977.5.......	0	0	0	5979.0	Si II	$4p^2P_{3/2}-5s^2S$
6045.8.......	0.7	0.6	0.5	....	....	....
6300.3.......	6.5	5.5	4.6	6300.3	[O I]	$2p^4\,^3P_2-2p^4\,^1D$
6312.0.......	16	14	11	6312.1	[S III]	$2p^2\,^1D-2p^2\,^1S$
6347.0.......	0.7	0.6	0.5	6347.5	Si II	$4^2S-4^2P_{3/2}$
6363.6.......	1.6	1.3	1.1	6363.8	[O I]	$2p^4\,^3P_1-2p^4\,^1D$

TABLE 9—*Continued*

Measured Wavelength (A)	*I* (obs)	*I* (0.5 cor)	*I* (1.0 cor)	Identification: Wavelength (A)	Element	Transition
6371.3.......	0.3	0.3	0.2	6371.3	Si II	$4^2S-4^2P_{1/2}$
6401.6.......	0	0	0	6401.5	[Ni III]	$a^3F_3-a^3P_1$
6518.0.......	0	0	0	....	....	....
6548.2.......	18	15	12	6548.1	[N II]	$2p^2\,^3P_1-2p^2\,^1D$
6563.4.......	350	287	231	6562.8	H I	2^2P-3^2D
6583.7.......	55	45	36	6583.4	[N II]	$2p^2\,^3P_2-2p^2\,^1D$
6590.6.......	0.3	0.2	0.2	....	....	....
6597.7.......	0	0	0	....	....	....
6678.0.......	25	20	16	6678.1	He I	2^1P-3^1D
6716.4.......	6	5	4	6716.4	[S II]	$2p^3\,^4S-2p^3\,^2D_{3/2}$
6730.2.......	8	6	5	6730.8	[S II]	$2p^3\,^4S-2p^3\,^2D_{3/2}$
7065.3.......	13	9.8	7.4	7065.3	He I	2^3P-3^3S
7135.8.......	17	13	9.5	7135.8	[Ar III]	$2p^4\,^3P_2-2p^4\,^1D$
7234.5.......	0.7	0.5	0.4	....	....	....
7254.4.......	0.7	0.5	0.4	....	....	....
7281.3.......	7	5	4	7281.3	He I	2^1P-3^1S
7319.9.......	8.6	6.2	4.5	7319.9	[O II]	$2p^3\,^2D_{5/2}-2p^3\,^2P$
7330.2.......	8.6	6.2	4.5	7330.2	[O II]	$2p^3\,^2D_{3/2}-2p^3\,^2P$
....	72	45	28	9069.0	[S III]	$3p^2\,^3P-3p^2\,^1D$
....	5.8	3.6	2.2	9229.0	H	3^2D-9^2F
....	181	110	67	9531.8	[S III]	$3p^2\,^3P-3p^2\,^1D$
....	8	5	3	9546.0	H	3^2D-8^2F
....	10	5.9	3.5	10049.4	H	3^2D-7^2F
....	70	39	22	10830.3	He I	$2^3S-2^3P^0$
....	20	11	6.4	10938.1	H	$3D^2-6^2F$

However, Sharpless (1952) has published the color excess E_y of 24 stars as a function of distance from θ^1 Ori A to 30′. The mean E_y for the Trapezium stars is not more than the average for the central value of the function. So, in addition to the error of ± 1.2 per cent quoted by Aller and Liller for all but the weakest intensities and the probability of large systematic errors in the infrared intensities because of the variable influence of water-vapor absorption in the Earth's atmosphere, it is likely that the correction for interstellar reddening is uncertain by an amount in the range between all and half the reddening of the Trapezium in magnitudes. Therefore, in Table 9 the intensity column is split into the observed datum, I(obs), the same datum corrected for half the smoothed reddening of the Trapezium, I(0.5 cor), and the observed intensity fully corrected for the smoothed reddening of the Trapezium, I(1.0 cor). The Stebbins, Whitford, and Mathis reddening curve extends from 3500 A to 10,100 A, so it needs very little extrapolation. As for the Balmer-line intensities outside the center of the nebula, Gurzadian (1955) has published data between Hβ and Hζ to about 16′ from the center.

Inspection of Table 9 shows that no constant multiplied by the $\Delta m(\lambda)$ reddening correction of Stebbins *et al.* will make either the Balmer decrement $I(\text{H}\alpha):I(\text{H}\beta):I(\text{H}\gamma):I(\text{H}\delta):I(\text{H}\epsilon):I(\text{H}\zeta)$ or the Paschen-Balmer ratios $I(n-3):I(n-2)$, where $n = 6, \ldots, 9$, match the predictions of the theory of Burgess (1958) or of Clarke (1965); see also chap. 9, Sec. 5. Seaton (1955) noticed the anomaly in the Balmer decrement and suggested that it could be explained by an appreciable optical depth in the Balmer lines and the degradation of high-energy quanta into low-energy quanta. Pottasch (1960*a*) has given a theory of the Balmer decrement in diffuse nebulae, taking account of appreciable optical depth in the early Balmer lines. However, it is impossible to fit the results of his theory, for any value of optical depth in Hα, to the observed Balmer intensities corrected by any amount for reddening. Part of the trouble may lie in peculiarities of the Trapezium stars, whose intrinsic colors in various passbands are certainly not well known, part in the difficulty of observing colors through diaphragms and filters that admit strong nebular radiations, and part in that the nebular light is being scattered by dust on a line of sight that certainly does not match the line of sight to the Trapezium. Pottasch also notes that approximations in his theory need to be examined, and Gershberg (1961) has discussed them. Peter Boyce (1963) measured Balmer-line intensities in Orion, and three other diffuse nebulae: M8, M16, and M17. He estimated reddening by comparing observed and theoretical Balmer decrements. The observed line ratios indicated that none of these nebulae shows strong self-absorption. His observations of the Hβ flux, when corrected for absorption according to the data of Whitford (1958), showed good agreement with the radio-frequency observations for these nebulae. Boyce concluded there was little evidence for the anomalous absorption law suggested for Orion.

Other recombination lines in the Orion spectrum are those of He I, Si II, O II, N II, and C II. The function U for helium, analogous to equation (17) for hydrogen, has been derived by Swihart (1952), by Aller (1956), and by Seaton (1960). A table of Swihart's results is reproduced by Mathis (1957). Except for low hydrogen abundance, for low effective temperature of the ionizing star(s), or for strong differential extinction of light below 912 A by dust in an H II region, the radius of ionization should be equal for helium and hydrogen. It is probable that emissions of these elements in the spectrum of NGC 1976 originate along the same line of sight. Thus the emissivity per cubic centimeter may be compared accurately if pairs of wavelengths are chosen close together, so that only a weak correction for interstellar reddening is necessary. Mathis (1962) has obtained $N_{\mathrm{He}}/N_{\mathrm{H}} = 0.117$ from the ratio $I(\lambda\,5876):I(\mathrm{H}\beta)$. Gurzadian (1955) has observed that the ratio $I(\lambda\,4471):I(\mathrm{H}\beta)$ decreases by about 18 per cent with increase of distance from the center to 9′. This is opposite to the expectation for decreased extinction at 9′, and seems to be a larger change than can be attributed to variations of ionization or electron temperature in this range.

For the theory of the collisional effects between electrons and other elements, such as O^{+}, O^{++}, N^{+}, and S^{+}, see, for example, Aller (1956, chap. 5), Seaton (1960), or chapter 9 (Aller and Liller) in this volume. The behavior is much as it is in planetary nebulae, except that the source(s) of ionization may not be as central and the ionized region not as symmetric. Internal dust also complicates the radiation field. Once again, only NGC 1976 has been adequately studied for the behavior of some ions as a function of position in the H II region, although some studies of the behavior in the 30 Doradus and η Carinae nebulae have been reported (Faulkner and Aller 1965). Osterbrock and Flather (1959) have found the intensity ratios of [O II] 3726 A, [O II] 3729 A, [O III] 4959–5007 A, and Hβ to a distance of 29′ in various directions from the Trapezium. These data have been corrected for central reddening in accordance with Mathis (1957), with an interpolation from center to 10′ in order to match the full Sharpless (1952) color excesses in the range $10' \leq \rho \leq 29'$. Table 10 summarizes the Osterbrock-Flather material. Deviations from the tabulated means usually seem to be correlated with fluctuations of intensity—hence density. At 2′, Table 10 ratios are I(3726–3729 A):I(4959–5007 A) = 0.50 and I(3726–3729 A): I(Hβ) = 0.71. These values should agree closely with the corresponding ratios of Table 9 entries under I(0.5 cor)—namely, 0.34 and 1.5. The differences between the photographic and photoelectric values probably reflect photographic errors or differences in the parts of the nebula spectrum admitted through the slits.

Unfortunately, no comparable information has been published about other ions. The spectrograms of Johnson (1953), for example, integrate over very large areas. They show only that the ratio I([S II] 6717–6731 A):I([N II] 6584 A) = 0.39 ± 0.06 (semirange), varying unsystematically to a radius of about 15′. White (1952) found I([N II] 6584 A): I(Hα) = 0.23 ± 0.11 (semirange) also varying unsystematically to a radius of about 15′. The measurements, however, are restricted to only six well-defined positions.

TABLE 10

INTENSITIES IN NGC 1976 (CF. TABLE 13)

Intensity	Projected Radius ρ or Volume Radius r(′)						
	0	4	8	12	16	20	24
I [O II] 3729 A/I [O II] 3726 A...	0.52	0.88	1.03	1.12	1.17	1.21	1.25
log I [O II] 3726–3729 A−log I [O III] 4959–5007 A..........	−0.75	+0.08	+0.49	+0.71	+0.76	+0.78	+0.78
log I [O II] 3726–3729 A−log I (Hβ)........................	−0.47	+0.11	+0.26	+0.32	+0.33	+0.34	+0.34

Further studies, such as O'Dell's and Hubbard's (1965), or Reitmeyer's (1965) photoelectric observations in seven wavelengths at 88 points in the nebula, should be made.

Osterbrock and Flather's use of their material to build a model of the Orion Nebula will be presented in Section 3.5.

3.3. The Bright-Line Spectra of Other Diffuse Nebulae

Table 11 is an index to spectrographic and filter-photographic observations in which some numerical estimate of relative line intensities has been made. The estimates vary widely in precision and in number of lines included. Known supernovae-shell and novae-shell spectra are excluded, but the range of types of nebulae is broad, from the Herbig-Haro prestellar objects to the probable supernovae shells IC 443 and NGC 6960–92. (Note that some shells, for example, NGC 6888, are probably not of supernova origin.) Qualitative estimates of relative line intensities may be found, for example, in Struve and Elvey's (1939) H II-region survey. The relatively few analyses of the intensity data available are summarized in Table 12. Quantities in parentheses have been assumed or separately established.

The ratio of two intensities in the spectrum of an H II region is generally a function of N_e, T_e, and the relative abundance of the elements. Aller (1954, 1956), Seaton (1954, 1955, 1960), Pronik (1957), and Parker (1962) have discussed these functions (see also chap. 9). Applications of plasma theory to diffuse nebulae have been few, apparently because of the faintness of the emissions except in NGC 1976. Treatment of the diffuse-nebula plasma is complicated by losses of energy by collisions involving grains (Spitzer 1949). A few stars, such as AE Aur, may be moving through their H II regions too fast for an equilibrium temperature to be reached (Spitzer and Savedoff 1950; Herbig 1958). The Balmer discontinuity is also affected by scattering by grains in diffuse nebulae. Boggess (1954) tried to derive T_e from the ratio of intensity in Hα to the flux density at 3200 Mc/s for NGC 6514, NGC 6523, NGC 6611, and NGC 6618. He found that uncertainties about extinction at Hα made the results ambiguous. Wade (1958)

TABLE 11

GALACTIC DIFFUSE NEBULAE (OTHER THAN NGC 1976) FOR WHICH THERE ARE NUMERICAL ESTIMATES OF BRIGHT-LINE INTENSITY RATIOS

Nebula	Ref.	Nebula	Ref.
NGC 281	17, 31	Struve-Elvey (1939) 44	17
IC 59	6, 17	Gum (1955) 17	18
IC 63	6, 17	NGC 3372	3, 18, 32
S 22	24	NGC 3603	18
IC 1805	6	NGC 6193	18
NGC 1499	6, 12, 16, 17, 23	IC 4628	18
T Tauri	22	NGC 6302	8
IC 405	12, 17	NGC 6357	18
IC 423	16	NGC 6514	1, 3, 30
λ Orionis	6, 17	NGC 6523	1, 3, 7, 13, 17, 27, 30, 33
NGC 1977	12, 23	NGC 6563	18
NGC 1982	12	NGC 6611	1, 16, 30, 33
S 147	24	NGC 6618	1, 3, 10, 11, 13, 28, 33
Herbig-Haro No. 1	2, 15, 22	Struve-Elvey (1939) 12	17
Herbig-Haro No. 2	15	NGC 6888	14, 24, 29
IC 434	6, 12, 16, 17, 23, 29	S 107, 108	29
NGC 2024	16, 23	Struve-Elvey (1939) 10	6
S 151, 155	29	IC 1318	6, 12
Struve-Elvey (1939) 41	23	NGC 6960-92	4, 5, 16, 19, 20, 21, 24, 26
IC 2144-8	18	NGC 7000	9, 12, 13, 16, 17, 25
NGC 2175	12, 16, 17	Struve-Elvey (1939) 12, Cederblad (1946) 190	17
IC 443	21, 24	Struve-Elvey (1939) 16	17
NGC 2237	12, 16, 17, 18, 23	10 Lacertae	17
NGC 2261	14	NGC 7538	17
NGC 2264	17, 29	NGC 7635	16, 17
IC 2177, Struve-Elvey (1939) 45–46	17		

References: (1) Boggess (1954); (2) Böhm (1956); (3) Campbell and Moore (1918); (4) Chamberlain (1953); (5) Code (1958); (6) Courtès (1960) also gives some inequalities for I([N II]/I(Hα); (7) Dufay (1954); (8) Evans (1959); (9) Gershberg and Pronik (1959*a*); (10) Gershberg and Pronik (1961); (11) Gershberg *et al.* (1961); (12) Greenstein and Henyey (1939*b*); (13) Gurzadian (1956*a*); (14) Gurzadian (1959); (15) Herbig (1951); (16) Hubble (1922*a*); (17) Johnson (1953); (18) Johnson (1960*b*); (19) Minkowski (1955*b*); (20) Minkowski (1958); (21) Osterbrock (1958*a*); (22) Osterbrock (1958*b*); (23) Page (1948); (24) Parker (1962); (25) Pikelner (1954*a*); (26) Pikelner (1954*b*); (27) Pronik (1960*a*); (28) Razmadze (1958); (29) Roshkovsky (1957*a*); (30) Thackeray (1950); (31) Hagen-Thorn (1962); (32) Faulkner and Aller (1965); (33) Boyce (1963).

went entirely into the radio-frequency spectrum and obtained $T_e = 8600°$ K in NGC 2237 from the ratio of flux densities at 85.5 Mc/s and 600 Mc/s. The error was estimated to be 30 per cent. In eight regions in the η Carinae Nebula, Faulkner and Aller (1965) found electron temperature between 9600° K and 10,400° K, while in eight other regions in 30 Doradus T_e lies between 10,300 and 10,800° K.

According to Miyamoto (1956), the effects of mass motions on collisional excitation are not observed in ordinary diffuse nebulae.

Little can be done to map the relative volumes occupied by various ions. A smaller volume of ion emission than that of Hα emission may merely mean that the former is too weak to be observed in regions farthest from the center of excitation—in the same way that Hβ is rarely observed to extend as far as Hα.

3.4. The Continuous Spectrum

The atomic theory of the continuous spectrum as applied to planetary nebulae may be also applied to diffuse nebulae if, in addition, scattering by grains is taken into account. Shain, Haze, and Pikelner (1954) observed 20 emission-type, 15 reflection-type, and 3 mixed-type diffuse (as well as three planetary) nebulae in passbands of 1400 A width centered at 5900 A and of 240 A width centered near Hα in order to measure the yellow continuum relative to Hα. After freeing the yellow continuum of Hα light leaked by its filter, and vice versa, they derived the corresponding surface brightnesses in magnitudes per square minute of arc, m_v and $m_{\mathrm{H}\alpha}$, from extrafocal images of the North Polar Sequence. Almost all nebulae, and different parts of given nebulae, obeyed the relation

$$m_v = m_{\mathrm{H}\alpha} + 2.79 \pm 0.09 . \tag{19}$$

This result led Shain, Haze, and Pikelner to believe that the yellow continuum was atomic. Of the possible bound-free, free-free, and two-quantum components, they sup-

TABLE 12

Physical Interpretations of Relative Line-Intensity Ratios in Galactic Bright Nebulae Excluding NGC 1976

Nebula	N_e (cm^{-3})	T_e (1000° K)	Note	Ref.
IC 443	590	(20)	I(3729 A)/I(3726 A) = 1.10	Osterbrock (1958*a*)
	350	(10–15)		Parker (1962)
NGC 6960–92	3×10^3 -5×10^4	80–200	Collisional excitation. No star.	Minkowski (1955*b*, 1958)
	40–500	40	$1.20\leq I$(3729 A)–I(3726 A) ≤1.47; I(3726–3729 A)/I(4959–5007 A) = 1.74	Osterbrock (1958*a*)
	220	(10)		Parker (1962)
	300–1000	400–450		Pikelner (1954*b*)
IC 405	....	7.6–9.0	Assumes N(O)/N(H) = 0.2×10^{-3}.	Pronik (1957)
NGC 6523	(30)	7.5–8.7	T_e highest at center.	Pronik (1960*a*)
	4×10^2–4×10^3	(10)	$0.53\leq I$(3729 A)/I(3726 A) ≤0.57 in region around Herschel 36.	Woolf (1961)
NGC 6888	400	(10)	WN star in supernovae shell?	Parker (1962)
NGC 7000	(10)	10	T_e same O$^+$ and O^{++} zones.	Gershberg and Pronik (1959*a*)
Herbig-Haro No. 1	1.3×10^4	7.5	Various forbidden lines; T_s = 24,000° K for $\lambda<353$ A.	Böhm (1956); cf. Haro and Minkowski (1960)
	3.7×10^3	(7.5)	I(3729 A)/I(3726 A) = 0.56; N(H$^+$)/N(H) = 0.5	Osterbrock (1958*b*)
T Tauri	3.7×10^3	(7.5)	I(3729 A)/I(3726 A) = 0.55	Osterbrock (1958*b*)
NGC 2261	(15)	....	N(H$^+$)/N(H) = 0.1; N(H) = 200 cm^{-3}.	Gurzadian (1959)
Rim of elephant trunk south of S Mon	200–320	(10)	$1.20\leq I$(3729 A)/I(3726 A) ≤1.29	Osterbrock (1957*b*)
NGC 3372	620	10	O II and other forbidden lines.	Faulkner and Aller (1965)

posed that the last was principally responsible. In discussing grains, they made the point that if gas and dust are present in nebulae, the intensity ratio of gas and dust will vary strongly as a function of density. Therefore, according to equation (19), the light scattered from grains would be a minor part of the continuum.

Other observations do not support the Russian data. Boggess (1954) has observed the ratio of continuous emission at 5800 ± 300 A to Hα emission in NGC 6514 and NGC 1976 to be very different from its value in the group NGC 6523, NGC 6611, and NGC 6618. Rozhkovsky (1955) has shown that the areal distribution of light in NGC 2237–9 is much more centralized in the wavelength region 4900–6200 A than in Hα, which radiates in a well-known ring structure. Wurm and Rosino (1956) have found a distinct lack of correlation between continuous emission at 5200 ± 90 A and Hα emission within NGC 1976, NGC 5067, and NGC 7000. Unpublished spectrograms taken by E. F. Carpenter at Steward Observatory and by Herbig at Lick Observatory show interesting changes in the continuum and monochromatic emissions in the Orion Nebula as the slits cross from the blue fringe near BD−5°1329 into the interior red area west of that star. The blue fringe has the stronger continuum in the blue, and the red area shows the Balmer discontinuity more distinctly. The Balmer lines smoothly increase in intensity from the blue to the red zones, but He I4471 A, [O III] 4959–5007 A, and [Ne III] 3869 A are relatively strong in the blue and weak in the red, and [N II] 6548–34 A and [S II] 4069 A are relatively weak in the blue, behaving somewhat like Balmer lines. It seems that the blue fringe is blue because of a strong continuum which is not proportional to Balmer intensities, and that conditions of excitation and ionization, especially ionization, are correlated with the intensity of the blue continuum. The ionizing radiation of the Trapezium may be selectively scattered before reaching different parts of the gas. Rozhkovsky (1956*a*) has shown that the red (western) fringe is polarized 5.8 per cent in the visual spectrum and the blue fringe is polarized 8.2 per cent.

Apart from the question as to whether equation (19) can be confirmed, it is not clear that observations of the yellow continuum have been freed of monochromatic emissions, particularly He I 5876 A, which is observed in NGC 1976 (Aller and Liller 1959) to be 2.6 mag fainter than Hα, or 2.8 mag fainter than Hα + [N II] 6548–84 A.

Processes other than atomic or scattering may contribute to the intensity in nebular continua. Struve and Swings (1948) have shown that the size of B10, a brightening in the dark nebula B7 associated with the RW Aurigae (RW_n) stars DD Tauri and CZ Tauri, is so large that the stars fall 7 mag off the Hubble relation, equation (1). The "Balmer continuum" of DD Tauri is so strong that it cannot be appreciably reddened by absorption. Struve and Swings suggest visible fluorescence of solid grains in B10, although Ambartsumian (1955) hypothesizes that matter is ejected from the stellar interiors and its energy converted into visual radiation. Dombrovsky (1958*a*) reports the presence of nonradial polarization in NGC 6618 as high as 50 per cent, and suggests an analogy with the Crab Nebula. This is not confirmed by radiofrequency observations (Little 1962).

The dependence of the intensity of the continuum of NGC 1976 on wavelength is rather well known. This cannot be said of any other diffuse nebula. The dependence is usually expressed in terms of a color temperature T_c and the Balmer discontinuity

$$D = \log \frac{I(\lambda < 3646)}{I(\lambda > 3646)}. \tag{20}$$

The best observations, all photographic, of regions near the center of NGC 1976 are summarized in Table 13. As stated in Section 3.2, corrections for reddening are not precisely known and are therefore not applied to the tabulated T_c. The uncorrected color temperatures are rather consistent at 10,000° or 12,000° K from 3100 A to 8700 A except for Barbier's anomalous $T_c = 23{,}500°$ K in the range 4500 to 6200 A. The infrared value agrees with the atomic continuum predicted for $T_e = 10{,}000°$ K by Spitzer and Greenstein (1951). Seaton (1955) discusses the Greenstein and Barbier observations as an example of an atomic continuum. The quantities D and T_c then depend upon the relative number of $2s \rightarrow 1s$ emission processes in the H II region and the number of such processes in the surrounding H I region. The agreement between the predicted values and the data of Table 13 is not very good, and Seaton notes that the discrepancies become greater if Balmer self-absorption and reddening are taken into account. The large

TABLE 13

THE CONTINUUM OF THE CENTRAL REGION OF NGC 1976

D (mag)	Observed T_c (° K)	Range of Wavelength (A)	Ref.	Note
0.64...	12000	3100–3646	1	Compared with θ^1 Ori C.
	11500	4000–5000	1	$12{,}000 < T_c < 20{,}000$ with correction for reddening.
0.52...	10300	3100–3646	2	Compared with β Ori.
	10000	4000–5000	2	No correction for reddening.
	23500	4500–6200	2	
....	10000	6000–8700	3	Compared with an extrapolation from α Leo and ϵ Hya. No correction for reddening.

(1) Greenstein (1946); (2) Barbier (1949); (3) Andrillat and Andrillat (1959).

Balmer discontinuity certainly means that at the center of NGC 1976 scattered light is a minor part of the continuum. A detailed study of processes in a dust-free H II region inside an H I envelope with dust leads Yada (1960) to conclude that the envelope does not play an important role in the spectrum of NGC 1976.

3.5. FORMS AND STRUCTURAL FEATURES

When nebular emissivity per unit volume j_λ is known as a function of the electron density and the density of ions concerned, the spectral intensity I_λ may be predicted. In a spherical model of radius R,

$$I_\lambda(\rho) = \int_{-\sqrt{(R^2-\rho^2)}}^{+\sqrt{(R^2-\rho^2)}} \frac{j_\lambda(r)}{4\pi}\,ds \qquad (21)$$

where ρ is the minimum distance from the line of sight to the center and $\rho^2 + s^2 = r^2$. Using the theoretical j_λ's given by several workers, Osterbrock and Flather (1959) found distributions of electrons and ions, summarized in Table 14, that agree with observed intensities when put into equation (21). From consideration of the emission measure (eq. [2]) and some radio flux densities (eq. [8]) predicted from their model, Osterbrock and Flather found that the quoted ion and electron densities could not be smooth functions of the radius out to $R = 24'$. Rough agreement with observations of flux densities was achieved with only one-thirtieth of the sphere filled with gas in separate clouds each with the Table 14 density as a function of r. Such piecemeal filling with gas reduces the flux but does not alter the relative intensities of the optical spectrum. Menon's (1961) and Pariiskii's (1961) radio observations show that the filling factor is a function of r, as though the density in the clouds falls off and the distance between clouds increases with increasing r. Radio observations give root-mean-square densities rather than cloud densities, and they lead to 100 $\mathfrak{M}_\odot$ (Menon's model), or as much as 156 $\mathfrak{M}_\odot$ (Pariiskii's model), of ionized gas, rather than 60 $\mathfrak{M}_\odot$ as in Osterbrock and Flather's model. Menon (1962) shows that the distribution of ionized gas closely resembles the distribution of the stars to apparent visual magnitude 18 (Johnson 1961). This common distribution is nearly that of an isothermal gas sphere. The sudden conversion of the gas sphere from neutral to ionized hydrogen, following the birth of a hot star such as θ^1 Ori C, would produce a rapid deformation from the isothermal distribution

TABLE 14

GAS AND STAR DENSITIES IN NGC 1976 (CF. TABLE 10)

ITEM	PROJECTED RADIUS ρ OR VOLUME RADIUS r (′)						
	0	4	8	12	16	20	24
$N(O^{++})/[N(O^+)+N(O^{++})]$ (degree of further ionization of O^+)....	0.74	0.53	0.30	0.17	0.14	0.13	0.13
log N_e (electrons cm^{-3} in clouds filling $\alpha=0.033$ of the volume)..	4.25	3.28	2.96	2.75	2.62	2.51	2.42
log d (stars pc^{-3})................	4.1:	2.08	1.49	1.08	0.74	0.32	−0.16

in as short a time as 10^4 years (Kahn and Menon 1961). This leaves us with an apparent paradox, for according to the cluster astrometry of Strand (1958) and according to stellar radial velocities (Wilson 1953, Herbig 1960, Johnson 1965), the velocity of escape in the system (Johnson 1961) is exceeded by the cluster motions. The use of Hayashi's (1961) improved theory of contracting stars for the recomputation of the cluster mass will only predict a reduction of the mass and a corresponding reduction of the velocity of escape. Thus the isothermal distribution of the stars and gas seems to be inconsistent with the positive energy of the system. Herbig (1962) has suggested that the positive energy of the present system is a result of the rapid expulsion of a considerable gaseous mass shortly after the formation of θ^1 Ori C—that the system hung together earlier because of the gravitational potential of the presently missing mass. However, it would be difficult to expel the mass in the Kahn-Menon time scale of 10^4 years and to expel much of the mass in any time scale without disturbing the isothermal distribution.

The fainter cluster stars around θ^1 Ori C are most easily photographed in the infrared, where they radiate much but where the gas emits little infrared radiation (see Plates 4 and 5).

Expansion is a natural characteristic of a hot gas mass imbedded in a cool gas. The imbedding is supposed to develop rather suddenly when a massive, hence hot, star contracts from the cool gas onto the main sequence in the time $T_{gc} = 6.32 \times 10^2 \mathfrak{M}^7 L^{-1} R^{-1}$ years (Sandage 1958), where $\mathfrak{M}$, L, and R are main-sequence values of mass, luminosity, and radius in solar units. During a short main-sequence lifetime of $T_N = 1.1 \times 10^{10} \mathfrak{M} L^{-1}$ years $\cong 10^2\ T_{gc}$ (Sandage 1958), the massive star energizes the gas to produce the nebula. The cooling of the surface of the star then de-energizes the gas and causes the nebula to disappear. Most H II regions seem to be denser than surrounding H I regions. This is expected if massive stars can contract only from the denser parts of the H I regions. Multiple centers of contraction within one dense region and within the time scale T_N account for Ambartsumian's (1949) stellar O-associations which, under Sharpless' definition, are components of H II regions. An O-association older than one generation, extending longer than T_N, would end after the expansion had rarefied the H II region too much to produce new massive stars.

Hot stars without an H II region but in a field of H I, as in h and χ Persei (Drake 1958), imply such dense nodules of H I that H II is evaporated from their surfaces and dispersed before becoming visible. The "rocket effect" of Oort and Spitzer (1955) might then explain why the mean H I radial velocity is -57 km/sec compared with the mean stellar velocity of -41 km/sec in h and χ Persei. An alternative explanation is Blaauw's (1961), involving supernovae explosions in massive and unstable double-star components with the explosions sweeping the field of interstellar gas.

Nebulae can take different forms during expansion. Shain and Haze (1951, 1953) described what they called peripheral nebulae, and supposed them to be a phenomenon of expansion and braking by the environs with the central parts being depleted of departing gases. It is not known whether the difference between centrally concentrated nebulae (for example, NGC 1976) and peripherally concentrated nebulae (for example, NGC 2237) is a result of difference in initial configuration, manner of expansion, or age (Kahn and Menon 1961). Perhaps the peripheral types are more uniform at the beginning or have smaller reserves of intermingling dense H I to be converted to H II near the central point. Westerlund (1960*b*) has discovered and described a most interesting series of peripheral nebulae centered on HD 148937: (*a*) the S-shaped nebulae NGC 6164–5, each about 3′ from the star, (*b*) an arc 13′.5 to the southeast of the star, (*c*) a nearly circular nebulosity extending 44′ to the northeast and 64′ to the west of the star, coinciding with NGC 6188 in the southeast, and (*d*) an outermost dark-cloud ring. The whole array is shown in Plate 6. According to Westerlund, HD 148937 is a most peculiar star of type B0 or earlier.

Bok and Reilly (1947) first drew attention to "globules"—small, symmetric, dark nebulae. They are usually seen projected against a bright diffuse nebula, and may be satellites of some types of bright nebulae since they are abundant near some, for example, NGC 2237 and NGC 6523, and absent near others, for example, NGC 1976. If, as Bok proposes, globules are protostars, a fundamental question is why some dust appears as small clouds in some H II regions and remains dispersed in others where the N_e may be greater. Leaving this evolutionary question unanswered, Pottasch (1956) put globules at one end of a sequence of nebulae that include the "elephant-trunk" structures with bright rims inside diffuse nebulae which are convex toward the exciting star. Next to globules in the sequence are dark trunks with the sharpest, shortest, and densest bright rims, which are found closest to the exciting star. This progression of forms and distances goes on to broad, long rims near the boundaries of diffuse nebulae which are sometimes concave toward the exciting star. These features are indistinguishable from some periph-

PLATE 1.—The η Carinae Nebula (left) and NGC 3199 (right) with a very distant H II region and radio source between them. The radio source is very similar to the η Carinae source at 1400 Mc/s. The plate was taken in Hα light with the Uppsala 20/26-inch Schmidt telescope at Mount Stromlo (Westerlund 1960*a*). The field is $2^\circ\!.6 \times 3^\circ\!.3$; N. top and E. left.

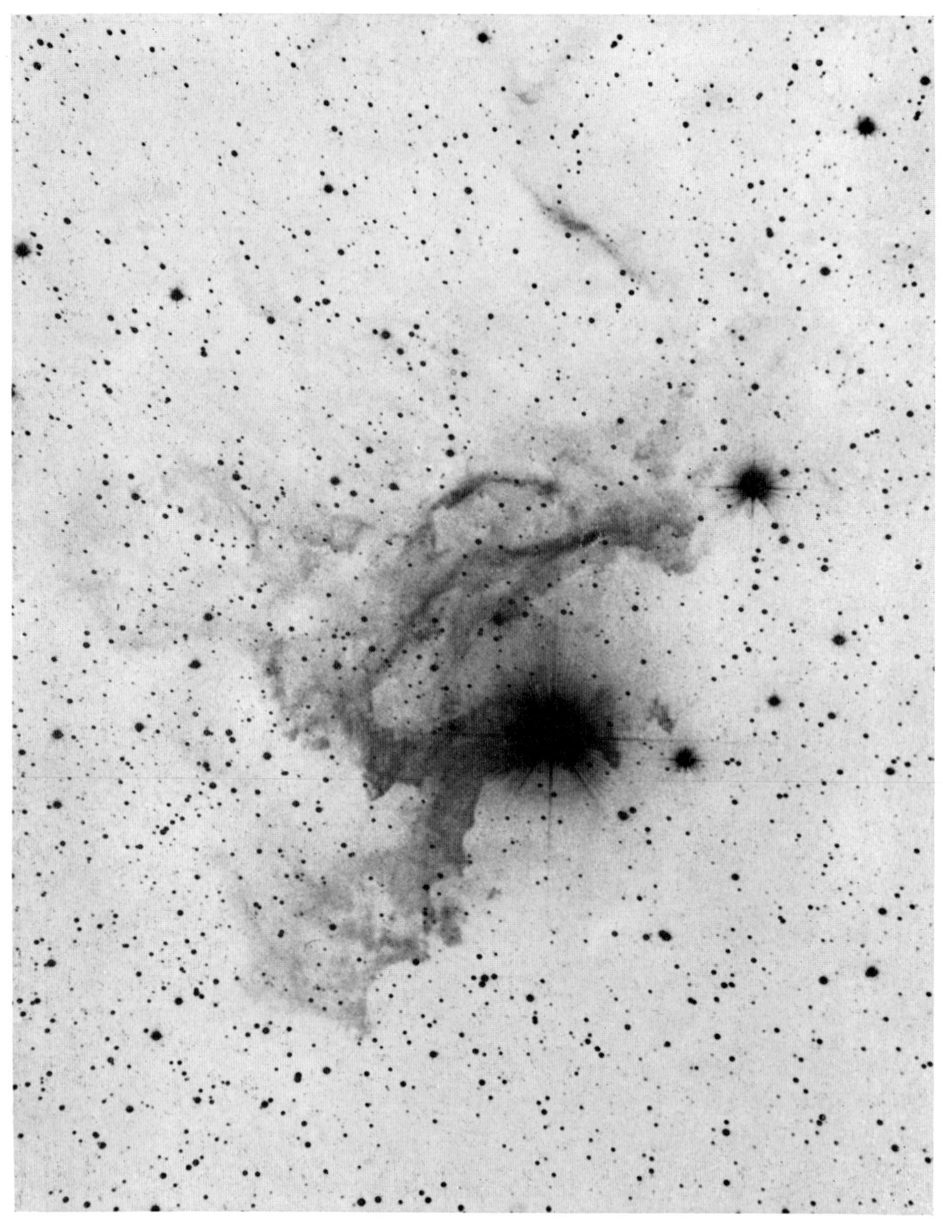

PLATE 2.—Negative print of IC 405 in the range 3800–5000 A; some filaments that are part of the emission nebulosity are also recorded (cf. Plate 3). AE Aurigae is the brightest star. The field is 26′ × 33′; N. top, E. left. Photographed with the Lick Crossley reflector by Herbig (1958).

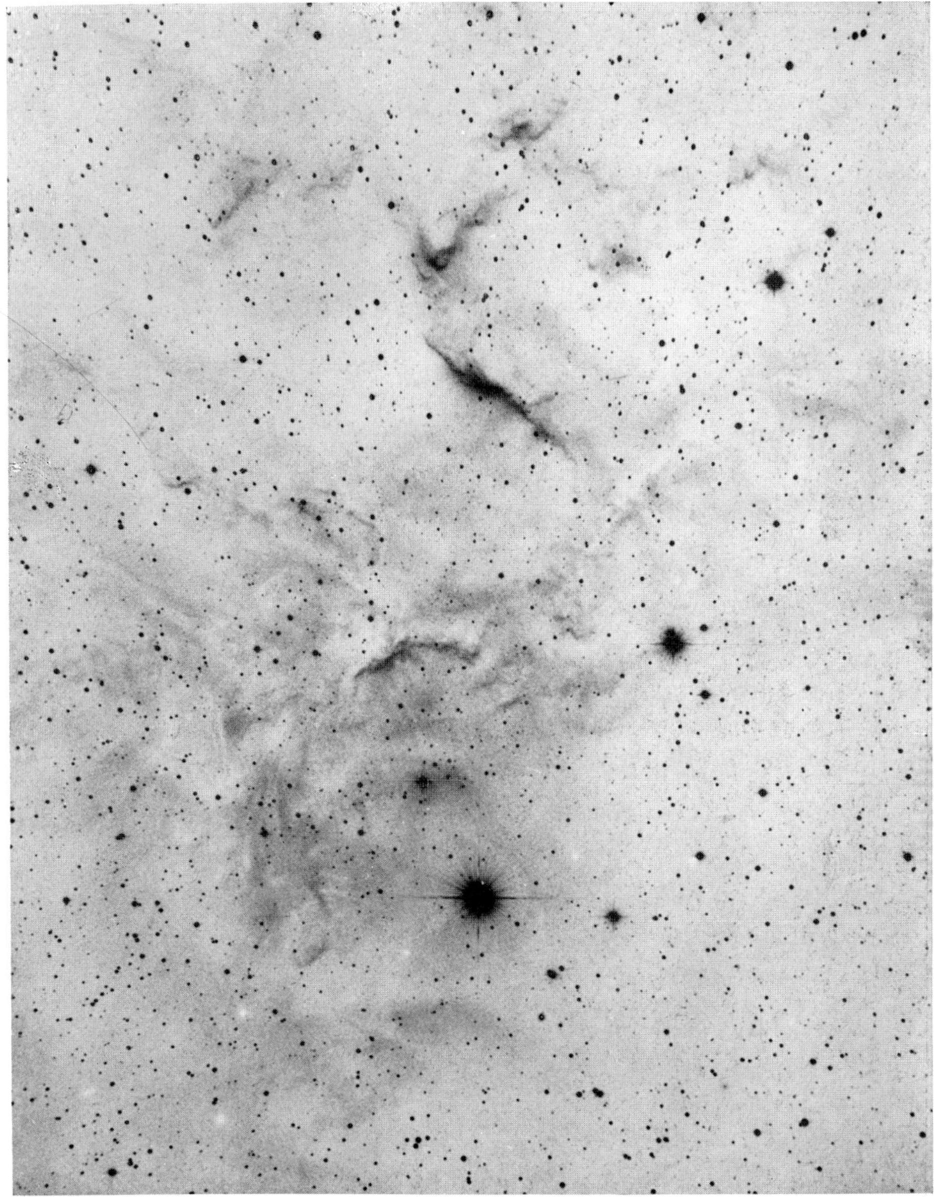

PLATE 3.—Negative print of IC 405 in emission, photographed with the Crossley reflector in the range 6300–6750 A to emphasize Hα. The field is centered north of Plate 2, but scale and orientation are the same.

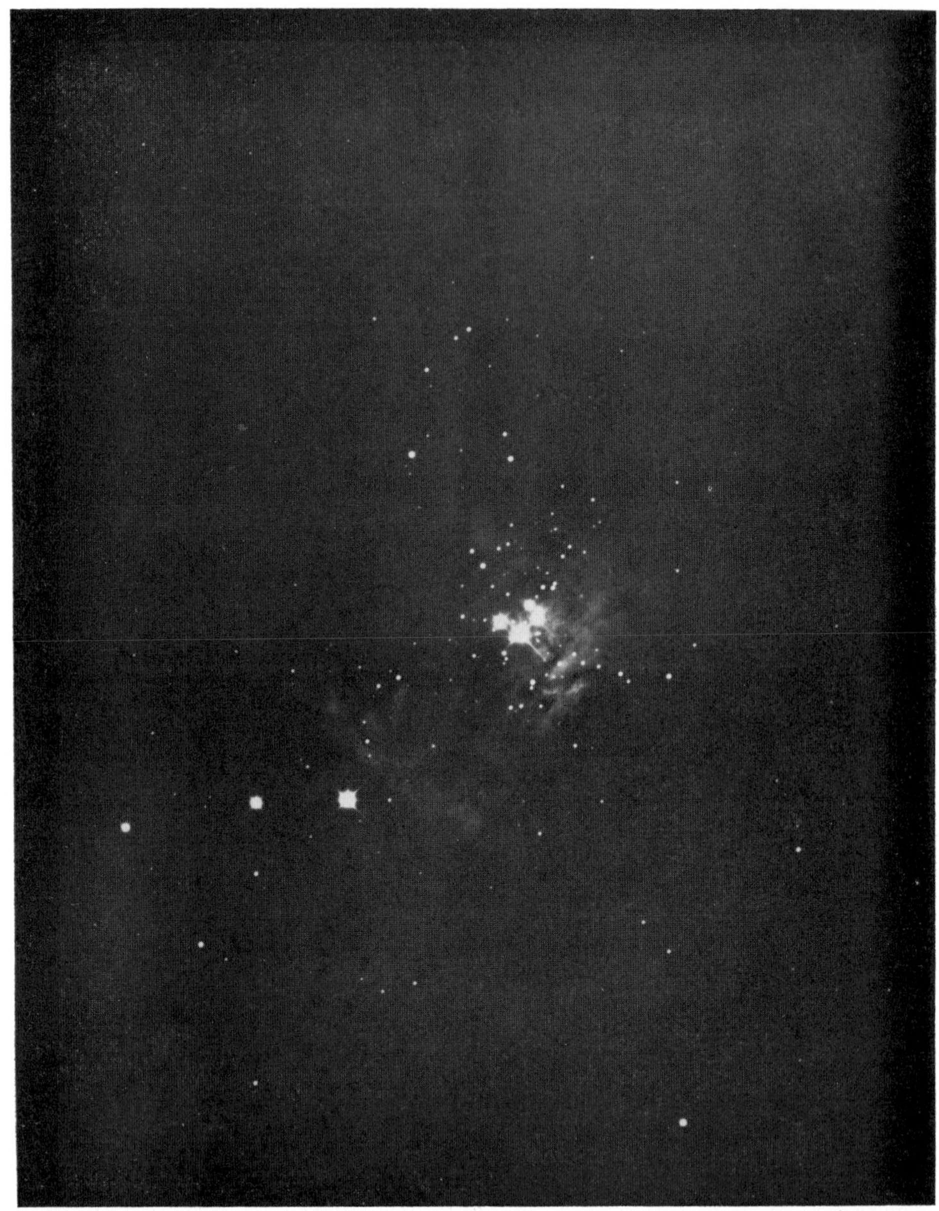

PLATE 4.—The Orion Nebula and cluster in the near infrared (7100–8700 A) taken by G. H. Herbig at the prime focus of the Lick 120-inch (exposure 3 min). Weak emissions of [Ar III], He I, and [O II] and some of the Paschen series and continuum contribute. The field is approximately 9′ × 11′; N. at top and E. at left.

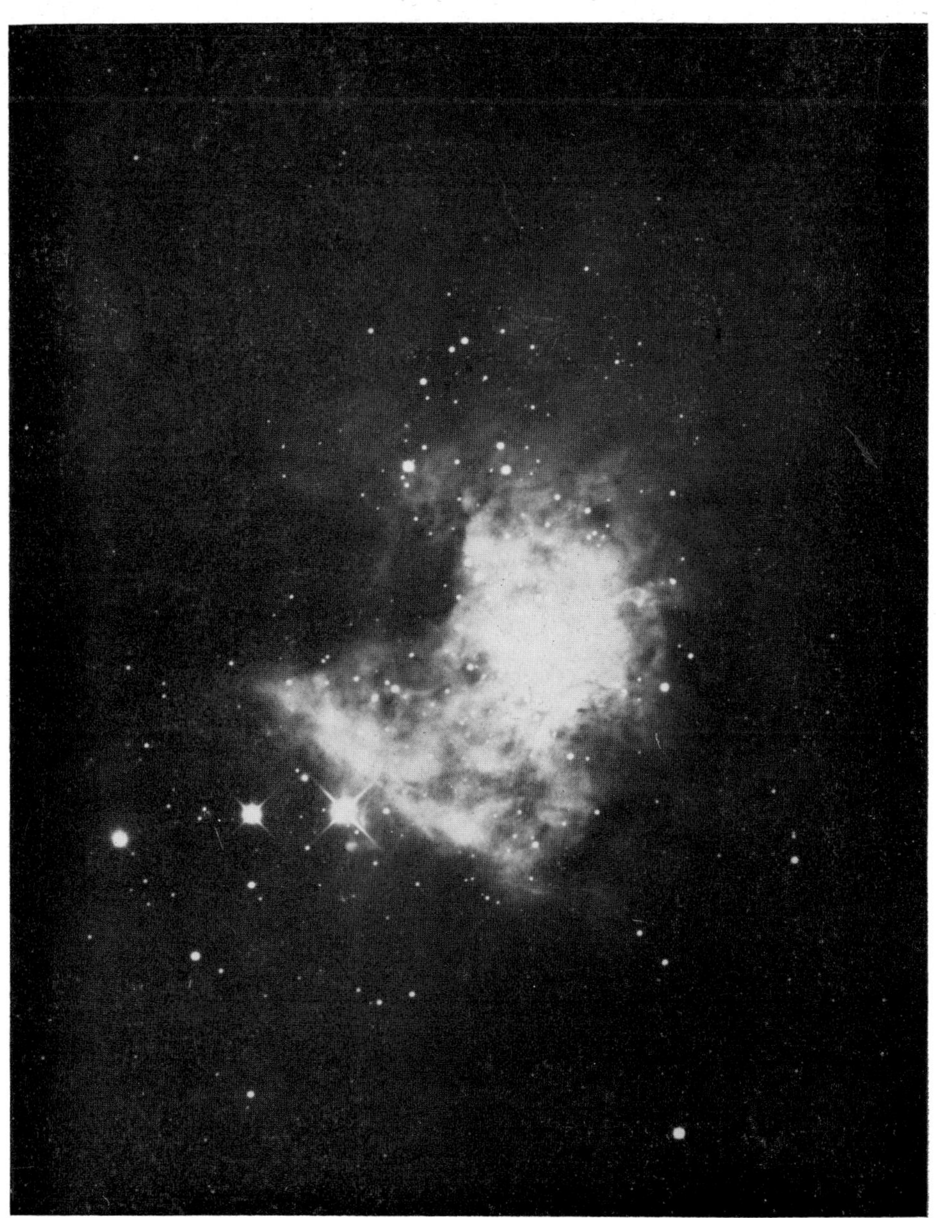

PLATE 5.—The same as Plate 4 but with a 20-minute exposure

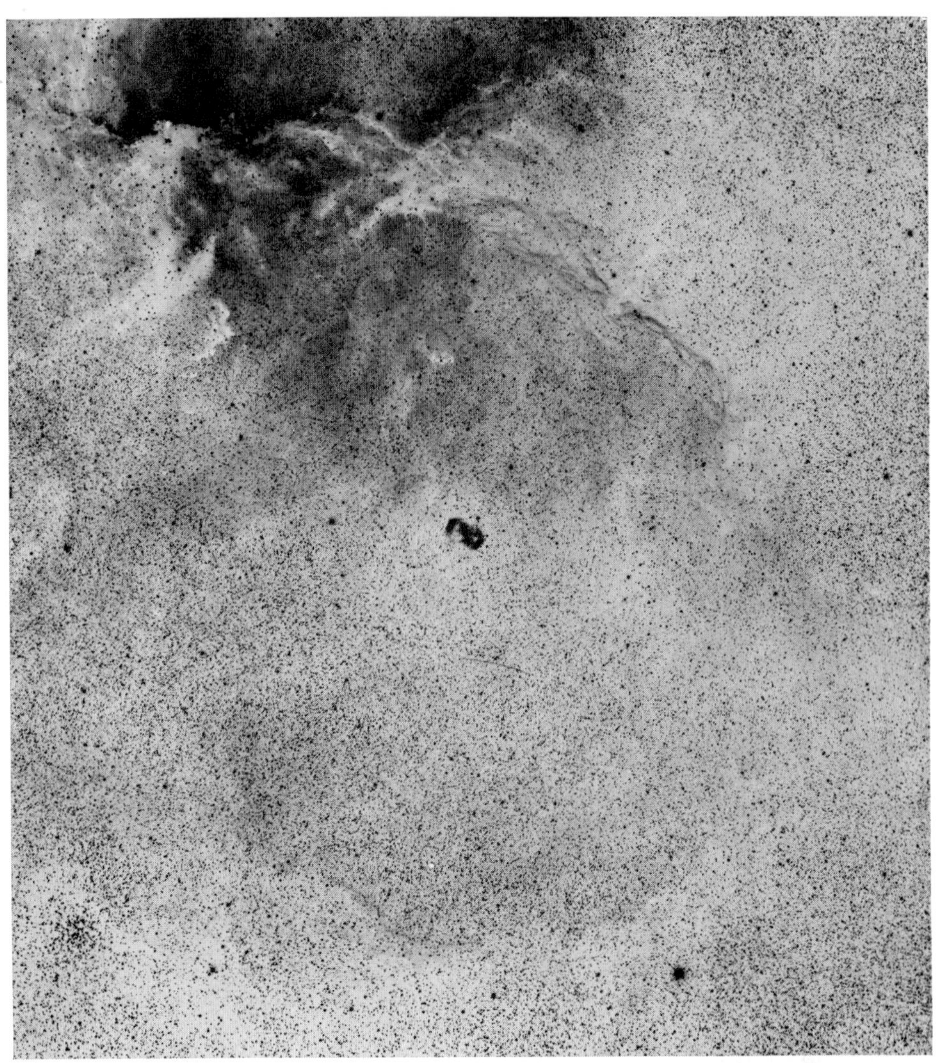

PLATE 6.—HD 148937 and the series of shells around it. NGC 6188 is the abutting bright region in the SE corner. This Hα plate was taken with the Uppsala 20/26-inch Schmidt telescope at Mount Stromlo (Westerlund 1960*b*). The field shown is approximately 2°.4 × 2°.6; N. at right, E. at top. A nearly straight streak west of center is a defect.

PLATE 7.—RS Puppis, a 41-day δ Cepheid uniquely surrounded by interstellar matter. The plate was taken by B. Westerlund at the Newtonian focus of the 74-inch Mount Stromlo reflector. Negative print; N. at top.

PLATE 8.—30 Doradus in the Large Magellanic Cloud. This Hα plate was taken at the Newtonian focus of the 74-inch Mount Stromlo reflector by L. H. Aller and D. J. Faulkner. The field is $29' \times 37'$; N. at top and E. at left.

PLATE 9.—The nebula Henize 11 (1956) in the Large Magellanic Cloud. Data as for Plate 8

Plate 10.—The nebula Henize 44 (1956) in the Large Magellanic Cloud. Data as for Plate 8

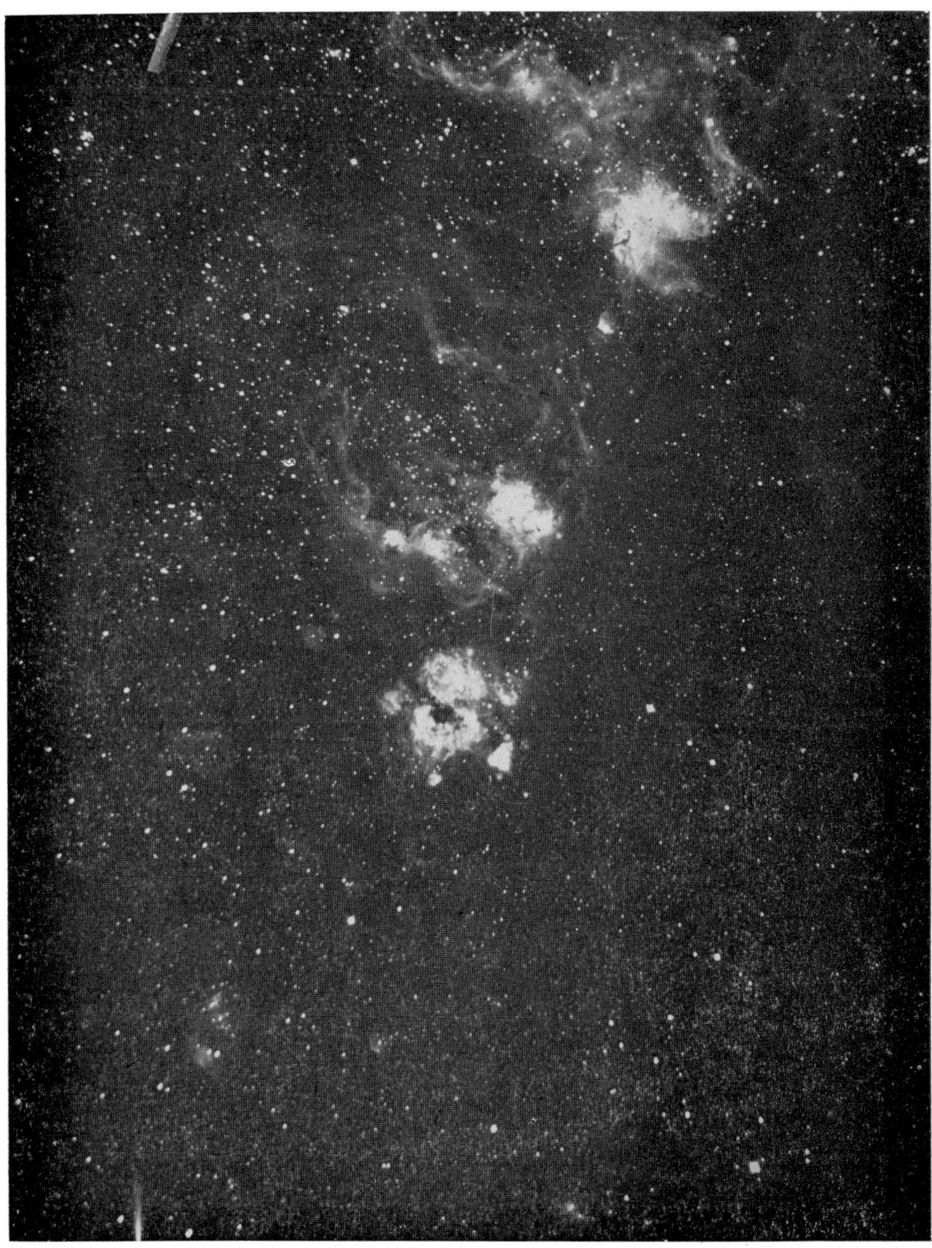

PLATE 11.—The nebula Henize 160 (1956) in the Large Magellanic Cloud. Data as for Plate 8

eral objects of Shain and Haze. The sequence displays the reaction of imbedded dense nodules to the ionizing radiation of the hot star and the expanding gas of the H II region, strikingly evident from the tendency of the elephant trunks to point toward the exciting star, in some cases the nearest of several exciting stars in the H II region. A nodule at the tip of a trunk consistently seems to be an impediment to the expansion of the rarer gas. The appearance of these features depends on geometry as well as physical factors. Objects with sharp, bright rims might be observed as either bright nodules or dark globules from directions at right angles to the actual line of sight. The bright rim is simply the H II skin of a mass too dense to be penetrated deeply by the ionizing light. The density of the nodule and the flux of ultraviolet radiation governs the rate of H II evaporation and the lifetime of the nodule. The one-dimensional shock-wave model of Kahn (1954) has been extended to three dimensions by Pottasch (1958), Goldsworthy (1961), and Axford (1961), and it has been compared satisfactorily with an observation of the brightness profile of a rim in NGC 1396 (Osterbrock 1957*b*). In the above outline no reference is made to Rayleigh-Taylor instability, earlier considered to be a likely explanation of elephant-trunk shapes (Spitzer 1954).

Minkowski (1942) explained comet shapes (prototype NGC 2261, Hubble's variable nebula, containing R Monocerotis) by means of the radiation pressure of involved giant stars. He said that T Tauri stars do not seem to form comet-like nebulae; in 1942 it was thought that they were dwarfs which would not generate much radiation pressure. Ambartsumian (1955), on the contrary, asserted that all known comet-like nebulae are connected with T Tauri stars. In the past there has been some confusion of terminology for "elephant-trunk" and "comet-like" nebulae, and it would be useful to reserve the former name for dark structures imbedded in H II regions, even when these have the comet-like bright rim. Although the tips of some elephant trunks contain no visible stars, others, such as that south of S Monocerotis, contain emission-line stars (Herbig 1954) that may become T Tauri stars. Dibai (1960) has proposed that comet-like nebulae may develop out of elephant trunks when a star forms in the tip of the trunk and illuminates the matter in it. Thus the comet shape is developed in relation to the expansion of an H II region which need not remain visible.

The complex field of H II regions in Monoceros (cf. Fig. 12 of Morgan *et al.* [1955] and Palomar Sky Survey charts E445 and E923) exhibits three overlapping rings: first, NGC 2237 centered on the O-association I Monocerotis at 1600 pc; second, an incomplete ring 4° in diameter centered on II Monocerotis at 900 pc; and third, a 3° ring linking the others in a chain. S Monocerotis and the elephant trunk south of it are in a central concentration well inside the 4° ring. NGC 2261 is also inside this ring, but the comet tail juts northward in the wrong way to have been influenced by S Monocerotis. Here, possibly, one of several O stars within 1° of the center of the 3° ring caused the ring and directed the NGC 2261 comet tail. However, Davies (1963) believes the 3° ring to be a supernova remnant. Another well-known comet-like nebula, NGC 6729, is not so easy to explain since it is rather far from any O-associations or known H II regions, at $l^{\mathrm{I}} = 327°$, $b^{\mathrm{I}} = -19°$. The amorphous nebulae IC 4812 and NGC 6726–7 are within a few minutes of arc of NGC 6729. The latter contains the B2 irregular variable star TY Coronae Austrinae, and the comet tail of NGC 6729 is about 30° out of line with the direction from TY to R Coronae Austrinae.

Although the scheme of Pottasch and Dibai is attractive, it does not easily explain the physical conditions in comet-like nebulae. If Gurzadian's (1959) value of $N_H = 200$ cm^{-3} is correct for NGC 2261, how can the shape remain cometary while the density is reduced by a factor of 500 from values estimated by Pottasch as typical of elephant-trunk regions? The excess luminosity of comet-like nebulae relative to their stars and the significance of their variability still need to be satisfactorily explained (Ambartsumian 1955, Gurzadian 1959).

Departure from sphericity may be equivalent to expansion in a preferred direction in order to conform to asymmetrical environs or equivalent to the presence of an ordered interstellar magnetic field. Johnson (1955) observed that the major axes of symmetric nebulae tend to lie in the galactic plane. Shain (1955) also found that dark nebulae are preferentially elongated in the galactic plane. They may be considerably more stable and have lifetimes longer than T_N, and they might then be subject to differential galactic rotation.

Filamentary structure has been treated by many authors. The usual approach is to find a rationale for the orientation, which is generally nonrandom. Filaments are found in both emission-type and continuous-type nebulae, but even the latter may be electrically conducting so that magnetic fields may govern the forms and motions. Some authors (for example, Shain and Haze 1952*a*, Fessenkov 1954) find that filaments tend to stretch parallel to the galactic equator; other authors (for example, Gershberg 1960) that the radiation field of nearby bright stars is important. Even in the latter situation, some regions are found (Osterbrock 1959) where the striae are perpendicular and others where they are parallel to the radiation front. In the eastern edge of the elliptical nebula around σ Orionis (near the Horsehead), two perpendicular systems of filaments interpenetrate, individual spikes being nearly parallel to the radiation of the star and a more braided kind at right angles. A finer scheme of classification is probably needed to make progress in understanding the nature of interstellar filaments. This is obvious if, as seems probable, different mechanisms operate to produce filaments out of interstellar matter in an unstressed state.

Westerlund (1960*c*) has discovered an interesting nebula around RS Puppis, currently a 41-day Cepheid with intermittently changing period. Despite the extreme Population I characteristics of both interstellar grains and galactic Cepheids, this is the only record of the illumination of grains by one of them. Moreover, the multiple rings shown in Plate 7 seem to be unique.

3.6. Radial Velocities

The radial velocities v of diffuse emission nebulae have been obtained with the spectrograph and the Fabry-Perot étalon. They give galactic motions, internal motions, and motions directly comparable with the exciting stars. Campbell and Moore (1918) list radial velocities for five diffuse nebulae, and radial velocties for seven diffuse nebulae are listed in Wilson's *General Catalogue of Radial Velocities* (1953). Only a very few spectrographic velocities, for example, Mayall's (1953, 1954) of NGC 1499 and IC 405, have been obtained since then. Most recent radial-velocity measures of diffuse nebulae have been made interferometrically by Courtès (1960), Courtès and Cruvellier (1960), and Courtès, Cruvellier, and Pottasch (1962). Excluding the very unusual velocity data for NGC 1976 (Wilson *et al.* 1959), most interstellar motions are known optically from interstellar absorption lines or the 21-cm line. Nothing as complete as the information about the internal and galactic kinematics of planetary nebulae has been gathered for diffuse nebulae.

Table 15 summarizes the material of Courtès (1960) from his Fabry-Perot plates. The Fabry-Perot étalon utilizes the interference phenomena of multiple waves produced by two plane-parallel semireflecting mirrors placed face to face. In Courtès' application, the radius of the Hα fringe was measured against a scale fringe in areas $10'' \times 10''$ at selected points in the image of an H II region. Columns 7 to 10 of the table are means of about ten or more measures—each on different plates. Detailed maps of v for many points in NGC 6523 and NGC 1976 are drawn by Courtès. The distribution of velocities of H II in galactic longitude is in excellent accord with the interstellar Ca II velocities (cf. Münch's chapter, this volume).

Lick spectrographic work on NGC 1976 (Campbell and Moore 1918) was analyzed much later by von Hoerner (1951). Meanwhile, the Fabry-Perot method had been applied (Fabry and Buisson 1911; Buisson, Fabry, and Bourget 1914; Baade, Goos, Koch, and Minkowski 1933; Minkowski 1934). The definitive work on NGC 1976, in selected parts to a radius of 10′, is that of Wilson *et al.* (1959). Radial velocities were measured at thousands of points in a network $1\overset{''}{.}3 \times 1\overset{''}{.}3$ defined by a multislit spectrograph working at 4.5 A/mm or 9.2 A/mm in outlying regions. The measures were made on the lines [O III] 5007 A, Hγ, and [O II] 3726 A with a probable error of less than 1 km/sec. In a small percentage of the areas measured there is a change from single to split lines ($\Delta v \leq 25$ km/sec) in distances as small as a few seconds of arc. This implies discontinuities in the flow produced by shock waves in a compressible medium (Münch 1958). In places, [O III] 5007 A splits into positive and negative components relative to the single-valued v. [O II] 3726 A then sometimes follows only one component, proving different ionization in the regions corresponding to the 5007 A components, since [Ne III] 3869 A behaves like [O III] 5007 A rather than like [O II] 3726 A, whereas the opposite would be expected if interstellar reddening had suppressed one branch of 3726 A. Correlations between the splitting and the geometry or brightness of the nebula have yet to be established. There are two areas in which both [O II] 3726 A and [O III] 5007 A show two such components. If we visualize these regions as bubbles in the gas, O^{++} is expanding faster than O^{+} by about 1 km/sec in one area and vice versa in the other. The mean rate of expansion is about 8 km/sec. If this were constant in time and if the dimensions of the split-line regions measure the size of the bubbles, their lifetimes could be 4 to 20×10^3 years.

A general expansion of the H II region into its surroundings may be inferred from a systematic velocity of approach of about 12 km/sec with respect to the Trapezium and about 6 km/sec with respect to H I (Menon 1958). The run of velocities along the projected radius is not actually that of an expanding sphere: the velocity drops by several kilometers per second between the center and a radius of 2′ or 3′, then returns about to the central value at 10′, the extreme point of measurement. In the same space, v ([O II] 3726 A) $- v$ (Hβ) drops from $+7$ km/sec to zero.

Wurm (1961) has attempted to relate the interstellar absorption-line velocities observed in the spectra of stars in NGC 1976 to the nebular emission-line velocities.

TABLE 15

RADIAL VELOCITIES OF H II REGIONS DETERMINED BY THE INTERFERENCE METHOD

COURTÈS (1951)	NEBULA	α (1950)	δ	H II RADIAL VELOCITY (km/sec)				H II MEAN RAD. VEL. (km/sec)	EXCITING STAR*	TYPE	STELLAR RADIAL VELOCITY (km/sec) WILSON (1953)	RADIAL VELOCITY OF INTERSTELLAR ABSORPTION LINES (km/sec) MÜNCH (1957)			DISTANCE (kpc)
OHP 109	...	0h 02m	+66°40′	−21.7				−21.7							
...	IC 1805	2 30	+61 17	−49.7	−44.2			−48.0	HD 15558	O5*f*	−50	−52	−31	−11	2.0
OHP 113	IC 1848	2 53	+61 15	−32.8				−32.8	HD 17505	O7	−17	−44		− 8	2.0
OHP 114	NGC 1499	4 02	+35 50	+ 6.5	+ 9.0	+ 3.8	+11.5	+ 7.7	HD 24912	O7	+70.1		+11.7		0.3
OHP 116	IC 405	5 13.5	+34 25	+16.2	+24.8	+16.9		+19.3	HD 34078	O9.5	+59.1		+15		0.6
OHP 118	IC 410	5 20	+33 25	+13.2	+ 5.7			+ 8.2							
OHP 128	λ Ori	5 36	+ 9 40	+31.2				+31.2	HD 34656	O7	0.0			+ 2.5	0.6
...	IC 434	5 38.7	− 1 28	+36.3	+37.7			+37.0	HD 37468	O9.5	+29.2		+14.5		0.4
...	Ori Loop	5 42.5	+ 2 04.3	+22.0				+22.0	HD 37742	O9.5 I*b*	+18.1		+19.6		0.4
OHP 140	NGC 2237	6 30	− 5 00	+32.3	+38.9	+37.3		+36.2	HD 46223	O5	+43.4		+18.5		1.4
OHP 147	NGC 2264	6 38.5	+ 9 50	+39.2				+39.2	HD 47839	O7	+33.2		+15.2		0.7
...	ζ Oph	16 34.4	−10 28	(+ 4)				(+ 4)	HD 149757	O5 V	−19		−16		
...	NGC 6514	17 59	−23 00	+ 1.2	+ 0.6			+ 0.9							
...	NGC 6523	18 03	−23 50	(− 8)				− 7.1	HD 165052	O6	+ 3			− 9	1.3
OHP 32	NGC 6823	19 41	+23 12	+26.3	+13.4			+19.8							
OHP 33	...	19 45.5	+28 05	+ 8.5				+ 8.5	HD 186943	O*a*	+10.0				
...	NGC 6871	20 02.7	+54 52	+ 4.1				+ 4.1							
OHP 54*b*	...	20 17	+46 00	(− 2)				(− 2)							
OHP 56	...	20 18	+39 00	+22.1				+22.1							
OHP 59	...	20 25	+42 00	−17.4				−17.4							
OHP 60	IC 1318	20 26	+39 50	−12.2				−12.2							
OHP 78	NGC 7000	20 58	+44 30	−13.9				−13.9	HD 199579	O*e*5	− 5.8		−11.7		
OHP 84	IC 1396	21 37	+57 00	−25.7	−14.5	−10.9	−17.7	−17.2	HD 206267	O6	− 7.8		−18.8		
OHP 92	Struve-Elvey (1939)	22 18.5	+56 00	−52.0				−52.0	HD 211853	WN6	−55				3.6
OHP 96	NGC 7380	22 44.7	+57 50	−47.9	−47.5			−47.7	HD 215835	O5*f*	−35.4	−41	−19		2.5
...	S 11	22 57.3	+58 30	−53.8				−53.8							
OHP 104	NGC 7635	23 18.5	+60 45	−54.6	−54.5			−54.6	HD 225160	O8*f*	−46	−60	−42	−11	2.8
OHP 106	...	23 59.0	+64 30	−53.8				−53.8							

* Johnson (1955).

In other nebulae, Courtès (1960) has determined the value of n in the relation $\Delta v = kl^n$ where l is the projected distance in minutes of arc between two points in the nebula. For the nebula around λ Orionis and two fields in IC 434, he gets $0.33 \leq n \leq 0.44$. In the Horsehead region of IC 434, he finds n indeterminate (about nil). In IC 443, NGC 2327, NGC 6960, NGC 6888, and in filaments near γ Cygni, he finds large internal motions: 40 km/sec $\leq \Delta v \leq$ 70 km/sec. Some of these filaments may be supernovae remnants (cf. Minkowski's chapter, this volume).

According to Table 15, Courtès has found that the following nebulae have mean radial velocities very near to those of the exciting stars: OHP 2, λ Orionis, NGC 2237, NGC 2264, IC 1805, OHP 33, and the Orion Loop. The following nebulae show differences of about 10 km/sec from their exciting stars: NGC 7000, NGC 6523, NGC 7635, NGC 7380, and IC 1396. Usually, Δv (star − nebula) is positive as in NGC 1976 where it is supposed that continuous absorption dims the far side of the gas (Wilson *et al.* 1959),

TABLE 16

FREQUENCY OF OCCURRENCE OF [O II] 3727 A AND FRACTION OF H I IN GALAXIES

Type	Per Cent with [O II] 3727 A	H I Sample	Mean Fraction of H I in Sample
E	12	M32, NGC 3115	0.001
S0-SB0	27	NGC 4111	0.015
Sa-SBa	45	....	....
Sb-SBb	62	Galaxy, M31, M81	0.036
Sc-SBc	68	M33, M51, M101, NGC 253, NGC 4236	0.085
Irr	94	LMC, SMC, IC 1613, NGC 6822	0.22

but in some instances this $\Delta v < 0$, as in NGC 2264, ζ Ophiuchi, and IC 434. IC 405 and NGC 1499 remain as interesting cases of very large Δv's in which it may be imagined that AE Aurigae and ξ Persei, respectively, are passing independently through the H II region. Plates 2 and 3 show AE Aurigae and IC 405. IC 405 appears to share the motion of AE Aurigae enough to sweep up the material of another nebula, S 126, according to direct photographs (Blaauw and Morgan 1953).

3.7. EMISSION NEBULAE IN EXTERNAL GALAXIES

In low-excitation nebulae [O II] 3726–29 A is the strongest emission on a blue-sensitive plate. Mayall (1958) gives the frequency of [O II] 3727 A (the blended doublet on low dispersion) in the spectra of external galaxies of different types, and Reddish (1961) gives the fraction of total mass of various galaxies in the form of interstellar H I. The progression in Table 16 is an integral part of the definition of stellar Populations I and II. Galaxies that contain much gas produce abundant young stars; the massive hot ones ionize the gas. Dust lanes commonly mark the equatorial planes of edgewise spiral galaxies; dust in elliptical galaxies is much rarer (Baade 1951). Most galaxies subtend small angles; separate H II regions in the images often seem to be integrated in the same way that dust clouds are fused into dust lanes. Besides (*a*) H II regions that may be supposed to be analogous to emission nebulae in the Sun's vicinity, there are (*b*) H II regions in the cores of external galaxies, although optical detection of similar regions in our own is not possible. Class (*b*) may be subdivided according to (*i*) normal and (*ii*) excessively widened and highly excited spectra. Hubble (1922*a*) made a comment on the "planetary" (high-excitation) spectra of NGC 1068 and NGC 4151, now placed in class (*ii*). Minkowski and Osterbrock (1959) and Osterbrock (1960) have studied class (*i*) spectra in elliptical galaxies. They conclude that the gas may be ejected from, and excited by, relatively few hot stars in the predominantly cool core populations. Seyfert (1943) observed a number of galaxies with class (*ii*) spectra. Woltjer (1959) has published an interpretation of class (*ii*) as massive, rapidly spinning or randomly stirred Doppler-broadened cores. In order to explain the special feature that class (*ii*) Balmer lines are broadened more than others, simple mass motions must be supplemented by change of excitation with nonrandom motions. Osterbrock and Parker (1965) conclude that the ionization and heating mechanisms in the nucleus of NGC 1068 are results of collisions between rapidly moving gas clouds. However, Dent and Haddock (1965) have found a new type of radio source spectrum in NGC 1275, another Seyfert (class *ii*) galaxy. Its high-frequency component of radiation appears to be the free-free emission of a gas at $N = 6 \times 10^4$ cm^{-3} and $T_e = 1.5 \times 10^6$ ° K. The interesting observations of [O II] 3726–3729 A in the core of M31 (Münch 1950*b*) put this galaxy in class (*b*, *i*) for the first time. Since his observation also belongs to galactic dynamics, it will not be discussed further here.

Type (*a*) spectra were known in the Large Magellanic Cloud before it was found to be extragalactic. This galaxy is so close that its H II regions are well resolved. When R. E. Wilson (1918) observed the spectra of 17 of them, he found a mean $v = +276$ km/sec and he found $I(\mathrm{H}\beta) \geq I(N_2)$ (here N_2 refers to the weaker "nebulium" line, 4959 A of [O III]). This situation is often encountered in diffuse nebulae but rarely in planetary nebulae, and this means that the level of excitation in a diffuse nebula is much lower than in an average planetary (see chapter 9 for further details).

Nail, Whitney, and Wade (1953) initiated the search for Hα-emission nebulae in the Large Cloud by direct photography in the two colors, used also for galactic searches (Table 1). Henize (1956) completed an Hα survey in both Clouds with an objective-prism-filter combination. He and Miller (1951) found nebulae and also emission-line stars in the Clouds. Lindsay (1956, 1961) and Koelbloed (1956) have studied planetary nebulae in the Small Magellanic Cloud, where planetaries are detectable with moderate telescopes, but in M31 planetaries are difficult to detect even with a large telescope (Baade 1955). Doherty, Henize, and Aller (1956) have catalogued absolute intensities and shapes of Large Cloud nebulae in Hα, and sample electron densities in the range 1 cm^{-3} $< N_e <$ 30 cm^{-3} were derived by Aller (1956) from absolute surface intensities.

Johnson (1959*c*) finds the spectrum of NGC 2070, the 30 Doradus Nebula, to be much like that of NGC 1976 in excitation (Aller class 3–4). With [O II] 3729/3726 A = 1.05 in NGC 2070, and with suitable scale factors applied to Osterbrock and Flather's (1959) model of NGC 1976, a central cloud density of 3600 cm^{-3} and a total interstellar mass of 5×10^6 $\mathfrak{M}_\odot$ follow. This mass is 10,000 times the model mass of H II regions in the neighborhood of the Sun (Fig. 6), suggesting that 30 Doradus is the nucleus of the Large Magellanic Cloud (Johnson 1959*a*). The dimensions are correspondingly large. A great loop of H II gas curves around 30 Doradus to the east, extending 52′ in radius toward the southeast (Johnson 1959*a*, Plates I and II). This is a radius of about 1 kpc, far larger than the largest known loop in the Galaxy—Gum's Nebula in Vela-Puppis, which may be about 160 pc in radius (Johnson 1959*b*). The 30 Doradus Loop is outside the field of Plate 8 but would be underexposed in that plate anyway. Its root is apparently the prominent arc starting northward in Plate 8. Another object called the "great loop" by Feast (1961) is a detail of the brighter structure of 30 Doradus. Mathis (1965*b*) has further discussed the physical characteristics of 30 Doradus but much more detailed studies have been made by Faulkner (thesis, 1964). In addition to using spectroscopic measurements, he analyzed isophotic contours from photometrically calibrated plates to construct a geometrical model for the nebula. He derives a mass considerably smaller than that obtained by Johnson. Excellent examples of the variety of H II regions in the Large Magellanic Cloud are shown in Plates 8 through 11. Mrs. Dickel (1965) gave detailed studies of the electron densities, masses, and isophotic contours of all of these objects except NGC 2070 (which was studied by Faulkner [thesis, 1964]). Dickel, Aller, and Faulkner (1964) reported photoelectric surface brightness measurements, S (Hβ), relative line intensities, electron densities, and ion abundances for 50 nebulosities in the Large Magellanic Cloud and 12 nebulosities in the Small Magellanic Cloud.

Aller (1942) has observed the relative line intensities of as many as nine emissions in each of 19 nebulae in M33. He has also (Aller 1956) published a map of about 160 nebulae in M33 and has estimated the electron densities of several of the brighter ones to be in the range of $10 \text{ cm}^{-3} < N_e < 40 \text{ cm}^{-3}$ by measuring absolute Hα intensities and assuming spherical, homogeneous volumes. The interstellar mass of NGC 604 is between 2 and 4×10^5 $\mathfrak{M}_\odot$ (Shain and Haze 1952*c*) according to an estimate that depends crucially on the degree of uniformity in the mass. Estimates of this kind also depend on the cube of the distance scale, and those made prior to 1952 should be increased by a factor of at least 10. The Burbidges (1962) have summarized their studies of ionized gas in spiral and irregular galaxies. The most striking result is a strong change in the conditions of ionization and excitation between the outer and inner parts of late spirals.

Following a comprehensive discussion of the distribution and motions of gaseous masses in spirals (Baade and Mayall 1951), Strömgren asked why the H II nebulae of M31 are fainter than those of M33 and other galaxies, although the range of their size and excitation seems to be similar in the nearby galaxies (NGC 2070 excepted). The emission spectra of NGC 604, NGC 588, and NGC 595 were observed in M33 a quarter of a century earlier than much fainter spectra of H II regions in M31 (Babcock 1939). The relative abundance of gas to stars in the two systems, Table 16, explains this. Unfortunately Baade and Mayall do not give H II data for M31 in a form which might be compared with the gross distribution of H I in M31 (van de Hulst, Raimond, and van Woerden 1957).

H II regions have also been observed spectroscopically in double galaxies, sometimes also extending continuously between the pair (Page 1952), and in a number of peculiar galaxies such as NGC 5128 (Burbidge and Burbidge 1959) where velocity fluctuations of between 100 and 200 km/sec may be correlated with structural features in the dust lane.

Table 2 gives the present data on the cosmic abundances of the elements in emission nebulae, the Crab Nebula, and the core of an elliptical galaxy. It also gives the same data about the Sun and local early-type stars for comparison.

3.8. Conclusions and Suggestions for Further Observations

When a comprehensive reliable theory for the Balmer decrement is established, it will be possible to obtain accurate corrections for interstellar reddening in H II regions. At present accurate comparisons of intensities cannot be made between parts of the spectrum of an H II region separated by more than 100 A. Nor can absolute optical intensities be compared with radio-frequency data until the extinction along an individual line of sight can be determined from comparison of theoretical with observed Balmer-line intensities. For a discussion of the Balmer decrement problem see chapter 9, also Kaler (1965) and Clarke (1965).

Too many data about nebulae have been given as integrations over the entire solid angle of the object, or without close specification of the part of the area observed. Except for electron temperature, most parameters in nebulae vary over too wide a range for an average to be meaningful. The problem of observing many points in a nebula is analogous to the problem of the stellar photometry of a cluster. The establishment of fundamental standards of surface brightness in various monochromatic emissions for H II regions along the Milky Way would be useful.

Knowledge of the stars in H II regions is extremely limited. Spectral types of even the brighter stars are frequently unknown or inaccurately known. Until we know what is the radiating source in an H II region, we cannot make accurate comparisons between observed nebular fluxes in the visible range and stellar model-atmosphere fluxes in the ultraviolet. Distinct from this astrophysical question are evolutionary questions: How is the stellar content related to the density, form, grain content, and other characteristics of the interstellar part of H II regions?

Observations should be directed toward understanding the structure during the evolution of an H II region from its initial configuration at the birth of a hot star. High-resolution 21-cm observations are needed for knowledge of the H I content and its sequence of configurations, and for a check of whether the seeming holes in H II are actually filled with H I.

The behavior of scattered light in the nebular continua could be more thoroughly investigated by colorimetry and polarimetry. The scattered continuum should be separated from the atomic continuum in emission nebulae. The relation between the scattering power of grains in H I and H II regions has yet to be checked observationally. What we know of interstellar grains has been deduced almost entirely from their effects on starlight in the line of sight to the star.

The dynamics of H II regions can be understood more completely only after many more observations of stellar radial velocities have been made. The Orion Nebula is one of few for which adequate proper-motion observations are possible. The Orion Nebula, because of its brightness and proximity (scale), is the object on which the most diverse work can be most easily correlated.

Diffuse nebula astronomy has never attracted as much attention as stellar astronomy. Astrometric measurements are difficult and require long time intervals to yield results of interest (e.g., Strand's study of motions of stars in Orion). Nebular problems are of interest to atomic and plasma physicists but less so to nuclear physicists, except that chemical compositions of nebulae shed some light on element synthesis processes in stars. Nebulae do not have the spherical symmetry characteristics of stellar structure, nor do they occur in such great numbers (as do stars) as to facilitate statistical studies. Nebulae are less amenable. Although 21-cm observations have revitalized interest in the interstellar medium, they have not told us much new about the optically bright parts of the medium. Most radio observations in the continuum have been directed at sources other than ordinary diffuse nebulae. New ideas may come from the data to be gathered from above the atmosphere; meanwhile, the existing framework needs to be made much more coherent.

I wish to thank everyone who contributed unpublished information and illustrations to this chapter.

This chapter was planned at the Mount Stromlo Observatory, written mainly at the Steward Observatory, and was further developed at the National Radio Astronomy Observatory and under the Lockheed Independent Research Program.

REFERENCES

Adams, W. S. 1949, *Ap. J.*, **109**, 354.
Aller, L. H. 1942, *Ap. J.*, **95**, 52.
———. 1954, *Nuclear Transformations, Stellar Interiors, and Nebulae* (New York: Ronald Press Co.).
———. 1956, *Gaseous Nebulae* (London: Chapman & Hall, Ltd.; New York: John Wiley & Sons, Inc.)
———. 1961, *The Abundance of the Elements* (New York: Interscience Publishers, Inc.).
———. 1964, *Astrophys. Norv.* **9**, 293.
Aller, L. H. and Faulkner, D. J. 1962, *Pub. A.S.P.*, **74**, 219.
Aller, L. H. and Liller, W. 1959, *Ap. J.*, **130**, 45.
Ambartsumian, V. A. 1949, *A.J.-U.S.S.R.*, **26**, 3.
———. 1955, *Mém. Soc. R. Sci. Liège*, Ser. 4, **15**, 458.
Andrillat, Y. and Andrillat, H. 1959, *Ann. d'ap.*, **22**, 104.
Arp, H. C. 1958, *Hdb. d. Phys.*, ed. S. Flügge (Berlin: Springer-Verlag), **51**, 75.
Axford, W. I. 1961, *Phil. Trans. R. Soc. London, A*, **253**, 301.
Baade, W. 1951, *Pub. U. Michigan Obs.*, **10**, 7.
———. 1955, *A.J.*, **60**, 151.
Baade, W., Goos, F., Koch, P. P., and Minkowski, R. 1933, *Zs. f. Ap.*, **6**, 355.
Baade, W. and Mayall, N. U. 1951, *Problems of Cosmical Aerodynamics*, eds. J. M. Burgers and H. C. van de Hulst (Dayton, Ohio: Central Air Documents Office), p. 165.
Babcock, H. W. 1939, *Lick Obs. Bull.*, **19**, 41.
Baker, J. G. and Menzel, D. H. 1938, *Ap. J.*, **88**, 52.
Barbier, D. 1944, *Ann. d'ap.*, **7**, 80.
———. 1949, *ibid.*, **12**, 6.
Barnard, E. E. 1895, *M.N.*, **55**, 442.
———. 1927, *A Photographic Atlas of Selected Regions of the Milky Way*, eds. E. B. Frost and M. R. Calvert (Washington: Carnegie Institution of Washington).
Bergh, S. van den. 1956, *Zs. f. Ap.*, **40**, 249.
Bigourdan, G. 1916, *C. R. Paris*, **162**, 489.
Blaauw, A. 1961, *B.A.N.*, **15**, 265.
Blaauw, A. and Morgan, W. W. 1953, *B.A.N.*, **12**, 76.
Böhm, K. H. 1956, *Ap. J.*, **123**, 379.
Boggess, A., III. 1954, University of Michigan. Dissertation.
———. 1961, *Space Astrophysics*, ed. W. Liller (New York: McGraw-Hill Book Co., Inc.), chap. 7.
Bok, B. J. 1932, *A Study of the η Carinae Region* (Groningen: Hoitsema Bros.); Harvard Reprint No. 77.
Bok, B. J., Bester, M. J., and Wade, C. M. 1955, *Proc. Amer. Acad. Arts Sci.*, **86**, 9.
Bok, B. J. and Reilly, E. F. 1947, *Ap. J.*, **105**, 255.
Bowen, I. S. 1927, *Pub. A.S.P.*, **39**, 295.
Boyce, P. 1963, University of Michigan. Thesis.
Buisson, H., Fabry, C., and Bourget, H. 1914, *Ap. J.*, **40**, 241.
Burbidge, E. M. and Burbidge, G. R. 1959, *Ap. J.*, **129**, 271.
———. 1962, *ibid.*, **135**, 694.
Burgess, A. 1958, *M.N.*, **118**, 477.
Byram, E. T., Chubb, T. A., and Friedman, H. 1964, *Ap. J.*, **139**, 1135.
Campbell, W. W., and Moore, J. H. 1918, *Pub. Lick Obs.*, **13**, 75.

Cannon, A. J. 1916, *Harvard Ann.*, **76**, 19.
Cederblad, S. 1946, *Lund Medd.*, Ser. 2, **12**, No. 119.
Chamberlain, J. W. 1953, *Ap. J.*, **117**, 399; cf. Minkowski (1955).
Chopinet, M. and Fehrenbach, C. 1961, *J. Observateurs*, **44**, 141.
Clarke, W. 1965, University of California, Los Angeles. Thesis.
Code, A. D. 1958, unpublished; cf. Minkowski (1958).
Collinder, P. 1931, *Ann. Lunds Obs.*, **2**, B43.
Collins, O. C. 1937, *Ap. J.*, **86**, 529; cf. Greenstein (1938).
Courtès, G. 1951, *C. R. Paris*, **232**, 795, 1283.
———. 1960, *Ann. d'ap.*, **23**, 115.
Courtès, G. and Cruvellier, P. 1960, *Ann. d'ap.*, **23**, 419.
Courtès, G., Cruvellier, P., and Pottasch, S. R. 1962, *Ann. d'ap.*, **25**, 214.
Curtis, H. D. 1919, *The Adolfo Stahl Lectures in Astronomy* (San Francisco: Astronomical Society of the Pacific), p. 98.
Davies, R. D. 1963, *Observatory*, **83**, 172.
Davis, L., Jr. and Greenstein, J. L. 1951, *Ap. J.*, **114**, 206.
Dent, W. A. and Haddock, F. T. 1965, *Nature*, **205**, 487.
Dibai, E. A. 1960, *A.J.-U.S.S.R.*, **37**, 16 (*Sov. Astr.-A.J.*, **4**, 13).
Dickel, H. 1965, *Ap. J.*, **141**, 1306.
Dickel, H. R., Aller, L. H., and Faulkner, D. J. 1964, *IAU-URSI Symposium No. 20* (Canberra: Australian Academy of Sci.), ed. Alex Rodger and Frank Kerr, p. 294.
Doherty, L., Henize K. G., and Aller, L. H. 1956, *Ap. J., Suppl.*, **2**, 345.
Dolidze, M. V. 1960, *Bull. Abastumani Obs.*, **25**, 105.
Dombrovsky, V. A. 1955*a*, *Doklady Akad. Nauk U.S.S.R.*, **102**, 907.
———. 1955*b*, *ibid.*, **105**, 924.
———. 1958*a*, *A.J.-U.S.S.R.*, **35**, 687 (*Sov. Astr.-A.J.*, **2**, 646).
———. 1958*b*, *Vestnik U. Leningrad*, No. 13.
Drake, F. D. 1958, Harvard University. Thesis; cf. (1959), p. 366.
———. 1959, *Paris Symposium on Radio Astronomy* (I.A.U. Symposium No. 9), ed. R. N. BRACEWELL (Stanford: Stanford University Press), p. 339.
Draper, H. 1880, *Amer. J. Sci.*, **20**, 433.
Dreyer, J. L. E. 1953, *New General Catalogue of Nebulae and Clusters of Stars (1888), Index Catalogue (1895), Second Index Catalogue (1908)* (London: Royal Astronomical Society).
Dufay, J. 1954, *Pub. Obs. Lyon*, **3**, No. 23.
Eddington, A. S. 1926, *Proc. R. Soc. London, A*, **111**, 424.
Elvius, A. 1960, *Ark. f. Astr.*, **2**, 309.
Evans, D. S. 1959, *M.N.*, **119**, 150.
Ewen, H. I., and Purcell, E. M. 1951, *Nature*, **168**, 356.
Fabry, C. and Buisson, H. 1911, *Ap. J.*, **33**, 406.
Faulkner, D. J. 1964, *IAU-URSI Symposium No. 20* (Canberra: Australian Acad. of Sci.), p. 310.
Faulkner, D. J. 1965, thesis Australian National University.
Faulkner, D. J. and Aller, L. H. 1965, *M.N.*, **130**, 121.
Feast, M. W. 1961, *M.N.*, **122**, 1.
Fessenkov, V. G. 1954, *Doklady Akad. Nauk U.S.S.R.*, **94**, 647.
———. 1955, *A.J.-U.S.S.R.*, **32**, 97.
Fessenkov, V. G. and Rozhkovsky, D. A. 1953, *Atlas of Gaseous-Dusty Nebulae* (Moscow: Acad. Sci. USSR).
Flather, E. and Osterbrock, D. E. 1960, *Ap. J.*, **132**, 18.
Gehrels, T. 1960, *Lowell Obs. Bull.*, **4**, 300.
Gehrels, T. and Teska, T. M. 1963, *Applied Optics*, **2**, 67.
Gershberg, R. E. 1960, *Izvest. Crimean Astr. Obs.*, **23**, 21.
———. 1961, *A.J.-U.S.S.R.*, **38**, 250; also **39**, 169 (*Sov. Astr.-A.J.*, **5**, 188; also **6**, 126).
Gershberg, R. E., Esipov, V. F., Pronik, V. I., and Shcheglov, P. V. 1961, *Izvest. Crimean Astr. Obs.*, **26**, 313.
Gershberg, R. E. and Metik, L. P. 1960, *Izvest. Crimean Astr. Obs.*, **24**, 148.
Gershberg, R. E. and Pronik, V. I. 1959*a*, *Izvest. Crimean Astr. Obs.*, **21**, 215.
———. 1959*b*, *A.J.-U.S.S.R.*, **36**, 902 (*Sov. Astr.-A.J.*, **3**, 876).
———. 1961, *Izvest. Crimean Astr. Obs.*, **26**, 303.
Gliese, W. and Walter, K. 1951, *Zs. f. Ap.*, **29**, 94.
Goldsworthy, F. A. 1961, *Phil. Trans. R. Soc. London, A*, **253**, 277.
Gould, R. J., Gold, T., and Salpeter, E. E. 1963, *Ap. J.*, **138**, 408.
Greenstein, J. L. 1938, *Ap. J.*, **87**, 581.
———. 1946, *ibid.*, **104**, 414.
———. 1948, *Harvard Obs. Monograph* No. 7, 19; *Ap. J.*, **107**, 375.
Greenstein, J. L. and Aller, L. H. 1947, *Pub. A.S.P*, **59**, 139.
Greenstein, J. L. and Henyey, L. G. 1939*a*, *Ap. J.*, **89**, 647.
———. 1939*b*, *ibid.*, **89**, 653.

Gum, C. S. 1955, *Mem. R.A.S.*, **67**, 155.
Gurzadian, G. A. 1955, *Comm. Burakan Obs.*, No. 16, 3.
———. 1956*a*, *ibid.*, No. 18, 3.
———. 1956*b*, *ibid.*, No. 20, 23.
———. 1959, *ibid.*, No. 27, 73.
Haddock, F. T. 1957, *Radio Astronomy* (I.A.U. Symposium No. 4), ed. H. C. van de Hulst (Cambridge: Cambridge University Press), p. 192.
Haddock, F. T., Mayer, C. H., and Sloanaker, R. M. 1954, *Ap. J.*, **119**, 456.
Hagen-Thorn, V. A. 1962, *Trans. Astr. Inst. Leningrad*, **19**, 166.
Hall, J. S. 1951, *J. Opt. Soc. Amer.*, **41**, 963.
———. 1955, *Mém. Soc. R. Sci. Liège*, Ser. 4, **15**, 543.
Hall, J. S. and Mikesell, A. H. 1950, *Pub. U.S. Naval Obs.*, Ser. 2, **17**, Part 1.
Hall, R. C. 1964, *Ap. J.*, **139**, 759.
———. 1965, *Pub. A.S.P.*, **77**, 158.
Haro, G. and Minkowski, R. 1960, *A.J.*, **65**, 490.
Hartmann, J. 1904, *Ap. J.*, **19**, 268.
Hayashi, C. 1961, *Pub. Astr. Soc. Japan*, **13**, 450.
Haze, V. F. and Shain, G. A. 1955, *Izvest. Crimean Astr. Obs.*, **15**, 11.
Henize, K. G. 1956, *Ap. J. Suppl.*, **2**, 315.
Henize, K. G. and Miller, F. D. 1951, *Pub. U. Michigan Obs.*, **10**, 75.
Henyey, L. G. 1936, *Ap. J.*, **84**, 609.
———. 1937, *ibid.*, **85**, 107.
Herbig, G. H. 1951, *Ap. J.*, **113**, 697.
———. 1954, *ibid.*, **119**, 483.
———. 1958, *Pub. A.S.P.*, **70**, 468.
———. 1960, *Ap. J. Suppl.*, **4**, 337.
———. 1962, *Ap. J.*, **135**, 736.
Hertzsprung, E. 1913, *A.N.*, **195**, 449.
Higgs, L. A., Broten, N. W., Medd, W. J., and Raghavao, R. 1964, *M.N.*, **127**, 367.
Hill, E. R., Slee, O. B., and Mills, B. Y. 1958, *Australian J. Phys.*, **11**, 530.
Hindman, J. V. and Wade, C. M. 1959, *Australian J. Phys.*, **12**, 258.
Hoerner, S. von. 1951, *Zs. f. Ap.*, **30**, 17.
Hoffleit, D. 1956, *Ap. J.*, **124**, 61.
Houten, C. J. van. 1961, *B.A.N.*, **16**, 1.
Hubble, E. P. 1920, *Pub. A.S.P.*, **32**, 155.
———. 1922*a*, *Ap. J.*, **56**, 162.
———. 1922*b*, *ibid.*, p. 400.
Huggins, W. 1864, *Proc. R. Soc. London, A*, **13**, 492.
———. 1882, *ibid.*, **33**, 425.
Hulst, H. C. van de, Raimond, E., and Woerden, H. van. 1957, *B.A.N.*, **14**, 1.
Johnson, H. M. 1953, *Ap. J.*, **118**, 370.
———. 1955, *ibid.*, **121**, 604.
———. 1956, *ibid.*, **124**, 90.
———. 1959*a*, *Pub. A.S.P.*, **71**, 301.
———. 1959*b*, *ibid.*, p. 342.
———. 1959*c*, *ibid.*, p. 425.
———. 1960*a*, *ibid.*, **72**, 10.
———. 1960*b*, *Mem. Mount Stromlo Obs.*, **3**, No. 15.
———. 1961, *Pub. A.S.P.*, **73**, 147.
———. 1965, *Ap. J.*, **142**, 964.
Joy, A. J. 1946, *Ap. J.*, **102**, 168.
Kahn, F. D. 1954, *B.A.N.*, **12**, 187.
Kahn, F. D. and Menon, T. K. 1961, *Proc. Nat. Acad. Sci.*, **47**, 1712.
Kaler, J. B. 1965, University of California, Los Angeles. Thesis.
Kaler, J. B., Aller, L. H., and Bowen, I. S. 1965, *Ap. J.*, **141**, 912.
Keeler, J. E. 1894, *Pub. Lick Obs.*, **3**, 161.
Keenan, P. C. 1936, *Ap. J.*, **84**, 600.
Khachikian, E. E. 1956, *Doklady Akad. Nauk Armenian SSR*, **23**, 49.
———. 1958, *Comm. Burakan Obs.*, No. 25, 67.
Khachikian, E. E. and Kallogian, N. L. 1962, *Comm. Burakan Obs.*, No. 30, 45.
Khavtassi, J. S. 1960, *Atlas of Galactic Dark Nebulae* (Tiflis: Acad. Sci. Georgian SSR).
Koelbloed, D. 1956, *Observatory*, **76**, 191.
Kopylov, I. M. 1958, *A.J.-U.S.S.R.*, **35**, 390 (*Sov. Astr.-A.J.*, **2**, 359).
Lindsay, E. M. 1956, *M.N.*, **116**, 649.
———. 1961, *A.J.*, **66**, 169.
Little, A. G. 1962, *Observatory*, **82**, 165.
Lynds, B. T. 1962, *Ap. J. Suppl.*, **7**, 1.

Martel, M.-T. 1958, *Ann. d'ap. Suppl.*, No. 7.
———. 1963, *Notes et Informations*, **14**, No. 1.
Mathis, J. S. 1957, *Ap. J.*, **125**, 328.
———. 1962, *ibid.*, **136**, 374.
———. 1965*a*, *Pub. A.S.P.*, **77**, 90.
———. 1965*b*, *ibid.*, p. 189.
Mayall, N. U. 1953, *Pub. A.S.P.*, **65**, 152.
———. 1954, *ibid.*, **66**, 132.
———. 1958, *Comparison of the Large-Scale Structure of the Galactic System with That of Other Stellar Systems* (I.A.U. Symposium No. 5), ed. N. G. Roman (Cambridge: Cambridge University Press), p. 23.
Mayall, N. U. and Oort, J. H. 1942, *Pub. A.S.P.*, **54**, 95.
Mendoza V., E. E. 1958, *Ap. J.*, **128**, 207.
Menon, T. K. 1958, *Ap. J.*, **127**, 28.
———. 1961, *Pub. Nat. Rad. Astr. Obs.*, **1**, No. 1.
———. 1962, *Ap. J.*, **136**, 95.
Menzel, D. H. 1926, *Pub. A.S.P.*, **38**, 295.
———. 1962, *Selected Papers on Physical Processes in Ionized Plasmas* (New York: Dover Publications, Inc.).
Meyer, W. F. 1920, *Lick Obs. Bull.*, **10**, 68.
Mills, B. Y., Little, A. G., and Sheridan, K. V. 1956, *Australian J. Phys.*, **9**, 218.
Minin, I. N. 1961, *A.J.-U.S.S.R.*, **38**, 641 (*Sov. Astr.-A.J.*, **5**, 487).
Minkowski, R. 1934, *Zs. f. Ap.*, **9**, 202.
———. 1942, *Pub. A.S.P.*, **54**, 190.
———. 1946, *ibid.*, **58**, 305.
———. 1947, *ibid.*, **59**, 257.
———. 1948, *ibid.*, **60**, 386.
———. 1949, *ibid.*, **61**, 151.
———. 1955*a*, *Gas Dynamics of Cosmic Clouds* (I.A.U. Symposium No. 2), eds. J. M. Burgers and H. C. van de Hulst (Amsterdam: North-Holland Publishing Co.), p. 9.
———. 1955*b*, *ibid.*, p. 106.
———. 1958, *Revs. Mod. Phys.*, **30**, 1048.
Minkowski, R. and Osterbrock, D. 1959, *Ap. J.*, **129**, 583.
Miyamoto, S. 1956, *Zs. f. Ap.*, **38**, 245.
Morgan, W. W., Sharpless, S., and Osterbrock, D. 1952, *A.J.*, **57**, 3.
Morgan, W. W., Strömgren, B., and Johnson, H. M. 1955, *Ap. J.*, **121**, 611.
Morgan, W. W., Whitford, A. E., and Code, A. D. 1953, *Ap. J.*, **118**, 318.
Münch, G. 1957, *Ap. J.*, **125**, 42.
———. 1958, *Revs. Mod. Phys.*, **30**, 1035.
———. 1960*a*, *A.J.*, **65**, 495.
———. 1960*b*, *Ap. J.*, **131**, 250.
Nail, V. M., Whitney, C. A., and Wade, C. M. 1953, *Proc. Nat. Acad. Sci.*, **39**, 1168.
O'Dell, C. R. and Hubbard, W. B. 1965, *Ap. J.*, **142**, 591.
Oort, J. H. and Hulst, H. C. van de. 1946, *B.A.N.*, **10**, 187.
Oort, J. H. and Spitzer, L., Jr. 1955, *Ap. J.*, **121**, 6.
Osterbrock, D. E. 1957*a*, *Pub. A.S.P.*, **69**, 227.
———. 1957*b*, *Ap. J.*, **125**, 622.
———. 1958*a*, *Pub. A.S.P.*, **70**, 180.
———. 1958*b*, *ibid.*, p. 399.
———. 1959, *ibid.*, **71**, 23.
———. 1960, *Ap. J.*, **132**, 325.
Osterbrock, D. and Flather, E. 1959, *Ap. J.*, **129**, 26.
Osterbrock, D. E. and Parker, R. A. R. 1965, *Ap. J.*, **141**, 892.
Osterbrock, D. E. and Stockhausen, R. E. 1960, *Ap. J.*, **131**, 310.
———. 1961, *ibid.*, **133**, 2.
Page, T. 1948, *Ap. J.*, **108**, 157.
———. 1952, *ibid.*, **116**, 63.
Pariiskii, Yu. N. 1961, *A.J.-U.S.S.R.*, **38**, 798 (*Sov. Astr.-A.J.*, **5**, 611).
Parker, R. A. R. 1962, California Institute of Technology. Thesis.
Parsamian, E. S. 1962, *Comm. Burakan Obs.*, No. 30, 51.
———. 1963, *ibid.*, No. 32, pp. 3, 17.
Pease, F. G. 1915, *Pub. A.S.P.*, **27**, 239.
Pickering, W. H. 1895, *Harvard Ann.*, **32**, 66.
Pike, E. M. and Drake, F. D. 1964, *Ap. J.*, **139**, 545.
Pikelner, S. B. 1954*a*, *Izvest. Crimean Astr. Obs.*, **11**, 8.
———. 1954*b*, *ibid.*, **12**, 93.

Plaskett, J. S. and Pearce, J. A. 1930, *M.N.*, **90**, 243.
Pottasch, S. R. 1956, *B.A.N.*, **13**, 77.
———. 1958, *Rev. Mod. Phys.*, **30**, 1053.
———. 1960*a*, *Ap. J.*, **131**, 202.
———. 1960*b*, *Ann. d'ap.*, **23**, 749.
———. 1960*c*, *Ap. J.*, **132**, 269.
Pronik, V. I. 1957, *Izvest. Crimean Astr. Obs.*, **17**, 14.
———. 1960*a*, *ibid.*, **23**, 3.
———. 1960*b*, *A.J.-U.S.S.R.*, **37**, 1001 (*Sov. Astr.-A.J.*, **4**, 935).
Razmadze, N. A. 1958, *Bull. Abastumani Obs.*, No. 23, 91.
Reddish, V. C. 1961, *Observatory*, **81**, 19.
Reitmeyer, W. L. 1965, *Ap. J.*, **141**, 1331.
Rishbeth, H. 1958, *M.N.*, **118**, 591.
Rodgers, A. W., Campbell, C. T., and Whiteoak, J. B. 1960, *M.N.*, **121**, 103.
Rodgers, A. W., Campbell, C. T., Whiteoak, J. B., Bailey, H. H., and Hunt, V. O. 1960, *An Atlas of H-Alpha Emission in the Southern Milky Way* (Canberra: Mount Stromlo Observatory).
Rozhkovsky, D. A. 1955, *Izvest. Astr. Inst. Acad. Sci. Kazakstan, S.S.R.*, **1**, 136.
———. 1956*a*, *ibid.*, **2**, 53.
———. 1956*b*, *ibid.*, **3**, 68.
———. 1957*a*, *ibid.*, **5**, 3.
———. 1957*b*, *ibid.*, p. 11.
———. 1960, *ibid.*, **10**, 15.
———. 1962, *ibid.*, **13**, 27.
Rozhkovsky, D. A., Glushkov, Y. I., and Jokusheva, K. G. 1962, *Izvest. Astr. Inst. Acad. Sci. Kazakstan S.S.R.*, **14**, 19.
Rozhkovsky, D. A. and Matiagin, V. S. 1956, *Izvest. Astr. Inst. Acad. Sci. Kazakstan, U.S.S.R.*, **2**, 64.
Ruskol, E. L. 1950, *A.J.-U.S.S.R.*, **27**, 341.
Sandage, A. 1958, *Stellar Populations*, ed. D. J. K. O'Connel (Amsterdam: North-Holland Publishing Co.; New York: Interscience Publishers, Inc.), p. 41.
Schalén, C. 1945, *Ann. Uppsala Obs.*, **1**, No. 9.
———. 1948, *ibid.*, **2**, No. 5.
———. 1953, *ibid.*, **3**, No. 9.
Schmidt, M. 1959, *Mém. Soc. R. Sci. Liège*, Ser. 4, **23**, 130.
———. 1962, *Symposium on Stellar Evolution*, ed. J. Sahade (LaPlata: Astr. Obs. Nat. University of LaPlata), p. 61.
Seaton, M. J. 1954, *M.N.*, **114**, 154.
———. 1955, *Mém. Soc. R. Sci. Liège*, Ser. 4, **15**, 462.
———. 1960, *Rept. Prog. Phys.*, **23**, 313.
Seyfert, C. K. 1943, *Ap. J.*, **97**, 28.
Shain, G. A. 1955, *A.J.-U.S.S.R.*, **32**, 110.
Shain, G. A. and Haze, V. F. 1951, *Izvest. Crimean Astr. Obs.*, **7**, 87.
———. 1952*a*, *ibid.*, **8**, 3.
———. 1952*b*, *ibid.*, p. 80; cf. Boggess (1954), p. 85.
———. 1952*c*, *ibid.*, **9**, 13.
———. 1952*d*, *Atlas of Diffuse Gaseous Nebulae* (Moscow: Acad. Sci. USSR).
———. 1953, *A.J.-U.S.S.R.*, **30**, 135; *Izvest. Crimean Astr. Obs.*, **10**, 210.
Shain, G., Haze, V. F., and Pikelner, S. B., 1954, *Izvest. Crimean Astr. Obs.*, **12**, 64; *A.J.-U.S.S.R.*, **31**, 105; English summary in *Mém. Soc. R. Sci. Liège*, Ser. 4, **15**, 441.
Shain, C. A., Komesaroff, M. M., and Higgins, C. S. 1951, *Australian J. Phys.*, **14**, 508.
Sharpless, S. 1952, *Ap. J.*, **116**, 251.
———. 1959, *Ap. J. Suppl.*, **4**, 257.
Sharpless, S. and Osterbrock, D. 1952, *Ap. J.*, **115**, 89.
Shklovsky, I. S. 1960, *Cosmic Radio Waves*, translation by R. B. Rodman and C. M. Varsavsky (Cambridge, Mass.: Harvard University Press; Toronto: S. J. Reginald Saunders & Co.), p. 196.
Slipher, V. M. 1912, *Lowell Obs. Bull*, **2**, 26.
———. 1918, *Pub. A.S.P.*, **30**, 63.
———. 1919, *ibid.*, **31**, 212.
Spitzer, L., Jr. 1941, *Ap. J.*, **93**, 369.
———. 1949, *ibid.*, **109**, 337.
———. 1954, *ibid.*, **120**, 1.
Spitzer, L., Jr. and Greenstein, J. L. 1951, *Ap. J.*, **114**, 407.
Spitzer, L., Jr. and Savedoff, M. P. 1950, *Ap. J.*, **111**, 593.
Stebbins, J. and Whitford, A. E. 1945, *Ap. J.*, **102**, 318.
Stecher, T. P. and Milligan, J. E. 1962, *Ap. J.*, **136**, 1.
Strand, K. Aa. 1958, *Ap. J.*, **128**, 14.
Strand, K. Aa. and Teska, T. M. 1958, *Ann. Dearborn Obs.*, **7**, 67.

Strömgren, B. 1939, *Ap. J.*, **89**, 526.
———. 1948, *ibid.*, **108**, 242.
———. 1951, *Problems of Cosmical Aerodynamics*, eds. J. M. Burgers and H. C. van de Hulst (Dayton, Ohio: Central Air Documents Office), p. 7.
Struve, O. 1941, *J. Washington Acad. Sci.*, **31**, 217.
Struve, O., Biesbroeck, G. Van, and Elvey, C. T. 1938, *Ap. J.*, **87**, 559.
Struve, O. and Elvey, C. T. 1939, *Ap. J.*, **89**, 119, 517; **90**, 301.
Struve, O., Elvey, C. T., and Keenan, P. C. 1933, *Ap. J.*, **77**, 274.
Struve, O., Elvey, C. T., and Roach, F. E. 1936, *Ap. J.*, **84**, 219.
Struve, O. and Swings, P. 1948, *Pub. A.S.P.*, **60**, 61.
Swihart, T. L. 1952, Indiana University. Master's degree thesis.
Thackeray, A. D. 1950, *M.N.*, **110**, 343.
———. 1956, *Observatory*, **76**, 154.
———. 1961, *ibid.*, **81**, 99.
Trumpler, R. J. 1930, *Lick Obs. Bull.*, **14**, 154.
Underhill, A. B. 1951, *Pub. Dom. Ap. Obs. Victoria*, **8**, 357.
Wade, C. M. 1958, *Australian J. Phys.*, **11**, 388.
———. 1959, *ibid.*, **12**, 418.
Walker, M. F. 1959, *Ap. J.*, **130**, 57.
Walter, K. 1941, *Zs. f. Ap.*, **20**, 256.
Weaver, H. F., Williams, D. R. W., Dieter, N. H., and Lum, W. T. 1965, *Nature*, **208**, 29.
Wesselink, A. J. 1962, *M.N.*, **124**, 501.
Westerhout, G. 1958*a*, *B.A.N.*, **14**, 215.
———. 1958*b*, *ibid.*, p. 261.
Westerlund, B. 1960*a*, *Ark. f. Astr.*, **2**, 419.
———. 1960*b*, *ibid.*, p. 467.
———. 1960*c*, *Pub. A.S.P.*, **73**, 72.
Weston, E. B. 1949, *Pub. A.S.P.*, **61**, 256.
———. 1952, *A.J.*, **57**, 28.
———. 1953, *ibid.*, **58**, 48.
White, M. L. 1952, *Ap. J.*, **115**, 71.
Whitford, A. E. 1958, *A.J.*, **63**, 201.
Whitney, W. T. and Weston, E. B. 1948, *Ap. J.*, **107**, 371.
Wilson, O. C., Münch, G., Flather, E. M., and Coffeen, M. F. 1959, *Ap. J. Suppl.*, **4**, 199.
Wilson, R. E. 1915, *Proc. Nat. Acad. Sci.*, **1**, 183.
———. 1918, *Pub. Lick Obs.*, **13**, 185.
———. 1953, *General Catalogue of Stellar Radial Velocities* (Washington: Carnegie Institution of Washington Pub. No. 601).
Wolf, M. 1923, *A.N.*, **219**, 109.
Woltjer, L. 1958, *B.A.N.*, **14**, 39.
———. 1959, *Ap. J.*, **130**, 38.
Woolf, N. J. 1961, *Pub. A.S.P.*, **73**, 206.
Wurm, K. 1961, *Zs. f. Ap.*, **52**, 149.
Wurm, K., and Rosino, L. 1956, *Mitt. Sternw. Hamburg-Bergedorf*, **10**, No. 103.
———. 1959, *Monochromatic Atlas of the Orion Nebula* (Padova: Osservatorio Astrofisico di Asiago).
Yada, B. 1960, *Pub. Astr. Soc. Japan*, **12**, 449.
Yang, K. S., and West, L. A. 1964, *A.J.*, **69**, 246.
Zanstra, H. 1926, *Phys. Rev.*, **27**, 644.

CHAPTER 3

Dark Nebulae

B. T. LYNDS
University of Arizona

1. DARK MARKINGS; EARLY HISTORY

THE ABSENCE of stars, or a marked reduction in star counts, in specific regions of the sky has been noted since the eighteenth century. It was not until the turn of the twentieth century, however, that these dark markings were recognized to be nonluminous, absorbing dust clouds in space.

The association of luminous nebulae with regions in which a noticeable decrease in star density existed had been explicitly pointed out much earlier by Sir William Herschel (1784). In fact, the frequency of association was such that Herschel related that if during his routine sweeps of the sky he came upon a region nearly devoid of stars he gave "notice to my assistant at the clock to prepare, since I expected in a few minutes to come at a stratum of [bright] nebulae, finding myself already (as I then figuratively expressed it) 'on nebulous gound.' "

Plate 1 shows an example of the association to which Herschel refers in the region near σ Orionis (NGC 2024), No. 28 in Herschel's *Catalogue of Second Thousand of New Nebulae and Clusters* of stars. Additional "starless fields" are given in Webb's *Celestial Objects for Common Telescopes* (1894, 1962), which was based on the visual observations of Sir John Herschel. These regions were later identified by Barnard as true dark nebulae.

In spite of the excellence of these diligent visual observers, the photographic technique was necessary in order to reveal the structural features of these dark regions. In this reconnaissance, the names of two astronomers are outstanding—E. E. Barnard and Max Wolf. Both skilled visual observers (in fact, both "comet seekers"), these two men essentially originated the study of dark nebulae per se, although the existence of these obscuring clouds was still somewhat in doubt even at the end of Barnard's career in 1923. Cautious to the end, Barnard seemed to prefer the noncommittal "dark markings" to the term "dark nebulae," although he personally seemed convinced that these objects were obscuring clouds, producing an effect analogous to that of small terrestrial clouds seen in silhouette against a dark, moonless sky.

With photography, the association of bright and dark nebulae was further emphasized by a few dark markings that were made visible against the background of a luminous nebula. One of the best examples is that of the Horsehead Nebula (Plate 2) described by Curtis (1918) as follows:

> The most striking feature of the region is a remarkable dark bay jutting into and bifurcating the long ray. It is impossible to look at the original negatives of these interesting objects and not be convinced that there is something dark between us and the general background of stars; I firmly believe that there are actually "dark nebulae" . . . that they are "holes" torn in the star fabric of the Milky Way by some rapidly rushing star cluster is difficult to believe when one studies carefully the sharply defined edges. If merely "holes" we must assume their age as of the order of hundreds of millions of years, in which time, as Dr. Campbell has pointed out, the random motions of the stars in the neighborhood would long since have obliterated the clean cut edge, if not the "hole" itself.

As is often true in the analysis of our own Galaxy, studies of the external galaxies provided strong arguments for the case of dark nebulae. Curtis' beautiful photographs of extragalactic systems showed many objects having dark bands. Curtis pointed out that an observer in one of these systems might well look up to a Milky Way like our own, crossed with bands of dark clouds. This point was also made by Wolf, who drew a comparison between his photograph of the Andromeda Nebula and our Milky Way (1908): "Auch diesen Nebel hat man vielfach als Analogen zu unserem Milchstrassen-system aufgefasst, besonders auch deshalb, weil er so schön die vielen dunkeln Höhlen erkennen lässt, wie wir sie in unserer Milchstrasse beobachten."

The definitive work of Hubble (1922) that established the reflecting properties of many luminous bright nebulae prompted Russell (1922) to postulate on the composition of the now-recognized dark objects, and he concluded that "it appears probable that the aggregate mass contained in one of these great obscuring clouds must be very considerable—probably sufficient to form hundreds of stars—and that a sensible fraction of the whole mass must be in the form of dust less than 0.1 mm in diameter."

2. CATALOGUES OF DARK NEBULAE

By 1930 it was generally accepted that nonluminous obscuring clouds existed in the Galaxy; in fact, Barnard had published a list of 349 such objects (Barnard 1927). This catalogue, based on his photographic survey of the Milky Way, lists the dark markings he had noted which he believed to be true dark nebulae. The catalogue contains the equatorial coordinates, an estimate of the size of the object, and a brief description of each. Barnard states explicitly that, although some of the dark markings listed "may be only vacancies among the stars," he had "tried to avoid such as much as possible." This latter statement emphasizes the intrinsic difficulty in identifying and cataloguing dark nebulosities. In contrast to the usual straightforward cataloguing of stars or luminous nebulae, when looking for dark objects, we are searching for the absence of stars rather than for the presence of any isolated object. Perhaps the best we can do is to assume that an obscuring cloud is present if the star density decreases by an amount greater than the random statistical fluctuations of the star numbers we would expect; that is, if the number of stars counted per unit area is less than $\sqrt{n_o}$, where n_o is the average number counted per equal area at the same galactic latitude, then we may assume the presence of an obscuring cloud in the field. Usually, then, the richer the star field, the easier it is to detect a dark nebula between the star field and the observer.

Plate 1.—Region near σ Ori showing NGC 2024, taken with the Crossley reflector of the Lick Observatory. The print has been dodged to bring out the nebulosity (Lick Observatory photograph).

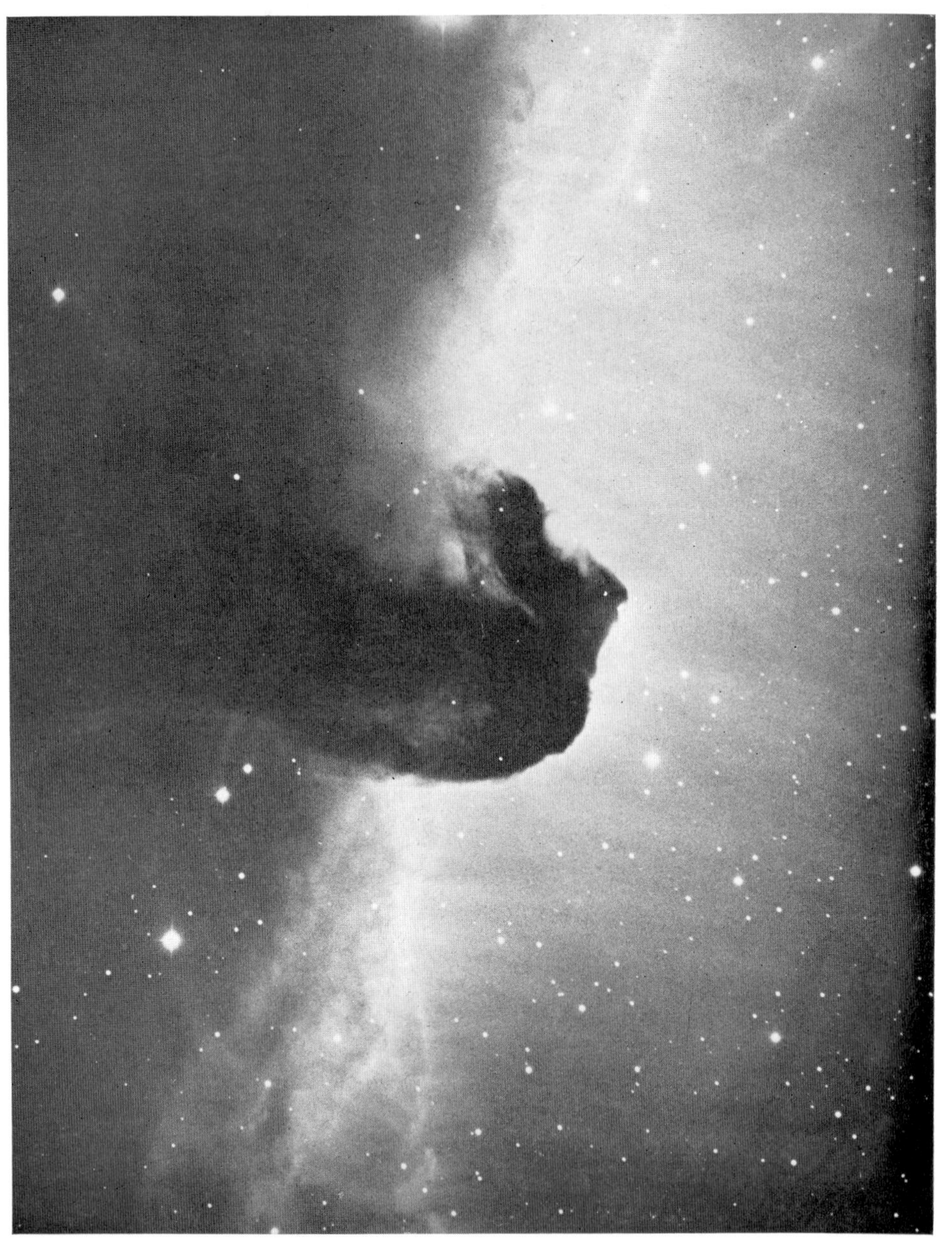

PLATE 2.—The Horsehead Nebula (IC 434, Barnard 33) in Orion, south of ζ Orionis, photographed in red light with the 200-inch Palomar reflector (Mt. Wilson and Palomar Observatories photograph).

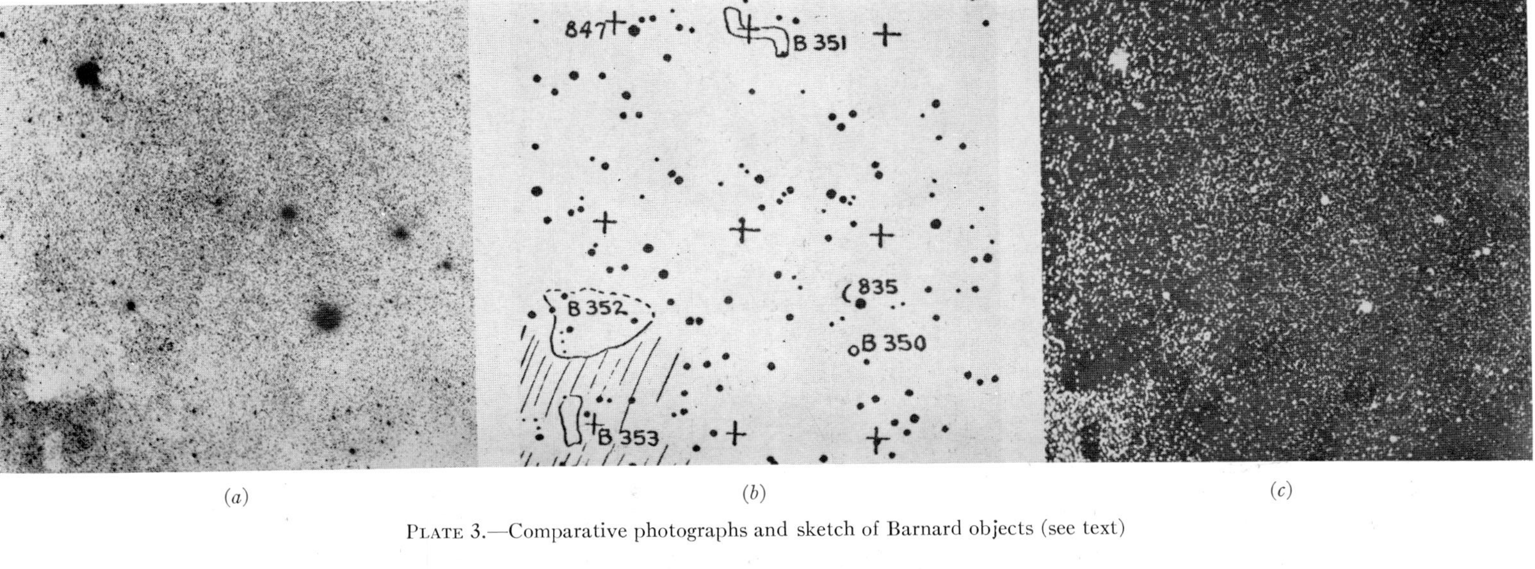

(a) (b) (c)

Plate 3.—Comparative photographs and sketch of Barnard objects (see text)

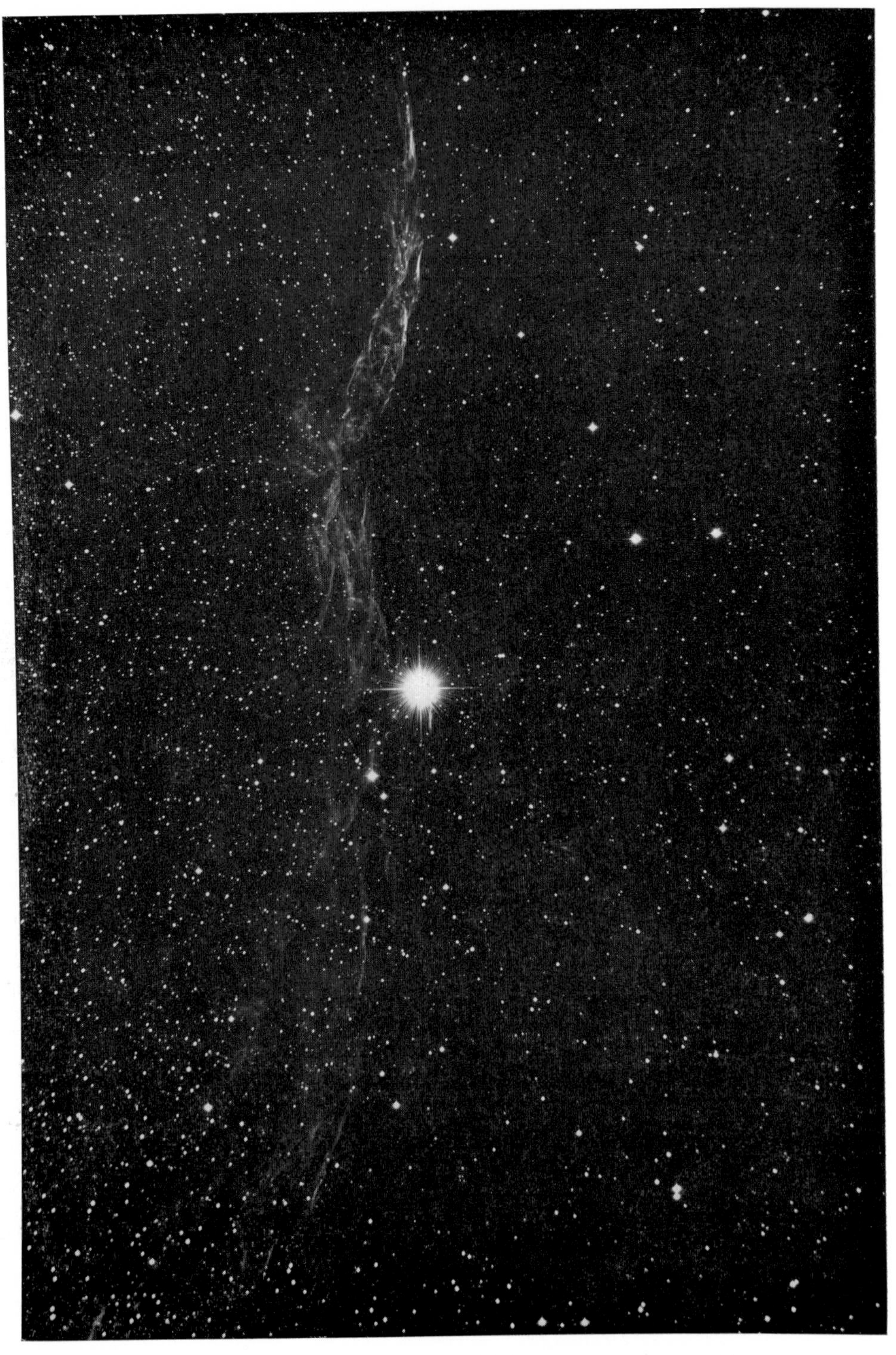

Plate 4.—The region of NGC 6960 (Lick Observatory photograph)

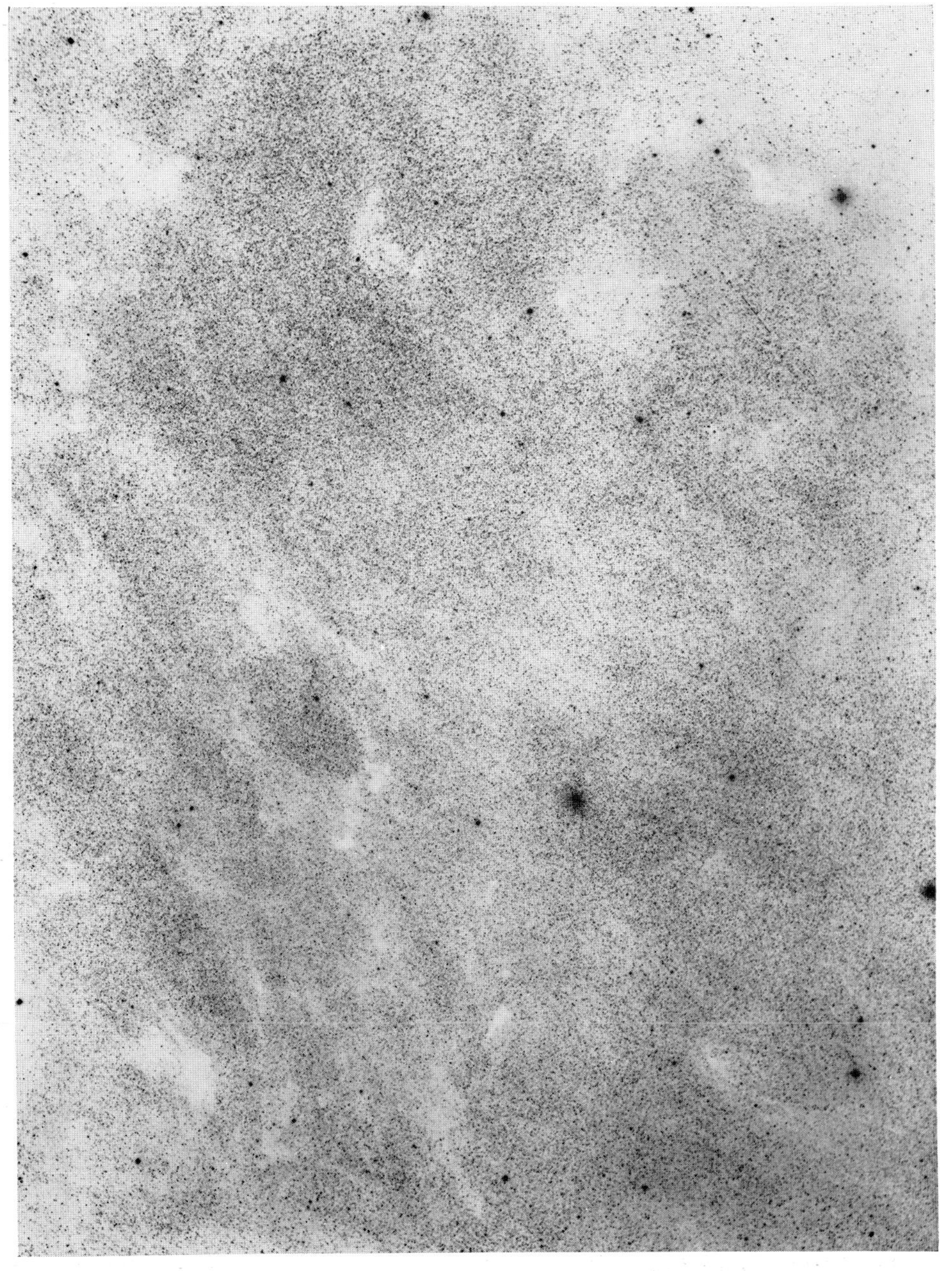

PLATE 5.—Typical dark nebulae on red Palomar negative No. 156 (Mt. Wilson and Palomar Observatories photograph).

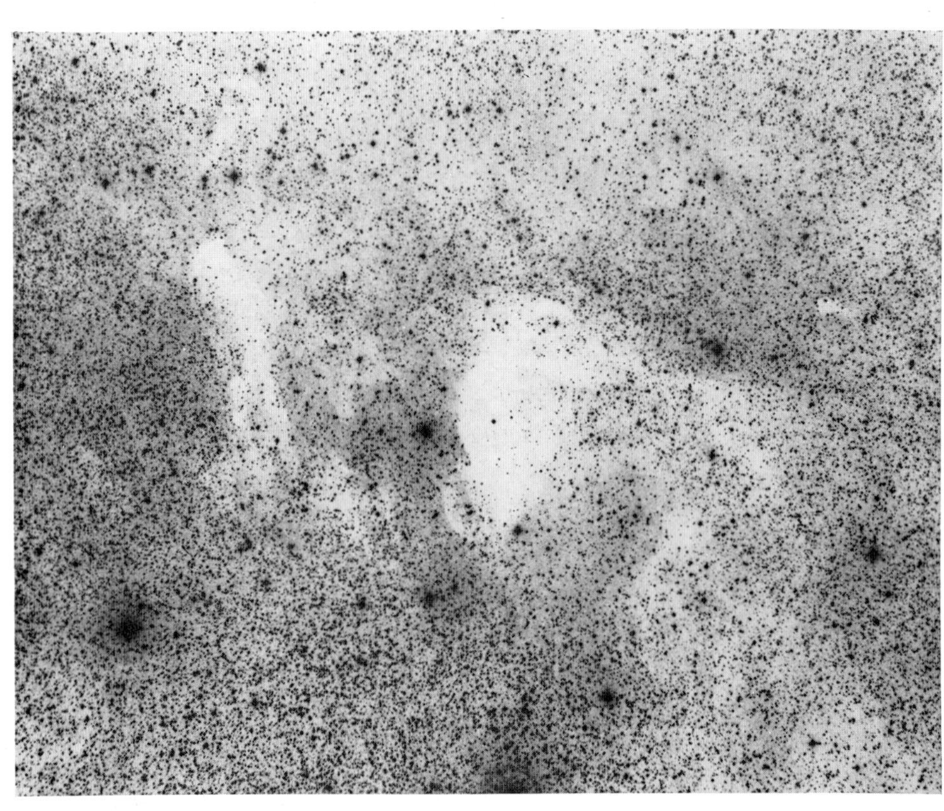

PLATE 6.—B92 (larger cloud) and B93 photographed in blue light with the Palomar Schmidt (Mt. Wilson and Palomar Observatories photograph).

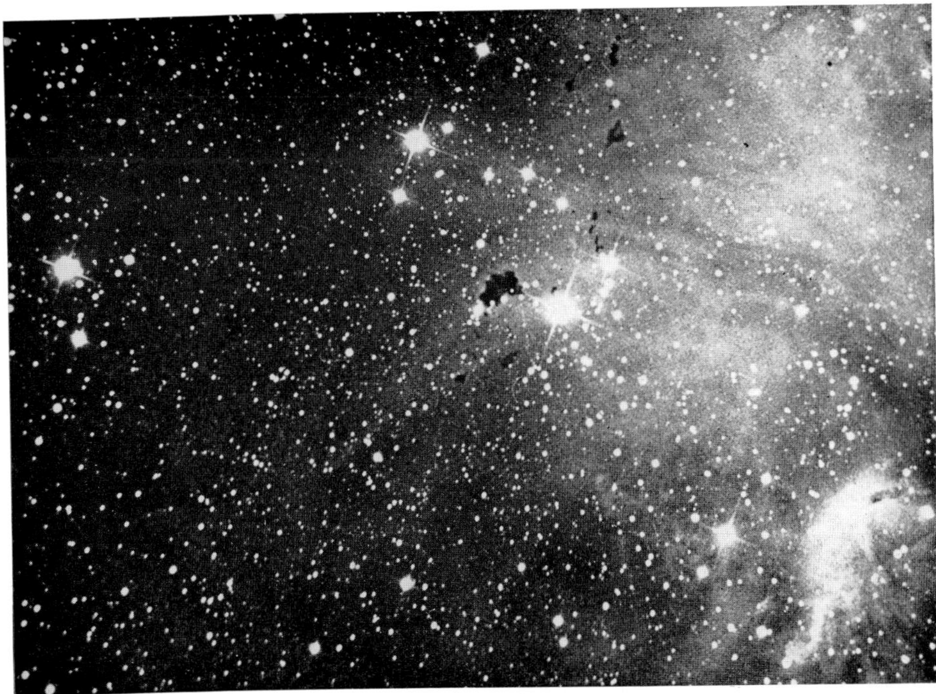

PLATE 7.—Field of dark clouds in IC 2944 taken at the Radcliffe Observatory (from Thackeray 1954).

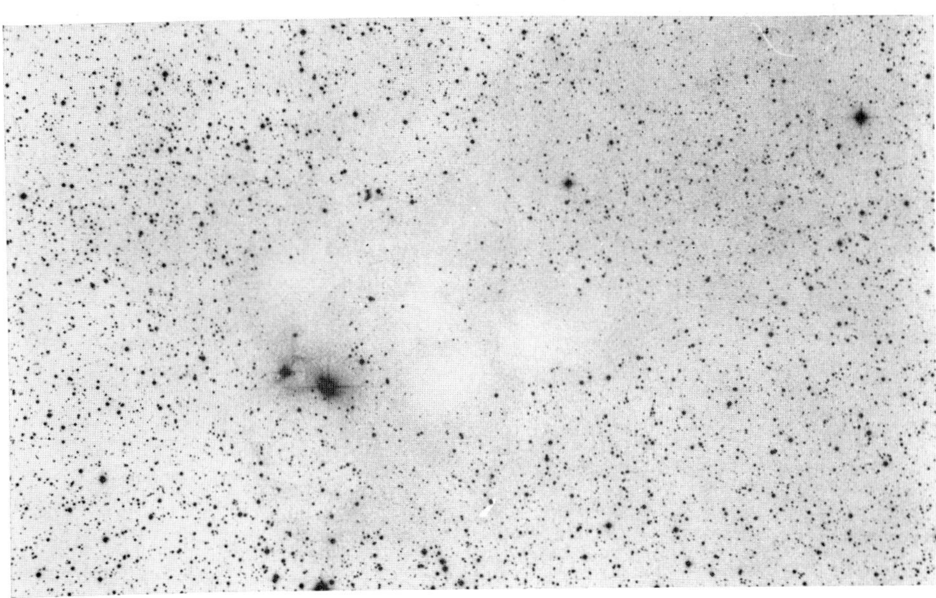

PLATE 9.—Dark nebula B27 associated with AB Aurigae (from the red negative of Palomar Schmidt No. 1314. Mt. Wilson and Palomar Observatories photograph).

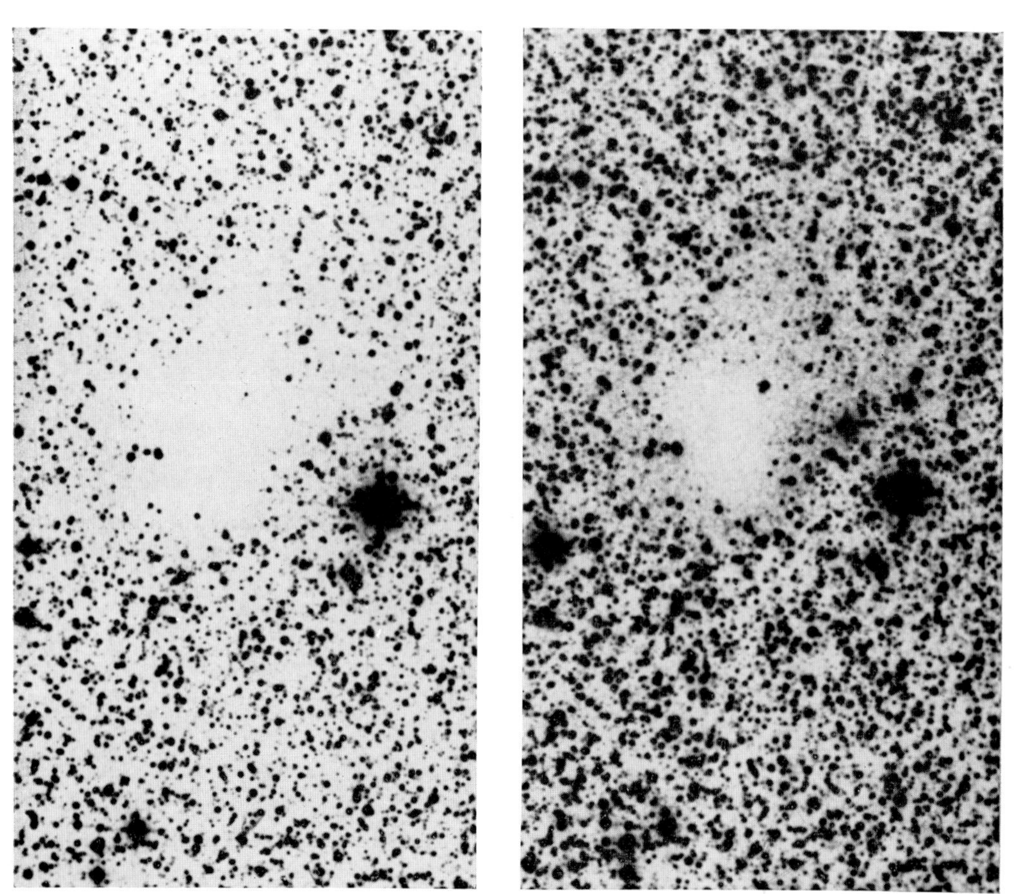

PLATE 8.—B338 photographed with the Palomar Schmidt in red (*right*) and blue (*left*). The scale is 9.3 seconds of arc/mm (Mt. Wilson and Palomar Observatories photograph).

[Plate 9 is on the previous page]

The ability to detect a dark nebula is also a function of the limiting magnitude of the survey. This is not a minor point, as evidenced by the Hagen Cloud controversy (Becker and Meurers 1956). The limiting magnitude problem occurs not only in visual observations but also in photographic studies. For example, several dark areas in Barnard's catalogue are not detectable on the Palomar prints. Plate 3*a* shows a region containing several of the Barnard objects as photographed in blue light by the 48-inch Palomar Schmidt (negative print, limiting magnitude 20). The sketch by Miss Calvert (Plate 3*b*) illustrates the location of Barnard's dark markings. Plate 3*c* is a reproduction of one of the Barnard and Ross photographs (positive print, limiting magnitude 17) of the same region. From the Palomar print, one would certainly not single out the objects B350 and B351 as isolated dark nebulae. The majority of the Barnard objects, however, can be easily identified on the Palomar prints; of the 349 objects, 22 are beyond the declination limit of the Schmidt survey, and only about 50 others are not clearly identifiable.

Shortly after the publication of Barnard's catalogue, Lundmark and Melotte undertook a search for dark nebulae on the one photographic survey available which covered the entire sky. This survey was made by Franklin-Adams and bequeathed to the Greenwich Observatory. The original negatives were examined by Lundmark and Melotte, and all regions where the star density dropped by a factor of 5 or 6 below the surrounding sky were called "dark nebulae." The Lundmark catalogue (1926) contains 1550 objects, together with their positions, sizes, and descriptions. Figure 1 shows the distribution of the nebulae listed in this catalogue; the size of the dots represents the areas of the dark regions on the scale used in the diagram for representation of the celestial sphere. This catalogue contains the results of the one search for dark nebulae which covers the entire sky.

The large complexes of dark nebulae that obscure the central regions of the Galaxy are the dominant features of the distribution in Figure 1. The Sco-Oph complex ($l^{\rm I} \sim 335$), lying about 20 degrees above the plane, and the Ori-Tau complex ($l^{\rm I} \sim 150$) at a comparable distance below the plane, are distinct. The region surrounding the north equatorial pole also contains areas of apparent obscuration. Most of the dark nebulae are within 30 degrees of the plane of the Galaxy. The existence of the smaller clouds in the diagram at distances as great as 75 degrees from the plane has not been confirmed. They may only represent the decrease in stellar density at higher galactic latitudes; associated statistical fluctuations in star-poor fields might produce the effect of a dark nebula.

As is apparent from the Lundmark-Melotte survey, the dark nebulae are for the most part confined to the belt of the Milky Way. Therefore, the Ross-Calvert photographic *Atlas of the Milky Way* (1934) contains a series of photographs of the region of the sky most likely to show dark clouds. Two catalogues of dark nebulae based on the *Atlas of the Milky Way* have been published.

The first survey was made by Khavtassi and is given in catalogue form (Khavtassi 1955) and also as an atlas (Khavtassi 1960). His *Atlas of Galactic Dark Nebulae* contains in graphical form the nebulae he recorded from the Ross-Calvert *Atlas of the Milky Way* and Hayden's *Photographic Atlas of the Southern Milky Way* (1952). Also represented schematically are the diffuse-emission nebulae, the galactic star clusters, and stars brighter than the fourth magnitude. Visual estimates of the opacity of the nebulae are

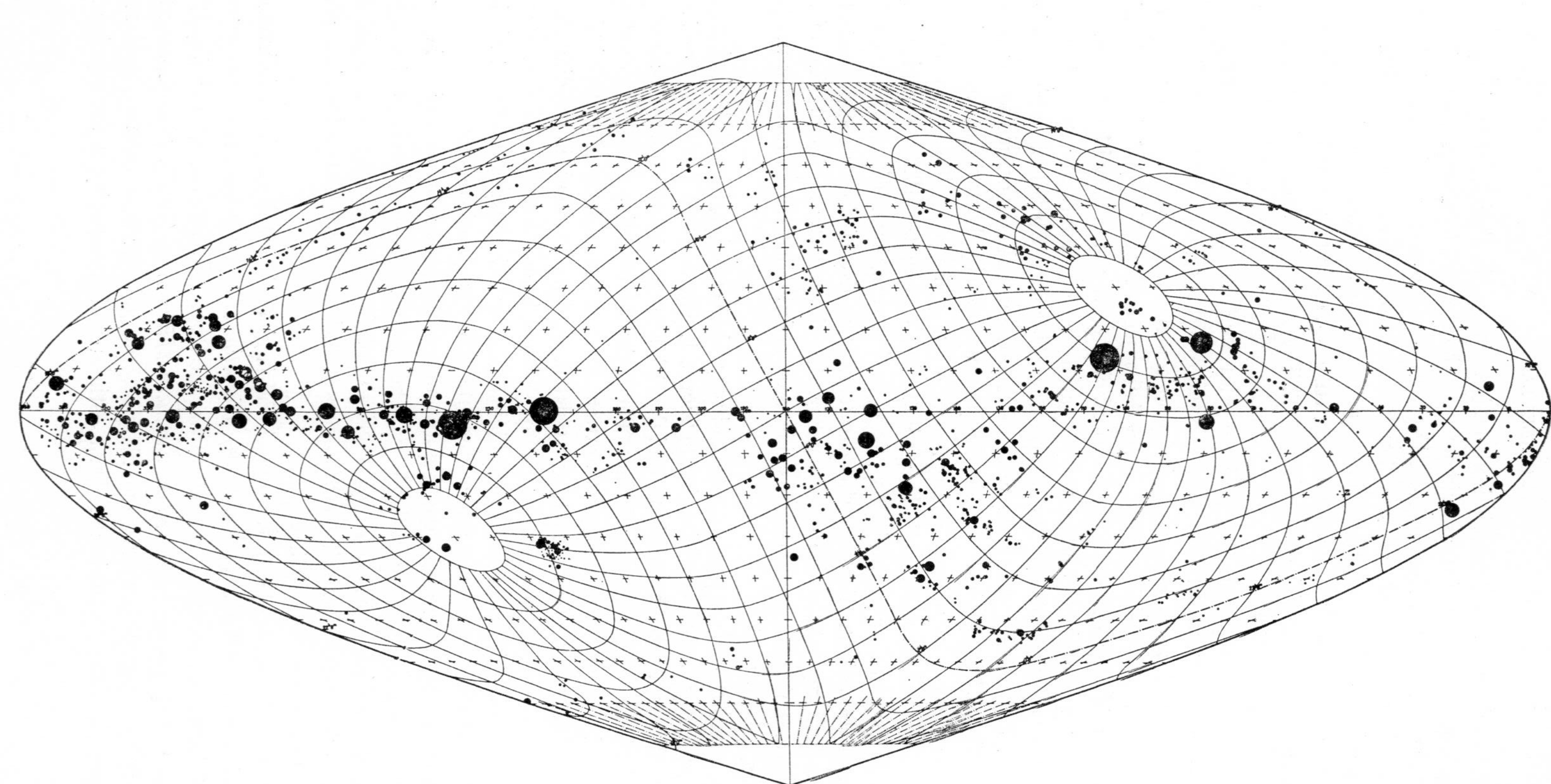

FIG. 1.—Distribution of dark nebulae from the Franklin-Adams charts (Lundmark and Melotte)

indicated by three degrees of shading in the figures. Khavtassi's catalogue lists 797 nebulae, together with their equatorial and galactic coordinates (l^{I}, b^{I}), their areas in square degrees, and an estimate of their opacity.

The second catalogue, based not only on the Ross-Calvert plates but also on the Lick photographs of the Milky Way (Barnard 1913*a*), was compiled by Schoenberg (1964). This catalogue lists 1456 dark nebulae, with equatorial and galactic coordinates (l^{I}, b^{I}), areas, shapes, and estimates of absorption for each object. Table 1, which is from this catalogue, gives the distribution of the observed nebulae as a function of their surface area.

The National Geographic Society–Palomar Observatory sky survey (see Minkowski and Abell 1963), photographed with the 48-inch Schmidt at Palomar, was the source of material for another catalogue of dark nebulosities (Lynds 1962). In this survey, both

TABLE 1

NUMBER OF OBSERVED DARK NEBULAE AS A FUNCTION OF THEIR SURFACE AREA

Surface Area (square minutes of arc)	Number of Nebulae	Surface Area (square minutes of arc)	Number of Nebulae
0–40	469	500–600	13
40–80	452	600–700	1
80–120	208	700–800	5
120–160	104	800–1000	5
160–200	86	1000–1200	4
200–240	41	1200–1400	3
240–280	38	1400–1600	1
280–320	20	1600–1800	3
320–360	12	2000–3000	2
360–400	11	3000–4000	3
400–500	17	4000–6000	1

After Schoenberg (1964).

the red and blue photographs were studied, and all dark nebulae detectable on these pairs of prints were recorded. The catalogue contains a list of 1802 condensations; most of the entries are condensations within the general obscuration of extensive dark-cloud complexes. Equatorial and galactic coordinates (l^{II}, b^{II}) are given for each condensation, together with the measured surface area on the charts and a visual estimate of the opacity of the cloud. It should be emphasized that measures of the areas of dark nebulae are extremely sensitive to the limiting magnitude of the survey and also to the wavelength sensitivity of the photographic plate. The outer regions of a dark nebula become essentially transparent on the red plate; therefore, the object appears to be smaller in the red. This "shrinking" effect is also present in two blue photographs, one having a limiting magnitude of 17, the other recording stars to the 20th magnitude. Here again, the very tenuous clouds are not easily seen in the richer star fields.

The 48-inch photographs have a larger scale and better resolution than do those in the Ross Atlas; therefore, the fine structure within an extended dark cloud is more apparent. It was found that dark nebulae cover some 1360 square degrees of the Palomar Survey plates; over most of this area, the clouds produce about one or two magnitudes absorption. The most opaque clouds cover a total area of only about 4 square degrees. However, clouds of greatest opacity increase in number to the resolving limit of the survey (see Table 1).

3. DISTRIBUTION OF DARK NEBULAE

3.1. Distances

Two methods are usually employed in order to estimate the distances of dark nebulae. If a dark nebula is known to have a luminous nebula associated with it, the distance of the bright nebula as determined from the spectroscopic parallax of the exciting star is

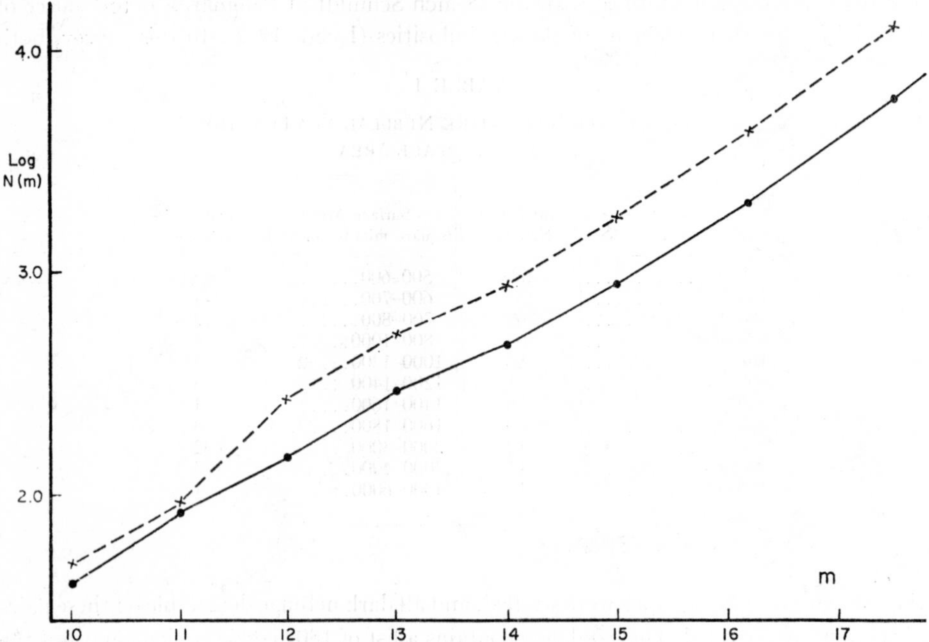

Fig. 2.—Star counts in NGC 6960 (after Wolf 1923). Solid line refers to counts in the obscured area, dotted line to a clear comparison region. The ordinate is log $n(m)$, the number of stars brighter than apparent magnitude m.

taken as the distance to the entire complex. In some cases, the distance to the bright nebula represents the upper limit of distance to the dark object. An example is M8, where the small globules are seen projected against the bright nebula. The dark markings cannot be at a distance greater than 1600 pc, which is that of the bright nebula; they may be closer and perhaps not be physically associated with the emission nebula at all (for further discussion of this point, see Razhkovsky 1955).

For more extended dark nebulosities, an estimate of their distances may be obtained from a comparison of star counts in the obscured region with those made in a nearby unobscured area. This approach to the problem of distance determinations was made by Wolf (1923). In photographs such as Plate 4 of the bright nebula NGC 6960, he noted a decrease in the star density to the west of this emission nebula and concluded that the nebula itself was obscuring the background stars. In order to illustrate this effect, he counted the number of stars of magnitude m, $A(m)$, per unit area in the clear region and compared this with similar counts made in the obscured area to the west of the nebula. Figure 2 is the Wolf diagram of the counts, with the solid line representing the cumulative counts in the clear region and the dashed representing those in the nebula. In this diagram, the counts for stars brighter than the eleventh magnitude are approximately the same, although around magnitude 11 the nebula begins to have an effect. For fainter stars the curves become parallel, indicating a total absorption of about $1^{\mathrm{m}}.0$ through the cloud. The distance to the cloud should be equal to the mean distance of the stars of the eleventh magnitude.

The graphical method developed by Wolf is a simplification of the analytical solution proposed by Pannekoek (1921), which includes the dispersive effects of the luminosity function.

In Pannekoek's notation, let $A(m)dm$ be the number of stars counted per unit solid angle of apparent magnitude between $m \pm \frac{1}{2}dm$. Then,

$$A(m) = \int_0^\infty F(r)\varphi(M)\,dr \tag{1}$$

where $F(r)dr$ is the number of stars per unit element of solid angle at a distance between $r \pm \frac{1}{2}dr$ and $\varphi(M)$ is the normalized distribution function of absolute magnitude.

The true space-density function $D(r)$ is related to $F(r)$ by

$$F(r) = r^2 D(r)\,. \tag{2}$$

If space were completely transparent, we would have

$$M = m + 5 - 5\log r \tag{3}$$

and

$$A(m) = \int_0^\infty r^2 D(r)\varphi(m + 5 - 5\log r)\,dr\,, \tag{4}$$

and if $\varphi(M)$ is known, the fundamental equation (4) may be used to find the space density.

It is necessary to modify the fundamental relation between M and r if space absorption is present. In this latter case, we may write

$$M = m + 5 - 5\log r - a(r) \tag{5}$$

where $a(r)$ is the space absorption up to a distance r. We may define a "fictitious" r_o such that

$$5\log r_o = 5\log r + a(r)\,. \tag{6}$$

In this notation, the basic equation retains the same form; that is,

$$A(m) = \int_0^\infty r_o^2 D_o(r_o)\varphi(m + 5 - 5\log r_o)\,dr_o \tag{7}$$

but now instead of the true space density being determined, the observations give only a solution of

$$D_o(r_o) = D(r)\{e^{3ba(r)}[1 + bra'(r)]\}^{-1} \tag{8}$$

where D_o is the so-called apparent density function and b is a constant of integration. (See also Trumpler and Weaver 1953, 1962.)

Malmquist (1943) has developed a method of correcting this observed density distribution for the effects of absorption by using the following expression for $N(r)$, the total number of stars per unit solid angle counted to a distance r:

$$N(r) = \int_0^r r^2 D(r)\,dr \tag{9}$$

and, if absorption is present, the same number of stars is also

$$N_o(r_o) = \int_0^{r_o} r_o^2 D_o(r_o)\,dr_o = N_o(r - 10^{0.2a(r)}) \tag{10}$$

or as Malmquist states, "Put into words this relation expresses the self-evident fact that the actual numbers of stars up to the true distance r is equal to the number of stars with the same space-cone up to the apparent distance $r_o = r - 10^{0.2a(r)}$, computed without regard to any space absorption." This statement is expressed simply in terms of the distance modulus, $y = m - M$, by

$$N(y) = N_o[y + a(y)] \tag{11}$$

where $a(y)$ is the total absorption at distance r, expressed in magnitudes.

Thus the method of determining the distance and absorption in a dark nebula reduces to the following steps: (1) Assume that a clear field (subscript c) near the dark nebula (subscript n) will supply the values of $A_c(m)$, and that the obscured region itself is also counted to determine $A_n(m)$. The solution of the fundamental equation gives $D_c(r)$ and $D_n(r)$. (2) Then $N_c(y)$ and $N_n(y)$ can be determined by equations (2) and (3). (3) A comparison of the two cases gives $a(y)$ from equation (11).

This analysis has neglected the general space absorption, which should be included as an additive term in both expressions for M; that is,

for a clear field,

$$M = m + 5 - 5 \log r + a^1(r)\,, \tag{12}$$

for a cloud field,

$$M = m + 5 - 5 \log r + a^1(r) + a(r) \tag{13}$$

and

$$\log N(y) = \log N_1[y + a(y)] = \log N_o[y + a^1(y) + a(y)] \tag{14}$$

where $N_1(y)$ and $N_o(y)$ are now the numbers of stars to the apparent distance $r = 10^{0.2y}$ in the comparison field and nebular field, respectively. In this manner, if $a^1(r)$ is known, the analysis of the star counts can be done as outlined and $a(r)$ determined for the nebula.

Figure 3, derived from a detailed analysis of the bright and dark regions in Cepheus by Wernberg (1941), is the graphical representation of step (2) in the Malmquist method. The four separate counts were made for stars of different spectral types.

The graphical representation of step (3) is illustrated in Figure 4, also from Wernberg (1941). The distance to the cloud is 160 pc; the extinction in the cloud is approximately $1^{m}.3$.

The dispersive influence of the luminosity function has been the major complication in analyses of star counts in clouded areas. The total absorption by a nebula can be easily determined by the graphical method of Wolf, but the radial extent of the nebula as evaluated for these star counts is greatly influenced by the spread in absolute magnitudes. If this difficulty is neglected, the data from star counts may lead to a misrepresentation of the nebula. Some of the early, uncritical applications of Wolf's method led

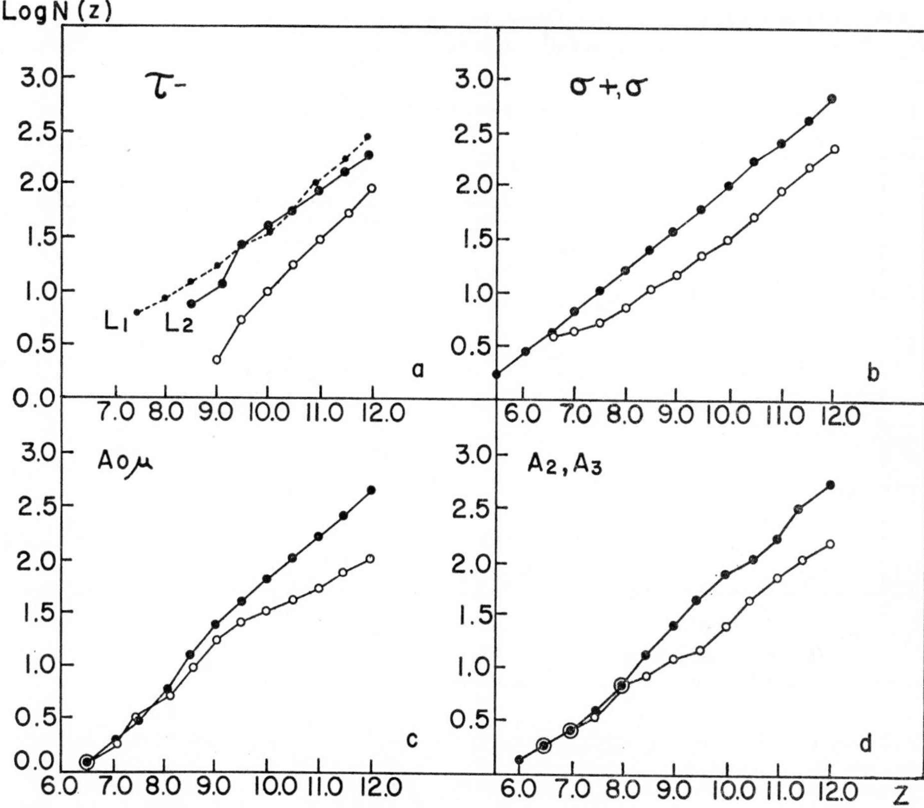

FIG. 3.—Malmquist diagrams from Wernberg (1941) for comparisons between regions L1 (*large filled circles*) or L2 (*small filled circles*) and an obscured region, M1, in Cepheus (*open circles*). The four separate diagrams refer to stars of the following different spectral and luminosity types (Uppsala luminosity classification): (*a*) B$\tau-$; (*b*) AOσ and AO$\sigma+$; (*c*) AOμ; (*d*) A2–A3.

researchers to conclude that dark nebulae of much greater radial extent than transverse were present, implying a system of, say, ellipsoidal nebulae all having their major axes lying in the line of sight. Better resolution is possible if the data on counts are confined to limited spectral classes having relatively small dispersion in absolute magnitude. This limitation, however, must decrease the number of stars counted and hence increases the statistical uncertainty in the analysis. Furthermore, the extensive data on spectral classification that are required are frequently not available.

In addition to the absorption of light from stars lying in and beyond a dark nebula, we also expect a reddening effect if the same type of particle produces the nebular obscuration as well as the general interstellar reddening. On this assumption, it is possible to estimate the distance of a cloud from multicolor measures of the stars in the vicinity of the nebula. One such method, based on a procedure used by Becker (1938) for general interstellar reddening estimates, was developed by Rodgers (1960). Let us restrict ourselves to the three-color system of Johnson and Morgan (1953). With this system, it is possible to represent the color difference $(U - B) - (B - V)$ as a function of one color, say $(B - V)$, for the entire spectral range of the main sequence for unreddened stars.

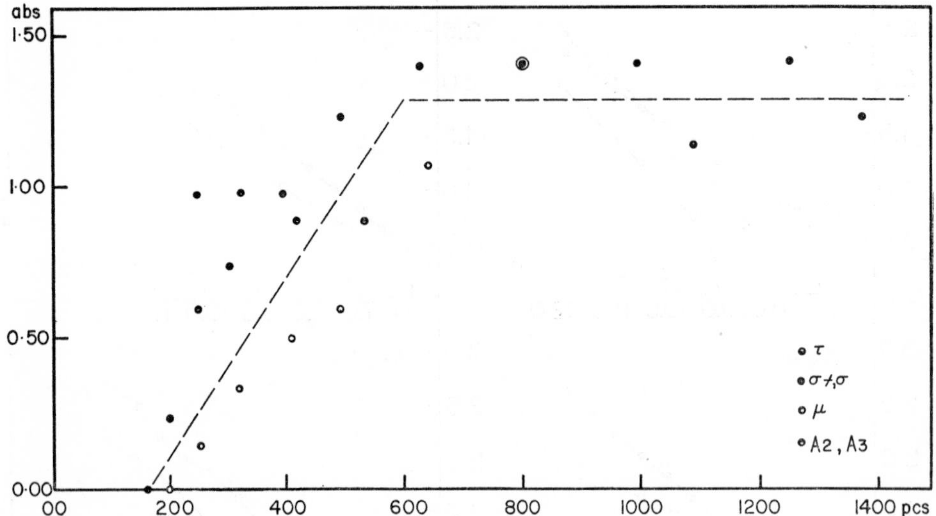

FIG. 4.—The relationship between distance and absorption in the dark Cepheus Nebula M1 as determined from the Malmquist diagrams (from Wernberg 1941).

Such a curve is shown (solid line) in Figure 5. If the stars are slightly reddened, the color excesses E are related by

$$E_{U-B} = (U-B)_{\text{obs}} - (U-B)_{\text{intrinsic}} \tag{15}$$

$$E_{U-B}/E_{B-V} = 0.76 \tag{16}$$

or

$$\frac{E_{U-B} - E_{B-V}}{E_{B-V}} = -0.24 \tag{17}$$

and the reddening lines are indicated in Figure 5 by the dashed straight lines of slope -0.24.

As in an idealized Wolf diagram, we assume that an isolated cloud is present, and in such a case there will be a discontinuity in the spectral sequence of the color-difference diagram at the point determined by the intrinsic colors of the stars at the same distance as the cloud. To illustrate this technique, consider the following example.

Let there be a discrete cloud with $m - M = 7.0$ and absorption of 1 mag. Consider stars of the ninth apparent magnitude. Then the color difference diagram would coincide with the intrinsic curve for main-sequence stars fainter than absolute magnitude +2.0, but stars brighter than this limit would lie behind the cloud and the resulting color-difference curve would be shifted slightly along the reddening lines. Thus the position of the discontinuity in the color-difference diagram enables us to determine the distance of the

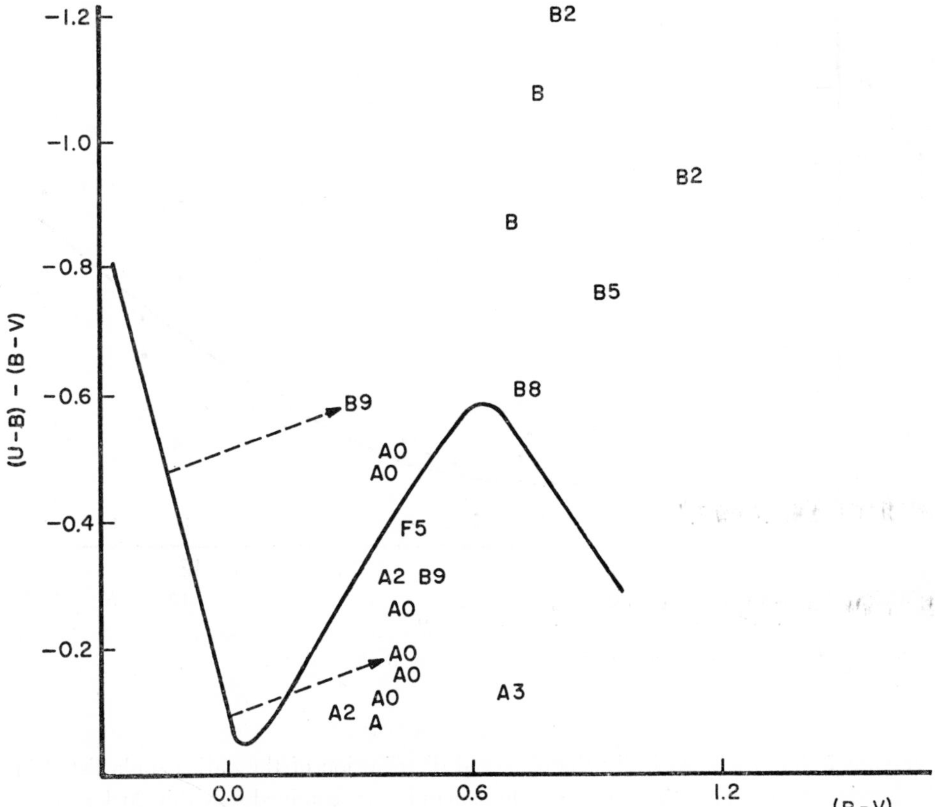

FIG. 5.—Color-difference diagram for stars on the *UBV* system. The spectral types refer to stars in the region of the Southern Coalsack (after Rodgers 1960). The solid line is the intrinsic color-difference relation; dotted lines indicate reddening lines.

nebula. The displacement of the curve beyond this discontinuity gives the total color excess, and the resulting total absorption may be found if the extinction ratio is known. Rodgers has applied this method to several regions of the Southern Coalsack.

The crosses plotted in Figure 5 represent the intrinsic color-difference diagram for one region of the Southern Coalsack for stars of magnitude V between 10.0 and 10.99. The solid line is the intrinsic color difference. The corresponding absorption is deduced by the displacement of the stars along the reddening lines given in Figure 5. Stars with V between 10.0 and 10.99 and with spectral types later than F appear to be unreddened; thus at a distance whose modulus $m - M = 6$, the cloud seems to begin to produce measurable absorption. The A-type stars of $10^{m}.5$ appear to be reddened by approximately $0^{m}.4$, which indicates a total absorption in the cloud of $1^{m}.2$. The total absorption in the cloud is $1^{m}.1$. Rodgers' final results for several regions analyzed in this manner indicate that there is a cloud at a mean distance of 174 pc having a photographic absorption of about $1^{m}.0$. Behind this nebula, there is a clear region extending to 800 pc. At greater distances, absorption again sets in, as illustrated in Figure 6.

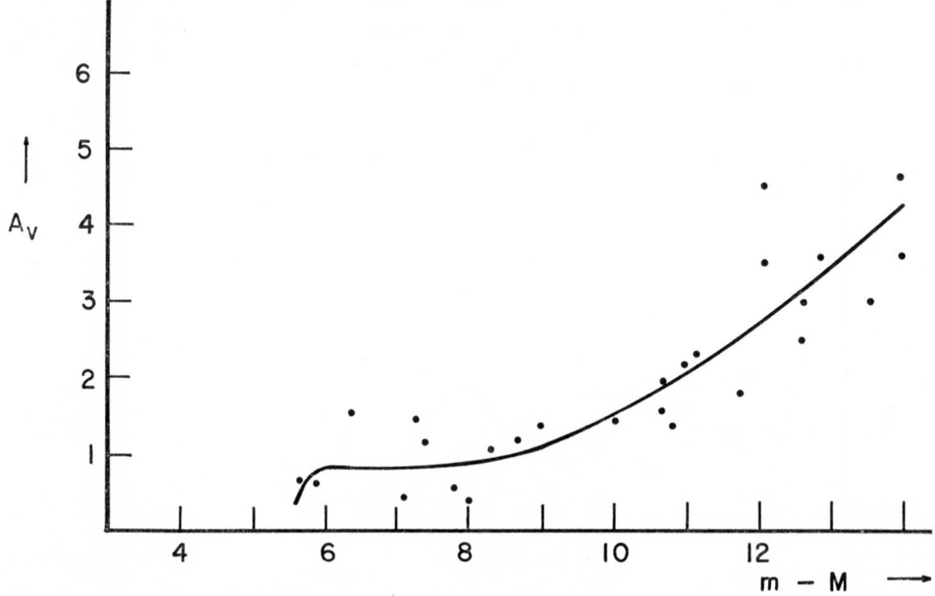

FIG. 6.—Visual absorption plotted against distance modulus for one region of the Coalsack. The plotted points are means for each magnitude interval (after Rodgers 1960).

3.2. Galactic Distribution

Figure 7 shows schematically the observed distribution of the dark nebula listed in the Lynds catalogue. For the most part, these nebulae are confined to within 20 degrees of the plane of the Galaxy. The systematic variations reflect the presence of Gould's Belt, or perhaps we should say that they are due to the Tau-Ori complex and the Sco-Oph clouds (see Bok 1937).

Table 2 lists the individual clouds or cloud complexes that have been studied in detail. Column 4 (distance estimates) indicates that we are observing to various depths in the plane. The very dark clouds of Taurus seem to be among the nearest, with distances less than 250 pc. The cloud complexes causing the Great Rift in Cygnus lie approximately 600 pc away, and the unique "tunnels" around θ Oph are about 200 pc distant.

An extensive survey of numerous regions of the Galaxy obscured by dark nebulae has been made by Khavtassi (1958), who compiled data on dark clouds determined by the method of star counts. Khavtassi points out that the regions studied by counts are

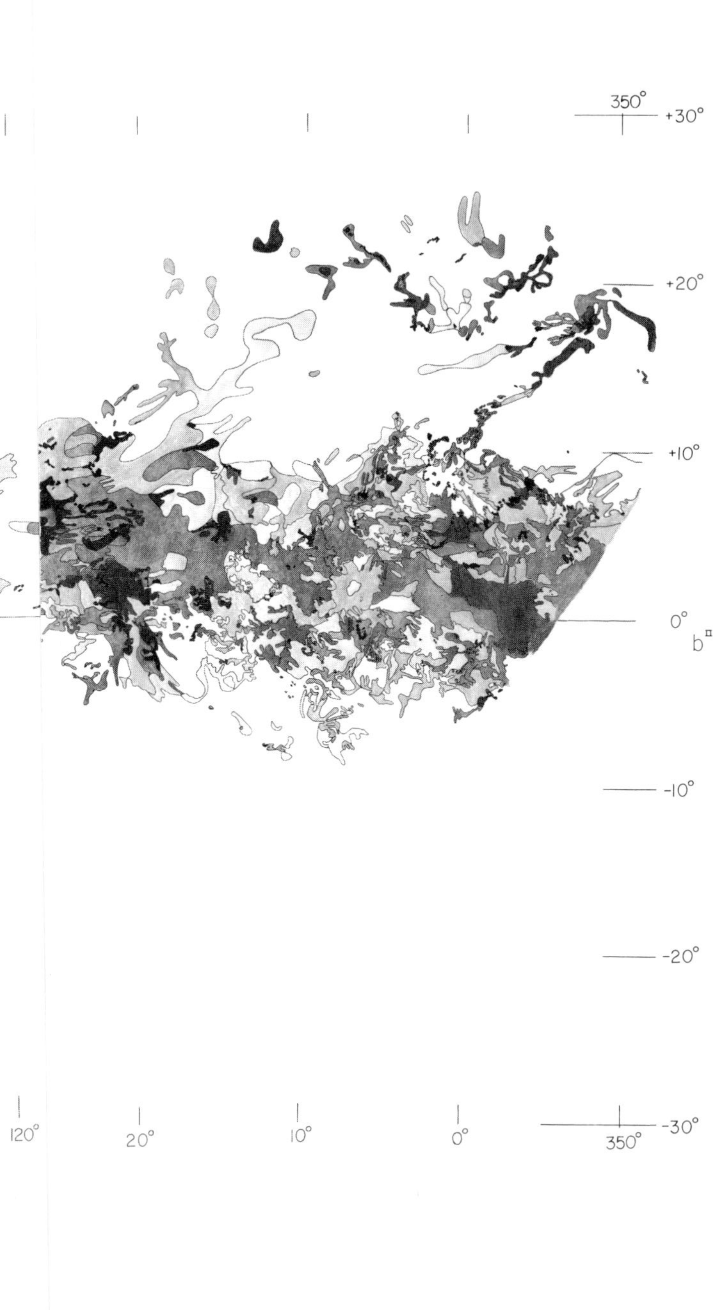

350°
+30°
+20°
+10°
0°
bII
-10°
-20°
120°
20°
10°
0°
350°
-30°

ordinarily divided into specific areas within which the visible density of stars is uniform. For each area, star counts are made and analyzed by Wolf diagrams or their equivalent, and the total absorption of light and the distance to the cloud is determined. Figure 8 (from Heeschen 1951) is a typical example of the representation of a clouded field by the method of star counts. Most of the references given in Table 2 are examples of this type of analysis.

According to Khavtassi, the star-count data on dark nebulae establish the presence of clouds of visible surface area extending from 1 square degree to 60 square degrees, the average being 14 square degrees. Obviously, the method of star counts fixes only the larger nebulae, since those of smaller dimensions do not contain a sufficient number of

TABLE 2

SELECTED DARK NEBULAE; POSITIONS AND ABSORPTION

Designation or Constellation	l^{II}	b^{II}	Distance Estimates (pc)	Absorption (mag)	Magnitude Limit of Survey	Reference
Scu	28°	− 2°	180–265	3.3	17	Wolf 1926
Aql	42	+ 0.5	110–300	4	14	Weaver 1949
	52	0	500	2	15	Baker 1941
Cyg	71	− 3	200, 500–700	2	14	Balanovsky and Haze 1935
52 Cyg	72	− 7	400–560	1	17	Wolf 1923
Cyg	75	− 2	600	1.5	11	Schalén 1934
Great Rift*	...	...	...	...	...	Miller 1937
NA Neb	85	+ 1	210, 560	<3	16	Wolf 1924*b*
Cyg	92	+ 3	250, 630	1.0, 1.0	11	Vanäs 1939
Cep	99	+ 3	150–250	0.5	12	Sticker 1937
Cep	100	+ 2	200–500	0.9	11	Schalén 1934
Cep	107	+ 2	200–600, 900	0.8, 1.7	11	Wernberg 1941
Cas	120	+ 5	500	2	15	Nantkes and Baker 1948
Cas	130	0	500, 800	1.5, 3	15	Baker and Nantkes 1944
Cas	132	+ 5	300	1.8	15	Nantkes and Baker 1948
Tau	165	−16	50, 175, 600	4, 1, 1.5	12	Adolfsson 1955
Aur	165	+ 4	130	0.6	15	Eklöf 1958
Aur	166	− 6	300	2.5	15	Heeschen 1951
Aur	167	− 6	115, 340	1.0, 1.1	11	Eklöf 1958
S Mon	202	+ 1	410–575 500–1000	2.1 1.5–2	17	Wolf 1924*a* Andrews 1933
Tau	204	−18	∼120	2–8	20	McCuskey 1941, Bok 1956
Ori	207	−21	300–400	1	13	Asklöf 1930
Vela	256	0	500–750	2	13	Greenstein 1936
η Car	286	0	800	0.8	15	Bok 1932
"Coalsack"	302	0	174	0.7–2.4	14	Rodgers 1960
ρ Oph	353	+17	200	2–8	17	Bok 1956
θ Oph	1	+ 5	250	2.0	12	Wallenquist 1939

* Details of Great Rift as follows:

Nebula	*Distance* (pc)	*Total Absorption*
East, south, and west of North America Nebula	550	4^m0
Absorption in NGC 7000 (NA Neb)	1200	1.5
Flare between NGC 7000 and Cyg star cloud	1300–2500	2.0
Rift, l^{II} from 73° to 79° (very uniform area)	1200	3.0
Rift, l^{II} from 65° to 73°	800–1600	2.0
Flare extending into negative latitudes between l^{II} 65° and 73°	1500	2.5
Rift, longitude l^{II} from 56° to 65°	600	2^m0 to 3^m0

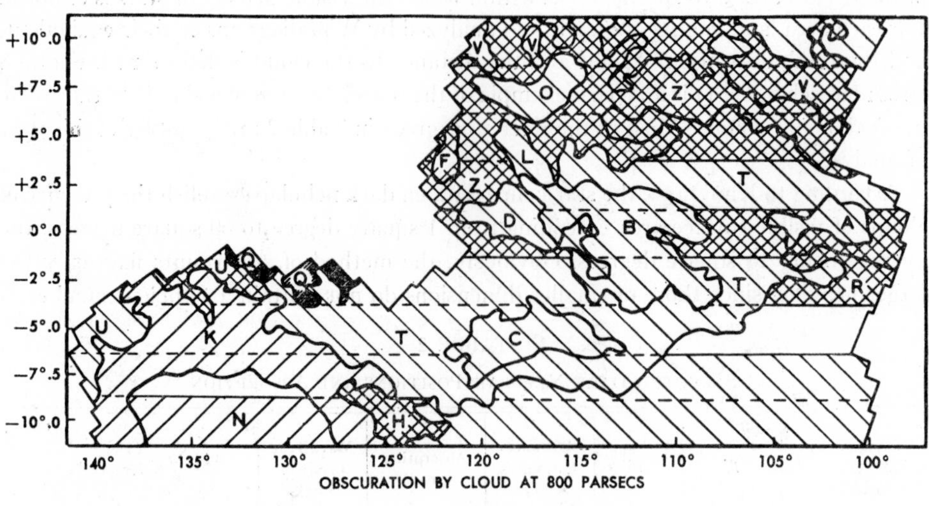

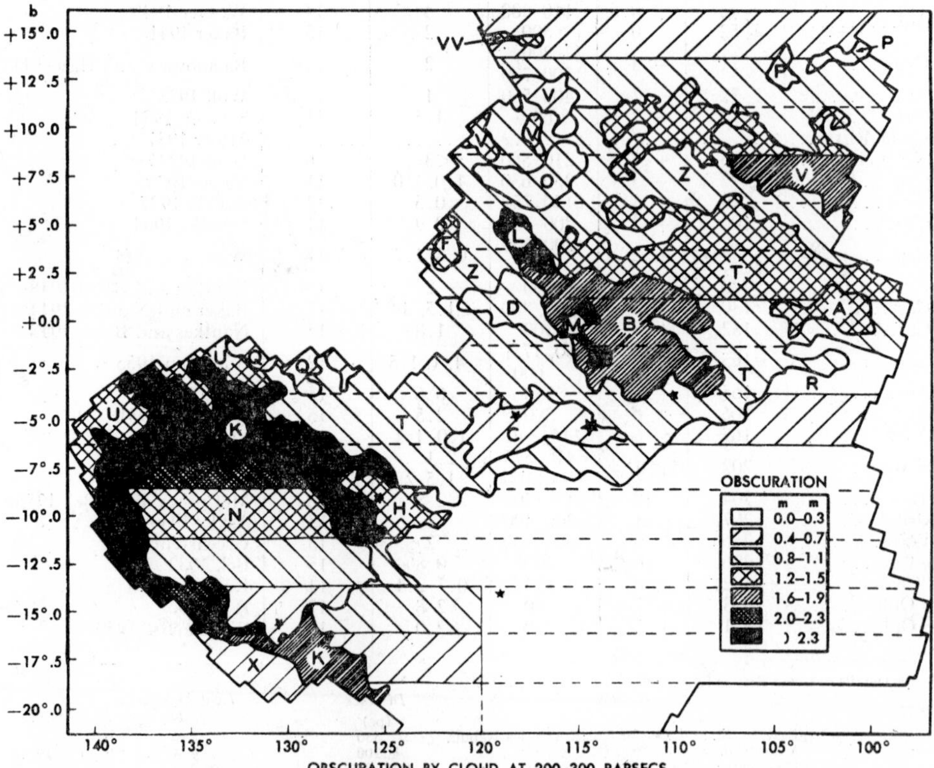

FIG. 8.—Divisions of the Perseus region showing obscuration from 200 to 300 pc and 800 pc; coordinates are (l^{I}, b^{I}) (from Heeschen 1951). The letters refer to Heeschen regions.

stars to make the counts statistically significant. Thus it is impossible to study the distribution of nebulae of small dimensions (<1 square degree), but small clouds compose more than 70 per cent of all known nebulae.

For the larger dark nebulae, Khavtassi points out a general trend. In almost all works that relate to star counts the presence of absorbing clouds in two distinct regions is indicated. The nebulae of the first region are at distances from 30 to 500 pc from the Sun, but nebulae of the second are 600 to 2500 pc distant. This is clearly seen in Figure 8, where Heeschen has plotted the obscuration produced by two clouds—one from 200 to 300 pc, the other at 800 pc. The presence of these is also apparent from Figure 6, where a dark nebula at a distance modulus ($m - M$) of 6 (160 pc) and another at about $m - M = 12$ (2500 pc) are clearly indicated.

With the exception of Orion, all nebulae listed in Table 2 lie within 100 pc of the plane of the Galaxy. The Ophiuchus nebulosity seems to be a spur of absorbing material starting about 20 pc above the plane and reaching to about 70 pc above the plane. The Orion absorbing material lies within 150 pc below the plane.

Although the galactic distribution of the dark nebulae of Table 2 does not represent that of a random sample of these objects, several regions that appear to be relatively free of dark clouds may be isolated. The galactic longitudes near 140°, and those from $l^{\text{II}} =$ 210° to 250° show little obscuration, as confirmed by the general surveys of Khavtassi and Lynds.

4. PHYSICAL PROPERTIES

4.1. Size and Shape

In attempting to categorize the sizes and shapes of dark nebulae, we should remember Herschel's comment that we are "on nebulous ground." No two clouds are identical, although it is perhaps not inappropriate to suggest some similarities among these dark markings. For the large obscuring areas found in the plane of the Milky Way, it is very tempting to comment on the similarity in appearance between them and our own terrestrial clouds, even to the extent of seeming to be wind-blown (Plate 5). The asymmetries of the smaller clouds have been analyzed by Schoenberg (1964). He interprets these clouds as flattened objects having the smallest extension in the direction toward the center of the Galaxy and the largest in the direction of galactic rotation (see also Ruskol 1950).

Some of the smaller dark nebulae (many of the Barnard objects, for example) have relatively sharp edges. Barnard states, "In many cases one side of a dark marking is very definite, while the other side is diffuse. This occurs so often that there must be some reason for it." In Plate 6, B92 and B93 are excellent examples of this type of nebulosity. Several of the sharp-edge clouds were measured for proper motion with base lines of one or two decades (Barnard 1919; Duncan 1937). Now that there are available pairs of photographs with a time lapse of more than half a century between them, it is worthwhile to look once more into the possibility of measurable proper motion of the clouds.

Some of the most striking dark nebulae have long rope-like appearances. Crescents, S-shapes, and arcs may be found among the Barnard objects. For any given longitude there seems to be a tendency for the structural features of dark nebulosities to align themselves with respect to each other (Fig. 7). The average position angle (measured from the direction of the galactic pole) does, however, vary with longitude. Behr (1955) and Tripp (1956) have shown that in some cases the direction of the parallel filaments is correlated with the plane of polarization and, hence, may be related to the influence of the galactic magnetic field (see Greenberg's and Davis' chapters, this volume).

Similar rope-like structures, on a much smaller scale, are also found associated with emission regions. These so-called elephant trunks seem to occur at the interface of an expanding H II region and a dense H I region (Osterbrock 1959). Plate 2 shows the elephant trunk of largest apparent size. The true physical association of these dark features and the H II region is evidenced by the brightening of the rims of these objects (for further details, see other chapters, this volume).

4.2. Globules

Many early observers of dark nebulae described small, essentially opaque, clouds that were usually circular or oval. They were called *globules* by Bok and Reilly (1947). These small masses are readily visible in projection against the luminous background of an emission nebula such as M8. Not all emission regions, however, have these objects associated with them. The Orion complex is singularly void of these features. Perhaps the richest field of such globules is in the region of η Car, recently studied by Thackeray (1950, 1954). Plate 7 shows a photograph of the region of IC 2944 taken with the Radcliffe reflector. The total absorption of these objects may be estimated either from star counts, if any stars are seen in the nebula field, or from an estimate of the absorption of the background emission, if the globule is seen projected against a bright nebula. In most cases, however, globules are almost totally opaque to the limiting magnitude of the photograph, so that often only lower limits to the total absorption may be reached.

These globules are almost always detected when they are projected against an emission field. Small dark nebulae, of the same apparent size, would be very difficult to detect if viewed against the general stellar background. If objects similar in physical structure to these globules were closer by about one order of magnitude, we might well be able to detect them in stellar fields. Bok and Reilly (1947) list several Barnard objects that seem to have the opacity and compact structure characteristic of the M8-type globules. The maximum distance of the globules found in M8 and η Car is about 1200 pc, whereas the Barnard nebulae 68, 69, 70, and 255, which may be similar in nature to the globules, very probably have distances comparable to that of the nearby Ophiuchus complex—200 pc.

If we define a globule as a small (radius <1 pc), nearly circular, opaque dark nebula, it is a relatively simple matter to search through the Palomar Schmidt photographs and select the nebulae that fit this definition. These objects are easily found by comparing red and blue photographs of the area. Most of the dark nebulae are much more opaque in the blue than in the red, although there is some indication that the extinction ratio may vary from cloud to cloud (Sears 1940; Bok and Warwick 1957). There are, however, a few small compact clouds that appear to be totally opaque to the limiting magnitudes of both the red and the blue photography. Plate 8 shows the dark globule B338. Approximately one hundred of these objects are detectable on the Palomar prints. An isolated globule, as shown in Plate 8, is not so common as the globules found as condensations within a cloud complex. It seems that, even in the larger cloud complexes, condensations exist which have the same apparent properties as the globules. Plate 9 is the dark nebula B27 associated with the star AB Aur. Note that within the dark nebula, several globular, opaque regions exist. Using 175 pc for the distance of AB Aur (Herbig 1960) as the distance of the cloud, we find sizes of about 0.2 pc for the diameters of these condensations.

4.3. Masses and Densities of Dark Nebulae

No direct measure of the mass of a nebula can be made from dynamical arguments. A fairly meaningful lower limit to the mass can be made with the data from star counts, that is, the distance to the cloud, its visible surface area, and the total extinction of starlight through the nebula. We shall follow the method described by Khavtassi (1958) in establishing the relations between the total mass and mean density of the nebula and the star count parameters.

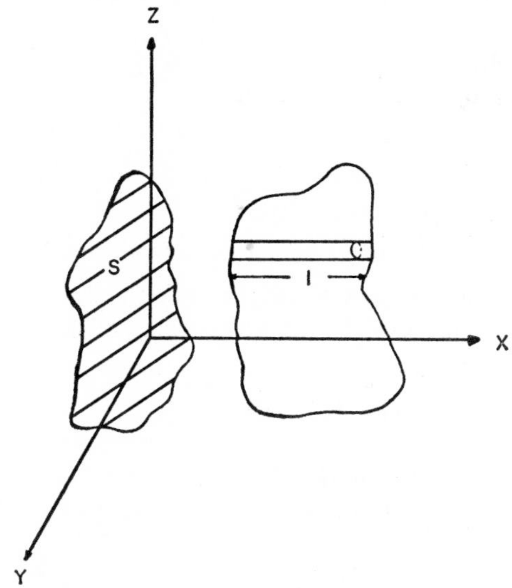

Fig. 9.—Schematic representation of absorption in a nebula. The x-axis is toward the observer; yz represents the plane of the sky.

In a nebula, consider a cylinder of cross-section of 1 cm², whose axis lies in the line of sight (Fig. 9). Let A be the absorption in magnitudes through the total length of the cylinder. Then we can write

$$A = (2.5 \log_{10}e)\pi a^2 QNl \tag{18}$$

where a is the geometrical radius of the nebula grains assumed to be spherical and of one size; Q is the efficiency factor of the grains; N is the number of grains per cubic centimeter (assumed constant throughout the nebula); and l is the length of the cylinder.

We may express Nl in terms of the mass of the cylinder $\mathfrak{M}_{cy}$ by

$$Nl = \frac{\mathfrak{M}_{cy}}{\frac{4}{3}\pi a^3 \delta} \tag{19}$$

where δ is the density of the grains (specific gravity).

Substituting in (18), we find

$$\mathfrak{M}_{cy} = 1.2 A \frac{a\delta}{Q} . \tag{20}$$

Now A will be a function of y and z (Fig. 9), but from star counts it is possible to determine only an average value of A for the entire nebula. This mean absorption Δm we will take as

$$\Delta m = \frac{\int\int A\, d\, y dz}{\int\int d\, y dz} . \tag{21}$$

Substituting for A from (18) and for Nl from (19), we have

$$\Delta m = \frac{0.81 \frac{Q}{a\delta} \int\int \mathfrak{M}_{cy} d\, y dz}{\int\int d\, y dz} . \tag{22}$$

But the integral of all of the cylinder masses is equal to the total mass of the nebula, $\mathfrak{M}$, and the integral of the projected area elements is the total projected surface area of the nebula S with

$$S = \text{angular surface area} \times \text{distance to nebula}$$

or

$$S = (\sigma \times 3.05 \times 10^{-4})(3.09 \times 10^{18} r)^2 = 2.9 \times 10^{33} \sigma r^2 \tag{23}$$

where σ is the surface area in square degrees and r is the distance to nebula in parsecs. Then, using the above relations and solving equation (22) for the mass of the nebula, we find

$$\mathfrak{M} = 1.8 \Delta m\, \sigma r^2 \frac{a\delta}{Q} \mathfrak{M}_\odot . \tag{24}$$

If we use the values of a and Q computed by Gehrels (1966) for the grains in reflection nebulae: Q (photographic) $= 1.3$, $a = 0.15 \times 10^{-4}$, and $\delta = 1$, we find

$$\mathfrak{M} = 2 \times 10^{-5} (\Delta m) \sigma r^2 \mathfrak{M}_\odot . \tag{25}$$

On the basis of equation (25), we can now estimate the mass of the extensive dark cloud illustrated in Figure 8, region K. Heeschen has calculated a distance of 200 pc and an absorption of 2.5 mag. We estimate the area to be approximately 50 square degrees; then,

$$\mathfrak{M} = 2 \times 10^{-5} \times 2.5 \times 50 \times (200)^2 \tag{26}$$

or

$$\mathfrak{M}_k = 100\, \mathfrak{M}_\odot . \tag{27}$$

If we use Rodgers' data for the Southern Coalsack, $r = 174$ pc, $\Delta m = 1.45$, and $\sigma = 19$ square degrees, the mass is found to be

$$\mathfrak{M}_c = 17\, \mathfrak{M}_\odot . \tag{28}$$

For one of the small globules seen projected against an emission nebula, we take Bok's estimate of 5 mag. as the extinction, the surface area as 10^{-4} square degrees, and 1600 pc as the distance, and find

$$\mathfrak{M}_{\text{glob}} = 0.03\, \mathfrak{M}_\odot . \tag{29}$$

We may also derive a relationship between the mean density of a nebula and the star-count parameters. If we substitute the value of A given in equation (18) into equation (21), we have

$$\frac{\int\int Nl\,dy\,dz}{\int\int dy\,dz} \tag{30}$$

and if we assume that N and l are independent, we may make the approximation

$$\Delta m = 2.5 \log_{10} e\pi a^2 Q\bar{N}\frac{\int\int l\,dx\,dy}{\int\int dx\,dy}$$

or

$$\Delta m = 2.5 \log_{10} e\pi a^2 A\bar{N}\bar{l}. \tag{31}$$

Let w equal the mass of one grain. Then

$$\Delta m = \tfrac{3}{4}(2.5 \log_{10} e)\frac{Q}{a\delta}(w\bar{N})\bar{l} \tag{32}$$

or

$$\Delta m = 0.81\frac{Q}{a\delta}\bar{\rho}\bar{l} \tag{33}$$

which is the usual formula of Russell (1922).

Using the values of Q, a, and δ given above, we find

$$\Delta m = 7 \times 10^{-4}\bar{\rho}\bar{l}. \tag{34}$$

Now if the nebula were spherical, $\bar{l}$ would be the average chord,

$$\bar{l} = \frac{4}{3\pi}\sqrt{S} \tag{35}$$

therefore, for a first approximation, we will use

$$\Delta m = 7 \times 10^{4}\bar{\rho} \times \frac{4}{3\pi}\sqrt{S} \tag{36}$$

or

$$\bar{\rho} = 6 \times 10^{-22}\frac{\Delta m}{r\sqrt{\sigma}}. \tag{37}$$

For the three examples used in the mass determination, we find

Heeschen region K	$\bar{\rho} = 1 \times 10^{-24}$ gm/cm³	(38)
Southern Coalsack	$\bar{\rho} = 1 \times 10^{-24}$ gm/cm³	(39)
M8 globule	$\bar{\rho} = 2 \times 10^{-22}$ gm/cm³	(40)

These densities are several orders of magnitude greater than the mean density of interstellar dust.

These values for the mass and density of a dark nebula are actually only a measure of the grain component, which is assumed to be identical with the interstellar grains. Until we have some estimate of the gas/dust ratio in these clouds, it is impossible to make an estimate of the total mass of the object. At present, the values for typical dark nebulae—$\bar{\rho} \sim 10^{-24}$ gm/cm³ and $\mathfrak{M} \sim$ 10–100 $\mathfrak{M}_\odot$—must be considered minimum values. The densities of the most opaque globules are at least 10^{-22} gm/cm^3, with minimum masses between 0.001 and 1 $\mathfrak{M}_\odot$.

Two questions raised since the dark nebulae were first identified are still unanswered. First, are some nebulae completely opaque to the background stellar radiation; and second, is there an intrinsic surface brightness associated with these objects? The answer to the first question must await future observations with large telescopes. Perhaps the most productive approach will be the use of direct photography in the far red and infrared.

Barnard (1919) commented that some of the dark nebulae are "not black except by contrast." Visual observations of several of his objects suggested the existence of "a dull, feebly luminous mass."

Struve and Elvey (1936), using a Fabry photometer on the 40-inch Yerkes refractor, found that in B15 the central core is blacker than the outer portion by about 0.08 mag. They concluded that "the increased brightnesses of the outer regions of the dark nebulae and of the Milky Way comparison fields is produced by interstellar scattering." Whether this faint luminosity is produced by scattered light from the Milky Way or by some intrinsic brightness of such nebulae is not yet known (Lynds 1967).

REFERENCES

Adolfsson, T. 1955, *Ark. f. Astr.*, **1,** 495.
Andrews, L. B. 1933, *Pub. A.A.S.*, **7,** 211.
Asklöf, S. 1930, *Upp. Obs. Medd.*, **51.**
Baker, R. H. 1941, *Ap. J.*, **94,** 493.
Baker, R. H., and Nantkes, E. 1944, *Ap. J.*, **99,** 125.
Balanovsky, I., and Haze, V. 1935, *Pulkovo Bull.*, **14,** 2.
Barnard, E. E. 1913*a*, *Pub. Lick Obs.*, **11.**
———. 1913*b*, *Ap. J.*, **38,** 496.
———. 1919, *ibid.*, **49,** 1.
———. 1927, *Photographic Atlas of Selected Regions of the Milky Way*, eds. E. B. Frost and M. R. Calvert (Washington: Carnegie Institution of Washington).
Becker, F., and Meurers, J. 1956, *Vistas in Astronomy*, ed. A. Beer (London and New York: Pergamon Press), **2,** 1069.
Becker, W. 1938, *Zs. f. Ap.*, **15,** 225.
———. 1950, *Sterne und Sternsysteme* (2d ed.; Dresden and Leipzig: T. Steinkopff).
Behr, A. 1955, *Mém. Soc. R. Sci. Liège*, Ser. 5, **15,** 547.
Bok, B. 1932, *Harvard Reprints*, No. 77.
———. 1937, *The Distribution of Stars in Space* (Chicago: University of Chicago Press).
———. 1948, *Harvard Monographs*, No. 7, p. 53.
———. 1956, *A.J.*, **61,** 309.
Bok, B., and Reilly, E. 1947, *Ap. J.*, **105,** 255.
Bok, B., and Warwick, C. 1957, *A.J.*, **62,** 323.
Curtis, H. D. 1918, *Pub. A.S.P.*, **30,** 65.
Duncan, J. C. 1937, *Ap. J.*, **86,** 496.
Eklöf, O. E. 1958, *Upp. Obs. Medd.*, No. 119.
Gehrels, T. 1966, private communication.
Greenstein, J. 1936, *Harvard Ann.*, **105,** 359.
Hayden, F. A. 1952, *Photometric Atlas of the Southern Milky Way* (Washington: Carnegie Institution of Washington).
Heeschen, D. S., 1951, *Ap. J.*, **114,** 132.
Herbig, G. 1960, *Ap. J. Suppl.*, **4,** No. 43, 337.
Herschel, W. 1784, *Phil. Trans. R. Soc. London,* **A74,** 437.
Hubble, E. 1922, *Ap. J.*, **56,** 162.
Johnson, H. L., and Morgan, W. 1953, *Ap. J.*, **117,** 313.
Khavtassi, J. 1955, *Bull. Abastumani Obs.*, No. 18.
———. 1958, *ibid.*, No. 23.
———. 1960, *Atlas of Galactic Dark Nebulae* (Abastumani Astrophysical Observatory).
Lundmark, K. 1926, *Upp. Obs. Medd.*, No. 12.

Lynds, B. T. 1962, *Ap. J. Suppl.*, **7**, No. 64, 1.

———. 1967, *Surface Brightness of Dark Nebulae*, in memorial volume for O. Struve (ed. M. Hack, Gauthier-Villars, in press).

Malmquist, K. G. 1943, *Upp. Obs. Medd.*, **1**, No. 7.

McCuskey, S. W. 1941, *Ap. J.*, **94**, 468.

Miller, F. D. 1937, *Proc. Nat. Acad. Sci.*, **23**, 405.

Minkowski, R., and Abell, G. 1963, *Basic Astronomical Data*, ed. K. Aa. Strand (Chicago: University of Chicago Press), p. 481.

Nantkes, E., and Baker, R. H. 1948, *Ap. J.*, **107**, 113.

Osterbrock, D. E. 1959, *Pub. A.S.P.*, **71**, 23.

Pannekoek, A. 1921, *Proc. Kon. Akad. Wetensch. Amsterdam*, **23**, No. 5.

Razhkovsky, D. A. 1955, *Contr. Alma-Ata*, **1**, Nos. 1–2, 136.

Rodgers, A. W. 1960, *M.N.*, **120**, 163.

Ross, F. E., and Calvert, M. R. 1934, *Atlas of the Milky Way* (Chicago: University of Chicago Press).

Ruskol, E. L. 1950, *A.J. (U.S.S.R.)*, **27**, No. 6, 341.

Russell, H. 1922, *Proc. Nat. Acad. Sci.*, **8**, 115.

Schalén, C. 1934, *Upp. Obs. Medd.*, Vol. **58**.

Schoenberg, E. 1964, *Veröffentl. Sternw. München*, **5**, No. 21.

Sears, F. 1940, *Pub. A.S.P.*, **52**, 80.

Sticker, B. 1937, *Veröffentl. Sternw. Bonn.*, Vol. **30**.

Struve, O., and Elvey, C. 1936, *Ap. J.*, **83**, 162.

Thackeray, A. 1950, *M.N.*, **110**, 524.

———. 1954, *Mém. Soc. R. Sci. Liège*, Ser. 4, **15**, 437.

Tripp, W. 1956, *Zs. f. Ap.*, **41**, 84.

Trumpler, R., and Weaver, H. 1953, *Statistical Astronomy* (Berkeley: University of California Press [2d ed.; 1962, New York: Dover Publications]).

Vanäs, E. 1939, *Upp. Obs. Ann.*, **1**, 1.

Wallenquist, A. 1939, *Ann. Bosscha-Sterr. Lembang (Java)*, Vol. **5**, p. E1.

Weaver, H. 1949, *Ap. J.*, **110**, 190.

Webb, T. W. 1894, *Celestial Objects for Common Telescopes* (London and New York: Longmans, Green and Co. [7th ed.; 1962, New York: Dover Publications]).

Wernberg, G. 1941, *Upp. Obs. Ann.*, Vol. **1**, No. 4.

Wolf, M. 1908, *Die Milchstrasse und die Kosmischen Nebel* (Leipzig: Barth).

———. 1923, *A.N.*, **219**, 109.

———. 1924*a*, *Probleme der Astronomie* (Berlin: J. Springer), p. 312.

———. 1924*b*, *A.N.*, **223**, 89.

———. 1926, *ibid.*, **229**, 1.

CHAPTER 4

*Flare Stars**

G. HARO
Tonantzintla Observatory, Mexico

1. INTRODUCTION

DURING the last two decades it has become well known that an extremely sudden increase in the integrated brightness of some stars takes place in an unpredictable, nonperiodic fashion. In a matter of minutes or even seconds, the total light of a star may be enhanced by anything from a few tenths of a magnitude up to 10 magnitudes, the increment being greatest in the ultraviolet. This extremely rapid light variation has been called a "flare" or "flash" and the stars in which the phenomenon occurs named flare or flash stars. In this chapter, to avoid confusion, we shall call these objects *flare stars* without implying that all such stars have the same spectroscopical or physical characteristics. On the contrary, as we will see, the flare phenomenon is observed in a wide range of stellar types.

The first flare stars found were in the vicinity of the Sun (Joy 1960; Oskanjan 1964), UV Ceti being the prototype. All known flare stars within 20 pc of the Sun are of low intrinsic luminosity with spectral types dM0 or later; usually they have emission lines mainly of H and Ca II during the "normal" minima. These features led astronomers to a rather restricted definition of the new class: that dMe features are shown during quiescence, that is, apart from the sudden outburst in light. Nevertheless, in the last 15 years, stars of a number of spectral types have been found to show extremely rapid light variation, similar in many respects to the outburst observed in the UV Ceti stars.

Most newly discovered flare stars, which are not necessarily restricted to the narrow criterion set by the nearby UV Ceti variables, are found in stellar aggregates and clusters of different ages (Haro and Chavira 1964) as, for example, the Orion Nebula, NGC 2264, the Pleiades, Coma, Praesepe, the Hyades, and so forth. In some of these aggregates, especially the youngest ones, conspicuous flares have been observed in stars also classified as T Tauri, RW Aurigae, or Orion type variables. These flare stars in stellar aggregates radically change the earlier picture based on the UV Ceti variables in the solar vicinity. Flares have been found to occur not only in Me dwarfs but also in stars of spectral type as early as K0, or earlier, and the absolute luminosities of these stars during their minima correspond to cool subgiants as well as to dwarfs.

* Manuscript received May, 1965.

Flares observed in W Ursa Majoris stars, in U Geminorum variables, in the close eclipsing binary UX Ursa Majoris, in old novae, and in a number of related objects (Joy 1957; Walker 1957; Kuhi 1964) further complicate the problem, as does Andrews' (1964) recent observation of a possible flare in the B8 star BD + 31° 1048.

Although several astronomers have insisted that rapid light outbursts in these stars are similar in nature to solar flares, it is not yet determined whether the kind of physical processes which causes the observed flare activity can be the same in such widely different types of stars, or whether fundamental differences in the flare phenomena could be caused by varying physical conditions over such a range of stellar types. This problem is far from being solved.

For reasons described below, current investigations suggest a connection between the UV Ceti stars of the solar neighborhood and the flare stars found in stellar aggregates; through this, a genetic link might be established between irregular variables of the T Tauri, RW Aurigae, and Orion types and most flare stars.

It is convenient to consider flare stars in this chapter under two headings without implying any conclusion about flare production mechanism in the two groups. These groups are (*a*) flare stars in stellar aggregates and the solar vicinity including stars of the UV Ceti type in the general field, and (*b*) binary stars of the U Geminorum and W Ursa Majoris types, novae and related objects, and Andrews' star.

This chapter deals particularly with flare stars of the first group, with incidental mention of those of the second.

2. FLARE STARS IN STELLAR AGGREGATES AND IN THE SOLAR VICINITY

Early discoveries of flare stars were in the vicinity of the Sun only, and the range of spectral features observed in them was restricted. As a result, for many years it was believed that only dMe stars with a rapid light burst could be considered as bona fide flare stars. Now, however, the evolution of flare stars may be described as starting with extremely rapid variables found in stellar aggregates (arranged approximately in order of increasing age) and ending with flare stars found in the solar vicinity and the general field. This procedure might be justified by the tendency of flare stars toward a decreasing space density from those in very young stellar aggregates to older ones, the solar vicinity and general field.

Basically, the unpredictable, short-lived outburst in integrated brightness distinguishes a flare star. Analyses of photometric and spectroscopic features indicate further common properties that, despite some differences in detail, make the class a rather homogeneous group of variable stars.

2.1. The Orion Aggregate

Joy (1949) first called attention to the sudden flares in dwarf stars such as the T Tauri variables, the bright-line stars of the Taurus clouds, and the faint dMe stars of large proper motion. Later Haro (1953), in his study of Hα emission stars in the Orion Nebula, called attention to the existence of several objects, such as his Hα star No. 70, strongly resembling flare-type variables; this suggestion was later confirmed (Haro and Chavira 1964). The first three flare stars recognized, Nos. 868, 886, and 905 in Brun's catalogue (1935), were found by Haro and Morgan (1953), who noted the similarities of their light curves to those of the UV Ceti stars in the solar neighborhood. These rapid variables in Orion were found to have mean luminosities, uncorrected for absorption, of $M_{pg} = +8$ to $+9$.

A systematic search in the Orion Nebula region by Haro and Terrazas (1954), Haro (1954), Haro and Chavira (1964), Rosino (1956), Rosino and Cian (1962), Dall'Olmo (1958, 1960, 1961), and Maffei (1963) led to the discovery of many flare stars. A list of all flare stars then known in the Orion Nebula region was presented by Haro and Chavira at the ONR Symposium held in Flagstaff, Arizona, in June, 1964. These 121 stars and information about them at that date are listed in Table 1A. For several stars, more than one flare has been observed. Additional flares observed by Haro and Chavira (1965), and by Rosino (1965) in stars of the Orion Nebula region during the winter months of 1964–65 are recorded in Table 1B. Tables 1A and 1B are of interest with respect to (*a*) frequency of recurrence of flares in the same star; (*b*) comparison of distinct flares in the same star; (*c*) the proportion of "new" flare stars in which the flare repeats compared with the ones previously known; (*d*) systematic data collection that eventually would permit statistical calculation of the probable total number of flare stars in a given region.

The observations by Haro and Chavira during the winter of 1964–65 produced a rather homogeneous series of plates with multiple exposures secured with the Tonantzintla Observatory Schmidt telescope (field 4° × 4°) only, reaching the apparent ultraviolet magnitude of 17.5 (103aO Eastman Kodak plates behind a Corning 9863 filter). (See Plates 1 and 2.) Long daily runs of observations were obtained, and in only 112 hours and 10 minutes of effective observation time, 573 different exposures were made and 62 flares detected in 60 stars. In contrast, identification of the 91 flare stars of Table 1A plus the 19 repeated flares found by Haro and co-workers took 529 hours and 5 minutes of observation (2356 exposures). However, this latter plate material was not homogeneous; it comprises exhaustively examined blue and ultraviolet multiple-exposure plates (1922 exposures were obtained during 344 hours of effective observational time), a large number of red spectral plates with one or two exposures, and many direct ultraviolet, blue, yellow, red, and infrared plates, with single and multiple exposures obtained with two different telescopes and not examined thoroughly (Haro and Chavira 1964).

Analysis of the data of Tables 1A and 1B led to the following conclusions and conjectures:

1. Most known flare stars in Orion follow the apparent distribution pattern of the T Tauri-like and Orion variables in that region (Haro 1962; Rosino and Cian 1962). Table 2 reproduces the data of Table 25 of Rosino and Cian's paper supplemented by all new flare-star data available up to the time of writing; a number of Hα emission objects and nebular variables which have shown flares are included.

Table 2 shows the general correlation between the apparent distribution of the nebular variables, the stars with Hα in emission, and the flare stars in the Orion region. In zone E (the densest part of the Orion Nebula), the number of flare stars observed is reduced by the bright nebulosity centered at the Trapezium, especially in the multiple-exposure plates. In zone A, where the number of flare stars is significantly greater than the number of known nebular variables, a dark lane emerging radially from zone E is conspicuous

TABLE 1A

FLARE STARS IN ORION UP TO JUNE, 1964

No.	Parenago No.*	Brun No.†	R.A. (1900)	Dec. (1900)	m_{vis}	Δm**	Sp. Type, eHα	Date of Flare	Ref.‡
1......			$5^h23^m.9$	−3° 58′	$14^m.5$	$0^m.8$		1959, 12/26	1
2......			5 24.6	5 37	16.5	2.4		1959, 12/23	1
3......			5 24.9	3 51	17.9	1.0		1963, 12/23	1
4......			5 25.8	7 8	13.3	1.2		1959, 12/26	1
5......			5 25.9	5 41	14.9	>1.3		1960, 1/20	2
6......			5 26.2	4 41	14.3	1.5		1959, 2/8	2
7......			5 27.5	5 7	19.5	>3.5		1960, 1/30	1
8......			5 27.5	4 4	16.5	1.8		1959, 12/26	1
						1.5		1960, 1/29	1
9......			5 27.7	3 19	14.4	2.0		1960, 1/2	1
10......			5 27.7	5 17	15.0	1.4		1954, 1/9	3
11......			5 28.0	3 11	14.2	2.6		1959, 12/23	1
12......			5 28.1	6 27	16.2	1.8	eHα	1959, 2/26	2, 4
						1.5		1955, 11/11	1
13......			5 28.3	5 4	16.5	4.0		1963, 12/13	1
14......			5 28.3	5 46	14.7	1.1	K, eHα	1953, 12/16	3, 4, 5
15......			5 28.3	5 4	15.5	2.5		1962, 11/25	1
16......			5 28.5	5 11	15.0	3.0	eHα:	1955, 2/22	6
						1.5		1956, 11/3	1
17......			5 28.5	4 7	15.1	3.1		1959, 12/26	1
18......	1218	112	5 28.6	5 6	15.3	1.5		1953, 12/15	1
						>2.0		1955, 12/10	1
19......			5 28.6	6 6	15.5	1.5		1959, 12/27	1
20......			5 28.7	5 38	16.1	1.4		1954, 12/14	3
21......			5 28.8	5 41	18.0		eHα (max.)	1963, 1/26	1
22......			5 28.8	5 35	14.8	1.4	K5–M0	1953, 12/16	3, 5
23......			5 28.9	6 18	14.2	1.9		1959, 2/25	2
24......			5 29.0	7 4	13.8	1.2		1963, 12/19	1
25......	1352	191	6 29.2	5 30	13.9	0.9		1959, 12/9	7
26......			5 29.3	6 52	16.5	2.0		1963, 12/15	1
27......			5 29.3	3 33	19.6	>4.0		1960, 1/30	1
28......			5 29.4	5 3	17.1	2.0		1956, 2/28	1
29......			5 29.5	4 37	16.2	2.1		1960, 1/27	1
30......			5 29.5	6 22	15.8	2.2		1959, 12/30	1
31......	1463	253	5 29.5	4 59	15.0	0.9		1960, 1/24	1
32......	1496	297	5 29.6	5 5	15.3	>1.0		1960, 1/26	1
						>1.0		1960, 1/29	1
33......			5 29.6	6 8	18.3	>2.5		1960, 1/25	1
34......	1485	284	5 29.6	5 51	13.8	0.9		1958, 1/16	8
35......			5 29.6	6 46	16.5	1.3		1958, 12/7	2
36......			5 29.7	6 6	16.5	2.0		1959, 11/6	1
37......			5 29.7	6 27	16.0	2.0		1959, 12/27	1
38......			5 29.8	6 32	15.5	1.5		1962, 11/26	1
39......	1576		5 29.8	5 49	16.8	2.0	eHα	1959, 12/5	1, 4
40......	1584	357	5 29.8	5 6	15.6	>2.0		1960, 2/1	1
41......		397	5 29.9	5 47	17.2	>2.0		1960, 1/29	1
42......	1614	404	5 29.9	5 46	15.1	1.0	eHα	1963, 12/23	1, 4
43......	1643	424	5 29.9	5 46	13.2	1.0		1958, 12/14	7
44......			5 30.0	3 6	14.1	1.6		1960, 1/25	1
45......	1648	422	5 30.0	5 7	14.3	0.8	G-Ke, eHα	1960, 1/27	1, 4, 9
46......			5 30.0	6 31	16.3	2.0		1962, 11/21	1
47......			5 30.0	5 32	15.2	1.8		1955, 11/8	2
48......			5 30.0	6 16	15.8	2.0		1960, 1/28	1
49......			5 30.0	5 10	16.6	2.2		1960, 11/21	2
50......	1656		5 30.0	5 13	16.0	1.0		1962, 11/22	1
51......	1669		5 30.0	6 28	15.8	1.0	eHα	1959, 12/26	1, 4
52......			5 30.0	4 38	19.5	>4.0		1962, 11/23	1
53......			5 30.1	−4 42	18.4	>3.0		1960, 1/31	1

TABLE 1A—*Continued*

No.	Parenago No.*	Brun No.†	R.A. (1900)	Dec. (1900)	m_{vis}	Δm**	Sp. Type, eHα	Date of Flare	Ref.‡
54......			$5^{h}30^{m}1$	−5°38′	$17^{m}2$	$1^{m}0$	eHα	1954, 10/30	6, 1
55......			5 30.4	5 44	16.2	2.5		1959, 1/29	2
56......			5 30.5	6 36	15.3	>1.5		1960, 1/30	1
57......			5 30.6	6 13	15.7	2.5		1959, 1/29	2
58......	2078	755	5 30.6	5 53	13.6	3.2	K2	1958, 1/28	8, 5
59......			5 30.7	5 42	17.3	2.9		1960, 1/3	1
60......			5 30.7	6 16	15.8	1.8		1959, 12/23	1
61......			5 30.7	6 22	17.6	1.5		1957, 1/28	10
62......			5 30.7	4 18	14.9	3.5		1959, 11/2	2
63......	2172	834	5 30.8	5 13	13.7	0.9		1961, 1/17	11
64......			5 30.8	5 37	15.3	1.8		1957, 3/1	2
65......			5 30.8	5 15	18.2	2.2		1960, 1/22	1
66......			5 30.9	4 28	14.3	2.1	K7e:	1954, 1/6	3, 5
						1.0		1960, 2/2	1
67......			5 30.9	5 26	16.2	2.1		1954, 10/30	6
68......	2210	868	5 30.9	5 45	16.1	0.5		1952, 1/30	12
69......	2228		5 30.9	5 57	16.9	1.0		1947, 1/19	1
70......			5 30.9	5 11	14.1	1.9		1958, 10/9	2
						0.7		1959, 12/5	1
71......	2245	886	5 30.9	5 19	14.8	1.0	K7	1952, 1/30	12, 5
72......	2246	889	5 30.9	5 20	13.7	2.7	K:(e), eHα		2, 4, 5
73......			5 30.9	6 37	16.0	0.7		1959, 12/27	1
74......	2270	905	5 31.0	5 31	15.8	0.8		1952, 12/21	12
75......			5 31.0	5 56	16.6	>2.0		1960, 1/29	1
76......			5 31.1	5 3	15.5	2.0	K3	1954, 1/4	3, 5
								1954, 12/18	2
77......	2305	935	5 31.1	5 21	13.4	0.6		1960, 2/24	11
78......			5 31.1	6 6	18.3	>2.0	eHα	1960, 1/29	1
						>1.0		1947, 1/19	1
						>1.0		1959, 12/22	1
						>1.0		1960, 2/2	1
						>1.0		1962, 11/22	1
						>1.0		1963, 12/18	1
79......	2318	947	5 31.1	5 13	14.2	0.6	eHα	1963, 12/13	1, 4
80......			5 31.1	6 19	14.5	0.9	eHα	1954, 10/25	13, 4
81......			5 31.2	4 33	17.2	>0.5		1961, 2/16	2
82......	2337	964	5 31.2	5 4	15.3	1.5	eHα	1959, 2/12	2, 4
83......			5 31.2	5 29	14.3	>1.0		1963, 12/15	1
84......			5 31.2	5 48	17.1	2.5		1956, 10/5	1
85......	2348	960	5 31.3	5 50	14.4	2.0		1955, 12/12	1
						2.5		1959, 12/22	1
86......	2347	965	5 31.3	5 24	13.7	3.0	eHα	1963, 12/11	1, 4
87......			5 31.3	5 40	16.2	1.5		1956, 11/10	1
88......			5 31.3	5 26	15.7	1.0		1953, 12/2	3
						2.0		1955, 12/19	1
						1.5		1963, 12/21	1
89......	2363	971	5 31.3	5 41	14.8	1.5		1956, 3/2	1
90......			5 31.3	4 28	14.1	1.7		1959, 2/5	2
91......			5 31.4	7 2	17.2	2.5		1959, 12/23	1
92......			5 31.5	5 11	15.3	>4.0	eHα	1963, 1/19	1
93......			5 31.6	7 10	15.0	0.9	eHα	1959, 12/24	1
94......			5 31.6	4 57	17.1	>1.5		1963, 12/18	1
						3.0		1963, 2/28	1
95......			5 31.6	7 51	15.5	0.7		1959, 12/28	1
96......			5 31.6	6 10	18.9	>4.0		1959, 12/5	1
97......			5 31.7	5 9	14.8	2.7	eHα	1960, 1/3	1
						1.2		1960, 1/27	1
						?	eHα	1951, 11/28	4
98......			5 31.7	5 52	18.0	>2.0		1960, 1/4	1
99......			5 31.8	−6 34	14.4	2.6		1957, 11/27	10

TABLE 1A—*Continued*

No.	Parenago No.*	Brun No.†	R.A. (1900)	Dec. (1900)	m_{vis}	Δm**	Sp. Type, eHα	Date of Flare	Ref.‡
100......			$5^h31^m.9$	$-5°\ 8'$	$15^m.5$	$>3^m.0$		1963, 12/21	1
						1.0		1963, 2/28	1
101......	2449	1038	5 31.9	5 12	14.8	1.8	eHα	1960, 1/26	1, 4
102......	2455	1039	5 32.0	5 4	14.4	1.0	M2	1960, 1/2	1, 14
103......			5 32.0	6 39	17.7	>2.0		1962, 11/27	1
104......	2570		5 32.1	5 27	14.2	2.5	eHα	1959, 12/27	1, 4
105......	2472	1047	5 31.1	5 29	16.2	3.0		1960, 2/1	1
106......			5 32.1	6 37	19.7	>5.0		1959, 12/26	1
107......			5 32.2	5 41	16.2	1.6		1959, 12/27	1
108......			5 32.3	6 34	18.0	1.7		1956, 1/17	2
109......			5 32.3	4 41	14.2	0.9		1959, 12/26	1
110......			5 32.4	3 37	18.0	>3.0		1959, 12/5	1
111......			5 32.7	5 56	16.2	0.8	eHα	1954, 1/8	13
112......			5 32.9	4 20	18.1	>4.0		1959, 12/26	1
113......			5 32.9	5 27	14.6	1.8		1959, 12/22	7
114......			5 33.1	6 22	15.0	1.4		1960, 1/31	1
115......			5 33.2	7 21	14.8	0.8	eHα	1959, 12/28	1, 4
116......			5 33.3	5 9	14.9	3.0		1963, 12/22	1
117......			5 33.7	5 42	17.9	>3.5		1963, 12/11	1
118......			5 34.6	4 30	14.4	3.9		1960, 1/25	1
119......			5 34.7	5 31	17.2	2.0		1962, 11/21	1
120......			5 35.6	3 29	18.1	>2.5		1959, 12/24	1
121......			5 37.3	−4 22	14.5	1.2		1959, 1/3	2

TABLE 1B

FLARE STARS IN ORION DISCOVERED IN WINTER OF 1964–1965

No.	Parenago No.*	Brun No.†	R.A. (1900)	Dec. (1900)	m_{vis}	Δm_U	Sp. Type: eHα	Date of Flare	Ref.‡
122......			$5^h22^m.9$	$-6°56'$	$17^m.1$	$>4^m.0$		1965, 2/7	15
123......			5 24.3	6 15	17.0	3.0		1965, 1/30	15
124......			5 25.1	4 28	17.7	4.0		1965, 1/8	15
125......			5 25.7	4 13	14.7	3.0		1965, 1/2	15
126......			5 25.9	6 57	15.3	>3.0		1965, 1/1	15
127......			5 26.0	4 18	14.8	0.7		1965, 1/27	15
128......			5 26.3	7 09	17.2	>3.0		1965, 1/7	15
129......			5 27.0	4 31	15.5	1.0		1965, 2/2	15
130......	987		5 27.8	5 44	14.4	4.0		1965, 1/1	15
131......			5 28.1	5 29	15.0	1.5		1964, 12/30	15
132......			5 28.1	5 04	15.3	3.0		1965, 1/27	15
133......			5 28.2	7 24	16.5	>2.0		1965, 1/5	15
134......	1191	99	5 28.6	5 24	14.8	0.7		1965, 1/26	15
135......			5 28.6	5 50	17.1	>3.0		1964, 12/31	15
136......			5 28.7	4 16	15.0	>2.5		1965, 1/5	15
137......			5 28.8	6 09	17.3	>3.0		1965, 1/5	15
138......			5 28.8	7 01	15.2	4.0		1965, 1/7	15
139......	1274	144	5 28.9	6 05	14.7	3.5		1964, 12/31	15
140......			5 29.1	4 49	16.7	>3.0		1965, 2/2	15
141......	1410	223	5 29.4	5 11	14.8	0.6		1964, 12/30	15
142......			5 29.6	6 26	15.6	1.5		1964, 12/29	15
143......	1502		5 29.6	6 43	15.0	0.8		1965, 1/9	15
144......	1530	319	5 29.7	6 10	15.2	0.9	eHα	1965, 1/27	15, 4
145......			5 29.7	5 07	17.3	>4.0		1965, 1/26	15
146......	1553	339	5 29.8	−5 44	12.4	1.3	K0:, eHα	1964, 12/27	15, 4

TABLE 1B—*Continued*

No.§	Parenago No.*	Brun No.†	R.A. (1900)	Dec. (1900)	m_{vis}	Δm_U	Sp. Type: eHα	Date of Flare	Ref.‡
147......	1571	362	5^h29^m8	−5°52′	15^m0	2^m0		1965, 1/27	15
148......			5 29.8	6 30	15.1	2.5	eHα	1964, 12/31	15, 4
149......			5 29.8	5 02	16.8	4.0		1964, 12/28	15
						1.5	eHα	1965, 1/26	15, 4
150......			5 29.9	6 25	15.7	1.5		1965, 1/31	15
151......	1625	423	5 29.9	5 50	14.7	1.4	eHα	1965, 1/19	16, 4
152......	1715	485a	5 30.1	6 00	15.2	2.5		1965, 1/1	15
153......	1741		5 30.2	6 09	14.2	3.0		1964, 12/30	15
						>5.0		1965, 1/26	15
154......			5 30.2	6 34	15.2	2.0		1965, 1/6	15
155......	1857	573	5 30.3	5 07	15.4	1.5		1965, 1/28	15
156......			5 30.4	7 24	14.7	1.0		1964, 12/30	15
157......	2039	709	5 30.5	6 06	17.6	2.0	eHα	1963, 2/25	16, 4
158......			5 30.6	6 30	14.7	0.5		1965, 1/26	15
159......	2112		5 30.7	5 33	14.6	2.5		1964, 1/1	16
160......			5 30.7	6 16	19.5	>4.0		1965, 2/5	15
68......	2210	868	5 30.9	5 45	16.1	2.0		1964, 1/8	16
74......	2270	905	5 31.0	5 31	15.8	2.0		1965, 1/30	15
161......			5 31.0	6 04	18.4	>3.0		1965, 1/7	15
78......			5 31.1	6 06	18.3	>2.0		1965, 1/8	15
82......	2337	964	5 31.3	5 04	15.3	1.0	eHα	1965, 1/26	15, 4
162......			5 31.2	5 19	17.5	>3.0	eHα	1965, 1/1	15, 4
85......	2348	960	5 31.3	5 50	14.4	1.5		1965, 1/9	15
86......	2347	965	5 31.3	5 24	13.7	2.0	K1:, eHα	1964, 12/31	15, 4
163......			5 31.3	5 18	15.0	4.0		1965, 1/1	15
92......			5 31.5	5 11	15.3	>3.0	eHα	1965, 1/7	15
164......			5 31.5	5 18	16.6	>3.0		1965, 1/2	15
93......			5 31.6	7 10	15.0	2.0		1965, 1/7	15
165......			5 31.7	6 46	15.7	1.2		1963, 1/19	16
166......			5 31.8	6 57	15.3	5.0		1965, 1/1	15
167......			5 31.9	4 53	15.4	1.5		1965, 1/7	15
168......			5 31.9	6 48	17.7	>4.0		1965, 1/30	15
100......			5 31.9	5 08	15.5	0.5		1964, 12/27	15
169......	2450		5 31.9	5 17	13.5	0.7	K1:	1964, 12/31	15
170......			5 32.2	4 42	16.8	>0.4		1964, 12/30	15
171......			5 34.4	6 52	15.4	3.0		1965, 1/1	15
172......			5 34.4	4 01	15.2	2.0		1965, 1/27	15
118......			5 34.6	4 30	14.4	3.5		1965, 1/5	15
173......			5 34.9	6 28	15.6	2.0		1965, 2/2	15
174......			5 37.1	5 59	12.7	3.0		1965, 1/6	15
175......			5 37.2	5 28	16.5	>4.0		1965, 1/1	15
176......			5 38.1	−7 21	14.6	5.0		1965, 2/2	15

* Parenago (1954).

† Brun (1935).

** Δm_U for all observations by Haro and Chavira in 1962 and 1963; for all others, Δm_{pg}.

‡ References: (1) Haro and Chavira (1964); (2) Rosino and Cian (1962); (3) Haro and Terrazas (1954); (4) Haro (1953); (5) Herbig (1962*a*); (6) Rosino (1956); (7) Dall'Olmo (1960); (8) Dall'Olmo (1958); (9) Herbig (1950); (10) Maffei (1963); (11) Dall'Olmo (1961); (12) Haro and Morgan (1953); (13) Haro (1954); (14) Blanco (1963); (15) Haro and Chavira (1965); (16) Rosino (1965).

§ Italics indicate repeated flare.

and a number of "normal" nebular variables may have been missed. In zones D and F there are more Hα emission stars than known nebular variables; this also indicates that the search may be incomplete. Plate 3 shows the distribution of flare stars in the Orion Nebula field.

2. The earliest spectral type of a flare star known in Orion is K0. The latest known is M2, although among the faint stars in this region some later type might be found. Since most of these flare stars (Tables 1A and 1B) are members of the Orion aggregate, their absolute magnitudes (without correction for interstellar absorption) should range from $M = +4.5$ to $+13$—that is, from subgiants to dwarfs.

3. Some well-known T Tauri-like and Orion variables occasionally show, superimposed on their typical irregular light, outstanding outbursts of the flare type.

4. Most flare stars have not been previously identified as nebular variables or as Hα emission objects. In most cases the star appears constant, at minimum (normal) light,

TABLE 2

DISTRIBUTION OF NEBULAR VARIABLES, Hα EMISSION OBJECTS, AND FLARE STARS IN ROSINO AND CIAN'S ORION ZONES

Zone	No. of Nebular Variables	No. of Stars with Hα in Emission	No. of Flare Stars	Nebular Variables That Have Shown Flares
A	8	4	14	
B	12	6	6	2
C	19	9	19	
D	69	78	30	4
E	239	119	57	16
F	51	67	34	4
G	7	6	7	
H	5	3	6	
I	4	5	3	

without emission lines on objective-prism records up to the time of the flare; but suddenly, in a few minutes or perhaps seconds, a sharp increase of brightness may take place with amplitudes ranging from tenths to more than 6 or 7 magnitudes. Whenever it has been possible to check the spectrum, bright emission lines have appeared almost simultaneously.

5. All flare stars in Orion have been observed photographically with exposures from 5 to 15 minutes, and the corresponding light curves are consequently smoothed; it can be surmised, however, from the shape of the light curve and the time of exposure for each point that if continuous photoelectric records were made during flares, the amplitude of the variation in the ultraviolet could reach 10 magnitudes, or perhaps more in the most intense outbursts. The same can be said of the outstanding flare stars in other aggregates and in the solar vicinity.

6. During outbursts, the amplitude of the variation in ultraviolet light is about one-third greater to twice as great as the amplitude in the blue, which is, in turn, considerably greater than in the yellow; in the near infrared (8400 A), no detectable variations have been noticed in the photographs; appearance or enhancement of spectral emission

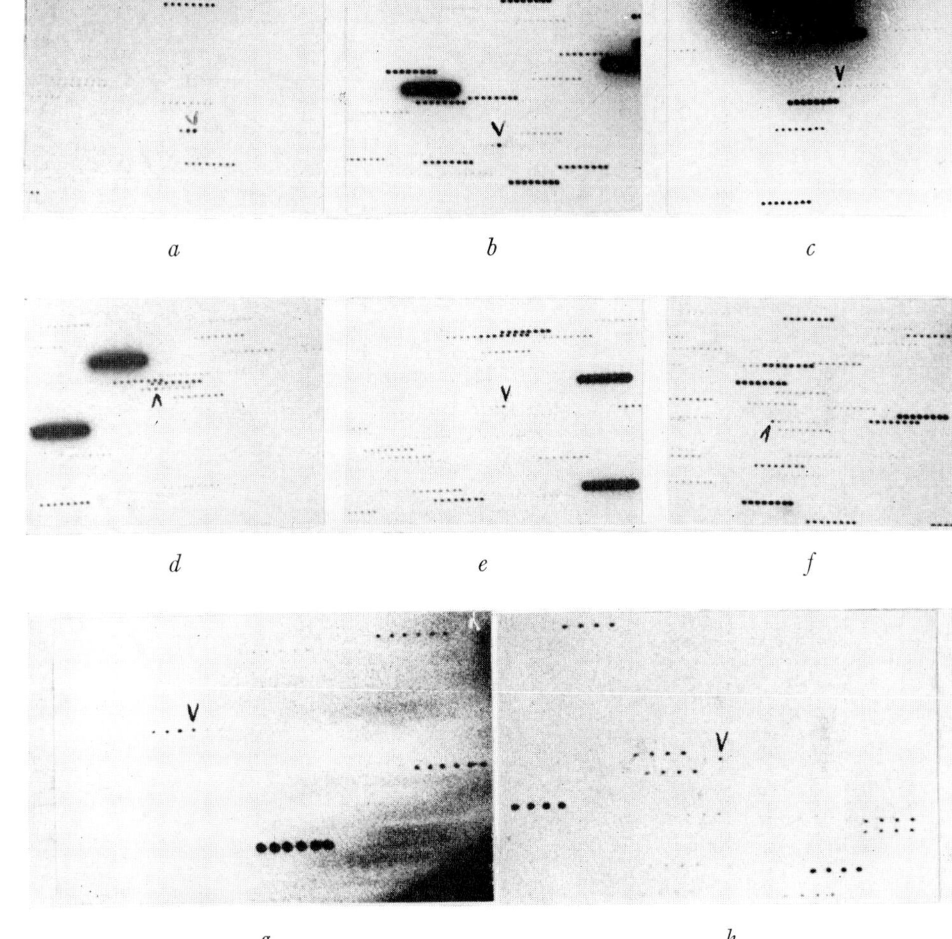

PLATE 1.—Multiple-exposure plates showing flares (marked with arrows) in some stars in the Pleiades and Orion regions. *a*, Pleiades No. 3; *b*, H_{II} 2411, a Hyades flare star not far from Alcyone; *c*, Pleiades No. 13, H_{II} 686; *d*, Pleiades No. 17, H_{II} 1306; *e*, Pleiades No. 20; *f*, Pleiades No. 30, H_{II} 3030; *g*, Orion No. 85, Parenago 2347; and *h*, Orion No. 7. Each exposure was 10^{m}, with 1 sec between exposures. The order of the exposures goes from right to left.

lines is conspicuous—especially for Hα, the most frequently observed feature during flares.

7. Observations in the red (6100 A to Hα) showed two extreme kinds of spectral change during flares in some Orion stars (Haro 1964): (*a*) the Hα emission line suddenly appears with great intensity or is strongly enhanced, but the nearby red continuum does not change appreciably, or (*b*) the red continuum is significantly intensified and the Hα emission line is relatively weaker and noticeable only after the maximum is reached.

In case (*a*), simultaneous direct observations in blue and ultraviolet light show that the rise to maximum is extremely rapid and the decline to minimum, although slower, is rapid also; at the end of the flare, the Hα emission usually disappears. Most Orion flare stars seem to belong to this group. In case (*b*), the rise to maximum is considerably slower than in (*a*), taking from 40 to 60 minutes to rise from normal (minimum) to maximum and about 5 to 6 hours to return to minimum. After minimum is again reached, the Hα line in emission is faintly visible for 1 or 2 days. Only four examples of this type are known in the Orion region (Nos. 66, 92, 149, and 153 of Tables 1A and 1B). Three of these "slow" flares (Nos. 66, 149, and 153) have also shown the "fast" flare features described in case (*a*).

Haro (1964) has stressed, following Ambartsumian's ideas (1954, 1957), that if the flare could be produced at different levels of the outer envelope of the star, there would be marked spectroscopic and photometric differences depending on the depth of the explosion. According to Ambartsumian, if the energy that causes changes in the star is liberated above the photospheric layers, a sudden increase of the nonthermal ultraviolet and blue continuous emission would occur, giving rise to a sharp light variation and, simultaneously, to an intense bright-line spectrum. When the energy is liberated in the photosphere, an increase in the thermal radiation as well as in the nonthermal continuous emission would be observed, with a lower relative intensity of the emission-line spectrum. This variation would not be as sharp as in the first case.

8. Of the 176 flare stars (Tables 1A and 1B) observed so far, only 23 have shown more than one outburst. This indicates a low frequency of repeated flares.

9. Qualitatively, the number of possible flare stars is estimated to be much larger than the known number of "normal" nebular variables in the Orion region. A quantitative and more detailed approach to this problem is desirable.

10. Although many flare stars in Orion lie above and to the right of the main sequence, a significant number lie on the main sequence or even below (Haro and Chavira 1964), for example, flare star No. 78 in Table 1A.

2.2. NGC 2264

The discovery of flare stars in the Orion Nebula region and their apparent close connection with the T Tauri-like variables led Haro (1954) to assume that wherever a group of T Tauri stars or related objects exists, flare stars may also be found. The young aggregate NGC 2264 (Walker 1956), among other T-associations, is a good place to test the suggestion.

All flare stars discovered in NGC 2264 by Haro, Rosino, and Wenzel are listed in Table 3 (see Haro and Chavira 1964). Only one (No. 1, Table 3 = Rosino's No. 1) has so far shown more than one flare.

No qualitative differences are known between the flare stars in Orion and those found in NGC 2264, although observations are not as extensive for NGC 2264 and no simultaneous spectral and direct observations are available for NGC 2264.

The concentration of the flare stars in a small area in NGC 2264 and the fact that two of them (LHα 67 and 69) were previously found and classified by Herbig (1954) as T Tauri stars indicate that they belong to the same stellar aggregate.

2.3. The Taurus Dark Clouds

Haro and Chavira (1955) obtained 138 direct blue plates centered at 4^h29^m, $+24°$ (1900), with a total of 690 different exposures in an effective observing time of 115 hours. On these plates, six flare stars, distributed much as are the T Tauri-like objects in the same region, were discovered. Usually during quiescence none of the flare stars shows Hα emission on objective-prism plates, although a faint emission line may sporadically be detected. In a nearby Taurus dark region, Petit (1958) gives a list of five additional flare stars. Table 4 contains data on these 11 stars.

TABLE 3

FLARE STARS IN NGC 2264

No.	Star	R.A. (1900)	Dec. (1900)	m_{pg}	Δm_{pg}	Sp. Type	Date of Flare	Ref.*
1.....	Rosino 1	$6^h34^m.5$	+ 9°53′	17.7	1.6		1955, 12/17	1
					1.4		1956, 4/7	1
2.....	Rosino 4	6 35.0	9 56	>17.5	>1.0		1955, 11/18	1
3.....	Rosino 7	6 35.7	10 01	17.0	1.0		1956, 3/9	1
4.....	Rosino 8	6 35.1	9 35	>18.0	>1.2		1955, 11/16	1
5.....	KSP 819	6 35.3	9 54	16.5:	0.7			2
6.....	Rosino 11	6 35.3	9 59	>18.0	>1.5		1955, 12/17	1
7.....	Rosino 16	6 35.5	9 56	15.4	0.7	K0	1956, 4/10	1, 3
8.....	Rosino 22	6 35.5	10 04	>17.5	>1.6		1956, 3/7	1
9.....	Haro 9	6 35.6	9 40	17.0	1.5		1954, 12/24	4
10.....	Haro 10	6 35.7	9 32	>19.0	>2.0		1952, 1/22	4
11.....	LHα67	6 35.8	9 45	17.5	1.0			2
12.....	LHα 69	6 35.8	9 32	18.0	2.0			2
13.....	Rosino 30	6 36.2	+ 9 39	>17.5	>1.3		1955, 11/9	1

* References: (1) Rosino *et al.* (1957); (2) Petit (1958); (3) Herbig (1959); (4) Haro and Chavira (1964).

In this particular nebulous area in Taurus, which is probably the seat of a very young T-association, some flare stars are of later M type than those observed in other young aggregates such as the Orion Nebula and NGC 2264. This could be either because the Taurus associations are relatively nearby stellar groups and, therefore, their members can be recorded to fainter absolute magnitudes—including late-type M stars—or because some of the flare stars observed are foreground objects. Because of the small number of flare stars in these particular regions and the lack of proper motions or radial velocities, no definite conclusions about membership in the Taurus T-associations can be drawn.

2.4. The Pleiades Region

Although the Pleiades region is one of the most intensively observed areas in the sky, few variable stars have been found there. Herbig (1962*a*) and Haro (1964) searched for possible T Tauri-like objects down to the sixteenth and seventeenth visual magnitudes, looking first for Hα emission stars in the Pleiades region. Their results were completely negative. When Johnson and Mitchell (1958) observed magnitudes and colors of stars considered, because of their proper motions, to be possible cluster members, they found that the discrepancy between two observations was sometimes large, being considerably greater for some stars fainter than $V = 12.0$ than for brighter ones. On this basis after observing a conspicuous flare in star H_{II} 1306, they concluded that many faint Pleiades members are flare-type variables.

Later, Haro (1964), Haro and Chavira (1964, 1965), and Rosino (1965) confirmed this conclusion. Sixty-one flare stars were found in a field of approximately 16 square degrees centered on Alcyone. These flare stars are listed in Table 5, in an arrangement similar to Table 1.

TABLE 4

FLARE STARS IN THE TAURUS DARK CLOUDS

No.	Star	R.A. (1900)	Dec. (1900)	m_{pg}	Δm_{pg}	Sp. Type	Ref.*
1	CY Tau	4^h11^m4	+28°06′	13.4	0.9	M2e	1
2	DF Tau	4 21.0	25 29	12.5	0.8	M0e	1
3	DK Tau	4 24.6	25 48	13.0	1.0	M0e	1
4	EY Tau	4 24.5	22 24	16.5	1.0	M3	2
5	EZ Tau	4 28.9	22 45	17.4	1.7	M5	2
6	FF Tau	4 29.1	22 42	15.8	1.4	M0–M1	2
7	DN Tau	4 29.4	24 02	13.5	0.8	K6e	1
8	FH Tau	4 30.6	22 50	15.9	1.1	M4–M5	2
9	FI Tau	4 30.8	23 25	17.5	1.2	M5	2
10	FK Tau	4 31.4	25 58	17.8	0.7	M6	2
11	DS Tau	4 41.5	+29 14	12.8	0.5	K6e	1

* References: (1) Petit (1958); (2) Haro and Chavira (1955).

A great deal of information on proper motions, photoelectric and photographic data, and spectroscopic surveys is available for relatively faint possible members of the Pleiades cluster. These data, with that summarized in Table 5, suggest the following:

1. Pleiades membership can be emphasized because in this cluster much work on proper motions and radial velocities has been done. A significant number of the flare stars found in the Pleiades region seem to be cluster members. For example, of 20 flare stars with known proper motions (Hertzsprung *et al.* 1947; van Maanen 1945*a*, *b*) 16, or 80 per cent, have been classed as probable members by Haro and Chavira (1965) and only four as possible nonmembers.

However, extrapolation from the given sample of flare stars with known proper motions to the fainter flare stars without known proper motions might not be valid, and a more restricted criterion for membership than the one used by Haro and Chavira could give different values. In Table 6, three criteria are applied to all Pleiades flare stars with known proper motions. In column 1, the Hertzsprung and van Maanen numbers are listed. Column 2 shows membership based on the Johnson-Mitchell criterion; of 20 flare stars, 14, or 70 per cent, are considered as probable cluster members and six as probable nonmembers. In column 3, the criterion of Ahmed, Lawrence, and Reddish (1965) is

TABLE 5

FLARE STARS IN THE PLEIADES

No.	Star	R.A. (1900)	Dec. (1900)	m_{pg}	Δm_U	Sp. Type, eHα	Date of Flare	Ref.*
1......		3^h33^m5	+24°39′	14.5	1.0	M3	1963, 11/11	1
2......		3 34.8	24 27	19.0	>3.0	M	1963, 2/17	2
					>3.5		1963, 11/17	1
3......		3 36.7	23 41	17.6	7.0		1963, 11/19	1
4......		3 37.5	22 18	20.9	>5.0		1963, 11/21	1
5......		3 37.6	25 06	18.6	>1.7		1963, 2/15	2
					>2.0		1963, 2/17	2
6......		3 37.7	22 53	20.0	>5.0		1963, 11/19	1
7......	H$_{II}$ 191	3 37.9	24 22	15.9	1.0	dK7	1963, 2/17	2, 3, 4
8......	H$_{II}$ 357	3 38.5	23 51	14.9	2.0	K5	1963, 11/10	1, 3, 4
					1.0		1963, 11/22	1
					≥0.5			5
9......		3 38.5	24 05	17.9	2.5		1963, 11/15	1
10......		3 38.5	25 43	17.5	1.5	M	1963, 11/21	1
11......		3 38.8	23 41	20.1	>2.5	M	1963, 11/12	1
12......		3 38.8	25 36	18.3	3.0	M	1963, 11/9	1
13......	H$_{II}$ 686	3 39.6	23 59	14.6	2.0		1963, 11/16	1, 3
14......	H$_{II}$ 906	3 40.2	24 22	15.9	3.0	M	1963, 2/15	2, 3
					2.0		1963, 2/16	2
					1.5		1063, 11/17	1
					1.0		1963, 11/19	1
					0.8		1965, 12/4	7
15......	vM 16	3 40.9	23.59	17.9	3.0	M	1963, 11/17	1, 6
16......	H$_{II}$ 1286	3 41.1	23 18	17.0	1.5	M	1963, 11/12	1, 3
17......	H$_{II}$ 1306	3 41.2	23 24	14.7	0.5	dK5(e)	1963, 11/19	1, 3, 4
					0.7		1963, 11/23	1
					3.7			5
					0.5		1965, 11/7	7
18......		3 41.6	22 02	17.2	2.0	M3–4, eHα	1963, 11/15	1
19......	H$_{II}$ 1531	3 41.7	23 40	14.4	0.4		1963, 11/17	1, 3
20......		3 41.8	24 19	20.2	>5.0		1963, 11/21	1
					>2.0		1947, 1/21	1
21......	H$_{II}$ 1653	3 42.0	24 25	14.7	0.8	K7	1963, 2/17	2, 3, 4
					0.4		1963, 11/17	1
22......		3 42.2	22 19	20.0	>2.5		1963, 2/26	1
23......		3 42.4	24 36	19.1	>2.0		1963, 11/10	1
24......		3 42.5	22 53	19.9	>3.0	M	1963, 11/21	1
25......		3 42.6	21 54	14.8:	3.0		1963, 11/19	1
26......		3 43.2	24 02	19.8	≥4.0		1963, 11/16	1
27......		3 44.1	23 33	19.0	>2.0	M	1963, 11/15	1
					>2.5		1964, 1/13	1
28......		3 44.2	23 37	17.9	2.0		1963, 11/9	1
29......		3 45.3	25 45	18.5	>2.5		1963, 11/10	1
30......	H$_{II}$ 3030	3 45.5	23 35	15.4	2.0	dK7	1963, 11/20	1, 3, 4
31......		3 45.6	24 18	18.6	3.5		1963, 2/18	2
32......		3 47.4	24 45	20.6	>5.0		1963, 11/22	1
33......		3 48.4	24 26	20.4	>5.0		1963, 11/18	1
34......		3 35.7	21 58	17.5	>5.0		1964, 11/3	7
35......		3 36.0	25 01	15.9	1.5	M	1964, 11/6	7
36......		3 36.2	23 46	17.1	2.5	M:	1964, 11/6	7
37......		3 36.7	23 44	>17.5	>2.0pg		1964, 11/26	8
38......		3 36.7	24 22	17.0	0.8		1964, 12/6	8
39......		3 37.1	24 30	16.7	5.0		1964, 11/7	7
40......		3 37.2	24 22	18.0	>3.0		1964, 11/3	7
41......		3 37.7	24 15	17.0	1.5		1964, 11/28	7
42......	H$_{II}$ 230	3 38.0	+23 16	15.4	0.7		1964, 12/7	7, 3

* References: (1) Haro and Chavira (1964); (2) Haro (1964); (3) Hertzsprung *et al.* (1947); (4) Herbig (1962*a*); (5) Johnson and Mitchell (1958); (6) van Maanen (1945*b*); (7) Haro and Chavira (1965); (8) Rosino (1965).

TABLE 5—*Continued*

No.	Star	R.A. (1900)	Dec. (1900)	m_{pg}	Δm_U	Sp. Type eHα	Date of Flare	Ref.*
43......		3^h38^m8	+23°53′	20.0	>6.0		1964, 11/6	7
44......	H_{II} 500	3 39.0	24 04	16.2	0.9		1964, 12/7	8, 3
45......	H_{II} 590	3 39.3	24 47	15.4	1.0		1964, 11/30	7, 3
46......	H_{II} 793	3 39.9	23 32	15.1	0.7		1964, 11/2	7, 3
47......	vM 6	3 40.5	23 51	17.4	2.3 pg	M	1964, 9/9	8, 6
48......	H_{II} 1061	3 40.6	23 48	15.2	1.5	dK5	1964, 11/8	7, 3, 4
49......		3 40.9	23 14	19.0	>4.0		1964, 11/8	7
50......		3 41.0	24 45	>17.2	>1.2 pg		1964, 10/15	8
51......	H_{II} 1827	3 42.4	23 40	15.9	2.0		1964, 12/7	7
52......		3 42.6	23 54	>17.5	>1.5 pg		1964, 9/9	8
53......		3 43.3	24 03	>18.5	>0.6		1964, 12/7	8
54......		3 43.6	23 37	>17.2	>1.3 pg		1964, 9/9	8
55......	H_{II} 2411	3 43.7	24 01	15.5	1.5	M4e	1963, 2/15	1, 3, 4
					0.5		1963, 11/10	1
					0.7		1963, 11/21	1
					3.0		1963, 11/23	1
					0.8		1964, 11/2	7
					0.9		1964, 12/8	8
					1.3		1964, 12/9	8
56......	H_{II} 2601	3 44.2	24 03	16.0	0.6	M2	1964, 11/6	7, 3, 4
57......	H_{II} 2879	3 44.9	23 25	15.6	0.9		1964, 12/6	8, 3
58......		3 46.1	25 04	21.5	>6.0		1964, 11/1	7
59......		3 46.3	23 02	15.5	0.7		1964, 12/7	8
60......		3 46.5	22 13	16.9	2.5		1964, 11/2	7
61......		3 48.1	+23 03	15.5	0.8		1964, 11/8	7

TABLE 6

MEMBERSHIP OF PLEIADES FLARE STARS WITH KNOWN PROPER MOTIONS

Hertzsprung/ van Maanen Nos.	Johnson and Mitchell	Ahmed, Lawrence, and Reddish	Haro and Chavira
H_{II} 191.........	PM	M	M
230.........	N-M	N-M	N-M
357.........	M	N-M	M
500.........	N-M	N-M	N-M
590.........	PM	M	M
686.........	M	N-M	M
793.........	PN-M	M	M
906.........	M	M	M
1061.........	M	N-M	M
1286.........	PM	M	M
1306.........	M	M	M
1531.........	M	PM	M
1653.........	M	PM	M
1827.........	M	M	M
2411.........	N-M	N-M	N-M
2601.........	M	M	M
2879.........	N-M	N-M	N-M
H_{II} 3030.........	M	N-M	M
vM 6...........	PM	....	M
vM 16...........	PN-M	....	M
% M and PM....	70	55	80

M = Member, N-M = Nonmember, PM = Probable member, PN-M = Probable nonmember.

applied to 18 flare stars (van Maanen Nos. 6 and 16 are excluded), and the resulting percentage of probable members is 55 per cent. In column 4 the membership, according to Haro and Chavira, would be 80 per cent.

To decide weights for these three values, the surface distribution of the flare stars of the Pleiades area was considered: the 51 rapid variables found by Haro and Chavira show a strong concentration toward the center of the cluster; in the inner area of 2×2 degrees, centered on Alcyone, 33 flare stars were found, an average of 8.2 flare stars per square degree.[1] In the surrounding 12 square degrees, only 18 flare stars were detected, that is, 1.5 rapid variables per degree. The 16 square degrees under consideration were exhaustively examined on all plates; the marked concentration of flare stars toward Alcyone cannot be due to any spurious instrumental or photographic effect. The flare stars discovered by Rosino have been excluded in this argument, since the total area searched by him was only approximately 4 square degrees.

Even with an over-all density of 1.5 field stars per square degree, in the central 4 square degrees surrounding Alcyone there would be $8.2 - 1.5 = 6.7$ cluster-member flare stars per degree; in other words, about 80 per cent of the flare stars found in the area searched for proper motions by Hertzsprung in the Pleiades can be tentatively considered as cluster members.

If we add to Haro and Chavira's 33 flare stars the 10 Rosino stars in the central 4 square degrees and deduct 80 per cent of possible cluster members in this area, no more than 9 flare stars are probable nonmembers. We have good evidence that among the 20 flare stars with known proper motion in the inner area covering 4 square degrees, four (H_{II} 230, H_{II} 500, H_{II} 2411, and H_{II} 2879) do not belong to the Pleiades cluster.

As Ahmed, Lawrence, and Reddish (1965) have commented, Johnson and Mitchell select as cluster members of the Pleiades only those stars which fall on or above a predefined narrow main sequence and which have the mean proper motion of the cluster. Thus, any field star included by the proper-motion criterion and lying in the main-sequence band might be spuriously accepted as a cluster member, while a real cluster member with the correct proper motion but lying below the predefined main sequence would be listed as a probable nonmember (good examples are H_{II} 793 and vM 16 in Johnson and Mitchell's lists). Ahmed, Lawrence, and Reddish adopted a more elaborate criterion by considering the distribution of all field stars on color-magnitude, two-color, proper-motion, and surface-density diagrams to the limiting magnitudes $U = 17.8$, $B = 17.5$, and $V = 16$, and by selecting each supposed member of the Pleiades according to its position on their diagrams. They investigated the effect of nebulous fog and, where necessary, made corrections, and they used the U magnitudes to give the $U - B$ colors as an additional criterion in separating faint field stars from the cluster stars.

In assessing the value of the Edinburgh astronomers' results, it should be noted that their photometric measurements seem to be affected by large errors. Some of the most obvious evidence is as follows: in Table *A* of their paper, two of the confirmed cluster members (H_{II} 624 and 890) with $B - V = +1.85$ have $U - B$ colors of 0.07 and -0.33, respectively, and another probable member (FC, H_{II} 2592) with $B - V = +1.32$ has $U - B = -0.73$. The star H_{II} 1061 (FF), classified by Herbig as dK5, has $B - V =$

[1] In an area of approximately 1.5 square degrees, centered on Alcyone, Haro and Chavira's data show 21 flare stars. This corresponds to 14 stars per square degree for the central region.

0.33 and $U - B = 0.72$. The star H_{II} 3030 (FF) of spectral type K7 has $B - V = 0.71$ and $U - B = 0.25$; and the star H_{II} 2548 (FF) of type M1 has $B - V = 0.60$ and $U - B = 0.89$. In Table *B* of the same paper, in the *Central Nebulous Area* alone, there are 27 stars with $B - V$ colors from $+1.0$ to $+2.75$ and $U - B$ colors from -0.02 to -1.74; the average value of the $B - V$ colors in these 27 stars is $+1.38$ and for the $U - B$ -0.37. In the same section of Table *B*, the star No. 73 (H_{II} 1538) has $V = 15.79$, $B - V = +0.97$ and $U - B = -3.52$, but obviously this must be a misprint.

The first thought about this and other very peculiar colors in the Ahmed *et al.* paper is that probably many stars were caught during a flare, but this is not reasonable because there are measurements on two plates in each color and it is improbable that so many stars would show flares of the same $B - V$ and, more especially, the same $U - B$ on different pairs of plates. The possibility of explaining the outstanding ultraviolet excesses as a result of dealing with extreme T Tauri variables is unacceptable because, as Herbig

TABLE 7

FLARE STARS IN COMA BERENICES

No.	R.A. (1900)	Dec. (1900)	m_{pg}	Δm_U	Sp. Type	Date of Flare	Ref.*
1.	$12^h20^m.2$	+26°18′	16.3	2.5	M	1964, 3/14	1
				1.5		1965, 4/6	2
				0.7		1965, 4/22	2
2.	12 19.5	+24 38	14.9	1.5	M	1964, 5/8	1
				3.5		1964, 5/15	1
3.	12 22.7	+27 35	16.8	1.3	M3–4	1964, 5/2	1
4.	12 17.1	+28 17	>17.5	>4.0	M:	1965, 4/24	2

* References: (1) Haro and Chavira (1964); (2) Haro and Chavira (1965).

and Haro have shown, to the seventeenth visual magnitude there are no Hα emission objects detectable on objective-prism plates. An alternative would be that the Pleiades central region is exceptionally rich in extremely peculiar stars, brighter than $V = 16.0$; but from other data, this does not seem to be true. The only conclusion which cannot be excluded, if the cases under consideration are taken as significant samples, is that the photometry of Ahmed *et al.* is seriously affected by errors.

2. The earliest spectral type of the flare stars known in the Pleiades is K5. If the star H_{II} 2407, where Johnson and Mitchell suspect flaring activity, is a real flare star, then according to Herbig (1962*a*) the earliest spectral type will be K3. There is no good spectroscopic data on the fainter flare stars but, from low-dispersion objective-prism red plates and the estimated colors, some of these rapid variables seem to be of spectral types later than M4.

3. Although part of the Pleiades cluster is imbedded in nebular matter (bright and dark), many flare stars seem to lie outside the nebulous regions; therefore, a straight correlation between interstellar material and flare stars, as found in the Orion Nebula region, in NGC 2264, and in the Taurus dark clouds, is not possible here.

2.5. THE COMA BERENICES CLUSTER

On plates centered at 12^h15^m, $+26° 30'$ (1900) and covering an area of 16 square degrees, Haro and Chavira (1964, 1965) found four flare stars (see Table 7). There is no conclusive evidence that these are cluster members. Their apparent magnitudes during minima and their average absolute magnitudes derived from the approximate spectral types indicate that their distance moduli agree with that of the Coma cluster; proper motions and, if possible, radial velocities would be necessary to settle the matter.

In the Coma cluster area, despite the relatively large number of observations obtained at the Tonantzintla Observatory, only four flare stars were detected—all of spectral type M and evidently not related to interstellar nebulous matter. The small number of flare stars in this region, in contrast to what has been found in Orion, NGC 2264, and the Pleiades, supports Haro's finding that flare stars are not distributed at random in our Galaxy.

2.6. Praesepe

Table 8 gives data on the 11 flare stars found at the Tonantzintla Observatory (Haro and Chavira 1964, 1965) in an area of 16 square degrees centered at 8^h34^m, $+19°50'$ (1900).

TABLE 8

FLARE STARS IN PRAESEPE

No.	R.A. (1900)	Dec. (1900)	m_{pg}	Δm_U	Sp. Type	Date of Flare	Ref.*
1.......	$8^h28^m.1$	$+19°48'$	18.1	>2.5	M:	1964, 2/18	1
2.......	8 31.6	20 18	17.5	2.0	M	1964, 2/14	1
				0.7		1965, 2/24	2
3.......	8 33.3	20 08	17.7	>4.0	M	1964, 2/14	1, 3
4.......	8 34.8	18 48	18.9	>2.5	M:	1964, 2/16	1
5.......	8 35.6	20 34	20.5	>5.0	M:	1964, 2/18	1
6.......	8 37.0	20 08	19.3	>3.0	M:	1964, 2/14	1
7.......	8 37.0	21 23	19.5	>4.0	M:	1964, 2/19	1
8.......	8 32.8	18 28	17.5	3.0	M	1965, 3/4	2
9.......	8 34.9	19 51	20.7	>5.0	M:	1965, 2/25	2
				>6.0		1965, 2/26	2
10.......	8 35.4	19 23	19.8	>4.0	M:	1965, 2/24	2
11.......	8 40.5	+19 09	17.6	3.0	M	1965, 3/6	2
				1.5		1965, 3/6	2

* References: (1) Haro and Chavira (1964); (2) Haro and Chavira (1965); (3) Klein-Wassink (1927).

As in Coma, all rapid variables in the Praesepe region are of M type and seem to be unrelated to interstellar clouds. Only one has a measured proper motion (No. 3 in Table 8 = No. 563 in the Klein-Wassink catalogue, 1927), and from this it can be considered as a cluster member.

Although the number of observations for Praesepe is less than for Coma (see Table 9), the number of flare stars found is significantly larger. The Praesepe cluster may be much richer in stellar content than the Coma cluster. Also, it is at lower galactic latitude; therefore, the foreground star-density per unit area is greater, and this may add a few flare stars to the total number observed.

2.7. The Hyades

As we have already seen (Table 5), the star H_{II} 2411, an M4e Hyades cluster member, projected not far from Alcyone, has shown several conspicuous flares.

The Tonantzintla observers (Haro and Chavira 1964) also found two flare stars in an area of 16 square degrees centered at 4^h15^m, $+15°20'$ (1900) in 35^h20^m of effective observational time (212 exposures). These have been measured for proper motion, and both seem to belong to the Hyades cluster. One, star No. 190 of Giclas, Burnham, and Thomas (1962), is also contained in Holmberg's (1944) list: Ho A 78; its spectral type, according to Herbig, is M3e. The other flare star, Giclas *et al.* No. 195, has been photoelectrically observed by Johnson, Mitchell, and Iriarte (1962) and is the faintest (No. 261) in their list of Hyades members; on the Tonantzintla Observatory infrared plates it seems to be an M4 star.

It is thus probable that these two flare stars are members of the Hyades cluster. Both have spectral types later than M2 and, according to Herbig (1962*a*), who obtained slit

TABLE 9

FLARE STARS IN DIFFERENT STELLAR AGGREGATES

Name of Aggregate (1)	Total No. of Known Flare Stars (2)	Stars with More Than One Flare (3)	% of Repeat Flares (4)	No. of Exposures (E) and Time of Obs. (5)	Sp. Types (6)	No. of Flares Observed per Hour (7)	Age of the Aggregate (years) (8)
Orion Nebula......	176	23	13	2495E–456^h* (573E–112^h)†	K0 to M	0.37 (0.55)	3×10^5–10^6
NGC 2264.........	13	1	8	326E–45^h‡	K0 to M:	0.24	10^6
Taurus Dark Cloud (Tonant.)........	6	None	—	690E–115^h*	M0 to M5	0.05	10^6:
Taurus Dark Cloud (Petit)...........	5	None	—	—	K5 to M5	—	10^6:
Pleiades...........	61	9	15	1112E–189^h*	K5 to M	0.33	2×10^7
Coma Berenices.....	4	2	50	803E–146^h*	M	0.05	3×10^8
Praesepe...........	11	3	28	539E–106^h*	M	0.13	4×10^8
Hyades............	2	1	50	212E–35^h*	M3–M5	0.08	4×10^8
Solar Neighborhood.	20	10:	50:	—	M0 to M6	—	4×10^8:

* Tonantzintla observations (total).
† Tonantzintla observations, winter 1964–1965.
‡ Rosino *et al.* and Tonantzintla observations.

spectrograms of H_{II} 2411 and Ho A 78, both variables show emission lines during minima. As in Coma and Praesepe, the flare stars in the Hyades do not seem to be associated with nebular clouds.

2.8. THE SOLAR VICINITY

In 1924, Hertzsprung, having noticed that the star DH Car had shown a rapid increase of about 2 mag on one occasion, wrote: "The supposition that a sudden outburst of unusually short duration has here occurred, seems to me to be the most plausible one. In that case, the star will be of exceptional interest." The history of further discoveries of flare stars within 20 pc of the Sun is well known due to Joy's (1960) excellent compilation of the relevant data reproduced in Table 10.

Data on flare stars in the solar neighborhood are more extensive than for those elsewhere and include good visual, photographic, photoelectric, and spectroscopic observations, interesting information on the kinematical properties of the UV Ceti stars and related objects, and radio observations. In summary:

1. All known flare stars in the solar vicinity are of low intrinsic luminosity (spectral types dM0 ~ dM6). Usually they exhibit strong, narrow emission lines of H and Ca II during minima. At the time of outburst, according to Joy and Humason (1949) and Joy (1960) who observed UV Ceti spectroscopically during two flares, the normal spectrum is heavily veiled by a strong continuous emission, especially enhanced in the ultraviolet region, indicating a temperature of at least 10,000° K. At the same time, the H emission lines are considerably widened and strengthened, and weak emission lines of He I (4026 and 4471 A) and He II (4685 A) appear. The emission lines of Ca II are slightly enhanced

TABLE 10

UV CETI FLARE STARS (JOY 1960)

Star and Constellation	Catalogue and Star Number	Normal Mag.	Max. Range	Spec.	M_v	Trig. Par.	Flare Obs.	Ref.*
Gr. 34 B-And	+43° 44 B	11^m0	....	dM4e	+13.1	0."28	Sp	Unpub.
vB. star-Aql	+4° 4048 B	17.9	....	dM5e	+19.0	.17	Sp	1
YZ CMi	Ross 882	11.8	1^m4	dM4.5e	+12.4	.15	D, Sp	2
DH Car	−B	12.2	1.8	....	....	......	D	3
Wolf 47-Cas	20C 70(B)	13.7	....	dM5e	+13.9	.11	Pe, Sp	4
V645 Cen	20C 861 (C)	13.4	1.0	dM4e	+15.4	.76	Sp, D	5
DO Cep	20C 1366 B	11.3	1.2	dM4.5e	+13.4	.25	D, Sp	6
UV Cet	L726–8 B	12.4	5.6	dM4.5e	+16.1	.37	Sp, D, Pe	7
Wolf 1130-Cyg	20C 1191 AB	11.9	....	dM3e	+10.8	.06	D, Sp	8
Furj. 54-Cyg	20C 1250 AB	12.2	....	dM3e	+11.5	.07	Sp	9
-Dra	+55° 1823	10.1	0.5	dM1.5e	+ 8.5	.05	D	10
GC25394-Dra	HD 234677	8.6	....	dM0e	....	......	Sp	11
EV Lac	20C 1382	10.2	2:	dM4.5e	+11.8	.20	D	12
AD Leo	20C 574 AB	9.4	0.5	dM4e	+10.9	.21	Pe	13
Wolf 359-Leo	20C 600	13.5	1	dM6e	+16.5	.40	D	14
-Mic	HD 196982 B	11.1	....	dM4.5e	+11.4	.11	Sp	15
V371 Ori	Wachmann AB	11.7	1.9	dM3e	....	......	Sp, D	16
EO Peg	+19° 5116 B	12.8	0.5	dM5.5e	+13.6	.14	Pe, Sp	17
V1216 Sgr	20C 1108	10.5	0.4	dM4.5e	+13.3	.35	Pe, Sp	18
WX UMa	20C 606 B	14.8	1.8	dM5.5e	+16.0	0.17	D	19

* References: (1) Herbig (1956); (2) van Maanen (1945*a*), Joy (1957); (3) Hertzsprung (1924); (4) Johnson and Morgan (1953), Bidelman (1954); (5) Thackeray (1950), Shapley (1951); (6) Van de Kamp and Lippincott (1951), Joy (1957); (7) Joy and Humason (1949), Luyten (1949), Roques (1953); (8) Sandig (1951), Joy (1947); (9) Luyten (1925); (10) Petit (1954); (11) Popper (1953); (12) Wagman (1953); (13) Gordon and Kron (1949); (14) Sandig (1951); (15) Luyten (1926), Joy (1957); (16) Wachmann (1939), Hoffleit (1952); (17) Roques (1954), Joy (1957); (18) Kron, Gascoigne, and White (1957), Joy (1957); (19) van Maanen (1940).

but no forbidden lines are apparent. The spectroscopic evidence of Joy and Humason indicates that the sudden increase in brightness arose mostly from the enhanced continuous spectrum and not from the strenghthened emission lines.

Although these characteristics have been observed only in UV Ceti, the general spectroscopic features described above are probably much the same for all flare stars because (*a*) the amplitude in the ultraviolet is always greater than in the blue, and the visual increment is the least conspicuous in all cases in which multicolor photometry has been obtained; and (*b*) Balmer emission lines always appear or are strengthened whenever spectroscopic evidence has been obtained during outbursts.

Joy (1960), comparing characteristics of solar flares with those of the UV Ceti-type outburst, pointed out that the times involved are much the same, but that there are spectroscopic differences. The solar flares have narrow emission lines mainly of H, Ca II, He I, Na, and Fe II, and the continuous emission is usually too weak to be observed although some intense solar flares have a continuous spectrum and produce radio fade-outs, indicating intense ultraviolet radiation.

2. Apart from the extremely rapid outburst, many flare stars behave as "normal" irregular variables of small amplitude. Roques (1958) and Oskanjan (1964), among others, have called attention to these "secondary" or "slow" changes in brightness. The same photometric behavior has been noticed by Haro and Chavira in some flare stars found in aggregates. It is often difficult to establish a sharp distinction between small amplitude flares and the "secondary" irregular variations. Oskanjan (1964) proposed that all brightness changes occurring faster than $0^m.05$/sec belong to the flare type, regardless of the amplitude of the change. This criterion only provides a first, restricted approach to the problem, being more suitable for dealing with variations of small amplitude, $\sim < 0^m.5$; otherwise, a flare of $1^m.2$ that takes 2 minutes to go from normal minimum to maximum—that is to say, with a velocity of brightening of only $0^m.01$/sec—would be rejected. Furthermore, an outburst of $3^m.0$ which takes 10 minutes to reach maximum, brightening at the rate of $0^m.005$/sec, undoubtedly should be accepted as a bona fide flare.

It seems to be safe, then, to adopt $0^m.005$/sec as the value for the minimum rate of brightening for the flare. This means that an eruptive variable which increases its total brightness by $0^m.3$/min can be classed as a flare star. Of course, the onset of a flare can be much faster, and Oskanjan has found, for UV Ceti itself, increments in the visual magnitude of the order of $0^m.25$/sec. The amplitude in the ultraviolet usually is greater (in many cases by a factor of more than 3) than in the visual; so the rate of increase at short wavelengths would be more than $0^m.5$/sec for very conspicuous flares.

In the light curve of a flare, the decline from maximum always takes place immediately, although at a rate significantly slower than the rise to maximum. The total time of variation from "normal" minimum to maximum and back to minimum in a flare star is, by far, the most rapid change observed in the integrated light in any intrinsic variable for a given amplitude of the variation. The relation in time between the ascending and descending branches on the light curve of a flare star is of the same order as that found in solar flares.

3. Joy (1960) found that of the 20 flare stars of Table 10, 14 are known binaries, and more recently, Kraft (1965) discovered that the flare star HD 234677 is a double-line spectroscopic binary of unknown period. Thus, of the 20 flare stars in the solar vicinity, 75 per cent are binaries. Three of these (Wolf 1130, V371 Ori, and HD 234677) are spectroscopic binaries; two (20C 1250 and AD Leo) are visual binaries with separations of less than 1 sec of arc, and of the remaining pairs five are so widely separated that little interaction can be expected between the components. Kraft's discovery of the spectroscopic binary nature of HD 234677, the brightest star in Table 10, leads to the question: How many more flare stars would show the same peculiarity if adequately observed? Herbig (1962*b*) pointed out that the dMe stars seem to lie systematically higher than the

dM's in a color-magnitude diagram and that there is a much greater proportion of close binaries among the dMe's than among the dM's; he interprets these data as indicating that the emission lines in M dwarfs may arise for either or both of the following reasons. Herbig states "(*a*) The star is still in its initial contractive phase or (*b*) The star is a member of a close binary system in which the emission lines are generated by some unknown phenomenon that may be of either dynamical or evolutionary origin."

This kind of reasoning could be applied to the flare phenomenon observed in stars.

4. Delhaye's study (1953) of the kinematic properties of 12 dMe stars, including 4 flare stars, in the vicinity of the Sun shows that there is a remarkably small velocity dispersion perpendicular to the galactic plane. These are probably young stars forming a fraction of a flat subsystem which has some of the structural and kinematical peculiarities of a T-association (Haro 1954, 1956).[2] Later Einasto (1954, 1955), Vyssotsky and Dyer (1957), and Gliese (1958) showed that the dMe stars have in all three coordinates a notably small velocity dispersion compared to that of the dM stars without emission.

5. The similarity of the outbursts in flare stars to flares in the solar chromosphere and the evidence that a nonthermal mechanism must operate in both events caused the British and Australian radio astronomers to observe nearby flare stars for radio emission. These observations were discussed by Sir Bernard Lovell (1964) in his Halley Lecture for that year. As expected, some stars simultaneously show radio bursts and optical flares. The Jodrell Bank and the Australian observations indicate conclusively that the radio wavelength emission is nonthermal.

According to Lovell, estimates of total energies involved in solar flares and in outbursts observed in the stars V371 Ori, UV Ceti, Ross 882, and EV Lac lead to the following conclusions. Lovell states "(*a*) The outburst of energy in the optical continuum for small flares of 1 mag or less is at least 10 to 100 times greater than that in some of the strongest solar flares recorded. (*b*) The output of energy in the radio spectrum shows an excess of 10^4 to 10^6 times on the large radio outbursts from the Sun. (*c*) Whereas the ratio of optical to radio energies for the large solar outbursts is about 10^5, the ratio for the flare stars is of the order of 10^2 to 10^3 for the 4 stars for which data are so far available."

Further radio studies of flare stars are needed for a better understanding of the flare phenomenon. So far, successful results have been obtained only for the dMe stars in the solar vicinity, but attempts to detect radio emission bursts in flare stars of considerably greater absolute luminosities in the Orion Nebula and in the Pleiades are under way.

[2] The high abundance of flare stars in regions where T Tauri stars are found (Orion, Taurus, Monoceros, etc.), that is, in regions characterized by clouds of interstellar matter, led Haro (1954) to suppose that the presence of flare stars implied the presence of interstellar clouds. Haro therefore suggested that the flare stars and emission objects in the solar neighborhood might be imbedded in clouds of interstellar matter and that some interaction exists between this group of objects and the nebular material. New data suggest that this is not correct but that T Tauri-like features, as well as the flare phenomenon, are intrinsic to particular groups of stars, depending on their initial conditions and subsequent evolutionary stages. However, some kind of small nebulous envelope not yet detected around flare stars might be present or, if these are close binaries, a transfer of mass might play an important role in the production of flares.

3. FURTHER DISCUSSION AND SUMMARY

1. Two criteria are necessary to classify variables as flare stars: (*a*) relative steadiness of the starlight during minima or "normal" phase; and (*b*) a sudden, unpredictable rise to maximum and a slower, but still rapid, decline toward minimum. There may also be characteristic spectroscopic changes but data for these are scarce or, in most cases, nonexistent.

Although most of the known flare stars remain most of the time at "normal" minima,[3] some, especially in young aggregates, can be classified as typical T Tauri, RW Aurigae, or Orion variables. In such cases, the flares can be distinguished only by taking into account the "normal" shape of the irregular light curve and the remarkable rapidity and conspicuousness of the superposed outbursts. Flare phenomenon can occur throughout the evolution of a late-type star, at least for stars of spectral types K0 or later, while it approaches the main sequence (Haro 1964). Then, the extremely young stars (with pronounced T Tauri, RW Aurigae, or Orion variable features) which show flares might be less numerous in our Galaxy than more highly evolved stars that have already lost their T Tauri-like characteristics and only show sporadic outbursts.

The rapidity of the outburst or the rise to maximum during a flare must now be defined in a more quantitative manner. In the solar vicinity, a minimum rate of increase of $0\overset{m}{.}005$/sec is a conservative estimate. This is undoubtedly correct for any flare star, even an irregular variable of the most pronounced T Tauri type, especially when the total flare amplitude is greater than $0\overset{m}{.}5$. But for stars characterized by steady and prolonged "minima," $0\overset{m}{.}0005$/sec may be provisionally accepted as the minimum rate of brightness increase if the total increase is greater than half a magnitude. Thus, for instance, a normally steady star, which suddenly increases its total brightness by 1.8 mag in about 1 hour ($0\overset{m}{.}0005$/sec), probably should be classified as a flare star.

If we consider the average light curve of a flare, the time interval corresponding to the ascending branch may be designated by A and that of the descending one by B. The ratio A/B for stellar flares gives values comparable to those obtained for solar flares.

If we accept that the outburst of energy in the optical continuum for a flare of 1 mag in UV Ceti is $\sim$10 times greater than in the strongest solar flares (Lovell 1964), then in the intrinsically brightest flare stars found—for instance, in the Orion Nebula—the outburst of energy in the optical-wavelength region for a flare of about 1 mag will be at least 10^5 to 10^6 greater than in the most conspicuous solar flares.

2. Notwithstanding the paucity of the spectroscopic data, it seems certain that the earliest spectral type of the brightest known flare stars is about K0. In very young stellar aggregates, such as Orion and NGC 2264, the brightest known flare stars are of late G or early K types. In older stellar groups, the brightest flare stars show progressively later "normal" spectral types. In the Pleiades the brightest flare stars now known are K5; and in Coma, Praesepe, the Hyades, and the solar neighborhood, all are of type M (Haro and Chavira 1964).

Although most flare stars do not show the Balmer lines in emission during the normal

[3] Many flare stars show irregular small variations with amplitudes from hundreds to a few tenths of a magnitude during the "normal" phase.

phase—at least on objective-prism plates—several have these spectroscopic features permanently present. Flare stars of dM spectral types tend to show H emission lines more than the others; examples are the dMe flare stars in the solar vicinity and the dMe rapid variables in the Hyades. During flares, all stars observed have shown strong Hα lines in emission and usually a conspicuous ultraviolet continuum appears.

Poveda (1964) found an interesting near-coincidence between Haro and Chavira's findings that the earliest known spectral type of a flare star in stellar aggregates lies near K0 and some theoretical results of Hayashi (1961, 1962; Hayashi and Nakano 1963). If the locus of the intersection of the purely convective tracks with the purely radiative ones for a range of stellar masses (curve B) is drawn on a theoretical HR diagram, the vertical line tangent to this curve corresponds to a surface temperature of 5012° K, or to a spectral type K1; according to Poveda, this would mark the earliest possible spectral type for a flare star since the flare stars should lie on the right side of curve B, where convection is complete. According to recent data, the spectral type for a surface temperature of 5050° K is K2 (Johnson 1964).

By analogy with solar flares, Poveda suggests that the cause of stellar flares is magnetic and that this, in turn, is a consequence of the interplay between convective motions in the outer layers of the stars and the differential rotation of these layers; therefore, wholly convective and rapidly rotating stars should show violent and frequent flares—the shorter the rotation period, the more frequent the flares. Also according to Poveda and based on Schatzman's (1962) theory of surface activity in stars and their loss of angular momentum with age, brighter stars above the main sequence should produce more flares per unit time than those nearer to the main sequence.

Several comments can be made on Poveda's schematic approach. The physical processes involved in production of solar flares are not well understood; so the hypothesis of similarity between the underlying mechanisms of stellar and solar flares does not solve outstanding questions about the mechanism of the stellar phenomenon (Severny 1964). There is no observational evidence that flare stars rotate more rapidly than nonflare stars of the same spectral type. Preliminary investigations by Kraft (1965) with the Palomar 200-inch coudé spectrograph at 9A/mm show that up to $v \sin i = 10$ km/sec no difference can be noticed between dMe flare stars and normal dM objects. The possibility of discovering a significant difference in rotational velocities for the dMe and dM stars is, however, not excluded if one goes to a higher dispersion than the one used by Kraft. Also, abnormally wide absorption lines have been observed in some T Tauri stars above the main sequence, but this line broadening is not necessarily due to axial rotation (Herbig 1962*c*). It would be desirable to measure the axial rotation of flare stars of the same spectral type and at different heights above the main sequence; the Orion Nebula flare stars are excellent candidates for this type of test. The frequency of flares in a given star seems to be higher for stars nearer the main sequence (see Table 9), although the absolute intensity of the outburst is related not only to the amplitude of the variation but also to the absolute brightness of the star. Flares of relatively small absolute intensity might therefore be detected in the intrinsically fainter stars and not in the bright ones. Among flare stars found in stellar aggregates and the solar vicinity, the earliest spectral type of known rapid variables is approximately K0 and thus not far from the Hayashi limit for the wholly convective stars; but if binary systems of the U Geminorum, W Ursa

Majoris, novae and related objects, and Andrews' star, in which rapid outbursts have been detected, are included, then the spectral type of these flare stars could be as early as B8, and the argument of complete or nearly complete convection as a fundamental parameter for the production of flares could not be maintained. Aside from the binary stars just mentioned (Andrews' star is not a known binary), 75 per cent of double stars, some of them close binaries, are among the well-observed dMe flare stars in the solar neighborhood; thus, the conspicuous flaring phenomenon observed in these stars might have something to do with their binary nature. It would be of interest to search spectroscopically for binary systems among flare stars in stellar aggregates. (These comments are intended only to emphasize the complexities of the problem.)

3. Apparent similarities between the nebular variables, the rapid irregular variables (dMe stars, T Tauri, RW Aurigae, and Orion variables), and the flare stars considered in this chapter are as follows:

A. There is a tendency to appear in groups, especially in young stellar aggregates in which interstellar matter is conspicuous.
B. Irregular variables and flare stars are not confined to regions containing interstellar matter, but some physical characteristics may be different for stars in clear regions from those of stars imbedded in interstellar clouds. For instance, no T Tauri stars (with the possible exception of RW Aurigae) are known outside the nebular regions, nor are flare stars of early K types. This may be an evolutionary effect rather than interaction with the nebular material.
C. Rapid irregular variables of all kinds can be found high above the main sequence, mainly in the very young aggregates. Yet, some irregular variables and flare stars also lie near or below the main sequence in the same stellar groups.
D. Typical T Tauri, RW Aurgae, and Orion variables may show the flare peculiarity; and, in turn, most flare stars which have been closely observed show a "normal" irregular variability of small amplitude.
E. All flare stars show T Tauri-like spectroscopic features during outbursts. Like most irregular variables, some show emission lines mainly of H and Ca II during minima; however, most do not.
F. The kinematic properties of irregular rapid variables, including the dMe stars and flare stars, seem to be similar.
G. The existence of flare stars can be expected in any group of rapid irregular variables, including dMe stars, and conversely.
H. The later the spectral type of the brightest flare star in a given volume of space, the fewer "normal" irregular variables or at least the smaller the amplitudes of their variations.

It seems reasonable to conclude that flare stars in the direction of different stellar aggregates and in the solar neighborhood belong to the same physical family and that diversities found among them are due to different stages of evolution.

Probably the last identifiable vestiges of irregular variability manifested by the T Tauri stars, the Orion and RW Aurigae variables, and the flare stars themselves during their first phase of evolution toward the main sequence may be found among the late dM-type flare stars of very low mass ($\leq 0.07\ \mathfrak{M}_\odot$) that, owing to internal degeneracy, cannot remain on the main sequence (Kumar 1963, 1964). If this is true, we would be unable to find any kind of rapid irregular variables in stellar groups older than about 10^9 years. This could be investigated by the study of stellar clusters such as NGC 752 or M67 (Haro and Chavira 1964).

4. In Table 9, a summary of observations of flare stars is presented. In column 1, the name of the aggregate is given; in column 2, the total number of flare stars known; in column 3, the number of stars with more than one flare; this number is also given as a percentage of the total in column 4; column 5 gives the total number of photographic exposures in each field and the effective total time of observation, when available; column 6 gives the approximate spectral range of the known variables; column 7 gives the average number of flares observed per hour; in computing the numbers in column 7, only flares occurring during the observations listed in column 5 have been considered. Column 8 gives the approximate age of the aggregates. Plates 1 and 2 show examples of flare stars in different aggregates.

The data of Table 9 are rather too limited for statistical use, but it seems evident that most flare stars belong to stellar clusters and have a nonrandom spatial distribution which follows the general pattern of the young ($\leq 10^9$ years) galactic aggregates. This conclusion is supported not only by simple statistics of the observational material involving the number of exposures, the total time of observation, and the number of flare stars found in the direction of each aggregate, but also by other considerations already presented in regard to the flare stars in the stellar groups studied.

The stellar groups of different ages listed in Table 9 suggest a spectroscopic sequence for the flare stars, starting with spectral types near K0 for the brightest objects—as in Orion and NGC 2264—passing through the Pleiades, where the earliest spectral type for the known flare stars is K5, and ending in the Hyades and in the solar vicinity with spectral types from M0 to M6. In the same direction, we find a notable decrease of the nebulous bright and dark clouds until, as in Coma, Praesepe, the Hyades, and the solar vicinity, nebulous material is practically absent. From extremely young to older aggregates, the T Tauri-like spectroscopic features of cluster members tend to diminish or disappear and the broadening of the main sequence, pronounced in very young clusters, decreases with the age of the stellar aggregates involved (Haro and Chavira 1964).

The above data indicate that a good part of the evolution of flare stars can be traced from those of a very young aggregate to those in the solar vicinity. Haro (1954, 1956) and Haro and Chavira (1955) pointed out, for the first time, that all such flare stars belong to the same physical class of variables and are members of the wide family of T Tauri-like stars. When they suggested ". . . the possibility that the flare stars belonging to T-associations of various ages disclose, by themselves, the evolutionary stage of the stellar aggregate to which they belong . . . ," it was not accepted by all astronomers. Now, a decade later, it seems that these objections have been partially removed.

REFERENCES

Ahmed, F., Lawrence, L. C., and Reddish, V. C. 1965, *Pub. R. Obs. Edinburgh*, **3**, No. 7, 187.
Ambartsumian, V. A. 1954, *Comm. Burakan Obs.*, No. 13.
———. 1957, *Non-Stable Stars* (*IAU Symposium No. 3*), ed. G. H. Herbig (London: Cambridge University Press), p. 177.
Andrews, A. D. 1964, *Contr. Armagh Obs.*, No. 46; *Irish A.J.*, **6**, No. 6, 212.

Bidelman, W. P. 1954, *Ap. J. Suppl.*, **1**, 214.
Blanco, V. 1963, *Ap. J.*, **137**, 513.
Brun, A. 1935, *Pub. Obs. Lyon*, **1**, Ser. 1, 12.
Dall'Olmo, U. 1958, *Coelum*, **26**, 3–4.
———. 1960, *ibid.*, **28**, 3–4.
———. 1961, *ibid.*, **29**, 3–4, 7–8.
Delhaye, J. 1953, *C. R. Paris*, **237**, 294.
Einasto, Ia. 1954, *Pub. Astr. Obs. Tartu*, **32**, 371.
———. 1955, *ibid.*, **33**, 57.
Giclas, H. L., Burnham, R., Jr., and Thomas, N. G. 1962, *Lowell Obs. Bull.*, **5**, No. 118, 13.
Gliese, W. 1958, *Zs. f. Ap.*, **45**, 293.
Gordon, K., and Kron, G. E. 1949, *Pub. A.S.P.*, **61**, 210.
Haro, G. 1953, *Ap. J.*, **117**, 73.
———. 1954, *Bol. Obs. Tonantzintla y Tacubaya*, **11**, 11.
———. 1956, *ibid.*, **14**, 3.
———. 1962, *Symposium on Stellar Evolution*, ed. J. Sahade (La Plata: Astr. Obs. Nat. University of La Plata), p. 37.
———. 1964, *IAU-URSI Symposium No. 20* (Canberra, 1962), eds. F. J. Kerr and A. W. Rodgers (Canberra: Australian Academy of Science), p. 30.
Haro, G., and Chavira, E. 1955, *Bol. Obs. Tonantzintla y Tacubaya*, **12**, 3.
———. 1964, *ONR Symposium, Flagstaff, Arizona.*
———. 1965, unpublished.
Haro, G., and Morgan, W. W. 1953, *Ap. J.*, **118**, 16.
Haro, G., and Terrazas, L. R. 1954, *Bol. Obs. Tonantzintla y Tacubaya*, **10**, 3.
Hayashi, C. 1961, *Pub. Astr. Soc. Japan*, **13**, 450.
———. 1962, *Prog. Theoret. Phys. Japan Suppl.*, No. 22.
Hayashi, C., and Nakano, T. 1963, *Prog. Theoret. Phys. Japan*, **30**, 460.
Herbig, G. H. 1950, *Ap. J.*, **111**, 15.
———. 1954, *ibid.*, **119**, 483.
———. 1956, *Pub. A.S.P.*, **68**, 531.
———. 1959, private communication.
———. 1962*a*, *Ap. J.*, **135**, 736.
———. 1962*b*, *Symposium on Stellar Evolution*, ed. J. Sahade (La Plata: Astr. Obs. Nat. University of La Plata), p. 45.
———. 1962*c*, *Advances in Astronomy and Astrophysics*, ed. Z. Kopal (New York: Academic Press, Inc.), **1**, 47.
Hertzsprung, E. 1924, *B.A.N.*, **2**, 87.
Hertzsprung, E., Sanders, C., Kooreman, C. J., *et al.* 1947, *Ann. Leiden Obs.*, **19**, No. 1A.
Hoffleit, D. 1952, *Harvard Bull.*, No. 921, p. 5.
Holmberg, E. 1944, *Lund. Medd.*, Ser. 2, No. 113.
Johnson, H. L. 1964, *Bol. Obs. Tonantzintla y Tacubaya*, **3**, No. 25, 305.
Johnson, H. L., and Mitchell, R. I. 1958, *Ap. J.*, **128**, 31.
Johnson, H. L., Mitchell, R. I., and Iriarte, B. 1962, *Ap. J.*, **136**, 75.
Johnson, H. L., and Morgan, W. W. 1953, *Ap. J.*, **117**, 313.
Joy, A. H. 1947, *Ap. J.*, **105**, 96.
———. 1949, *ibid.*, **110**, 424.
———. 1957, *Non-Stable Stars* (IAU Symposium No. 3), ed. G. H. Herbig (London: Cambridge University Press), p. 31.
———. 1960, *Stars and Stellar Systems*, ed. J. L. Greenstein (Chicago: University of Chicago Press), Vol. 6, 653.
Joy, A. H., and Humason, M. L. 1949, *Pub. A.S.P.*, **61**, 133.
Kamp, P. van de, and Lippincott, S. L. 1951, *Pub. A.S.P.*, **63**, 141.
Klein-Wassink, W. J. 1927, *Pub. Kapteyn Ap. Obs. Groningen*, No. 41.
Kraft, R. P. 1965, private communication.
Kron, G. E., Gascoigne, S. C. B., and White, H. S. 1957, *A.J.*, **62**, 205.
Kuhi, L. V. 1964, *Pub. A.S.P.*, **76**, 430.
Kumar, S. S. 1963, *Ap. J.*, **137**, 1121, 1126.
———. 1964, *Observatory*, **84**, No. 938, 18.
Lovell, B. 1964, *Observatory*, **84**, No. 942, 191.
Luyten, W. J. 1925, *Harvard Bull.*, No. 830.
———. 1926, *ibid.*, No. 835.
———. 1949, *Ap. J.*, **109**, 532.
Maanen, A. van. 1940, *Ap. J.*, **91**, 503.
———. 1945*a*, *Pub. A.S.P.*, **57**, 216.
———. 1945*b*, *Ap. J.*, **102**, 26.
Maffei, P. 1963, *Contr. Asiago Obs.*, No. 136.

Oskanjan, V. 1964, *Pub. Obs. Beograd*, No. 10.
Parenago, P. 1954, *Pub. Sternberg Astr. Inst.*, p. 25.
Petit, M. 1954, *Ciel et Terre*, **70**, 407.
———. 1958, *Contr. Asiago Obs.*, No. 95, p. 29.
Popper, D. M. 1953, *Pub. A.S.P.*, **65**, 278.
Poveda, A. 1964, *Nature*, **202**, No. 4939, 1319.
Roques, P. 1953, *Pub. A.S.P.*, **65**, 19.
———. 1954, *ibid.*, **66**, 256.
———. 1958, *ibid.*, **70**, 310.
Rosino, L. 1956, *Contr. Asiago Obs.*, No. 69.
———. 1965, private communication.
Rosino, L., and Cian, A. 1962, *Contr. Asiago Obs.*, No. 125.
Rosino, L., Grubissich, C., and Maffei, P. 1957, *Contr. Asiago Obs.*, No. 82.
Sandig, H.-U. 1951, *A.N.*, **280**, 39.
Schatzman, E. 1962, *Ann. d'ap.*, **25**, 18.
Severny, A. B. 1964, *Ann. Rev. Astr. Ap.*, **2**, 363.
Shapley, H. 1951, *Proc. Nat. Acad. Sci.*, **37**, 15.
Thackeray, A. D. 1950, *M.N.*, **110**, 45.
Vyssotsky, A. N., and Dyer, E. R. 1957, *Ap. J.*, **125**, 297.
Wachmann, A. A. 1939, *A.N.*, **21**, 25.
Wagman, N. E. 1953, *Harvard Ann. Card*, No. 1225.
Walker, M. F. 1956, *Ap. J. Suppl.*, **2**, 365.
———. 1957, *Non-Stable Stars* (IAU Symposium No. 3), ed. G. H. Herbig (London: Cambridge University Press), p. 46.

CHAPTER 5

Interstellar Extinction

H. L. JOHNSON

Lunar and Planetary Laboratory, University of Arizona, Tucson, Arizona

1. INTRODUCTION

SEVERAL methods of investigating interstellar extinction have been employed. Many years ago, Trumpler (1930) compared the diameters of galactic clusters with their distances determined from spectral types and photographic photometry. He noticed a systematic increase in the linear diameters of clusters with increasing distance as determined from the spectral types and photometry. From this observation, Trumpler deduced the existence of interstellar extinction. The method of estimating cluster distances from their apparent diameters is an effective tool in our investigation of interstellar extinction, because the apparent angular diameter of clusters, being in principle a linear function of only the true distance, should be independent of the amount of extinction.

Another important method is the variable-extinction method, which requires two-color photometric data and a means of estimating the absolute magnitudes and intrinsic colors of individual stars. The principle behind the method is illustrated by the following example. Suppose that we have observed a cluster of stars (membership in the cluster implies that the stars are all at the same distance from the Earth) and that all stars of the cluster have exactly the same intrinsic color and absolute magnitude. Let us suppose further that the reddening and extinction are variable across the face of the cluster. Then a graph of the observed magnitudes versus the observed colors of the stars exhibits a straight-line relationship, the slope of which is the ratio of total to selective extinction; in the *UBV* system, this ratio is $R = A_v/E_{B-V}$. The slope of the relationship between these two quantities is the ratio R, and the value of the ordinate for $E_{B-V} = 0$ is the true distance modulus of the cluster.

It is important to emphasize, however, that the data derived from the variable-extinction method apply *only to the component of the extinction that is variable across the cluster*. If there is additional extinction between the cluster and the Earth, we obtain from the analysis no information about this additional extinction and, in fact, we cannot even detect its existence. Thus, the extinction from the variable-extinction method is the minimum value for the extinction between the cluster and the Earth, and the value of R for the total extinction may be either greater or less than that derived, depending upon the characteristics of the undetected interstellar material. The first method, which depends upon cluster distances estimated from their angular diameters, does allow, at least in principle, evaluation of the total interstellar extinction between the clusters and the Earth.

The last method of investigating interstellar extinction that we will discuss here is the color-difference method. In its simplest form this is the comparison of the observed colors over as wide a range of wavelengths as possible of two identical stars, one of which is reddened and one which is not. In practice we first derive the intrinsic colors of all types of stars from observations of unreddened and little-reddened stars. Then we compare the colors of reddened stars with the derived intrinsic colors. The color-difference method permits us to determine the variation of interstellar extinction with wavelength but, since we almost never know with sufficient precision the true distances and absolute magnitudes of the stars, the deduction of the total extinction requires extrapolation to infinite wavelength ($1/\lambda = 0$) from the longest observed wavelength. In contrast to the variable-extinction method, the color-difference method produces data that apply to the total mass of interstellar material between the reddened stars and Earth.

Three methods of investigating the characteristics of interstellar extinction are considered here. Each provides only a part of the information we desire. We shall see, however, that combination of the results from the several methods provides us with fairly complete information about interstellar reddening and extinction.

2. PREVIOUS WORK AND RESULTS

Probably the earliest convincing demonstration of the existence of interstellar extinction is that of Trumpler (1930), who compared the distances of clusters estimated (without extinction corrections) from spectral types and photographic photometry, with distances estimated from apparent diameters of clusters. The comparison showed a systematic increase of cluster linear diameters with distance, unless a correction of about $0\overset{m}{.}7$/kpc was made to the photometric-spectroscopic data. Trumpler interpreted this finding as indicating the existence of interstellar extinction. He also showed that this extinction is accompanied by reddening.

Many investigations of reddening and extinction were made following Trumpler's publication. Among these is the work of Stebbins, Huffer, and Whitford (1940) on two-color photometry of B stars. They were able to make a crude estimate of the ratio of total-to-selective extinction, corresponding to approximately $R = 3.5$ on the *UBV* system.

Stebbins and Whitford's (1943, 1945) six-color photometry extended the range of wavelengths to about 1.0 μ. Among their results was the confirmation of the peculiar reddening in the Orion Nebula (Sword) region, first noticed by Baade and Minkowski (1937*b*). Whitford (1948) extended the range of wavelengths to 2.0 μ in the infrared. These multicolor data, ranging in wavelength from 0.35 μ to 2.0 μ, were interpreted by van de Hulst (1949) in terms of the physical characteristics of the particles producing the interstellar extinction. For example, he showed that the peculiar reddening in the Orion Trapezium (Sword) region can be explained as due to particles about 30 per cent larger than "normal."

Three years later, Sharpless (1952) derived, by the variable-extinction method, a value of $R = A_v/E_{B-V} = 6$ for the Orion Sword region. This determination was relatively weak, and other interpretations were possible. Divan (1954) made a spectrophotometric investigation of the interstellar reddening in the Orion and other regions over the range of wavelengths from 0.35 μ to 0.6 μ; the reddening curve was found to be essentially the same as in Perseus. Sharpless (1954) confirmed this result with *UBV* photometry.

Whitford (1958) reviewed the situation, adding several observations in the infrared and scanning-spectrometer data for shorter wavelengths. At that time, a value of $R \sim 3$

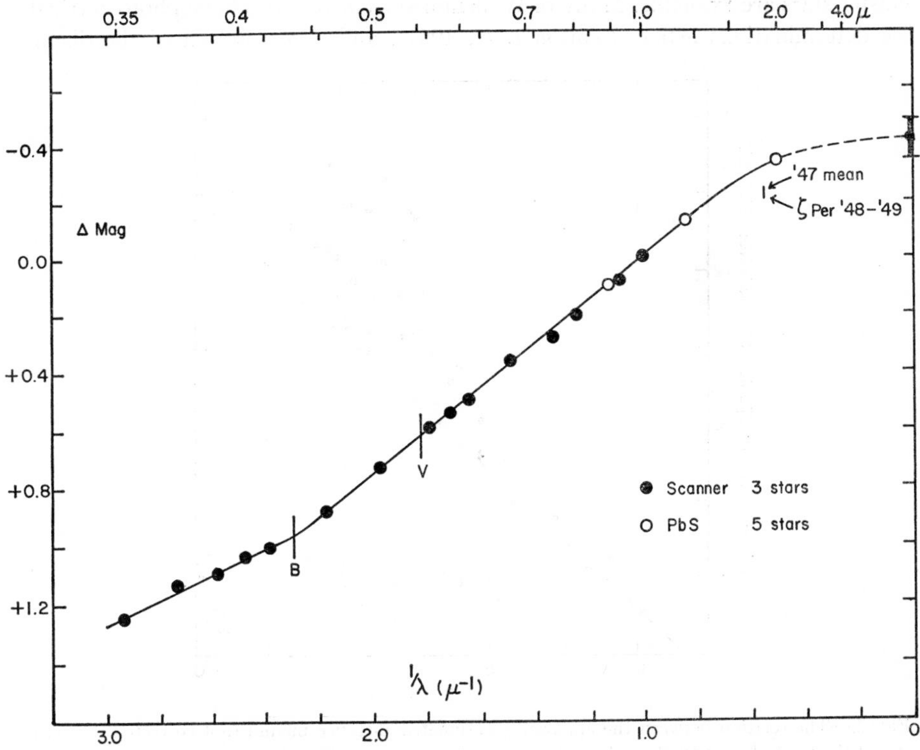

FIG. 1.—The interstellar-extinction law according to Whitford (1958)

seemed appropriate, with the ambiguous result for the Orion Sword the only outstanding discrepancy. Borgman's (1961) seven-color medium–narrow-band photometry confirmed the character of Whitford's curve in the *UBV* region. Whitford's graph of the law of interstellar extinction (Fig. 1) adequately summarizes our knowledge of the law of extinction as of the year 1961. We note, however, that Johnson and Morgan (1954) and Borgman (1961) found, even in the *UBV* spectral region, distinct variations in the reddening law.

Some of the earliest results of the infrared photometric program now being carried out at the University of Arizona were described by Johnson and Borgman (1963). Their interpretation of the new data confirmed the high value of R found by Sharpless for the Orion Sword region and also indicated significant variations of R with galactic longitude.

Later Arizona photometric work (Johnson 1965*a*) indicated a value of $R = 5.5$ in Cepheus. In another paper, Johnson (1965*b*) confirmed this result for Cepheus by the variable-extinction method and, from a combination of the variable-extinction and color-difference methods, derived $R = 6$ for NGC 2244 and the Orion Sword region.

3. THE CLUSTER-DIAMETER METHOD

In addition to Trumpler's (1930) cluster-diameter estimates, determinations by Wallenquist (1959) of cluster diameters from an analysis of the numbers and distributions of cluster stars are available. Many of Wallenquist's clusters also have photometric distance determinations listed (Johnson *et al.* 1961), and a comparison of the distance

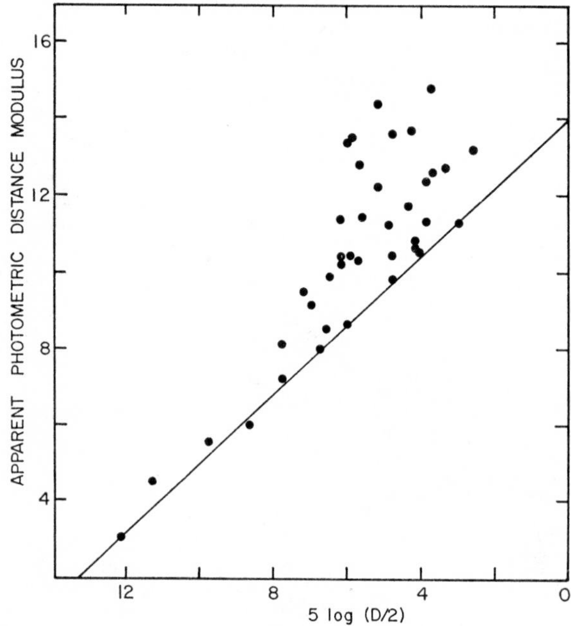

FIG. 2.—The relation between the apparent photometric distance moduli (not corrected for extinction) and the logarithm of Wallenquist's cluster diameters.

moduli from the two sources is illuminating. Figure 2 shows the relationship between apparent photometric distance moduli (not corrected for extinction) and the logarithm of Wallenquist's diameters. The scattering above the diagonal line clearly is due to interstellar extinction between the Earth and the more distant clusters. The height of a cluster point above the line is proportional to the extinction A_v between that cluster and the Earth.

Knowing the color excess E_{B-V} for each cluster from photometry, we plotted Figure 3, A_v versus E_{B-V}. The two lines in the figure correspond to $R = 3$ and $R = 6$. Note that a straight-line relation between A_v and E_{B-V} passes through the origin; the line in Figure 2 has been drawn to achieve this result, on the average, at all cluster diameters. We have, therefore, made the assumption that $A_v = 0$, if $E_{B-V} = 0$, and vice versa. If Wallenquist's diameters were entirely independent of distance, the diagonal line in Figure 2 would have an angle of slope of 45°; its deviation from this slope discloses that Wallenquist's diameters become systematically smaller with increase of cluster distance. The error is small and easily accounted for in the analysis.

Cluster-distance moduli were computed from Wallenquist's diameters by the equation

$$m - M = 14.0 - 4.5 \log \left(\frac{D}{2}\right),$$

which is the equation of the diagonal line in Figure 2. The resultant moduli are listed in the third column of Table 1. A similar analysis of Trumpler's (1930) diameters yielded the moduli in the fourth column of Table 1. The fifth column gives the mean of the two

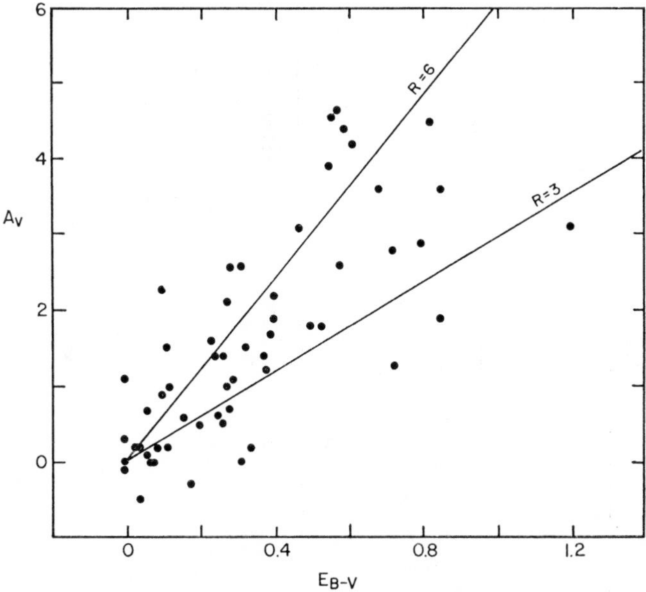

FIG. 3.—The relation between A_v and E_{B-V} for the clusters plotted in Figure 2

diameter modulus determinations; the sixth, the apparent photometric moduli (not corrected for extinction); the seventh, the color-excess E_{B-V}; the eighth, the extinctions A_v, defined as the difference between the photometric and diameter moduli. The ninth and tenth columns give the photometric distance moduli corrected for $R = 3.0$ and the difference between these corrected moduli and the diameter moduli. These differences, from the last column of Table 1, are plotted versus galactic longitude in Figure 4. A strong variation with longitude is present, and the extinction is higher than predicted from $R = 3.0$ in the region $100° < l^{II} < 220°$.

Part of the scatter in Figure 4 for $100° < l^{II} < 220°$ is due to the fact that the reddening of the clusters ranges from nearly zero to moderately high values. There is, therefore, a correlation of $\Delta(m - M)$ with E_{B-V}, a fact that is better shown in Figure 5. This figure is similar to Figure 3 except that only clusters for which $100° < l^{II} < 220°$ are plotted. The data for this part of the Milky Way plainly indicate a value of $R = A_v/E_{B-V} = 6$. Cer-

TABLE 1

COMPARISON OF CLUSTER-DISTANCE MODULI

Cluster	ℓ^{II}	m-M (dia)			m-M (App. Phot.)	E_{B-V}	A_v	m-M (R=3.0 Phot.)	Δ(m-M) Col 9-Col 5
		Wallenquist	Trumpler	Mean					
NGC 457	126°	10.6	10.6	10.6	13.7	0.47	3.1	12.3	+1.7
581	128	11.7	11.9	11.8	13.2	0.37	1.4	12.1	+0.3
663	129	9.8	10.4	10.0	14.4	0.85	4.1	11.8	+1.8
752	137	7.9	7.7	7.8	8.0	0.03	0.2	7.9	+0.1
869	133	8.7	8.8	8.8	13.5	0.56	4.7	11.8	+3.0
884	134	8.7	8.8	8.8	13.5	0.56	4.7	11.8	+3.0
IC 1805	133	----	9.6	9.6	14.1	0.82	4.5	11.6	+2.0
NGC 957	136	----	11.2	11.2	14.1	0.80	2.9	11.7	+0.5
1027	136	----	9.5	9.5	11.7	0.40	2.2	10.5	+1.0
1039	144	8.1	8.6	8.3	8.5	0.09	0.2	8.2	-0.1
IC 1848	137	----	9.3	9.3	13.5	0.61	4.2	11.7	+2.4
NGC 1245	146	----	11.9	11.9	12.6	0.28	0.7	11.8	-0.1
α Per	147	----	4.1	4.1	6.4	0.10	2.3	6.1	+2.0
Pleiades	167	5.2	5.6	5.4	5.6	0.04	0.2	5.5	+0.1
NGC 1528	152	----	9.3	9.3	10.4	0.29	1.1	9.5	+0.2
Hyades	179	3.0	3.0	3.0	3.0	0.00	0.0	3.0	0.0
NGC 1647	180	8.1	8.3	8.2	9.9	0.39	1.7	8.7	+0.5
1664	161	10.2	10.4	10.3	10.8	0.20	0.5	10.2	-0.1
1893	174	10.6	10.1	10.4	14.8	0.59	4.4	13.0	+2.6
1912	172	9.0	9.7	9.4	11.5	0.27	2.1	10.6	+1.2
NGC 1960	174°	9.6	10.0	9.8	11.2	0.24	1.4	10.5	+0.7
2099	177	8.5	9.1	8.8	11.4	0.31	2.6	10.5	+1.7
2158*	186	11.8	12.7	12.2	14.7	0.43	2.5	13.4	+1.2
2168	186	8.7	8.8	8.8	10.4	0.23	1.6	9.7	+0.9
2244	206	8.9	8.9	8.9	12.8	0.55	3.9	11.2	+2.3
2287	231	9.0	8.5	8.7	9.1	0.00	0.3	9.1	+0.4
2301	212	----	10.1	10.1	9.6	0.04	-0.5	9.5	-0.6
2323	221	10.2	10.0	10.1	10.6	0.26	0.5	9.8	-0.3
2324	213	10.9	11.5	11.2	12.7	0.11	1.5	12.4	+1.2
2353	224	----	9.5	9.5	10.5	0.12	1.0	10.1	+0.6
2362	238	11.3	(10.8)†	11.1	11.2	0.11	0.2	10.9	-0.2
2422	231	8.6	8.6	8.6	8.6	0.08	0.0	8.4	-0.2
2439	246	----	11.2	11.2	11.8	0.25	0.6	11.0	-0.2
2447	240	9.7	9.7	9.7	10.4	0.06	0.7	10.2	+0.5
2516	273	7.0	7.5	7.2	8.1	0.10	0.9	7.8	+0.6
2632	205	6.1	6.2	6.1	6.0	0.00	-0.1	6.0	-0.1
2682	216	9.7	9.7	9.7	9.8	0.06	0.1	9.6	-0.1
3330	284	----	11.5	11.5	11.2	0.18	-0.3	10.7	-0.8
4103	287	----	11.2	11.2	11.4	0.34	0.2	10.4	-0.8
Coma	228	3.8	(2.7)†	3.4	4.5	0.00	1.1	4.5	+1.1
NGC 4755	303	10.4	10.6	10.5	10.5	0.31	0.0	9.6	-0.9
6405	356	8.9	8.9	8.9	9.5	0.16	0.6	9.0	+0.1
6475	356	6.9	7.5	7.2	7.2	0.08	0.0	7.0	-0.2
6494	10	8.4	8.8	8.6	10.2	0.38	1.2	9.1	+0.5
6530	6	10.1	10.3	10.2	11.7	0.32	1.5	10.7	+0.5
NGC 6531	7°	10.5	(10.1)	10.3	11.3	0.27	1.0	10.4	+0.1
IC 4725	13	8.6	8.5	8.6	10.4	0.50	1.8	8.9	+0.3
NGC 6611	17	----	12.7	12.7	14.6	0.85	1.9	12.0	-0.7
6694	24	10.7	11.2	11.0	12.6	0.58	2.6	10.9	-0.1
6705	27	10.5	10.5	10.5	12.4	0.40	1.9	11.2	+0.7
Tr 35	28	----	12.5	12.5	15.6	1.20	3.1	12.0	-0.5
NGC 6830	50	----	11.0	11.0	12.8	0.53	1.8	11.2	+0.2
6834	65	----	11.8	11.8	14.6	0.72	2.8	12.4	+0.6
6940	70	8.9	8.9	8.9	10.3	0.26	1.4	9.5	+0.6
7086	94	----	11.6	11.6	12.9	0.72	1.3	10.7	-0.9
7654	112	9.7	10.4	10.0	13.6	0.68	3.6	11.6	+1.6
7789	115	9.4	9.7	9.6	12.2	0.28	2.6	11.4	+1.8

* Data for NGC 2158 from Arp and Cuffey (1962).

† Half weight because classified "poor" by Trumpler.

tainly, $R = 3$ is incompatible with the plotted points. On the other hand, Figure 6, which contains the remainder of the clusters, indicates $R = 3$ for other regions of the Milky Way.

The α Perseus cluster is within the range of galactic longitude of Figure 5. Its geometrical distance modulus is 5.4 mag, according to Morgan and Roman (1950), and 5.5 mag, according to Blaauw (1956); the photometric modulus is 6.1 mag (Mitchell 1960). The photometric modulus exceeds the geometric modulus by 0.6 or 0.7 mag. The diameter distance (Table 1) is also smaller than the photometric distance. Petrie (Eggen and

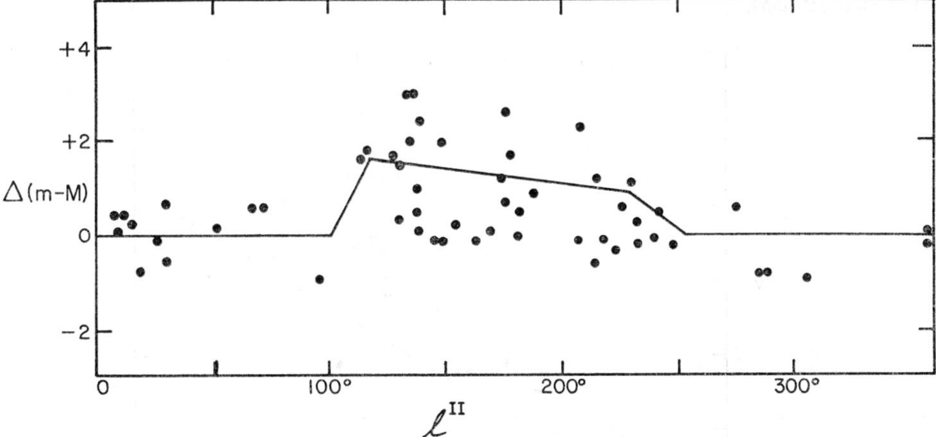

FIG. 4.—The differences between photometric distance modulus ($R = 3.0$) and diametric distance moduli $\Delta(m - M)$ for clusters at various galactic longitudes.

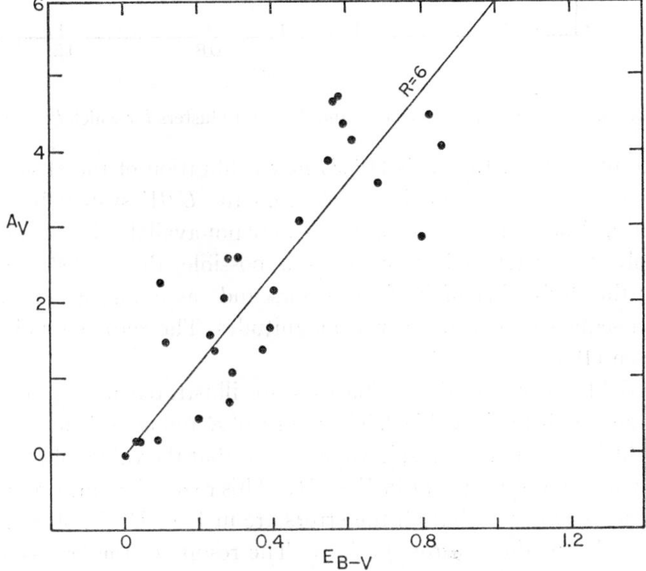

FIG. 5.—The relation between A_v and E_{B-V} for clusters for which $100° < l^{II} < 220°$

Herbig 1964, p. 51) called attention to the fact that a similar situation exists for the Pleiades. From the data listed by Petrie, the distance modulus of the Pleiades from trigonometric and dynamical parallaxes is about 0.9 mag smaller than that from photometric and spectroscopic data.

4. THE VARIABLE-EXTINCTION METHOD

Sufficient data on the *UBV* system are now available for 16 applications of the variable-extinction method to 14 different regions of the sky. The results for some of these regions have been discussed by Johnson and Hiltner (1956), Whitford (1958), and Johnson (1965*b*).

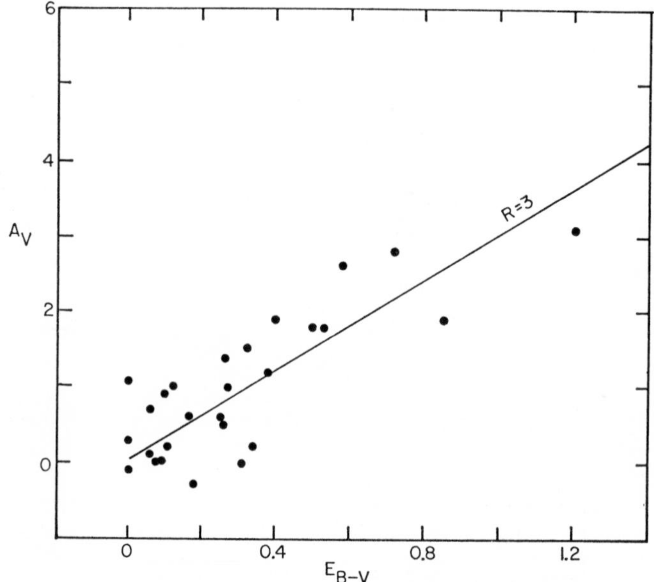

FIG. 6.—The relation between A_v and E_{B-V} for clusters for which $l^{II} < 100°$

In the present chapter, Blaauw's (1963) new calibration of the absolute magnitudes of the MKK system and the intrinsic colors on the *UBV* system by Johnson (1958, 1963) were used. Where MKK spectral types were not available but where identification of an "unevolved" cluster main sequence was possible, three-color *UBV* photometry was used for the derivation of intrinsic colors and (assuming all stars to lie on the zero-age main sequence) absolute visual magnitudes. The zero-age main sequence was that of Johnson (1963).

The 16 variable-extinction determinations are illustrated in Figures 7–22, and the numerical results are listed in Table 2. The values of R and $m - M$ are based upon least-squares straight-line fits to the data; it was assumed that the values of E_{B-V} are exact and that all observational errors appear in $V - M_V$. This regression line provides the smaller value of R. The assumption that all the errors are in $V - M_V$ is, of course, not correct since there are observational errors in E_{B-V}. The results of the least-squares solutions were therefore adjusted slightly toward the other regression line (all errors in E_{B-V}) in

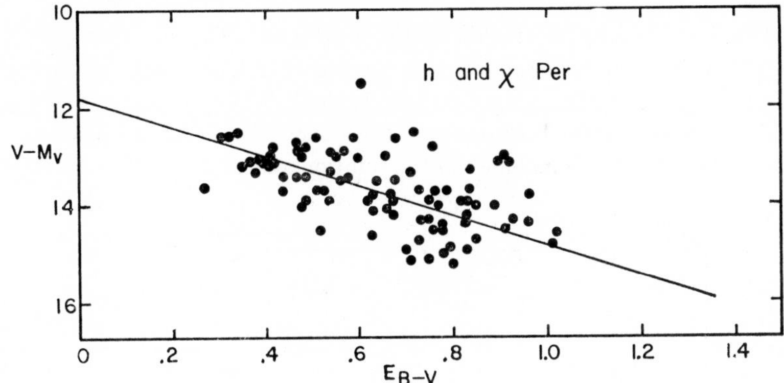

FIG. 7.—The variable-extinction diagram for h and χ Persei

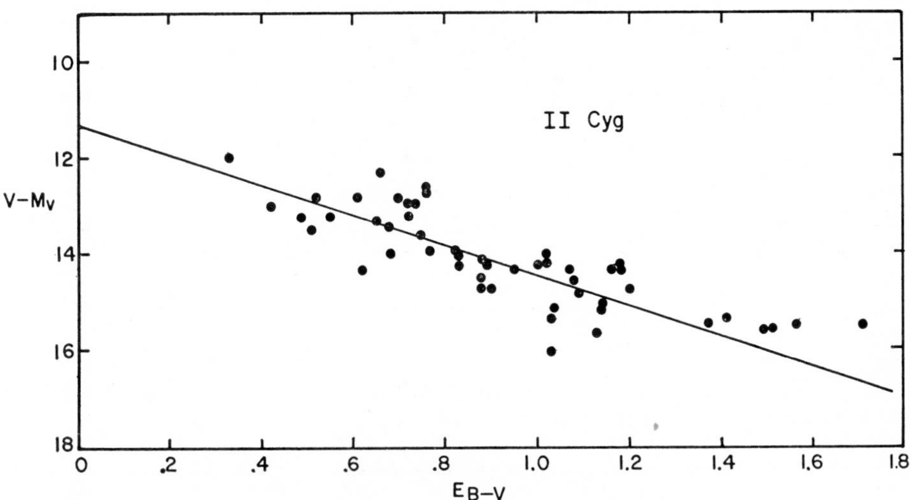

FIG. 8.—The variable-extinction diagram for II Cyg

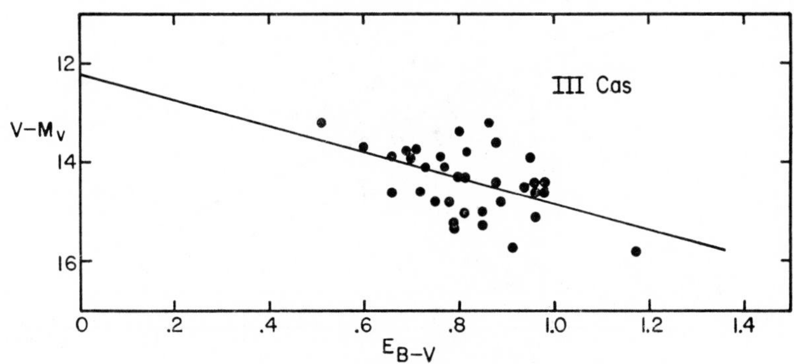

FIG. 9.—The variable-extinction diagram for III Cas

order to take into account the true distributions of errors. The quality of the values of R and $m - M$ in Table 2 can be judged from Figures 7–22. The sources of the data used in the computations are listed in the notes to Table 2. When Hiltner's (1956) data were used, all his stars within the boundaries of the association as defined by Morgan, Whitford, and Code (1953) were included in the solution and thus were assumed to be members.

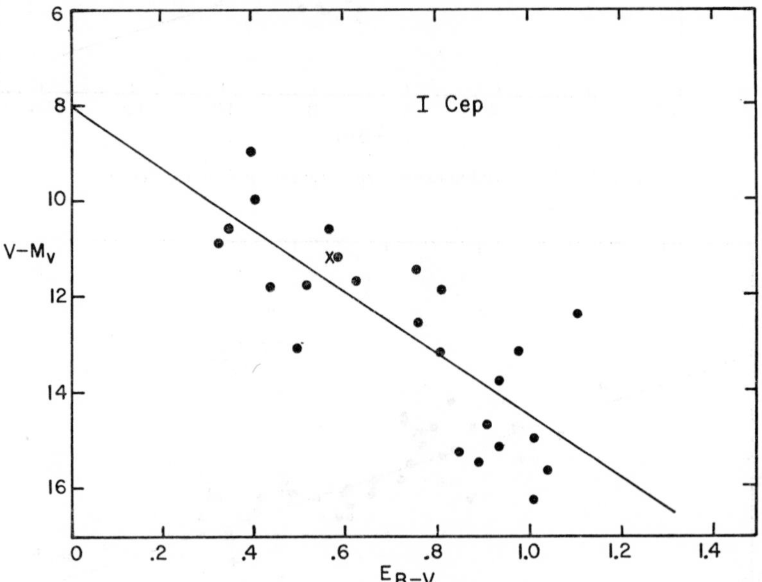

Fig. 10.—The variable-extinction diagram for I Cep. The cross represents μ Cep, on the assumption that it is a member of the association.

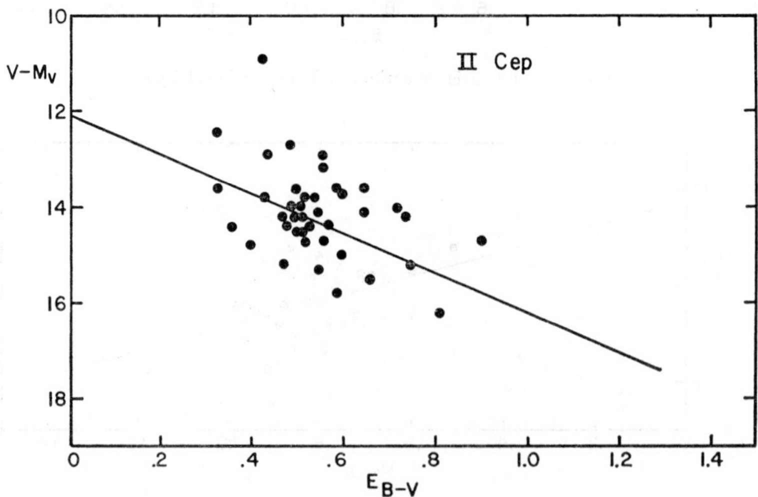

Fig. 11.—The variable-extinction diagram for II Cep

These computations were made under the assumption that all of the stars of a cluster or association are at exactly the same distance from the Earth. As mentioned by Sharpless (1952) and analyzed further by Walker (1962), the finite depth of a cluster can cause the variable-extinction method to produce a value of R that is larger than the true value. If the interstellar material that produces the extinction is intimately associated with the cluster stars, then the more distant stars are affected by more extinction and reddening

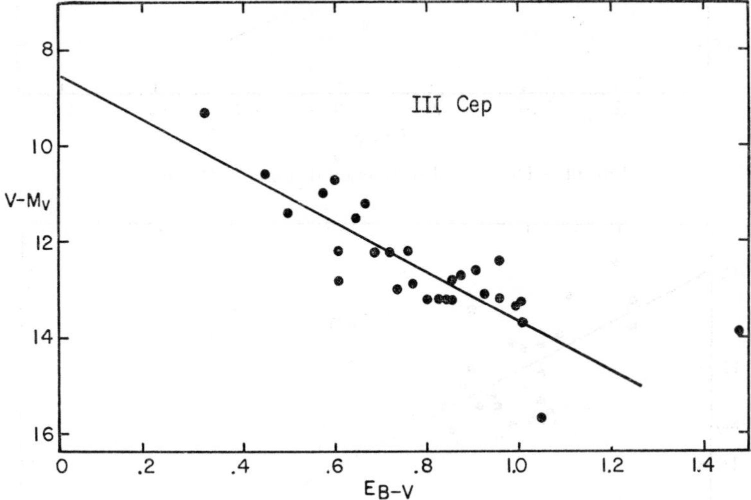

FIG. 12.—The variable-extinction diagram for III Cep. M_v was determined from MKK types and Blaauw's calibration (compare Fig. 13).

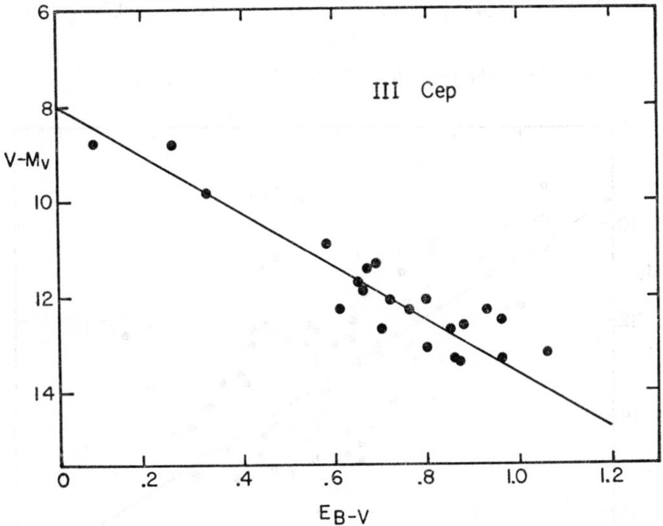

FIG. 13.—The variable-extinction diagram for III Cep. M_v was taken from Borgman and Blaauw's calibration of Borgman's seven-color photometry (compare Fig. 12).

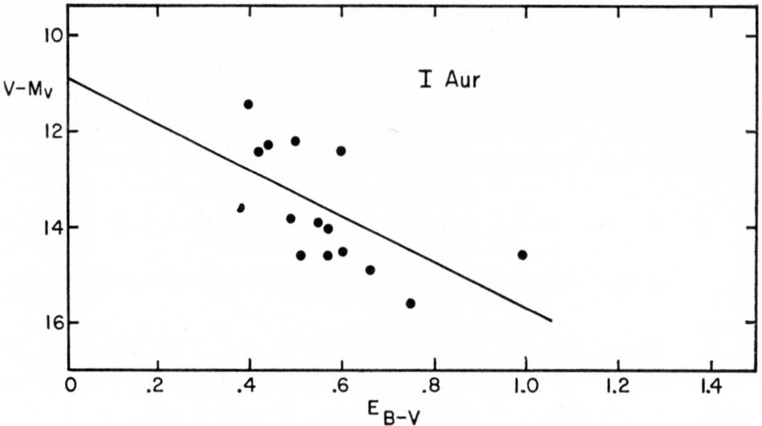

FIG. 14.—The variable-extinction diagram for I Aur

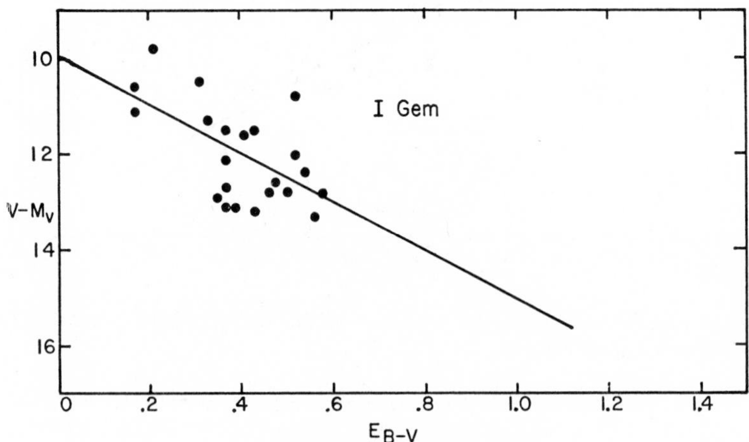

FIG. 15.—The variable-extinction diagram for I Gem

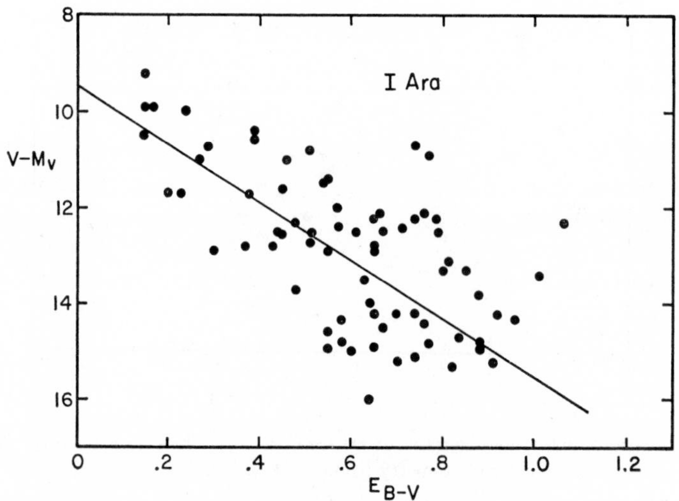

FIG. 16.—The variable-extinction diagram for I Ara

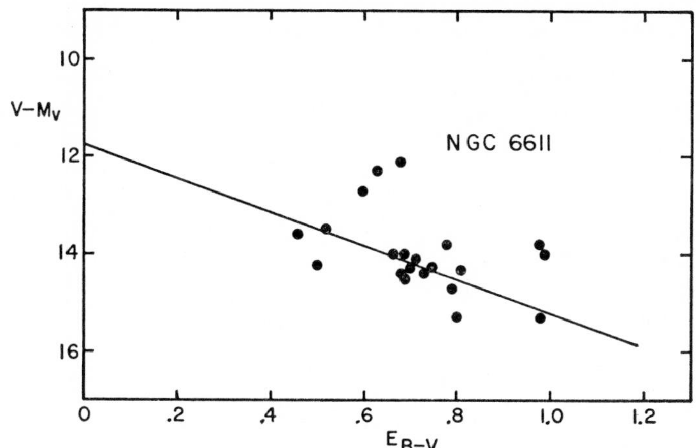

FIG. 17.—The variable-extinction diagram for NGC 6611

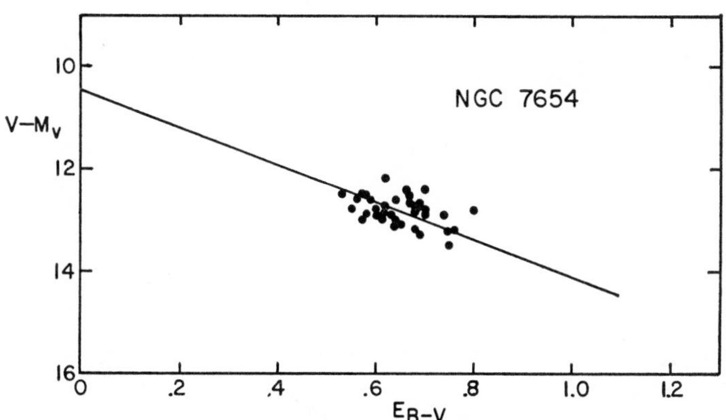

FIG. 18.—The variable-extinction diagram for NGC 7654

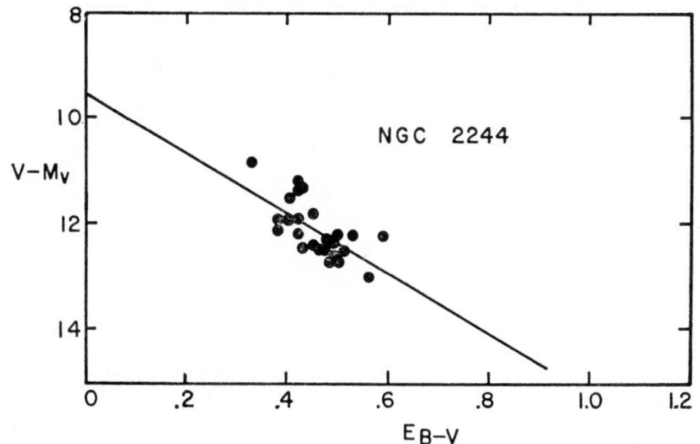

FIG. 19.—The variable-extinction diagram for NGC 2244

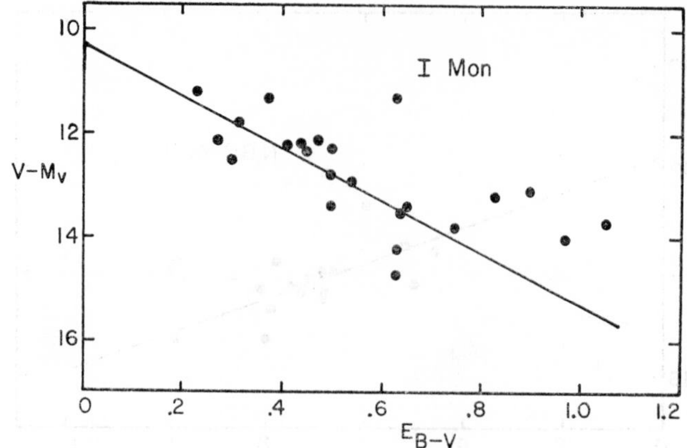

FIG. 20.—The variable-extinction diagram for I Mon

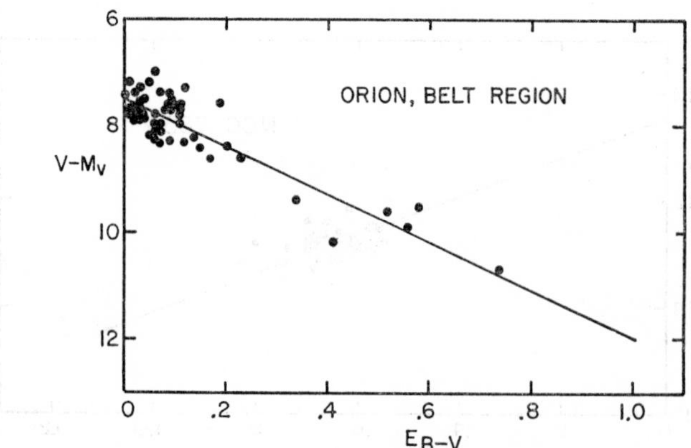

FIG. 21.—The variable-extinction diagram for the Orion Belt region

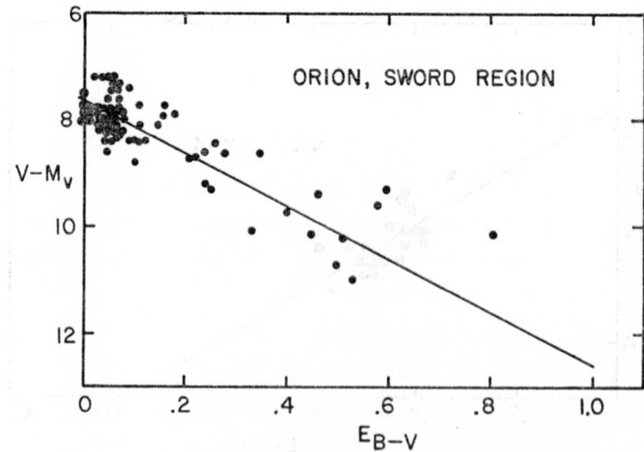

FIG. 22.—The variable-extinction diagram for the Orion Sword (Trapezium) region

than are the ones on the near side. In this case, the more distant stars are fainter because of both interstellar extinction and greater distance; since reddening is correlated only with extinction, not with distance, the variable-extinction method yields a spuriously large value of *R*. On the other hand, if the interstellar material is not connected with the cluster (being between it and the Earth), there is no systematic error in the computation.

As Walker (1962) showed, if one is willing to allow sufficient depth, the large values of *R* that are obtained by the variable-extinction method for some clusters and associations can be reconciled with $R = 3$. Sharpless' value of $R = 6$ for Orion is consistent

TABLE 2

RESULTS FROM THE VARIABLE-EXTINCTION METHOD

Cluster	Fig.	R p.e.		m-M p.e.		References UBV	MK
h, χ Per	7	3.0	± 0.3	11.8	± 0.2	1,2	1,2
II Cyg	8	3.1	0.3	11.3	0.3	3	3
III Cas	9	2.6	0.5	12.2	0.4	3	3
I Cep	10	6.5	0.7	8.0	0.6	3	3
II Cep	11	4.1	0.9	12.1	0.4	3	3
III Cep	12	5.2	0.4	8.5	0.3	4	4
III Cep	13	5.6	0.4	8.0	0.3	4	5
I Aur	14	4.8	1.2	10.9	0.7	3	3
I Gem	15	5.1	1.0	10.0	0.5	6	6
I Ara	16	6.0	0.5	9.5	0.3	12	12
NGC 6611	17	3.4	0.7	11.8	0.4	7	8
NGC 7654	18	3.6	0.6	10.5	0.3	9	8
NGC 2244	19	5.7	0.5	9.5	0.3	10	8
I Mon	20	5.0	0.5	10.3	0.4	3	3
Orion Belt	21	4.5	0.3	7.5	0.2	11	8
Orion Sword	22	5.0	± 0.3	7.7	± 0.2	11	8

References:

1. Johnson and Morgan (1954).
2. Johnson and Hiltner (1956).
3. Hiltner (1956).
4. Blaauw, Hiltner, and Johnson (1959).
5. M_v from Borgman and Blaauw (1964).
6. Crawford *et al.* (1955).
7. Walker (1961).
8. M_v from zero-age main sequence (Johnson 1963).
9. Pesch (1960).
10. Johnson (1962).
11. Sharpless (1952, 1954, 1962).
12. Whiteoak (1963).

with $R = 3$, if the depth of the cluster in the line of sight is about 200 pc. Hardie, Heiser, and Tolbert (1964) found a depth of 170 pc for the Orion Belt region, on the assumption that $R = 3$. But these depths, 170 and 200 pc, are nearly one-half the distance to the association, and are five times its linear diameter perpendicular to the line of sight. Walker also discussed the association III Cep, for which a depth of 400 pc (ten times the linear diameter perpendicular to the line of sight) is needed to produce $R = 3$. Similar extreme situations exist for I Ara (cf. Johnson 1965*b*), and other associations and clusters discussed here.

Petrie (Eggen and Herbig 1964, pp. 31–34) discussed his radial velocity and absolute magnitude data for I Lac, I Aur, and the double cluster. He assumed $R = 3$ and found three long "cigars" of stars, pointed at the Earth; he also commented that "the spread in distance cannot be explained as observational scatter, unless there is a gross error involved in assuming the extinction to be three times the colour excess." Petrie's data for the double cluster are especially interesting for they indicate that we are "looking down a pipe 100 parsecs wide and about 2000 parsecs long." The radial velocities "are all nearly the same, ranging from -41 ± 2 for the nearest [star] and -39 ± 2 km/sec for the farthest." As Blaauw commented at the time (Eggen and Herbig 1964, p. 34), the velocities can be taken to mean that the stars are actually all the same distance.

It is possible to assume that none of these associations is a true physically connected unit, but that we are merely looking through holes in the interstellar extinction. It is extremely difficult to accept this hypothesis for most of the associations; this is especially true for the double cluster and its radial-velocity data. It seems quite improbable that a large fraction of associations have their stars distributed in long, thin cigars pointed at the Earth. If all associations and clusters are spherical, in our most extreme case the value of R obtained from our procedure is too large by about 0.3; the average error is about 0.1, a negligible quantity.

TABLE 3

VALUES OF *R* IN CEPHEUS

Cluster	R	p.e.	m-M	p.e.	Method
I Cep	6.5	± 0.7	8.0	± 0.6	var.ext.
II Cep	4.1	± 0.9	12.1	± 0.4	var.ext.
III Cep	5.2	± 0.4	8.5	± 0.3	var.ext.
III Cep	5.6	± 0.4	8.0	± 0.3	var.ext.
NGC 7086	(1.8)	---	(11.6)	---	cl. dia.
NGC 7654	5.3	---	10.0	---	cl. dia.
NGC 7654	3.6	± 0.6	10.5	± 0.3	var.ext.
Mean	4.8	± 0.3			

The values of R for h and χ Persei (NGC 869 and 884) and for II Cyg are essentially the same as have been found before for these regions. III Cas is near the double cluster in the sky, and it also has about the same R. These clusters and associations yield the "normal" value, $R \sim 3$.

Table 2 shows three associations in Cepheus, and their data are collected in Table 3. There were two semi-independent solutions for III Cep. One used the MKK types of Blaauw, Hiltner, and Johnson (1959) and Blaauw's (1963) MKK calibration; the other used the values of M_v derived by Borgman and Blaauw (1964) from Borgman's seven-color photometry. The absolute-magnitude calibrations are, therefore, separate; furthermore, only some of the III Cep stars are common to both solutions. There are three more determinations of R in Cepheus, and seven values are listed in Table 3. The adopted mean value for Cepheus is 4.8 ± 0.3 (p.e.).

The cluster NGC 2244 appears to be a local condensation in the association I Mon and, as a result, we have two variable-extinction determinations for this region; these are shown in Table 2. In addition, Table 1 lists values of $m - M$ and A_v determined from the diameter of NGC 2244. All these data for I Mon are listed in Table 4; the value $R = 6.5$, listed in the third row, was computed from $A_v = 3^m\!.6$ and $E_{B-V} = 0^m\!.55$ from Table 1. The new distance for NGC 2244 (or I Mon) is 910 pc compared to 1660 pc (Johnson *et al.* 1961) derived with $R = 3$.

The cluster NGC 1893 is almost exactly centered in the association I Aur as defined by Morgan, Whitford, and Code (1953), but NGC 1912 and NGC 1960 are exactly at its edge. These four objects are listed in Table 5. The values for I Aur are taken from Table 2 and those for the other clusters from Table 1. Since the values of $m - M$ for all four objects agreed, within their errors, we assumed that the three clusters are physically associated with I Aur and computed the mean distance modulus of $10^{m}.2$ for the group. The mean value of R for Auriga is 6.3 ± 0.5 (p.e.). NGC 2099 might have been included in the group except that it is outside the boundary of I Aur as defined by Morgan, Whitford, and Code.

The two determinations for NGC 6611 are listed in Table 6. The value of R for this cluster, which is a local condensation in the association I Ser, is near the traditional value of 3.

TABLE 4

VALUES OF R IN MONOCEROS

Cluster	R	p.e.	m-M	p.e.	Method
I Mon	5.0	± 0.5	10.3	± 0.4	var.ext.
NGC 2244	5.7	± 0.5	9.5	± 0.3	var.ext.
NGC 2244	7.1	---	8.9	---	cl. dia.
Mean	5.7	± 0.4	9.7	± 0.3	

TABLE 5

VALUES OF R IN AURIGA

Cluster	R	p.e.	m-M	p.e.	Method
I Aur	4.8	± 1.2	10.9	± 0.7	var.ext.
NGC 1893	7.5	---	10.4	---	cl. dia.
NGC 1912	7.8	---	9.4	---	cl. dia.
NGC 1960	5.8	---	9.8	---	cl. dia.
Mean	6.3	± 0.5	10.2	± 0.3	

TABLE 6

VALUES OF R IN NGC 6611

Cluster	R	p.e.	m-M	p.e.	Method
NGC 6611	3.4	± 0.7	11.8	± 0.4	var.ext.
NGC 6611	2.3	---	12.7	---	cl. dia.
Mean	2.9	± 0.5	12.2	± 0.3	

There are three determinations in Gemini, the variable-extinction value for I Gem and the cluster-diameter values for NGC 2158 and NGC 2168. The three determinations, listed in Table 7, agree that $R \sim 6$, but the precision is low.

There are four determinations for the region covering the constellations Aquila, Vulpecula, and Cygnus, including the high-precision one for II Cyg. These are listed in Table 8, and the adopted mean value is 3.6 ± 0.2 (p.e.).

Now that I have discussed the regions where the data from the two methods agree, we must turn to Perseus and Cassiopeia, where there is disagreement. The data for this part of the sky, compiled in the same manner as for the others, are listed in Table 9. It is immediately evident that the value of R from the variable-extinction method is about one-half that from cluster diameters. This discrepancy arises because of the large angular diameters of the clusters in this region and their resulting smaller diameter distances. Thus, the extinction derived from cluster diameters is larger than that from variable extinction.

Perhaps this discrepancy occurs because the reddened clusters in Perseus and Cassiopeia are intrinsically larger by a factor of two than clusters elsewhere; however, as has been emphasized earlier, the extinction derived by the variable-extinction method is a minimum value for the extinction between the cluster and the Earth. The data of Table 7 are therefore not necessarily inconsistent if, in addition to the variable extinction, there exists a relatively nearby interstellar cloud that is essentially uniform across the region. We note again that the difference between the geometric and photometric distance moduli of the α Per cluster is 0.6 or 0.7 mag in the same direction, and that this difference in moduli for the Pleiades is 0.9 mag.

TABLE 7

VALUES OF *R* IN GEMINI

Cluster	R	p.e.	m-M	p.e.	Method
I Gem	5.1	± 1.0	10.0	± 0.5	var.ext.
NGC 2168	6.9	---	8.8	---	cl. dia.
NGC 2158	5.8	---	12.2	---	cl. dia.
Mean	5.9	± 0.5			

TABLE 8

VALUES OF *R* IN AQUILA, VULPECULA, AND CYGNUS

Cluster	R	p.e.	m-M	p.e.	Method
II Cyg	3.1	± 0.3	11.3	± 0.3	var.ext.
NGC 6830	3.4	---	11.0	---	cl. dia.
NGC 6834	3.9	---	11.8	---	cl. dia.
NGC 6940	(5.4)	---	8.9	---	cl. dia.
Mean	3.6	± 0.2			

TABLE 9

VALUES OF *R* IN PERSEUS AND CASSIOPEIA

Cluster	R	p.e.	m-M	p.e.	Method
h and χ Per	3.0	± 0.3	11.8	± 0.2	var.ext.
III Cas	2.6	± 0.5	12.2	± 0.4	var.ext.
NGC 457	6.6	---	10.6	---	cl. dia.
NGC 581	3.8	---	11.8	---	cl. dia.
NGC 663	4.8	---	10.3	---	cl. dia.
NGC 869	8.4	---	8.8	---	cl. dia.
NGC 884	8.4	---	8.8	---	cl. dia.
IC 1805	5.4	---	9.6	---	cl. dia.
NGC 957	3.6	---	11.2	---	cl. dia.
NGC 1027	5.5	---	9.5	---	cl. dia.
IC 1848	6.9	---	9.3	---	cl. dia.
NGC 1245	2.5	---	11.9	---	cl. dia.
NGC 7789	9.3	---	9.6	---	cl. dia.
Mean	5.6	± 0.4	from cluster diameters only.		

5. THE COLOR-DIFFERENCE METHOD

Most values of R listed in Tables 2–9 are larger than 3.0, the ratio that has usually been accepted as normal. For several of these regions we now have sufficient multicolor data to investigate the interstellar extinction by the color-difference method and to make a comparison of the results from the different procedures. In addition, we have data that allow us to apply the color-difference method to several other regions of the Milky Way. The photometric data that have been published (Johnson 1965*b*) are listed in Table 10. The necessary intrinsic colors have also been derived (Johnson 1964, 1965*b*) and are given in Table 11. Data for HD 37020, 37022, 37023, 37061, and 37903 were taken from Johnson and Borgman (1963); zero-point corrections for their data to the present system are +0.09 for $V - R$ and +0.05 for $V - I$.

The procedure was the same as before (Johnson 1965*b*), that is, comparison of the observed colors of reddened stars of known spectral type with the derived intrinsic colors. The resultant color excesses were normalized to $E_{V-K} = 1.00$. The averages for a given celestial region were then computed and renormalized to $E_{B-V} = 1.00$, so that direct comparison with the values of R in Tables 2–9 could be made. This comparison is important since these latter determinations of R provide data at $1/\lambda = 0$ in the Whitford (1948, 1958) type of diagram. The results of these computations are listed in Tables 13–26, and the mean interstellar-extinction curves are shown in Figures 24–37. The theoretical curve No. 15 of van de Hulst (1949, Table 14) is shown in Figure 23, and the corresponding color-excess ratios are given in Table 12. The theoretical data are included for comparison with the observational data. Figures 23–37 are printed to the same scale. Short discussions of the diagrams and tables are given blow. Note that mean values in the tables are weighted means, with weights dependent upon E_{B-V} and upon judgments of the quality of the data.

Perseus (*Fig. 24*). We have observed twenty-one stars in the general region of Perseus; the derived data are listed in Table 13. The spectral types range from 05*f* to M3 Iab. The early-type and late-type stars are separated in the table in order to facilitate comparison of the extinction curves derived from the two kinds of stars. There seems to be a small systematic error in the intrinsic $U - V$ colors of M supergiants (Johnson 1964), and the $U - V$ excess ratios in Table 13 reflect this error. The intermediate color-excess ratios for the M stars fall within the range of those found for the early-type stars, confirming the usefulness of late-type stars in investigation of interstellar extinction. There is a discrepancy between early-type and late-type stars for E_{V-L}/E_{V-K}, but comparison with the results for other stars in other parts of the sky (see below) indicates that the cause of the discrepancy might be in the early-type stars as easily as in the late-type stars.

The extinction curve in Figure 24 resembles rather closely the theoretical curve in Figure 23; the variable-extinction value of $R = 3.0$ agrees exactly with the extrapolated color-difference data. Out to L (3.4 μ), there is no indication that the extinction in Perseus is as high as is indicated by the cluster-diameter results in Table 9 (but see below).

Ophiuchus (*Fig. 25*). There is no observed value of R for this region, which is included

TABLE 10

THE OBSERVATIONAL DATA

	HD	Name	Sp	V	U-V	B-V	V-R	V-I	V-J	V-K	V-L	V-M	V-N
1	886	γ Peg	B2 IV	2.84	-1.10	- .23	- .07	- .25	- .49	- .60	----	----	----
2	2905	κ Cas	B1 Ia	4.16	- .67	+ .13	+ .14	+ .20	+ .16	+ .24	+ .23	----	----
3	10516	φ Per	B2pe	4.06	- .97	- .04	+ .17	+ .19	+ .22	+ .70	+1.20	+2.41	----
4	12533	γ And	K3 II+A	2.10	+2.13	+1.21	+ .94	+1.63	+2.06	+2.89	+2.95	+2.70	----
5	12953	BS 618	A1 Ia	5.70	+ .60	+ .60	+ .61	+1.11	+1.42	+1.80	----	----	----
6	14134		B3 Ia	6.55	+ .08	+ .45	+ .52	+ .87	+ .94	+1.09	----	----	----
7	14142		M2 Iab	8.45	+4.95	+2.33	+1.87	+3.51	+4.54	+5.79	----	----	----
8	14143		B2 Ia	6.66	+ .05	+ .50	+ .55	+ .92	+1.08	+1.32	----	----	----
9	14270		M3 Iab	7.84	+4.88	+2.28	+1.90	+3.55	+4.56	+5.85	+6.19	----	----
10	14322		B8 Ib	6.86	- .03	+ .31	+ .41	+ .65	+ .76	+ .97	+ .97:	----	----
11	14330		M1 Iab	7.96	+4.80	+2.25	+1.75	+3.24	+4.21	+5.41	+5.78	----	----
12	14404		M2 Ib	8.12	+4.90	+2.30	+1.88	+3.46	+4.49	+5.76	----	----	----
13	14433		A1 Ia	6.40	+ .59	+ .56	+ .60	+1.06	+1.24	+1.51	+1.63	----	----
14	14535		A2 Iap?	7.47	+ .83	+ .70	+ .73	+1.31	+1.58	+2.01	+2.10	----	----
15	----	BD+56°595	M0 Iab	8.13	+4.70	+2.25	+1.68	+3.14	+4.16	+5.37	+5.78	----	----
16	14580		M0 Iab	8.42	+4.92	+2.30	+1.76	+3.23	+4.17	+5.40	+5.73	----	----
17	14818	BS 696	B2 Ia	6.30	- .32	+ .30	+ .39	+ .57	+ .62	+ .78	+ .90	----	----
18	14826		M2 Iab	8.24	+4.76	+2.32	+2.16	+3.92	+4.93	+6.28	----	----	----
19	15570		O5f	8.11	+ .29	+ .69	+ .68	+1.24	+1.55	+1.66	----	----	----
20	17506	η Per	K3 Ib	3.79	+3.60	+1.70	+1.23	+2.12	+2.70	+3.64	----	----	----
21	17520		O8 V	8.26	- .36	+ .32	+ .30	+ .53	+ .61:	+ .75	----	----	----
22	20902	α Per	F5 Ib	1.80	+ .88	+ .48	+ .45	+ .78	+ .92	+1.24	+1.26	+1.30:	+1.30:
23	21291	BS 1035	B9 Ia	4.21	+ .17	+ .41	+ .37	+ .75	+1.01	+1.25	----	----	----
24	21389	BS 1040	A0 Ia	4.55	+ .46	+ .56	+ .51	+1.01	+1.30	+1.65	+1.75	----	----
25	22928	δ Per	B5 III	3.03	- .62	- .12	+ .04	- .07	- .22	- .33	----	----	----

TABLE 10—*Continued*

	HD	Name	Sp	V	U-V	B-V	V-R	V-I	V-J	V-K	V-L	V-M	V-N
26	23630	η Tau	B7 III	2.86	- .44	- .10	+ .02	- .01	- .07	- .10	----	----	----
27	24398	ζ Per	B1 Ib	2.85	- .67	+ .11	+ .15	+ .24	+ .21	+ .20	+ .20	----	----
28	24760	ε Per	B0.5 V	2.89	-1.18	- .18	- .04	- .20	- .48	- .61	----	----	----
29	24912	ξ Per	O7	4.05	- .91	+ .01	+ .14	+ .14	+ .11	+ .09	+ .09	----	----
30	26630	μ Per	G0 Ib	4.14	+1.59	+ .95	+ .78	+1.34	+1.64	+2.19	----	----	----
31	31398	ι Aur	K3 II	2.68	+3.31	+1.53	+1.07	+1.89	+2.38	+3.29	+3.47	+3.14	+3.88
32	32630	η Aur	B3 V	3.17	- .83	- .17	- .04	- .22	- .41	- .56	----	----	----
33	34085	β Ori	B8 Ia	.15	- .69	- .03	+ .03	.00	- .07	- .05	- .09	----	+ .01
34	35411	η Ori	B0.5 V	3.32	-1.09	- .17	- .08	- .29	- .43	- .58	----	----	----
35	35468	γ Ori	B2 III	1.64	-1.10	- .23	- .10	- .32	- .54	- .70	- .70	----	----
36	35497	β Tau	B7 III	1.66	- .62	- .13	.00	- .10	- .29	- .38	----	----	----
37	36389	119 Tau	M2 Ib	4.35	+4.27	+2.06	+1.75	+3.20	+4.04	+5.21	+5.59	----	----
38	36486	δ Ori	O9.5 II	2.20	-1.27	- .21	- .08	- .30	- .53	- .67	- .54	----	----
39	36512	υ Ori	B0 V	4.63	-1.33	- .26	- .08	- .33	- .65	- .83	----	----	----
40	36673	α Lep	F0 Ib	2.58	+ .47	+ .19	+ .22	+ .43	+ .54	+ .70	----	----	----
41	36861	λ Ori	O8	3.39	-1.18	- .18	- .03	- .19	- .38	- .53	----	----	----
42	37020-3	Trapezium	---	4.58	- .81	+ .05	+ .22	+ .43	+ .47	+ .86	+1.28	----	----
43	37041	θ^2 Ori	O9.5 Vp	5.06	-1.02	- .08	+ .15	+ .16	+ .04	+ .12	+ .79	----	----
44	37042		B1 V	6.38	-1.01	- .09	+ .16	+ .12	- .06	- .05	----	----	----
45	37043	ι Ori	O9 III	2.77	-1.29	- .23	- .07	- .25	- .51	- .71	- .66	----	----
46	37128	ε Ori	B0 Ia	1.70	-1.23	- .19	- .08	- .26	- .38	- .49	- .43	----	----
47	37742	ζ Ori	O9.5 Ia	1.74	-1.26	- .21	- .04	- .25	- .45	- .58	- .56	----	----
48	38771	κ Ori	B0.5 Ia	2.06	-1.20	- .18	- .01	- .18	- .41	- .53	- .57	----	----
49*	39801	α Ori	M2 Iab	.42	+3.96	+1.85	+1.64	+2.92	+3.42	+4.42	+4.72	+4.44	+5.09
50	41117	χ^2 Ori	B2 Ia	4.63	- .39	+ .29	+ .32	+ .54	+ .60	+ .77	----	----	----

TABLE 10—*Continued*

	HD	Name	Sp	V	U-V	B-V	V-R	V-I	V-J	V-K	V-L	V-M	V-N
51	46106		B0.5 V	7.91	- .60	+ .14	+ .17	+ .27	+ .46	+ .51	+ .83	----	----
52	46150		O6	6.80	- .70	+ .12	+ .24	+ .34	+ .30	+ .39	+ .77:	----	----
53	46223		O5	7.31	- .55	+ .22	+ .33	+ .49	+ .46	+ .70	+1.18	----	----
54	47839	15 Mon	O7	4.66	-1.31	- .24	- .08	- .30	- .64:	- .68:	----	----	----
55	53138	o^2 CMa	B3 Ia	3.01	- .95	- .11	+ .01	- .08	- .20	- .24	----	----	----
56	66811	ζ Pup	O5f	2.26	-1.39	- .28	- .10	- .32	- .50	- .67	----	----	----
57	87901	α Leo	B7 V	1.35	- .47	- .12	- .01	- .11	- .21	- .29	- .24	+ .23	+1.39
58	91316	ρ Leo	B1 Ib	3.85	-1.09	- .13	- .04	- .21	- .34	- .49	- .36:	----	----
59	106625	γ Crv	B8 III	2.59	- .45	- .10	- .02	- .12	- .16	- .23	----	----	----
60	108767	δ Crv	B9.5 Vn	2.94	- .14	- .05	- .05	- .09	- .06	- .13	----	----	----
61	109387	κ Dra	---	3.89	- .71	- .14	+ .05	- .03	- .10	+ .06	+ .43	----	----
62	116658	α Vir	B1 V	.96	-1.17	- .23	- .08	- .31	- .55	- .73	- .69	----	----
63	120315	η UMa	B3 V	1.86	- .86	- .19	- .09	- .28	- .36	- .49	- .44	----	----
64	135742	β Lib	B8 V	2.61	- .48	- .11	- .04	- .13	- .23	- .30	----	----	----
65	143275	δ Sco	B0 V	2.33	-1.01	- .11	- .01	- .15	- .23	- .43	- .48	----	----
66	144217-8	β Sco AB	B0.5V+B2 V	2.55	- .90	- .08	- .03	- .11	- .08	- .23	- .17	----	----
67	144470	ω^1 Sco	B1 V	3.98	- .86	- .04	+ .06	- .03	+ .04	0.00	- .05	----	----
68	145502	ν Sco	B2 IV-V	4.02	- .60	+ .04	+ .09	+ .13	+ .24	+ .27	+ .33	----	----
69	147165	σ Sco	B1 III	2.88	- .56	+ .14	+ .19	+ .31	+ .46	+ .51	+ .53	----	----
70	147394	τ Her	B5 IV	3.89	- .71	- .15	- .10	- .27	- .33	- .49	- .43	----	----
71	148478-9	α Sco	M1 Iab+B	.97	+3.10	+1.80	+1.59	+2.80	+3.37	+4.50	+4.81	----	+5.03
72	149757	ζ Oph	O9.5 V	2.56	- .84	+ .02	+ .10	+ .07	- .03	- .08	- .06	----	----
73	155763	ζ Dra	B6 III	3.17	- .53	- .11	- .06	- .18	- .23	- .31	----	----	----
74	163800		O8	7.02	- .39	+ .30	+ .44	+ .61	+ .64	+ .60	----	----	----
75	166734		O8f	8.42	+ .94	+1.07	+1.09	+1.95	+2.43	+3.01	+3.42:	----	----

TABLE 10—*Continued*

	HD	Name	Sp	V	U-V	B-V	V-R	V-I	V-J	V-K	V-L	V-M	V-N
76	167971		O8f	7.54	+ .44	+ .76	+ .86	+1.47	+1.83	+2.28	+2.59:	----	----
77	172167	α Lyr	A0 V	.03	.00	.00	- .04	- .06	.00	+ .01	+ .07	+ .07	+ .04
78	175588	δ^2 Lyr	M4 II	4.30	+3.32	+1.67	+1.78	+3.41	+4.33	+5.47	+5.70	----	+5.45
79	180809	θ Lyr	K0 II	4.37	+1.45	+1.25	+ .87	+1.46	+2.00	+2.68	+2.82	----	----
80	183143		B7 Ia	6.86	+1.39	+1.22	+1.12	+2.07	+2.73	+3.38	+3.97	----	----
81	186791	γ Aql	K3 II	2.73	+3.21	+1.53	+1.08	+1.83	+2.41	+3.29	+3.49	----	----
82	190603	BS 7678	B1.5 Ia	5.65	+ .07	+ .53	+ .53	+ .93	+1.15	+1.54	----	----	----
83	192876	θ^1 Cap	G3 Ib	4.26	+1.88	+1.08	+ .78	+1.30	+1.71	+2.32	+2.47	----	----
84	193237	P Cyg	P Cyg	4.81	- .16	+ .41	+ .54	+ .80	+1.01	+1.49	+1.92	----	----
85	194093	γ Cyg	F8 Ib	2.23	+1.21	+ .67	+ .49	+ .83	+1.07	+1.49	+1.53	----	----
86	195592		O9.5 Ia	7.08	+ .67	+ .87	+ .79	+1.48	+1.98	+2.44	----	----	----
87	195593	44 Cyg	F5 Iab	6.19	+1.74	+1.00	+ .86	+1.55	+2.07	+2.69	+2.94	----	----
88	197345	α Cyg	A2 Ia	1.26	- .14	+ .09	+ .11	+ .21	+ .24	+ .36	+ .50	----	----
89	200905	ξ Cyg	K5 Ib	3.70	+3.45	+1.65	+1.20	+2.10	+2.79	+3.79	+4.01	----	----
90	202109	ζ Cyg	G8 II	3.20	+1.75	+ .99	+ .69	+1.18	+1.54	+2.10	+2.21	----	----
91	202850	σ Cyg	B9 Iab	4.23	- .27	+ .12	+ .15	+ .29	+ .28	+ .45	----	-----	----
92	206165	9 Cep	B2 Ib	4.74	- .23	+ .30	+ .31	+ .49	+ .37	+ .59	+ .77	----	----
93*	206936	μ Cep	M2 Ia	4.17	+4.71	+2.26	+2.10	+3.86	+4.69	+5.82	+6.22	+6.20	+7.45
94	207198	BS 8327	O9 II	5.94	- .33	+ .31	+ .28	+ .45	+ .55	+ .60	----	----	----
95	207260	ν Cep	A2 Ia	4.29	+ .63	+ .51	+ .50	+ .94	+1.14	+1.43	+1.59	----	----
96	209750	α Aqr	G2 Ib	2.92	+1.77	+ .98	+ .66	+1.13	+1.43	+1.96	----	----	----
97	210839	λ Cep	O6f	5.04	- .49	+ .24	+ .28	+ .43	+ .41	+ .46	+ .69	----	----
98	214680	10 Lac	O9 V	4.88	-1.24	- .20	- .08	- .29	- .53	- .67	- .62	----	----
99*	217476	BS 8752	G0 Ia	5.13	+2.88	+1.55	+1.17	+2.02	+2.61	+3.33	+3.59	----	+4.18

TABLE 10—*Continued*

CLUSTERS

HD	Name	Sp	V	U-V	B-V	V-R	V-I	V-J	V-K	V-L	V-M	V-N
100* NGC 2024	No. 1	0	12.17	+1.70	+1.41	+1.80	+3.46	+4.79	+6.24	+7.09	----	----
101* NGC 6530	No. 7	05	5.97	- .90	.00	+ .25	+ .27	+ .18	+ .11	----	----	----
102* NGC 6530	No. 65	BOnne	7.46	- .69	+ 21	----	----	+ .71	+ .85	----	----	----
103* NGC 6611	No. 1	0	8.25	- .13	+ .47	+ .63	+1.04	+1.36	+1.59	+1.73	----	----
104* NGC 6910	No. 3	05	8.50	+ .71	+ .90	+ .87	+1.59	+2.07	+2.45	----	----	----
105* VI Cyg	No. 9	05f	10.77	+2.64	+1.90	+1.82	+3.08	+4.25	+5.33	+6.29	----	----
106* VI Cyg	No. 10	09 Ia	9.86	+1.89	+1.50	+1.44	+2.60	+3.38	+4.19	----	----	----
107* VI Cyg	No. 12	B8 Ia	11.48	+5.72	+3.22	+3.22	+5.54	+7.16	+8.82	+9.54	----	----

*NOTES:

No. 49. α Ori. Variable star; data for J. D. 2438400.0.

No. 93. μ Cep. Variable star; data for J. D. 2438300.0.

No. 99. BS 8752. Variable star; data for J. D. 2438300.0.

No. 100. Identification chart by Johnson and Mendoza (1964).

Nos. 101-102. Identification chart by Walker (1957). UBV data for No. 102 by Walker.

Nos. 103-104. Identification charts by Hoag, et al (1961).

Nos. 105-106. Identification chart by Johnson and Morgan (1954).

No. 107. Identification chart by Morgan, Johnson and Roman (1954).

as an example of the interstellar-extinction law in this part of the Milky Way. Within the accuracy of the observations (Table 14), the extinction law in Ophiuchus is identical with that in Perseus. Note, however, the correlation between E_{V-L}/E_{V-K} and E_{B-V}. This same correlation also exists among the early-type data of Table 13 and suggests that the intrinsic $V-L$ colors (Johnson 1965*b*) should be bluer by 0.04 mag. Such a change would, in fact, reduce the $V-L$ discrepancy between the early-type and the late-type stars of Table 13 to almost zero. It would, however, have other consequences, which will be discussed later.

TABLE 11

THE INTRINSIC COLORS OF EARLY-TYPE STARS

Luminosity Classes III, IV and V

Sp	U-V	B-V	V-R	V-I	V-J	V-K	V-L
O5-7	-1.46	- .32	- .15	- .43	- .73	- .94	- .92
O8-9	-1.44	- .31	- .15	- .43	- .73	- .94	- .92
O9.5	-1.40	- .30	- .14	- .42	- .73	- .94	- .92
B0	-1.38	- .30	- .13	- .41	- .70	- .93	- .91
B0.5	-1.29	- .28	- .12	- .39	- .66	- .90	- .88
B1	-1.19	- .26	- .11	- .36	- .62	- .83	- .81
B2	-1.10	- .24	- .10	- .32	- .53	- .71	- .69
B3	- .91	- .20	- .08	- .27	- .44	- .59	- .57
B5	- .72	- .16	- .06	- .22	- .36	- .47	- .45
B6	- .63	- .14	- .06	- .19	- .31	- .41	- .39
B7	- .54	- .12	- .05	- .17	- .27	- .35	- .33
B8	- .39	- .09	- .03	- .12	- .18	- .24	- .22
B9	- .25	- .06	- .02	- .07	- .09	- .12	- .10
A0	.00	.00	.00	- .02	- .02	.00	+ .02

Luminosity Classes Ia and Ib

Sp	U-V Ia	U-V Ib	B-V	V-R	V-I	V-J	V-K	V-L
O8-9	-1.41	-1.41	-.29	-.15	-.43	-.73	-.94	-.92
O9.5	-1.37	-1.36	-.27	-.13	-.42	-.68	-.86	-.84
B0	-1.31	-1.29	-.24	-.11	-.36	-.59	-.77	-.75
B0.5	-1.26	-1.23	-.22	-.09	-.30	-.50	-.66	-.64
B1	-1.19	-1.15	-.19	-.07	-.25	-.45	-.58	-.56
B2	-1.13	-1.08	-.17	-.04	-.18	-.34	-.42	-.40
B3	-1.00	- .95	-.13	-.01	-.12	-.25	-.31	-.29
B5	- .87	- .81	-.09	+.01	-.06	-.18	-.20	-.18
B6	- .80	- .74	-.07	+.01	-.03	-.14	-.16	-.14
B7	- .73	- .67	-.05	+.02	-.01	-.11	-.10	-.08
B8	- .62	- .55	-.02	+.03	+.02	-.06	-.03	-.01
B9	- .56	- .48	.00	+.04	+.05	-.01	+.04	+.06
A0	- .47	- .41	+.01	+.05	+.08	+.04	+.11	+.13
A1	- .35	-----	+.03	+.06	+.11	+.07	+.17	+.19
A2	- .23	-----	+.05	+.07	+.14	+.12	+.22	+.24

NGC 6611 (*Fig. 26*). This is a young cluster (Walker 1961), which is embedded in emission nebulosity. It is in the same general part of the sky as Ophiuchus. The extrapolated color-difference datum for $1/\lambda = 0$ is a little larger than R from Table 6.

NGC 6530 (*Fig. 27*). This is another young cluster (Walker 1957), which is also associated with emission nebulosity. Unfortunately, there are no observations at wavelengths longer than 2.2 μ and no determination of R. There is, however, a clear indication that the extinction law for NGC 6530 differs from those for the previously discussed regions; compare the color-excess ratios in the tables. The differences are undoubtedly significant.

Capricornus (*Fig. 28*). This region is represented by only one relatively little-reddened star, but the shape of the derived extinction curve is sufficiently different from the previously discussed curves to be of interest. Curves similar to that in Figure 28 will be shown later.

Cygnus (*Fig. 29*). The extinction curve for this important region is well determined since several highly reddened stars have been observed. The stars in this region are, in fact, among the most highly reddened stars known. The value of R from Table 8 is 3.6. The extrapolation of the color-difference data to $1/\lambda = 0$ fits well with this value of R, even with the turn-up from K to L. This turn-up is well established, as Table 18 shows. Compare the Cygnus data with the theoretical ones of van de Hulst in Table 12. The small dip at K certainly is not expected from the theoretical analysis.

Aquila (*Fig. 30*). The extinction curve for Aquila is almost identical with that for Cygnus, including the turn-up from K to L. The value of R is from Table 8. The spectral types of the stars observed in Cygnus and Aquila range from O5 to K3 II, with no appreciable changes of the color-excess ratios with spectral type. Furthermore, except for $V - L$, the extinction curves for Aquila and Cygnus are almost identical with the curves for Perseus and Ophiuchus. This is true even in $V - L$, if we accept the data from the late-type Perseus stars and make the correction to intrinsic $V - L$ that is discussed under Ophiuchus. We would then have a turn-up from K to L in Perseus and Ophiuchus.

Cepheus (*Fig. 31*). This region is, in several respects, one of the best observed. There are seven extinction determinations (cf. Table 3) with a mean of $R = 4.8 \pm 0.3$ (p.e.). This is the only region where the photometry of a reddened star has been carried out to wavelength region O (11.5 μ). The color-difference curve shown in Figure 31 is very different from van de Hulst's theoretical curve in Figure 23; note, however, that the turn-up from K to L is almost identical with that found in Cygnus and Aquila. The intermediate color-excess ratios appear to be a little different in Cepheus, but the difference is relatively small (and in the opposite direction) compared with the ratio of NGC 6530.

The color-excess ratios for μ Cep in Table 20 differ somewhat from those computed earlier (Johnson 1965*a*, *b*), partly because the comparison star for the ratios in Table 20 is α Ori alone. The modified procedure, in which the data were first normalized to $E_{V-K} = 1.00$, also has some effect upon the result. Nevertheless, the general shape of the curve is preserved throughout and is confirmed by the data for BS 8752 and, especially, by the well-determined value of $R = 4.8$. The water-vapor bands found by Danielson, Woolf, and Gaustad (1965) in the spectrum of μ Cep, and which presumably also exist in the double-cluster M supergiants, apparently do not significantly affect the interstellar-extinction data. But then we would not expect these bands to be a significant factor, because the same bands are much stronger in the atmosphere of the Earth, and the interference filters used in the photometry are necessarily designed to avoid them as much as possible. The O filter (11.0 μ and 12.0 μ) avoids water-vapor bands.

NGC 2244 (*Fig. 32*). Another region in which an extinction curve similar to that in Cepheus exists is NGC 2244 (or I Mon). The value of $R = 5.7 \pm 0.4$ (p.e.) seems to be rather well determined. The turn-up from K to L is much exaggerated in comparison with the other curves. The shape of the curve in Figure 32, in the intermediate part, is similar to that in Figure 28.

Auriga (*Fig. 33*). In Figure 33 is another curve similar to that found in Cepheus.

TABLE 12

COLOR-EXCESS RATIOS FOR VAN DE HULST'S CURVE NO. 15*

Curve	$\frac{E_{U-V}}{E_{V-K}}$	$\frac{E_{B-V}}{E_{V-K}}$	$\frac{E_{V-R}}{E_{V-K}}$	$\frac{E_{V-I}}{E_{V-K}}$	$\frac{E_{V-J}}{E_{V-K}}$	$\frac{E_{V-K}}{E_{V-K}}$	$\frac{E_{V-L}}{E_{V-K}}$	$\frac{E_{V-M}}{E_{V-K}}$	$\frac{E_{V-N}}{E_{V-K}}$	$\frac{E_{V-O}}{E_{V-K}}$	$\frac{A_V}{E_{V-K}}$
No. 15	0.62	0.36	0.29	0.58	0.83	1.00	1.04	1.06	1.08	1.09	1.10
Curve	$\frac{E_{U-V}}{E_{B-V}}$	$\frac{E_{B-V}}{E_{B-V}}$	$\frac{E_{V-R}}{E_{B-V}}$	$\frac{E_{V-I}}{E_{B-V}}$	$\frac{E_{V-J}}{E_{B-V}}$	$\frac{E_{V-K}}{E_{B-V}}$	$\frac{E_{V-L}}{E_{B-V}}$	$\frac{E_{V-M}}{E_{B-V}}$	$\frac{E_{V-N}}{E_{B-V}}$	$\frac{E_{V-O}}{E_{B-V}}$	$\frac{A_V}{E_{B-V}} = R$
No. 15	1.71	1.00	0.80	1.62	2.30	2.78	2.91	2.95	3.01	3.02	3.05

* van de Hulst, 1949.

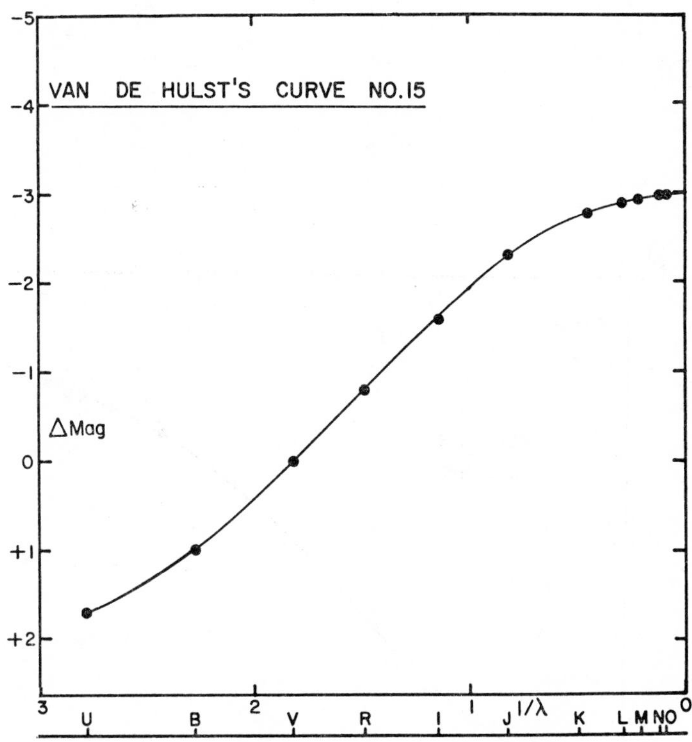

FIG. 23.—van de Hulst's theoretical curve No. 15

TABLE 13

COLOR-EXCESS RATIOS IN PERSEUS

Star	Sp	$\frac{E_{U-V}}{E_{V-K}}$	$\frac{E_{B-V}}{E_{V-K}}$	$\frac{E_{V-R}}{E_{V-K}}$	$\frac{E_{V-I}}{E_{V-K}}$	$\frac{E_{V-J}}{E_{V-K}}$	$\frac{E_{V-K}}{E_{V-K}}$	$\frac{E_{V-L}}{E_{V-K}}$	E_{B-V}
HD 2905	B1 Ia	.59	.39	.26	.55	.74	1.00	.96	.32
HD 12953	A1 Ia	.58	.35	.34	.61	.83	1.00	----	.57
HD 14134	B3 Ia	.66	.41	.38	.71	.85	1.00	----	.58
HD 14143	B2 Ia	.62	.39	.34	.63	.82	1.00	----	.67
HD 14322	B8 Ib	.52	.33	.38	.63	.82	1.00	.98	.33
HD 14433	A1 Ia	.70	.40	.40	.71	.87	1.00	.98	.53
HD 14535	A2 Iap	.59	.36	.37	.65	.82	1.00	1.04	.65
HD 14818	B2 Ia	.67	.39	.36	.62	.80	1.00	1.08	.47
HD 15570	O5 f	.67	.39	.32	.64	.88	1.00	----	1.01
HD 17520	O8 V	.64	.37	.27	.57	.79	1.00	----	.63
HD 21291	B9 Ia	.60	.34	.27	.58	.84	1.00	----	.41
HD 21389	A0 Ia	.60	.36	.30	.60	.82	1.00	1.05	.55
HD 24398	B1 Ib	.62	.38	.28	.63	.85	1.00	.97	.30
HD 24912	O7	.53	.32	.28	.55	.82	1.00	.98	.33
Mean, Early Types		.614	.370	.325	.620	.825	1.000	1.005	
HD 14142	M2 Iab	.78	.37	.31	.58	.86	1.00	----	.65
HD 14270	M3 Iab	.90	.42	.28	.58	.86	1.00	1.10	.60
HD 14330	M1 Iab	.85	.40	.32	.60	.88	1.00	1.14	.57
HD 14404	M2 Iab	.76	.39	.32	.56	.85	1.00	----	.68
HD 14580	M0 Iab	.89	.42	.33	.61	.85	1.00	1.11	.65
HD 17506	K3 Ib	.83	.43	.32	.64	.80	1.00	----	.38
BD +56° 595	M0 Ib	.76	.39	.28	.56	.86	1.00	1.16	.60
Mean, Late Types		.824	.403	.309	.590	.851	1.000	1.128	
		$\frac{E_{U-V}}{E_{B-V}}$	$\frac{E_{B-V}}{E_{B-V}}$	$\frac{E_{V-R}}{E_{B-V}}$	$\frac{E_{V-I}}{E_{B-V}}$	$\frac{E_{V-J}}{E_{B-V}}$	$\frac{E_{V-K}}{E_{B-V}}$	$\frac{E_{V-L}}{E_{B-V}}$	
Perseus Mean (all types)		1.80	1.00	0.84	1.60	2.19	2.62	2.75	

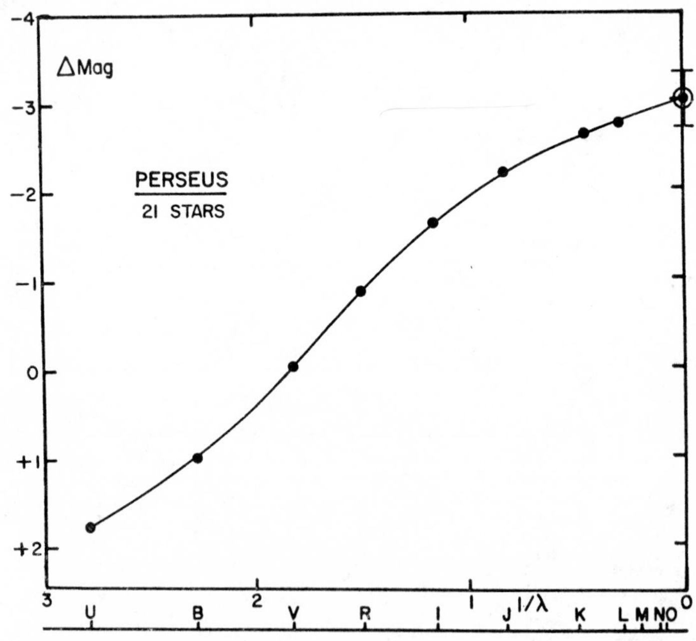

Fig. 24.—The interstellar-extinction curve for Perseus

TABLE 14

COLOR-EXCESS RATIOS IN OPHIUCHUS

Star	Sp	$\frac{E_{U-V}}{E_{V-K}}$	$\frac{E_{B-V}}{E_{V-K}}$	$\frac{E_{V-R}}{E_{V-K}}$	$\frac{E_{V-I}}{E_{V-K}}$	$\frac{E_{V-J}}{E_{V-K}}$	$\frac{E_{V-K}}{E_{V-K}}$	$\frac{E_{V-L}}{E_{V-K}}$	E_{B-V}
ζ Oph	09.5 V	.65	.37	.28	.57	.81	1.00	1.00	.32
HD 163800	08	.68	.40	.38	.68	.89	1.00	----	.61
HD 166734	08 f	.60	.35	.31	.60	.80	1.00	1.10	1.38
HD 167971	08 f	.58	.33	.31	.59	.80	1.00	1.09	1.07
Mean		.62	.36	.32	.61	.82	1.00	1.08	
		$\frac{E_{U-V}}{E_{B-V}}$	$\frac{E_{B-V}}{E_{B-V}}$	$\frac{E_{V-R}}{E_{B-V}}$	$\frac{E_{V-I}}{E_{B-V}}$	$\frac{E_{V-J}}{E_{B-V}}$	$\frac{E_{V-K}}{E_{B-V}}$	$\frac{E_{V-L}}{E_{B-V}}$	
Ophiuchus Mean		1.72	1.00	0.89	1.69	2.28	2.78	3.00	

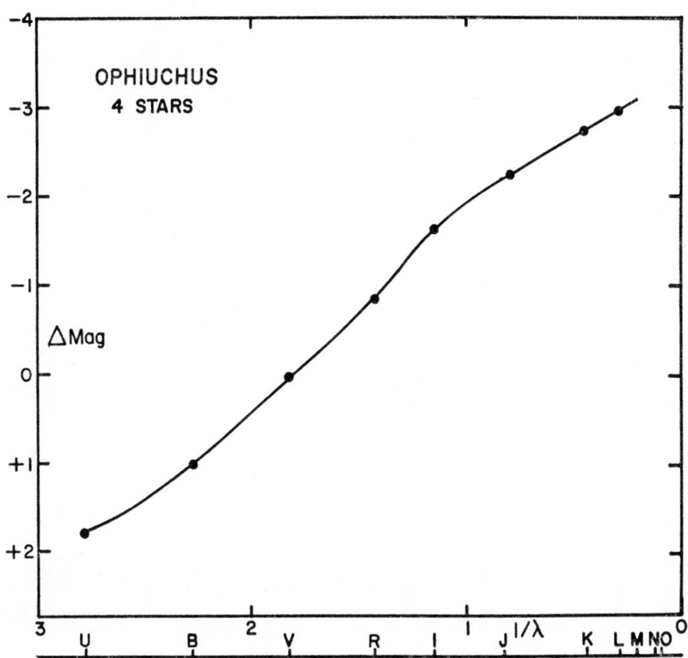

FIG. 25.—The interstellar-extinction curve for Ophiuchus

TABLE 15

COLOR-EXCESS RATIOS IN NGC 6611

Star	Sp	$\frac{E_{U-V}}{E_{B-V}}$	$\frac{E_{B-V}}{E_{B-V}}$	$\frac{E_{V-R}}{E_{B-V}}$	$\frac{E_{V-I}}{E_{B-V}}$	$\frac{E_{V-J}}{E_{B-V}}$	$\frac{E_{V-K}}{E_{B-V}}$	$\frac{E_{V-L}}{E_{B-V}}$	E_{B-V}
No. 1	O	1.68	1.00	.99	1.86	2.65	3.20	3.35	.79
No. 4	O	1.70	1.00	.93	1.80	----	----	----	.74
No. 5	O9.5 V	1.66	1.00	.92	1.66	----	----	----	.59
No. 7	O5	1.74	1.00	.95	1.80	----	----	----	1.22
No. 9	O	1.74	1.00	.96	1.80	----	----	----	.94
NGC 6611 Mean		1.70	1.00	.95	1.78	2.65	3.20	3.35	

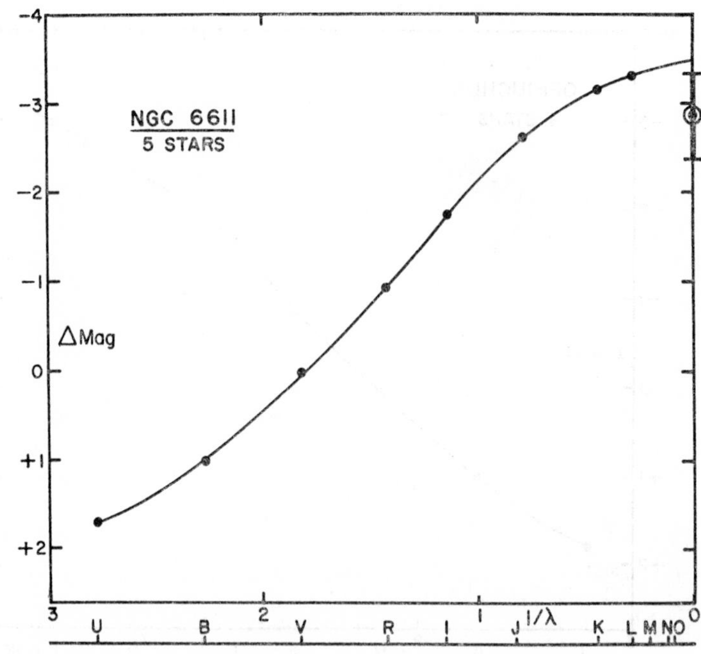

FIG. 26.—The interstellar-extinction curve for NGC 6611

TABLE 16

COLOR-EXCESS RATIOS IN NGC 6530

Star	Sp	$\frac{E_{U-V}}{E_{B-V}}$	$\frac{E_{B-V}}{E_{B-V}}$	$\frac{E_{V-R}}{E_{B-V}}$	$\frac{E_{V-I}}{E_{B-V}}$	$\frac{E_{V-J}}{E_{B-V}}$	$\frac{E_{V-K}}{E_{B-V}}$	E_{B-V}
No. 7	05	1.75	1.00	1.25	2.19	2.84	3.28	.32
No. 9	B0 V	1.71	1.00	1.25	2.11	----	----	.28
No. 65	B0nne	----	1.00	----	----	2.76	3.49	.51
No. 118	07	1.77	1.00	1.18	2.10	----	----	.39
NGC 6530 Mean		1.74	1.00	1.23	2.13	2.80	3.38	

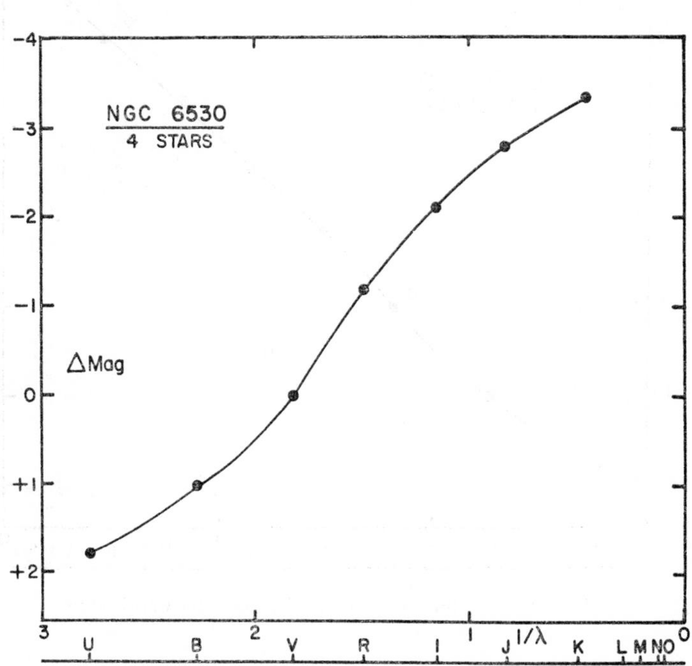

FIG. 27.—The interstellar-extinction curve for NGC 6530

TABLE 17

COLOR-EXCESS RATIOS FOR α^1 CAPRICORNI

Star	Sp	$\frac{E_{U-V}}{E_{B-V}}$	$\frac{E_{B-V}}{E_{B-V}}$	$\frac{E_{V-R}}{E_{B-V}}$	$\frac{E_{V-I}}{E_{B-V}}$	$\frac{E_{V-J}}{E_{B-V}}$	$\frac{E_{V-K}}{E_{B-V}}$	$\frac{E_{V-L}}{E_{B-V}}$	E_{B-V}
α^1 Cap	G3 Ib	1.74	1.00	.87	1.53	2.34	3.00	3.48	.18

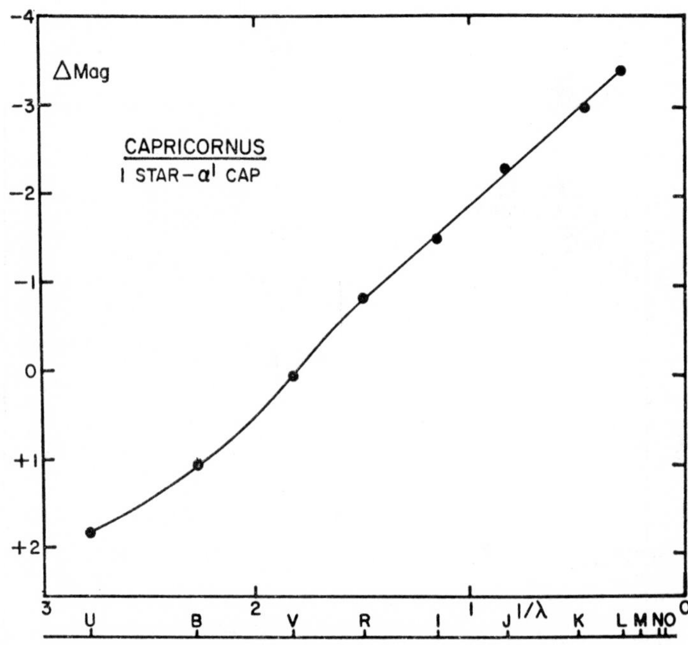

FIG. 28.—The interstellar-extinction curve for Capricornus

TABLE 18

COLOR-EXCESS RATIOS IN CYGNUS

Star	Sp	E_{U-V}/E_{V-K}	E_{B-V}/E_{V-K}	E_{V-R}/E_{V-K}	E_{V-I}/E_{V-K}	E_{V-J}/E_{V-K}	E_{V-K}/E_{V-K}	E_{V-L}/E_{V-K}	E_{B-V}
HD 195592	O9.5 Ia	.62	.34	.28	.58	.81	1.00	----	1.14
44 Cyg	F5 Iab	.59	.36	.28	.53	.79	1.00	1.16	.52
NGC 6910 No.3	O5	.64	.36	.30	.60	.83	1.00	----	1.22
VI Cyg No. 9	O5 f	.65	.35	.31	.56	.79	1.00	1.15	2.22
VI Cyg No. 10	O9 Ia	.64	.35	.31	.59	.80	1.00	----	1.79
VI Cyg No. 12	B8 Ia	.72	.37	.36	.62	.82	1.00	1.09	3.25
Mean		.64	.36	.31	.58	.81	1.00	1.13	
		E_{U-V}/E_{B-V}	E_{B-V}/E_{B-V}	E_{V-R}/E_{B-V}	E_{V-I}/E_{B-V}	E_{V-J}/E_{B-V}	E_{V-K}/E_{B-V}	E_{V-L}/E_{B-V}	
Cygnus Mean		1.81	1.00	.87	1.63	2.26	2.81	3.18	

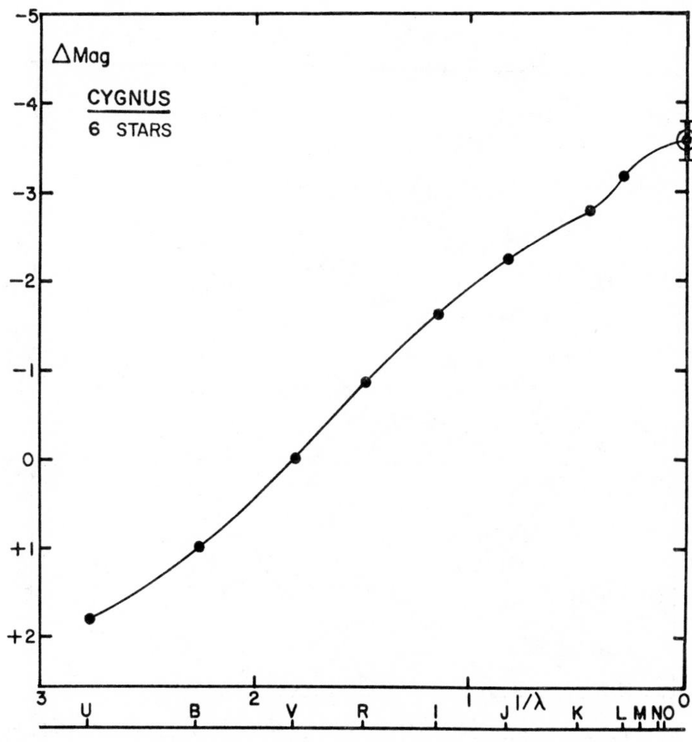

FIG. 29.—The interstellar-extinction curve for Cygnus

TABLE 19

COLOR-EXCESS RATIOS IN AQUILA

Star	Sp	$\frac{E_{U-V}}{E_{V-K}}$	$\frac{E_{B-V}}{E_{V-K}}$	$\frac{E_{V-R}}{E_{V-K}}$	$\frac{E_{V-I}}{E_{V-K}}$	$\frac{E_{V-J}}{E_{V-K}}$	$\frac{E_{V-K}}{E_{V-K}}$	$\frac{E_{V-L}}{E_{V-K}}$	E_{B-V}
γ Aql	K3 II	.64	.40	.25	.51	.77	1.00	1.13	0.21
HD 183143	B7 Ia	.61	.36	.32	.60	.82	1.00	1.16	1.25
Mean		.62	.37	.31	.59	.81	1.00	1.16	
		$\frac{E_{U-V}}{E_{B-V}}$	$\frac{E_{B-V}}{E_{B-V}}$	$\frac{E_{V-R}}{E_{B-V}}$	$\frac{E_{V-I}}{E_{B-V}}$	$\frac{E_{V-J}}{E_{B-V}}$	$\frac{E_{V-K}}{E_{B-V}}$	$\frac{E_{V-L}}{E_{B-V}}$	
Aquila Mean		1.68	1.00	0.84	1.59	2.19	2.72	3.14	

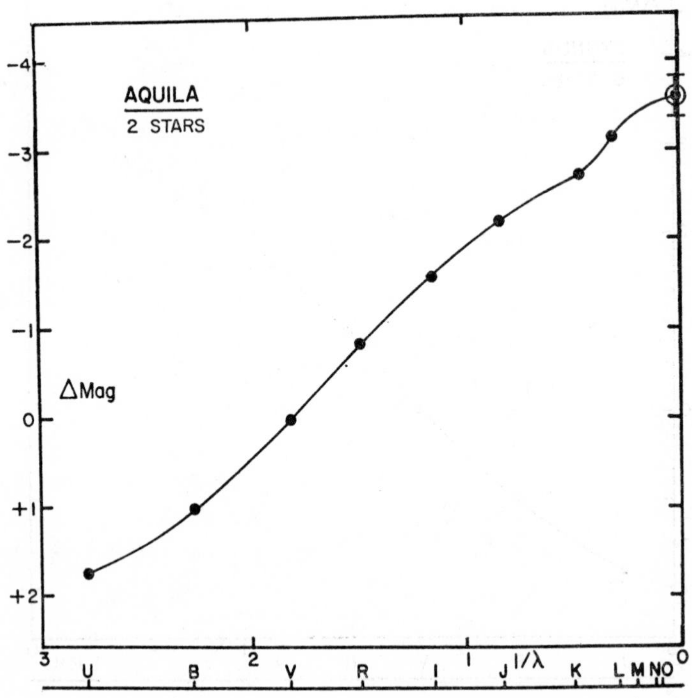

Fig. 30.—The interstellar-extinction curve for Aquila

TABLE 20

COLOR-EXCESS RATIOS IN CEPHEUS

Star	Sp	$\frac{E_{U-V}}{E_{V-K}}$	$\frac{E_{B-V}}{E_{V-K}}$	$\frac{E_{V-R}}{E_{V-K}}$	$\frac{E_{V-I}}{E_{V-K}}$	$\frac{E_{V-J}}{E_{V-K}}$	$\frac{E_{V-K}}{E_{V-K}}$	$\frac{E_{V-L}}{E_{V-K}}$	$\frac{E_{V-M}}{E_{V-K}}$	$\frac{E_{V-N}}{E_{V-K}}$	$\frac{E_{V-O}}{E_{V-K}}$	E_{B-V}
μ Cep	M2 Ia	.54	.29	.33	.67	.91	1.00	1.07	1.26	1.69	1.72	.41
9 Cep	B2 Ib	.84	.47	.35	.66	.70	1.00	1.16	----	----	----	.47
HD 207198	O9 II	.71	.40	.28	.57	.83	1.00	----	----	----	----	.61
ν Cep	A2 Ia	.71	.38	.36	.66	.84	1.00	1.12	----	----	----	.46
λ Cep	O6 f	.69	.40	.31	.61	.81	1.00	1.15	----	----	----	.56
BS 8752	G0 Ia	.89	.43	.34	.60	.86	1.00	1.12	----	1.45	----	.82
Mean		.73	.40	.33	.63	.82	1.00	1.12	1.26	1.61	1.72	
		$\frac{E_{U-V}}{E_{B-V}}$	$\frac{E_{B-V}}{E_{B-V}}$	$\frac{E_{V-R}}{E_{B-V}}$	$\frac{E_{V-I}}{E_{B-V}}$	$\frac{E_{V-J}}{E_{B-V}}$	$\frac{E_{V-K}}{E_{B-V}}$	$\frac{E_{V-L}}{E_{B-V}}$	$\frac{E_{V-M}}{E_{B-V}}$	$\frac{E_{V-N}}{E_{B-V}}$	$\frac{E_{V-O}}{E_{B-V}}$	
Cepheus Mean		1.82	1.00	0.82	1.58	2.05	2.50	2.80	3.15	4.02	4.30	

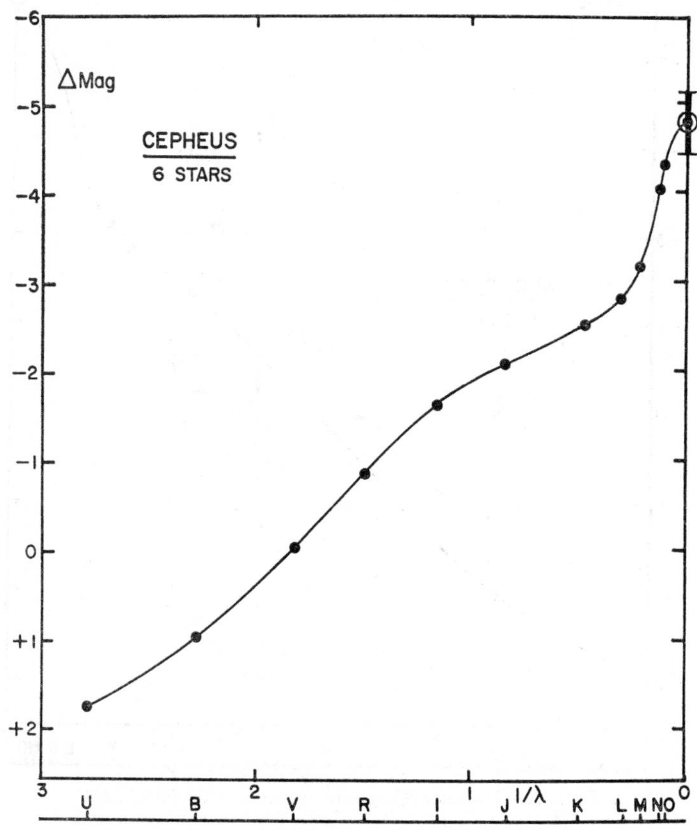

FIG. 31.—The interstellar-extinction curve for Cepheus

TABLE 21

COLOR-EXCESS RATIOS IN NGC 2244

Star	Sp	$\frac{E_{U-V}}{E_{B-V}}$	$\frac{E_{B-V}}{E_{B-V}}$	$\frac{E_{V-R}}{E_{B-V}}$	$\frac{E_{V-I}}{E_{B-V}}$	$\frac{E_{V-J}}{E_{B-V}}$	$\frac{E_{V-K}}{E_{B-V}}$	$\frac{E_{V-L}}{E_{B-V}}$	E_{B-V}
HD 46106	B0.5 V	1.64	1.00	.69	1.57	2.67	3.36	4.07	.42
HD 46149	O8	1.71	1.00	.81	1.62	----	----	----	.48
HD 46150	O6	1.73	1.00	.89	1.75	2.34	3.02	3.84	.44
HD 46202	O9 V	1.78	1.00	.78	1.61	----	----	----	.49
HD 46223	O5	1.69	1.00	.89	1.70	2.20	3.04	3.89	.54
NGC 2244 Mean		1.71	1.00	.81	1.65	2.40	3.14	3.93	

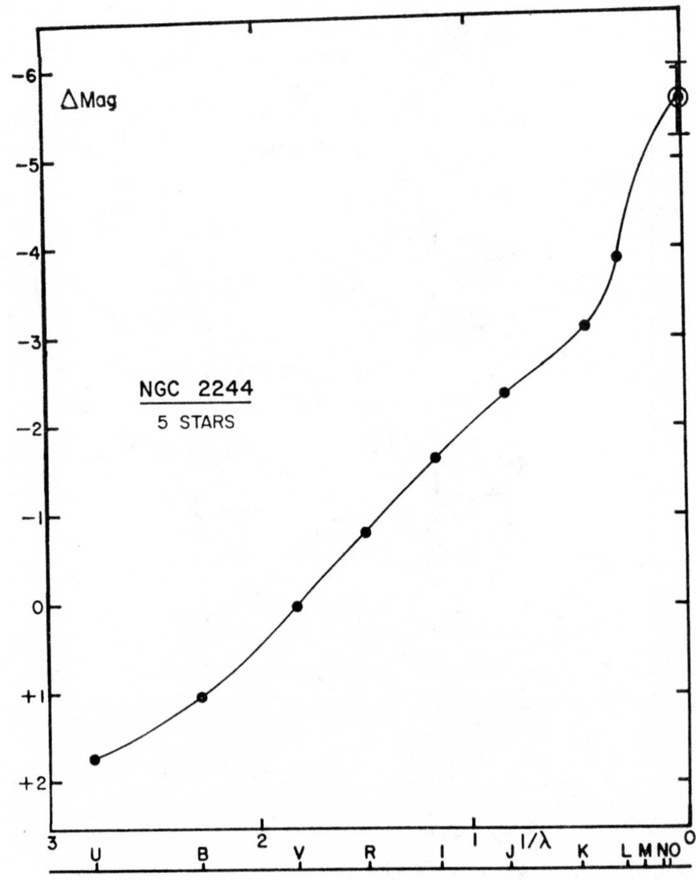

FIG. 32.—The interstellar-extinction curve for NGC 2244

TABLE 22

COLOR-EXCESS RATIOS FOR ι AURIGAE

Star	Sp	$\frac{E_{U-V}}{E_{B-V}}$	$\frac{E_{B-V}}{E_{B-V}}$	$\frac{E_{V-R}}{E_{B-V}}$	$\frac{E_{V-I}}{E_{B-V}}$	$\frac{E_{V-J}}{E_{B-V}}$	$\frac{E_{V-K}}{E_{B-V}}$	$\frac{E_{V-L}}{E_{B-V}}$	$\frac{E_{V-M}}{E_{B-V}}$	$\frac{E_{V-N}}{E_{B-V}}$	E_{B-V}
ι Aur	K3 II	2.10	1.00	0.57	1.57	1.81	2.52	2.76	2.86	5.6	0.21

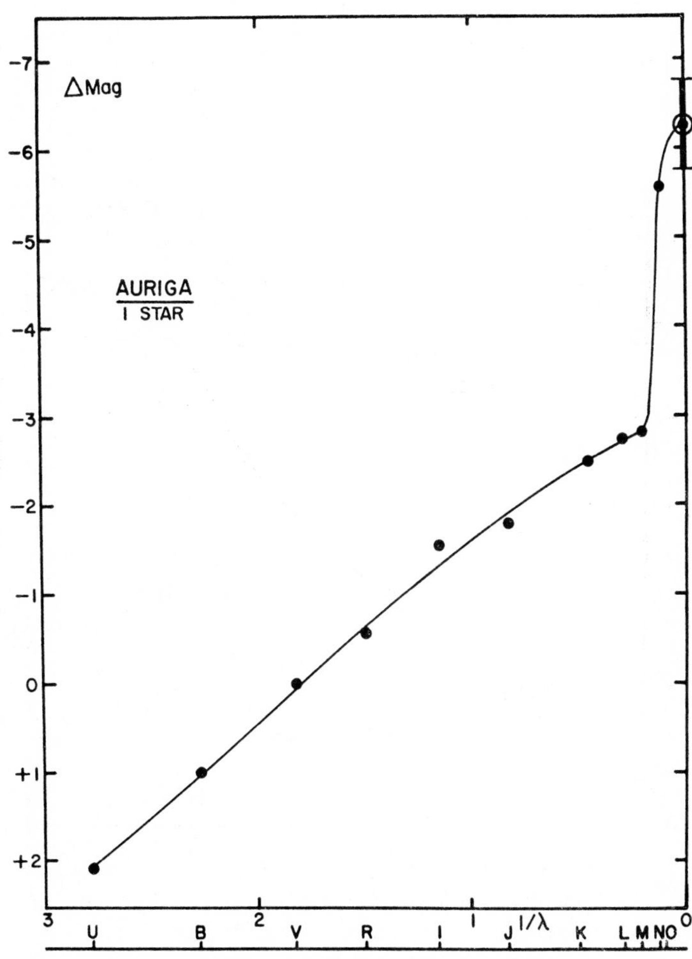

FIG. 33.—The interstellar-extinction curve for Auriga

TABLE 23

COLOR-EXCESS RATIOS FOR SCORPIUS

Star	Sp	$\frac{E_{U-V}}{E_{B-V}}$	$\frac{E_{B-V}}{E_{B-V}}$	$\frac{E_{V-R}}{E_{B-V}}$	$\frac{E_{V-I}}{E_{B-V}}$	$\frac{E_{V-J}}{E_{B-V}}$	$\frac{E_{V-K}}{E_{B-V}}$	$\frac{E_{V-L}}{E_{B-V}}$	E_{B-V}
δ Sco	B0 V	1.95	1.00	.63	1.37	2.47	2.63	2.47	.19
β Sco AB	B0.5V + B2V	1.61	1.00	.44	1.39	3.00	3.33	3.78	.18
ω^1 Sco	B1 V	1.50	1.00	.77	1.50	3.00	3.77	3.64	.22
ν Sco	B2 IV-V	1.79	1.00	.68	1.61	2.75	3.50	3.79	.28
σ Sco	B1 III	1.58	1.00	.75	1.68	2.70	3.35	3.45	.40
Mean		1.69	1.00	.67	1.55	2.77	3.35	3.48	

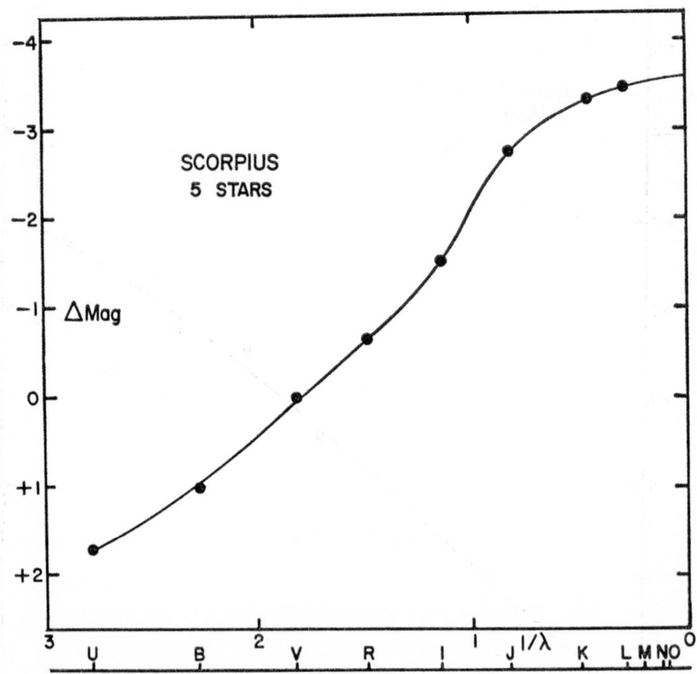

Fig. 34.—The interstellar-extinction curve for Scorpius

TABLE 24

COLOR-EXCESS RATIOS FOR THE BORDER REGION BETWEEN TAURUS AND ORION

Star	Sp	$\frac{E_{U-V}}{E_{V-K}}$	$\frac{E_{B-V}}{E_{V-K}}$	$\frac{E_{V-R}}{E_{V-K}}$	$\frac{E_{V-I}}{E_{V-K}}$	$\frac{E_{V-J}}{E_{V-K}}$	$\frac{E_{V-K}}{E_{V-K}}$	$\frac{E_{V-L}}{E_{V-K}}$	$\frac{E_{V-M}}{E_{V-K}}$	E_{B-V}
χ^2 Ori	B2 Ia	.62	.39	.30	.61	.79	1.00	----	----	.46
119 Tau	M2 Ib	.59	.32	.35	.61	.86	1.00	1.17	1.10	.38
Mean		.60	.36	.32	.61	.82	1.00	1.17	1.10	
		$\frac{E_{U-V}}{E_{B-V}}$	$\frac{E_{B-V}}{E_{B-V}}$	$\frac{E_{V-R}}{E_{B-V}}$	$\frac{E_{V-I}}{E_{B-V}}$	$\frac{E_{V-J}}{E_{B-V}}$	$\frac{E_{V-K}}{E_{B-V}}$	$\frac{E_{V-L}}{E_{B-V}}$	$\frac{E_{V-M}}{E_{B-V}}$	
Taurus - Orion Mean		1.67	1.00	.89	1.69	2.28	2.78	3.25	3.06	

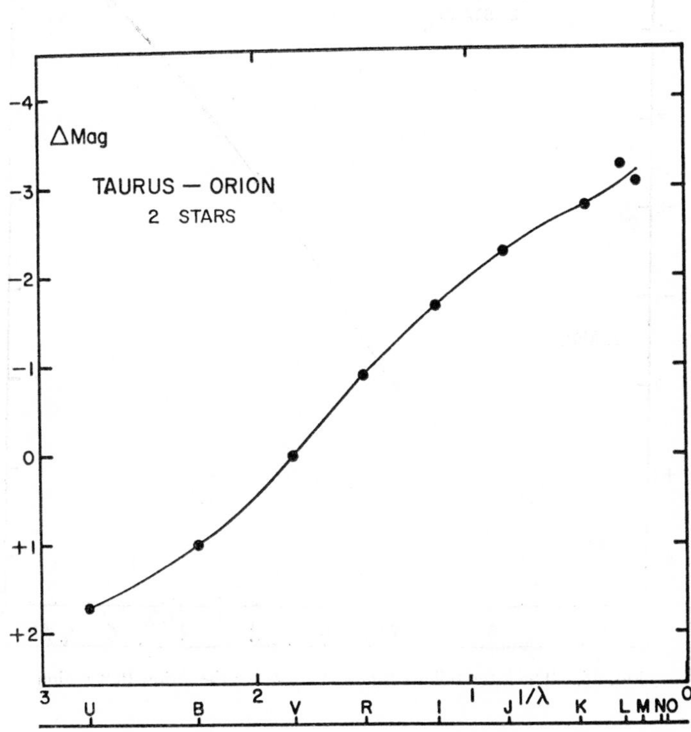

Fig. 35.—The interstellar-extinction curve for the border region between Taurus and Orion

TABLE 25

COLOR-EXCESS RATIOS FOR THE ORION BELT REGION

Star	Sp	$\frac{E_{U-V}}{E_{B-V}}$	$\frac{E_{B-V}}{E_{B-V}}$	$\frac{E_{V-R}}{E_{B-V}}$	$\frac{E_{V-I}}{E_{B-V}}$	$\frac{E_{V-J}}{E_{B-V}}$	$\frac{E_{V-K}}{E_{B-V}}$	$\frac{E_{V-L}}{E_{B-V}}$	E_{B-V}
NGC 2024 No. 1	(B1 Ia)	1.81	1.00	1.17	2.32	3.28	4.26	4.78	1.60
HD 37903	B1.5 V	1.81	1.00	.99	2.03	----	----	----	.36

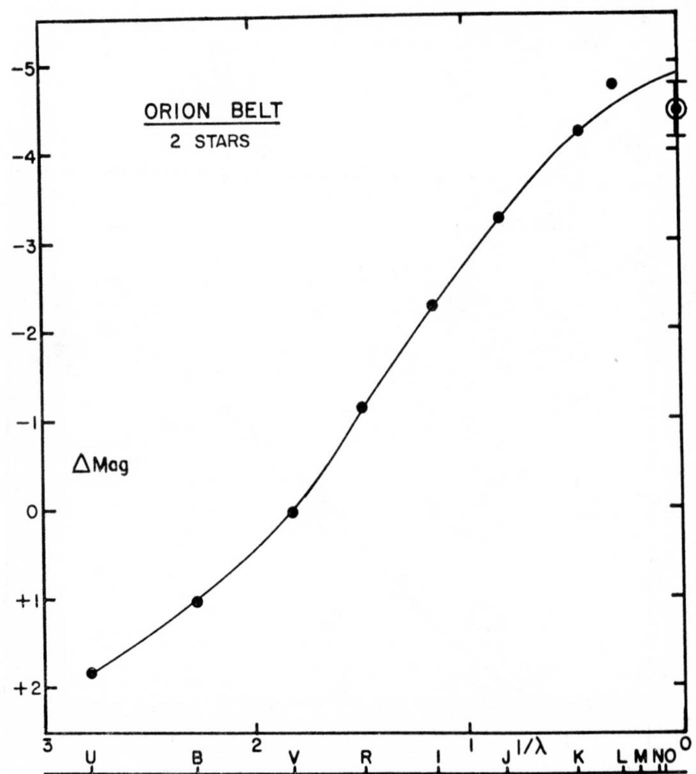

FIG. 36.—The interstellar-extinction curve for the Orion Belt region

TABLE 26

COLOR-EXCESS RATIOS FOR THE ORION SWORD REGION

Star	Sp	$\frac{E_{U-V}}{E_{V-K}}$	$\frac{E_{B-V}}{E_{V-K}}$	$\frac{E_{V-R}}{E_{V-K}}$	$\frac{E_{V-I}}{E_{V-K}}$	$\frac{E_{V-J}}{E_{V-K}}$	$\frac{E_{V-K}}{E_{V-K}}$	$\frac{E_{V-L}}{E_{V-K}}$	E_{B-V}
Trapezium	(O9.5)	.33	.19	.20	.47	.67	1.00	1.22	.35
θ^2 Ori	O9.5 Vp	.36	.21	.27	.55	.73	1.00	1.61	.22
HD 37042	B1 V	(.21)	.19	.31	.55	.77	1.00	----	.17
Mean		.34	.20	.25	.51	.71	1.00	1.37	----
		$\frac{E_{U-V}}{E_{B-V}}$	$\frac{E_{B-V}}{E_{B-V}}$	$\frac{E_{V-R}}{E_{B-V}}$	$\frac{E_{V-I}}{E_{B-V}}$	$\frac{E_{V-J}}{E_{B-V}}$	$\frac{E_{V-K}}{E_{B-V}}$	$\frac{E_{V-L}}{E_{B-V}}$	E_{B-V}
Orion Sword Mean		1.70	1.00	1.25	2.55	3.55	5.00	6.86	----
HD 37020*	B0.5 Vp	1.50	1.00	1.61	3.32	----	----	----	.28
HD 37022*	O6 p	1.65	1.00	1.23	2.71	----	----	----	.32
HD 37023*	B0.5 Vp	1.50	1.00	1.19	2.50	----	----	----	.36
HD 37061	B1 V	1.57	1.00	1.09	2.25	----	----	----	.53

* Individual stars of the Trapezium.

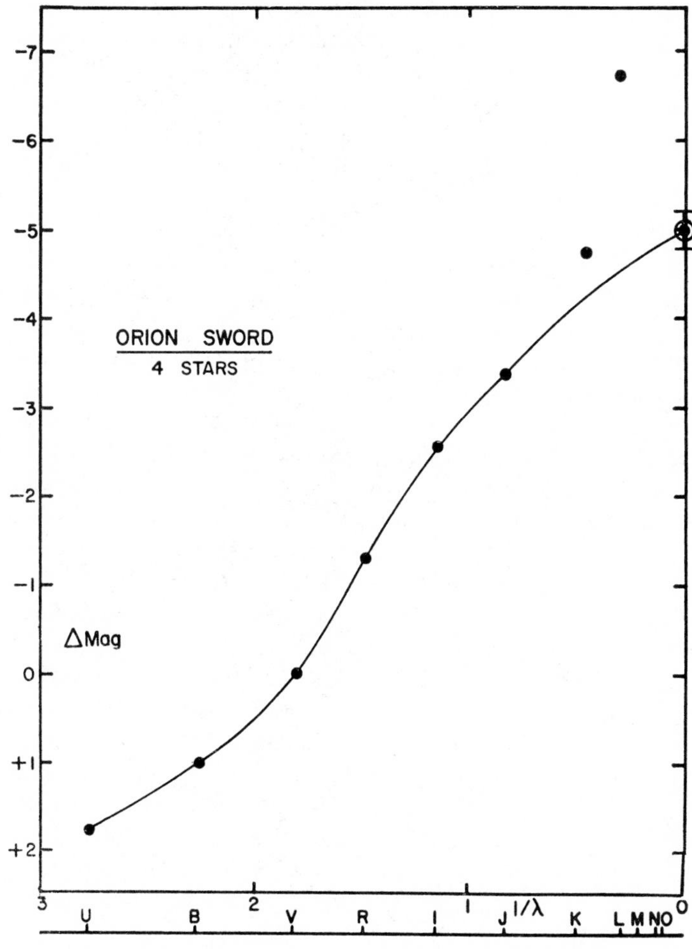

FIG. 37.—The interstellar-extinction curve for the Orion Sword (Trapezium) region

The data—both the color differences and the value of $R = 6.3 \pm 0.5$ (p.e.)—are, however, more uncertain. Nevertheless, the extrapolation of the color-difference data does agree well with the independently determined value of R. If this curve is correct, there is no large turn-up at L, or even at M, and there is a sudden rise at N. This is another curve which has, in the intermediate part, some similarity with that in Capricornus (Fig. 28).

Scorpius (*Fig. 34*). We have observed five stars in Scorpius and their color-excess ratios are listed in Table 23. The observational data for these stars depend upon a minimum of three, and as many as eight, accordant observations and have high weight. The color excesses of the Scorpius stars are rather small, however, and the ratios in Table 23 are less precise than those derived, for example, from the less well-observed but highly-reddened Cygnus stars. Nevertheless, it seems plain that the interstellar-extinction curve in Scorpius differs significantly from those found elsewhere. The extrapolated value of R is 3.6, a deviation from "normal" also suggested by Borgman and Blaauw (1964).

Taurus-Orion (*Fig. 35*). The first of three extinction curves for the region of Orion is shown in Figure 35 for the border of Taurus and Orion. The extinction curve seems similar to those for Perseus and Cygnus. $V - L$ for 119 Tau is well determined from several accordant observations, and would indicate a turn-up from K to L if the M data (a single observation) did not exist. In fact, comparison of Tables 24 and 18 shows that the color-excess ratios from $B - V$ to $V - L$ are almost identical. Perhaps the single M observation is in error.

Orion Belt (*Fig. 36*). Another type of extinction curve is found in the Belt region of Orion. Two stars have been measured here, but only one has been measured out to L (3.4 μ). Actually, the curve in Figure 36 is somewhat like those in Perseus and Cygnus, except that the extrapolated value of R is approximately 5, a value which is confirmed by the variable-extinction value of 4.5. The reddening of NGC 2024 No. 1 is so great that the color-excess ratios must be quite precise. The spectral type given for this star is only a rough estimate but, as shown by Johnson and Mendoza (1964), the color-excess ratios are not sensitive to the assumed spectral type. The color-difference data have been extrapolated to $1/\lambda = 0$ to agree with the value of R; this extrapolation does, however, leave a discrepancy at L that is large compared with the observational error.

Orion Sword (*Fig. 37*). This is the region in which Baade and Minkowski (1937*b*) first found evidence of "abnormality" in the interstellar-extinction law. All succeeding photometric work that included the infrared has shown the same result. There has been, from time to time, a suspicion that the excess infrared radiation, which we interpret here as due to interstellar extinction, might be due to unknown K-type companions of the early-type stars. There are, however, three arguments against this interpretation. (1) Essentially the same extinction curves are obtained from the integrated Trapezium, θ^2 Ori, HD 37042, and HD 37061, even though E_{B-V} differs among these stars by a factor of 3. It is improbable that all these stars have exactly the proper K-type companions to produce this result. Also, contrary to Underhill's (1964) assertion, the data of Johnson and Borgman (1963) for three of the individual Trapezium stars confirm, out to I (0.90 μ), the large infrared excess compared to that expected for a "normal" extinction law. (2) Although the excess radiation at I is significant, Sharpless (1963) reported that infrared spectrograms show no evidence for the existence of a late-type spectrum. (3) The results of two variable-extinction determinations, those of Sharpless (1952) and of Figure 22, confirm the high value of R obtained by extrapolation of the color-difference data. In view of the new data that have revealed many types of extinction curves for other regions, the curves for Orion no longer seem peculiar.

The color-difference data have been extrapolated to $1/\lambda = 0$ without regard to the K and L points. This extrapolation implies the assumption that $R = 5.0$, and the excess radiation at K and L must be explained in another way. This is, to be sure, only an interpretation of the data, but it is interesting to see where it leads. The excess radiation in the Trapezium region, above the extinction curve drawn in Figure 37, is

$$\text{Excess } K = 3.0 \times 10^{-16} \text{ watt cm}^{-2}\,\mu^{-1}$$

$$\text{Excess } L = 1.6 \times 10^{-16} \text{ watt cm}^{-2}\,\mu^{-1}\,.$$

The blackbody temperature computed from these energies is about 1300° K. It has been known for some time (Baade and Minkowski 1937*a*) that there is a clustering of very red stars surrounding the Trapezium. Dr. Haro and I have recently re-examined his infrared and visual plates of the Orion Sword region, and we agree that the I magnitudes of the red stars that were included in the (36 sec of arc) diaphragm average about 13 or 14, and that the visual magnitudes are much fainter. It is perhaps not a coincidence that the V and I magnitudes for 1300° K blackbodies that together produce the K and L energies are in accord with the photographic estimates. The corresponding excesses for θ^2 Ori are about one-tenth of those for the Trapezium and indicate a blackbody temperature between 800° and 1000° K. The probable errors of all these excesses are high—possibly as large as 50 per cent. Efforts to measure L for HD 37042, which is fainter than θ^2 Ori by $\Delta V = 1.4$ mag, were unsuccessful; however, if this star had an L-excess similar to those of the other objects, we would have detected it.

This interpretation that $R = 5.0$ and that the extra radiation at K and L comes from other objects, such as those found by Baade and Minkowski in the neighborhood of the Trapezium and θ^2 Ori, carries the implication that the interstellar-extinction law for the Orion region does not vary with distance from the Trapezium; such variation has been suggested earlier (Hallam 1959; Sharpless 1962, 1963; Johnson and Borgman 1963). With our present interpretation, interstellar extinction with $R = 5.0$ exists in the entire Orion region (including both the Sword and Belt regions), with excess K and L radiation in the region of the Trapezium, θ^2 Ori, and, probably, NGC 2024 No. 1. Haro and Moreno (1954) found that faint red stars, like those of Baade and Minkowski (1937*a*), also are associated with NGC 2024.

α Leonis (*Fig. 38*). The color-excess curve (not normalized) for α Leonis is shown in Figure 38. The UBV data indicate that, although the spectral type of α Leo is B7 V, the color type is nearer to B7.5 V; the latter type was used in computing the color excess in Table 27 and Figure 38. There seems to be no appreciable excess for α Leo out to L (3.4 μ), but there is a rapidly rising excess at longer wavelengths. If we interpret this far-infrared excess as due to a cool companion of α Leo, this companion must have a blackbody temperature of only about 300° K. Another possible interpretation is in terms of visually neutral extinction, perhaps caused by particles about 5 μ in diameter. There are, of course, other curves (Figs. 31–33) that might receive a similar interpretation.

α Leo is about 20 pc distant (Jenkins 1952), but this fact does not make improbable the existence of an interstellar cloud between the star and the Earth. Münch and Unsöld (1962) found, from observations of interstellar calcium lines, an interstellar gas cloud in front of α Oph. Our (unpublished) photometry shows the star to be slightly reddened. α Oph is also about 20 pc from the Earth.

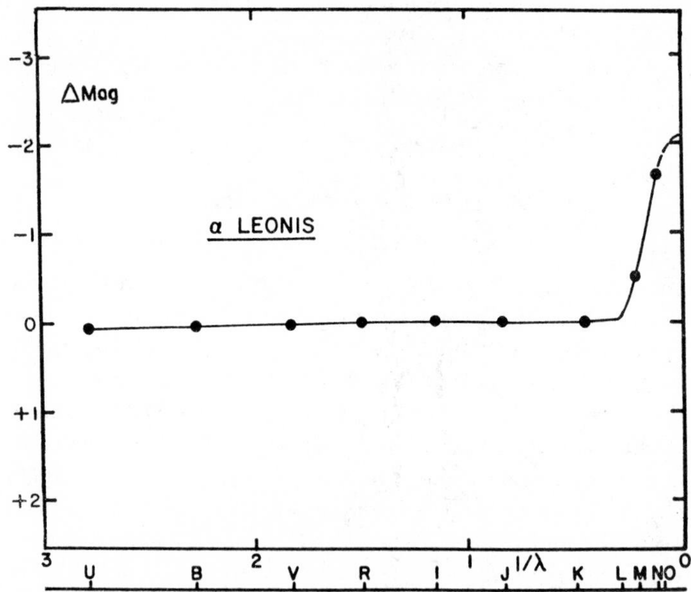

FIG. 38.—The color-excess curve for α Leonis. This curve differs from all other similar curves in this chapter; since $E_{B-V} = 0.00$ for α Leo, the un-normalized color excesses are plotted.

TABLE 27

COLOR EXCESSES FOR α LEO AND φ PER

Star	Sp	E_{U-V}	E_{B-V}	E_{V-R}	E_{V-I}	E_{V-J}	E_{V-K}	E_{V-L}	E_{V-M}	E_{V-N}	Comparison Type
α Leo	B7 V	- .01	- .01	+ .03	+ .03	+ .01	+ .01	+ .04	+ .51	+1.67	B7.5 V
φ Per	B2pe	+ .13	+ .20	+ .27	+ .51	+ .75	+1.41	+1.89	+3.10	----	B2 V
φ Per	B2pe	+ .16	+ .10	+ .12	+ .22	+ .32	+ .64	+ .77	----	----	κ Dra

TABLE 28

COLOR-EXCESS RATIOS FOR φ PERSEI

(Comparison Star: κ Draconis)

Star	E_{U-V}/E_{B-V}	E_{B-V}/E_{B-V}	E_{V-R}/E_{B-V}	E_{V-I}/E_{B-V}	E_{V-J}/E_{B-V}	E_{V-K}/E_{B-V}	E_{V-L}/E_{B-V}	E_{B-V}
φ Per	1.60	1.00	1.20	2.20	3.20	6.40	7.70	0.10

ϕ Persei (*Fig. 39*). Another notable curve is exhibited by ϕ Persei (Tables 27, 28, and Fig. 39). The peculiarities of this extinction curve, in particular the rapid rise at the longer wavelengths, were noticed first by Stebbins and Kron (1956). Hiltner's (1947) spectrograms show that, in the region of the *I* filter (0.90 μ), the spectra of ϕ Per and κ Dra are almost identical. According to our unpublished data, the latter star is bluer, a fact that also is shown by the density gradients of Hiltner's spectrograms. There is no evidence in the spectrum of ϕ Persei for the existence of a late-type companion—only a rise in the continuous spectrum relative to that of κ Draconis. If, in accord with the spectra, we interpret the differences in color-indices between the two stars as due to

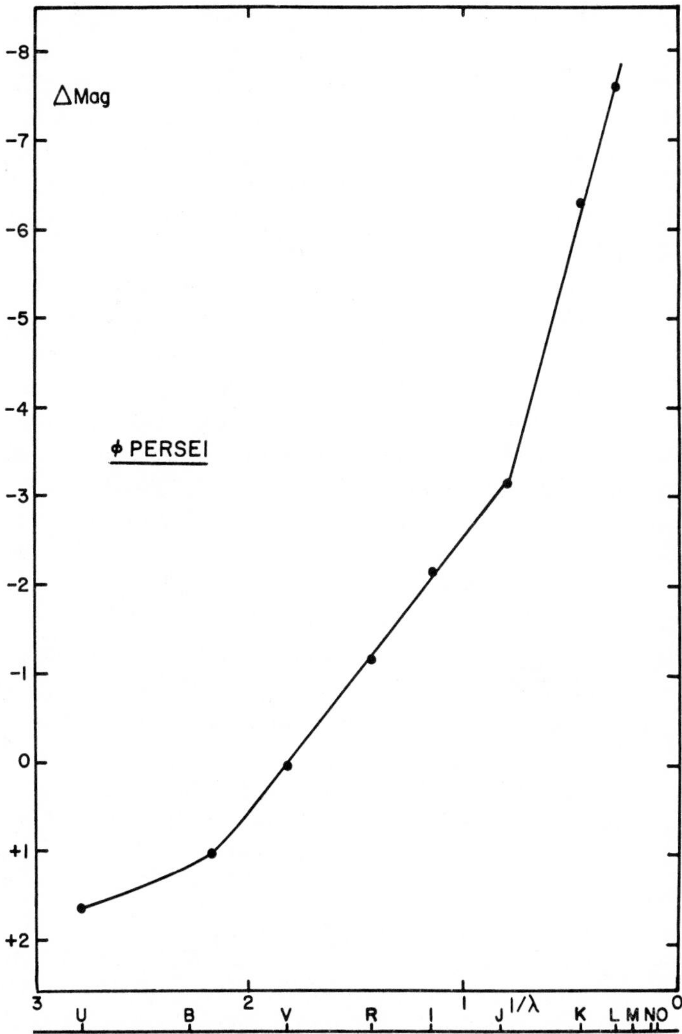

Fig. 39.—The interstellar-extinction curve for ϕ Persei with κ Draconis as comparison star

interstellar reddening, the color-excess ratios in Table 28 are the result. Out to L (3.4 μ) they are similar to those derived (Johnson 1965*b*) using a B2 V comparison star, indicating a large value of R for ϕ Persei.

6. DISCUSSION

More extensive investigation of interstellar reddening and extinction has again shown that the results of the color-difference method are in close agreement with those of the variable-extinction method. We have also found that the use of cluster distances determined from their apparent diameters strengthens the extinction determinations. We have not attempted to analyze the interstellar-extinction curves now known; instead, the reader is invited to re-examine the data and the figures for himself.

The subject of the extinction in the region of Perseus requires disquisition. The cluster diameters indicated a larger extinction in this region than did the variable-extinction method, even though the two methods agreed in the Milky Way regions on both sides of Perseus. The color-difference data in Perseus tended to agree with the variable-extinction result, but at the cost of discordance between the values of E_{V-L} for early-type and late-type stars in Perseus and the introduction of a correlation between E_{V-L}/E_{V-K} with E_{B-V} in Ophiuchus and Perseus. This correlation could be demonstrated to exist elsewhere, and suggests that the intrinsic $V - L$ (Table 11) should be made bluer by 0.04 mag. When we make this change (which reduces greatly the $V - L$ discordance between early-type and late-type stars in Perseus), however, we find a turn-up from K to L in Perseus and Ophiuchus similar to that observed in Aquila, Cygnus, Cepheus, and elsewhere. In some regions where this turn-up has been observed (Cepheus, NGC 2244), it has been associated with high extinction and large values of R. Thus, this modification of intrinsic $V - L$ may possibly be interpreted as bringing the color-difference data into agreement with the extinction determined from cluster diameters in Perseus. It has already been emphasized that the variable-extinction method determined only the minimum extinction and that, therefore, these different results are not necessarily contradictory.

10 Lacertae (*Fig. 40*). In this connection there is another star whose data should be mentioned. Table 29 contains the color-excess ratios for 10 Lacertae; the values of E_{V-L} were computed from the corrected intrinsic $V - L$ of -0.96. Figure 40 shows the extinction curve for this star; the short dotted line represents the curve before the correction of $V - L$.

At the time that the intrinsic colors were being derived, it was noticed that 10 Lac would show this turn-up from K to L unless the intrinsic $V - L$ were set at -0.92, a fact that had some weight in my decision to adopt this value. On the other hand, the use of -0.92 has resulted in values of E_{V-L}/E_{V-K} for ζ Oph and the early-type stars in the double cluster that seem too small. Perhaps we should adopt the bluer value and accept the implication of higher extinction in Perseus and, from Figure 40, in Lacerta. The extinction and reddening are so great in Aquila and Cygnus that their turn-ups from K to L are unaffected by this change. The turn-up in Cepheus is increased from its already high value.

7. CONSEQUENCES OF THE VARIATION IN INTERSTELLAR EXTINCTION

7.1. Range of Values of *R*

One conclusion seems inescapable: that there is no unique interstellar-extinction law. Even after the examination of much observational evidence, we still find no values of R significantly smaller than 3, although there are many larger ones. Thus, $R \approx 3$ seems to be the minimum value, with the maximum around 6—if we ignore the color-excess curves for α Leo and ϕ Per in Figures 38 and 39. The common practice of blanket application of $R = 3$ to all regions of the sky almost certainly is incorrect and undoubtedly has resulted in the derivation of a distance scale for our Galaxy that is too large. It may well be that the present photometric distance scale should be reduced by an average of from 25

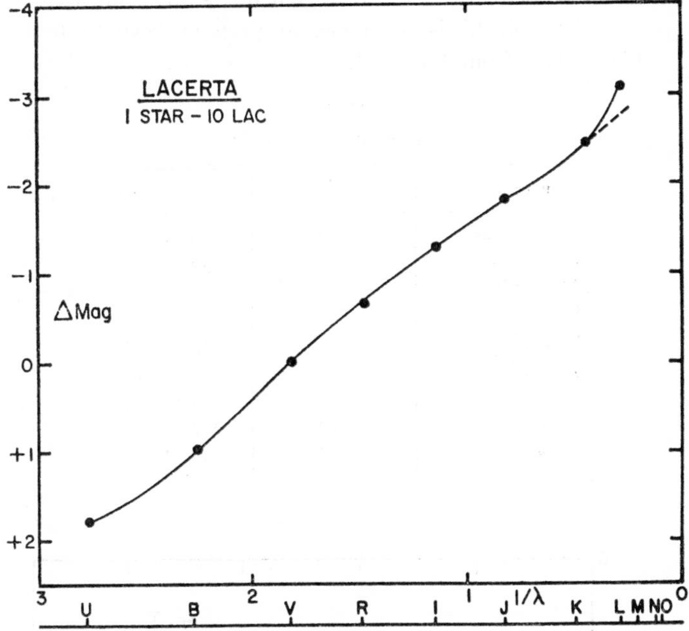

Fig. 40.—The interstellar-extinction curve for 10 Lacertae

TABLE 29

COLOR-EXCESS RATIOS FOR 10 LACERTAE

Star	Sp	$\frac{E_{U-V}}{E_{V-K}}$	$\frac{E_{B-V}}{E_{V-K}}$	$\frac{E_{V-R}}{E_{V-K}}$	$\frac{E_{V-I}}{E_{V-K}}$	$\frac{E_{V-J}}{E_{V-K}}$	$\frac{E_{V-K}}{E_{V-K}}$	$\frac{E_{V-L}}{E_{V-K}}$	E_{B-V}
10 Lac	09 V	0.74	0.41	0.26	0.52	0.74	1.00	1.26	0.11
		$\frac{E_{U-V}}{E_{B-V}}$	$\frac{E_{B-V}}{E_{B-V}}$	$\frac{E_{V-R}}{E_{B-V}}$	$\frac{E_{V-I}}{E_{B-V}}$	$\frac{E_{V-J}}{E_{B-V}}$	$\frac{E_{V-K}}{E_{B-V}}$	$\frac{E_{V-L}}{E_{B-V}}$	
10 Lac	09 V	1.82	1.00	0.64	1.27	1.82	2.46	3.09	

to 30 per cent. As Howard and Kirk (1964) have emphasized, this would cause an approximately equal percentage change in Oort's constant, A (which would be increased to around 20 km/sec/kpc), and the distance to the galactic center, R_o (which would be reduced to 7 or 8 kpc).

The analyses of the data in this chapter have resulted in a number of determinations of $R = A_v/E_{B-V}$. The selected, most precise values that can be obtained are listed in Table 30 and are plotted in Figure 41. The larger value of R for the Cas-Per region is represented by a filled circle, whereas the smaller value is represented by an open circle. The error bars correspond to the probable errors listed in Table 30.

A strong variation of R with galactic longitude is evident in Figure 41. This diagram (solid line) is similar to that in Figure 4, which was plotted from the cluster-diameter data. A value of $R = 5.6$ for the Cas-Per region ($l^{II} \sim 130°$) fits well with the trend of the other points, although this is, of course, not proof of its correctness. The range in R exhibited in Figure 41 is from 3.6 to 6.1.

TABLE 30

THE BEST VALUES OF R

Region	ℓ^{II}	R	p.e.
Sagittarius	5°	3.8	± 0.4
Serpens	17°	3.6	0.3
Aquila-Cygnus	50° - 70°	3.6	0.2
Cepheus	90° - 110°	4.8	0.2
Cassiopeiea-Perseus	120° - 135°	3.0 5.6	0.3 0.4
Auriga	170° - 180°	6.3	0.4
Gemini	188°	5.9	0.5
Monoceros	207°	5.7	0.4
Orion	207°	5.0	± 0.2

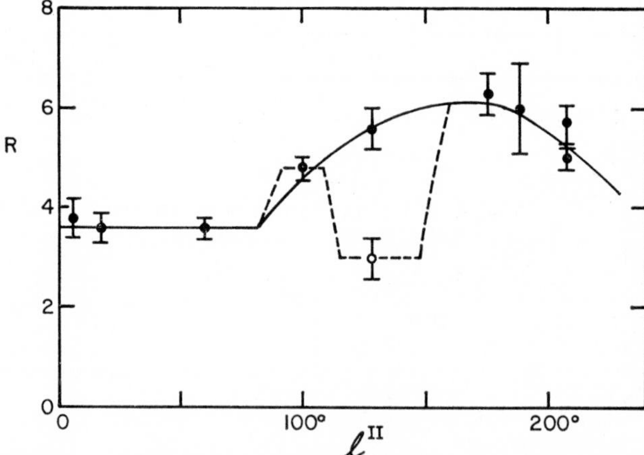

FIG. 41.—The variation of $R = A_v/E_{B-V}$ with galactic longitude. The solid line represents the relationship if $R = 5.6$ in Perseus; the dotted line, if $R = 3.0$ (see text).

Note added in proof.—A more comprehensive investigation of the extinction curve in the region of the double cluster in Perseus has recently been completed (Johnson and Mendoza V. 1966). New observations, including measures at 5 μ (magnitude M), on M supergiants in this region yielded from the color-difference method an interstellar-extinction curve similar to that found in Cepheus (Fig. 31). The change from the curve in Figure 24 is due almost entirely to the great brightnesses of the reddened stars at 5 μ compared with those expected if R were 3. Furthermore, the new M-star observations gave the same extinction curve from 0.35 μ to 3.4 μ (U to L) as the early-type stars in the same region, again confirming the usefulness of M supergiants for this purpose. According to this new interpretation, in Perseus $R = A_v/E_{B-V}$ is greater than 3.0, the (minimum) value obtained from the variable-extinction method, and may be as high as 5.6, the value obtained from cluster diameters.

7.2. Distances of Clusters and the Galactic Center, and the Oort Constant A

If Figure 41 is correct, then it is necessary to revise the distances of many galactic clusters, especially those with galactic longitudes between 100° and 200°. In order to explore the consequences of the new and larger values of R upon galactic structure and rotation, we have revised the distances of approximately one hundred galactic clusters. These distances are based upon the photometric distance moduli of Johnson *et al.* (1961) with interstellar-extinction corrections computed from R as indicated by the solid line in Figure 41 (or for $l^{II} > 230°$, estimated on the basis of Figure 4). The individual values of R and the revised distances are listed in Table 31. These distances are not identical with those listed in earlier tables of this chapter, because those in Table 31 are the final product of the analysis, whose several stages are illustrated by the other tables.

The positions of these clusters projected upon the galactic plane are shown in Figure 42. They appear to define three "arms," which are remarkably similar to those described by Morgan, Sharpless, and Osterbrock (1952) and Morgan, Whitford, and Code (1953). The diagrams of Becker (1964) and Schmidt-Kaler (1964) are also similar. The distance scale between arms in Figure 42 is about three-fourths of that found by the other investigators; this scale is a product of the larger extinction corrections that we have made. The two branches in the solar arm, first detailed by Morgan, Whitford, and Code, are rather well defined in Figure 42, although the diagrams of Becker and Schmidt-Kaler are not so clear on this point.

A number of clusters in the Perseus region, among them the double cluster, have transferred their allegiance to the solar arm; others, NGC 457 and NGC 581, for example, have remained in the outer arm. According to this analysis, NGC 1893 is not associated with I Aur (cf. Table 5) but is an inhabitant of the outer arm. It is noteworthy that this outer arm no longer appears to terminate abruptly at $l^{II} = 140°$, as, for example, in the diagrams of Becker; it may be traced in Figure 42 at least as far as $l^{II} = 180°$. The inner arm, at a distance of about 1000 pc, is rather well defined and appears to be curved, perhaps in the manner indicated by Schmidt-Kaler.

Although the distance revisions were made following Figures 4 and 41, which indicate $R = 3.6$ in the half of the sky toward the galactic center, it is not safe to assume that this value may be applied everywhere in the region. For example, I Ara has a well-determined value of $R = 6.0$ (cf. Fig. 16), which was used in deriving its distance for Table 31.

TABLE 31

REVISED CLUSTER DISTANCES

Cluster	R	Distance (parsecs)	Cluster	R	Distance (parsecs)
NGC 103	5.4	1820	NGC 2324	5.2	2700
129	5.4	870	2353	4.8	950
225	5.4	460	2362	4.0	1440
457	5.5	1660	2422	4.5	460
581	5.6	1660	2439	3.8	1440
654	5.6	520	2447	3.8	1090
663	5.6	830	2516	3.6	350
744	5.7	870	2632	---	160
752	5.8	390	2682	5.0	790
869	5.7	1150	3330	3.6	1320
884	5.7	1150	4103	3.6	1090
957	5.8	790	4755	3.6	760
1027	5.8	760	6087	3.6	870
1039	5.9	380	6405	3.6	600
1245	5.9	1590	6475	3.6	240
1342	6.0	380	6494	3.6	600
1444	6.0	380	6530	3.6	1260
1502	5.9	320	6531	3.6	1150
1528	6.1	520	6611	3.6	2000
1545	6.0	480	6633	3.6	300
1647	6.0	320	6694	3.6	1260
1662	5.8	260	6705	3.6	1590
1664	6.1	830	6709	3.6	830
1778	6.1	830	6755	3.6	1450:
1893	6.1	1740	6802	3.6	870:
1907	6.1	790	6823	3.6	1320
1912	6.1	910	6830	3.6	1520
1960	6.1	910	6834	3.6	2500
2099	6.0	830	6866	3.7	1150
2129	5.8	870	6871	3.6	1520
2158	5.9	2750	6882/5	3.6	600
2168	5.8	660	6910	3.6	1260
2169	5.7	870	6913	3.6	870
2215	5.0	910	6940	3.6	760
2244	5.3	910	7031	4.3	660:
2251	5.4	1250	7062	4.3	1090
2264	5.4	690	7063	3.9	600
2287	4.5	660	7067	4.3	1740:
2301	5.2	760	7086	4.5	830
2323	4.8	720	7092	4.3	250

TABLE 31—*Continued*

Cluster	R	Distance (parsecs)	Cluster	R	Distance (parsecs)
NGC 7128	4.6	1090	I Ara	6.0	790
7142	4.8	870:	I Aur	6.0	1090
7160	4.9	630	I Cas	5.7	1000
7209	4.5	830			
7235	4.8	1380	III Cas	5.7	760
			I Cep	6.0	400
7261	4.9	460	II Cep	4.5	2620
7380	5.0	1200	III Cep	4.8	440
7510	5.1	1150	Coma	---	80
7654	5.1	1090			
7789	5.2	1450	II Cyg	3.6	1650
			I Gem	6.0	1000
7790	5.3	2090	Hyades	---	40
IC 1805	5.7	760	I Lac	4.5	520
1848	5.8	1000	I Ori	5.4	380
4665	3.6	320			
4725	3.6	520	α Per	6.0	145
			Pleiades	6.1	115
4996	3.6	1260	Tr 2	5.8	380
5146	3.6	870	Tr 35	3.6	1820

As a further test of the consequences of the revision of R, we repeated the galactic-rotation computation of Johnson and Svolopoulos (1961) using the new distances from Table 31. We obtained $A = 20$ km/sec/kpc instead of the 15 computed from the old distance scale. If $AR_o = 156$ km/sec (Schmidt 1956), the distance to the galactic center R_o is 7.8 kpc.

Although it may seem a retrogression to obtain such values for A and R_o, it will not be entirely a surprise to workers in the field of galactic rotation. For example, Kraft and

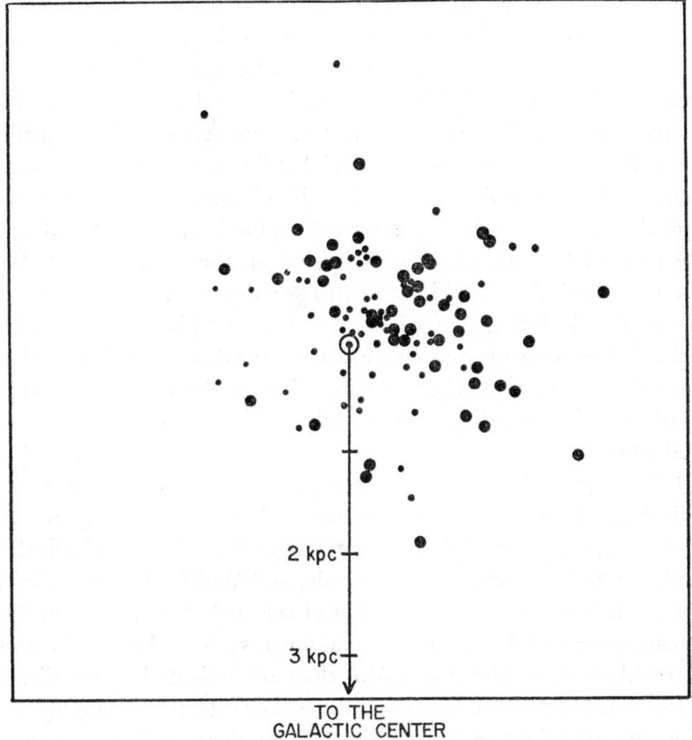

FIG. 42.—The positions in the galactic plane of one hundred galactic clusters in Table 30. The large spots designate young clusters, whose earliest type stars are B2 or earlier; the small spots, all other clusters; ⊙ indicates the Sun. Galactic longitude increases anticlockwise.

Schmidt (1963) say: "All one can do is keep in mind that evidence from proper motions would tend toward fainter absolute magnitudes, smaller distances and a larger value of A." They state further that their Cepheid distance scale (which is the same as that of galactic clusters with $R = 3.0$) "is 1.27 times that implied by the proper motions." This point is discussed also by Stibbs (1964), who obtained 1.29 as the ratio of the photometric (Cepheid and galactic cluster) distance scale to the proper-motion distance scale. The new cluster-distance scale (Table 31) is smaller than the old one by almost exactly this factor. Münch and Münch (1964) found a discrepancy between the rotational velocities of stars and of interstellar hydrogen, and suggest that possible departures from the " 'standard' reddening-extinction law" could be the source. There is no contradiction of Arp's (1965) distance determinations to the galactic center; for example, if we assume that $R = 4.1$, a value completely in agreement with Figure 41 for $l^{II} = 0$, we obtain $R_o = 7.9$ kpc from Arp's data. This emphasizes the strong dependence of the galactic distance scale upon R.

8. CONCLUSION

Much further work will be necessary to clarify the interstellar-extinction picture. At first glance, the multiplicity of interstellar-extinction curves and the resultant values of $R = A_v/E_{B-V}$ seem only to complicate matters. As we have seen, however, this seeming complication may help to resolve some outstanding discrepancies.

In concluding this chapter, I wish to emphasize that I have not intended to present a consistent picture of the law of interstellar extinction; indeed, such a picture cannot be put together from the observational material at hand. The single firm conclusion that can be drawn from the discussion presented in this chapter is that the interstellar-extinction law is not the same everywhere. It seems also quite likely that $R = 3$ is the minimum value of the ratio of total-to-selective extinction on the *UBV* system. We cannot go beyond these two conclusions at this time with any degree of certainty, although, as the discussion in Section 7 shows, high values of R in many places in the Milky Way may be acceptable. It has been my purpose to bring together such observational evidence as can be brought to bear upon the problem of interstellar extinction and to show that the analysis of these data by the direct procedures does not confirm the existence of an universal extinction law.

Perhaps different procedures of interpretation and analysis are needed, and it is by no means certain that the results of this chapter will be confirmed by future work. For example, we have assumed that all of the early-type stars that lie within the boundaries of the associations as defined by Morgan, Code, and Whitford are physical members of the associations. It is obvious that rejection of selected stars from membership in the associations can make variable-extinction diagrams, such as those of Figures 10, 12, 13, and 16, compatible with smaller values of R than are indicated by the diagrams as they stand. But the boundaries of these associations were defined originally because there appear to be concentrations of early-type stars in these regions, compared with the regions outside the boundaries. We cannot reject many stars from association membership without running the risk of saying that some associations are merely background stars viewed through holes in the interstellar material. This conflict illustrates the uncertainty of the analysis of the presently available data. If I were to hazard a guess at this time, I would say that interstellar extinction generally may be characterized by a value of $R = A_v/E_{B-V}$ around 3.5, with local deviations to higher values, mostly in regions near hot stars—a conclusion that Borgman and I have already suggested.

It is plain that much more observational work in the infrared is necessary before we will be able to reach a real understanding of the nature of interstellar extinction and the material producing it. The situation in Cepheus must be explored much more thoroughly, because the extinction curve indicated by the available data is quite unexpected. We must observe more stars, of different spectral types, out to very long wavelengths. We must also observe many other stars elsewhere in the Milky Way. Unfortunately, these additional data will be forthcoming slowly, since the stars are very faint in the long wavelengths and infrared detectors are not as sensitive as we would like. We intend to continue our instrumentation and observational programs at the University of Arizona, and it is to be expected that additional observational data bearing on these matters will become available fairly soon.

REFERENCES

Arp, H. C. 1965, *Ap. J.*, **141**, 43.
Arp, H. C., and Cuffey, J. 1962, *Ap. J.*, **136**, 51.
Baade, W., and Minkowski, R. 1937*a*, *Ap. J.*, **86**, 119.
———. 1937*b*, *ibid.*, p. 123.
Becker, W. 1964, *Zs. f. Ap.*, **58**, 202.
Blaauw, A. 1956, *Ap. J.*, **123**, 408.
———. 1963, *Basic Astronomical Data*, ed. K. Aa. Strand (Chicago and London: University of Chicago Press), p. 383.
Blaauw, A., Hiltner, W. A., and Johnson, H. L. 1959, *Ap. J.*, **130**, 69; and Erratum, *Ap. J.*, **131**, 527.
Borgman, J. 1961, *B.A.N.*, **16**, 99.
Borgman, J., and Blaauw, A. 1964, *B.A.N.*, **17**, 358.
Crawford, D., Limber, D. N., Mendoza V., E. E., Schulte, D., Steinman, H., and Swihart, T. 1955, *Ap. J.*, **121**, 24.
Danielson, R. E., Woolf, N. J., and Gaustad, J. E. 1965, *Ap. J.*, **141**, 116.
Divan, L. 1954, *Ann. d'ap.*, **17**, 456.
Eggen, O. J., and Herbig, G. H. 1964, *Royal Obs. Bull.*, No. 82.
Hallam, K. L. 1959, Ph.D. thesis, University of Wisconsin.
Hardie, R. H., Heiser, A. M., and Tolbert, C. R. 1964, *Ap. J.*, **140**, 1472.
Haro, G., and Moreno, A. 1954, *Bol. Obs. Tonantzintla y Tacubaya*, **2**, No. 7, 15.
Hiltner, W. A. 1947, *Ap. J.*, **105**, 212.
———. 1956, *Ap. J. Suppl.*, **2**, 389.
Howard, W. E. III, and Kirk, J. G. 1964, *A.J.*, **69**, 544.
Hulst, H. C. van de. 1949, *Rech. Astr. Obs. Utrecht*, **11**, Part 1, 1.
Jenkins, L. F. 1952, *General Catalogue of Trigonometric Stellar Parallaxes* (New Haven: Yale University Observatory).
Johnson, H. L. 1958, *Lowell Obs. Bull.*, **4**, 37.
———. 1962, *Ap. J.*, **136**, 1135.
———. 1963, *Basic Astronomical Data*, ed. K. Aa. Strand (Chicago and London: University of Chicago Press), p. 204.
———. 1964, *Bol. Obs. Tonantzintla y Tacubaya*, **3**, 305.
———. 1965*a*, *Vistas in Astronomy* (London: Pergamon Press).
———. 1965*b*, *Ap. J.*, **141**, 923.
Johnson, H. L., and Borgman, J. 1963, *B.A.N.*, **17**, 115.
Johnson, H. L., and Hiltner, W. A. 1956, *Ap. J.*, **123**, 267.
Johnson, H. L., Hoag, A. A., Iriarte, B., Mitchell, R. I., and Hallam, K. L. 1961, *Lowell Obs. Bull.*, **5**, 133.
Johnson, H. L., and Mendoza V., E. E. 1964, *Bol. Obs. Tonantzintla y Tacubaya*, **3**, 331.
———. 1966, *Ann. d'ap.*, **29**, 525.
Johnson, H. L., and Morgan, W. W. 1954, *Ap. J.*, **119**, 344.
Johnson, H. L., and Svolopoulos, S. 1961, *Ap. J.*, **134**, 868.
Kraft, R. P., and Schmidt, M. 1963, *Ap. J.*, **137**, 249.
Mitchell, R. I. 1960, *Ap. J.*, **132**, 68.
Morgan, W. W., Johnson, H. L., and Roman, N. G. 1954, *Pub. A.S.P.*, **66**, 85.
Morgan, W. W., and Roman, N. G. 1950, *Ap. J.*, **111**, 426.
Morgan, W. W., Sharpless, S., and Osterbrock, D. 1952, *A.J.*, **57**, 3.
Morgan, W. W., Whitford, A. E., and Code, A. D. 1953, *Ap. J.*, **118**, 318.
Münch, G., and Münch, L. 1964, *Ap. J.*, **140**, 162.
Münch, G., and Unsöld, A. 1962, *Ap. J.*, **135**, 711.
Pesch, P. 1960, *Ap. J.*, **132**, 689.
Schmidt, M. 1956, *B.A.N.*, **13**, 15.
Schmidt-Kaler, Th. 1964, *Zs. f. Ap.*, **58**, 217.
Sharpless, S. 1952, *Ap. J.*, **116**, 251.
———. 1954, *ibid.*, **119**, 200.
———. 1962, *ibid.*, **136**, 767.
———. 1963, *Basic Astronomical Data*, ed. K. Aa. Strand (Chicago and London: University of Chicago Press), p. 225.
Stebbins, J., Huffer, C. M., and Whitford, A. E. 1940, *Ap. J.*, **91**, 20; *Ap. J.*, **92**, 193.

Stebbins, J., and Kron, G. E. 1956, *Ap. J.*, **123**, 440.
Stebbins, J., and Whitford, A. E. 1943, *Ap. J.*, **98**, 20.
———. 1945, *ibid.*, **102**, 273.
Stibbs, D. W. N. 1964, *Observatory*, **84**, 108.
Trumpler, R. J. 1930, *Lick Obs. Bull.*, **14**, 154.
Underhill, A. B. 1964, *Observatory*, **84**, 35.
Walker, G. A. H. 1962, *Observatory*, **82**, 52.
Walker, M. F. 1957, *Ap. J.*, **125**, 636.
———. 1961, *ibid.*, **133**, 438.
Wallenquist, A. 1959, *Ann. Upp. Obs.*, Vol. **17**, No. 7.
Whiteoak, J. B. 1963, *M.N.*, **125**, 105.
Whitford, A. E. 1948, *Ap. J.*, **107**, 102.
———. 1958, *A.J.*, **63**, 201.

CHAPTER 6

Interstellar Grains

J. MAYO GREENBERG
Rensselaer Polytechnic Institute, Troy, New York

1. INTRODUCTION

ALTHOUGH IT IS now well established that an appreciable amount of matter in the form of gas and solid particles exists between the stars, as recently as forty years ago the question of whether dark markings in the Milky Way were holes between the stars or were due to obscuring matter was still open. When it became known that obscuration existed not only in the visible dark clouds, but also in apparently clear regions, the problem of estimating its amount became important in theories of galactic structure. The apparent brightness of stars, even in relatively clear regions, may be considerably reduced by interstellar extinction, and distance measurements would consequently be modified.

Trumpler's discovery (1930) of color excesses provided the first definitive proof that interstellar reddening is characteristic of the extinction. Estimates of the total extinction on the basis of reddening were, and still are, useful in correcting distances. The basic assumption that the wavelength dependence of the extinction is invariant is, however, clearly not valid (see, e.g., Baade and Minkowski 1937, and chapter by Harold Johnson, this vol.), although large variations may be restricted to small regions of interstellar space.

About thirty years ago (Hall 1937; Stebbins, Huffer, and Whitford 1939), investigations into wavelength dependence of the extinction led to the discovery that over a fairly wide spectral range the extinction varied inversely as the wavelength (the "λ^{-1} law"). Later work showed that the λ^{-1} law is only a rough approximation of the actual law of reddening and is certainly not even valid in the spectral region from 3000 to 20,000 A.

The following questions remain: (1) How uniform is the extinction law? (2) How accurately can we determine the photographic extinction (ratio of total to selective extinction) from the amount of reddening? (3) Do the observed variations of the extinction law follow some pattern and/or can they be correlated with some other observational phenomenon?

Various proposals about the nature of the mechanism of interstellar extinction have been made. An early hypothesis that extinction is produced by the scattering of starlight by particles such as very small meteors or micrometeorites, perhaps fragments of larger bodies, led to the consideration of metallic grains. It has since become evident that meteors and meteorites are not of interstellar origin but are localized within our Solar System.

The early work of Schalén (1936) and Greenstein (1938) was mostly concerned with grains of a metallic character; otherwise, there is considerable overlap between their work and that of much more recent origin.

In 1935 Lindblad investigated the possibility of grain growth by condensation from the interstellar gas; this investigation was taken further by Oort and van de Hulst (1946). Many present theories of formation of interstellar (including metallic) grains are based on these investigations, or at least apply some of the principles which were considered at that time.

The discovery of interstellar polarization (Hall 1949; Hiltner 1949) raised further questions, since the interstellar grains undoubtedly play a role. According to present theories, polarization is produced by aligned grains, the alignment presumably being due to magnetic fields which are galactic; thus the grains are important for modern theories of galactic structure.

In addition, the effect of the grains on the temperature balance of the interstellar gas, the possibility of molecule formation (e.g., of H_2 or OH) on the grains, and their role in star formation gives them an important place in astrophysics, although grains constitute only a small fraction, no more than about 10^{-3} to 10^{-4}, of the total galactic mass.

Several possible explanations for the extinction and polarization exist, and at present we can only establish some preferences. It has been generally accepted that the grains are dielectric (mostly ices), but both metallic and graphite particles have been proposed to explain the rather high degree of stellar polarization. Qualitative theoretical considerations had led some workers to doubt that dielectric grains could account for the maximum observed ratio of polarization to extinction. Primarily for this reason, Cayrel and Schatzman (1954), and subsequently Hoyle and Wickramasinghe (1962), proposed the existence of graphite flakes in interstellar space because of their inherently enormous potential for polarizing radiation. Very small particles (large molecules or clumps of molecules) consisting of free radicals must be considered also since growth processes of the dielectric grains are still not fully understood.

Since growth and formation of grains, grain alignment, and polarization by grains cannot all be quantitatively explained at the same time by any present theory, all models so far proposed should be considered as fully as possible. If a choice is not completely defined, new observations may aid in the selection. Regardless of whether or not the conclusions reached in this chapter turn out to be correct, it is hoped that at least most of the theoretical tools have been amply explained and that the important physical processes involving grains have been mentioned so that workers in the field can use this chapter as a "do it yourself" monograph.

2. OBSERVATIONAL EVIDENCE FOR GRAINS

Extinction and polarization of light from stars in distant parts of the Galaxy are the best indicators of the presence of interstellar particles. The existence of reflection nebulae and the scattering and polarization of light in other galaxies also suggest the presence of grains, but scattering effects are more difficult to interpret quantitatively.

In this section we attempt to summarize observations in a form convenient for theoretical interpretation; observational methods are discussed only insofar as they have a bearing on the interpretations.

2.1. Extinction

Within the wavelength region most completely investigated ($\sim$3000 to 10,000 A), absorption of starlight increases with decreasing wavelength and thus gives rise to *reddening*. A considerable amount of new work on extinction in the infrared as far as $\lambda \approx 11.5\ \mu$ is reported in the chapter (this volume) by Harold Johnson. The ultraviolet has now been investigated as far as $\lambda \approx 1200$ A. We shall treat the two regions, $\lambda <$ 3000 A and $\lambda >$ 10,000 A, separately from the region 3000 A $\lesssim \lambda \lesssim$ 10,000 A, first, because they are not as well known, and second, because of theoretical reasons which will be explained later. The absorption at any wavelength may be found from a comparison of radiation from a reddened star with that from an unreddened star of the same spectral and luminosity class. Because of their brightness and nearly featureless spectra, O and B stars have been the principal test objects. However, see the chapter by Harold Johnson for the types of stars which are used to study the far-infrared spectral region.

Let $m(\lambda)$ be the apparent magnitude of an unreddened star, $M(\lambda)$ its absolute magnitude, and r its distance expressed in parsecs. Then

$$m(\lambda) = M(\lambda) + 5 \log r - 5 . \tag{1}$$

If we denote by primes the same quantities for a reddened star and let $A(\lambda)$ be its extinction, then

$$m'(\lambda) = M'(\lambda) + 5 \log r' - 5 + A(\lambda) . \tag{2}$$

The extinction is given by

$$A(\lambda) = m'(\lambda) - m(\lambda) - [M'(\lambda) - M(\lambda)] - 5 \log (r'/r) . \tag{3}$$

The *color excess* E_{ij} of the reddened star is defined as the difference between the extinctions at two wavelengths, λ_i and λ_j, $(\lambda_i < \lambda_j)$. Thus

$$E_{ij} \equiv A(\lambda_i) - A(\lambda_j) \equiv \Delta m'_{ij} - \Delta m_{ij} , \tag{4}$$

where it has been assumed that the reddened and unreddened comparison stars are of the same spectral type, and where Δm_{ij} and $\Delta m'_{ij}$ are the color indices defined by

$$\Delta m_{ij} \equiv m(\lambda_i) - m(\lambda_j) . \tag{5}$$

To compare the wavelength dependence of extinction directly for different stars, it is customary to normalize the extinction curves (see Fig. 1 of chap. by Harold Johnson, this volume). Whitford (1958) normalizes to a standard color excess $E_{ij} = 1.00$ mag, where $\lambda_i^{-1} = 2.40\ \mu^{-1}$ ($\lambda_i = 4170$ A) and $\lambda_j^{-1} = 0.99\ \mu^{-1}$ ($\lambda_j = 10{,}100$ A). The extinction at each wavelength λ_a is then obtained by finding the ratio E_{aj}/E_{ij}. This arbitrarily makes the extinction at 10,100 A equal to 0.0 mag. The high degree of uniformity of almost all extinction curves plotted in this way between $\lambda^{-1} = 1\ \mu^{-1}$ and $\lambda^{-1} = 3\ \mu^{-1}$ led to the suggestion of a universal reddening law (see, for example, Divan 1954). A high degree of uniformity has important theoretical implications both from the point of view of the character of the grains, as well as in the application to stellar distance corrections.

But numerous deviations now appear to be well established (see below, this section; Sharpless 1963; also chapter by Harold Johnson).

The important characteristics of the Whitford curve are (1) the very small positive slope (perhaps zero) for $\lambda = \infty$ ($\lambda^{-1} = 0$),[1] (2) the roughly linear region (λ^{-1} law) in the visible ($1\ \mu^{-1} \leq \lambda^{-1} \leq 2\ \mu^{-1}$), and (3) the decrease in slope in the ultraviolet.

If we denote the normalized extinction of a star by $A^{(n)}$ (see Johnson's Fig. 1), then the absolute value of the extinction is given by

$$A(\lambda) = \alpha_1[A^{(n)}(\lambda) - \alpha_2] \tag{6}$$

where α_2 is the value of $A^{(n)}(\lambda)$ at $\lambda^{-1} = 0$, and α_1 is the actual difference in the extinctions at $\lambda^{-1} = 2.4\ \mu^{-1}$ and $\lambda^{-1} = 1.0\ \mu^{-1}$. Thus $A(\lambda) = 0$ at $\lambda^{-1} = 0$, which is theoretically justified since the extinction by any kind of particle becomes zero when λ goes to ∞.

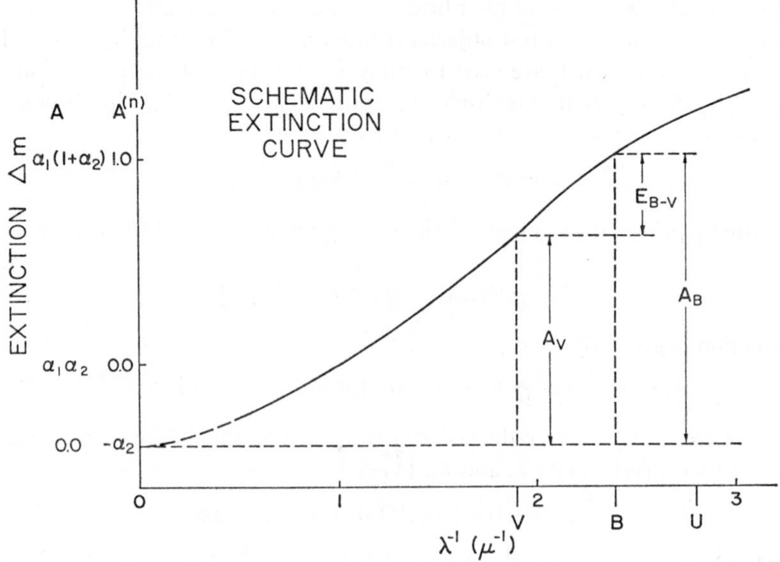

FIG. 1.—Schematic extinction curve

The schematic extinction curve (Fig. 1) conveniently illustrates some important astrophysical terms to which we shall refer. Note that Figures 1 in this and in the Johnson chapter are inverted with respect to each other and that, for theoretical reasons, the abrupt change in the slope[2] in the blue in Johnson's Figure 1 is replaced by the smooth

[1] Johnson (1961) indicates that by extrapolation the extinction curve may sometimes have a non-zero slope at $\lambda = \infty$. See also chapter by Harold Johnson, this volume. There are important theoretical consequences of this result if true. Note that the extinction curve for electron scattering would have a finite slope at $\lambda^{-1} = 0$.

[2] There appear to be several possible theoretical explanations of the abrupt change in slope at $\lambda^{-1} = 2.2\ \mu^{-1}$ which Whitford mentions, and Nandy (1964*a*) has also found evidence for the slope discontinuity at $\lambda^{-1} = 2.2\ \mu^{-1}$. However, the importance of this observation is still open to question. See the section on cylindrical dielectrics, and on graphite particles for possible explanation.

variation of the curve in Figure 1 here. The locations of the inverse wavelengths corresponding to the *UBV* system of Johnson and Morgan (1953) are roughly estimated to show on what portions of the extinction curve they occur. The quantities A_v and A_p are the total extinctions in the photovisual (5290 A) and the photographic (4250 A) regions.

It can be seen from equation (2) that the absolute value of $A(\lambda)$ is needed to determine M or r. Direct determination of $A(\lambda)$ is, however, difficult. Since the extinction must vanish at $\lambda = \infty$, direct evaluation of $A(\lambda)$ involves measurements at extremely long wavelengths. An alternative indirect evaluation involves extrapolations beyond the region readily accessible for extinction observations. The dotted portion of the extinction curve in Harold Johnson's Figure 1 denotes extrapolation beyond the point $\lambda = 2\ \mu$, which was the longest wavelength studied by Whitford (1948).

If we assume a reasonable uniformity of the wavelength dependence of extinction, we can determine $A(\lambda)$ indirectly. (See, however, Johnson and Borgman 1963; chapter by Harold Johnson, this volume.)

The constancy of the reddening law would imply that the ratio of the total extinction at one wavelength to the difference between the extinctions at two wavelengths may be given uniformly as

$$\frac{A(\lambda_1)}{A(\lambda_2) - A(\lambda_1)} = \frac{A^{(n)}(\lambda_1)}{A^{(n)}(\lambda_2) - A^{(n)}(\lambda_1)} = R \tag{7}$$

where R is called the ratio of total to selective extinction.[3]

Since we can measure $A(\lambda_2) - A(\lambda_1)$, it is only necessary to have a value of the ratio R to determine an absolute scale for the extinction. Average values of this ratio can be obtained by star counts and by observation of the dependence on galactic latitude of the number of external galaxies. The presently much used value $R = A_V/E_{B-V} = 3.2$ is in good agreement with the extrapolation of the reddening curve given in Figure 1. An average extinction of about 1.6 mag/kpc in the neighborhood of the Sun has been obtained by the same method used to obtain R.

The determination of R is important in the theory of the physical properties of the interstellar grains, which depends on accurate extrapolation of the reddening curve beyond the infrared, e.g., whether the slope is zero or not for $\lambda^{-1} = 0$. Also, of course, the shape of the extinction curve between $\lambda^{-1} = 0$ and $1\ \mu^{-1}$ is needed to determine R. The "double S" curve shown by Johnson for specific cases in Figure 2 clearly indicates a wide range of variability of extinction in this wavelength region. This will be discussed further in Sections 7 and 8.

We must know the wavelength dependence of the extinction to find the intrinsic intensity of a reddened star as a function of wavelength. Where observations of the extinction are not yet extensive, as in the far ultraviolet where rocket and satellite measurements are now being made, theoretical extrapolations of the extinction must be used because the intrinsic colors of high-temperature stars in this region are not known (see, for example, Boggess 1961).

[3] In some systems, the wavelength in the numerator differs from both wavelengths in the denominator. It should also be noted that the width of the wavelength band must be taken into account in evaluating the quantities in equation (7). This is particularly important in theoretical studies.

Although the reddening law seems to be fairly uniform over the whole sky in the visual regions, clear deviations exist in some wavelength regions, mainly the ultraviolet and infrared. Johnson and Morgan (1955) noted a small but distinct difference in the *UBV* plot of the heavily reddened association VI Cygni when compared with stars in the sector from Cepheus to Monoceros. The difference lies in the direction of greater absorption in the ultraviolet for stars in the Rift and in the smaller decrease in the slope of

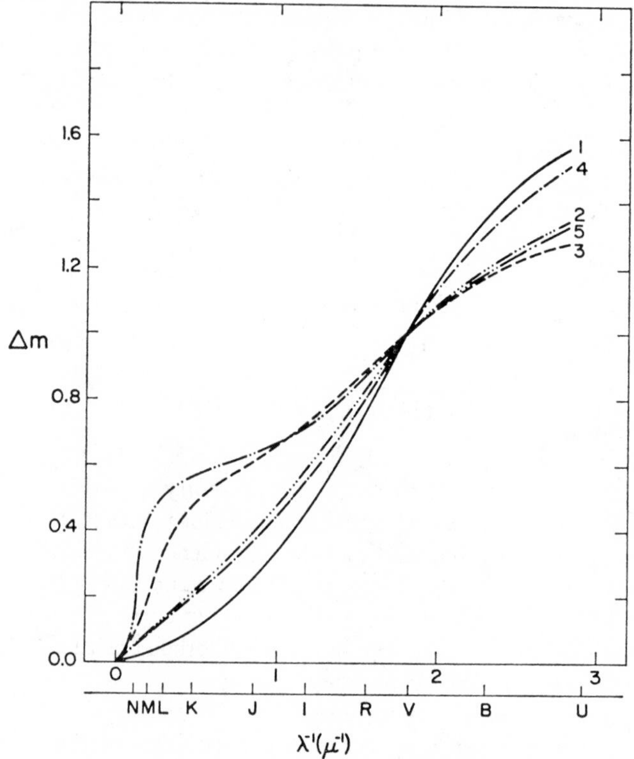

FIG. 2.—Smoothed extinction curves for different regions: (1) Perseus, (2) Orion Belt, (3) NGC 2244, (4) Cygnus, and (5) Cepheus. Note particularly the extremely rapid rise of the extinction curves (2) and (3) in the far infrared. (Johnson 1965.) Effective wavelengths in the infrared have been defined by Harold Johnson as follows: $J = 1.1–1.4\ \mu$, $K = 1.9–2.5\ \mu$, $L = 3.2–4.1\ \mu$, $M = 4.4–5.5\ \mu$.

the reddening line in the blue for these stars. This also can be expressed as a greater value of E_{U-B}/E_{B-V}. Several deviations from a mean extinction law may be seen in Figure 3 (Borgman 1961). Wampler (1961) has found a systematic variation with galactic longitude of the ratio of slopes in the blue to those in the visible region (see Fig. 4). Recent observations by Borgman and Johnson (1962) extend the wavelength range of comparison to obtain the ratio of slopes in the infrared to those in the blue and ultraviolet. The results are shown in Figure 5. Müller (1955) has reported an anomaly in the sense of too little ultraviolet absorption for stars in the cluster NGC 654 in the Perseus Arm. The difference between the extinction curves in Cygnus and Perseus has been carefully investigated by Nandy (1964*a*), and a comparison of these is shown in Figure 6. Such observations of variations may provide fundamental information for determination of the nature of the interstellar grains.

Undoubtedly, variations in the extinction in the far ultraviolet are at least as common as variations in the visual and near ultraviolet. We reproduce a summary of results in Figure 7, as reported by Stecher (1965). The points for which $\lambda^{-1} \leq 3$ will later be used to

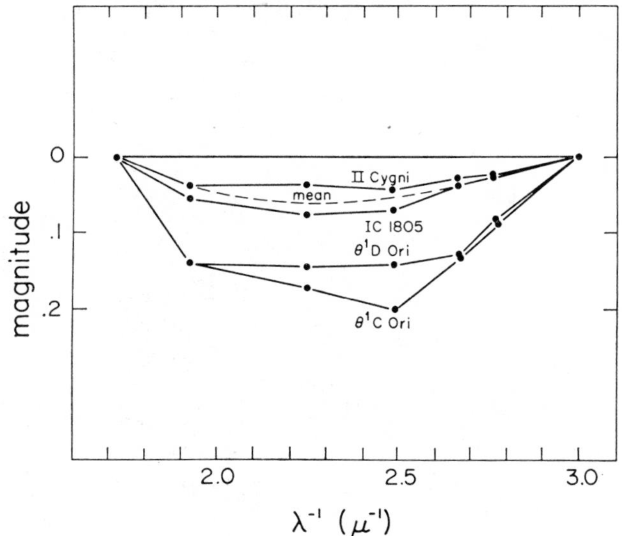

FIG. 3.—Deviations from a linear λ^{-1} absorption law normalized to an absorption difference of 1 magnitude between the reciprocal wavelengths $\lambda^{-1} = 1.729$ and $\lambda^{-1} = 3.005$. The broken curve refers to a mean absorption law normalized as indicated above. (Borgman, 1961.)

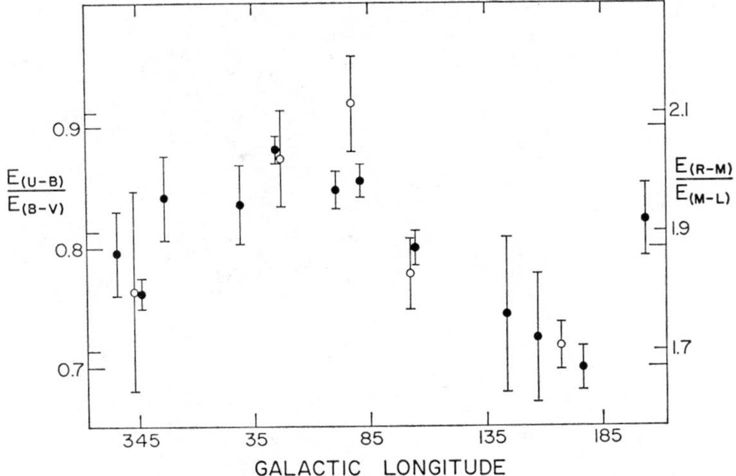

FIG. 4.—Variation of wavelength dependence of extinction with galactic longitude. (Wampler, 1961.)

represent a "standard" extinction curve for comparison with some of the theoretical calculations.

The infrared extinction as reported by Johnson (this volume; 1966) exhibits curious anomalies, at least from a theoretical point of view. In Figure 2 are presented a set of idealized curves representing what, from theoretical considerations, are the cogent points of interest. Perhaps related to some of the variations in the infrared is the reported (Neckel 1965) work on the magnitude and variation of the ratio of total to selective extinction.

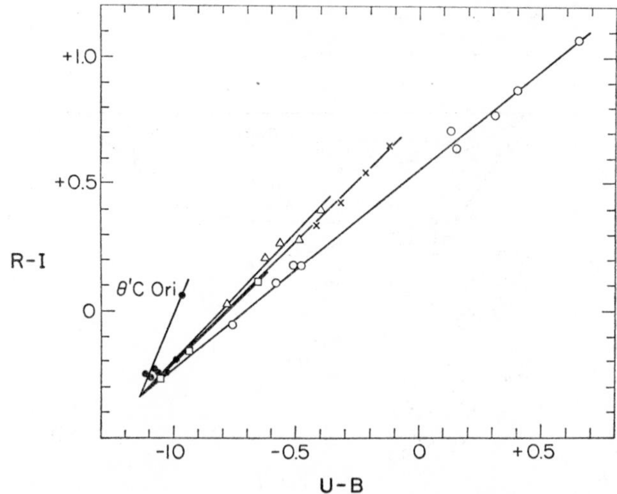

FIG. 5.—The relations between $R-I$ and $U-B$ for O stars in several constellations. Symbols in this diagram refer to the constellations in which the stars are situated: open circles: Cygnus; squares: Monoceros; crosses: Ophiuchus-Scorpio; triangles: Cassiopeia-Perseus; filled circles: Orion and unreddened blue stars not belonging to any of these constellations. (Borgman and Johnson, 1962.)

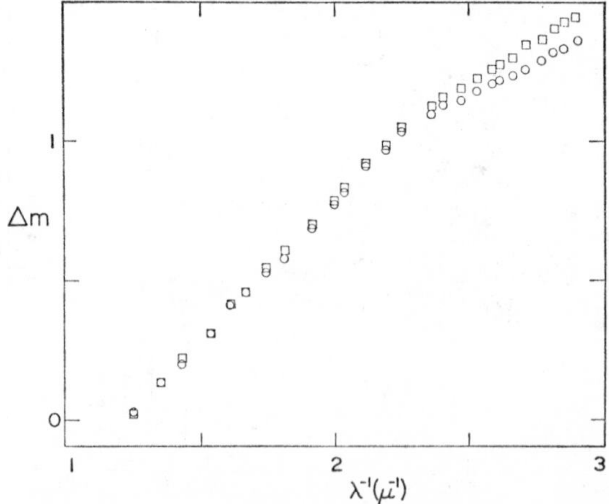

FIG. 6.—Wavelength dependence of extinction in Cygnus (squares) and Perseus (circles). Note slope discontinuity at $\lambda^{-1} = 2.2\ \mu^{-1}$. (Nandy, 1964*b* 1965.)

Comparison of the densities of gas (measured by interstellar line intensities) and dust (measured by reddening) is important in any theory of formation and growth of grains. The distribution of gas, as determined either by radio or optical methods, is known to be quite irregular, as are also regions of obscuration. On a broad scale there appears to be strong correlation between the gas and dust in the spiral arms of our Galaxy, and extinction has not been found in regions of sufficiently low gas density. The problem is under investigation. Spitzer (1961) gives references to work (Davies 1960; Kurochkin 1957; Lambrecht and Schmidt 1958; and K. H. Schmidt 1958) which seems to confirm

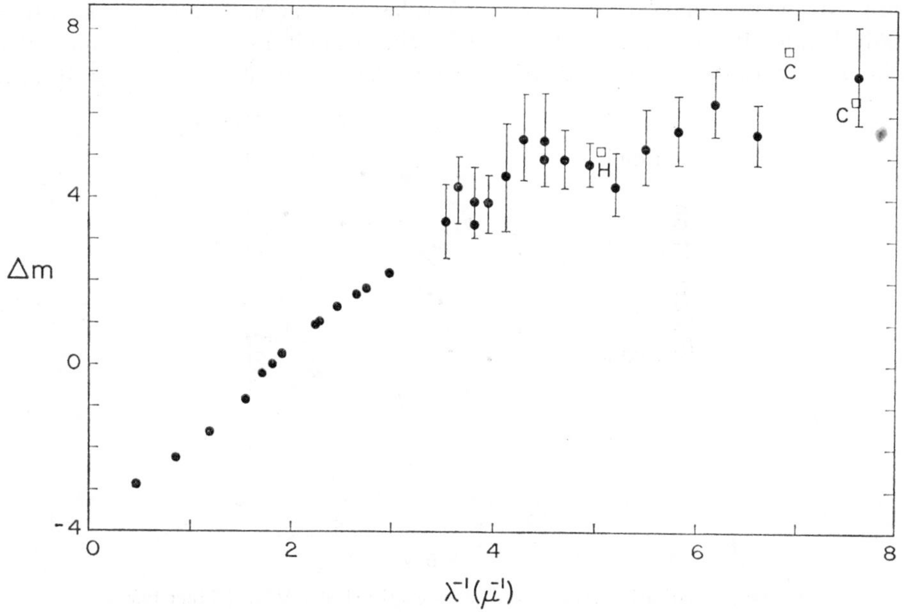

FIG. 7.—Observed interstellar extinction in the ultraviolet. Points up to $\lambda^{-1} = 3$ from Boggess and Borgman (1964). Points marked C are from Chubb and Byram (1963). The point marked H is from Alexander *et al.* (1964). The points with error bars are mean values for up to five pairs of stars. All values are normalized to $B - V = 1$ and $V = 0$. (From Stecher 1965.)

the correlation between dust and neutral hydrogen in the interstellar medium. For further details, see the *Proceedings of the Third Symposium on Cosmical Gas Dynamics* (1958). More recent work on the distribution of gas and dust is referred to by Strömgren (1965). By a statistical analysis, Zonn (1962) finds that the extinction by a single cloud ($R \sim 1.5$ pc) is $0^{m}_{.}19$ (using a ratio of total-to-selective extinction of 3.2). This would give $1^{m}_{.}9$ extinction/kpc if there were 10 clouds/kpc (see Sec. 4).

2.2. POLARIZATION

Partial linear polarization has been observed for many stars, often showing (1) association with extinction, (2) the existence of a maximum ratio of polarization relative to extinction, (3) wavelength dependence, and (4) correlation in position angle from star to star over large areas of the sky.

The polarization P may be defined by

$$P=\frac{I_1-I_2}{I_1+I_2} \tag{8}$$

where I_1 is the intensity in the plane of polarization and I_2 is the intensity in the perpendicular plane. Theoretically, it is more convenient to express the polarization as a magnitude difference,

$$\Delta m_P=2.50 \log \frac{I_1}{I_2}=2.1717\left(P+\frac{P^3}{3}+\ldots\right) \text{ with } P<1\,. \tag{9}$$

The maximum ratio of polarization to extinction has been variously stated as $\Delta m_P/A_V = 0.061$ (Hiltner 1956) and $\Delta m_P/A_V = 0.065$ (Th. Schmidt 1958). Schmidt finds that $\Delta m_P/A_V$ decreases with increasing extinction for galactic latitudes less than $b = 10°$, and

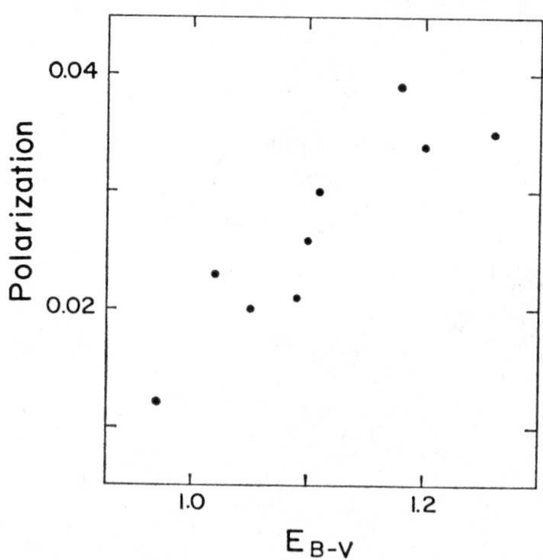

FIG. 8.—Polarization versus color excess for the cluster M29. (Hiltner 1956.)

is smaller where the dispersion of polarization alignment is greater. The correlation between polarization and extinction within the star cluster M29 is shown in Figure 8. The position angles of the planes of vibration are similar for all of the stars. These effects indicate that polarization of starlight is an interstellar phenomenon associated with extinction. An interpretation of polarization as a stellar phenomenon (Thiessen 1961) has been shown (Behr 1961) to be based on faulty selection of data. Furthermore, Thiessen's assumption that polarization is due to synchrotron radiation in stellar atmospheres raised formidable theoretical problems (Struve 1961). However, Serkowski (1966*b*) has recently investigated the polarization in Mira variables and has shown that at least some of their polarization is not of interstellar origin.

The variation of interstellar polarization with wavelength has been investigated by Behr (1959), Gehrels (1960), Hiltner (as communicated to van de Hulst 1957), Treanor (1963), Serkowski (1965), Serkowski *et al.* (1965), and Visvanathan (1965). The observations by Treanor with eight narrow-band filters distributed between 3400 A and 6600 A for 10 stars have confirmed the decrease of polarization toward the ultraviolet. Serkowski *et al.* (1965) and Serkowski (1966*a*), in addition to examining the linear polarization, showed that the degree of ellipticity does not exceed 0.02 per cent. This could be a very significant result in determining the optical characteristics of the grains. Both Treanor and Gehrels believe that the polarization angle is not constant for different wavelengths, but Visvanathan, who has studied λ-dependence of polarization in 26 stars (some of which are the same as those observed by Gehrels and Treanor), is not convinced that this variation is beyond observational uncertainty. There is no doubt that there is a general wavelength dependence of polarization (which is generally similar to that shown in Fig. 9), just as there is a general wavelength dependence of extinction. In other words, all observers find variations from star to star and from region to region to be clearly indicated, although the intercorrelations are not yet known. The work of

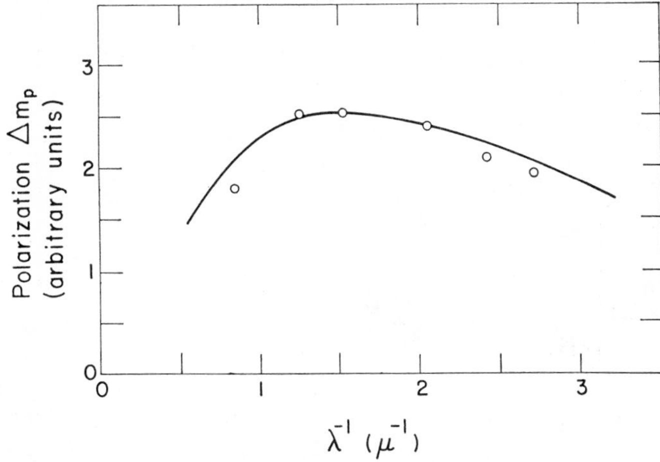

FIG. 9.—Observed percentage polarization, average of eight stars, as a function of inverse wavelength. Open circles are observed points. Solid curve is theoretical (see Fig. 83 and Gehrels, T. 1960, *A.J.*, **65**, 470).

Gehrels (1960), averaged over the results for 8 stars, indicates a distinct maximum in the red (6500 A) (see Fig. 9), while that of Hiltner shows only a decrease toward the infrared. Behr, using three colors, found that the polarization in the yellow (5150 A) is larger than in the blue (4500 A) for some stars and smaller for others, and always smaller in the ultraviolet (3310 A) than in the blue. So far, correlation of these variations with other physical phenomena has been inconclusive except perhaps in the Cygnus region, where Serkowski finds a shift of the polarization maximum to the red; in this region it is well known that the *R* value is anomalous. Visvanathan has considered the possibility that the wavelength dependence of the polarization might be influenced by whether the polarization occurs in H I or H II regions, but no significant difference has been shown between the polarization values in stars in the different regions (see Table 20). Visvanathan generally finds that the polarization maximum occurs at the visual filter. In the work reported by Coyne and Gehrels (1966), about half of the shapes of the polarization curves seem to be rather similar to those of Visvanathan. Furthermore, the maxima are shifted toward larger values of 1/λ than in Gehrels' earlier work (Fig. 9). The Coyne-Gehrels-Visvanathan averages of the polarization are shown in Table 19 and also in

Figure 10. There is remarkable agreement between these. In a later section, these averages are also compared with a sample computation.

Hall (1958) has compiled a vast amount of data on polarization and has constructed a set of maps showing the degree and orientation of the polarization (see also Hall and Serkowski 1963). Over extended regions of galactic longitude (but within a narrow band of latitude from $b^{\text{II}} = -5°$ to $b^{\text{II}} = 5°$), the plane of polarization is uniformly oriented parallel to the galactic plane (see Figs. 11 and 12). In some regions, however, the polarization is randomly oriented and is generally smaller (Fig. 13). These observations are consistent with the picture of the local spiral structure in that regions of high polarization and low dispersion correspond to viewing across a spiral arm, and regions of low polarization and high dispersion correspond to viewing along a spiral arm. An attempt

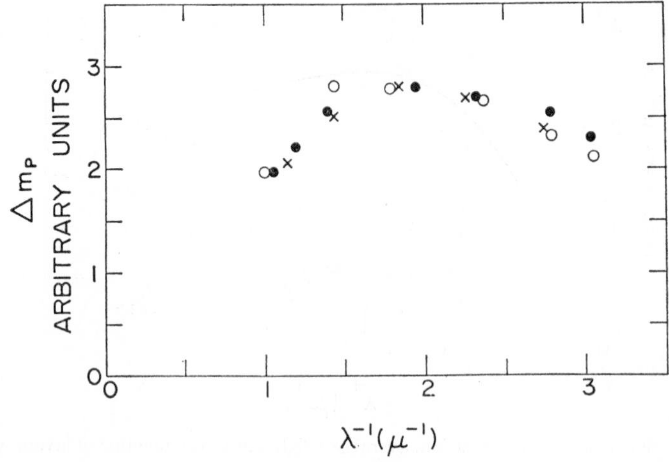

FIG. 10.—Averages of various observed polarizations as a function of wavelength. Open circles, Gehrels (1960); filled dots, Coyne and Gehrels (1966); crosses, Visvanathan (1965). The average maximum for the majority of stars appears to be at $\lambda^{-1} \approx 1.8$.

by Lodén (1965) to find a statistical correlation between degree of polarization and galactic longitude, however, gives inconclusive results. He finds that the degree of polarization statistically depends more upon the stellar distance.

Polarization associated with more detailed structural features in the interstellar medium is also evident. Behr (1955) and Behr and Tripp (1955) found that polarization tends to be parallel to nebular streamers; this was also noted by Shain (1956). The tendency of the electric vector to lie along fine filamentary structures has been observed by Hall (1954, 1955).

Grain alignment by polarization and possible control of the local features of spiral-arm structure and the more detailed galactic features by magnetic fields in space are discussed in detail in the chapter by Davis and Berge, this volume. Confirmation of preliminary data on the correlation of the polarization of 75-cm synchrotron radiation and the optical polarization (Westerhout, Seeger, Brouw, and Tinbergen 1962) would provide apparently direct proof of grain alignment by magnetic fields. According to theories of synchrotron radiation and the presently accepted grain orientation mechanism, the directions of the optical and radio polarizations should be orthogonal to each other, as indicated by the data. Recent work at high radio frequencies, when Faraday rotation may be neglected, firmly reinforces the conviction that it is the magnetic fields which align the dust grains (Brouw 1966, private communication).

The direction of polarization of synchrotron radiation is perpendicular to the plane defined by the magnetic field and the line of sight as can be seen to arise as follows. (1) Highly relativistic electrons spiralling about the magnetic field radiate in a small cone

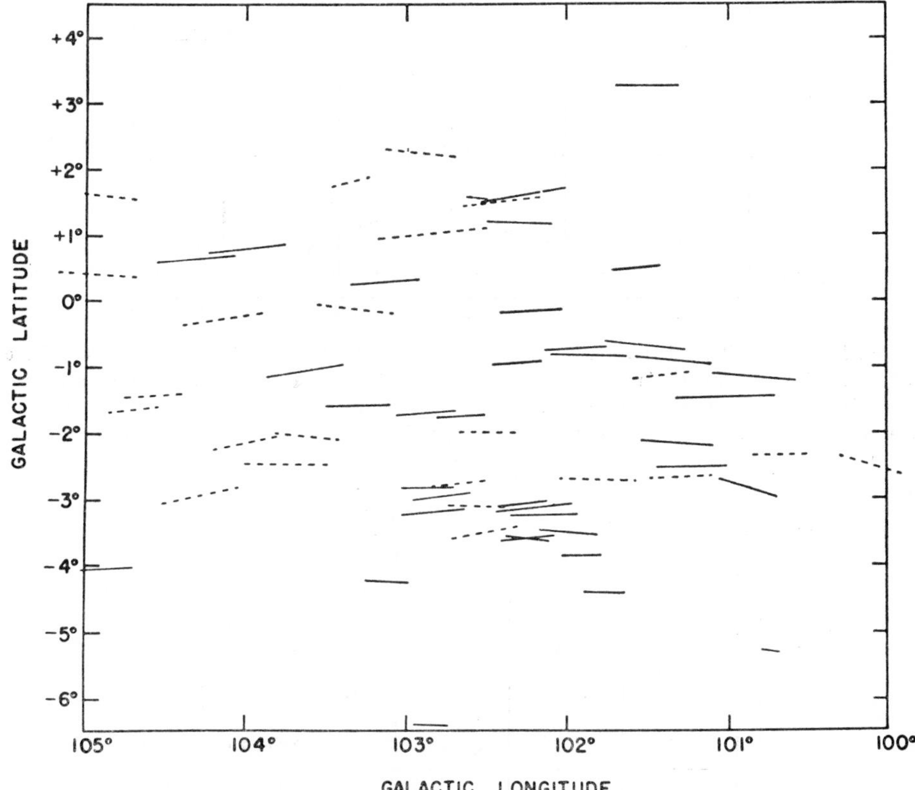

FIG. 11.—Polarization of stars near the longitude of minimum dispersion in the plane of vibration. Stars whose photometric distances exceed 2.5 kpc are denoted by solid lines. Stars closer than 1.6 kpc are indicated by dashed lines. (Hall 1958.)

pointing along the direction of electron motion; (2) an observer sees the radiation from each electron as the direction of motion of the electron is along the line of sight; (3) at the moment of the pulse (at the electron) the electron is just moving perpendicular to the magnetic field–line of sight plane and therefore the electric vector of the radiation is polarized perpendicular to the magnetic field. This direction of polarization is the same as that produced by a nonrelativistic electron moving in a circular orbit about the magnetic field. If the magnetic fields are predominantly in the galactic plane, the syn-

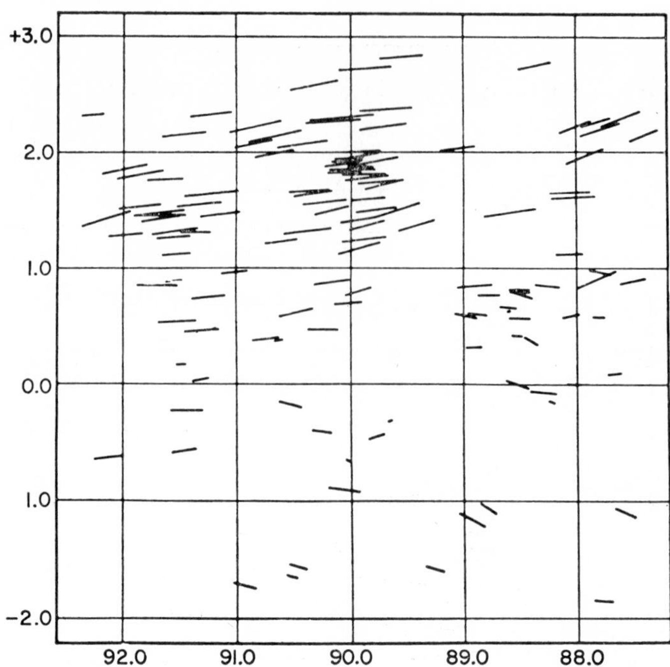

FIG. 12.—Interstellar polarization in Cassiopeia. The coordinates are galactic longitude (abscissae) and latitude in degrees. Each line represents a star. The length of the line is proportional to the amount of polarization, and the position angle of the line corresponds to the position angle of the plane of vibration. Note that in general the plane of vibration is approximately parallel to the galactic plane. (Hiltner 1956.)

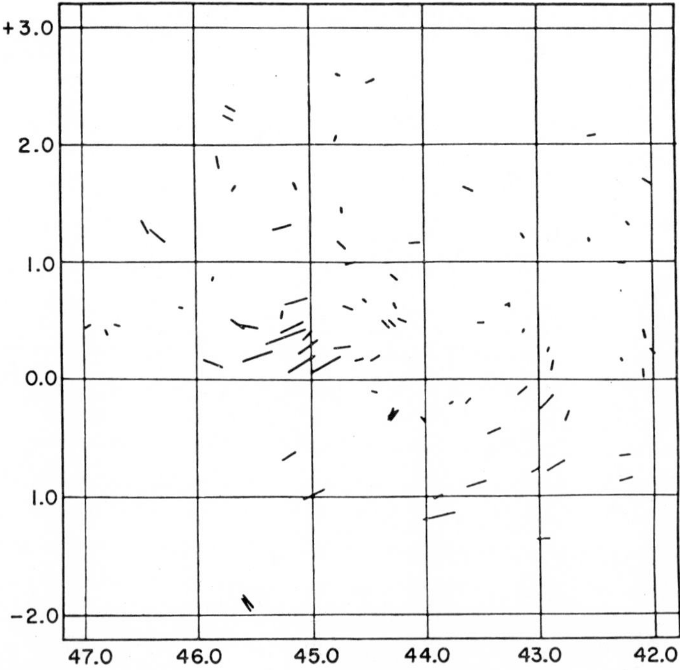

FIG. 13.—Interstellar polarization in Cygnus. Compare Figure 12. Note that the planes of vibration have a rotationally ordered but not preferred orientation and that in general the amount of polarization is smaller. (Hiltner 1956.)

chrotron radiation is polarized perpendicular to the galactic plane and this direction is orthogonal to the optical polarization (interstellar) as shown in Figures 11 and 12.

The assumption that oriented particles are responsible for both the extinction and the polarization seems well borne out by the correlations which have been mentioned.

2.3. Reflection Nebulae; Scattering and Extinction in Other Galaxies

Within the Galaxy, scattering of light in reflection nebulae seems to be due to interstellar grains. Henyey and Greenstein's (1941) interpretation of the nature of diffuse nebulae and galactic background light led to the tentative conclusion that the grains have a high albedo and a forward scattering phase function. On the other hand, Schalén (1945) interpreted the colors of reflection nebulae using grains which do *not* have a forward phase function and which have a low albedo.

Dust and corresponding extinction exist in external galaxies. Hiltner (1958), in his work on the association of polarization of globular cluster in M31 with reddening, suggested that the electric vector has a preferential direction along the major axis of the Galaxy.

Further observational studies of reflection nebulae and extragalactic observations (see, for example, Elvius 1951, 1956*a*, *b*) are needed to determine the scattering characteristics of the interstellar grains. The questions of the albedo and the phase function are not yet resolved (van Houten 1961), and we shall therefore consider mainly theoretical interpretations of extinction and polarization within the Galaxy. A simple theory of the colors of reflection nebulae and the interpretation of a rather limited amount of data are developed in Section 9.

3. GENERAL FORMULATION OF THE THEORY OF EXTINCTION AND POLARIZATION BY GRAINS

Each grain in the line of sight between a star and the observer reduces the starlight by a combination of scattering and absorption. The *extinction cross-section* of a grain is defined by

$$C_{\text{ext}} = C_{\text{sca}} + C_{\text{abs}} = \frac{\text{Total radiant energy scattered and absorbed per unit time}}{\text{Incident radiant energy per unit area per unit time}} \quad (10)$$

where C_{sca} and C_{abs} are, respectively, the scattering and absorption cross-sections. To take account of polarization we must postulate that (1) the extinction of each grain is a function of its orientation, and (2) the grains are preferentially oriented.

We denote by I_x and I_y the intensities (fluxes F_x and F_y per unit area) of two incoherent beams of radiation propagated along the z-direction and polarized along the x- and y-directions, respectively.

For simplicity we consider particles with axial symmetry and define $C_x(\theta, \varphi)$ and $C_y(\theta, \varphi)$ as the single-particle extinction cross-sections (the subscript *ext* is omitted) for radiation polarized along the x- and y-directions (see Fig. 14). Then the reduction in each flux by the single particle is given by

$$\begin{aligned} F_x &= -I_x C_x(\theta, \varphi) \\ F_y &= -I_y C_y(\theta, \varphi) . \end{aligned} \quad (11)$$

The extinction by a grain will, in general, be a function of its size (a characteristic linear dimension a), shape (elongation e, that is, ratio of length to width), and the wavelength of the radiation. If the scattering is classical, the index of refraction can be incorporated into the functional dependence on a and λ. We define

$n(z, a, e)da\,de$ = number of grains of size a in da and elongation e in de per unit volume at z.

$f(z, \theta, \varphi, a, e)$ = angular distribution function of the particles at z.

$\left.\begin{matrix} C_x(\theta, \varphi, a, e, \lambda) \\ C_y(\theta, \varphi, a, e, \lambda) \end{matrix}\right\}$ = extinction cross-sections.

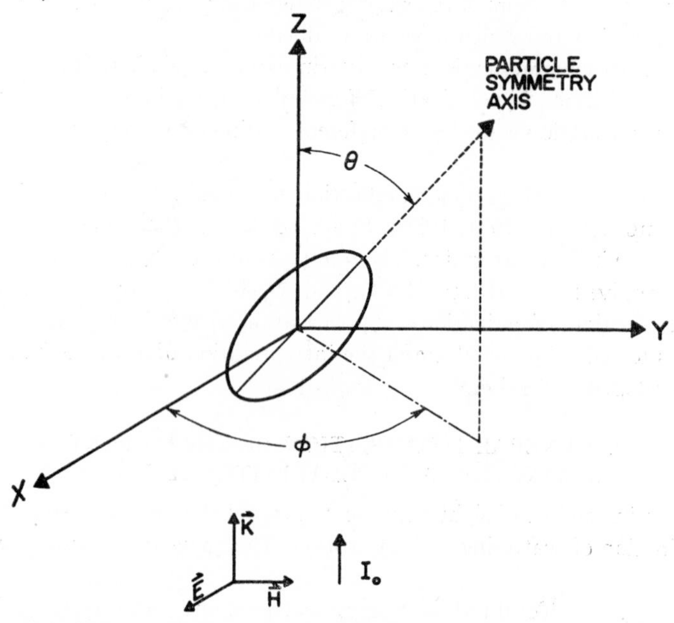

Fig. 14.—Schematic illustration of a nonspherical particle arbitrarily oriented with respect to incident unpolarized radiation of intensity I_0. The incident radiation may be divided into two incoherent beams of intensity $I_0/2$ polarized along the x- and y-directions.

The definition of the angular distribution function is such that

$$\int_0^{2\pi}\int_0^{\pi} f(z, \theta, \varphi, a, e) \sin\theta \, d\theta \, d\varphi = 1 .$$

Consider a star at a distance D from the observer and of intrinsic intensity $I_0(\lambda)$ at some wavelength λ. Integrating equation (11) (generalized in accordance with the above definitions) for the residual intensity of starlight polarized along the two directions, we obtain

$$I_x(\lambda) = \frac{I_0}{2} e^{-\tau_x(\lambda)}$$
$$I_y(\lambda) = \frac{I_0}{2} e^{-\tau_y(\lambda)} \qquad (12)$$

where

$$\tau_{x,y}(\lambda) = \int_0^D dz \int_0^\infty da \int_0^\infty de \int_0^\pi \sin\theta d\theta \int_0^{2\pi} d\varphi n(z, a, e) \times f(z, \theta, \varphi, a, e) C_{x,y}(\theta, \varphi, a, e, \lambda) \tag{13}$$

are the optical depths for the two directions of polarization. In practice we shall use a much simpler form for τ_x and τ_y.

We now relate the interstellar extinction and polarization to these optical depths.

The total intensity of received light is given by

$$I = I_x + I_y = \frac{I_0}{2}(e^{-\tau_x} + e^{-\tau_y}) \tag{14}$$

and the extinction expressed in magnitudes is

$$\Delta m = -2.5 \log \frac{I}{I_0} = -1.086 \ln \left(\frac{e^{-\tau_x} + e^{-\tau_y}}{2}\right). \tag{15}$$

This can be written as

$$\Delta m = -1.086 \ln \left[\exp\left(-\frac{\tau_x + \tau_y}{2}\right) \cosh\left(\frac{\tau_x - \tau_y}{2}\right)\right]. \tag{16}$$

It is generally sufficiently accurate to replace $\cosh(\tau_x - \tau_y)/2$ by unity. Then

$$\Delta m = 1.086 \frac{\tau_x + \tau_y}{2}. \tag{17}$$

The extinction is then linearly related to the cross-sections of the grains. In terms of the optical depths, the polarization in magnitudes is simply

$$\Delta m_P = 1.086 (\tau_y - \tau_x), \tag{18}$$

if τ_y and τ_x represent the maximum and minimum absorption, respectively.

The total mass of the grains per unit area along the line of sight is

$$\mu = \rho \int_0^D dz \int_0^\infty da \int_0^\infty n(z, a, e) V(a, e) de \tag{19}$$

where ρ is the mass density and $V(a, e)$ is the volume of the particle.

Equations (13) and (19) can be simplified in some special cases. For example, suppose all grains to be spherical and of the same radius a. Then

$$\tau_x = \tau_y = \tau = N C_{\text{ext}}(a, \lambda) = N Q_{\text{ext}}(a, \lambda) \pi a^2 \tag{20}$$

where N is the total number of grains per unit area in the line of sight and Q_{ext} is the extinction efficiency of the grain. From equations (17) and (20) the wavelength dependence of the interstellar extinction is shown to be the same as the wavelength dependence of the extinction cross-section of the particle of radius a.

Assume that the extinction cross-section for the particle at some wavelength can be approximated by its geometrical cross-section ($Q \sim 1$). Then the interstellar extinction at this wavelength is

$$\Delta m = 1.086 N \pi a^2. \tag{21}$$

Similarly the total mass $\mathfrak{M}$ along the line of sight per unit sky area due to grains of density s is

$$\mathfrak{M} = \tfrac{4}{3}\pi a^3 N s , \qquad (22)$$

and for an extinction $\Delta m/D$ per unit length we obtain the average space density ρ_g of the grains as

$$\rho_g = \tfrac{4}{3} a s \frac{\Delta m}{D} \times \frac{1}{1.086} . \qquad (23)$$

In summary, interstellar extinction and polarization phenomena can be defined in terms of the optical depths (determined by extinction by individual grains), τ_x and τ_y should follow the reddening curves in all varieties (for average purposes we shall use either the Whitford or the Boggess-Borgman curve), the particles are sufficiently non-isotropic and well oriented that $\tau_y/\tau_x \approx 1.06$ is possible, and $\tau_y - \tau_x \rightarrow 0$ as $\lambda \rightarrow \infty$ and as $\lambda \rightarrow 0$. (In detail, $\tau_y - \tau_x$ should follow the trends shown in Fig. 9.)

4. PHYSICAL AND CHEMICAL PROPERTIES OF THE INTERSTELLAR MEDIUM

Physical conditions in the general interstellar medium are considered in detail elsewhere in this volume, but it is convenient to summarize here some basic quantities that are important for the existence and growth of the grains which constitute a small fraction of the medium.

The abundances according to Aller (1967) of some elements are given in Table 1. The relative abundances of the elements seem to vary from region to region in the Galaxy (see Aller 1961). There is reason to believe that the metal-to-hydrogen ratio is higher in the spiral arms (where the dust is concentrated) than in the galactic center. Table 2 gives values of parameters for gas clouds where dust may also be present. The values for N and R have been obtained by assuming 8 clouds/kpc^3 and by assuming that the faction of space occupied by clouds may be estimated at 0.1. (See *Proceedings of the Third Symposium on Cosmical Gas Dynamics;* also, in particular, van de Hulst 1958, and Spitzer chapter, this vol.)

Galactic magnetic fields appear to play a crucial role in the orientation of the interstellar grains. The smallest quoted value of the magnetic field (on a galactic scale) in spiral arms is given as the upper bound of the field strength for which the Zeeman effect on the 21-cm line is observable; this value is $B < 0.6\ \Gamma(\Gamma = 1 \times 10^{-5}$ gauss). The highest estimate for the magnetic field is $B = 3\Gamma$, based on an argument involving the confinement of cosmic rays in the Galaxy. Other observational-theoretical results group around $B = 1\Gamma$. The temperature of 10,000° K and dilution factor of 10^{-14} for radiation in the average region in space correspond to an energy density of 7.5×10^{-13} erg/cm^3, which lies within the range of 5 to 12×10^{-13} erg/cm^3 computed by Dunham (1939).

5. PHYSICS AND CHEMISTRY OF GRAINS

5.1. Grain Characteristics

The principal physical and chemical characteristics of the grains to be determined are size, nonsphericity (irregularity of shape), chemical constitution, internal temperature, optical and magnetic properties, and, most important, the possibilities for growth and continued existence.

Many physical and chemical processes involving the grains are very general in application, regardless of the specific grain model chosen. The main differences are in interpretation and, more particularly, in the values of the parameters one chooses to consider as correct. We summarize here the formulation of the appropriate fundamental physical processes to which all kinds of grains are subjected. Unless otherwise specified, the grain shape is assumed to be spherical for simplicity.

5.1.1. *Nucleation.*—A problem that has plagued theories of interstellar grain formation has been that of the production of condensation nuclei out of the interstellar gas.

TABLE 1

RELATIVE ABUNDANCES OF SOME OF THE LIGHTER ELEMENTS

Element	Relative Number of Atoms	Relative Mass	Notes*
H	1	1	
He	1.2×10^{-1}	4.8×10^{-1}	1
C	2.5×10^{-4}	3.0×10^{-3}	1, 2, 4
N	1.3×10^{-4}	1.7×10^{-3}	1, 2, 4
O	7.9×10^{-4}	1.3×10^{-2}	1, 2, 4
Ne	6.3×10^{-4}	1.2×10^{-2}	1, 2
Na	1.3×10^{-6}	2.9×10^{-5}	3
Mg	4.5×10^{-5}	1.1×10^{-3}	2, 3, 4
Al	1.8×10^{-6}	4.8×10^{-5}	2, 3, 4
Si	2.5×10^{-5}	7.0×10^{-4}	2, 3, 4
P	3.2×10^{-7}	9.8×10^{-6}	2, 3
S	2.8×10^{-5}	9.0×10^{-4}	1, 2, 4
Cl	1.0×10^{-6}	3.5×10^{-5}	1, 2
Ar	7.6×10^{-6}	3.0×10^{-4}	1, 2, 4
K	7.9×10^{-8}	3.1×10^{-6}	3
Ca	1.6×10^{-6}	6.4×10^{-5}	3
Ti	5.0×10^{-8}	2.4×10^{-6}	3
Cr	8.9×10^{-8}	4.7×10^{-6}	3
Mu	6.3×10^{-8}	3.5×10^{-6}	3
Fe	2.5×10^{-5}	1.4×10^{-3}	4
	4.0×10^{-6}	2.2×10^{-4}	3
Co	5.4×10^{-8}	3.2×10^{-6}	3
Ni	1.1×10^{-6}	6.6×10^{-5}	4

* Abundances derived from (1) diffuse nebulae, particularly Orion (Aller and Liller 1959, see also Johnson, chap. 2, this vol.); (2) stars recently formed from interstellar medium (cf. Aller, 1961, p. 115); (3) sun, stars whose compositions fit that of diffuse nebulae, and young stars when compositions can be compared (cf. Aller 1961; Goldberg, Müller, and Aller 1960; Müller and Mutschlecner 1964; Aller 1965; O'Mara 1967); and (4) analysis of spectrum of τ Scorpii (Scholz 1967).

Table modified from compilation by L. H. Aller (private communication).

TABLE 2

REPRESENTATIVE PHYSICAL PARAMETERS OF GAS CLOUDS

Mean H density in the galactic plane	0.5–3 cm^{-3}
Average radius, R	10 pc
Number per kpc along line of sight ($\pi R^2 N$)	8
Number per kpc^3 (N) in galactic plane	3×10^4
Root-mean-square cloud velocity, V	12 km/sec
Fraction of galactic space occupied	0.1
Temperature	100° (H I), 10,000° (H II)
Radiation Field	$T_R=10{,}000°$ K; dilution factor $W=10^{-14}$

The early work (ter Haar 1943; Kramers and ter Haar 1946) was not entirely convincing to all workers. A source of nuclei now receiving attention is in the atmospheres of cool stars. The major work in this direction was initiated by Hoyle and Wickramasinghe (1962), particularly with regard to the formation of graphite particles in sufficient number and of sufficient size to explain interstellar extinction (see Section 8).

The details of the physical processes involved in the formation of carbon particles are under continuing investigation (Donn *et al.* 1966). The basic physical considerations of the process of formation of graphite grains in stellar atmospheres or circumstellar envelopes are (1) nucleation—either via direct molecule formation or condensation on ions or other molecules, (2) subsequent condensation because of the high partial pressure—exceeding the saturation vapor pressure of carbon atoms, and (3) ejection of grains by radiation pressure. These conditions seem to be realized in N stars and perhaps also in M giants. In any case, there is a strong theoretical argument in favor of the formation of carbon particles by such a mechanism. The important point is that the crystalline nature and size of the carbon particles are dependent on the growth conditions and that whiskers, platelets, or polycrystalline carbon (soot) may be formed, depending mostly on the degree of supersaturation of carbon vapor.

Similar ideas have been considered by Kamijo (1963), leading to the theoretical possibility of the formation of other solid or liquid particles in the circumstellar envelopes of the M-type long-period variable stars. The conclusion is that particles primarily consisting of SiO_2 and whose size is of the order of 2×10^{-7} cm are formed in large numbers. These particles, though small, are large enough to be considered as condensation nuclei in the interstellar medium.

Where such processes take place in stellar atmospheres, they may be thought of as leading either to the production of nuclei serving as condensation points in interstellar space or to grains of a sufficient size and number to be optically significant in interstellar space. By optically significant, we mean that they contribute a substantial part of interstellar extinction with or without subsequent formation of mantles in interstellar space.

5.1.2. *Growth.*—Growth of a grain will occur if there exist atoms in the interstellar gas which may stick to the grain. Let m_A = mass of the atom, V_A = mean atom speed, γ = sticking probability (related to the accommodation coefficient), r = grain radius, N_A = number density of condensible atoms, and s = grain density. Then the rate of increase of size of a spherical grain is given by

$$\frac{dr}{dt} = \frac{\gamma N_A m_A V_A}{4s}. \quad (24)$$

5.1.3. *Destruction.*—Partial or complete destruction of a grain may be accomplished by inelastic collision of grains with grains, grains with gas atoms, grains with low-energy cosmic rays (suprathermal particles), and by evaporation produced by elevated grain temperatures. Excluding cosmic-ray destruction, all of these mechanisms are associated with the dynamics and motions of gas clouds. In particular, one must consider the possibility of clouds colliding with clouds and clouds colliding with O-B associations and the accompanying changes in the physical conditions within the clouds. Assuming all clouds have a radius of R, a spatial density N (number of clouds per unit volume), and a random mean speed V, it may be shown that the time τ_{cc} between cloud-cloud collisions is given approximately by

$$\frac{1}{\tau_{cc}} = (4NV\pi R^2). \tag{25}$$

if grazing encounters are included.

In spatially fixed O-B associations, or Strömgren spheres, represented by a radius R_I and spatial density N_I, the frequency of collisions between an individual cloud and an ionizing region is given by

$$\frac{1}{\tau_{co}} = N_I V \pi (R+R_I)^2 . \tag{26}$$

The temperature of a grain is reached by a balance between the rates at which energy is absorbed and emitted. Energy may be absorbed from the electromagnetic radiation, from inelastic gas atom and ion collisions, or from chemical release (molecule formation). Energy may be emitted as electromagnetic radiation, or by evaporation (emission of molecules). In general, the dominant energy absorption and emission processes involve electromagnetic radiation. The absorption (emission) cross-section of a grain is a function of its size or radius a, the wavelength λ, and the complex index of refraction m; we may write $m = m' - im''$. It may be shown (van de Hulst 1949, and later in this section) that to a high order of approximation the energy balance is dominated by emission and absorption of radiation. The equation defining the grain temperature is then

$$\int_0^\infty C_{\text{abs}}(a, \lambda) R(\lambda) d\lambda = \int_0^\infty C_{\text{abs}}(a, \lambda) B(\lambda, T_g) d\lambda \tag{27}$$

where $R(\lambda)$ is the wavelength distribution of radiation in space and $B(\lambda, T_g)$ is the Planck blackbody radiation at a temperature T_g. If T_g is sufficiently high, evaporation is possible (see eq. [31]).

If the radiation incident on a grain is not isotropic, it will not only raise the grain temperature, but also exert a force due to radiation pressure. The radiation force is equal to the rate of transfer of momentum to the grain, which is equal in turn to the net rate of loss of momentum in the incident beam of electromagnetic radiation. It is readily shown (see eqs. [66], [67], below, and [10]) that this force F per unit incident intensity of radiation is given by

$$cF = C_{\text{abs}} + (1 - \langle\cos\theta\rangle) C_{\text{sca}} . \tag{27a}$$

where c = velocity of light and

$$\langle\cos\theta\rangle = \frac{1}{4\pi}\int_0^{2\pi}\int_0^{\pi}\left[\frac{C_{\text{sca}}(\Theta, \Phi)}{C_{\text{sca}}}\right]\cos\theta \sin\theta \, d\theta \, d\varphi . \tag{27b}$$

A process that may lead to a modification of the size distribution of grains without destruction is the differential effect of the radiation pressure on small and large grains that are in the neighborhood of hot bright stars.

The only practical possibility for colliding grains to have enough relative energy to cause evaporation is when two clouds, each containing grains, collide. If two grains of mass m collide, then the total maximum energy available is $\frac{1}{4}mV^2$. Total evaporation may then occur if

$$\tfrac{1}{4}mV^2 > 2mL \tag{28}$$

where V is relative speed of the clouds (or grains) and L is the heat of vaporization. (For ice, $L \cong 10^{10}$ erg gm^{-1}; for graphite, $L \cong 5 \times 10^{11}$ erg gm^{-1}.)

Evaporation may also occur during the collision of two clouds as a result of sputtering produced by the atoms and ions of one cloud striking the grains in the other cloud. According to Wickramasinghe (1963, quoting other sources), H atoms with 2 ev energy have a 0.25 chance of releasing a lattice molecule from an ice crystal. When two clouds collide at a relative speed of 10 km/sec, an H atom or an atom of weight A has energy relative to a grain of .5 or $A/2$ volts, respectively; i.e., all atoms of helium, or heavier atoms, have a good chance of causing sputtering. An important limiting factor for this process is the time of deceleration of the clouds. The sputtering rate due to atom A is

$$\frac{dr}{dt} \approx \frac{-\varpi_A N_A V m_{mol}}{4s}$$

where ϖ_A is the probability that atom A sputters off a molecule, N_A is number density of atoms A, V is cloud velocity, s is the grain density as before and m_{mol} is an average molecular mass.

Finally, when two clouds collide, according to Kahn (1955), the collision will be at least partially inelastic, thus causing a considerable increase in cloud temperature. For a totally inelastic collision this conversion of energy is $\frac{1}{2}m_{\rm H} V^2 = \frac{3}{2}kT$, which for relative cloud speeds of 10 km/sec gives rise to a temperature of 4000° K. Assuming a Boltzmann distribution of atom energies, a fraction $e^{-E/kT}$ has an energy greater than E. Let $\varpi_{\rm H}$ be the probability that an H atom of energy E is able to cause a molecule to sputter off. Then the rate of decrease of radius of the grain is given by

$$\left(\frac{dr}{dt}\right)_{sp} = \varpi_{\rm H}\, e^{-E/kT} \frac{N_{\rm H}}{4s} (3 m_{\rm H} kT)^{1/2} \left(\frac{m_{mol}}{m_{\rm H}}\right). \tag{29}$$

The vapor pressure of a grain is an extremely sensitive function of its temperature; i.e., it varies essentially exponentially to the power T_g^{-1}. At low temperatures, the vapor pressure of a solid may be given by

$$\log P = -\frac{A}{T_g} + B \log T_g + C. \tag{30}$$

(For ice, $A = 2480$, $B = 4$, $C \sim 4.06$.) When the vapor pressure exceeds the partial pressure of the external medium, the grain will evaporate. For example, at $T_g = 100°$ K, $P \cong 1.6 \times 10^{-12}$ dynes cm^{-2}. The partial pressure of O atoms in the interstellar gas at $T = 100°$ K is given by $P = nkT = 1 \times 4 \times 10^{-16}$ dynes cm^{-2}, where we have let the number density N_O of O atoms be 10^{-2} cm^{-3}. One would certainly expect grain evaporation to take place under these conditions. The rate of decrease of the radius is just the same as the increase by growth in a medium with the appropriate partial pressure for atom A (e.g., oxygen) for equilibrium; i.e.,

$$\frac{dr}{dt} = \frac{N_A}{4s} (3 m_A k T_g)^{1/2} = \frac{P}{4s} \left(\frac{3 m_A}{k T_g}\right)^{1/2} \frac{m_{mol}}{m_{\rm H}}. \tag{31}$$

Integrating equation (31), one readily finds that the time required for evaporation of a grain of radius a is given by

$$t \cong \frac{4sa}{P} \left(\frac{kT_g}{3 m_A}\right)^{1/2} \tag{31a}$$

when the grain vapor pressure is sufficiently larger than that of the ambient medium.

Low-energy cosmic rays ($E < 10$ Mev) could be effective in evaporating grains. The basic question is whether there are enough such particles to make a significant contribution to grain destruction. Following Kimura (1962), let $p(a,E)$ be the probability that a grain of radius a is entirely decomposed by a bombarding proton of energy E. Then

$$p(a, E) = \frac{\zeta(-dE/dx)\bar{x}}{mL} \tag{32}$$

where ζ is the fraction of proton energy loss available for decomposition, dE/dx is the stopping power of the grain, $\bar{x}$ is the stopping length, and L is roughly given by the heat of vaporization; i.e., mL is the total energy required for evaporation.

The rate $\phi(a)$ at which the grains of radius a are decomposed is then given by

$$-\frac{dn(a)}{dt} = \phi(a) = \pi a^2 \int p(a, E)\, j(E)\, dE \tag{33}$$

where $j(E)$ is the energy spectrum of the suprathermal protons. The effect of alpha particles is estimated as equivalent to 1.6 times that of the protons. Substituting from equation (32) into equation (33) and replacing $\bar{x}/a$ by its approximate value 1, we get

$$\phi(a) = \frac{3}{4sL} \int \zeta \left(-\frac{dE}{dx}\right) j(E)\, dE . \tag{33a}$$

In the remainder of this section we shall show how, by appropriate choice of the parameters involved in the growth and destruction mechanisms discussed above and by judicious auxiliary physically plausible reasoning, it is possible to predict the existence of each of the several kinds of models of interstellar grains which have been postulated. This procedure is not to be construed as cynical but rather to point out where the uncertainties and plausibilities lie and also to indicate the problems which warrant further investigation.

We shall now attempt to summarize the present extent of our knowledge concerning each grain type which has been proposed to date.

5.2. Dirty Ice

The conclusion that grains grow from the interstellar gas is based on the correlation between gas clouds and regions of obscuration. Regions of very low gas density ($\lesssim 0.1$ H atoms/cm^3, as between the spiral arms) exhibit no measurable extinction. Furthermore, the existence of interstellar diatomic molecules (CH, CN) demonstrates the possibility of the formation of larger molecules which may act as nuclei for grain growth. The possibilities of formation of larger molecules was investigated theoretically by ter Haar (1943) and Kramers and ter Haar (1946). The suggestion by Lindblad (1935) that small solid particles in the galactic system were formed by condensation from the interstellar gas was subjected to further analysis (ter Haar 1943; van de Hulst 1943; Oort and van Woerkom 1943). These authors found that with reasonable assumptions the grains could grow to sizes well beyond those which would account for the interstellar extinction. A consequence of their theory was, however, that for various initial gas densities the grains would either cause a practically complete disappearance of all condensable gases or would not form at all. Coexistence of gas and solid particles would then be rare, whereas observations indicate that the densities of grains and condensable gas are, on the average, about the same. The later theory of growth and formation of interstellar grains developed by Oort and van de Hulst (1946) introduces a process which prevents the total sublimation or condensation of the condensable gases. The problem is treated in three basic steps: (1) condensation nuclei (large molecules) form out of the gas; (2) the condensable gases (excluding H and He which are shown not to condense) strike and generally stick to the nuclei, the growth of a grain being characterized by $\dot{r}$, the rate of increase of its radius; and (3) separate clouds impinge on each other, causing sufficiently energetic collisions between the grains to produce occasional grain evaporation. Major problems associated with steps (1) and (3) remain. The continuing and present formation of condensation nuclei by growth of large molecules have never been satisfactorily worked out. However, the problem may be sidestepped by requiring only that such nuclei exist now by virtue of, say, large-grain destruction processes which produced the seeds for future grain growth. In other words, the question of formation of nuclei is pushed back to primordial time and formation may well have taken place at a very slow initial rate. Experiments on grain-grain collisions can conceivably be performed in the laboratory with high-energy particle accelerators. The theory that nuclei for grain condensation are emitted in the form of small particles from cool stars is a quite reasonable one. Thus, although the reality of all three processes can be questioned, it cannot be stated at present that they are qualitatively incorrect.

An important consequence of the Oort–van de Hulst procedure is that it leads to an equilibrium distribution of particle sizes, consistent with the general uniformity of the reddening law. Let us consider steps (2) and (3) qualitatively. An estimate of the scale of particle sizes may be made by comparing the rate of growth and the average time between cloud collisions and the probability that a grain will evaporate during such a collision. The rate of increase of the radius of a grain is obtained from equation (24). Assuming that the velocity of the gas atoms corresponds to the kinetic temperature of the medium, we may rewrite equation (24) in the form

$$\dot{r} = \frac{\gamma N_A}{4s}(3m_A kT)^{1/2} \approx f\gamma\sqrt{(TA)}\,N_{\mathrm{H}} \times 2.6 \times 10^{-20}\ \mathrm{cm/sec} \qquad (34)$$

where f is the fraction of condensable gases, A their atomic weight, T the kinetic temperature, and N_{H} the number density of hydrogen atoms. Following Oort and van de Hulst, we take $\gamma \approx 1$, $s = 1$ (as for ice), $A = 16$ (corresponding to oxygen), $f = 10^{-3}$ (see oxygen in Table 1), $T = 10{,}000°$ K, $N_{\mathrm{H}} = 2$ atoms per cm^3, and we obtain $\dot{r} \approx 10^{-7}$ cm/10^6 years. In H I regions, although the temperature is less, the density is generally higher, so that this value of $\dot{r}$ is still a reasonable estimate. This rate of grain growth implies that a sizable fraction of the grains would have a radius of the order of 10 μ (assuming the age of the Galaxy to be of the order of 10×10^9 years and assuming a sufficient supply of oxygen). However, such particles would exhibit effects which are not observed.[4]

With the values of N, V, and R from Table 2 in equation (25), the time τ_{cc} between

[4] Diffraction by particles of size of the order of 10 μ would produce a halo of angular dimension 3°. Assuming that the intervening clouds contain a large number of such particles and also span a sufficiently large solid angle, such a halo could be observed. Another effect would be rapid rise of interstellar extinction in the form of a rough λ^{-1} law for $\infty > \lambda > 10\ \mu$. See chapter by Harold Johnson.

cloud collisions is given approximately by $\tau_{cc} = 10^7$ years. The average lifetime of a grain is the time between cloud collisions divided by the probability that a grain will be vaporized inside a cloud. This probability of evaporation has been estimated by Oort and van de Hulst to be approximately 0.1 for grains whose radius[5] is 10^{-5} cm. The average lifetime of a grain is then about 10^8 years; during this time the radius grows to about 0.1 μ, in qualitative agreement with the initial assumption. The lifetime of a grain is about 1/100 the age of the Galaxy, and the processes of growth and evaporation occur with sufficient frequency to make the concept of an equilibrium state reasonable.

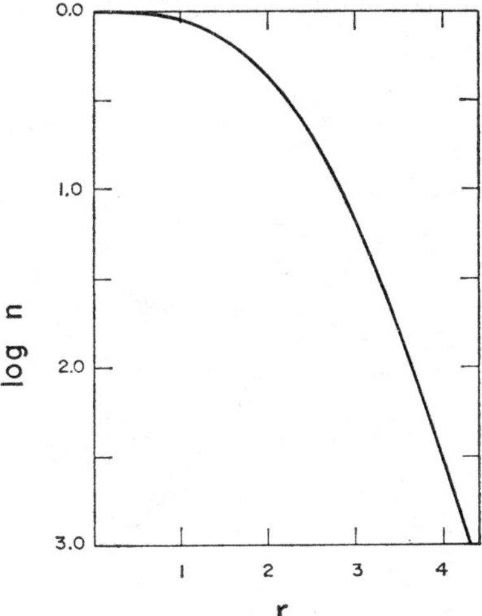

FIG. 15.—The Oort–van de Hulst distribution function of the radii. (Oort and van de Hulst 1946.)

TABLE 3

THEORETICAL GRAIN SIZE DISTRIBUTION FUNCTION

r	$n(r)$	r	$n(r)$	r	$n(r)$
0	1.00	0.5	0.67	1.0	0.064
0.1	0.99	.6	.53	1.1	.029
0.2	0.96	.7	.37	1.2	.011
0.3	0.90	.8	.233	1.3	.004
0.4	0.80	0.9	0.128	1.4	0.002

Source: van de Hulst 1949.

The more quantitative calculations of Oort and van de Hulst lead to an equation whose solution, representing the distribution in sizes of the particles, has been obtained in numerical form. Figure 15 and Table 3 show it graphically and as tabulated by van de Hulst (1949). The dimensionless quantities n and r of Figure 15 are given by

$$n = \frac{n(r)}{n(O)}, \qquad r = \frac{r}{a_O} \tag{35}$$

[5] The number density of the grains in the clouds must be known to calculate the probability of evaporation. The value which has been obtained is for dielectric grains and an extinction of 1 mag/kpc. Thus, the calculation makes use of the final results and is a measure of the internal consistency of the theory.

where a_O is a scale factor which is a measure of the size to which grains may grow, and $n(O)$ is a measure of the total number of grains, which is related to the rate of formation of condensation nuclei. An admittedly simplified derivation of an Oort–van de Hulst type of distribution function (Greenberg 1966) is based on the assumption that the destruction probability is proportional to the grain area. The equation for steady-state growth and destruction is

$$\dot{r}\frac{dn(r)}{dr}+P(r)n(r)=0 \tag{36}$$

where $P(r)$ is the destruction probability, assumed to be a function of r alone. Letting $P(r)$ be proportional to r^2, one obtains

$$n(r) \propto e^{-Ar^3}\,. \tag{37}$$

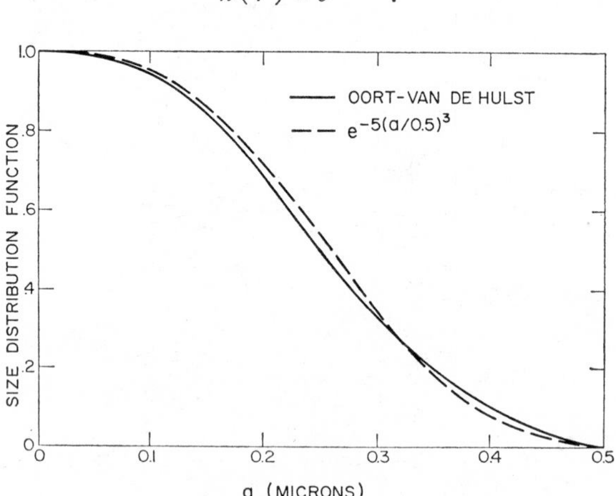

FIG. 16.—Comparison of the Oort–van de Hulst size-distribution function with a distribution function derived on the basis of grain destruction proportional to grain area.

In Figure 16 we compare an appropriately normalized Oort–van de Hulst distribution for spheres with $n(a) = e^{-5(a/0.5)^3}$ grain radius a in microns. The value of this size-distribution function is that it allows further analytical work to be carried through conveniently. If we let $P(r) = D\pi r^2$ where D is a constant, then the coefficient A in equation (37) is given by $A = D\pi/3\dot{r}$. If, furthermore, we use the probability per unit time of evaporation of a grain of 10^{-5} cm radius as given by Oort and van de Hulst, then $D\pi r^2 = 10^{-8}\ \text{yr}^{-1}$. It is interesting to note that if we combine this with the value of $\dot{r} = 10^{-7}$ cm/10^6 yr, then $A = 300\ \mu^{-3}$; whereas the value of A which almost reproduces the Oort–van de Hulst distribution as shown in Figure 16 is $A = 40$. This is equivalent to reducing the effective cutoff (or characteristic) radius from $a_i = 0.5\ \mu$ to $a_i = (7.5)^{-1/3} \times 0.5\ \mu \approx 0.25\ \mu$. If the rate of destruction has been overestimated and/or the rate of growth underestimated, then, of course, it is not too difficult to raise the value of a_i by a factor of 2.

The average spherical radius and sphere area corresponding to the size distribution function $e^{-5(a/a_i)^3}$ are easily shown to be given by

$$\langle a \rangle = \frac{5^{-1/3}\Gamma(\frac{2}{3})}{\Gamma(\frac{1}{3})a_i} = 0.298a_i$$

$$\langle a^2 \rangle = \frac{5^{-2/3}}{\Gamma(\frac{1}{3})\, a_i^2} = 0.127 a_i^2$$

where $\Gamma(\frac{2}{3})$ and $\Gamma(\frac{1}{3})$ are factorial (gamma) functions. Thus the effective single-size sphere equivalent to the distribution of spheres has a radius $a \approx 0.3a_i$, which gives $a \approx 0.15\ \mu$ for $a_i = 0.5\ \mu$.

The chemical composition of particles formed in this manner has been considered further by van de Hulst (1949). He concludes that most oxygen atoms frozen onto a grain may combine with impinging H atoms to form water molecules which, because of their

TABLE 4

SUGGESTED COMPOSITION OF THE INTERSTELLAR PARTICLES

100 molecules H_2O	$m = 1.31$
30 molecules H_2	1.10
20 molecules CH_4	1.26
10 molecules NH_3	1.32
5 molecules MgH, etc.	complex

Source: van de Hulst 1949.

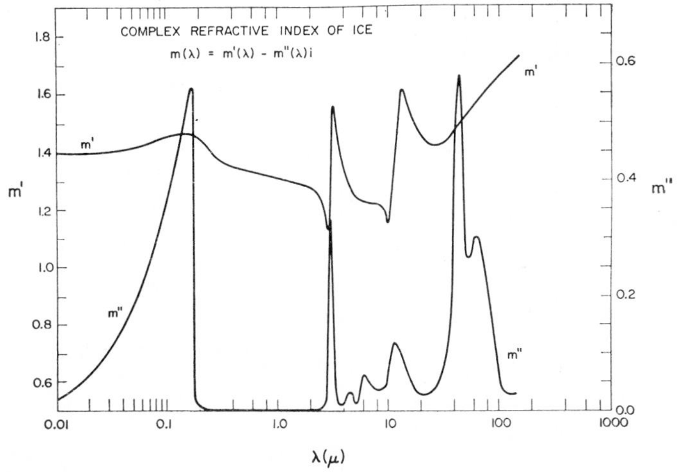

FIG. 17.—Complex refractive indices of ice as a function of wavelength according to a recent compilation by Irvine and Pollack (unpublished; see Table 5).

stability, become the major constituent of the grain. Molecules containing the less abundant metallic elements would exist in the grain roughly in proportion to their relative abundances in space. The grain composition given in Table 4 suggests a dielectric particle whose refractive index is close to $m = 1.25$ with an additional small imaginary part. For most purposes, it is reasonable to use the complex indices for ice itself for the optical properties of these grains as deduced from a number of sources (Dressler and Schnepp 1960; Dorsey 1940; Irvine and Pollack 1966; Kislovskii 1959). Generally speaking, all dielectric substances will have absorptivities in the infrared and the ultraviolet at least roughly similar to the absorptivity of ice. In Figure 17 we present the complex index of refraction of ice as a function of the wavelength. It should be noted that, in the visible, even dirty ice is possibly almost perfectly transparent ($m'' \ll 1$).

Let us assume that ice itself is perfectly transparent ($m'' = 0$) and that only impurities contribute to absorption. If the impurities are atomic or molecular and have a number of absorption lines, then the total effect on the absorption is given by

$$2m'^2\kappa = \frac{4e^4N\lambda^3}{3m^2c^4}\,\omega\sum_k \frac{f_k\lambda_k^2}{(\lambda^2-\lambda_k^2)^2},$$

TABLE 5

INDICES OF REFRACTION FOR ICE

$m_{\text{ice}} = m' - im''$

(As shown in Figure 17)

$\lambda(\mu)$	m'	m''
.5000 (−2)	1.307	0.0008
.1000 (−1)	1.308	.0141
.2000 (−1)	1.311	.0655
.3000 (−1)	1.314	.1090
.6000 (−1)	1.324	.2070
.8000 (−1)	1.331	.2810
.1000	1.339	.3779
.1100	1.346	.4101
.1200	1.353	.4414
.1300	1.360	.4727
.1400	1.370	.5040
.1500	1.380	.5030
.1600	1.392	.0316
.1700	1.404	.0005
.1800	1.417	~0
.2000	1.389	|
.2500	1.346	|
.3000	1.328	|
.4000	1.316	|
.5000	1.307	|
.6000	1.306	|
.7000	1.305	|
.8000	1.304	~0
.9000	1.303	
.1000 (+1)	1.302	
.2000 (+1)	1.291	.0016
.2800 (+1)	1.155	.0123
.3000 (+1)	1.130	.2273
.3200 (+1)	1.557	.1562
.3400 (+1)	1.490	.0307
.4000 (+1)	1.327	.0124
.5000 (+1)	1.247	.0133
.6000 (+1)	1.235	.0617
.7000 (+1)	1.221	.0491
.8000 (+1)	1.219	.0369
.9000 (+1)	1.210	.0365
1.0000 (+1)	1.152	.0413
2.0000 (+1)	1.455	.0255
3.0000 (+1)	1.427	.0525
4.0000 (+1)	1.461	.242
5.0000 (+1)	1.520	.325
6.0000 (+1)	1.559	.275
7.0000 (+1)	1.585	.285
8.0000 (+1)	1.615	.225
9.0000 (+1)	1.638	.150
10.0000 (+1)	1.65	0.100

where $m = m' - im'' = m'\ (1 - i\kappa)$, N = number of atoms or molecules per unit volume, λ_k = wavelength at center of absorption line, λ = observed wavelength, $\omega = 2\pi\lambda/c$, f_k = oscillator strength, and e, m = electron charge, mass. Choosing $\lambda = 0.5\ \mu$, $\lambda - \lambda_k = 100$ A, all f_k's equal, and assuming 10 lines, each given by $f_k = 0.1$, one finds that

$$m'^2\kappa \approx 3 \times 10^{-28} N .$$

A reasonable value of the number of water molecules per cm³ is 4×10^{22}, so that even if the impurities are in the proportion of Fe to O, one has $N \approx 10^{21}$ cm^{-3}, which gives a negligibly small value of $m'\kappa \approx 10^{-7}$. If one considers the Fe atoms to aggregate into

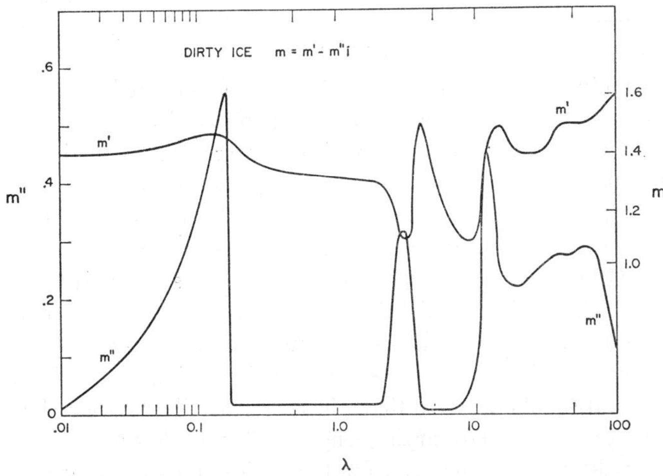

FIG. 18.—Dirty ice complex refractive indices are used in model computations unless otherwise specified. Note the arbitrary imposition of $m'' = 0.02$ in the region $\lambda = 0.17\ \mu$–$2.8\ \mu$.

metallic clumps of perhaps 10^3 atoms/particle, one has 10^{18} such particles/cm³ $= N_p$. Assuming that these particles are sufficiently large to exhibit metallic properties, we treat them as small absorbing spheres of radius $a_p = 2 \times 10^{-7}$ cm. If we use the absorption cross-section as given by equation (68), we find that

$$m'' = -2\pi a_p^3 N_p I m \left(\frac{m^2 - 1}{m^2 + 2}\right) \cong 0.05 .$$

We are thus left with a range of justifiable values of m'' for dirty ice in the visible wavelength region between $m'' \approx 10^{-7}$ and $m'' \approx 0.05$.

Figure 18 shows the form of the complex indices used in earlier computations (e.g. Lichtenstein and Greenberg 1962) involving spherical dirty ice grains.

A rather interesting consequence of the existence of the dielectric grains is that although the *average* atomic composition of space is preserved, a relatively larger proportion of the higher atomic number elements exists in dust clouds than elsewhere. This will be discussed further in the section on interpretation.

In the foregoing theory, particles are assumed to grow by random accretion, and their shape would therefore be roughly spherical. Nonspherical grains might result from collisions which are not sufficiently energetic to cause total evaporation but lead to melting and subsequent fusion between grains. Van de Hulst (1949) has considered fusion processes in relation to the formation of nonspherical particles and has estimated their number. Amalgamation of the colliding grains leads to a modification in the size distribution function which we neglect. The question of selective (or partial) evaporation has also been taken up by Spitzer and Tukey (1951) and is discussed below for metallic grains.

Formation of elongated ice grains has been treated by Kahn (1952) and Piotrowski (1962). Kahn considers the growth of regular ice crystals to be strongly influenced by an electrostatic potential produced by the electric dipole moments of the water molecules. Polarized ice crystals tend to grow in the form of long needles, since the field at the positive end holds free H atoms which can combine with an impinging O atom to form a water molecule, and this then lines up with the existing crystal. Although Kahn's theory suggests a mechanism for the growth and existence of elongated grains, it does not predict a size distribution. Kahn's estimate of the maximum size attainable by growth with a regular structure is at least consistent with the interstellar extinction. This line of investigation is highly significant and should be re-examined in the light of modern work on crystal growth, such as the formation of ice whiskers. A comment on the question of increasing the elongation of grains of a nonspecific crystal character has been made by Piotrowski (1962). The suggested mechanism is that negatively charged grains (Spitzer 1961) which are already slightly nonspherical will tend to attract positive ions toward the ends and thus size growth is accompanied by increased elongation. This has not yet been adequately investigated. As a matter of some importance in the later section on interpretation, it should be noted that whiskers generally have a rather narrow range of characteristic diameters, but varying lengths. This would imply a spread of elongations but an almost uniform diameter for the grains. Finally, regarding the chemical composition of dielectric grains, it should be noted that an attempt to detect an ice absorption band was unsuccessful (Danielson, Woolf, and Gaustad 1965).

We calculate the temperature of a grain as the temperature it will assume in radiative balance with a radiation field. The internal temperature of dielectric grains is important relative to orientation and grain growth by accretion. Heat gains and losses by collisions with atoms and ions will be shown to be unimportant—certainly for dielectric grains. The average interstellar radiation field is replaced by a diluted blackbody radiation field of temperature $T = 10{,}000^\circ$ K and a dilution factor $W = 10^{-14}$. The equation of energy balance becomes (see eq. [27])

$$W\int_0^\infty C_{\text{abs}}(\lambda, a, T_g)B(\lambda, T_{\text{R}})\,d\lambda = \int_0^\infty C_{\text{abs}}(\lambda, a, T_g)B(\lambda, T_g)\,d\lambda \quad (38)$$

where T_g is the temperature of the grain, C_{abs} is the absorption cross-section, and $B(\lambda, T)$ is the radiation of a blackbody at temperature T, according to Planck's law. The Planck function is given by $\pi B = C_1\lambda^{-5}/(e^{C_2/\lambda T} - 1)$, where $C_1 = 3.74 \times 10^{-5}$ erg cm^2 sec^{-1} and $C_2 = 1.439$ cm degree. If C_{abs} is independent of λ, equation (38) becomes

$$W\sigma T_{\text{R}}^4 = \sigma T_g^4\,.$$

Thus, if the grain is a blackbody, its equilibrium temperature would be $T_g = 10^{1/2} = 3.16^\circ$ K for the postulated form of the radiation field.

To use equation (38) correctly, the actual absorption cross-section (Sec. 7) and the size range (Sec. 8) must be found. An estimate of the absorptive properties of dielectric materials at all wavelengths is also needed. Van de Hulst assumes that most of the absorption by dielectric grains is due to the existence of proper frequencies of a crystal in the infrared. The particles would then be small compared with the wavelength. The absorption cross-section of small grains with complex indices of refraction is proportional to λ^{-1} (see eq. [68]). If the index of refraction is complex and roughly constant over a band of wavelengths from ∞ to λ_0 and real for $\lambda < \lambda_0$, we find $T_g \leq W^{1/5}T$ (the maximum value occurring for $\lambda_0 = 0$). Using the values of T and W given in Table 2, we get $T_g \leq 16°$ K. Van de Hulst has considered other possibilities for absorption (such as the presence of metallic impurities which give rise to a small imaginary part of the index of refraction in the visible and ultraviolet regions) and different forms for the absorption

TABLE 6

ENERGY ABSORBED (ergs/sec) IN VARIOUS WAVELENGTH BANDS
(Normalized to Dilution Factor $W = 1$;
$m'' = 0.02$ from $\lambda^{-1} = 0.17\mu$–2.8μ)

T (° K)	$a(\mu)$	$\lambda(\mu)$				
		0–0.19	.19–.30	.30–1.0	1.0–5.2	5.2–∞
10,000	0.1	29.7	31.2	38.9	2.1	101.9
	1.0	34×10^2	136×10^2	325×10^2	26×10^2	521×10^2
50,000	0.1	4.91×10^5	$.06\times10^5$	$.01\times10^5$	$.001\times10^5$	4.99×10^5
	1.0	3.73×10^7	$.25\times10^7$	$.10\times10^7$	$.002\times10^7$	4.09×10^7
100,000	0.1	8.09×10^6	$.01\times10^6$	....	....	8.10×10^6
	1.0	6.49×10^8	$.08\times10^8$	$.02\times10^8$	....	6.59×10^8

cross-section. His calculations give a wide range of grain temperature. If the absorption occurs at sufficiently high wavelength ($\lambda_0 > 10\ \mu$), then grain temperatures of the order of $T_g = 10°$ to $20°$ K appear most likely. Calculations by Lichtenstein and Greenberg (1962), using rigorous Mie theory for spheres and the complex indices of Figure 18, indicate that dielectric grain temperatures are probably somewhat lower—perhaps $T_g = 10°$ K is a fair upper limit. Using Table 6 along with Figures 19–26, one may refine this estimate under varying physical conditions. For standard physical conditions, we find temperatures as given in Tables 7 and 8.

Table 6 shows the relative proportions of the total energy absorbed from a blackbody radiation field. In the event that the actual radiation field differs from this appreciably (which may well be the case), it is possible to determine the appropriate modifications needed to improve the grain temperature estimate. Furthermore, it should be noted that in the $\Delta\lambda$ range 0.30 to 1.0 μ, we have assumed that the imaginary part of the refractive index for dirty ice grains is $m'' = 0.02$, and if it turns out that this is an overestimate, which it may well be, we see that the grain temperature estimate in average interstellar regions (H I) would have to be reduced by as much as 20 to 30 per cent because the total energy absorbed is reduced by about one-half. Of particular note is the fact

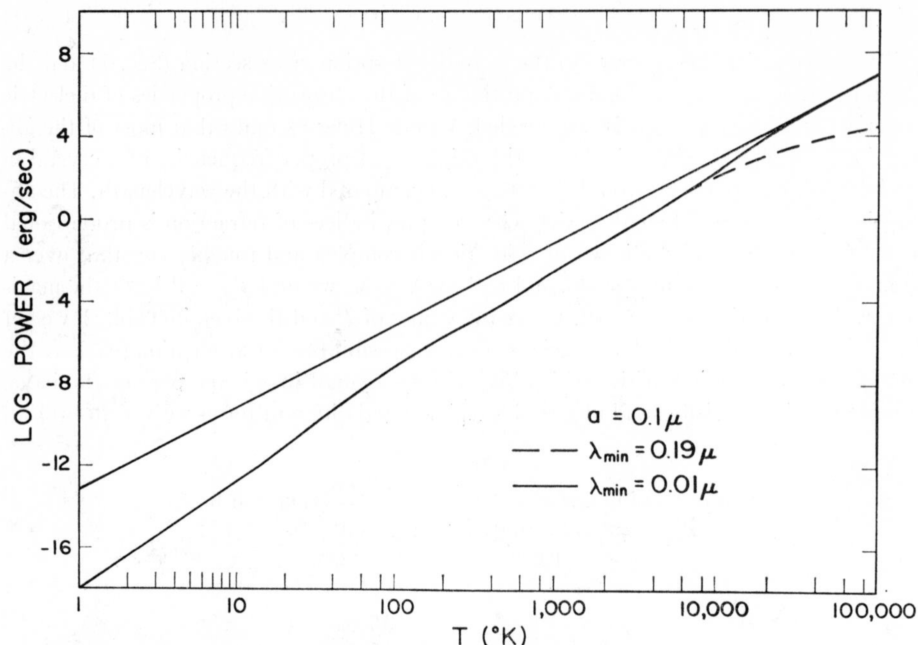

FIG. 19.—Energy absorbed or emitted by a dirty ice grain of 0.1 μ radius. Straight line corresponds to a perfect blackbody (T^4 law). Solid curved line and dashed line refer to cutoff wavelengths for the radiation fields of 0.01 μ and 0.19 μ, respectively.

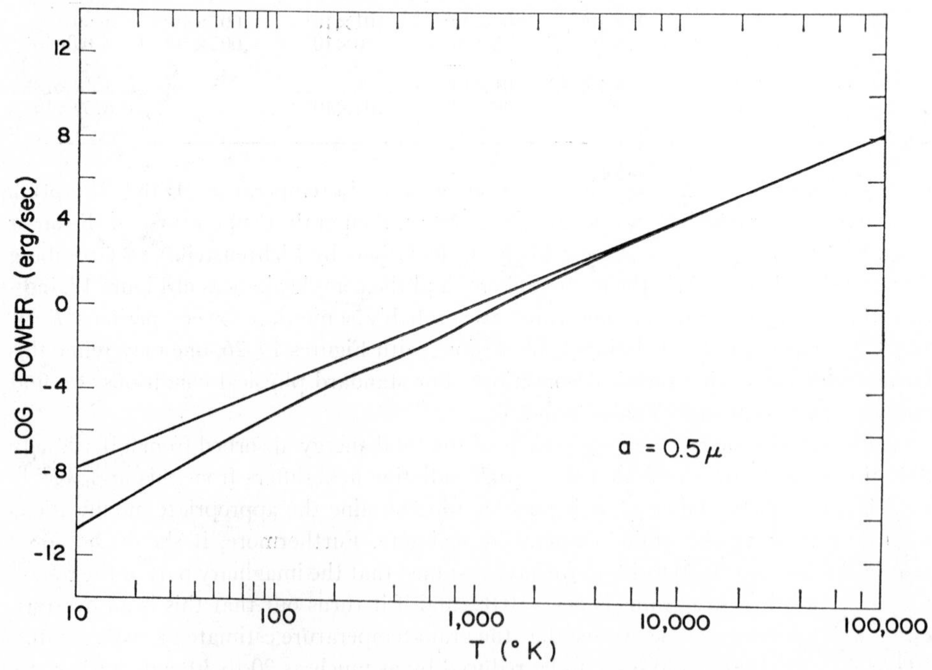

FIG. 20.—Solid lines same as Figure 19, but for 0.5 μ radius

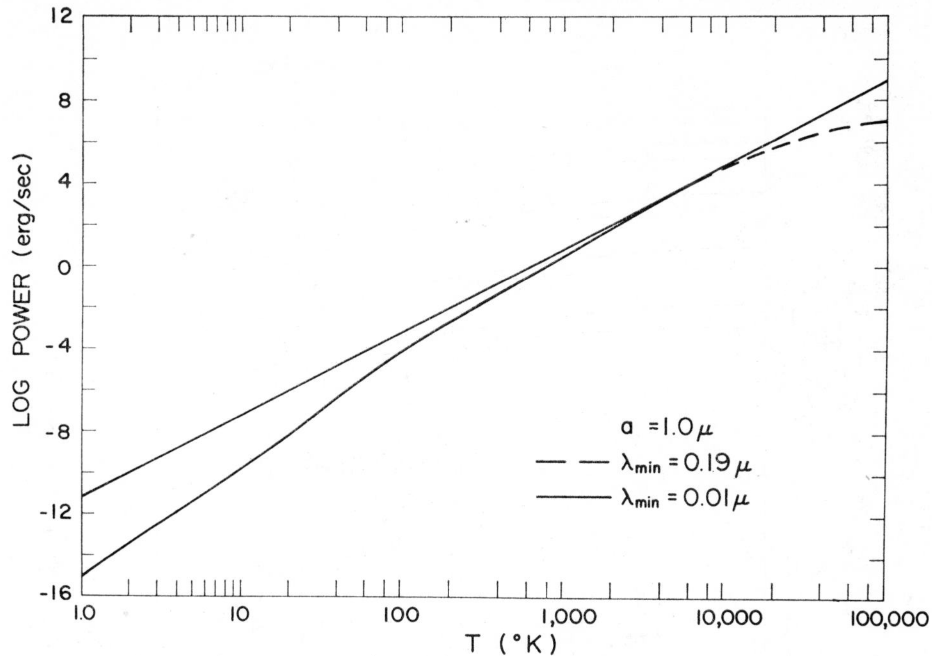

FIG. 21.—Same as Figure 19, but for 1.0 μ radius

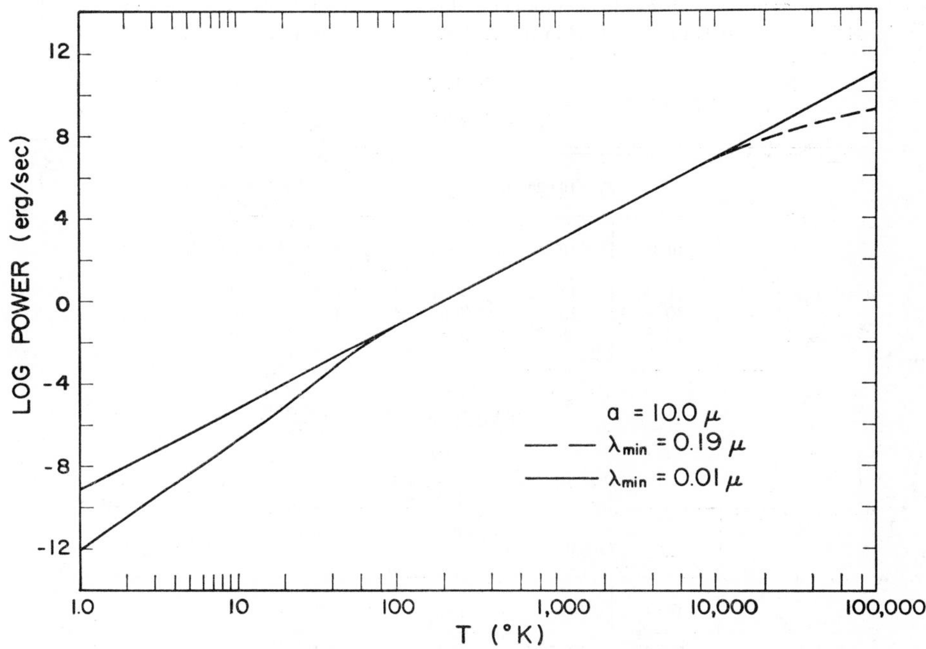

FIG. 22.—Same as Figure 19, but for 10.0 μ radius

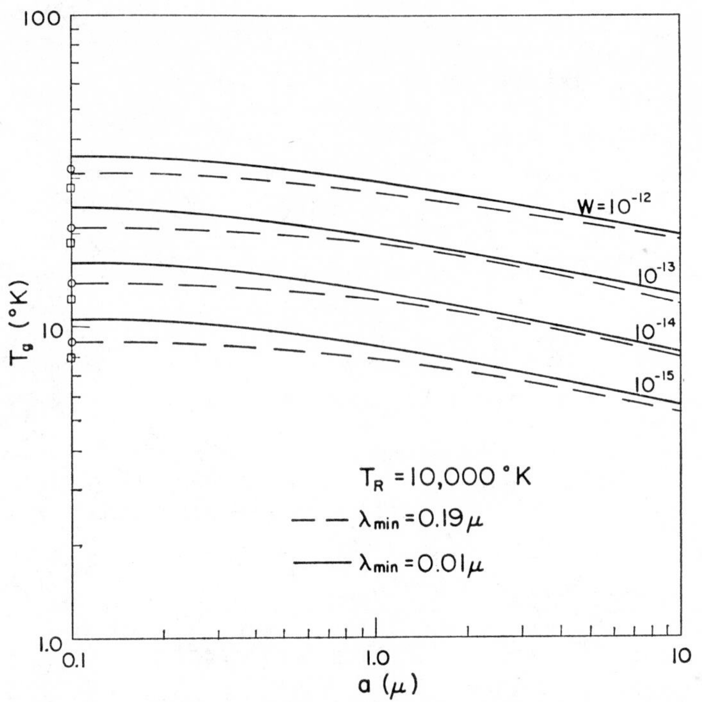

FIG. 23.—Temperatures of dirty ice grains as a function of size. Solid and broken curves are for a blackbody radiation field with 0.01 μ and 0.19 μ wavelength cutoff. The circles and squares denote the modification of grain temperatures for the former and latter radiation fields when the grain absorption is reduced from $m'' = 0.05$ to $m'' = 0.02$ in the visible. The values of W are the dilution factors.

TABLE 7

$m'' = .02$

W	$T_R = 10,000°$ K			$T_R = 50,000°$ K		
	10^{-13}	10^{-14}	10^{-15}	10^{-13}	10^{-14}	10^{-15}
$a = 0.1\ \mu$	21° K	14° K	9° K	95° K	62° K	40° K
$1\ \mu$	20	13	8.6	60	40	26
$10\ \mu$	12.5	8.2	5.5	38	26	18

TABLE 8

$m'' = .02$

$B(\lambda) = 0,\ \lambda < 1905$ A

W	$T_R = 10,000°$ K			$T_R = 50,000°$ K		
	10^{-13}	10^{-14}	10^{-15}	10^{-13}	10^{-14}	10^{-15}
$a = 0.1\ \mu$	19° K	12.5° K	8° K	45° K	30° K	20° K
$1\ \mu$	10	12.5	8	39	26	18
$10\ \mu$	12	8	5.3	26	18	11

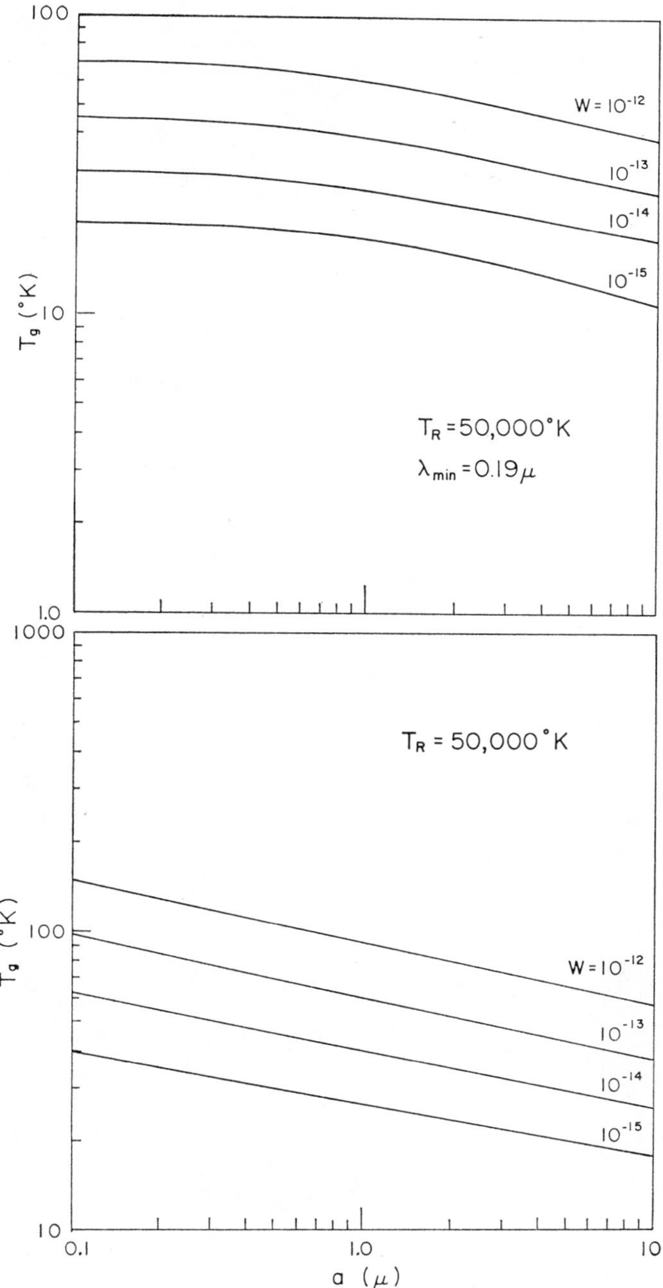

FIG. 24.—Dirty ice ($m'' = 0.02$ in the visible) grain temperatures for modified (upper diagram) and unmodified (lower diagram, $\lambda_{min} = 0.01\ \mu$) blackbody radiation fields at 50,000° K with various dilution factors.

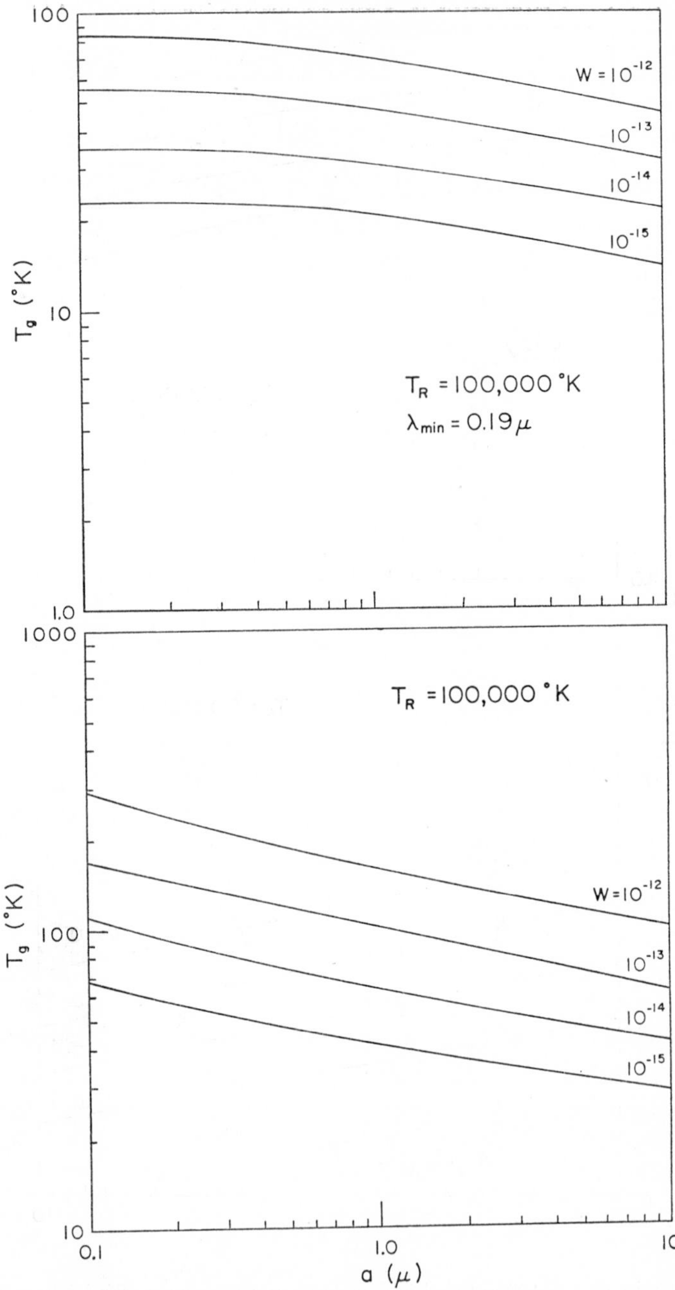

FIG. 25.—Same as Figure 24, but for radiation field temperature T_R = 100,000° K

that if there is indeed a large deficit of radiation in interstellar space below $\lambda \approx 2000$ A, a significant reduction in estimated grain temperature must result.

In Figures 19, 20, and 21, we show the curves for total energy absorbed by a spherical blackbody (upper curve) of radii $a = 0.1\ \mu$, $1.0\ \mu$, and $10.0\ \mu$, respectively, and also the total energies absorbed by the model dirty ice grains of the same size. Both of these curves are normalized to a dilution factor of unity. Because this is the correct factor for emission, we arrive at a grain temperature by the following procedure: (1) Choose a radiation field temperature, (2) drop down from the blackbody curve by the log of the dilution factor, and (3) carry this point horizontally to the left and read off the temperature at the point of intersection with the grain radiation curve. In this manner, we have

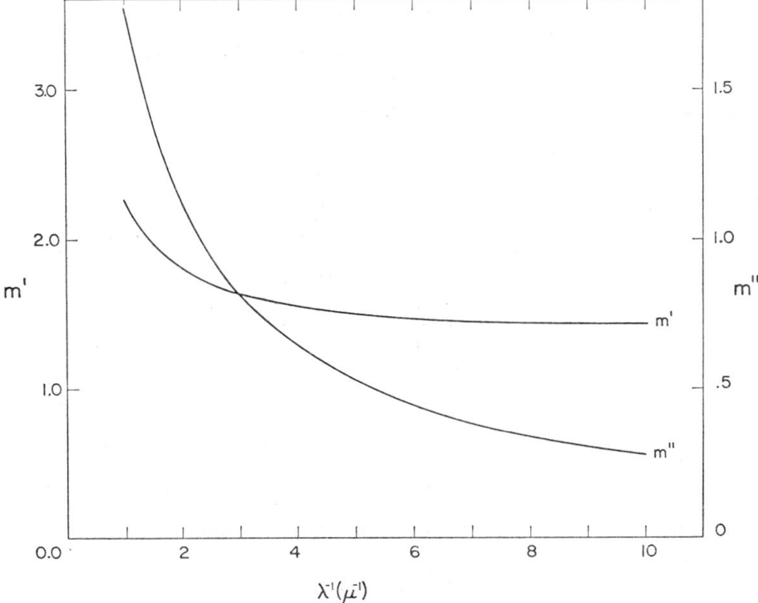

FIG. 26.—Schematic graphite complex indices of refraction from $m^2 = 2 - 8i\lambda$ with λ in microns

generated the curves shown in Figures 22, 23, 24, and 25, which show how grain temperature depends on grain size, radiation field temperature, and dilution factor. Where points are shown separated from the curves, we see the effect of decreasing the absorptive part of the index of refraction in the visible from $m'' = 0.05$ to $m'' = 0.02$.

As has already been stated elsewhere, it may be easily verified that the maximum possible rate of kinetic energy input to a grain by collisions of gas atoms is generally small compared with the rate of absorption of radiant energy. For example, assuming that the total energy per gas atom is given to the grain upon impact, we find that this would be given by

$$\frac{dE}{dt} = n\pi a^2 v_{\mathrm{H}} E_{\mathrm{H}} \approx n\pi a^2 \frac{(kT)^{3/2}}{m^{1/2}} \tag{39}$$

where n = hydrogen atom density, E_H and v_H are the energy and velocity of the atoms, and m is the atomic mass. Letting $n = 10\ cm^{-3}$, $m = m_H$, $T = 300°$ K, and $a = 10^{-4}$ cm, we get $dE/dt \approx 10^{-15}$ erg/sec. This, for example, may be compared (from Table 6) with the value of $5.21 \times 10^4 \times W = 5.21 \times 10^{-10}$ erg/sec as the rate of energy input from a blackbody radiation field of 10,000° K diluted by a factor of 10^{-14}. Even in H II regions (using a gas temperature $T = 10{,}000°$ K), the collisional energy input is only increased to 10^{-13} erg/sec.

Another significant fact regarding grain temperature which has not been sufficiently considered concerns the basic definition of temperature. The equilibrium temperature of a system may be defined only if the fluctuations of system energy are small compared with the average. We should compare the average energy per absorbed photon (or the

TABLE 9

RADIATION FORCE IN DYNES ON DIRTY ICE GRAINS
DILUTION FACTOR $W = 1$
(POWERS OF 10)

BLACKBODY TEMPERATURE (° K)	RADIUS a (μ)		
	0.1	1.0	10.0
1,000	.289 (−13)	.445 (−10)	.732 (−8)
2,500	.138 (−11)	.283 (− 8)	.260 (−6)
5,000	.753 (−10)	.497 (− 7)	.391 (−5)
10,000	.459 (− 8)	.773 (− 6)	.594 (−4)
25,000	.310 (− 6)	.257 (− 4)	.224 (−2)
50,000	.505 (− 5)	.381 (− 3)	.354 (−1)
100,000	.745 (− 4)	.586 (− 2)	.565

rate of absorption of radiant energy) with the total heat content of the grain. The heat content of a grain is given by (see, for example, Kittel [1956], p. 130)

$$Q \approx 60N(kT_g)(T_g/\Theta)^3 \qquad (40)$$

where Θ is the Debye temperature and N is the total number of atoms (or molecules) in the grain. Using $\Theta = 300$ and $N = 4 \times 10^{10}$ (for radius $a = 10^{-4}$ cm), we get $Q = 9 \times 10^{-8}$ erg for $T_g = 10°$ K. This is approximately 100 times the rate of radiant energy input (and output) of 5.2×10^{-10} erg/sec. For $a = 10^{-5}$ cm, $Q \approx 9 \times 10^{-11}$ erg and $(dE/dt)_{rad} \approx 10 \times 10^{-13}$ erg/sec, and the temperature is still definable. For $dE/dt = 10^{-12}$ erg/sec, this is equivalent to one photon of wavelength $\lambda = 2\ \mu$ per second. In view of the fact that $Q/(dE/dt) \propto aT_g^4$, it is possible to find conditions under which this ratio is of the order of unity, and the concept of a grain temperature is then not meaningful.

The radiation pressure exerted on ice grains (spherical, radii 0.1 μ, 1 μ, and 10 μ, and with complex index of refraction as given in Fig. 17, with $m'' = 0.02$ in the visible) is given in Table 9. These computations are performed using Mie theory in equation (27*a*).

The magnetic properties of dielectric grains have been discussed by Davis and Greenstein (1951). We limit ourselves to a statement of their results on the paramagnetic susceptibility of grains which contain a small amount of ferromagnetic impurities (assumed to be mostly Fe atoms). The effect of trapped free radicals could be expected to be similar to that of Fe atoms (see Sec. 5.5 on free radicals).

If a substance is put in a sinusoidally varying magnetic field

$$B = B_0 \cos \omega t\,, \tag{41}$$

then its magnetization is given by

$$M = B_0(\chi' \cos \omega t + \chi'' \sin \omega t) \tag{42}$$

where χ' and χ'' are the real and imaginary parts of the complex magnetic susceptibility. When $\omega = 0$, then $\chi'' = 0$ and $\chi' = \chi_0$ (the static magnetic susceptibility) which is reasonably well given by

$$\chi_0 = 7.3\,\frac{n}{T_g} \tag{43}$$

where n is number of Fe atoms per cubic Angstrom and T_g the grain temperature. Davis and Greenstein estimate χ'' to be

$$\chi'' = 2.5 \times 10^{-12}\,\frac{\omega}{T_g} \tag{44}$$

limited by the condition that

$$10^{-2} > n > 6 \times 10^{-21}\left(\frac{T}{a^5 \rho}\right)^{1/2} \tag{45}$$

where ω is assumed to be the angular velocity of the grain that is imparted by collisions with the gas atoms (see eq. [46]) and where T is the kinetic temperature of the ambient medium, a is a characteristic linear dimension of the grain, and ρ is the density of the grain. The lower limit to n varies from 6×10^{-4} ($a = 10^{-6}$ cm, $T = 10{,}000°$ K) to 1.8×10^{-7} ($a = 10^{-5}$ cm, $T = 100°$), and even for the larger of these two lower limits the amount of Fe required is well within the limit prescribed on the basis of the cosmic abundances given in Table 1.

5.3. Metallic Grains

In this group we include grains with a high fractional content of metallic compounds. It is unlikely that purely metallic grains are formed directly out of the interstellar medium because metal atoms are relatively rare. However, the heavy atom content in a grain may be progressively increased by selective evaporation of lighter atoms and molecules from initially predominantly dielectric grains which may collide in the manner discussed in Section 5.1 (Spitzer and Tukey 1951). Spitzer and Tukey estimate the range of relative grain velocities within which a solid residue (core) of Fe (and Mg) compounds will form as between 2.0 and 5.2 km/sec. Since interstellar clouds collide at speeds higher than this, many grains will be completely vaporized, but a certain fraction are slowed down by collisions with gas atoms so that the selective evaporation process can operate. Arbitrarily assuming that only half of the mass of two grains remains after collision because of evaporation of the more volatile constituents, it is possible to conclude that the number of secondary grains is a considerable fraction of the total number of grains. The disagreement with van de Hulst (1949) on the relative importance of the secondary grains appears to be due to a difference in the assumption regarding the size of the fused particles. The larger the particle, the shorter its life, because the collision cross-section which affects further interaction with other grains is proportional to the area of the grain.

The formation of small grains of high metallic content may be pictured in the following way. Assume that instead of a total destruction of grains every 10^8 years—using the figures of the last section—there remains a fractional residue f of low vapor pressure elements. After n generations, one would find nuclei with nfN atoms, where N is the number of atoms in a typical nonmetallic grain. The total volume of such a grain would be of the order of nf times the original volume, and the radius would be $(nf)^{1/3}$ times the original radius. Using a typical original radius of 0.5 μ, $n = 100$, and $f = 10^{-5}$ (from cosmic abundance data), we get a nuclear radius of 0.05 μ.

An analysis of all chemical and physical reactions which could result from high-speed collisions of grains is beyond the scope of this chapter. Spitzer and Tukey conclude that a ferromagnetic particle consisting of metallic iron and various iron oxides is likely to be the end product of such a collision. In any case, as previously pointed out, the need for small nuclei to be created could well be satisfied by this process.

If selective evaporation is important, then an appreciable fraction of grains formed by deposition of lighter elements (ice) on the metallic (or metallic oxide) nuclei should be present, along with the primary dielectric grains. Accretion would occur by the process outlined in the previous section. The optical properties of such a coated grain could be characteristic of a core with a complex or high refractive index and a shell with a real refractive index similar to that of ice.

The only work on the grain temperatures applicable to ferromagnetic grains is that of van de Hulst (1946). His calculation, based on the conductivity of purely metallic substances, is not applicable to metallic compounds. The conductivity, σ, defines the imaginary part of the index of refraction through the relation

$$m^2 = \epsilon - \frac{i\sigma}{\omega}. \tag{46}$$

For typical pure metallic (iron) particles, we shall use as a complex index in the infrared $m = (1 - i)(3 \times 10^{-4}\lambda/\rho)^{1/2}$, where λ is in microns and ρ is in ohm cm. At $\lambda = 1$ μ, one finds $m = 2.45\ (1 - i)$. In the visible and ultraviolet, the complex index of refraction is not as given by equation (46) but as in Table 10.

If the particle size is assumed to be $a \leq 0.1$ μ, a fair approximation for the absorption cross-section in equation (27) would be given by equation (70). The absorption is then proportional to λ^{-2} for large wavelengths. Using this form and a range of possible specific resistivities, van de Hulst concludes that metallic particles ($a = 0.1$ μ) reach equilibrium temperatures at least ten times higher than the blackbody temperature (3.2° K). The range of plausible temperatures obtained by van de Hulst is 50° to 100° K for $a = 0.1$ μ and higher for smaller particles. The conductivity of graphite given in the next section is close enough to that of iron so that we may use the calculated graphite grain temperature as a good approximation to the temperature of iron grains; namely, $T_g \approx 30°$. A calculation of the equilibrium temperatures of particles of a general type requires appropriate absorption properties. Such particles as are made of metallic oxides and hydrides may act like dielectrics but with high real m, in which case their temperatures would again be in the 10° to 20° K range.

The magnetic properties of a grain consisting of a ferromagnetic core and a mantle of ice are determined by the core. The ferromagnetic property used in relation to the Davis-Greenstein orientation mechanism depends not only on the material but also on the shape of the grain.

5.4. Graphite

Cayrel and Schatzman (1954) postulated that graphite flakes comprise a small component of the total mass of interstellar grains. The original justification for their theory was an attempt to account easily for the maximum observed interstellar polarization. Cayrel and Schatzman were concerned more with the stability of graphite in interstellar

TABLE 10

Complex Indices of Refraction for Iron

$m = m' - im''$

$\lambda(\mu)$	m'	m''
0.39	1.12	1.24
0.42	1.19	1.34
0.45	1.26	1.42
0.48	1.32	1.48
0.49	1.34	1.50
0.51	1.38	1.52
0.52	1.40	1.55
0.55	1.46	1.60
0.58	1.52	1.66
0.60	1.56	1.70
0.61	1.58	1.71
0.80	1.94	2.08
0.91	2.34	2.34
1.00	2.45	2.45
2.00	3.46	3.46
2.99	4.24	4.24
4.19	5.01	5.01
10.5	7.92	7.92
104.7	25.06	25.06

space than with its production. Hoyle and Wickramasinghe (1962) proposed a mechanism for the formation of graphite in the atmospheres of N-type stars, and because even these may not produce an adequate supply, Wickramasinghe has proposed that other stars may be required to produce graphite grains in their atmospheres. The Hoyle-Wickramasinghe mechanism has been studied further (Donn *et al.* 1966). The conditions for the production of whiskers, platelets, and polycrystalline carbon particles are investigated, and it is tentatively concluded that in the highly supersaturated atmospheres of carbon (N) stars, the smaller particles would tend to be polycrystalline and spherical and only a small proportion of relatively large platelets would be formed (Donn *et al.* 1966). It is estimated that 10^5 N-stars in the Galaxy (assumed a constant number in time) may be sufficient to produce the required grain density (in the form of carbon alone) to account for the observed interstellar extinction. Actually one would need less than this amount of carbon if one considers the additional extinction produced as a result of dielectric mantle formation on small carbon particles.

Wickramasinghe has reinvestigated the growth-destruction mechanisms for production of dirty ices in interstellar space, with particular application to the possibility of graphite grains being surrounded by dielectric mantles. He considers the destruction of mantles by grain-grain collision, sputtering by high-temperature gas atoms during cloud-cloud collision, and thermal evaporation during cloud encounter with O and B stars. Using equation (29) with $\varpi_H \approx 0.25$, $n_H \approx 40\ \text{cm}^{-3}$, $T = 3000°$ K, $E = 2$ ev, and $s \approx 2$, one finds that

$$\frac{dr}{dt} \approx 10^{-12}\ \text{cm/yr}. \tag{47}$$

This rate is 10 times faster than the growth rate, so that mantles of $\sim 10^{-5}$ cm outer radius have a low probability of existence on the time average.

Uncertainty in the parameters leading to the value given in equation (47) lies in the use of $E = 2$ ev for the fractional atom-to-ion yield of 0.25. This value is based on the extrapolation of results of metal sputtering to consideration of ice crystals. It is not clear whether the required sputtering energy is overestimated or underestimated. The number of effective atoms for sputtering is given by the exponential $e^{-E/kT}$ and is therefore a rather sensitive function of both E and T. The cloud temperature may decrease below, say, 2000° K to 3000° K in a time short compared with the time required for grain evaporation if we invoke an efficient cooling process like that involving the H_2 molecule. In such a case, the sputtering process would become considerably less effective. Using equation (36) with $P(r) = (dr/dt)/r_{\max}$ where $r_{\max}$ is the peak size and where dr/dt is as in equation (47), we get

$$n(r) = e^{-(P/\dot{r}_{\text{growth}})r} \tag{48}$$

with $r > r_0 =$ graphite core radius and where

$$\frac{P}{\dot{r}_{\text{growth}}} \approx \frac{10^{-12}\ \text{cm per yr}}{r_{\max} 10^{-13}} \approx \frac{10}{r_{\max}}.$$

Thus if $r_{\max}$ is governed by standard growth for 10^8 yrs, the coefficient of r in the exponent is $\sim(5 \times 10^{-6})^{-1}$ times this value and we have an exponential falloff giving maximum mantle radius no more than perhaps 2 times core radius r_0 if r_0 is $\sim 5 \times 10^{-6}$ cm.

The optical properties, as well as the shape, of a graphite crystal are extremely anisotropic. Graphite, a hexagonal layer-latticed crystal, occurs in the form of flakes; electrons are free to move within, but not perpendicular to, the layers. In other words, a graphite flake at low temperatures acts like a metal (complex refractive index) or a dielectric (the refractive index having at most a small imaginary part), depending on whether the electric field is perpendicular to or parallel to the axis of the hexagon. Dutta (1953) found that the conductivities perpendicular to and parallel to the hexagonal axis of a thin (0.011 cm) graphite crystal (at $T = 80°$) are $\sigma_\perp = 1.5 \times 10^4\ \text{ohm}^{-1}\ \text{cm}^{-1}$, and $\sigma_\parallel = 0.8\ \text{ohm}^{-1}\ \text{cm}^{-1}$. The literature on graphite has been followed intensively (Wickramasinghe and Guillaume 1965), and the known available complex indices (homogenized) for graphite appear to be as shown in Figure 27. Also shown are the complex indices as assumed in the original paper by Hoyle and Wickramasinghe; namely, $m^2 = 2 - 8i\lambda$ (λ in microns). For very long wavelengths (the infrared up to perhaps $\lambda = 1\ \mu$), the complex indices are well determined by the conductivities. In Table 11 we give a set of proposed values of the conductivities of hexagonal graphite for various temperatures and the subsequent values of m^2 to be compared with each other. We note that there are significant temperature effects which could affect the scattering properties of these grains. In considering the optics of nonspherical graphite—the more realistic form—we will use the strong anisotropy of the conductivity ($\approx 100:1$); for the electric field directed on the hexagonal axis, we will consider the imaginary part of the index to be much less than that shown in Figure 26.

TABLE 11

GRAPHITE CONDUCTIVITIES AND COMPLEX INDICES IN THE INFRARED AND NEAR VISIBLE (FROM A. I. P. HANDBOOK 1957)

λ	$m^2 = 4 - 10i\lambda$*		$\sigma = 2\times10^4 \parallel c$-axis 300° K	$\sigma = 6\times10^4 \parallel c$-axis 15° K	$\sigma = 2.6\times10^6 \perp c$-axis 300° K	$\sigma = 2.5\times10^7 \perp c$-axis 15° K
	m'	m''	$m' = m''$	$m' = m''$	$m' = m''$	$m' = m''$
2.094	3.56	2.94	1.12	1.94	12.78	39.63
2.416	3.77	3.20	1.20	2.09	13.72	42.57
2.992	4.13	3.62	1.34	2.32	15.27	47.37
3.306	4.32	3.83	1.41	2.44	16.06	49.80
3.696	4.54	4.07	1.49	2.58	16.97	52.64
4.488	4.95	4.53	1.64	2.84	18.7	58.01

GRAPHITE COMPLEX INDICES IN THE VISIBLE AND ULTRAVIOLET*

$\lambda(\mu)$	m'	m''	$\lambda(\mu)$	m'	m''
.1499	1.127	0.266	0.4016	2.456	1.452
.2000	0.786	1.272	0.5181	2.456	1.452
.2202	0.849	1.649	0.5464	2.456	1.425
.2597	1.480	2.365	0.6410	2.456	1.425
.3322	2.252	1.665	0.8403	2.579	1.629
.3597	2.425	1.443	1.1627	2.852	2.034
.3731	2.456	1.452			

* From Wickramasinghe and Guillaume (1965).

Using the complex m^2 based on a constant conductivity ($m^2 = 2 - 8i\lambda$), one arrives at the radiation-absorption and radiation-emission curves as shown in Figure 28. Using the standard (H I) radiation field, one arrives at a grain temperature for $a = 0.05\ \mu$ of 30° K, which is significantly but not greatly different from that for the dielectric grains.

The radiation pressure exerted on spherical graphite grains of radius $a = 0.05\ \mu$ whose complex index is given by $m^2 = 2 - 8i\lambda$ is shown in Table 12 for various blackbody models of stars. As before, these are given for a dilution factor of unity.

The magnetic properties of graphite can be explained (Ganguli and Krishnan 1941) on the basis of the free motion of electrons perpendicular to the hexagonal axis. The specific diamagnetic susceptibilities parallel to and perpendicular to the c-axis are in cgs units; that is, $\chi_{\parallel} \approx 21.5 \times 10^{-6}$, $\chi_{\perp} \approx -0.5 \times 10^{-6}$, and at low temperatures, $\chi_{\parallel} - \chi_{\perp} \approx -30 \times 10^{-6}$. For a small flake (diameter $<0.2\ \mu$)$\chi_{\parallel} \approx -1 \times 10^{-6}$. Cayrel and Schatzman have estimated the imaginary part of $\chi_{\parallel}(\chi_{\parallel} = \chi_{\parallel}' + i\chi_{\parallel}'')$ of which the

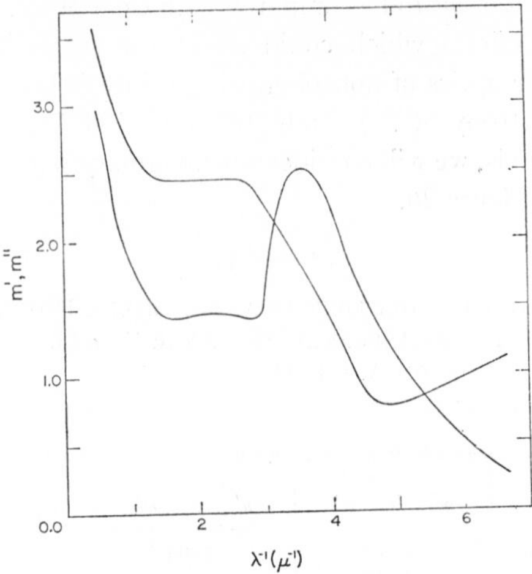

FIG. 27.—Complex refractive indices for graphite as given by Wickramasinghe and Guillaume (1965). The peaked curve is for m''.

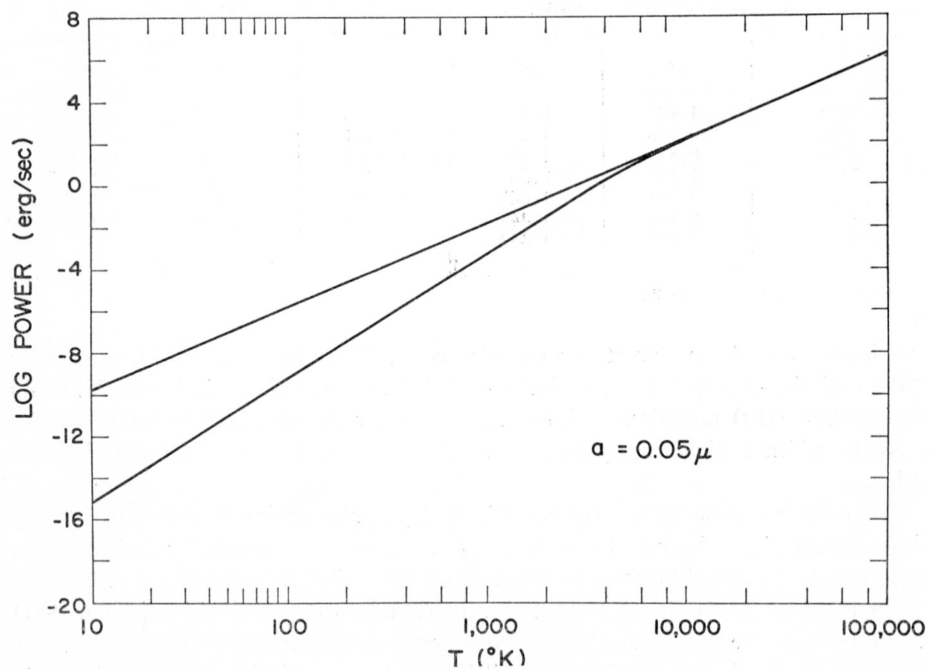

FIG. 28.—Rate of absorption of radiant energy by a graphite sphere ($m^2 = 2 - 8i\lambda$) of radius $a = 0.05\ \mu$. Straight line corresponds to a perfect blackbody.

real part is roughly the one given above) by a calculation involving the relaxation time of electron collisions in the plane perpendicular to the c-axis. This gives[6]

$$\frac{\chi_{\parallel}''}{\chi_{\parallel}} \approx 8 \times 10^{-7} l\omega$$

where l, the mean free path of the electrons, decreases rapidly with increasing grain temperatures ($l \propto T_g^{-2}$). Using the value of $\chi_{\parallel}$ for small grains, we get

$$\chi'' \approx 8 \times 10^{-13} l\omega$$

$$\approx 8 \times 10^{-13} \frac{\omega}{T_g^2} .$$

TABLE 12

RADIATION FORCE ON SMALL SPHERICAL GRAPHITE GRAINS

$a = 0.05\ \mu;\ m^2 = 2 - 8i\lambda\ (\mu)$

Temperature (° K)	Force [Dynes (powers of 10)]
1,000	0.667 (−14)
2,500	0.152 (−11)
5,000	0.723 (−10)
10,000	0.208 (− 8)
25,000	0.904 (− 7)
50,000	0.138 (− 5)
100,000	0.211 (− 4)

A more recent evaluation of the diamagnetic susceptibility by Wickramasinghe (1962) gives (in our notation)

$$\frac{\chi''}{\chi_{\parallel}} \approx 6 \times 10^{-6} \frac{\omega}{T_g}$$
$$\chi'' \approx 12 \times 10^{-11} \frac{\omega}{T_g} , \qquad (49)$$

which is larger than the Cayrel-Schatzman estimate (as modified here) by a factor greater than 100 and furthermore is quite a bit less temperature dependent. The earlier estimate of χ'' may be more reliable.

5.5. Free Radicals

Platt (1956) first suggested that a small particle grown by random accretion from the interstellar gas and containing many ions and radicals is probably no more chemically stable or electrically neutral than its constituents, and that its energy bands may not be filled (as would be true for a dielectric). Platt then derived a set of plausible physical characteristics for such particles whose chemical constituents could be expected to follow normal cosmic composition.

[6] The value of $\chi_{\parallel}' / \chi_{\parallel}$ quoted here is a factor of ten less than that given by Cayrel and Schatzman (1954).

Kimura (1962) has approached the production of very small particles (free radicals) from the point of view that the rate of destruction of dielectric particles by low-energy cosmic rays would not allow such particles to become large ($\sim 10^{-5}$ cm). Assuming the simplified growth and destruction differential equation (eq. [36]), but with $P(r)$ replaced by $\varphi(r)$ (eq. [33*a*]), one finds

$$n(a) = e^{-(\varphi/\dot{r})a} . \tag{50}$$

Kimura estimates φ as $\approx 2.6 \times 10^{10}\zeta/L \gtrsim 10^{-14}$ sec^{-1} on the basis of a cosmic-ray flux (Hayakawa, Nishimura, and Takayanagi 1961) which would be required to maintain the mean H I cloud temperatures (neglecting the cloud-cloud collision process as considered by Kahn). Using the value of $\dot{r}$ in equation (34), we find an exponential factor of $(0.2 \times 10^7)^{-1}$ cm $= 5 \times 10^{-7}$. This corresponds to particle size (at the e^{-1} point in the particle size distribution) of $a = 50$ A, which is characteristic of the Platt-type particle.

The degree of shape anisotropy estimated by Platt is based on statistically random growth by accretion, to be given by a relative root-mean-square variation in any dimension as $N^{-1/2}$ where N is the number of atoms in the grain. Because of the existence of unpaired electrons, the particles should be quasi-metallic. Theoretically, the longest wavelength absorbed by a one-electron jump between the highest filled and lowest empty Fermi levels of a particle of length L is about $\lambda = 400\,L$. This means that particles about 10 A in diameter will absorb and re-emit radiation at wavelengths shorter than 4000 A. This type of scattering is obviously quantum mechanical rather than classical. Platt shows qualitatively that a three-dimensional structure would have an extinction cross-section roughly constant and equal to the geometrical cross-section into the ultraviolet (see Fig. 69).

If we assume that a finite absorption cross-section exists with the same form as the extinction cross-section, we may estimate the internal temperature of the grain. The absorption associated with the quantum-mechanical scattering process is due to the fact that a re-emitted photon has slightly less energy than the incident photon. Let C_{abs} be constant for $\lambda < \lambda_0$ and zero for $\lambda > \lambda_0$. This is essentially opposite to the type of absorption characteristic of classical dielectric grains, but is in some ways similar to that of classical metallic grains. For $\lambda_0 = 0.5\ \mu$, equation (28) gives the internal temperature as $T_g \approx 1000°$ K. If no *actual* absorption were associated with the quantum-mechanical scattering process, the internal temperature would be undefined in terms of radiation equilibrium. We should then investigate the effect of collisions with the gas molecules in the medium. It may be necessary to incorporate the effect of the collisions into the total energy balance of the Platt particles to determine the equilibrium temperature. If 1000° K is the correct order of magnitude for the temperature of the Platt grains, such grains could not exist and the growth process would follow the sequence appropriate to dielectric particles where at all stages the grains consist of chemically stable materials.

A grain containing trapped free radicals would be paramagnetic to about the same degree as the dielectric particles with ferromagnetic impurities. However, to our knowledge no estimate has yet been made of the imaginary part of the susceptibility for the complex aggregates envisioned by Platt.

6. ORIENTATION MECHANISM

Of all the types of orientation mechanisms thus far proposed, only one seems plausible. The interstellar polarization is roughly parallel to the galactic plane; this implies that the axis of elongated interstellar particles should be, on the average, perpendicular to the galactic plane (disk-like particles being thus seen edge on from a point in the plane). Any satisfactory orientation mechanism must take this into account.

Gold (1952*a*, *b*) noted that if the velocity of a dust grain relative to the gas is supersonic, the angular momentum imparted to a grain by collisions with the gas tends to be normal to the relative velocity. Since the elongated axis of a grain tends to be normal to its angular momentum, the relative motion of the grain and gas must be perpendicular to the galactic plane. Since relative motions of clouds are the source of the relative motion of grains and gas, they must be predominantly normal to the galactic plane because relative cloud motions in the galactic plane would produce orientation in the wrong direction. Such anisotropy of cloud motions seems unlikely. Furthermore, it is difficult to explain why, over wide ranges of galactic longitude, the optical polarization should be either uniform or quite random, the implication being that the relative cloud motions are at right angles in one region and random in another (e.g., Perseus vs. Cygnus).

A more quantitative objection to Gold's theory has been raised by Davis (1955). Davis calculates the average penetration of grains into a cloud within which the relative velocity of the grain and gas atoms is greater than the thermal velocities of the gas atoms. He finds that the Gold mechanism would be effective only in a thin layer at the edge of a gas cloud, the average effect throughout the interior of the cloud being much smaller.

A further objection has been raised by Zirin (1952) who points out that the Gold mechanism applies only to elongated grains, and that if the grains are as likely to be flattened as elongated, the fraction of grains effective in producing optical polarization is further reduced. Zirin has presented a qualitative alternative picture of grains sputtering out atoms along the direction of relative motion of cloud-cloud collisions and thus becoming thinner along the direction of motion. This theory is highly speculative; it also requires that relative motions of the clouds should be predominantly *in* the galactic plane to give the observed general direction of polarization, and this is not apparently borne out. Zirin was led to his consideration of gas-dynamical orientation by an apparent correlation between the velocities of gas clouds and the degree of polarization. Using semiquantitative magnetohydrodynamic ideas to relate cloud velocities to magnetic field strengths, we could perhaps also state that the magnetic fields are correlated with the degree of polarization, but there is too little evidence to pursue this further. Perhaps the relative motions of clouds could tend to stretch magnetic lines of force.

Existence of large-scale galactic magnetic fields can be accepted without question. Whether they are strong enough to orient the grains sufficiently to produce the observed amount of polarization is not yet clear, but magnetic orientation appears to be the simplest explanation at present. Spitzer and Tukey (1951) proposed a static theory of magnetic orientation which applies only to ferromagnetic grains. We may picture the particles as magnetic needles spinning under bombardment by the gas atoms; the magnetic moments of the grains tend to lie along the direction of the magnetic field. The degree of alignment is defined by a state of equilibrium and is high if the Boltzmann factor is much greater than unity, i.e., the ratio of the magnetic potential energy to the kinetic energy of rotation, $IVB/kT > 1$, where IV is the magnetic moment of the grain. Assuming $V \approx 10^{-14}$ cm³ ($a \approx 10^{-5}$ cm), $T = 100°$, and $I = 2000$ gauss (pure iron), we find that $B \gtrsim 10^{-3}$ gauss, an impossibly high value, even under favorable conditions. It should be noted that the Spitzer-Tukey mechanism implies that the magnetic fields are generally perpendicular to the galactic plane.

The most satisfactory dynamical theory of grain orientation is that developed by Davis and Greenstein (1951) for paramagnetic grains, generalized to include ferromagnetic grains by Henry (1958). The details of the Davis-Greenstein theory are described in the original paper, with a later modification by Davis (1958). A more realistic thermodynamic application of the Davis-Greenstein mechanism has been made by Jones and Spitzer (1967). A simplified and approximate, but physically correct, treatment of their work is given in this section.

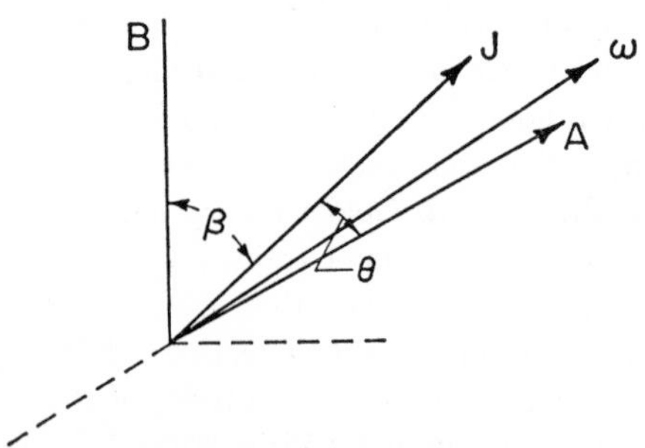

FIG. 29.—Geometric configuration of a particle spinning with angular velocity ω and angular momentum J in a magnetic field B. A is along the particle axis.

The motion of an individual nonspherical particle is affected by collisions with gas atoms and to the torque generated by its own motion in an external magnetic field. The gas atoms play a dual role in this process: (1) They set the particle spinning, creating a dynamical situation in which there is an aligning torque due to paramagnetic relaxation, and (2) the random collisions tend to disalign the rotation axes of the particles. As we shall later see, the grains would spin even *without* collision.

We examine first the dynamics of a spheroidal paramagnetic grain in a uniform magnetic field B. By the equipartition theorem, the mean kinetic energy of rotation of a grain on which no torques act is equal to the mean kinetic energy of the atoms of the interstellar gas. Thus if I is the average moment of inertia,

$$\tfrac{1}{2}I\omega^2 = \tfrac{3}{2}kT \qquad (51)$$

where ω is the root-mean-square angular velocity. Figure 29 shows a geometric configuration of the angular velocity ω, the angular momentum J, and the symmetry axis A of a grain with respect to the magnetic field B. With respect to axes fixed in the grain, the external magnetic field appears to vary sinusoidally with time. If $\chi'' = \chi'\omega t_0$ (where $t_0 \ll 1$ and is constant), $\chi' \neq \chi'(\omega)$, and the grain is assumed isotropic so that the magnetization can be obtained by superposition, the magnetization of the grain at the time t has the direction that B had at $t - t_0$ (Davis 1958). Thus $\boldsymbol{M} \approx \chi'\boldsymbol{B}(t - t_0)$. The direction of $B(t - t_0)$ is given by the direction of $B(t) + [\omega t_0 \times B(t)]$ if ωt_0 is an infinitesimal rotation. The torque on the grain is then given by

$$\dot{J} = L = VM \times B \approx V\left(\frac{\chi''}{\omega}\right)(\omega \times B) \times B \tag{52}$$

and the rate of change of the rotational energy with respect to time is given by

$$\dot{R} = -V\left(\frac{\chi''}{\omega}\right)[\omega^2 B^2 - (\omega \cdot B)^2] \tag{53}$$

where V is the volume of the grain. The negative sign in equation (53) implies energy dissipation associated with the paramagnetic relaxation.

Davis and Greenstein use equations (52) and (53) to obtain the average rate of change of θ and β over one rotation of A and ω about J (nutation), where θ is the angle between the angular momentum and the magnetic field and β is the angle between the symmetry axis and the magnetic field. Thus

$$\langle \dot{\beta} \rangle = -\frac{\chi''}{\omega}\frac{VB^2}{I\gamma} \sin\beta \cos\beta (\gamma \cos^2\theta + \sin^2\theta)$$

and (54)

$$\langle \dot{\theta} \rangle = \frac{\chi''}{\omega}\frac{VB^2}{I\gamma}(\gamma - 1)\sin\theta \cos\theta (1 - \tfrac{1}{2}\sin^2\beta)$$

where γ is the ratio of the moments of inertia perpendicular to and along the symmetry axis of the grain. These expressions have been rederived by Cugnon (1963) for the more general case in which the magnetic susceptibility is anisotropic, as it is for ferromagnetic particles and for graphite. One obtains

$$\langle \dot{\beta} \rangle = -\frac{\chi_T''}{\omega}\frac{VB^2}{I\gamma} \sin\beta \cos\beta [\gamma \cos^2\theta + (1 + r\sin^2\theta)]$$

(55)

$$\langle \dot{\theta} \rangle = -\frac{\chi_T''}{\omega}\frac{VB^2}{I\gamma} \sin\theta \cos\theta \left[1 - \tfrac{1}{2}\sin^2\beta \left(1 + \frac{2r}{\gamma - 1}\right)\right]$$

where $r = (\chi_A'' - \chi_T'')/2\chi_T''$ and χ_A'' and χ_T'' are the imaginary parts of the magnetic susceptibilities along and transverse to the particle symmetry axis, respectively.

For isotropic susceptibilities, one has $r = 0$ and equations (54) reduce to equations (55), obtained by Davis and Greenstein. For the ferromagnetic case, one has $\chi_A'' = 0$ and $r = -\frac{1}{2}$. Since $\dot{\beta}$ is negative, the angular momentum of a grain tends to line up with the direction of the magnetic field. For elongated particles, $\gamma > 1$ and θ approaches $\pi/2$, which means that the long axis spins about the magnetic vector. For flat particles, $\gamma < 1$ and θ approaches zero. Generally, both elongated and flattened particles, when perfectly oriented by the Davis-Greenstein mechanism, line up with their short dimension along the magnetic field. The long dimension is then perpendicular to the magnetic field. The general direction of observed polarization supports this theory if the galactic magnetic fields lie along the direction of the spiral arms.

The presence of a torque tending to orient the grains does not necessarily mean that the grains will be oriented. Random bombardment by the gas atoms acts against such order, and if the system were in thermal equilibrium, there would be no orientation at all. The assumption that the system is not in thermal equilibrium, but is in a steady state, leads to an estimate of the conditions needed for orientation. In the absence of a torque, assume the spinning of an elongated (later considered to be spheroidal for simplicity) body to be produced by collision with the gas atoms at a temperature T. Let I = moment of inertia about a symmetry axis of γI = moment of inertia about the transverse axis. Then by equipartition we have the distribution functions

$$F_B(J_B) = (2\pi I kT)^{-1/2} \exp\left(-\frac{J_B^2}{2IkT}\right)$$
$$F_A(J_A) = (2\pi\gamma I kT)^{-1/2} \exp\left(-\frac{J_A^2}{2I\gamma kT}\right) \tag{56}$$

where $F_B(J_B)$ = fraction of particles with angular momentum about the B axis between J_B and $J_B + dJ_B$, etc.

Using the body-fixed system of coordinates and defining θ as the angle between $\boldsymbol{J}$ and $\boldsymbol{b}$, one may find that the fraction of spheroids from which the principal axis $\boldsymbol{b}$ lies within a range of solid angle $d\Omega$ at an angle θ with respect to $\boldsymbol{J}$ is given by

$$f_\theta(\theta) = \frac{\gamma^{1/2}}{4\pi}(\gamma \cos^2\theta + \sin^2\theta)^{-3/2} \tag{57}$$

where $\gamma = \frac{1}{2}[(b^2/a^2) + 1]$.

The full treatment of the orientation of nonspherical particles is not simple; so we shall consider (as in Jones and Spitzer 1967) the orientation of spherical particles and then estimate qualitatively the effects of nonsphericity.

Let the external magnetic field $\boldsymbol{B}$ be directed along the z-axis. In the absence of the magnetic field, the distribution of angular momenta about each separate axis is Maxwellian (eq. [56]). Since for angular momentum along the z-axis there will be no magnetic torques, we still find

$$F_z(J_z) = (2\pi I kT)^{-1/2} \exp\left(-\frac{J_z^2}{2IkT}\right). \tag{58}$$

However, the distribution of angular momentum about a transverse axis is now influenced by the torque as well as by gas-atom collisions.

We denote the relaxation time for approach to equilibrium via gas atoms alone by τ_c (c = collision) and by torque alone as τ_o (o = orientation). The temperature associated with the collisional equilibrium is T and the temperature associated with the magnetic torque is the grain internal temperature T_g. It should be noted here that even in the absence of gas-atom collisions the non-zero internal temperature would produce random magnetization which would cause the grain to spin about the z-axis with an angular momentum distribution corresponding to its internal temperature. Then the distribution function about the x-axis is controlled by two rates; namely, T/τ_c and T_g/τ_o. The steady-state equivalent (or average) temperature T_{av} is obtained by averaging these rates

$$T_{av} = \frac{(T/\tau_c)+(T_g/\tau_o)}{(1/\tau_c)+(1/\tau_o)}$$
$$\frac{T_{av}}{T} = \frac{1+\delta(T_g/T)}{1+\delta} \tag{59}$$

where $\delta = \tau_c/\tau_o$ is the ratio of the collisional to the torque relaxation times.

We may now write for the distribution function of angular momentum about the x- (or y-)axis

$$F_x(J_x) = (2\pi I kT_{\text{av}})^{-1/2} \exp\left(-\frac{J_x^2}{2IkT_{\text{av}}}\right). \tag{60}$$

The fraction of *spheres* whose angular momentum vectors make an angle β with $\boldsymbol{B}$ is then given by

$$f_0(\beta) = \frac{(T_{\text{av}}/T)^{1/2}}{4\pi}\left(\frac{T_{\text{av}}}{T}\cos^2\beta + \sin^2\beta\right)^{-3/2}. \tag{61}$$

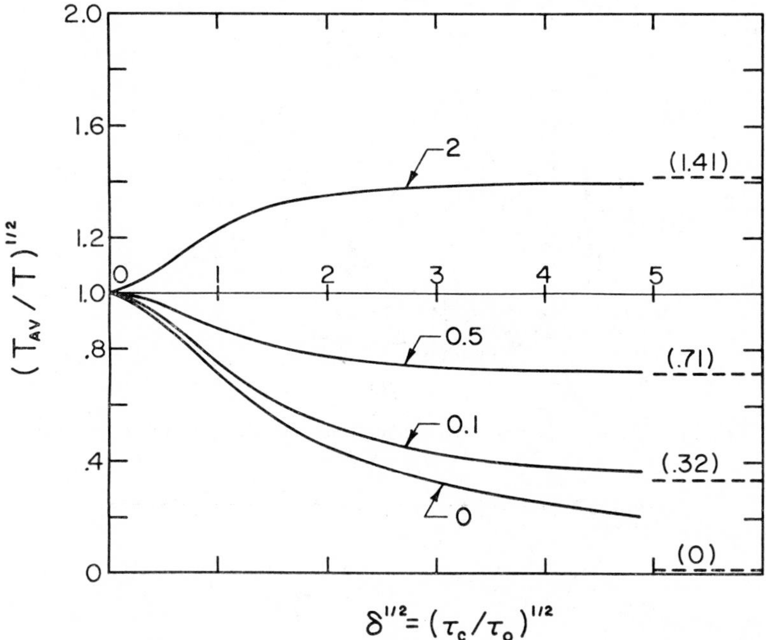

FIG. 30.—Orientation parameter $\xi = (T_{\text{av}}/T)^{1/2}$ for several ratios of grain temperature to gas temperature. For $\xi < 1$, orientation is such that particles tend to spin about an axis directed along the magnetic field. When grain temperature is greater than gas temperature the opposite kind of orientation occurs no matter how strong the magnetic field; i.e., the particles spin about axes perpendicular to the magnetic field.

We see, then, that if the grain temperature equals the gas temperature, the distribution will be uniform and that only if the grain temperature is less than the ambient gas temperature will the Davis-Greenstein mechanism be effective. Thus the degree of orientation is determined by the two (not entirely independent) parameters δ and (T_{av}/T). In Figure 30 we show this dependence. It is interesting to note that orientation is possible even for $\delta < 1$.

The function in equation (61) will be peaked toward $\beta = 0$ as $T_{\text{av}}/T \to 0$. It is pictured in Figure 31 for various values of T_{av}/T. For later reference, we show in Figure 32 the function $f^c(\theta)$ which describes how flat ($\gamma < 1$) and elongated ($\gamma > 1$) particles

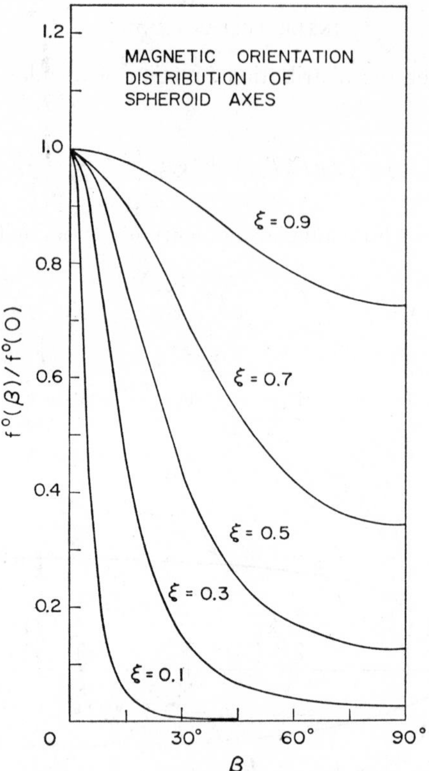

FIG. 31.—Relative orientation of sphere (or spheroid) angular momentum axes and magnetic field direction.

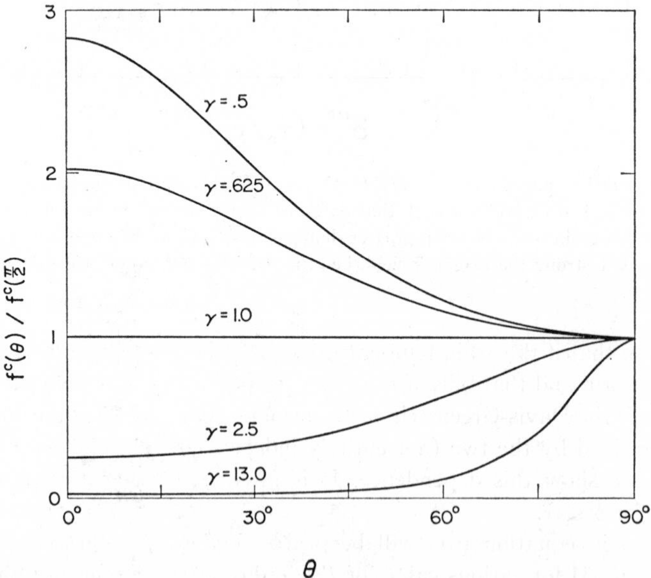

FIG. 32.—Relative orientation of spinning spheroid symmetry axis about the angular momentum direction: $\gamma > 1$ for elongated (prolate) spheroids; $\gamma < 1$ for flattened (oblate) spheroids; $\gamma = \infty$ for a needle; $\gamma = 0.5$ for a disk.

tend to spin in collisional equilibrium, i.e., the distribution of symmetry axes about the angular momentum when only collisions with gas atoms cause the spinning.

The relaxation time for orientation τ_O is estimated from equation (52) by calculating the time it takes for the torque to change the angular momentum by an amount comparable to its initial value; i.e., $1 = \Delta J/J \approx V\chi'' B^2 \tau_O/J$, which gives

$$\tau_O \approx \frac{\omega I}{\chi'' V B^2}. \qquad (62)$$

The collision or disorientation relaxation time is defined similarly as the time needed for the cumulative effect of impacts by the gas atoms to produce an equivalent change in the angular momentum. The average angular momentum imparted to the grain by a hydrogen atom that sticks to the grain is

$$\Delta J = d m_{\mathrm{H}} V_{\mathrm{H}} \qquad (63)$$

where d is the average collision radius.

Since the total angular momentum is given roughly by $J = I\omega$, where I is an average moment of inertia, the change in direction of the angular momentum due to one collision is

$$\Delta\beta = \frac{\Delta J}{J} \approx \left(\frac{m_{\mathrm{H}}}{m_g}\right)^{1/2} \qquad (64)$$

where m_g is the mass of the grain and m_{H} is the mass of a hydrogen atom.

Because the collisions are random, the root-mean-square change in angle for N collisions is $N^{1/2}\Delta\beta = (N m_{\mathrm{H}}/m_g)^{1/2}$, and the number of collisions required to rotate the angular momentum vector through one radian is $N = m_g/m_{\mathrm{H}}$. The time required for this number of collisions is the relaxation time for collisions τ_c. The rate at which hydrogen atoms hit the grain is $A v_{\mathrm{H}} n_{\mathrm{H}}$ where A is the average cross-sectional area of the grain and v_{H} and n_{H} are the average velocity and the number density of the hydrogen atoms, respectively. We then get

$$\tau_c = \frac{m_g}{m_{\mathrm{H}} A v_{\mathrm{H}} n_{\mathrm{H}}}. \qquad (65)$$

For most cases of interest, $\tau_c \approx 10^4$ to 10^6 years (see Table 13).

The fundamental question of whether the dissipative torque due to paramagnetic relaxation can lead to preferential orientation of the grains so that the system is not in a state of thermal equilibrium now appears to be answered. If the temperature of the grain is the same as the kinetic temperature of the surrounding gas, the system will be in thermal equilibrium, and spontaneous fluctuations in the magnetization would, in combination with the gas atom collisions, establish a random angular distribution of the grains. If the grain temperature is higher than the external temperature, spontaneous fluctuation in the magnetization alone will establish orientation opposite to the Davis-Greenstein sense.

Using the value of the appropriate parameters given in Section 5 for different types of grains, we obtain qualitative conditions for magnetic orientation of grains given in Table 13. The higher and lower values given for the minimum magnetic field requirement for $\delta = 1$ are for gas temperatures of 100° K (H I) and 10,000° K (H II), and in all cases the value of n_{H} is taken to be 10 (a high estimate).

The question of the degree of orientation and its polarizing effect on the starlight will be further discussed in the next two sections. We will need to involve the detailed optical properties of nonspherical particles in order to consider the quantitative question of degree of orientation by magnetic fields.

7. GRAIN OPTICS

All grains, except those suggested by Platt, scatter electromagnetic radiation by a classical process involving application of Maxwell's equation and for which the particles are defined by size, shape, and index of refraction. Although the classical methods are completely understood, numerical results for the range of applications in the interstellar dust problem are few. They are summarized below. The scattering by Platt particles is a quantum mechanical process and will be discussed separately.

Table 14 lists the classical scattering investigations considered in this section. Exact

TABLE 13

MINIMUM MAGNETIC FIELDS NEEDED FOR GRAIN ORIENTATION

	χ'' (cgs)	T_g (° K)	B^2 (gauss)^2	a (10^{-5} cm)	B (gauss)	τ_c (years)
Dielectric particles	$2.5\times10^{-12}\,\omega/T_g$	10	$10^{-8}aT^{1/2}n_{\rm H}T_g$	2	1.4×10^{-5} 4.5×10^{-5}	6×10^5 6×10^4
Ferromagnetic "	$3\times10^{-7}\,\omega$	100	$10^{-13}\,aT^{1/2}n_{\rm H}$	0.4	0.6×10^{-8} 2.0×10^{-8}	6×10^5 6×10^4 ($\rho\approx5$)
Ferrite "	$2\times10^{-10}\,\omega$	(10)	$10^{-10}\,aT^{1/2}n_{\rm H}$	0.4	2×10^{-7} 6×10^{-7}	3×10^5 3×10^4 ($\rho\approx3$)
Graphite "	$8\times10^{-13}\,\omega/T_g^2$	100	$3\times10^{-9}\,aT^{1/2}n_{\rm H}T_g^2$	0.4	1×10^{-4} 3×10^{-4}	1×10^5 1×10^4
Platt "	?	(1000)	?	0.01	?	

TABLE 14

TOTAL SCATTERING CROSS-SECTIONS

Size (ka)	Shape	Refractive Index	Method
$\ll1$	Sphere	Real, complex	Th.
	Ellipsoid	Real, complex	Th.
Arbitrary	Sphere	Real	Th., Exact (Mie theory)
		Complex	Th., Exact (Mie theory)
		$\|m-1\|\ll1$	Th., Approximate
	Infinite cylinder	Real	Th., Exact
		Complex	Th., Exact
		$\|m-1\|\ll1$	Th., Approximate
	Spheroid	Real	Exp., Microwave analog
		$\|m-1\|\ll1$	Th., Approximate
	Finite cylinder	Real	Exp., Microwave analog

Th. = theoretical; Exp. = experimental.

numerical results for scattering cross-sections for arbitrary-size particles are presently available only for a sphere and an infinite cylinder. Approximate and analog methods have been used to extend the range of results. The microwave analog method (Greenberg, Pedersen, and Pedersen 1961; Greenberg, Lind, Wang, and Libelo 1963; Greenberg, Libelo, Lind, and Wang 1963; Lind, Wang, and Greenberg 1965; Greenberg, Lind, Wang, and Libelo 1967) discussed below can, in principle, supply all information on scattering by nonspherical particles needed for interstellar application which cannot be obtained by any other method at this time.

Figure 33 illustrates the geometry of a particle in an incident radiation field of intensity I_0 (ergs/cm^2 sec) which scatters radiation of intensity I in direction Θ, Φ. We follow the notation[7] of van de Hulst (1957) and define angular scattering function $F(\Theta, \Phi)$ by

$$I = \frac{I_0 F(\Theta, \Phi)}{k^2 r^2} \tag{66}$$

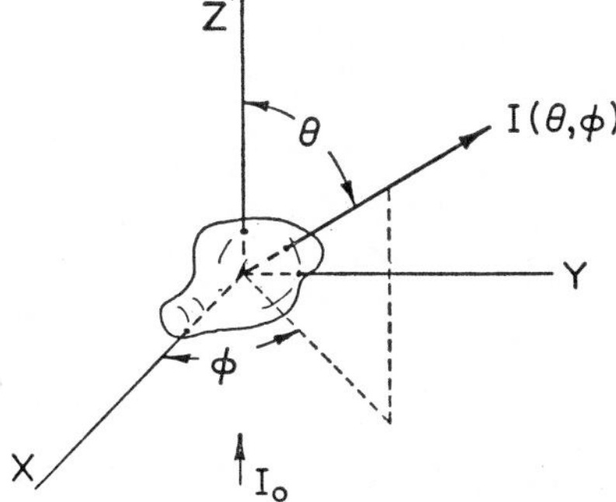

FIG. 33.—Definition of scattering angle. Radiation incident along the positive z-direction

where r is the distance (large) from the scatterer and $|\boldsymbol{k}| = k$ is the wave number defined by $k = 2\pi/\lambda$, where λ is the wavelength of the radiation. The total scattering cross-section $C_{\rm sca}$ is obtained from

$$C_{\rm sca} = \frac{1}{k^2} \int\int F(\Theta, \Phi) \sin \Theta d\Theta d\Phi . \tag{67}$$

If the particle absorbs as well as scatters energy, we may define an absorption cross-section $C_{\rm abs}$. The total (or extinction) cross-section is again given by equation (10).

Other frequently used terms are defined as follows: G = geometrical cross-section, $Q_{\rm sca} = C_{\rm sca}/G$ = scattering efficiency, $Q_{\rm abs} = C_{\rm abs}/G$ = absorption efficiency, $Q_{\rm ext} = C_{\rm ext}/G$ = extinction efficiency, $Q_{\rm sca}/Q_{\rm ext}$ = albedo, and $x = ka$, where a = characteristic linear dimension of a particle.

Although it is unreasonable to expect interstellar grains to have any simple shape, it is useful to consider scattering by spherical and nonspherical particles of high symmetry if

[7] Unfortunately there is no uniformity in notation in scattering problems. The quantity $F(\Theta, \Phi)/k^2$ is often referred to as the differential scattering cross-section and is then denoted by $\sigma(\Theta, \Phi)$. Similarly, C is denoted by σ. Furthermore, the quantity $\sigma(\Theta, \Phi)/\sigma$ is called the phase function.

only because systematic effects can be made apparent. Since refracting properties vary with grain type, we shall consider particles with real, complex, and nonisotropic indices of refraction.

7.1. Pure and Impure Dielectric Grains

For dielectric grains we generally take $m = n(1 - i\delta)$ and $0 \leq \delta \ll 1$.

Although the index of refraction of dielectric materials is wavelength-dependent, it generally does not vary greatly over the visual range. In the infrared and ultraviolet, the value of δ is of the order of unity, and, of course, in the neighborhoood of an absorption band, the real and imaginary parts of the index vary appreciably.

We first consider $a \ll \lambda$. Such particles may be visualized as being at every instant in the uniform (but time-dependent) electric field of an incident plane-polarized wave. The induced dipole moment, $p = \alpha E_0$ (α is the polarizability), then varies with time as $p \exp(i\omega t)$ and radiates according to classical electromagnetic theory. The only nontrivial case in which the polarizability of a particle can be computed by elementary means is that of homogeneous, but not necessarily isotropic, ellipsoids. The rate of radiation of energy by the oscillating dipole may then be used to define the scattering cross-section. For an isotropic sphere the polarizability is

$$\alpha = \frac{(m^2 - 1)}{(m^2 + 2)} a^3 \tag{68}$$

where a is the radius of the sphere and the scattering efficiency is (van de Hulst 1957)

$$Q_{\text{sca}} = \tfrac{8}{3}\pi k^4 |\alpha| \frac{2}{\pi a^2} = \tfrac{8}{3} x^4 \left| \frac{m^2 - 1}{m^2 + 2} \right|^2 . \tag{69}$$

If m is real, the absorption is zero, and thus the extinction efficiency varies as λ^{-4} (Rayleigh scattering). If m is complex, absorption takes place, and the absorption efficiency is given by (van de Hulst 1957)

$$Q_{\text{abs}} = -4x\,Im\left(\frac{m^2 - 1}{m^2 + 2}\right) \tag{70}$$

which varies with the wavelength as λ^{-1} and will therefore dominate over the scattering at sufficiently long wavelengths.

Ellipsoidal particles may be similarly treated, the only difference being that the polarizability now depends on the geometrical configuration of the axes with respect to the electric and propagation vectors of the incident radiation. Consider the prolate spheroid. Define C_E and C_H as the extinction cross-sections for the spheroid whose axis of symmetry is parallel to the electric and magnetic vectors, respectively. Then for an arbitrary orientation angle between the electric vector and the symmetry axis, the extinction is given by

$$C_{\text{ext}} = C_E \cos^2 \chi + C_H \sin^2 \chi . \tag{71}$$

We show some typical numerical results in Table 15.

Except for isotropic spheres, the ratio of cross-sections C_E/C_H is not equal to unity; polarization of incident unpolarized radiation by nonspherical particles is possible if they are oriented.

For dielectric particles, the dominant sizes are of the order of visual wavelengths. A discussion of the scattering properties of particles for which $a \gtrsim \lambda$ should therefore be included.

Before considering the results of exact methods which apply to particles of arbitrary indices of refraction, it is useful to demonstrate the nature of the wavelength dependence of extinction cross-sections by means of theoretical approximations, valid when the index of refraction is close to unity. The method of van de Hulst (1946) and Schiff (1956) has been used to calculate the extinction by a variety of shapes (Greenberg 1960*a*). We shall here present the results for spheroids. This approximation applies only to scattering

TABLE 15

EXTINCTION CROSS-SECTIONS FOR SMALL ($ka \ll 1$) PROLATE AND OBLATE SPHEROIDS WITH REAL AND COMPLEX INDICES OF REFRACTION*

b/a		5.025	3.203	1.667	1	.600	.312	.199
$m=\sqrt{2}$	$C_H/\frac{8}{3}\pi a^2(ka)^4(b/a)^2$	0.0513	0.0528	0.0571	0.0625	0.0697	.0756	0.0849
	C_E/C_H	1.945	1.738	1.328	1	0.732	0.505	0.398
$m=\sqrt{2}(1-i)$	$C_H/4\pi a^2(ka)(b/a)$	0.346	0.376	0.464	0.600	0.809	1.143	1.304
	C_E/C_H	4.08	3.65	2.18	1	0.422	0.169	0.105

* Since equations (69) and (70) are equally valid for real and complex indices of refraction (δ not $\ll 1$), we include the results for both cases for later reference.

Source: Davis and Greenstein 1951.

of scalar waves, and no polarization effects are shown. The extinction cross-section for a spheroidal scatterer (see Fig. 34) is given by

$$C_{\text{ext}} = 4\pi AB\, Re\, \{\tfrac{1}{2} - (ia/C\rho^*) \exp(-iC\rho^*/a) + (a/C\rho^*)^2 [1 - \exp(iC\rho^*/a)]\} \quad (72)$$

where

$$A^2 = b^2 \sin^2\chi + a^2 \cos^2\chi \qquad \rho^* = \rho(1 - i\delta) = \rho(1 - i\tan\beta)$$

$$B^2 = a^2 \qquad \rho = 2ka(m' - 1)$$

$$C^2 = (ab)^2/A^2 \qquad m = m'(1 - i\delta) = m' - im''$$

and b and a are semimajor and semiminor axes of the spheroid.

Figure 34 shows the effect of orientation on the wavelength dependence of the total cross-section. The two orientations considered here have nothing to do with the polarization and must not be confused with those corresponding to C_E and C_H above. The quantities C_E and C_H both correspond to the single curve for $k \perp b$. Furthermore, the values shown in Table 15 would apply (where $|m - 1| \ll 1$) in the range of very small particles; namely, for $\rho \ll 1$. Equation (72) is a more general form of that derived by van de Hulst (1946) for spheres. Van de Hulst's equation, useful in describing the important features of scattering by spheres even for real values of m as large as 2, is shown in Figure 35 where the extinction curves (exactly calculated from Mie theory) for $m = 1.5$, $m = 1.33$, $m = 0.93$, $m = 0.80$ are shown. The results of equation (72) for spheres are plotted with a common scale for ρ. Equation (72) should be at least qualitatively valid for spheroids. The discussion of the microwave analog method shows that this assumption is reasonable, but with stricter limitations than apply to spheres. The wavelength dependence of the extinction for a spheroid oriented with its symmetry axis perpendicular to the direction of propagation of the incident radiation differs from that when the axis is parallel to it. This seems to be true for the more general case of any elongated or flattened particle characterized by a real (or slightly complex) index of refraction. This can be seen if the parameter ρ of equation (72) is interpreted as a measure of the phase lag suffered by the central ray that passes through the particle. When this phase lag is large, one approaches the geometrical limit in which the extinction cross-section is twice the geometrical cross-section. This factor of 2 has been amply discussed elsewhere (for example, by van de Hulst 1957). Thus, an elongated particle seen end-on has, for each value of the wavelength, a larger phase lag than the same particle seen sideways. Consequently the end-on extinction curve will show characteristics similar to that for the sideways view, but shifted to longer wavelengths. The asymptotic limits for each orientation will be twice the geometrical cross-section for that orientation.

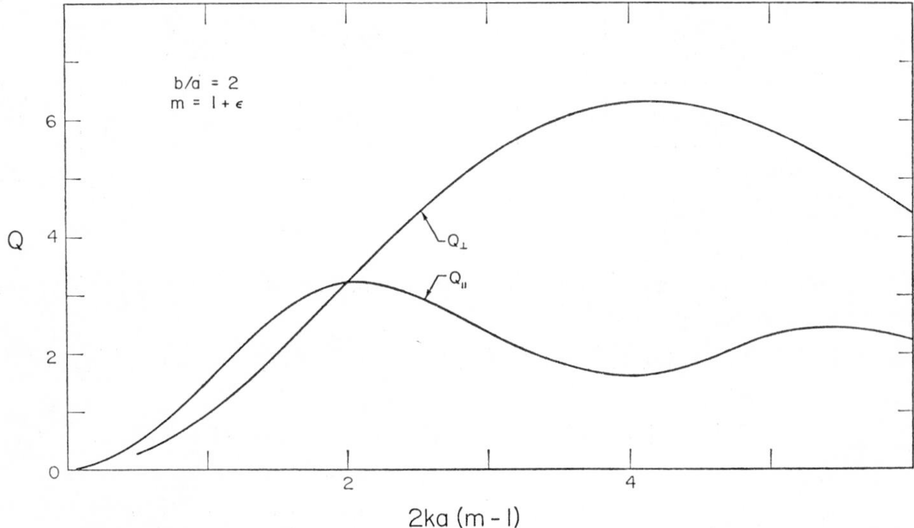

FIG. 34.—Total cross-sections of spheroidal particles calculated from the approximation in equation (72). (Greenberg, 1960*a*.)

Equation (72) is also useful with respect to the effect of an imaginary part (absorption) of the index of refraction. Figure 36 shows the effect of changes in the absorption coefficient on the extinction curves for spheres. As in the approximate treatment of particles small compared with the wavelength (eqs. [69] and [70]), the absorption cross-section dominates the scattering cross-section for sufficiently small particles; the slopes of the extinction curves at $\rho = 0$ are non-zero only for particles with complex indices of refraction; this slope increases with the imaginary part of the index of refraction.[8]

As already indicated in Figure 35, exact calculation for scattering by homogeneous spheres of arbitrary size and index of refraction can easily be made, particularly with the aid of high-speed computers.

Much information is available on this subject (for example, van de Hulst 1957; Deir-

[8] This is true for a moderate imaginary part of the index. However, as the real or imaginary part of the refractive index approaches infinity, the slope tends to zero at $\lambda = \infty$.

mendjian, Clasen, and Viezee 1961; Giese 1961), and we shall not present the computations involved in the application of the general electromagnetic equations to this problem.

No exact solution for scattering by simple refracting spheroids is presently available. An approach to an exact numerical solution (Greenberg *et al.* 1967) so far includes only computations for scattering of scalar waves axially incident on spheroids. The scattering from infinite cylinders can be handled analytically, and some exact computations of the extinction cross-sections (per unit length) for circular cylinders with various indices of refraction, are shown in Figures 37–40. The curves labeled Q_E and Q_H are for the electric vector and the magnetic vector of the incident radiation parallel to the symmetry axis; the ordinates in the two cases are analogous to the quantities labeled C_E

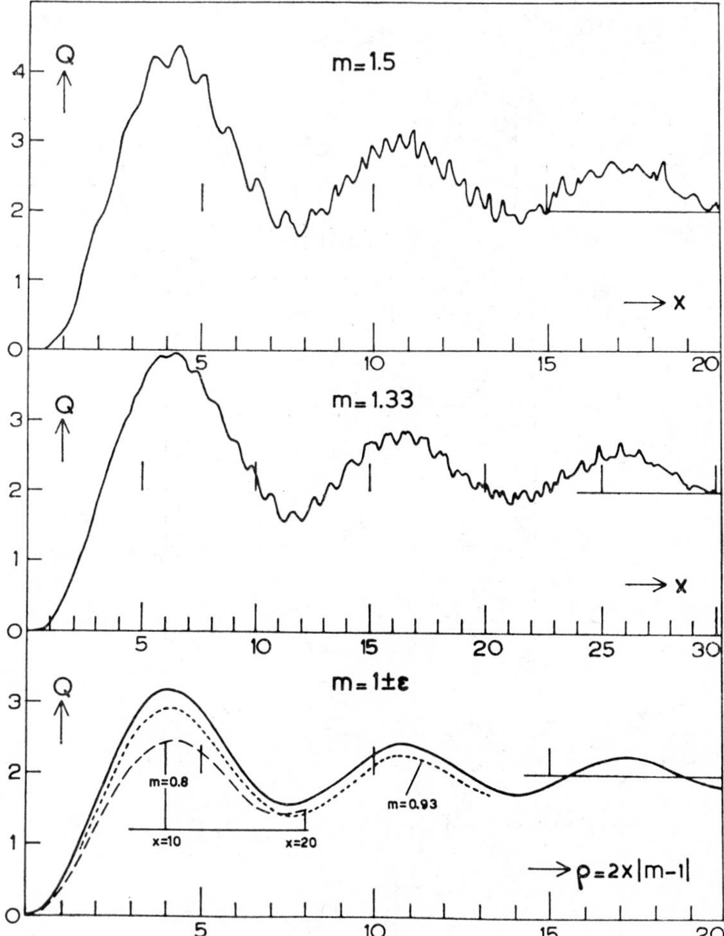

FIG. 35.—Extinction curves computed from Mie's formulae for $m = 1.5, 1.33, 0.93$, and 0.8. The scales of x have been chosen in such a manner that the scale of $\rho = 2x\,|m - 1\,|$ is common to these four curves and to the extinction curves and to the extinction curve for $m = 1 + \epsilon$. (Van de Hulst, 1957.)

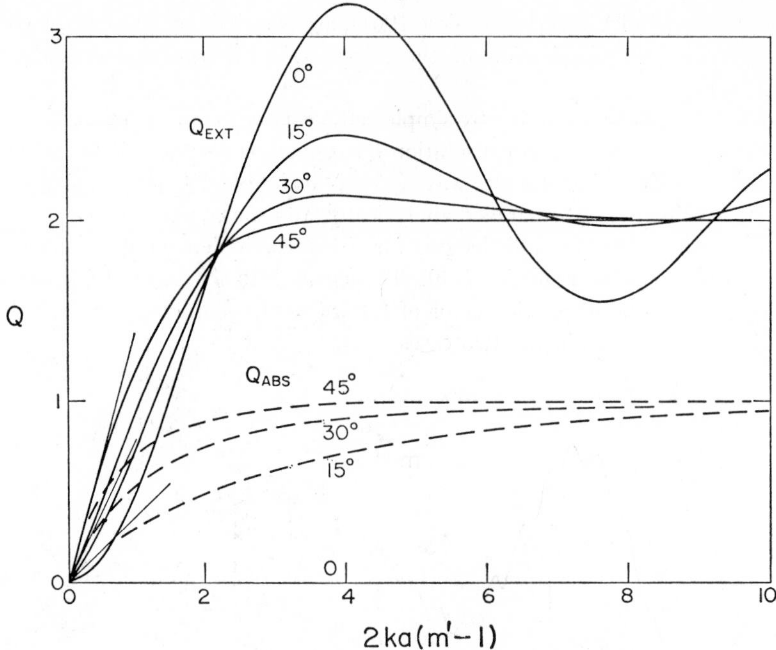

FIG. 36.—Influence of an imaginary term in the refractive index upon the extinction curve for m close to 1: $m = m' - im''$. The refractive index is $1 + \epsilon - i\epsilon \tan \beta$ (ϵ small). (Van de Hulst, 1957.)

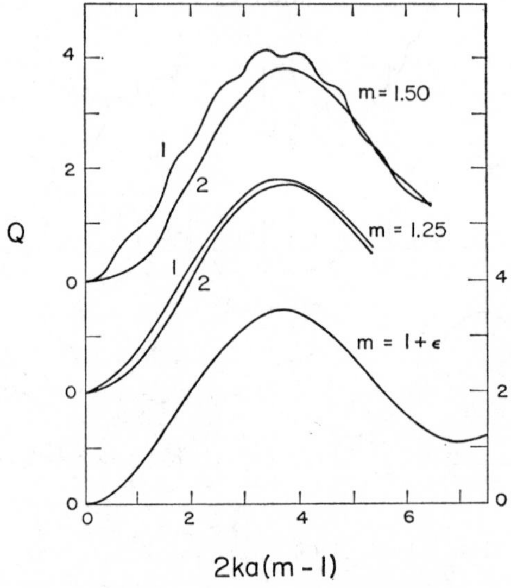

FIG. 37.—Extinction curves of very long cylinders for Case 1 (E axis) and Case 2 (H axis) computed for three different values of the refractive index. (Van de Hulst, 1957.)

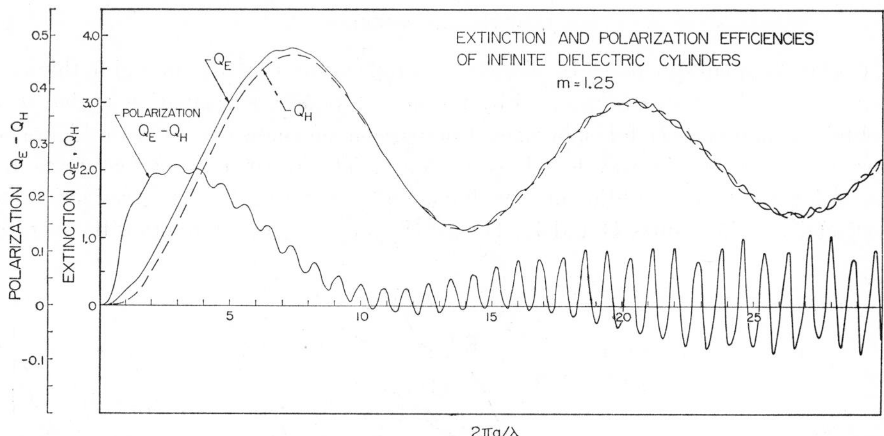

FIG. 38.—Extinction efficiencies for dielectric ($m = 1.25$) cylinders oriented normal to incident radiation: Q_E for $E\|$ axis, Q_H for $H\|$ axis. Polarization efficiency is also shown as $Q_E - Q_H$. The maximum polarization efficiency occurs at a value of a/λ significantly less than one-half that of the position of maximum extinction efficiencies. Note also the interesting oscillation (positive and negative) of the polarizations for large values of a/λ.

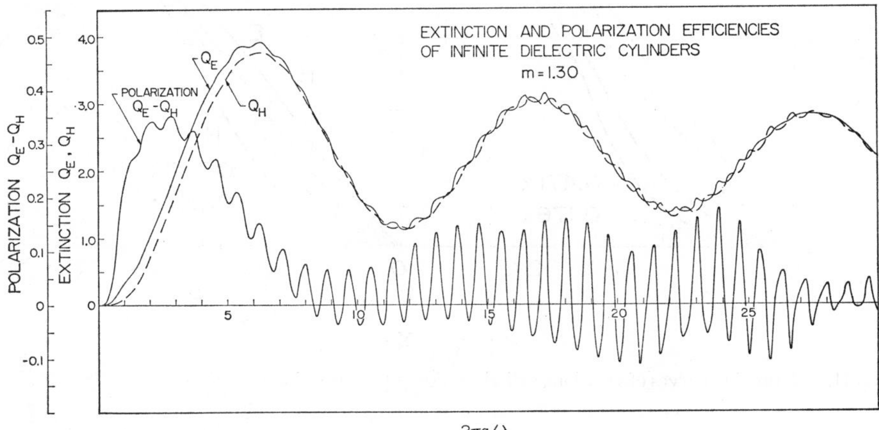

FIG. 39.—Same as Figure 38, but $m = 1.30$

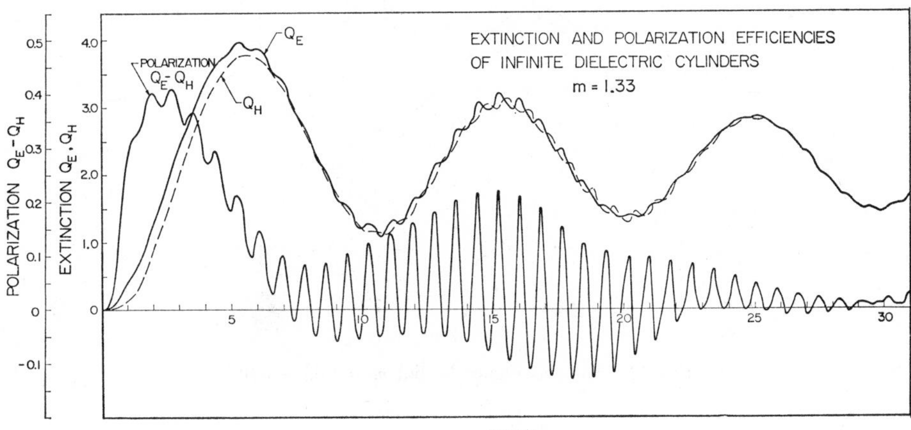

FIG. 40.—Same as Figure 38, but $m = 1.33$

and C_H for the small spheroids. In all cases the propagation vector is normal to this axis. The curve for $m = 1 + \epsilon$ is obtained by the same type of approximation as that used to obtain equation (72) for spheroids. Comparison of these curves with Figure 35 shows the similarity in wavelength dependence. The distinguishing feature is the polarization effect. The addition of a small imaginary part to the index of refraction has the effect shown in Figures 41 and 42. The principal change for all shapes is the increase

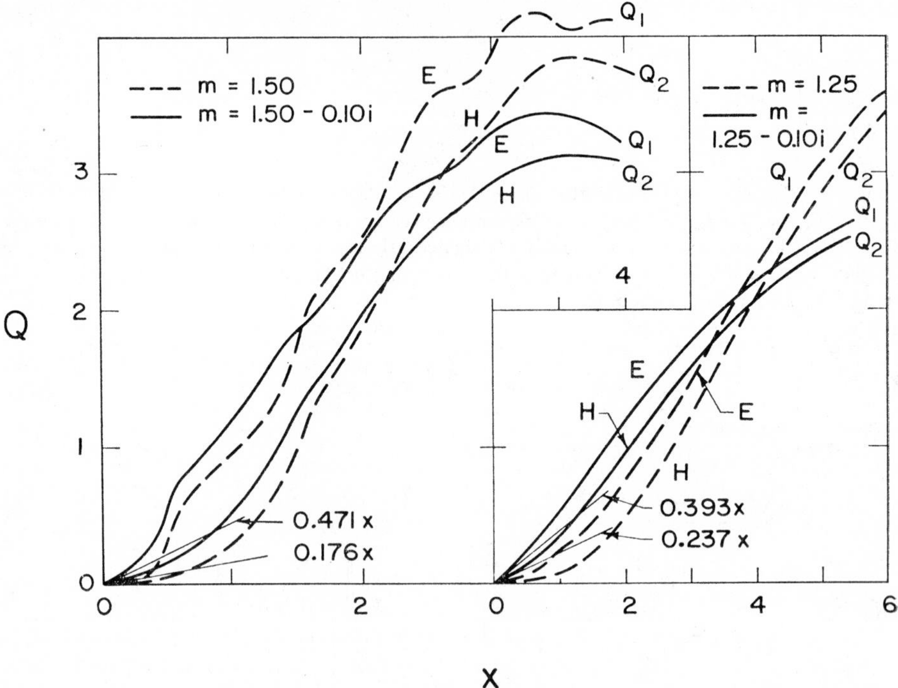

FIG. 41.—Extinction curves of very long cylinders of slightly absorbing material. (Van de Hulst, 1957.)

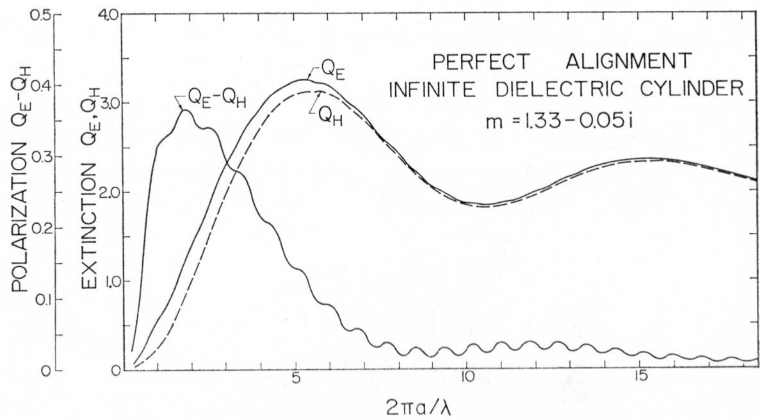

FIG. 42.—Same as Figure 38, but $m = 1.33 - 0.05i$

of the slopes for small a from zero to a positive value. In finding the polarization produced by aligned cylinders, we discover the difference, $Q_E - Q_H$, increases with the index of refraction and always $\rightarrow 0$ for small diameters, although the ratio Q_E/Q_H will remain finite as was shown for the small spheroids. For sufficiently short wavelengths, the curves for Q_E and Q_H approach each other and may even cross over, thus reversing the sign of the polarization. A region intermediate between long and short wavelengths is most effective for polarization produced by long aligned cylinders.

Although an infinite cylinder seems to be an unrealistic representation of the shape of interstellar particles, it is the only nonspherical shape for which analysis is available in the range $ka \sim 1$. The justification for the infinite cylinder model lies in the fact that

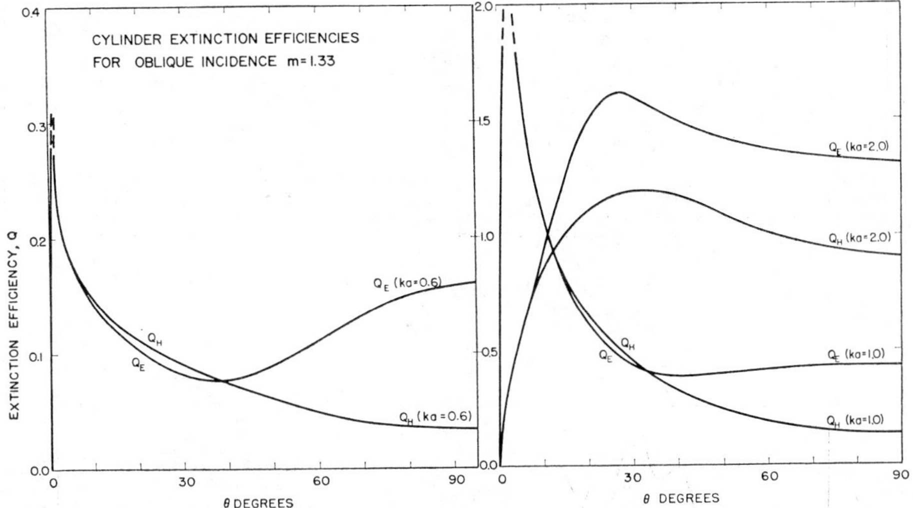

FIG. 43.—Extinction efficiencies *versus* obliqueness angle θ for tilted circular cylinder. Normal incidence (Q_E and Q_H of Figure 40) is given by $\theta = 90°$. The peak value of Q for $ka = 0.6$ is 3.33 and for $ka = 1.0$ is 3.1. Note the crossing of the curves for $Q_E(\theta)$ and $Q_H(\theta)$ and consequent polarization reversal as the obliqueness increases (θ decreases). This produces a reduction in polarization efficiency.

the extinction and polarization by infinite cylinders differ quantitatively but not qualitatively from the extinction and polarization produced by finite elongated particles. Furthermore, if the grains are indeed needle-like (or whisker-like) in structure or perhaps consist of loose agglomerations of needles, this representation would not be far from realistic. Cylinder extinction efficiencies for normal incidence and $m = 1.33$ are shown in Figure 38. Varying angles of incidence, limited to selected sizes, are shown in Figures 43 and 44. For comparison, we present equivalent results for $m = 1.6$ in Figures 45, 46, and 47. The dashed curves are computed from equation (72); the thin line is $2\times$ Mie theory extinction for sphere of radius a. Thin curves are exact computations for infinite cylinders with appropriately normalized extinction efficiencies.

Appropriate cross-sections for an arbitrarily shaped particle at all orientations with respect to the incident radiation are needed even for semiquantitative work. The microwave analog method is now being used to determine extinction cross-sections and angular scattering distributions for particles with a wide range of optical properties. Details

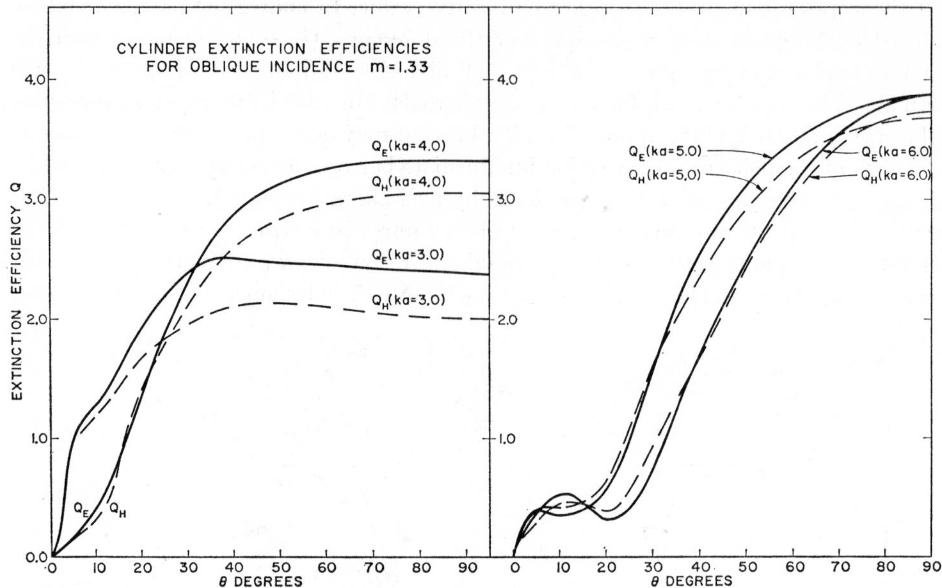

FIG. 44.—Same as Figure 43

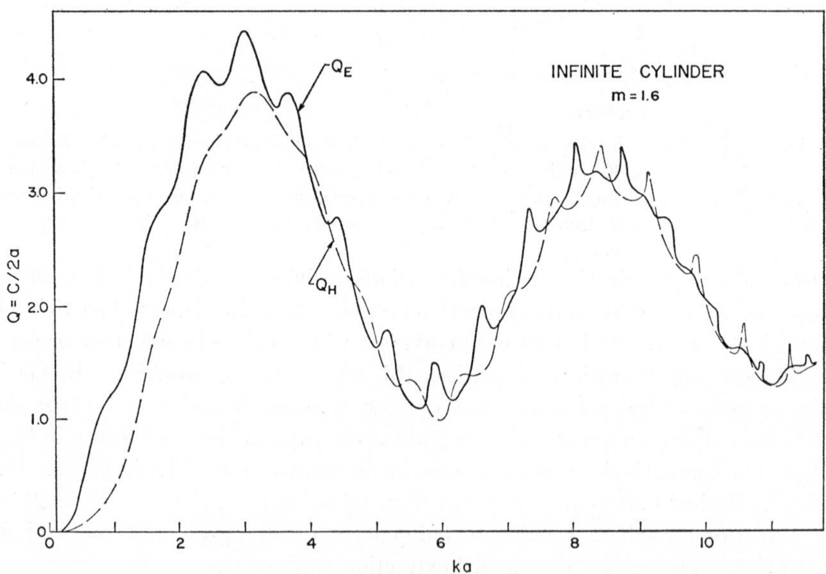

FIG. 45.—Extinction efficiencies for cylinders with a high real refractive index $m = 1.6$ oriented normal to the radiation.

of the method are beyond the scope of this section, but a description of the essential features involved in simulating possible optical characteristics of interstellar grains is useful. In the experiment the radiation has a wavelength of 3.18 cm. The scaling of particle sizes is simple; for example, a particle whose characteristic linear dimension is 0.5 μ and which scatters radiation whose wavelength is 0.5 μ has, in the experiment, a characteristic linear dimension of 3 cm. The creation of optical properties (indices of refraction) in the microwave region analogous to those for a particular substance in the

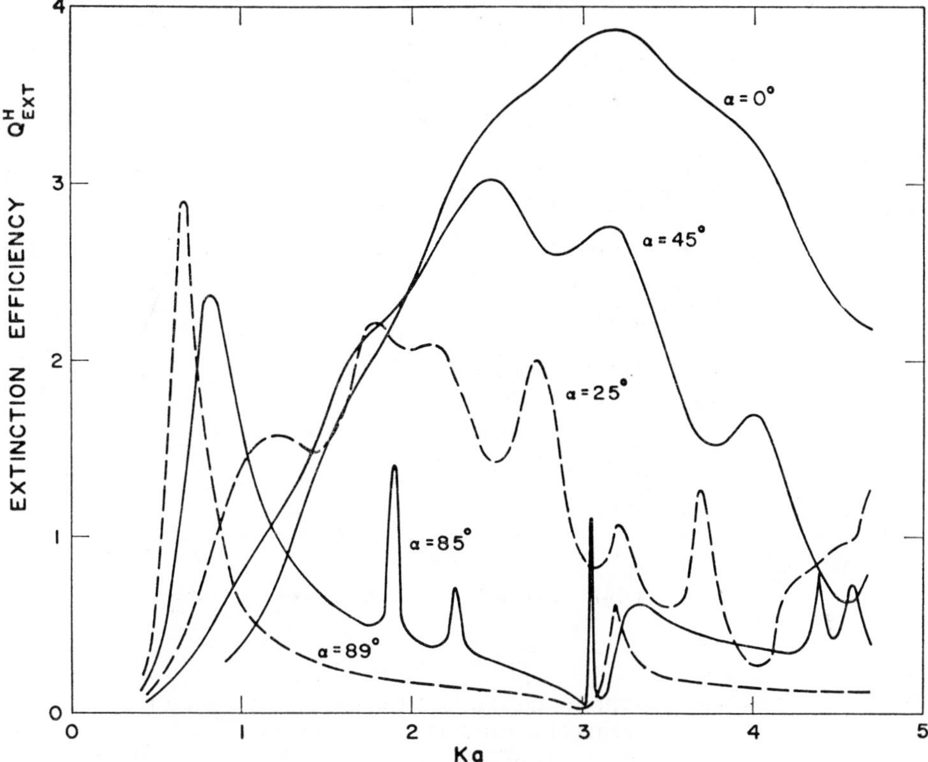

FIG. 46.—Extinction efficiencies *versus* ka for cylinders with $m = 1.6$ tilted at various angles α; $\alpha = 0°$ is normal incidence (corresponding to $\theta = 90°$ in Figures 43 and 44). $H \parallel$ axis. (Correction: $\alpha = 25°$ and $\alpha = 45°$ should be interchanged in figure.)

optical region is more subtle. Since the index of refraction always varies with wavelength, materials different from the actual particles must be used for the analog particles. Variously expanded plastics give a range of real refractive indices from $m = 1 + \epsilon$ (ϵ small) to $m = 1.6$. The imaginary part can be varied by the inclusion of different amounts of absorbing material, such as carbon dust. Finally, it is, in principle, possible to make particles with nonisotropic refractive properties. Figures 48 to 59 and Tables 16 and 17 show the type of results obtained by the microwave analog method. Since the wavelength in the experiment is fixed, different values of ρ are obtained by changing the particle sizes, characteristics such as shape and index of refraction being held constant.

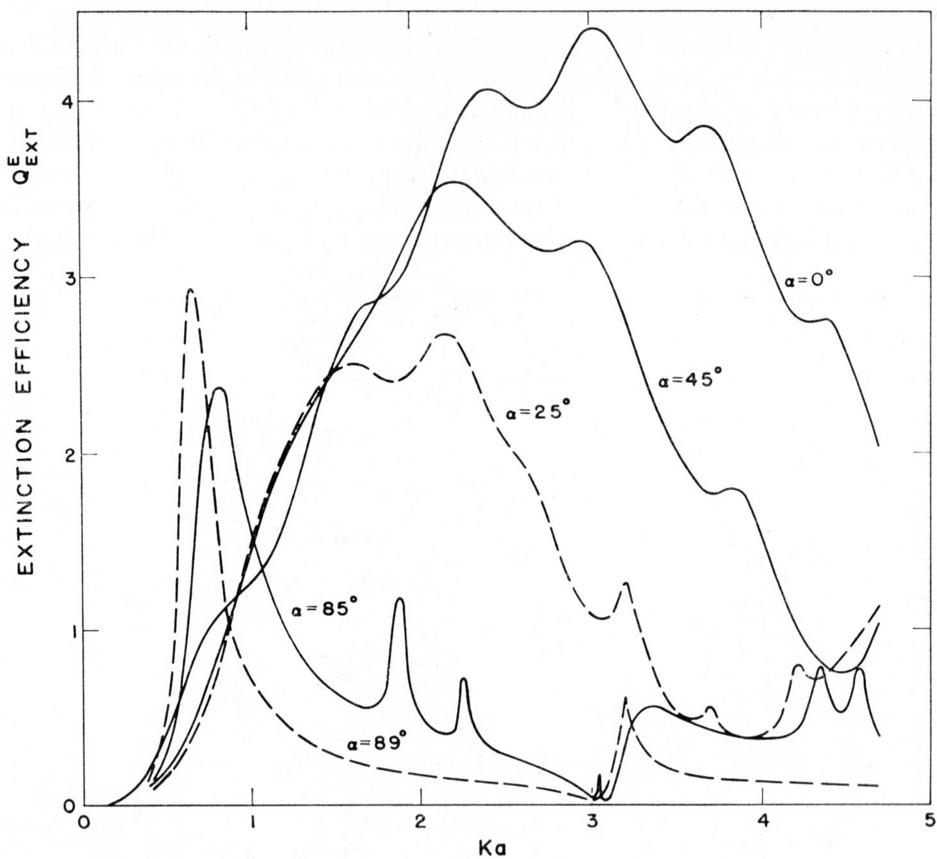

FIG. 47.—Same as Figure 46, but $E \parallel$ axis

TABLE 16

COMPARISON OF THE POLARIZING ABILITIES OF ALIGNED REFRACTING SPHEROIDS AND INFINITE CYLINDERS

ELONGATED Q_E/Q_H
FLAT Q_H/Q_E

ρ	SPHEROIDS			∞ CYLINDERS	
	$m=1.6$	$m=1.33-.05i$		$m=1.3$	$1.33-0.05i$
	$b/a=2$	$b/a=2$	$b/a=0.5$		
1.5	....	1.14	....	....	....
1.8	....	1.10	1.46	1.19	1.21
1.9	....	1.15	....	1.15	....
2.0	1.12	1.11	....	1.17	1.15
2.1	....	1.11	1.19	1.14	....
2.2	....	1.08	1.16	1.12	1.12

TABLE 17

EFFECT OF GRAIN SPINNING ON AMOUNT OF POLARIZATION

PROLATE SPHEROID $m = 1.33 - 0.05i$
ELONGATION $b/a = 2$

ρ	$\frac{\langle Q_E(x)-Q_H(x)\rangle}{Q_E-Q_H}$
1.52	0.808
1.67	.853
1.80	.671
1.94	.692
2.07	.628
2.23	.610
2.36	.582
2.52	.536
2.64	.589
2.77	.424
2.94	.379
3.32	0.368

Note that $\langle Q_E - Q_H\rangle/(Q_E - Q_H) = \frac{1}{2}$ is the Rayleigh value.

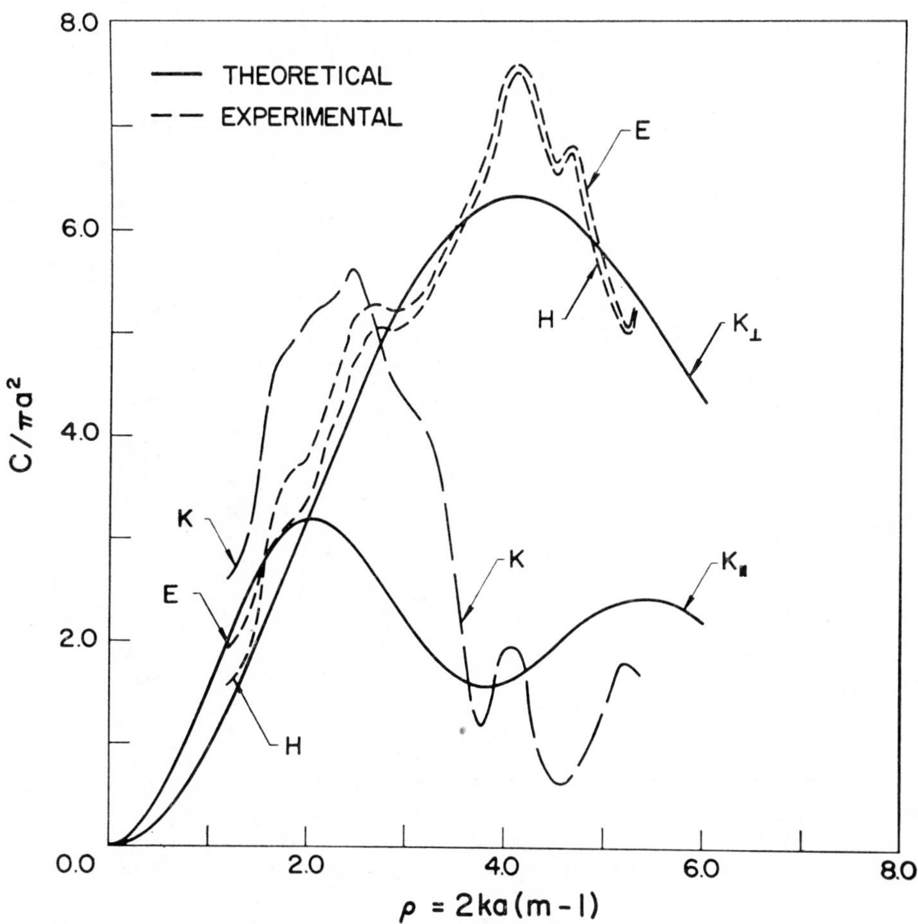

FIG. 48.—Experimental extinction efficiencies for three orthogonal orientations of prolate spheroids. Elongation $b/a = 2$, index of refraction $m = 1.26$. E, H, K refer respectively to particle axes parallel to the electric, magnetic, and propagation vectors of the incident radiation. The terms $k_{\|}$ and $k_{\perp}$ define two cases for the scalar wave approximation in which the particle axis is parallel or perpendicular to the propagation direction.

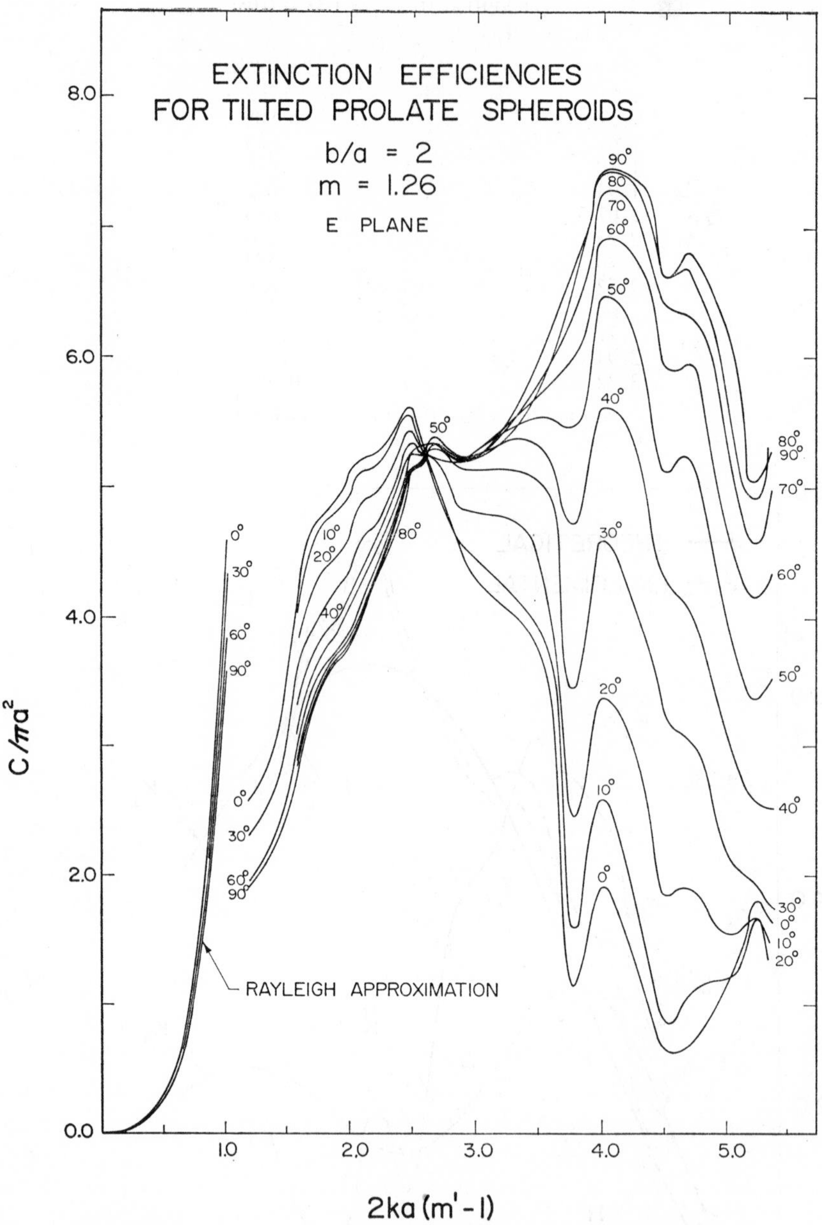

FIG. 49.—Experimental extinction efficiencies for prolate spheroids obliquely oriented in the plane containing the particle axis and the electric vector of the incident radiation (elongation $b/a = 2$, index of refraction $m = 1.26$). Each curve is labeled according to the angle between the particle axis and the propagation vector. Calculated values according to the Rayleigh approximation are shown for small x.

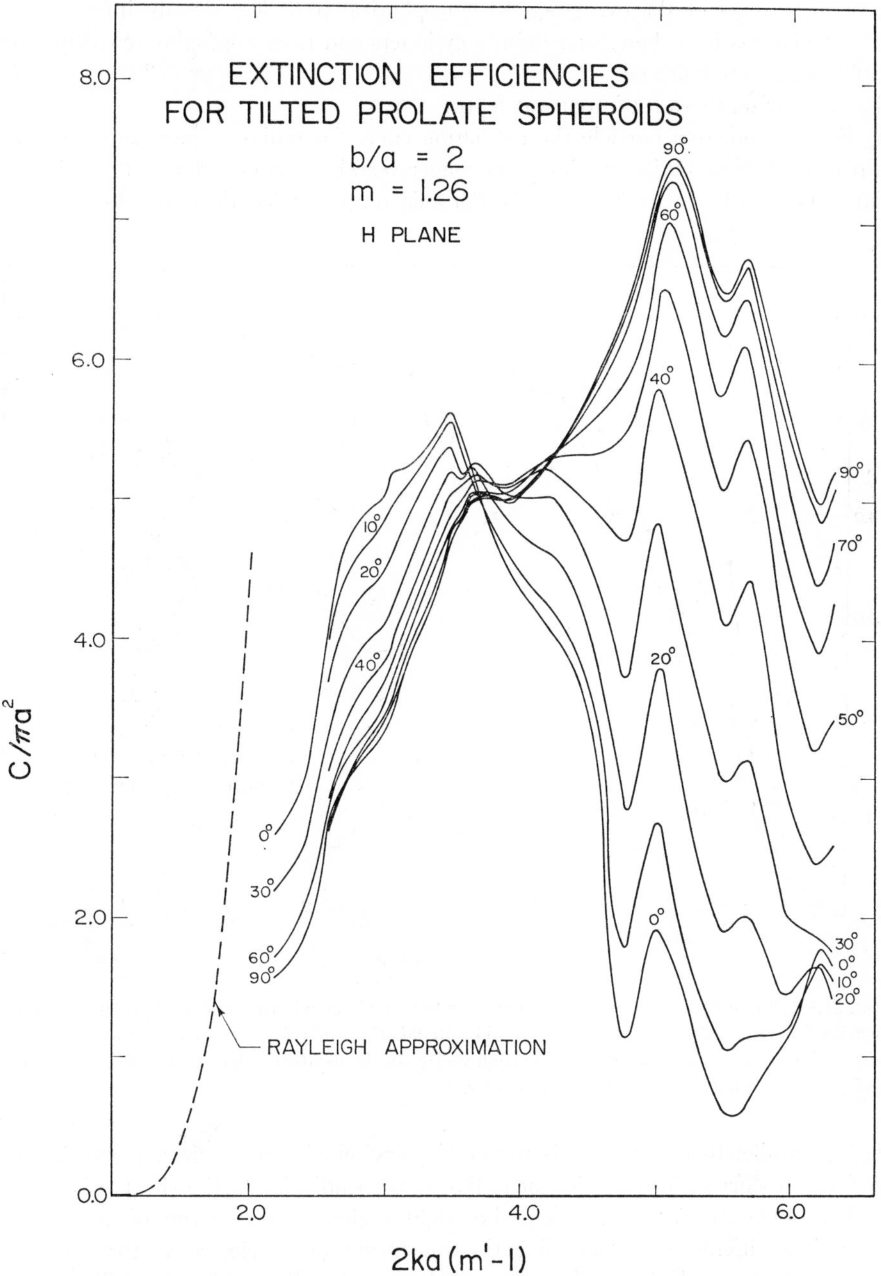

FIG. 50.—Experimental extinction efficiencies for prolate spheroids obliquely oriented in the plane containing the particle axis and the magnetic vector of the incident radiation (elongation $b/a = 2$, index of refraction $m = 1.26$). Each curve is labeled according to the angle between the particle axis and the propagation vector. Calculated values according to the Rayleigh approximation are shown for small x.

The main features of the extinction by nonspherical particles, already inferred from exact calculations for spheres and infinite cylinders and from approximate calculations for spheroids, are confirmed.

To summarize, the most important features are

1) For an elongated particle the extinction curve for radiation parallel to the long axis reaches its first major maximum at longer wavelengths (smaller ρ) than that for radiation perpendicular to the long axis. For a flattened particle the extinction curve for

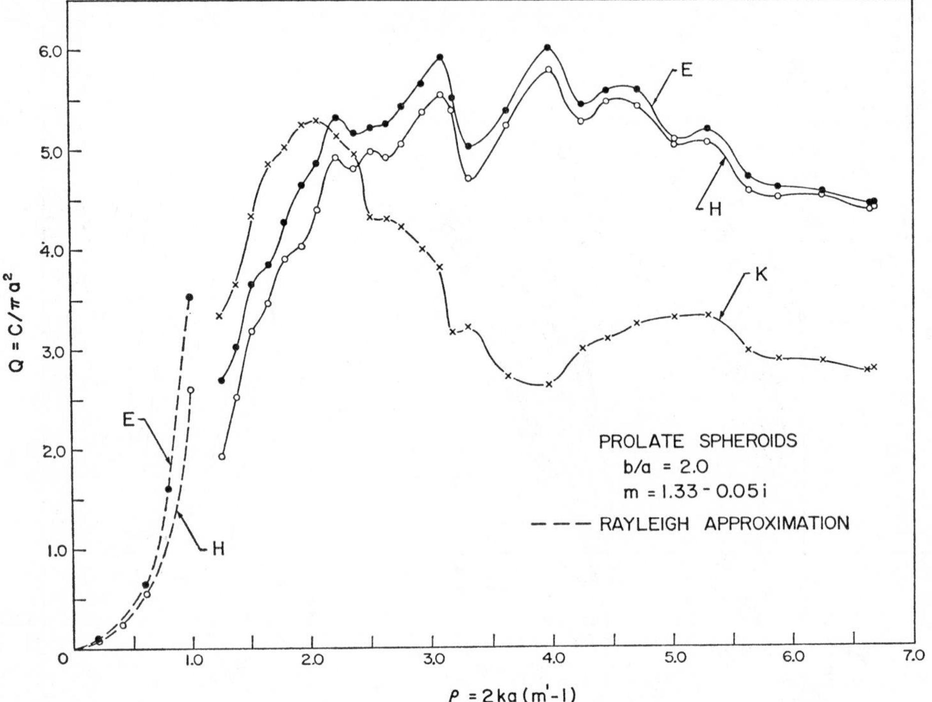

FIG. 51.—Experimental extinction efficiencies for three orthogonal orientations of prolate spheroids. Elongation $b/a = 2$, index of refraction $m = 1.33 - 0.05i$. E, H, K refer, respectively, to particle axes parallel to the electric, magnetic, and propagation vectors of the incident radiation. Calculated values of Q_E and Q_H are shown as broken curves for small x.

radiation parallel to the short axis reaches its first maximum at shorter wavelengths than does the curve for extinction of radiation perpendicular to the short axis.

2) For an elongated particle oriented at right angles to the direction of the incident radiation, the difference between the extinction curves for the electric vector parallel to and perpendicular to the long axis is *generally largest* (there are numerous oscillations) at some wavelength between that corresponding to $\rho = 0$ and that corresponding to a value of ρ near the first major maximum of the extinction curves. An analogous statement holds for flattened particles.

3) The polarizing ability of aligned finite elongated particles is not necessarily much

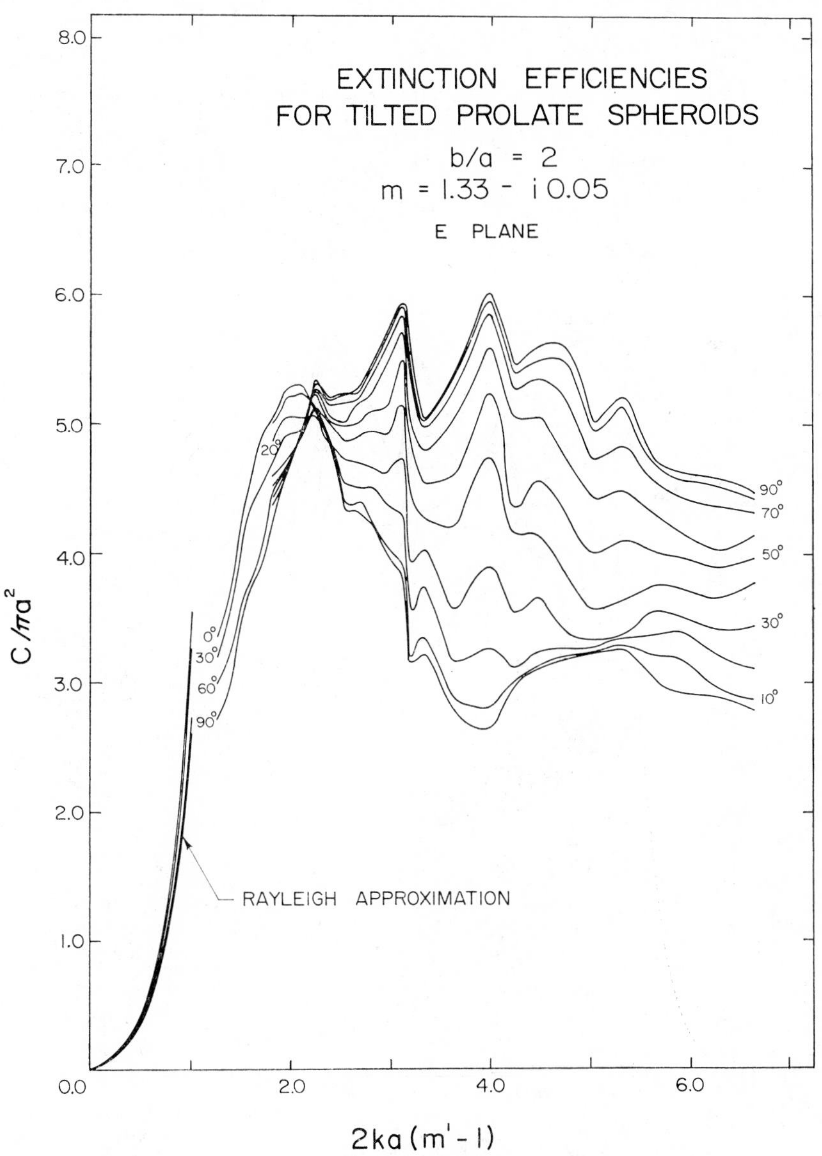

FIG. 52.—Same as Figure 49, but $m = 1.33 - 0.05i$

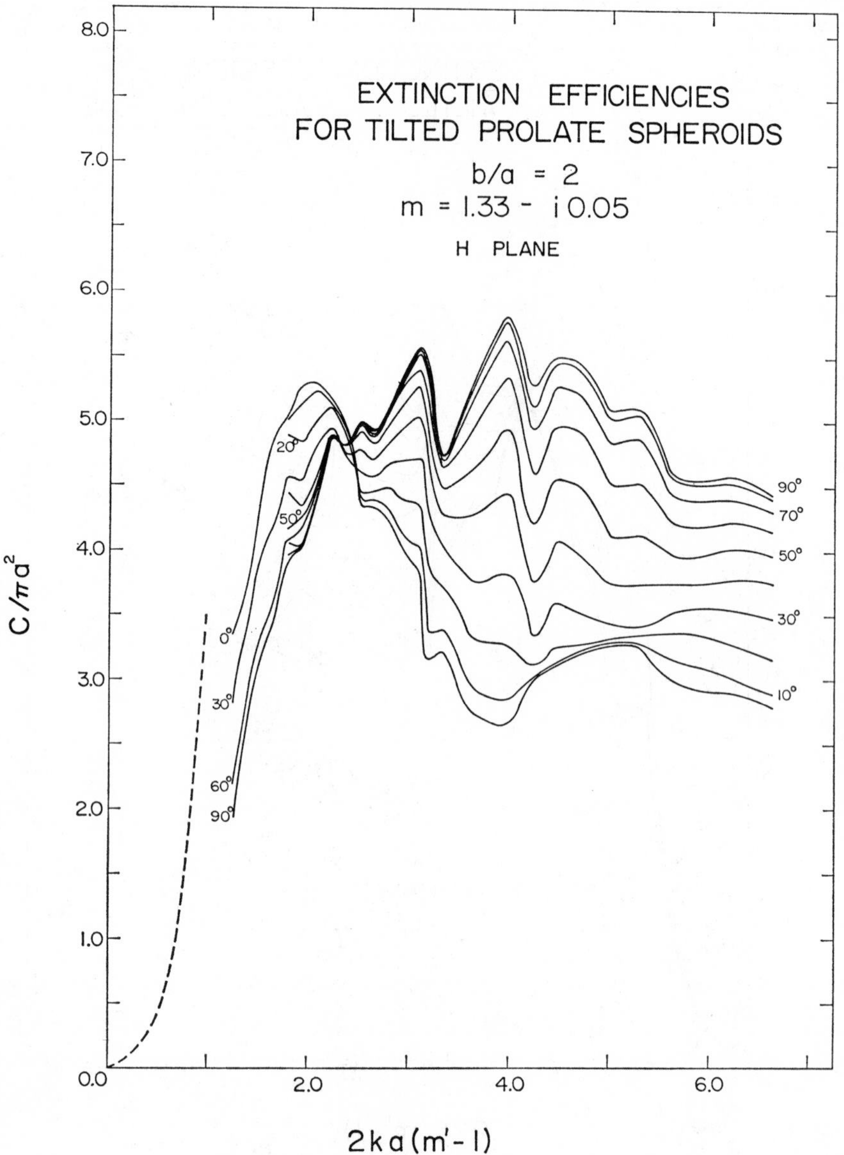

FIG. 53.—Same as Figure 50, but $m = 1.33 - 0.05i$

less than that of infinite cylinders (see Tables 16 and 17, and Fig. 59). A particle of finite length may be a more effective polarizer (i.e., may give a higher polarization to extinction ratio) than one of infinite length and the same axial geometric cross-section.

4) The addition of a small imaginary part in the index of refraction does not greatly modify the *general* characteristics of the extinction curves, except for large $\lambda (\rho \to 0)$, where the slope of the extinction curve is changed from zero ($Q \sim \lambda^{-4}$ for m real) to a non-zero amount ($Q \sim \lambda^{-1}$ for m complex). Two quantitative changes are important:

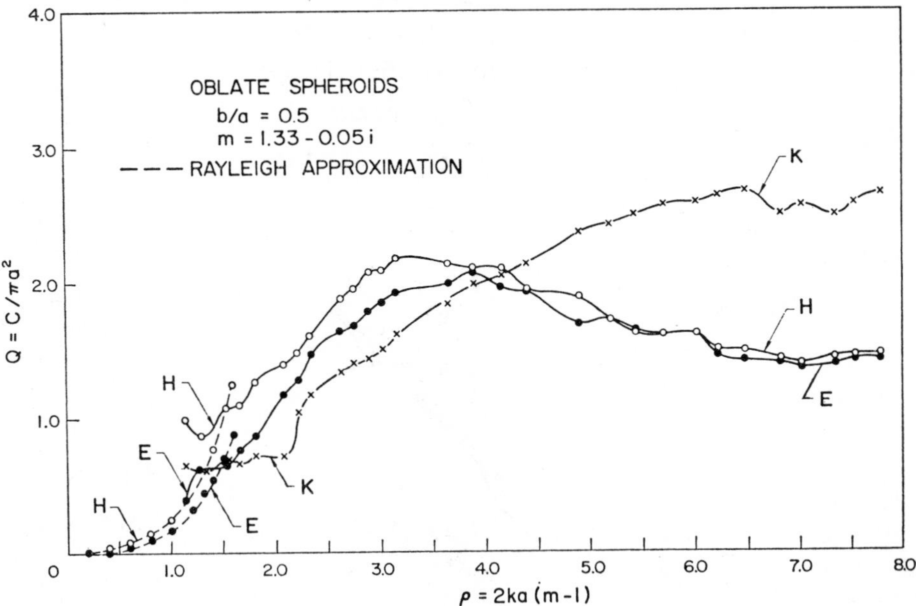

FIG. 54.—Same as Figure 51, but "oblate" replaces "prolate." Open and filled circles on dashed curves are calculated.

(a) a general increase in the value of $C_E - C_H$ for wavelengths up to that for which C_E or C_H reach their first maximum and (b) a reduction of the height of the first maximum for $C_{k\|}$ relative to that for C_E.

7.2. Metallic Particles

For metallic particles, we take $m = n(1 - i\delta)$, $\delta \gtrsim 1$, $n \sim 1$ to $\gg 1$.

For spheres, infinite cylinders and small spheroids ($ka \ll 1$), there is a similar amount of information on scattering by metallic particles as for the dielectric particles.

Scattering by metallic spheres is shown in Figure 60; the index of refraction $m = 1.27 - 1.37i$ is for iron at $\lambda = 0.42\ \mu$. The extinction curve is quite linear up to $x \approx 1$ and is due largely to absorption up to that value of x. The calculations of Figure 60 are for a constant index of refraction.

Indices of refraction for metallic particles vary strongly with wavelength. In particular, both the real and imaginary parts of the index of refraction in the visible region are of the order of unity, but become extremely large in the infrared; metallic particles there-

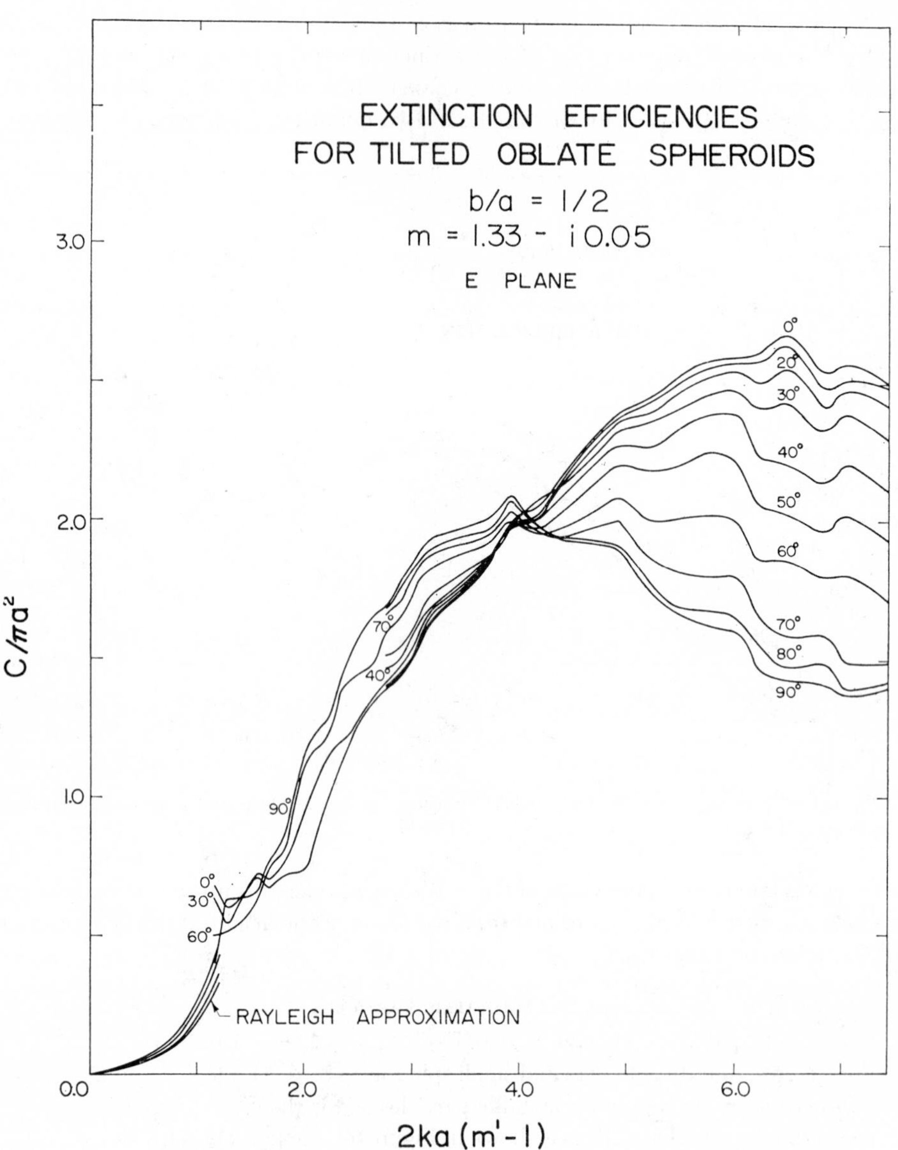

FIG. 55.—Same as Figure 49, but "oblate" replaces "prolate"

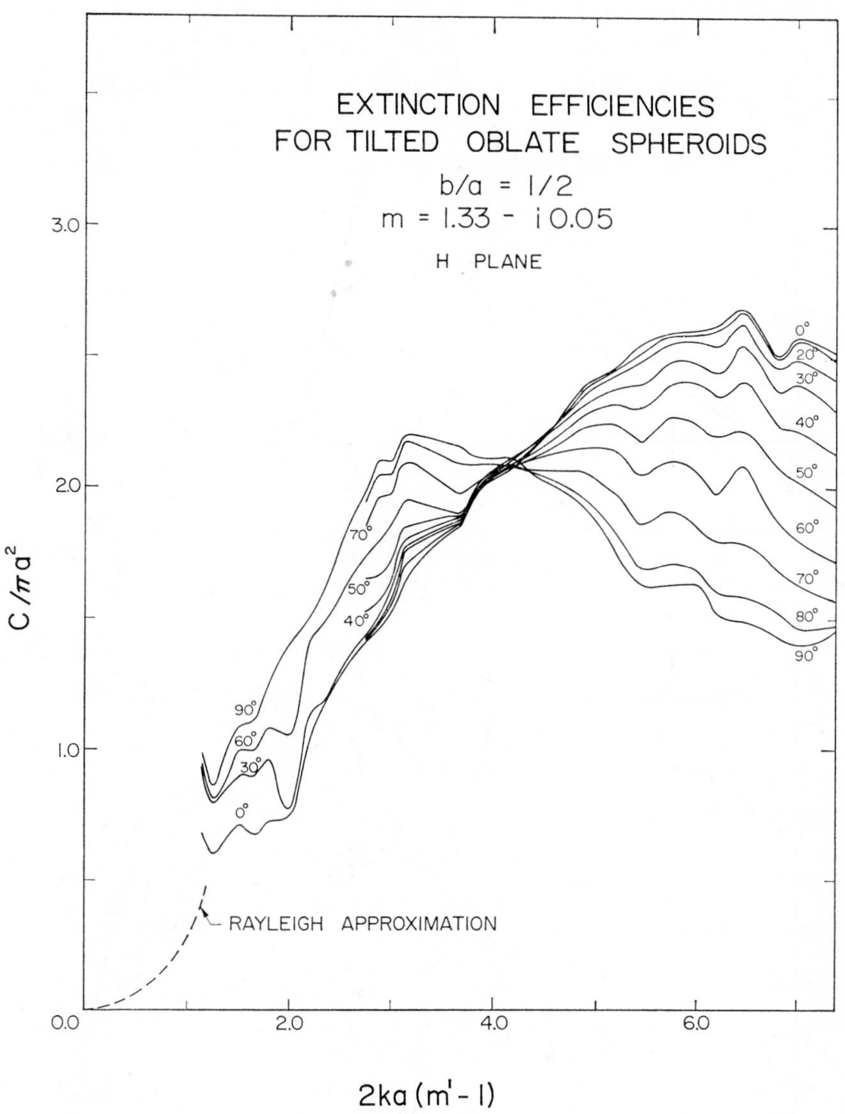

FIG. 56.—Same as Figure 50, but "oblate" replaces "prolate"

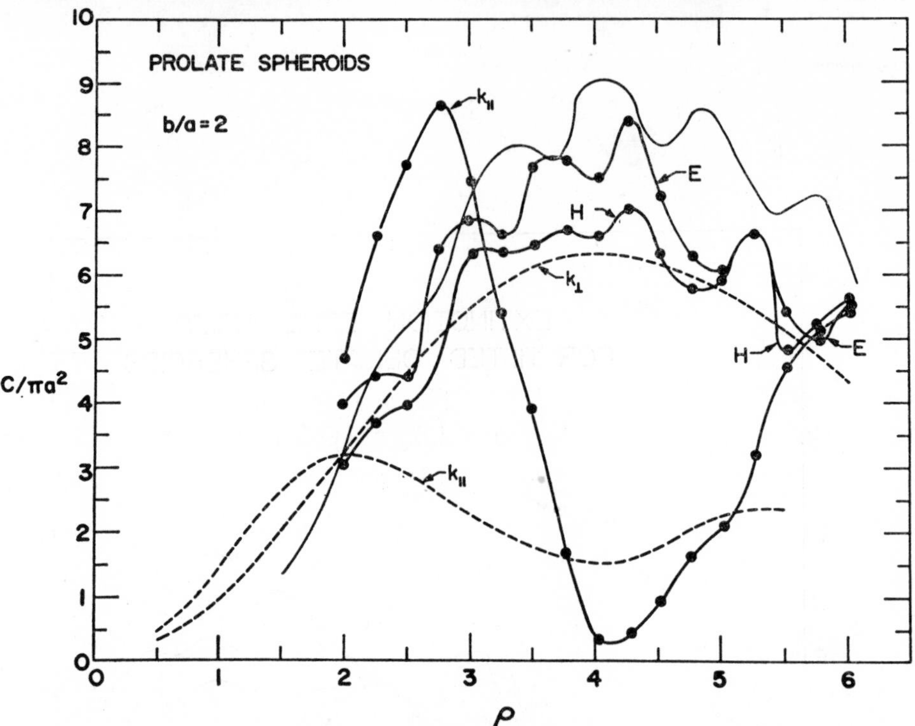

FIG. 57.—Total cross-sections divided by axial geometric cross-sections for prolate spheroids with an elongation of $b/a = 2$. The terms K, E, and H define the direction of the propagation vector and the electric and magnetic vector of the incident radiation with respect to the spheroid axis. (Greenberg *et al.* 1961.) Note: The index of refraction is $m = 1.61$.

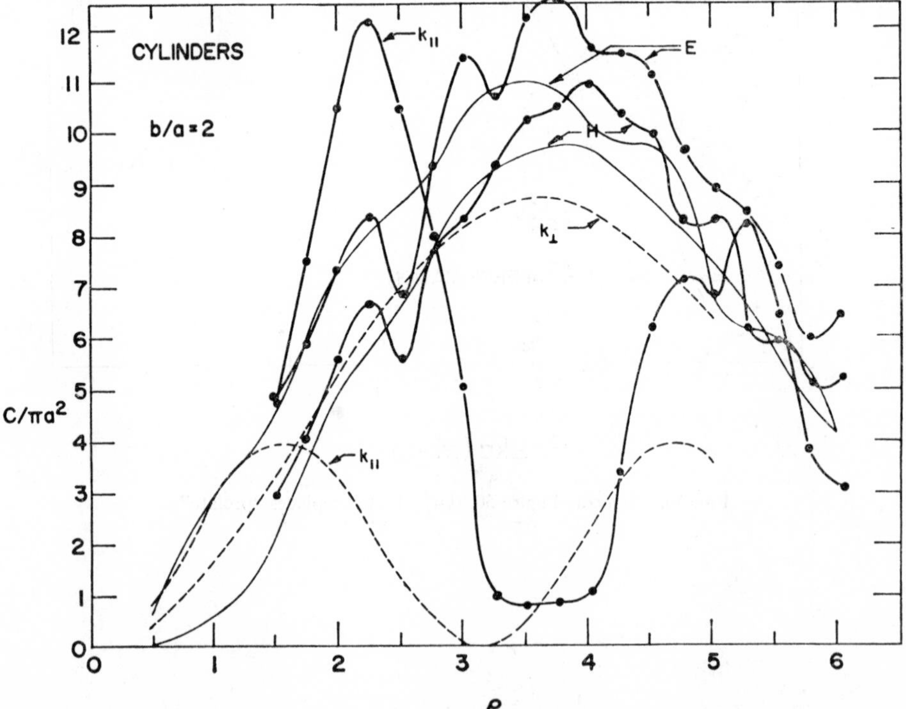

FIG. 58.—Total cross-sections divided by axial geometric cross-sections for circular cylinders with an elongation of $b/a = 2$. The terms K, E, and H define the direction of the propagation vector and the electric and magnetic vector of the incident radiation with respect to the cylinder axis. Comparison is shown with appropriately normalized extinction efficiencies of infinite cylinders. (Greenberg *et al.* 1961.) Note: The index of refraction is $m = 1.61$.

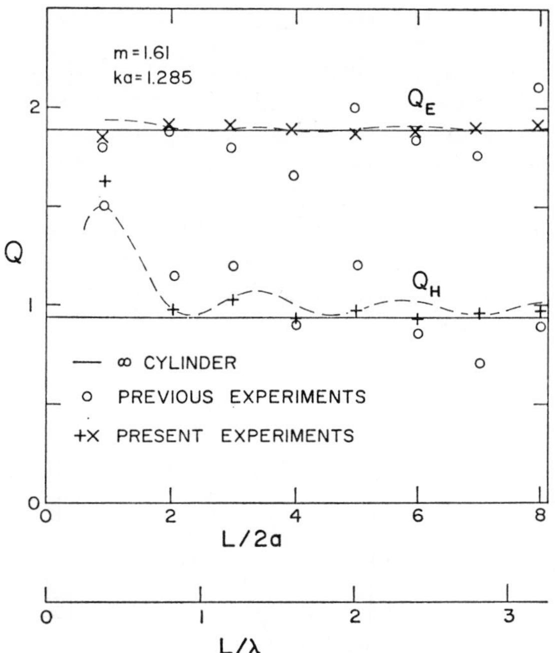

FIG. 59.—Total cross-sections divided by transverse geometric cross-sections for cylinders of fixed diameter as a function of elongation. The horizontal lines represent appropriately normalized theoretical total cross-sections for an infinite cylinder of the same diameter as the finite cylinders. (Greenberg, Pedersen, and Pedersen 1961; Lind, A. C., Ph.D. Dissertation 1966.)

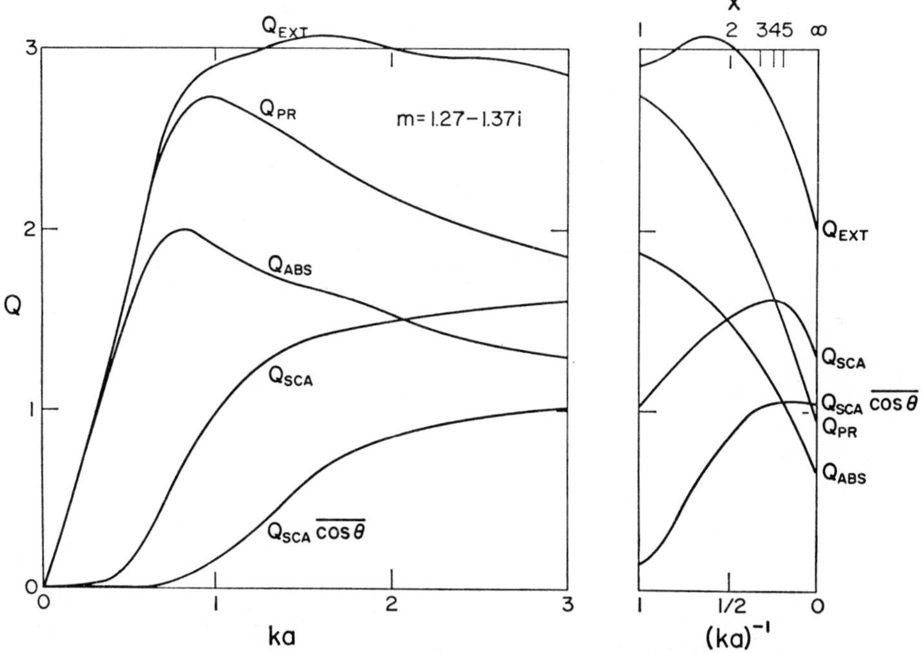

FIG. 60.—Efficiency factors for extinction, radiation pressure, absorption, and scattering for m = 1.27–1.37i. (Van de Hulst, 1957.)

fore act like perfect reflectors (conductivity ∞) in this region. Extinction curves for spheres approaching this condition are shown in Figure 61. For sufficiently long wavelengths, the slopes tend to zero, as for dielectrics. Figures 62 to 66 show extinction efficiencies for variously sized graphite simulating spheres and also metallic spheres where the variation of the index of refraction with wavelength has been included.

The extinction by very small metallic spheroidal particles was included in the discussion of dielectric particles (Sec. 7.1). Numerical values of C_E/C_H are given in Table 8 for $m = \sqrt{2}\,(1 - i)$. Comparison with values for a pure dielectric, for which $m = \sqrt{2}$,

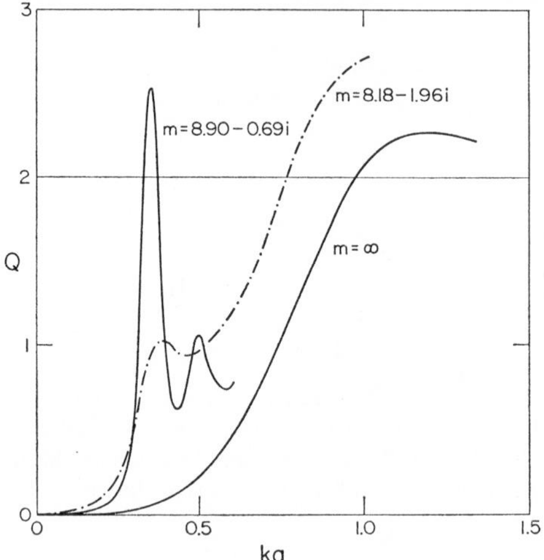

FIG. 61.—Extinction curves for three very large values of the refractive index. (Van de Hulst, 1957.)

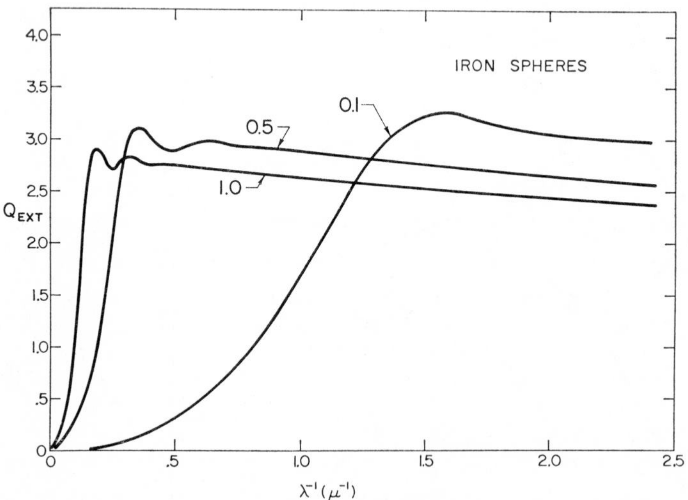

FIG. 62.—Extinction curves for iron spheres of radii 0.1, 0.5, and 1.0 μ. The indices of refraction depend on wavelength according to $m = 2.45\,\lambda^{1/2}\,(1 - i)$.

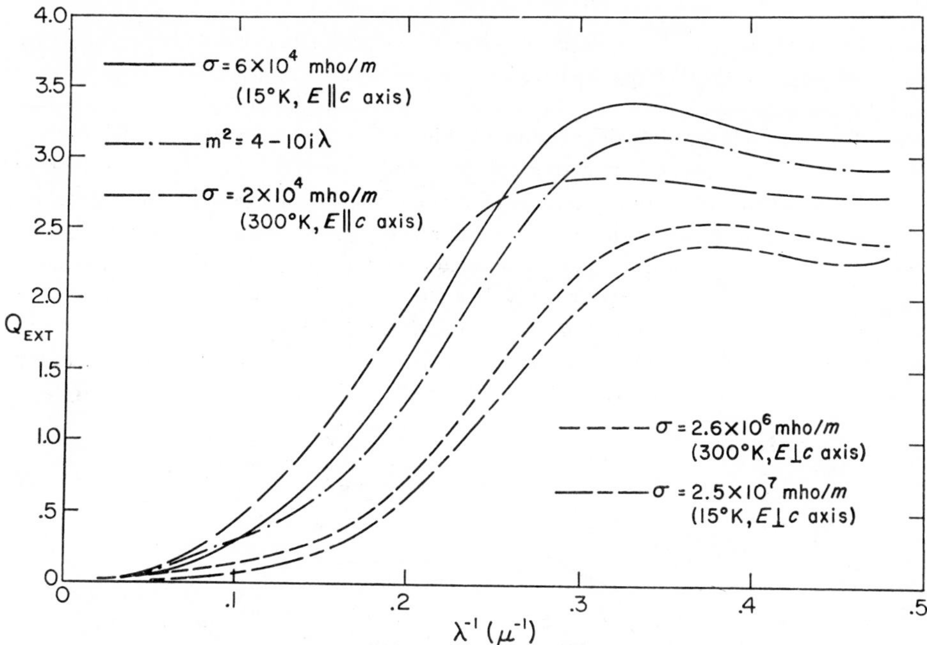

Fig. 63.—Extinction curves for graphite spheres of radius $a = 0.5\ \mu$. The complex indices for each curve are obtained from conductivities corresponding to varying grain temperatures.

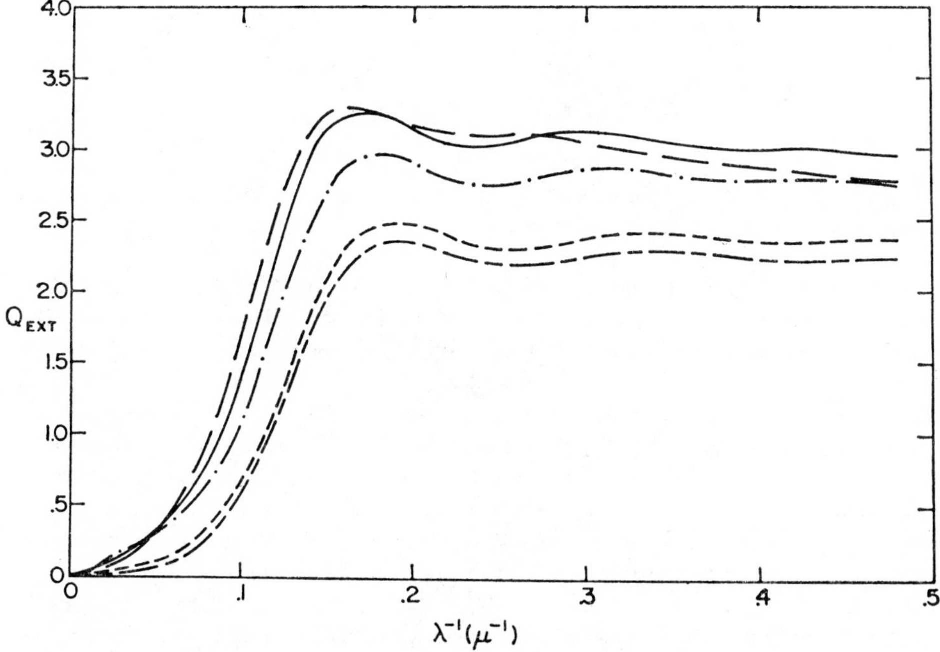

Fig. 64.—Extinction curves for graphite spheres of radius $a = 1.0\ \mu$. The complex indices are obtained from conductivities corresponding to varying grain temperatures. See Fig. 63.

suggests that small metallic particles are more effective polarizers than small dielectric particles, at least for visible wavelengths and for an index of refraction similar to that in the example chosen.

Exact calculations for absorbing cylinders (Figs. 65 and 66; compare Fig. 37) suggest that both extinction curves start with a non-zero slope. The difference $Q_E - Q_H$ is larger for dielectric cylinders for the same value of x when $x \ll 1$. For $x >$ about 0.5, this does not appear to hold for comparisons of the difference $Q_E - Q_H$ at $m = 1.5$ with that at $m = \sqrt{2}\ (1 - i)$. The addition of an absorption term to the index of refraction has a more pronounced effect on the polarizing ability of small spheroids than on that of thin cylinders. The ratio Q_E/Q_H for thin infinite cylinders decreases from 4.5 for $m = \sqrt{2}$ to 4.3 for $m = \sqrt{2}\ (1 = i)$; this ratio for spheroids is always greater for complex than for real indices of refraction (see Table 17).

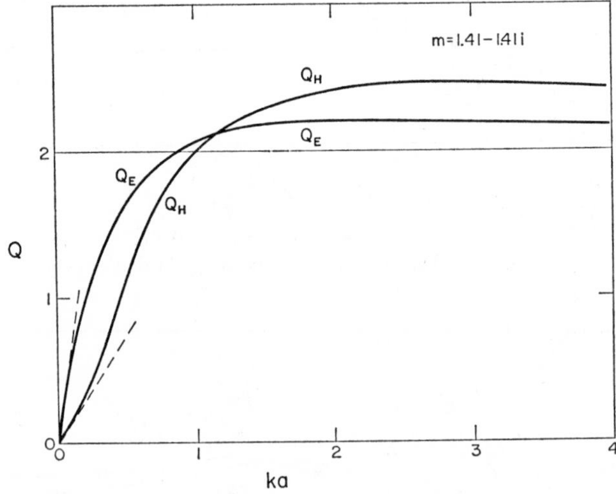

FIG. 65.—Extinction curves of very long metallic cylinders with $m = 1.41\text{–}1.41i$. Dashed lines: approximate values for $ka \ll 1$. (Van de Hulst 1957.)

As m becomes large, elongated particles become enormously effective polarizers, as seen by comparisons of the two extinction curves for infinite cylinders in Figure 67. The quantity C_E is infinite for infinitely thin cylinders. The slope of the extinction curve for very elongated conducting particles with axes distributed at random may differ from zero, the value for perfectly conducting spheres.

An extension to the theory of scattering by very small particles was attempted by Stevenson (1953). However, it has been shown (Greenberg, Libelo, Lind, and Wang 1963) that calculations of the total cross-sections of spheres (with both real and complex refractive indices) and spheroids are unreliable with this theory (often not as good as those obtained from the simple Rayleigh approximations of eqs. [69] and [70] and Table 8).

The differences in the extinction characteristics of dielectric and metallic particles may be summarized as follows: (1) Metallic particles are generally better polarizers and

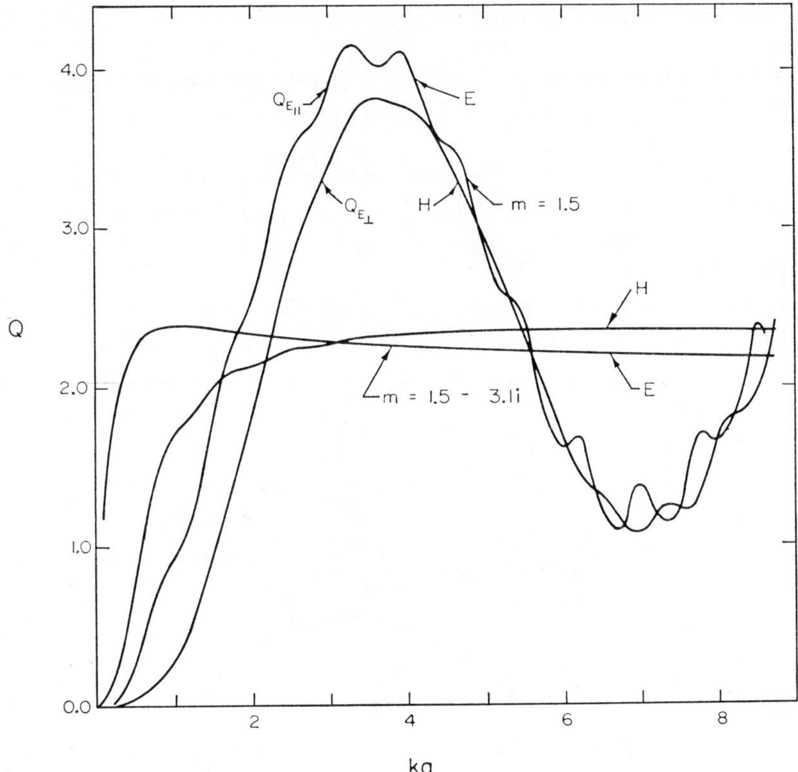

FIG. 66.—Comparison of extinction curves for long dielectric ($m = 1.5$) and metallic ($m = 1.5–3.1i$) cylinders. Note again, as in Figure 65, the complete reversal of polarization by metallic cylinders at about the peak of the extinction curves.

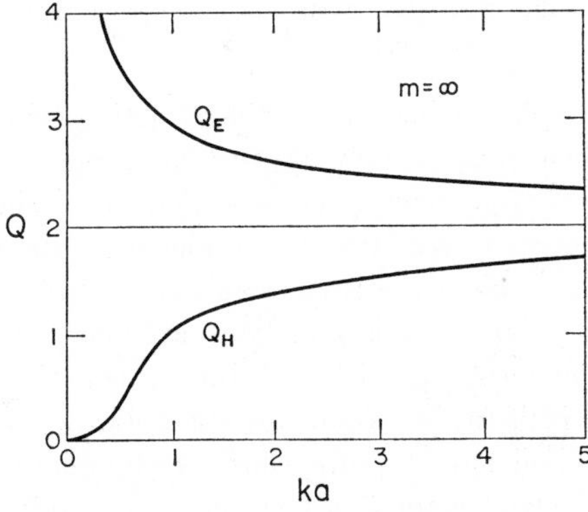

FIG. 67.—Extinction curves of very long, totally reflecting cylinders for the two states of linear polarization of the incident light. (Van de Hulst 1957.)

(2) the extinction curve rises more rapidly as a function of a/λ for metals than for dielectrics.

The refractive index for metals $\rightarrow \infty$ with λ and the difference between metallic and dielectric refractive properties in the infrared are not so pronounced as in the visual region.

7.3. Particles with Non-Isotropic Optical Characteristics

Little is known about scattering by birefringent particles. The microwave analog method is capable of supplying such information, and it is being used for this purpose at the time of writing.

For particles for which $a \approx \lambda$, some semiquantitative inferences can be drawn on the basis of the theory of particles with low indices of refraction.

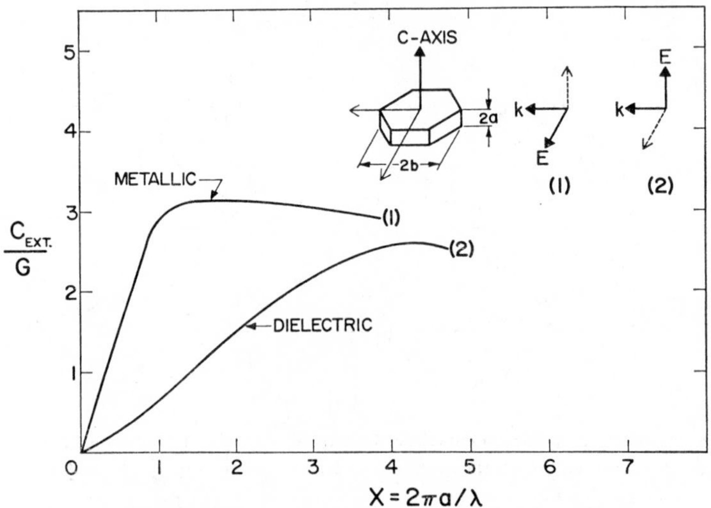

Fig. 68.—Hypothetical representation of the difference in the scattering properties of graphite flakes oriented with (1) electric vector of the radiation in the plane of the flake and (2) electric vector perpendicular to the plane of the flake (along the c-axis). The area normalizing parameter G is proportional to b^2 and to ab in cases (1) and (2), respectively.

Consider a birefringent sphere whose indices of refraction for two directions of polarization are $n_{\parallel}$ and $n_{\perp}$. Then, qualitatively, $C_E \approx C_s(\rho_{\parallel})$ and $C_H \approx C_s(\rho_{\perp})$, where $\rho_{\parallel} = 4\,a(n_{\parallel} - 1)/\lambda$ and $C_s(\rho_{\parallel})$ is the extinction cross-section for a sphere whose index of refraction is $n_{\parallel}$. The extinction curves (as a function of ρ) for the two directions of polarization are shifted laterally with respect to each other; consequently, oriented birefringent spheres will polarize radiation. This argument can also be applied to spheroids; the polarization of the particle will be greatest if the birefringent axis is properly oriented with respect to the axes of the spheroid.

For particles optically similar to graphite, sufficient difference exists in the refractive indices for waves whose electric vector $\boldsymbol{E}$ lies parallel and perpendicular to the c-axis for a certain asymmetry in the scattering properties to be predictable. The metallic character of graphite for $\boldsymbol{E}$ perpendicular to the c-axis is shown by the shape of the upper curve in Figure 68 with its rapid rise at long wavelengths. The lower curve is indicative of the dielectric (with some absorption) character of graphite for $\boldsymbol{E}$ parallel to the c-axis. The predicted high polarization by oriented graphite has been confirmed by experiment (Cayrel and Schatzman 1954) and has been calculated qualitatively by Wickramasinghe (1962) and Greenberg (1966).

For ellipsoids of dimensions $\ll \lambda$, Jones (1945) has generalized the method discussed in Section 7.1 to include the effect of anisotropy of the material. The polarizability is a complicated function of the orientations of the ellipsoids and of the crystals, but can be used in the same way to determine the extinction cross-section.

We assume that the tensor representing the anisotropic complex index of refraction of graphite has the same principal axes as the spheroid. Following the notation of van de Hulst (1957), we may describe the polarizability along the a-(transverse) and c-(symmetry) axes (for an oblate spheroid) by

$$\alpha_a = \frac{a^2 c}{3} \frac{m_a^2 - 1}{L_a(m_a^2 - 1) + 1}$$
$$\alpha_c = \frac{a^2 c}{3} \frac{m_c^2 - 1}{L_c(m_c^2 - 1) + 1} \tag{73}$$

where m_a^2 and m_c^2 are the complex dielectric constants along the a- and c-directions and the L_a and L_c are numerical factors depending on the rato a/c (see definition of L_a and L_c in van de Hulst 1957).

The various absorption and scattering cross-sections are given by

$$C^a_{\text{abs}} = 4\pi k Re\,(i\alpha_a)$$
$$C^c_{\text{abs}} = 4\pi k Re\,(i\alpha_c) \tag{74}$$

$$C^a_{\text{sca}} = \tfrac{8}{3}\rho k^4 |\alpha_a|^2$$
$$C^c_{\text{sca}} = \tfrac{8}{3}\pi k^4 |\alpha_c|^2 \tag{75}$$

where the superscripts a and c correspond to the electric vector of the incident radiation being along a- and c-directions, respectively.

7.4. Free Radicals

Optical properties of large free radicals can be explained theoretically in terms of a free-electron approximation like that for metals (Platt 1956). Platt shows that the longest wavelength λ of strong allowed absorption for the electric vector of the radiation parallel to the largest diameter d of such a particle will probably be of the order of gd, where g is some small multiple of 137 (the fine structure constant). With $g = 400$, a roughly spherical particle of about 6 A radius containing about 500 atoms would be large enough to absorb intensely in the visible wavelength region. For such a particle size, scattering is by a quantum-mechanical process of absorption by electron transition and subsequent reradiation. Greenberg (1960*b*) showed that the extinction cross-section is of the order of the geometrical cross-section or, more specifically, the square of the diameter parallel to the electric vector. Absorption by a small isolated particle is followed by reradiation at the same frequency with a lifetime of the order of 10^{-9} seconds for strong allowed transitions in the visible. This reradiation, equivalent to random reflection, gives the particles a high albedo,[9] although the exact form of the extinction curve for such particles is obscure. Figure 69 shows a hypothetical model. The form of the curve is based on the consideration that in a solid particle each layer absorbs at a different peak wavelength and that the longest peak wavelength corresponds to the largest diameter. Thus the cross-section might be expected to be zero down to some wavelength and be roughly constant for wavelengths shorter than this, at least over a broad band of wavelengths.

7.5. Core-Mantle Particles

The simplest core-mantle particle is a sphere of one material of radius r_1 surrounded by a shell of another material of outer radius r_2. In the context of this chapter, one should consider the core material as metallic (including graphite) and the mantle material as some generalized dielectric substance accumulated in space. Figures 70 through 72 present extinction efficiencies for several combinations of core and mantle radii and core and mantle optical properties. As there is little significant difference between iron and graphite cores, only the results for graphite are shown. The qualitatively important feature is that in the far infrared (sufficiently long wavelength) the metallic core dominates the extinction independent of the size of the mantle (assuming it to have a pure real index as used here). The departure from the pure core extinction curve occurs for shorter wavelengths as the mantle size is increased. If the mantle has a complex index, it is evident that the core will be shielded for even longer wavelengths. It is not necessary for our purposes to consider core radii still larger than $r_1 = 0.1\ \mu$ because for this size the extinction curve is (except for detailed oscillations) almost independent of the mantle size through the wavelength region of interest in the sense that there is a rapid rise in extinction in the infrared dominated by the core and a subsequent relatively flat extinction—certainly not increasing with increasing λ^{-1}.

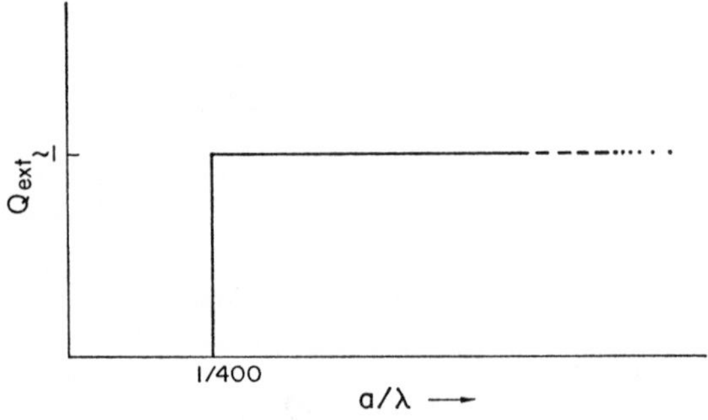

Fig. 69.—Hypothetical extinction curve for a small spherical particle of the type envisaged by Platt (1956).

The possibility that the cores are plate-like particles or whiskers should not be ex-

[9] It seems reasonable to assume that coupling of the electromagnetic field with the lattice would cause a slight reduction in the energy of the outgoing photon and, consequently, a small real absorption of energy in the particle.

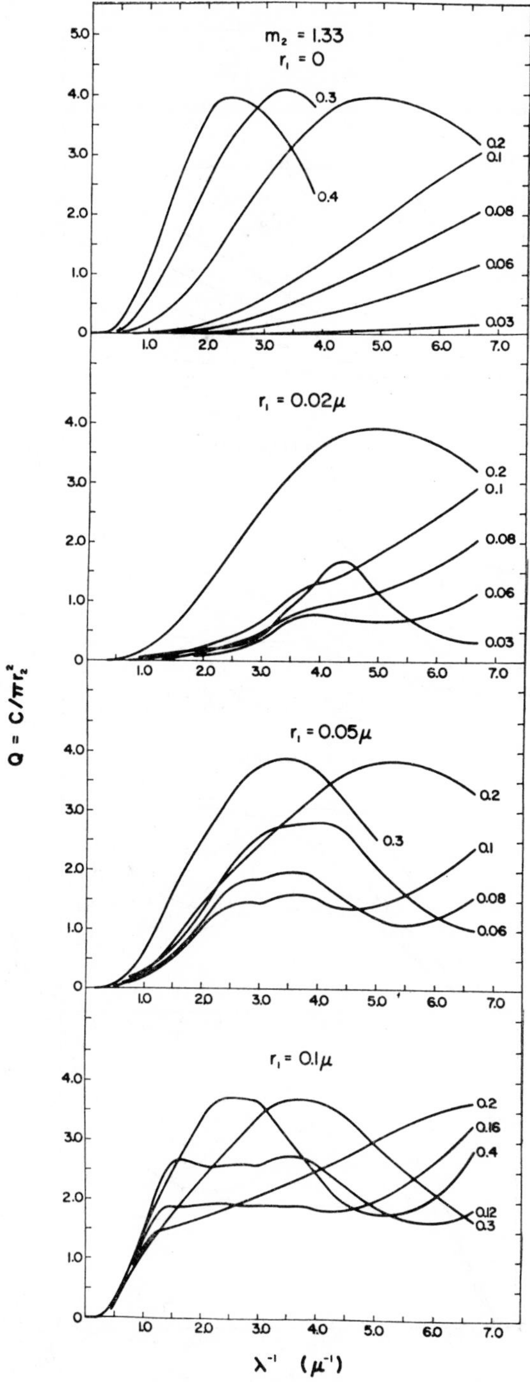

FIG. 70.—Extinction curves for graphite core–dielectric mantle spheres. Inner radii $r_1 = 0$, $0.02\ \mu$, $0.05\ \mu$, and $0.1\ \mu$. Graphite index as shown in Table 11. Radii r_2 of dielectric mantles as shown by labeling of each curve. Refractive index of mantle $m = 1.33$.

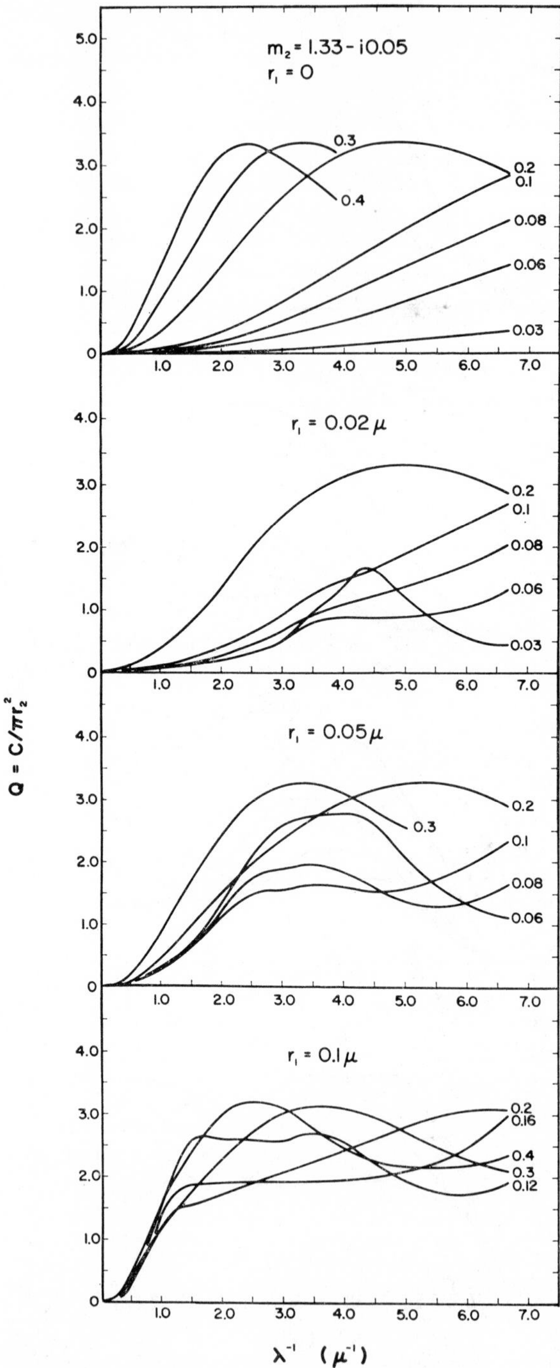

FIG. 71.—Same as Figure 70, but refractive index of mantle $m = 1.33–0.05i$

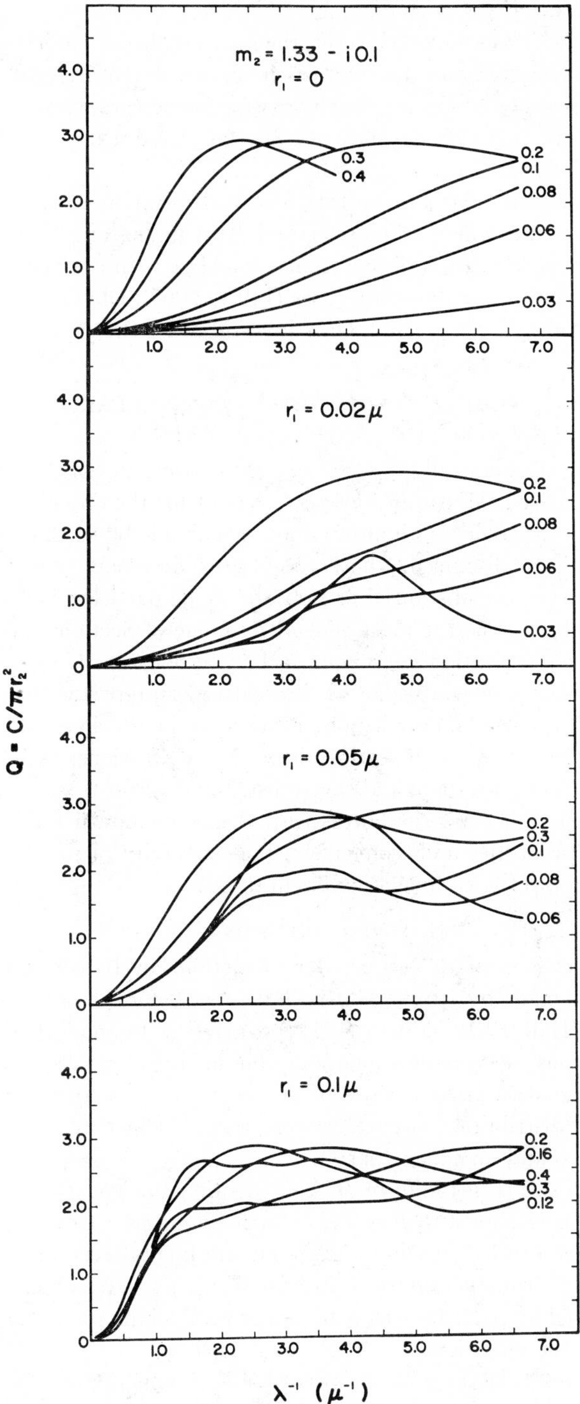

FIG. 72.—Same as Figure 71, but refractive index of mantle m = 1.33–0.1i

cluded from our consideration. For such cores, the mantles also would likely take on the same general shape characteristics, i.e., flat or elongated. Scattering calculations have already been done for propagation incident normally to the core-mantle cylinder axis. Now the arbitrarily oriented cylinder may also be considered using the new computational techniques (Lind and Greenberg 1966).

An exact treatment of the flat core and flat or spherical mantle is not available. Of course, the microwave analog method may be used. Also an analytical solution for axial incidence could be found for particles that are not too large with the method mentioned as previously applied only to scalar wave cases (Libelo 1965). Obvious approximations present themselves, but they have not been investigated with sufficient rigor to be presented here.

8. INTERPRETATIONS OF INTERSTELLAR EXTINCTION AND POLARIZATION

In this section the astronomically observable optical properties of the different types of grains will be considered. Mixtures may best account for the observations, but here mainly the degree of consistency with observations attainable by considering each type of grain separately will be discussed. The Davis-Greenstein model, assumed to be sufficiently effective to produce substantial orientation of the particles, is used.

It would be desirable to use the most general formulae of Section 3 to derive theoretical extinction and polarization curves for each type of grain. However, the aforementioned limitations in our knowledge of the scattering properties (even of the dielectric particles) and the lack of detailed observations make this impossible at present. Simplified models are used in the following theoretical discussions, and the effects of size distribution, index of refraction, and shape distribution on extinction and polarization are considered separately. We shall investigate the degree to which the observations determine the grain properties and, conversely, the sensitivity of predicted results to modifications of grain properties (including orientation).

8.1. Dielectric Grains

A detailed interpretation of the interstellar extinction law based on scattering by dielectric spheres was made by van de Hulst (1949) just before the discovery of interstellar polarization (Hall 1949; Hiltner 1949). Scattering by nonspherical particles modifies the calculations; nevertheless, many of van de Hulst's qualitative conclusions are still valid. Among these are estimates of (1) average size of dielectric grains, (2) effects of varying the distribution of particle sizes, and (3) effects of varying the index of refraction (both real and imaginary parts).

A clue to the grain size is in the form of the extinction efficiencies of pure dielectric particles. These have been shown to vary as λ^{-4} for small grains, to go through an interval roughly proportional to λ^{-1}, reach a maximum, and finally level off asymptotically to twice the geometrical cross-section (see Figs. 35 and 36). Thus, the observed extinction law cannot be produced by particles which are either very small or very large compared to the wavelength of the radiation.

Van de Hulst showed that the gross properties of the extinction curve in the visible region could be reproduced by a variety of size distribution functions (see Fig. 73). The form of the size distribution function is not completely defined by the observations and will probably have to be obtained from a theory of grain growth. We are therefore justified in making a limited selection of distribution functions to be used in the following analysis. We shall generally use the Oort–van de Hulst distribution $n_{\mathrm{OH}}(u)$ (see Fig. 15, Table 3, and function g of Fig. 73), the e^{-a/a^3} distribution (Fig. 16), and the δ-function (all particles of the same size and function a of Fig. 73). By considering the derivative of the extinction curve with respect to λ^{-1}, one might expect to obtain a finer method of distinguishing between various theoretical grain models. In Figure 73 the derivatives of extinction curves are compared. Because the index of refraction of the grains is allowed here to be independent of wavelength, it is possible to use a common scale for the

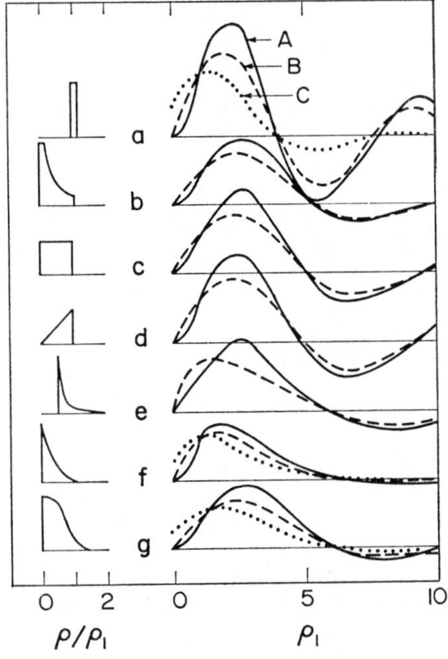

FIG. 73.—Extinction gradients, $\delta A/\delta\rho_0$, for the distribution functions of the radii sketched at the left side of the figure. The curves of A itself would look very similar to each other except when drawn on a large scale. The maximum extinction is reached where the gradient curves cross the zero line. (A) $m = 1.33$; (B) $m = 1 + \epsilon$; (C) $m = 1 + \epsilon - i\epsilon \tan 15°$. For ρ_1 read ρ_0 in figure. (Van de Hulst 1949.)

abscissae, namely, $\rho_0 = 4\pi(a_0/\lambda)(m' = 1)$, where m' is defined by $m = m' - im''$ and where a_0 is the particle size scale factor. The particle size scale factor is defined such that for the δ-function distribution all particles are of size $a = a_0(\rho/\rho_0 = 1)$.

We choose the function $n_{\mathrm{OH}}(u)$ because there is some physical basis (growth mechanism) for its representing the actual state of affairs. The δ-function is physically unrealistic but simplifies calculations and, as shown below, leads to reasonable qualitative predictions. In the case of long cylindrical particles, the δ-function distribution could even have physical significance if used as a representation of "whisker" growth particles.

For spheres, the optical depth (eq. [13]) reduces to the form

$$\tau(\lambda) = T\int_0^\infty u^2 n(u) Q(\rho_0 u, m)\, du(= \Delta m/1.086 \qquad (76)$$

where the dimensionless quantities u and ρ_0 are defined by $u = a/a_0$, $\rho_0 = 4\pi(m - 1)a_0/\lambda$; a_0 is an adjustable scale length and T is a numerical factor measuring the total extinction. We have here let the extinction efficiency Q be a function both of $\rho_0 u$ and m, which is the general case. With $n(u) = n_{\text{OH}}(u)$ or $n(u) = \delta(u - 1)$, certain sets of values for a_0 and m fit $\tau(\lambda)$ best to the shape of the reddening curve. Table 18 shows some typical results for relative values (no attempt has been made to normalize these in the standard way) of the extinction as a function of wavelength as obtained by van de Hulst. Comparison of extinction values across the table is, for our purpose, not significant.

We find that particle sizes of the order of $a_0 = 0.3\ \mu$ are characteristic of the dielectric grains. This can be seen more directly by calculating relative contribution of particles of various sizes, namely, the form of the integrand of equation (76) for $n = n_{\text{OH}}$. Assuming

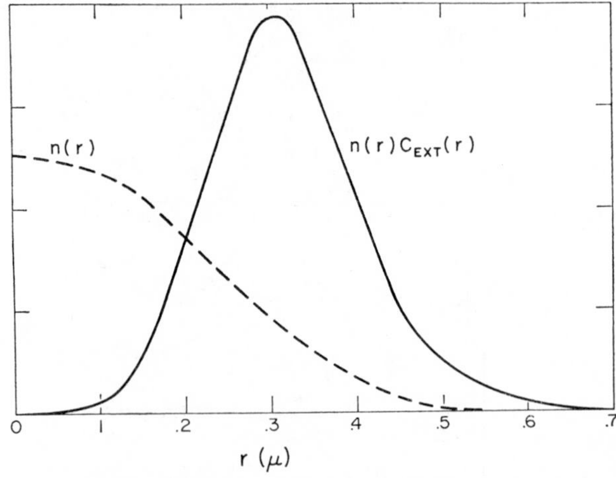

FIG. 74.—Relative contribution of particles of various sizes to the total photographic scattering (full curve). The dashed curve shows the distribution function of the radii. (Van de Hulst 1946*b*.)

$m = 1.33$, we find the result shown in Figure 74; the rather sharp peak around $a = 0.3\ \mu$ makes $a_0 = 0.3$ a good choice for the δ-function distribution of Table 18 justifying the use of the δ-function distribution as a means of arriving at qualitative inferences concerning properties of dielectric grains.

The data of Table 18 are significant for the dependence of the reddening law on the nature of the grains. The bottom two rows of Table 18 illustrate the effect of varying size and index of refraction on the ratio of total to selective extinction and on the decrease of slope of the extinction curve in the ultraviolet relative to the visual.

Columns 3 through 7 show that the ratio of total to selective extinction increases with the particle size and that the rate of decrease of the slope of the reddening curve in the ultraviolet is more rapid for the larger particle sizes. Columns 1 and 2 show that the presence of a slight amount of absorption increases the ratio of total to selective absorption. This increase may occur because for grains with a complex index of refraction, the theoretical reddening curve does not approach $\lambda^{-1} = 0$ with zero slope, whereas the rest of the curve is only slightly modified (see Fig. 73).

TABLE 18

THEORETICAL EXTINCTION VALUES AS A FUNCTION OF WAVELENGTH

	(1)	(2)	(3)	(4)	(5)	(6)*	(7)*	(8)
Size distribution function	n_{OH}	n_{OH}	n_{OH}	n_{OH}	n_{OH}	n_{OH}	n_{OH}	δ
Index of refraction†	$1+\delta$	$1+\delta-i\delta \tan 15°$	$1.25-0.03i$	$1.25-0.03i$	$1.25-0.03i$	$1.25-0.03i$	$1.25-0.03i$	1.33
$\rho_0\lambda$	1.60	1.30	1.60	1.70	1.80	1.70	1.78	1.25
$a_0(\mu)$	0.42	0.34	0.51	0.54	0.57	0.54	0.57	0.30
λ^{-1}	Relative extinction values $A^{(r)}(\lambda^{-1})$							
0	0	0	0	0	0	0	0	0
1.5	0.63	0.81	0.61	0.67	0.74	0.64	0.68	0.48
2.0	0.95	1.11	0.92	1.01	1.08	0.98	1.02	0.82
2.5	1.21	1.38	1.21	1.27	1.32	1.24	1.26	1.11
$A^{(r)}(2.0)/[A^{(r)}(2.5)-A^{(r)}(2.0)]$	3.65	4.12	3.17	3.89	4.50	3.77	4.26	2.8
$\frac{A^{(r)}(2.5)-A^{(r)}(2.0)}{A^{(r)}(2.0)-A^{(r)}(1.5)}$	0.81	0.90	0.93	0.72	0.71	0.72	0.71	0.86

Adapted from van de Hulst 1949.

* Columns (6) and (7) were computed with a slowly varying dependence of m on the wavelength which, for our purposes, can be ignored.

† Assignment of an index of refraction of the form $1 + \delta$ or $1 + \delta - i\delta \tan 15°$ means that the approximate scattering formula (eq. [72]) applied to spheres has been used. The value of δ can then be chosen arbitrarily so long as $\delta < 0.3$.

If we consider the same arguments for particles of one size only, we may refer directly to the extinction curves of Section 7. Further, we may assume that the extinction coefficient Q is a function of ρ only (eq. [72] and Fig. 35, lower curve). Since $\alpha = 4\pi a_0(m - 1)/\lambda$, the extinction curve will be unchanged as long as $a(m - 1) =$ constant. Change in a implies change in m for constant α, although it seems more likely that the sizes and not the refractive indices of the interstellar grains may vary from region to region. If m is assumed constant, then increasing a is equivalent to decreasing λ^{-1} in the same proportion to keep α constant. Therefore, if $a_2 > a_1$, then $Q(a_2)$ is shifted to smaller values of λ^{-1} relative to $Q(a_1)$.

In Figure 75 the schematic extinction curve is obtained from the efficiency curve for

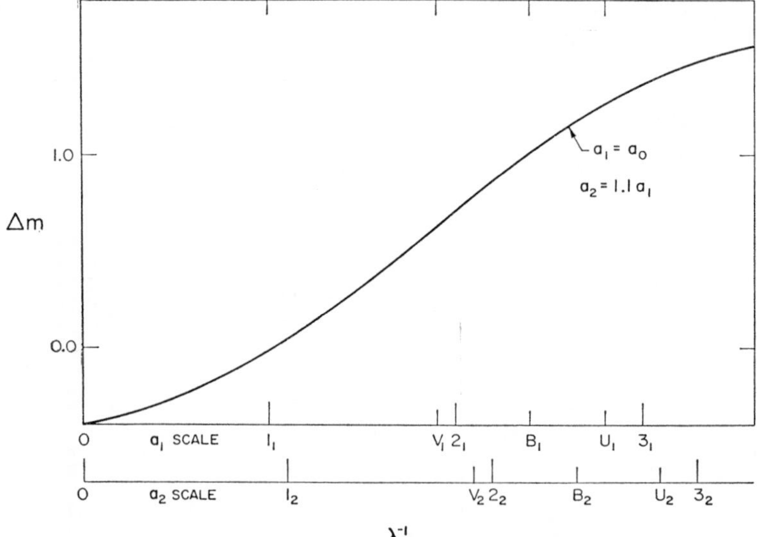

FIG. 75.—Effect of particle size on extinction curve. The two wavelength scales are adjusted according to the particle sizes instead of adjusting the normalized extinction curve.

spheres of low refractive indices (Fig. 35, lower curve) by assigning a wavelength scale corresponding to $a_0 = 0.30$ and $m = 1.33$ (Table 18, col. 8). If $a_1 = a_0$ and $a_2 = 1.1\ a_0$, we obtain two different wavelength scales for the extinction curve. The gradually decreasing slope toward shorter wavelengths shows that increase in particle size increases the difference $A_B - A_V$ less than it increases A_V; i.e., $(A_{B_2} - A_{V_2})/(A_{B_1} - A_{V_1}) < A_{V_2}/A_{V_1}$.

Table 18 shows that the ratio of total to selective extinction is larger for larger particles, equivalent to the above relation. Similarly, the ratio E_{U-B}/E_{B-V} should decrease with increasing particle size.[10]

In Figure 76 we present a set of extinction curves computed using exact Mie theory for seven size distributions of dirty ice grains. The size distributions are of the form $n = e^{-5(a/a_i)^3}$ where $a_i = 0.1, 0.2, \ldots, 0.7\ \mu$. Throughout the visible the form with $a_i = 0.5$

[10] In this calculation, the *form* of the size-distribution function is assumed constant. If the average grain size differs because of differences in the form of the distribution function, it may be necessary to modify the conclusions.

fits very closely to the Whitford points. In Figure 77, two attempts are made to get a best-size distribution by a least-squares analysis. The agreement with the Whitford points is better than that with a single parameter, but there may still be difficulty in getting such purely dielectric grains to fit correctly such curves as have been found by Johnson (see Fig. 2), especially those with sharp dropoff in the infrared. These are further discussed in the section on core-mantle particles, where it is shown that even there the difficulties remain.

In the most extreme curve (number 5 in Fig. 2: Cepheus), the initial rise at $\lambda^{-1} \approx 0.1$ reaches $\Delta m \approx 0.5$ at $\lambda^{-1} \approx 0.2$ if we assume that the remaining portion of the extinction is normal. The extinction may be produced by two size distributions of particles: large

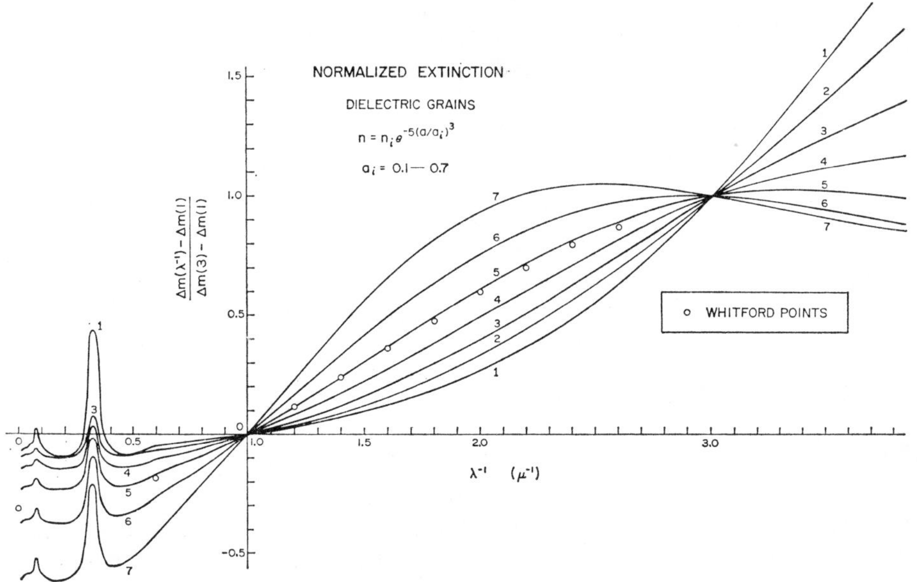

FIG. 76.—Normalized extinction curves computed for spherical grains of dirty ice. Each curve is labeled according to the appropriate distribution of sizes. Comparison is made with the Whitford curve.

particle (l), total extinction $\Delta m_l = 0^m.5$; and normal particle (n), total extinction (in the ultraviolet) $\Delta m_n = 1^m.0$. If we assume spherical grains of (constant) $m = 1.33$, the radius of the larger grains can be shown to be $a_l = 5\ \mu$ calculated from the apparent extinction maximum at $\lambda^{-1} = 0.2\ \mu^{-1}$, letting $4\pi a_l \lambda^{-1}(m - 1) = 4$ for $\lambda^{-1} = 0.2\ \mu^{-1}$ (see Fig. 35). Radius of normal grains is assumed to be $a_n = 0.3\ \mu$. The assumed constant index of refraction implies that the relative number of ordinary to large particles, N_n/N_l, is inversely proportional to the ratio of the square of their radii and is also proportional to their relative contribution to extinction. Thus,

$$\frac{N_n}{N_l} = \frac{m_n}{m_l} \left(\frac{r_l}{r_n}\right)^2 = 5.5 \times 10^2 ,$$

and the mass of a larger particle is consequently *eight times larger* than that of a smaller particle.

Insofar as the ultraviolet is concerned, it is significant that the three-parameter size-distribution function in Figure 77 extrapolates rather closely to the measurements of extinction up to $\lambda^{-1} \approx 4\ \mu^{-1}$ ($\lambda = 2600$ A) by Boggess and Borgman (1964) and by Stecher (1965). Generally one requires relatively more smaller particles to achieve such extinction in the ultraviolet, thus implying that perhaps the e^{-a^3} distribution is a poor initial choice.

The strong ice absorption at $\lambda^{-1} \approx 0.3\ \mu^{-1}$ is larger by a factor of $\sim$3 to 4 than the upper limit on its intensity established by observation (Danielson, Woolf, and Gaustad 1965).

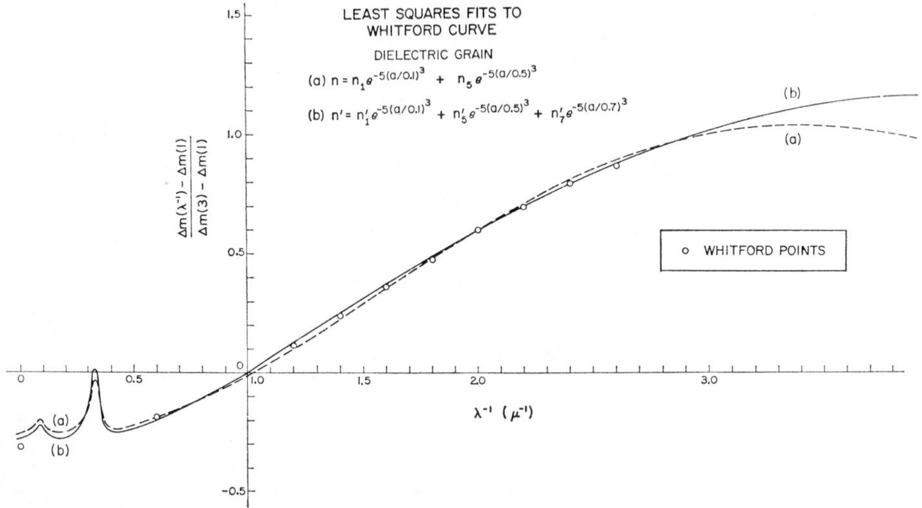

FIG. 77.—The curves labeled (*a*) and (*b*) are the results of a least-squares analysis of, respectively, two-parameter and three-parameter size-distribution functions for best fits to the Whitford curve. The values of the parameters are, respectively: $n_1 = 215$, $n_5 = 16$, $n_1' = 3450$, $n_5' = 12.2$, $n_7' = 0.88$.

In Figure 78 another way of representing the difficulties inherent in a one-parameter size distribution is shown. Here we plot calculated values of normalized extinction at selected wavelengths versus a_i [in $n = e^{-5(a/a_i)^3}$]. A perfect fit for all wavelengths would exist if the observed points were to fall on a vertical line. However, as the longer wavelengths are considered, one needs larger particles than those required for the shorter wavelengths. In other words, this single-parameter size-distribution function will not give the observed extinction curve. Although a basic size-distribution function [$e^{-5(a/a_i)^3}$ with $a_i = 0.5\ \mu$] gives excellent fit in the visible, one needs more small and more large particles to increase the ultraviolet and the infrared extinction, respectively, and this seems to be an inconsistency in the model. However, perhaps the extension to consideration of nonspherical particles should be made before arriving at a definite conclusion. The increase of the ratio of total to selective extinction due to the addition of a slight absorption term to the refractive index can be deduced from the curves of Figure 36, where in the linear region the slope of the curve for $\beta = 15°$ is less than that for $\beta = 0$, whereas its ordinate is greater.

When two or more parameters are changed simultaneously, a quantitative treatment is perhaps not too meaningful, unless nonspherical particles are included. The need to consider these introduces additional complications.

For a first approximation, let all particles be of one size and $m = 1.33$. By comparing Figure 35 with the Whitford curve (Fig. 1 of the Harold Johnson chapter), we may infer that $\rho = 2.0$ corresponds approximately to $\lambda^{-1} = 2.0$. This gives $a_0 = .24\ \mu$. At $\rho = 2.0$ the cross-section is about $1.5\ \pi a_0^2$. Taking the factor 1.5 into account, equation (23) then gives $\rho_g = 1 \times 10^{-26}$ gm/cm³, which is about 1 per cent of the mean hydrogen density in the galactic plane (see Table 2) and of the same order as the density of the condensable gases.

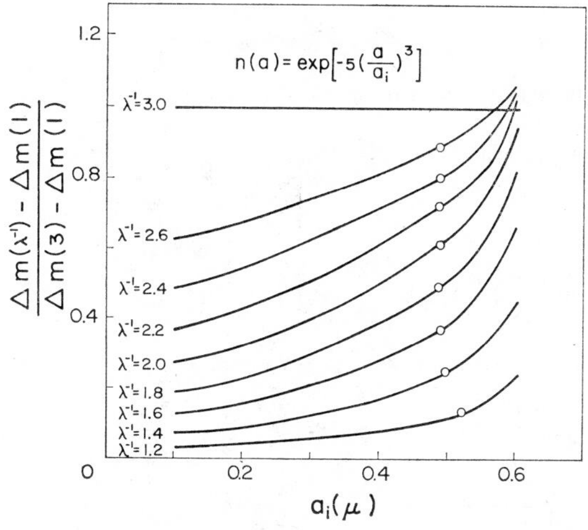

FIG. 78.—Curves of normalized extinction versus a_i, the size-distribution parameter in $n(a) = \exp[-5(a/a_i)^3]$. The normalization is such that at $\lambda^{-1} = 3.0$ and $\lambda^{-1} = 1$, the curves are horizontal lines. Observed Whitford points are shown on each wavelength designated curve. If the given size-distribution function were correct, the points would fall on a vertical line for all wavelengths.

If we assume that the dust is unevenly distributed so that it exists primarily in the clouds, then its density within the clouds is increased by a factor of 10, that is to say, 10^{-25} gm/cm³. However, if the gas and dust densities are correlated, then the hydrogen density may also be increased by the same factor so that within the cloud the mass of condensable atoms (C, O, N) bound in the grains is still of the order of the mass of free condensable atoms. This assumes, of course, that the relative abundances by mass of condensable atoms is as given in Table 1; namely, of the order of 1 per cent. Thus, local exhaustion of condensable atoms is not likely on the average. On the other hand, as has already been noted, the degree of correlation between gas and dust is not yet firmly established. It is often assumed that the lack of high-density H atoms in regions of *very* high extinction is due to the formation of H_2 molecules (still undetectable). Although this appears to be borne out by the detection of the OH molecule, nevertheless, more observational work is needed to confirm this assumption. That so little mass can produce so much extinction (total mass per unit area in 1 kpc is only $\approx 3 \times 10^{-5}$ gm) is due to the high extinction efficiency for particles whose size is of the order of the wavelength.[11] The average number density of the grains is of the order of 100/km³.

As a matter of convenience, it is worth noting that the approximate extinction formula, equation (72) (for spheres), lends itself, when combined with some forms of the size-distribution function, to analytical formulae for properties related to the grains. We present some of these results for handy reference. For spheres, the cross-section simplifies to

$$\frac{C}{4\pi}=\frac{a^2}{2}-Re\left(\frac{ia}{\alpha}\,e^{-i\alpha a}-\frac{1}{\alpha^2}+\frac{1}{\alpha^2}\,e^{-i\alpha a}\right) \tag{77}$$

where $\alpha = 2k(m' - 1)(1 - i\delta)$ and $m = m' - im'' = m'(1 - i\delta)$.

Define the three-size distribution functions by

$$n_1 = e^{-a_1 a}, \qquad n_2 = e^{-a_2^2 a^2}, \qquad n_3 = e^{-a_3^3 a^3}. \tag{78}$$

The total extinction is readily computed for the first two of these to be

$$\frac{C_1}{4\pi}=a_1^{-3}-Re\left\{i\alpha^{-1}(i\alpha+a_1)^{-2}-\alpha^{-2}a_1^{-1}+\alpha^{-2}(i\alpha+a_1)^{-1}\right\}$$

$$[\alpha = \text{real or complex}]$$

$$\begin{aligned}\frac{C_2}{4\pi}= \sqrt{\pi}(2a_2)^{-3}-Re\Big\{&i(2\alpha a_2^2)^{-1}+\sqrt{\pi}(2a_2^3)^{-1}\exp(-\alpha^2/4a_2^2)\\ &-i a_2^{-3}\int_0^\infty \exp(-x^2)\sin(\alpha x/a_2)\,dx-(\sqrt{\pi}/2)\alpha^{-2}a_2^{-1}\\ &+(\sqrt{\pi}/2\alpha^2 a_2)\exp(-\alpha^2/4a_2^2)-(i/\alpha^2 a_2)\int_0^\infty e^{-x^2}\sin(\alpha x/a_2)\,dx\Big\}\end{aligned} \tag{79}$$

$$[\alpha = \text{real}].$$

The integrals may be evaluated approximately when necessary by expanding s in $(\alpha x/a_2)$ to obtain

$$\int_0^\infty e^{-x^2}\sin(\alpha x/a_2)\,dx=\tfrac{1}{2}\left[\frac{\alpha}{a_2}-\frac{(\alpha/a_2)^3}{3!}+\frac{2(\alpha/a_2)^5}{5!}-\frac{3!}{7!}(\alpha/a_2)^7+\ldots\right].$$

For the size distribution which we most often have used in the numerical computation, namely, $e^{-5(a/a_i)^3}$, only two of the various integrals may be put in analytical form. However, the integral which is a measure of the total surface area of the grains is in analytical form and is useful in problems relating to the availability of surfaces for the formation of interstellar molecules.

The surface areas and volumes of grains appropriate to each of these size distribution functions are

$$a_1 = 8\pi a_1^{-3}, \qquad a_2 = \pi^{3/2}a_2^{-3}, \qquad a_3 = a_3^{-3}/3, \tag{80}$$

$$V_1 = 8\pi a_1^{-4}, \quad V_2 = \frac{2\pi}{3}\,a_2^{-4}, \quad V_3 = 0.4\pi a_3^{-4}. \tag{81}$$

[11] A simple demonstration of this fact can be made by shining the light from a slide projector on a screen and blowing a puff of smoke in the light beam. The dramatic decrease in light reaching the screen is considerably greater than that given by interposition of a fairly large obstacle of mass much greater than that of the smoke. Analytically, the function Q/a^3 (extinction per unit volume) must have a maximum between $\lambda^{-1} = 0$ and $\lambda^{-1} = \infty$.

Although polarization complicates the theoretical analysis of interstellar grain phenomena, consistency of a physical theory with the polarization data as well as that of extinction would increase its probability of correctness.

First, theoretical arguments based on the existence of polarization lead to additional important possibilities for varying the extinction curve. Starlight viewed through a cloud of elongated particles whose short axes are perpendicular to the line of sight will be polarized; extinction is produced by an average of the cross-sections for particles seen sideways and end-on (in Fig. 35, the mean of $C_{k\parallel}$ and $C_{k\perp}$; in Fig. 57, the appropriate

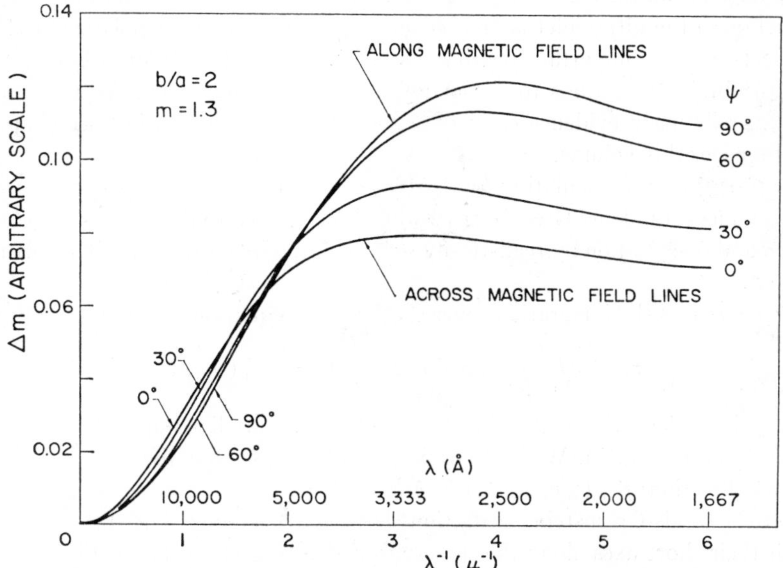

FIG. 79.—Effect of the orientation on wavelength dependence of extinction. The curves for $\psi = 0°$ and 90° are the same as those obtained previously by Greenberg and Meltzer (1960). Calculations are based on the scalar wave approximation.

mean of C_E, C_H, $C_{k\parallel}$). The cross-section $C_{k\perp}$ may be thought of as a mean value of C_E and C_H. The same cloud of particles whose short axes are aligned along the line of sight (direction of magnetic field = line of sight) will produce no polarization,[12] and the extinction will be produced only by $C_{k\perp}$, which will have a dependence on wavelength different from the average of $C_{k\parallel}$ and $C_{k\perp}$. For the uniform size model (choosing $\rho = 2.0$ to correspond to the visual region), unpolarized radiation will have higher extinction in the ultraviolet relative to the visual than will the polarized radiation. The same will hold for flattened particles. If one uses the Oort–van de Hulst size distribution for a (the shorter axis) of prolate spheroids of index of refraction $m = 1.3$, $b/a = 2$, and the approximate formula, equation (72), for scattering by such particles, the results shown in Figure 79

[12] There are many orientation distributions which would produce zero polarization and for which wavelength dependence of extinction would differ from that considered here. If two equal clouds of grains along the line of sight have magnetic fields perpendicular to each other and also to the line of sight, there would be no difference between the wavelength dependence of extinction (with no polarization) and that produced by the first of the above cases (with polarization).

are obtained. The upper and lower curves (in the ultraviolet) correspond to the situations described above. The others are for intermediate cases: $\pi/2 - \eta \rightarrow \psi$ is the angle between the spin axis and the line of sight. This set of curves illustrates qualitatively some anticipated effects of orientation on the wavelength dependence of extinction. Thus (1) E_{U-B}/E_{B-V} is greater when the radiation is polarized than when the radiation is unpolarized (Greenberg and Meltzer 1960; Wilson 1960) (with possible, but improbable, exceptions as seen in the footnote); (2) A/E, the ratio of total-to-selective extinction, is a function of polarization; whether its value is greater for unpolarized than for polarized radiation depends on the correct wavelength scale for Figure 79 as well as other factors; (3) extinction in the ultraviolet is greater for unpolarized than for polarized radiation. These effects have an important bearing on whether a standard Whitford curve should be used to obtain corrections to the magnitudes of reddened stars. As was shown in Section 2, preliminary evidence seemed to suggest that it should, but see chapter by Harold Johnson, this volume.

A new procedure for calculating total cross-sections for arbitrarily oriented dielectric circular cylinders (Lind and Greenberg 1966) may be applied to the problem of obtaining extinction and polarization by realistically oriented interstellar particles (Greenberg and Shah 1966).

We first suppress the integrations over z, a, and e in equation (13) to obtain the form

$$\tau_{x,y} = \int_0^{\pi} \int_0^{2\pi} d\varphi \sin\theta \, d\theta \, f(\theta, \varphi) C_{x,y}(\theta, \varphi). \tag{82}$$

The simplest orientation is that in which all particles are lined up in the same direction (picket fence = P.F.). We will consider for this case that the particles all point along the x-direction: $f_{\text{P.F.}}(z, \theta, \varphi, a, e) = \delta(\cos\theta)\delta(\varphi)$.

For perfect Davis-Greenstein orientation (Davis-Greenstein = D.G.) the particles all spin with their short axes along the direction of the magnetic field. In Figure 80, $\boldsymbol{B}$ is defined by the angles Θ, Φ with respect to the z- and x-axes. If we let $\boldsymbol{b}$ be a unit vector along $\boldsymbol{B}$ and $\boldsymbol{a}$ be a unit vector along the particle symmetry axis, then the plane of rotation of the axis is defined by $\boldsymbol{a}\cdot\boldsymbol{b} = 0$. If the symmetry axis makes the angles θ, φ with respect to the x-y-z frame and the angle α with respect to some fixed direction $\boldsymbol{n}$ in its plane of rotation, we may write $\boldsymbol{b}\cdot\boldsymbol{n} = 0$, $\boldsymbol{a}\cdot\boldsymbol{n} = \cos\alpha$, $\boldsymbol{n}\cdot\hat{\boldsymbol{z}} = \sin\Theta$ where $\hat{\boldsymbol{z}}$ is a unit vector along the z-axis.

Using the appropriate vector relations, we then show that

$$\cos\theta = \sin\Theta \cos\alpha \tag{83}$$

$$\cot\varphi = \frac{\cos\Theta \cos\Phi \cos\alpha - \sin\Phi \sin\alpha}{\cos\Theta \sin\Phi \cos\alpha + \cos\Phi \sin\alpha}. \tag{84}$$

For any axially symmetric particle we may derive the cross-sections for arbitrary orientations with respect to the direction and state of polarization of the incident radiation from two basic cross-sections, $C_E(\theta)$ and $C_H(\theta)$, where θ is the angle between the incident radiation and the symmetry axis.

We may then show that for the configuration in Figure 80 with incident unpolarized radiations

$$C_x = C_E(\theta)\cos^2\varphi + C_H(\theta)\sin^2\varphi \tag{85}$$

$$C_y = C_E(\theta)\sin^2\varphi + C_H(\theta)\cos^2\varphi. \tag{86}$$

The extinction and polarization are derived from the sum and difference of these two quantities given by

$$C_x + C_y = C_E(\theta) + C_H(\theta) \tag{87}$$

$$C_x - C_y = [C_E(\theta) - C_H(\theta)] \cos 2\varphi \,. \tag{88}$$

Substituting equations (87) and (88) into equation (82), and using the appropriate forms for the angular distribution function f, we obtain

Perfect alignment

$$\frac{2}{1.086}(\Delta m)_{\text{P.F.}} = C_E\left(\frac{\pi}{2}\right) + C_H\left(\frac{\pi}{2}\right) \tag{89}$$

$$\frac{1}{1.086}(\Delta m_p)_{\text{P.F.}} = C_E\left(\frac{\pi}{2}\right) - C_H\left(\frac{\pi}{2}\right) \tag{90}$$

Perfect Davis-Greenstein alignment

$$\frac{2}{1.086}(\Delta m)_{\text{D.G}} = \frac{1}{\pi}\int_0^{\pi}[C_E(\theta) + C_H(\theta)]d\alpha \tag{91}$$

$$\frac{1}{1.086}(\Delta m_p)_{\text{D.G.}} = \left\{\frac{1}{\pi}\int_0^{\pi}[C_E(\theta) - C_H(\theta)]\cos 2\varphi d\alpha\right\}_{\max} \tag{92}$$

where θ and φ are defined by equations (83) and (84).

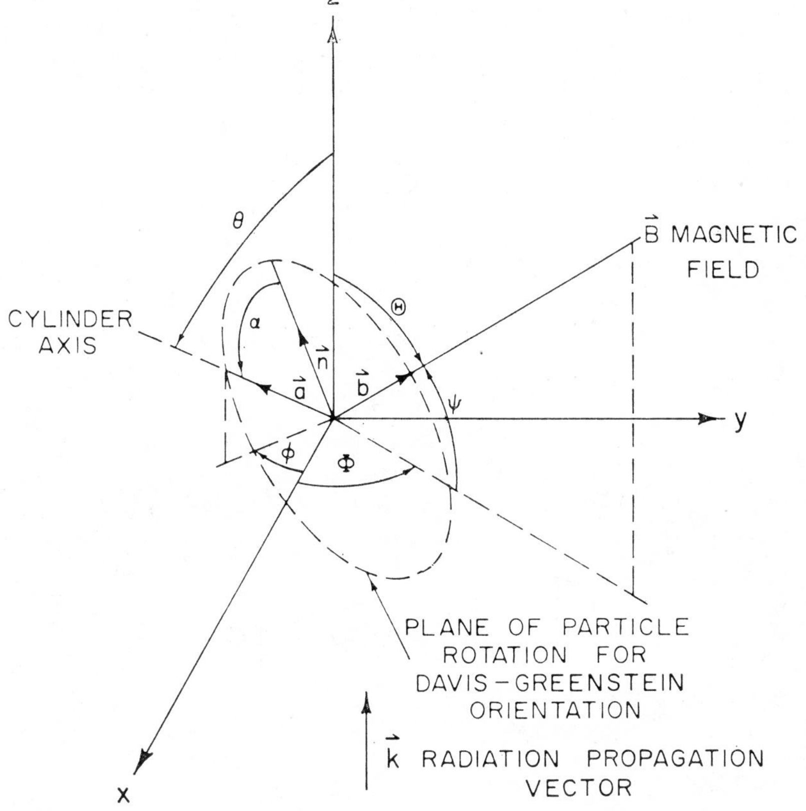

FIG. 80.—Geometrical configuration for axially symmetrical, elongated particle spinning about its short axis which is aligned with the magnetic field direction. Radiation is propagated along the z-direction.

It is perhaps instructive to denote "perfect alignment" by "perfect *static* alignment"; i.e., all of the particles are fixed in space. In this case "perfect Davis-Greenstein alignment" may be distinguished as "perfect *spinning* alignment."

The maximum value of the integrand in equation (92) is, for symmetry reasons, obviously obtained for $\Phi = \pi/2$ (**B** in the *yz*-plane in Fig. 80). Under more general conditions, such as would hold when the radiation passes through more than one cloud in each of which the orientation is different, it is convenient to define for each cloud the quantities

$$C = \frac{1}{\pi}\int_0^{\pi} [C_E(\theta) - C_H(\theta)] \cos 2\varphi d\alpha \tag{93}$$

$$S = \frac{1}{\pi}\int_0^{\pi} [C_E(\theta) - C_H(\theta)] \sin 2\varphi d\alpha \tag{94}$$

in terms of which the polarization would be given by

$$(1.086)^{-1}(\Delta m_p)_{\text{D.G.}} = (C^2 + S^2)^{1/2} . \tag{95}$$

For very small particles satisfying the condition for validity of the Rayleigh approximation, we may readily obtain the various polarization and extinction measures analytically. For an axially symmetric particle, let us define $C_{\|}$ and $C_{\perp}$ as the extinction cross-sections when the electric vector is parallel to the symmetry axis and perpendicular to the symmetry axis, respectively. For spheroids, the formulae for $C_{\|}$ and $C_{\perp}$ may readily be obtained (see eq. [71] and van de Hulst 1957).

For arbitrary orientation we find that

$$\begin{aligned} C_E(\theta) &= C_{\|} \sin^2\theta + C_{\perp} \cos^2\theta , \\ C_H(\theta) &= C_{\perp} \end{aligned} \tag{96}$$

and

$$\begin{aligned} C_E + C_H &= (C_{\|} + C_{\perp}) + (C_{\perp} - C_{\|}) \cos^2\theta \\ C_E - C_H &= (C_{\|} - C_{\perp}) \sin^2\theta . \end{aligned} \tag{97}$$

For very small spheroidal particles satisfying the Rayleigh approximation, we may readily obtain the various polarization and extinction measures analytically (eqs. [69] and [70]). Substituting equation (97) into equations (89) to (92), performing the required integration, and denoting the appropriate quantities by the superscript R (for Rayleigh), we have

Perfect alignment

$$\frac{2}{1.086}(\Delta m)^{\text{R}}_{\text{P.F.}} = C_{\|} + C_{\perp} \tag{98}$$

$$\frac{1}{1.086}(\Delta m_p)^{\text{R}}_{\text{P.F.}} = C_{\|} - C_{\perp} \tag{99}$$

$$\left(\frac{\Delta m_p}{\Delta m}\right)^{\text{R}}_{\text{P.F.}} = 2\,\frac{C_{\|} - C_{\perp}}{C_{\|} + C_{\perp}} \tag{100}$$

Perfect Davis-Greenstein alignment

$$\frac{2}{1.086}(\Delta m)^{\mathrm{R}}_{\mathrm{D.G.}} = C_{\parallel}\left(1-\frac{\cos^2\psi}{2}\right)+C_{\perp}\left(1+\frac{\cos^2\psi}{2}\right) \quad (101)$$

$$\frac{1}{1.086}(\Delta m_p)^{\mathrm{R}}_{\mathrm{D.G.}} = (C_{\parallel}-C_{\perp})\frac{\cos^2\psi}{2} \quad (102)$$

$$\left(\frac{\Delta m_p}{\Delta m}\right)^{\mathrm{R}}_{\mathrm{D.G}} = 2\,\frac{(C_{\parallel}-C_{\perp})\frac{\cos^2\psi}{2}}{C_{\parallel}\left(1-\frac{\cos^2\psi}{2}\right)+C_{\perp}\left(1+\frac{\cos^2\psi}{2}\right)}. \quad (103)$$

The analogous results for infinite cylinders whose index of refraction is $m = 1.33$ are contained in Figures 81 and 82. The upper curve in Figure 82 is the wavelength dependence of polarization if the cylinders are in picket fence alignment and is obtained by taking the difference between the extinction efficiencies Q_E and Q_H as given in Figure 40.

By comparing the results of Figures 81 and 82 with the observed extinction and

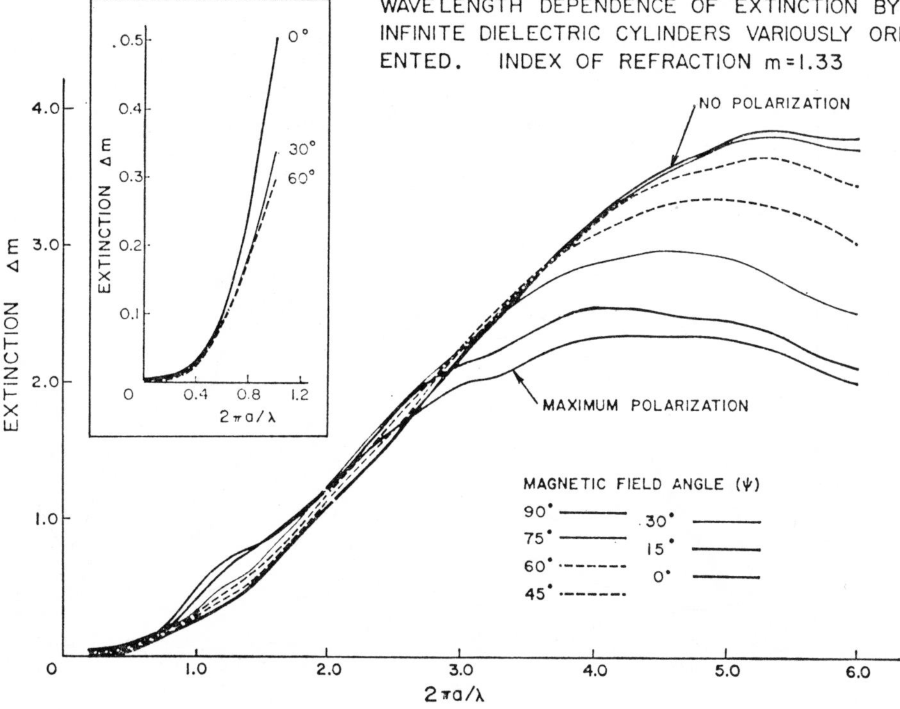

Fig. 81.—Variation with $ka = 2\pi a/\lambda$ of extinction produced by infinite dielectric cylinders spinning in planes which make various angles ψ with respect to the direction of incident radiation. The angle $\psi = 0°$ corresponds to the case in which the spinning plane contains the propagation vector and thus gives rise to the maximum degree of polarization. The vertical scale for extinction is chosen to be that given directly by evaluating the integral

$$\frac{1}{\pi}\int_0^{\pi}[Q_E(\theta)+Q_H(\theta)]\,d\alpha$$

(see equation [91]).

polarization curves, we see that, even for a single size particle (all cylinders of the same radius), there is a fair resemblance between theory and observation. The scale to be chosen corresponds roughly to an association of an inverse wavelength, $\lambda^{-1} = 1$, with a value of ka in the range $1.1 \leq ka \leq 1.25$. Noting that the calculations are for a pure real index of refraction as well as for a single size of particle, we should not yet make more than qualitative inferences, particularly in the infrared and ultraviolet where dielectric particles may be expected to be absorbing. See Greenberg and Lind (1967), who compare calculations for $a = 0.095\ \mu$, $0.127\ \mu$, $0.19\ \mu$, $0.254\ \mu$ (Fig. 81) with the observed extinction curve (Fig. 7) to $\lambda^{-1} \approx 6\ \mu^{-1}$.

As a preliminary illustration of the effect of an extended distribution of sizes, the wavelength dependence of polarization was calculated (Libelo 1962) for perfectly aligned cylinders (Oort–van de Hulst distribution for the radii) whose index of refraction is $m = 1.3$, and this result was compared with that for an appropriate single size (Fig. 83). The scale factor in $n_{\mathrm{OH}}(a)$ was chosen so that spheres with this distribution would give approximately the Whitford extinction curve. We note that the single size $a_0 = 0.2\ \mu$ gives $ka = 1.36$ at $\lambda^{-1} = 1$, which is about 10 per cent higher than that which

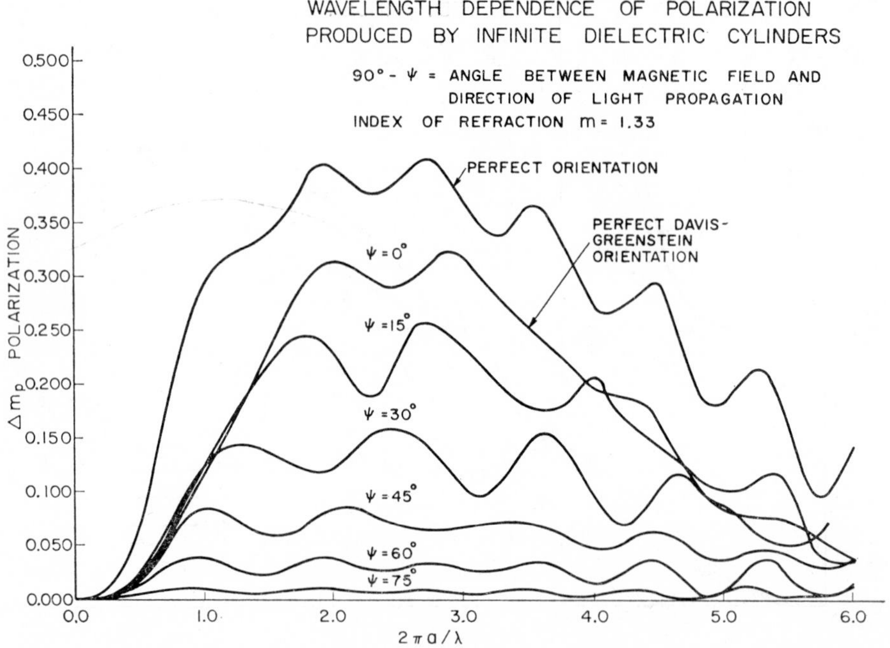

FIG. 82.—The uppermost curve is the variation of polarization with $ka = 2\pi a/\lambda$ for perfectly aligned, infinite dielectric cylinders ($Q_E - Q_H$). The other curves give the variation of the wavelength dependence of polarization produced by cylinders spinning in planes which make various angles ψ with respect to the direction of incident radiation. $\psi = 0°$ corresponds to the case in which the spinning plane contains the propagation vector. The vertical scale for polarization is chosen to be that given directly by evaluating the integral

$$\left\{\frac{1}{\pi}\int_0^{\pi}[Q_E(\theta) - Q_H(\theta)]\cos 2\varphi\, d\alpha\right\}_{\max} .$$

one needs for $m = 1.33$ [remembering, as for spheres, that it is the value of $\rho = (4\pi a/\lambda)(m - 1)$ which is significant].

The scattering properties of individual grains make it seem certain that the wavelength dependence of the polarization produced by dielectric grains should have a maximum somewhere between the far infrared and far ultraviolet.

We shall now explore a possible explanation (other than particle size) for variation of the wavelength dependence of polarization. We shall assume a distribution of particle sizes of the Oort–van de Hulst type, and we shall assume that the Davis-Greenstein mechanism is less effective for large particles than for small particles (and can therefore

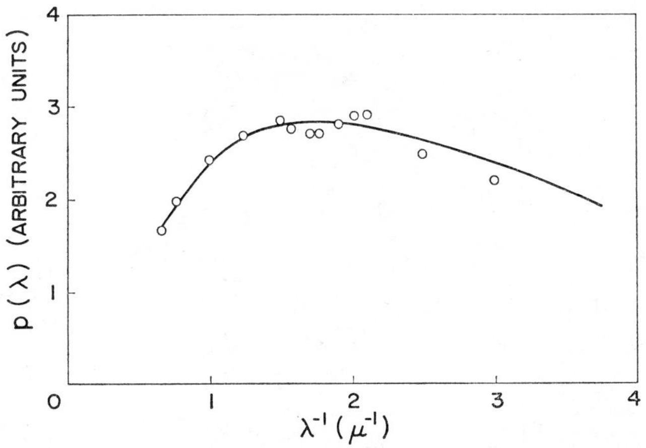

FIG. 83.—Theoretical wavelength dependence of polarization as calculated for perfectly aligned, infinite cylinders of index of refraction $m = 1.3$. The solid curve is calculated using the Oort–van de Hulst distribution of radii with $l = 0.4\ \mu$. This scale factor corresponds to an effective single particle size of $a_0 \approx 0.33$ for extinction. The circles refer to a single cylinder size, $a_0 = 0.20\ \mu$ (δ-function distribution). The effective single size representing the wavelength dependence is considerably smaller than that which is representative of the wavelength dependence of extinction.

be represented by a function, $f(a)$, with negative gradient). Let $p_\lambda(a)$ denote the polarization at the wavelength λ produced by a particle of size a. Then we may express the polarization by

$$\Delta m_p(\lambda) \sim \int_0^\infty n(a) f(a) p_\lambda(a) da . \qquad (104)$$

The form of the functions in the integrand is shown schematically in Figure 84. The shape of $p_{\lambda 2}(a)$ is given by $(\lambda_2/\lambda_1) p_{\lambda 1}(\lambda_1 a/\lambda_2)$.

If $f(a)$ decreases with increasing particle size (as for an insufficiently strong magnetic field), the polarization will tend to be less at longer wavelengths, so that the maximum can be expected to be shifted toward the shorter wavelengths. The relative contribution to the polarization by particles of various sizes (see Fig. 85) will not be the same, either in form or effectiveness, as the extinction contribution. Qualitative effects of changes in the size distribution can also be determined from the nature of the curves in Figure 84. Increasing the length parameter in the size distribution increases the polarization at the longer wavelengths, tending to shift the polarization maximum to the red. These results agree with those predicted by Davis (1959) on the basis of a somewhat different type of analysis. Similar methods may be used for other parameters, such as index of refraction, elongation, etc., to demonstrate the effect of variations on wavelength dependence or polarization.

In the range of ka (or λ^{-1}) corresponding to high polarization ($2 \lesssim ka \lesssim 3$ in Fig. 82), it is interesting to note that all extinction curves in Figure 36 are similar in magnitude

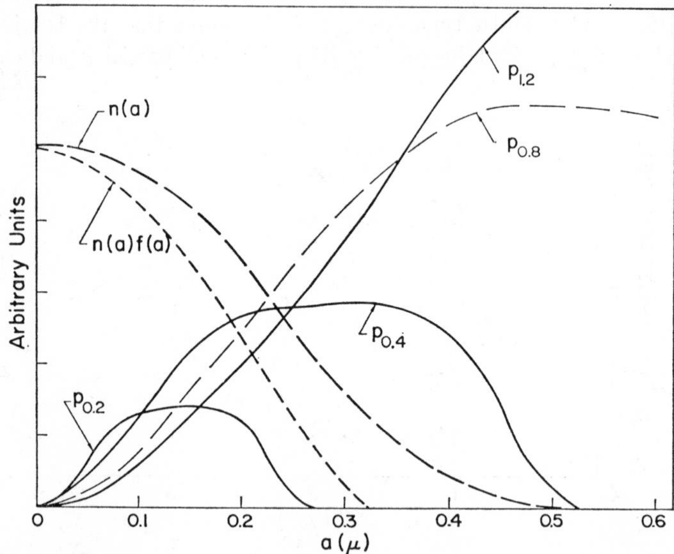

FIG. 84.—Schematic curves from which the effect of size distribution and degree of orientation on polarization may be obtained. The function $n(a)$ is the same as that of Figure 74. The curves for $p_\lambda(a)$ are obtained from the scattering curves for infinite cylinders whose index of refraction is $m = 1.3$.

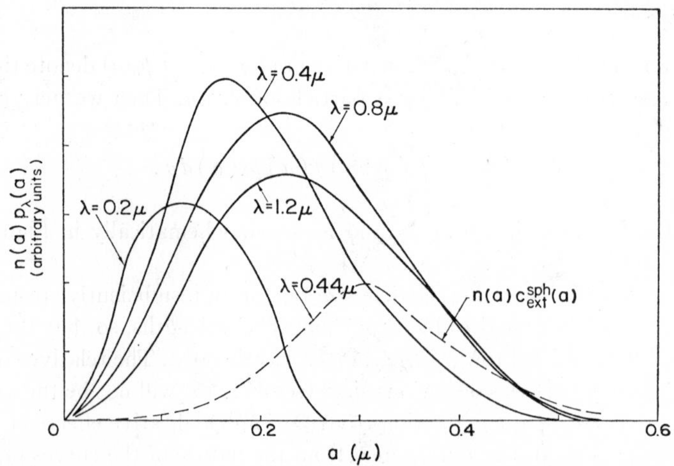

FIG. 85.—Relative contribution of particles of different sizes to the polarization and extinction. The dashed curve is taken from Figure 74.

and that, as a consequence, the ratio of polarization to extinction varies almost entirely according to the amount of polarization. The oscillations in the polarization curves are due to the fact that we are considering only one size of particle. These oscillations will almost completely disappear when a distribution of sizes is considered.

According to Figures 81 and 82, the obvious consequences of decreasing the angle between the magnetic field and the line of sight (increasing ψ) are (1) reduction of the amount of polarization, (2) broadening of the wavelength dependence of polarization, and (3) increasing extinction in the ultraviolet relative to the visible (assuming a reasonable scale of λ^{-1} relative to ka). Item (3) has already been predicted on the basis of various approximations (Greenberg and Meltzer 1960; Wilson 1960).

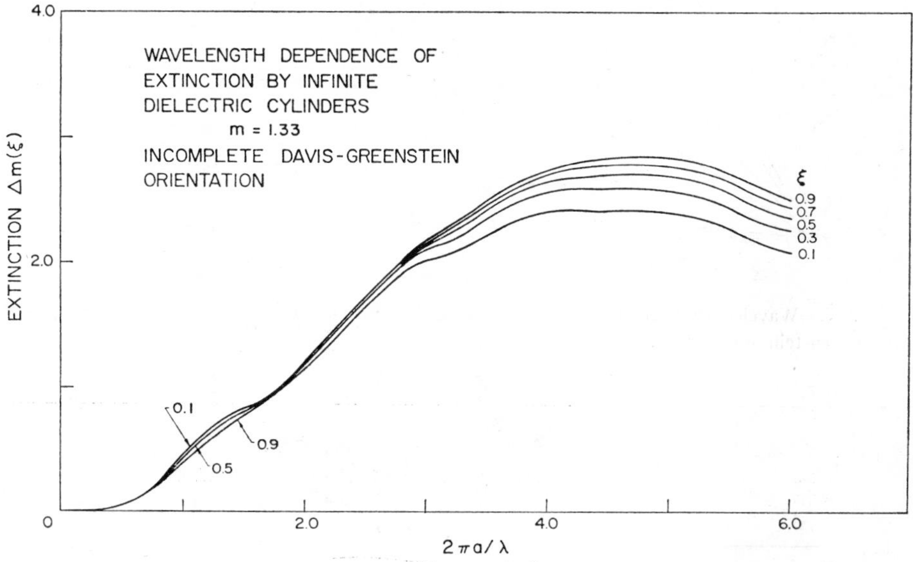

FIG. 86.—Wavelength dependence of extinction for incomplete orientation

Even more realistic results may be obtained by considering the distribution function of the angular momentum as given in equation (61) and plotted in Figure 31 and substituting this into equation (82). These are shown in Figures 86–89 where an indication is given of the functional relationship between the degree of polarization and the strength of the magnetic field.

We define the orientation parameter ξ by

$$\xi^2 = \frac{T_{\text{av}}}{T} = \frac{1+\delta\,(T_g/T)}{1+\delta} \tag{105}$$

(see eqs. [59] and [61]). Using τ_0 from equation (62) and τ_c from equation (65), we can show that

$$\delta = \frac{\chi'' B^2}{2a\omega N_{\text{H}}}\left(\frac{2\pi}{m_{\text{H}} kT}\right)^{1/2}. \tag{106}$$

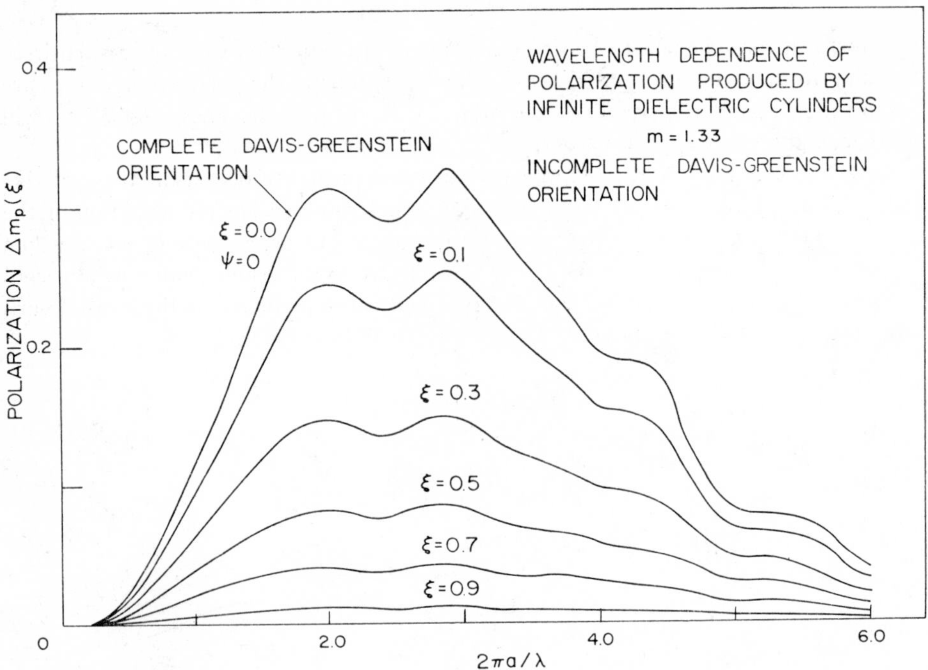

Fig. 87.—Wavelength dependence of polarization for incomplete ($\xi > 0$) and complete ($\xi = 0$) Davis-Greenstein orientations.

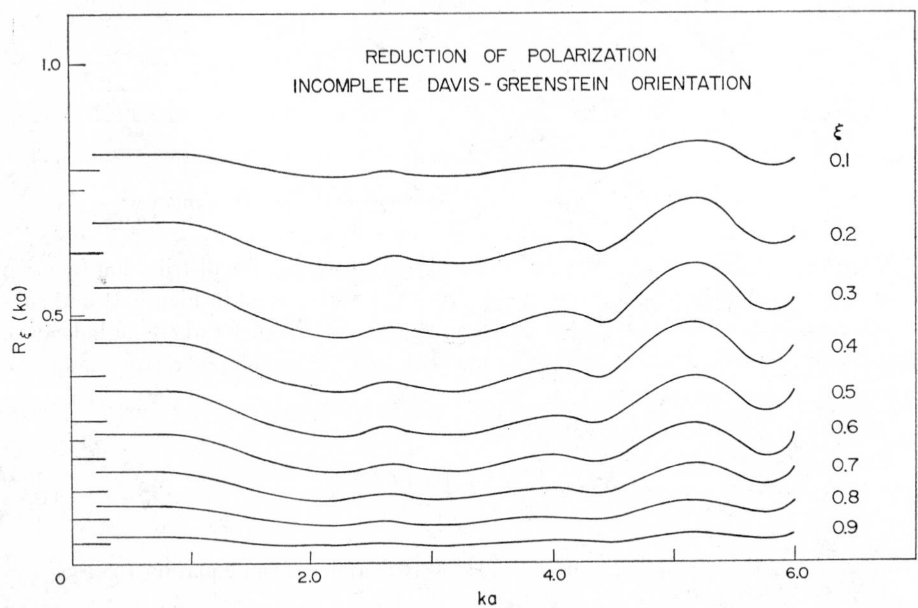

Fig. 88.—Reduction of degree of polarization as a function of cylinder size (ka) for different degrees of disorientation (increasing ξ). Marks on the ordinate correspond to reduction of degree of polarization for incompletely oriented small spheroids satisfying conditions for the Rayleigh approximation.

Using $\chi'' = 2.5 \times 10^{-12}\, \omega/T_g$, $T_g = 10°$ K, $T = 100°$ K, $n_{\rm H} = 10$, $a = 0.2\ \mu$, we obtain $\delta \approx 10^{10} B^2$. Thus $\delta = 1$ when $B = 10^{-5}$ gauss; and when $T_g/T \ll 1$, we can write for equation (105)

$$\xi^2 \approx \frac{1}{1+B^2} \tag{107}$$

where B is in units of 10^{-5} gauss. Equation (107), while not general, is nevertheless a convenient and useful simplification.

We see in Figure 87 how the wavelength dependence and the degree of polarization are modified downward by decreasing the degree of orientation from perfect Davis-Greenstein alignment.

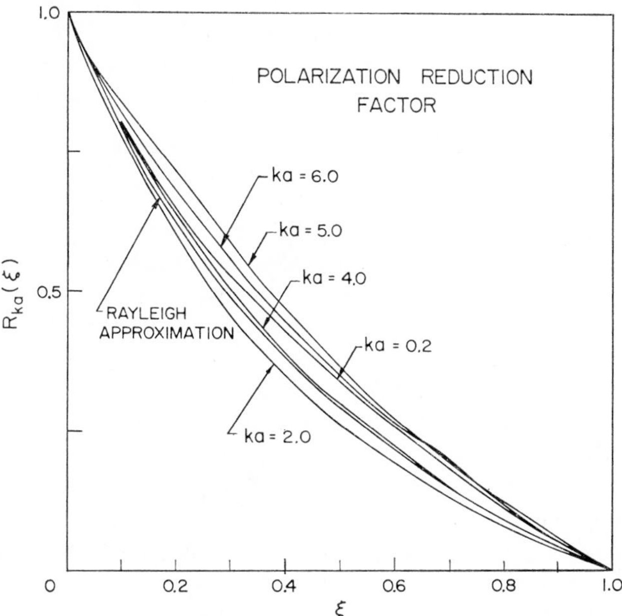

FIG. 89.—Reduction of polarization for several cylinder sizes and for the small spheroids as a function of the orientation parameter ξ.

To visualize the interstellar grains as they appear when incompletely oriented, we may consider only elongated particles and limit ourselves to the case in which the magnetic field is perpendicular to the line of sight. In this configuration, perfect Davis-Greenstein alignment implies that the particle spins in a plane whose normal is along the magnetic field direction. Averaged over one rotation, the spinning grain looks like a flat circular plate seen edge-on. Imperfect Davis-Greenstein alignment implies that the plate wobbles about its axis. The more the plate wobbles, the larger is the cone swept out by its axis, and the time-average picture of the particle becomes thicker and thicker. For completely random motion, the time-average of the spinning particle is exactly spherical and the polarization becomes exactly zero.

Let us define this reduction ratio of polarization for imperfect to perfect Davis-

Greenstein orientation by the factor R. In Figure 88 we have plotted the reduction factor as a function of ka for different degrees of orientation defined by different values of ξ. $\xi = 0$ means perfect D.G. orientation, and $\xi = 1$ means spherical angular distribution of particle axes. We see that R is fairly constant up to rather large values of ka. An equally interesting and important point is that the values of R obtained in Rayleigh approximation for spheroids are very close to those for the cylinders. These values are shown on the ordinate of Figure 88. In Figure 89 the dependence of R on ξ is plotted for the Rayleigh approximation and for several selected values of ka. Thus although the Rayleigh approximation may neither be used to infer the absolute magnitude of the polarization (or polarization relative to extinction) nor to find the wavelength dependence of polarization, it may be used to infer the dependence of polarization on degree of orientation. Merely by combining the basic orientation dependence of extinction on particle orientation as given in equation (82) with the distribution of particle axes given in equation (61), we may show that the ratio of polarization for imperfect orientation to polarization at perfect Davis-Greenstein orientation is given by the reduction factor

$$R^2 = \frac{3}{2(1-\xi^2)} - \frac{1}{2} - \frac{3\xi}{2} \frac{\sin^{-1}\sqrt{(1-\xi^2)}}{(1-\xi^2)^{3/2}}. \qquad (108)$$

This, of course, applies to any kind of grain, dielectric or otherwise, as long as it is spheroidal. In Table 21 we show explicitly the dependence of the ratio of polarization to extinction on the magnetic field strength (using the approximate relation [107]). We see that if the field is perpendicular to the line of sight and uniform, we need a value of $\xi = 0.5$ and $B = 1.7 \times 10^{-5}$ gauss to get the maximum observed ratio of polarization to extinction, and that to get the polarization to extinction ratio of $\gtrsim 0.025$ characteristic of regions of low dispersion of position angle, such as in Figure 9 (Behr 1965), we require a magnetic field $B \gtrless 1 \times 10^{-5}$ gauss. This result agrees with that previously given by various authors, but it is now on a much firmer theoretical foundation.

It is well to note that to a good approximation it is possible to calculate further the reduction factor introduced by tangling the magnetic field. Let the field $\boldsymbol{B} = (B_x, B_y, B_z)$ be characterized by a spheroidal distribution given by a Gaussian probability distribution for B_x, B_y, B_z,

$$F(\boldsymbol{B}) = \exp -\left(\frac{B_x^2 + B_z^2}{a^2} + \frac{B_y^2}{c^2}\right)$$

where $c > a$ means the field is predominantly directed along the y-axis (see Fig. 80). We may define $X = a/c = \langle B_y \rangle / \langle B_x \rangle$. Then the field-tangling reduction factor is defined as $R(X)$, where R is the same function as given in equation (108). The total reduction is given by the product $R(\xi)R(X)$.

Again using single size particles but taking the wavelength dependence of the refractive index of ice, we find the polarization and extinction by cylinders of radius $0.10\ \mu$ (Figures 90 and 91) and by cylinders of radius $0.25\ \mu$ (Figures 92 and 93) for varying magnetic field strengths.

We see in Figures 94 and 95 how a distribution of sizes—in both of these cases $n = e^{-5(a/0.4)^3}$—smooths out the wavelength dependence of polarization. We define the "constant elongation" case for infinite cylinders by first making the assumption that a finite but sufficiently elongated particle acts per unit length like an infinite cylinder. If a particle of radius a has a length $2b$ where $e = b/a =$ constant for all a, the extinction cross-section is given by $c = 4baQ_{\text{cyl}} = 4ea^2Q_{\text{cyl}}$, where the cylinder extinction efficiency Q_{cyl} is, as usual, defined as the cross-section per unit length divided by the diameter $2a$. The "constant length" case is obtained by assuming that all particles are so long that the

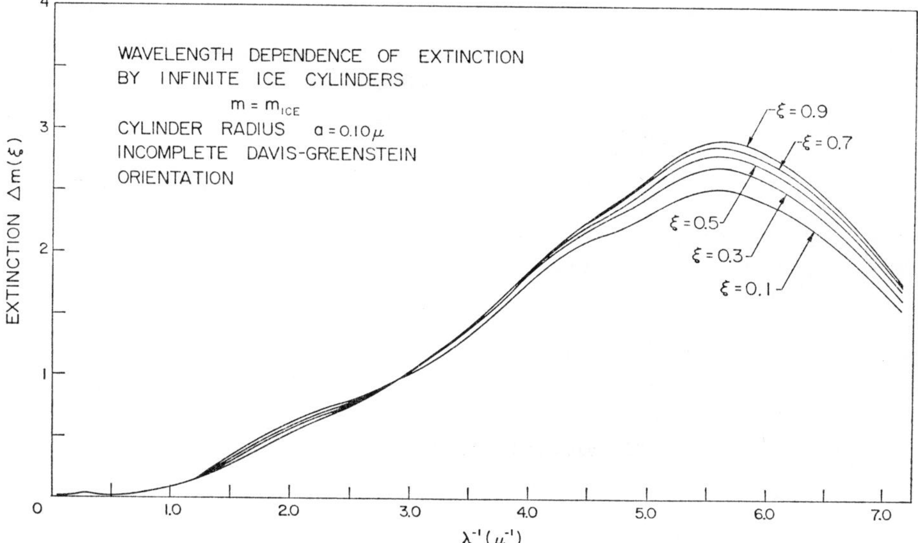

Fig. 90.—Extinction by ice cylinders of 0.10 μ radius for varying degrees of orientation. Complex index of refraction is as given in Figure 17.

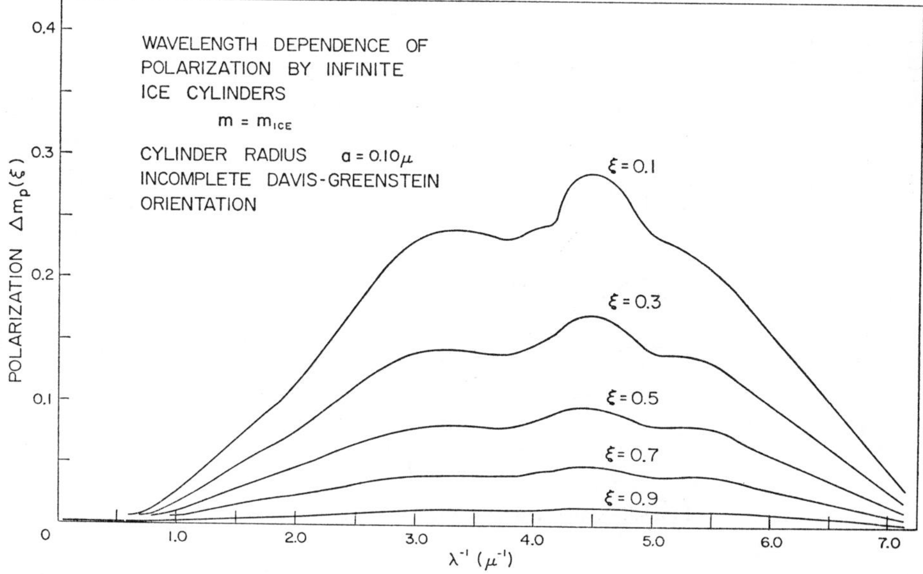

Fig. 91.—Polarization by ice cylinders of 0.10 μ radius for varying degrees of orientation; m(ice) is as given in Figure 17.

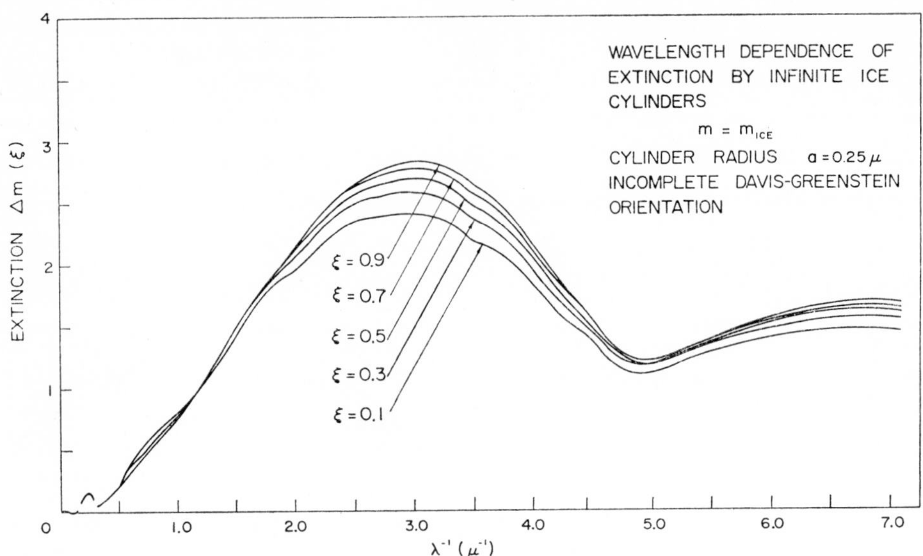

Fig. 92.—Same as Figure 90, but radius $a = 0.25\ \mu$

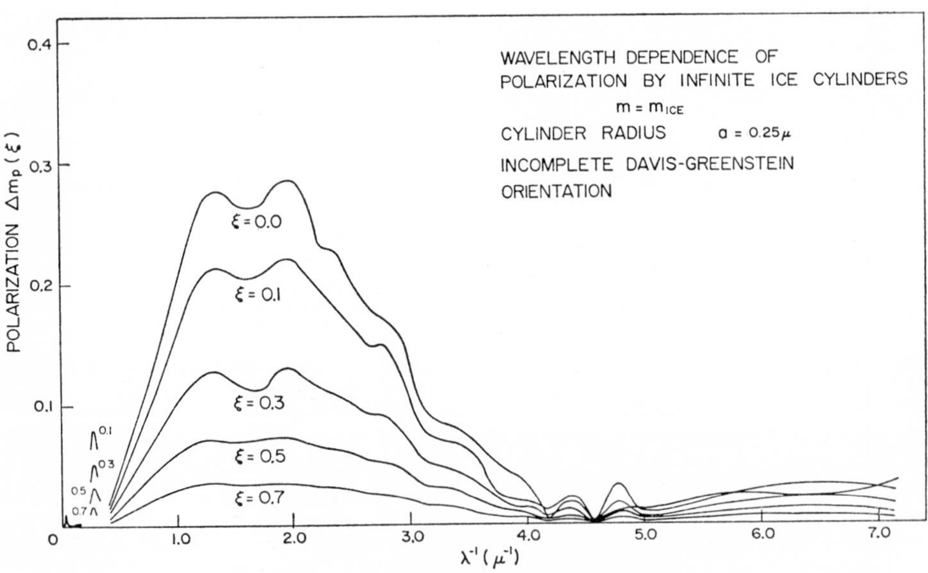

Fig. 93.—Same as Figure 91, but radius $a = 0.25\ \mu$

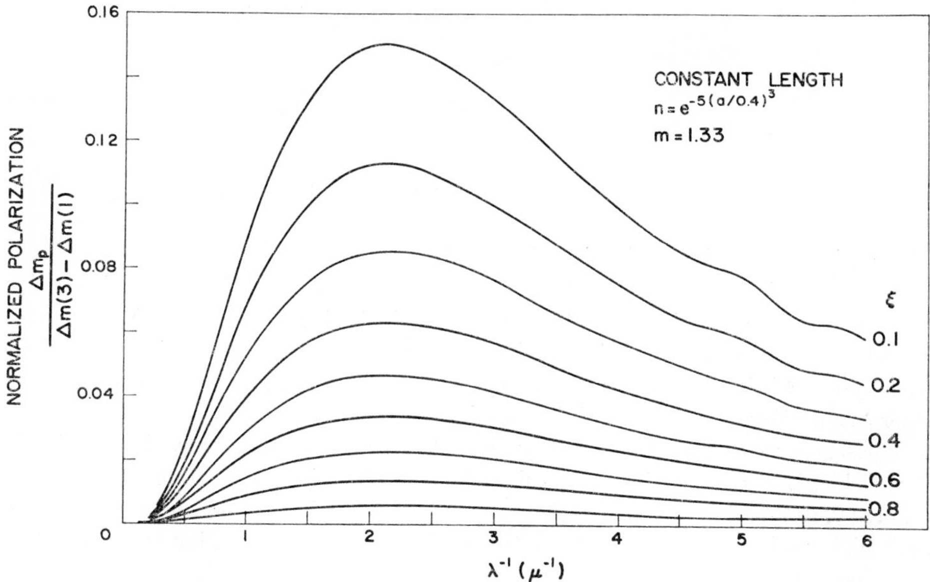

Fig. 94.—Polarization by a size distribution of cylinders of constant length for varying degrees of orientation.

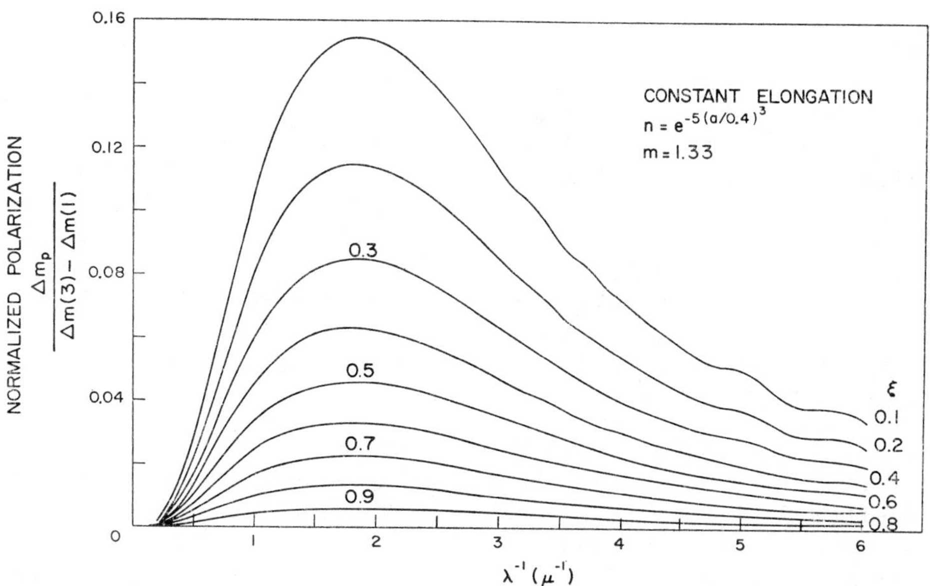

Fig. 95.—Polarization by a size distribution of cylinders of constant elongation for varying degrees of orientation.

elongation is always $\gg 1$, even for the largest diameter particles in the size distribution; i.e., $c \sim aQ_{\text{cyl}}$, as for the infinite cylinder. Thus (from eq. [13]) the two cases are given by

$$\text{constant elongation} \quad \tau_{x,y} = \int_0^{\pi} \int_0^{2\pi} d\varphi \sin\theta \, d\theta \, d\varphi f_{\xi}(\theta, \varphi) a^2 n(a) Q_{x,y}(\theta, \varphi) \tag{109}$$

$$\text{constant length} \quad \tau_{x,y} = \int_0^{\pi} \int_0^{2\pi} d\varphi \sin\theta \, d\theta \, d\varphi f_{\xi}(\theta, \varphi) a n(a) Q_{x,y}(\theta, \varphi) \tag{110}$$

where f_{ξ} is the axis distribution function for the orientation parameter ξ.

The results in Figures 94 and 95 are normalized and give an immediate indication that the ratio $\Delta m_p / \Delta m$ is similar to that for the single size model shown in Figure 87.

In Figures 96 and 97 we consider variation of sizes, but the orientation is of the picket-fence type. This is done for computational convenience as well as for comparisons with older calculations, particularly of the polarization.

The size parameter $a_i \approx 0.4$ perhaps most closely matches the observed extinction in the visible, but $a_i \approx 0.3$ is better suited for the ultraviolet extinction, as shown in Figure 99. Comparing the polarization curve for $a_i = 0.4$ in Figure 97 with that for the most highly oriented case ($\xi = 0.1$) in Figure 95, we see that spinning orientation has a tendency to narrow and peak the wavelength dependence relative to picket fence, but that as ordering decreases, the dependence again flattens. We shall refer to this again. The set of Figures 98 through 101 are computed polarization and extinction results for cylinders with a small absorption ($m = 1.33 - 0.05i$). In the constant elongation cases (Figures 99 and 101) are shown the observed extinction (Boggess and Borgman 1964) and the observed average polarization (Tables 19, 20). For both extinction and polarization, there is a fair (but not too good for the extinction in the visible) correspondence for rather similar particle sizes. The faster drop of the observed average polarization toward the ultraviolet and infrared relative to the best-fitting theoretical curve could be perhaps related to the phenomena already noted—that correctly oriented (spinning) particles tend to show the same effect relative to the picket-fence case treated here. In Figure 99, the observed point at $\lambda^{-1} \approx 1.6$ is clearly discrepant and perhaps the effective inverse wavelength should be shifted to the left because of bandwidth effects (see also Fig. 7).

Finally, we note that the microwave data for finite spheroidal particles, as shown in Figures 48 through 56, are fairly complete. With some theoretical extrapolation toward the Rayleigh region, we may begin to carry out—if warranted—calculations as detailed as those for infinite cylinders.

8.2. Metallic Grains

Güttler (1952) has shown that the Whitford curve can be reasonably well represented by a distribution of metallic spheres consisting of a mixture of two sizes. He uses an index of refraction similar to that for the curves of Figure 60 for the region $1 \leq \lambda^{-1} \leq 3$ and lets it approach infinity for very long wavelengths, thus insuring that the extinction curve has zero slope in this region. Reasonable values may be obtained for the ratio A/E. The work of Schalén (1939) and Güttler indicates that the dominant size of metallic particles is roughly one-tenth the radius of the dielectric particles; namely $a \approx 0.05\ \mu$. This value is obtained by letting $ka = 1$ in Figure 60 correspond to $\lambda = 0.3\ \mu$.

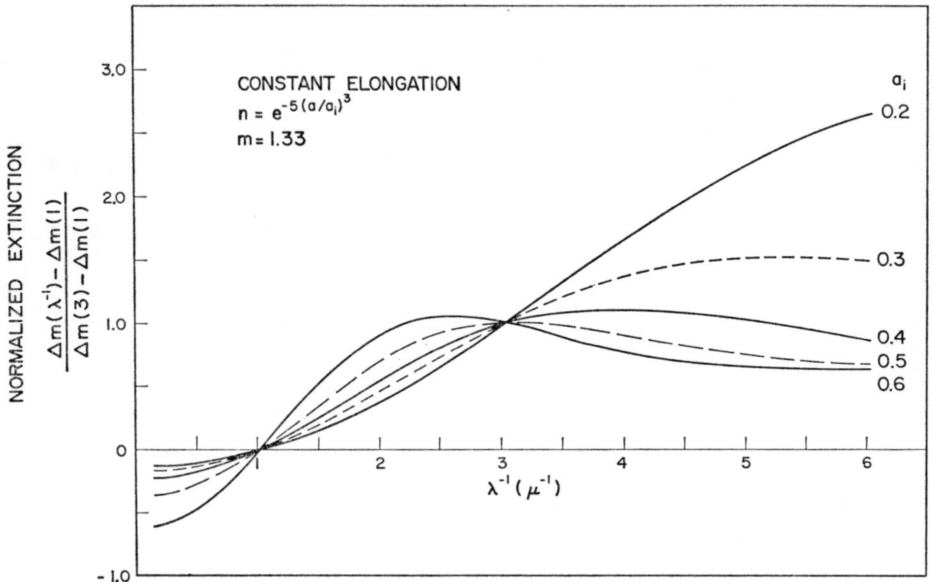

FIG. 96.—Extinction for several size distributions of perfectly oriented dielectric cylinders

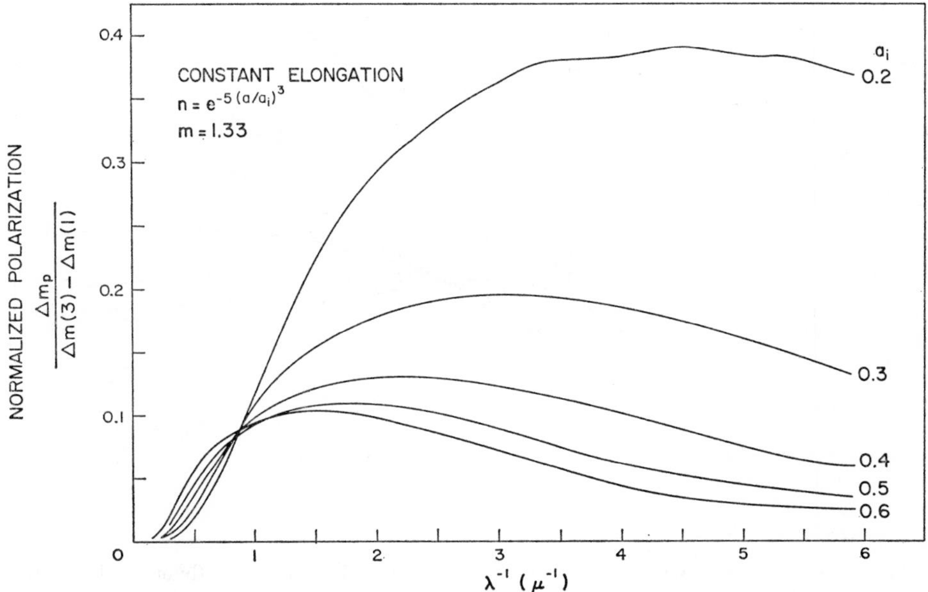

FIG. 97.—Polarization for several size distributions of perfectly oriented dielectric cylinders

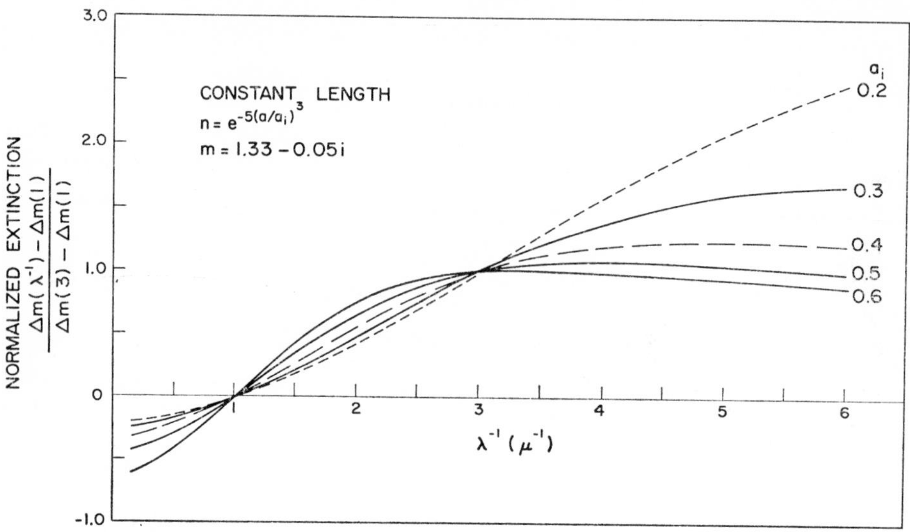

FIG. 98.—Extinction by several size distributions of slightly absorbing, perfectly oriented dielectric cylinders of the same length.

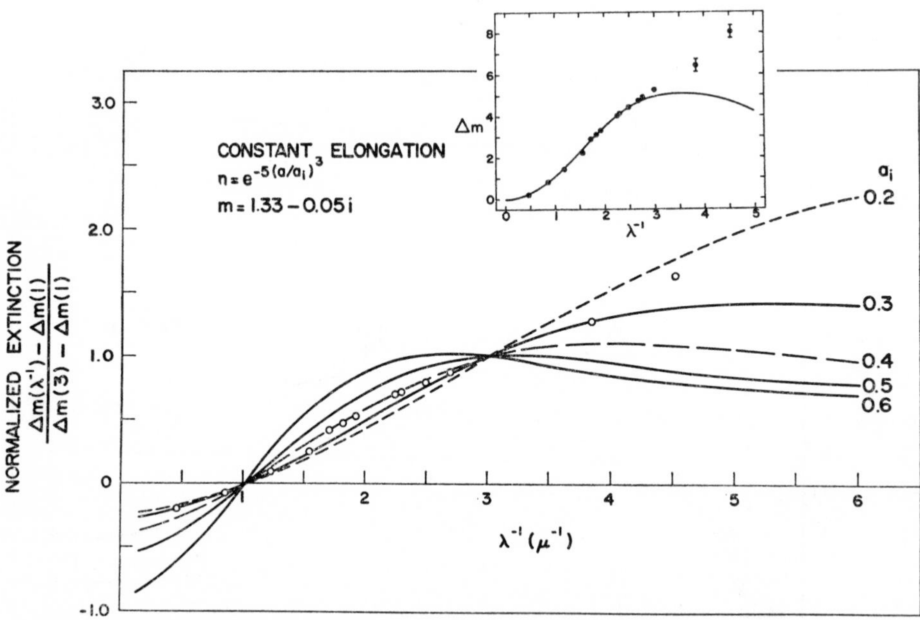

FIG. 99.—Extinction by several size distributions of slightly absorbing, perfectly oriented dielectric cylinders of constant elongation. Comparison is made with observations as given by Boggess and Borgman (open circles). Insert is comparison of the same observation with curve 15 of van de Hulst.

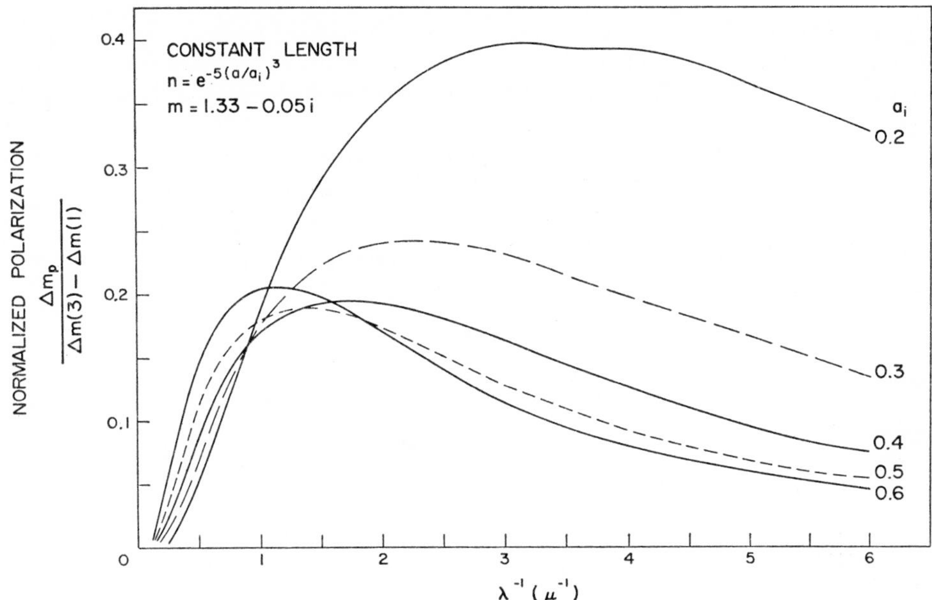

FIG. 100.—Polarization by several size distributions of slightly absorbing, perfectly oriented dielectric cylinders of constant length.

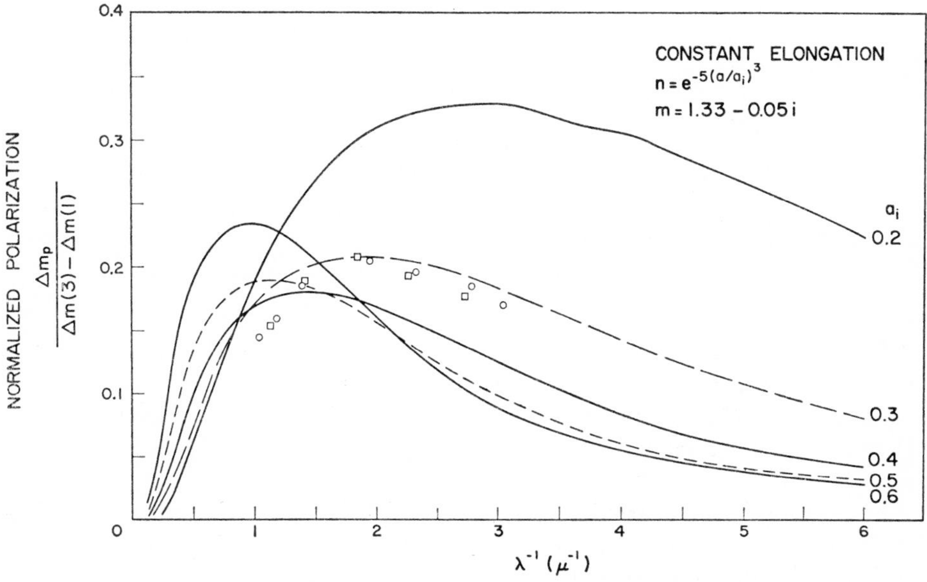

FIG. 101.—Polarization by several size distributions of slightly absorbing, perfectly oriented dielectric cylinders of constant elongation. Comparison is made with the averages of observations given in Figure 10. Circles from Coyne and Gehrels (1966); squares from Visvanathan (1966).

Equation (23), $\rho_g = \frac{4}{3} sa \, \Delta m/L$, can now be used to obtain the mass density of metallic grains. For iron (density 7.9 gm/cm^3), ρ_g (Fe) $= 2.0 \times 10^{-26}$ gm/cm^3. This calculated total metallic mass in interstellar space is inconsistent with the data of Table 1, by at least a factor of 10.

With regard to polarization, Zimmermann (1957) made the only systematic attempt to use the scattering properties of nonspherical metallic particles with ka not extremely small, but the scattering approximation he uses (Stevenson 1953) is not valid. Zimmermann also used the Spitzer-Tukey orientation mechanism rather than the Davis-Greenstein mechanism.

TABLE 19

AVERAGED WAVELENGTH DEPENDENCE OF POLARIZATION P_{C-G} FOR 30 SELECTED STARS FROM COYNE AND GEHRELS (1966)

λ^{-1}..............	1.05	1.19	1.39	1.95	2.33	2.79	3.04
P_{C-G}............	1.44	1.61	1.86	2.04	1.96	1.86	1.69
λ^{-1}..............	1.15	1.43	1.85	2.25	2.76		
P_V*............	1.52	1.87	2.07	1.93	1.76		

*P_V = Average of Visvanathan data in H I and H II (see Table 20).

TABLE 20

POLARIZATION IN H I AND H II REGIONS

FILTER	λ^{-1}	NO. OF STARS		WEIGHTED MEANS (P)	
		H I	H II		
UV......	2.76	11	14	0.879±.014	0.877±.012
B........	2.25	11	15	0.972 .005	0.963 .004
V........	1.85	11	15	1.031 .005	1.035 .003
R........	1.43	8	8	0.938 .019	0.930 .019
IR.......	1.15	7	8	0.807 .017	0.724 .026

From Visvanathan (1965).

The questions of the albedo and phase function are answered in Figure 60, where at $ka = 0.5$, $Q_{\text{sca}}/Q_{\text{ext}} \approx 0.1$ and the average value of the cosine of the scattering angle is $\langle \cos \theta \rangle$ is <0.1. These values are outside the limits required by Henyey and Greenstein. This argument is not very strong because the value of the observed albedo is open to question.

Thus, clouds of metallic particles may produce the observed extinction and polarization. They would not, however, produce the significant differences in the reddening law associated with different orientations as indicated by the variations of polarization values over the sky.

Fick (1954, 1955) has investigated the polarizing properties of grains consisting of Fe_2O_3. These are likely to be dielectric (with a large refractive index $m \approx 2$) rather than metallic. Fick considers orientation only by the Spitzer-Tukey mechanism. No useful calculations on wavelength dependence of extinction have been made for this type of grain although, of course, they are now possible, at least for cylinders in the correct size range.

8.3. Graphite Flakes

In this section we consider only graphite particles and graphite spheres devoid of dielectric mantles. Core-mantle particles are treated separately.

Cayrel and Schatzman (1954) concluded that a small number of oriented graphite flakes added to the main constituent consisting of dielectric particles (not assumed well oriented) could account for the observed degree of polarization. The number of graphite particles required to produce the polarization may have an appreciable effect on the wavelength dependence of the extinction. This effect, if it exists, would show up as a small increase in the value of A/E.

We present below a simplified picture of the Cayrel-Schatzman hypothesis. Some of the formulae used in the Cayrel-Schatzmann paper are valid only for extremely small spheroids, but the treatment given here does not depend on the sizes of the particles. We assume perfect orientation of the graphite flakes and postulate a scale for Figure 68 corresponding to metallic grains for $E \perp C$ and to dielectric grains for $E \parallel C$.

We assume that the extinction consists of two parts: (1) τ^D, due to non-oriented dielectric grains, and (2) $\tau^G_{x,y}$, due to oriented graphite grains. We may then write

$$\tau_x = \frac{\tau^D}{2} + \tau_x^G\,, \qquad \tau_y = \frac{\tau^D}{2} + \tau_y^G\,, \qquad \tau \approx \frac{\tau_x + \tau_y}{2}\,. \tag{111}$$

Let $\tau_y^G - \tau_x^G$ give the correct amount of polarization and be roughly constant in the range $1 \leq \lambda^{-1} \leq 3$. This means that

$$\frac{2(\tau_y^G - \tau_x^G)}{\tau^D + (\tau_x^G + \tau_y^G)} = 0.06\,. \tag{112}$$

According to Cayrel and Schatzman, $\tau_y^G/\tau_x^G \cong 3$. This gives $\tau_y^G\tau \approx 0.05$. If we assume that the contribution to the extinction by the graphite grains is essentially of a metallic character in the observed range (Q = constant), then the slope of the extinction curve is not modified, and only the extinction at $\lambda^{-1} = 0$ is affected. The net result is to change the ratio of the total to selective extinction expected on the basis of spherical dielectric grains alone by only about 5 per cent. We may estimate the density of the graphite grains as $\rho_g^G \approx 0.02 \times 10^{-26}$ gm/cm^3, well within the acceptable limit.

A semiquantitative estimate of the wavelength dependence of polarization by graphite flakes may be based on a Rayleigh approximation calculation.

It can be shown that for oblate particles whose scattering is given adequately by the Rayleigh approximation, the ratio of polarization to extinction is

$$\left(\frac{\Delta m_p}{\Delta m}\right)_{\rm calc} = \text{constant}\; \frac{C^{(a)}_{\rm ext}/C^{(c)}_{\rm ext} - 1}{2C^{(a)}_{\rm ext}/C^{(c)}_{\rm ext} + 1} \tag{113}$$

where the subscript "calc" is used to denote the theoretically calculated value and where the constant is dependent on, among other factors, the degree and kind of orientation of the particles.

We do not use the quantity $(\Delta m_p)_{\text{calc}}$ directly to estimate the wavelength dependence of polarization. A somewhat better estimate of this quantity is obtained by multiplying equation (113) by the observed wavelength dependence of extinction to give

$$\Delta m_p(\lambda) = \left(\frac{\Delta m_p}{\Delta m}\right)_{\text{calc}} (\Delta m)_{\text{obs}} . \tag{114}$$

We present the following table as a reasonable representation (see Wickramasinghe and Guillaume 1965) of the complex dielectric constants of graphite

$\lambda^{-1}(\mu^{-1})$	m_a^2	m_c^2
.9	$4-10i$	$4-10\beta i$
1.2→2.8	$4-\ 7i$	$4-\ 7\beta i$

where β is the ratio of conductivities along the c-direction and the a-direction and is expected to be less than unity. In order to span a range of possibilities, we will consider, in turn, $\beta = 10^{-2}$, 10^{-1}, and 1. According to our data, $\beta \approx 10^{-2}$ is quite likely.

The size of the graphite flake is taken to be such that $a = 0.05\ \mu$, which corresponds to the radius for spherical graphite particles that roughly give the extinction curve. Graphite flakes are quite flat, and we shall consider two values of the degree of oblateness: $a/c = 5$ and $a/c = \infty$ (disk). A summary of the numerical results is presented in Tables 21 and 22. We see from the tables that in equation (113) we may effectively replace $C_{\text{ext}}^{(a)}/C_{\text{ext}}^{(c)}$ by $C_{\text{abs}}^{(a)}/C_{\text{abs}}^{(c)}$. Furthermore, it can be seen that the value of $(\Delta m_p/\Delta m)_{\text{calc}}$ is a constant within 5 per cent over the entire visible spectrum. Consequently, to this approximation, the wavelength dependence of the polarization is just that of the extinction; i.e., it rises monotonically from the far infrared to the ultraviolet. This dependence

TABLE 21

CROSS-SECTIONS FOR OBLATE SPHEROIDS
$a/c = 5$, $L_c = .75$, $L_a = .125$

λ^{-1}/β	$C_{\text{abs}}^{(a)}/C_{\text{abs}}^{(c)}$: 10^{-2}	10^{-1}	1
0.9	305	32	19
1.2–2.8	396	41	14

λ^{-1}/β	$C_{\text{abs}}^{(c)}/C_{\text{sca}}^{(c)}$: 10^{-2}	10^{-1}	1
0.9	11.0	99	91
2.8	.3	2.8	4.5

λ^{-1}	$C_{\text{abs}}^{(a)}/C_{\text{sca}}^{(a)}$
.9	40
2.8	4.5

TABLE 22

CROSS-SECTIONS FOR FLAT DISKS
$a/c \to \infty$, $L_c = 1.0$, $L_a = 0$

λ^{-1}/β	$C_{\text{abs}}^{(a)}/C_{\text{abs}}^{(c)}$: 10^{-2}	10^{-1}	1
0.9	1600	170	116
1.2–2.8	1600	165	65

λ^{-1}/β	$C_{\text{abs}}^{(c)}/C_{\text{sca}}^{(c)}$: 10^{-2}	10^{-1}	1
0.9	1.25 a/c	12.5 a/c	17.2 a/c
2.8	.03 a/c	.32 a/c	8 a/c

λ^{-1}	$C_{\text{abs}}^{(a)}/C_{\text{sca}}^{(a)}$
.9	18 a/c
2.8	.9 a/c

of polarization on wavelength has also been predicted (see next section) for the Platt-type (Platt 1956) particles. However, it is generally observed that the polarization is rather flat, at least through the range $1.3 \leq \lambda^{-1} \leq 3$ with perhaps, on the average, a maximum in the green (see Table 19).

In Figure 102 are shown the observations of Behr (1963), a solid theoretical curve (Greenberg 1962; Greenberg, Lind, Wang, and Libelo 1963) which fits Gehrels' (1960) observations, and a dotted theoretical curve based on the result that the polarization by small graphite flakes follows the extinction curve. The solid theoretical curve is normalized to unity at $\lambda^{-1} = 2$ and is exceedingly close to 1 at $\lambda^{-1} = 1.5$, which we have for convenience chosen as the point of normalization for the dotted curve. It seems

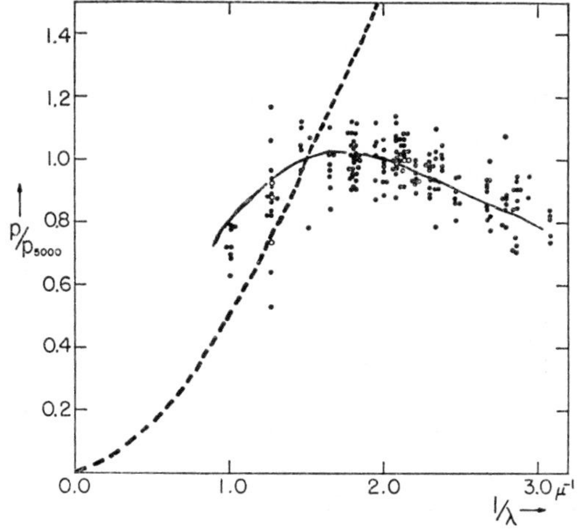

FIG. 102.—Polarization by small graphite flakes (broken curve). Solid curve is theoretical (see Fig. 9). Experimental points are those of A. Behr for stars whose polarization-to-extinction ratio is greater than 0.026.

that even where the Rayleigh approximation should be valid, the pure graphite model does not give a good representation of the polarization. The values of ka at $\lambda^{-1} =$ 0.9 and 2.8 are $ka_{0.9} = .28$ and $ka_{2.8} = .88$, and one would expect the approximation to be valid up to $ka \sim 0.5$.

All of the above calculations have assumed uncoated graphite flakes. If graphite flakes do indeed exist in interstellar space, it is to be expected that they would accrete dielectric (dirty ice) mantles. If the mantles are not too thick (relative to the core thickness), the grain would probably still preserve a reasonable degree of oblateness. Since the ratio $C_{\text{ext}}^{(a)}/C_{\text{ext}}^{(c)}$ would, at least for the longer wavelengths, still be determined primarily by core absorption (a Rayleigh approximation result), we should find that the wavelength dependence of polarization would still have the tendency to increase, although perhaps not so steeply. This statement is admittedly rather qualitative, but when taken with the quantitative results for pure graphite flakes, it leads one to conclude that, at the very least, the graphite flakes do not *readily* lend themselves to a theoretical explanation of the wavelength dependence of polarization.

Insofar as extinction is concerned, the work of Wickramasinghe and Guillaume (1965),

Greenberg (1966), Stecher and Donn (1966), and Nandy (1964) comparing extinction by small graphite spheres with observation affords proof that the extinction in a limited spectral region can be reasonably well matched by pure graphite particles. See Section 8.5 for further modifications produced by dielectric mantles on the graphite. An interesting point is the correlation between a specific refractive index effect of graphite and the slope discontinuity in the extinction curves, as determined by Nandy and noted originally by Whitford at $\lambda^{-1} - 2.2\ \mu^{-1}$ (see Fig. 103). In Figure 104, the extinction by small graphite spheres with an Oort–van de Hulst size distribution giving an e^{-1} size value of .056 μ is compared with observed extinction from the far infrared to the far ultraviolet. In the visible, the match is fair, but not good in the curvature sense.

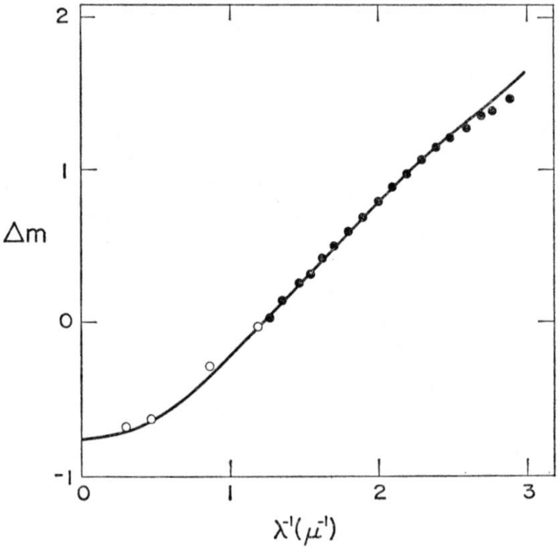

FIG. 103.—Theory (solid curve) and observation for stars in Perseus. Theory is obtained for pure graphite spheres. (From Nandy 1964*a*.)

The calculated and observed values of extinction up to $\lambda^{-1} = 4\ \mu^{-1}$ rise in a roughly similar way, but the calculated values drop subsequently and do not show the rise (this may still be open to question) for the observations to $\lambda^{-1} \approx 7\ \mu^{-1}$. Figure 105 shows qualitatively how various size graphite spherical particles must be included to give a match with the observed extinction in various wavelength regions. According to these curves, one needs some particles with $a > 0.11\ \mu$ in order to fit the far infrared point. The calculations shown in Figures 103 through 105 make use of the recent index of refraction measurements for graphite (see Table 11). Using the index of refraction $m^2 = 2 - 8i\lambda$ (as given in Hoyle and Wickramasinghe 1962), one obtains a rather poor fit in the ultraviolet with grains of .05-μ radius (see Fig. 107).

Further studies of graphite are shown in the section on core-mantle particles.

8.4. FREE RADICALS

With the simplified scattering curves of Figure 69 it is possible to derive some extinction and polarization properties of clouds consisting of Platt particles.

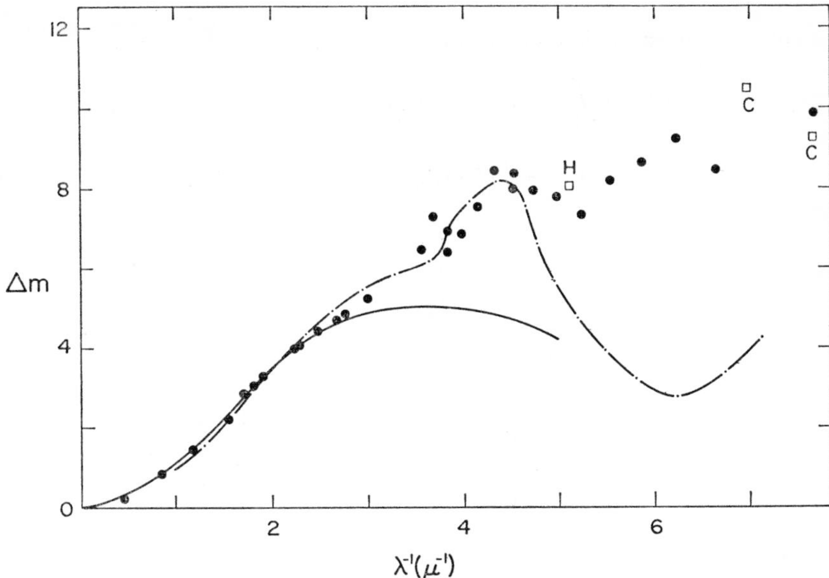

FIG. 104.—Comparison of van de Hulst curve 15 (solid line) and theoretical calculation for pure graphite spheres (broken line) with observations in the visible and ultraviolet (see Fig. 6). (From Stecher 1965.)

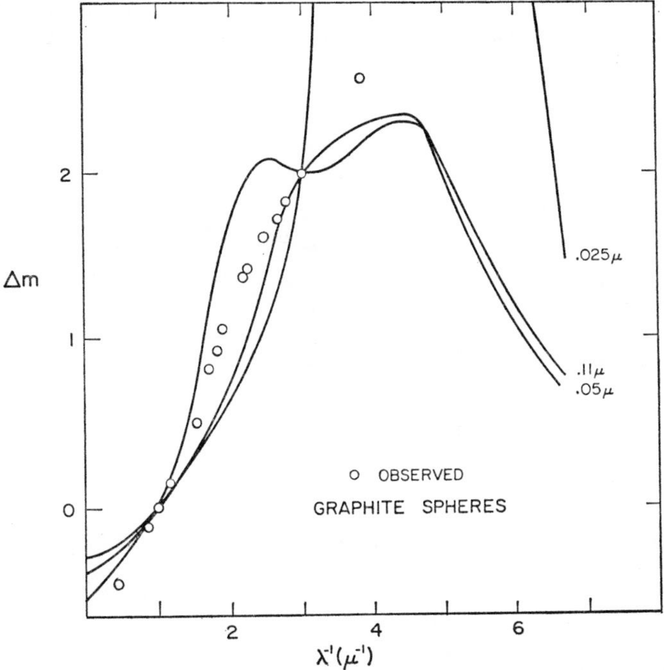

FIG. 105.—Comparison of observed extinction with extinction by three sizes of graphite spheres. (From Wickramasinghe 1963.)

First consider the extinction for spherical particles, then

$$\begin{aligned}\tau(\lambda) &= \int_0^\infty \sigma(a, \lambda) n(a) da \\ &= \int_{\lambda/200}^\infty \pi a^2 n(a) da = \int_{c\lambda}^\infty \pi a^2 n(a) da\,.\end{aligned} \tag{115}$$

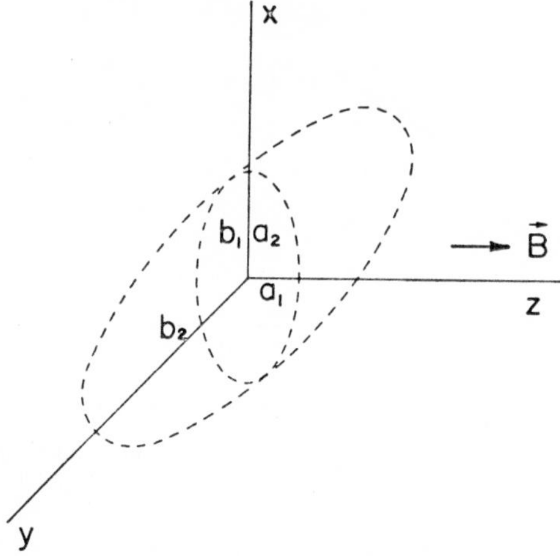

FIG. 106.—Schematic illustration of different sizes of particles which scatter radiation directed along the y-axis by resonance absorption.

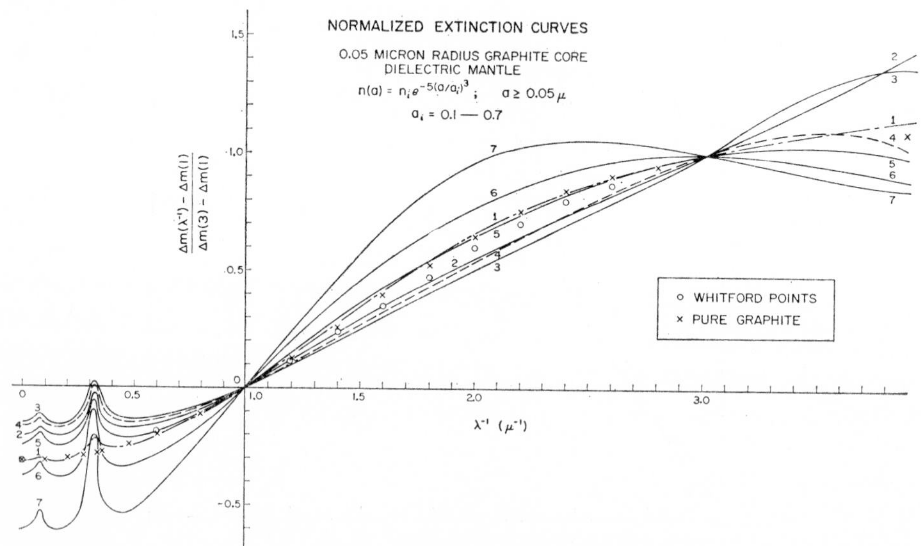

FIG. 107.—Theoretical extinction curves computed for spherical graphite ($m^2 = 2 - 8i\lambda$) of 0.05 μ radius and for graphite core plus dirty ice mantles of various distributions of outer radii. Comparison is made with the Whitford curve.

For $n(a) = a^{-(n+3)}$, we find that $\tau(\lambda) \approx \lambda^{-n}$. Thus if $\tau \approx \lambda^{-1}$, $n(a) \sim a^{-4}$. The extinction curve defines $n(a)$ through the above integral equation, and a reasonable conclusion appears to be that its shape would be similar to that of Figure 15.

With linear size resonance along the direction of polarization of the radiation if all of the particles are spheroidal with the same elongation ($e = 1 + \epsilon$, $\epsilon \ll 1$ [see Fig. 106]) and perfect spinning orientation (with the long axes of one-half of the particles oriented along each of the x, y axes), we obtain

$$\begin{aligned} \tau_y(\lambda) &= \int_{ec\lambda}^{\infty} \pi a b n(a) da + \int_{ec\lambda}^{\infty} \pi a^2 n(a) da \\ \tau_x(\lambda) &= \int_{c\lambda}^{\infty} \pi a b n(a) da + \int_{ec\lambda}^{\infty} \pi a^2 n(a) da . \end{aligned} \tag{116}$$

We find then that

$$\Delta m_p \sim \tau_y - \tau_x \approx \pi (c\lambda)^3 n(c\lambda) \epsilon = \epsilon \frac{\lambda^{-1} d\tau}{d(\lambda^{-1})} \tag{117}$$

and

$$\frac{\Delta m_p}{\Delta m} \cong 2 \frac{\tau_y - \tau_x}{\tau_y + \tau_x} \approx \frac{1}{2\lambda} \frac{d\tau/d(\lambda^{-1})}{\tau(\lambda)} \epsilon . \tag{118}$$

For constant ϵ, these equations imply that, in the region where $\Delta m \sim \lambda^{-1}$,

$$\frac{\Delta m_p}{\Delta m} \approx \tfrac{1}{2}\epsilon , \qquad \Delta m_p \approx \tfrac{1}{2}\epsilon\lambda^{-1} . \tag{119}$$

These results are similar to those obtained (Greenberg 1960*a*) with a different form for the extinction curve of scattering by the Platt particles. Furthermore, for ϵ = constant, it seems that the polarization would be at a maximum at some value of λ^{-1} greater than that for which the extinction curve has decreasing slope. In other words, the polarization maximum would not be in the visible, contrary to the observations. Actual scattering functions for small particles could differ considerably from those used in this section, but the calculated total mass density is not critically dependent on the scattering mechanism, and it is of the order of 10^{-2} less than that required for the dielectric grains.

Greenberg (1960*b*) has shown that variation of the extinction curve due to orientation can be expected to be negligible. The Platt particles would have at least as high an albedo as the dielectric grains but would scatter more or less uniformly in all directions.

8.5. Core-Mantle Particles

Extensive computations on extinction by core-mantle particles have been made by Greenberg and by Wickramasinghe. The original motivation for the work of Greenberg was an attempt to find a theoretical fit for the infrared extinction curves of Johnson. We see in Figures 62 and 63 that no reasonable size core can produce the bend in the extinction curve at the observed small value of λ^{-1}. As a matter of fact, we see that a metallic or graphite core of at least 1μ radius is needed to produce this knee (see Figs. 62, 63). For such a core, the extinction throughout the visible and ultraviolet parts of the spectrum would be essentially independent of wavelength (gray extinction). Limiting ourselves to attributing extinction to grains (of any type whatever), it seems that the only way to achieve the extinction in the infrared as given by Johnson *and* the more or less uniform wavelength dependence ($\sim\lambda^{-1}$) in the visible region is to attribute the extinction to *two* distributions of grains—one having very large sizes (1μ radius metallic or 5μ radius dielectric [$m = 1.33$]), and the other having the "typical" sizes (0.05 to 0.10μ radius metallic or graphite of 0.3 to 0.5μ radius dielectric). However, to fit a generalized nonexotic curve as given by Borgman and Johnson, Greenberg (1966) has obtained an excellent fit by an appropriate size distribution of mantles on a metallic core (Fig. 108).

In Figure 107 are shown the theoretical extinction curves to be expected for various distributions of ice mantle sizes on small graphite cores ($a_{\text{core}} = 0.5\ \mu$). In Figure 109,

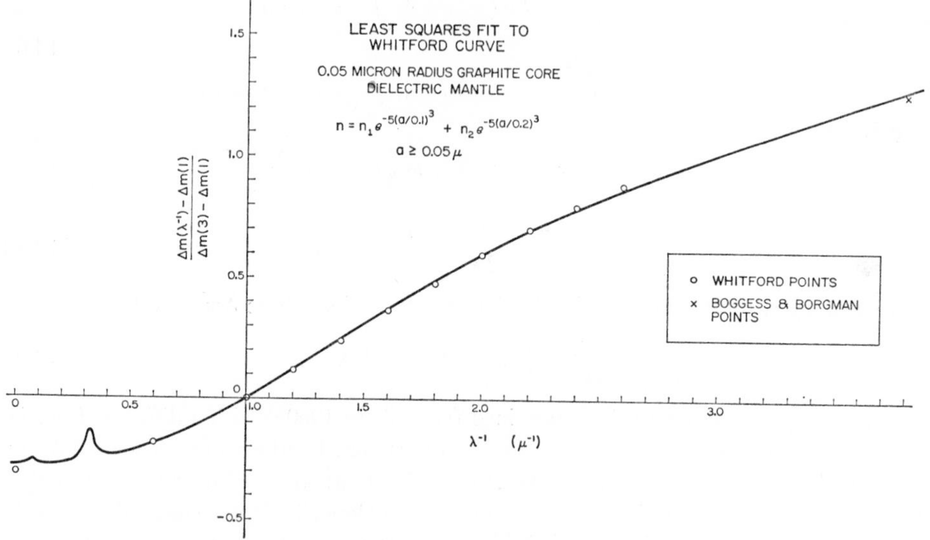

FIG. 108.—Theoretical extinction curve obtained from a least-squares fit of a two-parameter size-distribution function of spherical cores of graphite ($m^2 = 2 - 8i\lambda$) surrounded by mantles of dirty ice. The values of the parameters are $n_1 = 4{,}462$, $n_5 = 273.4$.

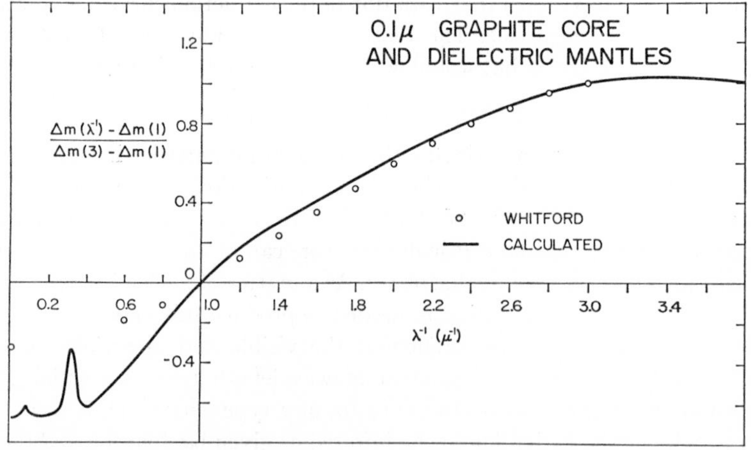

FIG. 109.—Theoretical extinction curve for 0.1 μ radius graphite ($m^2 = 2 - 8i\lambda$) surrounded by dirty ice mantles distributed according to $n = e^{-5(a/0.5)^3}$, $a > 0.1\ \mu$.

the core radius is increased to $a = 0.1\ \mu$. The curves in Figure 107 and those for pure ice are, as expected, almost identical for mantle radii of the order of $0.3\ \mu$. This is because as the mantle size is increased, the total extinction cross-section is dominated by the scattering by the mantle—the absorption by the core becoming negligible (see Figs. 70–73). In the ultraviolet, where ices (dielectrics) themselves are absorbing, the effect of the core is negligible even for small thickness of mantles.

Wickramasinghe has considered mantles as probably growing on graphite cores and has indicated the modification in the extinction curve given by Nandy in Cygnus may be met by appropriately varying the mantle thickness (Fig. 110).

The wavelength dependence of polarization by small graphite flakes surrounded by spherical and oblate spheroidal mantles has been investigated (Wickramasinghe, Donn,

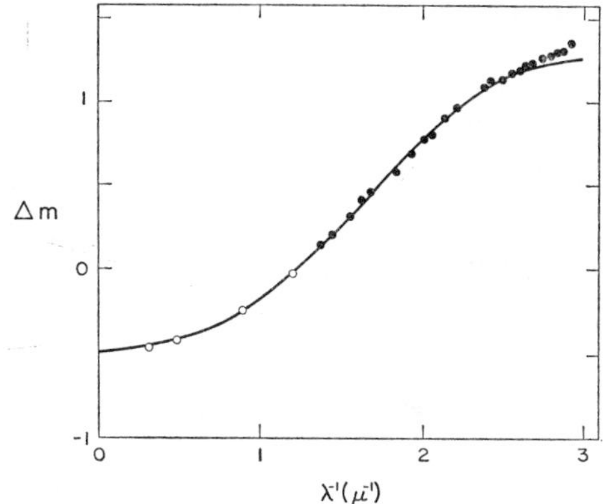

FIG. 110.—Theoretical fit of observations in Cygnus by core-mantle particles whose core radii are less than $0.06\ \mu$ and whose mantles are distributed according to $e^{-\alpha a}$ where $a > 0.06\ \mu$ and $\alpha = 33.3\ \mu^{-1}$. Open circles are from Johnson and Borgman (1963), and dots are observations of Nandy. (From Nandy and Wickramasinghe 1965.)

Stecher and Williams 1966; Swamy and Wickramasinghe 1967). The scattering theory required for such particles is not well developed, and at best we are limited to obtaining qualitative results.

Extinction curves calculated for spherical particles with core ($m^2 - 2 = 8i\lambda$) and mantle ($m = m_{\text{ice}}$) are shown in Figure 111. Compare extinction curves of Whitford (1958) (crosses) and Boggess and Borgman (1964) (open circles).

Following Krishna Swamy and Wickramasinghe (1967 private communication), we let the graphite core be of radius $0.056\ \mu$ and of thickness $0.01\ \mu$, and the mantle radius be $0.16\ \mu$. The value of the imaginary part of the susceptibility used may be too large.

The most important and perhaps amusing consequence (from a historical point of view) of the requirement of mantles on graphite flakes is that the degree of orientation required for such particles imposes the need for galactic magnetic field strengths of the same order as those needed for elongated dielectric particles. We apply the theory of Section 6. The orientation relaxation time is given by $\tau_0 \approx \omega I / \chi_T'' V_{\text{core}} B^2$ where I is the moment of inertia of the grain (essentially that of the mantle $I = \frac{8}{15}(\pi a^3 s)$, V_{core} = volume of core = $\pi\,(.056)^2(.01)\mu^2$, ω is the angular velocity given by $\frac{1}{2}I\omega^2 = (3/2)\,kT$. The collisional relaxation time is as given in equation (65) for spheres. Letting $a_m = 0.16$, $\chi_T'' = 12 \times 10^{-11}\ \omega/T_g$ (eq. [49]), $n_H = 10$, $T_g = 20°$ K, and $T = 100°$ K, we obtain $\tau_c/\tau_0 = \delta \approx 0.1B^2$ in units of Γ. Note that T_g is taken as less than that for pure graphite. Assuming the maximum (perfect D.G.) $\Delta m_p/\Delta m \approx 0.3$, we are allowed a minimum reduction factor $R = 0.2$. This implies (see Table 23) $\xi = 0.6$ and $\delta = 1.8$, and finally $B = \sqrt{18} \times 10^{-5}$ gauss $\approx 4 \times 10^{-5}$ gauss.

9. REFLECTION NEBULAE

We reproduce here some results of recent calculations dealing with reflection nebulae (Greenberg and Roark 1967).

The question we pose here is, Can the colors of reflection nebulae be used to determine clearly and precisely the chemical and physical nature of interstellar grains? The colors are those of the *UBV* system (Johnson and Morgan 1953).

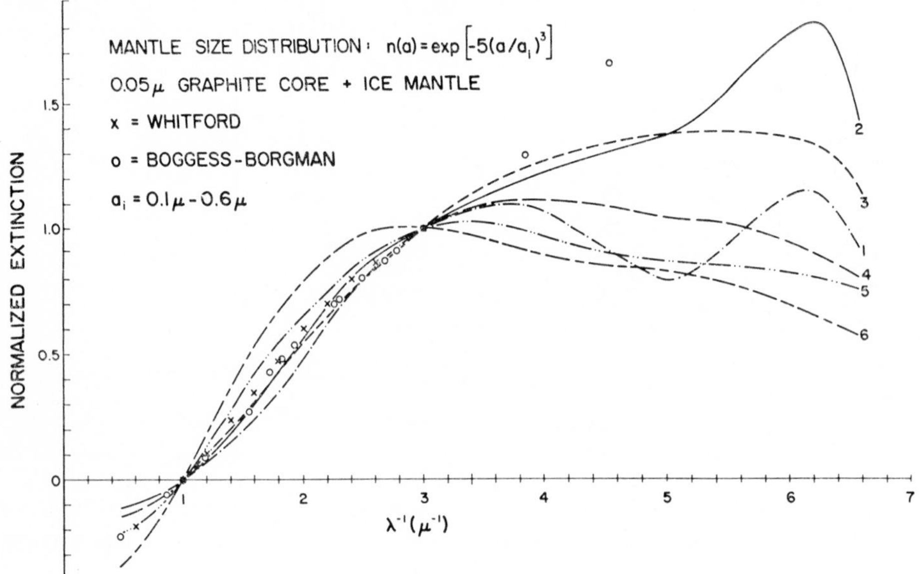

FIG. 111.—Extinction curves for spherical particles (see text)

TABLE 23

POLARIZATION REDUCTION FACTORS

ξ	$\Delta m_p/\Delta m$	R^R	$B(\Gamma)$	δ
0.1	0.15	.772	10	99
.2	....	.624	5	24
.3	0.09	.492	3.3	10.1
.4	....	.382	2.2	5.25
.5	0.06	.291	1.7	3.00
.6	....	.213	1.3	1.78
.7	0.025	.152	1.02	1.04
.8	....	.094	....	0.56
0.9	0.01	.047	....	0.23

The theoretical situation would be ideal if, between metallic and dielectric grains, one type of grain could be excluded on very general grounds; i.e., the colors predicted by that type of grain would be universally different from the colors observed for a variety of nebulae.

Although this is not generally possible, nevertheless, we can demonstrate rather significant and reasonably consistent differences between the colors from certain models of reflection nebulae.

We define the intensity of light scattered at wavelength λ from a single spherical particle of radius a and index of refraction m by the equation

$$I_R(\Theta, \lambda) = I_0(\lambda)\frac{F(\Theta, \lambda, a, m)}{k^2R^2} \tag{120}$$

where I_0 is the light intensity at wavelength λ at the particle, $k = 2\pi/\lambda$, R = distance from the particle to the point of observation, and $F(\Theta, \lambda, a, m)$ is the angular scattering factor defined in the notation of van de Hulst by

$$F(\Theta, \ldots) = \tfrac{1}{2}[i_1(\Theta) + i_2(\Theta)] \tag{121}$$

where Θ is the angle between the incident and the scattered radiation. The scattering functions $i_1(\Theta)$ and $i_2(\Theta)$ correspond to polarization of the incident radiation along and perpendicular to the plane of scattering. We assume here that the incident light is unpolarized, and we ignore polarization effects in the scattered radiation. The generalization to include polarization is straightforward and is certainly important, but we do not include it here.

The basic cloud model considered here is a plane-parallel slab of homogeneously distributed spherical scattering particles. Figure 112 shows the appropriate configuration if the star is situated between the front face of the nebula and the observer. Light from the star at S is scattered by particles at B in the direction of the observer. We assume that the total contribution to the nebular brightness is obtained by adding all such contributions along the line of sight. We only partially include the effect of multiple scattering by considering the attenuation of the starlight at B as a result of extinction by particles in that part of the path which passes through the nebula. In some of our calculations, the effect of this attenuation is implicitly evaluated by comparisons with results in which this attenuation is completely neglected.

Neglecting this internal attenuation, the scattered light contributed by the grains of size a in the range da in an element of volume ΔV at B is given by

$$dI'_N(\lambda) = I_S(\lambda)\frac{\gamma^2}{4r^2}\frac{F(\Theta, \lambda, a, m)}{k^2R^2}n(a)\,da\,dV \tag{122}$$

where $I_S(\lambda)$ = surface light intensity of the star, γ = stellar radius, r = distance from the star to the scattering element of volume dV, and $n(a)da$ = number of grains per unit volume of size a in da. If the extinction per unit length within the nebula is given by $\kappa(\lambda)$, then the starlight and the subsequent scattered light in equation (122) are each attenuated, and we incorporate this in equation (122) to give

$$dI_N(\lambda) = \exp[-\kappa(\lambda)L_S - \kappa(\lambda)L_N]\cdot dI_N(\lambda) \tag{123}$$

where L_S = that portion of the ray trajectory between the star and the scattering volume which lies within the nebula (AB in Fig. 112) and L'_N = that portion of the ray trajectory between the scattering volume and the observer which lies within the nebula (BP in Fig. 112). The element of volume dV is shown in expanded form in Figure 112 as a truncated cone whose solid angle $d\omega$ is that subtended by the observer. In all actual cases the viewing angle φ is small ($\varphi < 30'$) and the nebular thickness T is small compared to the nebular distance R. Under these conditions, the volume element is given by

$$dV = R^2 \, d\omega \sec a \, dz \, . \tag{124}$$

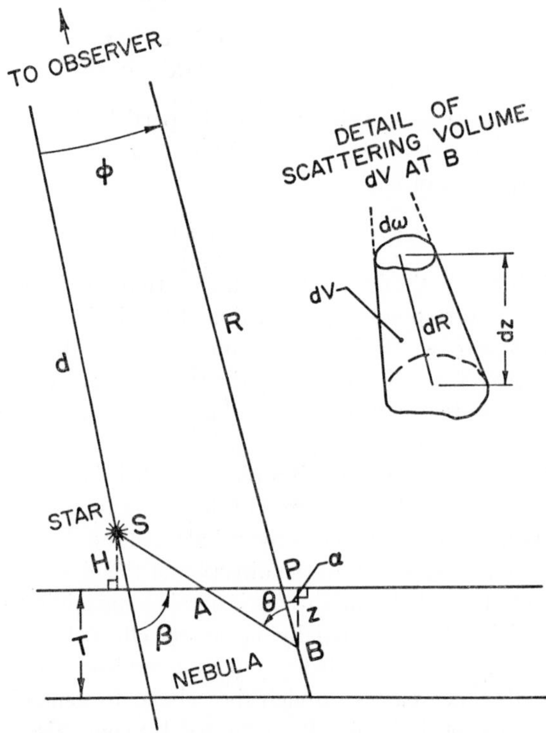

FIG. 112.—Geometrical configuration of a plane-parallel model of a reflection nebula. The case pictured here is that in which the star is between the observer and the front surface of the nebula. The perpendicular distance between the star and the nebula is H, the nebular thickness is T, the distance to the star is d, the variable distance to a nebular volume element dV is R, the nebular tilt angle is β, the scattering angle for nebular particles is $\Theta = \pi - \theta$ and the observation or telescope offset angle is ϕ. The solid angle subtended by the telescope is $d\omega$.

This expression is independent of the shape of the telescope aperture. Combining equations (122) through (124) and integrating the scattering contributions of all nebular particles along the line of sight, we obtain

$$I_N(\varphi, \lambda) = I_S(\lambda) \frac{\gamma^2}{4k^2} \sec a \, d\omega \times \int_0^T \int_{a_1}^{a_2} \frac{1}{r^2} F(\Theta, \lambda, a) n(a) \, da \, dz \exp[-\kappa(\lambda)(L_S + L_N)] \, . \tag{125}$$

The basic nebular parameters for all cases are defined as follows: d = the distance from the star to the observer, H = the normal distance from the star to the *front* surfaces of the nebula, T = the thickness of the nebula, β = the tilt of the nebula, and φ = the angle between the star and the portion of the nebula observed.

For most of our calculations we have chosen the following range of values for these parameters: $d \sim 100$ pc, $H < T$, $T \sim 1$ pc, $0 < \varphi < 30'$. The choice of a typical nebular distance d is based on the accepted Pleiades distance $d = 126$ pc (Allen 1963). If we assume that the nebulosity illuminated by Merope is reasonably uniform dimensionally, then its thickness is of the same order as its width. The angular width of about $30'$ at 126 pc implies a linear width of 1 pc which, in turn, leads us to choose a thickness of $T \simeq 1$ pc.

There is no a priori reason why the solid particles in reflection nebulae should be essentially different from grains in the more general regions of interstellar space. However, because reflection nebulae constitute a rather limited sample of interstellar space, it is also not necessarily true that the particles of the reflection nebulae should have the same characteristics as the average of those which produce extinction and polarization. One difference which stands out is that the grain density in the reflection nebulae seems to be considerably higher than that in the average interstellar cloud (see the end of this section).

Because we are more interested in intercomparisons of theoretical models than in comparisons with observations, the restriction to the spherical shape for the grains and a limited selection of types of size distribution are well justified.

The first restriction is reasonable because we assume that the degree of parallel orientation of the grains in the reflection nebula is small and, furthermore, we are here not considering polarization effects. Consider first a size distribution in which all particles are the same size. The other size distribution considered is defined by $n = e^{-5(a/a_i)^3}$. Unless otherwise noted, we use the extinction coefficient $\kappa(\lambda)$ in our calculations of nebula colors in the form given observationally by Boggess and Borgman, and we shall hereafter refer to this as the empirical extinction. In a few cases we shall calculate the extinctions specifically for the scattering particles and these will be referred to as calculated extinctions. $\kappa(\lambda)$ will be chosen throughout to be defined by $\kappa(5540) = 1.2$ mag/pc, a large value compared with that associated with normal interstellar clouds but not inconsistent with the observational data of Mendoza (1965).

Mendoza has measured the extinction of the Pleiades star HZ 1 = 371 behind the nebulosity illuminated by Merope. He arrives at a color excess of 0.36 mag and a ratio of total to selective extinction of $A_V/E_{B-V} = 3.6$ to 4.2. This implies a total extinction of 1.3 to 1.5 mag which, if caused entirely by the nebula assumed to be approximately 1-pc thick, implies an extinction coefficient of $\kappa \simeq 1$ mag/pc. The grain density implied by this extinction is of the order of 10^3 times that of the average in the spiral arms and approximately 10^2 times that of an average interstellar dust cloud.

Reflection nebulae are not very bright; so it is customary to use wide band filters in defining their colors. For this reason, the calculations presented here are in the *UBV* system (Johnson and Morgan 1953). In this chapter, only results on $B - V$ colors are given because our calculations on $U - B$ do not seem to add any information at this qualitative level. They become more important in detailed discussion of particular nebulosities.

We define the integrated light in each band by

$$I(\lambda)_i = \int I(\lambda)Q_i(\lambda)d\lambda \tag{126}$$

where $Q_i(\lambda)$ is the transmissivity function for the color $i(i \equiv U, B, V)$ and $I(\lambda)$ is the light intensity received by the observer at the wavelength λ. The difference between the intrinsic $B - V$ color of the star and the $B - V$ color of the selected nebular region is then given by

$$\text{Color Difference} = 2.5 \log \frac{I_S(\lambda)_V}{I_S(\lambda)_B} - 2.5 \log \frac{I_N(\lambda)_V}{I_N(\lambda)_B}. \tag{127}$$

The observed color difference between star and nebula must take into account the extinction of the starlight in those cases where the star is within or behind the nebula. To include this general case, we define the quantities $I'_S(\lambda)_i$ by

$$I'_S(\lambda)_i = \int I_S(\lambda)\exp[-\kappa(\lambda)L_0]Q_i(\lambda)d\lambda \tag{128}$$

where L_0 is the pathlength in the nebula between star and observer. The observed color difference between star and nebula is then given by

$$\text{C.D.} = 2.5 \log \frac{I'_S(\lambda)_V}{I'_S(\lambda)_B} - 2.5 \log \frac{I_N(\lambda)_V}{I_N(\lambda)_B} \tag{129}$$

which reduces to equation (127) when the star is in front of the nebula.

TABLE 24

REPRESENTATIVE SPECTRAL DISTRIBUTION (NORMALIZED AT 5500 A) OF THE NEBULA-ILLUMINATING STAR USED IN THE COLOR-DIFFERENCE COMPUTATIONS

λ	I_S	λ	I_S	λ	I_S
3000.......	1.639	4500.......	1.577	5500.......	1.000
3300.......	1.539	4800.......	1.367	5800.......	0.876
3600.......	1.442	4900.......	1.267	6100.......	0.730
3900.......	2.132	5100.......	1.199		
4200.......	1.838	5200.......	1.135		

The colors, as defined in equations (126) through (129), depend on the transmissivity functions Q_i, and the calculated and observed colors must therefore use identical transmissivity functions. Furthermore, even with this proviso, it is not generally correct to use the same "effective" monochromatic wavelengths for the star and the nebula, unless their spectral intensity distributions are quite similar.

Most reflection nebulae are illuminated by stars of MK spectral types from about B3 to A0. The nebulae in the Pleiades cluster receive their illumination from stars of type B6 or B7, and of these stars we shall be principally concerned with Merope ($B - V$, -0.06, spectral type B6 IV). The values of I_S used in the computations (see Table 24) are based on Willstrop's (1965) observations of α Eri and HD 188350. The wavelengths listed are those used in the numerical integrations appropriate to equations (126) and (129). The values of $Q_i(\lambda)$ appropriate to B and V are taken from Allen (1963).

Figures 113–118 show that, in general, small particles ($r \leq 0.05\ \mu$) produce a bluer and more uniform color than large particles.

Figures 113–115 show that the color of a nebula consisting of very small particles is generally bluer than that consisting of standard size dielectric particles. Within the framework of the models calculated using the modified single scattering form (including extinction within the nebula), we find this conclusion seems quite generally valid, regardless of whether the star is in front of (Fig. 113), within (Fig. 114), or behind (Fig. 115) the nebula.

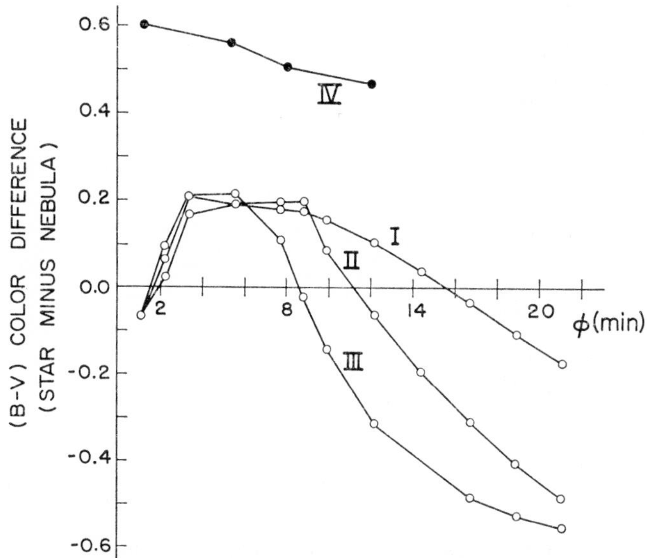

FIG. 113.—Computed $B - V$ colors of reflection nebulae consisting either of small iron grains (filled circles) or dielectric grains (open circles). Effect of varying the nebular tilt angle β for star in front with $T = 1.0$ pc, $H = 0.2$ pc, $d = 126$ pc. Radii of the iron and dielectric grains distributed according to $n = \exp[-5(a/a_i)^3]$, with $a_i = 0.05\ \mu$ for iron and $a_i = 0.5\ \mu$ for dielectric grains. Empirical extinction κ (5540 A) = 1.2 mag/pc for both grain types. For the iron grains tilt angle effect is negligible (IV: $\beta =$ 70°, 80°, 90°). For dielectric grains curves I: $\beta = 80°$; II: $\beta = 90°$; III: $\beta = 100°$.

Figure 116 shows that the dominant grain quality in the determination of nebular colors is the size rather than the optical properties. The difference between dielectric and graphite curves for the same (small) particle radius and the differences between curves for different size graphite spheres seem to be negligible.

Figure 117 shows the contribution to the nebular color by a range of different particle sizes, for the pure single scattering approximation. Of particular note is the curve for the very small ($a = 0.05\ \mu$) dielectric particles which exhibit the usual greater blueness and exhibit small variation with offset angle, characteristic of small particles under an extremely wide variety of nebular models and approximations.

The colors of reflection nebulae consisting of very small ($a \sim 0.05\ \mu$) grains, whether dielectric or metallic, show a high degree of insensitivity with variable telescope offset angle and with reasonable variations of the geometrical parameters defining the star-nebula system as long as the star is not deeply imbedded in the nebulosity. If the star is close to the front surface of the nebula and the offset angle is kept small, the effects of detailed nebular geometry such as shape are reduced.

As a consequence, it appears to be necessary only to find a reflection nebula whose illuminating star is unreddened or only very little reddened in order to apply the appropriate criterion. If it is observed that the color of this nebula is significantly less blue

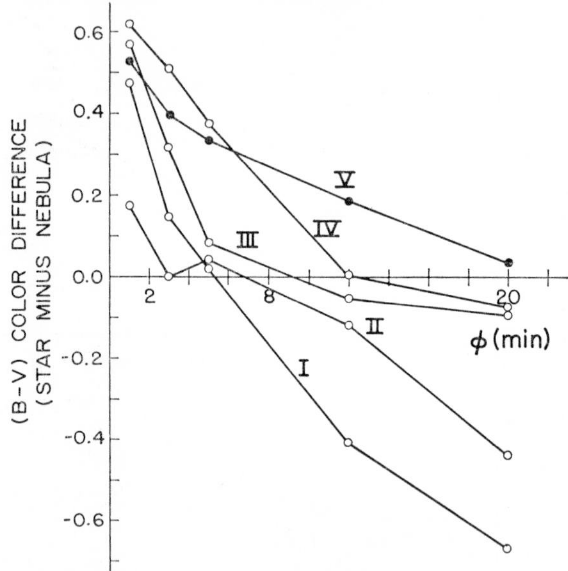

FIG. 114.—Computed $B - V$ colors for varying star-to-surface distances. Dielectric grains (open circles), small (0.05 μ) graphite grains (filled circles). Star within nebula, $\beta = 90°$, $T = 1.0$ pc, $d = 160$ pc. Dielectric grains distributed according to $n = \exp[-5(a/0.50)^3]$. κ (5540 A) = 1.2 mag/pc. Curves I and V: $H = 0.01$ pc; II: $H = 0.10$ pc; III: $H = 0.49$ pc; IV: $H = 0.99$ pc.

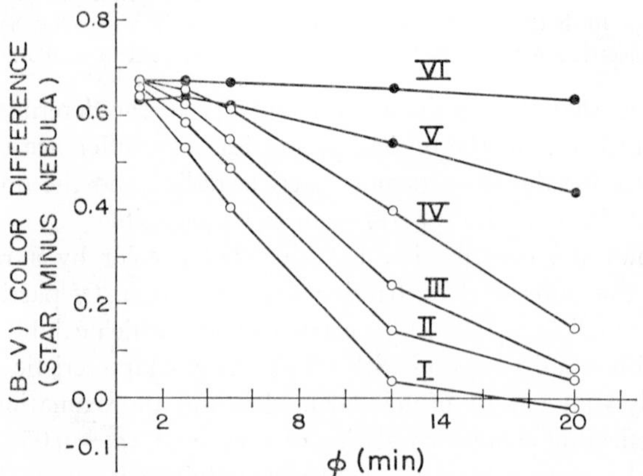

FIG. 115.—Same as Figure 114, except star is behind the nebula. Curves I and V: $H = 1.01$ pc; II: $H = 1.25$ pc; III: $H = 1.50$ pc; IV and VI: $H = 2.00$ pc.

near the star than C.D. = 0.6 and that the color exhibits significant change with offset angle, we may conclude that the small scattering grains may be excluded. It is to be noted that in none of the cases calculated for star in front and for dielectric grains that are not small has the color difference for small offset angles turned out to be greater than C.D. = 0.5.

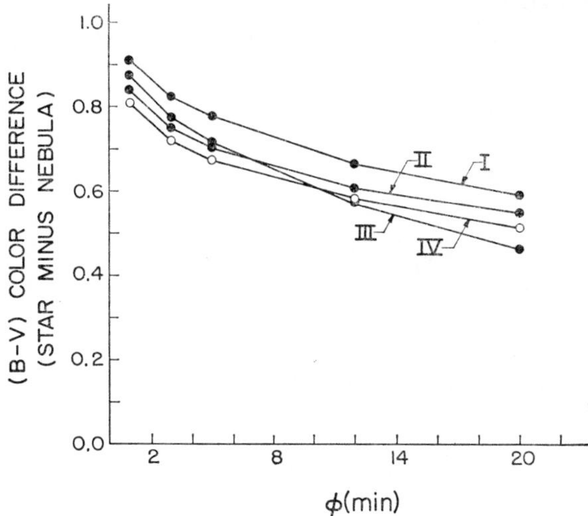

FIG. 116.—Computed $B - V$ colors of reflection nebulae consisting either of small dielectric (open circles) or small graphite (filled circles) grains. Star in front with $\beta = 90°$, $T = 1.0$ pc, $H = 0.01$ pc, $d = 160$ pc. Extinction κ (5540 A) = 1.2 mag/pc. Curve I: single-size ($a = 0.015\ \mu$) graphite spheres; II: graphite spheres with $a = 0.05\ \mu$; III: graphite spheres distributed according to $n = \exp[(-5(a/0.05)^3]$; IV: single-size ($a = 0.05\ \mu$) dielectric spheres.

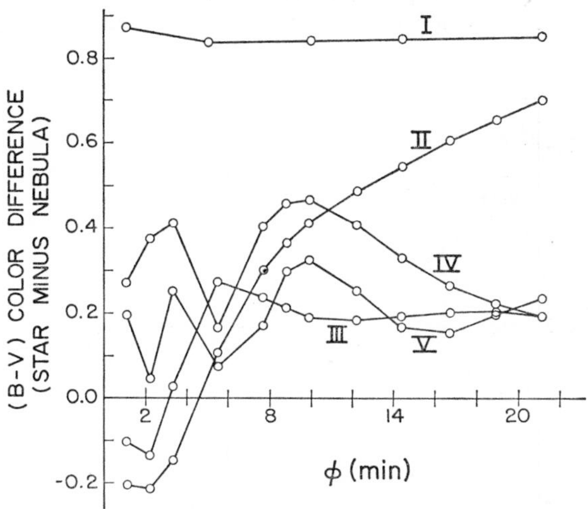

FIG. 117.—Computed $B - V$ colors of reflection nebulae consisting of single-size spherical dielectric grains with pure single scattering formulation ($\kappa = 0.0$). Star in front with $\beta = 90°$; $T = 1.0$ pc, $H = 0.0001$ pc, $d = 126$ pc. I: $a = 0.05\ \mu$; II: $a = 0.20\ \mu$; III: $a = 0.30\ \mu$; IV: $a = 0.40\ \mu$; V: $a = 0.60\ \mu$.

Figures 118 and 119 show, respectively, the $B - V$ and $U - B$ color differences between the Pleiades star Merope and its attendant nebulosity. It can be seen that the observed color difference curves are qualitatively and quantitatively unlike those predicted for model nebulae composed of small metallic grains. A model for the Merope nebula has been constructed using dielectric grains and using a geometrical configuration in which the illuminating star is situated very close to the front surface of the nebulosity. This configuration is compatible with the possibility that Merope may be

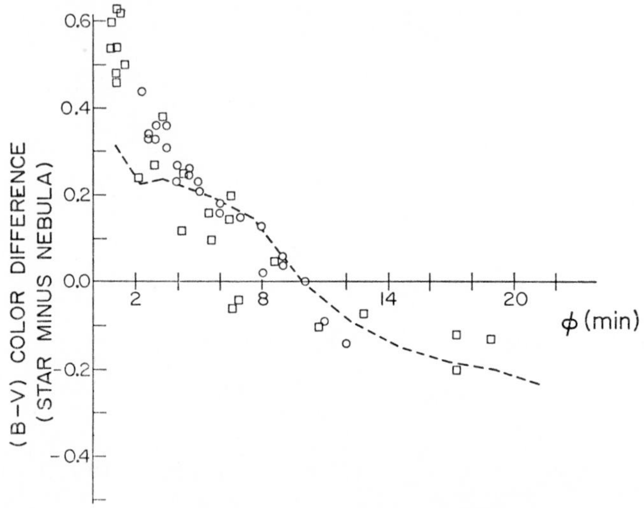

FIG. 118.—Observed $B - V$ color differences in the Merope nebula. Circles from Greenberg and Roark (1967); squares are those of Hall and Elvius (1965). Broken line represents computed color difference for $\beta = 90°$, $T = 1.0$ pc, $H = 0.04$ pc, $d = 126$ pc. Star in front of nebula and dielectric nebular grains distributed according to $n = \exp[-5(a/0.6)^3]$ with κ (5540 A) = 1.2 mag/pc.

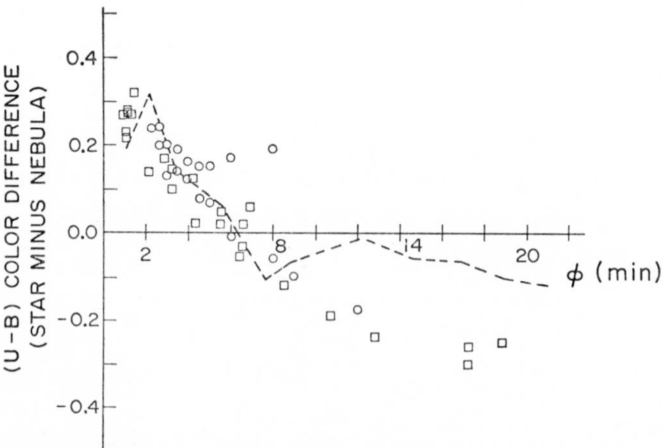

FIG. 119.—Same as Figure 118 for $U - B$ color differences in Merope

entirely unreddened. The color-difference curves predicted by this model are indicated in the figures by a broken line. Fair agreement between model and observation is obtained for $B - V$ color differences at offset values greater than $\varphi = 3'$ and for $U - B$ color differences at φ less than about 8 min of arc. These predicted curves are based on a grain size distribution $n = e^{-5(a/a_i)^3}$ but with $a_i = 0.60\ \mu$ rather than the usual $a_i = 0.50\ \mu$. Although the latter value gives a good representation of the average wavelength dependence of interstellar extinction, no nebular models based on this value give as good agreement as we obtain for the larger a_i value.

10. DIFFUSE INTERSTELLAR LINES

It has long been recognized that the strength of the 4430 A absorption band is strongly correlated with the interstellar extinction (Greenstein and Aller 1950). The ratio of 4430 A to color excess seems to be reduced in high-velocity clouds and/or in the presence of H II regions; i.e., grains in the vicinity of hot bright stars may be modified in the sense of reduced efficiency in the production of 4430 A. The exact mechanism for this reduced efficiency depends on the manner in which the 4430 A line is produced by the grains. In addition to the 4430 A line, about twenty other diffuse lines in the range of 4000 to 6000 A have been detected which are in excellent correlation with each other (Herbig 1966). Herbig shows that the diffuse lines remove about six times as much stellar flux as do the lines identified as atomic or molecular. We are obviously dealing with an abundant substance or substances. The observation (Routly and Spitzer 1952) that the ratio of Na/Ca is correlated with 4430 A in the sense that 4430 A is reduced where interstellar Ca is increased has led to the conjecture that perhaps Ca atoms that are bound to grains are responsible for 4430 A and that, consequently, where Ca atoms are released, the diffuse line strength is decreased. Although the resonance line of Ca is close enough (4227 A) to suggest this, there still remains—among other questions—an accounting for the diffuse lines other than 4430 A.

In spite of various attempts (Herbig 1963; Unsöld 1964), the mystery of the diffuse lines remains with us. It is to be hoped that a satisfactory explanation will also be helpful in determining the nature of the interstellar grains.

We limit ourselves to a few qualitative remarks concerning the optics of the 4430 A band, assuming it to be produced by some as yet unidentified absorbing substance within or on the surface of the grains. As was first pointed out by van de Hulst (1947), it can be stated categorically that any constituent which is dispersed through the dielectric grain material as an atom or molecule with a resonant absorption will produce a highly asymmetric line (or dispersion curve) for sufficiently large grains because the extinction by absorbing particles whose size is of the order of the wavelength is not a monotonic function of increasing absorptivity. The complex index of refraction of an ice grain with a suspension of atoms with a resonance line whose strength is determined by the proportion of the relative abundance of Ca to O would be as shown in Figure 120. In Figures 121 and 122 are shown the shapes of the line for particle size distributions $e^{-5(a/a_i)^3}$ for $a_i = 0.1\ \mu$ and $0.2\ \mu$, respectively. We see that even for $a_i = 0.1\ \mu$ (particles of average size $\sim 0.03\ \mu$), the line is not symmetrical and that for the larger particle sizes the line actually exhibits emission properties toward shorter wavelengths. Since very small particles have a small effect on the shape of the extinction curve in the visible,

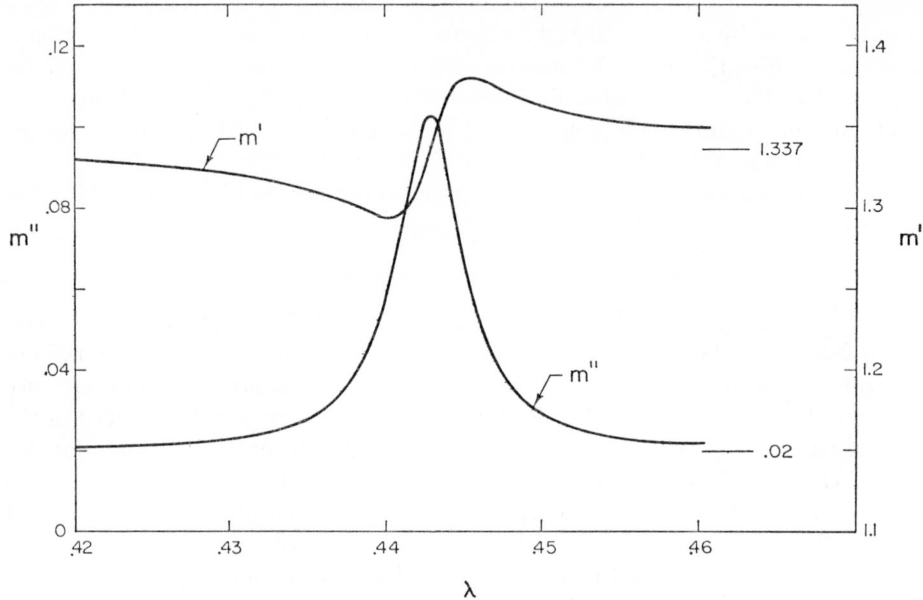

FIG. 120.—Computed complex index of refraction for small amount of resonant absorbing material in dirty ice grains. Real and imaginary parts of the index of refraction relative to that of the ambient medium.

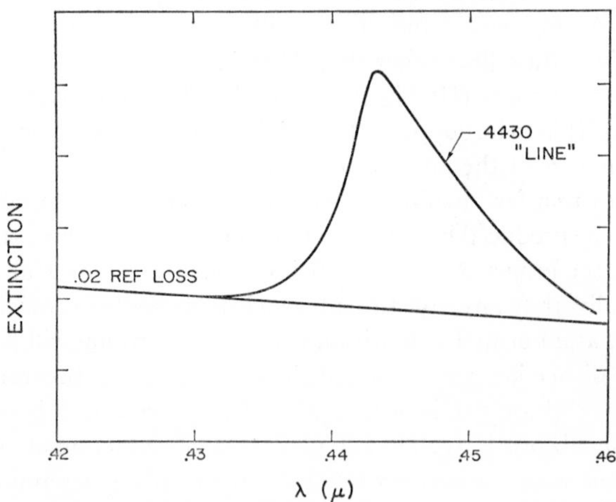

FIG. 121.—Wavelength dependence of extinction computed around the resonant absorption for grains distributed according to $n = \exp[-5(a/0.1)^3]$. Notice that the extinction "line" is asymmetric even for these very small particles.

it might be questioned whether they could be added to a standard size distribution in sufficient number to mask the dispersion line produced by large particles. Apparently this is not possible (see Fig. 123). An alternative suggested by Herbig (1966) is the ad hoc assumption that *only* small particles are diffuse line carriers, these small particles appearing in space both as separate entities and as cores within larger particles. This scheme would be optically feasible.

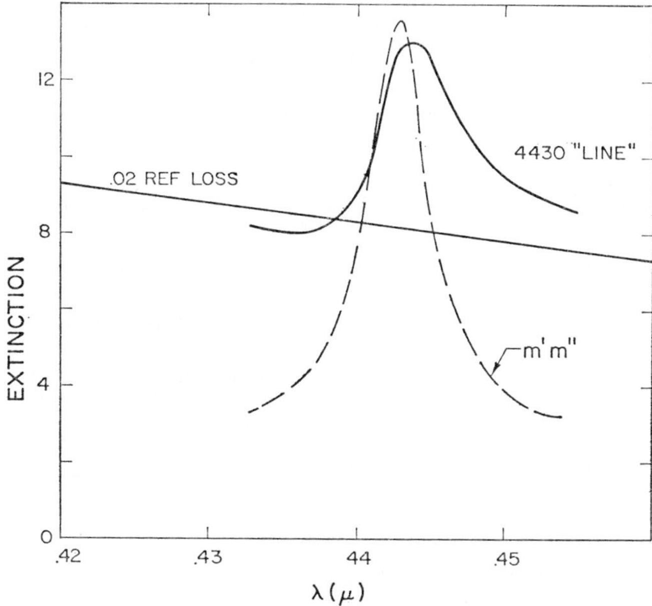

FIG. 122.—Wavelength dependence of extinction computed around the resonant absorption for grains distributed according to $n = \exp[-5(a/0.2)^3]$. Note the appearance of dispersion (emission) character of the line short of absorption center. Comparison is made with the absorption by extremely small (Rayleigh size) particles as given approximately according to the broken line.

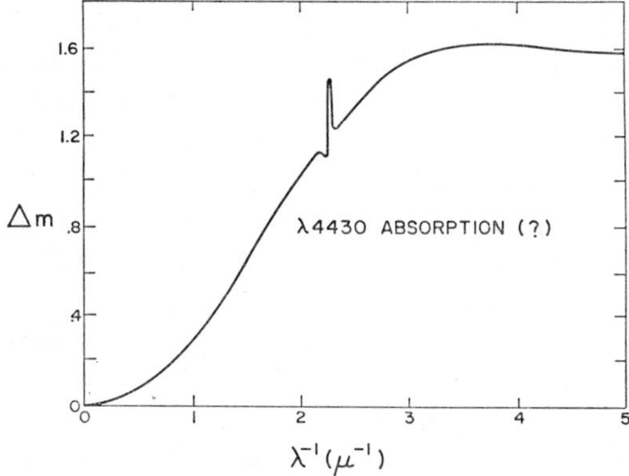

FIG. 123.—Dispersion character of "absorption line" for a two-parameter size distribution according to $n = n_1 \exp[-5(a/0.1)^3] + n_5 \exp[-5(a/0.5)^3]$ with $n_1/n_5 \approx 50$.

Finally we mention another optically attractive possibility—that the diffuse line carriers are distributed on grain surfaces.

The number of line carriers needed to give the observed ratio of 4430 A to extinction is simply obtained. At resonance, an atomic or molecular absorption coefficient has the form $\alpha_{\nu_0} = (\pi e^2/mc)(4f/\Gamma)$ where f is the oscillator strength and Γ is a measure of the width of the line in frequency units; that is, $\Gamma/4\pi = \Delta\nu$ where $\alpha_{\nu_0 \pm \Delta\nu} = \alpha_0/2$. The total optical depth at resonance is then given by

$$\tau_{\nu_0} = N_{\nu_0}\left(\frac{e^2}{mc}\right)\left(\frac{f}{\Delta\nu}\right) \tag{130}$$

where N_{ν_0} = number of 4430 A "molecules" per unit area along the line of sight. At the visible or photographic wavelengths, the ratio

$$\left[\int_0^\infty C_{\text{ext}} e^{-5(a/0.5)^3} da\right] \Big/ \left[4\pi \int_0^\infty a^2 e^{-5(a/0.5)^3} da\right] \approx 1 .$$

We may therefore, at least for approximate purposes, represent the grain extinction by

$$\Delta m_{\text{phot}} \approx 4\pi N \int_0^\infty a^2 e^{-5(a/0.5)^3} da = \text{total grain area}, \tag{131}$$

because the average extinction efficiency of the grain is ≈ 1.

The extinction efficiency of the 4430 A molecules is

$$Q_{\text{ext}}^{\nu_0} = \frac{\left(\frac{e^2}{mc^2}\right)\lambda_0 f\left(\frac{\nu_0}{\Delta\nu}\right)}{\pi a_{4430}^2} \approx 1 \tag{132}$$

where we have used $f = 1$, $\Delta\nu/\nu_0 = 10^{-2}$, and $a_{4430} = 2 \times 10^{-8}$ cm.

Thus, we are assuming that a ratio of 4430 A to interstellar extinction of $\sim$0.05 implies the necessity of covering about 1/20 of the available grain area by the 4430 A carrier.

11. SUMMARY AND CONCLUSIONS

As was stated in the introduction, the subject of the interstellar grains supplies us with many unanswered questions and even some real mysteries and puzzles. The types of physical reasoning and the calculational procedures required to arrive at the proper answers have been illustrated by numerous analytical and computational examples.

In Table 25 I attempt to summarize the results of the preceding sections and to evaluate the present status of the various theories. The entries, +, 0, and − in the table mean, respectively, *possible*, *uncertain*, and *unlikely*. If an entry is enclosed by parentheses it is understood that some element of the underlying theory or process is itself uncertain. In some cases the theoretical principle may be well understood but the parameters which one uses to arrive at a result are subject to various interpretations. In such cases I prefer not to use parentheses except for cases in which the parameters needed are at wide variance with generally accepted values.

None of the grain models can account for the extremely anomalous extinction reported by Johnson in the infrared for NGC 2244 and Cepheus and illustrated in Figure 2. Certainly no modification of a normal size distribution (one which accounts for both visible

TABLE 25

Grain Type	Growth (by Accretion or Otherwise)	Existence (According to Abundance of Elements and Total Mass)	Physics for Growth as Non-spherical Particle	Magnetic Orientation	Wavelength-Dependence of Extinction			Variations of λ-dependence of Extinction[d]	λ-dependence of Polarization	Amount of Polarization Relative to Extinction[e]	Colors of Reflection Nebulae[f]
					Infrared[a] $\lambda^{-1}<1\mu^{-1}$	Middle[b] $1\leq\lambda^{-1}\leq 3$	Ultraviolet[c] $3\leq\lambda^{-1}\leq 5$				
Dielectric (dirty ice)	+	+	0	+	+	+	0	+	+	+	+
Metallic (iron)	0	−	0	+	+	+	0	−	(+)	+	−
Graphite (pure)	+	+	+	+	+	+	0	−	−	+	−
Ferrite	0	0	+	+	0	+	0	−	(0)	+	−
Core+mantle (graphite+dirty ice)	+	+	+	0	+	+	+	+	(+)	(−)	0
Large free radicals	+	+	−	+		(+)		−	−	(0)	0

[a] Normal or small deviations in the infrared (but not grossly anomalous as in Cepheus and NGC 2244 [Fig. 7]) are considered here.

[b] Because extinction in the visible is the best known, we assume that this is matched before attempting to reproduce the infrared and ultraviolet.

[c] Extinction observed at $\lambda^{-1} = 8$ is not yet well established. If it were included it might well be that all types would be − in this category.

[d] No extreme variations are considered here such as λ-independent extinction or extremely rapid increase in extinction in the infrared, as in NGC 2244 and Cepheus (Fig. 7).

[e] The amount of polarization as produced by a magnetic field $B \approx 10^{-5}$ gauss.

[f] The assumption is that reflection particles are similar in size to interstellar particles.

and near ultraviolet extinction) by physical interactions can produce these rapid rises in extinction at or below $\lambda^{-1} = 0.4\ \mu^{-1}$. As a matter of fact the particle sizes needed to account for such λ-dependence are not present within a standard size distribution. The fact that there may be separate local clouds of very large particles (10 times larger than normal) should not be excluded. An additional tentative solution to this problem could be to invoke the presence of an additional mode of producing extinction which is wavelength-independent. Free electron scattering has this property. The more normal variations of extinction are usually accounted for by size-distribution modifications or by orientation effects. One possibility for further investigation which applies to dielectric materials is the modification of the imaginary part of the index of refraction by bombardment of high energy photons or particles in the neighborhood of very hot stars.

The discontinuity in slope of the extinction curve at $\lambda^{-1} = 2.25$ as noted by Whitford (1958) and more recently by Nandy (1965, 1966) has not been included in the table. If it were included, it would tend to favor grains containing graphite because of graphite's index-of-refraction change in this region. Because of the possibility of variation of graphite's optical properties with temperature and other physical modification, this point seemed to be of somewhat lesser weight than the others and was therefore not included in the table. The association of the slope discontinuity in the extinction curve with the short wavelength limit (λ 4430) of the unidentified diffuse lines may or may not be associated with carbon. The diffuse lines may yet provide a way to unlock the puzzle of the definite identification of the grains.

Note that not all of the considerations in Table 25 are of equal weight. Particular emphasis should be given to any negative entry. A combination of graphite and graphite core-mantle grains is on an equal footing with the dielectric dirty-ice grains except for the negative entries having to do with scattering properties and with degree of magnetic orientation. Probably the more critical of these two is the degree of orientation and the polarizing ability of graphite core–ice mantle particles. The latter should be investigated more carefully by the microwave method.

In future observational and theoretical work the obvious choice will be to try to find results which can be used to exclude as many of the vast variety of models as possible. One observational test may be the wavelength dependence of polarization in the ultraviolet. Complete information for nonspherical dielectric particles indicates that the polarization should decrease, but at a rather slow rate, toward shorter wavelengths. On the other hand, metallic and graphite particles would exhibit a more rapid decrease of polarization and perhaps even a polarization reversal (rotation through 90°) in the ultraviolet.

Detailed simultaneous observations of the wavelength dependence of polarization and extinction by individual stars will be extremely important in fixing on particular grain models. So far the results are not conclusive, because similar general statements about variations of polarization and extinction may be made for competing models. One probably needs more refined calculating procedures. Some of these have been indicated as they apply to elongated dielectric particles. The same can and should be done for other grain models.

One of the calculations which probably requires further work is that concerning the extension of the Davis-Greenstein theory as shown in the section on orientation.

Variations of the wavelength dependence of extinction with degree of orientation is still a possible method of distinguishing between dielectric and metallic (graphite) grains, but its effects are obscured by other quantities—such as size—producing similar extinction variations. A slight possibility exists that the wavelength dependence of polarization can be used to determine the particle sizes (according to the maximum in the polarization) and that this parameter may thus be isolated.

Acknowledgments.—A first draft of this chapter was completed in 1961 while the author was a Senior Visiting Fellow at Leiden Observatory. The author is most grateful to Professor J. H. Oort for his hospitality and to Professor H. C. van de Hulst who was extremely helpful in the early stages of this work. The author would also like to express his thanks to Dr. A. C. Lind, R. T. Wang, G. Shah, and to Dr. P. R. Lichtenstein whose contributions to this chapter go beyond the references to their published work. Assistance in obtaining some of the microwave scattering data was given by Y. K. Minn, P. F. Faix, and W. Harrison. A major portion of the chapter was rewritten during the author's stay at the Institute for Advanced Study in 1965. For this opportunity grateful acknowledgment is made to Professor J. Robert Oppenheimer and to Professor Bengt Strömgren, who also helped the author through many clarifying discussions.

Much of the material appearing here has not been published elsewhere. This work has been supported by grants from the National Aeronautics and Space Administration and the National Science Foundation.

REFERENCES

Some useful general references are:

American Institute of Physics (*A.I.P.*) *Handbook*, 1957 (York, Pa.: Maple Press Co.), pp. 5–162.

Lowell Observatory Bulletin No. 105, **4**, 264–321. A conference on *Polarization of Starlight by the Interstellar Medium* (1960). A discussion of the observational and theoretical problems of polarization.

"*Les particules solides dans les astres*" (Rept. Liège Symposium, 1954), *Mém. Soc. R. Sci. Liège*, ser. 4, **15**, 687 (1955).

Proceedings of the Third Symposium on Cosmical Gas Dynamics (Cambridge, Mass., June 24–29, 1957), I.A.U. Symposium No. 8, Smithsonian Symposium Publ. No. 1, eds. J. M. Burgers and R. N. Thomas, *Rev. Mod. Phys.*, **30**, 905–1108, 1958.

Alexander, J. D., Bowen, P. J., Gross, M. J., and Heddle, D. W. O. 1964, *Proc. Roy. Soc. London, A*, **279**, 510.

Allen, C. W. 1963, *Astrophysical Quantities* (London: Athlone Press).

Aller, L. H. 1961, *The Abundance of the Elements* (New York: Interscience Publishers).

———. 1965, *Advances in Astronomy and Astrophysics*, **3**, 1.

———. 1967, Private communication.

Aller, L. H., and Liller, W. 1959, *Ap. J.*, **130**, 45.

Baade, W., and Minkowski, R. 1937, *Ap. J.*, **86**, 123.

Behr, A. 1955, *Mém. Soc. R. Sci. Liège*, ser. 4, **15**, 547.

———. 1959, *Zs. f. Ap.*, **47**, 54.

———. 1961, *ibid.*, **53**, 95.

———. 1963, private communication.

———. 1965, *IAU Colloquium on Interstellar Grains* (*RPI*).

Behr, A., and Tripp, W. 1955, *Naturwiss.*, **42**, 9.

Boggess, A. 1961, *A.J.*, **66**, 279.

Boggess, A., and Borgman, J. 1964, *Ap. J.*, **140**, 1636.

Borgman, J. 1961, *B.A.N.*, **16**, No. 512, 99.

Borgman, J., and Johnson, H. L. 1962, *Ap. J.*, **135**, 306.

Brown, H. 1949, *Rev. Mod. Phys.*, **21**, 625.

Burgers, J. M., and Thomas, R. N. 1958, *Rev. Mod. Phys.*, **30**, 905–1108 (I.A.U. Symposium No. 8: *Third Symposium on Cosmical Gas Dynamics; Smithsonian Symp. Pub. No. 1*).

Cayrel, R., and Schatzman, E. 1954, *Ann. d'ap.*, **17**, 555.

Chubb, T. A., and Byram, E. T. 1963, *Ap. J.*, **138**, 617.

Coyne, G., and Gehrels, T. 1966, *A.J.*, **71**, 355.

Cugnon, P. 1963, *Bull. Soc. R. Sci. Liège*, **32**, 228–39.
Curtis, H. D. 1918, *Lick Obs. Pub.*, **13**.
Danielson, E. E., Woolf, N. J., and Gaustad, J. E. 1965, *Ap. J.*, **141**, 116.
Davies, R. D. 1960, *M.N.*, **120**, 483.
Davis, L., Jr. 1955, *Vistas in Astronomy*, ed. A. Beer (London and New York: Pergamon Press), **1**, 336.
———. 1958, *Ap. J.*, **128**, 508.
———. 1959, *Zs. f. Ap.*, **47**, 59.
Davis, L., Jr., and Greenstein, J. L. 1951, *Ap. J.*, **114**, 206.
Deirmendjian, D., Clasen, R., and Viezee, W. 1961, *J. Opt. Soc. Amer.*, **51**, 620.
Divan, L. 1954, *Ann. d'ap.*, **17**, 456.
Donn, B. 1955, *Mém. Soc. R. Sci. Liège*, ser. 4, **15**, 571.
Donn, B., Wickramasinghe, N. C., Hudson, J. P., and Stecher, T. P. 1966, in production.
Dorsey, N. E. 1940, *Properties of Ordinary Water Substance* (New York: Reinhold).
Dressler, K., and Schnepp, O. 1960, *J. Chem. Phys.*, **33**, 270.
Dunham, T., 1939, *Proc. Am. Phil. Soc.*, **81**, 277.
Dutta, A. K. 1953, *Phys. Rev.*, **90**, 187.
Elvius, A. 1951, *Stockholms Obs. Ann.*, **17**, No. 4.
———. 1956*a*, *ibid.*, **18**, No. 9.
———. 1956*b*, *ibid.*, **19**, No. 1.
Elvius, A., and Hall, J. S. 1966, *Lowell Obs. Bull.*, **6**, 257.
Fick, E. 1954, *Zs. f. Phys.*, **138**, 183.
———. 1955, *ibid.*, **140**, 308.
Ganguli, N., and Krishnan, K. S. 1941, *Proc. R. Soc. London, A*, **177**, 168.
Gehrels, T. 1960, *A.J.*, **65**, 470.
Giese, R. H. 1961, *Zs. f. Ap.*, **51**, 119.
Gold, T. 1952*a*, *Nature*, **169**, 322.
———. 1952*b*, *M.N.*, **112**, 215.
Goldberg, L., Müller, E. A., and Aller, L. H. 1960, *Ap. J. Suppl.*, **5**, 1.
Greenberg, J. M. 1960*a*, *J. App. Phys.*, **31**, 82.
———. 1960*b*, *Ap. J.*, **132**, 672.
———. 1963, in *Ann. Rev. Astr. and Ap.*, **1**, 267.
———. 1966, Proceedings of I.A.U. Symposium No. 24 (Spectral Classification), p. 291.
Greenberg, J. M., Libelo, L. F., Lind, A. C., and Wang, R. T. 1963, *Electromagnetic Theory and Antennas*, Part I, ed. E. C. Jordan (New York: Macmillan Company), p. 81.
Greenberg, J. M., and Lichtenstein, P. R. 1963, *A.J.*, **68**, 74.
Greenberg, J. M., Lind, A. C., Wang, R. T., and Libelo, L. F. 1963, *Proc. ICES*, I (London: Pergamon Press), p. 123.
———. 1967, *Proc. ICES*, II (New York: Gordon and Breach), in press.
Greenberg, J. M., and Meltzer, A. S. 1960, *Ap. J.*, **132**, 667.
Greenberg, J. M., Pedersen, N. E., and Pedersen, J. C. 1961, *J. App. Phys.*, **32**, 233.
Greenberg, J. M., and Roark, T. 1967, *Ap. J.*, **147**, 917.
Greenberg, J. M., and Shah, G. 1966, *Ap. J.*, **145**, 63.
Greenstein, J. L. 1938, *Harvard Obs. Circ.*, No. 422.
Greenstein, J. L., and Aller, L. H. 1950, *Ap. J.*, **111**, 328.
Güttler, A. 1952, *Zs. f. Ap.*, **31**, 1.
Haar, D. ter. 1943, *B.A.N.*, **10**, No. 361, 1.
———. 1944, *Ap. J.*, **100**, 288.
Haar, D. ter, van de Hulst, H. C., Oort, J. N., and van Woerkom, A. J. J. 1943, *Ned. v. Naturkunde*, **10**, 238.
Hall, J. S. 1937, *Ap. J.*, **85**, 145.
———. 1949, *Science*, **109**, 166.
———. 1954, *A.J.*, **59**, 364.
———. 1955, *Mém. Soc. R. Sci. Liège*, 4th ser., **15**, 543.
———. 1958, *Pub. U.S. Naval Obs.*, **17**, 275.
Hall, J. S., and Serkowski, K. 1963, *Basic Astronomical Data*, ed. K. Aa. Strand (Chicago: University of Chicago Press), p. 293.
Hayakawa, S., Nishimura, S., and Takayanagi, K. 1961, *Pub. Astr. Soc. Japan*, **13**, 184.
Henry, J. 1958, *Ap. J.*, **128**, 497.
Henyey, L. G., and Greenstein, J. L. 1938, *Ap. J.*, **88**, 580.
———. 1941, *ibid.*, **93**, 70.
Herbig, G. H. 1966, paper presented at I.A.U. Symposium No. 31 (Radio Astronomy and the Galactic System).
———. 1967. *Ap. J.*, in press.
Hiltner, W. A. 1949, *Science*, **109**, 165.
———. 1956, *Vistas in Astronomy*, ed. A. Beer (London and New York: Pergamon Press), **2**, 1086.
———. 1958, *Ap. J.*, **128**, 9.
Houten, C. J. van. 1961, thesis, Leiden University.

Hoyle, F., and Wickramasinghe, N. C. 1962, *M.N.*, **124**, 417.
Hulst, H. C. van de. 1943, *Ned. Tijdschr. v. Natuurkunde*, **10**, 251.
———. 1946, *Rech. Astr. Obs. Utrecht*, **11**, Part 1.
———. 1949, *ibid.*, Part 2.
———. 1955, *Mém. Soc. R. Sci. Liège*, ser. 4, **15**, 393.
———. 1957, *Light Scattering by Small Particles* (New York: J. Wiley & Sons, Inc.; London: Chapman & Hall, Ltd.).
———. 1958, *Rev. Mod. Phys.*, **30**, 913.
Irvine, W. M., and Pollack, J. B. 1966, Private communication.
Johnson, H. L., 1961, unpublished private communication by A. E. Whitford.
———. 1965, *Ap. J.*, **141**, 923.
———. 1967, *ibid.*, **147**, 912.
Johnson, H. L., and Borgman, J. 1963, *B.A.N.*, **17**, 115.
Johnson, H. L., and Morgan, W. W. 1953, *Ap. J.*, **117**, 313.
———. 1955, *ibid.*, **122**, 142.
Jones, R. C. 1945, *Phys. Rev.*, **68**, 93, 213.
Jones, R. V., and Spitzer, L., Jr. 1967, *Ap. J.*, **147**, 943.
Kahn, F. D. 1952, *M.N.*, **112**, 518.
———. 1955, *Gas Dynamics of Cosmic Clouds*, ed. H. C. van de Hulst and J. M. Burgers (Amsterdam: North-Holland Pub. Co.), p. 115.
Kamijo, F. 1963, *Pub. Astr. Soc. Japan*, **15**, 440.
Kimura, H. 1962, *Pub. Astr. Soc. Japan*, **14**, 374.
Kislovskii, L. D. 1959, *Opt. Spek.*, **7**, 201.
Kittel, C. 1956, *Solid State Physics* (New York: John Wiley & Sons), p. 130.
Kramers, H. A., and Haar, D. ter. 1946, *B.A.N.*, **10**, No. 371, 137.
Kurochkin, N. E. 1957, *A.J.-U.S.S.R.*, **34**, 31.
Lambrecht, H., and Schmidt, K. H. 1958, *A.N.*, **284**, 71; *Mitt. Univ.-Sternw. Jena*, No. 32; *Irish A.J.*, **5**, 70.
Libelo, L. 1962, *U.S. Naval Ordnance Laboratory*, NOLTR 62–142, 62–157, 62–161.
———. 1965, unpublished thesis, Rensselaer Polytechnic Institute.
Lichtenstein, P. R., and Greenberg, J. M. 1962, Report to the A.A.S. Meeting, Cambridge, Massachusetts.
Lind, A. C. 1966, thesis (Rensselaer Polytechnic Institute).
Lind, A. C., and Greenberg, J. M. 1966, *J. Ap. Phys.*, **37**, 3195.
Lind, A. C., Wang, R. T., and Greenberg, J. M. 1965, *Applied Optics*, **4**, 1555.
Lindblad, B. 1935, *Nature*, **135**, 133.
Lodén, L. O. 1965, *Stockholms Obs. Ann.*, **22**, No. 8.
Mendoza, V, E. E. 1965, *Bull. Tonantzintla y Tacubaya*, **4**, 3.
Mie, G. 1908, *Ann. Phys.*, **25**, 377.
Morgan, W. W., Harris, D. L., and Johnson, H. L. 1953, *Ap. J.*, **118**, 92.
Müller, E. A. 1955, *Zs. f. Ap.*, **38**, 110.
Müller, E. A., and Mutschlecner, J. P. 1964, *Ap. J. Suppl.*, **9**, 1.
Nandy, K. 1964*a*, *Pub. Roy. Obs. Edinburgh*, **4**, 57.
———. 1964*b*, *ibid.*, **3**, 142.
———. 1965, *ibid.*, **5**, 13.
———. 1966, *ibid.*, p. 233.
Nandy, K., and Wickramasinghe, N. C. 1965, *Pub. Roy. Obs. Edinburgh*, **5**, No. 3.
Neckel, H. 1965, *Zs. f. Ap.*, **62**, 180.
O'Mara, B. J. 1967, Thesis (University of California, Los Angeles).
Oort, J. H., and Hulst, H. C. van de. 1946, *B.A.N.*, **10**, No. 376, 187.
Piotrowski, S. L. 1962, *Acta Astronomica*, **12**, 221.
Platt, J. R. 1956, *Ap. J.*, **123**, 486.
Routly, P., and Spitzer, L., Jr. 1952, *Ap. J.*, **115**, 227.
Schalén, C. 1936, *Medd. Uppsala Astr. Obs.*, vol. 64.
———. 1939, *Uppsala Ann.*, **1**, No. 2, 51.
———. 1945, *ibid.*, No. 9.
Schiff, L. I. 1956, *Phys. Rev.*, **103**, 443.
Schmidt, K. H. 1958, *A.N.*, **284**, 73.
Schmidt, Th. 1958, *Zs. f. Ap.*, **46**, 145.
Schmidt-Kaler, Th. 1966, unpublished paper presented at Noordwijk Symposium.
Scholz, M. 1967, *Zs. f. Ap.*, **65**, 1.
Serkowski, K. 1962, *Advances in Astronomy and Astrophysics*, ed. Z. Kopal (New York: Academic Press), Vol. 1.
———. 1965, *Ap. J.*, **141**, 1340.
———. 1966*a*, *Lowell Obs. Bull.*
———. 1966*b*, *Ap. J.*, **144**, 857.

Serkowski, K., Chojnacki, W., and Rucinski, S. 1965, in *Proc. I.A.U. Rensselaer Colloquium* (in press, 1967).
Shain, G. A. 1956, *A.J.-U.S.S.R.*, **33**, 469.
Sharpless, S. 1963, in *Basic Astronomical Data*, ed. K. Aa. Strand (Chicago: University of Chicago Press), chap. 12.
Spitzer, L. 1961, I.A.U. Draft Report of Commission 34.
Spitzer, L., Jr., and Tukey, J. W. 1951, *Ap. J.*, **114**, 187.
Stebbins, J., Huffer, C. M., and Whitford, A. E. 1939, *Ap. J.*, **90**, 209.
Stecher, T. P. 1965, *Ap. J.*, **142**, 1683.
Stecher, T. P., and Donn, B. 1965, *Ap. J.*, **142**, 1681.
Stevenson, A. F. 1953, *J. App. Phys.*, **24**, 1143.
Stoeckly, R., and Dressler, K. 1964, *Ap. J.*, **139**, 240.
Strömgren, B. G. 1965, Report on I.A.U. Commission 34, *Trans. I.A.U.*, **XII A**, 557.
Struve, O. 1961, *Sky and Tel.*, **21**, 317.
Suess, H. E., and Urey, H. C. 1956, *Rev. Mod. Phys.*, **28**, 53.
Swamy, K., and Wickramasinghe, N. C. 1967, to be published.
Thiessen, G. 1961, *Astr. Abh. Hamburg*, **5**, No. 9.
Treanor, P. 1963, *A.J.*, **68**, 185.
Trumpler, R. J. 1930, *Lick Obs. Bull.*, **14**, 154.
Visvanathan, S. 1965, unpublished thesis, Mount Stromlo Observatory.
Wampler, E. J. 1961, *Ap. J.*, **134**, 861.
Westerhout, G., Seeger, Ch. L., Brouw, W. N., and Tinbergen, J. 1962, *B.A.N.*, **16**, No. 518, 187.
Whiteoak, J. B. 1966, *Ap. J.*, **144**, 305.
Whitford, A. E. 1948, *Ap. J.*, **107**, 102.
———. 1958, *A.J.*, **63**, 201.
Wickramasinghe, N. C. 1962, *M.N.*, **125**, 87.
———. 1963, *ibid.*, **126**, 99.
Wickramasinghe, N. C., Donn, B., Stecher, T. P., and Williams, D. A. 1966, *Ap. J.*, **145**, 949.
Wickramasinghe, N. C., and Guillaume, C. 1965, *Nature*, **207**, 366.
Willstrop, R. V. 1965, *Mem. R.A.S.*, **69**, 83.
Wilson, R. 1959, *Observatory*, **79**, 169.
———. 1960, *M.N.*, **120**, 51.
Zimmermann, O. 1957, *Zs. f. Naturforsch.*, **12a**, 647.
Zirin, H. 1952, *Bull. Harvard Coll. Obs.*, No. 921, pp. 19–27.
Zonn, W. 1962, *Acta Astron.*, **12**, 142.

CHAPTER 7

Interstellar Absorption Lines

GUIDO MÜNCH

Mount Wilson and Palomar Observatories, Carnegie Institution of Washington, California Institute of Technology

1. INTRODUCTION

THE EXISTENCE of interstellar gas was first suggested by Hartmann (1904) following his observation of the constant wavelengths of the Ca II resonance lines (H and K) in the spectrum of the spectroscopic binary δ Orionis. The presence of H- and K-lines in the spectra of other high-temperature stars, with depth, form, and radial velocity different from those of lines generally observed in such stars, was noticed also in the early days of stellar spectroscopy. In 1909, V. M. Slipher discussed the problems raised by the "stationary" H- and K-lines, suggesting an interstellar origin for such features and predicting the existence of similar lines due to Na I. Slipher's contribution seems to have been overlooked, for it was not until ten years later that Miss Heger (1919) actually discovered the "stationary" D-lines in δ Orionis and other stars. Shortly thereafter the weaker ultraviolet resonance doublet of Na I was photographed by Miss Heger (1921). The conception of the stationary lines as truly interstellar in origin and thus independent of the stars in which they were observed, however, seems to have evolved gradually. It was stated by Eddington (1926) in his Bakerian Lecture, in which the physical conditions that may prevail in the interstellar medium were first analyzed. This important theoretical investigation answered previously unsettled questions and profoundly stimulated subsequent development of the subject. The work of Struve (1927, 1928) and Gerasimovitsch and Struve (1929) explaining the observed relation between interstellar line intensities and the location of the stars in which they were observed generally confirmed the ideas advanced by Eddington (1926). As the complexity of the problem of line intensities and velocities became evident, systematic observations of higher precision were carried out with intermediate dispersions at the Dominion Astrophysical Observatory (Plaskett and Pearce 1933) and at the Mount Wilson Observatory (Merrill, Sanford, Wilson, and Burwell 1937). The work of Plaskett and Pearce (1933) established firmly the effects of galactic rotation on interstellar matter. From the radial velocities of the Ca II lines in 314 early-type stars, the galactic-rotation shift of the interstellar lines was found to be nearly one-half that of the stars in which they were observed; this was interpreted as indicating a nearly uniform density distribution of the absorbing matter throughout the volume surveyed. However, it soon became apparent that this uniformity could not hold strictly both for density and velocity, since the line intensities, or widths, did not seem to depend markedly on galactic longitude (Beals 1936), as would be expected if differential galactic rotation were the main agent of line broadening (Unsöld, Struve, and Elvey 1930; Eddington 1934). Until this time, the line intensities, except for the few obtained by Williams (1934), were eye estimates, and their inherent uncertainty made the results insufficiently conclusive. For this reason, the spectrophotometric and radial velocity work done at Mount Wilson (Merrill *et al.* 1937) had transcendental importance. Analysis of this material by Wilson and Merrill (1937) led to an understanding of the seemingly discrepant data regarding the galactic-rotation effect on interstellar line intensities. These authors showed, by applying their Doublet-Ratio (*DR*) method (see Sec. 3.2), that "the observations can be accounted for by making the hypothesis that interstellar sodium occurs in discrete aggregations or clouds, which, while participating in the general galactic rotation, have in addition considerable random motions." The observation by Beals (1936) of the K-line in the spectra of ϵ Orionis, ζ Orionis, and ρ Leonis, which showed contours with two clear minima, provided direct confirmation for the inference of Wilson and Merrill.

Further advance resulted from the great increase in speed and resolving power achieved by Adams (1941) and Dunham (1956) in the development of the grating coudé spectrograph of the Mount Wilson Observatory. The discovery of faint interstellar lines due to Ti II (Dunham and Adams 1937; Dunham 1937*a*), Ca I (Dunham 1937*b*), K I (Dunham 1937*b*), and Fe I (Adams 1943) was a direct result of this instrumental development. Some additional absorption features with unmistakable interstellar appearance were detected (Dunham 1937*b*) but could not be identified with any resonance line of atom or ion. The suggestion by Swings and Rosenfeld (1937) that the line observed at 4300.4 A might be due to CH, led McKellar (1940) and Douglas and Herzberg (1941, 1942) to identify the unexplained lines as transitions arising from the lowest rotational levels of the CH, CN, and CH^+ radicals.

Interstellar Ca II and molecular lines in 300 bright O and B stars were studied at coudé dispersions by Adams (1943, 1949), who found that occurrence of multiple components was a general phenomenon. This extensive material provided a basis for the modern study of the distribution and physical state of the interstellar gas.

Studying the spectra of distant stars under medium resolving power, Merrill (1934) discovered a number of absorption features with apparent interstellar origin. Unlike the identified atomic and molecular lines, however, these features are broad and shallow. Understanding their origin remains still the most challenging problem of stellar spectroscopy.

The considerable increase in light-gathering power achieved in the coudé spectrograph of the Hale telescope on Palomar Mountain has allowed the study of interstellar lines in very distant stars. The relation between the large-scale distribution of the interstellar gas and the structural features or spiral arms of the galactic system was thereby established (Münch 1953, 1957).

The search for additional molecular lines in the interstellar spectrum has been resumed recently by Herbig at the Lick Observatory. According to a preliminary report (Herbig 1960), the line at 3143 A, detected by Spitzer and Field (1955) in the spectrum of ζ Persei, is also present in the spectrum of ζ Ophiuchi, together with lines at 3137 A and 3146 A. All of these lines lie (Feast 1955) at the position of CH lines, as predicted by McKellar (1941).

2. THE DATA OF OBSERVATIONS

2.1. Identification

The interstellar absorption features known at present are listed in Table 1. The 2^3S–3^3P absorption line of He I, observed in the spectra of stars imbedded in dense emission nebulae, is definitely interstellar in nature, but because the conditions required for its formation are different from those for other atomic lines given in Table 1, it should be considered separately and will be distinguished by the name *nebular absorption line.* The presence of a particular line in the interstellar spectrum is dependent on the abundance of the ion or molecule concerned and by the population of the lower state of the transition involved. Without calculating the state of ionization and excitation prevalent in interstellar space, we can see that all atomic and molecular lines of Table 1 arise from ground levels. The Ti II ion deserves special comment, since the $^4F_{5/2}$ level lies only 0.0165 ev above the $^4F_{3/2}$ ground state. No lines are observed arising from $^4F_{5/2}$, despite their relatively large transition probabilities. This means that the radiation downward transitions $^4F_{5/2}$–$^4F_{3/2}$, which occur only through magnetic-dipole interaction with the low probability of 10^{-4} sec^{-1}, are sufficient to keep the population of the $^4F_{5/2}$ sublevel insignificant (Burgess, Field, and Michie 1960). A similar remark applies to the sublevels J = 3, 2, and 1 of the ^5D ground term of Fe I. More striking still is the case offered by the molecular lines. The lowest rotational state of CH, for example, is double, with a separation of 0.0022 ev. The $R_2(1)$ line in the (0,0) $X^2\Pi \rightarrow A^2\Delta$ band has as its initial state the lower of these two sublevels, and it is observed in the interstellar spectrum at 4300.32 A. The $R_1(1)$ line at 4309.9 A arises from the upper sublevel, but it is not an interstellar feature. A different situation arises for interstellar CN, for which the population of an excited level is detectable. The interstellar feature at 3874.00 A is the R(1) line of the (0,0) band in the $^2\Sigma$–$^2\Sigma$ system and, therefore, arises in the first rotational state, which lies 3.78 cm^{-1} = 4.7×10^{-4} ev above the ground state. The relative intensity of this line to that of the $R(0)$ line at 3874.60 A, expressed in terms of the temperature that would characterize a Boltzmann population of the levels, provides a value of 2.3° K for such rotational temperature (McKellar 1941).

The absence from Table 1 of particular ions with resonance lines in the accessible region may be in general ascribed to ionization conditions, although in some cases they cannot be computed accurately (see Sec. 3.5). The lines of Li I at 6707.74 A and 6707.89 A have been searched for unsuccessfully by Spitzer (1949 *b*). The lines of Be II at 3130.42 A and 3131.06 A have not been detected either (Spitzer 1949*b;* Spitzer and Field 1955). The absence of Al I at 3944 A (Rogerson, Spitzer, and Bahng 1959) is more conspicuous, since aluminum has about the same cosmic abundance as calcium or sodium. The calculations of the state of inoization show, however, that the intensity of the Al I 3944 A line in a star with strong interstellar lines, as χ^2 Orionis, should be below the limits of detectability of existing equipment (Burgess *et al.* 1960; Münch and Zirin 1961).

The conditions necessary for a molecule to be a possible interstellar absorber are much more difficult to specify than those for atoms, because little is known at present

TABLE 1

INTERSTELLAR ABSORPTION FEATURES

Atomic Lines			Molecular Lines			Diffuse Lines	
Atom or Ion	λ (A)	Transitions	Molecule	λ (A)	Transitions	λ (A)	Int.
Na I.	3302.34	$3^2S_{1/2}$–$4^2P^0_{3/2}$	CH	4300.31	$A^2\Delta \leftarrow X^2\Pi$ (0,0) $R_2(1)$	4430.6	10
	2.94	–$4^2P^0_{1/2}$		3890.23	$B^2\Sigma^- \leftarrow$. . . . . $^PQ_{12}(1)$	4760.	. . . .
	5889.95	–$3^2P^0_{3/2}$		86.39	. . . . $Q_2(1)+{}^QR_{12}(1)$	5780.5	3
	95.92	–$3^2P^0_{1/2}$		78.77	. . . . $R_2(1)$	5797.1	1
				3146.01	$C^2\Sigma^+ \leftarrow$. . . . . $^PQ_{12}(1)$	6203.0	1
K I.	7664.91	$4^2S_{1/2}$–$4^2P^0_{3/2}$		43.15	. . . . $Q_2(1)+{}^QR_{12}(1)$	6270.0	1
	98.98	–$4^2P^0_{1/2}$		37.53	. . . . $R_2(1)$	6283.9	6
						6613.9	2
Ca I.	4226.73	4^1S_0 –$4^1P^0_1$	CN	3874.61	$B^2\Sigma^+ \leftarrow X^2\Sigma^+$ (0,0) R(0)		
				75.77	. . . . P(1)		
Ca II.	3933.66	$4^2S_{1/2}$–$4^2P^0_{3/2}$		74.00	. . . . R(1)		
	68.47	–$4^2P^0_{1/2}$					
			CH⁺	4232.58	$A^1\Pi \leftarrow X^1\Sigma^+$ (0,0) R(0)		
Ti II.	3072.97	$a^4F_{3/2}$–$2^4D^0_{1/2}$		3957.74	. . . . (1,0) R(0)		
	3229.19	–$2^4F^0_{5/2}$		3745.33	. . . . (2,0) R(0)		
	41.98	–$2^4F^0_{3/2}$					
	83.76	–$2^4G^0_{5/2}$					
Fe I.	3719.94	a^5D_4 –$2^5F^0_5$					
	3859.91	–$2^5D^0_4$					

about the processes which lead to molecule formation. A valuable list of all lines from diatomic molecules that could be of interest in the identification of or search for new molecular interstellar lines has been compiled by McKellar (1941). The laboratory data on radicals and molecular ions, however, are rather meager, as shown by the difficulties encountered in the identification of lines eventually attributed by Douglas and Herzberg (1941) to CH^+. An interstellar line at 3934.3 A, measured by Dunham (1937*b*) in the spectrum of χ^2 Orionis and HD 190603, was at one time suggested as due to NaH. This line is now thought to be most likely the K-line of a cloud moving with a radial velocity of +55 km/sec. The only unidentified sharp interstellar line is thus 3579.04 A.

Problems of identification of diffuse absorption features differ from those encountered with atomic or molecular lines. Several authors (Beals and Blanchet 1938; Greenstein and Aller 1950; Duke 1951) have shown that the intensity or depth of the strongest of these "bands," at 4430 A, is rather well correlated with color excess. It has been suggested that they might be formed by gases absorbed in the solid particles producing the reddening; however, this correlation with color excess may not mean more than that of the general correlation of all interstellar features with distance in the galactic plane. As will be explained later, the intensity of sharp interstellar features is only loosely correlated with distance (or color excess) because these features are strongly saturated. Were the 4430 A and other unidentified bands unsaturated absorption features, they should exhibit a closer correlation with distance. An alternative suggestion for the origin of the diffuse bands is that they are absorbed by a homonuclear diatomic molecule, although it has been argued that if the diffuseness of the bands were due to unresolved rotational structure, the molecule involved would have to consist of two atoms heavier than neon, which could be expected to have a rather low abundance in interstellar space (Herzberg 1955). Still, if the diffuseness were due to predissociation, it would require a low dissociation energy (less than 1.5 ev), compared with the rather high dissociation energies of all expected diatomic molecules. Thus it seems that we are as far now from identifying the diffuse interstellar features as we were at the time of their discovery a quarter of a century ago.

2.2. Catalogues

A useful list of all publications containing observational data pertaining to the interstellar absorption lines has been given by Binnendijk (1952). This list should be supplemented by a few recent publications. Beals and Oke (1953) have given the measures of velocities and intensities of interstellar Na I and Ca II lines that were obtained at the Dominion Astrophysical Observatory up to 1946. The almost complete lack of data about interstellar lines in stars of the southern hemisphere has been partially met by the list of velocities of the Ca II lines published by Feast, Thackeray, and Wesselink (1955, 1957). Supplementing Adams' material (1949), recent results obtained with the coudé spectrographs of the Mount Wilson and Palomar Observatories have been given by Münch (1957) and Münch and Zirin (1961). The photoelectric measures of the interstellar D-lines and of CH^+ 4232 A in nine stars by Rogerson, Spitzer, and Bahng (1959) should also be noted.

Recently a few observations of interstellar lines with extremely high spectral resolving power have been made. Using the Czerny spectrograph of the McMath solar telescope at Kitt Peak, Livingston and Lynds (1964) have photographed the D-lines in a few stars with an image intensifier tube. Under a resolving power of 250,000, a D-line in 1b Scorpii (m_v = 4.8) has been recorded with a 30-minute exposure.

At the Lick Observatory, photoelectric observations of the D2-line with a stack of Fabry-Perot etalons have been carried out by Hobbs (1965). Two of the interstellar components on the D2-line in α Cygni observed with this PEPSIOS spectrometer appear double, with an apparent separation closely corresponding to the hyperfine structure splitting. Further work is required to prove convincingly that hyperfine structure splitting is actually being observed and that the internal velocity dispersion in the clouds giving rise to the components is less than 0.64 km/sec. The time required to obtain a scan 0.7 A wide in α Cygni with this instrument is 25 minutes.

At the 100-inch Mount Wilson coudé spectrograph, Münch and Vaughan (1966) have used a pressured scanned Fabry-Perot interferometer to obtain photoelectric high-resolution profiles of the interstellar D-lines. The use of a fairly large entrance diaphragm (0.5 $\times$ 0.5 mm) instead of a narrow slit, allows the scanning with resolution of 0.02 A, a spectral range of 1 A around the D-lines in a 5.0 magnitude star in about 1 hour. The gain of this interferometer system over conventional photographic methods is thus near ten, and it compares in efficiency with the image-tube work of Livingston and Lynds (1964).

2.3. Structure of the Interstellar Absorption Lines

It is appropriate to describe here the main results obtained by Adams (1949) in regard to the structure of the Ca II lines. Among the 300 stars observed by this author, 87 had lines that could be measured as double, 17 as triple, and 4 as quadruple. In another 40 stars, the lines appeared sufficiently broad or showed enough incipient resolution to indicate almost certain complexity. Thus essentially one in two of the observed stars showed complex lines, and it should be kept in mind that the finite resolving power of the spectrograph is the limiting factor for resolution of weaker components. We may illustrate the point by comparing the structure of the K-line observed by Adams in ϵ Orionis in spectra with 2.9 A/mm dispersion to the result of the observation under a dispersion of 1.14 A/mm (Oke 1961; Münch 1961). In the spectra showing multiple components of the K-line and the Na I D-lines reproduced in Plate 1, components at +3.0, +11.8, +18.3, +25.5, and +28.2 km/sec may be measured, although at 2.9 A/mm only components at +1.2, +10.6, and +26.5 km/sec could be measured by Adams. The multiplicity of the K-line suggests that the interstellar gas is concentrated in discrete units or clouds. Since the separation between components is sometimes complete, in the sense that no trace of absorption may be detected between them, the internal motions in each cloud in general must be smaller than the relative motions between clouds. From the detailed analysis of the extensive and homogeneous material obtained by Adams (1949), we may draw the following two conclusions:

1. The radial velocity of the stronger component of a line always agrees with the velocity expected from galactic rotation—at a point half the distance to the star—to a closer extent than that of any other component. These close agreements may be illustrated by the root-mean-square deviation of the stronger components in 84 stars of a mean distance of 320 pc from a double sine wave of conveniently adjusted amplitude, which amounts to less than 1 km/sec.

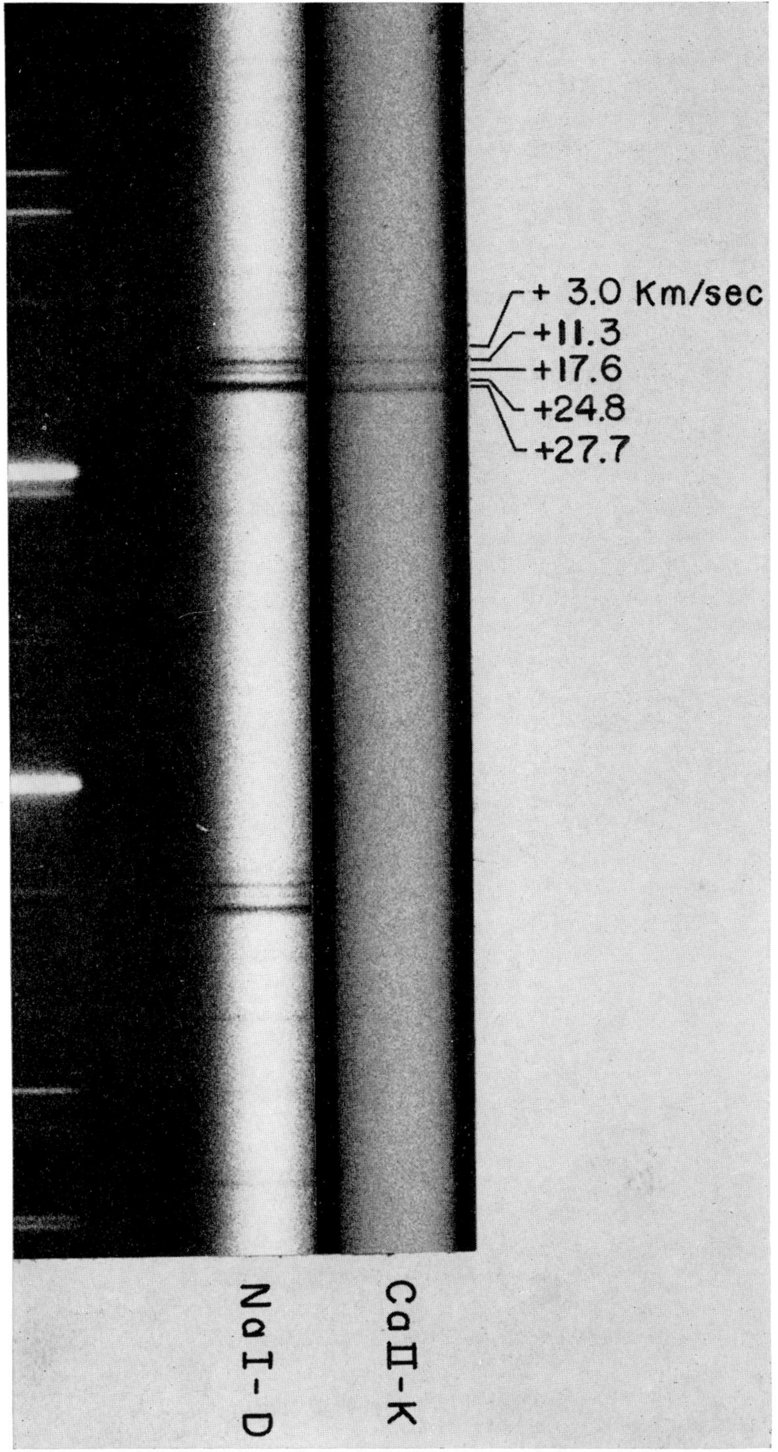

PLATE 1.—Interstellar absorption lines showing multiple components of Ca II K and Na I D_1 and D_2 in ϵ Orionis. The original spectrogram of the K-line was obtained by J. B. Oke with the 100-inch coudé in a IIa-O emulsion at a dispersion of 1.14 A/mm. The spectrum of the D-lines was obtained by the author with the same instrument in a 103a-D emulsion at 1.7 A/mm dispersion. The many sharp lines in the yellow spectrum are telluric. D1 components are left of center, D2 at right.

2. The occurrence of multiple lines in spectra of stars of a particular region of the Milky Way does not depend on the galactic longitude of the region. Highly complex lines are observed in Sagittarius and Cygnus, where galactic-rotation shifts vanish at the small distances concerned. The mean separation of the weaker components from the stronger one, with or without regard to sign, does not depend on galactic longitude either, although some degree of asymmetry exists between the numbers of components with positive and negative high velocities.

The decomposition of the velocities of interstellar clouds into a systematic component, arising from galactic rotation, and a random component representing peculiar mass

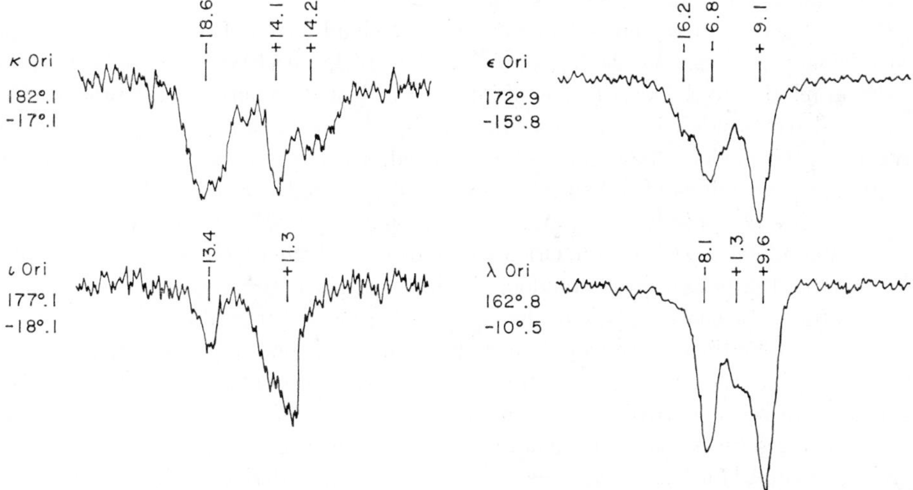

FIG. 1.—Microphotometer tracings of the K-line in four members of the Orion association, from 2.7 A/mm plates. The intensity scale is about the same for the two stars on the left, which were obtained in 103a-O emulsion. The spectra on the right, with higher contrast and less granularity, were obtained in Cramer-Contrast plates. The various components are labeled by their residual radial velocities (in km/sec), as measured by Adams (1949).

motion, thus seems to be physically significant. The two components cannot be separated unambiguously since there is no way to determine the distance to a particular cloud. Such a separation may be carried out only statistically for a group of stars. The apparent angular scale over which the structure of the interstellar lines changes is too small to delineate the boundaries of individual stars in the sky. Attempts have been made to derive the angular extent of individual clouds (Whipple 1948) from observed data, but their usefulness is questionable because the structure of the lines may change radically over very small areas. For example, we may consider the structure of the K-line in κ, ϵ, ι, and λ^1 Orionis as observed by Adams (1949) and illustrated in Figure 1. The same lines are observed in the various members of several double stars or multiple systems, but it is practically impossible to extend the observations to a close network of, say, a few degrees extent because of the scarcity of suitable background objects for observation.

In other words, the structure of the interstellar lines changes over distances smaller than those between stars in which they may be observed. It is therefore necessary to characterize the structure of the medium statistically as a whole. The system of interstellar clouds is then described by the following parameters:

A. The mean number of clouds, ν kpc^{-1}, intercepted by a line of sight in the galactic plane.

B. The mean intensity of the lines produced in one cloud. As equivalent, we may give the mean number of atoms contained in one cloud and their internal velocity distribution.

C. The fractional volume of space, α, occupied by the clouds and the relative density of atoms inside and outside the clouds.

Essentially these parameters define the discrete-cloud model of the interstellar medium originally proposed by Ambartsumian and Gordeladse (1938) and Ambartsumian (1940) in relation to the distribution of interstellar dust. In Chapter 1 the properties of this model are considered in various contexts. The description of the interstellar medium in terms of this model is, however, highly idealized. The mean number of clouds per unit length, ν, has meaning only when the clouds are distributed at random. This is not true on a galactic scale (spiral structure), but considering only small distances from the Sun, for example <500 pc (within the Orion spiral arm), we have no observational evidence to contradict the assumption of random distribution. The determination of the parameters and functions characteristic of the discrete-cloud model from data related to interstellar absorption lines alone would be a simple matter if lines produced by individual clouds were always completely separated. The resolving power of the 2.9-A/mm material is about 8 km/sec, and, depending on the relative intensity of the components, the effective limit of resolution between them may be even higher. Statistics of interstellar line components in Adams' catalogue have led Blaauw (1952) to derive a value for ν between 8 and 12 kpc^{-1}; he allows for the finite resolving power, but disregards the different intensities of the components. Blaauw finds that the observed frequencies of the velocity components (see Chap. 1, Fig. 1) may be accounted for in terms of an isotropic velocity distribution function of the form

$$\psi_1(v) = \frac{1}{\sigma_1\sqrt{2}}\, e^{-|v|\sqrt{2}/\sigma_1} \tag{1}$$

where $\sigma_1 = 7$ km/sec. A Gaussian velocity distribution

$$\psi_2(v) = \frac{1}{\sigma_2\sqrt{(2\pi)}}\, e^{-v^2/2\sigma_2^2} \tag{2}$$

fits the observed facts less well than $\psi_1(v)$, since the former fails to account for the high observed frequencies of high-velocity components ($|v| > 24$ km/sec).

This simple statistical description of the motions of the interstellar clouds can be criticized on the grounds that it makes no distinction between the strong low-velocity components and the weak highly displaced ones. As we shall see later, there seem to be intrinsic differences in the physical properties of these two kinds of clouds. The kinematical properties of high-velocity clouds are further discussed by Spitzer in this volume.

3. METHODS OF ANALYSIS

3.1. Curves of Growth for Single Clouds

The profile of an interstellar absorption line depends on the number of absorbers along the line of sight and their radial velocity components. Let us suppose that the line of sight passes through one cloud of linear extent L and constant number density N of the relevant atom or ion. The absorption coefficient at wavelength λ of an undisturbed atom at rest is

$$k_\lambda = \frac{\lambda_o^4}{8\pi^2 c} \frac{g_j}{g_i} a_{ji} \frac{\gamma_j}{\gamma_j^2 + (\lambda - \lambda_o)^2} \tag{3}$$

where λ_o is the wavelength of the line center, which is assumed to arise between two states of statistical weights g_j and g_i and which has a transition probability a_{ji}. The radiation damping constant γ_j, expressed in λ-units, is

$$\gamma_j = \frac{\lambda^2}{4\pi c} \sum_{r<j} a_{jr} . \tag{4}$$

Now let $\psi(y)$ be the normalized distribution of velocity components v along the line of sight, measured from their mean, v_o, in units of their dispersion σ:

$$\int_{-\infty}^{+\infty} \psi(y)\,dy = 1 , \qquad \text{with} \qquad y = \frac{v - v_o}{\sigma} . \tag{5}$$

In general, $\psi(y)$ will depend on the distance r through v_o or σ, and we would have to express the line intensity as an integral over distance. Restricting ourselves to clouds of linear dimensions small enough to make galactic-rotation effects negligible, we write for the optical depth

$$\tau_\lambda = \frac{\lambda_0^4}{8\pi c} \frac{g_j}{g_i} \frac{a_{ji}}{b_\lambda} NL \frac{a}{\pi} \int_{-\infty}^{+\infty} \frac{\psi(y)\,dy}{a^2 + (v - y)^2} \tag{6}$$

where $b_\lambda = \lambda_o \sigma / c$ is the dispersion measured in λ-units, a is the damping constant in units of b_λ, and

$$v = \frac{1}{b_\lambda} \left[\lambda - \lambda_o \left(1 + \frac{v_o}{c} \right) \right] \tag{7}$$

is the wavelength shift from the line center at the cloud. Since re-emission effects are negligible, we write for the intensity within the line, in units of the background continuum,

$$r_\lambda = e^{-\tau_\lambda} , \tag{8}$$

and the equivalent width is

$$W = \int_{-\infty}^{+\infty} (1 - e^{-\tau_\lambda})\,d\lambda . \tag{9}$$

It is convenient to write W in units of b_λ:

$$\frac{W}{b_\lambda} = \int_{-\infty}^{+\infty} \{ 1 - \exp[- \tau_o H(a, v)] \} \, dv \tag{10}$$

where

$$\tau_o = \frac{\lambda^4}{8\pi c} \frac{g_j}{g_i} \frac{a_{ji}}{b_\lambda} NL\psi(0) \tag{11}$$

is the optical depth at the line center for vanishing damping and

$$H(a,v)=\frac{a}{\pi\psi(0)}\int_{-\infty}^{+\infty}\frac{\psi(y)\,dy}{a^2+(v-y)^2}. \tag{12}$$

For interstellar lines, a is very small, about 10^{-3} or less. For a Maxwellian velocity distribution, when $\psi(v)$ is of the form (2), Strömgren (1948) has verified that the values of W computed for $a = 10^{-3}$ do not differ sensibly from those derived with $a = 0$ until τ_o reaches a value of about 10^3. The following calculations may then be carried out with $a = 0$:

$$H(a,v)=\frac{\psi(v)}{\psi(0)}. \tag{13}$$

For the Maxwellian case (2), $W/b_\lambda\sqrt{2}$ has been tabulated by Strömgren (1948) as a function of $\tau_o/\sqrt{2}$. For small values of τ_o we have

$$\frac{W}{b_\lambda}=\sqrt{(2\pi)}\sum_{n=1}^{\infty}(-1)^{n-1}\frac{\tau_o^n}{n!\sqrt{n}} \tag{14}$$

but for $\tau_o \gg 1$, we have the asymptotic expansion

$$\frac{W}{b_\lambda\sqrt{2}}=2(\log\tau_o)^{1/2}\left[1+\frac{0.2886}{\log\tau_o}-\frac{0.1335}{(\log\tau_o)^2}+\ldots\right]. \tag{15}$$

If the velocity distribution were of the form (1), we have explicitly (Münch 1957)

$$\frac{1}{b_\lambda\sqrt{2}}W(\tau_o)=\log\gamma\tau_o+E_1(\tau_o) \tag{16}$$

where $\log\gamma = 0.5772\ldots$ is Euler's constant and $E_1(\tau_o)$ is the first exponential integral of τ_o. The limiting behavior of equation (16) is described by the power series

$$\frac{1}{b_\lambda\sqrt{2}}W(\tau_o)=\sum_{n=1}^{\infty}(-1)^{n-1}\frac{\tau_o^n}{n!n} \tag{17}$$

and by the asymptotic series ($\tau_o \gg 1$)

$$\frac{1}{b_\lambda\sqrt{2}}W(\tau_o)=\log\gamma\tau_o+e^{-\tau_o}\sum_{n=0}(-1)^n\frac{n!}{\tau_o^{n+1}}. \tag{18}$$

Equation (16) would apply to situations with one line that arises by absorption through many clouds with negligible internal motions but with a distribution of radial velocities of the form of equation (1).

3.2. The Doublet-Ratio Method

We have now shown how the equivalent widths of an interstellar absorption line depend on the number of atoms N in the ground state and on the Doppler widths. These parameters cannot both be determined from the strength of one line. But when the strength of both components of a doublet are available, such as K and H of Ca II, or D1 and D2 of Na I, we may evaluate both parameters, since N is the same for the two lines of the doublet and the ratio between the transition probabilities of the two lines is known. The intensity ratio *DR* between the members of the Na I or Ca II doublets,

$$DR(\tau_o)=\frac{W(2\tau_o)}{W(\tau_o)}=\frac{W_2}{W_1}, \tag{19}$$

is thus a function of τ_o alone. A value of *DR* from observations fixes the corresponding values of τ_o and W_1/b_λ. The observed W_1 then determines b_λ, and from equation (11), rewritten in the form

$$b_\lambda\frac{\tau_o}{W_1}=\frac{\lambda^4}{8\pi c}\frac{g_j}{g_1}a_{j1}\frac{NL}{W_1}\psi(0), \tag{20}$$

we evaluate *NL* for the particular line considered.

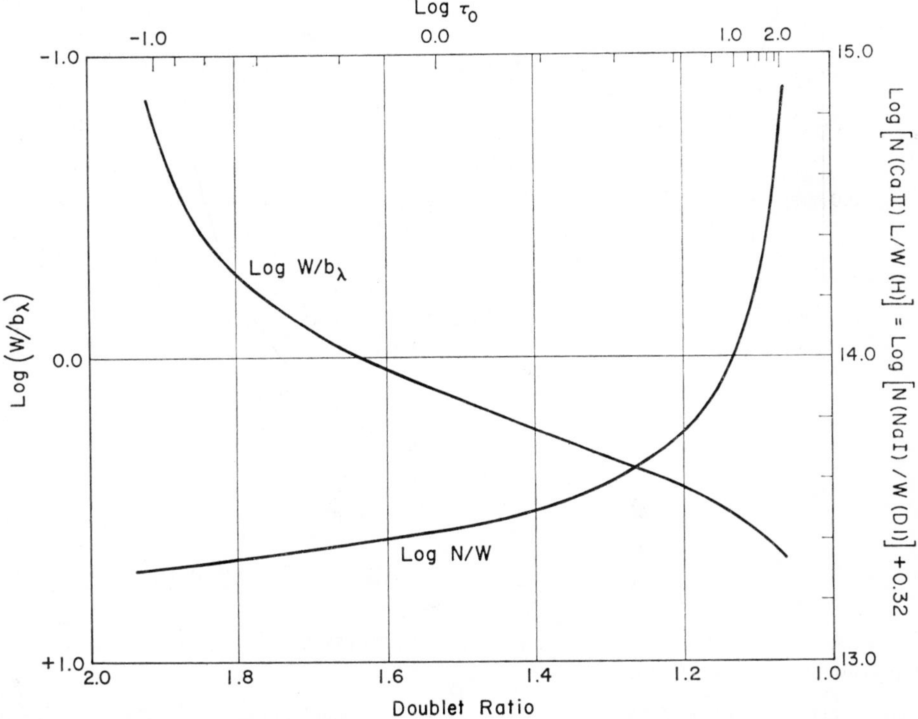

FIG. 2.—Theoretical curve of growth for the interstellar Na I D-lines and Ca II H- and K-lines, for a Gaussian velocity distribution.

A convenient table giving log N/W_1 and log W_1/b_λ in numerical form as function of *DR* or τ_o for a Gaussian ψ and for both the Na I and Ca II doublets has been published by Strömgren (1948) and is graphically represented in Figure 2. The curve of growth for an exponential ψ has been given by Münch (1957) and is shown in Figure 3.

For weak lines, the intensity of a line is determined by the number of atoms, independently of b_λ. Let W_{01} be the equivalent widths of a completely unsaturated line corresponding to given *NL;* then

$$W_{01}=\frac{\lambda^4}{8\pi c}\frac{g_j}{g_1}a_{j1}NL. \tag{21}$$

The actual line strength W_1 may then be written

$$W_1 = W_{01}\, S(DR) \tag{22}$$

where $S(DR) \leq 1$. The saturation function S may be expressed as a power series in $1 - DR/2$ from equations (19) and (14) or (19) and (17). For a Gaussian ψ, we have

$$S_G(DR) = \frac{DR}{2} - \left(\frac{8}{3\sqrt{3}} - 1\right)\left(1 - \frac{DR}{2}\right)^2 \tag{23}$$

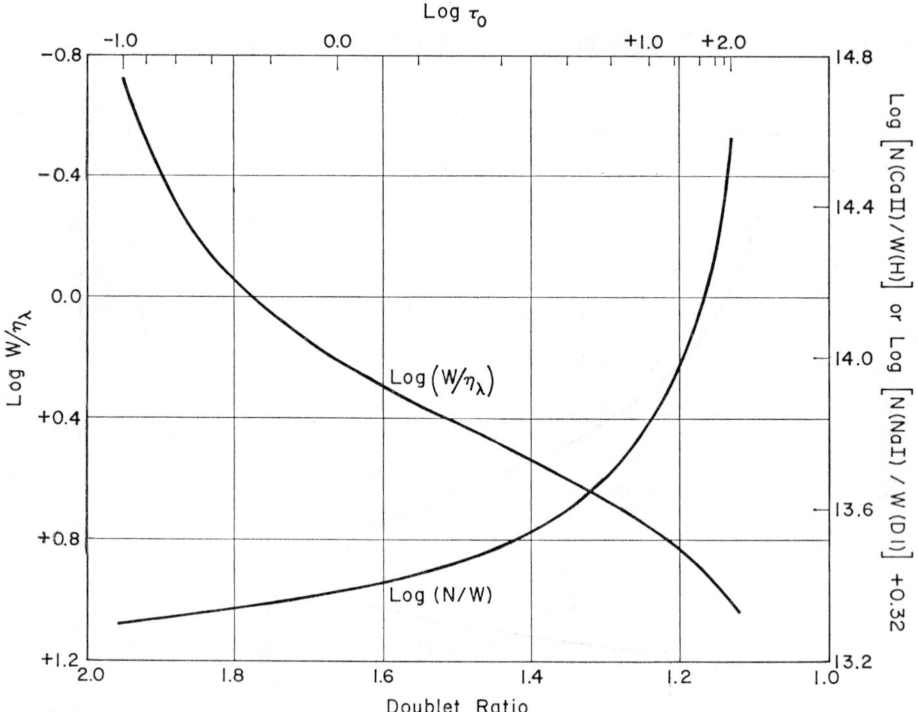

FIG. 3.—Theoretical curve of growth for the interstellar Na I D-lines and Ca II H- and K-lines, for an exponential velocity distribution.

where terms of order higher than $(1 - DR/2)^2$ are neglected. To the same order of approximation we have for an exponential ψ

$$S_E(DR) = \frac{DR}{2} - \tfrac{7}{9}\left(1 - \frac{DR}{2}\right)^2. \tag{24}$$

When the doublet ratio approaches unity, the line strengths become insensitive both to NL and to the velocity dispersion parameter. From the first terms of the asymptotic expansions (15) and (16), which are poor approximations in practice, we obtain

$$b_\lambda^{(2)} = W_1 \left(\frac{DR - 1}{2 \log 4}\right)^{1/2} \tag{25}$$

and

$$b_\lambda^{(1)} = W_1 \frac{DR - 1}{2^{1/2} \log 4}. \tag{26}$$

The corresponding values of NL could then be obtained from equation (20).

The failure of the doublet-ratio method for strong lines imposes serious limitations to its application. With Na I lines, however, when the D-lines become saturated, the ratio of their intensity to that of the doublet between 3302 and 3303 A may be used, as Dunham (1939) has done for χ^2 Orionis. The ratio between the transition probabilities of the two Na I doublets is 21.8 (Filippov and Prokofjef 1929), and therefore the ratio D1/λ 3303 between the strengths of their two fainter lines may be calculated as a function of $N/W(D_1)$ by the same procedure followed for the two yellow lines. The values of D1/λ 3303 for a Gaussian ψ have been calculated by Strömgren (1948).

3.3. Curves of Growth for More than One Cloud

When an absorption line is formed through a small number n of clouds moving with respect to each other and with nonvanishing internal motions, the considerations of Section 3.1 must be extended. Let us assume, following Strömgren (1948), that each of the n clouds is characterized by an internal Gaussian velocity distribution with dispersion b_i and that their relative radial mass motions are large enough that no significant overlap occurs in their internal velocity distributions. The total number of absorbing atoms in each cloud is then NL/n, and the optical thickness of each is τ_o/n. The total equivalent width is then $nW(\tau_o/n)$. Considering the relation between W and NL for the n clouds, then, we see that it is the same as that for one cloud when the Doppler constant is taken equal to nb_i. In the doublet-ratio method, the relation between DR and NL/W, after b has been eliminated, remains unchanged.

When the number of clouds along the line of sight is large and their velocity distribution is Gaussian with parameter b_e, the resulting velocity distribution of all the absorbing atoms is also Gaussian with parameter $(b_e + b_i)^{1/2}$. The relation between DR and NL/W, again, remains unchanged.

The case of a small number of clouds producing partially overlapping interstellar lines is rather difficult to handle. Consider the case of two equal clouds with the mean radial velocities v_o' and v_o''. We should take

$$2\psi(v) = \frac{1}{\sigma\sqrt{(2\pi)}}\, e^{-(v-v_o')^2/2\sigma^2} + \frac{1}{\sigma\sqrt{(2\pi)}}\, e^{-(v-v_o'')^2/2\sigma^2}. \tag{27}$$

With $a = 0$, as in Section 3.1, we now have

$$\frac{\tau_\lambda}{\tau_o} = \tfrac{1}{2} e^{-(v-\alpha/2)^2} + \tfrac{1}{2} e^{-(v+\alpha/2)} \tag{28}$$

where

$$\alpha = \frac{v_o' - v_o''}{\sigma\sqrt{2}} \tag{29}$$

is the velocity difference between the two clouds measured in units of the Doppler width. Introducing equation (28) in equation (10), W/b_λ may be found by numerical integration as a function of α and τ_o. Values of W/b_λ for $\alpha = 1$, 2, and 3, evaluated by Strömgren (1948), can be reproduced rather closely by the relation valid for one cloud with effective Doppler constants b^* with values 1.2 b, 1.5 b, and 1.8 b, respectively. The relation between NL/W and DR, which is explicitly independent of b, seems to be unaffected by changes in α. The doublet-ratio method for one cloud with a Gaussian distribution of internal radial velocities is therefore of wide application. Although NL always has the same meaning, the effective Doppler constant varies between the limits b_e and $(b_e^2 + b_i^2)^{1/2}$ and generally increases with the number of clouds.

3.4. Statistical Curve of Growth

If the interstellar clouds are spatially distributed at random, in the sense of Poisson's law, and in velocity according to a definite probability density function, the relation between the strength W of an interstellar line and distance R can be specified only as a probability. That is, for a given R the probability of occurrence of any value W may be predicted. The calculation of such a statistical curve of growth is difficult and has been carried out only for the case in which the line produced in any cloud is rectangular in profile. Let $r_j(\lambda)$ be the residual intensity of the line produced in the jth cloud:

$$r_j(\lambda) = \begin{cases} 1 & \text{when} \quad |\lambda - \lambda_j| > Fb_i \\ 0 & \text{when} \quad |\lambda - \lambda_j| \leq Fb_i \end{cases} \tag{30}$$

where F is a constant, b_i measures the velocity dispersion of the absorbing atoms in one cloud, and λ_j is the central wavelength of the line. The total equivalent widths formed through n clouds is

$$W_n = \int_0^\infty d\lambda (1 - r_1 r_2 \ldots r_j). \tag{31}$$

In this expression $n = n(R)$ should be considered as a random variable taking integral values with the probability

$$p_n(R) = e^{-\nu R} \frac{(\nu R)^n}{n!}. \tag{32}$$

The values of W_n are also dependent on the probability density $\phi(\lambda_j)$ of the λ_j's. The mean value $\langle W_n \rangle$ over the λ_j's,

$$\langle W_n \rangle = \int_0^\infty d\lambda \left[1 - \prod_{j=1}^{n} \int d\lambda_j \phi(\lambda_j) r_j(\lambda - \lambda_j) \right], \tag{33}$$

has been obtained by Spitzer (1948) for the case in which λ_j is uniformly distributed around the line center λ_o:

$$\phi(\lambda_j) = \begin{cases} \dfrac{1}{2Fb_e}, & \text{when} \quad |\lambda_j - \lambda_o| < Fb_e \\ 0, & \text{when} \quad |\lambda_j - \lambda_o| > Fb_e \end{cases}. \tag{34}$$

Here b_e measures the dispersion of radial velocities among the clouds themselves. Equation (33) may then be written

$$\langle W_n \rangle = \int_0^\infty d\lambda \{ 1 - [1 - B(\lambda)]^n \} \tag{35}$$

where

$$B(\lambda) = \frac{1}{2Fb_e} \int_{\lambda_o - Fb_e}^{\lambda_o + Fb_e} r_j(\lambda - \lambda_j) d\lambda_j. \tag{36}$$

If we average $\langle W_n \rangle$ over all values of n, we find

$$\langle W \rangle = \sum_{n=0}^{\infty} p_n(R) \langle W_n \rangle = \int_0^\infty d\lambda [1 - e^{-\nu R B(\lambda)}]. \tag{37}$$

To evaluate $B(\lambda)$ we notice in equation (36) that when $|\lambda - \lambda_o| > F(b_e + b_i)$, $r_j = 1$ for all λ in the range $(\lambda_o - Fb_e < \lambda < \lambda_o + Fb_e)$ and thus $B = 0$. If $|\lambda - \lambda_o| < F(b_e - b_i)$ over the range of integration, then $\lambda - \lambda_j$ will sweep over the entire line width $2Fb_i$ for a single cloud. Thus, for a range $2Fb_i$ of λ_j, the residual intensity equals the central intensity r_o instead of unity. Accordingly we have

$$B(\lambda) = \begin{cases} (1 - r_o)\dfrac{b_i}{b_e}, & \text{if} \quad |\lambda - \lambda_o| \leq F(b_e - b_i) \\ 0 \quad , & \text{if} \quad |\lambda - \lambda_o| > F(b_e - b_i) \end{cases} . \tag{38}$$

For $F(b_e - b_i) < |\lambda - \lambda_o| < F(b_e + b_i)$, we may take $B(\lambda)$ to vary linearly between the two extremes established in equation (38), in the form

$$B(\lambda) = (1 - r_o)\frac{F(b_e + b_i) - |\lambda - \lambda_o|}{2Fb_e} . \tag{39}$$

Equation (37) may now be integrated to give

$$\langle W \rangle = 2Fb_e(1 - e^{-\nu Rq})\left[1 + \frac{b_i}{b_e} h(\nu Rq)\right] \tag{40}$$

where

$$q = (1 - r_o)\frac{b_i}{b_e} \tag{41}$$

and

$$h(x) = \frac{1 + e^{-x}}{1 - e^{-x}} - \frac{2}{x} . \tag{42}$$

Noticing that $h(x)$ is uniformly bounded by unity and that for small x it behaves as $x/6$, we may neglect the term depending on $h(\nu Rq)$ in equation (40). Of practical interest is the case in which $b_i < b_e$. When $h(x)$ is of importance, the assumption of a rectangular line profile would be unrealistic. We may thus write

$$\langle W \rangle = 2Fb_e(1 - e^{-\nu Rq}) . \tag{43}$$

The dispersion around this mean curve of growth should be measured by the mean square value of equation (31), taken over both the fluctuations in the λ_j-distribution and in $n(R)$. For simplicity we take only the mean square value of $\langle W_n \rangle$ in equation (35) over $n(R)$:

$$\langle W^2 \rangle = \sum_{n=0}^{\infty} p_n(R) \langle W_n \rangle^2 . \tag{44}$$

From equations (32), (35), (39), and (43) we find

$$\sigma_w = (\langle W^2 \rangle - \langle W \rangle^2)^{1/2} = W(e^{\nu Rq^2} - 1)^{1/2} . \tag{45}$$

When galactic rotation is taken into account, an additional dependence of W and σ_w on distance arises. The radial velocity shift produced by galactic rotation will increase the velocity spread among the clouds. Considering distances small enough that the double sine wave approximation holds, we may write

$$b_e(R) = b_e(0) + \frac{AR\lambda}{c} \sin 2(l - l_o) \tag{46}$$

for stars near the galactic plane, where A is Oort's constant, $l - l_o$ is the galactic longitude from the center, and c the velocity of light. Equation (46) must then be used with equations (43) and (45) when analyzing data in different galactic longitudes.

3.5. Ionization Equilibrium in Interstellar Space

To obtain information about the abundances of the elements in the interstellar medium, it is necessary to study their degree of ionization. The steady state requires that the ionization rate equal the recombination rate. Denote by N' and N'' the number densities of some ion in two successive stages of ionization, and by N_e the electron density. The radiative recombination rate $n(\text{rec})$ per unit volume is the same as in thermodynamic equilibrium and has a value

$$n(\text{rec}) = \alpha N'' N_e \tag{47}$$

where the recombination rate α is a function of the electron temperature T_e given by

$$\alpha = \frac{8\pi}{c^2}(2\pi m k T_e)^{-3/2} \sum_j \frac{g_j'}{2 g_1''} e^{\chi_j/kT_e} \int_{\chi_j}^{\infty} (h\nu)^2 a_j(\nu)\, e^{-h\nu/kT_e} d(h\nu) \tag{48}$$

where g_j' denotes the statistical weight, χ_j the ionization potential, $a_j(\nu)$ the continuous absorption cross-section of the jth excited state of the atoms in the lower state of ionization, and g_1'' the statistical weight of the ground state of the next state of ionization. If it is assumed that ionization is produced only by radiation of intensity $I(\nu)$, the photoionization rate is

$$n(\text{ion}) = 4\pi N' \int_{\nu_1}^{\infty} a_1(\nu) \frac{I(\nu)}{h\nu} d\nu = N'\Gamma \tag{49}$$

where $a_1(\nu)$ is the atomic absorption coefficient from the ground state of the low state of ionization and ν_1 is the ionization limit. We may thus write the ionization equation in the form

$$\frac{N'' N_e}{N'} = \frac{\Gamma}{\alpha}. \tag{50}$$

To show the relation between this equation and the one corresponding to thermodynamic equilibrium, we introduce equations (48) and (49) into equation (50) to obtain

$$\frac{N'' N_e}{N'} = \frac{2 g_1''}{g_1'} \frac{(2\pi m k T_r)^{3/2}}{h^3} e^{-\chi_1/kT_r} \left(\frac{T_e}{T_r}\right)^{1/2} w \tag{51}$$

where T_r is some temperature characteristic of the radiation $I(\nu)$ and the dilution factor w is defined by

$$w = \frac{g_1' \int \nu^2 a_1(\nu) G(\nu)\, e^{-(h\nu-\chi_1)/kT_r} d(h\nu/kT_r)}{\sum_j g_j' \int \nu^2 a_j(\nu)\, e^{-(h\nu-\chi_j)/kT_e} d(h\nu/kT_e)} \tag{52}$$

with

$$G(\nu) = \frac{c^2}{2 h\nu^3} I(\nu)\, e^{h\nu/kT_r}. \tag{53}$$

Under equilibrium conditions and when $h\nu_1 > kT_r$, equation (51) reduces to Saha's equation when the effect of captures in excited states is neglected.

The recombination coefficients α for a number of ions of importance have been computed by Strömgren (1948), Seaton (1951), and Weigert (1955). In Table 2 we reproduce Seaton's results for values of $T_e = 10^2, 10^3$, and 10^4 ° K. To evaluate the photoionization rates, the intensity of the radiation field must be specified. The radiation field usually adopted is that produced by all stars at the Sun, as evaluated by Dunham (1939), who expressed the net radiation field as the sum of blackbody intensities, each characterized by some temperature T_j and a dilution factor $\delta(T_j)$, in the form

$$I(\nu) = \sum_j \delta(T_j) B_\nu(T_j). \tag{54}$$

The ionization limits of all ions with which we are concerned in practice fall in the inaccessible ultraviolet. The stars which have more weight in determining the radiation field at these wavelengths have spectral types B and O, but the energy curves of these

TABLE 2

RECOMBINATION COEFFICIENTS $\alpha(T_e)$
(in units of 10^{-12} cm³ sec⁻¹)

Final Ion	Final State	T_e (° K)			Final Ion	T_e (° K)		
		10^4	10^3	10^2		10^4	10^3	10^2
H I........	1s	0.16	0.50	1.6	C I..........	0.33	1.7	7.6
	2s	.020	.077	0.24	Na I.........	0.17	1.2	5.9
	2p	.057	.22	0.69	K I..........	0.16	1.1	5.6
	3	.044	.20	0.62	Ca I.........	0.031	0.10	0.33
	4	.032	.17	0.54	Na II........	1.6	7.9	34
	$n\geq 5$	0.095	0.83	4.9	K II.........	2.0	9.1	38
					Ca II........	1.7	8.1	35
	Total	0.41	2.0	8.6				

stars may depart considerably from blackbody radiation at their effective temperatures. Further uncertainty arises from the effects of interstellar continuous extinction. Not only do we still lack information about the values of the absorption below 3000 A, but also we would expect the energy density of ionizing radiation to vary considerably from point to point because of the highly irregular distribution of interstellar dust. For these reasons, Seaton (1951) has calculated ionization rates $\gamma(T_j)$ for various blackbody intensities in the form

$$\gamma(T_j) = 10^{27} \int_0^\infty a_1(\epsilon)(\chi_1 + \epsilon)^2 e^{-\epsilon/kT_j} d\epsilon \tag{55}$$

where $a_1(\epsilon)$ is in cm² with χ_1 and $\epsilon = h(\nu - \nu_1)$ in rydbergs. The interstellar ionization rate then is given by

$$\Gamma = \sum_j \delta(T_j)\gamma(T_j). \tag{56}$$

The values of $10^{-8}\, e^{\chi/kT_j}\gamma\,(T_j)$ calculated by Seaton (1951) for a number of ions and various temperatures T_i are given in Table 3. In some cases a cutoff at the ionization potential of hydrogen has been introduced to account for the different conditions in

H I and H II regions (see Sec. 3.6). Table 4 gives the values of Γ/α obtained for the dilution factors $\delta(T_j)$ estimated by Dunham (1939), without taking into account interstellar continuous extinction.

The physical state of interstellar matter, in general, is determined by the degree of ionization of hydrogen, because of the large cosmic abundance of this element. Strömgren (1939) first pointed out that in the neighborhood of high-temperature stars interstellar hydrogen should be essentially ionized out to some distance from the star (H II regions). At larger distances, the radiation field will be depleted below 912 A, and the hydrogen will be neutral (H I regions). Next we consider the state of ionization in a pure hydrogen medium of constant density as a function of distance from a star of effective temperature T_r and radius R. Because of the simple structure of the hydrogen atom, we may discuss its ionization on the basis of equation (51), with w expressed as a simple function of distance. For low densities the population of excited states may be disregard-

TABLE 3

VALUES OF $10^{-8} e^{\chi/kT_j} \gamma(T_j)$ in $\sec^{-1}$ FOR BLACKBODY RADIATION

Ion	χ (ryd)	T_i (° K)				
		50000	25000	15000	10000	5000
H I	1.000	16.0	9.0	5.6	3.8	2.0
C I*	0.828	12.0	9.4	7.0	5.1	2.6
Na I	0.378	.0081	0.0067	.0056	.0046	.0031
K I*	0.319	.42	0.11	.026	.0071	.00091
Ca I	0.449	10.4	6.1	4.0	2.7	1.6
Na II	3.47	330.0	150.0	86.0	56.0	28.0
K II	2.34	640.0	300.0	170.0	120.0	57.0
Ca II	0.873	0.57	0.21	0.11	.067	0.031
Ca II*	0.873	0.13	0.095	0.073	.055	0.030

* The radiation field is cut off at λ = 912 A.

TABLE 4

IONIZATION EQUILIBRIUM Γ/α in cm^{-3} FOR DUNHAM'S STELLAR RADIATION FIELD

Ion	Γ (10^{-12} $\sec^{-1}$)	T_e (° K)		
		10^4	10^3	10^2
H I	43	105	22	5.0
C I*	110	320	63	14
Na I	4.0	24	3.3	0.68
K I*	42	260	38	7.5
Ca I	1400	8200	1200	230
Na II	0.62	0.39	0.079	0.018
K II	27	13	3.0	0.71
Ca II	2.5	1.5	0.31	0.071
Ca II*	0.86	0.51	0.11	0.025

* The radiation field is cut off at λ = 912 A.

ed. If the star is assumed to radiate as a blackbody of temperature T_r and we neglect recombinations taking place in the ground state, then

$$w = \frac{R_*^2}{4s^2} e^{-\tau} \tag{57}$$

where τ is some mean optical depth in the Lyman continuum. The factor $e^{-\tau}$ takes account of the loss to the radiation field of every quantum-producing ionization, which is assumed to be converted upon recombination into soft quanta that escape freely from the nebula. Let N be the total number density, and

$$N_{\rm H}^+ = N_e = xN, \text{ and } N_{\rm H} = (1-x)N\,. \tag{58}$$

The ionization equilibrium equation (51) is then written

$$\frac{x^2}{1-x} N = C \frac{1}{s^2} e^{-\tau(s)} \tag{59}$$

where

$$C = \frac{(2\pi m k T_r)^{3/2}}{4h^3 R_*^2} e^{-\chi/kT_r} \left(\frac{T_e}{T_r}\right)^{1/2}. \tag{60}$$

Let a_o be an effective mean value of the absorption cross-section in the Lyman continuum. Then

$$d\tau = (1-x) a_o N ds\,. \tag{61}$$

Putting

$$y = e^{-\tau} \quad \text{and} \quad z^{1/3} = \frac{s}{s_o} = \left(\frac{a_o N^2}{3C}\right)^{1/3} s\,, \tag{62}$$

we obtain $y = y(s)$ by solving the differential equation

$$\frac{dy}{dz} = -x^2 \tag{63}$$

obtained from (61) and (62) under the boundary condition $y(z=0) = 1$ simultaneously with the equation

$$\frac{1-x}{x^2} = \alpha \frac{z^{2/3}}{y}. \tag{64}$$

Here

$$\alpha = \frac{N s_o^2}{C} = \left(\frac{9}{a_o^2 N C}\right)^{1/3} \tag{65}$$

is a pure number, which in practice is small compared to unity. We notice that for small z, $x \approx 1$ and from equations (63) and (64)

$$y = 1 - z + \tfrac{6}{5}\alpha z^{5/3} + \dots\,. \tag{66}$$

For large z, from equation (64) we see that $x^2 \approx y/\alpha z^{2/3}$, and thus from equation (63) we obtain

$$y = A e^{-3z^{1/3}/\alpha}\,, \tag{67}$$

where A is a constant. Since $\alpha \ll 1$, from equations (66) and (67) we see that x will remain of order unity as long as y does not become comparable to α, and that when

y becomes smaller than $az^{2/3}$, x will tend rapidly to zero. This happens near $z = 1$. The value of s corresponding to $z = 1$,

$$s = s_o = \left(\frac{3C}{a_o N^2}\right)^{1/3}, \tag{68}$$

defines the linear dimensions of the region where the ionization is nearly complete, and it is called the Strömgren radius. The numerical integration of equations (63) and (64) leads to a dependence of x on s, for $s \approx s_o$, as shown in Table 5. The width of the transition zone, or the range of s over which x changes from about 1 to about 0, is seen in equation (67) to be measured by

$$\Delta s = \frac{a\, s_o}{3} = \frac{1}{a_o N} \tag{69}$$

which is independent of the source of ionization (the star) and represents the mean free path of an ultraviolet quantum in the neutral medium with density N. The dependence

TABLE 5

RADII OF H II REGIONS

Sp	T_i (° K)	$S_o\, N_H^{2/3}$ (pc)	Sp	T_i (° K)	$S_o\, N_H^{2/3}$ (pc)	Sp	T_i (° K)	$S_o\, N_H^{2/3}$ (pc)
O5.....	79000	140	O9.......	32000	46	B3.......	18600	7.2
O6.....	63000	110	B0.......	25000	26	B4.......	17000	5.2
O7.....	50000	87	B1.......	23000	17	B5.......	15500	3.7
O8.....	40000	66	B2.......	20000	11	A0.......	10700	0.5

Strömgren (1948).

on distance of the degree of ionization in the transition region may be found explicitly by setting $z = 1$. Eliminating y between equations (63) and (64), we find

$$-\frac{2-x}{x(1-x)}\frac{dx}{dz} = \frac{2}{3z} + \frac{1-x}{az^{2/3}}. \tag{70}$$

Since $a \ll 1$, for $z \approx 1$ we may neglect the term $2/3z$, and the equation can then be integrated to give

$$3z^{1/3} = B - a\left(\frac{1}{1-x} + 2\log_e \frac{x}{1-x}\right) \quad \text{with} \quad B = \text{const}. \tag{71}$$

The radii of the Strömgren spheres associated with main-sequence stars of various spectral classes (Strömgren 1939, 1948) calculated on the basis of the temperature scale established by Kuiper (1938) are given in Table 5.

The effective radius of the ionized sphere produced by a group of stars with different spectral types is computed by adding together the volumes of the H II regions that would be produced by each star.

When the spatial distribution of the interstellar hydrogen is nonuniform, determination of the extent of the H II region becomes more complicated. In principle it may be found given the distribution of ionizing stars and of the gas. An example of practical interest is that of a homogeneous cloud at distance d from the star but not surrounding it. The thickness S_o of the ionized shell is again found by requiring that its volume equal that of the corresponding Strömgren sphere, in the form

$$S_o = \frac{s_o^3}{3d^2} \quad \text{for} \quad S_o \ll d\,. \tag{72}$$

So far the calculations have been made using a number of simplifying assumptions. First, possible differences between the electron and the radiation temperatures have been disregarded. Actually, the temperature of H II regions is always close to 10,000° K. The value of C for an O5-type star (T = 80,000° K), according to equation (60), should be divided by $\sqrt{8}$ because of this, and the corresponding value of s_o should be divided by $\sqrt{2}$, a decrease of about 30 per cent. The numerical values of s_o given in Table 5, on the other hand, have been obtained with the value of a_o at the Lyman limit instead of with a mean value in the Lyman continuum. If a Lyman mean were used, the values of s_o would be increased by less than 20 per cent.

The neglect of recombinations in the ground state which give rise to a diffuse field of ultraviolet radiation have more important physical effects. According to Table 2, at $T_e = 10^4$ ° K, the relative frequency of recombinations in the ground state is $\varpi_o = 16/41$, which is not negligible. The extent of H II regions has been investigated by Jefferies and Pottasch (1959) and Münch (1961), who took into consideration the diffuse field of ionizing ultraviolet radiation. Without entering into detail, for $\varpi_o = 0.4$ and $a = 10^{-2}$, the radius of the H II regions is 1.20 s_o, with s_o defined by equation (68).

3.6. Excitation in Interstellar Space

Consider an atom or ion with a level m of excitation $\chi_m = h\nu$ above the ground state n. Let the Einstein probability coefficients B_{nm} and B_{mn} for absorption and induced emission, respectively, be expressed in terms of A_{mn}, the probability for spontaneous decay, through the standard relations

$$B_{mn} = \frac{c^2}{2h\nu^3} A_{mn} \quad \text{and} \quad B_{nm} = \frac{g_m}{g_n}\frac{c^2}{2h\nu^3} A_{mn} \tag{73}$$

where the g's are the statistical weights. Also let $\sigma_{mn}(v)$ and $\sigma_{nm}(v)$ be the cross-sections for deactivation and excitation to and from the ground state by collisions with particles of number density $\mathfrak{N}$ moving with relative velocities v. The transition collision rates then satisfy the reversibility relation

$$\langle v\sigma_{nm}(v)\rangle = \frac{g_m}{g_n} e^{-\chi_m/kT} \langle v\sigma_{mn}(v)\rangle \tag{74}$$

where the averages indicated by the angular brackets are taken through the distribution of velocities of the colliding particles. The steady-state condition on the occupancy numbers N_m and N_n of the levels is expressed by

$$N_n[B_{nm}I + \mathfrak{N}\langle v\sigma_{nm}(v)\rangle] = N_m[A_{mn} + B_{mn}I + \mathfrak{N}\langle v\sigma_{mn}(v)\rangle] \tag{75}$$

where I is the specific intensity of the radiation field at the frequency $\nu = \nu_{nm}$. If there is an emission line at this frequency, I in equation (75) is a mean value over the effective width of the line; that is,

$$I = \frac{1}{\langle\Delta\nu\rangle}\int I(\nu)\,d\nu\,. \tag{76}$$

Let the actual population of the level m be expressed in terms of its Boltzmann value $N_m^{(o)}$ and a "dilution factor" b_m be defined by

$$N_m = b_m \frac{g_m}{g_n} N_n e^{-\chi_m/kT} = b_m N_m^{(o)} . \tag{77}$$

From equations (73), (74), (75), and (77), we find

$$b_m = \frac{K + \mathfrak{J} e^{h\nu/kT}}{1 + K + \mathfrak{J}} \tag{78}$$

where K measures the relative importance of collisional radiative deactivations,

$$K = \mathfrak{N} \frac{\langle v \sigma_{mn}(v) \rangle}{A_m} \tag{79}$$

and

$$\mathfrak{J} = \frac{c^2}{2h\nu^3} I . \tag{80}$$

As $K \to \infty$, for fixed $\mathfrak{J}$ we see in equation (78) that $b_m \to 1$. When the intensity I has its equilibrium value, we see that equation (78) also implies a Boltzmann distribution for all K. In general, the correct value of I is given by the solution of the equation of transfer for radiation in the (m,n) transition:

$$\langle \Delta\nu \rangle \frac{4\pi}{h\nu} \frac{dI}{ds} = N_m (A_{mn} + B_{mn} I) - N_n B_{nm} I \tag{81}$$

where it is assumed that absorption of radiation takes place only in the transition $n \to m$. With the mean optical depth τ corrected for induced emissions and defined by

$$\frac{d\tau}{ds} = -N_n \frac{h\nu}{4\pi} \frac{B_{nm}}{\langle \Delta\nu \rangle} (1 - b_m e^{-h\nu/kT}) \tag{82}$$

as an independent variable, we may rewrite equation (81) in the form

$$\frac{d\mathfrak{J}}{d\tau} = \mathfrak{J} - (b_m^{-1} e^{h\nu/kT} - 1)^{-1} . \tag{83}$$

This equation may be formally integrated along the line of sight through the total optical depth τ_o of the medium:

$$\mathfrak{J} = \int_0^{\tau_o} e^{-\tau} (b_m^{-1} e^{h\nu/kT} - 1)^{-1} d\tau . \tag{84}$$

Equation (84) solved simultaneously with equation (78) gives the general solution to the excitation problem. This solution applies to permitted or forbidden transitions in the optical, infrared, or radio-frequency regions of the spectrum. We shall now show how more restricted expressions, used in the past for each of these spectral regions separately, are embodied in equation (84).

When the temperature is not space dependent, we have

$$\mathfrak{J} = (1 - e^{-\tau_o})(b_m^{-1} e^{h\nu/kT} - 1)^{-1} . \tag{85}$$

Then equation (78) becomes

$$b_m^{-1} = 1 + \frac{e^{-\tau_o}}{K} . \tag{86}$$

In the optical range, induced emissions are negligible ($h\nu \gg kT$). If the total optical depth of the medium is small ($\tau_o < 1$), from equation (83), the radiation intensity is related to the Planck intensity $B_\nu(T)$ by

$$I = \tau_o \frac{K}{1+K} B_\nu(T) . \tag{87}$$

If L is the net linear extent of the emitting region, the total optical depth when $h\nu \gg kT$ is

$$\tau_o = L \frac{N_m}{4\pi \langle \Delta\nu \rangle b_m} h\nu \frac{A_{mn}}{B_\nu(T)} \tag{88}$$

and the total energy emitted in the $m \rightarrow n$ transition is

$$\langle \Delta\nu \rangle I = \frac{L}{4\pi} N_m A_{mn} h\nu . \tag{89}$$

This expression has been used to compute forbidden line strengths in the optical region. However, some forbidden lines of astrophysical interest, not necessarily in the radio-frequency range, may be saturated, as is the $2p^2P_{1/2}$–$2p^2P_{3/2}$ transition of C II (Seaton 1955).

For small optical depth and negligible induced emissions, let us consider the emission coefficient Γ per atom in the ground state and unit density of collision agents as

$$\Gamma = \frac{N_m A_{mn} h\nu}{\mathfrak{N} N_n} . \tag{90}$$

Expressing N_m in terms of $N_m^{(o)}$ and b_m (eqs. [77] and [86]), we may rewrite

$$\frac{1}{\Gamma} = \frac{\mathfrak{N}}{\Gamma_1} + \frac{1}{\Gamma_2} \tag{91}$$

where

$$\Gamma_1 = \frac{N_m^{(o)} A_{mn} h\nu}{N_n} \tag{92}$$

is the specific emission rate when collisional deactivations are much more frequent than radiative decays. The emission rate

$$\Gamma_2 = \langle v\sigma_{nm}(v) \rangle h\nu \tag{93}$$

would correspond to the situation in which the radiative downward transitions are much more frequent than collisional deactivations, but the population of the upper state remains Boltzmannian. Equation (91) thus provides justification, in a two-level system, for a general interpolation formula as suggested by Spitzer (1949*a*).

Consider finally a transition in the radio-frequency range (21-cm line of hydrogen) for which $b_m = 1$ and $h\nu \ll kT$, but allow for variations in temperature along the line of sight. If the intensity, given by equation (84), is expressed in terms of a mean brightness temperature T_b defined by

$$I = (1 - e^{-\tau_o}) B(T_b) , \tag{94}$$

we have

$$(1 - e^{-\tau_o}) T_b = \int_0^{\tau_o} T(\tau) e^{-\tau} d\tau . \tag{95}$$

Introducing as an independent variable, an optical depth t at the constant temperature T_b such that $t_o = \tau_o$, by integrating by parts, we obtain

$$\frac{1}{T_b}\int_0^{\tau_o} \exp\left(-\int_0^t \frac{T_b}{T}\,dt\right) = \int_0^{\tau_o} \frac{1}{T} \exp\left(-\int_0^t \frac{T_b}{T}\,dt\right) dt. \qquad (96)$$

The brightness temperature is thus a harmonic mean of the temperature along the line of sight, with a weight function which is nearly a negative exponential of the mean optical depth. This result, in slightly different form, was derived by Kahn (1955).

4. INTERPRETATION OF OBSERVATIONS

4.1. Doublet-Ratio Method

The interpretation of interstellar line intensities by the doublet-ratio method encounters the difficulty of allowing properly for the effects of systematic motions arising from galactic rotation. As noted in Section 1, the mean radial velocities of the interstellar lines show a galactic-rotation shift closely corresponding to a distance equal to one-half that of the star in which they are observed. However, it is not possible to establish from the somewhat limited data such a clear-cut dependence of total intensities on galactic longitude. The observable interstellar lines at distances small enough to be unaffected by large-scale inhomogeneities in the spatial distribution of the gas have widths exceeding the spread expected from galactic rotation. This and the intrinsically random distribution of the gas tends to obscure the effect of galactic rotation on line widths. We therefore analyze the mean values of the doublet ratios observed in nearby stars, disregarding galactic rotation, to derive mean values of the velocity-dispersion parameters and of the density; for example, we can use the mean relation between the doublet ratio of Na I as a function of D1, as given by Binnendijk (1952) for weak lines. We find for each pair (*DR*, D1) a unique pair (σ, NL) for an assumed velocity distribution (Sec. 3.1). The result of the analysis is shown in Table 6 and further illustrated in Figures 4 and 5, where the loci of constant NL or σ are drawn in (*DR*, D1) coordinates, both for the Gaussian ψ (Fig. 4) and for an exponential ψ (Fig. 5). For D1 < 0.2 A, the points representing the observations fall almost along a line with $b/\sqrt{2} = 4.2$ km/sec, or along the locus $\eta\sqrt{2} = 4.7$ km/sec. For larger values of D1, the observed points seem to depart from a curve of constant σ, the effect being more pronounced for a Gaussian ψ. The mean distances corresponding to those stars are large enough that galactic rotation must cause some line broadening. The differences expected at different galactic longitudes are, however, not pronounced for stars nearer than 1 kpc (see Sec. 4.2; also Jentzsch and Unsöld 1948). At greater distances, observational difficulties with faint stars hinder verification of the theory. Nevertheless, measured values of *DR* seem to be more consistent with an exponential velocity distribution than with a Gaussian one. If the velocity dispersion *is* Gaussian, the velocity distribution seems to increase with the distance from the Sun in all directions. Since the statistics for single clouds provide velocity-distribution functions showing an excess of large velocities over a Gaussian function (Blaauw 1952), the former point of view would seem to describe the true physical phenomenon more accurately.

The result of a similar analysis for the Ca II doublet is given in Table 6 and Figures 6 and 7. Again, we have used the mean values given by Binnendijk (1952) and the theoretical curves of growth given in Section 3.1. For weak lines, the observed points follow theoretical curves with $b/\sqrt{2} = 7.8$ km/sec and $\eta\sqrt{2} = 8.5$ km/sec, values almost twice as large as those found for the Na I lines. Accurate measurement of doublet ratios for Ca II is more difficult than for Na I, because of the position of the H-line with respect to Hϵ. The difference between the velocity-dispersion parameters for lines of Ca II and Na I, however, is well established. This difference can be understood (see below) in view of the different behavior of the Na I and Ca II lines with high velocity.

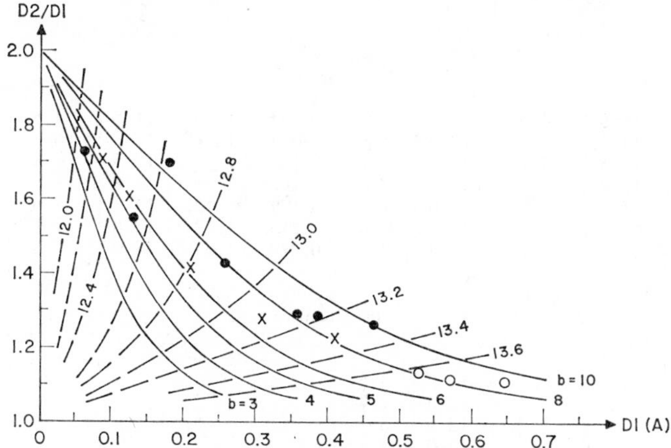

FIG. 4.—Doublet-ratio curves of Na I for a Gaussian velocity law of dispersion $\sigma_1 = b/\sqrt{2}$. Continuous lines are theoretical loci for constant b; the values of b (in km/sec) are given at right. Broken lines represent loci for a constant number of atoms along the line of sight (NL); corresponding values of $\log_{10} NL$ are shown at left. Crosses denote observed mean values from nearby stars; open circles represent lines formed through the full width of the Orion Arm; filled circles refer to lines formed in the Perseus Arm.

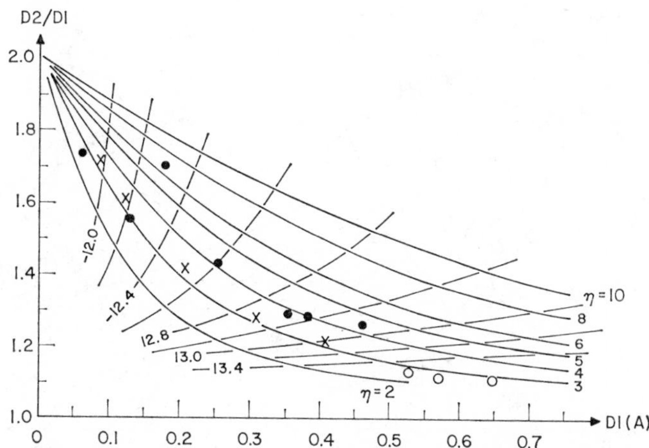

FIG. 5.—Doublet-ratio curves of Na I for an exponential velocity law of dispersion $\sigma_2 = \eta\sqrt{2}$. Continuous lines are theoretical loci for constant η; values of η (in km/sec) are given at right. Broken lines represent loci of constant number of atoms along the line of sight (NL); corresponding values of $\log_{10} NL$ are shown at left. Crosses denote observed mean values from nearby stars; open circles represent lines formed through the full width of the Orion Arm; filled circles refer to lines formed in the Perseus Arm.

TABLE 6

ANALYSES OF DOUBLET RATIOS*

Na I						Ca II					
D1 (A)	D2/D1	log N_1L	$b/\sqrt{2}$ (km/sec)	log N_2L	log $\eta\sqrt{2}$	H	K/H	log N_1L	$b/\sqrt{2}$ (km/sec)	log N_2L	log $\eta\sqrt{2}$
Orion Arm											
0.055..	1.80	11.8	4.0	11.8	4.7	0.033..	1.85	11.8	4.8	11.8	5.7
.124..	1.61	12.2	4.2	12.3	4.8	.074..	1.78	12.2	6.8	12.2	8.5
.226..	1.42	12.5	4.7	12.5	4.7	.131..	1.66	12.5	7.8	12.5	8.8
.312..	1.28	12.9	4.8	12.8	4.2	.162..	1.59	12.6	7.8	12.6	8.5
.410..	1.22	13.2	5.7	13.0	4.8	.194..	1.52	12.7	8.0	12.7	8.5
.524..	1.13	....	5.8	13.9	3.6	.213..	1.42	12.8	7.1	12.9	7.0
0.646..	1.11	....	6.7	14.4	3.8	0.279..	1.31	13.0	7.3	13.1	6.8
Perseus Arm											
0.065..	1.76	....	....	11.8	4.5						
.162..	1.66	....	....	12.3	7.4						
.258..	1.42	....	....	12.5	5.7						
.382..	1.28	....	....	13.0	5.7						
0.468..	1.25	....	....	13.2	6.2						

* Münch (1957).

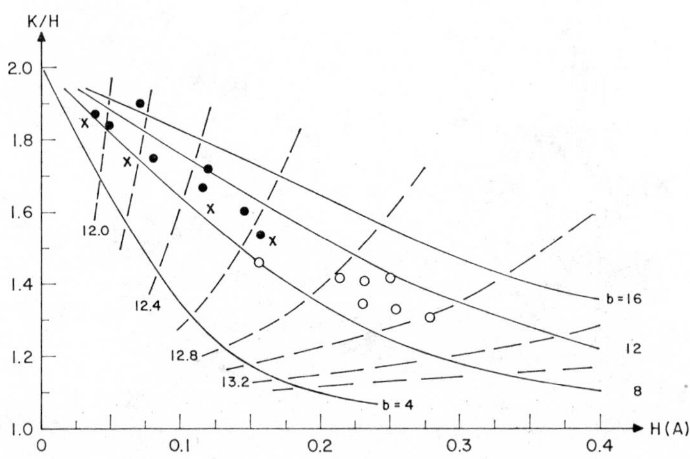

FIG. 6.—Doublet-ratio curves of Ca II for a Gaussian velocity law of dispersion $\sigma_1 = b/\sqrt{2}$. Data and symbols are as in Figure 4.

The doublet-ratio method, as noted above, provides inaccurate data if the lines are strong, especially for the number of atoms along the line of sight. The practical limitation of finding stars with strong lines within our spiral arm is more serious in the study of the velocity-dispersion parameters at small doublet ratios. In the northern hemisphere, the only direction relatively free from nearby obscuration is near the maximum of galactic rotation at $l^{\mathrm{I}} = 100°$. The appearance of the interstellar lines in distant stars near this direction suggests that the large-scale distribution of the gas is not random, but that two large-scale structural features of the Galaxy (spiral arms) exist. With sufficiently high spectral resolution, the components of the line arising in the local arm may be separated from that arising from the distant arm (Münch 1953, 1957). The component with

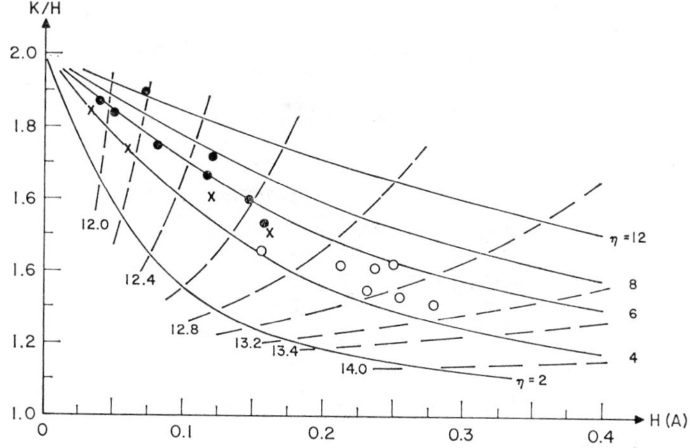

FIG. 7.—Doublet-ratio curves of Ca II for an exponential velocity law of dispersion $\sigma_2 = \eta\sqrt{2}$. Data and symbols are as in Figure 5.

small displacement arises, then, through the entire width of the local spiral arm. Observed intensities of these nearby components have been analyzed by the doublet-ratio method, with the results shown in Figures 4 and 5, (Na I) and 6 and 7 (Ca II). These points represent the continuation of a pattern set by values obtained from nearer stars. However, values for the components in the Perseus Arm suggest velocity-dispersion parameters (Münch 1957) systematically higher than those found in the Orion Arm by factors between 1.2 and 1.5.

4.2. INTENSITY-DISTANCE RELATIONS

The intensities of interstellar lines have frequently been used to determine the distances of individual stars. In one of the most recent studies, van Rhijn (1946) found, empirically, a mean error of about 40 per cent for such a distance determination. The mean-intensity–distance relations for different directions have been given by Binnendijk (1952). Because of the importance of these determinations, we shall re-examine the dependence of interstellar line intensities on distance using the statistical curve of growth constructed in Section 3.2. Specifically, we shall analyze the total intensities of the K-line as measured by Spitzer, Epstein, and Li Hen (1950) from Adams' coudé spectrograms. Stellar distances are known from the spectroscopic distance moduli corrected for interstellar extinction (Ramsay 1950; Morgan, Code, and Whitford 1955). The observational relations are given graphically in Figures 8 and 9 showing separately the data for stars within $\pm 15°$ of the null points of galactic rotation and for stars within $\pm 15°$ of the maxima. At the small distances concerned, departures from the double wave for the galactic-rotation shifts are negligible. To form a theoretical curve of growth from equation (43), we adopt the value $\nu = 8$ kpc^{-1} derived by Blaauw (1952). Since the value of the velocity dispersion is known from the analysis of the doublet ratio (cf. Sec. 4.1), the only parameters we are free to choose for a fit between theory and observations

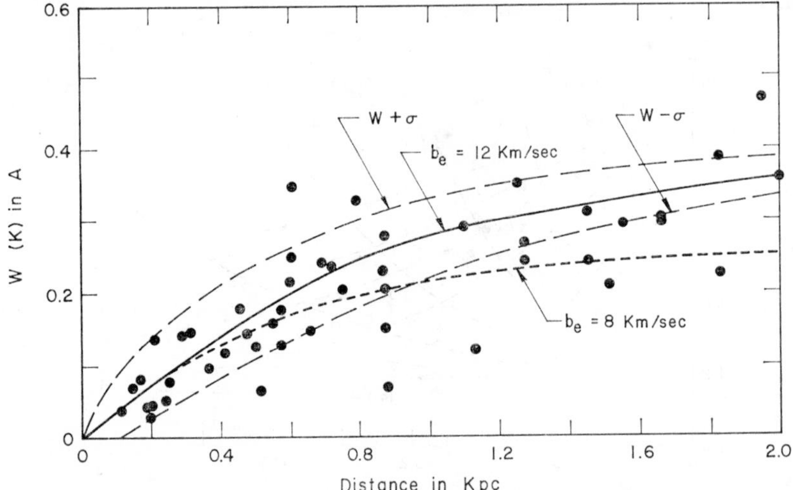

FIG. 8.—Relation between the strength of interstellar K of Ca II and distance near the null points of galactic rotation. The points represent observed values at high dispersion. The full line represents the theoretical relation for $b_e = 12$ km/sec. The dotted line shows the theoretical relation for $b_e = 8$ km/sec. The dashed lines show the limits $W \pm \sigma$.

are b_i, the internal velocity dispersion in a cloud, and r_o, the central intensity of the line formed in a single cloud. The pure number F is related to the shape of the internal velocity distribution. If it is assumed to be Gaussian, then for small r_o, $2F = 2.5$. In Figure 8, for the nulls of galactic rotation we have drawn the curves W for $b_e = 12$ km/sec, with $q = 0.145$ chosen by trial and error. The values of $\langle W \rangle \pm \sigma_w$, computed from equation (45), have been shown as dashed lines around the average relation to indicate the mean error to be expected from the randomness of the cloud distribution. The internal velocity dispersion b_i would then be 1.7 km/sec. To show the effect of the external velocity distribution, we have drawn a curve W for $\nu = 8$ kpc^{-1}, $b_e = 8$ km/sec, and $q = 0.218$ or $b_i = 1.7$ km/sec. This curve differs appreciably from the relation obtained for $b_e = 12$ km/sec only for $r > 1$ kpc.

To illustrate the effect of galactic rotation, we have represented (Fig. 9) the two curves $b_e = 12$ km/sec and $b_e = 8$ km/sec obtained with the parameters ν, r_o, and b_i as before. The effect of galactic rotation is appreciable for $r > 1$ kpc, and the dispersion is larger than at the null points but decreases very slowly for $r > 0.6$ kpc.

If σ_w is taken as a measure of the mean error in distance corresponding to a fixed equivalent width, at 1 kpc the mean percentage error in a distance determination would be 35 per cent, essentially the same at the maxima and zeros of galactic rotation. At smaller distances, the mean error increases rapidly, being 50 per cent at 400 pc and nearly 100 per cent at 200 pc. These results essentially agree with the empirical mean errors in distance determinations from interstellar line strengths found by van Rhijn (1946).

From the observational points included in Figures 8 and 9, it would hardly be possible to determine all the parameters required to form the theoretical curves. Nevertheless, a larger number of observations, for example, from Na I intensities, would lead to similar results. In particular, the value $b_i = 1.7$ km/sec, derived from statistical arguments, is

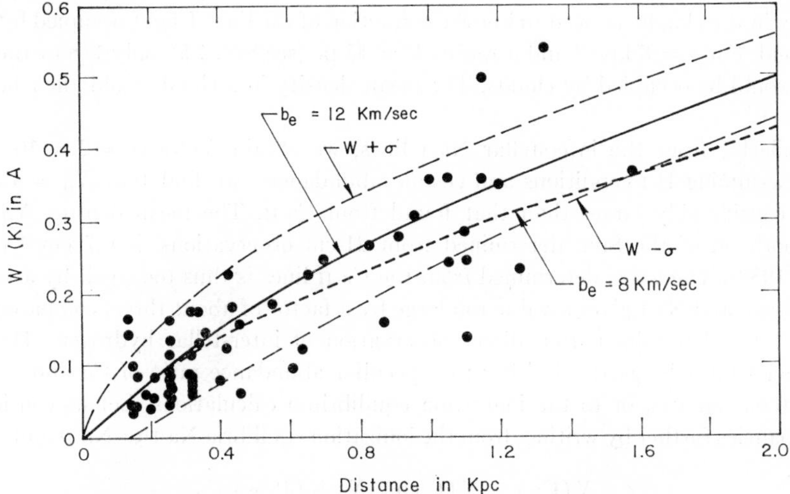

FIG. 9.—Relation between the strength of the interstellar K-line and distance, near the maxima of galactic rotation. The points represent observed values at high dispersion. The full line represents the theoretical relation for $b_e = 12$ km/sec. The dotted line shows the theoretical relation for $b_e = 8$ km/sec.

in good agreement with the values of velocity dispersions in individual clouds determined from 21-cm absorption profiles (Muller 1957; Clark, Radhakrishnan, and Wilson 1962). The values of the velocity dispersions measured for highly displaced optical components, which probably arise in single clouds, are about twice as large (Spitzer and Skumanich 1952). Thus there may be an intrinsic difference between the velocity dispersion in high-velocity clouds and that in the slow-moving ones.

4.3. The Mean Space Densities of Ca II and Na I

By combining the results of the two preceding sections, we may derive values for the average spatial densities of Na I and Ca II. Using mean values for the number of atoms along the line of sight for suitably unsaturated lines ($DR > 1.7$), we find from Figure 7, for example, that log $NL = 12.4$ when $DR = 1.64$ or $K = 0.17$ A. The mean distance r corresponding to this intensity is 500 pc, and $N(\text{Ca II}) = 2 \times 10^{-9}$ cm^{-3} is then the mean density along the line of sight. To derive the values of the total density, we must consider the ionization equilibrium of calcium, and we therefore require the electron density. Suppose that the line of sight is entirely in an H I region. The electrons, then, are provided by atoms with ionization potentials of less than 13.5 ev (mainly carbon), and if their abundance is the same as that found in the Sun, $N_e = 3 \times 10^{-4} N_{\rm H}$. We then write

$$N(\text{Ca II}) \approx N(\text{Ca}) N_e \left(\frac{a}{\Gamma}\right)_{\text{Ca II}} \tag{97}$$

and, assuming that the abundance of calcium has the general cosmic value $N(\text{Ca}) = 1.4 \times 10^{-6} N_{\rm H}$, we find

$$N_{\rm H}^2 \approx \frac{10^4}{3} N(\text{Ca II}) \frac{N_{\rm H}}{N(\text{Ca})} \left(\frac{\Gamma}{a}\right)_{\text{Ca II}} . \tag{98}$$

From the values of Γ/a given in Table 4 it follows that $N_{\rm H}^2 = 0.1$ cm^{-6}. To find the density in the clouds we need to know the fraction of the line of sight occupied by a typical cloud. For $\nu = 8$ kpc^{-1} and a radius $R = 15$ pc (see Sec. 4.5), only 1/8 of the line of sight would be occupied by clouds. The mean density in a cloud would then be $N_{\rm H}' = 2.5$ cm^{-3}.

Similarly, from the interstellar Na I lines, we obtain $N(\text{Na I}) = 2 \times 10^{-9}$ cm^{-3}. Again assuming H I conditions and cosmic abundances, we find that $N_{\rm H}^2 = 3$ cm^{-6}, a value considerably larger than that found from Ca II. The mean density $N_{\rm H}$ in the neighborhood of the Sun, determined from 21-cm observations, is 0.7 cm^{-3} (van de Hulst 1958). The value determined from the Ca II lines is thus too small by a factor of about two, and Na I gives a value too large by a factor of about three, compared to the mean value determined from direct observation of interstellar hydrogen. These discrepancies must be ascribed either to a peculiar abundance ratio of Ca and/or Na in the interstellar gas, or to the ionization equilibrium calculations. Let us consider the discrepancies further by writing, from the ionization equilibria Na I ⇄ Na II and Ca II ⇄ Ca III,

$$\frac{N(\text{Ca})}{N(\text{Na})} = \frac{N(\text{Ca II})}{N(\text{Na I})} \frac{N_e + (\Gamma/a)_{\text{Ca II}}}{N_e + (\Gamma/a)_{\text{Na I}}} . \tag{99}$$

From the mean values of the densities of Ca II and Na I found above and the ionization equilibria calculated for H I conditions by Seaton,

$$\frac{N(\text{Ca})}{N(\text{Na})} = 0.87 \frac{N_e + 0.025}{N_e + 0.68} . \tag{100}$$

If $N_e = 3 \times 10^{-4}$, then $N(\text{Ca})/N(\text{Na}) = 0.036$, a value about twenty times smaller than the abundance ratio observed in the Sun. This result is independent of electron density and electron temperature. For an H II region we would obtain $N(\text{Ca})/N(\text{Na}) = 0.17$, but then the Na/H and Ca/H ratios would be inadmissibly large. Possible inaccuracies in the atomic data used are now certainly too small to account for the discrepancy. An interstellar Ca/Na ratio much less than the cosmic value must then be accepted, or significant modifications in the radiation field assumed in the ionization calculations must be introduced. There is some evidence in favor of the second alternative, since the Ca II and Na I ionization rates depend on different spectral regions (2413–1800 A for Na I and 1044–912 A for Ca II). For Na I, 80 per cent of the ionization is produced by stars with $T_i \leq 16{,}000°$ K, while for Ca II, only 20 per cent of the ionization comes from these stars. The possibility that the energy curves of cool stars are depressed in the region 2000 A with respect to the blackbody curve at their effective temperatures has been demonstrated by recent rocket observations. At 2200 A, the solar flux is observed to be ten times smaller than that of the blackbody at 6000° K (Malitson, Purcell, Tousey, and Moore 1960). On the other hand, model atmospheres for the early-type–star calculations of the fluxes immediately to the red of the Lyman jump have shown (Underhill 1950; Traving 1957) that the radiation may exceed by factors of four or more that expected from blackbodies. An increase of the ionization rate of Ca II by a factor of 4, together with a decrease in the ionization rate of Na I by a factor of 10, would bring the Ca/Na abundance ratio derived from interstellar line strengths into better agreement with its cosmic value.

4.4. Correlation between Interstellar Line Strengths and Reddening

The discrete-cloud model was originally formulated from data related to the continuous extinction of starlight by interstellar solid particles. The existence of discrete clouds of interstellar gas was independently suggested by absorption profiles of interstellar absorption lines. The similarity of some parameters pertaining to the two distributions suggested that the dust clouds might be identical with the gas clouds. Still, a proof of the identity of the two kinds of clouds was desirable. Since color excess is a random variable, additive with respect to distance, and interstellar lines are mostly saturated, we should not expect a one-to-one correspondence between interstellar line strength and reddening. Many objects with strong reddening and small interstellar line strengths are known (Greenstein and Struve 1939; Morgan 1939). Arguments have also been raised against any general association of dust and gas (Evans 1941). In a thorough analysis of the correlations between line strengths of interstellar Na I and color excess, however, Spitzer (1948) has shown that the clouds of dust and gas can indeed be regarded as identical. The color excess produced by the average cloud is $0^{m}.024$ in the E_1 system (Stebbins, Huffer, and Whitford 1940), corresponding to an equivalent width $D = \frac{1}{2}(D1 + D2) = 0.15$ A. The relation (D, E_1) and its dependence on distance and galactic longitude found by Spitzer essentially agrees with the discrete-cloud model. However, for small values of E_1, the D-lines appear stronger than predicted by the theory, indicating the presence of clouds more tenous than the average or, alternatively, the presence of a nearby uniform intercloud medium. From the analysis of doublet ratios in little-reddened stars, Spitzer (1948) found that in an average star at 1-kpc distance, about three-fourths of the 2×10^{13} cm^{-2} Na I atoms in the line of sight are in typical clouds, and the other fourth may be in the intercloud medium. Since the pathlength through the cloud averages only about one-eighth of the total distance (see Sec. 4.5), the intercloud density must be less than that in the clouds by a factor of at least 20. Spitzer's (1948) results conflict with the earlier results of Evans (1941) and Schildt (1947). The apparent disagreement has been shown by Spitzer (1948) to arise from the failure of the earlier workers to account properly for the random effects of line saturation.

4.5. The Intercloud Medium

Analysis of the total strengths of the interstellar D-lines suggests the existence of an intercloud medium considerably less dense than the clouds. From a study of complex interstellar line profiles with well-separated components, it is also possible to say that there is no absorption greater than a few per cent between them. If the relative pathlengths in and between the clouds are known, we can evaluate the ratio between the corresponding densities. The average "radius" of a cloud cannot be determined from interstellar line data alone, but is generally determined from the fractional volume $\alpha = 0.05$ occupied by the clouds. This number follows from statistics of reflection nebulae (Ambartsumian and Gordeladse 1938) and of diffuse emission nebulae (Oort and van de Hulst 1946). The average cloud radius determined in this way is about 15 pc, in agreement with a recent direct determination from the absorption components observed at 21-cm (Clark *et al.* 1962). The density in the intercloud medium cannot therefore be higher than one-twentieth of that in the clouds. This estimate is extremely rough, so it is important to obtain further information regarding the gas density between the clouds. In principle, such information can be derived from interstellar lines in stars at small distances from the Sun, although observation of very faint lines is difficult with existing techniques. The nearest B-type stars, α Vir and η UMa, in which weak interstellar lines have been observed by Dunham (1941), have been used for this purpose by Strömgren (1948). The Ca II and Na I lines in these stars are so weak that it may be assumed that the line of sight to either of the stars does not pass through any cloud and that the lines are produced in the intercloud medium. The possibility that the column between observer and star contains a small part of a cloud cannot, of course, be excluded. The density derived with the assumption that the lines arise entirely in the intercloud gas should therefore be considered an upper limit. This assumption has been strengthened by the discovery of a rather strong interstellar K-line in α Oph (Münch and Unsöld 1962), a star at a smaller distance than α Vir or η UMa. For the D1-line in α Vir, Dunham (1941) has given $W = 0.007$ A as an upper limit. Such a line is so weak that it may be considered unsaturated. The corresponding number of Na I atoms along the line of sight is then $N(\text{Na I})\, L < 6 \times 10^{10}\ \text{cm}^{-2}$. Finding the density of all Na atoms requires the degree of ionization and, hence, knowledge of whether the line arises in an H I or H II region. Assume first that H II conditions prevail throughout the whole pathlength ($L \approx 90$ pc). Since Na would be essentially singly ionized, from the ionization equilibrium given in Table 4 we find, for $N_e = N_{\text{H}}$, $T_e = 10^4\,{}^\circ$ K, that

$$N(\text{Na})\, N_{\text{H}} < \frac{6 \times 10^{10}}{L} \left(\frac{\text{T}}{\alpha}\right)_{\text{Na I}} = 5 \times 10^{-9}\ \text{cm}^{-6}. \tag{101}$$

The usual cosmic value for the Na/H abundance ratio then gives $N_{\text{H}} < 0.04\ \text{cm}^{-3}$; this result depends only on the square root of the assumed abundance ratio or ionization rate.

Alternatively, the column between α Vir and the Sun may not be ionized by the effect of the general galactic radiation. We have to consider the ionizing effect of α Vir itself, however. For a star of type B1.5V, the radius of the Strömgren sphere is 13 $N_{\text{H}}^{-2/3}$ pc (cf. Table 6). Even for the maximum value set by Oort's limit $N_{\text{H}} = 5\ \text{cm}^{-3}$, the fractional pathlength within the H II region would be more than 10 per cent. The contribution of the H II column to the line strength will be considerably larger than that of the H I part, since $N_e\,(\text{H II}) > 10\, N_e\,(\text{H I})$. We can then neglect the H I column. Assuming, again, a cosmic abundance for sodium, we write

$$N_{\text{H}}^{4/3} < \frac{N(\text{Na I})L}{3.9 \times 10^{19}} \left(\frac{\text{T}}{\alpha}\right)_{\text{Na I}} \frac{N(\text{H})}{N(\text{Na})} \tag{102}$$

or $N_{\rm H} < 0.07$ cm^{-3}. In this context, note that the minimum density $N_{\rm H}$ needed to produce measurable emission in the Balmer lines is $N_{\rm H} = 15$ cm^{-3}. For such a high density, the radius of the Strömgren sphere would be only about 2 pc and the angular semi-diameter 1°5. At such small distances from α Vir, however, the ionization of Na I would be determined by α Vir itself and would be enhanced with respect to that produced by the general galactic radiation, to which the ionization calculations of Table 4 refer and the preceding calculation would not apply. Making the rather extreme assumption that the observed Na I arises in an H I region, we would obtain

$$N_{\rm H}^2 < \frac{N(\text{Na I})L}{L} \frac{N_{\rm H}}{N(\text{Na})} \left(\frac{\rm T}{\alpha}\right)_{\text{Na I}} = 0.51 \text{ cm}^{-6}, \qquad (103)$$

a value 20 times smaller than the mean value found from the strong Na I lines.

If the K-line in α Vir (Dunham 1941), with intensity $W = 0.0035$ A, is analyzed in the same manner, we obtain

$$N(\text{Ca II})\, L = 3.7 \times 10^{10} \text{ cm}^{-2}\,.$$

If we have H II conditions through the whole pathlength, we have

$$N_{\rm H} = 0.004 \text{ cm}^{-3} \qquad (\text{H II}, L = 90 \text{ pc}).$$

If only the H II region around α Vir need be considered,

$$N_{\rm H} = 0.009 \text{ cm}^{-3} \qquad (\text{H II}, L = 13\, N_{\rm H}^{-2/3} \text{ pc}),$$

but if the line of sight were entirely H I,

$$N_{\rm H} = 0.04 \text{ cm}^{-3} \qquad (\text{H I}, L = 90 \text{ pc}).$$

The results from the K-line are, thus, about ten times smaller than those obtained from the D1-line. Considering that the K-line was actually measured and the D1-line is only an estimated upper limit, it would seem that the smaller values provided by the K-line are more reliable. In comparison with the mean densities obtained in Section 4.3 from Ca II, we infer that densities in the intercloud medium are at least twenty times less than those in the typical cloud.

4.6. The Chemical Composition of the Interstellar Gas

The spectrum of the star χ^2 Ori contains every atomic interstellar line known, and their intensities have been given by Dunham (1941). From the strengths of the four Na I lines, Dunham (1939) constructed an empirical curve of growth for the interstellar lines in this star. We shall make use of the Na I strengths first to determine the Doppler width and the number of atoms on the line of sight. The D-lines are so saturated here that the *DR* method fails; but using the D1/λ 3303 ratio, Strömgren (1948) has found $NL = 10^{14}$ cm^{-2} and $\sigma_1\sqrt{2} = 4.8$ km/sec. Comparing this result with the average values derived before (see Sec. 4.1) for the distance of χ^2 Ori (1.2 kpc), we find that the mean density seems abnormally high by a factor of about 10, and the Doppler width is smaller by a factor of two. This suggests that the absorption lines are mainly formed in a single cloud of relatively high density. The K-line observed at high dispersion provides an additional basis for this hypothesis since it consists of a single strong line with a component 15 km/sec to the red that is less than one-tenth as strong. We shall assume, accordingly, a pathlength L corresponding to the dimension of one cloud. From the number of atoms along the line of sight we can derive the number densities of the various elements, applying the ionization corrections given in Table 4 for the H I and H II conditions. The results of the calculation are given in Table 7, which is essentially from Seaton (1951). For an H I region it has been assumed that all the titanium is singly ionized. If the titanium density is then set equal to the observed Ti II density and cosmic abundances are assumed, we obtain $N_{\rm H} = 70$ cm^{-3} and $N_e = 0.02$ cm^{-3}. Using this electron density, we compute the number densities for Na, Ca, and K. The results given in Table 7 are in general agreement with the mean cosmic values, which are fitted to agree with titanium. For a path of 25 pc the agreement is almost perfect. The same agreement would be obtained when the pathlength $L = 15$ pc and the ionization rates are increased by 5/3.

If the cloud were an H II region, most titanium would be in the form of Ti III. The electron density may be fixed as the maximum value that would produce minimum

TABLE 7

ABUNDANCES $N \times 10^6$ FROM THE INTERSTELLAR ABSORPTION LINES IN χ^2 ORIONIS

Ion	NL (cm^{-2})	Atom	H I				H II	
			D = 15pc	Cosmic	D = 25pc	Cosmic	D = 15pc	Cosmic
....	....	H	....	70×10^6	....	40×10^6	....	6×10^6
Na I......	1.0×10^{14}	Na	60	90	60	50	10	8
K I.......	1.5×10^{12}	K	10	14	10	8	4	1
Ca I......	1.0×10^{11}	Ca	40	130	60	80	4	10
Ca II.....	5.4×10^{12}	Ca	....	....	....	....	....	....
Ti II......	2.3×10^{14}	Ti	5	5	3	3	75	0.4

Seaton (1951).

detectable emission at Hβ, or an emission measure $EM = 1000$. For $L = 15$ pc, we would obtain $N_{\rm H}^2 = N_e^2 = 6$ cm^{-6}. The abundances of Na, Ca, and K given in Table 7 are then obtained. Agreement with the cosmic values is not as satisfactory as it was for the H I hypothesis.

The possibility that the Na and Ca and H- and K-lines originate in an H II region and the Ti lines in an H I region was considered by Strömgren (1948). However, it seems that an H I region of sufficient extent to account for the observed Ti II lines would also suffice to account for all other lines.

An observational determination of the ratio between the number densities of Ca II and Ca I would lead to a direct determination of the electron density, independent of any assumptions regarding the pathlength and the chemical composition. Unfortunately, although the Ca I lines are barely measurable, the Ca II lines are strongly saturated. In χ^2 Ori, additional uncertainty is produced by the presence of Ca II stellar lines. The Ca II interstellar intensities measured by Spitzer, Epstein, and Li Hen (1950) are unreliable because of this. If $DR = 1.40$, then $W(\rm H) = 0.173$ A would imply $NL = 5.4 \times 10^{12}$ cm^{-2}, a number far too small to be compatible with the numbers derived from Na I, K I, and Ca I, and from a cosmic abundance for the element providing the free electrons.

Considering the evidence provided by interstellar absorption lines as a whole, it seems fairly well established that the abundances of some elements in the interstellar gas are of the same order of magnitude as in the Sun and stars. This result refers directly to the metals Na, K, Ca, and Ti and indirectly to the elements supplying the free electrons in H I regions, mostly carbon. The agreement is better if the average cloud is an H I region than if it is an H II region. The observed anomalous abundance ratio Ca II/Na I probably results from the incorrect radiation field currently used to determine photo-ionization rates. The mean values of the hydrogen density derived on the assumption of cosmic abundance are in good agreement with the direct determination from the 21-cm hyperfine structure line.

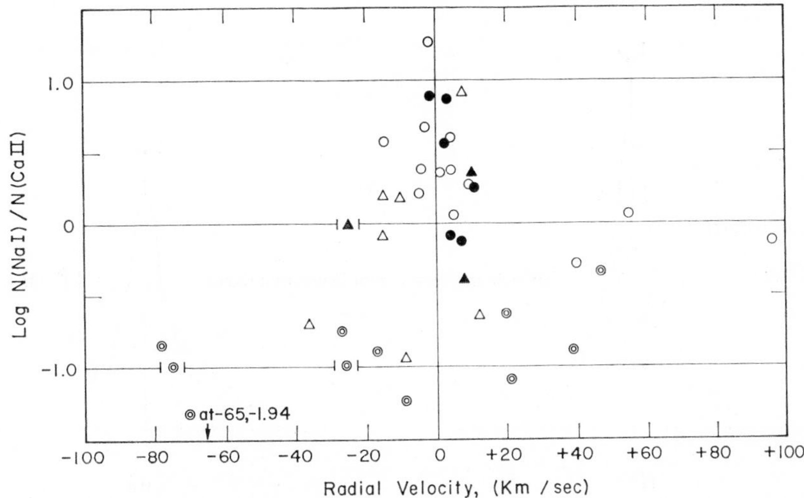

FIG. 10.—Relation between log N(Na I)/N(Ca II) and residual velocity. The solid symbols refer to blends. Open circles represent well-determined values, the triangles altered values, and the double circles upper limits. After Routly and Spitzer (1952).

4.7. The High-Velocity Clouds

Following the discovery of multiple interstellar lines of Ca II, Merrill and Wilson (1947) obtained spectra of some stars in the D-line region with similar high dispersion. Comparing the D-lines with the Ca II lines, these authors found that although the strongest component of Na I is always stronger than the corresponding Ca II component, the weaker Na I components tend to appear weaker, if they appear at all. The phenomenon is clearly illustrated in Plate 1 by the relative intensities of the various components of Ca II and Na I observed in ϵ Ori. A quantitative study of the relation between the components of Na I and Ca II as a function of their intensity has been carried out by Routly and Spitzer (1952). By measuring the intensities of corresponding separate components of the Ca II and Na I lines, these authors applied the doublet-ratio method to each component. The result of their analysis is represented graphically in Figure 10, where the values of log N(Na I)/N(Ca II) for individual components are plotted against the radial velocity of the component in the local standard of rest. It is apparent from the figure that in high-velocity clouds ($|v| > 20$ km/sec) the Na I concentration is about one-tenth that of Ca II. The large scatter of the points with $v \simeq 0$ weakens the correlation between $N(\text{Na I})/N(\text{Ca II})$ and v, but if the small values of this abundance ratio are correlated with space velocities, then those points near $|v| = 0$ with low ratios $N(\text{Na I})/N(\text{Ca II})$ may be due to clouds having large space velocities but small radial components. Routly and Spitzer (1952) also showed that the effect is independent of cloud size, since the Na I/Ca II ratio seems uncorrelated with $N(\text{Na I})L$ or $N(\text{Ca II})L$.

The relation between the kinematics of a cloud and its composition and/or ionization equilibrium seems well established and is an important problem for further study. Routly and Spitzer (1952) have related this phenomenon to the result mentioned earlier that the velocity dispersion parameters b derived from strong lines of Ca II are about 1.5

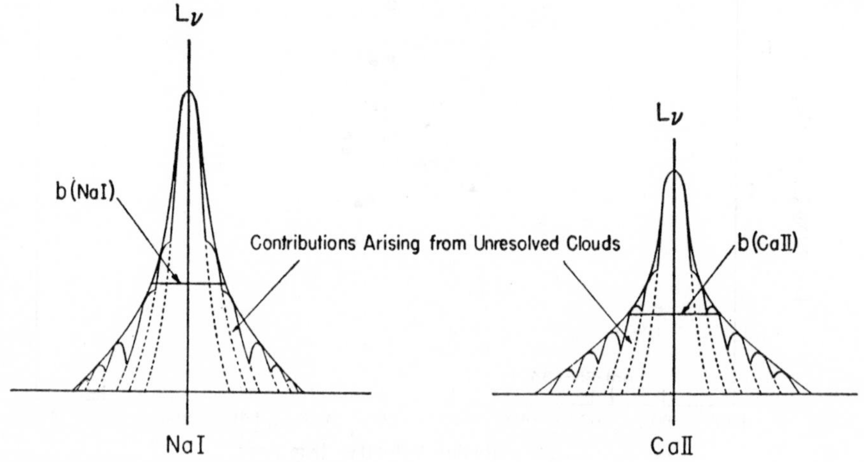

FIG. 11.—Qualitative illustration of the absorption coefficient L_ν for strong lines of Na I and Ca II, showing why $b(\text{Ca II}) > b(\text{Na I})$ for saturated lines. After Routly and Spitzer (1952).

times as large as those derived from Na I (Sec. 4.3). Since the b's represent a mean velocity of the absorbing atoms, they govern the frequency dependence of the absorption coefficient. Let L_ν denote the net line-absorption coefficient, including the effects of all clouds. Evidently L_ν for a strong line has a highly complex structure. The central part of L_ν is due to low-velocity clouds having more Na I than Ca II; consequently, the central height of L_ν for Na I is greater than for Ca II. On the other hand, the wings of L_ν arise from high-velocity clouds, where Ca II is more abundant than Na I; thus the wings of L_ν for Ca II are wider than the wings of L_ν for Na I. The net result is a sharply peaked L_ν for Na I and a lower but broader L_ν for Ca II. The effect is illustrated schematically in Figure 11 (from Routly and Spitzer 1952). The variation of $N(\text{Na I})/N(\text{Ca II})$ is explained by Routly and Spitzer (1952) by the increased ionization of Na I, relative to Ca II, produced by electron collisions. The increased electron temperature in a high-velocity cloud, in turn, is explained by collisions with intercloud atoms. Routly and Spitzer showed that at an electron temperature of 5000° K, the collisional ionization of Na I could account for the observed effect. Whether the electron temperature in a high-velocity cloud may rise to 5000° K has not been definitely established, owing to the complexity of the gas dynamics involved.

An alternate qualitative explanation considered by Routly and Spitzer (1952) is the possibility that Ca and Na atoms adhere to, or evaporate from, solid particles at different rates, depending on the nature of their chemical bonds. If the solid grains are concentrated in low-velocity clouds and are nearly absent in high-velocity clouds, a differential effect is produced. No clear-cut observational evidence on this matter exists at present, except that some stars at high galactic latitudes show interstellar line components with high-velocity characteristics but appear unreddened to any appreciable extent (Münch and Zirin 1961).

Variations in the relative concentration of the elements in the interstellar gas clouds should be studied further with a view to detecting true abundance variations or varying ionization conditions. The greater efficiency of photoelectric detectors compared with photographic plates should in principle enable us to extend the limits of detectability of faint interstellar absorption lines by an order of magnitude, with increased accuracy also in the intensity measurements.

REFERENCES

Adams, W. S. 1941, *Ap. J.*, **93**, 11.
———. 1943, *ibid.*, **97**, 105.
———. 1949, *ibid.*, **109**, 354.
Ambartsumian, V. A. 1940, *Bull. Abastumani Obs.*, **4**, 17.
Ambartsumian, V. A., and Gordeladse, Sh. G. 1938, *Bull. Abastumani Obs.*, **2**, 37.
Beals, C. S. 1936, *M.N.*, **96**, 661.
Beals, C. S., and Blanchet, G. H. 1938, *M.N.*, **98**, 398.
Beals, C. S., and Oke, J. B. 1953, *M.N.*, **113**, 530.
Binnendijk, L. 1952, *Ap. J.*, **115**, 428.
Blaauw, A. 1952, *B.A.N.*, **11**, 459.
Burgess, A., Field, G. B., and Michie, R. W. 1960, *Ap. J.*, **131**, 529.
Clark, G. B., Radhakrishnan, V., and Wilson, R. W. 1962, *Ap. J.*, **135**, 151.
Douglas, A. E., and Herzberg, G. 1941, *Ap. J.*, **94**, 381.
———. 1942, *Canadian J. Res.*, *A*, **20**, 71.
Duke, D. 1951, *Ap. J.*, **113**, 100.
Dunham, T., Jr. 1937*a*, *Nature*, **139**, 246.
———. 1937*b*, *Pub. A.S.P.*, **49**, 26.
———. 1939, *Proc. Amer. Phil. Soc.*, **81**, 277.
———. 1941, *Pub. A.A.S.*, **10**, 50.
———. 1956, *Vistas in Astronomy*, ed. A. Beer (London and New York: Pergamon Press), Vol. **2**, 1223.
Dunham, T., Jr., and Adams, W. S. 1937, *Pub. A.A.S.*, **9**, 5.
Eddington, A. S. 1926, *Proc. R. Soc. London*, *A*, **111**, 424.
———. 1934, *M.N.*, **95**, 2.
Evans, J. W. 1941, *Ap. J.*, **93**, 275.
Feast, M. W. 1955, *Observatory*, **75**, 182.
Feast, M. W., Thackeray, A. D., and Wesselink, A. J. 1955, *Mem. R.A.S.*, **67**, 51.
———. 1957, *ibid.*, **68**, 1.
Filippov, A., and Prokofjef, W. 1929, *Zs. f. Phys.*, **56**, 458.
Gerasimovitsch, B., and Struve, O. 1929, *Ap. J.*, **69**, 7.
Greenstein, J. L., and Aller, L. H. 1950, *Ap. J.*, **111**, 328.
Greenstein, J. L., and Struve, O. 1939, *Ap. J.*, **90**, 625.
Hartmann, J. 1904, *Ap. J.*, **19**, 268.
Heger, M. L. 1919, *Lick Obs Bull.*, **10**, 59.
———. 1921, *ibid.*, **10**, 141.
Herbig, G. 1960, *A.J.*, **65**, 491.
Herzberg, G. 1955, *Les particules solides dans les astres*, Sixième Colloque International d'Astrophysique, Liège (Cointe-Sclessin, Belgique: Institut d'Astrophysique), p. 291.
Hobbs, L. M. 1965, *Ap. J.*, **142**, 160.
Hulst, H. C. van de. 1958, *Rev. Mod. Phys.*, **30**, 913.
Jefferies, J., and Pottasch, S. 1959, *Ann. d'ap.*, **22**, 318.
Jentzsch, C., and Unsöld, A. 1948, *Zs. f. Ap.*, **125**, 370.

Kahn, F. D. 1955, *Gas Dynamics of Cosmic Clouds* (Amsterdam: North-Holland Publishing Co.; New York: Interscience Publishers, Inc.), chap. 12.
Kuiper, G. P. 1938, *Ap. J.*, **88**, 429.
Livingston, W. C., and Lynds, C. R. 1964, *Ap. J.*, **140**, 818.
McKellar, A. 1940, *Pub. A.S.P.*, **52**, 187.
———. 1941, *Pub. Dom. Ap. Obs. Victoria*, **7**, 251.
Malitson, H. H., Purcell, J. D., Tousey, R., and Moore, C. E. 1960, *Ap. J.*, **132**, 746.
Merrill, P. W. 1934, *Pub. A.S.P.*, **46**, 206.
Merrill, P. W., Sanford, R. F., Wilson, O. C., and Burwell, C. G. 1937, *Ap. J.*, **86**, 274.
Merrill, P. W., and Wilson, O. C. 1947, *Pub. A.S.P.*, **59**, 132.
Morgan, W. W. 1939, *Ap. J.*, **90**, 632.
Morgan, W. W., Code, A. D., and Whitford, A. E. 1955, *Ap. J. Suppl.*, **2**, 41.
Münch, G. 1953, *Pub. A.S.P.*, **65**, 179.
———. 1957, *Ap. J.*, **125**, 42.
———. 1961, unpublished.
Münch, G., and Unsöld, A. 1962, *Ap. J.*, **135**, 711.
Münch, G., and Vaughan, A. 1966, in press.
Münch, G., and Zirin, H. 1961, *Ap. J.*, **133**, 11.
Muller, C. A. 1957, *Ap. J.*, **125**, 830.
Oke, J. B. 1961, unpublished.
Oort, J. H., and Hulst, H. C. van de. 1946, *B.A.N.*, **10**, 187.
Plaskett, J. S., and Pearce, J. A. 1933, *Pub. Dom. Ap. Obs. Victoria*, **5**, 167.
Ramsay, J. 1950, *Ap. J.*, **111**, 434.
Rhijn, P. J. van. 1946, *Pub. Kapteyn Astr. Lab. Groningen*, No. 50.
Rogerson, J. B., Spitzer, L., Jr., and Bahng, J. D. 1959, *Ap. J.*, **130**, 991.
Routly, P. McR., and Spitzer, L., Jr. 1952, *Ap. J.*, **115**, 227.
Schildt, J. 1947, *A.J.*, **52**, 209.
Seaton, M. J. 1951, *M.N.*, **111**, 368.
———. 1955, *Ann. d'ap.*, **18**, 188.
Slipher, V. M. 1909, *Lowell Obs. Bull.*, **2**, 1.
Spitzer, L., Jr. 1948, *Ap. J.*, **108**, 276.
———. 1949*a*, *ibid.*, **109**, 337.
———. 1949*b*, *ibid.*, **109**, 548.
Spitzer, L., Jr., Epstein, I., and Li Hen. 1950, *Ann. d'ap.*, **13**, 147.
Spitzer, L., Jr., and Field, G. 1955, *Ap. J.*, **121**, 300.
Spitzer, L., Jr., and Skumanich, A. 1952, *Ap. J.*, **116**, 452.
Stebbins, J., Huffer, C. M., and Whitford, A. E. 1940, *Ap. J.*, **91**, 20.
Strömgren, B. 1939, *Ap. J.*, **89**, 526.
———. 1948, *ibid.*, **108**, 242.
Struve, O. 1927, *Ap. J.*, **65**, 163.
———. 1928, *ibid.*, **67**, 353.
Swings, P., and Rosenfeld, L. 1937, *Ap. J.*, **86**, 483.
Traving, G. 1957, *Zs. f. Ap.*, **41**, 215.
Underhill, A. 1950, *Ap. J.*, **111**, 203.
Weigert, A. 1955, *Wiss. Zs. Schiller V.-Jena*, **4**, 435.
Whipple, F. L. 1948, *Harvard Monographs*, No. 7, p. 109.
Williams, E. G. 1934, *Ap. J.*, **79**, 280.
Wilson, O. C., and Merrill, P. W. 1937, *Ap. J.*, **86**, 44.

CHAPTER 8

Atomic Processes with Special Application to Gaseous Nebulae

S. J. CZYZAK
General Physics Research Laboratory, Aerospace Research Laboratories, Wright-Patterson Air Force Base, Ohio

1. INTRODUCTION

GALACTIC nebulae which show an emission-line spectrum are of two types, that is, diffuse (Type I population) and planetary nebulae (Type II population). The diffuse nebulae are normally irregular, often of low density and surface brightness, and sometimes extensive; whereas, planetary nebulae are generally symmetrical and compact and usually have a higher surface brightness and density than the diffuse bright nebulae. The spectra of these gaseous nebulae, as well as those of certain stars, including the Sun, show strong emission lines owing to allowed and forbidden transitions of a number of "heavy" elements (atoms and ions other than H and He), such as N, O, Ne, S, Cl, Ar, P, Fe, Ca, Mn, Cr, V, Co, and Ni in various stages of ionization. The allowed transitions are accounted for by electric-dipole radiation; whereas, the forbidden transitions are due to magnetic-dipole and/or electric-quadrupole radiation. The forbidden emission lines result from collisional excitation of the metastable levels and were first identified in gaseous nebulae by Bowen (1928).

The physical nature of the excitation of the line spectra of gaseous nebulae seems to be well understood, but the continuous spectra of normal, purely gaseous nebulae need further study. Recombination of hydrogen and helium ions with electrons and double-photon emission is a generally satisfactory explanation, but additional observational data are still required for verifying various theoretical concepts.

Interpretation of gaseous nebulae requires proper evaluation of their geometrical structure, physical state, and chemical composition. One must be able to interpret the spectra quantitatively in terms of physical parameters and chemical composition of the emitting medium. Accurate transition probabilities and collisional cross-sections for al-

Dr. Czyzak is now at the Astronomy Department, The Ohio State University, Columbus, Ohio.

lowed and forbidden transitions, and thus accurate atomic wave functions, are necessary for proper interpretation of the properties of gaseous nebulae. Wave functions best suited for this purpose are the Hartree-Fock type. For a comprehensive exposition of the physical theory, the reader is referred to the textbooks by Hartree (1957) and Slater (1960).

Atomic wave functions now used in the calculation of transition probabilities and collision cross-sections are usually obtained by the Hartree-Fock self-consistent field method (HFSCF). Two basic procedures are used, one employing analytical techniques and the other using numerical techniques. The HFSCF method of calculating the wave functions still gives the best one-electron representation of many-electron atomic configurations. Within the past few years, many of these functions have appeared in the literature and more recently, programs for calculating them have become available. Iterative programs have been developed by Piper (1956, 1957, 1959, 1961), Worsley (1958), Douglas (1954), Mayers and Hirsh (1963), Froese (1963*a*, *b*), and Herman and Skillman (1963). Analytical programs have been developed by Boys and Price (1954), Nesbet (1955*a*), Nesbet and Watson (1960), Roothaan (1960), and Watson and Freeman (1961*a*, *b*). Further work is necessary, since the majority of the programs do not take into account configuration interaction. Those that do have been used only for the light elements and have given more accurate results for these. In addition, very little work has been done on atoms requiring the inclusion of relativistic effects. However, recently Froese (1966) and Mayers and O'Brien (1967*a*) have developed programs that now include configuration interaction, and the latter two investigators (1967*b*) have developed a program which takes into account relativistic effects. In transition probability calculations, the wave functions are needed for determining the value of $\sigma^2(nl,n'l')$ for allowed transitions and $\overline{r^2}$ and s_q, i.e., the quadrupole moment, in forbidden transitions.

In his classical paper, Bowen (1928) identified the so-called nebulium lines (specifically, N_1 and N_2, or the forbidden green lines of [O III] found in the spectra of nebulae and novae) as being due to doubly ionized atoms which had been excited to metastable levels by electron collision and which then escaped by emitting forbidden quanta. The discovery that many other nebular lines were due to forbidden transitions of atoms or ions involving p^2, p^3, and p^4 configurations and not to transitions involving some unknown light element greatly stimulated study of forbidden-transition probabilities. Calculations of these transition probabilities were undertaken shortly thereafter by Condon (1932, 1934), Rubinowicz (1932), Stevenson (1932), and Boyce, Menzel, and Payne (1933) followed by a comprehensive study by Pasternack (1940) and by Shortley, Aller, Baker, and Menzel (1941). These last two classical papers served for many years as basic references in the study of forbidden transitions and remain of great importance in astrophysics today.

As a consequence of Bowen's discovery, further study and investigation of the higher approximations for the interaction between matter and the radiation field was undertaken. To a first approximation, the selection rules allow radiation due to electric-dipole transitions only; that is, only the first approximation of the Hamiltonian is considered. If, now, we consider the next higher approximation for the interaction of matter and the radiation field, two additional types of radiation come into play: that owing to the magnetic dipole and that owing to electric quadrupole. It is these latter two that account for the forbidden transitions.

To estimate correctly the physical conditions in and chemical compositions of stars and nebulae, it is necessary to have accurate data on the transition probabilities and collision cross-sections for both allowed and forbidden lines. The general theory for the allowed and forbidden-transition probability calculations has been thoroughly described by Condon and Shortley (1935) in *The Theory of Atomic Spectra*.

Quite reliable results for allowed-transition probability calculations have been obtained, largely owing to the important work of Bates and Damgaard (1949). These authors found that in calculating the transition integrals, it was permissible to neglect the departure of the potential of an atom or ion from its asymptotic Coulomb form. Others, for example, Biermann and Lubeck (1948, 1949), Biermann and Trefftz (1949), Trefftz and Biermann (1952), and Trefftz (1949, 1950, 1952) also investigated allowed transitions but used the Hartree-Fock wave functions with exchange rather than the Coulomb functions of Bates and Damgaard. They also obtained good results. Further, they, as well as Trees (1951, 1952), Green (1949), and Boys (1950, 1953*a*, *b*), took into account configuration interaction and showed, after some extensive study of light elements, that this would lower the allowed-transition probabilities in some instances by as much as 25 per cent.

For forbidden transitions, the theory of magnetic-dipole radiation has been worked out explicitly for LS-coupling. Both Pasternack (1940) and Shortley (1940) give the necessary formulae. The theory of quadrupole radiation in LS-coupling is more complicated and the electron configuration of the atom must be considered. Shortley extended this theory and showed that many general methods used for electric-dipole calculations could also be used for the electric-quadrupole calculations. In particular, he developed methods for calculating transitions involving intermediate coupling and absolute line strengths. As a result of these developments, detailed computations for atoms and ions of the p^q configuration which are of astrophysical importance were performed by Pasternack (1940) and Shortley *et al.* (1941). This work was further extended by Aller, Ufford, and Van Vleck (1949), Osterbrock (1951), Garstang (1951, 1952*a*, *b*, 1956), Seaton and Osterbrock (1957), Rohrlich (1956, 1959*a*, *b*), Naqvi and Talwar (1957), and others. Some of the methods originally devised by Shortley did not, however, lend themselves readily to the solution of problems involving the more complex atoms, and this led Garstang (1957, 1958) to reformulate the procedure for calculating the quadrupole intensity by employing Racah's powerful theory (1942*a*, *b*, 1943).

The problem of calculating collision cross-sections due to electron excitation is still far from being solved. Here, too, accurate radial wave functions are necessary prerequisites; but with the wave functions available today, it is reasonable to expect better than order-of-magnitude accuracy in collision cross-section results. In this work, the methods devised by Seaton (1953*a*, *b*, 1955*a*, *b*, 1958, 1961) for calculating the collision strengths by the exact resonance and distorted wave methods are the best available.

In the following sections, the theory for calculating wave functions, transition probabilities, and collision cross-sections is described. In addition, an example shows the mathematical manipulations involved in such calculations. The ion Fe XVI was chosen for this purpose because its relatively simple electronic configurations ($1s^2\,2s^2\,2p^6\,3s$) simplified the calculations for the wave functions, transition probabilities, and collision cross-sections without detracting from the demonstration of the general procedure for such computations. In Section 5, results obtained and existing problems are discussed.

2. ATOMIC STRUCTURE CALCULATIONS

2.1. Wave Functions for Hydrogen

The motion of an electron in a central force field serves as the basis for the theoretical treatment of atomic states (such as calculation of the various states for hydrogen) where the electron moves in the field of force of a proton and obeys Coulomb's law of attraction. For hydrogen, the calculations can be carried out with mathematical rigor.

The time-independent Schrödinger wave equation,

$$H\phi = E\phi\,, \tag{1}$$

where $H = -\frac{1}{2}\nabla^2 + V$, E = the energy, and ϕ = the wave function of the electron, may be written in atomic units for an electron in a central field in the following manner:

$$\nabla^2\phi + 2[E - V(r)]\phi = 0\,. \tag{2}$$

$V(r)$ is the potential energy of the electron, a function depending on r only; ∇^2 is the Laplacian operator in spherical coordinates; and $\phi = \phi(r, \theta, \varphi)$. Expression (2) is solved in the usual way by the method of separable variables, that is, by writing $\phi(r, \theta, \varphi) = R(r)U(\theta, \varphi)$. An expression equating a function of r to a function of θ and φ is obtained; this can only hold in general if each side of the equation is equal to the same constant or eigenvalue λ, where $\lambda = -l(l+1)$ and l can only be zero or a positive integer. The solutions of the equation for θ and φ, which is the spherical harmonics equation, are the products of the associated Legendre and exponential functions, $P_l^{|m|}(\cos\theta)e^{\pm im\varphi}$. The quantum number m occurs as a result of the further separation of the equation for θ and φ, and it takes on $(2l+1)$ values from $-l$ to l. The solutions of the radial equation are the associated Laguerre functions, and with the boundary conditions imposed, the form is $x^l e^{-x/2} L_{n+l}^{2l+1}(x)$. The wave functions for hydrogen may then be written explicitly in the following manner:

$$\phi_{nlm}(r, \theta, \varphi) = A x^l e^{-x/2} L_{n+l}^{2l+1}(x) P_l^{|m|}(\cos\theta)\, e^{im\varphi} \tag{3}$$

where A is the numerical normalizing factor and $x = (2N/n)r$. N is the atomic number, and n is the principal quantum number which is the result of the imposed boundary condition.

The part of the solution of the Schrödinger equation involving only the angular dependence on θ and φ is the same for all atoms as for hydrogen. The radial wave function is not the same for atoms of different elements. It is desirable to rewrite the radial equation for hydrogen,

$$\frac{1}{r^2}\frac{d}{dr}\left(r^2\frac{dR}{dr}\right) + \left\{2[E - V(r)] - \frac{l(l+1)}{r^2}\right\}R = 0\,, \tag{4}$$

in the following form, by setting $R = P(nl; r)/r$

$$\left\{\frac{d^2}{dr^2} + 2[E - V(r)] - \frac{l(l+1)}{r^2}\right\}P(nl; r) = 0\,, \tag{5}$$

where for hydrogen $V(r) = -1/r$, and $P(nl; r)$, a function of r only, is the solution; for hydrogen these $P(nl; r)$ are the associated Laguerre functions with the boundary conditions as before: $P(r = 0) = P(r = \infty) = 0$. They are normalized in the sense that the

$$\int_0^\infty P^2(nl; r)\, dr = 1\,,$$

where $P^2(nl; r)dr$ represents the probability of the electron being found between r and $r + dr$; that is, $P^2(nl; r)$ is the probability per unit radius. Equation (5) can be rewritten in a form suitable for ready application to the Hartree method. The expression $H\phi = E\phi$ can be written in the following form:

$$\phi_{nlm} = \frac{1}{E} H\phi$$

where $\phi_{nlm} = CP_l^{|m|}(\cos\theta)e^{im\varphi}[P(nl; r)]/r$, and for the radial function, $P(nl; r)$, the operator will be

$$\frac{d^2}{dr^2} + \frac{2N}{r} - \frac{l(l+1)}{r^2}\,,$$

where $2N/r$ is the potential energy of one electron around the nucleus of charge N. Hence, the expression for ϕ_{nlm} can be written as follows:

$$\phi_{nlm}(r, \theta, \varphi) = A'P_l^{|m|}(\cos\theta)\, e^{im\varphi} \frac{1}{r}\left[\frac{d^2}{dr^2} + \frac{2N}{r} - \frac{l(l+1)}{r^2}\right] P(nl; r)\,, \quad (6)$$

where A' is the normalizing factor for θ and φ and includes the $1/E$ eigenvalue.

2.2. Wave Functions for Atoms and Ions Other than Hydrogen

For calculating the wave functions for atoms and ions other than hydrogen, rigorous mathematical solutions are not possible, but very good simplifying approximations have been developed. The approximations, although still based on the central-field model, depart from the inverse-square law; the representation is a more general central field in that the force is directed toward the center of attraction and is a general function of distance only. These are known as the central-field approximations.

In the central-field approximation on the Hartree self-consistent field approximations, the fundamental assumptions are as follows:

1. Each electron is acted upon by the average charge distribution of each of the other electrons. When this is summed over all the electrons, a nearly spherical charge distribution is obtained. If, now, a spherical average is calculated, then the potential from this charge distribution and nucleus is also spherical; hence, we have a central field.

2. Since each electron moves in a central field because of the spherically symmetrical potential, the angular wave functions will be the same as those for hydrogen, that is, the spherical harmonics. Only the radial part, $P(nl; r)$, will differ.

3. In the central-field model, the energy is independent of the magnetic quantum numbers m_l and m_s; therefore, the energy for each of the $2(2l + 1)$ states permitted by the Pauli exclusion principle is degenerate.

Consider an n-electron atom for which the Hamiltonian operator is

$$H = -\sum_{i=1}^{n} \nabla_i^2 - \sum_{i=1}^{n} \frac{2N}{r_i} + \sum_{i<j}^{n}\sum^{n} \frac{2}{r_{ij}}\,, \quad (7)$$

where the summation over i is over all n-electrons and that over ij is once for each pair with $i = j$ excluded. The

$$\sum_i \nabla_i^2$$

is the kinetic energy of the electrons,

$$\sum_i 2N/r_i$$

is the potential energy of the electrons in the field of the nucleus, and

$$\sum_{i<j}\sum 2/r_{ij}$$

is the Coulomb repulsion and is the term in assumption (1), that is, the term over which a spherical average must be made.

As a first approximation in the Hartree self-consistent field approximation (1928, 1948, 1957), each electron is regarded as moving independently in an average potential field caused by the other electrons. Such motion would correspond to a product of one-electron wave functions, ϕ_a; that is, the radial wave function ψ may be written

$$\psi = \phi_1(r_1)\phi_2(r_2)\phi_3(r_3)\ldots\phi_n(r_n)\,. \tag{8}$$

In order for ψ to approach the true wave function of the Hamiltonian as closely as possible, the variational principle is used to calculate the ϕ_a's. Essentially, this principle states that the closer ψ approaches the true wave function of the Hamiltonian, the lower the energy E becomes; that is,

$$2\,E \leq \int\psi^*H\psi d\tau/\int\psi^*\psi d\tau\,. \tag{9}$$

The set of n Hartree wave equations would then have the following form:

$$\nabla_i^2\phi_a(r_i) + 2[E - V_i(r_i)]\phi_a(r_i) = 0 \tag{10}$$

where $V_i(r_i)$ is the potential energy involving the latter two terms of equation (7) with the requirement that the last term is spherically averaged (Hartree 1957; Slater 1960). To obtain the Hartree equations, however, one starts with the expression for the expectation value of H, which may be written in the following way:

$$\sum_i\Big[\int\phi_a^*(r_i)(-\nabla_i^2)\phi_a(r_i)d\tau_i - \int\phi_a^*(r_i)\Big(\frac{2N}{r_i}\Big)\phi_a(r_i)d\tau_i + \sum_{j\neq i}{}' \int\phi_a^*(r_i)\phi_\beta^*(r_j)\frac{2}{r_{ij}}\phi_a(r_i)\phi_\beta(r_j)d\tau_i d\tau_j\Big] \geq 2E \tag{11}$$

where the ϕ's are normalized. Although the potential is not actually spherically symmetrical, it can be made so by averaging over all directions for a particular r_i. The second and third integrals represent the expectation value of the electrostatic potential due to the other electrons and the nucleus.

The energy integrals can be calculated in the following manner. For the kinetic energy and the potential energy of the electrons in the field of the nucleus, the expression for that part of the operator is multiplied by ϕ_a^* and integrated over all the coordinates, where the wave function for ϕ_a is of the form of equation (6). This yields

$$\int_0^\infty \phi_a^*(r_i)\left(-\nabla_i^2-\frac{2N}{r_i}\right)\phi_a(r_i)\,d\tau_i$$
$$= -\int_0^\infty P(n_il_i;\, r_i)\left[\frac{d^2}{dr_i^2}+\frac{2N}{r_i}-\frac{l_i(l_i+1)}{r_i^2}\right]P(n_il_i;\, r_i)\,dr_i, \qquad (12)$$

the right side usually being denoted by $2I(n_il_i)$. After the integrations over θ_i and φ_i, only the radial part remains.

The calculation of the Coulomb interaction energy, which is represented by the third term of equation (11), requires the assumptions made earlier in this section. This term is the Coulomb interaction between the charge distributions $\phi_a^*(r_i)\phi_a(r_i)$ and $\phi_\beta^*(r_j)\phi_\beta(r_j)$. To evaluate this term one can first calculate the electrostatic potential of the jth electron in the ith position, which is $\int\phi_\beta^*(r_j)\phi_\beta(r_j)2/r_{ij}d\tau_j$. If this is then multiplied by the charge distribution of the ith electron, $\phi_a^*(r_i)\phi_a(r_i)$, and integrated over $d\tau_i$, the result is the third term of equation (11). Now a spherical average of the potential of the charge distribution must be found; this is equal to the potential of the spherically averaged charge density. For the spherical average of the charge density of the jth electron, one considers either the total charge enclosed in a volume between τ_j and $\tau_j + d\tau_j$ or the volume element, $d\tau_j = r_j^2 \sin\theta_j dr_j d\theta_j d\varphi_j$. If $\phi_\beta^*(r_j)\phi_\beta(r_j)d\tau_j$ is integrated over θ_j and φ_j, then the charge located between r_j and $r_j + dr_j$ is $P^2(n_jl_j;\, r_j)dr_j$, where $P^2(n_jl_j;\, r_j)$ is the charge density. Now the potential energy of the ith electron in the presence of the jth electron is $2/r$, where r is the distance between the two. Thus the potential energy of an electron at distance r_i from the origin in the presence of the jth electron of charge $P^2(n_jl_j;r_j)$ is $(2/r_i)P^2(n_jl_j;r_j)dr_j$ for $r_i > r_j$ and $(2/r_j)P^2(n_jl_j;r_j)dr_j$ for $r_j > r_i$. The expression for the total potential energy becomes

$$\frac{2}{r_i}\left[\int_0^{r_i}P^2(n_jl_j;\, r_j)\,dr_j + r_i\int_{r_i}^\infty \frac{P^2(n_jl_j;\, r_j)}{r_j}\,dr_j\right]. \qquad (13)$$

Following Hartree, we define the term in the brackets as $Y^0(n_jl_j;\, n_jl_j;\, r_i)$. For the third-term integral of equation (11), we then have the following expression:

$$\int\phi_a^*(r_i)\phi_\beta^*(r_j)\frac{1}{r_{ij}}\phi_a(r_i)\phi_a(r_j)\,d\tau_i d\tau_j$$
$$= \int_0^\infty P^2(n_il_i;\, r)\frac{2}{r_i}\,Y^0(n_jl_j;\, n_jl_j;\, r_i)\,dr_i \qquad (14)$$

with the integral on the right-hand side defined as $F^0(n_il_i;\, n_jl_j)$, which is the same as that defined by Slater (1960). The total energy E is then

$$E = \sum_{i=1}^{n} 2I(n_il_i) + \tfrac{1}{2}\sum_{i\neq j} F^0(n_il_i;\, n_jl_j), \qquad (15)$$

which follows from the fact that the energy is the expectation value of the Hamiltonian for the particular state of the system represented by the wave function ψ; that is,

$$E = \frac{\int\psi^* H\psi\, d\tau}{\int\psi^*\psi\, d\tau}. \qquad (16)$$

Next we employ the variational principle in order to determine the lowest value of E and also the best possible wave function $P(nl; r)$. To do this it is necessary to obtain δE, the variation of E, as a function of the variations $\delta P(n_il_i; r_i)$. At the minimum, δE must be zero for arbitrary δP. At the same time, the wave functions must be orthonormal. For this to be true, the $P(nl; r)$ must satisfy the Hartree equations. These variations are considered separately; that is, $P(n_il_i; r_i)$ is the wave function to be varied and $P(n_jl_j; r_j)$ with $i \neq j$ are the others. The contributions to E depending on $P(n_il_i; r_i)$ consist of contributions from the integrals $I(n_il_i)$ and $F^0(n_il_i; n_jl_j)$ where $(n_il_i) \neq (n_jl_j)$. The method of undetermined multipliers is used with the requirement that

$$\delta\left[E+\Sigma\lambda_{n_il_i}\int_0^\infty P^2(n_il_i; r_i)\,dr_i\right]=0$$

(Hartree 1948, 1957; Slater 1960). The variation leads to the final result,

$$\left[\frac{d^2}{dr^2}+\frac{2}{r}\,Y(nl; r)-\epsilon_{nl}-\frac{l(l+1)}{r^2}\right]P(nl; r)=0, \qquad (17)$$

where the subscripts i and j have been dropped with the understanding that $(n_il_i) \equiv (nl)$ and $(n_jl_j) \equiv (n'l')$; ϵ_{nl} is the Lagrange multiplier, $-\lambda_{n_il_i}$, and is also the eigenvalue of the problem; and

$$Y(nl; r)=N-\sum_{n'l'} q(n'l')\,Y^0(n'l', n'l'; r)+Y^0(nl, nl; r), \qquad (18)$$

where $q(n'l')$ represents the number of electrons in each shell. Equation (17) thus represents the Hartree Self-Consistent Field Equation Without Exchange and does not take into account the effect of spin; that is, the spin-orbit interaction term was not included in the Hamiltonian in equation (7).

An improvement of the self-consistent field method was made by Slater (1929) and independently by Fock (1930). Since the Pauli exclusion principle must be satisfied for these calculations, antisymmetric functions must be used, and the simplest way of accomplishing this is by writing the wave function Ψ as a determinant. It is essential to include spin right from the start. If the method of Slater (1929, 1937, 1951, 1960) is followed, the wave functions can be written in determinant form:

$$\Psi=\frac{1}{\sqrt{n!}}\begin{vmatrix} \psi_1(\rho_1) & \psi_1(\rho_2)\ldots \\ \psi_2(\rho_1) & \psi_2(\rho_2)\ldots \\ \cdots & \cdots \\ \cdots & \cdots \\ \cdots & \cdots \\ \cdots & \cdots \end{vmatrix}. \qquad (19)$$

The individual electron wave function is $\psi_r(\rho_i) = \phi_a(r_i)\chi_\eta(\sigma_i)$ where χ_η represents the spin function, η the spin state, and σ_i the spin coordinate of the ith electron.

The expectation value of $E = \int\psi^*H\psi d\tau/\int\psi^*\psi d\tau$, is the same as in the Hartree method, and the best wave functions are obtained by varying the ϕ_a's to get the minimum E.

This will lead to a set of equations known as the Hartree-Fock self-consistent field equations. One begins by substituting equation (19) into equation (9) to get

$$\sum_i \left\{ \int \phi_\alpha^*(r_i)[-\nabla_i^2]\phi_\alpha(r_i)\,d\tau_i - \int \phi_\alpha^*(r_i)\frac{2N}{r_i}\phi_\alpha(r_i)\,d\tau_i \right\}$$
$$+\sum_{j\neq i} \int \phi_\alpha^*(r_i)\phi_\beta^*(r_j)\frac{2}{r_{ij}}\phi_\alpha(r_i)\phi_\beta(r_j)\,d\tau_i d\tau_j \qquad (20)$$
$$-\sum_{j\neq i}{}'' \int \phi_\alpha^*(r_i)\phi_\beta^*(r_j)\frac{2}{r_{ij}}\phi_\alpha(r_j)\phi_\beta(r_i)\,d\tau_i d\tau_j \geq 2E\,.$$

The kinetic energy, potential energy of the electrons in the field of the nucleus, and the Coulomb interaction energy appear here just as in equation (11). The new term is the well-known exchange term, and the Σ'' denotes that the summation takes place only over parallel spins; a detailed discussion of its physical significance has been given by Slater (1960). The procedure for evaluating the energy integral and performing the variation is the same as in the Hartree method (1928). Besides $F^0(nl, n'l')$ we also have terms $F^k(nl, n'l')$ with $k \neq 0$, where $F^k(nl, n'l') = \int_0^\infty [P^2(nl; r)/r]\,Y^k(n'l', n'l'; r)\,dr$; and, in addition, a term $G^k(nl, n'l')$ is obtained which takes into account the exchange effect,

$$G^k(nl, n'l') = \int_0^\infty P(nl; r)P(n'l'; r)\frac{1}{r}\,Y^k(nl, n'l'; r)\,dr\,. \qquad (21)$$

For a discussion of F^k and G^k terms see Condon and Shortley (1935), Pauling and Wilson (1935), Hartree (1948, 1957), and Slater (1937, 1951). The total energy can be written in the following manner (Hartree 1948, 1957):

$$E = E^0 + \Sigma E' + \Sigma E''$$

where

$$E^0 = \sum_{nl} q(nl)I(nl) + \tfrac{1}{2}\sum_{nl}[q(nl)-1]\,q(nl)F^0(nl; nl)$$
$$+\sum_{nl\neq n'l'} q(nl)\,q(n'l')F^0(nl; n'l') - \sum_{nl,k} A_{lk}F^k(nl; nl) - \sum_{nl,n'l',k} B_{ll'k}G^k(nl; n'l')$$

= total energy of complete groups;

$$E' = q(nl)I(nl) + q(nl)\sum_{n'l'} q(n'l')F^0(nl; n'l')$$
$$-\frac{q(nl)}{2(2l+1)}\sum_{n'l'k} B_{ll'k}G^k(nl; n'l')$$ = energy of each incomplete group; and

$E'' = -A'_{lk}F^k(nl; nl)$ = energy when one or more incomplete groups exist.

The coefficients $-A_{lk}$, $-A'_{lk}$, and $-B_{ll'k}$ are the coefficients of F^k and G^k in the energy expansion. If the variation is carried out in the same manner as in the Hartree method, a general wave equation of the form

$$\left[\frac{d^2}{dr^2} + \frac{2}{r}\,Y(nl; r) - \epsilon_{nl,nl} - \frac{l(l+1)}{r^2}\right]P(nl; r)$$
$$= X(nl; r) + \sum_{nl\neq n'l} \epsilon_{nl,n'l}P(n'l; r) \qquad (22)$$

is obtained, where

$$Y(nl;\, r) = N - \sum_{n'l'} [\, q(n'l') - \delta_{nln'l'}\,]\, Y^0(n'l',\, n'l';\, r) + \sum_{k\neq 0} a_{lk} Y^k(nl,\, nl;\, r); \tag{23}$$

$$X(nl;\, r) = -\frac{2}{r} \sum_{k,n'l'\neq nl} \beta_{ll'k} Y^k(nl,\, n'l';\, r) P(n'l';\, r)\cdot \tag{24}$$

and

$$Y^k(nl,\, n'l'\ \ r) = \frac{1}{r^k}\int_0^r \rho^k P(nl;\, \rho) P(n'l';\, \rho)\, d\rho + r^{k+1}\int_r^\infty \frac{1}{\rho^{k+1}} P(nl;\, \rho) P(n'l';\, \rho)\, d\rho$$

with $a_{lk} = 2A_{lk}/q(nl)$ and $\beta_{ll'k} = B_{ll'k}/q(nl)$. Equations (23) and (24) clearly indicate the additional terms that come into being when exchange is included in calculation of the wave functions. The actual calculations involving exchange become more complex and tedious; however, the results are more accurate in that they more closely approximate the true wave functions; that is, eigenfunctions of the Hamiltonian. Equation (22) is the Hartree-Fock self-consistent field equation. These equations can be solved approximately either by analytical or numerical methods.

The analytical method for solving the HFSCF equations has been developed by Boys (1950, 1953*a*, *b*), Löwdin (1953, 1955*a*, *b*, *c*, 1962, 1963), Nesbet (1955*a*, *b*, 1958*a*, *b*, 1960, 1961*a*, *b*, 1963), Roothaan (1951, 1960), Slater (1929, 1937, 1951, 1960), and Nesbet and Watson (1958, 1960). Basically, there are three ways of obtaining orthonormal and analytic wave functions. Since many wave-function calculations have been reported for the $3p^q$ and $4p^q$ using the method developed by Watson and Freeman (1961*a*, *b*), which employs a version of Nesbet's symmetry and equivalence restrictions and which represents a compromise between the very accurate calculations of Roothaan and Weiss (1960) and those by Watson (1959) for the iron series, the general procedure is here briefly described. In the analytic procedure, matrix techniques originally formulated by Roothaan are employed to obtain orthonormal wave functions. The radial wave functions $P_i(r)$ have the following form:

$$P_i(r) = \sum_j c_{ij} R_j(r) \tag{25}$$

with the requirement that

$$\int_0^\infty |P_i(r)|^2 dr = 1, \tag{26}$$

where

$$R_j(r) = N_j r^{(l+A_j+1)} e^{-Z_j r} \tag{27}$$

and

$$N_j = \left[\frac{(2Z_j)^{2l+2A_j+3}}{(2l+2A_j+2)!}\right]^{1/2} \tag{28}$$

is the normalization constant; the Z_j's (the screening constants) and the A_j's are the parameters which define the basis functions R_j. The size of the basis sets affects the accuracy of the solutions if they are not too much alike, and it is essential to take this into account in the calculation of the wave functions. Once the sets have been chosen, a "best choice" of the individual R_j's can be made by varying the Z_j parameters. Given the R_j's, one then solves the Hartree-Fock equation for the eigenvectors c_{ij} and their eigenvalues by matrix techniques.

Numerical procedures have been developed for the IBM 650, 1620, 7090, and other computers. Recently, Froese (1963*a*, *b*) and Mayers and Hirsh (1963) have developed procedures for the IBM 7090 computer. A detailed bibliography of atomic wave functions by Knox (1957) lists the published wave functions and methods of calculating them. This bibliography tabulates the published work up to March, 1957. Since that time, additional work has been reported in the literature by Piper (1959, 1961), Roothaan and Weiss (1960), Roothaan, Sachs, and Weiss (1960), Watson and Freeman (1961*a*, *b*), Czyzak (1962), Douglas (1954, 1956), Douglas and Garstang (1962), Douglas, Hartree, and Runciman (1955), Clementi (1962, 1963*a*, *b*), Herman and Skillman (1963), Mayers (1963),[1] and Froese (1963*a*). With numerical procedures for high-speed computers, atomic wave functions will be much more readily available.

In order to give the reader, not familiar with these types of calculations, some appreciation of the way they are performed, let us use Fe XVI as an example. Although so far of limited astrophysical value, this ion has a relatively simple configuration. The calculation of the $1s^2 2s^2 2p^6 3s$ ground configuration for Fe XVI using only the Hartree method (1928, 1948, 1957) will be developed, since it is considerably more simple to handle than the Hartree-Fock method and yet will illustrate the technique without loss of generality. The additional conditions which must be taken into account when the Hartree-Fock method is employed are discussed below.

The equations (17), including the definition (18) for the $1s^2 2s^2 2p^6 3s$ configuration, which are used in the Hartree self-consistent field method, are written explicitly in the following manner:

$$\frac{d^2}{dr^2}P(1s;r)+\left\{\frac{2}{r}[26-Y^0(1s1s;r)-2Y^0(2s2s;r)-6Y^0(2p2p;r) - Y^0(3s3s;r)]-559.33\right\}P(1s;r)=0; \tag{29}$$

$$\frac{d^2}{dr^2}P(2s;r)+\left\{\frac{2}{r}[26-Y^0(2s2s;r)-2Y^0(1s1s;r)-6Y^0(2p2p;r) - Y^0(3s3s;r)]-98.38\right\}P(2s;r)=0; \tag{30}$$

$$\frac{d^2}{dr^2}P(2p;r)+\left\{\frac{2}{r}[26-5Y^0(2p2p;r)-2Y^0(1s1s;r)-2Y^0(2s2s;r) - Y^0(3s3s;r)]-89.60-\frac{2}{r^2}\right\}P(2p;r)=0; \tag{31}$$

and

$$\frac{d^2}{dr^2}P(3s;r)+\left\{\frac{2}{r}[26-2Y^0(1s1s;r)-2Y^0(2s2s;r) - 6Y^0(2p2p;r)]-35.65\right\}P(3s;r)=0. \tag{32}$$

[1] Computing Laboratory, University of Oxford, Oxford, England.

The calculation is begun by first choosing a set of trial functions and electron energies ϵ_{nl} for the various shells of the configuration. The trial values of ϵ_{nl} are in the above equations. The ϵ_{nl}'s may be estimated from experimental data (Slater 1955), or they may be interpolated from known values of other atoms or ions along the isoelectronic sequence. When it is not possible to obtain the ϵ_{nl}'s by either of these methods, it is necessary to establish them from the trial wave functions (Froese 1963*a*, *b*). Normally, the speed and success in obtaining final self-consistent wave functions depends to a large degree on the accuracy of the original estimates. Various methods have been proposed for arriving at the desired set of trial functions by using known wave functions. Hartree (1955, 1957) and Froese (1957, 1958, 1959, 1963*b*) have developed an excellent interpolating scheme, which will be discussed here. All interpolating methods developed to date rely on defining a function which varies slowly with atomic number N so that the interpolation can be performed between known functions with reasonable accuracy. Several procedures can be employed: pure scaling, interpolation based on pure scaling, interpolation from the variation of reduced wave function, and variation of screening constants with $\bar{r}$ or N. Pure scaling provides a method for estimating both radial wave function and the contributions to $Z^0(nl, n'l'; r)$. In pure scaling, it is assumed that normalized $P(nl; r)$'s for atoms of different N are geometrically similar, and that on this basis they are related in the sense that $\bar{r}^{1/2}P(nl; r)$ is a function of $r/\bar{r}$ only; that is, two atoms of atomic numbers N and N' are related in the following manner:

$$P_N(nl; r) = \left(\frac{\bar{r}_N}{\bar{r}_{N'}}\right)^{1/2} P_{N'}\left(nl; \frac{\bar{r}_{N'}}{\bar{r}_N} r'\right) \tag{33}$$

and

$$Z_N^0(nl, nl; r) = Z_{N'}^0\left(nl, nl; r' \frac{\bar{r}_{N'}}{\bar{r}_N}\right) \tag{34}$$

where

$$\bar{r} = \int_0^\infty r P^2(nl; r)\, dr .$$

Also, if $\bar{r}_H$ is the value of $\bar{r}$ for the wave function of hydrogen then in the Coulomb field of a point charge $(N - \sigma)$, the $\bar{r}$, which is the mean radius, can be written as

$$\bar{r} = \frac{\bar{r}_H}{(N - \sigma)}, \quad \text{or} \quad \sigma = N - \frac{\bar{r}_H}{\bar{r}}, \tag{35}$$

which expresses the effect of the screening of the nucleus by the electrons. It has been shown that for complete (nl) groups, σ will generally vary slowly with N, so that for neighboring atoms N and N'

$$\frac{\bar{r}_N}{\bar{r}_{N'}} = \frac{N' - \sigma}{N - \sigma}. \tag{36}$$

We get

$$P_N(nl; r) = \left(\frac{N - \sigma}{N' - \sigma}\right)^{1/2} P_{N'}(nl; r'), \tag{37}$$

and

$$Z^k(nl, n'l'; r) = \int_{\rho=0}^{r} \left(\frac{\rho}{r}\right)^k P(nl; \rho) P(n'l'; \rho)\, d\rho ,$$

which is part of the $Y^k(nl, n'l'; r)$ expression

$$Z_N^0(nl, nl; r) = Z_{N'}^0(nl, nl; r') \tag{38}$$

where $r' = (N - \sigma)r/(N' - \sigma)$ and, for neighboring atoms, the scaling ratio $(N - \sigma)/(N' - \sigma)$ is not very sensitive to the adopted value of σ. A better method than the pure-scaling procedure is to calculate the desired wave functions from scaled wave functions of two or several neighboring atoms of different N and then to interpolate to get the values of the desired atom. This takes care of the departures from pure scaling. Hartree has developed a very convenient and accurate method for interpolating atomic wave functions with respect to N. In his procedure, it is necessary to take into account $\sigma(nl)$, $\bar{r}(nl)$, and the reduced wave function $P^*(nl; s)$ which Hartree defines as follows:

$$P^*(nl; s) = \bar{r}^{1/2}P(nl; r); \qquad s = \frac{r}{\bar{r}}. \tag{39}$$

Hartree (1957) and Froese (1957, 1958) have made plots of self-consistent field results with exchange and have shown that, for a fixed s, both $\sigma(nl)$ and $P^*(nl; s)$ vary almost linearly with respect to $\bar{r}(nl)$ over a large range of N's. Plots of $\sigma(nl)$ vs. $\bar{r}$ for different N's are shown in Figures 1, 2, and 3. The results of such interpolation may be sufficiently accurate for the determination of certain parameters, so that they can be used directly; or if not, they serve as excellent initial estimates for self-consistent field calculations. Froese and Hartree (1957) have shown that the results of such interpolations may be accurate enough to use as they stand, and Czyzak (1962) has shown this to be the case for ions of the $3p^q$ configuration.

For large atoms, very few data are available for purposes of interpolation; and, therefore, extrapolation has to be employed. The above general procedure is applicable for making initial estimates, as interpolation between known results and those for $N = \infty$ is made instead of an extrapolation. To make the interpolation possible, the limit $\sigma_0(nl)$ of the screening number $\sigma(nl)$ is determined for each shell as well as the limiting value of $[\partial P^*(nl; s)]/\partial\bar{r}$ as $N \rightarrow \infty$. Froese has described this procedure in great detail. Once the trial functions (i.e., the initial estimates) have been made, one can proceed to make the self-consistent field calculations.

To illustrate the procedure, we shall perform the calculations to obtain the $P(1s; r)$ wave functions for one iteration. In Tables 1, 2, and 3, the terms of equation (29) that have to be calculated have been tabulated. Table 1 represents the trial wave functions; Table 2, the $Y^0(1s,1s; r)$; and Table 3, the term required to calculate the $g(1s; r)$. In order to calculate the $Y^0(nl, nl; r)$ and $Y^0(n'l', n'l'; r)$, the expressions are broken into two parts as shown in equation (13) and the quadrature for computing the terms is done by using a modification of Simpson's rule:

$$\int_{r+h}^{r+2h} P^2(1s; \rho)\,d\rho$$
$$= \tfrac{1}{12}h[-P^2(1s; r) + 8P^2(1s; r+h) + 5P^2(1s; r+2h)] \tag{40}$$

and

$$\int_{r}^{r-h} \frac{P^2(1s; \rho)}{\rho}\,d\rho = -\tfrac{1}{12}h\left[\frac{5P^2(1s; r)}{r} + \frac{8P^2(1s; r-h)}{r} - \frac{P^2(1s; r-2h)}{r}\right] \tag{41}$$

where h is the interval width. These results are then added to give the $Y^0(1s1s; r)$, or column 6 of Table 2. This also is done for the $2s$, $2p$, and $3s$ shells and the terms in the bracket; that is,

$$g(1s; r) = \left\{\frac{2}{r}\left[26 - Y^0(1s1s; r) - 2Y^0(2s2s; r) - 6Y^0(2p2p; r) - Y^0(3s3s; r)\right] - 559.33\right\} \quad (42)$$

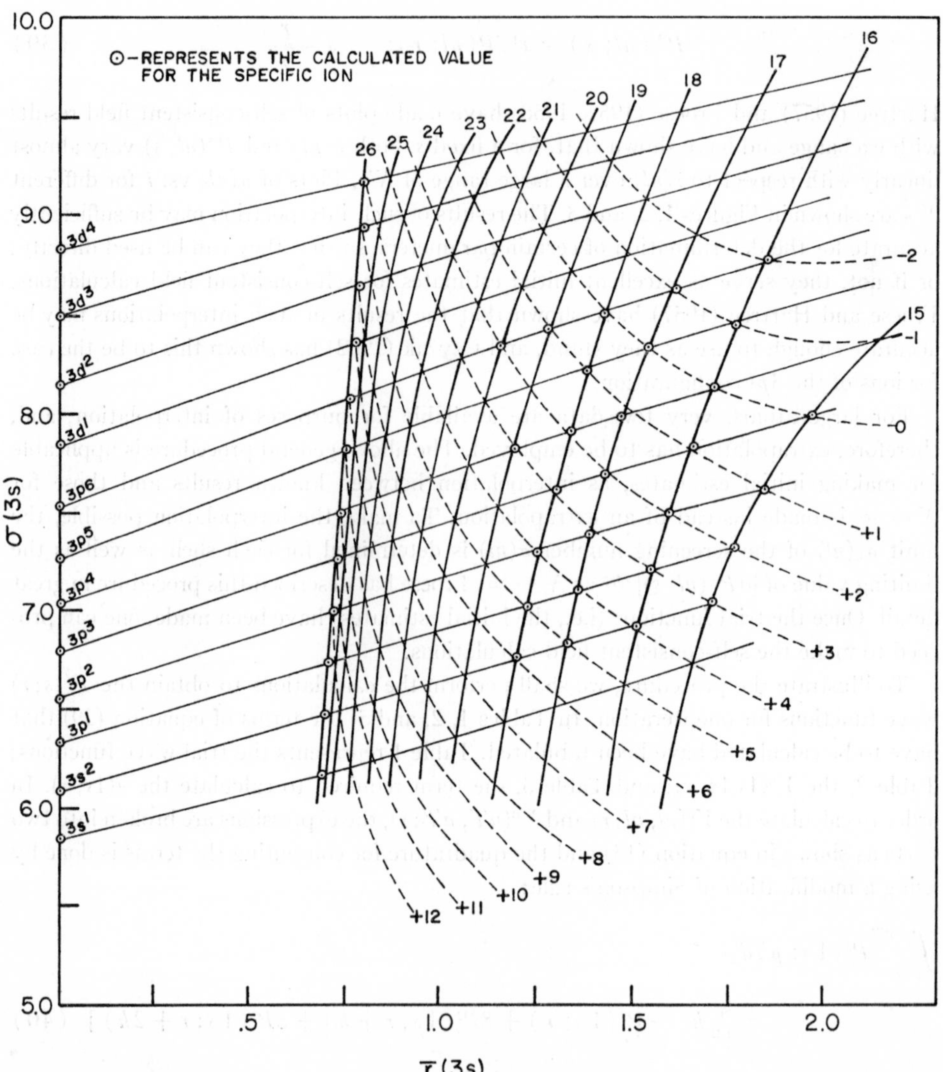

FIG. 1.—Screening numbers σ_{3s} for $3s$ wave functions vs. $\bar{r}_{3s}$. Numbers associated with near-vertical solid lines refer to z; those associated with dashed lines refer to degree of ionization.

is computed for each value of r as shown in Table 3. With these data, the homogeneous differential equation,

$$\frac{d^2}{dr^2}P(1s;\,r)+g(1s;\,r)P(1s;\,r)=0\,, \tag{43}$$

can be solved by Numerov's method (see Hartree 1958); that is, Numerov's difference equation is used to solve equation (43). This is done in the following way. The associated function,

$$b(r_i)=[1+\tfrac{1}{12}h^2g(nl;\,r_i)]P(nl;\,r_i) \tag{44}$$

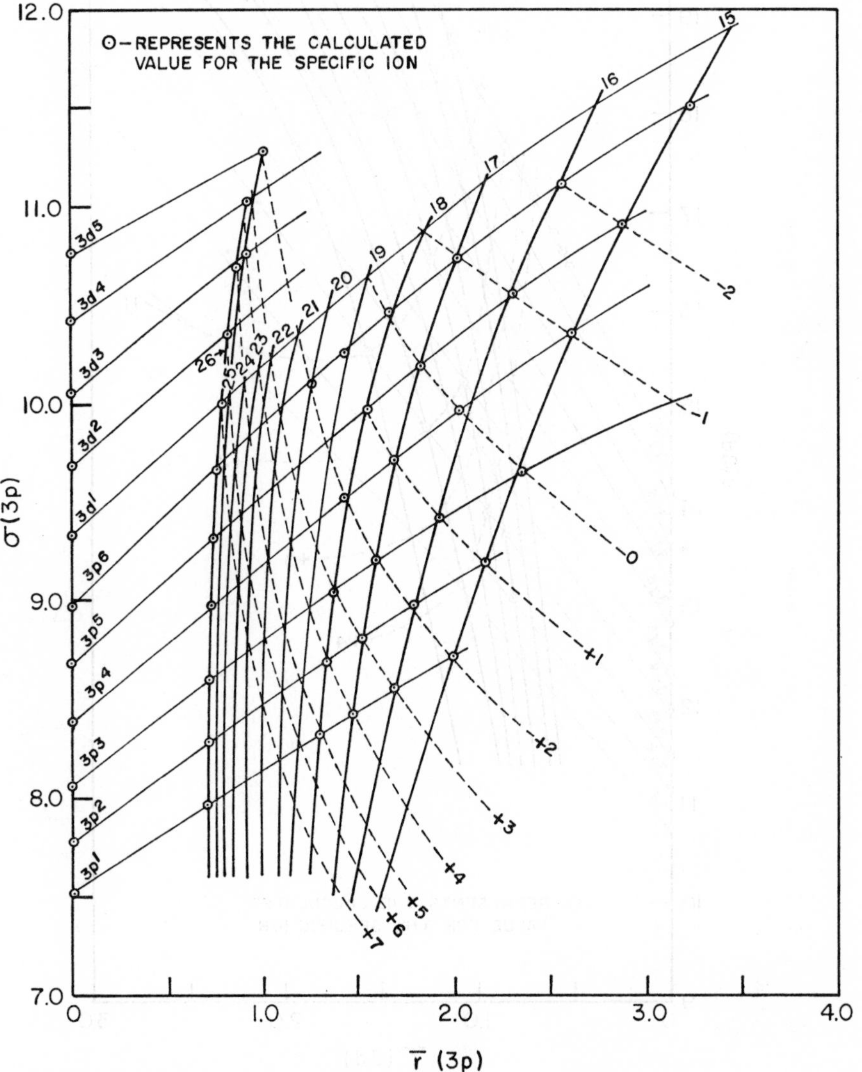

FIG. 2.—Screening numbers σ_{3p} for $3p$ wave functions vs. $\bar{r}_{3p}$. Numbers associated with near-vertical solid lines refer to z; those associated with dashed lines refer to degree of ionization.

is used to carry out the forward integration,

$$b(r_{i+1}) = [2 - G(r_i)]\,b(r_i) - b(r_{i-1}) \tag{45}$$

where

$$G(r_i) = \frac{h^2 g(nl;\, r_i)}{1 + \frac{1}{12} h^2 g(nl;\, r_i)}. \tag{46}$$

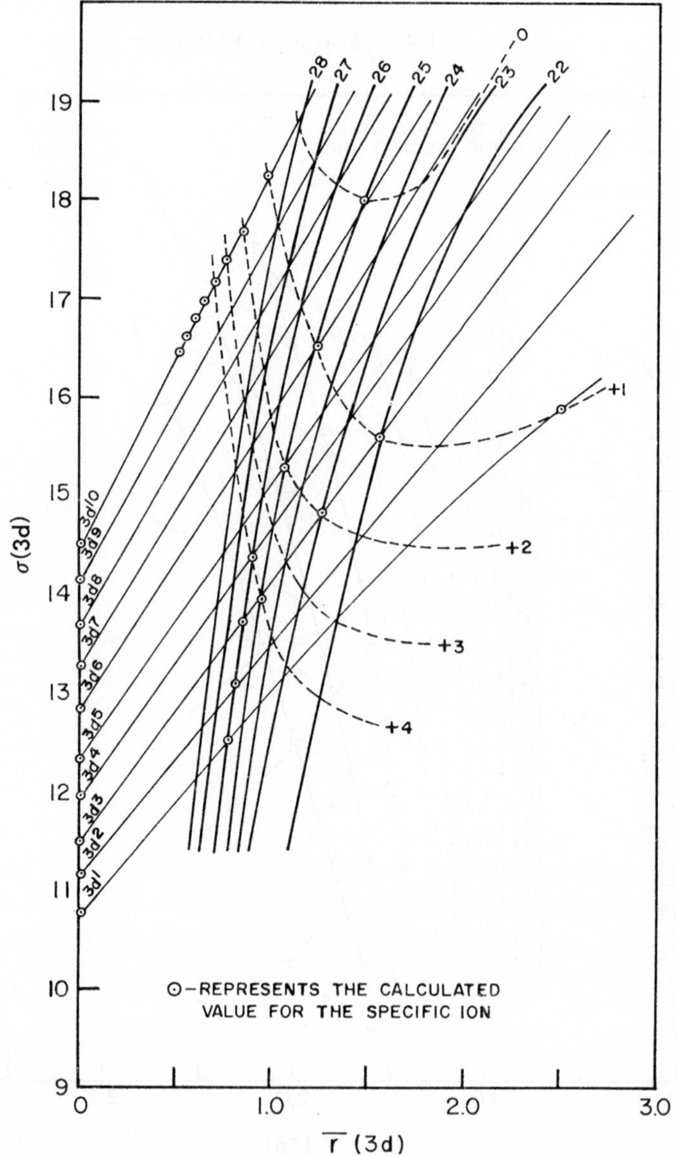

FIG. 3.—Screening numbers σ_{3d} for $3d$ wave functions vs. r_{3d}. Numbers associated with near-vertical solid lines refer to z; those associated with dashed lines refer to degree of ionization.

The b is a function of the interval size, so it is adjusted at each change of the interval. The integration is started at $r = 0$, and for s and p functions, respectively, we have

$$b(r_0) = \tfrac{1}{6}h^2 N \lim_{r\to 0} \frac{P(ns;\, r)}{r} \tag{47}$$

and

$$b(r_0) = -\tfrac{1}{6}h^2 \lim_{r\to 0} \frac{P(np;\, r)}{r^2}\,; \tag{48}$$

TABLE 1

HFSCF Wave Functions with Exchange for Fe XVI

r	P(1s;r)	P(2s;r)	P(2p;r)	P(3s;r)	P(3p;r)	P(3d;r)
+ 0.0000	+ 0.0000	+ 0.0000	+ 0.0000	+ 0.0000	+ 0.0000	+ 0.0000
0.0020	0.4942	+ 0.1514	0.0021	+ 0.0700	+ 0.0010	0.0000
0.0040	0.9383	+ 0.2873	0.0081	+ 0.1328	+ 0.0039	0.0000
0.0060	1.3362	+ 0.4085	0.0177	+ 0.1887	+ 0.0086	0.0001
0.0080	1.6916	+ 0.5159	0.0307	+ 0.2383	+ 0.0149	0.0001
0.0100	2.0076	+ 0.6104	0.0467	+ 0.2818	+ 0.0226	0.0003
0.0120	2.2875	+ 0.6929	0.0655	+ 0.3197	+ 0.0318	0.0005
0.0160	2.7501	+ 0.8245	0.1107	+ 0.3800	+ 0.0536	0.0011
0.0200	3.1000	+ 0.9164	0.1644	+ 0.4216	+ 0.0796	0.0020
0.0240	3.3551	+ 0.9738	0.2250	+ 0.4468	+ 0.1088	0.0034
0.0280	3.5309	+ 1.0011	0.2913	+ 0.4579	+ 0.1407	0.0052
0.0320	3.6406	+ 1.0025	0.3618	+ 0.4567	+ 0.1745	0.0075
0.0360	3.6955	+ 0.9815	0.4357	+ 0.4448	+ 0.2098	0.0103
0.0440	3.6786	+ 0.8855	0.5894	+ 0.3954	+ 0.2826	0.0177
0.0520	3.5425	+ 0.7350	0.7463	+ 0.3202	+ 0.3560	0.0274
0.0600	3.3322	+ 0.5478	0.9014	+ 0.2278	+ 0.4273	0.0395
0.0680	3.0801	+ 0.3376	1.0511	+ 0.1249	+ 0.4946	0.0541
0.0760	2.8089	+ 0.1153	1.1929	+ 0.0170	+ 0.5565	0.0710
0.0840	2.5342	- 0.1106	1.3249	- 0.0916	+ 0.6119	0.0902
0.1000	2.0127	- 0.5496	1.5550	- 0.2985	+ 0.7007	0.1351
0.1160	1.5602	- 0.9455	1.7368	- 0.4775	+ 0.7581	0.1875
0.1320	1.1884	- 1.2816	1.8706	- 0.6194	+ 0.7844	0.2462
0.1480	0.8934	- 1.5521	1.9597	- 0.7204	+ 0.7813	0.3098
0.1640	0.6650	- 1.7578	2.0089	- 0.7808	+ 0.7517	0.3770
0.1800	0.4912	- 1.9033	2.0237	- 0.8027	+ 0.6990	0.4467
0.2120	0.2636	- 2.0412	1.9717	- 0.7477	+ 0.5388	0.5885
0.2440	0.1395	- 2.0244	1.8441	- 0.5913	+ 0.3282	0.7278
0.2760	0.0733	- 1.9070	1.6735	- 0.3694	+ 0.0914	0.8584
0.3080	0.0385	- 1.7322	1.4837	- 0.1134	- 0.1523	0.9762
0.3400	0.0203	- 1.5314	1.2915	+ 0.1523	- 0.3885	1.0784
0.3720	0.0108	- 1.3258	1.1075	+ 0.4093	- 0.6069	1.1633
0.4360	0.0032	- 0.9483	0.7861	+ 0.8517	- 0.9664	1.2795
0.5000	0.0010	- 0.6455	0.5382	+ 1.1635	- 1.2075	1.3274
0.5640	0.0004	- 0.4300	0.3586	+ 1.3401	- 1.3350	1.3171
0.6280	0.0003	- 0.2790	0.2340	+ 1.4001	- 1.3672	1.2616
0.6920		- 0.1784	0.1501	+ 1.3718	- 1.3275	1.1743
0.7560		- 0.1131	0.0950	+ 1.2831	- 1.2389	1.0674
0.8840		- 0.0451	0.0368	+ 1.0175	- 0.9896	0.8330
1.0120		- 0.0184	0.0138	+ 0.7363	- 0.7276	0.6125
1.1400		- 0.0079	0.0051	+ 0.4997	- 0.5043	0.4302
1.2680		- 0.0036	0.0018	+ 0.3235	- 0.3343	0.2915
1.3960		- 0.0015	0.0006	+ 0.2020	- 0.2140	0.1918
1.5240				+ 0.1225	- 0.1334	0.1233
1.7800				+ 0.0421	- 0.0484	0.0480
2.0360				+ 0.0135	- 0.0164	0.0176
2.2920				+ 0.0041	- 0.0053	0.0061
2.5480				+ 0.0012	- 0.0016	0.0021
2.8040				+ 0.0003	- 0.0005	0.0007
3.0600				+ 0.0001	- 0.0001	0.0002

but for $l > 1$, $b(r_0) = 0$. For the Fe XVI example involving the 1*s* shell, we begin by determining the $b(r_0)$ and $b(r_1)$ from equations (44) and (47) using the trial function value for $P(1s; r)/r$ at $r = 0$ and $P(1s; r)$ at $r = 0.002$. From these two values, the b values for all the other r values are determined. The first two values $b(r_0)$ and $b(r_1)$ are not as accurate as they should be since the trial $P(1s; r)$ wave-function values were used for the first two points; these can be redetermined from the equation

$$P(1s; r) = r^{l+1}(A + Br + Cr^2) \tag{49}$$

TABLE 2

Calculation of $Y^0(nl, nl; r)$

r	$P^2(1s;r)$	$\frac{P^2(1s;r)}{r}$	$\int_0^r P^2(1s;\rho)d\rho$	$r\int_r^\infty \frac{P^2(1s;\rho)d\rho}{\rho}$	$Y^0(1s,1s;r)$
0.0000	0.0000	000			0.00000
0.0020	0.2440	122.0	0.00066	0.05066	0.05132
0.0040	0.8800	220.0	0.00220	0.09995	0.10215
0.0060	1.7850	298.0	0.00506	0.14679	0.15185
0.0080	2.8630	358.0	0.00846	0.19046	0.19892
0.0100	4.0320	403.0	0.01535	0.23044	0.24579
0.0120	5.2350	436.0	0.02461	0.26643	0.29104
0.0160	7.5630	460.0	0.05022	0.32628	0.37650
⋮	⋮	⋮	⋮	⋮	⋮
0.0440	13.535	308.0	0.38934	0.38317	0.77251
⋮	⋮	⋮	⋮	⋮	⋮
0.1000	4.0520	40.52	0.88092	0.09438	0.97530
⋮	⋮	⋮	⋮	⋮	⋮
0.5000	0.0^510	0.0^520	0.99919	0.00000	0.99919
⋮	⋮	⋮	⋮	⋮	⋮
0.6280	0.0^790	0.0^6143	0.99919	0.00000	0.99919

r	$P^2(2s;r)$	$\frac{P^2(2s;r)}{r}$	$\int_0^r P^2(2s;\rho)d\rho$	$r\int_r^\infty \frac{P^2(2s;\rho)d\rho}{\rho}$	$Y^0(2s2s;r)$
0.0000	0.0000	0.0000	0.00000	0.00000	0.00000
0.0020	0.026	11.45	0.00000	0.01106	0.01106
0.0040	0.0825	2.0625	0.00007	0.02204	0.02211
0.0060	0.1169	27.8167	0.00039	0.03276	0.03315
0.0080	0.2662	33.2750	0.00114	0.04318	0.04432
0.0100	0.3726	37.2600	0.00253	0.05325	0.05578
0.0120	0.4801	40.0083	0.00477	0.06295	0.06772
0.0160	0.6798	42.4875	0.00709	0.08130	0.08839
⋮	⋮	⋮	⋮	⋮	⋮
0.1000	0.3021	3.0210	0.03552	0.39492	0.43044
⋮	⋮	⋮	⋮	⋮	⋮
0.5000	0.4206	0.8412	0.96124	0.02664	0.98788
⋮	⋮	⋮	⋮	⋮	⋮
1.3960	0.0^522	0.0^516	0.99120	0.00000	0.99193

TABLE 2—*Continued*

r	$P^2(2p;r)$	$\frac{P^2(2p;r)}{r}$	$\int_0^r P^2(2p;\rho)d\rho$	$r\int_r^\infty \frac{P^2(2p;\rho)}{\rho}d\rho$	$Y^0(2p2p;r)$
0.0000	0.00000	0.00000			0.00000
0.0020	$0.0^{5}441$	0.00220	0.0000000	0.01096	0.01096
0.0040	0.00007	0.01650	0.0000000	0.02193	0.02193
0.0060	0.00030	0.05220	0.0000003	0.03289	0.03289
0.0080	0.00090	0.11780	0.0000010	0.04385	0.04385
0.0100	0.00220	0.21810	0.0000040	0.05481	0.05481
0.0120	0.00430	0.35750	0.0000100	0.06576	0.06577
0.0160	0.01230	0.76560	0.0000400	0.08765	0.08769
⋮	⋮	⋮	⋮	⋮	⋮
0.1000	2.40250	24.02500	0.0762700	0.44192	0.51819
⋮	⋮	⋮	⋮	⋮	⋮
0.5000	0.28970	0.57940	0.9775200	0.01850	0.99602
⋮	⋮	⋮	⋮	⋮	⋮
1.3960	$0.0^{6}360$	$0.0^{6}200$	0.9990600	0.00000	0.99906

r	$P^2(3s;r)$	$\frac{P^2(3s;r)}{r}$	$\int_0^r P^2(3s;\rho)d\rho$	$r\int_r^\infty \frac{P^2(3s;\rho)}{\rho}d\rho$	$Y^0(3s3s;r)$
0.0000	0.0000	0.0000			0.00000
0.0020	0.0049	2.4500	0.000004	0.00424	0.00424
0.0040	0.0176	4.4000	0.000030	0.00845	0.00848
0.0060	0.0356	5.9333	0.000080	0.01262	0.01270
0.0080	0.0568	7.1000	0.000170	0.01672	0.01689
0.0100	0.0794	7.9400	0.000300	0.02074	0.02104
0.0120	0.1022	8.5167	0.000490	0.02470	0.02519
0.0160	0.1444	9.0250	0.000980	0.03236	0.03334
⋮	⋮	⋮	⋮	⋮	⋮
0.1000	0.0891	0.8910	0.008950	0.17876	0.18771
⋮	⋮	⋮	⋮	⋮	⋮
0.5000	1.3537	2.7074	0.182830	0.57006	0.75289
⋮	⋮	⋮	⋮	⋮	⋮
1.1400	0.2497	0.2129	0.967360	0.02894	0.99630
⋮	⋮	⋮	⋮	⋮	⋮
1.3960	0.0408	0.0292	0.996930	0.00305	0.99998
⋮	⋮	⋮	⋮	⋮	⋮
3.0600	$0.0^{7}10$	$0.0^{8}33$	1.000000	0.00000	1.00000

by using the values for the wave functions computed at positions $r = 0.004$, 0.006, and 0.008 to evaluate the three constants, A, B, and C. Once these are known, the value for $P(1s; r)$ at $r = 0.002$ can be recomputed. After all $P(1s; r)$ values (Table 4) have been computed, it is essential to examine the general shape and tail of the wave function.

If a shape and tail such as is indicated by curve I for $P(1s; r)$ in Figure 4 is obtained, the ϵ_{nl} is then the best available eigenvalue for the wave functions at this stage of iteration. It is then necessary to renormalize $P(1s; r)$. If a shape and tail is of the type

TABLE 3

Calculation of g(nl; r)

r	$Y^0(1s,1s;r)$	$2Y^0(2s2s;r)$	$6Y^0(2p2p;r)$	$Y^0(3s3s;r)$	Σ	g(1s;r)
0.000	0.00000	0.00000	0.00000	0.00000	0.00000	0.00
0.002	0.05132	0.02212	0.06756	0.00424	0.14344	25297.23
0.004	0.10215	0.04422	0.13158	0.00847	0.28642	12297.50
0.006	0.15185	0.06630	0.19734	0.01270	0.42819	7963.75
0.008	0.19892	0.08864	0.26310	0.01689	0.56755	5798.78
0.010	0.24579	0.11156	0.32886	0.02104	0.70725	4499.22
0.012	0.29104	0.13544	0.39462	0.02519	0.84629	3633.79
0.016	0.37650	0.17678	0.52614	0.03334	1.11276	2551.58
0.020	0.45486	0.19678	0.65760	0.04136	1.35060	1905.61
0.024	0.52491	0.23550	0.78822	0.04924	1.59787	1474.10
0.028	0.58760	0.27296	0.91932	0.05787	1.83775	1166.58
0.032	0.64322	0.30930	1.05018	0.06460	2.06730	936.75
0.036	0.69224	0.34466	1.17948	0.07212	2.28850	758.08
.	.	.	.	.	.	.
0.100	0.97530	0.86088	3.10914	0.18771	5.13303	-141.99
.	.	.	.	.	.	.
0.500	0.99919	1.97576	5.97612	0.75289	9.70396	-501.85
.	.	.	.	.	.	.
1.140	1.00000	1.98384	5.99316	0.99630	9.97330	-531.23
.	.	.	.	.	.	.
1.396	1.00000	1.98386	5.99436	0.99998	9.97818	-559.33

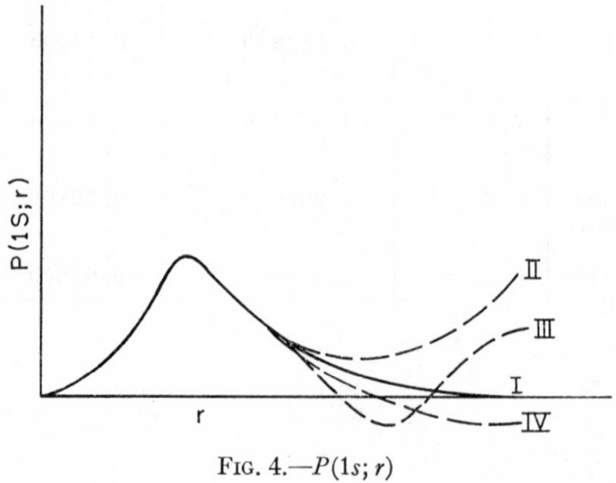

FIG. 4.—$P(1s; r)$

indicated by curves II, III, or IV, the parameter ϵ_{nl} is not sufficiently close to the proper value. A new ϵ_{nl} is therefore chosen and another iteration is made. This procedure is continued for all the wave functions in the configuration until self-consistency is achieved. Self-consistency is obtained when the difference in ϵ_{nl} and wave-function values between the nth and $(n-1)$st iterations is negligible. Normally, however, the procedure would be started with the outermost shell; that is, the iteration in this case would begin with the $P(3s; r)$ because the estimates of the outer shells are very likely to be the least accurate. Once the Hartree wave functions have been calculated, the other necessary parameters can be determined for both the transition probability and collision cross-sections. Before proceeding to determine the necessary parameters, it should be borne in mind that the Hartree method of calculating the wave functions does not include the exchange effects. The values for the parameters necessary for the determination of the various spectral properties would, therefore, be normally less accurate than those determined from the Hartree-Fock method, since the latter takes the exchange into account. It can be seen when the Hartree-Fock equations are written out explicitly that the difference in the two methods is significant. The Hartree-Fock equations written out explicitly are as follows:

$$\left[\!\left[\frac{d^2}{a\,r^2}+\frac{2}{r}\left\{26-\sum_{n'l'}^{3s}[q(n'l')-\delta_{1sn'l'}]\,Y^0(n'l',n'l';r)\right\}\right]\!\right]P(1s;r)$$
$$+\frac{2}{r}[Y^0(1s2s;r)P(2s;r)+Y^1(1s2p;r)P(2p;r) \quad (50)$$
$$+\tfrac{1}{2}Y^0(1s3s;r)P(3s;r)]=\sum_{n''=1}^{3}\epsilon_{1sn''s}P(n''s;r);$$

$$\left[\!\left[\frac{d^2}{d\,r^2}+\frac{2}{r}\left\{26-\sum_{n'l'}^{3s}[q(n'l')-\delta_{2sn'l'}]\,Y^0(n'l',n'l';r)\right\}\right]\!\right]P(2s;r)$$
$$+\frac{2}{r}[Y^0(2s1s;r)P(1s;r)+Y^1(2s2p;r)P(2p;r) \quad (51)$$
$$+\tfrac{1}{2}Y^0(2s3s;r)P(3s;r)]=\sum_{n''=1}^{3}\epsilon_{2sn''s}P(n''s;r);$$

$$\left[\!\left[\frac{d^2}{d\,r^2}+\frac{2}{r}\left\{26-\frac{1}{r}-\sum_{n'l'}^{3s}[q(n'l')-\delta_{2pn'l'}]\,Y^0(n'l',n'l';r)\right.\right.\right.$$
$$\left.\left.\left.+\tfrac{2}{5}Y^2(2p2p;r)\right\}\right]\!\right]P(2p;r)+\frac{2}{3r}[Y^1(2p1s;r)P(1s;r) \quad (52)$$
$$+Y^1(2p2s;r)P(2s\ \ r)+\tfrac{1}{2}Y^1(2p3s;r)P(3s;r)]=\epsilon_{2p2p}P(2p;r);$$

and

$$\left\{\frac{d^2}{d\,r^2}+\frac{2}{r}\left[26-\sum_{n'l'}^{2p}q(n'l')\,Y^0(n'l',n'l';r)\right]\right\}P(3s;r)$$
$$+\frac{1}{r}[Y^0(3s1s;r)P(1s;r)+Y^0(3s2s;r)P(2s;r) \quad (53)$$
$$+Y^1(3s2p;r)P(2p;r)]=\sum_{n''=1}^{3}\epsilon_{3sn''s}P(n''s;r).$$

If these equations are compared to equations (29)–(32), the difference is obvious. The wave functions for Fe XVI listed in Table 1 were calculated by the Hartree-Fock method. Although the data of Table 1 were used as trial wave functions for purposes of illustrating the computational procedure, they are actually the final results for Fe XVI wave functions with exchange. Therefore, it should not be surprising that the first iteration of $P(1s; r)$ wave functions (Table 4) are literally the same as the trial functions. Furthermore, very little difference, if any, would be observed for the 1*s* wave functions whether calculated with or without exchange. The maximum difference would be observed at the peak points and would in the case of Fe XVI 1*s* wave function not exceed 0.001. The difference would, of course, increase with the quantum number, *n*, and generally becomes significant beyond the 2*s* shell. It has been shown, however, that for very highly ionized

TABLE 4

Calculation of $P(nl; r_i)$

r	$h^2g(1s;r)$	$[1+\frac{1}{12}h^2g(1s;r_i)]$	$\frac{h^2g(1s;r_i)}{1+\frac{1}{12}h^2g(1s;r_i)}$	$b(r_i)$	$P(1s, r_i)$
0.0000	0.00000	0.00000	0.00000	0.00451	0.000
0.0020	0.10119	1.00840	0.10035	0.49829	0.494
0.0040	0.04919	1.00408	0.04899	0.94207	0.938
0.0060	0.03186	1.00264	0.03178	1.33970	1.336
0.0080	0.02320	1.00193	0.02316	1.69475	1.692
0.0100	0.01800	1.00149	0.01797	2.01055	2.008
0.0120	0.01454	1.00121	0.01452	2.29022	2.288
0.0160	0.04083	1.00339	0.04069	2.75950	2.750
0.0200	0.03049	1.00253	0.03041	3.10818	3.100
0.0240	0.02359	1.00196	0.02354	3.36234	3.355
.	.	.	.	.	.
.	.	.	.	.	.
0.0600	0.01223	1.00102	0.01222	3.33632	3.332

atoms, exchange becomes relatively unimportant. Both the analytical and numerical methods give generally satisfactory results, and neither method has any particular advantage over the other in the long run. Further work is still necessary for both analytical and numerical methods for the following reasons.

Although atomic wave-function calculations today would, as a matter of course, include exchange interaction, configuration interaction and correlation effects are not normally taken into account. This may cause part of the discrepancy between calculated and observed spectral line-strength results. If configuration interaction is to be included in the calculations, then it must be included as an integral part of the wave-function calculation, and since only very few atomic wave functions have had this included, it is not possible at this time to make a generalization of this effect. It would appear that when configuration interaction is taken into account, *f*-values and transition probabilities could be lowered by as much as 25%. Trees (1951, 1952), Green (1949), and Boys (1950, 1953*a*, *b*), as well as Biermann and Trefftz (1949), Trefftz and Biermann (1952), and Trefftz (1949, 1950, 1952), have made extensive studies on the effects of configuration interaction on the energy levels of various atoms. Recently, Burke (1963) and Weiss (1963) have made extensive calculations of the wave functions of light elements.

Burke has included correlation effects and Weiss, configuration interaction. Each obtained an improved value for the total energy of the atom and ions investigated.

With reference to the correlation effects (if the quality of ψ as determined by the HFSCF method with exchange is regarded as the criterion), it has been found that the total calculated energy is negative and at least 90% of the observed value. (Although these deviations are small compared to total energy, they are large compared to the energy difference between various energy levels.) As it was pointed out earlier in this section, the HFSCF method takes care of only the interelectronic repulsion in the Schrödinger equation, that is, ψ is a spherically symmetrical function in a central field. This is a good approximation, but it is not good enough to eliminate completely differences with respect to the observed energy. The correlation energy is then the difference between the experimental energy, corrected for relativistic effects, and the HFSCF total energy. The observed total energy is known for atoms up to ten electrons only, although the HFSCF energy is known for many more atoms. We shall consider the significance of the correlation effect in the case of Li I. The total observed energy is 14.96 rydbergs and the HFSCF energy is 14.81 rydbergs—a difference of 0.15 rydbergs; the energy difference between the $2s$ and $2p$ level is 0.14 rydbergs (see Burke 1963; Weiss 1963).

It should be borne in mind that the HFSCF-type atomic wave functions are the best one-electron representations for many-electron atomic configurations and are themselves sufficiently accurate to permit accurate calculation of line and collision strengths for the determination of the chemical abundances in celestial bodies. Today, atomic wave functions can be readily calculated for ground state and excited state configurations. Due to the work of Roothaan *et al.*, Piper, Watson, Douglas, Froese, Mayers, Clementi, Herman and Skillman, and Czyzak and Krueger, many wave functions have already been determined. Froese[2] and Worsley[3] have been collecting available wave functions and those can be obtained upon request. Krueger, McDavid, and Czyzak (1965) are at present computing wave functions for various configurations and compiling them for general use.

Once a set of wave functions for a particular atom or ion has been determined, the calculation of certain parameters, of value in transition probability and collision cross-section calculations, becomes relatively easy. The following parameters are useful:

$$\bar{r} = \int_0^\infty r P^2(nl;\, r)\, dr\,, \tag{54}$$

$$\overline{r^2} = \int_0^\infty r^2 P^2(nl;\, r)\, dr\,, \tag{55}$$

$$\sigma = N - (\bar{r}_H / \bar{r})\,, \tag{56}$$

and

$$\sigma^2(nl, n'l') = \sum_m \left| \frac{1}{\left(\sum_{nl} q(nl)\right)!} \sum_j \sum_p \sum_{p'} (-1)^{\pi+\pi'} \times \int p\psi_{\mathrm{I}} \left[\left(\frac{4\pi}{3}\right)^{1/2} r_j U_{1m}(\theta_j \varphi_j) \right] p'\psi_{\mathrm{II}} d\tau \right|^2 , \tag{57}$$

[2] Mathematics Department, University of British Columbia, Vancouver, B.C., Canada.

[3] McLennan Computing Laboratory, University of Toronto, Toronto, Ontario, Canada.

where p and p' are the permutations and π and π' represent the respective parities. The $\psi_I = \prod_a \phi_a(r_i)$, where ϕ_a is of the same form as in equation (6) with the spin function included and where the range of a includes all quantum numbers of the electron of State I. For State II we have a similar product. The spherical harmonics $U_{1m}(\theta_j\varphi_j)$ are normalized.

$$s_q = \int_0^\infty r^2 P(nl; r) P(n'l'; r)\, dr \tag{58}$$

$$\zeta_{\text{calc}} = \frac{e\hbar^2}{2m^2c^2} \int_0^\infty \left(-\frac{1}{r}\frac{\partial V}{\partial r}\right) P^2(nl; r)\, dr \tag{59}$$

$$\eta_{\text{calc}} = \frac{e^2\hbar^2}{4m^2c^2} \int_0^\infty \int_0^{r_2} \frac{1}{r_2^3} P^2(nl; r_2) P^2(nl; r_1)\, dr_1 dr_2 . \tag{60}$$

Table 5 gives the results for the $1s^2\, 2s^2\, 2p^6\, 3s\, {}^2S_{1/2}$ and $1s^2\, 2s^2\, 2p^6\, 3p\, {}^2P_{1/2}$ configurations and levels of Fe XVI.

TABLE 5

Atomic Parameters for Fe XVI Ground and Excited Levels

$\lim_{r\to 0} P(nl; r)/r^{l+1}$	1s 260.	2s 79.8	2p 531.	3s 36.9	3p 258.	3d 289.
$\bar{r}$	.0591	.2644	.2323	.6811	.6705	
$\overline{r^2}$	.0047	.0821	.0658	.5252	.5155	
σ	.6193	3.3071	4.4761	6.1719	7.3572	
ζ_{calc}					13500.	1531.
η_{calc}					6.90	6.41
$\epsilon_{nl,nl}$	559.33	98.38	89.60	35.65	33.17	29.74
	${}^2S - {}^2P$	${}^2P - {}^2D$	${}^2S - {}^2D$			
$\bar{\sigma}^2(nl, nl)$	.1506	.0244				
s_q			.2125			

Finally, in the Hartree-Fock method of calculating wave functions, it is presumed that the total wave function for the states that are involved can be expressed as a single determinant. If this should not be the case; that is, if the total wave function is a sum of the determinants, then it may not be possible to orthogonalize two open shells with the same l-value and the same number of electrons in each l-shell. Froese (1965) has given a detailed explanation for this possible inconsistency; Fock (1930) was aware of it when dealing with the $1s\ 2s$ configuration of helium.

3. THEORY OF TRANSITION PROBABILITIES

Radiation is regarded as a spatial flow of energy basically governed by the following Maxwell equations:

$$\nabla \cdot H = 0 , \tag{61}$$

$$\nabla \times E + \frac{1}{c}\frac{\partial H}{\partial t} = 0\,, \tag{62}$$

$$\nabla \cdot E = 4\pi\rho\,, \tag{63}$$

$$\nabla \times H - \frac{1}{c}\frac{\partial E}{\partial t} = \frac{4\pi}{c} I\,. \tag{64}$$

In order to discuss the theory of radiation, we consider a charged particle in an electromagnetic field. The classical description of the electromagnetic radiation field may be conveniently expressed by the vector potential A and the scalar potential φ rather than H or E where

$$H = \nabla \times A\,, \tag{65}$$

and

$$E = -\frac{1}{c}\frac{\partial A}{\partial t} - \nabla\varphi\,, \tag{66}$$

which will satisfy Maxwell's equations. If we then consider a particle of mass m and charge e moving with a velocity v in an electromagnetic field, it will be subjected to a force,

$$F = e\left(E + \frac{1}{c} V \times H\right). \tag{67}$$

The equations of motion can be written using Newton's second law and the relations (65) and (66) in (67). The equations of motion can also be derived from the Lagrangian function, which can be readily transformed into a Hamiltonian function by canonical equations. The Hamiltonian function in terms of coordinates and momenta is

$$H = \frac{1}{2m}\left(P - \frac{e}{c}A\right)^2 + e\varphi\,. \tag{68}$$

The construction of the Hamiltonian operator is performed in the usual way, and the momentum P is replaced by the operator $-i\hbar\nabla$. If the operation is performed on the wave function ψ, an expression is obtained from which the Hamiltonian operator can be extracted and which in vector notation has the form (Schiff 1955):

$$H = \frac{i\hbar}{2m}\left(i\hbar\nabla^2 + \frac{e}{c}\nabla \cdot A + \frac{2e}{c} A \cdot \nabla + \frac{e^2}{i\hbar c^2}|A|^2\right) + e\varphi\,. \tag{69}$$

For an electromagnetic field involving light waves, $\nabla \cdot A = 0$, $\varphi = 0$, and $(e^2/c^2)|A|^2$ is small. We then have for the Hamiltonian,

$$H \approx -\frac{\hbar^2}{2m}\nabla^2 + \frac{i\hbar e}{mc} A \cdot \nabla\,. \tag{70}$$

For a particle or system of particles with a potential energy V the Hamiltonian becomes

$$H = -\sum \frac{\hbar^2}{2m}\nabla_i^2 + V + \sum \frac{i\hbar}{m}\frac{e}{c} A_i \cdot \nabla_i$$

or

$$H = H_0 + H_1\,. \tag{71}$$

H_0 is the Hamiltonian for the system without the electromagnetic field and is time-independent; whereas H_1 is of particular importance in the emission and absorption of radiation, that is, the atomic system is subject to the time-dependent perturbations of H_1 in the electromagnetic field. Briefly reviewing the perturbation procedure, we begin with the usual expression,

$$H\psi = i\hbar \frac{\partial}{\partial t}\psi , \tag{72}$$

and for an unperturbed system, we can write

$$H_0\psi^0 = i\hbar \frac{\partial}{\partial t}\psi^0 , \tag{73}$$

where ψ^0 represents the unperturbed wave functions. It is also assumed that wave functions are of the form $\psi^0 = \psi^0(r, t) = \varphi^0(r)e^{-iEt/\hbar}$, that is, satisfy equation (73). By expanding ψ in terms of the unperturbed ψ_n's with coefficients as a function of time only and such that $|a|^2$ = probability of finding the system in the state with the energy E_n, we can express $\psi_n(r, t)$ in the following manner:

$$\psi_n(r, t) = \Sigma a_n(t)\psi_n^0(r, t), \tag{74}$$

and equation (72) becomes

$$\Sigma a_n(t) H_0\psi_n^0(r, t) + \Sigma a_n(t) H_1\psi_n^0(r, t) = i\hbar \sum \frac{da_n(t)}{dt}\psi_n^0(r, t) + i\hbar \sum a_n(t) \frac{\partial\psi_n^0(r, t)}{\partial t}. \tag{75}$$

But $\Sigma a_n(t)H_0\psi_n^0(r, t) = i\hbar\Sigma a_n(t)[\partial\psi_n^0(r, t)/\partial t]$ satisfies the unperturbed equation, and we are left with the expression

$$\sum a_n(t) H_1\psi_n^0(r, t) = -i\hbar \sum \frac{da_n(t)}{dt}\psi_n^0(r, t). \tag{76}$$

If multiplied by $\psi_m^{0*}(r, t)$ and integrated, this gives the matrix elements, that is,

$$\sum a_n(t) \int \psi_m^{0*}(r, t) H_1\psi_n^0(r, t) d\tau = i\hbar \sum \frac{da_n(t)}{dt} \int \psi_m^{0*}(r, t)\psi_n^0(r, t) d\tau ;$$

but

$$\int \psi_m^{0*}(r, t)\psi_n^0(r, t) d\tau = \delta_{mn} ;$$

therefore,

$$\frac{da_m(t)}{dt} = -\frac{i}{\hbar} \sum a_n(t) \int \psi_m^{0*}(r, t) H_1\psi_n^0(r, t) d\tau , \tag{77}$$

or

$$\dot{a}_m(t) = -\frac{i}{\hbar} \sum a_n(t) e^{i(E_m - E_n)t/\hbar} \int \varphi_m^{0*}(r) H_1\varphi_n^0(r) dr . \tag{78}$$

If at $t = 0$ the system is in an unperturbed state k then $a_n(0) = 0$, except for $n = k$, $a_k(0) = 1$. In equation (78) the time-dependent wave functions, $\psi_n^0(r, t)$, have been replaced by the product $\varphi_n^0(r)e^{iE_nt/\hbar}$. If the field is considered as a plane-polarized light wave, then the vector potential $\boldsymbol{A}$ will have only one component; that is, A_x, with $A_y = A_z = 0$. The electromagnetic field operator H_1 will be of the form:

$$H_1 = -\frac{e}{c}\frac{\hbar}{i} \sum_j \frac{1}{m_j} A_{xj}P_{xj} , \tag{79}$$

and

$$\int \psi_m^{0*}(r,t) H_1 \psi_n^0(r,t)\, d\tau = \left(\psi_m^0 \,\middle|\, -\frac{e\hbar}{mci} A_x \sum \frac{\partial}{\partial x_j} \,\middle|\, \psi_n^0\right)$$
$$= -\frac{e}{mc}\frac{\hbar}{i} A_x e^{2\pi i \nu_{mn} t} \left(\varphi_m^0 \,\middle|\, \sum \frac{\partial}{\partial x_j} \,\middle|\, \varphi_n^0\right), \qquad (80)$$

where

$$P_{xj} = \frac{\partial}{\partial x_j} \quad \text{and} \quad \nu_{mn} = (E_m - E_n)/h .$$

It can be shown that

$$\left(\varphi_m^0 \,\middle|\, \frac{\partial}{\partial x_j} \,\middle|\, \varphi_n^0\right) = -\frac{2\pi m \nu_{mn}}{\hbar}(\varphi_m^0 \,|\, x_j \,|\, \varphi_n^0)$$

in the following way: The φ_m^{0*} and φ_n^0 satisfy the wave equations,

$$\frac{\partial^2 \varphi_m^{0*}}{\partial x_j^2} + \frac{2m}{\hbar}[E - V(x_j)]\varphi_m^{0*} = 0$$

$$\frac{\partial^2 \varphi_n^0}{\partial x_j^2} + \frac{2m}{\hbar}[E - V(x_j)]\varphi_n^0 = 0 .$$

The first expression is multiplied by $x_j\varphi_n^0$ and the latter by $x_j\varphi_m^{0*}$; the results are subtracted and the resultant expression is integrated. One obtains

$$\int_{-\infty}^{\infty}\left(x_j\varphi_n^0 \frac{\partial^2}{\partial x_j^2}\varphi_m^{0*} - x_j\varphi_m^{0*}\frac{\partial^2}{\partial x_j^2}\varphi_n^0\right)dx_j = \frac{2m}{\hbar^2}(E_m - E_n)\int_{-\infty}^{\infty}\varphi_m^{0*}x_j\varphi_n^0\,dx_j .$$

Integrating the left-hand integrals by parts gives

$$\int_{-\infty}^{\infty}\varphi_n^0\frac{\partial}{\partial x_j}\varphi_m^{0*}\,dx_j - \int_{-\infty}^{\infty}\varphi_m^{0*}\frac{\partial}{\partial x_j}\varphi_n^0\,dx_j = \frac{2m}{\hbar^2}(E_m - E_n)\int\varphi_m^{*}x_j\varphi_n^0\,dx_j .$$

However,

$$\int_{-\infty}^{\infty}\varphi_n^0\frac{\partial}{\partial x_j}\varphi_m^{0*}\,dx_j = -\int_{-\infty}^{\infty}\varphi_m^{0*}\frac{\partial}{\partial x_j}\varphi_n\,dx_j ,$$

which can be readily obtained upon integrating the expression by parts. Therefore,

$$\left(\varphi_m^0 \,\middle|\, \frac{\partial}{\partial x_j} \,\middle|\, \varphi_n^0\right) = -\frac{m}{\hbar^2}(E_m - E_n)(\varphi_m^0 \,|\, x_j \,|\, \varphi_n^0) = -\frac{2\pi m \nu_{mn}}{\hbar}(\varphi_m^0 \,|\, x_j \,|\, \varphi_n^0);$$

then

$$\int \psi_m^{0*} H_1 \psi_n^0\, d\tau = -\frac{2\pi i \nu_{mn}}{c} A_x \mathfrak{a}_{mn} e^{2\pi i \nu_{mn} t}, \qquad (81)$$

where

$$\mathfrak{a}_{mn} \equiv (\varphi_m^0 \,|\, e\Sigma x_j \,|\, \varphi_n^0). \qquad (82)$$

The time dependence of A_x for light of frequency ν can be written as

$$A_x = \tfrac{1}{2}A_x^0(e^{2\pi i \nu t} + e^{-2\pi i \nu t}),$$

where A_x is a function of frequency and coordinates. Since atomic dimensions are of the order of 10^{-3} times the wavelength of light, A_x^0 in the above equation is treated as an invariant over the atom; that is, $A_x^0 = A_x^0(\nu_{mn})$. On the basis of equation (79), the Hamiltonian operator H_1 may then be written as

$$H_1(r, t) = H_1' \cdot \tfrac{1}{2}(e^{2\pi i\nu t} + e^{-2\pi i\nu t}). \tag{83}$$

From equation (78) and considering $a_m(0) = \delta_{mn}$, one obtains for $t > 0$,

$$a_m(t) = \int \varphi_m^{0*} H_1' \varphi_n^0 d\,r \left\{ \tfrac{1}{2} \left[\frac{1 - e^{2\pi i(\nu_{mn}+\nu)t}}{h\nu_{mn} + h\nu} + \frac{1 - e^{2\pi i(\nu_{mn}-\nu)t}}{h\nu_{mn} - h\nu} \right] \right\}, \tag{84}$$

provided

$$t \ll |\nu_{mn} \pm \nu|^{-1}. \tag{85}$$

If the transitions are such that

$$E_m - E_n \approx h\nu, \tag{86}$$

then the first term in the braces will be relatively small. Thus, the probability of finding the system in state m at time t will be, provided we neglect the first term of the braces,

$$|a_m(t)|^2 = |(\varphi_m^0 | H_1' | \varphi_n^0)|^2 \left\{ \frac{1 - \cos[2\pi(\nu_{mn} - \nu)t]}{(h\nu_{mn} - h\nu)^2} \right\}. \tag{87}$$

Since there is more than one frequency to be considered, equation (87) should be averaged with respect to ν in the region where relation (86) holds. Now $|a_m(t)|^2$ will be small except in this region; therefore, it is satisfactory to integrate from $-\infty$ to $+\infty$, and this gives the following expression:

$$|a_m(t)|^2 = \frac{\pi^2 \nu_{mn}^2}{c^2 \hbar^2} |A_x^0(\nu_{mn})|^2 |\alpha_{mn}|^2 t, \tag{88}$$

considering $A_x^0(\nu_{mn})$ as slowly varying in this region.

With the conditions (85) and (86), the probability of finding the system in state m due to a transition from state n can be approximated to

$$|a_m(t)|^2 \approx \frac{1}{\hbar^2} |(\varphi_m^0 | H_1' | \varphi_n^0)|^2 t^2. \tag{89}$$

This, then, is the approximate result that is obtained for the special case, that is, for plane-polarized light.

Equation (88) contains only the A_x component, so that for the general case which would contain the A_y and A_z components, the expression becomes

$$|a_m(t)|^2 = \frac{\pi^2 \nu_{mn}^2}{c^2 \hbar^2} (|A_x^0(\nu_{mn})|^2 |\alpha_{mn}|^2 + |A_y^0(\nu_{mn})|^2 |\beta_{mn}|^2 + |A_z^0(\nu_{mn})|^2 |\gamma_{mn}|^2) t \tag{90}$$

where

$$\beta_{mn} \equiv (\varphi_m^0 | e\Sigma y_j | \varphi_n^0)$$

and

$$\gamma_{mn} \equiv (\varphi_m^0 | e\Sigma z_j | \varphi_n^0).$$

For isotropic radiation, we have

$$|A_x^0(\nu_{mn})|^2 = |A_y^0(\nu_{mn})|^2 = |A_z^0(\nu_{mn})|^2 = \tfrac{1}{3} |A^0(\nu_{mn})|^2. \tag{91}$$

The electric field E can be expressed in terms of the $|A^0(\nu_{mn})|^2$,

$$E(\nu_{mn}) = -\frac{1}{c}\frac{\partial A}{\partial t} = -\frac{1}{c}\frac{\partial}{\partial t}[A^0(\nu_{mn})\cos 2\pi\nu_{mn}t],$$

or

$$E(\nu_{mn}) = \frac{2\pi\nu_{mn}}{c} A^0(\nu_{mn})\sin 2\pi\nu_{mn}t. \tag{92}$$

If the average of the square of equation (92) is taken, the result is

$$\overline{E^2}(\nu_{mn}) = \frac{2\pi^2\nu_{mn}^2}{c^2}|A^0(\nu_{mn})|^2; \tag{93}$$

and from electromagnetic theory, the radiation density $\rho(\nu_{mn})$ is equal to $\overline{E^2}(\nu_{mn})/4\pi$ or

$$|A^0(\nu_{mn})|^2 = \frac{2c^2}{\pi\nu_{mn}^2}\rho(\nu_{mn}). \tag{94}$$

Equation (90) can now be rewritten in the following form, taking advantage of equations (91) and (94).

$$|a_m(t)|^2 = \frac{2\pi}{3\hbar^2}\rho(\nu_{mn})(|\alpha_{mn}|^2 + |\beta_{mn}|^2 + |\gamma_{mn}|^2)t. \tag{95}$$

Quite obviously, at $t = 0$ the probability that the system is in state m is zero. On the other hand, where t is not zero, the probability that a transition resulting in the absorption or emission of energy from the electromagnetic field from state n to m or m to n will take place in unit time is

$$B_{nm}\rho(\nu_{mn}) = \frac{2\pi}{3\hbar^2}\rho(\nu_{mn})(|\alpha_{mn}|^2 + |\beta_{mn}|^2 + |\gamma_{mn}|^2) \tag{96}$$

or

$$B_{mn}\rho(\nu_{mn}) = \frac{2\pi}{3\hbar^2}\rho(\nu_{mn})(|\alpha_{mn}|^2 + |\beta_{mn}|^2 + |\gamma_{mn}|^2). \tag{97}$$

These last two equations give the Einstein transition probabilities, B_{nm} and B_{mn} being the coefficients for absorption and induced emission. This discussion applies to states of unit weight, that is, Zeeman states. The absorption probability is assumed to be proportional to the radiation density as in equation (96). For emission which may be induced and/or spontaneous, however, expression (97) only accounts for the former. Since radiation can be emitted in the absence of an electromagnetic field, as is the case of a transition from an excited level m to a lower level n, expression (97) is incomplete and, therefore, requires an additional term to take into account the spontaneous emission. Thus the expression for the emission of radiation can be written as

$$A_{mn} + B_{mn}\rho(\nu_{mn}).$$

For emission then, just as in the case of absorption, we have a term which is proportional to the radiation density and another which is independent of it. A_{mn} is Einstein's coefficient for spontaneous emission and B_{mn} the one for induced emission. The quantum mechanical calculation of A_{mn} is somewhat tedious and complicated; however, the A_{mn}'s may indirectly be found quite readily from the B_{mn}'s (see eq. [206]). Einstein (1917) determined its value by considering the equilibrium between two states of different energy.

The A_{mn}'s and B_{mn}'s are defined in terms of energy density rather than intensity. From Planck's radiation law, the radiation density can be expressed as

$$\rho(\nu_{mn}) = \frac{16\pi^2\hbar\nu_{mn}^3}{c^3}\frac{1}{e^{h\nu_{mn}/kT}-1}. \tag{98}$$

If the number of the system is N_m and N_n for the states m and n, then the number of states undergoing transitions from n to m is $N_n B_{nm}\rho(\nu_{mn})$, and those undergoing transitions from m to n are $N_m[A_{mn} + B_{mn}\rho(\nu_{mn})]$. At equilibrium these quantities are equal, or

$$\frac{N_n}{N_m} = \frac{A_{mn}+B_{mn}\rho(\nu_{mn})}{B_{nm}\rho(\nu_{mn})}. \tag{99}$$

The Boltzmann distribution law requires for equilibrium that

$$\frac{N_n}{N_m} = e^{-(E_n-E_m)/kT} = e^{h\nu_{mn}/kT}. \tag{100}$$

From these last two equations, the radiation density can be determined; that is,

$$e^{h\nu_{mn}/kT} = \frac{A_{mn}+B_{mn}\rho(\nu_{mn})}{B_{nm}\rho(\nu_{mn})}$$

or

$$\rho(\nu_{mn}) = \frac{A_{mn}}{B_{nm}e^{h\nu_{mn}/kT}-B_{mn}}. \tag{101}$$

To satisfy Planck's equation (98), the Einstein coefficients must be related in the following way:

$$B_{nm} = B_{mn} \tag{102}$$

and

$$A_{mn} = \frac{16\pi^2\hbar\nu_{mn}^3}{c^3}B_{mn}. \tag{103}$$

Since it was originally assumed that $\boldsymbol{A}$, the vector potential, and thus the field was constant over the atom, $A_{mn}^{(E)}$ depends only on the matrix elements for the electric-dipole moment,

$$A_{mn}^{(E)} = \frac{32\pi^2\nu_{mn}^2}{3\hbar c^3}|(\varphi_m^0|\boldsymbol{P}|\varphi_n^0)|^2, \tag{104}$$

where

$$|(\varphi_m^0|\boldsymbol{P}|\varphi_n^0)|^2 = (|\alpha_{mn}|^2+|\beta_{mn}|^2+|\gamma_{mn}|^2)$$

and

$$B_{mn} = \frac{2\pi}{3\hbar^2}|(\varphi_m^0|\boldsymbol{P}|\varphi_n^0)|^2.$$

If, however, consistent with the true state of affairs, the constancy of the field over the atom is not assumed, then A_{mn} will have additional terms. In equation (104) only electric-dipole radiation is taken into account; whereas, if the original assumption were not made, there would be magnetic-dipole, electric-quadrupole, magnetic-quadrupole, electric-octupole, etc. terms. The first two terms are very important in the transition probability calculation of forbidden lines; whereas, the others are of no importance in optical spectroscopy. The expansion of $\boldsymbol{A}$ may be represented by a Taylor series,

$$\boldsymbol{A} = (\boldsymbol{A}_i)_0 + r_j\left(\frac{\partial}{\partial r_j}\boldsymbol{A}_i\right)_0 + \frac{r_j r_k}{2!}\left(\frac{\partial^2}{\partial r_j \partial r_k}\boldsymbol{A}_i\right)_0 + \frac{r_j r_k r_l}{3!}\left(\frac{\partial^3}{\partial r_j \partial r_k \partial r_l}\boldsymbol{A}_i\right)_0 + \ldots , \tag{105}$$

where r_j, r_k, r_l are components of the vector $\boldsymbol{r}$ and the suffix zero indicates the value at $\boldsymbol{r} = 0$. The repetition of the index denotes summation.

If the expanded vector potential is introduced in the electromagnetic-field operator H_1, it will include not only the electric-dipole radiation terms but also two additional terms, that is, the magnetic-dipole and electric-quadrupole radiation terms that are the second and third terms of expression (105). Thus

$$H_1 = -\frac{e}{mc}\frac{\hbar}{i}\left[(\boldsymbol{A}_i)_0 \cdot \boldsymbol{P}_i + r_j\left(\frac{\partial}{\partial r_j}\boldsymbol{A}_i\right)_0 \cdot \boldsymbol{P}_i + \tfrac{1}{2}\left(\frac{\partial^2}{\partial r_j \partial r_k}\boldsymbol{A}_i\right)_0 \cdot \boldsymbol{P}_i(r_j r_k) + \ldots\right]. \tag{106}$$

This expression can be written as

$$H_1 = -\frac{e}{mc}\frac{\hbar}{i}[\boldsymbol{A}_0 \cdot \boldsymbol{P} + \tfrac{1}{2}(\nabla \times \boldsymbol{A}_0) \cdot (\boldsymbol{r} \times \boldsymbol{P}) + \ldots]. \tag{107}$$

The procedure for obtaining the transition probabilities for the magnetic-dipole and electric-quadrupole radiations is the same as for electric-dipole radiation.

If the detailed calculations are carried out, the transition probabilities due to magnetic-dipole and electric-quadrupole radiation will have the following forms:

$$A^{(M)}_{mn} = \frac{32\pi^3 \nu^3_{mn}}{3\hbar c^3} |(\varphi^0_m|\boldsymbol{M}|\varphi^0_n)|^2 \tag{108}$$

and

$$A^{(Q)}_{mn} = \frac{16\pi^5 \nu^5_{mn}}{5\hbar c^5} |(\varphi^0_m|\mathfrak{Q}|\varphi^0_n)|^2 \tag{109}$$

where $\boldsymbol{M}$ and $\mathfrak{Q}$ are the magnetic and quadrupole moments;

$$(\varphi^0_m|\boldsymbol{M}|\varphi^0_n) = \int \varphi^{0*}_m \left(\frac{-e}{2mc}\sum_i (\boldsymbol{L}_i + 2\boldsymbol{S}_i)\right)\varphi^0_n d\boldsymbol{r}$$

and

$$(\varphi^0_m|\mathfrak{Q}|\varphi^0_n) = \int \varphi^{0*}_m \left(-e\sum_i (\boldsymbol{r}_i \boldsymbol{r}_i - \tfrac{1}{3}r_i^2 I)\right)\varphi^0_n d\boldsymbol{r};$$

so that when there is a variation of the field over the atom, the expression for the transition probability includes the above two terms, or

$$A_{mn} = \frac{32\pi^3 \nu^3_{mn}}{3\hbar c^3}\left[|(\varphi^0_m|\boldsymbol{P}|\varphi^0_n)|^2 + |(\varphi^0_m|\boldsymbol{M}|\varphi^0_n)|^2 + \frac{3}{10}\frac{\pi^2 \nu^2_{mn}}{c^2}|(\varphi^0_m|\mathfrak{Q}|\varphi^0_n)|^2\right].$$

The superscripts (M) and (Q) identify the transition probabilities due to the magnetic dipole and electric quadrupole, respectively. For a detailed exposition of radiation theory, the reader is referred to textbooks on quantum mechanics, particularly those by Eyring, Walter, and Kimball (*Quantum Chemistry*, 1944), Pauling and Wilson (*Introduction to Quantum Mechanics*, 1935), and Condon and Shortley (*The Theory of Atomic Spectra*, 1935).

The actual emitted intensity due to spontaneous emission resulting from a transition between a particular pair of states ($m \rightarrow n$); that is, the component of a line in erg/sec is

$$I_{mn} = N_m h\nu_{mn} A_{mn} \,. \tag{110}$$

This equation can be used only when the radiation density is sufficiently low that the induced emission is negligible. It is difficult to make absolute comparisons with experiments for emission intensity because the N_m is generally not known; therefore, it is common practice to work with relative intensities of related lines or groups of lines. The conditions for the excitation of atoms are such that isotropy can be assumed; at least it is an excellent approximation. It can also be assumed that the number of atoms in different states of the same level are equal. This is very often referred to as natural excitation. If, on the other hand, the excitation is not isotropic, large departures from natural excitation will occur and problems arising from such effects are numerous. These have been reviewed in great detail by Mitchell and Zemansky (1961).

The total intensity of a line is the sum of the intensities of its components, that is, all the states between two levels, and the natural excitation for a line from level J to J' is

$$I(J, J') = \frac{32\pi^3}{3\hbar} \sum_{mn} h\nu_{mn} N_m \bar{\nu}^3_{mn} |(\varphi^0_m | \boldsymbol{P} | \varphi^0_n)|^2 \tag{111}$$

for electric-dipole radiation. The sum of the squared matrix is defined as the strength of the line, $S^{(E)}(J, J')$, and $S^{(E)}_{mn} = |(\varphi^0_m | \boldsymbol{P} | \varphi^0_n)|^2$ is the strength of the component involving a transition from state m to state n. Also useful are the following partial sums introduced by Condon and Shortley (1935):

$$S^{(E)}(m, J') = \sum_n |(\varphi^0_m | \boldsymbol{P} | \varphi^0_n)|^2 ,$$

$$S^{(E)}(J, n) = \sum_m |(\varphi^0_m | \boldsymbol{P} | \varphi^0_n)|^2 ,$$

$$S^{(E)}(J, J') = \sum_m S^{(E)}(m, J') = \sum_n S^{(E)}(J, n)$$

and

$$S^{(E)}(J, J') = \sum_{mn} S^{(E)}_{mn} \,.$$

From equation (111) and from the above expressions, it can be seen that the line intensity is proportional to the number of atoms in any one of the initial states N_m to the fourth power of the wave number $\bar{\nu}$ and $S(J, J')$. If the approximations $N_m = N(J)$ (a constant) and $h\nu_{mn} = h\nu$ (a constant) are made, then these terms may be factored out of the sum in equation (111). The equation for $I(J, J')$ also can be written as

$$I(J, J') = N(J) h\nu A(J, J') \,. \tag{112}$$

But $N(J) = (2J+1)N_m$. We have for electric-dipole radiation

$$(2J+1)N_m h\nu A^{(E)}(J, J') = \frac{32\pi^3\bar{\nu}^3}{3\hbar} N_m h\nu \sum_{mn} |(\varphi_m^0|\boldsymbol{P}|\varphi_n^0)|^2$$

or

$$A^{(E)}(J, J') = \frac{32\pi^3\bar{\nu}^3}{3\hbar} \frac{S^{(E)}(J, J')}{2J+1} \tag{113}$$

and

$$A^{(M)}(J, J') = \frac{32\pi^3\bar{\nu}^3}{3\hbar} \frac{S^{(M)}(J, J')}{2J+1} \tag{114}$$

for the magnetic-dipole radiation. Similarly, to obtain the quadrupole transition probability we have

$$I(J, J') = \frac{16\pi^5}{5\hbar} \sum_{mn} h\nu_{mn} N_m \bar{\nu}_{mn}^5 |(\varphi_m^0|\mathfrak{Q}|\varphi_n^0)|^2$$

or

$$I(J, J') = \frac{16\pi^5}{5\hbar} h\nu N_m \bar{\nu}^5 \sum_{mn} |(\varphi_m^0|\mathfrak{Q}|\varphi_n^0)|^2$$

and

$$I(J, J') = N(J) h\nu A^{(Q)}(J\ J').$$

We define

$$\sum_{mn} |(\varphi_m^0|\mathfrak{Q}|\varphi_n^0)|^2 \equiv S^{(Q)}(J, J'),$$

so that we get

$$A^{(Q)}(J, J') = \frac{16\pi^5\bar{\nu}^5}{5\hbar} \frac{S^{(Q)}(J, J')}{2J+1} \tag{115}$$

and $\mathfrak{Q}$ is a dyad or $\mathfrak{Q} = -e\sum_j (\boldsymbol{r}_j\boldsymbol{r}_j - \frac{1}{3}r_j^2 I)$ where $I = \boldsymbol{ii} + \boldsymbol{jj} + \boldsymbol{kk}$. In order to employ equations (113), (114), and (115), we change to convenient dimensions, and the expressions are written in the form,

$$A^{(E)}(J, J') = \frac{2.662 \times 10^9\bar{\nu}^3}{2J+1} S^{(E)}(J, J')\,\text{sec}^{-1}, \tag{116}$$

$$A^{(M)}(J, J') = \frac{3.532 \times 10^4\bar{\nu}^3}{2J+1} S^{(M)}(J, J')\,\text{sec}^{-1}, \tag{117}$$

and

$$A^{(Q)}(J, J') = \frac{2.648 \times 10^3\bar{\nu}^5}{2J+1} S^{(Q)}(J, J')\,\text{sec}^{-1}, \tag{118}$$

where $\bar{\nu}$ is the wave number of the line expressed in rydbergs, J and J' are the upper and lower levels, respectively, and the line strengths [$S^{(E)}$, $S^{(M)}$, and $S^{(Q)}$], are expressed in atomic units.

In addition to the transition probability A and the line strength S, another very useful parameter in the calculation of line intensities is the oscillator strength, or the so-called f-value.

In classical electromagnetic theory, accelerated charges radiate; so, if these were a system of moving charges, they would produce a radiation field which might be viewed as originating from an electric-dipole, magnetic-dipole, electric-quadrupole, or higher order of multipoles. As has been previously indicated, multipoles of higher order than electric quadrupole are of no interest in optical atomic spectra. Classically then, if dipole oscillators of frequency ν_0 are exposed to a beam of monochromatic radiation ν, they undergo a forced vibration of frequency ν; that is, they attempt to follow the rapidly varying electric field and oscillate with the frequency of the radiation. This causes the beam of radiation to undergo refraction and absorption. The classical theory of the absorption by a dipole, which is basic to the development of the expressions for the complex dielectric constant, complex index of refraction, and the absorption coefficient, is described in detail by Aller (1963).

The intensity varies according to the following relation:

$$I = I_0 e^{-\kappa_\nu \rho x} \tag{119}$$

where $\kappa_\nu \rho$ is the absorption coefficient per unit volume and is written as

$$\kappa_\nu \rho = \left(N_0 \frac{e^2}{m c} \right) \frac{\gamma}{4\pi} \left[\frac{1}{(\nu - \nu_0)^2 + (\gamma/4\pi)^2} \right], \tag{120}$$

where N_0 = the number of oscillators of frequency ν_0; γ = the classical damping constant $(8\pi^2/3mc^2)\nu^2$; ν = the natural frequency of the monochromatic radiation; and e, m, and c are, respectively, the electron charge, electron mass, and the velocity of light. The absorption coefficient per atom $\kappa_\nu \rho / N_0$ is

$$\alpha_\nu = \left(\frac{e^2}{m c} \right) \frac{\gamma}{4\pi} \left[\frac{1}{(\nu - \nu_0)^2 + (\gamma/4\pi)^2} \right]. \tag{121}$$

Since the quantum theory of radiation differs fundamentally from the classical theory, it might be expected that the absorption coefficient, as derived from quantum mechanics, would not bear any resemblance to the classical one. Using quantum mechanics, Weisskopf and Wigner (1930) derived the shape of the line absorption coefficient. Their formula turns out to be analogous to the classical one if γ is replaced by a quantum mechanical Γ and N_0 is replaced by $N_0 f$; so

$$\kappa_\nu \rho = \left(\frac{\pi e^2}{m c} \right) N_0 f \left(\frac{\Gamma}{4\pi} \right) \left[\frac{1}{(\nu - \nu_0)^2 + (\Gamma/4\pi)^2} \right], \tag{122}$$

and per atom it becomes

$$\alpha_\nu = \left(\frac{\pi e^2}{m c} \right) f \left(\frac{\Gamma}{4\pi} \right) \left[\frac{1}{(\nu - \nu_0)^2 + (\Gamma/4\pi)^2} \right], \tag{123}$$

where f is the quantum mechanical value for the oscillator strength. The quantity f is not an integer (as would be required by classical theory) and Γ may be expressed in the following way:

$$\Gamma_{mn} = \Gamma_m + \Gamma_n , \tag{124}$$

where

$$\Gamma_n = \sum_n \frac{A_{mn}}{(1 - e^{h\nu_{mn}/kT})} + \sum_j \frac{\bar{\omega}_j}{\bar{\omega}_n} \frac{A_{jn}}{(e^{h\nu_{jn}/kT} - 1)}. \tag{125}$$

The first summation is taken over n levels lower than m, and the second summation over the levels j above m. For any level, Γ_n is defined in general as

$$\Gamma_n = \Sigma A_{mn} + \Sigma B_{mn} I(\nu_{mn}) + \Sigma B_{nk} I(\nu_{nk}) . \tag{126}$$

The f-values are related to the Einstein coefficients and the line strengths, respectively, in the following way:

$$A(J, J') = \frac{8\pi^2 e^2 \bar{\nu}}{m c^3} \frac{\bar{\omega}_{J'}}{\bar{\omega}_J} f(J, J') \sec^{-1} \tag{127}$$

and

$$f(J, J') = \frac{4\pi m c}{3\hbar e^2} \frac{\bar{\nu}}{\bar{\omega}_{J'}} S(J, J'), \tag{128}$$

where the $\bar{\omega}$'s represent the statistical weight of the appropriate levels, that is, $\bar{\omega}_{J'} = 2J' + 1$ and $\bar{\omega}_J = 2J + 1$. $\bar{\nu}$ is the wave number of the line in rydbergs. The expressions for $f(J, J')$ for electric-dipole, magnetic-dipole, and electric-quadrupole radiation in terms of line strength $S(J, J')$ become:

$$f^{(E)}(J, J') = \frac{3.336 \times 10^{-1}}{2J' + 1} \bar{\nu} S^{(E)}(J, J'), \tag{129}$$

$$f^{(M)}(J, J') = \frac{4.426 \times 10^{-6}}{2J' + 1} \bar{\nu} S^{(M)}(J, J'), \tag{130}$$

and

$$f^{(Q)}(J, J') = \frac{3.318 \times 10^{-7}}{2J' + 1} \bar{\nu}^3 S^{(Q)}(J, J'). \tag{131}$$

The validity of the theoretical f-values depends basically upon deviation from LS-coupling and the evaluation of σ^2. Since equations (129), (130), and (131) presume LS-coupling, the validity of the theoretical f-values computed according to the scheme described depends on the following:

1. The accuracy with which the necessary radial quantum integrals can be evaluated, that is, $\sigma^2(nl, n'l')$ (see eqs. [136], [137], or [138]).

2. Deviation from LS-coupling. This means that the relative line strengths within a multiplet are not noticeably distorted. For heavy atoms, these deviations from LS-coupling may be considerable.

3. If the matrices of the spin-orbit, spin-spin, and spin-other-orbit interactions are diagonalized to take into account deviations from LS-coupling, the line strength S can be calculated, although by somewhat more cumbersome expressions than those for LS-coupling. The resulting f-values should then be as accurate as the radial quantum integrals permit. A description of the theory and application of f-values is given by Aller (1963).

An important relationship concerning atomic f-values is the Thomas-Kuhn sum rule, discovered independently by Thomas (1925) and Kuhn (1925). Oscillator strengths in atoms obey this rule; that is, if f is summed over all possible transitions between all possible configurations, the result should give the number of electrons N in the atom, or

$$\Sigma f(nl, n'l') = N . \tag{132}$$

If the electrons in the inner shell are bound tightly enough, the interactions between the inner and outer (valence) electrons may be neglected, and equation (132) may be written as

$$\Sigma f(nl, n'l') = q\,, \tag{133}$$

where q represents the number of valence electrons, and summation is carried out only over the levels to which transitions occur, or

$$\sum_{m} f_{nm} - \sum_{k<n} \frac{\bar{\omega}_k}{\bar{\omega}_n} f_{kn} + \int_0^{\infty} f_{nk} d\,k = q\,. \tag{134}$$

The first term of equation (134) represents the absorption term from level n, and the summation is over all levels with $m > n$. The second term represents downward transitions, that is, the emissions with summation over all levels with $n > k$. The integral term represents the bound-free absorptions in which the atom loses the electron.

TABLE 6

Selection Rules for Allowed and Forbidden Transitions

Electric-Dipole	Magnetic-Dipole	Electric-Quadrupole
$\Delta J = 0, \pm 1$ ($0 \nleftrightarrow 0$ forbidden)	$\Delta J = 0, \pm 1$ ($0 \nleftrightarrow 0$ forbidden)	$\Delta J = 0, \pm 1, \pm 2$ ($0 \nleftrightarrow 0$; $1/2 \nleftrightarrow 1/2$, $0 \nleftrightarrow 1$) forbidden
$\Delta M = 0, \pm 1$	$\Delta M = 0, \pm 1$	$\Delta M = 0, \pm 1, \pm 2$
Parity changes $(-1)^{\Sigma l_i}$	No parity change $(-1)^{\Sigma l_i}$	No parity change $(-1)^{\Sigma l_i}$
One electron jump	No electron jump	One or no electron jump
$\Delta l = \pm 1$	$\Delta l = 0$ $\Delta n = 0$	$\Delta l = 0, \pm 2$
$\Delta S = 0$	$\Delta S = 0$	$\Delta S = 0$
$\Delta L = 0, \pm 1$ ($0 \nleftrightarrow 0$ forbidden)	$\Delta L = 0$	$\Delta L = 0, \pm 1, \pm 2$ ($0 \nleftrightarrow 0$, $0 \nleftrightarrow 1$) forbidden

If the transitions are due to electric-dipole radiation (eq. [116]), the other types of transitions can be neglected. When the radiation conditions are such that the electric-dipole radiation is missing, however, those due to magnetic-dipole and electric-quadrupole radiations become important. The symmetry properties of the wave functions are such that the matrix elements for electric-dipole, magnetic-dipole, and electric-quadrupole radiation vanish unless certain conditions, that is, the selection rules, are satisfied. These selection rules for atomic spectra are shown in Table 6.

Table 6 furnishes the selection rules for allowed and forbidden transitions. The forbidden transitions are not only those due to magnetic-dipole and electric-quadrupole radiation but also those, not discussed here, due to two quantum processes, electric-dipole radiation caused by (1) external perturbations and (2) the atomic nucleus. When magnetic-dipole and electric-quadrupole transitions are absent, the rules hold rigorously provided that there is only negligible configuration interaction and the coupling is LS. Otherwise, only the first three rules apply rigorously. Transitions are usually regarded as forbidden if

1. The first three selection rules are violated.
2. The last three selection rules are violated.
3. The atom is subjected to external perturbations.
4. Appreciable nuclear perturbations occur.
5. Two quantum processes occur.

Normally, however, transitions involving condition (2) without any other violations are not considered to be forbidden. The rules of Table 6 apply strictly for LS-coupling. The interesting cases, however, are the ones that involve intermediate coupling, for example, ${}^1D_2 - {}^3P_2$ in [O III], etc., which clearly violate magnetic-dipole rule $\Delta S = 0$ for pure LS-coupling. For example, a transition $3s^2\,{}^1S_0 - 3s3p\,{}^3P_1$, although excluded in LS-coupling, is not considered as forbidden. Transitions that violate the Laporte parity rule (1924) $\Delta l = \pm 1$ are considered forbidden transitions. Hence, magnetic-dipole and electric-quadrupole transitions will be considered "forbidden transitions" in this context.

The allowed-transition probabilities and the oscillator strengths are determined from equations (116) and (129), respectively. It is, however, first necessary to calculate the absolute line strength $S^{(E)}(J, J')$. The absolute line strength is defined by the following relation:

$$S^{(E)} = \mathfrak{S}(\mathfrak{L})\,\mathfrak{S}(\mathfrak{M})\,\sigma^2(nl, n'l') = \mathfrak{S}(\mathfrak{R})\,\sigma^2(nl, n'l'), \tag{135}$$

where $\mathfrak{S}(\mathfrak{L})$ is the relative line strength and $\mathfrak{S}(\mathfrak{M})$ is the relative multiplet strength. The expression $\mathfrak{S}(\mathfrak{R}) = \mathfrak{S}(\mathfrak{L})\mathfrak{S}(\mathfrak{M})$ represents relative strength for the transition involved, that is $(J \rightarrow J')$; $\sigma^2(nl, n'l')$ has been defined by equation (57). Generally, however, $\sigma^2(nl, n'l')$ is defined as

$$\sigma^2(nl, n'l') = \frac{1}{4l_>^2 - 1}\left[\int_0^\infty rP(nl; r)P(n'l'; r)\,dr\right]^2, \tag{136}$$

where $l_>$ is the greater of the two azimuthal quantum numbers. This expression is satisfactory if the changes in the orbitals of the passive electrons are not taken into account; otherwise, the more complicated formula has to be employed in order to include the involved configurations properly. With reference to $\sigma^2(nl, n'l')$, it should be mentioned that Chandrasekhar (1945) was the first to point out that, in addition to the expression (136), two other expressions can be used for calculating the transition integral. These two equations are obtained by the proper transformation of equation (136) (see Schiff [1955], pp. 25, 131) and are

$$\sigma^2(nl, n'l') = \frac{1}{4l_>^2 - 1}\left\{\frac{2}{E_{n'l'} - E_{nl}}\int_0^\infty P(nl; r) \times\left[\frac{1}{r}P(n'l'; r) - \frac{d}{dr}P(n'l'; r)\right]dr\right\}^2 \tag{137}$$

and

$$\sigma^2(nl, n'l') = \frac{1}{4l_>^2 - 1}\left[\left(\frac{2}{E_{n'l'} - E_{nl}}\right)^2 \int_0^\infty P(n'l'; r)\frac{dV}{dr}P(nl; r)\,dr\right]^2, \quad (138)$$

where $E_{n'l'} - E_{nl}$ is the energy difference between the two levels and V is the potential of the field in which the electron moves. The expressions of (136), (137), and (138) are known as the dipole length, dipole velocity, and dipole acceleration, respectively. For exact wave functions, the transition integral value would be the same regardless which of these equations is used.

When approximate wave functions, for example, the Hartree-Fock types, are used, however, one may obtain somewhat different results (Weiss 1963). Which of the three formulae gives the best result depends to a large extent on the accuracy of the wave functions over the radial distance. Since the largest contribution to the integral for equation (136) comes from the region of large r, for equation (137) from the region of the intermediate r, and for equation (138) from small r, it would appear that for Hartree-Fock–type wave functions, the dipole acceleration equation would give very poor results, due to the fact that these functions are most inaccurate near the origin. According to Chandrasekhar, the best choice would be the dipole velocity equation because, in calculating the wave functions by minimizing the energy, the region of intermediate r would be the most accurate. The difference between the results for transition probabilities, when determined by using the dipole length and dipole velocity, is of the order of 10 per cent, however, and one cannot always be sure that one of these is better than the other or even that the correct result lies somewhere between the two.

Many of the relative line strengths and relative multiplet strengths have been reported in the literature by Goldberg (1935, 1936) and Allen (1963). Rohrlich (1959*a*, *b*), however, developed general formulae for all the allowed-transition arrays of astrophysical interest, and these arrays are

(*a*) $l^n \cdot l' - l^n \cdot l''$,

(*b*) $l^n - l^{n-1} \cdot l'$,

(*c*) $l^n \cdot l' - l^{n-1} \cdot l'^2$,

(*d*) $l^n l' \cdot l'' - l^n \cdot l'^2$,

and

(*e*) $l^n l'' \cdot l' - l^n \cdot l'^2$,

where the symbol to the right of the dot denotes the jumping electron and that on the left denotes the electrons not in closed shells. The expression for the relative line strength $\mathfrak{S}(\mathfrak{L})$ is the same for all the above transition arrays and following Rohrlich (1959*a*, *b*) is

$$\mathfrak{S}(\mathfrak{L}) = \frac{(2J+1)(2J'+1)}{2S+1} W(LJL'J'; S1), \quad (139)$$

where $W(LJL'J'; S1)$ represents the Racah coefficient and where the primed and unprimed numbers refer to the final and initial states, respectively. The relative multiplet strength $\mathfrak{S}(\mathfrak{M})$ for transition (*a*) is:

$$\mathfrak{S}[l^n(\alpha_1 S_1 L_1)l'SL; l^n(\alpha_1' S_1 L_1)l''SL'] = \delta(\alpha_1, \alpha_1')g(SL)[(2L'+1)l_> (4l_>^2 - 1)W^2(l'Ll''L'; L_1 1)], \quad (140)$$

where $g(SL) = (2S+1)(2L+1)$.

For transition (*b*),

$$\mathfrak{S}[l^n\alpha SL;\, l^{n-1}(\alpha_1S_1L_1)l'SL] = n[l^n\alpha SL\{\!|l^{n-1}(\alpha_1S_1L_1)lSL]^2\mathfrak{S}[l^{n-1}(\alpha_1S_1L_1)lSL;\, l^{n-1}(\alpha_1S_1L_1)l'SL'] \tag{141}$$

where $[l^n\alpha S^nL\{\!|l^{n-1}(\alpha_1S_1L_1)lSL]$ = fractional parentage coefficient as defined by Racah.

For transition (*c*),

$$\begin{aligned}\mathfrak{S}[l^n(\alpha_1S_1L_1)l'SL;\, l^{n-1}(\alpha_1'S_1'L_1')l'^2(S_2'L_2')SL'] &= 2n[l^n\alpha_1S_1L_1\{\!|l^{n-1}(\alpha_1'S_1'L_1')lS_1L_1]^2g(SL)\,g(S_1L_1)\,g(S_2'L_2')\\ &\times[(2L'+1)l_>(4l_>^2-1)W^2(S_1'\tfrac{1}{2}S\tfrac{1}{2};\, S_1S_2')]\\ &\times X^2(L_1'L_1l;\, L'L1;\, L_2'l'l'),(S_2'+L_2'=\text{even})\end{aligned} \tag{142}$$

where

$$X(abe;\, cdf;\, ghk) = \sum_\lambda (2\lambda+1)W(ackh;\, g\lambda)W(bhfc;\, d\lambda)W(fhba;\, e\lambda)$$

is a 9_j coefficient of Wigner.

For transition (*d*),

$$\begin{aligned}\mathfrak{S}[l^n(\alpha_1S_1'L_1')l'S_1L_1,\, l''SL;\, l^n(\alpha_1'S_1'L_1')l'^2(S_2'L_2')SL'] &= 2l_>(4l_>^2-1)\,g(SL)\,g(S_1L_1)\,g(S_2'L_2')(2L'+1)\\ &\times W^2(S_1'\tfrac{1}{2}S\tfrac{1}{2};\, S_1S_2')W^2(l'l'L'L_1';\, L_2'L_1)\\ &\times W^2(L'l'Ll'';\, L_11)(l_>=\text{larger of } l'l'').\end{aligned} \tag{143}$$

For transition (*e*),

$$\begin{aligned}\mathfrak{S}[l^n(\alpha_1'S_1'L_1')l''S_1L_1,\, l'SL;\, l^n(\alpha_1'S_1'L_1')l'^2(S_2'L_2')SL'] &= 2l_>(4l_>^2-1)\,g(SL)\,g(S_2'L_2')(2L'+1)\\ &\times[W^2(S_1'\tfrac{1}{2}S\tfrac{1}{2};\, S_1S_2')X^2(L_1'L_1l'';\, L'L1;\, L_2'l'l')].\end{aligned} \tag{144}$$

With the necessary line and multiplet strengths evaluated from these equations and $\sigma^2(nl;\, n'l')$ calculated from the wave functions, the theoretical absolute line strengths, transition probabilities, and oscillator strengths can be determined. Using Fe XVI as an illustration, allowed transitions, involving the ground configuration ($3s\,^2S_{1/2}$) and two of the excited levels ($3p\,^2P_{1/2\,3/2}$ and $3d\,^2D_{3/2\,5/2}$) are listed in Table 7 and shown in Figure 5. The $\sigma^2(nl,\, nl')$ were calculated using the wave functions listed in Table 1, and the results for ^2S–^2P and ^2P–^2D transitions are shown in Table 5. The value for $\mathfrak{S}(\mathfrak{L})$ is computed from equation (139) and for $\mathfrak{S}(\mathfrak{M})$ from the appropriate equation of the set (140) to (144). For example, for the transition $^2S_{1/2}$–$^2P_{1/2}$ in Fe XVI, these values are determined in the following way:

a) Using equation (139), we get

$$\mathfrak{S}(\mathfrak{L}) = \frac{[(2)(\frac{1}{2})+1][(2)(\frac{1}{2})+1]}{2} W^2(0\tfrac{1}{2}1\tfrac{1}{2};\tfrac{1}{2}1)$$

$$\mathfrak{S}(\mathfrak{L}) = 2\left[\frac{(1-\frac{1}{2}+\frac{1}{2}+1)(1+\frac{1}{2}-\frac{1}{2})}{(1)(2)(2)(3)}\right]$$

$$\mathfrak{S}(\mathfrak{L}) = (2)\left[\frac{(2)(1)}{12}\right]$$

$$\mathfrak{S}(\mathfrak{L}) = \tfrac{1}{3}.$$

b) Using equation (140) we get

$$\mathfrak{S}(\mathfrak{M}) = (1)(2)(1)[(2+1)(1)(4-1)W^2(0011;01)],$$

where

$$W^2(0011;01) = \frac{1}{(1)(3)} = \tfrac{1}{3}$$

TABLE 7

Allowed Transition Probabilities for Fe XVI (3s, 3p & 3d configurations)

Transitions	λ(Å)	$\bar{\nu}$ (ryd.)	$\sigma^2(nl,nl')$	$\mathfrak{S}(\mathfrak{M})$	$\mathfrak{S}(\mathfrak{L})$	$S^{(E)}(J, J')$	$f^{(E)}(J, J)$	$A^{(E)}(J, J')$
$^2P_{\frac{1}{2}}$ - $^2S_{\frac{1}{2}}$	361.66	2.5197	.1506	6	1/3	3012	.12616	6.42×10^9
$^2P_{\frac{3}{2}}$ - $^2S_{\frac{1}{2}}$	336.17	2.7108	.1506	6	2/3	.6024	.27109	7.99×10^9
$^2D_{\frac{3}{2}}$ - $^2P_{\frac{1}{2}}$	249.73	3.6490	.0244	60	1/3	.4874	.29484	1.58×10^{10}
$^2D_{\frac{3}{2}}$ - $^2P_{\frac{3}{2}}$	263.53	3.4579	.0244	60	1/15	.09748	.02790	2.68×10^9
$^2D_{\frac{5}{2}}$ - $^2P_{\frac{3}{2}}$	261.57	3.4838	.0244	60	3/5	.8773	.25296	1.65×10^{10}

Erratum: Heading, col. 8, for (J, J) read (J, J′).

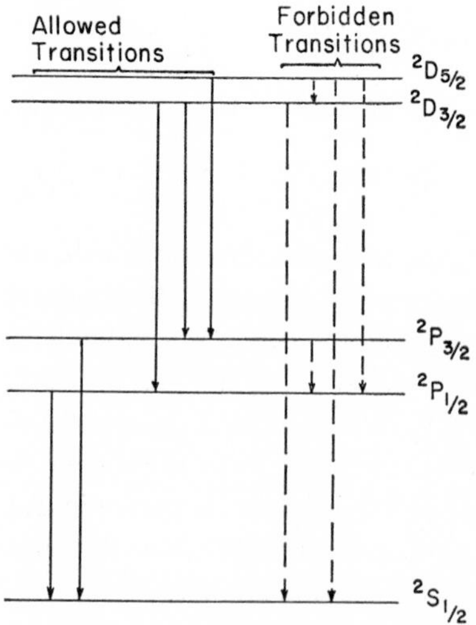

FIG. 5.—Allowed transition probabilities for Fe XVI (3*s*, 3*p*, and 3*d* configurations)

or

$$\mathfrak{S}(\mathfrak{M}) = 2[(3)(1)(3)(\tfrac{1}{3})]$$

$$\mathfrak{S}(\mathfrak{M}) = 6\,.$$

In a similar manner, with the appropriate equations (139) to (144), the $\mathfrak{S}$'s for the remainder of the allowed transitions were calculated and are listed in Table 7. With these data, the absolute line strengths $S^{(E)}(J, J')$, the oscillator strengths $f^{(E)}(J, J')$, and transition probabilities $A^{(E)}(J, J')$ can be calculated from equations (116), (129), and (135). Table 7 lists the results for the transitions.

The calculation of the forbidden transitions due to magnetic-dipole and electric-quadrupole radiations for LS-coupling may be worked out explicitly by using the formulae (117) and (118), respectively. For transitions involving the magnetic dipole, Pasternack (1940) and Shortley (1940) developed the necessary formulae for calculating the absolute line strengths $S^{(M)}(J, J')$. The equation for a $J \to J' = J+1$ transition is

$$S^{(M)}(SLJ,\ SLJ+1) = \frac{(J-S+L+1)(J+S-L+1)(J+S+L+2)(S+L-J)}{4(J+1)}\,, \tag{145}$$

and for a $J \to J' = J$ transition,

$$S^{(M)}(SLJ,\ SLJ) = \frac{2J+1}{4J(J+1)}[S(S+1) - L(L+1) + 3J(J+1)]^2\,. \tag{146}$$

Since the magnetic-dipole strengths do not satisfy the ordinary rule within a multiplet, the formula obtained is

$$\sum_{J'} S^{(M)}(SLJ,\ SLJ') = (2J+1)[2J(J+1) + 2S(S+1) - L(L+1)]\,. \tag{147}$$

This sum rule is also useful for checking purposes. Once the $S^{(M)}(J, J')$ has been calculated from the appropriate equation above, the transition probability $A^{(M)}(J, J')$ can be evaluated from equation (117).

Calculation of the transition probabilities, $A^{(Q)}(J, J')$ due to electric-quadrupole radiation is a little more complicated. In this calculation, the electron configuration of the atom must be taken into account. As in the case of the dipole radiation, it is necessary first to determine the absolute line strength $S^{(Q)}(J, J')$, which is the matrix

$$\sum_{mn} |(\varphi_m^0\ |Q|\varphi_n^0)|^2$$

and has been defined in connection with equation (115). It should be noted that the absolute square of the dyad is equal to the sum of the absolute squares of the nine elements. Garstang (1957, 1958), in developing the procedures for the computation of the quadrupole line strengths, applied the Racah (1942*a*, *b*, 1943) techniques. Here we follow Garstang's (1957) detailed procedure and method.

Garstang obtained the following reduced matrix:

$$S^{(Q)}(J, J') = \tfrac{2}{3}|(\alpha J||T^{(2)}||\alpha J')|^2\,. \tag{148}$$

This is a typical matrix element in Racah's theory and is the fundamental formula for the strength of an electric-quadrupole line expressed in terms of tensor operators. For the relative line strength in an electric-quadrupole multiplet, Garstang obtained the following expression:

$$\frac{(SLJ\|T^{(2)}\|SL'J')}{(SL\|T^{(2)}\|SL')} = (-1)^{S-L-J'}[(2J+1)(2J'+1)]^{1/2}W(LJL'J';S2). \quad (149)$$

If this expression is inserted in equation (148), the relative strengths of lines in an electric quadrupole are obtained. Equations (148) and (149) are equivalent to Rubinowicz' formulae (1932). When the Racah orthogonality relation for W is used, one obtains

$$\sum_{J'}\left|\frac{(SLJ\|T^{(2)}\|SL'J')}{(SL\|T^{(2)}\|SL')}\right|^2 = \frac{2J+1}{2L+1}, \quad (150)$$

which is a verification of the J sum rule. Summation over J gives

$$\sum_{JJ'}\left|\frac{(SLJ\|T^{(2)}\|SL'J')}{(SL\|T^{(2)}\|SL')}\right|^2 = 2S+1; \quad (151)$$

hence, for the total multiplet strength one gets

$$S^{(Q)}(SL, SL') = \tfrac{2}{3}(2S+1)\,|(SL\|T^{(2)}\|SL')|^2. \quad (152)$$

Racah's method permits the evaluation of matrix elements for complex transition arrays. The matrix element in equation (152) can be further reduced by means of fractional parentage to two-electron–like matrix elements of the form $(S_1L_1l_nSL\|T_n^{(2)}\|S_1L_1l_n'S'L')$ where $T_n^{(2)}$ is a one-electron operator and by application of Racah's formula gives

$$(S_1L_1l_nSL\|T_n^{(2)}\|S_1L_1l_n'S'L') = (L_1l_nL\|T_n^{(2)}\|L_1l_n'L');$$

then

$$\frac{(L_1l_nL\|T_n^{(2)}\|L_1l_n'L')}{(l_n\|T_n^{(2)}\|l_n')} = (-1)^{L_1-l_n-L'}[(2L+1)(2L'+1)]^{1/2} \times W(l_nLl_n'L';L_12). \quad (153)$$

Equations (150) and (151) can be expressed in the following appropriate forms:

$$\sum_{L'}\left|\frac{(L_1l_nL\|T_n^{(2)}\|L_1l_n'L')}{(l_n\|T_n^{(2)}\|l_n')}\right|^2 = \frac{2L+1}{2l_n+1}$$

and

$$\sum_{LL'}\left|\frac{(L_1l_nL\|T_n^{(2)}\|L_1l_n'L')}{(l_n\|T_n^{(2)}\|l_n')}\right|^2 = 2L_1+1.$$

It should be understood that these equations are a restatement of the well-known rules for relative strengths of multiplets.

It is now necessary to evaluate the expression $(l_n\|T_n^{(2)}\|l_n')$. These are the one-electron elements which can be evaluated in terms of radial integrals, that is, equation (58), which calculates the s_q parameter. The evaluation of $(l_n\|T_n^{(2)}\|l_n')$ has been done essentially by Gaunt (1929), Racah (1942*b*), Shortley (1940), and Seaton (1955*a*). Values are

$$(nl\|T_n^{(2)}\|n'l) = \left[\frac{l(l+1)(2l+1)}{(2l-1)(2l+3)}\right]^{1/2} s_q(nl, n'l) \qquad (154)$$

$$(nl\|T_n^{(2)}\|n'l+1) = 0 \qquad (155)$$

$$(nl\|T_n^{(2)}\|n'l+2) = -\left[\frac{3(l+1)(l+2)}{2(2l+3)}\right]^{1/2} s_q(nl, n'l+2), \qquad (156)$$

where n need not be equal to n'. These expressions will apply in all possible cases.

To illustrate the use of these expressions in the calculation of the magnetic dipole and electric quadrupole, let us consider the ${}^2D_{3/2}$–${}^2D_{5/2}$ transition in Fe XVI.

For determining the magnetic-dipole line strength $S^{(M)}$ equation (145) is used, which upon proper substitution of J, L, and S values gives

$$S^{(M)}\left(\tfrac{1}{2}2\tfrac{3}{2};\tfrac{1}{2}2\tfrac{5}{2}\right) = \frac{\left(\frac{3}{2}-\frac{1}{2}+2+1\right)\left(\frac{3}{2}+\frac{1}{2}-2+1\right)\left(\frac{3}{2}+\frac{1}{2}+2+2\right)\left(\frac{1}{2}+2-\frac{3}{2}\right)}{4\left(\frac{3}{2}+1\right)}$$

$$S^{(M)}\left(\tfrac{1}{2}2\tfrac{3}{2};\tfrac{1}{2}2\tfrac{5}{2}\right) = 2.4\,,$$

and the magnetic-dipole contribution to the transition probability (see eq. [117]) is

$$A^{(M)}\left(\tfrac{3}{2},\tfrac{5}{2}\right) = \frac{35320\left(\frac{2840}{109737}\right)^3(2.4)}{6}$$

$$A^{(M)}\left(\tfrac{3}{2},\tfrac{5}{2}\right) = 0.245\,.$$

In order to calculate the electric-quadrupole line strength $S^{(Q)}$ equations (149), (153), and (154) are inserted into equation (148), or

$$S^{(Q)}(J, J') = \tfrac{2}{3}\{[(-1)^{S-L-J'}(2J+1)(2J'+1)W^2(LJL'J'; S2)]$$

$$\times[(-1)^{L_1-l_n-L'}(2L+1)(2L'+1)W^2(l_nLl_n'L'; L_12)]\}\frac{l(l+1)(2l+1)}{(2l-1)(2l+3)}s_q^2$$

$$S^{(Q)}(J, J') = \tfrac{2}{3}\{[(1)(6)(4)W^2(2\tfrac{3}{2}2\tfrac{5}{2};\tfrac{1}{2}2)]$$

$$\times[(1)(5)(5)W^2(2222; 02)]\}\frac{(2)(3)(5)}{(3)(7)}(\overline{r^2})^2$$

$$S^{(Q)}\left(\tfrac{3}{2},\tfrac{5}{2}\right) = [(\tfrac{2}{3})(6)(4)(\tfrac{1}{120})][(5)(5)(\tfrac{1}{25})](\tfrac{10}{7})(.1871)$$

$$S^{(Q)}\left(\tfrac{3}{2},\tfrac{5}{2}\right) = 0.0356\,.$$

The electric-quadrupole contribution to the transition probability (see eq. [118]) becomes

$$A^{(Q)}\left(\tfrac{3}{2},\tfrac{5}{2}\right) = \frac{2648\left(\frac{2840}{109737}\right)^5(.0356)}{6}$$

$$A^{(Q)}\left(\tfrac{3}{2},\tfrac{5}{2}\right) = 1.83\times10^{-7}\,.$$

Table 8 lists the magnetic-dipole and electric-quadrupole contributions to the transition probability for the various forbidden transitions in Fe XVI.

Equations (148) and (154) are equivalent to the following Shortley (1940) formulae for the line strength and total multiplet strength, respectively,

$$S^{(Q)}(J, J') = (2J + 1)H(J, J')\,|\,(J\|N\|J')\,|^2 \tag{157}$$

and

$$S^{(Q)}(SL, SL') = (2S + 1)(2L + 1)H(L, L')\,|\,(SL\|N\|SL')\,|^2, \tag{158}$$

where $H(J, J')$, $H(L, L')$, and the matrix elements are as defined in his paper.

In retrospect it should be pointed out that Shortley gave explicit formulae for the magnetic-dipole line strengths in LS-coupling which are valid for all configurations, so that for this coupling scheme, the theoretical study can be regarded as being complete.

TABLE 8

Forbidden Transition Probabilities for Fe XVI (3s, 3p & 3d configurations)

	$\lambda(\mathring{A})$	$\bar{\nu}$	s_q	$S^{(M)}$	$S^{(Q)}$	$A^{(M)}$	$A^{(Q)}$
$^2D_{\frac{3}{2}}$ - $^2D_{\frac{5}{2}}$	35211.	2840.	.1871	2.400	.0356	2.45×10^{-1}	1.83×10^{-7}
$^2P_{\frac{1}{2}}$ - $^2P_{\frac{3}{2}}$	4768.7	20970.	.5155	1.333	.1417	27.38	2.39×10^{-2}
$^2S_{\frac{1}{2}}$ - $^2D_{\frac{5}{2}}$	147.11	679770.	.2125	0	.1700	0	6.84×10^{5}
$^2S_{\frac{1}{2}}$ - $^2D_{\frac{3}{2}}$	147.73	676930.	.2125	0	.1134	0	6.71×10^{5}
$^2P_{\frac{1}{2}}$ - $^2D_{\frac{5}{2}}$	0	0	0	0	0	0	0

Shortley also formulated a general theory for quadrupole line strengths in which he showed that many methods developed for the electric dipole could be extended to the quadrupole case. The method is, however, a tedious procedure when applied to complex configuration transitions. Racah's powerful techniques, on the other hand, simplify this considerably, as has been so well demonstrated by Garstang, for example, as in the work on the transitions between d and s electrons.

These procedures work very well for LS-coupling, or at least where departures from LS-coupling are small. There is, of course, a high percentage of atoms and ions whose spectra do indicate a sizable departure from LS-coupling; this is certainly the case for many atoms and ions of astrophysical interest. These departures are attributed to configuration interaction and intermediate coupling.

The effect of configuration interaction on line strengths for allowed and forbidden transitions has been investigated and reported in a series of important papers. Biermann and Trefftz (1949), Trefftz and Biermann (1952), Trees (1951), Trefftz (1949, 1950, 1952), Green (1949), Boys (1950, 1953*a*, *b*), and others have made extensive studies on the effect of configuration interaction on the energy levels of various atoms. Garstang (1956), Rohrlich and Pecker (1963), Weiss (1963), Burke (1963), and others have investigated configuration interaction for certain specific configurations. The calculations of the transition probabilities which take into account configuration interaction, however, have in the past been performed for one atom at a time or for atoms along a particular isoelectronic sequence, so a satisfactory generalization is yet to be made. In some instances, transition probabilities have been lowered by as much as 30 per cent by including configuration interaction in the calculation. Much detailed investigation is still necessary, and to date the emphasis has been primarily on the relatively light atoms or ions.

An important theoretical scheme for describing atomic spectra that takes into account configuration interaction and is different from the usual presentation was proposed by Layzer in 1959. In this method the nuclear charge enters explicitly into all predictions in the structure of term spectra; that is, they apply to all atoms along an isoelectronic sequence. The spectroscopic terms are no longer assigned to definite configurations in the first approximation, but instead are considered in terms of "complex" configurations specified by a set of radial quantum numbers and a definite parity value. For example, the "complex" designated as $(1^2 2^4)$ contains all the states belonging to $1s^2 2p^4$ and $1s^2 2s^2 2p^2$, but not $1s^2 2s 2p^3$ configurations. The first two configurations have terms with the same S, L and parity; and, therefore, there is a configuration interaction involving these two. Further, the theoretical predictions are expressed in terms of screening parameters, so that screening takes on a quantitative meaning. In a series of papers by Layzer (1963), Layzer and Bahcall (1962), Layzer, Horak, Lewis, and Thompson (1964), and Layzer, Varsavsky, and Froese (1964), the theoretical scheme has been extended and applied to various problems. Other investigators (Varsavsky 1961; Froese 1964, 1965) have also applied this scheme to transition probability calculations of a whole series of ions and expanded on the theory. The new scheme in many ways is a simple one and in some respects, the predictions, primarily for the light elements investigated, have been more accurate.

In intermediate coupling, the Hamiltonian has to be altered to include terms due to electron spin. Thus, in any calculation of transition probabilities based on intermediate coupling, the spin-orbit (ζ) and the spin-spin, and spin-other-orbit (η) interactions should be included. Of the calculations performed so far, it has been sufficient to include only the spin-orbit interactions; that is, the empirical values from observed or laboratory-measured lines and theoretical values for multiplet splitting were in relatively good agreement. To a large extent, however, agreement depends on the configuration and complexity of the atom investigated. For example, in highly ionized Fe atoms for which ζ is very large, the effect of η is relatively insignificant. For light atoms ($Z < 18$), the η can significantly alter the results; and thus, the discrepancy between the empirical and theoretical values is large.

The general theory of line strength for intermediate coupling was developed by Condon and Shortley. Shortley extended the work by developing specific computation methods for magnetic-dipole and electric-quadrupole transitions. It has been shown that magnetic-dipole strengths do not satisfy the ordinary J sum rule and one obtains equation (147); however, in intermediate coupling, only the J-group sum rule is obeyed. This rule states that:

The sums of the strengths of the lines connecting all the levels of a given J' in the fina configuration is independent of the coupling.

If the initial and final configurations are different, then the J-file sum rule applies.

For any coupling, the strengths of the J files referring to the levels of the initial (final) configuration are proportional to $2J + 1$, if the jumping electron is not equivalent to any other in the final (initial) configuration.

Forbidden transitions for $2p^q$, $3p^q$, and $3d^q$ configurations involving intermediate coupling were extensively investigated by Pasternack (1940), Shortley *et al.* (1941), Aller *et al.* (1949), Garstang (1951), and Seaton and Osterbrock (1957). By expressing the wave functions (primed) due to intermediate coupling in terms of pure LS-coupling, for example, for a p^3 configuration, we get

$$\begin{aligned}
\psi(^2D'_{3/2}) &= a\psi(^2P_{3/2}) + b\psi(^4S_{3/2}) + c\psi(^2D_{3/2})\\
\psi(^4S'_{3/2}) &= a'\psi(^2P_{3/2}) + b'\psi(^4S_{3/2}) + c'\psi(^2D_{3/2})\\
\psi(^2P'_{3/2}) &= a''\psi(^2P_{3/2}) + b''\psi(^4S_{3/2}) + c''\psi(^2D_{3/2})\\
\psi(^2P'_{1/2}) &= \psi(^2P_{1/2})\\
\psi(^2D'_{5/2}) &= \psi(^2D_{5/2})
\end{aligned}$$

and solve for the coefficients. It is also necessary to obtain the solution of the secular determinant; that is,

$$\begin{vmatrix}
E_0(^2D) + \frac{111}{10}\eta - E & \frac{1}{2}\zeta\sqrt{5} & -\frac{6}{\sqrt{5}}\eta\\
\frac{1}{2}\zeta\sqrt{5} & E_0(^2P) - \frac{5}{2}\eta - E & \zeta\\
-\frac{6}{\sqrt{5}}\eta & \zeta & E(^4S) - E
\end{vmatrix} = 0$$

which gives the following approximate equations:

$$\begin{aligned}
E(^4S_{3/2}) &= E(^4S) - \frac{\zeta^2}{(PS)}\\
E(^2D_{5/2}) &= E(^2D) - \tfrac{37}{5}\eta\\
E(^2D_{3/2}) &= E(^2D) + \tfrac{111}{10}\eta - \tfrac{5}{4}\frac{\zeta^2}{(PD)}\\
E(^2P_{3/2}) &= E(^2P) - \tfrac{5}{2}\eta + \tfrac{5}{4}\frac{\zeta^2}{(PD)} + \frac{\zeta^2}{(PS)^2}\\
E(^2P_{1/2}) &= E(^2P) + 5\eta
\end{aligned}$$

where

$$\begin{aligned}
E(^2P) - E(^2D) &\equiv (PD)\\
E(^2D) - E(^4S) &\equiv (DS)\\
E(^2P) - E(^4S) &\equiv (PS)\,.
\end{aligned}$$

Detailed computational procedures were developed by these investigators. Later, Garstang (1957, 1958), in addition to using the perturbation method which was used by all the investigators mentioned, developed a more simplified method for calculating the quadrupole line strengths from Racah's theory, which is most useful for the heavy ions or elements (see Garstang 1962*a*, *b*, *c*). Garstang points out that the line strength as given by equation (148) applies for any coupling. He represents the transformation coefficients from LS-coupling to intermediate coupling by the expression $(\beta J|\alpha J)$. They are independent of the magnetic quantum number, so the matrix elements of $T_q^{(k)}$ transform as follows:

$$(\beta JM|T_q^{(k)}|\beta' J'M') = \sum_{\alpha\alpha'} (\beta J|\alpha J)(\alpha JM|T_q^{(k)}|\alpha' J'M')(\alpha' J'|\beta' J').$$

TABLE 9

$3p^2$ Configuration

Transition	Type	P II	S III	Cl IV	A V
$^1D_2$-$^1S_0$	λ	7864.5	6312.1	5323.3	4625.5
	e	1.95	2.54	3.15	3.78
$^3P_2$-$^1S_0$	λ	4736.6	3796.7	3203.2	2784.4
	e	0.0063	0.0163	0.0382	0.0811
$^3P_1$-$^1S_0$	λ	4669.5	3721.8	3118.3	2691.4
	m	0.22	0.85	2.61	6.82
$^3P_2$-$^1D_2$	λ	11898.2	9532.1	8045.6	7005.7
	(m>>e)	0.0169	0.0638	0.195	0.513
$^3P_1$-$^1D_2$	λ	11483.2	9069.4	7530.5	6435.1
	(m>>e)	0.00627	0.0246	0.0797	0.221

Using Racah's procedure (his eq. [7]) to separate the dependence of matrix elements on M and M', one obtains

$$(\beta J\|T^{(2)}\|\beta' J') = \sum_{\alpha\alpha'} (\beta J|\alpha J)(\alpha J\|T^{(2)}|\alpha' J')(\alpha' J'|\beta' J'). \qquad (159)$$

This permits the calculation of the intermediate-coupling line strength; care must be exercised to use the correct phases of the matrix elements in this equation.

As an example of the use of intermediate coupling in the calculation of transition probabilities, let us consider the $^2D_{5/2}$–$^4S_{3/2}$ and $^2D_{3/2}$–$^4S_{3/2}$ transitions for [O II], [S II], and [A IV]. Let us also examine the effect of the wave functions and, hence, some of the parameters on transition probability results. The necessary coefficients, a, b, c, a', b', c', a'', b'', and c'', which take into account the departure from LS-coupling, that is, give the necessary correction for intermediate coupling, are determined; and, with these coefficients, the $S^{(M)}(J, J')$ and $S^{(Q)}(J, J')$ for intermediate coupling can be calculated.

In Table 12 we have tabulated the transition probabilities for the three ions for which intermediate coupling applies. If we examine ${}^2D_{5/2}-{}^4S_{3/2}$ and ${}^2D_{3/2}-{}^4S_{3/2}$ transitions, we immediately notice that for the ${}^2D_{5/2}-{}^4S_{3/2}$ transition there seems to be a large discrepancy between the results of Pasternack and the others; whereas, for the ${}^2D_{3/2}-{}^4S_{3/2}$ transition, this is not the case except for [O II]. Offhand, it would seem that this could be attributed primarily to the type and accuracy of the wave functions used; however, if we further analyze these results by examining the magnetic-dipole and electric-quadrupole contributions to the total transition probability as well as the ζ and η parameters for S II, we observe the following from Table 13:

TABLE 10

$3p^3$ Configuration

Transition	Type	P I	S II	Cl III	A IV
${}^2P_{\frac{3}{2}}-{}^2D_{\frac{5}{2}}$	λ	13565.6	10320.6	8481.6	7236.0
	m	0.0190	0.0602	0.1690	0.4440
	e	0.0941	0.1540	0.1950	0.2260
${}^2P_{\frac{3}{2}}-{}^2D_{\frac{3}{2}}$	λ	13538.4	10287.1	8433.7	7169.1
	m	0.0341	0.1080	0.3060	0.8140
	e	0.0404	0.0665	0.0844	0.0981
${}^2P_{\frac{1}{2}}-{}^2D_{\frac{3}{2}}$	λ	13585.7	10338.8	8501.8	7263.3
	m	0.0211	0.0665	0.1860	0.4880
	e	0.0801	0.1310	0.1650	0.1900
${}^2P_{\frac{1}{2}}-{}^2D_{\frac{5}{2}}$	λ	13613.0	10372.6	8550.5	7332.0
	e	0.0530	0.0865	0.1080	0.1220
${}^2D_{\frac{5}{2}}-{}^4S_{\frac{3}{2}}$	λ	8787.6	6717.3	5517.2	4711.4
	m	$0.0^{5}905$	$0.0^{4}363$	$0.0^{3}146$	$0.0^{3}610$
	e	$0.0^{3}185$	$0.0^{3}429$	$0.0^{3}861$	$0.0^{2}162$
${}^2D_{\frac{3}{2}}-{}^4S_{\frac{3}{2}}$	λ	8799.1	6731.5	5537.7	4740.3
	m	$0.0^{3}177$	0.00156	0.00651	0.02730
	e	$0.0^{3}119$	$0.0^{3}274$	$0.0^{3}545$	0.00101
${}^2P_{\frac{3}{2}}-{}^4S_{\frac{3}{2}}$	λ	5332.4	4068.5	3342.7	2854.8
	m	0.0108	0.3410	0.9630	2.550
	e	$0.0^{6}327$	$0.0^{5}130$	$0.0^{5}486$	$0.0^{4}156$
${}^2P_{\frac{1}{2}}-{}^4S_{\frac{3}{2}}$	λ	5339.7	4076.5	3353.4	2869.1
	m	0.0426	0.1340	0.3740	0.9720
	e	$0.0^{5}467$	$0.0^{4}138$	$0.0^{4}397$	$0.0^{3}119$

The discrepancies in the transition probabilities arising from the difference in the magnetic-dipole contribution for S II are of the order of 20 per cent. For the $3p^3$ ions, this variation is from 5 to 25 per cent and can be shown to be due to the ζ and η values used. For example, the expressions for the line strengths $S^{(M)}$ for $^2D_{5/2}$–$^4D_{3/2}$ and $^2D_{3/2}$–$^4S_{3/2}$ transitions are as follows:

$$S^{(M)}(\tfrac{5}{2},\tfrac{3}{2}) = 12c'^2, \tag{160}$$

where

$$c' = \frac{6\eta}{\sqrt{5}}\frac{1}{(DS)} + \frac{\sqrt{5}}{2}\frac{\zeta^2}{(PD)(DS)};$$

and

$$S^{(M)}(\tfrac{3}{2},\tfrac{3}{2}) = \tfrac{4}{15}(10aa' + 15bb' + 6cc')^2, \tag{161}$$

where

$$a = -\frac{\sqrt{5}}{2}\frac{\zeta}{(PD)} - \frac{34\zeta\eta}{(PD)^2}; \qquad a' = -\frac{\zeta}{(PS)} - \tfrac{5}{2}\frac{\zeta\eta}{(PS)^2};$$

and

$$b = -c';\ b' = c = 1;$$

and the quantities (DS), (PS), and (PD) are the unperturbed energy level separation; that is $(DS) = E_0(^2D) - E_0(^4S)$, etc.

TABLE 11

$3p^4$ Configuration

Transition	Type	S I	Cl II	A III
$^1D_2$-$^1S_0$	λ	7724.7	6152.9	5191.4
	e	1.78	2.29	3.10
$^3P_2$-$^1S_0$	λ	4506.9	3583.0	3005.1
	e	0.00731	0.01800	0.04250
$^3P_1$-$^1S_0$	λ	4589.0	3675.0	3109.0
	m	0.35	1.34	4.02
$^3P_2$-$^1D_2$	λ	10819.8	8579.5	7135.8
	(m>>e)	0.0275	0.1005	0.3214
$^3P_1$-$^1D_2$	λ	11305.8	9125.8	7751.0
	(m>>e)	0.00802	0.02926	0.08303

In calculating the $A^{(M)}$'s, Czyzak and Krueger (1963*a*) used ζ_{obs} and η_{obs}; Garstang (1951) used ζ_{obs} and η_{calc}; and Pasternack (1940) used only ζ_{obs}, which had been determined by Robinson and Shortley (1937). Equations (160) and (161) show that a small difference in ζ accounts for a large portion of the difference in the $A^{(M)}$'s. Although η_{obs} and η_{calc} differ by a factor of two, the effect on the transition probability is small because the η's have only a secondary effect in the calculation.

The discrepancy in the quadrupole transitions is due primarily to the difference in the quadrupole moment s_q. The quadrupole-moment value depends on the way in which it is determined. From the results obtained for O II and A IV, it can be seen that the hydrogenic wave functions give an s_q that is roughly four times greater than that determined from the Hartree-Fock method with exchange. If the expressions for $S^{(Q)}$ for the two transitions are examined, it will be seen that the significant quantity in the equation is the quadrupole moment.

TABLE 12

Comparison of Transition Probabilities

Ion		Pasternack[1] (Hydrogenic)	Aller, Ufford & Van Vleck[2] (HFSCF)	Garstang[3] (HFSCF with Ex)
O II:	$^2D_{\frac{5}{2}} - {}^4S_{\frac{3}{2}}$	8.52×10^{-5}	5.5×10^{-5}	4.0×10^{-5}
	$^2D_{\frac{3}{2}} - {}^4S_{\frac{3}{2}}$	6.80×10^{-5}	14.0×10^{-5}	13.1×10^{-5}
	s_q	2.25	.70	.59
		Pasternack (Hydrogenic)	Garstang (HFSCF)	Czyzak & Krueger[4] (HFSCF with Ex)
S II:	$^2D_{\frac{5}{2}} - {}^4S_{\frac{3}{2}}$	11.3×10^{-4}	6.3×10^{-4}	4.65×10^{-4}
	$^2D_{\frac{3}{2}} - {}^4S_{\frac{3}{2}}$	19.7×10^{-4}	17.4×10^{-4}	18.3×10^{-4}
	s_q	6.93	2.10	1.72
		Pasternack (Hydrogenic)	Naqvi & Talwar[5] (HFSCF)	Czyzak & Krueger (HFSCF with Ex)
A IV:	$^2D_{\frac{5}{2}} - {}^4S_{\frac{3}{2}}$	4.64×10^{-3}	2.59×10^{-3}	2.23×10^{-3}
	$^2D_{\frac{3}{2}} - {}^4S_{\frac{3}{2}}$	26.6×10^{-3}	27.0×10^{-3}	28.3×10^{-3}
	s_q	3.49	.97	.87

[1] Pasternack (1940).
[2] Aller, Ufford & Van Vleck (1949).
[3] Garstang (1951).
[4] Czyzak & Krueger (1963<u>a</u>).
[5] Naqvi & Talwar (1957).

The expressions for the $S^{(Q)}$'s for these transitions are

$$S^{(Q)}(\tfrac{5}{2},\tfrac{3}{2}) = 14a'^2 s_q^2 \tag{162}$$

and

$$S^{(Q)}(\tfrac{3}{2},\tfrac{3}{2}) = 6(ac'+ca')^2 s_q^2. \tag{163}$$

TABLE 13

Magnetic Dipole & Electric Quadrupole Contributions to the $^2D_{\frac{5}{2}} - {}^4S_{\frac{3}{2}}$ and $^2D_{\frac{3}{2}} - {}^4S_{\frac{3}{2}}$ Transitions of S II

Transition	Type	Pasternack[1] (Hydrogenic)	Garstang[2] (HFSCF)	Czyzak & Krueger[3] (HFSCF with Ex)
$^2D_{\frac{5}{2}} - {}^4S_{\frac{3}{2}}$	$A^{(M)}$	.28 x 10^{-4}	.31 x 10^{-4}	.36 x 10^{-4}
	$A^{(Q)}$	11.0 x 10^{-4}	6.01 x 10^{-4}	4.29 x 10^{-4}
$^2D_{\frac{3}{2}} - {}^4S_{\frac{3}{2}}$	$A^{(M)}$	13.0 x 10^{-4}	13.5 x 10^{-4}	15.6 x 10^{-4}
	$A^{(Q)}$	6.7 x 10^{-4}	3.85 x 10^{-4}	2.70 x 10^{-4}
	ζ_{obs}	538.	538.	553.
	ζ_{calc}	-	459.	444.
	η_{obs}	-	-	0.42
	η_{calc}	-	0.23	0.26

NOTE: The subscript "obs" indicates values of ζ and η obtained from the secular determinant in which observed or experimental values for the energy levels were used. The subscript "calc" refers to ζ and η obtained from eqs. (59) and (60).

[1] Pasternack (1940).
[2] Garstang (1951).
[3] Czyzak & Krueger (1963).

The differences between the HFSCF s_q's and the HFSCF with exchange s_q's are sufficiently small that the results of the $A^{(Q)}$'s obtained by Garstang and by Czyzak and Krueger are reasonably close. The difference can be attributed primarily to the exchange effect, which also accounts for the small difference in the ζ's and η's. Thus, it seems that in this type of transition, the results are very sensitive to the manner in which the quadrupole moment is obtained; that is, it depends on the type and accuracy of the wave functions used. Although the difference is mainly due to the quadrupole moment, the ζ and η contribution should not be overlooked. The effect of η on the results depends on the size of the atom; that is, for light atoms or ions this appears more significant than for the heavier atoms or ions (see Krueger and Czyzak [1964]). Thus, in determining the $S^{(M)}$'s and $S^{(Q)}$'s, the ζ_{obs} and η_{obs} were used instead of ζ_{calc} and η_{calc} because in calculating the former, observed or experimental energy level values are used in the energy expressions, so that configuration interaction or correlation energy is taken into account implicitly, at least in part; whereas, the present solutions of the Hartree-Fock equations do not include these effects.

For the $3p^2$ and $3p^4$ configurations, the agreement between the semiempirically and integral-determined ζ's, and η's is quite good (≤ 5 per cent) (see Czyzak and Krueger [1963*a*]); and, therefore, either method for obtaining the ζ's, and η's may be used. The relatively poorer agreement of the $3p^3$ configuration may be attributed in part to the fact that for half-filled *p*-shells the η's play a more significant role in the calculation of multiplet splitting.

Garstang has shown that although energy levels may be appreciably affected by relatively small changes in the parameters (ζ and η), there is little change in transition probabilities. In Tables 9, 10, and 11, the latest theoretical transition probabilities for the $3p^2$, $3p^3$, and $3p^4$ configurations have been compiled.

4. CALCULATION OF COLLISION CROSS-SECTIONS DUE TO ELECTRON IMPACT

Since particle collisions involve rather complex phenomena, it is generally necessary to use approximate methods in treating problems. The accuracy of the approximation depends on conditions involved in the collision, so special methods have to be devised for each particular problem. As yet, no general method has been developed. Particle collisions may be loosely classified into four general categories:

(*a*) atom-molecule collisions;

(*b*) atom-atom collisions or molecule-molecule collisions;

(*c*) atom–heavy-particle collisions; and

(*d*) atom- or molecule-electron collisions.

In these collisions there may be a direct energy exchange without any particle transfer upon impact or there may be a particle exchange involving a rearrangement or redistribution in the total system.

An important type of collision in problems of astrophysical interest is that between atoms or ions and electrons. This type of collision is involved in transitions observed in nebulae and is the one that will be discussed here.

In atom-electron collisions, a fast electron is defined as one in which the speed of the incident electron is very much higher than the orbital speed of the atomic electron; otherwise, the collision electron is called slow. A general theory for both elastic and inelastic scattering due to fast electrons is the Born and Coulomb-Born approximation. Although the Born method is a high-energy approximation, it may be used to obtain low-energy results which are not inaccurate by more than a factor of two, except for very low energies, and even for these the results are within an order of magnitude. Results are usually better for forbidden transitions not involving a spin change than for allowed transitions. Where spin-change transitions occur, the Born approximation cannot be used. In such processes the overlap between the atomic wave functions and that of the colliding electron come into play and, therefore, the treatment requires greater refinement than is available in the Born approximation. Consequently, Oppenheimer's modification of the Born approximation, which takes into account exchange, is of little help. It also fails for collisions between positive ions and slow electrons, since cross-section values with incorrect energy dependence are obtained. In general, if cross-sections due to excitation by electron impact near the threshold energy are required, then other schemes using different assumptions must be made. Two schemes developed by Seaton (1953 to 1962) and Seaton and Peach (1962) and used successfully for the $p^q(q = 2, 3, 4)$ configurations are the exact resonance and distorted wave methods. The exact resonance method has been applied to problems involving electron excitation of various terms in the ground state configuration of ions or atoms. In this method the wave function must satisfy the Pauli exclusion principle and the wave functions and energies for the various states are assumed to be equal. For this reason it is referred to as the exact resonance approximation. The distorted wave method, on the other hand, although still requiring the Pauli exclusion principle to hold, does not assume the wave functions of the various states of the ground configurations and their energies to be the same. The text by Mott and Massey (1949) as well as the texts and reviews by Bates *et al.* (1950), Massey and Burhop (1952), and Massey (1956*a*, *b*) gives comprehensive expositions on the theory of atomic collisions. In addition, Seaton (1958, 1962) has reviewed recent developments on the theory of excitation and ionization by electron impact.

Fundamentally, the collision between an atom and an electron is a many-body problem, the simplest of which is the collision of electrons with hydrogen atoms. Collision with hydrogen atoms has been investigated quite thoroughly, both theoretically and experimentally, by several investigators.

For a beam of electrons colliding with an atom or ion, the expression for the total cross-section Q may be written as follows:

$$Q(\nu \rightarrow \nu') = \int I_{\nu\nu'}(\theta)\, d\omega \,, \tag{164}$$

where $I_{\nu\nu'}(\theta)$ has the dimensions of area and θ is the angle through which the electrons are scattered into an element of solid angle $d\omega$. The integrand is the differential cross-section, and ν and ν' are the initial and final states, respectively. Semiclassically, two collision mechanisms of producing atomic transitions are possible. One may be thought of as having the electron create a variable electronic field in the neighborhood of the atom due to its passage near the atom with a resulting definite probability that a radiative transition can take place. In this mechanism the velocity of the incident electron must be much larger than that of the atomic electrons. In the other mechanism, the incident electron is captured by the atom and the atomic electron is ejected. Here, if the spins of the captured and ejected electrons are different, the atomic spin will change. The former mechanism is effective at large separations of the atom and incident electron, whereas the latter would be at distances of the order of atomic dimensions. The angular momentum according to quantum theory is $L_\nu = \hbar\sqrt{[l_\nu(l_\nu + 1)]}$ and classically, $L = mv_b r$. Here v_b represents the velocity of the incident electron at a large distance from the atom or ion, and r is the distance of closest approach if the incident electron were undeflected. The total cross-section $Q(\nu \rightarrow \nu')$ based on the probability that such a transition can take place is

$$Q(\nu \rightarrow \nu') = \int_0^\infty P_r(\nu, \nu')\, 2\pi r\, dr \,, \tag{165}$$

where $P_r(\nu, \nu')$ is the transition probability and

$$\sum_{\nu \neq \nu'} P_r(\nu, \nu') \leq 1$$

must be satisfied for at least all values of r which give a significant contribution to the total cross-section integral. The angular momentum is quantized by putting

$$(mv_b r)^2 \simeq l_\nu(l_\nu + 1)\hbar^2 ,$$

which upon differentiation yields

$$2(mv_b)^2 r dr \approx \hbar^2(2l_\nu + 1)\delta l_\nu ;$$

and if the integral in equation (165) is replaced by a summation and δl_ν is set equal to one, then

$$Q(\nu \rightarrow \nu') = \sum_{l_\nu} \frac{P_{l_\nu}(\nu, \nu')\pi(2l_\nu + 1)\hbar^2}{(mv_b)^2} .$$

However, $k_b = mv_b/\hbar$. If we define $\Omega^{l_\nu}(\nu', \nu) = (2l_\nu + 1)P_{l_\nu}(\nu, \nu')$ which, summed over all l_ν, becomes

$$\Omega(\nu', \nu) = \sum_{l_\nu} \Omega^{l_\nu}(\nu', \nu)$$

we get for the collision cross-section

$$Q(\nu \rightarrow \nu') = \frac{\pi}{k_b^2} \Omega(\nu', \nu)$$

To account for transitions between the levels n and n' whose statistical weights are $\omega_n = (2L_n + 1)(2S_n + 1)$ and $\omega_n' = (2L_{n'} + 1)(2S_{n'} + 1)$, the cross-section must be summed over the final states and averaged over the initial states. Thus,

$$Q(n \rightarrow n') = \frac{\pi}{\omega_n k_b^2} \Omega(n', n), \tag{166}$$

where $\Omega(n', n)$ is the collision strength. Detailed balancing requires that $\Omega(n', n) = \Omega(n, n')$. Also, $\Sigma P_r(\nu, \nu') \leq 1$ corresponds to

$$\sum_{n'} \frac{\Omega(n, n')}{(2J_n + 1)} \leq (2l_n + 1)$$

in accord with the conservation theorem of Mott, Bohr, Peierls, and Placzek (see Mott and Massey 1949). In the calculation of the collision cross-section, the dimensionless parameter must be determined. The difficulty lies in the numerical procedure to be employed in solving the appropriate wave equations to obtain the collision strength.

The wave equation for an atom or ion and a colliding electron is

$$\Big[\underbrace{\sum_i \Big(-\tfrac{1}{2}\nabla_i^2 + \frac{Ze^2}{r_i}\Big) - \sum_{i<j} \frac{e^2}{r_{ij}}}_{\text{atomic electrons}} \underbrace{- \tfrac{1}{2}\nabla_b^2 + \frac{Ze^2}{r_b}}_{\text{incident electron}} - \sum_i \frac{e^2}{r_{ib}}\Big]\psi(r_a, r_b) = E\psi(r_a, r_b), \tag{167}$$

where $\psi(r_a, r_b)$ is the wave function for the complete system, r_a refers to the atomic electrons, r_b to the colliding electron, and the subscript i to summation over the atomic electrons. The total energy E which is conserved during collision is given by

$$E = E_n + \tfrac{1}{2}k_b^2 = E_{n'} + \tfrac{1}{2}k_b'^2 , \tag{168}$$

where E_n and $E_{n'}$ are the atomic energies before and after collision, and k^2 is numerically equal to the kinetic energy (in rydbergs) of the colliding electron. The wave function $\psi(r_a, r_b)$ can be expanded; that is,

$$\psi(r_a, r_b) = \Sigma\varphi_m(r_a)F_m(r_b)\,, \tag{169}$$

where the summation is over the discrete levels, the integration is over the continuous states, and $\varphi_m(r_a)$ is the wave function for the various atomic states.

If equation (169) is substituted into equation (167) and advantage is taken of the equation

$$\Big[\sum_i\Big(-\tfrac{1}{2}\nabla_i^2+\frac{Ze^2}{r_i}\Big)-\sum_{i<j}\frac{e^2}{r_{ij}}-E_n\Big]\varphi_n(r_a)=0\,, \tag{170}$$

the following expression for the wave equation (167) is obtained after multiplying by $\varphi_{n'}^*(r_a)$ and integrating over the configuration space,

$$\begin{aligned}\int\varphi_{n'}^*\Big[\sum_i\Big(-\tfrac{1}{2}\nabla_i^2+\frac{Ze^2}{r_i}\Big)-\sum_{i<j}\frac{e^2}{r_{ij}}\Big]\sum_m\varphi_mF_m d\tau&\\+\int\varphi_{n'}^*\big(-\tfrac{1}{2}\nabla_b^2\big)\sum\varphi_mF_m d\tau+\int\varphi_{n'}^*\Big(\frac{Ze^2}{r_b}-\sum_i\frac{e^2}{r_{ib}}\Big)\sum_m\varphi_mF_m d\tau&\\=E\int\varphi_{n'}^*\sum_m\varphi_mF_m d\tau\,.&\end{aligned} \tag{171}$$

Integrating the first integral in equation (171), involving the atomic electrons only and taking into account the relation (170), gives

$$\begin{aligned}\int\varphi_{n'}^*\Big[\sum_i\Big(-\tfrac{1}{2}\nabla_i^2+\frac{Ze^2}{r_i}\Big)-\sum_{i<j}\frac{e^2}{r_{ij}}\Big]\sum_m\varphi_mF_m d\tau&\\=\int\varphi_{n'}^*\sum_m E_m\varphi_mF_m d\tau=E_{n'}F_{n'}&\end{aligned} \tag{172}$$

and the second integral becomes

$$\int\varphi_{n'}^*\big(-\tfrac{1}{2}\nabla_b^2\big)\sum_m\varphi_mF_m d\tau=-\tfrac{1}{2}\nabla_b^2F_{n'}\,. \tag{173}$$

The integral on the right-hand side of equation (171) is simply

$$E\int\varphi_{n'}^*\sum_m\varphi_mF_m d\tau=EF_{n'}\,. \tag{174}$$

Thus, equation (171) may be rewritten as follows:

$$-\tfrac{1}{2}\nabla_b^2F_{n'}-(E-E_{n'})F_{n'}=-\int\varphi_{n'}^*\Big(\frac{Ze^2}{r_b}-\sum_i\frac{e^2}{r_{ib}}\Big)\sum_m\varphi_mF_m d\tau \tag{175a}$$

or

$$(\nabla_b^2+k_b'^2)F_{n'}=2\int\varphi_{n'}^*\Big(\frac{Ze^2}{r_b}-\sum_i\frac{e^2}{r_{ib}}\Big)\sum_m\varphi_mF_m d\tau\,, \tag{175b}$$

where

$$k_b'^2=2(E-E_{n'})\,.$$

The solutions of equations (175) are well-behaved functions of asymptotic form,

$$F_{n'}(r_b) \sim \delta_{nn'} e^{ik_b \cdot r_b} + \frac{1}{r} f_{nn'}(\theta)\, e^{ik_b' r_b} \tag{176}$$

for which the differential cross-section for the $n \rightarrow n'$ transition may be written as

$$I_{nn'}(\theta) = \frac{k_b'}{k_b} | f_{nn'}(\theta) |^2 d\omega \,. \tag{177}$$

To calculate the collision cross-section for a particular transition, it is necessary to solve equations (175*a* or *b*) by numerical techniques in which specific assumptions are made.

For a Born approximation where the kinetic energy of the incident electron is much greater than the energy of the atomic electrons; that is, the interaction of incident electron with the atom is small, $F_{n'}(r_b)$ can be set equal to zero for $n' \neq n$ on the right-hand side of equations (175*a*, *b*). Then

$$(\nabla_b^2 + k_b'^2) F_{n'}(r_b) = 2F_n(r_b) \int \varphi_{n'}^* \left(\frac{Ze^2}{r_b} - \sum_i \frac{e^2}{r_{ib}} \right) \varphi_n d\tau \,, \tag{178}$$

and if it is assumed that F_n on the right side of equation (178) is a plane wave, the following expression is obtained:

$$(\nabla_b^2 + k_b'^2) F_{n'}(r_b) = 2\, e^{ik_b \cdot r_b} \int \varphi_{n'}^* \left(\frac{Ze^2}{r_b} - \sum_i \frac{e^2}{r_{ib}} \right) \varphi_n d\tau \,, \tag{179}$$

so that

$$f_{nn'}(\theta) = -\frac{1}{4\pi} \int 2\, e^{ik_b \cdot r_b} \left\{ \int \varphi_{n'}^* \left[\frac{Ze^2}{r_b} - \sum_i \frac{e^2}{r_{ib}} \right] \varphi_n d\tau \right\} e^{-ik_b \cdot r_b} d r_b \tag{180}$$

and

$$\Omega(n, n') = k_b k_b' \frac{\omega_n}{\pi} \int | f_{nn'}(\theta) |^2 d\omega \,. \tag{181}$$

In attempting to calculate the collision cross-section due to electron excitation by the exact resonance method (Seaton 1953*b*), the wave function of the complete system, that is, the atom or ion and the colliding electron, is represented as sum of the terms, each of which is an antisymmetric sum of the products of the wave function of a term of the ground state configuration with the corresponding wave function of the scattering electron. The object is to obtain the asymptotic behavior of the scattering electron wave function. In order to accomplish this, the formal total wave function is used with the wave equation for the system to obtain a series of coupled integro-differential equations for the scattering-electron wave function. These integro-differential equations resemble the Hartree-Fock equations. By assuming equal energies and using the Hartree-Fock self-consistent field wave functions with exchange for the terms of the configuration of the atom or ion and by neglecting certain nonspherical terms in the potential, it is possible to obtain uncoupled equations. The solutions of these equations may be expressed in terms of two functions $\mathfrak{F}$ and $\mathfrak{G}$. With the procedure and notation of Seaton, the relations satisfied by these two functions are

$$[\mathfrak{H}^P + V - \tfrac{1}{2}k^2]\,\mathfrak{F} = \mathfrak{J}(\mathfrak{F}) \tag{182}$$

and

$$[\mathfrak{H}^P + V - \tfrac{1}{2}k^2]\,\mathfrak{G} = -q\,\mathfrak{J}(\mathfrak{G}) \,, \tag{183}$$

where $\mathfrak{H}^P$ is a modified Hamiltonian, V is a spherical potential due to all atomic electrons, k^2 is the energy in rydbergs of the incoming electron, and $\mathfrak{J}(\mathfrak{F})$ and $\mathfrak{J}(\mathfrak{G})$ represent the exchange effects. The equations (182) and (183) are obtained from the general expressions of Seaton's equations (1953*b*), in which the energy differences and the terms Y_2 and R_2 have been omitted.

Explicitly the terms in equations (182) and (183) are as follows:

$$\mathfrak{H}^P = -\tfrac{1}{2}\left(\frac{d^2}{dr^2}+\frac{2}{r}\right)-\frac{Z}{r},$$

$$V = \Sigma q(nl)\,Y_0(nl, nl),$$

$$\mathfrak{J}(F) = [\,Y_0(PF)+\lambda(PF)\,]P, \qquad \text{where} \qquad F = \mathfrak{F} \text{ or } \mathfrak{G}, \qquad (184)$$

$$Y_0(nl, nl) = \frac{1}{r}\int_0^r P_{nl}P_{nl}\,dr' + \int_r^\infty \frac{1}{r'}P_{nl}P_{nl}\,dr',$$

$$Y_0(PF) = \frac{1}{r}\int_0^r P_{nl}F\,dr' + \int_r^\infty \frac{1}{r'}P_{nl}F\,dr',$$

$P_{nl} \equiv P(nl; r)$ the atomic wave function for the nl shell, and F = the wave function representing the scattering electron. The $\lambda(P\mathfrak{F})$ and $\lambda(P\mathfrak{G})$ are adjusted to make $\mathfrak{F}$ and $\mathfrak{G}$ satisfy $\Delta(P\mathfrak{F}) = 0$ and $\frac{1}{2}(\epsilon + k^2)\Delta(P\mathfrak{G}) = R_0(P^3\mathfrak{G})$, where ϵ is the ϵ_{nlnl} parameter; and

$$\Delta(PF) = \int_0^\infty PF\,dr', \qquad R_0(P^3F) = \int_0^\infty PP\,Y_0(PF)\,dr', \qquad \text{where} \qquad F = \mathfrak{F} \text{ or } \mathfrak{G}.$$

Linear combinations of the uncoupled equations give approximate solutions of the coupled equations. The uncoupled equations are solved by iterative techniques similar to those used in solving the Hartree-Fock equations. In this case, however, each iteration contributes linearly to the general solution. The correct linear combination is found by using the variational principle. It is possible to correct for the assumption of equal term energies and the omitted parts of the potential, at least in part, by using the general solution and the perturbation theory.

Although it is true for a number of situations that the p-waves are responsible for most of the interaction between the scattering electron and the target and hence, in the exact resonance method, that the other waves have been neglected, this is not always the case. A specific example is in the calculation of $\Omega(2, 3)$ for S II. When such a situation arises, it is necessary to depart from this scheme and use the distorted wave method.

In the distorted wave method (see Seaton [1955*a*, *b*, 1956]), the total wave function of the system is written as an antisymmetric linear combination of products of the wave functions of states of the target with wave functions representing the incoming electron. As in the case of the exact resonance method, the total wave function must satisfy the Pauli exclusion principle to allow for exchange of the scattering electron with those in the target. The total wave function is put into the wave equation, and one obtains a series of coupled integro-differential equations, as in the exact resonance method. At this point, however, a different set of assumptions is used to simplify the problem. First, the energies and wave functions of the various terms of a configuration are not assumed to be the same as were assumed in the exact resonance method. Next, it is assumed that certain of the integrals, that is, those involving products of atomic wave functions with the interaction potential, are negligible. As a result of these assumptions, a series of equations is obtained which may be solved successively. The first equations give the wave function $F_n(r)$ for the elastic scattering. This function is then used in the remaining equations to give the various wave functions $F_{n'}(r)$ describing inelastic scattering. Omitting the exchange terms for the sake of simplicity, the expressions for elastic and inelastic scattering, respectively, may be written as follows:

$$\left[\nabla^2 + k_n^2 - \frac{2M}{\hbar^2} V_{nn}\right] F_n(r_b) = 0$$

and

$$\left[\nabla^2 + k_{n'}^2 - \frac{2M}{\hbar^2} V_{n'n'}\right] F_{n'}(r_b) = \frac{2M}{\hbar^2} V_{nn'} F_n(r_b), \tag{185}$$

where ∇^2 is the usual Laplacian operator, k_n^2 and $k_{n'}^2$ are the energies in rydbergs, V_{nn} and $V_{n'n'}$ are the interaction potentials between the incoming electron and the ion or atom, and $V_{nn'}$ is the coupling term. Explicitly, these terms are written in the following way:

$$V_{nn} = \int \varphi_n^* \left(\frac{Ze^2}{r_b} - \sum_i \frac{e^2}{r_{ib}}\right) \varphi_n d\tau ; \qquad V_{n'n'} = \int \varphi_{n'}^* \left(\frac{Ze^2}{r_b} - \sum_i \frac{e^2}{r_{ib}}\right) \varphi_{n'} d\tau ;$$

$$V_{nn'} = \int \varphi_n^* \left(\frac{Ze^2}{r_b} - \sum_i \frac{e^2}{r_{ib}}\right) \varphi_{n'} d\tau .$$

Here, too, the asymptotic behavior of these functions supplies the necessary information for the particular case of electron excitation. For transitions involving a spin change in p^q configurations, it is found in some instances that the distorted wave method may give results which are as much as two orders of magnitude greater than the theoretical upper limit. Hence, the exact resonance method and the distorted wave method should not be regarded as equivalent for obtaining collision cross-section results.

If Born's approximation and Seaton's method for strong coupling are applied to collisions between electrons and atomic configurations with one electron outside the closed shells, then the general procedure for determining the collision strength Ω for allowed transitions may be obtained in the following manner. Let us rewrite equation (175*b*) in the form

$$(\nabla_b^2 + k_b'^2) F_{n'}(r_b)$$
$$= 2\left[\int \varphi_{n'}^* \frac{Ze^2}{r_b} \sum_m \varphi_m F_m d\tau - \int \varphi_{n'}^* \left(\sum_i \frac{e^2}{r_{ib}}\right) \sum_m \varphi_m F_m d\tau\right]. \tag{186}$$

This equation is then analogous to van Regemorter's (1960) expression, $L(\alpha\alpha') = 2(\alpha|1/r_{cb} + V(r_b)|\alpha')$ for $\alpha \neq \alpha'$, where α and α' refer to the initial and final states. The $\alpha = nLMklm$ and the $\alpha' = n'L'M'k'l'm'$ are the appropriate quantum numbers, and the subscripts c and b refer to the atomic and colliding electrons, respectively. The first integral on the right-hand side of equation (186) is simply $(Ze^2/r_b)F_n'$. For inelastic collisions the expression is

$$\left(\nabla_b^2 + k_b'^2 - 2\frac{Ze^2}{r_b}\right) F_{n'}(r_b) = -2\left\{\left[\int \varphi_{n'}^* \left(\sum_i \frac{e^2}{r_{ib}}\right) \varphi_{n'} d\tau\right] F_{n'}(r_b) + \left[\int \varphi_{n'}^* \left(\sum_i \frac{e^2}{r_{ib}}\right) \varphi_n d\tau\right] \mathfrak{F}_n(r_b)\right\}, \tag{187}$$

where the approximation has been made that $F_m = 0$ for $m \neq n$ or n' and $F_n = \mathfrak{F}_n$ with $\mathfrak{F}_n$ satisfying the equation $(\nabla_b^2 + k_b^2 - 2Ze^2/r_b)\mathfrak{F}_n = 0$. If the first integral on the right side is combined with the potential energy term on the left side, that is,

$$-2\left[\frac{Ze^2}{r_b} - \int \varphi_{n'}^* \left(\sum_i \frac{e^2}{r_{ib}}\right) \varphi_{n'} d\tau\right],$$

then we can approximate this as $-2ze^2/r_b$, where z is the ionic charge. Equation (187) becomes

$$\left(\nabla_b^2 + k_b'^2 - \frac{2ze^2}{r_b}\right) F_{n'}(r_b) = -2\mathfrak{F}_n(r_b)\int \varphi_{n'}^* \frac{e^2}{r_{cb}} \varphi_n d\tau, \qquad (188)$$

where the subscript c in r_{cb} refers to the one electron outside the atomic closed shells and $\Sigma(e^2/r_{ib})$ has been put approximately equal to its largest term e^2/r_{cb}. Since $f_{nn'}(\theta)$ is necessary for the calculation of the collision parameter Ω, and hence to the determination of the cross-section, it must be obtained first. It can be written as follows:

$$f_{nn'}(\theta) = -\frac{2}{4\pi}\int \mathfrak{F}_{n'}(r_b, \pi - \Theta)\left(\varphi_{n'}^* \frac{e^2}{r_{cb}} \varphi_n d\boldsymbol{r}_c\right)\mathfrak{F}_n(r_b, \theta_b) d\boldsymbol{r}_b,$$

where

$$\cos\Theta = \cos\theta\cos\theta_b + \sin\theta\sin\theta_b\cos(\varphi - \varphi_b).$$

By choosing the axis in a convenient way, one can set

$$\mathfrak{F}_n(r_b, \theta_b) = A_l(r_b) Y_{l0}(\theta_b \varphi_b),$$

and if $1/r_{cb}$ is expanded; that is,

$$\frac{1}{r_{cb}} = 4\pi \sum_{\lambda\mu} \gamma_\lambda Y_{\lambda\mu}^*(\theta_c \varphi_c) Y_{\lambda\mu}(\theta_b, \varphi_b)/2\lambda + 1$$

where

$$\gamma_\lambda(r_b, r_c) = \begin{cases} r_b^\lambda / r_c^{\lambda+1} & \text{for } r_b < r_c \\ r_c^\lambda / r_b^{\lambda+1} & \text{for } r_c < r_b, \end{cases}$$

then $f_{nn'}(\theta)$ can be solved in the following way:

$$f_{nn'}(\theta) = -\frac{2e^2}{4\pi}\int \mathfrak{F}_{n'}(r_b, \pi - \Theta)\left[Y_{L'M'}^* P_{n'L'}\left(4\pi \sum_{\lambda\mu} \frac{\gamma_\lambda Y_{\lambda\mu}^* Y_{\lambda\mu}}{2\lambda + 1}\right) Y_{LM} P_{nL} d\boldsymbol{r}_c\right] \times A_l Y_{l0} d\boldsymbol{r}_b$$

$$= -\sum_{\lambda\mu} \frac{2e^2}{2\lambda + 1}\int Y_{L'M'}^* P_{n'L'} \gamma_\lambda Y_{\lambda\mu}^* Y_{LM} P_{nL} d\boldsymbol{r}_c \int \mathfrak{F}_{n'} Y_{\lambda\mu} A_l Y_{l0} d\boldsymbol{r}_b$$

$$= -\sum_{\lambda\mu} \frac{2e^2}{2\lambda + 1}\int Y_{L'M'}^* Y_{\lambda\mu}^* Y_{LM} \sin\theta_c d\theta_c d\varphi_c \int \gamma_\lambda P_{n'L'} P_{nL} d r_c \times \int \mathfrak{F}_{n'} Y_{\lambda\mu} A_l Y_{l0} d\boldsymbol{r}_b.$$

If we write

$\mathfrak{F}_{n'} = B_{l'}(r_b) P_{l'}(\cos\Theta)$, where $P_{l'}(\cos\Theta)$

$$= (4\pi)/(2l' + 1)\Sigma Y_{l'm'}(\theta, \varphi) Y_{l'm'}^*(\theta_b \varphi_b),$$

then

$$f_{nn'}(\theta) = -\sum_{\lambda\mu m'} \frac{8\pi e^2}{(2\lambda+1)(2l'+1)} \left(\int Y^*_{L'M'} Y^*_{\lambda\mu} Y_{LM} d\hat{r}_c \right.$$

$$\left. \times \int Y^*_{l'm'} Y_{\lambda\mu} Y_{l0} d\hat{r}_b \int A_l B_{l'} d r_b \right) Y_{l'm'} \int \gamma_\lambda P_{n'L'} P_{nL} d r_c$$

or

$$f_{nn'}(\theta) = -\sum_{\lambda\mu m'} \frac{8\pi e^2 Y_{l'm'}}{(2\lambda+1)(2l'+1)} \left[\int Y^*_{L'M'}(\theta_c\varphi_c) Y^*_{\lambda\mu}(\theta_c\varphi_c) Y_{LM}(\theta_c\varphi_c)\right.$$

$$\times d\hat{r}_c \int Y^*_{l'm'}(\theta_b\varphi_b) Y_{\lambda\mu}(\theta_b\varphi_b) Y_{l0}(\theta_b\varphi_b) d\hat{r}_b \quad (189)$$

$$\left. \times \int A_l(r_b) B_{l'}(r_b) d r_b \int \gamma_\lambda P_{n'L'}(r_c) P_{nL}(r_c) \; d r_c \right],$$

where $d\hat{r}_c = \sin\theta_c d\theta_c d\varphi_c$ and $d\hat{r}_b = \sin\theta_b d\theta_b d\varphi_b$. Equation (189) is now analogous to van Regemorter's equations (47); that is, the expressions for the R matrix are shown as angular and radial integrals in his equations (48) and (49). Following the procedure of van Regemorter, we obtain for Ω the expression

$$\Omega(nL, n'L') = \sum_{l m l' m'} |T(nLMlm; n'L'M'l'm')|^2, \quad (190)$$

where T is the transmission matrix and nLM and $n'L'M'$ are the quantum numbers of the atomic configuration and lm and $l'm'$ are the ones for the colliding electron before and after collision, respectively. Since the total angular momentum is conserved during collision and the wave functions may be expressed in terms of $L^T = L + l$ and $M^T = M + m$, then the transmission matrix, which is diagonal in L^T and M^T as well as being independent of M^T, may be expressed as

$$T = (nLlL^TM^T; n'L'l'L^{T'}M^{T'}) = T(nLlL^T, n'L'l'L^T)\,\delta(L^TL^{T'})\,\delta(M^TM^{T'}).$$

Parity conservation requires $T = 0$ unless $(-1)^{L+l} = (-1)^{L'+l'}$, so that the expression for the collision strength Ω becomes

$$\Omega(nL, n'L') = \sum_{Ll'L^T} (2L^T+1)\,|T(nLlL^T; n'L'l'L^T)|^2. \quad (191)$$

The transmission matrix T^2 is analogous to $\int |f_{nn'}(\theta)|^2 d\omega$ and is obtainable for either plane or Coulomb waves. The reciprocity and conservation conditions require T to be symmetric and the scattering matrix $S(=1 - T)$ to be unitary. For both the Born and Coulomb-Born approximation (usually noted as Born I and CB I), the conservation conditions are not necessarily satisfied; however, it has been shown by Seaton that satisfactory results may be obtained if $T \ll 1$. When T is large, then the Born II or CB II approximation must be used. The difference between the Born I and Born II and hence between CB I and CB II is that for the former $T = -2iR$, whereas for the latter, $T = -2iR/(1 - iR)$, the reactance matrix R being hermitian. Seaton (1955*b*) showed that the Born II and the CB II are superior to the Born I and CB I when the interactions are strong.

Van Regemorter (1960), in developing the scheme for the determination of the collision cross-section between singly ionized positive ions and one electron outside the closed shells, took into account the long-range Coulomb distortion, so that the solutions of the radial wave function were Coulomb functions $F_{kl}(r_b)$ instead of spherical Bessel functions; that is,

$$\left[\frac{d^2}{dr_b^2}-\frac{l(l+1)}{r_b^2}+\frac{2z}{r_b}+k_b^2\right]F(k_b lz;\ r_b)=0\,, \tag{192}$$

where

$$F(k_b lz;\ r_b)\sim\frac{1}{k^{1/2}}\sin\left[k_b r_b-\tfrac{1}{2}l\pi+\frac{z}{k_b}\ln(2kr)+arg\Gamma\left(l+1-\frac{iz}{k_b}\right)\right]. \tag{193}$$

Here and in the following equations z is the ionic charge. For inelastic collisions, the $1/r_{cb}$ may be written as a multipole expansion so that the integral for the $R_{aa'}$ matrix can be separated into two parts, that is, the angular and radial integrals, and $R_{aa'}$ can be expressed in the form

$$R_{aa'}=-2\sum_{\lambda}f_\lambda\Re_\lambda(nl_1kl_2;\ n'l_1'k'l_2'), \tag{194}$$

where

$$\begin{aligned}\Re_\lambda=&\int_0^\infty P(nl_1;\ r_1)P(n'l_1';\ r_1)\,r_1^\lambda dr_1\int_0^\infty F(kl_2z;\ r_2)F(k'l_2'z;\ r_2)\frac{dr_2}{r_2^{\lambda+1}}\\&+\int_0^\infty F(kl_2z;\ r_2)F(k'l_2'z;\ r_2)Z_\lambda(nl_1n'l_1';\ r_1)\,dr_2\,,\end{aligned} \tag{195}$$

$$\begin{aligned}Z_\lambda(nl_1,\ n'l_1';\ r_1)=&\ r_2^\lambda\int_{r_2}^\infty P(nl_1;\ r_1)P(n'l_1';\ r_1)\frac{1}{r_1^{\lambda+1}}dr_1\\&-\frac{1}{r_2^{\lambda+1}}\int_{r_2}^\infty P(n_1l_1;\ r_1)P(n'l_1';\ r_1)\,r_1^\lambda dr_1\,.\end{aligned} \tag{196}$$

Also,

$$\begin{aligned}f_\lambda=&(-1)^{L+L'-L^T}(2\lambda+1)^{-1}[(2l+1)(2l'+1)(2L+1)(2L'+1)]^{1/2}\\&\times C_{000}^{ll'\lambda}C_{000}^{LL'\lambda}W(LlL'l';\ L^T\lambda),\end{aligned} \tag{197}$$

and $F(klz; r)$ = asymptotic form and is the same as in expression (193), $C_{000}^{ll'\lambda}$ and $C_{000}^{LL'\lambda}$ = Clebsch-Gordan coefficients, and $W(LlL'l';\ L^T\lambda)$ = Racah coefficients. For the coefficients f_λ, tables have been given by Percival and Seaton (1957). For a singly ionized atom and an electron, $z = 1$, and we have

$$\begin{aligned}F_{kl}(r_b)=&\left(\frac{\pi}{2}\right)^{1/2}\frac{2^{l+1}}{(2l+1)!}\left(\frac{A_{kl}}{1-e^{-2\pi/k}}\right)^{1/2}r_b^{l+1}e^{ikr_b}\\&\times F\left(l+1-\frac{i}{k},\ 2l+2,\ -2ikr_b\right)\end{aligned} \tag{198}$$

with $A_{kl} = (1 + k^2l^2)[1 + k^2(l - 1)^2]\ldots(1 + k^2)$ and $A_{k0} = 1$. In equation (198) F is the confluent hypergeometric function. Since it is first necessary to evaluate the radial term $\Re_\lambda$ (eq. [195]), the method outlined by van Regemorter (1960) will be used.

The term $\Re_\lambda$ can be conveniently written as the sum of two parts, namely, long- and short-range contributions, shown in equation (195). Van Regemorter's original procedure for handling a practical case was to divide the computation into the following five parts: (1) evaluate the multipole length integral, $\int_0^\infty P(nl_1;\, r)P(n'l_1';\, r)\, r^\lambda dr$ and $Z_\lambda(nl_1, n'l_1'; r)$; (2) evaluate the dipole ($\lambda = 1$) short-range integrals, $1/z^2 \int_0^\infty F(Kl_2;\, \rho)$ $F(K'l_2';\, \rho) Z_1(n_1l_1,\, n_1'l_1';\, \rho)\, d\rho$ with $l_2' = l_2 \pm 1$; (3) evaluate the dipole long-range integrals at threshold, that is, $\int_0^\infty F(Kl_2;\, \rho) F(K'l_2';\, \rho) \frac{d\rho}{\rho^2}$ with $l_2' = l_2 \pm 1$; (4) evaluate the dipole long-range integrals for $K' \neq 0$; and (5) evaluate the monopole and quadrupole integrals: $1/z^2 \int_0^\infty F(Kl_2;\, \rho) F(K'l_2';\, \rho) Z_0(n_1l_1,\, n_1'l_1';\, \rho) d\rho$ for $l_2' = l_2$, (monopole short range), $1/z^2 \int_0^\infty F(Kl_2;\, \rho) F(K'l_2';\, \rho) Z_2(n_1l_1,\, n_1'l_1';\, \rho)\, d\rho$ for $l_2' = l_2$, $l_2 \pm 2$ (quadrupole short range), and $1/z \int_0^\infty F(Kl_2;\, \rho) F(K'l_2';\, \rho)\, d\rho/\rho^3$ for $l_2' = l_2$, $_2 \pm 2$ (quadrupole long range).

Here, the original expression for $F(klz; r)$ has been defined as $1/z^2 F(kl; \rho)$ independent of the parameter z with K defined by $K = k/z$ and the transformation variable $\rho = zr$. Let us now apply this procedure to determine the collision strength Ω for the $3s \rightarrow 3p$ transition of Fe XVI. In this example, we shall consider the threshold case $E/\Delta E = 1$ and $K' = 0$, where ΔE is the separation energy of the $3s$ and $3p$ levels. For the ionic charge $z = 15$ and $\lambda = 1$, the two expressions for $R_{aa'}$ are

$$R_{aa'} = -2f_1\mathfrak{R}_\lambda$$

or

$$R_{12} = -2f_1(l, l-1)\Big[\int_0^\infty P(3s;\, r)\, rP(3p;\, r)\, dr \int F(Kl;\, \rho) \times F(K', l-1;\, \rho)\frac{d\rho}{\rho^2} + \frac{1}{z^2}\int_0^\infty F(Kl;\, \rho) F(K', l-1;\, \rho) Z_1 d\rho\Big] \tag{199}$$

$$R_{13} = -2f_1(l, l+1)\Big[\int_0^\infty P(3s;\, r)\, rP(3p;\, r)\, dr \int_0^\infty F(Kl;\, \rho) \times F(K', l+1;\, \rho)\frac{d\rho}{\rho^2} + \frac{1}{z^2}\int_0^\infty F(Kl;\, \rho) F(K', l+1;\, \rho) Z_1 d\rho\Big] \tag{200}$$

where for Fe XVI, $3s \rightarrow 3p$ transition $K^2 = 0.01163$, $K'^2 = 0$, and $z^2 = 225$. Thus, for this calculation we employ parts 1, 2, and 3. The contributions from part 5 have been neglected, since they are negligible. In Table 14 the data for the integrals in equations (199) and (200) and the sum in the brackets are tabulated. The next step requires that the appropriate $f_1(l, l')$ values be determined; and these, for this example, have been obtained from the tables of Percival and Seaton (1957), which are listed in Table 15. This now permits calculation of the R matrix. The terms of the R matrix, given by equation (194) for $\lambda = 1$ are

$$\begin{aligned} \alpha &= R(3sl, 3pl-1) & \epsilon &= R(3pl+1, 3dl) \\ \beta &= R(3sl, 3pl+1) & \zeta &= R(3pl+1, 3dl+2) \\ \gamma &= R(3pl-1, 3dl) & \nu &= R(3pl, 3dl+1) \\ \delta &= R(3pl-1, 3dl-2) & \mu &= R(3pl, 3dl-1)\,. \end{aligned}$$

Taking into account coupling with a *d*-level for the elements of the T matrix for an $ns \rightarrow np$ transition gives

$$T(n\,sl,\, n\,pl-1) = \frac{2i\alpha(1+\zeta^2+\epsilon^2) - 2i\beta\gamma\epsilon}{D} \tag{201}$$

$$T(n\,sl,\, n\,pl+1) = \frac{2i\beta(1+\delta^2+\gamma^2) - 2i\beta\gamma\epsilon}{D}, \tag{202}$$

TABLE 14

Calculations of Integrals in Equations (200) and (201) for $\frac{E}{\Delta E} = 1.0$

		$\int_0^\infty P(3s;r)\, rP(3p;r) = -.6729$		
$l \rightarrow l-1$		$\int F(kl;\rho)\, F(k', l-1;\rho)\, d\rho$	$\frac{1}{z^2}\int F(kl;\rho)\, F(k'\, l-1) Z_1 d\rho$	$\Re_1$
1	0	0.16515	0.16491	0.05378
2	1	0.16070	0.13940	0.03127
3	2	0.17250	0.09814	-0.01793
4	3	0.17550	0.03791	-0.08019
5	4	0.16875	0.00918	-0.10438
6	5	0.15304	0.00291	-0.10007
7	6	0.13073	0.00036	-0.08833
8	7	0.10517	0.00001	-0.07076
9	8	0.07970	0.00000	-0.05363
10	9	0.05694		-0.03832
11	10	0.03840		-0.02584
12	11	0.02448		-0.01647
13	12	0.01477		-0.00994
14	13	0.00846		-0.00569
15	14	0.00460		-0.00309
16	15	0.00238		-0.00160
17	16	0.00117		-0.00079
18	17	0.00055		-0.00037
19	18	0.00025		-0.00017
$l \rightarrow l+1$				
0	1	0.12052	0.15594	0.07484
1	2	0.07654	0.12658	0.07507
2	3	0.05770	0.08750	0.04867
3	4	0.04200	0.03305	0.00479
4	5	0.02953	0.00772	-0.01214
5	6	0.02003	0.12658	-0.01116
6	7	0.01308	0.08750	-0.00905
7	8	0.00823	0.03305	-0.00553
8	9	0.00497	0.00772	-0.00334
9	10	0.00288	0.00231	-0.00194
10	11	0.00160	0.00025	-0.00108
11	12	0.00086	0.00000	-0.00058
12	13	0.00044		-0.00029
13	14	0.00022		-0.00014
14	15	0.00010		-0.00007
15	16	0.00005		-0.00003
16	17	0.00002		-0.00001
17	18	0.00001		-0. 00000
18	19	0.00000		

where

$$D = 1 + \alpha^2 + \beta^2 + \gamma^2 + \delta^2 + \epsilon^2 + \zeta^2 + \delta^2\zeta^2 + \alpha^2\zeta^2 + \gamma^2\zeta^2 + \beta^2\gamma^2 + \epsilon^2\delta^2 + \alpha^2\epsilon^2 + \beta^2\delta^2 - 2\alpha\beta\gamma\epsilon .$$

If we consider the Fe XVI $3s \to 3p$ transition at threshold, for which there is no coupling with the $3d$ level, then the expressions for T become

$$T(3sl, 3pl-1) = \frac{2i\alpha}{1+\alpha^2+\beta^2} \tag{203}$$

$$T(3sl, 3pl+1) = \frac{2i\beta}{1+\alpha^2+\beta^2}. \tag{204}$$

TABLE 15

f_λ Values

l	$\to l+1$	f_λ	l	$\to l-1$	f_λ
0	1	-0.577350	1	0	0.333333
1	2	-0.471404	2	1	0.365148
2	3	-0.447213	3	2	0.377964
3	4	-0.436436	4	3	0.384900
4	5	-0.430330	5	4	0.389249
5	6	-0.426401	6	5	0.392232
6	7	-0.423659	7	6	0.394405
7	8	-0.421636	8	7	0.396059
8	9	-0.420084	9	8	0.397359
9	10	-0.418853	10	9	0.398409
10	11	-0.417855	11	10	0.399274
11	12	-0.417028	12	11	0.400000
12	13	-0.416333	13	12	0.400616
13	14	-0.415739	14	13	0.401147
14	15	-0.415227	15	14	0.401609
15	16	-0.414780	16	15	0.402015
16	17	-0.414387	17	16	0.402374
17	18	-0.414039	18	17	0.402693
18	19	-0.413728	19	18	0.402980
19	20	-0.413449			

A similar set of expressions can be obtained for a $3p \to 3d$ transition. After the $\mathfrak{R}_1$ values have been determined, then the R_{12} and R_{13} quantities can be calculated. The R_{12} and R_{13} can then be introduced into the following equation for CB I:

$$T = -2iR_{\alpha\alpha'}, \text{ where } R_{12} = \alpha \text{ and } R_{13} = \beta ; \tag{205}$$

and if

$$\Omega_l = (2L^T + 1)|2i\alpha|^2 + (2L^T + 1)|2i\beta|^2 , \tag{206}$$

then

$$\Omega(n, n') = \Sigma\Omega_l , \tag{207}$$

where α represents the values of $R_{\alpha\alpha'}$ for $l \to (l-1)$ and β for $l \to (l+1)$. The Ω_l and Ω results with $(E/\Delta E) = 1$ for the CB I approximation are shown in Table 16 with the $(E/\Delta E) = 1.5$, 2.0, and 4.0. The Ω_l and Ω for the CB II approximation for $E/\Delta E =$

1.0, 1.5, 2.0, and 4.0, using equations (203) and (204) are shown in the same table for comparison. In addition, in Table 17 the results are shown for the $3p \rightarrow 3d$ transition. These results have been published by Tully (1963) and Bely, Tully and van Regemorter (1963) as well as Krueger and Czyzak (1964). This example, for which the van Regemorter schemes for the CB I and CB II methods without exchange were used, shows how the collision strength Ω can be determined for the collision between an electron and a particular type of configuration, that is, for one electron outside the closed shells. The Born I, CB I, Born II, and CB II may, of course, be applied to other atomic configurations; however, appropriate assumptions and approximations must be made to fit the particular case.

TABLE 16

Collision Strengths, Ω, for Fe XVI 3s→3p Transition

$E/\Delta E$ / l	1.0		1.5		2.0		4.0	
	CB I	CB II	CB I	CB II	CB I	CB II	CB I	CB II
0	0.030	0.029	0.042	0.041	0.041	0.040	0.036	0.035
1	0.075	0.075	0.082	0.081	0.078	0.077	0.065	0.064
2	0.048	0.048	0.045	0.045	0.042	0.042	0.033	0.033
3	0.006	0.006	0.005	0.005	0.005	0.005	0.004	0.004
4	0.141	0.140	0.131	0.130	0.122	0.121	0.096	0.095
5	0.295	0.290	0.274	0.270	0.256	0.253	0.202	0.201
6	0.324	0.320	0.302	0.299	0.284	0.281	0.230	0.228
7	0.293	0.290	0.284	0.281	0.276	0.273	0.254	0.252
8	0.214	0.213	0.218	0.217	0.219	0.218	0.209	0.208
9	0.138	0.138	0.155	0.155	0.165	0.164	0.175	0.175
10	0.078	0.078	0.102	0.102	0.118	0.117	0.144	0.143
11	0.039	0.039	0.062	0.062	0.080	0.080	0.116	0.116
12	0.017	0.017	0.036	0.036	0.052	0.052	0.092	0.092
13	0.007	0.007	0.019	0.019	0.033	0.033	0.073	0.073
14	0.002	0.002	0.010	0.010	0.020	0.020	0.057	0.057
15	0.001	0.001	0.005	0.005	0.012	0.012	0.044	0.044
16			0.002	0.002	0.007	0.007	0.034	0.034
17			0.001	0.001	0.004	0.004	0.026	0.026
18					0.002	0.002	0.020	0.020
19					0.001	0.001	0.016	0.016
Ω	1.708	1.693	1.775	1.761	1.817	1.802	1.926	1.916

In dealing with collision cross-sections for forbidden transitions, three general approximation methods have been developed: the Coulomb-Born-Oppenheimer wave method (as developed by Hebb and Menzel), the exact resonance method (as developed by Seaton), and the distorted wave method (as developed by Seaton). The main objection to the Hebb and Menzel method has been that it violates charge conservation conditions (see Bates *et al.* [1950] regarding effect on results). While this is not always the case, it can, at times, lead to serious errors. Also, while this method is not considered particularly satisfactory for $l < 3$, it may be quite useful for larger values of l. The distorted wave method appears to be best for *dd* and *sd* configurations, but not particularly good for the

TABLE 17

Collision Strengths, Ω, for Fe XVI 3p→3d Transition

$E/\Delta E$	1.0		1.5		2.0		4.0	
l	CB I	CB II	CB I	CB II	CB I	CB II	CB I	CB II
0	0.156	0.152	0.248	0.238	0.239	0.230	0.205	0.199
1	0.175	0.172	0.180	0.177	0.170	0.168	0.137	0.135
2	0.087	0.086	0.081	0.081	0.076	0.076	0.059	0.059
3	0.021	0.021	0.019	0.019	0.017	0.017	0.013	0.013
4	0.267	0.265	0.243	0.241	0.222	0.221	0.164	0.163
5	0.491	0.485	0.449	0.444	0.415	0.410	0.314	0.311
6	0.494	0.489	0.462	0.457	0.433	0.428	0.343	0.341
7	0.408	0.405	0.409	0.406	0.406	0.403	0.393	0.390
8	0.264	0.263	0.289	0.288	0.302	0.301	0.310	0.309
9	0.148	0.148	0.189	0.189	0.216	0.215	0.255	0.254
10	0.071	0.071	0.114	0.114	0.146	0.145	0.205	0.205
11	0.030	0.030	0.064	0.064	0.094	0.094	0.163	0.163
12	0.011	0.011	0.034	0.034	0.059	0.059	0.129	0.128
13	0.004	0.004	0.017	0.017	0.036	0.036	0.100	0.100
14	0.001	0.001	0.008	0.008	0.021	0.021	0.078	0.078
15			0.004	0.004	0.012	0.012	0.060	0.060
16			0.002	0.002	0.007	0.007	0.046	0.046
17			0.001	0.001	0.004	0.004	0.035	0.035
18					0.002	0.002	0.027	0.027
19					0.001	0.001	0.021	0.021
Ω	2.628	2.603	2.813	2.784	2.878	2.850	3.057	3.037

pp-type, whereas the exact resonance method is particularly good for pp-type configurations. However, very recent results reported by Seaton [see Billings, Czyzak, Krueger, Saraph, Seaton, and Shimming (1965)] show that the agreement between the exact resonance and distorted wave methods is quite close for the pp-type waves in a $2p^2$ configuration. For O III this is a difference of approximately 15% for Ω (1, 2) and Ω (1, 3) and 25% for Ω (2, 3); but for the latter the contribution from the pp-waves is very small.

Hebb and Menzel (1940) worked out a method for determining the collisional excitation and de-excitation of the metastable levels of (O III) by electron impact. This method uses the Coulomb-Born-Oppenheimer approximation. It represents both the incoming and outgoing electrons as solutions of the wave equation in the Coulomb field of the ion. The total wave function of the system is constructed to satisfy the Pauli exclusion principle. This allows exchange of the scattering electron with either of the outer electrons of the ion. The exchange effect is important for low-energy electrons interacting with light atoms. The scattering electron is treated as though it were moving in a potential resulting from the two outer electrons plus that due to a point charge equal to 2 located at the center of the atom. The essential information relating to the transition caused by electron impact is contained in the integral of the product of this potential V with the total wave functions ψ_n and $\psi_{n'}$ representing the initial and final states of the system. The first order differential cross-section may be written in the following way:

$$Q(n \to n')\,d\omega = \frac{d\omega}{4\pi^2}\frac{p}{p'}(n'|V|n)(n|V|n'), \tag{208}$$

where p and p' represent the incident and scattered momentum; $d\omega$ is the solid angle; and

$$(n'|V|n) = \int \psi_{n'}^* V \psi_n \, d\tau , \tag{209}$$

where $n = JLSMkm_s$, $n' = J'L'S'M'k'm_{s'}$, and the integration is over the space coordinates of the several electrons concerned. The complete wave function may be written as $\psi_n = \varphi_n \chi_n$ with φ_n referring only to the ion and χ_n only to the colliding electron. This procedure is suitable for both neutral atoms and ions; however, for neutral atoms, plane waves represent χ_n, the wave function for the colliding electron, and for ions, Coulomb functions are used. In order to satisfy the Pauli exclusion principle and allow for exchange, the simple product function $\varphi_n \chi_n$ cannot be used, but instead ψ_n must be written as a sum of the terms of $\varphi_n \chi_n$ in which the arguments of the electrons are permuted. With these conditions satisfied, equation (209) may be written more explicitly as follows:

$$\begin{aligned}(n'|V|n) = 2[&\int \varphi_{n'}^*(1, 2)\varphi_n(1, 2)V(1, 3)\chi_{n'}^*(3)\chi_n(3)\,d\tau \\ &+\int \varphi_{n'}^*(2, 3)\varphi_n(1, 2)V(1, 3)\chi_{n'}^*(1)\chi_n(3)\,d\tau \\ &+\int \varphi_{n'}^*(3, 1)\varphi_n(1, 2)V(1, 3)\chi_{n'}^*(2)\chi_n(3)\,d\tau].\end{aligned} \tag{210}$$

Here the second and third integrals are the exchange integrals, and the Pauli exclusion principle is satisfied by requiring that $\varphi_n(1, 2) = -\varphi_n(2, 1)$. Since the total collision cross-section $Q_T(JLSk \to J'L'S'k')$ is desired, equation (208) is integrated over all directions of p', then averaged over the initial values of M and m_s and summed over the final values of M' and $m_{s'}$, or

$$Q_T(JLSk \to J'L'S'k') = \frac{1}{2(2J+1)} \sum_{MM'm_s m_{s'}} \int Q(n \to n')\, d\omega . \qquad (211)$$

The radial wave functions for the ion or atom are determined from the procedures outlined in Section 2. The wave functions for the incoming and outgoing electrons, on the other hand, are the solutions for the wave equation in the Coulomb field of the ion or atom. Each matrix element (eq. [209]) must be evaluated to obtain the cross-sections, and the integrations are over the space coordinates of the electrons. To perform these integrations, the method of partial cross-sections is used. The final matrix then becomes

$$(n'|V|n) = \frac{2\pi}{(pp')^{1/2}} \sum_{l'=0}^{\infty} C_{l'}(n', n) P_{l'm'}(\cos\theta)\, e^{im'\varphi} , \qquad (212)$$

where $C_{l'}$ is complex, depending on the phases in χ; and $P_{l'm'}$ (cos θ) are the usual Legendre functions. The expression for the final cross-section then becomes

$$Q_T(JLSp \to J'L'S'p') = \frac{\pi}{(2J+1)p^2} \sum_{MM'm_s m_{s'}} C_{l'}(n'; n) C_{l'}(n; n') \qquad (213)$$

or

$$Q_T(JLSk \to J'L'S'k') = \frac{\pi}{\omega_n k^2} \sum_{l'} \Omega_{l'} = \frac{\pi}{\omega_n k^2} \Omega(JLSk \to J'L'S'k') \qquad (214)$$

where $\omega_n = 2J + 1$, k = momentum of incoming electron, and

$$\Omega_{l'} = \sum_{MM'm_s m_{s'}} C_{l'}(n'; n) C_{l'}(n; n') .$$

Table 18 shows the results reported in the literature for the collision strengths of various p^2 ions. The results obtained by this method have uncertain limits of accuracy. It has been shown that for every transition involved in [O III], this method gives results that exceed the theoretical upper limit by factors varying from 1.2 to 2.8. On the other hand, for [N II], some of the individual transitions did not exceed the upper limit. That these results may be questionable is best illustrated when one compares the total [O III] obtained by the Hebb and Menzel method (Table 18) with collision strengths determined by Seaton using the distorted wave method where the charge conservation condition is not violated. Seaton's optimum values [(1958), obtained by taking the mean of the results obtained by variational and semi-empirical methods] were Ω (1, 2) = 1.59, Ω (1, 3) = .220, and Ω (2, 3) = .64. It becomes quite apparent that the failure of the Hebb and Menzel method to satisfy the conservation theorem can lead to results which may be higher by as much as an order of magnitude. How serious the situation is for the $3p^q$ configuration cannot be ascertained at this time, since additional study and investigation into the various methods are still required.

Particularly successful have been the methods devised by Seaton for calculating the electron excitation of allowed and forbidden lines. The theory and methods (exact resonance and distorted wave) have been extensively described by Seaton (1953*a*, *b*, 1955*a*, *b*, 1956, 1958, 1961, 1962, Seaton and Peach, 1962). He has applied them primarily to the $2p^q$ configuration. Here the exact resonance method for determining the cross-sections is described for the $3p^q$ configuration as used by Czyzak and Krueger (1963*b*) and based on detailed procedure suggested to them by Seaton (1963). This method differs from Seaton's original procedure (eqs. [182] and [183]) in that exchange with the core electrons is included and orthogonalization with the $2p$ wave function is required.

TABLE 18

Collision Strengths, Ω, for p^2 Configurations Determined by the CBO Method as Devised by Hebb & Menzel (1940) for Forbidden Transitions

Configuration	Ion	Ω(1, 2)	Ω(1, 3)	Ω(2, 3)	Reference
$2p^2$	N II	9.54	1.68	1.55	Aller & White (See Bates, et. al. 1950)
	O III	19.08	3.37	3.11	Hebb & Menzel (1940)
$3p^2$	P II	114.57	16.21	13.7	Czyzak & Krueger (1964)
	S III	3.25	.003	.537	Czyzak & Krueger (1963b)
	Cl IV	1.59	.040	.136	"
	A V	.255	.00011	.160	"
	Fe XIII	.346	.075	.126	Czyzak & Krueger (1964)

The exact resonance equations for a $3p^q$ configuration atom or ion may be written more explicitly than in equations (182) and (183). Following the notation of Seaton, we have

$$(\mathfrak{H}+k^2)F_n = 2r\,Y_0(P_{3p}F_n)P_{3p}+\eta P_{3p}+\varphi F_n+\epsilon_{kp3p}P_{2p} \tag{215}$$

where

$$\mathfrak{H}=\frac{d^2}{dr^2}-\frac{2}{r}+\frac{2Z}{r}-2V\,,$$

$$V=\sum_{nl} q(nl)\,Y_0(P_{nl}P_{nl})\,,\qquad \text{with}\qquad q(nl)=\text{number of electrons in shell,}$$

$$Y_0(P_{nl}P_{nl})=\frac{1}{r}\Big(\int_0^r P_{nl}P_{nl}\,dr'\Big)+\int^{\infty}\frac{1}{r'}\,P_{nl}P_{nl}\,dr'\,,$$

and $P_{nl} \equiv P(nl; r)$ = wave functions for the nl shell. The exchange expression φF_n may be written in the following manner:

$$\varphi F_n = -2Y_0(P_{2p}F_n)P_{2p} - \tfrac{4}{5}Y_2(P_{2p}F_n) - \tfrac{2}{3}\sum_{n=1}^{3} Y_1(P_{ns}F_n)P_{ns}$$

$$Y_1 = \frac{1}{r^2}\int_0^r P_{ns}F_n r' dr' + r\int_r^\infty P_{ns}F_n \frac{1}{r'^2} dr' \qquad (216)$$

$$Y_2 = \frac{1}{r^3}\int_0^r P_{2p}F_n r'^2 dr' + r^2\int_r^\infty P_{2p}F_n \frac{1}{r'^3} dr' .$$

The P_{nl}'s are wave functions of the atom or ion for the appropriate shells, and the F_n's are collision wave functions and are the sum of the incident and scattered waves. The expression for the η parameter is $\eta = -q(nl)[(k^2 + \epsilon_{3p3p})\Delta(P_{3p}F_n)]$ and is such that

$$1 + q(nl)](k^2 + \epsilon_{3p3p})\Delta(P_{3p}F_n) = 2(1+\tau)R_0(P_{3p}^3F_n), \qquad (217)$$

where

$$\Delta(P_{3p}F_n) \equiv \int_0^\infty P_{3p}F_n dr$$

$$R_0(P_{3p}^3F_n) \equiv \int_0^\infty P_{3p}P_{3p}Y_0(P_{3p}F_n)dr.$$

In this method, the energy differences as well as the $Y_2(P_{3p}P_{3p})$ and $R_2(P_{3p}^3F_n)$ terms in the P_{3p} wave functions are neglected; however, a good approximation can still be expected. The solution of equation (215) is expressed in terms of two functions $\mathfrak{F}_n$ and $\mathfrak{G}_n$ which satisfy the conditions:

(*a*) $\tau = -1$, $\Delta(P_{3p}F_n) = 0$ for $F_n = \mathfrak{F}_n$;

(*b*) $\tau = q(nl)$, $(k^2 + \epsilon_{3p3p})\Delta(P_{3p}F_n) = 2R_0(P_{3p}^3F_n)$ for $F_n = \mathfrak{G}_n$; and

(*c*) the parameter ϵ_{kp2p} is adjusted so that $\Delta(F_nP_{2p}) = 0$ and is imposed at each iteration.

One proceeds to solve the equation by iterative processes, and it is convenient to rewrite equations (215) and (217) in the following form:

$$\mathfrak{L}F_n = \omega F_n + \eta P$$

$$\Delta(\theta F_n) = 0 ,$$

where

$$\mathfrak{L} = \mathfrak{H} + k^2$$

$$\omega F_n = 2\tau Y_0(P_{3p}F_n)P_{3p} + \varphi F_n$$

$$\theta = \{[1 + q(nl)][k^2 + \epsilon_{3p3p}] - 2(\tau + 1)Y_0(P_{3p}P_{3p})\}P_{3p} .$$

The iterative process is continued until a satisfactory set of functions $F_0, F_1, F_2 \ldots$ is obtained which satisfy the following conditions:

$$(\theta F_n) = 0,\ n \geq 0,$$

$$\mathfrak{L}F_0 = \eta_0 P_{3p}, \text{ and}$$

$$\mathfrak{L}F_n = \omega F_{n-1} + \eta_n P_{3p},\ n \geq 1 ,$$

and the collision functions are

$$F_0 \sim k^{-1/2} \sin (k + \delta) + a_0 k^{-1/2} \cos (k + \delta);\ a_0 = \eta_0 \alpha$$

or

$$F_n \sim a_n k^{-1/2} \cos (k + \delta);\ a_n = A_n + \eta_n \alpha\ ,$$

where A_n and η_n are the coefficients and α is a constant. Once these have been obtained, the Kohn variational method is used to calculate the scattering parameter a and the collision strengths are obtained from the following equation:

$$\Omega(n \rightarrow n') = C_q w_n w_{n'} |X|^2$$

where

$$X(\text{E.R.}) = \sin (\xi - \zeta)\ .$$

The ξ and ζ are the phases of $\mathfrak{F}$ and $\mathfrak{G}$, and $C_q w_n w_{n'}$ are coefficients which properly weight the levels involved. Tables 19 and 20 show the values obtained for several of

TABLE 19

Collision Strengths, Ω, for p^1 Configurations Using the Distorted Wave and ER Methods as Originally Devised by Seaton (1953a, b)

Configuration	Ion	Ω	Reference
p^1	Si II	7.095	Blaha (1964)
	P III	5.494	"
	S IV	4.096	"
	Fe XIV	0.249	" (1962, 1964)

the ions. Compare results from the earlier method of Hebb and Menzel (see Table 18). The results obtained by using the exact resonance approximation should be considered only as preliminary data. In Table 19 the collision strengths for the $3p^1$ configuration are shown.

The calculation of the collision strengths by the distorted wave method as presented is that developed by Seaton (1963, 1965) and follows his procedure. The wave function for the complete system, i.e., the ion and the colliding electron, may be written in the usual notation in the following manner:

$$\psi_{i'} = \sum_a \Big\{ \sum_i \psi(p^q(S_iL_i)l_aSL) + c_{i'}\delta(l_a, 1)\psi(p^{q+1}SL) \Big\} \qquad (218)$$

where $c_{i'}$ is non-zero only when $\psi_{i'}$ contains p-waves and SL is allowed in p^{q+1} and $q = 1, 2, 3, 4$, or 5. The wave function contains the radial wave function for the colliding electron $F_{ii'}$ which has the asymptotic form:

$$F_{ii'} \sim k^{-1/2} \sin (\chi_i + \tau_i)\delta_{ii'} + k^{-1/2} \cos (\chi_i + \tau_i) R^{(0)}_{ii'} \qquad (219)$$

where $\chi_i = k_i r - \frac{1}{2} l_i \pi + 2zk_i^{-1} \ln (2k_i r) + \arg \Gamma(l_i + 1 - izk_i^{-1})$, $z = Z - q$, τ_i is the reference phase and may be taken to have any convenient value, and $k_i^2 = 2[E - E_q(i)]$, with E the energy of the total system and $E_q(i)$ is the energy of ion in state i. $R^{(0)}_{ii'}$ is the reactance matrix. Thus, here we have the complete wave function expanded in terms of vector-coupled antisymmetric products. The radial functions for the colliding electron are taken to be orthogonal to the atomic radial wave functions.

If we define

$$L_{i'j'} = (i'|H_{q+1} - E|j') = \int \psi^*_{i'}(H_{q+1} - E)\psi_{j'}d\tau \tag{220}$$

then the variational principle gives:

$$\delta(R - 2L) = 0 \tag{221}$$

where R is the reactance matrix and L is defined in (220). The coefficients $c_{i'}$ and $c_{j'}$ in $\psi_{i'}$ and $\psi_{j'}$ are obtained from

$$\frac{\partial L}{\partial c_{i'}} = \frac{\partial L}{\partial c_{j'}} = 0\,.$$

It is more convenient in performing a distorted wave approximation, for example for a $3p^q$ configuration, to let

$$(\mathfrak{L}_{l_i} - \tfrac{1}{2}k_i^2)\mathfrak{F}_i = -\varphi\mathfrak{F}_i \tag{222}$$

where

$$\mathfrak{L}_{l_i} = -\tfrac{1}{2}\left[\frac{d^2}{dr^2} - \frac{l_i(l_i+1)}{r^2}\right] - \frac{Z}{r} + \sum_{nl} q(nl)\,Y_0(P_{nl}P_{nl})$$

$$-\frac{1}{2l_i+1}\left\{\sum_{n=1}^{3} Y_{l_i}(P_{ns}\mathfrak{F}_i)P_{ns} + \frac{3l_i}{2l_i-1}Y_{l_i-1}(P_{2p}\mathfrak{F}_i)P_{2p} + \frac{3l_i+3}{2l_i+3}\,Y_{l_i+1}(P_{2p}\mathfrak{F}_i)P_{2p}\right\}$$

$$\mathfrak{F}_i \sim k_i^{-1/2}\sin\chi_i + k_i^{-1/2}\cos\chi_i R^{(0)}_{ii'} \tag{223}$$

$$\varphi\mathfrak{F}_i = \frac{1}{2l_i+1}\left\{\sum_{n=1}^{3} Y_{l_i}(P_{ns}\mathfrak{F}_i)P_{ns} + \frac{3l_i}{2l_i-1}\,Y_{l_i-1}(P_{2p}\mathfrak{F}_i)P_{2p} + \frac{3l_i+3}{2l_i+3}\,Y_{l_i+1}(P_{2p}\mathfrak{F}_i)P_{2p}\right\}$$

$$Y_{l_i}(AB) = \frac{1}{r^{l_i+1}}\int_0^r ABr'^{l_i}dr' + r^{l_i}\int_r^\infty AB\frac{1}{r'^{l_i+1}}\,dr$$

and A and B represent the appropriate wave function. $P_{nl} \equiv P(nl;\,r)$ = atomic wave function for the (nl) shell, and $q(nl)$ = number of electrons in the (nl) shell. Thus, in equation (222) the exchange interactions with the core are not included in the calculation of $\mathfrak{F}_i$ and are obtained separately through the expression $\varphi\mathfrak{F}_i$, i.e., if one were to take $\varphi = 0$ then the exchange interaction with the core would have been taken into account.

Defining

$$F_{ii'} = \delta(ii')G_i$$

$$G_i = \mathfrak{F}_i + \sum_{n=1}^{2} a_i(ns)P_{ns} + \sum_{n=2}^{3} a_i(np)P_{np}$$

and

$$a_i(ns) = -\delta(l_i0)(\mathfrak{F}_i|P_{ns})$$

$$a_i(np) = -\delta(l_i1)(\mathfrak{F}_i|P_{np}), \qquad \text{where} \qquad (\mathfrak{F}_i|P_{np}) = \int_0^\infty \mathfrak{F}_i P_n\,dr',$$

then $F_{ii'}$ will also satisfy

$$\delta(l_i 0)(P_{ns}|F_{ii'}) = 0 \qquad \text{and} \qquad \delta(l_i 1)(P_{np}|F_{ii'}) = 0\,.$$

By performing the necessary algebraic manipulations $L_{i'j'}$ comes out to be:

$$L_{i'j'} = A_{i'j'} + c_{i'}\delta(l_a 1)B_{j'} + c_{j'}\delta(l_\beta 1)D_{i'} + c_{i'}c_{j'}Q\delta(l_a 1)\delta(l_\beta 1) \tag{224}$$

where l_a and l_β are the incoming and outgoing angular momenta associated with states i' and j' respectively. The terms in the above equation are defined as follows:

$$A_{i'j'}(l_a l_\beta) = \delta(l_a l_\beta)\,\delta(i'j')\,[G_{i'}(l_a)\,|\,\mathfrak{L}_{l_a} - \tfrac{1}{2}k_{i'}^2|G_{i'}(l_a)]$$
$$+ f_2(L_{i'}l_a L_{j'}l_\beta;L)R_2[P_{3p}G_{i'}(l_a)P_{3p}G_{j'}(l_\beta)]$$
$$- \sum_\lambda g_\lambda(L_{i'}l_a L_{j'}l_\beta;SL)R_\lambda[P_{3p}G_{i'}(l_a)G_{j'}(l_\beta)P_{3p}],$$

$$B_{j'}(l_a) = (q+1)^{1/2}\{\delta(l_\beta 1)\,A_{q+1}(SL,\,j')R_0[P_{3p}^3G_{j'}(l_\beta)] + d_2R_2[P_{3p}^3G_{j'}(l_\beta)]\},$$

$$D_{i'}(l_\beta) = (q+1)^{1/2}\{\delta(l_a 1)\,A_{q+1}(SL,\,i)R_0[P_{3p}^3G_{i'}(l_a)] + d_2R_2[P_{3p}^3G_{i'}(l_a)]\},$$

$$Q = -(E - E_{q+1}) = \tfrac{1}{2}(\epsilon_{3p3p} + k^2) - R_0(P_{3p}^4) - f_2(p^{q+1}SL)R_2(P_{3p}^4),$$

$$k^2 = k_i^2 + 2f_2(p^{q+1}SL)R_2(P_{3p}^4),$$

ϵ_{3p3p} = one electron energy for the $3p$-shell of the ion, $f_2({}^{q+1}SL)$ = coefficient of $R_2(P^4)$ in the energy expression for the ${}^{2S+1}L$ term of the configuration p^{q+1},

$$c_i' = -\frac{D_{i'}(1)}{Q}, \qquad c_j' = \frac{B_{j'}(1)}{Q},$$

and

$$R_\lambda(EFGH) = \int_0^\infty EG\,Y_\lambda(FH)\,dr' = \int_0^\infty FH\,Y_\lambda(EG)\,dr'.$$

The f_λ, g_λ, and d_λ are special coefficients multiplying electrostatic interaction integrals:

$$f_\lambda(L_il_iL_jl_j;L) = \delta(S_iS_j)\,q[(2L_i+1)(2L_j+1)]^{1/2}\sum_k A_q(i,\,k)\,A_q(j,\,k)$$
$$\times \sum_{\mathfrak{L}} (2\mathfrak{L}+1)W(L_k1Ll_i;L_i\mathfrak{L})W(L_k1Ll_j;L_j\mathfrak{L})f_\lambda(1l_i1l_j;\mathfrak{L})$$

$$g_\lambda(L_il_iL_jl_j;SL) = q[(2S_i+1)(2L_i+1)(2S_j+1)(2L_j+1)]^{1/2}$$
$$\times \sum_k A_q(i,\,k)\,A_q(j,\,k)W(S_j\tfrac{1}{2}\tfrac{1}{2}S_i;SS_k)$$
$$\times \sum_{\mathfrak{L}} (2\mathfrak{L}+1)W(L_k1Ll_i;L_i\mathfrak{L})W(L_k1Ll_j;L_j\mathfrak{L})g_\lambda(1l_i1l_j;\mathfrak{L}),$$

$$d_\lambda(L_il_iSL) = \sum_j A_{q+1}(SL,\,j)\,f_\lambda(L_ilL_j1;L).$$

The expression $A_q(i, j) \equiv (p^{q-1}(S_jL_j)pS_iL_i]\!]p^qS_iL_i)$ is a fractional parentage coefficient; see Racah (1942, 1943). $A_q(SL, j)$ is the same coefficient written for cases in which the spin and angular momentum of the final state are not subscripted. The $f_\lambda(1l_i1l_j; \mathfrak{L})$ and $g_\lambda(1l_i1l_j; \mathfrak{L})$ have been tabulated by Percival and Seaton (1957) for some of the simpler cases.

From the variational principle (eq. 221), the best estimate for the R-matrix becomes

$$R_{ij} = \delta(ij)R_{ii}^{(0)} - 2L_{ij} . \tag{225}$$

From the R-matrix we obtain the $\mathfrak{R}$-matrix and T-matrix:

$$\mathfrak{R} = [(\sin \tau) + (\cos \tau)R][(\cos \tau) - (\sin \tau)R]^{-1} \tag{226}$$

where $(\sin \tau)_{ij} = \delta_{ij} \sin \tau_i$, etc. Then

$$T = 2i\mathfrak{R}\{1 - i\mathfrak{R}\}^{-1} \tag{227}$$

from which the Ω's may be calculated directly:

$$\Omega_{ij}(l_il_j) = \tfrac{1}{2}\sum_{SL}(2S+1)(2L+1)\,|T(L_il_iSL, L_jl_jSL)|^2$$

or

$$\Omega_{ij} = \sum_{l_il_j}\Omega_{ij}(l_il_j) ,$$

and finally

$$Q(i \to j) = \frac{\pi\Omega_{ij}}{k_i^2\omega_i} .$$

The collision strengths for the $3p^q$ configuration, as determined by the exact resonance and distorted wave methods, are shown in Tables 19 and 20.

The collision strengths Ω calculated by the exact resonance method as outlined in this section are for those that did not take into account the following: (1) correction for the energy differences, and (2) Y_2- and R_2-coupling terms. The iterative processes to obtain the set of functions F_n were calculated to $n = 4$. This appears to be sufficient for neutral or not too highly ionized atoms. The S II calculated by Czyzak and Krueger differs significantly from that of Seaton for Ω (2, 3) because Seaton considered contributions from other waves in addition to the p-waves. Seaton's p-wave value for $\Omega(2, 3) = 1.34$ can be compared with Czyzak-Krueger's value of 1.74. It is apparent that to consider p-waves only is insufficient and can lead to differences of as much as an order of magnitude for the Ω (2, 3) transition. The differences in Ω (1, 2) and Ω (1, 3) appear to be due to the technique employed in solving the uncoupled equations. Thus the data obtained by the exact resonance method described here is satisfactory to the degree that only p-waves are considered: the Ω (2, 3) values cannot be expected to be regarded as reliable unless the other wave contributions are taken into account for the $2 \to 3$ transition. Seaton (1955*a*, 1965) showed that the additional contributions were significant and therefore had to be included. This has been described in detail in Seaton's papers (1955*a*, *b*, 1956) on the distorted wave method and in a forthcoming paper in which the latest developments using this method will be reported.

In Table 20 results of collision strengths $\Omega(i, j)$ obtained by the distorted wave and exact resonance methods are shown. For the results which were obtained by the exact resonance method with or without conditions (1)–(3), referred to above, only *p*-waves

TABLE 20

Collision Strengths, Ω, for p^q(q = 2, 3 & 4) Configurations Using the ER and the DW Methods

Configuration	Ion	$\Omega(1, 2)$	$\Omega(1, 3)$	$\Omega(2, 3)$	Method	Reference
p^2	N II	2.39	0.22	0.46	DW	Seaton (1953b)
		2.17	0.31	0.50	Average of DW, ER, etc.	Seaton (1958)
		2.14	0.43	0.24[a]	ER	Czyzak & Krueger (1964)
		2.14	0.17	0.24[a]	ER with Cond. (1), (2) & (3) Included	" "
	O III	1.73	0.20	0.61	DW	Seaton (1953b)
		1.59	0.22	0.64	Average of DW, ER, etc.	Seaton (1958)
		1.68	0.34	0.19[a]	ER	Czyzak & Krueger (1964)
		1.63	0.19	0.19[a]	ER with Cond. (1), (2) & (3)	" "
	Fe XIII	0.07	0.015	0.01[a]	EK	" "
		0.06	0.006	0.007[a]	ER with Cond. (1), (2) & (3)	" "
p^3	O II	1.44	0.22	1.92	DW	Seaton (1953b)
		1.28	0.58	2.12	Average of DW, ER, etc.	Seaton (1958)
		0.99	0.60	1.49[a]	ER	Czyzak & Krueger (1964)
		1.25	0.23	1.49[a]	ER with Cond. (1), (2) & (3)	" "
	S II	2.02	0.38	12.7[b]	DW	Seaton (1953b)
		0.96	0.58	1.45[a]	ER with Cond. (1), (2) & (3)	Czyzak & Krueger (1964)
	A IV	0.38	0.23	0.56[a]	ER	" "
		0.45	0.14	0.67	ER with Cond. (1), (2) & (3)	Krueger & Czyzak (1965)
	Fe XII	0.05	0.03	0.08[a]	ER	Czyzak & Krueger (1964)
		0.05	0.02	0.07	ER with Cond. (1), (2) & (3)	Krueger & Czyzak (1965)
p^4	Ne III	0.76	0.08	0.27	DW	Seaton (1953b)
	Fe XI	0.10	0.02	0.01	ER	Czyzak & Krueger (1964)
	Ca XIII	0.06	0.01	0.006	ER	" "
		0.05	0.006	0.006	ER with Cond. (1), (2) & (3)	Krueger & Czyzak (1965)

[a] Only p-p waves considered for this calculation.

[b] Seaton's value for p-wave contribution only is 1.34.

were considered. Since the contributions for the other waves are not considered in this method, a large discrepancy for the Ω (2, 3) exists for the ions, as has been pointed out above for the $2p^3$ and $3p^3$ cases, i.e., O II and S II.

When conditions (1)–(2) are included in the calculations, better agreement is obtained for Ω (1, 2) and Ω (1, 3) with the earlier results of Seaton (1953*b*) who also included the same conditions in his calculations. However, the Ω (2, 3) result is helped very little by the inclusion of these conditions. Thus, this further indicates the necessity for considering other waves in addition to the *p*-waves. Since the Ω's play such an important role in the calculation of the electron temperatures (T_ϵ), densities (N_ϵ) and collision excitation rates, one still must regard the existing values for T_ϵ and N_ϵ obtained

TABLE 21

PARTIAL WAVE RESULTS FOR O III AND S II

	O III			S II		
l l'	Ω(1,2)	Ω(1,3)	Ω(2,3)	Ω(1,2)	Ω(1,3)	Ω(2,3)
1 1	1.664	0.209	0.012	0.232	0.037	0.268
2 2	0.718	0.125	0.090	2.753	1.240	2.700
3 3	0.003	0.000	0.047	0.006	0.001	0.569
4 4	0.000	0.000	0.009	0.000	0.000	0.086
5 5	....	....	....	0.000	0.000	0.011
2 0, 0 2	0.001	0.000	0.091	0.070	0.000	1.198
3 1, 1 3	0.004	0.000	0.048	0.000	0.000	0.893
4 2, 2 4	0.001	0.000	0.005	0.004	0.000	0.182
5 3, 3 5	....	....	....	0.000	0.000	0.311
Total	2.391	0.335	0.301	3.065	1.278	6.218

from forbidden lines as not fully definitive, i.e., the present Ω values preclude their use for such calculations.

The recent results on O III reported by Seaton (see Billings, Czyzak, Krueger, Saraph, Seaton, and Shemming [1965] and Saraph, Seaton, and Shemming [1966]) and on S II reported by Czyzak and Krueger (1967) show that all the previously reported collision strengths of the various ions (see Table 20) will have to be redone. This is particularly true of the very important Ω(2, 3). Saraph *et al.* (1966) report the following values: Ω(1, 2) = 2.391, Ω(1, 3) = 0.335, and Ω(2, 3) = 0.301 for O III collision strengths. The newest S II values obtained by Czyzak and Krueger, using the latest distorted wave method devised by Seaton and modified by them, for the $3p^q$ configurations are: Ω(1, 2) = 3.065, Ω(1, 3) = 1.278, and Ω(2, 3) = 6.218. Table 21 shows the contributions of the various waves for O III and S II; it is apparent that the contributions from the different waves can be quite significant, so that it cannot now be assumed, as was done earlier, that the *pp* waves always predominate.

5. SUMMARY

The method of determining Hartree-Fock self-consistent field wave functions with exchange has been developed to a high degree, and the majority of the wave functions reported in the literature do not include configuration interaction or correlation effects. Only for the lighter elements has configuration interaction been taken into account. This is usually done by applying an expansion method to the wave functions wherein each term of the expansion represents another configuration. The most recent work by Burke (1963), Weiss (1963), and Froese (1964, 1965) gives results for the energy levels ($2s^q$ and $2p^q$ configurations) that are in very good agreement with observation. For the heavier elements, however, a satisfactory technique for including configuration interaction as an integral part of the wave function calculation still remains to be found. So far, techniques are more limited by computer size than by the procedure itself. As of this writing work is in progress to include configuration interaction and relativistic effects for the heavier atoms. The wave functions, however, are quite satisfactory for problems of astrophysical interest such as transition probability and collision cross-section calculations for allowed and forbidden lines. Wave functions for various types of configurations are available and may be obtained from any of the following: Froese (1965); Mayers (1963); Krueger *et al.* (1965); and Worsley (1964).

In general, transition probability calculations are in a relatively good state. Good wave functions are available, and this is particularly important in determining the $\sigma^2(nl, n'l')$ and $s_q(nl, n'l')$ parameters. It is true, of course, that the energy levels may be appreciably affected by relatively small changes in the parameters ζ and η; and, hence, the accuracy of the wave functions is very important here. There is, however, little change with ζ, η in the transition probability values. The calculations of transition probabilities have been pursued intensively for $2p^q$, $3p^q$, and $3d^q$ configurations, but much work still remains to be done for the heavier atoms and ions.

Relatively few collision cross-sections of forbidden lines have been calculated, and the most reliable results are those obtained by the methods developed by Seaton (1953, 1966). When the results of Table 20 are examined, it becomes apparent that most are questionable. Recent calculations on O III and S II (Table 21) show that the earlier values are in considerable doubt. The latest results are accurate to approximately ± 10 per cent. The results in Table 18, calculated by the earlier method of Hebb and Menzel (1940), are highly questionable because their method violates the conservation theorem (see Bates *et al.* [1950]). In general, only relatively few collision cross-section calculations have been made, so that there is very little to go on. With the latest developments in the distorted wave and quantum-defect methods of calculation, it now appears that good results for the $2p^q$ and $3p^q$ configurations will be available shortly. However, one should not overlook the fact that one of the major difficulties in evaluating the accuracy of theoretical work is the severe lack of experimental data that are so necessary to the verification of the calculated results.

The author is indebted to Mr. T. K. Krueger, Dr. Froese, Dr. Seaton, and the editors for their very helpful comments, suggestions, and critical reading of the manuscript.

REFERENCES

Allen, C. W. 1963, *Astrophysical Quantities* (2d ed.; London: Athlone Press).
Aller, L. H. 1956, *Gaseous Nebulae* (New York: John Wiley & Sons, Inc.).
———. 1963, *Astrophysics—The Atmospheres of the Sun and Stars* (2d ed.; New York: Ronald Press Co.).
Aller, L. H., Ufford, L. W., and Vleck, J. H. van. 1949, *Ap. J.*, **109**, 42.
Bates, D. R., and Damgaard, A. 1949, *Phil. Trans. R. Soc. London*, **A242**, 101.
Bates, D. R., Fundaminsky, A., Leech, J. W., and Massey, H. S. W. 1950, *Phil. Trans. R. Soc. London*, **A243**, 93.
Bely, O., Tully, J. A., and Van Regemorter, H. 1963, *Ann. d. Phys.*, **8**, 303.
Biermann, L., and Lubeck, K. 1948, *Zs. f. Ap.*, **25** 325.
———. 1949, *ibid.*, **26**, 43.
Biermann, L., and Trefftz, E. 1949, *Zs. f. Ap.*, **26**, 213.
Billings, A., Czyzak, S. J., Krueger, T. K., Saraph, H. E., Seaton, M. J., and Shemming, J. 1965, *Fourth International Conference on Physics of Electronic and Atomic Collisions.*
Blaha, M. 1962, *Bull. Astr. Inst. Czech.*, **13**, 81.
———. 1964, *ibid.*, **15**, 33.
Bowen, I. S. 1928, *Ap. J.*, **67**, 1.
Boyce, J., Menzel, D. H., and Payne, C. 1933, *Proc. Nat. Acad. Sci.*, **19**, 581.
Boys, S. F. 1950, *Proc. R. Soc. London*, **A201**, 125.
———. 1953*a*, *ibid.*, **A217** 136.
———. 1953*b*, *ibid.*, **A217**, 235.
Boys, S. F., and Price, V. E. 1954, *Phil. Trans. R. Soc. London*, **A246**, 451.
Burke, E. A. 1963, *Phys. Rev.*, **130**, 1871.
Chandrasekhar, S. 1945, *Ap. J.*, **102**, 223.
Clementi, E. 1962, *J. Chem. Phys.*, **36**, 33.
———. 1963*a*, *ibid.*, **38**, 2248.
———. 1963*b*, *ibid.*, **39**, 175.
Condon, E. U. 1932, *Phys. Rev.*, **41**, 759.
———. 1934, *Ap. J.*, **79**, 217.
Condon, E. U., and Shortley, G. H. 1935, *The Theory of Atomic Spectra* (Cambridge: Cambridge University Press).
Czyzak, S. J. 1962, *Ap. J. Suppl.*, **7**, 53.
Czyzak, S. J., and Krueger, T. K. 1963*a*, *M.N.*, **126**, 177. Corrections in *M.N.*, **129**, 103.
———. 1963*b*, *3d International Conference on Physics of Electronic and Atomic Collisions*, p. 213.
———. 1964, unpublished data.
———. 1965, in press.
———. 1967, *Proc. Phys. Soc.*, **90**, 623.
Douglas, A. S. 1954, *Computer Methods and Programs*, unpublished.
———. 1956, *Proc. Cambridge Phil. Soc.*, **52**, 687.
Douglas, A. S., and Garstang, R. H. 1962, *Proc. Cambridge Phil. Soc.*, **58**, 377.
Douglas, A. S., Hartree, D. R., and Runciman, W. A. 1955, *Proc. Cambridge Phil. Soc.*, **51**, 486.
Einstein, A. 1917, *Phys. Zs.*, **18**, 121.
Eyring, H., Walter, J., and Kimball, G. E. 1944, *Quantum Chemistry* (New York: John Wiley & Sons, Inc.).
Fock, V. 1930, *Zs. f. Phys.*, **61**, 126.
Froese, C. 1957, *Proc. R. Soc. London*, **A239**, 311.
———. 1958, *ibid.*, **A244**, 390.
———. 1959, *Phys. Rev.*, **116**, 900.
———. 1963*a*, *Canadian J. Phys.*, **41**, 1895.
———. 1963*b*, *ibid.*, **41**, 50.
———. 1964, *Ap. J.*, **140**, 361.
———. 1965, *ibid.*, **141**, 1206 and 1557.
———. 1966, private communication.
Froese, C., and Hartree, D. R. 1957, *Proc. Cambridge Phil. Soc.*, **53**, 663.
Garstang, R. H. 1951, *M.N.*, **111**, 115.
———. 1952*a*, *Ap. J.*, **115**, 506.
———. 1952*b*, *ibid.*, **115**, 569.
———. 1956, *Proc. Cambridge Phil. Soc.*, **52**, 107.
———. 1957, *ibid.*, **53**, 214.
———. 1958, *ibid.*, **54**, 383.
———. 1962*a*, *M.N.*, **124**, 321.
———. 1962*b*, *Ann. d'ap.*, **25**, 109.
———. 1962*c*, *Atomic and Molecular Processes*, ed. D. R. Bates (New York and London: Academic Press, Inc.), 1–46.
Gaunt, J. A. 1929, *Phil. Trans. R. Soc. London*, **A228**, 195.
Gottschalk, W. H. 1948, *Ap. J.*, **108**, 326.

Goldberg, L. 1935, *Ap. J.*, **82**, 1.
———. 1936, *ibid.*, **84**, 11.
Green, L. C. 1949, *Ap. J.*, **109**, 289.
Hartree, D. R. 1928, *Proc. Cambridge Phil. Soc.*, **24**, 89 and 111.
———. 1948, *Rept. Progr. Phys.*, **11**, 113.
———. 1955, *Proc. Cambridge Phil. Soc.*, **51**, 684.
———. 1957, *The Calculation of Atomic Structures* (New York: John Wiley & Sons, Inc.).
———. 1958, *Numerical Analysis* (Oxford: Clarendon Press).
Hebb, M., and Menzel, D. 1940, *Ap. J.*, **92**, 408.
Herman, F., and Skillman, S. 1963, *Atomic Structure Calculations* (Englewood Cliffs: Prentice-Hall, Inc.).
Knox, R. S. 1957, *Bibliography of Atomic Wave Functions Solid State Physics* (New York: Academic Press, Inc.).
Krueger, T. K., and Czyzak, S. J. 1965, *Mem. R.A.S.* **69**, Part 4, p. 145.
Krueger, T. K., McDavid, W. L., and Czyzak, S. J. 1965, *ARL Tech. Rept. ARL* 65-10.
Kuhn, W. 1925, *Zs. f. Phys.*, **33**, 408.
Laporte, O. 1924, *Zs. f. Phys.*, **23**, 135.
Layzer, D. 1959, *Ann. of Phys.*, **8**, 271.
———. 1963, *Phys. Rev.*, **132**, 735.
Layzer, D., and Bahcall, J. 1962, *Ann. of Phys.*, **17**, 177.
Layzer, D., Horak, Z., Lewis, M. N., and Thompson, D. P. 1964, *Ann. of Phys.*, **29**, 101.
Layzer, D., Varsavsky, C. M., and Froese, C. 1964, *Ann. of Phys.* (in press).
Löwdin, P. O. 1953, *Phys. Rev.*, **90**, 120.
———. 1955*a*, *ibid.*, **97**, 1474.
———. 1955*b*, *ibid.*, **97**, 1490.
———. 1955*c*, *ibid.*, **97**, 1509.
———. 1962, *Rev. Mod. Phys.*, **34**, 80.
———. 1963, *J. Mol. Spectr.*, **10**, 12.
Massey, H. S. W. 1956*a*, *Hdb. d. Phys.*, ed. S. Flügge (Berlin: Springer-Verlag), Vol. 36.
———. 1956*b*, *Rev. Mod. Phys.*, **28**, 199.
Massey, H. S. W., and Burhop, E. H. S. 1952, *Electronic and Ionic Impact Phenomena* (London: Oxford University Press).
Mayers, D. 1963, private communication and unpublished results.
Mayers, D., and Hirsh, A. 1963, *C.E.I.R. Rept. to Aerospace Research Laboratories*, Wright-Patterson Air Force Base, Ohio (unpublished).
Mayers, D., and O'Brien, F. 1947*a*, unpublished results.
———. 1947*b*, unpublished results.
Mitchell, A. C. G., and Zemansky, M. W. 1961, *Resonance Radiation and Excited Atoms* (Cambridge: Cambridge University Press).
Mott, N. F., and Massey, H. S. W. 1949, *The Theory of Atomic Collisions* (2d ed.; Oxford: Clarendon Press).
Naqvi, A. M., and Talwar, S. P. 1957, *M.N.*, **117**, 463.
Nesbet, R. K. 1955*a*, *Proc. R. Soc. London*, **A230**, 312.
———. 1955*b*, *Phys. Rev.*, **100**, 228.
———. 1958*a*, *ibid.*, **109**, 1632.
———. 1958*b*, *Ann. of. Phys.*, **4**, 87.
———. 1960, *Phys. Rev.*, **119**, 658.
———. 1961*a*, *ibid.*, **122**, 1497.
———. 1961*b*, *Rev. Mod. Phys.*, **33**, 28.
———. 1963, *ibid.*, **35**, 552.
Nesbet, R. K., and Watson, R. E. 1958, *Phys. Rev.*, **110**, 1073.
———. 1960, *Ann. of Phys.*, **9**, 260.
Osterbrock, D. E. 1951, *Ap. J.*, **114**, 469.
Pasternack, S. 1940, *Ap. J.*, **92**, 129.
Pauling, L., and Wilson, E. B. 1935, *Introduction to Quantum Mechanics* (New York: McGraw-Hill Book Co., Inc.).
Percival, I. C., and Seaton, M. J. 1957, *Proc. Cambridge Phil. Soc.*, **53**, 654.
Piper, W. W. 1956, *Trans. Amer. Inst. Elec. Engr.*, **75**, Part I, 152.
———. 1957, *Bull. Amer. Phys. Soc.*, **2**, 132A, 265A.
———. 1959, *Gen. Elec. Rept. No. 59-RL-2242G* (Schenectady, N.Y.: General Electric Laboratory).
———. 1961, *Phys. Rev.*, **123**, 1281.
Racah, G. 1942*a*, *Phys. Rev.*, **61**, 186.
———. 1942*b*, *ibid.*, **62**, 438.
———. 1943, *ibid.*, **63**, 367.
Regemorter, H. van. 1960, *M.N.*, **121**, 213.
Robinson, H. H., and Shortley, G. H. 1937, *Phys. Rev.*, **52**, 713.
Rohrlich, F. 1956, *Ap. J.*, **123**, 521.
———. 1959*a*, *ibid.*, **129**, 441.
———. 1959*b*, *ibid.*, **129**, 449.

Rohrlich, F., and Pecker, C. 1963, *Ap. J.*, **138**, 1246.
Roothaan, C. C. J. 1951, *Rev. Mod. Phys.*, **23**, 69.
———. 1960, *ibid.*, **32**, 179.
Roothaan, C. C. J., Sachs, L. M., and Weiss, A. W. 1960, *Rev. Mod. Phys.*, **32**, 186.
Roothaan, C. C. J., and Weiss, A. W. 1960, *Rev. Mod. Phys.*, **32**, 194.
Rubinowicz, A. 1932, *Ergeb. Exakt. Naturwiss.*, **11**, 170.
Saraph, H. E., Seaton, M. J., and Shemming, J. 1966, *Phys. Soc. Proc.*, **89**, 27.
Schiff, L. I. 1955, *Quantum Mechanics* (2d ed.; New York: McGraw-Hill Book Co., Inc.).
Seaton, M. J. 1953*a*, *Phil. Trans. R. Soc. London*, **A245**, 469.
———. 1953*b*, *Proc. R. Soc. London*, **A218**, 400.
———. 1955*a*, *ibid.*, **A231**, 37.
———. 1955*b*, *Phys. Soc. Proc.*, **68**, 457.
———. 1956, *Contr. Inst. Astrophys. Paris*, *B*, No. 150, 289.
———. 1958, *Rev. Mod. Phys.*, **30**, 979.
———. 1960, *Rept. Prog. Phys.*, **23**, 314.
———. 1961, *Phys. Soc. Proc.*, **77**, 174.
———. 1962, *Atomic and Molecular Processes*, ed. D. R. Bates (New York and London: Academic Press, Inc.), 374–420.
———. 1963, private communication.
———. 1965, private communications.
Seaton, M. J., and Osterbrock, D. E. 1957, *Ap. J.*, **125**, 66.
Seaton, M. J., and Peach, G. 1962, *Phys. Soc. Proc.*, **79**, 1296.
Shortley, G. H. 1940, *Phys. Rev.*, **75**, 225.
Shortley, G. H., Aller, L. H., Baker, J. G., and Menzel, D. H. 1941, *Ap. J.*, **93**, 178.
Slater, J. C. 1929, *Phys. Rev.*, **34**, 1293.
———. 1937, *ibid.*, **51**, 846.
———. 1951, *ibid.*, **92**, 603.
———. 1955, *ibid.*, **98**, 1039.
———. 1960, *Quantum Theory of Atomic Structure* (New York: McGraw-Hill Book Co., Inc.), Vols. I and II.
Stevenson, A. F. 1932, *Proc. R. Soc. London*, **A137**, 298.
Thomas, W. 1925, *Naturwiss.*, **13**, 627.
Trees, R. E. 1951, *Phys. Rev.*, **83**, 756.
———. 1952, *ibid.*, **85**, 382.
Trefftz, E. 1949, *Zs. f. Ap.*, **26**, 240.
———. 1950, *ibid.*, **28**, 67.
———. 1952, *ibid.*, **29**, 287.
Trefftz, E., and Biermann, L. 1952, *Zs. f. Ap.*, **30**, 275.
Tully, J. A. 1963, Ph.D. thesis, Physics Dept., University College, London, England.
Varsavsky, C. M. 1961, *Ap. J.*, *Suppl.*, **6**, 75.
Watson, R. E. 1959, *Tech. Rept. No. 12, Solid State and Molecular Theory Group* (Massachusetts Institute of Technology).
Watson, R. E., and Freeman, A. J. 1961*a*, *Phys. Rev.*, **123**, 521.
———. 1961*b*, *ibid.*, **123**, 2027.
Weiss, A. W. 1963, *Ap. J.*, **138**, 1262.
Weisskopf, V., and Wigner, E. 1930, *Zs. f. Phys.*, **63**, 54; **65**, 18.
Worsley, B. H. 1958, *Canadian J. Phys.*, **36**, 289.

CHAPTER 9

Planetary Nebulae

L. H. ALLER
Department of Astronomy, University of California, Los Angeles
AND
WILLIAM LILLER
Harvard College Observatory

1. INTRODUCTION

AMONG the most intriguing objects in the galactic system are the planetary nebulae, so designated because their images often resemble the disks of the remote planets, Uranus and Neptune. They first attracted considerable interest because of the development of an extensive, relatively simple, physical theory of the excitation of their spectra. More recently, they have attracted renewed attention because of the role they may play in stellar evolution.

The central stars of planetary nebulae represent an advanced stage in the life history of some kind of star. Our evidence indicates that they evolve into white dwarfs, but we do not yet know whether they represent an intermediate stage for most stars or not. Neither do we know from what specific kinds of stars they may evolve.

Planetary nebulae are of great significance to galactic structure studies because (1) they are recognizable to great distances from the Sun and (2) their radial velocities may be measured in a straightforward way from their emission lines. This aspect has been reviewed in Volume V of this series by R. Minkowski.

Quantities measurable for individual planetaries are (1) angular diameter, (2) surface brightness (or in some instances calibrated isophotic contours), (3) relative brightnesses of the principal emission lines, and (4) often but not always, the brightness of the central star, and (5) sometimes a measurable radio-frequency flux.

The greatest difficulty in studies of planetary nebulae has been and remains that of getting reliable distances. Furthermore, it is not easy to obtain the interstellar extinction intensities of the Balmer lines, and/or relative intensities of certain Balmer and Paschen lines in their spectra.

1.1. SURVEY OF THE DATA

Identification or recognition of a planetary nebula at a great distance may prove difficult. Some galactic nebulae, for example, IC 1470 or NGC 7635, that were included as planetaries in older lists are now recognized as diffuse galactic nebulae or Strömgren spheres. Combination variables (cf., for example, Sahade 1960), such as Z Andromedae, show [O III], [N III], and [S II] lines characteristic of many planetaries, but such stars can often be identified by their spectral changes and by the appearance of an underlying M-type spectrum. A technique used by Minkowski (1946, 1947, 1948, 1951) and later by Henize (1951), Haro (1952), Perek (1960), Kohoutek (1964, 1965), and Pik-Sin The (1962) is to photograph fields in the light of Hα with an objective prism on the telescope. To distinguish highly reddened planetaries from Be stars, Minkowski secured direct photographs with the 60-inch and 100-inch telescopes. If the planetary is not highly reddened, its spectrum will show the green nebular lines of [O III] or the [O II] pair at 3727 A.

Star-like images produced by faint, highly reddened objects with bright central stars are, however, very difficult to distinguish from those of Be stars.

The classical discussion of planetary nebulae of *Lick Observatory Publications*, Volume 13, lists 78 objects, most of which were found by visual methods, although a few were detected by their spectra. By 1934, Vorontsov-Velyaminov listed 134 planetaries. Intensive search programs initiated by Minkowski in the early forties and extended by Henize, Haro, The, Kohoutek, and Perek as noted above have now increased the number of known planetaries to about 1034.

The statistics are far from complete for a number of reasons. The surveys certainly do not reach a fixed limiting magnitude as do many stellar programs. Near the galactic center, the crowded background of stars discriminates against objects of low surface brightness, and furthermore, the effects of space absorption are quite marked in these regions; the number of planetaries in the central region of the Milky Way is probably ten times the number actually observed.

The Palomar Sky Survey allowed Abell (1955, 1966) to identify many new objects of low surface brightness at great distances from the galactic plane. Explicit searches for new planetaries near the galactic center have been carried out with more powerful instruments than elsewhere. Recently Kohoutek (1965) used the Hamburg 80-cm Schmidt to extend the suvey into regions away from the Milky Way. Use of an objective prism permits detection of stellar planetaries which would be missed on the Palomar survey. The survey should be extended to other parts of the sky, particularly to the southern hemisphere.

Hence, the total number of planetaries is very difficult to estimate. After allowing for the effects of observational selection and absorption, Minkowski suggested that the total number would be between 3,000 and 4,000. Shklovsky (1956) suggested about 6,000; O'Dell (1963*a*), revised in Seaton (1966), estimates that the total number of planetaries with radii less than 0.7 pc is 8,500. Much higher estimates have been given by others. On the basis of present statistics, we would not expect to find more than one planetary nebula in a globular cluster. The only such object known is in M15.

1.2. Catalogues

Earlier data are contained in Vorontsov-Velyaminov's catalogue (1950), but the most comprehensive catalogue is the one recently prepared by Perek and Kohoutek (1967). This list gives 1034 objects and supplies finding charts for many nebulae which would otherwise be difficult to identify. Little other than their approximate positions is known about most of the nebulae. The tables give for each nebula (1) its 1950 position; (2) the precession in α and δ; (3) the galactic longitude and latitude on the new (II) system; (4) the descriptive type according to the classification by Vorontsov-Velyaminov (1934*a*); (5) the discoverer; and (5) data (when available) on dimension, radial velocity, magnitude, surface brightness, and spectrum. They list references to further descriptions. Radio data are listed when available. Surface brightnesses are expressed in terms of ergs cm^{-2} sec^{-1} in the monochromatic radiations of strong lines, for example, at 4955 A, 5007 A, or Hβ = 4861 A. They also give the surface brightness in terms of a circle of 1′ of arc diameter: H(mag/circle 1′) = H(mag/square 1′) + 0.26. A rough conversion from one system to the other may be made (see, for example, Aller 1963, p. 143):

$$H(\text{mag/circle } 1') = 7.57 - 2.5 \log \mathfrak{F} \quad \text{where } \mathfrak{F} = \text{flux through nebular surface in ergs cm}^{-2}\text{ sec}^{-1}\,. \tag{1}$$

Perek and Kohoutek's valuable catalogue includes references to data on spectral line broadening, doubling and splitting, internal motions, isophotes, excitation, and so forth. The last table in their series gives data for the central stars, the parallax and its probable error, the proper motion, the stellar magnitude, and the data pertaining to the spectrum.

The galactic concentration of planetary nebulae indicate that they belong to the Type II disk population (Minkowski and Abell 1963), but Type I planetaries cannot be excluded. They occur not only in the disk population but also in the extreme population Type II as represented by globular clusters. The average distance of a planetary nebula from the galactic plane has been estimated as 187 pc (Parenago 1946), 217 pc (Vorontsov-Velyaminov 1950), 280 pc (O'Dell 1963*a*), and 325 pc (Minkowski 1965*a*); the last estimate takes into account the objects of low surface brightness. According to Schmidt (1963), the average of the absolute value of the distance, z, of a main-sequence star from the plane of the Galaxy depends smoothly on the mass, in the sense that the more massive stars are more tightly concentrated toward the plane of the Galaxy. If his calibration is correct, the z-dispersion of the planetaries would indicate a mass of about 1.2 solar masses (1.2 $\mathfrak{M}_\odot$) (O'Dell 1963*a*). Minkowski's data, however, would suggest that the masses of the central stars would be 0.8 to 1.0 $\mathfrak{M}_\odot$. Hence the central star of a typical planetary nebula would be much older than the Sun, although some may be younger.

A plot of radial velocity against galactic longitude shows a very great range of line-of-sight velocities in the direction of the galactic center and a pronounced, but somewhat smaller, scatter at other longitudes. We must conclude that most planetary nebulae do not follow the usual law of galactic rotation. Minkowski (1965*a*) notes that most of them appear to move in highly elongated orbits and concludes "It is not established that, apart from those in the central region and direction of the center where the radial motions are obvious, there is a system with pure differential rotation. If such a system exists, the rotation doesn't differ much from that of the gas and Population I."

Hence, no methods in which galactic rotation effects have been used to establish distances give realistic results. Eventually, however, enough may be learned about the kinematics of planetary nebulae so that proper motion and radial velocity data may be combined to give a detailed picture of the system on the one hand and clues to individual distances on the other.

2. DISTANCES OF PLANETARY NEBULAE

Any adequate discussion of the physics of gaseous nebulae requires knowledge of their distances. This problem is not unduly severe for diffuse galactic nebulae such as Orion or M8 which are associated with stars of known luminosity or with star clusters. The situation with planetary nebulae is much more difficult. Although their galactic distribution and kinematical properties can be found, distances are much more difficult to establish.

Statistical methods based on the assumption that all nebulae are alike or differ from one another in some simple fashion that can be handled by a single, easily evaluated parameter bog down because pronounced evolutionary effects occur. We cannot usefully compare objects like NGC 40 (Plate 1) with NGC 7027, nor Campbell's hydrogen-envelope star (BD + 30°3639) or IC 4997 with the giant nebula in Aquarius, NGC 7293. See also, for example, Minkowski (1966), Aller (1954, p. 196, 1956, chap. 7), Minkowski and Osterbrock (1960).

Initially, a planetary nebula is a physically small, optically thick envelope surrounding a hot core. Only the inner portion is ionized as a Strömgren sphere. As the shell expands, the radius of the ionized sphere grows even more rapidly until finally the Strömgren sphere meets the outer boundary of the nebula, which continues to expand as a tenuous shell. As the nebula fades away, the central star likewise seems to decline in brightness (see discussions by Shklovsky 1956, Abell 1966, O'Dell 1963*a*), although the rates are not yet determined quantitatively.

In view of the evolutionary factors involved, it is clear that such a quantity as a mean parallax $\langle\pi\rangle$ or a mean absolute magnitude for the nebulae $(\bar{M}_n)$ or for central stars $(\bar{M}_s)$ would be of very limited usefulness.

2.1. Trigonometric Parallaxes

Trigonometric parallaxes are at best hopelessly small and often negative because the supposed background stars are closer than the nebula itself (see, for example, Aller 1956, p. 49).

2.2. Proper Motions

Proper motion data combined with radial velocity data still provide the soundest basis for attempts to effect a fundamental calibration of all distance methods. The older data are being superseded by modern determinations using long focal-length instruments in which positions are ultimately tied to external galaxies. Although the accuracy of individual proper motion data is greatly increased, we still face the problem of interpreting τ- and υ-components of proper motion in terms of a kinematical model for these objects.

Furthermore, we cannot usually obtain individual distances in this way. Also, objects with well-defined central stars are favored observationally, as are objects of low surface brightness.

2.3. Parallaxes of Individual Objects

In some cases it is possible to secure good determinations of the distances of individual objects. Among the special characteristics which have been employed are (1) the rate of growth of expanding shells; (2) the association of a planetary nebula with the globular cluster M15 = NGC 7078; (3) the association of a planetary nucleus with a "normal" star in a binary system, that is, the nucleus of NGC 246; (4) the presence of planetaries in the central bulge of the Galaxy; (5) the presence of planetary nebulae in external galaxies.

In the expanding-shell methods we compare the angular rate of expansion dr/dt in seconds of arc per year with the expansion velocity V_R (km/sec). The method is valid, provided the actual expansion of the shell is being observed and it has also been employed to get distances of novae (McLaughlin 1960, p. 587). If the radius of the ionized region increases at a greater (or smaller) rate than the physical radius of the shell, the method leads to spurious results (see Sec. 12).

A single planetary nebula is found in the globular cluster M15 = NGC 7078. At the generally accepted distance of this cluster, 13 kpc, its major axis of 1″ corresponds to 0.06 pc and the photographic absolute magnitude of the central star is −1.9. O'Dell, Peimbert, and Kinman (1964) conclude that the nebula is subluminous and has an abnormal chemical composition.

The central star of the faint, large planetary nebula, NGC 246, is a remarkable high-excitation object showing emission lines of O VI. The companion is similar to τ Ceti, with an absolute visual magnitude of +6.3. From the magnitude difference between the two stars, Baum and Minkowski infer that the primary has an absolute visual magnitude between +3.6 and +3.9.

It has long been known that the planetaries show a high concentration toward the central bulge of our Galaxy. Thus from measurements of their brightnesses and angular diameters, it might be possible to draw some conclusions concerning their spatial distributions and actual size (Minkowski 1951). Most of them fall in crowded regions with high space absorption, and it would be necessary to determine accurate values of the integrated magnitude and space absorption in each case. This project would be well worth undertaking.

The brightest planetaries are sufficiently luminous to be observable in the nearest of external galaxies. Baade (1955) observed five planetaries in M31, and their magnitudes have been revised by Miss Swope (1963). It is difficult to understand what B and V magnitudes actually mean for these objects in view of the monochromatic nature of their radiation. The mean value $M_B = -2.55$ must refer to the very brightest objects observed. Henize and Westerlund (1963) have found a number of small, unresolved emission nebulae that are probably true planetary nebulae in the Small Magellanic Cloud. They range upward in luminosity from $M_{pg} = -0.7$ to -3 which appears to represent an upper limit to the luminosities of these objects. Unfortunately, these data obtained from external galaxies do not seem capable of giving mean values or dispersions, at least with available equipment.

2.4. Statistical Methods for Distance Determinations

As mentioned above, the methods proposed are based on the idea of some parameter that remains reasonably constant from one object to another, for example, the absolute photographic nebular magnitude, M_n (assumed constant except perhaps for a small temperature correction). Theoretically this implies an optically thick nebula surrounding a hot central star, essentially all of whose radiation contributes to the photographic brightness of the nebula. If all these central stars are assumed to have about the same bolometric magnitude, and if all their radiation is converted to nebular emission in the photographic range, M_n would show only a small dispersion. If, however, the nebular shell is optically thin, so that a substantial portion of the radiation from the central star escapes, this assumption leads to distances that are much too large! This hypothesis, originally proposed by Zanstra (1931*b*), has since been used by Vorontsov-Velyaminov (1934*b*), Berman (1937), Parenago (1946), and others.

$\mathfrak{M}$ = *Mass of nebular shell is constant.* As the nebula expands with time and becomes more and more attenuated, it becomes optically thin and all the nebular gas becomes visible. If we assume the mass of the shell does not exceed some limiting value or is constant from one object to another, we can obtain a distance scale (Minkowski and Aller 1954; Shklovsky 1956).

There are also methods which attempt to take some account of nebular evolution in a more or less orderly way. For example, Kohoutek (1960, 1961) assumed that, at least to a first approximation,

$$\Delta M = M_n - M^*_{\text{bol}} = \text{constant}$$

and is independent of the evolution of the system. The most reasonable model envisages an orderly development of the emitting nebula from a compact Strömgren sphere surrounded by a shell of neutral gas to an attenuated ionized envelope.

Let us describe briefly some of the principal results that have been obtained. Where $\bar{M}$ denotes a mean value, Zanstra obtained $\bar{M}_n = -0.5$, $\bar{M}^*_p = 3.7$, $\bar{M}^*_{\text{bol}} = 0.0$, while Vorontsov-Velyaminov obtained $\bar{M}_n = +0.20$. Berman's elaborate treatment (1937) gave a distance scale which seemed to be of the right order for many objects, even though a few of his basic assumptions are known to be untenable. Berman's upper limit $M_n = -3.1$ for the brightnesses of planetary nebulae in the Galaxy is comparable with Henize and Westerlund's result for the Magellanic Clouds.

Since planetaries evolve from population Type II stars, their masses cannot exceed 1 $\mathfrak{M}_\odot$, and may actually be less than about 0.2 $\mathfrak{M}_\odot$. In any event, we can establish an upper limit for the distance of any nebula. These considerations form the basis of the so-called Shklovsky Method (1956), first employed by Minkowski and Aller (1954) to fix limits on the distance of the Owl Nebula, NGC 3587.

The essential argument is the following: In any particular recombination line, the emission per unit volume is proportional to the product of the ion and electron densities, that is to the square of the gas density. Hence the total emission from an emitting nebula of mass $\mathfrak{M}$ and volume V is

$$L \propto \frac{\mathfrak{M}^2}{V} \propto \frac{\mathfrak{M}^2}{d^3(r'')^3} \qquad (2)$$

where d is the distance and r'' is the angular radius of the nebula. The constancy of proportionality involves the geometrical shape of the nebula, which in simpler situations may be a shell or a sphere. If F is the flux (ergs cm^{-2} sec^{-1}) in the line, measured at the Earth, then $L \propto Fd^2$. By eliminating L and solvlng for d, we have

$$d = \text{const}\ \frac{\mathfrak{M}^{2/5}}{F^{1/5}(r'')^{3/5}} . \tag{3}$$

The mass $\mathfrak{M}$ must be assumed, but it enters only to the 2/5 power. The observed quantities F and r'' enter with the 1/5 and 3/5 powers respectively. A real uncertainty arises if the emitting gas is actually concentrated in dense filaments and we assume that it is spread out evenly.

Let the monochromatic flux in Hβ received at the top of the Earth's atmosphere from a spherical nebula of radius r (cms) and angular radius r'' be F(Hβ) ergs cm^{-2} sec^{-1}. The corresponding flux through the surface of the nebula itself will be

$$\mathfrak{F}(\mathrm{H}\beta) = F(\mathrm{H}\beta)\left(\frac{206,265}{r''}\right)^2 = \frac{\frac{4}{3}\pi r^3 E(\mathrm{H}\beta)}{4\pi r^2} = \tfrac{1}{3}E(\mathrm{H}\beta)\,r , \tag{4}$$

where E(Hβ) is the volume rate of emission in Hβ in ergs cm^{-3} sec^{-1}. We may write

$$E(\mathrm{H}\beta) = 10^{-25} E^0_{4,2}(T_e) N_i N_e \tag{5}$$

where $E^0_{4,2}(T_e)$ has been tabulated by Clarke (1965). (See Table 6.) Here N_i and N_e are the numbers of protons and electrons per cm^3, and T_e is the gas kinetic electron temperature, respectively.

Suppose that the nebular shell is completely ionized and that the ratio of helium to hydrogen is 0.17 by numbers of atoms. Then, if M_{H} is the mass of a hydrogen atom, the nebular mass will be

$$\mathfrak{M} = 1.68 \times \tfrac{4}{3}\pi r^3 N_i M_{\mathrm{H}} . \tag{6}$$

Noting that

$$r(\mathrm{cm}) = d(\mathrm{pc})r'' \times 1.496 \times 10^{13}\ \mathrm{cm},\ r(\mathrm{a.u.}) = d(\mathrm{pc})r'' ,$$

we have, since $N_e = 1.17 N_i$,

$$d = 1.72 \times 10^4 [\mathfrak{F}(\mathrm{H}\beta)]^{-1/5} \mathfrak{M}^{0.4} (r'')^{-1} [E^0_{4,2}(T_e)]^{1/5}, \tag{7}$$

where d is pc, $\mathfrak{M}$ is in solar masses, $\mathfrak{F}$ is the flux in ergs cm^{-2} sec^{-1} through the surface, r'' is the radius of the nebula in seconds of arc and $E^0_{4,2}(T_e)$ is given by Table 6.

As an illustrative example, consider NGC 7662. We adopt $\mathfrak{M} = 0.2\ \mathfrak{M}_\odot$ (Goldberg and Aller [1943, p. 177]; Shklovsky 1956), $T_e = 10{,}000^\circ$ K, $E^0_{4,2}\ (T_e) = 1.241$. $\mathfrak{F}(\mathrm{H}\beta) = 0.16$ erg cm^{-2} sec^{-1}, $r'' = 15''.2$. Then we find $d = 893$ pc. If the nebula is regarded as a shell rather than a sphere, d will be increased. If the mass of the emitting volume is less than 0.2 $\mathfrak{M}_\odot$, d will be smaller than the calculated value.

Even if one adopts the premise that the luminous shells of all planetary nebulae have exactly the same mass, difficulties in the application of the method still remain. Actual planetaries are not homogeneous. Only part of the volume may be filled by an emitting plasma, while the electron density N_e and the electron temperature T_e may vary from point to point. Furthermore, the fraction of the volume that is filled with radiating matter ϵ varies from one nebula to another. (For example, compare the nearly homogeneous Owl Nebula with an object such as NGC 40 or NGC 1501; see Aller 1956, pp. 245, 256.) See Minkowski (1964).

Kohoutek (1960, 1961, 1962) employed an iterative process in an attempt to improve Shklovsky's distance scale. He found nebular masses and distances that tended to be smaller than those found by other investigators and may be subject to large errors for individual objects. The essential spirit of his investigation that evolutionary effects must be considered in setting up a distance scale for planetary nebulae should underlie all future work on this problem.

For the Palomar survey planetaries, Abell (1966) has derived distances that are probably actually superior to those found for the better-known objects. Except for the postulate of some average mass for the shells, they do not depend on unknown assumptions, for example, the value of ϵ, unless most of the light comes from unresolved filaments of much higher than average density. From measurements of the *UBV* colors of the central stars, Abell was able to evaluate the amount of interstellar extinction.

One of the best sets of distances so far published is that given by O'Dell. In place of our equation (5), he uses an expression for $E(\mathrm{H}\beta)$ given by Menzel and his co-workers (Menzel and Aller 1941*c*, eq. [11]; Aller 1953, eq. [3]) and defines

$$f(T_e) = 10^6 T_e^{-3/2} \exp\left(\frac{9800}{T_e} - 0.98\right). \tag{8}$$

With the mass of the shell given in terms of $\mathfrak{M}_\odot$, N_e in units of cm^{-3}, $F(\mathrm{H}\beta)$ in ergs cm^{-2} sec^{-1}, the radius of the nebula R and its distance, d, in parsecs, O'Dell obtains

$$\log \overline{\mathfrak{M}} = 51.60 + 3 \log F(\mathrm{H}\beta) - 6 \log (r'') - 5 \log N_e - 2 \log \epsilon - 3 \log f(T_e) \tag{9}$$

$$\log R = 17.53 + \log F(\mathrm{H}\beta) - 2 \log (r'') - 2 \log N_e - \log \epsilon - \log f(T_e) \tag{10}$$

$$\log d = 22.85 + \log F(\mathrm{H}\beta) - 3 \log (r'') - 2 \log N_e - \log \epsilon - \log f(T_e)\,. \tag{11}$$

Eliminating N_e from these expressions, O'Dell finds

$$d = 149\, K_1^{-1} (r'')^{-3/5} F(\mathrm{H}\beta)^{-1/5}\,, \tag{12}$$

where $K_1 = [\mathfrak{M}^2 f(T_e)/\epsilon]^{-1/5}$. Another relation between $f(T_e)$, ϵ, and $\mathfrak{M}$ is given by

$$\mathfrak{F}(\mathrm{H}\beta) = 2.68 \times 10^{-7} f(T_e)\, \epsilon^{2/3} \mathfrak{M}^{1/3} N_e^{5/3} \text{ ergs cm}^{-2}\text{ sec}^{-1}\,. \tag{13}$$

Now one may estimate N_e from the 3726A/3729A [O II] lines ratio (cf. Sec. 7). Then, by plotting the observed quantity $\log \mathfrak{F}(\mathrm{H}\beta)$ against $\log N_e$, O'Dell established the value of the constant

$$K_2 = f(T_e)\epsilon^{2/3}\mathfrak{M}^{1/3} = 0.394\,. \tag{14}$$

He obtained the constant K_1 from the τ- and υ-components of proper motion, finally establishing a distance scale

$$d(\mathrm{pc}) = 75\, (r'')^{-3/5} F(\mathrm{H}\beta)^{-1/5} \tag{15}$$

whose average uncertainty he regards as within 50 per cent. $F(\mathrm{H}\beta)$ is defined by equation (4).

If T_e is taken as 10,000°, $\overline{\mathfrak{M}} = 0.14$ and $\bar{\epsilon} = 0.7$ give a good representation of the data.

There are difficulties. The non-isotropy of the nebular space motions and the great distances of many of these objects complicates the interpretation of the τ- and υ-components. The mass of the nebular shell and the factor ϵ must certainly vary from one object to another.

For our illustrative calculations we have adopted the distance scale given by Minkowski (1964). He proposes that one calculate the distance d_M on the assumption that the absolute magnitude is constant, and the distance $d_{\mathfrak{M}}$ on the alternate hypothesis that the mass of the luminous shell is constant. If the nebula is optically thick, d_M will approach the correct value whereas $d_{\mathfrak{M}}$ will be too large; that is, $d_M/d_{\mathfrak{M}} < 1$. On the other hand, for an optically thin nebula, $d_{\mathfrak{M}}$ will be the more nearly correct, d_M will be too large, and $d_M/d_{\mathfrak{M}} > 1$. Thus one uses the ratio $d_M/d_{\mathfrak{M}}$ as the discriminant between the two methods.

One must still choose the proper values of M_n and the mass $\mathfrak{M}$. Minkowski calibrated his distance scale by using some of the data for individual objects we have mentioned earlier. He assumed that M_n depends on the magnitude difference between the star and the nebula $(m_s - m_n)$ and finally adopted

$$M_n = -1.1 - 0.27\,(m_s - m_n)\,. \tag{16}$$

Actually we should write

$$m_n = -1.1 - \mathfrak{f}(T)(m_s - m_n)\,, \tag{16a}$$

where $\mathfrak{f}(T)$ has been calculated as a function of T and $m_s - m_n$ by O'Dell (private communication) as follows

$T_{\text{star}}/10^3$	30	41.5	61	87.5	125
$M_s - M_n$	−0.1	+1.7	+3.3	+4.2	+5.2
$\mathfrak{f}(T)$	0.80	0.63	0.56	0.40	0.20

In all of these discussions one must correct the observed fluxes for the effects of interstellar extinction. If measurements of both Balmer and Paschen lines are available (see Sec. 5), one can calculate the space extinction directly. For other nebulae Minkowski used data from the work of Parengo (1946), who determined local values of the photographic extinction A_{pg}/kpc on the assumption that the density of the absorbing medium depended on the distance Z from the galactic plane, as $\exp(-Z/Z_0)$.

Seaton (1966) attempted to obtain improved distances as a by-product of his evolutionary studies of planetaries and their central stars. He divided the nebulae into two groups: (*a*) those with He II lines (for which $M(\mathrm{H}^+) \sim M(\mathrm{H})_{\text{total}}$) and (*b*) those without He II lines, which are mostly optically thick. Then, using electron densities deduced from forbidden-line intensities (see Sec. 7), the Hβ surface brightness, and making due allowances for the filling factor, ϵ, he derived the radius of each nebula r(cm) from which its distance d can be found at once since the angular radius r'' is known. The total mass of the shell is adopted as $0.6\mathfrak{M}_\odot$.

In conclusion, we must admit that except for certain special objects, the distance of any given planetary is subject to considerable uncertainty. Nevertheless, the situation is greatly improved over that which existed just a few years ago when older methods gave implausibly large distances for many objects.

3. METHODS OF OBSERVATION OF PLANETARY NEBULAE

Our knowledge of any celestial body is limited by the kinds of observations that can be made. Since planetary nebulae present finite surfaces and emit radiation that is mostly monochromatic, they present a number of problems and opportunities not found in other astronomical objects.

The principal methods of observation of planetary nebulae may be summarized as follows: (a) visual, (b) direct photography, (c) spectroscopic (including special devices), (d) polarization, (e) photoelectric photometry, and (f) radio-frequency.

3.1. Visual and Direct Photographic Observations

Nearly all of the bright extended planetary nebulae were discovered by visual methods. Stellar objects or objects of low surface brightness evade discovery by visual methods. The lowest surface brightness which the eye can detect depends on the degree of dark adaptation, the absence of background illumination, the color of the object, and the observer.

In meaningful photographic studies of planetary nebulae, we take account of the circumstance that their radiation is primarily monochromatic. We use filters to isolate, for example, Hα, the green nebular lines, the 3727 A [O II] emission, or the nebular continuum. See Plates 7–12. Particularly useful are interference filters with which narrow-passband photographs may be obtained (see, e.g., Aller, 1956, pp. 12–14).

3.2. Spectroscopic Methods

Spectroscopic methods include the following techniques: (a) conventional slit spectrograph with a fast camera, with supplementary devices such as multislits, image slicers, and electronic cameras; (b) slitless spectrograph; and (c) Fabry-Perot etalon used with or without an auxiliary spectrograph.

If the nebula is of small angular size, the distribution of radiation within the monochromatic images may be studied most conveniently by means of a slitless spectrograph (Berman 1930). The technique has been used most successfully by O. C. Wilson (1958) at the coudé focus of the 100-inch telescope, employing a compensating image rotator so that details of the nebular image are not obliterated by the rotation of the image in the course of the exposure.

When a slit spectrum is to be recorded on a photographic plate, the speed of the camera is important. Thus we may use a fast camera at the coudé focus or a nebular spectrograph at the prime focus. With conventional equipment at the coudé focus, it is possible to observe only the strongest lines in the spectrum of a planetary nebula.

On the other hand, the disadvantage of going to the prime focus or to a very fast system is that the scale is greatly reduced, so that fine structure in the nebula may be lost in the plate grain. We may overcome this difficulty by working at the coudé focus with an electronic camera such as the Lallemand tube (Lallemand 1962, Hiltner 1962, Chopinet 1963, Aller and Walker 1965). In contrast to the ordinary photographic process, the relation between light intensity and plate density can be a linear one, provided proper types of developers and emulsions are used. The sensitivity varies over the surface of the cathode so that we must also photograph the spectrum of a comparison star, but this step is necessary in almost all photometry. The image converter is particularly effective in the green region of the spectrum where most emulsions are slow and, of course, in the infrared where all photographic plates are notoriously insensitive. The Lallemand image converter technique requires considerable experimental skills. Peter Boyce notes that the RCA two-stage cascaded image intensifier as distributed by the Department of Terrestrial Magnetism of the Carnegie Institution is simple to operate and is very promising.

One disadvantage of using a conventional slit spectrograph on an extended surface is that the slit covers only a part of the image and most of the nebular light goes to waste. Bowen (1938) sought to overcome this difficulty by using an image slicer, a device involving a series of mirrors that added strips of the spectrum side by side on the final plate. An alternate procedure for reaching the same result might be to use fiber optics (Kapany 1957). No such devices have yet been employed in astronomical work.

For studies of internal motions in a small nebula (or small regions of a large nebula), we may use a multislit spectrograph (Wilson 1950). In this, the single slit of a conventional spectrograph is replaced by a series of slit images. Each slit image corresponds to a different traverse of the nebula. Since the light of a gaseous nebula is concentrated in a relatively small number of strong lines, often well separated at the coudé focus, there is usually little confusion between images at different wavelengths. Hence differential motions can be studied from one part of the nebula to another in different monochromatic images. See Plate 6. Osterbrock, Miller, and Weedman (1966) have measured internal motions in a number of nebulae from emission line profiles.

The Fabry-Perot interferometer has been applied extensively to studies of internal motions in extended gaseous nebulae (Fabry and Buisson 1911; Baade, Goos, Koch, and Minkowski 1933; Courtès 1958), but its versatility also permits it to be used on planetaries with various combinations and arrangements (see especially Geake, Ring, and Woolf 1959; Davies, Ring, and Selby 1964). For example, to attain a very high resolving power, we may use a Fabry-Perot interferometer in series with a grating; this combination limits the range of the spectrum as would an interference filter. In effect, we may use an entrance aperture a hundred times larger than the slit area of a conventional grating instrument of similar size with the same spectral resolution. In the technique used by Davies *et al.*, the spectrograph is replaced by interference filters that isolate strong lines to be studied. The wavelength dependence of the transmission function is controlled by varying the pressure between the etalon plates, hence by changing the pressure at a known rate, one may sweep over short intervals of the spectrum, for example, over the 3726 A and 3729 A [O II] lines. Insofar as one may wish to investigate particular line ratios and so forth, the Fabry-Perot technique has very great advantages.

3.3. Polarization Measurements

The measurement of polarization in planetary nebulae is subject to the considerable limitations imposed by their low surface brightnesses, a difficulty that frustrated the efforts of W. F. Meyer (1920) many years ago. Since his time, the development of Polaroid and increases in the sensitivity of other techniques has made it possible to measure low levels of polarization over extended surfaces. Although Gurzadian and Razmadze (1959) found a large polarization for NGC 7026, interstellar reddening is high and the polarization is probably of interstellar origin. Measurable polarization has been found in the small nebula immediately surrounding η Carinae (Thackeray 1956, Visvanathan 1965). The most extensive and interesting polarization measurements are those obtained for the Crab Nebula, a nonthermal source.

3.4. Photoelectric Methods

Extensive photoelectric photometry of planetary nebulae has been undertaken both with filters and with spectrum scanners. Since radiations of a planetary nebula are primarily monochromatic, measurements made with conventional *UBV* filters supply only photometric gibberish, unless supplemented by careful spectrophotometry. By a proper choice of narrow-passband interference filters, it is possible to make meaningful and readily interpretable measurements (Liller 1955; see also Aller 1956, pp. 31–36).

Table 1 gives the surface brightness of a number of planetary nebulae in the light of the green nebular "N_1" line of [O III] at 5007 A. We tabulate the logarithm of the emergent flux in ergs cm^{-2} sec^{-1} on the basis of the diameter in seconds of arc as adopted in the second column. The measurements were made with the 60-inch and 100-inch telescopes at Mount Wilson in 1954.

In place of interference filters, we may use an objective prism (Liller and Aller 1954, MacRae and Stock 1954) or a scanner or some combination of interference filters and scanning (Capriotti and Daub 1960, Collins, Daub, and O'Dell 1961, O'Dell 1962, 1963*c*). Photoelectric measurements of either the green nebular line or Hβ or both now exist for about 100 planetaries, but much more work remains to be done.

A spectrum scanner has many advantages over a conventional photoelectric photometer for nebular work. The entire spectrum can be scanned or the instrument can be used as a monochromator at pre-selected wavelengths. Two extensive investigations of planetary nebulae using a spectrum scanner covering both the ordinary spectral regions and the infrared have been carried out at Mount Wilson. In 1966, using a spectrum scanner designed for the purpose by Liller (1957), Liller and Aller (1963) observed about 20 planetaries. A second series of measurements carried out during 1962 and 1963 by O'Dell (1963*b*) with the same telescope, but different scanners and photocells, concentrated on more detailed measurements of fewer objects. Other observers have studied many additional, fainter objects. During 1960 and 1961, photoelectric observations in the range of 3300 A to 5100 A were secured for 14 southern planetaries (Aller and Faulkner 1964). A basic limitation of photoelectric scanning is that although strong lines can be measured with accuracy, poor results are often obtained for weak lines, partly because of blending and partly because of the difficulty in estimating the position of the continuum on the tracings. Hence, spectral scans should be supplemented with spectrograms obtained with a conventional spectrograph.

Spectrophotometric studies of planetary nebulae have been published by Berman (1930, 1936), Page (1936, 1942) Aller (1941, 1951), Liller and Aller (1954, 1963, 1966), Andrillat (1954, 1955), Minkowski and Aller (1956), Collins, Daub, and O'Dell (1961), Osterbrock, Capriotti, and Bautz (1963), O'Dell (1963*b*), Chopinet (1963), Aller and Faulkner (1964), Aller and Kaler (1964*a*, *b*, and *c*), Aller and Walker (1965), Vorontsov-Velyaminov *et al.* (1966), Czyzak *et al.* (1966), Aller, Kaler, and Bowen (1966).

3.5. Summary

In summary, our observational data include:

a) Direct photographs (preferably in monochromatic radiations), photometrically calibrated so that isophotic contours may be derived.

b) Slit spectrograms secured for selected points in the nebula, preferably made with a fast spectrograph and ideally with dispersion sufficiently high so that the lines are not lost in the background continuum. A multislit can be used to measure internal motions in the stronger monochromatic radiations.

TABLE 1

PHOTOELECTRIC MEASUREMENTS OF PLANETARY NEBULAE

LOG S (λ5007) (ergs cm^{-2} sec^{-1})

Nebula	Diam.	Log S_1	Nebula	Diam.	Log S_1	Nebula	Diam.	Log S_1	Nebula	Diam.	Log S_1	α (1950)	δ (1950)
NGC 650	64″	−2.10	NGC 6778	16″	−1.65	IC I 289	45″	−2.96	VV5	5″	−0.68	0m25s.5	+55° 38′
NGC 1501	56	−2.34	NGC 6781	106	−2.78	IC I 351	7	−0.82	M1-1	2	−0.71	1 34.2	+50 13
NGC 2022	20	−1.57	NGC 6790	2	+0.84	IC II 1747	13	−1.37	M1-4	4	−0.69	3 38.0	+52 07
NGC 6058	25	−2.86	NGC 6803	5	−0.30	IC II 2003	5	−0.34	M2-2	10	−1.82	4 09.2	+56 49
NGC 6309	15	−1.23	NGC 6804	63	−2.76	IC II 3568	24	−1.28	J320	5	−1.10	5 02.8	+10 38
NGC 6369	28	−2.78	NGC 6807	1	+0.87	IC II 4634	10	−0.63	M2-9	40	−1.65	17 02.9	−10 04
NGC 6439	5	−0.75	NGC 6833	1	+1.14	IC II 4732	1	+0.84	M1-22	2	−1.08	17 32.2	−18 32
NGC 6445	34	−1.80	NGC 6879	5	−0.68	IC II 4776	8	−0.37	M3-11	4	−1.58	17 32.4	−20 55
NGC 6537	5	−0.75	NGC 6881	5	−1.14	IC II 4846	1	−0.99	M1-25	2	−0.66	17 35.5	−22 07
NGC 6565	10	−0.92	NGC 6884	8	−0.50	IC II 4997	1.5	+1.18	VV232	5	−0.99	17 38.8	−24 40
NGC 6567	8	−0.51	NGC 6886	6	−0.45	IC II 5117	1	+1.01	−29°13998	3	−0.10	17 44.7	−29 59
NGC 6578	8.5	−1.29	NGC 6891	10	−0.54	IC II 5217	6	+0.42	M4-9	20	−2.64	18 11.6	− 5 00
NGC 6620	5	−0.78	NGC 6894	44	−2.58				M3-25	2	−1.04	18 12.5	−10 11
NGC 6629	15	−1.18	NGC 6905	40	−1.99				M1-46	10	−2.33	18 25.1	−15 35
NGC 6644	2.5	+0.66	NGC 7008	40	−2.76				M3-27	2	−0.79	18 25.5	+14 27
NGC 6741	8	−0.65	NGC 7026	12	−0.86				M3-28	5	−1.07	18 29.9	−10 08
NGC 6751	21	−1.76	NGC 7048	60	−2.71				M2-44	5	−1.38	18 35.6	− 3 04
NGC 6772	75	−3.07	NGC 7354	20	−1.78				VV458	2	+0.40	18 47.6	+20 47
									−32°14673	4	−0.77	18 52.0	−32 20
									M1-66	2	−0.18	18 55.8	− 1 08
									M4-13	2	−1.17	19 11.4	+14 54
									M4-14	2	−1.12	19 18.6	+ 7 21
									M3-34	2	+0.20	19 24.4	− 6 41
									M1-71	2	−0.43	19 34.2	+19 36
									VV505	1	+0.86	19 36.9	+15 50
									M1-73	2	−0.48	19 38.9	+14 50
									M1-74	20	−0.06	19 40.0	+15 02
									M4-17	2	−0.73	20 07.4	+43 35
									M3-35	2	−0.73	20 19.1	+32 20
									VV553	5	−1.91	21 31.1	+39 25
										1	+1.02	22 29.6	+47 32
									M1-80	5	−1.84	22 54.2	+56 53
									VV576	1	−1.21	23 23.4	+57 55
									M2-55	30	−2.83	23 29.7	+70 06

c) Integrated intensities in the monochromatic images may be measured best by photoelectric photometry, although much earlier work was done photographically. A complete description of a monochromatic image consists of an isophotic map calibrated in terms of surface brightness in absolute units and supplemented by multislit radial velocity measurements which give the internal motions in the line of sight. Such detailed data are available for very few objects.

d) Polarization data (usually negative).

e) Radio-frequency measurements, preferably at several frequencies to isolate nonthermal radiations that may occur. See Section 9.

4. ORIGIN OF SPECTRA OF PLANETARY NEBULAE

The principal spectroscopic features of planetary nebulae have been described in many articles. (For a succinct description and references to earlier literature, see e.g., Aller 1956, pp. 57–107.) The spectral lines arise from the following mechanisms.

4.1. Primary Mechanisms

An atom or ion may be photo-ionized by far-ultraviolet quanta. When the electrons are recaptured in highly excited levels, they may cascade to lower levels with the emission of familiar permitted lines. The prominent hydrogen and helium lines are produced in this manner, as are the permitted lines of carbon, nitrogen, oxygen, and neon, which are much weaker because of the lower abundances of these elements.

4.2. Collisional Excitation

The strongest lines in the spectra of most planetary nebulae are the so-called *forbidden* lines, originating in magnetic-dipole and/or electric-quadrupole transitions between terms of the ground configuration of a relatively abundant ion. Bowen suggested that the great strength of these lines in gaseous nebulae could be understood if electrons excited the ions to low-lying metastable levels by inelastic collisions and if, as the ions cascaded back to lower levels, they emitted these forbidden transitions. The forbidden lines in gaseous nebulae attain great strength as compared with those in laboratory emissions, not because the metastable levels are de-excited by collisions with the walls in the latter situation, but rather because the radiating volume of the nebula is so very large compared with the laboratory source.

Nonetheless, it must be emphasized that planetary nebulae are essentially transparent to even the strongest forbidden transitions, such as the green nebular lines of [O III] (see Aller 1956, p. 80).

Except in the "satellite" ultraviolet, $\lambda < 2900$ A, few permitted lines can be excited in this way because for the relevant higher levels of all ions that are likely to be present their upper values lie so far above the ground level. One exception is 4571 A $3^1S - 3^3P$ transition of Mg I, whose upper level at 2.70 ev is almost certainly excited by electron collisions. This intercombination line is a permitted transition of very low transition probability, with $f = 24 \times 10^{-7}$ (see Allen 1963).

4.3. Bowen's Fluorescent Mechanism

In the spectra of many high-excitation planetaries certain permitted lines of O III and N III appear with considerable intensity, while other lines, equally strong (or even stronger) under conditions of laboratory excitation or by recombination, are missing.

Bowen (1934) noticed that all the O III lines could be produced by atoms cascading from the $2p3d\ ^3P_2$ level, and that there was a close coincidence between the frequency of the transitions from the $2p^2\ ^3P_2$ level of O III to the $2p3d\ ^3P_2$ level on the one hand, and that of the resonance Lyα transition of ionized helium on the other. Hence, in high-excitation planetaries where helium is doubly ionized and the Lyα of ionized helium attains great strength, O III ions could easily be excited to the $2p3d\ ^3P_2$ level from which they could cascade, emitting the characteristic "fractional multiplets" observed in planetaries. An equally surprising coincidence is that the final transition of the O III fluorescent cycle, $2p3s\ ^3P_1^0$–$2p^2\ ^3P_2$, has a wavelength 374.436 A so nearly coincident with a resonance line 374.442 A ($2p\ ^3P_{3/2}$–$3d\ ^2D_{3/2}$) of N III that it produces a somewhat similar fluorescent cycle in this ion.

Detailed analysis shows that Bowen's fluorescent mechanism works not only qualitatively, but also quantitatively (Menzel and Aller 1941*a*, Hatanaka 1944, Unno 1955, Burgess and Seaton 1960*a*). Since the wavelength of the He II Lyα line, 303.780 A, differs sensibly from the O III 303.799 A line, the efficiency of the mechanism might be expected to be low. Actually, since a given He II quantum may be scattered several times, the resultant efficiency of the O III fluorescence cycle may build up to as high as 40 to 50 per cent in some nebulae (Seaton 1960*a*). The N III cycle may show an even higher resultant efficiency.

4.4. The Continuum

In addition to discrete-line radiation, planetaries emit a characteristic continuous emission which originates from the following physical mechanisms:

a) Recombination of electrons to excited levels of hydrogen and helium (see Sec. 5).

b) Free-free transitions involving kinetic energy losses of free electrons in the electrostatic fields of ions. These free-free emissions are responsible for the thermal radiation observed in the radio-frequency spectra of the planetaries.

c) The 2-quantum emissions produced by hydrogen atoms cascading from the 2*s*-level to the ground level (Mayer 1931, Spitzer and Greenstein 1951).

Other possible sources of a nebular continuum, for example, emission of the negative hydrogen ion, electron scattering, and fluorescence of small particles have been shown to be negligible. Electron scattering cannot be significant because the total amount of energy from the central star scattered by electrons per unit volume is small compared to the continuum radiation produced by other processes. Presumably, planetary nebular shells contain no solid particles that can produce a continuum by fluorescence. Some planetaries sometimes have been suspected to be nonthermal sources of radio-frequency emissions, but it is probable that most if not all of such objects are confused with nearby nonthermal sources.

The most striking, immediately evident feature of the continuous spectra of planetary nebulae is the intensity jump at the head of the Balmer series, produced by recombinations of electrons and ions in the second level of hydrogen. A similar, but smaller, break at the Paschen limit is also observed.

4.5. Differences between the Spectra of Different Objects

Differences from one object to another in the spectra of planetary nebulae may be ascribed almost entirely to differences in excitation conditions, although chemical composition may play a secondary role. Low-excitation planetaries such as NGC 40 are characterized by a strong [O II] pair at 3727 A, prominent Balmer lines, and weak green nebular lines. See Plate 3. As the level of excitation increases, the green nebular lines at 5007 A and 4959 A dominate the spectrum and the pair at 3727 A becomes less important; lines of [Ne III] and eventually [Ne V] become prominent. In objects of the highest excitation, the He II line at 4686 A becomes comparable with Hβ, [Ne V] is strong, and the lines of the Bowen fluorescent mechanisms become conspicuous. Excitation classes may be assigned on the basis of these criteria (Fig. 1; see also Aller 1956, p. 66; Page 1942).

It must be emphasized that stratification effects within a nebula have a profound influence on the appearance of the spectrum. A high-excitation inner region may be surrounded by a low-excitation envelope, a circumstance which is readily explained by the

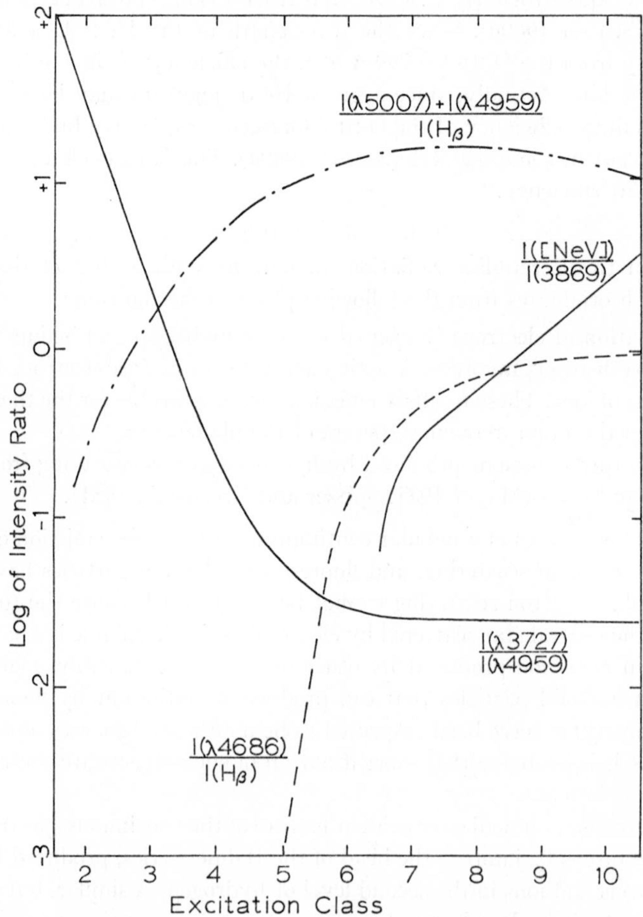

FIG. 1.—Intensity ratios used to define the excitation class of a gaseous nebula. Notice that the $I(3727)/I(4959)$ ratio provides an excellent criterion at low excitation levels; $I(\lambda\ 4686)/I(H\beta)$ is valuable at intermediate to moderately high excitation, whereas $I([Ne\ V])/I(3869 = [Ne\ III])$ is useful at the very highest excitations.

systematic absorption of the highest energy quanta in the inner region of the nebula (Bowen 1928). Optically thick nebulae are more often adequately described by a single excitation class than are optically thin nebulae (for example, the Ring Nebula, NGC 6720 or NGC 7008 [see Plate 5]) which often have strong 4686 A in the inner region surrounded by a shell with strong 3727 A [O II] (Minkowski 1942, Page 1942). These circumstances emphasize the need for spectroscopic descriptions and line-intensity measurements that refer to specific identifiable regions in planetary nebulae.

Earlier spectroscopic studies made by slitless spectrographs and low-dispersion slit spectrographs have been extended as new equipment has been developed. Since the classical work of Bowen and Wyse (1939) and Wyse (1942) with the image slicer, fast modern nebular spectrographs have become available. The brightest planetaries can be studied with coudé spectrographs, supplemented sometimes with electronic cameras. NGC 7027 (Plate 8) has the richest spectrum in number of lines, but it is less satisfactory for theoretical studies than a number of others. NGC 7662 is the most thoroughly studied of high-excitation planetaries; it is very favorably placed for observation (see Plate 1). IC 418 is the brightest of low-excitation planetaries; its spectrum rather closely resembles that of the Orion Nebula (Kaler, Aller, and Bowen 1965), whereas NGC 7009 (see Plate 2) shows a wealth of recombination lines of oxygen and nitrogen. Since many of the most interesting planetaries have low surface brightnesses and therefore intrinsically very faint lines, the importance of such techniques as the electronic camera (especially as applied to nebular spectrographs and so forth) cannot be overemphasized.

4.6. The Ultraviolet and Infrared Spectra of Planetary Nebulae

The extremely large red shifts of certain quasi-stellar objects (QSO's) cause emission lines that fall normally in the inaccessible ultraviolet to be displaced into the visible region (Schmidt 1965). The observed transitions include λ 2326 C II, λ 1909 C IV, λ 1550 C IV, and even Lyα λ 1216.

Although conditions in a QSO may differ considerably from those existing in a planetary nebula (Bahcall 1966), these lines are probably important there also. Further, since space telescopes will soon be available, it is of interest to predict the ultraviolet and infrared spectra of planetary nebulae. Such predictions have been made by Code (1960), Aller (1961), and most thoroughly by Osterbrock (1963). Of recombination lines, only those of helium are important but a large number of collisionally excited lines are likely to be significant; these include transitions of C II, C III, C IV, N II, N III, N V, Mg II, Si II, Si III, Si IV, S II, S III, S IV, S VI, Ar II, and Ar III. The infrared spectral region probably contains lines of [Mg IV] (4.492 μ), [S IV] (10.58 μ), [Ne III] (15.38 μ) and [Ne V] (14.33 μ), while the [Ne II] line at 12.8 μ may be quite strong in low-excitation objects such as IC 418 (Gould 1966).

5. THE PRIMARY MECHANISM: THE RECOMBINATION LINES OF HYDROGEN AND HELIUM

5.1. The Hydrogen Lines in Optically Thin Nebulae

As an initial model of a planetary nebula, we consider a thin shell of attenuated gas surrounding a hot star. The density of the gas, which is mostly hydrogen and helium, is about $10^2 - 10^5$ atoms/cm^3. Since the radius of the shell is very large compared with that of the illuminating star, the radiation field is greatly attenuated. If the star radiates as a blackbody of temperature T_1 beyond the Lyman limit, the geometrical dilution factor will be

$$W_g = \frac{I_\nu}{B_\nu(T_1)} = \frac{\pi R_s^2 B_\nu(T_1)}{4\pi r^2 B_\nu(T_1)} = \frac{1}{4}\left(\frac{R_s}{r}\right)^2, \quad (17)$$

where R_s is the radius of the star and r is the radius of the nebular shell. If $R_s = R_\odot = 6.96 \times 10^{10}$ cm and $r = 10{,}000$ a.u. $= 1.494 \times 10^{17}$ cm, then $W_g = 5.4 \times 10^{-14}$. Consequently, a volume element in the shell is bathed by radiation whose energy *distribution* corresponds to a high temperature (for example, 35,000 to 100,000° K), although the energy *density* corresponds to a very low equivalent temperature. The dilution factor may be modified further by radiative processes occurring within the nebula.

Thus, as was pointed out long ago by Zanstra (1926) and independently by Menzel (1926), photo-ionization may occur only from the ground level, whereas recaptures may occur on *all* levels. If the nebula is optically thick, so that Lyman quanta are repeatedly re-absorbed and scattered, each Lyman line or continuous quantum is eventually degraded into a Balmer plus perhaps higher series quanta and a Lyα quantum. For example, a recapture on level 4 can result in emission of Lyγ or Hβ + Lyα. If Lyγ is emitted, there is a finite chance that upon the next absorption it will be broken down into Paschen α + Hα + Lyα or Paschen α + Lyβ or Hβ + Lyα. Eventually every Lyman quantum yields one Balmer quantum and one Lyα quantum. That is, for a thick nebula, the number of ultraviolet stellar quanta beyond the Lyman limit must equal the number of Balmer quanta emitted by the nebula. Hence, by comparing the number of Balmer quanta with the number of stellar quanta emitted at the same frequencies in the visual region, we may determine the color temperature of the central star. In this way Zanstra was able to show that the nuclei of planetary nebulae often had temperatures in the range of 25,000 to 100,000° K.

We postulate that the nebula is in a steady state, that is, that the number of atoms in a given level does not change with time. On the basis of this condition of statistical equilibrium, we may calculate the population in each of the excited levels and predict the intensities of recombination lines. In particular we may calculate the relative intensities of the hydrogen lines and predict the Balmer and Paschen intensity decrements.

Early work by Plaskett (1928), Carroll (1930), and Cillié (1932, 1936) was followed by extensive calculations by Menzel and Baker (1937, 1938), whose results were revised and corrected by Seaton (1959*a*). In these computations, all levels of a given n were grouped together and the dependence of the population on the azimuthal quantum number l was not investigated. Let $F_{n'n}$ denote the number of radiative excitations cm^{-3} sec^{-1} from a lower level n' to a higher level n; $F_{nn'}$, the number of radiative de-excitations from level n to n'; and $F_{n\kappa}d\nu$, the number of photo-ionizations from level n by absorption of quanta between ν and $\nu + d\nu$. Conversely, let $F_{\kappa n}d\nu$ denote the number of radiative recaptures on level n with the emission of quanta between ν and $\nu + d\nu$. Because of the extreme dilution of the radiation, excitations and ionizations from levels higher than the first can be neglected. The condition of statistical equilibrium for level n expresses the fact that the number of atoms entering level n by direct recombination from the continuum, cascading from higher levels, n'', and by excitation by line radiation from the ground level, 1, must equal the total number cascading from level n to all lower levels, n'.

$$\int_{\nu_n}^{\infty} F_{\kappa n} d\nu \quad + \quad \Sigma F_{n''n} \quad + \quad F_{1n} \quad = \sum_{1}^{n-1} F_{nn'} . \quad (18)$$

Recombination to level n	Cascades from higher levels	Excitation from ground level	Cascade to lower levels

In this treatment, collisional processes are neglected, although they can be taken into account when necessary. See page 521.

Menzel (1937) showed that the calculations would be simplified if we referred the populations of the atomic levels and the emission rates to the gas kinetic electron temperature of the gas, that is, we take advantage of the fact that the velocity distribution of the free electrons corresponds very closely to the Maxwellian distribution for some temperature, T_e, defined as the electron temperature (Bohm and Aller 1947).

The population of atoms in a level n may be expressed in terms of the ionic density N_i, the electron density N_e, and the electron temperature by the combined Boltzmann and Saha equations, viz.,

$$N_n = b_n N_i N_e \frac{h^3}{(2\pi m k T_e)^{3/2}} \frac{\varpi_n}{2} e^{X_n}, \quad (19)$$

where

$$X_n = \frac{hRZ^2}{n^2 k T_e} = \frac{158{,}000}{n^2 T_e} . \quad (20)$$

Here ϖ_n is the statistical weight of the level n; that is, $2n^2$, R is Rydberg's constant; h, m, and k have their usual meanings; and b_n is the factor that measures the degree of departure from thermodynamic equilibrium at temperature T_e.

The intensity of the radiation I_ν may be written in the form

$$I_\nu = W_\nu \frac{2h\nu^3}{c^2} \frac{1}{\exp(h\nu/kT_1) - 1} . \quad (21)$$

Noting that the oscillator strength $f_{n'n}$ for a transition between levels n' and n in hydrogen is given by (Menzel and Pekeris 1935)

$$f_{n'n} = \frac{2^6}{3\sqrt{3}\pi} \frac{1}{\varpi_{n'}} \frac{R^3 Z^6}{\nu^3_{n'n}} \left| \frac{1}{n^3} \frac{1}{n'^3} \right| g_{nn'} \quad (22)$$

and making use of the well-known relationship between the Einstein absorption coefficient, B, and the f-value (see, for example, Aller 1963, p. 175), we have

$$\begin{aligned} F_{n'n} &= N_{n'} B_{n'n} I_\nu = 4\pi N_{n'} I_\nu \frac{\pi\epsilon^2}{mc} f_{n'n} \frac{1}{h\nu_{n'n}} \\ &= N_i N_e W_\nu \frac{KZ^4}{T_e^{3/2}} \frac{b_{n'} g_{n'n}}{n'} \frac{e^{h\nu/kT_e}}{e^{h\nu/kT_1} - 1} \frac{2e^{X_n}}{n(n^2 - n'^2)}, \end{aligned} \quad (23)$$

where we have used equations (19), (20), (21), and (22). Here K is a coefficient whose numerical value is 3.26×10^{-6}. The number of downward jumps cm^{-3} sec^{-1} from level n to n' is

$$F_{nn'} = N_n A_{nn'} = N_i N_e \frac{KZ^4}{T_e^{3/2}} b_n \frac{g}{n'} \frac{2}{n} \frac{e^{X_n}}{(n^2 - n'^2)}, \quad (24)$$

where we have used the relation between the Einstein coefficient of spontaneous emission, $A_{nn'}$, and the f-value, viz.,

$$A_{nn'}=\frac{\varpi_{n'}}{\varpi_n}\frac{8\pi^2\epsilon^2\nu^2}{m c^3}f_{n'n} \tag{25}$$

and equations (19), (20, (21), and (22). It may be shown that the number of radiative recaptures cm^{-3} sec^{-1} on a level n is given by

$$\int_{\nu_n}^{\infty}F_{\kappa n}d\nu=N_iN_e\frac{KZ^4}{T_e^{3/2}}\frac{\langle g_{\rm II}\rangle}{n^3}S_n, \tag{26}$$

where

$$S_n=e^{X_n}E_1(X_n) \quad \text{and} \quad E_1(x)=\int_1^{\infty}\frac{e^{-yx}}{y}dy. \tag{27}$$

There $\langle g_{\rm II}\rangle$ is an approximate mean value of the bound-free Gaunt factor.

The population of a level n will depend on the character of the radiation field both in the lines and in the continuum. In Menzel and Baker's case A, the line radiation that comes from the star or that is generated within the nebula itself by radiative transformations may be neglected.

The equation of statistical equilibrium then becomes

$$\sum_{n''=n+1}^{\infty}F_{n''n}+\int_{\nu_n}^{\infty}F_{\kappa n}d\nu=\sum_{n'=1}^{n-1}F_{nn'}. \tag{28}$$

Now put

$$u_{n''n}=\frac{2n^2}{n''(n''^2-n^2)}, \qquad t_n=\sum_{1}^{n-1}-u_{n'n}g_{nn'}. \tag{29}$$

Each level yields an equation of the form

$$N_iN_e\frac{KZ^4}{T_e^{3/2}}\left(\sum_{n''=n+1}^{\infty}b_{n''}e^{X_{n''}}g_{n''n}u_{n''n}+\langle g_{\rm II}\rangle S_n-b_ne^{X_n}t_n\right)=0, \tag{30}$$

which must be solved by some systematic procedure for the b_n's.

In Menzel and Baker's case B, the nebula is presumed to be optically thick, and excitation by Lyman-line absorption is now important since Lyman continuum quanta are broken down into Lyman-line radiation and quanta in subordinate lines and continua which escape directly from the nebula. We assume the optical thickness is sufficient so that the number of absorptions from level 1 to n is exactly compensated by the number of emissions from level n to 1.

Since $F_{n1}=F_{1n}$,

$$\sum_{n''=n+1}^{\infty}F_{n''n}+\int_{\nu_n}^{\infty}F_{\kappa n}d\nu=\sum_{2}^{n-1}F_{nn'}. \tag{31}$$

By redefining t_n as

$$t_n=\sum_{2}^{n-1}-u_{n'n}g_{nn'}, \tag{32}$$

equation (30) may be solved exactly as before. Menzel and Baker took advantage of the the fact that $b_n \to 1$ as $n \to \infty$, and calculated b_n by a method of successive approximations.

Seaton (1959*b*) generalized a procedure proposed by Plaskett to calculate the rate at which atoms enter a level n. Consider case A in which atoms enter a level n either by direct capture of electrons from the continuum on that level or by capture on some higher level n'' and subsequent cascade to n. If a cascade coefficient $C_{n''n}$ is defined as the probability that capture on a level n'' is followed by a transition to n when all different modes of cascade are considered, atoms will enter level n at a rate

$$F_n = N_i N_e \sum_{n''>n} \alpha_{n''}(T_e) C_{n''n}, \tag{33}$$

where $\alpha_{n''}(T_e)$ is the recombination coefficient for level n. Clearly $C_{nn} = 1$. Seaton defines

$$A_n = \sum_{n'=1}^{n-1} A_{nn'}, \qquad P_{nn'} = A_{nn'}/A_n, \tag{34}$$

where $A_{nn'}$ is given by equation (25). Then

$$C_{n''n} = \sum_{m=n}^{n''-1} P_{n''m} C_{mn}. \tag{35}$$

Note that the cascade coefficients do not depend on the temperature and may be calculated once and for all. The number of quanta emitted in the $n - n'$ transition cm^{-3} sec^{-1} will be

$$Q_{nn'} = P_{nn'} F_n = N_i N_e P_{nn'} \sum_{n''>n} \alpha_{n''} C_{n''n} = N_i N_e \alpha_{nn'}, \tag{36}$$

where Seaton defines

$$\alpha_{nn'} = P_{nn'} \sum_{n''>n} \alpha_{n''} C_{n''n} \tag{37}$$

as the effective recombination coefficient for the $n \to n'$ transition. For case B we may employ the same procedure except that $n' = 2$ replaces $n' = 1$ as the lower limit in the expression for A_n.

The analysis by Menzel and Baker is oversimplified in that no account is taken of the dependence of the factor b as on l well as on n. Their solution amounts to requiring that l-states are populated according to their statistical weights, viz.,

$$N(n, l) = \frac{2l+1}{n^2} N(n) \cdot \qquad N(n) = \sum_{l=0}^{n-1} N(n, l). \tag{38}$$

The problem of the l-dependence has been considered by Burgess (1958), by Searle (1958), by Seaton (1959*b*), by Pengelly (1964) and by Clarke (1965). In place of equation (31) we now have

$$\sum_{n''=n+1}^{\infty} \sum_{l''=l\pm1} N(n'', l'', T_e) A(n''l''; nl) + N_i N_e \alpha(nl; T_e) = N(nl; T_e) A(n, l), \tag{39}$$

where $\alpha(nl;T_e)$ denotes the radiative recombination coefficient for level (n,l). Following Pengelly, in analogy to equation (34) we define

$$A(n,l) = \sum_{n'=n_0}^{n-1} \sum_{l'=l\pm 1} A(nl;n'l'); \qquad P(n''l'';nl) = \frac{A(n''l;nl)}{A(n''l'')}, \tag{40}$$

where $n_0 = 1$ for case A and $n_0 = 2$ for case B. The equation of statistical equilibrium may be written in the form

$$N(n,l) = \frac{N_i N_e}{A(nl)} \sum_{n''=n}^{\infty} \sum_{l''=0}^{n''-1} \alpha(n''l'')C(n''l'';nl) \tag{41}$$

where

$$C(n''l'';nl) = \sum_{n'=n}^{n''-1} \sum_{l''=l\pm 1} P(n''l'';n'l')C(n'l';nl) \tag{42}$$

(compare eq. [35]).

The solution of equation (38) requires calculation of the $A(nl;n'l')$'s and $\alpha(n,l)$'s which turns out to be a formidable numerical task, chiefly because of cancellation between nearly equal terms of opposite sign. Burgess (1964) has discussed the problem of the recombination coefficient in detail. Clarke found that by applying recursion relations he could overcome these cancellation difficulties and calculate values up to $n = 80$.

Equation (39) expresses the condition of statistical equilibrium for a level (n,l) under conditions of strict radiative equilibrium. Collisions may modify this situation in several ways. At high electron temperatures, inelastic collisions will excite atoms directly from the ground level to any level n and produce collisional ionizations as well. At lower electron temperatures typical of most planetary nebulae, collisions can act to reshuffle the atoms among levels of different l and even levels of different n. Menzel and Baker's (1937) and Seaton's (1959*b*) hypotheses clearly require enough collisions to insure that each l-level is populated according to its statistical weight (eq. [38]). According to calculations by Seaton (1964) and Pengelly and Seaton (1964), collisions between excited atoms and protons can be effective in shuffling atoms from a level (n,l) to $(n,l \pm 1)$. If $T_e \sim 10{,}000°$ K and $N_e \sim 10^4$ per cm^3 and $n = 15$, collisional depopulations are comparable with radiative transitions. Hence the levels tend to be populated according to their statistical weights. If n is about 38, collisional transitions of the type $\mathrm{H}(n) + e \rightarrow \mathrm{H}(n+1) + e$ compete in significance to the sum of all radiative escapes from level n. Hence $b(n)$ approaches $b(n+1)$. If n is as large as 82, collisional ionization and 3-body recombination become as important as all possible downward transitions forcing $b(n)$ to 1. Since Balmer lines may be observed in some planetaries to n-values as high as 35 (Aller, Bowen, and Wilson 1963), we must take into account the influence of collisions upon the populations of the excited levels.

Clarke divided the energy levels into four "domains." For n between two (case A) or three (case B) and n_{coll} = typically 20, $b(n,l;\,T_e)$ was calculated by the radiative theory. Then between n_{coll} (the value of n at which collisions become important) and $n_{\max}$ (where b_n approaches 1; typically $n_{\mathrm{coll}} = 20$ and $n_{\max} = 60$) he computed b_n by the "Menzel-Baker Hypothesis," that is, assuming the collisional effects cause the atoms to be distributed among the l-levels according to their statistical weights. Then beyond

$n = 60$, he assumed $b_n = 1$. At what point is the calculation to be truncated; that is, what is the largest possible value of $n = n_{\mathrm{lim}}$? If we use Griem's (1964) modification of the Inglis-Teller formula, viz.,

$$\log 2N = 22.82 - 7.5 \log n_m, \tag{43}$$

we find that for $N = 10^4$, $n_m = 290$. Transitions between $n = 110$ and $n' = 109$ have been observed by Hoglund and Mezger (1965). Clarke (1965) considered a variety of situations in addition to case A and case B as noted above—the pure Menzel-Baker case (that is, $n_{\mathrm{coll}} = 0$), which agrees very well with Seaton's (1959*b*) results, and the pure radiative case where no collisions are considered up to $n = 60$.

In the standard case A where b depends on l up to $n = 20$, the $l = 1$ levels are underpopulated because of the large transition probability to the ground level. In case B, $l = 0,1$ levels are underpopulated because of direct radiative transitions to $n = 2$ yielding Balmer quanta.

The Balmer decrement is given by

$$\frac{I(n,2)}{I(4,2)} = \frac{e^{X_n}\nu(n,2)}{e^{X_4}\nu(4,2)} \frac{\sum_{l'=0}^{1} \sum_{l=l'\pm 1} (2l+1)\, b(n,l;T_e)\, A(nl;2l')}{\sum_{l'=0} \sum_{l=l'\pm 1} (2l+1)\, b(4,l;T_e)\, A(4l;2l')}. \tag{44}$$

Corresponding expressions may be derived for the Paschen series and for the Pickering series of He II. Note, however, that the temperature T_e corresponding to an He II decrement corresponds to $\frac{1}{4}T_e$ for hydrogen.

Clarke's (1965, Table 3) results for the Balmer decrement are presented in Table 2, "Comparison of the Balmer Decrements Obtained by Various Methods." Column 1 gives n; columns 2 and 3 the Menzel-Baker method $b(n)$ decrement; columns 4 and 5 the $b(n,l)$ values for the combined $b(n,l)$ method with $n_{\mathrm{coll}} = 10$, $n_{\mathrm{max}} = 60$, $n_{\mathrm{lim}} = 700$; columns 6 and 7 give similar values for $n_{\mathrm{coll}} = 20$; and columns 8 and 9 give the decrements obtained using only the $b(n,l)$ method (that is, $n_{\mathrm{coll}} = 60$). The decrement is *not* sensitive to the upper limit n_{lim}.

The results of the calculations are quite informative. For the Menzel-Baker $b(n)$ methods, the decrements show considerable differences between cases A and B in that the B decrement is the steeper. Note, however, that the pure $b(n,l)$ decrements are steep and that there is *no* significant difference between cases A and B. The decrements which we consider to be more realistic, that is, those in which collisional effects are invoked for the higher levels, lie between these two extremes. A hiatus appears where collisional redistribution is suddenly "turned on." Lines arising from levels below the changeover point fall close to the radiative gradient, whereas those arising from levels where a collisional redistribution has been postulated show a more gentle decrement. The decrements obtained by using the pure $b(n,l)$ method follow closely those obtained by Pengelly (1964); there is no detectable difference between cases A and B. It does not make much difference if n_{lim} is taken as 80 or 700, or even if b_n is set equal to 1 at 40 rather than 60.

How well do the predicted decrements agree with the observations? Let us lay aside for the moment the problem of the Hα:Hβ:Hγ ratio and consider the higher members of the series. Observations by Berman (1936) and by Aller (1951) suggested that case B

TABLE 2

COMPARISON OF THE BALMER DECREMENTS OBTAINED BY VARIOUS METHODS

(T_e = 10,000° K; n_{max} = 60; n_{lim} = 700)

	$b(n)$ METHOD		COMBINED $b(n, l)$ AND $b(n)$ METHOD				$b(n, l)$ METHOD	
	$n_{coll}=0$		$n_{coll}=10$		$n_{coll}=20$		$n_{coll}=60$	
	A	B	A	B	A	B	A	B
n 3	1.9092E 02	2.7054E 02	2.6437E 02	2.7823E 02	2.7619E 02	2.8405E 02	2.8543E 02	2.8838E 02
4	1.0000E 02	1.0000E 02	1.0000E 02	1.0000E 02	1.0000E 02	1.0000E 02	1.0000E 02	1.0000E 02
5	5.8882E 01	5.0555E 01	4.8254E 01	4.7206E 01	4.7594E 01	4.6928E 01	4.7304E 01	4.6809E 01
6	3.7814E 01	2.9744E 01	2.7131E 01	2.6179E 01	2.6595E 01	2.5969E 01	2.6369E 01	2.5879E 01
7	2.5841E 01	1.9188E 01	1.6851E 01	1.6097E 01	1.6459E 01	1.5954E 01	1.6298E 01	1.5892E 01
8	1.8497E 01	1.3180E 01	1.1219E 01	1.0634E 01	1.0932E 01	1.0536E 01	1.0816E 01	1.0492E 01
9	1.3725E 01	9.4804E 00	7.8635E 00	7.4041E 00	7.6483E 00	7.3372E 00	7.5616E 00	7.3049E 00
10	1.0482E 01	7.0651E 00	5.7346E 00	5.3675E 00	5.5683E 00	5.3213E 00	5.5021E 00	5.2970E 00
11	8.1972E 00	5.4159E 00	5.7688E 00	5.3862E 00	4.1842E 00	3.9856E 00	4.1324E 00	3.9668E 00
12	6.5380E 00	4.2486E 00	4.6012E 00	4.2253E 00	3.2261E 00	3.0644E 00	3.1846E 00	3.0495E 00
13	5.3029E 00	3.3977E 00	3.7319E 00	3.3791E 00	2.5411E 00	2.4080E 00	2.5074E 00	2.3959E 00
14	4.3636E 00	2.7621E 00	3.0709E 00	2.7470E 00	2.0381E 00	1.9272E 00	2.0101E 00	1.9173E 00
15	3.6360E 00	2.2772E 00	2.5588E 00	2.2647E 00	1.6601E 00	1.5668E 00	1.6367E 00	1.5585E 00
16	3.0633E 00	1.9006E 00	2.1558E 00	1.8901E 00	1.3706E 00	1.2913E 00	1.3506E 00	1.2843E 00
17	2.6062E 00	1.6034E 00	1.8341E 00	1.5946E 00	1.1449E 00	1.0770E 00	1.1277E 00	1.0709E 00
18	2.2367E 00	1.3656E 00	1.5741E 00	1.3582E 00	9.6638E-01	9.0777E-01	9.5136E-01	9.0248E-01
19	1.9346E 00	1.1731E 00	1.3615E 00	1.1666E 00	8.2329E-01	7.7229E-01	8.1005E-01	7.6764E-01
20	1.6852E 00	1.0154E 00	1.1860E 00	1.0099E 00	7.0726E-01	6.6256E-01	6.9548E-01	6.5846E-01
21	1.4775E 00	8.8507E-01	1.0398E 00	8.8022E-01	1.0071E 00	8.8246E-01	6.0157E-01	5.6906E-01
22	1.3029E 00	7.7630E-01	9.1692E-01	7.7205E-01	8.8813E-01	7.7402E-01	5.2386E-01	4.9517E-01
23	1.1551E 00	6.8482E-01	8.1293E-01	6.8106E-01	7.8741E-01	6.8280E-01	4.5900E-01	4.3356E-01
24	1.0292E 00	6.0729E-01	7.2430E-01	6.0396E-01	7.0156E-01	6.0551E-01	4.0444E-01	3.8177E-01
25	9.2116E-01	5.4116E-01	6.4827E-01	5.3819E-01	6.2792E-01	5.3956E-01	3.5820E-01	3.3792E-01
26	8.2796E-01	4.8438E-01	5.8268E-01	4.8173E-01	5.6439E-01	4.8296E-01	3.1876E-01	3.0054E-01
27	7.4711E-01	4.3536E-01	5.2578E-01	4.3298E-01	5.0927E-01	4.3408E-01	2.8491E-01	2.6849E-01
28	6.7661E-01	3.9281E-01	4.7617E-01	3.9066E-01	4.6122E-01	3.9165E-01	2.5570E-01	2.4084E-01
29	6.1486E-01	3.5569E-01	4.3271E-01	3.5374E-01	4.1913E-01	3.5464E-01	2.3035E-01	2.1686E-01
30	5.6054E-01	3.2316E-01	3.9448E-01	3.2138E-01	3.8210E-01	3.2220E-01	2.0824E-01	1.9597E-01
31	5.1254E-01	2.9452E-01	3.6071E-01	2.9291E-01	3.4938E-01	2.9365E-01	1.8888E-01	1.7768E-01
32	4.6999E-01	2.6921E-01	3.3075E-01	2.6774E-01	3.2037E-01	2.6842E-01	1.7185E-01	1.6159E-01
33	4.3211E-01	2.4676E-01	3.0410E-01	2.4541E-01	2.9455E-01	2.4604E-01	1.5681E-01	1.4740E-01
34	3.9829E-01	2.2678E-01	2.8030E-01	2.2554E-01	2.7150E-01	2.2611E-01	1.4347E-01	1.3482E-01
35	3.6798E-01	2.0893E-01	2.5897E-01	2.0778E-01	2.5084E-01	2.0831E-01	1.3161E-01	1.2363E-01
36	3.4076E-01	1.9293E-01	2.3981E-01	1.9187E-01	2.3228E-01	1.9236E-01	1.2102E-01	1.1364E-01
37	3.1623E-01	1.7855E-01	2.2255E-01	1.7758E-01	2.1556E-01	1.7803E-01	1.1153E-01	1.0471E-01
38	2.9407E-01	1.6560E-01	2.0695E-01	1.6469E-01	2.0045E-01	1.6511E-01	1.0302E-01	9.6685E-02
39	2.7399E-01	1.5389E-01	1.9282E-01	1.5305E-01	1.8677E-01	1.5344E-01	9.5348E-02	8.9462E-02
40	2.5577E-01	1.4328E-01	1.8000E-01	1.4250E-01	1.7435E-01	1.4286E-01	8.8421E-02	8.2941E-02
$E(4, 2)$	5.5956E-26	1.2374E-25	7.9511E-26	1.2442E-25	8.2088E-26	1.2411E-25	8.2718E-26	1.2354E-25

After W. Clarke (1965).

gave the best representation of the data. Later, Aller, Bowen, and Minkowski (1955) found that in NGC 7027 the higher members of the series showed marked deviations in the sense that the observed intensities were too high. These observations were confirmed by a new series obtained by Aller, Bowen, and Wilson (1963). Subsequently, Kaler (1964) made an intensive study of the Balmer predictions using the theories of Burgess (1958), Seaton (1959*b*), and Pengelly (1964). Some nebulae, such as IC 2165, showed very good agreement with theory; others—most notoriously NGC 7027—showed strong deviations. Kaler also found that the intensities of lines of various series of H and He II which arise from levels with the *same* principal quantum number n show the same percentage deviation. Furthermore, the intensities of lines of various series of H, He I, and He II which arise from levels of the same statistical weight show the same percentage deviation. The deviation did not appear to depend on temperature, density, character of the spectrum of the central star, or uniformity of the nebula, although there appeared to be a slight tendency for very inhomogeneous nebulae to show stronger deviations.

In order to compare theoretical with observed decrements, we must first correct the observed intensities for interstellar reddening. On the assumption that the reddening law is the same for all planetary nebulae studied,[1] we may write (see Berman 1936)

$$\log I(\lambda)_c = \log I(\lambda)_0 + Cf(\lambda) . \tag{45}$$

Here $I(\lambda)_0$ is the observed intensity, $I(\lambda)_c$ is the intensity corrected for space absorption, $f(\lambda)$ is the space absorption function (Whitford 1948, 1958, Wampler 1961, Johnson 1966) which must be known accurately from 3,500 to 11,000 A. The constant C depends on the amount of absorbing material between the nebula and the observer.

Assuming that the electron temperature is known (see Sec. 7), we may determine C either (a) by forcing the observed decrement to agree with the predicted decrement (cf., for example, Aller 1951) or (b) by comparing Paschen and Balmer lines arising from the same upper level (cf. Minkowski and Aller 1956), taking into account, however, that different sets of l-states are involved in the two instances. Care must be taken also to insure that the observational data are free from wavelength-dependent errors. In practice one proceeds by an iterative process which converges rapidly.

Following Clarke, we now compare the predicted and observed decrements for NGC 7662 for which Kaler found substantial disagreements with theory. With $C = 0.25$, best agreement is obtained for $n_{\text{coll}} = 20$. We have smoothed the theoretical decrements arbitrarily over the discontinuity at $n = 20$, where collisional redistribution was imposed. In Table 3, notice that the lines still remain substantially stronger than the predicted values, although the departure is less marked. See Figure 2. NGC 7027 shows deviations which cannot be explained on our present assumptions. Possibly collisional redistributions among the excited levels may be more important than we have assumed. If this is true, the discordance between theory and observation would be less than we supposed.

To summarize, in some nebulae, for example, NGC 1535, NGC 2392, NGC 6741,

[1] But see chapter by Harold Johnson (this volume), who shows that there are differences in the absorption law in different directions in the Galaxy. The known deviant law for Orion makes no difference for planetaries.

IC 351, and Anon 21^h31^m, there is fairly good agreement between theory and observation. In others, such as IC 418, there are substantial deviations. These departures are always in the sense that the observed lines are *too* bright as compared to Hβ; that is, the theoretical intensities always seem to be *lower limits.*

Significantly, in the situation most likely to be encountered in planetary nebulae, there is no appreciable difference between cases A and B, even though the physical situation is quite different.

The recombination theory for helium is discussed below. As Goldberg (1941) and Mathis (1957) emphasized, for this element the dependence of the *b*-factor on *l* must be taken into account explicitly, even in the first approximation. As we successively con-

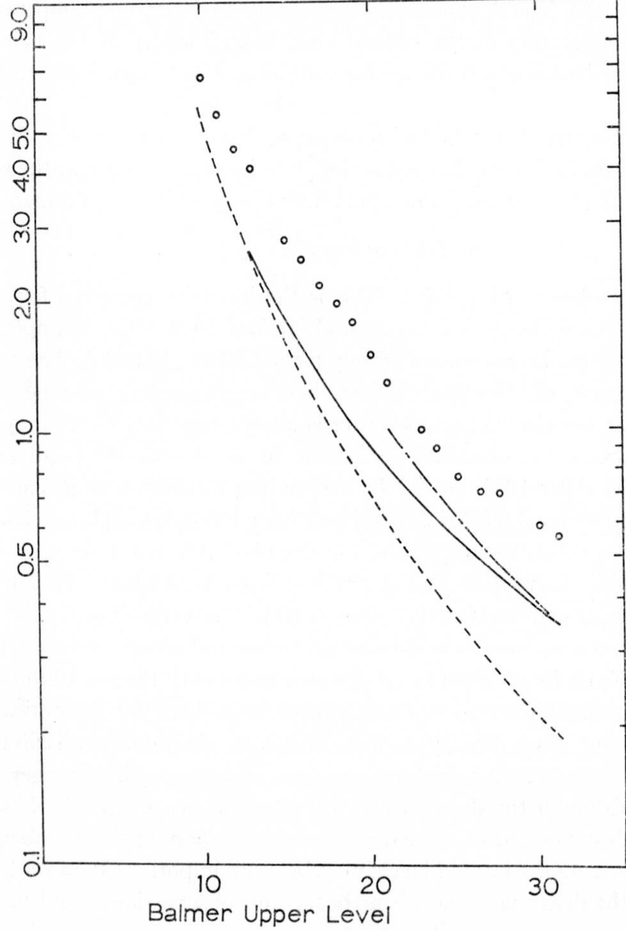

FIG. 2.—Comparison of the observed Balmer decrement for NGC 7662 with theoretical decrements. The circles give the observed intensities (Aller, Kaler, and Bowen 1966); the dashed line represents the purely radiative decrement for $n_{coll} = 60$ and $n_{max} = 60$; the dot-dash line represents radiative processes plus collisions with $n_{coll} = 20$ and $n_{max} = 60$ (Clarke 1965); and the solid curve gives the (interpolated) theoretical decrement. The curves are normalized at Hβ ($n = 4$) with $I = 100$. See Table 3 and text.

sider carbon, nitrogen, oxygen, and neon in various stages of ionization, further complications are added by the multiplicity of terms, the presence of intercombination lines, and the necessity of calculating radial quantum integrals either by the Bates-Damgaard (1949) method or the Hartree-Fock method (see chapter by Czyzak, this volume; also Burgess and Seaton 1960*a*).

5.2. The Hydrogen Lines in Optically Thick Nebulae

Until now we have considered the recombination theory using the hypothesis that the atoms cascaded back to the ground level without delay, the concentration of atoms in any of the higher levels at any one time being assumed extremely small. Actually, this

TABLE 3

Comparison of Observed and Predicted Balmer Decrement for NGC 7662

n	I_0	I_i	I_{Rad}	I_{Int}	n	I_0	I_i	I_{Rad}	I_{Int}
3	330	279	285	285	18	1.7	2.0	0.95	1.13
4	100	100	100	100	19	1.6	1.8	.81	1.00
5	43	46	47	47	20	1.4	1.5	.70	0.90
6	19.3	21	26	26	21	1.1	1.3	.60	0.80
7	14.7	17	16	16	22	1.1	1.3	.52	0.73
8	(14.7)	(17)	11	11	23	0.86	1.0	.46	0.66
9	7.0	8.1	7.6	7.6	24	0.79	0.91	.40	0.61
10	5.9	6.8	5.5	5.6	25	0.67	0.78	.36	0.56
11	4.8	5.5	4.1	4.2	26	0.73	0.83	.32	0.51
12	4.0	4.6	3.6	3.7	27	0.63	0.72	.28	0.47
13	3.6	4.1	2.6	2.6	28	0.61	0.71	.26	0.44
14	3.7	4.2	2.0	2.1	29	0.67	0.78	.23	0.40
15	2.5	2.8	1.6	1.8	30	0.53	0.60	.21	0.38
16	2.2	2.5	1.4	1.55	31	0.49	0.56	0.19	0.35
17	2.0	2.2	1.1	1.30					

I_0 = Observed line intensity (Aller, Kaler, and Bowen 1966).
I_i = Observed intensity corrected for space absorption.
I_{Rad} = Predicted intensity for pure radiative processes to $n = 60$.
I_{Int} = Interpolated theoretical intensities to allow for gradual onset of collisional redistribution at n near 20.

condition may not always be fulfilled. We must examine the consequences of two factors: (1) the great optical thickness of the nebula in Lyα and (2) the influence of the metastability of the 2*s*-level in H (in He I, the 2^3S-level is highly metastable).

The Lyα photons build up to a high intensity, but how high? An upper limit to the number of scatterings can be obtained by assuming that they are perfectly coherent (Ambartsumian 1933). Since the number of free paths required for a quantum to diffuse out of the nebula by a random-walk process would be t_0^2, an optical thickness of $t_0 = 10^5$ to 10^6 would require 10^{10} to 10^{12} scatterings before an escape. Zanstra (1934) argued that such a situation was unlikely; a radiation pressure of sufficient size would be produced to cause the nebular shell to expand. More recent work suggests that radiative pressure might not be as effective as previously supposed, and Lyα quanta may be eliminated by other mechanisms.

Occasionally, a Lyα photon may be degraded to the 2-photon continuum, but the process is inefficient. An atom which has been excited to a 2*p*-level collides with a proton or an electron and is shifted to a 2s-level, whence it decays via a 2-photon continuum. This process has a probability

$$p_{2p-2s} = \frac{C_{2p-2s}}{A_{2p-1s}} = 2.7 \times 10^{-13} N_e \qquad (T_e = 12{,}000^\circ \mathrm{K}) \tag{46}$$

(cf. Seaton 1955*b*). Thus at $N_e = 4{,}000$ per cm³, 10^9 scatterings are required to convert an average Lyα photon into two photons of the 2-quantum continuum.

5.2.1. *Lyα Scattering.*—Let Q denote the number of scatterings an average photon undergoes before it escapes from the nebula; Q is the ratio of the total number of photons emitted or scattered per unit time in the nebula to the total number escaping.

The number of Lyα photons created per second by recombination is given by the total number reaching the $2p$-level. In Menzel and Baker's case B, every recapture on the second or higher level ultimately leads to a transition to $n = 2$, and hence produced a Balmer quantum. Thus we may write for the total rate of emission of all Balmer photons

$$\sum_{n=3}^{\infty} F_{n,2} + \int F_{\kappa 2} d\nu = \sum_{n=2}^{\infty} \int F_{\kappa n} d\nu = N_i N_e \alpha_{\mathrm{B}} , \tag{47}$$

where (see eq. [26])

$$\alpha_{\mathrm{B}} = \frac{K Z^4}{T_e^{3/2}} \sum_{2}^{\infty} \frac{e^{X_n}}{n^3} \int_{\nu_n}^{\infty} g_{\mathrm{II}} \exp(-h\nu / kT_e) \frac{d\nu}{\nu} . \tag{48}$$

Following Seaton, we define α_{B} as the coefficient for recombination for all levels $n \geq 2$. In Menzel and Baker's case B, two-thirds of these reach the $2p$-level and one-third reach the $2s$-level.

Each atom reaching the $2p$-level by recapture and/or cascade creates a Lyα quantum which is scattered Q times before it escapes. Hence the total number of atoms actually reaching the $2p$-level cm^{-1} sec^{-1} will be $\frac{2}{3}\alpha_{\mathrm{B}} N_e N_i(\mathrm{H}^+) Q$, which will equal the number leaving the level each second by radiative transitions for which the Einstein coefficient of spontaneous emission is $A(2p \rightarrow 1s) = 6.25 \times 10^8$ sec^{-1}. Thus,

$$\tfrac{2}{3}\alpha_{\mathrm{B}} N_e N_i(\mathrm{H}^+) Q = N_{2p}(\mathrm{H}) A(2p \rightarrow 1s) . \tag{49}$$

If $t_0 = 370{,}000$, $Q = 2.4 \times 10^6$ (Osterbrock 1962). And $\alpha_{\mathrm{B}} = 2.3 \times 10^{-13}$ cm^{-3} sec^{-1} at $T = 12{,}000^\circ$ K; so

$$N_{2p}(\mathrm{H}) = 5.9 \times 10^{-16} N_e N_i . \tag{50}$$

Thus if $N_e \sim N_i = 10^4$, $N_{2p}(\mathrm{H}) \sim 6 \times 10^{-8}$.

Actually, the scattering cannot be coherent; in fact, it must be largely noncoherent. Zanstra suggested that the photons were re-emitted with a Gaussian thermal distribution about the rest frequency of the line, and that no correlation exists between the frequency of the photon before and after the scattering process. Within the framework of these assumptions, we may calculate Q by the following elementary argument due to Osterbrock (1962).

After each scattering process, the probability that the frequency ν of the photon lies between $\Delta\nu = x\Delta\nu_D$ and $\Delta\nu + d\Delta\nu = (x + dx)\Delta\nu_d$ is

$$p(x) = \frac{1}{\sqrt{\pi}} \exp(-x^2) . \tag{51}$$

Let $x_1 = (\Delta\nu_1/\Delta\nu_D)$ denote that value of x for which $t(x_1) = 1$. (Here $\Delta\nu_D = \nu/cV$, where V is the most probable velocity of the radiating atoms.) Thus

$$t_1 = t_0 \exp(-x_1^2) = 1, \qquad x_1 = \sqrt{(ln t_0)}. \tag{52}$$

The probability that the frequency emitted differs from the rest frequency by an amount greater than x_1 is

$$w(x_1) = 2\int_{x_1}^{\infty} p(x)\,dx = 1 - \frac{2}{\sqrt{\pi}}\int_0^{x_1} \exp(-y^2)\,dy. \tag{53}$$

We assume that the photon escapes directly if the frequency shift from the rest frequency much exceeds x_1. The number of scatterings the photon suffers before it escapes is

$$Q = \frac{1}{w(x_1)}. \tag{54}$$

For example, if $t_0 = 10^4$, $Q = 5.6 \times 10^4$. The method is satisfactory if $t_0 < 20{,}000$, but fails for much larger values of t_0.

When the photon is scattered, ν_{initial} and ν_{final} are correlated; so the assumption of complete redistribution is not correct. For optical depths exceeding about 20,000, it is necessary to consider the effects of damping wings. Spitzer and Greenstein (1951) developed a physical theory for the rate at which Lyman alpha photons escape from a nebula. In a somewhat similar problem in a solar context, Unno showed that frequency redistribution leads to an escape of photons in the far wings, where the optical depth is small, but Osterbrock gave a more complete treatment for the optically thick nebula. We can put

$$Q = \mathfrak{Q}(t_0)t_0, \tag{55}$$

where $6 < \mathfrak{Q} < 12$ for $7{,}500 < t < 10^6$, while for $t_0 > 10^6$,

$$Q \to 6 \times 10^{-6} t_0^2. \tag{56}$$

We may regard the frequency redistribution in the wings as a diffusion problem in frequency; that is, the flow of photons along the frequency displacement axis can be treated as a diffusion process. Except in the very far wings, however, we must consider an inherent bias in the scattering which tends to shift the photons preferentially back toward the center of the line. Near the center of the line itself, the redistribution takes place in a more or less random fashion. It is not even necessary that a photon reach a place where the optical depth is less than 1 before it can escape from the nebula. Photons may get out of the nebula as a consequence of several passages through the wings; in each of which passage they are scattered several times and find their way to the surface. We assume that the boundary of the ionized zone is the boundary of the nebula. Suppose there is a surrounding shell of neutral gas. Then a Lyα photon will be thrown back into the nebula and will be repeatedly scattered before it escapes or is converted into 2-photon emission. O'Dell (1965) has presented arguments that many planetaries are surrounded by H I regions.

Because the mean free path of a Lyman α quantum is so very short, the influence of the expansion of the nebula cannot be very great. On the other hand, we should consider the possible role of turbulence which could act to broaden the Doppler profile and facilitate the escape of quanta.

5.2.2. *The role of the metastable 2s-level.*—The total number of captures in this level will be very nearly

$$\alpha(2s)N_eN(\mathrm{H}^+) = \tfrac{1}{3}\alpha_{\mathrm{B}}N_eN(\mathrm{H}^+)\,. \tag{57}$$

Atoms may escape from this metastable level by the following processes

a) Photo-ionization to the continuum. We can neglect the contribution of the stellar continuum because of the extreme dilution of the stellar radiation. Under most circumstances we can probably also neglect the influence of the photo-ionization produced by the Lyman α radiation.

b) By 2-photon "2*q*" emission, the Einstein coefficient for the spontaneous emission of 2-quanta being $A_{2q} = 8.227\ \mathrm{sec}^{-1}$. Spitzer and Greenstein (1951).

c) Collisional transitions to the 2*p*-level. The rate coefficients for this process have been computed by Seaton (1955*a*).

Then

$$\tfrac{1}{3}\alpha_{\mathrm{B}}N_eN(\mathrm{H}^+) = N(2s)[A_{2s,1s} + N(\mathrm{H}^+)q_p(2s-2p) + N_eq_e(2s-2p)]\,. \tag{58}$$

Putting in numerical values, for $T = 12{,}000°$ K we get

$$N(2s)[8.2 + N(\mathrm{H}^+) \times 4.65 \times 10^{-4} + N_e \times 0.54 \times 10^{-4}]. \tag{59}$$

At the limit of extremely low densities, we have

$$N(2s) = 9.3 \times 10^{-15}N_eN(\mathrm{H}^+)\,, \tag{60}$$

whereas at intermediate densities

$$N(2s) \cong \frac{1.5\times10^{-10}N_e}{1+\dfrac{1.57\times10^4}{N_i(\mathrm{H}^+)}}, \qquad T\cong12{,}000°\mathrm{K}. \tag{61}$$

We may also calculate the ratios:

$$\frac{N(2s)}{N(2p)} = \frac{A(2p-1s)}{2Q[A_{2s-1s}+N_i(\mathrm{H}^+)q_p+N_eq_e]} \cong \frac{3.8\times10^7}{Q[1+0.63\times10^{-4}N_i(\mathrm{H}^+)]}, \tag{62}$$

where we have used $A(2p - 1s) = 6.25 \times 10^8\ \mathrm{sec}^{-1}$ (Bethe and Salpeter 1957). Note that if $Q \approx 10^7$ and $N_e \approx 5 \times 10^4$, then $N(2s) \approx N(2p)$.

5.2.3. *Effect of optical thickness.*—We may now estimate the optical thickness of the nebular shells in the principal Balmer lines. Compare Osterbrock (1964). As an example, consider a nebula where $N(\mathrm{H}^+) = 2 \times 10^4$, $N_e = 2.4 \times 10^4$, $r = 1.5 \times 10^{17}$ cm, and $Q = 2.4 \times 10^6$. Then $N(2s) = 2.0 \times 10^{-6}$ and $N(2s)/N(2p) = 7$. Using $f(2s - 3p) = 0.425$ and $f(2p - 3d) = 0.694$, we find that $\alpha_0(2s - 3p) = 3.0 \times 10^{-13}$ and $\alpha_0(2p - 3d, 3s) = 4.9 \times 10^{-13}$ are the absorption coefficients at the line center (cf. for example, Aller 1963, p. 323). We find also $t(\mathrm{H}\alpha) \approx 0.11$. Similarly, $t(\mathrm{H}\beta) \approx 0.015$ and $t(\mathrm{H}\gamma) = 0.006$, so that self-reversal effects are not important for the hydrogen lines. For a dense planetary such as IC 4997, for which $N(\mathrm{H}^+) \approx 3 \times 10^5$ and $R \approx 0.8 \times 10^{17}$ cm, these effects can become significant.

Note, however, that if Q is greatly enhanced, for example, by a surrounding shell of neutral gas to throw back the quanta, it can attain a value of 10^8 or more (at which point degradation to 2-photon emission can occur) and appreciable optical depths in Hα and even in Hβ might be produced. Such effects could be important in objects such as the Orion Nebula, where there is an extensive surrounding cloud of neutral hydrogen gas.

The self-reversal problem was considered by Pottasch (1960*a*, *b*), who treated a simplified model in the details of the line profile, and handled thermal motions in a schematic fashion. He used a plane-parallel atmosphere rather than a spherical one. Capriotti (1964*a*, *b*) discussed emission in the Balmer and Paschen lines for an idealized 20-level hydrogen atom. His model nebulae consisted of spherical shells in which both electron temperature and density were constant. Upward transitions can occur not only from the 1*s*-level, but also from the 2*p*-level. The absorbing atoms are assumed to have a Maxwellian velocity distribution; the line-emission function is assumed to have a Doppler profile.

When scattering occurs, the redistribution in frequency is assumed to be complete. In solving the transfer equation in integral form, Capriotti assumes that emission and scattering are isotropic. He tabulated Balmer and Paschen decrements for $T_e =$ 10,000° K and 20,000° K for his 20-level atom for $t(\mathrm{H}\alpha) = 1.0$, 25, 63, 125, 250, and for values of $R = R_1/R_2 = 0, \frac{1}{4}, \frac{1}{2}$, and $\frac{3}{4}$, where R_1 was the inner radius of the nebula and R_2 was its outer radius. The Balmer decrement depends not only on the optical depth in Hα, but also on R. Furthermore, for a given combination of $t(\mathrm{H}\alpha)$ and R, absorption from the 2*s*-level has a more pronounced effect than that from the 2*p*-level. For most gaseous nebulae, the value of $t(\mathrm{H}\alpha)$ is such that the population in the 2*s*-level is greater than that in the 2*p*-level.

Following Pottasch, Osterbrock, Capriotti, O'Dell, and others, we may plot the two Balmer-line ratios log $I(\mathrm{H}\beta)/I(\mathrm{H}\gamma)$ and log $I(\mathrm{H}\alpha)/I(\mathrm{H}\beta)$ as ordinates and abscissae, respectively (see, for example, Capriotti 1964*a*, Fig. 1). If all nebulae conformed to the same theory and if there were no space absorption, the plots would all fall on the same point in the diagram at least to within observational errors. Space absorption tends to steepen the decrement, that is, increase the ratio $I(\mathrm{H}\alpha):I(\mathrm{H}\beta):I(\mathrm{H}\gamma)$, and spread the nebulae out along a line passing through the theoretical point. If measurements of both Paschen and Balmer lines are available, we can correct all the measured intensities for space absorption. Effects of self-absorption, however, will tend to increase the Hα/Hβ ratio at the same time as the Hβ/Hγ ratio is decreased.

No planetaries appear to fall near such a self-absorption curve, which corresponds to high optical thickness and small space absorption.

A nebula affected by both space absorption and high optical thickness which leads to self-reversal effects would fall between these two lines. Such a nebula is the small, very dense planetary VV8, which has been studied extensively by O'Dell.

5.3. Collisional Excitation

In addition to effects of self-reversal and reddening on the Balmer decrement, we must also consider the influence of collisional excitations (Miyamoto 1938, Chamberlain 1953, Parker 1964). Chamberlain showed that if the nebular gas is excited both by inelastic collision processes and by stellar radiation, the b_n factors are given by

$$b_n = (1 - \mathcal{H})b_n^R + \mathcal{H}b_n^c\,, \tag{63}$$

where b_n^R is the factor calculated for a radiative situation, for example, Menzel and Baker's case B, and where b_n^c is the factor calculated for pure collisional excitation. Also

$$\mathcal{H} = 1 - \frac{b_1 e^{X_1}}{G(T_e)} \int_{y_1}^{\infty} \frac{W_\nu g\, dy}{y(e^y - 1)} = \frac{q_c(T_e) T_e^{3/2} N_1}{G(T_e) K N_e}, \tag{64}$$

where $q_c(T_e)$ is the rate coefficient for collisional ionizations. In practice we can anticipate that $\mathcal{H}$ will be very small in all planetaries. For pure collisional excitation $\mathcal{H} = 1$, but for pure radiative excitation $\mathcal{H} = 0$. Collisions always tend to steepen the Balmer decrement. For example, both IC 418 and NGC 7027 contain neutral oxygen, and since oxygen and hydrogen have the same ionization potential, the presence of O I would imply the presence of H I. In NGC 7027, Seaton (1960*a*) estimates $N(H^0)/N(H^+) = 0.35$, whereas Minkowski and Aller (1956) estimated that this ratio could be as high as 0.50. Osterbrock (1964) estimated that this ratio could be as high as 0.50. Osterbrock (1964) concludes that collisional excitation of the $n = 3$ level may well be important in IC 418 and NGC 7027.

In Figure 3, we plot the ratio log $I(H\beta)/I(H\gamma)$ against log $I(H\alpha)/I(H\beta)$ corrected for space absorption. The observational data are from O'Dell (1963*b*) supplemented with some measurements by Liller and Aller (1963) and by Osterbrock, Capriotti, and Bautz (1963). The theoretical curves are for self-absorption (Capriotti 1964*a*) and for collisional excitation (Parker 1964). The plotted points correspond to $t(H\alpha) = 25$, 63, and to $\mathcal{H} = 0.03$, 0.10, and 0.20, interpolated for $T = 12{,}000°$ K. The theoretical points corresponding to cases A and B (Clarke 1965) are indicated. Note that many of the nebulae fall to the left, that is, have too small a ratio of Hα to Hβ (cf. see also O'Dell 1964).

The conclusions that may be drawn are that some imperfections remain in the theory: In some nebulae the higher members of the Balmer series appear to be abnormally strong. Also, as we have seen, collisional and self-reversal effects must be taken into account. Finally, it may yet be necessary to solve the transfer equations for the Lyman lines, rather than to consider only the limiting cases A and B. Capriotti (1966) used a somewhat schematic procedure to estimate the consequences of a partial leakage of Lyman-line radiation upon the Balmer-line spectrum. He found derivations from the conventional decrements of such a magnitude as to warrant a more detailed study of Lyman-line radiation.

5.4. The Helium Spectrum in Planetary Nebulae

Ambartsumian (1932) was the first to recognize that because of the metastability of the 2^3S- and 2^1S-levels of helium, the spectrum of this atom might show some interesting features in gaseous nebulae. The problem was considered subsequently by Goldberg (1941), who gave the first quantitative theory of the statistical equilibrium of the levels, and by Mathis (1957), Seaton (1960*b*), Pottasch (1962), Münch and Wilson (1962), Osterbrock (1964), Münch (1964), and O'Dell (1965).

Menzel and Aller (1945) suggested that collisional excitations from the high metastable 2^3S- and 2^1S-levels to higher permitted levels could have a profound effect on the line intensities predicted on any purely radiative theory. Furthermore, these effects would be strongly density sensitive (at least at a certain range of densities); so fluctuations in the spectrum might be pronounced.

Consider first the triplet levels. Let us first write down the equation of statistical equilibrium for the 2^3S-level. On the assumption that these recombinations are hydrogen-like, with three times as many atoms arriving at the triplet level as at the singlet level, we may write (following Burgess and Seaton 1960)

$$(\text{number entering } 2^3S\text{-level}) = \tfrac{3}{4}\alpha_B N_e N(\mathrm{He}^+)\,. \tag{65}$$

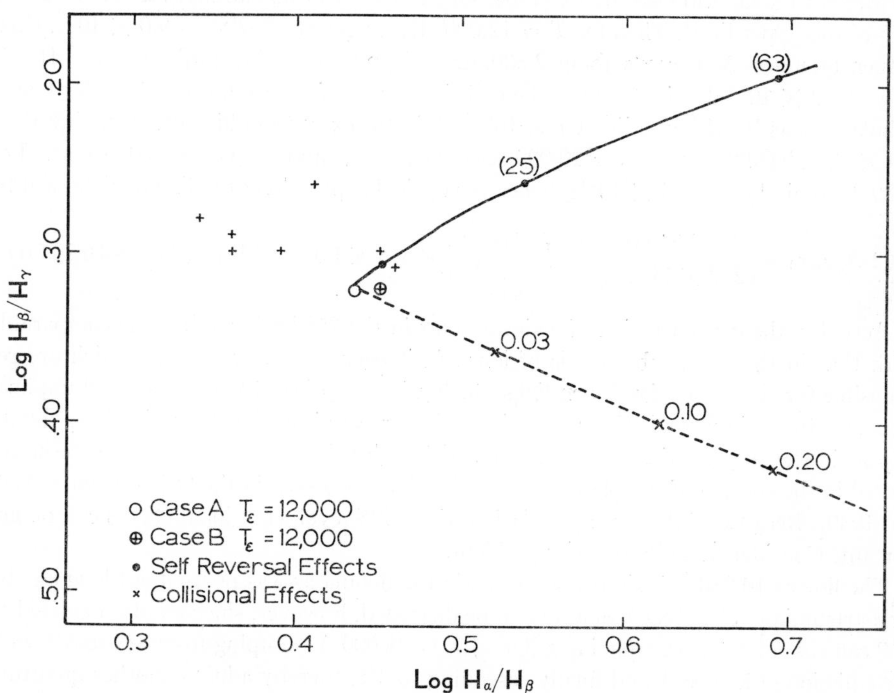

FIG. 3.—The Balmer ratios Hα/Hβ and Hβ/Hγ. The observed points (*crosses*) correspond to IC 2149' NGC 6826, NGC 7662, NGC 6543, IC 418, and NGC 7027. All ratios have been corrected for interstellar absorption.

The upper solid curve displays the effects of self-reversal as calculated by Capriotti (1964–). The points corresponding to the optical depth in Hα, $t(\mathrm{H}\alpha) = 25$, and $t(\mathrm{H}\alpha) = 63$ are indicated. The lower (*dotted*) curve gives the predictions for collisional effects (Parker 1964). Points are plotted corresponding to the values, 0.03, 0.10, and 0.20 of the parameter $\mathcal{H}$ (see eq. [64]).

The number leaving the 2^3S-level will depend on the spontaneous 2-photon emission for which $A(3^3S\text{–}1^1S) = 2.2 \times 10^{-5}$ sec Mathis (1957), collisions to the 2^1S-level for which the rate factor is $q(2^3S\text{–}2^1S)N_e$, and collisions to higher triplet levels (which can be neglected if all that happens is excitation to higher levels from whence the atoms cascade back to 2^3S).

Significant, however, are collisional excitations to higher singlet levels from whence atoms can escape to ground state or direct collisional de-excitations $\Sigma q[2^3S\text{–}n^1(L)]N_e$. We can neglect photo-ionizations by quanta from the central star, although photo-ionizations by Lyα quanta which occur at a rate n_a may be important (cf. Münch 1964). The equilibrium equation is then

$$N(2^3S)\{[A(2^3S\text{–}1^1S) + q(2^3S\text{–}2^1S)N_e] + \Sigma q[2^3S\text{–}n^1(L)]N_e + n_a\} = \tfrac{3}{4}\alpha_B N_e N(\mathrm{He}^+) . \quad (66)$$

In a first rough examination of the problem, following Pottasch (1962) and Osterbrock (1964), we may neglect the influence of the ionization by Lyman α and collisions to singlet states other than 2^1S. For this latter collisional process we review the discussions by Marriot (1955) and Osterbrock (1964) and adopt a collision strength Ω amounting to 0.75 of the upper limit. Then for $T = 12{,}000°$ K, we get $q(2^3S\text{–}2^1S) = 9.0 \times 10^{-9}$. Consequently, unless N_e exceeds about 2,500, even the extremely low radiative rate $A(2^3S\text{–}1^1S) = 2.2 \times 10^{-5}$ dominates over the collisional rate of de-excitation. Consider now a nebula such as IC 418 of radius 1.7×10^{17} cm. At the extreme of high density, $N(2^3S) = 1.9 \times 10^{-5} N(\mathrm{He}^+)$ cm^3, $N_e = 20{,}000$ electrons/cm^3, and $N(\mathrm{He}) = 3{,}000$/cm^3. For 3889 A, we find $\alpha_0 = 5.3 \times 10^{-14}$; hence the optical depth along the line of sight will be

$$N(2^3S)\alpha_0 r_0 = \frac{1.72 \times 10^{-13} N_e N(\mathrm{He}^+)}{(2.2 \times 10^{-5} + 9 \times 10^{-9} N_e)} \times 5.3 \times 10^{-14} \times 1.7 \times 10^{17} = 465 . \quad (67)$$

Actually, the concentration of helium atoms in the 2^3S-level can become comparable with that in the metastable levels of ions of nitrogen, neon, or oxygen, which are responsible for the strong forbidden lines. Such a large optical depth has several implications: (*a*) the self-reversal effects are going to be large and (*b*) the absorption lines should appear in the spectra of the central stars. Although this phenomenon has not been observed in the spectra of the planetary nuclei, it has been seen in the Orion Nebula (Wilson 1939). Because of the large population of the 2^3S-level, collisional effects can be important. Consider first the transfer problem.

The line at 10,830 A is simply scattered in the nebula as was pointed out long ago by Ambartsumian. The 3889 A quanta can be degraded, however, since an atom excited to 3^3P can cascade to 2^3P, emitting a line at $\lambda \approx 21{,}000$ A, jumping from 3^3S to 3^3S with the emission of λ 7065 A and finally cascading to 2^3S, thereby adding another quantum of λ 10,830 A to the radiation field.

The reverse cycle is not possible; such a Rosseland-type cycle can go only in one direction. A similar process occurs with 3187 A; this line is not subject to interference from the hydrogen lines—but, alas, falls so far in the ultraviolet that photometric difficulties are terrible.

Since the n^3P, n^3D, n^1P, n^1D, and other terms lie just a few volts above the 2^3S-level, collisional excitation to them can be important (Menzel and Aller 1945). Detailed calculations for the triplet levels have been made by Osterbrock (1964); see also Seaton (1967). Under high-density conditions the number of collisional excitations from 2^3S to 2^3P at $T = 12{,}000°$ K is

$$F(2^3S\text{–}2^3P) = 5.1 \times 10^{-12}\, N(\mathrm{He}^+)N_e , \quad (68)$$

while the effective recombination rate for λ10830 is $1.27 \times 10^{-13}\, N(\mathrm{He}^+)N(\mathrm{H})$; so for every atom entering the 2^3P-level by direct recapture or by cascade, about 40 enter it by collisions.

These considerations predict a terrifically strong line at 10,830 A. At 12,000° K, the number of quanta emitted/cm^{-3} sec^{-1} in Hβ will be

$$F_{4\rightarrow 2} = 2.8 \times 10^{-14} N_i N_e \,. \tag{69}$$

Hence the ratio of *intensities* will be

$$\frac{E(10{,}830\mathrm{A})}{E(\mathrm{H}\beta)} = 81 \frac{N(\mathrm{He}^+)}{N(\mathrm{H}^+)} \,. \tag{70}$$

Since hydrogen and helium are completely ionized in the regions considered, $N(\mathrm{He}^+)/N(\mathrm{H}^+) \approx N(\mathrm{He})/N(\mathrm{H}) = 0.17$ (O'Dell 1964; Aller 1964*a*, *b*). It follows that I(10,830 A) $\approx 14\, I(\mathrm{H}\beta)$. After correction for space extinction we find the following observed values for two relatively dense nebulae (O'Dell 1963*b*, Liller and Aller 1963): I(10,830 A)/I(Hβ) = 1.17 for IC 4997 and 0.70 for IC 418. These ratios are considerably smaller than the predicted values.

TABLE 4

THEORETICAL INTENSITIES OF HELIUM TRIPLETS

t_0 (3889 A)

λ(A)	Transition	0	0.3	1.0	3.0	10.0	30.0
10,830....	2S–2P	1.58	1.59	1.61	1.69	1.84	1.91
3889....	2S–3P	1.08	1.06	0.99	0.75	0.32	0.10
3188....	2S–4P	0.451	0.448	0.436	0.381	0.223	0.082
7065....	2P–3S	0.183	0.199	0.240	0.375	0.668	0.823
4713....	2P–4S	0.032	0.032	0.034	0.038	0.052	0.068
4121....	2P–5S	0.0105	0.0105	0.0105	0.0104	0.0106	0.0096
5876....	2P–3D	1.0	1.0	1.0	1.0	1.0	1.0
4471....	2P–4D	0.394	0.393	0.392	0.390	0.384	0.385
4026....	2P–5D	0.210	0.209	0.208	0.207	0.199	0.191

Pottasch (1962).

Before we discuss this discordance, let us first examine some predictions of the purely radiative theory of the helium triplets (Pottasch 1962) which might be applied as the limiting situation with low densities. Table 4 gives Pottasch's predicted values of the relative intensities of the helium lines as a function of the optical depth in 3889 A on the scale I(5876 A) = 1.0.

Table 5 gives the measured intensities of the various helium lines. From I(3889 A) and the second row of Table 4 we estimate t_0(3889 A), read off the predicted values of I(10,830 A), and tabulate the ratio of observed and predicted values, that is, I_0/I_p. Although the line at 10,830 A is considerably (~2 to 4.4 times) stronger than a purely radiative theory would predict, it is weaker than a collisional theory would indicate. The predicted intensities at 7065 A are in good agreement with O'Dell's observations. Note that the 4471 A line is systematically stronger than the predicted value. Kaler has found deviations for the He I lines similar to those appearing for the hydrogen lines. They appear at a lower value of n; λ 4471 may be affected by this process.

We still have to ask why the 10,830 A line is not as strong as Osterbrock's calculations would suggest. An explanation is perhaps to be found in a recent calculation by O'Dell, who adopts Münch's point of view that helium atoms escape from the 2^3S-level mainly by the absorption of Lyα. He then uses the observed intensities of the helium lines to investigate the transfer problem for Lyα before the quanta escape from the nebula. If the parameter f denotes the ratio of the distance that a Lyα photon travels before it is lost by escape or destruction inside the nebula to the radius of the nebula, O'Dell finds that the photoelectric line ratios indicate that f ranges from nearly 100 to more than 3,000, although theoretical discussions of density-bounded nebulae indicate that f should never exceed about 40. He explained this discordance by postulating a thin outer shell of neutral hydrogen which is optically thick to Lyα. This shell acts as an effective barrier and scatters the photons back again into the ionized interior of the nebula rather than allowing them to escape. This process would serve to increase the mean number of

TABLE 5

OBSERVED INTENSITIES OF HELIUM TRIPLETS

λ (A)	IC 2149	NGC 6543	IC 418	IC 4997
10830	3.32	1.23	6.4	7.8
3889	0.85	0.71	0.51	0.45
3188	...	...	0.53	0.13
7065	0.34	0.42	0.47	0.59
4713	...	...	0.055	0.10
4121	...	...	0.029	0.027
5876	1.00	1.00	1.00	1.00
4471	0.46	0.56	0.50	0.48
4026*	...	0.32	0.22	0.36 (0.23)
t_0 (3889)	2.0	3.2	5.6	6.6
Pred. 10830	1.65	1.70	1.76	1.78
Ratio I_0/I_p	2.0	...	3.6	4.4
N	2×10^4	1.1×10^4	1.9×10^4	15×10^4
Pred. I(7065A)	0.32	0.41	0.54	0.58
I_0(4471A)$/I_p$	1.18	1.44	1.30	1.24

O'Dell (1963*b*), Liller and Aller (1963), and Aller and Kaler (1964 *b*, *c*). (Corrections for space reddening and blending of λ 3889 with hydrogen have been applied.)

* For IC 418 the photoelectric and photographic intensity measurements of I(4026A) are in good agreement; for IC 4997, the photographic value (indicated in parentheses) is closer to the theoretical predictions and, indeed, to our expectations.

scatterings to the point where the Lyα photons would be destroyed by the 2-quantum emission in hydrogen. The difficulty with this hypothesis is that so much momentum would be transferred to the shell by the scattered photons that it would be blown away rapidly (Capriotti, private communication).

The problem of the He/H ratio is examined in Section 10. Lines of He I, He II, and H all arise from recombination, and interpretation of the data is complicated by the fact that helium can exist in two states of ionization, but hydrogen only in one, but more particularly interpretation is complicated by the fact that the radius of the hydrogen ionization zone may be larger than that of the first ionization zone of helium. In some planetaries, all hydrogen is ionized and all helium is ionized at least once.

6. THE THERMAL BALANCE OF A GASEOUS NEBULA

6.1. General Considerations

If we can neglect dissipation of mechanical energy, shock waves, and so forth, the amount of energy absorbed in a volume element must be equal to the amount of energy radiated. Energy is absorbed by the photo-ionization of hydrogen and helium atoms. It is radiated away as electrons are recaptured on discrete levels by free-free emissions and by collisional excitation of atomic levels which are followed by the re-emission of energy —particularly in forbidden lines—although collisional excitation of hydrogen lines can become important at higher electron temperatures or in regions where the density of neutral hydrogen is high. We follow the treatment of Menzel and Aller (1941*b*) with modifications for the collisional excitation of hydrogen as given by Hummer (1963).

6.2. Statistical Equilibrium in the Continuum

First consider the equation of statistical equilibrium for the continuum,

$$\int_{\nu_1}^{\infty} F_{1\kappa} d\nu + \int \mathfrak{F}_{1\kappa} d\nu = \sum_{n=1}^{\infty} \int F_{\kappa n} d\nu , \tag{71}$$

where the first term on the left corresponds to the photo-ionizations from the ground level and the second term includes the collisional ionizations. The right-hand side gives the number of recombinations on all levels. Three-body collisions can be neglected. Following Menzel (1937) and Baker, Menzel, and Aller (1938), we see that the radiative ionizations are given by

$$\int_{\nu_1}^{\infty} F_{1\kappa} d\nu = N_i N_e \frac{KZ^4}{T_e^{3/2}} e^{X_1} b_1 \int_{y_1}^{\infty} W_\nu \frac{g_{\mathrm{II}}}{e^y - 1} \frac{dy}{y} , \tag{72}$$

where X_n is defined by equation (20) and

$$y = h\nu / kT_1 . \tag{73}$$

T_1 is the temperature of the central star, and W_ν is the dilution of the radiation field, which includes not only the geometrical factor (defined in eq. [17]) but also modifications produced by radiative transfer in the nebula. The g_{II} correction factor to Kramers' formula has been calculated by Menzel and Pekeris (1935). Among more recent discussions and calculations, we mention those by Grant (1958), Karzas and Latter (1961), Brussaard and van de Hulst (1962), and Clarke (1965). Burgess (1958) writes the expression for g_{II} in the following form:

$$\begin{aligned} g_{\mathrm{II}}(n;\nu/\nu_1) = {} & 1 + 0.1728 n^{-2/3} (u+1)^{-2/3} (u-1) \\ & - 0.0496 n^{-4/3} (u+1)^{-4/3} (u^2 + \tfrac{4}{3} u + 1) + \ldots , \end{aligned} \tag{74}$$

where $u = n^2(\nu/\nu_1)$, n being the principal quantum number and ν_1 the frequency at the Lyman limit. The rate of recaptures on any level n is given by equation (26). Assuming that collisional ionizations may be neglected, we employ equations (26), (27), (71), and (72) to obtain the equation of statistical equilibrium, viz.,

$$e^{X_1} b_1 \int_{y_1}^{\infty} W_\nu \frac{g_{\mathrm{II}}}{(e^y - 1)} \frac{dy}{y} = \sum_{n=1}^{\infty} \frac{\langle g \rangle S_n}{n^3} = G(T_e) , \tag{75}$$

which defines $G(T_e)$, the explicit values of g_{II} being taken into account. In Spitzer's (1948) notation, $\beta = X_1 = h\nu_1/kT_e$ and $\phi(\beta)/\beta$ is our $G(T_e)$.

6.3. Radiative Equilibrium

We may now write the equation of radiative equilibrium for the continuum. The total energy supplied by photo-ionization from the ground level equals the total amount of energy liberated by recapture on discrete levels and by subsequent cascade to the ground level in free-free emissions and in collisional excitation of forbidden lines and discrete levels in H_1. Thus,

$$\int_{\nu_1}^{\infty} F_{1\kappa} h\nu_{1\kappa} d\nu = \sum_1^{\infty} \int_{\nu_n}^{\infty} F_{\kappa n} h\nu_{\kappa n} d\nu + \sum_2^{\infty} \int F_{\kappa n} h\nu_{n1} d\nu + \int_0^{\infty} \int_0^{\infty} F_{\kappa\kappa'} h\nu_{\kappa\kappa'} d\nu_\kappa d\nu_{\kappa'} + E_{\text{coll}}. \quad (76)$$

Putting in explicit expressions (see Menzel 1937, 1963; Aller 1956, pp. 119 *et seq.*), we have

$$\int_{\nu_1}^{\infty} E_{1\kappa} d\nu = N_i N_e \frac{KZ^4}{T_e^{3/2}} kT_1 b_1 e^{X_1} \int_{y_1}^{\infty} \frac{W_\nu g_{\text{II}}}{(e^y - 1)} dy\,; \quad (77)$$

$$\int_{\nu_n}^{\infty} E_{\kappa n} d\nu = N_i N_e \frac{KZ^4}{T_e^{3/2}} \frac{e^{X_n}}{n^3} \int g \exp(-h\nu/kT_e) h d\nu = N_i N_e \frac{kKZ^4}{T_e^{1/2}} \frac{\langle g \rangle}{n^3}\,; \quad (78)$$

$$\sum_2^{\infty} h\nu_n \int_\nu^{\infty} F_{\kappa n} d\nu = \Sigma hRZ^2 \left(1 - \frac{1}{n^2}\right) \int_{\nu_n}^{\infty} F_{\kappa n} d\nu = hRZ^2 N_i N_e \frac{KZ^4}{T_e^{3/2}} \left(G^*_{T_e} - \Sigma \frac{S_n^*}{n^5}\right); \quad (79)$$

$$\int_0^{\infty} \int_{\nu_{\kappa'}=0}^{\infty} E_{\kappa\kappa'} d\nu_\kappa d\nu_{\kappa'} = N_i N_e \frac{Kk^2 T_e^{1/2} Z^2}{2hR} \ll g_{\text{III}} \gg . \quad (80)$$

Here $E_{ij} = F_{ij} h\nu_{ij}$ is the energy released per unit time and volume; the * indicates that the Gaunt factor is taken into account in computing $G(T_e)$ and S_n. The $\ll \gg$ around g_{III}, the Gaunt factor for free-free transitions, indicates that the mean value of g_{III} has to be found by a double integration over a Maxwellian distribution of velocities.

6.4. Collisional Excitation of Forbidden Lines

We must now consider the collision terms. The energy emitted in the forbidden lines simply escapes from the nebula. If $I(FL)$ is the intensity of a forbidden line and $I(\text{H}\beta)$ that of $\text{H}\beta$, then the energy dissipation per unit volume will be

$$E_1(\text{coll}) = E(\text{H}\beta) \Sigma \frac{I(FL)}{I(\text{H}\beta)}, \quad (81)$$

where

$$E(\text{H}\beta) = F_{4,2} h\nu_{4,2} = N_i N_e \frac{KZ^4}{T_e^{3/2}} b_4(T_e) e^{X_4} \frac{g_{4,2} 2hRZ^2}{(2)^3 (4)^3} = E^0_{4,2} N_i(\text{H}) N_e \times 10^{-25}$$
$$= 22.8 \times 10^{-20} N_i N_e b_4(T_e) e^{X_4} / T_e^{3/2} \quad (82)$$

is the emission in $\text{H}\beta$ in ergs cm^{-3} sec^{-1}. Here $E^0_{4,2}(T_e)$ may be given most conveniently in the form of Table 6 from Clarke (1965).

The summation is taken over all lines that are collisionally excited. Some of these "cooling" transitions fall in the inaccessible regions of the spectrum. Some correspond to excitation of permitted lines, for example, the Mg II 2800 A doublet in planetaries of low excitation, and certain resonance lines of C IV and C V in objects of higher excitation (Schmidt 1965).

Transitions between fine-structure levels of ground terms of Ne II, S IV, and Mg IV giving lines at 12.8 μ, 10.5 μ, and 4.49 μ may also play important cooling roles (Gould 1966).

One must then estimate the cross-section or collision strength and calculate the energy dissipation

$$E = \Sigma \mathfrak{F}_{AB} h\nu_{AB} \tag{83}$$

over all these possible excitations. Such calculations can be carried out provided we know the likely concentrations of relevant ions. In most planetary nebulae, the bulk of the energy dissipation by the collisonal excitation of forbidden lines seems to involve lines falling in the normally observable wavelength ranges.

TABLE 6

VALUES OF EMISSIVITY COEFFICIENTS FOR Hβ

$E^{0}_{4,2} = E(\text{H}\beta)/N_e N_i \times 10^{25}$ (ergs cm^{-3} sec^{-1})

Case	5000	10,000	15,000	20,000	40,000
A.......	1.525	0.821	0.5570	0.4182	0.2024
B.......	2.222	1.241	0.8629	0.6596	0.3327

Clarke (1965).

6.5. Collisional Excitation of Discrete Levels of Hydrogen

The dissipation of energy in the excitation of discrete levels of hydrogen and in collisional ionization of hydrogen is

$$E_2(\text{coll}) = \Sigma \mathfrak{F}_{1n} h\nu_{1n} + \int \mathfrak{F}_{1\kappa} h\nu_{1\kappa} d\nu \,, \tag{84}$$

which may be written in the form (Hummer 1963)

$$E_2(\text{coll}) = N_e N_{\text{H}} hR[\Phi(T_e) + q(T_e)] \,, \tag{85}$$

where hR corresponds to 13.60 volts, while

$$\Phi(T_e) = \sum_{n=2}^{\infty} \left(1 - \frac{1}{n^2}\right) q_n(T_e) \tag{86}$$

gives the rate coefficient for collisional excitations to discrete levels and $q(T_e)$ gives the rate coefficient for collisional ionizations.

If $Q_n(T_e)$ denotes the collisional inelastic cross-section for the nth level, the corresponding collisional rate coefficient is given by

$$q_n(T_e) = \sqrt{\left(\frac{8kT}{\pi m_e}\right)} \int_{E=En}^{\infty} \left(\frac{E}{kT_e}\right) Q(E) \exp(-E/kT_e) d(E/kT_e) \,, \tag{87}$$

which Hummer expresses in the form:

$$q_n(T_e) = C_n \left(\frac{T}{D_n}\right)^d \exp\left(-\frac{D_n}{T}\right), \tag{88}$$

where $d = 0.12$ for hydrogen and the constants C_n and D_n are given in Table 7 for a few values of n. Finally, if we set $\lambda = 157{,}890/T_e$, we get Hummer's formula,

$$\Phi(T_e) = \exp(-0.12 \ln \lambda)[39 \exp(-0.75\,\lambda) + 7.7 \exp(-0.889\,\lambda) + 2.8 \exp(-0.938\,\lambda) + 1.4 \exp(-0.96\,\lambda) + 3.8 \exp(-\lambda)] \times 10^{-9}\,. \quad (89)$$

For ionizations, Hummer finds that $q(T) = 5.3 \times 10^{-11} \sqrt{T} \exp(-157{,}900/T)$.

For optically permitted transitions (for example, 2^3S–n^3P) in helium, Osterbrock (1964) writes equation (88) in the form:

$$q(i \rightarrow j) = 3.0 \times 10^{-6} \frac{f}{E_0^{3/2}} \left(\frac{T}{11{,}600 E_0}\right)^{0.2} 10^{-5040\,E_0/T}\,, \quad (88a)$$

where E_0 is the excitation energy of the triplet level above 2^3S (expressed in ev) and f is the oscillator strength. For example, for the 10,830 A line, $E_0 = 1.14$ and $f = 0.546$.

TABLE 7

PARAMETERS FOR COLLISIONAL EXCITATION

n	$C_n \times 10^9$	D_n
2	51	118,400
3	8.5	140,400
4	3.0	148,040
5	1.44	151,600

Courtesy, D. G. Hummer, *M.N.R.A.S.*, **125**, 437, 1963.

We now put in the explicit expressions for the terms in equation (76), viz., equations (77), (78), (79), and (80). Thus we obtain

$$N_i N_e \frac{KZ^4}{T_e^{3/2}} kT_1 b_1 e^{X_1} \int_{y_1}^{\infty} \frac{W g_{II} dy}{(e^y - 1)} = N_i N_e \frac{KZ^4}{T_e^{3/2}} \left(\frac{k^2}{2hRZ^2}\right) g_{III} T_e^2 + N_i N_e \frac{KZ^4}{T_e^{3/2}} k \Sigma \frac{\langle g \rangle}{n^3} T_e + N_i N_e \frac{KZ^4}{T_e^{3/2}} 2hRZ^2 \left\{\tfrac{1}{2} G^*_{T_e} - \tfrac{1}{2} \sum_1^{\infty} \frac{S_n^*}{n^5}\right\} + (E_{\text{coll}})\,. \quad (90)$$

Now we divide equation (90) by $2hRZ^2$ and $N_i N_e KZ^4 T_e^{-3/2}$, and by the equation of statistical equilibrium:

$$\frac{\left(\dfrac{kT_1}{2hRZ^2}\right) \displaystyle\int_{y_1}^{\infty} W_\nu \frac{g_{II} dy}{e^y - 1}}{\displaystyle\int_y^{\infty} \frac{W_\nu g_{II} dy}{y(e^y - 1)}} = \frac{\left(\dfrac{k}{2hRZ^2}\right)^2 T_e^2 \ll g_{III} \gg + \left[\left(\dfrac{k}{2hRZ^2}\right) \displaystyle\sum_1^{\infty} \frac{g_{II}}{n^3}\right] T_e + \tfrac{1}{2} G^*_{T_e} - \tfrac{1}{2} \Sigma \dfrac{S_n^*(T_e)}{n^5}}{G^*_{T_e}} + \frac{T_e^{3/2}}{N_i N_e 2hRZ^2 G^*_{T_e} KZ^4} \Big[E_1(\text{coll}) + E_2(\text{coll})\Big]\,. \quad (91)$$

We assume that although energy may be dissipated in the collisional excitation of hydrogen lines, the number of collisional ionizations is negligible compared with the number of radiative ionizations. Otherwise we must include the collisional term in equation (71), and equation (75) is replaced by a more complex expression (see, for example, Chamberlain 1953). In practice, however, it would appear from the work of Hummer that collisional ionization is usually negligible compared with photo-ionization.

The left-hand side of equation (75) depends on W_ν—the dilution factor which is modified by the transfer of radiation through the nebula—and on T_1—the temperature (parameter) used to characterize the energy distribution of the central star. A particularly simple situation is provided when W_ν is independent of ν, so that the left-hand side depends only on T_1.

Thus, if collisional ionization is negligible, we may write the equation in the form (Menzel and Aller 1941*a*, eq. [16])

$$f_1(T_1) = f_2(T_e) + C(T_e)\,, \tag{92}$$

where $f_2(T_e)$ is the first term on the right-hand side of equation (91) and depends only on T_e. Note that $\langle g_{\rm II}\rangle$ and $\langle\langle g_{\rm III}\rangle\rangle$ also depend on T_e and must be evaluated accurately. Finally, substituting for $E(\mathrm{H}\beta)$ from equation (32) and equation (85), we get

$$C(T_e) = 1.61\times 10^{-3}\,\frac{b_4(T_e)\,e^{X_4}}{G_{T_e}}\,\Sigma\,\frac{I(FL)}{I(\mathrm{H}\beta)} + \frac{T_e^{3/2}}{2KZ^6}\,\frac{N_{\rm H}}{N(\mathrm{H}^+)}\,\frac{[\,q(T_e)+\Phi(T_e)\,]}{G_{T_e}}\,. \tag{93}$$

Equation (92) simplifies to

$$f_1(T_1) = f_2(T_e) + \mathcal{A}(T_e)\,\Sigma\,\frac{I(FL)}{I_{\mathrm{H}\beta}} + 1.53\times 10^5\,\frac{T_e^{3/2}}{G_{T_e}}\,\Phi(T_e)\,\frac{N_{\rm H}}{N(\mathrm{H}^+)}\,, \tag{94}$$

where $\mathcal{A}(T_e)$ has been tabulated (Aller 1956, p. 137), and $f_2(T_e)$ increases from 0.531 to 0.600 as T_e increases from 10,000° K to 30,000° K.

The dissipation of thermal energy by excitation of forbidden lines suffices to depress the electron temperature to something less than 20,000° K (Zanstra 1931*b*, Menzel and Aller 1941*b*). Hummer (1963) showed that even for a pure hydrogen nebula, the electron temperature could not rise above 20,000° K because of collisional excitation of discrete levels. An alternate treatment of the energy balance problem was given by Spitzer (1948). Osterbrock (1965*b*) has discussed the influence of fine-structure transitions in the lowest-lying terms of nitrogen, oxygen, neon, and so forth upon the cooling of planetary nebulae.

7. THE FORBIDDEN LINES; ELECTRON TEMPERATURES AND DENSITIES

7.1. Theoretical Considerations

A considerable body of theoretical work has been accumulated to substantiate Bowen's (1928) conclusion that the forbidden lines in the spectra of nova shells and gaseous nebulae are produced by downward transitions from metastable levels to which atoms and ions have been excited by inelastic collisions.

Suppose that an electron traveling with a velocity between v and $v + dv$ hits an atom in a state A and excites it to a state B. If $\sigma(v)$ is the cross-section for collisional excitation (that is, if the electron hits within the target of area $\sigma(v)$, it will surely excite the atom), then the total number of collisional excitations/cm^{-3} sec^{-1} will be $N_A N_e v\sigma(v)f(v)dv$, since a column $v\sigma(v)$ is swept up each second by a particle moving with a velocity v. In each cm^3 there are N_A atoms in level A and $N_e f(v)dv$ electrons with velocity between v and $v + dv$. Here

$$f(v)dv = 4\pi\left(\frac{m}{2\pi kT_e}\right)^{3/2} v^2 \exp(-mv^2/2kT)\,dv \tag{95}$$

is the Maxwellian distribution of velocities. Integrating over all velocities greater than the threshold velocity v_0, and noting

$$\tfrac{1}{2}mv_0^2 = \chi_B - \chi_A = \chi_{AB}\,, \tag{96}$$

where χ_{AB} is the difference in energy between levels B and A, we obtain

$$\mathfrak{F}_{AB} = N_A N_e \int_{v_0}^{\infty} v\sigma(v) f(v)\,dv = N_A N_e q_{AB} \tag{97}$$

as the total number of collisional excitations cm^{-3} sec^{-1}. Here q_{AB} is sometimes called the *activation coefficient*. Under conditions of thermodynamic equilibrium, detailed balancing holds, and

$$N_A N_e q_{AB} = N_B N_e q_{BA} \tag{98}$$

where q_{BA} is called the *deactivation coefficient*. These two coefficients are related by

$$q_{AB} = q_{BA}\frac{\varpi_B}{\varpi_A}e^{-\chi_{AB}/kT_e}\,, \tag{99}$$

since in thermodynamic equilibrium, N_B and N_A are related by Boltzmann's law. Furthermore, a relation of the type (98) must hold for all velocities

$$\varpi_A v_A^2 \sigma(A \to B) = \varpi_B v_B^2 \sigma(B \to A)\,. \tag{100}$$

Menzel and Hebb (1940) introduced a quantity Ω called the *collision strength* such that

$$\sigma(A \to B) = \frac{\pi}{k^2}\Omega(A\ B) \tag{101}$$

where

$$k = mv/\hbar\,. \tag{102}$$

In practice each level J will consist of $2J + 1$ Zeeman states. Let there be a such states in level A and b such states in level B. Then

$$\sigma(A - B) = \frac{\pi}{k^2}\frac{\Omega(A, B)}{2J_A + 1}, \quad \text{where} \quad \Omega(A, B) = \sum_{a,b}\Omega(a, b)\,. \tag{103}$$

From equation (100) the collision strengths are such that

$$\Omega(A \to B) = \Omega(B \to A)\,. \tag{104}$$

Calculations of collision strengths have been carried out by a variety of methods including the Born-Oppenheimer approximation and what Seaton (1951, 1953*a*, 1955*c*,

1957, 1958) calls the exact resonance method and the distorted wave method. Czyzak's chapter (this volume) gives a discussion of this difficult problem.

For collisions between electrons and neutral atoms, $\Omega(A - B) \rightarrow 0$ at the threshold energy. Accordingly, Seaton calculates the deactivation coefficient:

$$q_{BA} = \frac{8.63 \times 10^{-6}}{\varpi_B \sqrt{T_e}} \left[\int_0^\infty \Omega(B \rightarrow A) \exp\left(-\frac{E}{kT}\right) dE/kT \right] \text{cm}^3 \text{ sec}^{-1}, \quad (105)$$

where ϖ_B is the statistical weight of the upper level, $E = \frac{1}{2}mv^2$, that is, the kinetic energy of the free electron. Then q_{AB} may be calculated from equation (99). Seaton (1958) gives the deactivation coefficients for oxygen and nitrogen listed in Table 8.

TABLE 8

COLLISIONAL DEACTIVATION COEFFICIENTS FOR OXYGEN AND NITROGEN

T	Oxygen (Calculations)			T	Nitrogen (Estimates)		
	$10^9 q_{21}$	$10^9 q_{31}$	$10^9 q_{32}$		$10^9 q_{21}$	$10^9 q_{31}$	$10^9 q_{32}$
1000	1.6	1.2	0.6	1000	0.8	0.6	1.5
5000	5.0	3.5	1.9	5000	2.6	1.8	4.4
10,000	6.7	5.1	3.0	10,000	3.5	2.7	6.9
50,000	7.6	7.2	5.6	50,000	4.0	3.7	13

Courtesy, M. J. Seaton, *Reviews of Modern Physics*, **30**, 986, 1958 (July).

Because of the attractive field of the ion, an impinging electron can pick up considerable speed in its immediate neighborhood, even if the velocity at infinity is small. Hence Ω will remain finite at the energy threshold. We can write from equation (101)

$$\sigma_{AB} = \frac{1}{2J_A + 1} \frac{h^2}{4\pi m^2} \frac{\Omega(A \rightarrow B)}{v^2} = \frac{4.17}{2J_A + 1} \frac{\Omega(A \rightarrow B)}{v^2} (\text{cm}^2). \quad (106)$$

Using equations (95), (97), and (106), we get

$$\mathfrak{F}_{AB} = N_A N_e \frac{\Omega(A \rightarrow B)}{2J_A + 1} \frac{h^2}{2\pi m^2} \left(\frac{m}{2\pi kT}\right)^{1/2} \exp\left(-\frac{\chi_{AB}}{kT_e}\right) \quad (107)$$

or

$$\mathfrak{F}_{AB} = 8.63 \times 10^{-6} \frac{N_A N_e}{\sqrt{T_e}} \frac{\Omega(A \rightarrow B)}{2J_A + 1} \exp\left(-\frac{\chi_{AB}}{kT_e}\right) \quad (108)$$

or

$$\mathfrak{F}_{BA} = 8.63 \times 10^{-6} \frac{N_B N_e}{\sqrt{T_e}} \frac{\Omega(B \rightarrow A)}{2J_B + 1}. \quad (109)$$

From the calculations by Seaton, Czyzak, Krueger, and their associates, we compiled Table 9, giving the preferred Ω-values which will be used in further calculations.

We may now write down the equations of statistical equilibrium for each level. Consider first the O^{++} [O III] ion. For the ^1S level 3 we have (Menzel, Hebb, and Aller 1941)

$$N_1 N_e q_{13} + N_2 N_e q_{23} = N_3(N_e q_{32} + N_e q_{31} + A_{32} + A_{31}) \quad (110)$$

number of collisional excitations = number of collisional de-excitations + number of radiative de-excitations

Similarly, for the second or $^1D_2$ level,

$$N_1N_eq_{12}+N_3N_eq_{32}+N_3A_{32}=N_2N_e(q_{23}+q_{21})+N_2A_{21}\,. \tag{111}$$

The equation for the ground term, here considered as a single level, could be written down or derived from these two.

We introduce, in the usual way, the factors b_1, b_2, and b_3 to express the deviation of the population from thermodynamic equilibrium. Thus

$$\frac{N_2}{N_1}=\frac{b_2}{b_1}\frac{\varpi_2}{\varpi_1}\exp(-\chi_2/kT_e)\,;\qquad \frac{N_3}{N_1}=\frac{b_3}{b_1}\frac{\varpi_3}{\varpi_1}\exp(-\chi_3/kT_e)\,. \tag{112}$$

For convenience we introduce Seaton's variables

$$x=\frac{10^{-4}N_e}{t^{1/2}}\quad\text{and}\quad t=\frac{T_e}{10{,}000}\,. \tag{113}$$

TABLE 9

TARGET AREA PARAMETERS FOR SELECTED IONS

	O II	O III	F IV	Ne III	Ne IV	Ne V	S II	S III	Cl III	Cl IV	Ar III	Ar IV
Ω(1,2)......	1.43	2.39	1.93	1.26	1.04	1.38	3.07	4.97	3.19	1.99	4.75	1.43
Ω(1,3)......	0.428	0.335	0.279	0.164	0.427	0.218	1.28	1.07	1.97	0.33	0.724	0.645
Ω(2,3)......	1.70	0.310	0.237	0.188	1.42	0.185	6.22	0.96	6.64	1.08	0.665	4.92

Billings, Czyzak, Krueger, Seaton, Saraph, Shemming, and de Martins (1967).

We can solve for the ratios b_2/b_1 and b_3/b_1 (see Menzel, Hebb, and Aller 1941; Aller 1956, pp. 192–93). Defining

$$C=8.63\times10^{-6}\frac{N_e}{\sqrt{T_e}}=8.63\times10^{-4}x\,, \tag{114}$$

we now obtain

$$\frac{b_3}{b_2}=\frac{\left(1+\dfrac{\Omega_{23}}{\Omega_{12}}\dfrac{e^{-\chi_{23}/kT_e}}{(1+\Omega_{23}/\Omega_{13})}+\dfrac{A_{21}\varpi_2}{C\Omega_{12}(1+\Omega_{23}/\Omega_{13})}\right)}{\left[1+\dfrac{(A_{32}+A_{31})\varpi_3}{\Omega_{13}C(1+\Omega_{23}/\Omega_{13})}+\left(\Omega_{32}+\dfrac{A_{32}\varpi_3}{C}\right)\dfrac{e^{-\chi_{23}/kT_e}}{\Omega_{12}(1+\Omega_{23}/\Omega_{13})}\right]}\,. \tag{115}$$

Note that as N_e increases, C increases and $b_3/b_2\rightarrow1$—as indeed it must. We also have

$$\frac{b_2}{b_1}=\frac{[\Omega_{12}+\Omega_{13}d\,\exp(-\chi_{23}/kT_e)]}{\left[\left(\Omega_{12}+A_{21}\dfrac{\varpi_2}{C}\right)+\Omega_{23}(1-d)\exp(-\chi_{23}/kT_e)\right]}\,, \tag{116}$$

and

$$d=\frac{\Omega_{32}+\dfrac{A_{32}}{C}\varpi_3}{\Omega_{13}+\Omega_{23}+(A_{32}+A_{31})\dfrac{\varpi_3}{C}}\,. \tag{117}$$

As an illustration, let us consider the determination of the electron temperature T_e from the intensity ratio of the green nebular lines. We define

$$\eta_2=\frac{b_2}{b_1}\,;\qquad \eta_3=\frac{b_3}{b_1}\,. \tag{118}$$

Then the emission per unit volume in the green nebular lines is

$$E(\lambda 5007) + E(\lambda 4959) = N_2 A_{21} h\nu_{21} = \eta_2 N_i \frac{\varpi_2}{\varpi_1} \exp(-\chi_2/kT_e) A_{21} h\nu_{21}$$

$$E(\lambda 4363) = N_3 A_{32} h\nu_{32} = \eta_3 N_i \frac{\varpi_3}{\varpi_1} \exp(-\chi_{23}/kT_e) A_{32} h\nu_{32}. \quad (119)$$

We use the recommended [O III] cross-sections and $A_{32} = 1.6$, $A_{31} = 0.23$, and $A_{21} = 0.028$.

TABLE 10

TEMPERATURES AND DENSITIES DERIVED FROM FORBIDDEN-LINE DATA

Nebula	Assumed Radius	d/A	$D \times 10^{-17}$	C	I_n/I_a	T_e	Log $\mathfrak{F}$ (Hβ)	Log N_e from	
								Hβ Flux	[O II]
IC 351.......	3″5	0.2	0.66	0.00	94.2	12,500	−1.90	3.41	3.85
NGC 1535....	9″	0.2	0.83	.15	73.3	14,000	−1.49	3.58	3.67
NGC 2022....	10″	0.2	2.1	.32*	60.0	15,100	−2.08	3.11	3.70
NGC 6309....	7″5	1.0	3.2	.05	100	12,200	−2.25	2.89	3.62
NGC 6720....	35″	0.2	1.9	.19	62.7	14,900	−2.32	3.01	2.89†
NGC 6741....	4″	1.0	0.76	.19	131	11,100	−1.72	3.47	3.53
VV$_{\text{I}}$ 267......	2″5	1.0	1.4	0.14	49	16,700	−1.50	3.50	....

* Minkowski's estimate (personal communication).

† Seaton and Osterbrock (1957), *VV*$_{\text{I}}$ 267 = Anon 21^{h}29^{m} = *VV*$_{\text{II}}$ 553 α = 21^{h}31^{m}1, δ = +39°25′ (1950).

Taking advantage of the fact that for most planetary nebulae $x \approx 1$, $\exp(-\chi_{23}/kT_e) \approx 0.1$ and that the Ω-values are subject to large uncertainties, we find

$$\frac{b_2}{b_1} \approx \frac{1}{1+67/x}, \qquad \frac{b_3}{b_2} \approx \frac{1+0.028x}{93}. \quad (120)$$

Then

$$\frac{I(N_1+N_2)}{I(4363\text{A})} \approx \frac{a10^{1.43/t}}{1+bx} = \frac{a10^{14,300/T_e}}{1+0.01bN_e/\sqrt{T_e}}, \quad (121)$$

where $a = 7.1$ and $b = 0.028$ with the above choice of target areas.

As an example of the application of these equations, we consider six nebulae of moderate to high excitation; spectroscopic data are from Aller and Walker (1965). First, we estimate the degree of space absorption, preferably by comparing Balmer and Paschen lines—otherwise from a comparison of predicted and observed Balmer decrements. Column 5 of Table 10 gives the correction parameter C for space absorption for each nebula.

$$C = \log F(\text{H}\beta)/F_0(\text{H}\beta), \quad (122)$$

where $F(\text{H}\beta)$ is the 4861 A flux corrected for space absorption and $F_0(\text{H}\beta)$ is the observed flux.

Next, we make an estimate of T_e from the intensity ratio (5007 A + 4959 A)/4363 A (column 6). Equation (121) with $x = 1$ gives a very good approximation (column 7) for the nebulae under consideration. If the nebula can be approximated as a shell of thickness d and outer radius A, we define

$$D = 3d\left(1 - \frac{d}{A} + \frac{d^2}{3A^2}\right). \quad (123)$$

The emission in Hβ per unit volume, $E(\text{H}\beta) = 10^{-25} N_i N_e E^0_{4,2}$, is related to the flux in ergs cm^{-2} sec^{-1} through the surface by

$$\mathfrak{F}(\text{H}\beta) = \tfrac{1}{3} N_i N_2 E^0_{4,2} D 10^{-25} = \tfrac{1}{3} E(\text{H}\beta) D \,. \qquad (124)$$

If $N_i(\text{H}) = 0.85 N_e$ then

$$N_e^2 = 3.52\, \mathfrak{F}(\text{H}\beta) D^{-1} E^{0\,-1}_{4,2} 10^{25} \,. \qquad (125)$$

The second column of Table 10 gives the outer radius of the nebula in seconds of arc (see Curtis 1918); the third column gives the value of d/A for those nebulae that can be regarded approximately as shells.

To calculate $A(\text{cm}) = A'' \times r \times 1.494 \times 10^{13}$ where r is the distance in parsecs, and ultimately D(cm) (see column 4), we employ the distances as given by Minkowski (private communication). Column 8 gives log $\mathfrak{F}(\text{H}\beta)$ (in ergs cm^{-2} sec^{-1}) as deduced from photoelectric measurement (mostly by the present authors) or by O'Dell (1963*c*) combined with measurements of the 5007 A/Hβ ratio.

Using Clarke's data for $E^0_{4,2}$ (Table 6) for Menzel and Baker's case B, we now calculate log N_e (column 9) which can be compared with the log N_e values from the [O II] lines in column 10. For the objects listed in Table 6 which are of only moderate surface brightness, the electron density is about 1×10^3 to 5×10^3 electrons/cm^3. For objects of higher densities, we could recalculate T_e with improved values of x and then obtain a new estimate of N_e from the surface brightness.

For N$^+$ we employ $A_{21} = 0.0041$, $A_{32} = 1.08$, $A_{31} = 0.034$, and $\Omega(1,2) = 3.13$, $\Omega(1,3) = 0.342$, and $\Omega(2,3) = 0.38$.

$$\frac{I(6548\text{A}) + I(6584\text{A})}{I(5755\text{A})} = \frac{8.5 \times 10^{+10,900/T}}{1 + 0.0029 N_e/\sqrt{T_e}} = \frac{8.5 \times 10^{+1.09/t}}{1 + 0.29x} \,. \qquad (126)$$

These coefficients differ somewhat from those given by Seaton (1960*a*).

7.2. Electron Temperatures and Densities from Observed Intensity Ratios

We assume throughout that the densities are such that collisional transitions between the members of the ground term dominate over the radiative transitions.

Notice that equations (121) and (126) each connect an observed ratio $I_{\text{neb}}/I_{\text{aur}}$ with a function of T_e and N_e. Hence, as was pointed out by Aller and White (1949), if we are justified in assuming that the [N II] and [O III] radiations are emitted by the same strata in the nebula, the measured intensity ratios can be used to obtain both T_e and N_e. The electron densities obtained in this way seemed to be greater than those found from surface-brightness measurements, suggesting that much of the emission might occur in filaments of higher than average density. Subsequent work (Seaton 1954; Aller 1954*a*, 1956, pp. 193–99, 1957; Seaton and Osterbrock 1957) with improved target areas has substantiated this conclusion.

In addition to [N II] and [O III] lines, we may use transitions of certain other ions for which reliable Ω-values have been computed, for example, [Ne III], [S II], [S III], and [Cl III]. Unfortunately, for most of these ions, the relevant lines fall in regions of the spectrum that are difficult to observe, are very weak, or are blended. The [S II] lines show a particular anomaly; the transauroral lines, 4068 A and 4076 A, always tend to be enhanced in regions of high density; hence they are particularly valuable for assessing density fluctuations in an object such as NGC 6543.

7.3. The 3729 A/3726 A Ratio of [O II]

The ground configuration of the O^+ ion is $2p^3$ with a $^4S_{3/2}$ term as the lowest level. Transitions from the $^2D_{3/2}$, $^2D_{5/2}$ levels to this ground $^4S_{3/2}$ level produce the 3726–3729 A pair of emissions which are prominent in the spectra of many diffuse and low-density planetary nebulae. Pasternack (1940) predicted an intensity ratio, $I(3729\text{ A})/(3726\text{ A}) = 1.9$, and the theory by Shortley *et al.* (1941) gave 1.64 (Aller and Menzel 1945). These ratios contradict the observations. The discrepancy was clarified by a refined calculation of the transition probabilities for these [O III] lines in which account was taken of the spin-spin and spin-other orbit interactions (Aller, Ufford, and van Vleck 1949). In comparing theoretical predictions with observations secured at the McDonald Observatory, they noticed that the ratio

$$r = \frac{I(3729\text{ A})}{I(3726\text{ A})} = \frac{I(^4S_{3/2} - ^2D_{5/2})}{I(^4S_{3/2} - ^2D_{3/2})} \qquad (127)$$

was not constant, but varied from one nebula to another in such a way as to indicate a dependence on the density. At low densities, collisional processes do not suffice to maintain a Boltzmann distribution among the ^2D levels.

The exact theoretical value of the ratio, therefore, depends on the electron density, the electron temperature, and the cross-section for collisional transitions between $^2D_{3/2}$, $^2D_{5/2}$, and $^2P_{3/2}$, $^2P_{1/2}$. The ratio r will be larger the smaller the electron density, while with increasing electron density, the intensity ratio should approach the theoretical value (predicted by radiative de-excitations alone). The large value of the ratio for NGC 40 (Plates 1, 3, 4), which presumably has the lowest density, and the small value for IC 4997, which is known to be the most dense of the nebulae studied, are in qualitative agreement with expectations.[2]

Reliable calculations of the target-area parameters enabled Seaton (1953*a*, *b*), and later Seaton and Osterbrock (1957), to obtain a precise expression relating r to temperature and density. If x is defined by equation (113) and ϵ is given by

$$\epsilon = \exp(-1.96t) , \qquad (128)$$

Seaton and Osterbrock obtain

$$r = 1.50 \left[\frac{1 + 0.33\epsilon + 2.30x(1 + 0.75\epsilon)}{1 + 0.40\epsilon + 9.9x(1 + 0.84\epsilon)}\right]. \qquad (129)$$

If the newer target areas are used, the expression for r is changed slightly but not by enough to require modification of equation (129). For a temperature of 10,000° K and low densities,

$$x = (1.49/r - 1)/(10.45 - 3.64/r) . \qquad (130)$$

Expressions may be derived for the ratio

$$r' = I(7320\text{ A})/I(7330\text{ A}) = 0.810 \qquad (131)$$

[2] Aller, Ufford, and van Vleck (1949).

to within the precision of the theory, and for the ratio $r'' = [I(3726\text{ A}) + I(3729\text{ A})]/$ $I(7320\text{ A}) + I(7330\text{ A})]$

$$r'' = \frac{5.5}{\epsilon}\left[\frac{1+0.36\epsilon+5.3x(1+0.82\epsilon)}{1+13.8x(1+0.38\epsilon)+38.4x^2(1+0.78\epsilon)}\right]. \tag{132}$$

Note that equation (129) gives the density in filaments, and so forth, where the [O II] lines actually originate. These densities often differ from those found from the surface brightnesses which yield a sort of average of N_iN_e over the radiating volume. The 3726 A/3729 A ratio is sensitive to density only over a limited range where the rates of radiative and collisional depopulations of the ^2D term are comparable. At lower densities, radiative processes dominate and r approaches a limiting value of about 1.5 for $t = 1.0$. At very high densities, the ratio again becomes insensitive to N_e, approaching the limiting value 0.35. At other density ranges, we may use lines of other ions such as [S II].

7.4. Comparison of Ionic Densities

Since all forbidden lines arise from the same physical mechanism, it is convenient to refer the corresponding ionic densities to the density of O^{++} ions. Let the subscript (or superscript) "O" denote quantities pertaining to the O^{++} ion; for example, $N_O = N(O^{++})$ and $I_O = I(5007\text{ A}) + I(4979\text{ A})$. From equations (118), (119), and (120), we may write for the nebular-type transitions, for the ions of the jth type

$$\frac{N_j}{N_O} = \left(\frac{I_n}{I_O}\,\frac{5}{9}\,\frac{\lambda}{\lambda_O}\,\frac{\varpi_1}{\varpi_2}\,\frac{A_O}{A}\right)\frac{\overset{O}{\eta_2}}{\eta}\,10^{(\chi-\chi_o)\theta} \tag{133}$$

$$\frac{N_j}{N_O} = C_\lambda\,\frac{\overset{O}{\eta_2}}{\eta_\circ}\,10^{(\chi-\chi_o)\theta}\,\frac{I_n}{I_O} = P_j(T_e,\,N_e)\frac{I_n}{I_O}, \tag{134}$$

where C_λ is defined as the term in parantheses in equation (133). Most of the quantities referring to the other ion are not designated with any subscript. Here A is the sum of the A-values for the nebular-type transitions; I_n is the intensity of the nebular transitions, corrected for space absorption; $\bar{\lambda}$ is the mean wavelength; $\theta = 5040/T$ and χ is the excitation potential of the upper term of the nebular transition. We calculate $P_j(T_e, N_e)$ as a function of T_e, N_e and tabulate it for ranges of astrophysical interest, provided we have sufficiently accurate target areas.

Here $\eta_2^0 = (1 + 67/x)^{-1}$, $\lambda_0 = 4990$ A, $A_0 = 0.0281$, $\chi = 2.48$ ev. Similar expressions can be derived for the auroral lines. To get the actual ionic densities $N_{i/j}$, we must now determine the ratio $N(O^{++})/N(H^+)$. Noting that

$$\frac{E(O^{++})}{E(H\beta)} = \frac{E(5007\text{A})+E(4959\text{A})}{E(H\beta)} = \frac{I(5007\text{A})+I(4959\text{A})}{I(H\beta)} = \frac{I_O}{I(H\beta)}, \tag{135}$$

and since, equation (5),

$$E(H\beta) = N_iN_eE^0_{4,2}\times 10^{-25} \tag{136}$$

$$\begin{aligned} E(5000\text{A})+E(4959\text{A}) &= \tfrac{5}{9}N(O^{++})\frac{10^{-1.25/t}}{1+67/x}A_O h\nu \\ &= 6.18\times 10^{-14}\,N(O^{++})\frac{10^{-1.25/t}}{1+67/x}, \end{aligned} \tag{137}$$

we have

$$N(\mathrm{O}^{++})/N(\mathrm{H}^{+}) = 0.161 \times 10^{-11} E^{0}_{4,2} 10^{12,500/T_e} (N_e + 6700\sqrt{T_e}) \frac{I_{\mathrm{O}}}{I(\mathrm{H}\beta)} \quad (138)$$

or

$$\frac{N(\mathrm{O}^{++})}{N(\mathrm{H}^{+})} = \mathfrak{J}(T_e, N_e) \frac{I_{\mathrm{O}}}{I(\mathrm{H}\beta)}, \quad (139)$$

wherein the dependence of $\mathfrak{J}$ on N_e is very slight unless $N_e \gg 10^4$. For the lower densities, we calculated $\mathfrak{J}$ from $E^0_{4,2}$ for case B as given in Table 6. See Table 11.

TABLE 11

COEFFICIENT FOR CALCULATION OF $N(\mathrm{O}^{++})/N(\mathrm{H}^{+})$

T_e (° K)	log $\mathfrak{J}$ (N_e, T_e)
5,000	−4.270
10,000	−4.625
15,000	−5.105
20,000	−5.374
40,000	−5.834

8. CONTINUOUS SPECTRA OF THE PLANETARY NEBULAE

The most striking, immediately evident feature of the continuous spectra of planetary nebulae is the intensity jump at the head of the Balmer series, produced by recombinations to the second level of hydrogen (Fig. 4; Plate 4). The amount of energy emitted $\mathrm{cm}^{-3}\ \mathrm{sec}^{-1}$ in the Balmer continuum between frequencies ν and $d\nu$ will be (Menzel 1937, eq. [23])

$$E_{\kappa 2} d\nu = N_i N_e \frac{hKZ^4}{T_e^{3/2}} e^{X_2} \frac{g_{\mathrm{II}}(\nu)}{2^3} \exp(-h\nu/kT_e)\, d\nu. \quad (140)$$

The frequency dependence of $g_{\mathrm{II}}(\nu)$ is relatively low (Menzel and Pekeris 1935, Burgess 1958, Clarke 1965).

In principle, we could use this equation to obtain the electron temperature of the gas, provided we knew the energy distribution in the underlying continuum. In practice, the latter is not known, and there are great difficulties in accurately measuring the ultraviolet spectral energy distribution. Space absorption must also be determined accurately.

The flux emergent from the nebula at the Balmer limit may be used to get the elec tron density, provided we have some estimate of the electron temperature. Let $\mathfrak{F}_{\mathrm{Bac}}(\Delta\lambda)$ be the Balmer continuum flux in ergs $\mathrm{cm}^{-2}\ \mathrm{sec}^{-1}$ in a wavelength interval $\Delta\lambda$ at the Balmer limit. Suppose that the nebula is a spherical shell of inner and outer radius r_i and r_o, respectively. Then, putting in numerical values

$$\mathfrak{F}_{\mathrm{Bac}}(\Delta\lambda) = 5.5 \times 10^{-4} \frac{N_i N_e}{T^{3/2}} r_o \left[1 - \left(\frac{r_i}{r_o}\right)^3\right] \Delta\lambda \text{ ergs cm}^{-2}\text{ sec}^{-1}, \quad (141)$$

where r_o and r_i are in parsecs and $\Delta\lambda$ is in angstroms.

Measurements of $\mathfrak{F}_{\mathrm{Bac}}(\Delta\lambda)$ compared with $\mathfrak{F}(\mathrm{H}\beta)$ (Page 1942, Aller 1941) are in reasonable accord. The main difficulty lies in estimating the position of the background continuum. Now from equation (82),

$$\mathfrak{F}(\mathrm{H}\beta) = 0.236 \frac{N_e N(\mathrm{H}^{+})}{T_e^{3/2}} b_4(T_e) e^{X_4} r_o \left[1 - \left(\frac{r_i}{r_o}\right)^3\right] \text{ergs cm}^{-2}\text{ sec}^{-1}. \quad (142)$$

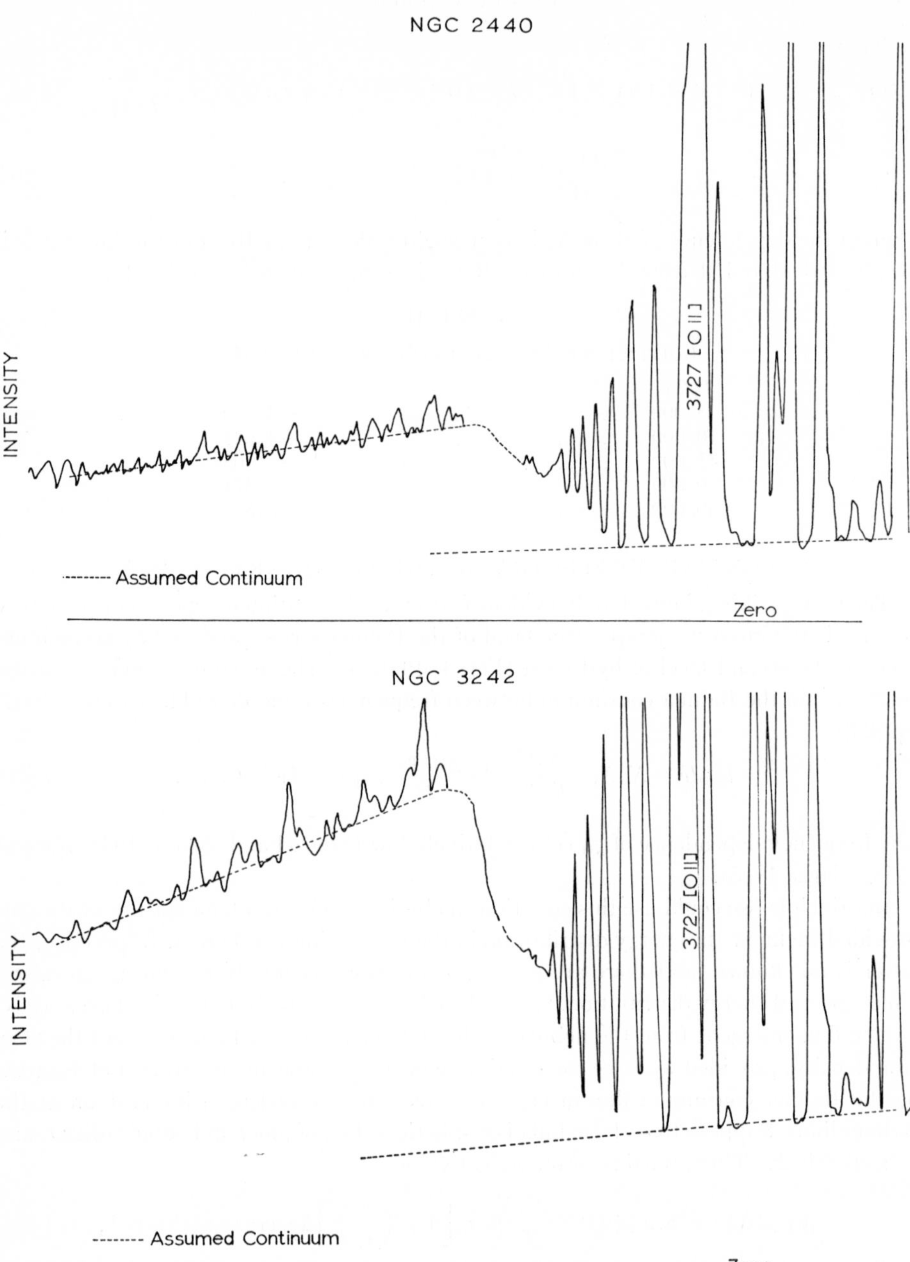

FIG. 4.—The Balmer discontinuity in NGC 2440 and NGC 3242. These tracings (on an intensity scale) show the region near the Balmer limit. The [O II] lines 3726 and 3729 are blended together as one line denoted as λ 3727. The line immediately to the right of this pair is λ 3734 (H13). Several helium lines are prominent on the short wavelength side of the Balmer jump. The dotted lines show the estimated position of the continuum.

Hence, the ratio

$$\frac{\mathfrak{F}(\mathrm{H}\beta)}{\mathfrak{F}_{\mathrm{Bac}}(20\mathrm{A})}=21.4\,b_4(T_e)\,e^{X_4} \tag{143}$$

where $\mathfrak{F}_{\mathrm{Bac}}$ (20A) is the flux from a 20A interval at the Balmer limit, should be dependent on the temperature and on the theory for the Balmer decrement which determines $b_4(T_e)$. Analyses by Seaton (1955*d*) and by Aller (1956, p. 148) show good agreement with predictions at least to within the errors of observation.

We can use either equation (141) or (142) to determine the product $N_eN(\mathrm{H}^+)$, provided that the distance of the nebula is known. Because of the observational difficulty in measuring $\mathfrak{F}_{\mathrm{Bac}}(\Delta\lambda)$, it is usually more practical to determine $N_eN(\mathrm{H}^+)$ from the surface brightness in the Hβ line.

Recombinations to the second level contribute to the continuum only at wavelengths shorter than 3650 A. Recombinations to the third and higher levels contribute throughout the visible region and near-infrared. Some contribution is also made by recombinations to excited levels of helium and in high-excitation planetaries by recombinations in ionized helium for which the formal theory is similar to that of hydrogen (see, for example, Seaton 1960*a*, *b*). The recombination emission spectrum of He I has been observed by Menzel in NGC 6543.

We must also consider the free-free emissions. These quanta contribute the thermal radiation of the planetaries in the radio-frequency region (see Sec. 9). The energy emitted in the frequency range ν to $\nu + d\nu$ from bound-free recombinations and free-free transitions in hydrogen will be (see Menzel 1937, 1962; Aller 1956, pp. 118 *et seq.*, p. 152; Seaton 1960*a*)

$$\begin{aligned}E_{\mathrm{H}}d\nu&=N_iN_e\left\{\frac{K}{T_e^{3/2}}\exp(-h\nu/kT_e)\left[\sum_{n_m}^{\infty}\frac{\langle g_{\mathrm{II}}\rangle}{n^3}\exp(X_n)+\frac{kT_e}{2hR}\langle g_{\mathrm{III}}\rangle\right]\right\}h\,d\nu\\&=N_iN_eh\gamma\,d\nu\,,\end{aligned} \tag{144}$$

where γ, the quantity in brackets, has been tabulated by Seaton (1960*a*). Here the lower limit of the summation n_m is given by $\nu > \nu_m = hRZ^2/n_m^2$. The $\langle g_{\mathrm{II}}\rangle$ and $\langle g_{\mathrm{III}}\rangle$ factors vary slowly with frequency.

The 2-quantum emission (Mayer 1931) is produced by hydrogen atoms escaping from the 2*s* metastable level. The atom jumps from the 2*s*-level to a fictitious *p*-level between it and the ground level, and thence to the 1*s* ground level. Hence the condition on the photons emitted is that $\nu_1+\nu_2\sim 2\nu$ (Lyα). Spitzer and Greenstein (1951) showed that this process could make an important contribution to the continuous spectra in planetaries. Let X denote the probability that a recombination in the second or higher level will result in the atom reaching the 2*s*-level. Then the 2-photon emission in the interval ν to $\nu+d\nu$ will depend on the energy of the transition, $h\nu$, the probability of emission in the interval involved, $\psi(\nu)$, and X multiplied by the sum of recaptures on all levels except the first. There is also a factor of 2 because two quanta are emitted in every 2-photon emission. Thus (cf. Aller 1956, p. 152; Seaton 1960*a*),

$$E_{2q}d\nu=N_iN_e\left\{\frac{K}{T_e^{3/2}}\left[\sum_{2}^{\infty}\frac{e^{X_n}}{n^3}g_{\mathrm{II}}E_1(X_n)\,2X\nu\psi(\nu)\right]\right\}h\,d\nu=\gamma_o(2q)\,N_iN_eh\,d\nu\,, \tag{145}$$

provided the density is so low that essentially all atoms in the 2*s*-level escape from it by radiation rather than by collisions.

When the density is high, atoms tend to escape from the 2*s*-level by collision; hence the 2-photon emission will be reduced. Seaton (1955*d*) finds the rate of collisional deactivation to be determined primarily by electron impacts. For $T_e \approx 10{,}000^\circ$ K, he obtains

$$\gamma(2q)=\frac{\gamma_0(2q)}{1+0.6\times10^{-4}N_e}. \qquad (146)$$

This correction becomes important in nebulae with dense filaments, for example, NGC 7027.

TABLE 12

DATA FOR THE CALCULATION OF CONTINUOUS EMISSION IN PLANETARY NEBULAE

λ (A)	$\nu\times10^{14}$	10,000		12,000		15,000		20,000	
		[1]	[2]	[1]	[2]	[1]	[2]	[1]	[2]
9100	3.29	7.75	2.06	7.65	1.80	7.41	1.53	7.02	1.24
Paschen limit	3.65	6.55	2.42	6.62	2.08	6.60	1.77	6.43	1.46
	3.65	17.5	2.4	15.0	2.08	12.5	1.77	10.3	1.46
6750	4.45	12.1	3.2	11.1	2.79	9.88	2.33	8.61	1.92
5700	5.26	8.36	4.05	8.11	3.52	7.72	3.0	7.13	2.43
4560	6.59	4.46	5.46	4.85	4.76	5.11	4.05	5.25	3.29
Balmer limit = 3650	8.22	2.09	7.23	2.56	6.31	3.05	5.36	3.57	4.36
	8.22	37.6	7.23	29.6	6.31	22.5	5.36	16.1	4.36
3310	9.05	25.8	8.12	21.6	7.07	17.5	6.00	13.4	4.88
3040	9.86	17.7	8.97	15.85	7.83	13.7	6.66	11.2	5.41

Adapted from Seaton (1960*a*), courtesy, Institute of Physics and the Physical Society.

Table 12 (adapted from Seaton 1960*a*) gives data for calculation of continuous emission by a hydrogen nebula from 3040 A to 9100 A. The first two columns give the wavelength and frequency. Then, for each temperature, are tabulated two quantities:

$$[1]=10^{14}\gamma=\frac{E_{\rm H}\times10^{14}}{N_iN_eh}; \qquad [2]=10^{14}\gamma^0_{2q}=\frac{E_{2q}\times10^{14}}{N_iN_eh}. \qquad (147)$$

Figure 5 displays the sum of the quantities [1] and [2], log $10^{14}E(\mathrm{H}+2q)/N^1N_eh$. Note that the effect of the 2-quantum emission is to cut down the magnitude of the Balmer jump and to flatten the energy distribution when it is plotted on a 1/λ scale.

Finally, we may add the helium contribution, noting that it may be regarded as essentially hydrogenic insofar as the higher levels and free-free transitions are concerned (cf., for example, Burgess and Seaton 1960*c;* Goldberg 1939). The continuous absorption coefficients $a_j(\nu)$ may be connected to recombination coefficients $\sigma_j(\nu)$ with the aid of Milne's relation:

$$\frac{a_j(\nu)}{\sigma_j(\nu)}=\frac{m^2v^2c^2}{\nu^2h^2}\frac{\varpi_\epsilon\varpi_i}{2\varpi_j}, \qquad (148)$$

where $\varpi_\epsilon = 2$ is the statistical weight of the electron, ϖ_i is the weight of the ion, and ϖ_y is the weight of level *j*. Much more observational work needs to be done on the continuous spectra of planetaries. Measurements obtained by various observers (Aller 1941, Page

1942, Barbier and Andrillat 1954, Minkowski and Aller 1956) are not in sufficiently good agreement. One difficulty arises from the extreme faintness of these continua in most planetaries. Observers are tempted to use wide slits; then weak nebular lines contribute to the observed "continuum." Measurement of the Balmer "jump" is complicated by the influence of the hydrogen lines.

The theory of the nebular continuum has important applications in the context of radio-frequency observations. The observed values of the Balmer discontinuity (Table 13) (secured from plates obtained at Lick Observatory) are compared with values pre-

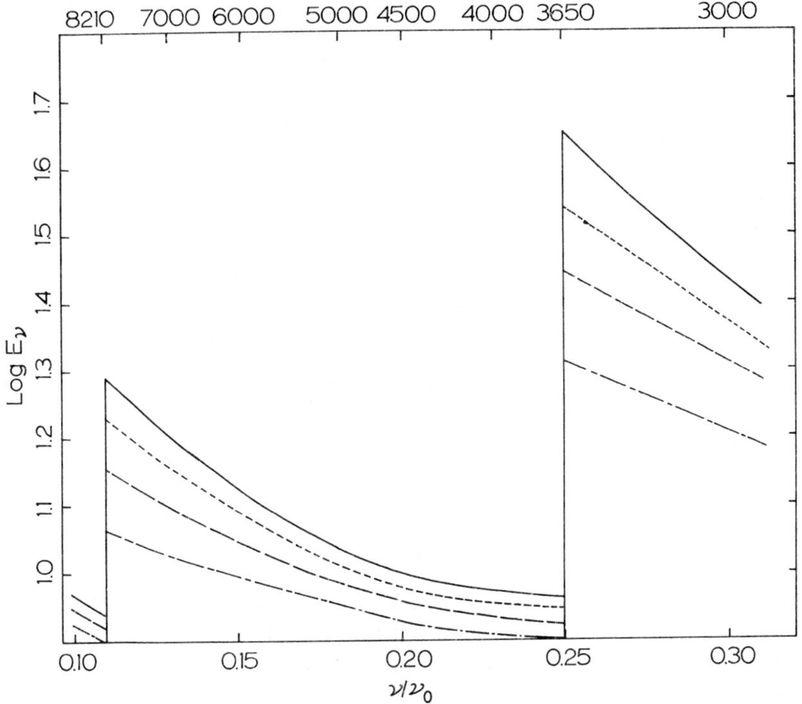

FIG. 5.—The continuous emission in hydrogen. We plot $\log E_\nu = \log 10^{14}E(\mathrm{H} + 2q)/N_iN_ehKT_e^{-2/3}$ vs. ν/ν_0 where ν_0 is the frequency of the Lyman limit. The upper scale gives the wavelength in angstroms. *Solid line*, $T = 10{,}000°$ K; *short dashes*, $T = 12{,}000°$ K; *long dashes*, $T = 15{,}000°$ K; *dash-dot*, $T = 20{,}000°$ K.

TABLE 13

BALMER DISCONTINUITY IN PLANETARY NEBULAE

Nebula	T_e (° K)	log N_e	D (Predicted)	D (Observed)
NGC 2440.......	13,400	3.47	0.55	0.45
NGC 3242.......	11,500	4.00	.80	.58
IC II 4634.......	12,400	3.40	.63	.14
NGC 6309.......	14,500	3.00	.53	.52
NGC 7009.......	11,300	3.80	.76	.77
NGC 7026.......	10,300	3.93	0.68	0.67

dicted from Table 12 and equation (146) with the aid of electron temperatures and densities as given in columns 2 and 3. Here $D = \log I(\lambda < 3650\text{ A}) - \log I(\lambda > 3650\text{ A})$. The measurement is difficult because we must extrapolate the underlying continuum from beneath a host of strong lines through the confluence of the Balmer series. Observed and predicted values of D are in fairly good agreement for NGC 6309, NGC 7009, and NGC 7026, but NGC 3242, IC II 4634, and the very inhomogeneous nebula NGC 2440 give too small a discontinuity.

9. RADIO-FREQUENCY EMISSION

9.1. Theoretical Considerations

An incandescent ionized gas will emit radiation in the radio frequency as well as in infrared, optical, and possibly X-ray regions. If the gas has a "normal" chemical composition, recombination and discrete transitions dominate the optical regions, but except for transitions between very high discrete levels the emission in the radio-frequency region arises entirely from free-free transitions. Let k_ν denote the mass absorption coefficient for these free-free transitions. Then the emission per unit volume over all angles will be

$$E_\nu d\nu = \rho j_\nu d\nu = 4\pi\rho k_\nu B_\nu(T) d\nu\,. \tag{149}$$

In the far-infrared and particularly in the radio-frequency region, the Planckian function $B_\nu(T)$ may be replaced by the Rayleigh-Jeans approximation; that is,

$$B_\nu(T) = \frac{2\nu^2 kT}{c^2}. \tag{150}$$

The absorption coefficient per unit volume (for a derivation, see, for example, Aller 1963, p. 187) may be written in the form

$$\rho k_\nu = \frac{8\pi\epsilon^6 Z^2}{3\sqrt{(6\pi)}\, c\,(mk)^{3/2}} \frac{N_i N_e}{T_e^{3/2}} \frac{g_{\text{III}}}{\nu^2} = 0.01771 \frac{N_i N_e}{T_e^{3/2}} \frac{g_{\text{III}}}{\nu^2}, \tag{151}$$

with the same notation as above; g_{III} is the Gaunt correction factor for free-free transitions. The Gaunt factor has been discussed by Menzel and Pekeris (1935), by Grant (1958), by Karzas and Latter (1961), by Oster (1961), and by Brussaard and van de Hulst (1962).

We employ here, however, a formula by Elwert (1948), which involves asymptotic expressions valid for the radio-frequency region. The asymptotic form to be used depends on the value of the discriminant

$$d = \frac{8.1 \times 10^{-6}\nu}{T_e^{1/2} N_e^{1/3}}. \tag{152}$$

If $d \ll 1$,

$$g_{\text{III}} = \frac{\sqrt{3}}{\pi} \log_e \frac{420 T_e}{Z N_i^{1/3}}, \tag{153}$$

whereas if $d > 1$,

$$g_{\text{III}} = \frac{\sqrt{3}}{\pi} \log_e \left(\frac{(2k)^{3/2}}{4.22\pi m^{1/2}\epsilon^2} \frac{T_e^{3/2}}{Z\nu} \right) \tag{154}$$

or

$$g_{\text{III}} = 0.5513 \log_e \left(4.97 \times 10^7 \frac{T_e^{3/2}}{Z\nu} \right). \tag{155}$$

Consider as an example the thermal radiation from a planetary for which $N_e = 10^4$ and $T_e = 17{,}000°$ K for $\lambda = 4$ cm (that is, $\nu = 7.49 \times 10^9$ sec^{-1}). We find $d \approx 21$. Hence

$$g_{\rm III} \approx -2.76 + 1.904 \log_{10} T_e \tag{156}$$

and

$$\rho k_\nu = 0.315 \times 10^{-21} \frac{N_i N_e}{T_e^{3/2}} g_{\rm III} . \tag{157}$$

Consider now a homogeneous sphere of radius R and uniform temperature T and emissivity. Consider a column of length $l(\theta)$ parallel to the line of sight to the Earth which intersects the surface of the sphere at a latitude θ. The intensity of the emergent beam will be

$$I_\nu(0,\theta) = B_\nu(T) \int_0^{\tau_{\rm max}} e^{-t} dt = B_\nu(T)[1 - e^{-\tau_{\rm max}(\theta)}] . \tag{158}$$

In the usual way the flux emergent through the surface is obtained by calculating

$$\mathfrak{F}_\nu(0) = 2\pi B_\nu(T) \int_0^{\pi/2} (1 - e^{-\tau_{\rm max}(\theta)}) \sin\theta \cos\theta \, d\theta , \tag{159}$$

whereas the flux received per unit area at the Earth's distance r from the nebula is

$$F_\nu(r) = 2\pi B_\nu(T) \frac{R^2}{r^2} \int_0^{\pi/2} (1 - e^{-\tau_{\rm max}(\theta)}) \sin\theta \cos\theta \, d\theta . \tag{160}$$

Here

$$\tau_{\rm max} = \rho k_\nu l(\theta) = \rho k_\nu 2R \cos\theta . \tag{161}$$

For a shell of optical thickness

$$\tau_0 = \rho k_\nu 2R , \tag{162}$$

we have for the flux received at the Earth's surface

$$\mathfrak{F}_\nu(r) = \frac{\pi l^2}{2 r^2} B_\nu(T) \left[\tfrac{1}{2} - \frac{1}{\tau_0^2} + e^{-\tau_0} \left(\frac{1}{\tau_0^2} + \frac{1}{\tau_0} \right) \right] . \tag{163}$$

For an optically thin sphere we have

$$\mathfrak{F}_\nu(r) = \tfrac{4}{3} \pi \frac{R^3}{r^2} \rho k_\nu B_\nu(T) \tag{164}$$

whereas in the limiting case of an optically thick nebula,

$$\mathfrak{F}_\nu(r) = \pi B_\nu(T) \frac{R^2}{r^2} . \tag{165}$$

If the nebula is a spherical shell of inner radius R_i and outer radius R_o, we have

$$\mathfrak{F}_\nu(r) = \tfrac{4}{3} \frac{\pi R_o^3}{r^2} \rho k_\nu B_\nu(T) \left[1 - \left(\frac{R_i}{R_o} \right)^3 \right] \tag{166}$$

if the nebula is optically thin. For a shell of arbitrary optical thickness we have to derive a somewhat more complicated expression.

The difficulty with radio-frequency observation of planetaries as compared with diffuse objects such as Orion has been that, although the electron density is high, the angular size is extremely small so that the total flux is small. The detectability of a planetary nebula by a radio dish may be discussed as follows (Upton 1959).

If an antenna is receiving radiation from some uniform source at a temperature T_B, and if an additional flux of radiation $\mathfrak{F}_\nu$ is added, the increase in energy received per second by the antenna is $\Delta E_\nu = \frac{1}{2}\mathfrak{F}_\nu A_0$; the factor $\frac{1}{2}$ comes from the fact that we can observe only one plane of polarization at a time. Here A_0 is the effective area of the antenna. The flux causes a rise in antenna temperature by an amount $\Delta T_A = \Delta E_\nu/k = (A_0/2k)\mathfrak{F}_\nu$, where k is Boltzmann's constant. Consider an 85-foot antenna whose effective area is 0.65 times the geometrical area. Then $A_0 = 3.42 \times 10^2$ cm², and the temperature rise will be $\Delta T_A = 1.241 \times 10^{22}\, \mathfrak{F}_\nu$ if $\mathfrak{F}_\nu$ is in cgs units and ΔT_A is in °K. For some typical nebulae, the predicted rise in antenna temperature for such a device is (for $\lambda = 4$ cm)

NGC	3587	6572	6803	6818	7009	7027	7662
ΔT_A	0.02	0.17	0.008	0.03	0.10	0.71	0.10

If a gain of 0.01° K can be measured, all of these objects except NGC 6803 could be detected with an 85-foot dish, although meaningful measurements would require a large antenna.

In recent years, measurements of radio-frequency radiation from planetaries have been made by a number of workers (Lynds 1961, Menon and Terzian 1965, Davies *et al.* 1965, Slee and Orchiston 1965, Terzian 1966, Thompson *et al.* 1967, Ehman [private communication], Khromov *et al.* 1965). Most of these observations were made in the range from 3,000 to 620 megacycles/sec. Slee and Orchiston used the 210-foot Parkes dish to measure the emission of about 70 planetaries, mostly at λ 11 cm, although some data were secured at 21 and 48.5 cm. Terzian used the Arecibo 1,000-foot radio telescope to observe accessible planetaries in the declination range $-2°$ to $+38°$ at 430 megacycles. Stanley, Colvin, and Thompson, working at 2,891 mc, used the two 90-foot dishes at the Owens Valley Radio Observatory to obtain both flux and positions and widths of the sources. Their results are generally in good agreement with those of Slee and Orchiston but only in fair agreement with those of Menon and Terzian and the Jodrell Bank group. The Russian observations were obtained at 32.5 cm.

9.2. Comparison with Observations

If a gaseous nebula is optically thin, we see (from eqs. [150, 151, and 164]) that the flux will be independent of frequency except for the slow dependence through the $g_{\rm III}$ factor. Note that since the absorption coefficient varies inversely as the frequency squared, a nebula that is optically thin at a high frequency may be optically thick at a low frequency.

Furthermore, from the nebular Hβ flux (see eqs. [4, 5, and 82]), one may predict the emission in the radio-frequency region (eqs. [150, 151, and 164]). In general, the results are in satisfactory agreement, thus substantiating the interpretation of the hydrogen spectra of these nebulae as arising from recombination and free-free emissions.

In some instances the observed radio-frequency fluxes are greater than the predicted values, indicating that the Hβ fluxes are affected by interstellar extinction. From a comparison of observed and predicted fluxes, reasonable values for the interstellar extinction may be derived for a number of planetaries (Menon and Terzian 1965).

Using equations (149) through (166), we may now predict the emission from a number of planetaries for various wavelengths of interest. One must have starting values of the radius, density, temperature, and angular size. In order to simplify the calcuations, we assume that the planetaries are either shells or are spherically symmetrical. We must also distinguish between optically thin and optically thick objects, noting that the same object may be optically thin at one frequency (for example, at 4 cm) and optically thick at a lower frequency (for example, at 40 cm).

Table 14 gives the results of the calculations. Column 1 gives the object; column 2 the angular diameter d'' (in most instances essentially as given by Curtis 1918); column 3 the assumed linear diameter in units of 10^{17} cm, based on column 2; and the distance as supplied by Minkowski. Next, on the basis of the measured Hβ fluxes and electron temperatures T_e based on forbidden-line ratios, we adopt initial values of $T_e/10^3$ and log N_e (cols. 4 and 5).

The remainder of the table gives the radio-frequency data. First we consider the 11-cm radiation. The g_{III} factor is first calculated from equations (152)–(155); it varies slowly with frequency, ranging from 5.0 at 4 cm to around 6.5 at 40 cm. Column 6 gives the volume absorption coefficient ρk_ν in units of 10^{-18} cm^2/gram calculated by equation (151); column 7 gives the optical depth τ_0, and column 8 the predicted flux—all for 11-cm radiation in units of 10^{-26} watts (m)$^{-2}$ (c/s)$^{-1}$. Column 9 gives the observed flux, mostly from the data of Slee and Orchiston (1965) and of Stanley, Colvin, and Thompson (private communication); values indicated in parentheses. Note that the predicted flux is usually larger than the observed flux.

These predicted values depend, of course, on our assumed model for the planetaries, on the assumed distances (particularly for the optically thin ones), and especially on the adopted temperatures and electron densities. The temperatures depend on the oxygen-line intensities; the electron densities quoted depend on the measured surface brightness in Hβ, the temperature, assumed thickness of the radiating layers (that is, on the distance and angular diameter), and on the adopted value for the space absorption. Estimates of the latter are taken from Minkowski's compilation. For most of these nebulae, the two sets of values can be brought into closer agreement by assuming that the amount of space absorption has been overestimated. Column 10 gives this revised absorption, and column 11 the corresponding revised log N_e.

Next we tabulate values of the volume absorption coefficient, in terms of 10^{-18} cgs units for 4-cm and 21-cm radiation, τ_0 ($\lambda = 21$ cm), and finally the corresponding fluxes in units of 10^{-26} watts (m)$^{-2}$ (c/s)$^{-1}$ in columns 15 and 16. The starred values indicate nebulae for which the optical thickness effects are important. A number of planetaries tend to become optically thick at the very lowest frequencies (cf. Terzian 1966, Khromov 1965, Ehman [thesis]).

In accordance with expectation, flat spectral energy distributions are frequently found at higher frequencies among the planetaries. Some, such as IC 418, NGC 6572, and NGC 7027, show a steep decline of flux with wavelength as one would predict for optically thick objects. In such instances it is possible to estimate the electron temperature. Thus Menon and Terzian (1965) find $T_e = 13{,}300^\circ$ K for NGC 6572 as compared with $T_e = 11{,}500^\circ$ K derived from forbidden-line intensities. Terzian (1966) finds an electron temperature of 16,300° K for NGC 6720, as compared with $T_e = 15{,}700^\circ$ K from spectroscopic data with Minkowski's estimate for space absorption.

In comparing predicted and observed fluxes, one must keep in mind that not only is

TABLE 14

RADIO FLUXES FROM PLANETARY NEBULAE

(1) Nebula	(2) Diameter	(3) l_0 10^{17} (cm)	(4) $T_e/10^3$	(5) Assumed Log N_e	(6) (10^{-18}) ρk_ν	(7) τ_0	(8) 11-cm Flux Predicted	(9) 11-cm Flux Observed	(10) Revised Absorp.	(11) Revised Log N_e	(12) $\rho k_\nu/10^{-18}$ 4 cm	(13) $\rho k_\nu/10^{-18}$ 21 cm	(14) τ_0 21 cm	(15) 4-cm Flux	(16) 21-cm Flux
IC 418.......	12″4	3.27	12.5	4.09	1.43	0.47	2.14*	1.29 (1.34)	......	3.94	0.008	2.82	0.92	1.16	0.99*
NGC 1535....	18.	3.4	16	3.39	0.04	.014	0.20	0.21	......	3.39	.0050	0.161	0.05	0.19	0.22
NGC 2792....	13.	5.9	15	3.41	0.05	.30	0.21	0.13	0.88	3.30	.0036	0.115	0.07	0.11	0.133
IC II 2448....	8.	2.3	14.5	3.98	0.72	.17	0.44	0.09	0.14	3.63	.017	0.55	0.13	0.08	0.092
NGC 2867....	12.	2.4	12	3.78	0.376	.09	0.44	0.28	0.44	3.58	.17	0.55	0.13	0.15	0.18
NGC 3132....	69	7.1	10.5	2.95	0.0098	.07	0.94	0.18 (0.22)	......	2.80	.00058	0.019	0.013	0.42	0.50
NGC 3242....	36.	5.4	15	3.38	0.043	.23	1.30	0.81 (0.72)	0.10	3.18	.0020	0.066	0.036	0.47	0.54
NGC 3587....	200.	15.0	13.6	2.00	0.00083	.002	0.19	(0.10)	......	2.00	10^{-5}	0.00033	....	0.174	0.208
NGC 4361....	42	5.75	20	2.84	0.00246	.014	0.14	0.20 (0.21)	0.29	2.90	.00038	0.01225	0.070	0.162	0.187
IC II 4634....	8.4	4.21	13	3.74	0.278	.117	0.29	0.12 (0.15)	0.46	3.54	.013	0.425	0.18	0.105	0.116
NGC 6572....	14.4	1.96	13	4.19	2.19	.43	2.80*	1.07 (0.92)	0.15	3.95	.085	2.76	0.54	0.95	0.92
NGC 6803....	5.2	2.8	12	4.00	1.03	.26	0.26	0.08	0.41	3.74	.036	1.19	0.33	0.068	0.068
NGC 6818....	18.	3.7	20	3.86	0.286	.10	1.77	0.34 (0.34)	0.00	3.47	.005	0.17	0.06	0.27	0.31
NGC 7009....	28	3.65	12	3.55	0.120	.044	1.14	0.79 (0.62)	0.14	3.48	.011	0.353	0.13	0.78	0.93
NGC 7027....	14	3.0	16	4.21	1.75	.53	3.8*	(3.53)	0.96	4.27	.28	9.25	2.8	5.7	2.34
NGC 7662....	15.2	2.35	14	3.86	4.26	0.10	0.91	(0.65)	......	3.86	0.052	1.65	0.4	0.84	0.83

* Nebulae for which optical thickness effects are important.

the space absorption correction uncertain but structural complexities and dense filaments may cause a nebula (for example, NGC 7027) to be optically much thicker than a homogeneous planetary of the same mass (see also Osterbrock 1965*a*).

Most planetaries lie below the limit of angular resolution of available radio telescopes, although the radio diameters of some may be measurable with an interferometer. An exception is the giant, presumably nearby, planetary in Aquarius, NGC 7293, which Slee was able to resolve with an 11-cm beam at the 210-foot Parkes dish. The radio brightness distribution resembles the optical picture if the latter is "smeared" by the finite resolution of the radio beam.

When isophotic contours are available, one may compare predicted and observed radio-frequency fluxes, determine space absorption, and estimate electron temperatures. Although in most instances the electron temperatures derived from optical and radio data are in good agreement, Thompson (1967) finds that in some nebulae, e.g., NGC 6572, his radio data indicate a lower temperature in contradiction to Menon and Terzian's results. Electron density and temperature fluctuations within the nebula may partly explain the radio-optical discrepancies.

9.3. Nonthermal Radio-Frequency Radiation

Are nonthermal sources included among planetary nebula? In principle, it might appear easy to answer this question. A thermal source weakens as the wavelength increases, whereas the flux from a nonthermal source rises with increasing wavelength.

In practice, reliable identification of nonthermal sources among planetaries is difficult. All of them are relatively weak as radio-frequency sources go; hence they may be confused with background nonthermal emitters such as supernova remnants or radio galaxies which have not been identified optically. Near the galactic plane or center where steep gradients of nonthermal emission may occur, further complications may exist.

Several possible nonthermal planetaries have been suggested. One candidate was NGC 3242, for which detailed spectroscopic studies in the blue spectral region (Czyzak *et al.* 1966) failed to reveal any features which could be attributed to a unique nonthermal cause. Subsequent work (Terzian 1967) showed that the excess long-wavelength radiation arose from a nebulous wisp 10′ away. Similarly, abnormalities attributed to NGC 6781 could be due to a complex, nearby nebulosity that has the appearance of a nonthermal source.

10. CHEMICAL COMPOSITION OF PLANETARY NEBULAE

10.1. Interpretation of Data; Special Difficulties

Presumably, the chemical composition of a planetary nebula reflects that of the outer envelope of the highly evolved giant star from which it originated. To what extent does this composition represent that of the material from which the star was formed, and to what extent has it been influenced by nuclear processes occurring within the star followed by mixing with the outer layers?

At first sight, a planetary nebula might appear to offer distinct advantages for the determination of element abundances. All parts of the nebula are accessible to observations. The physical processes underlying the formation of the spectral lines and continua are all believed to be reasonably well understood. We can compare the spectra originating from different shells and filaments.

On the other hand, the difficulties which beset such an enterprise exceed those normally encountered in analyses of stellar atmospheres, at least for most elements. Since we have to use detailed mechanisms to interpret spectral-line intensities, we must know not only the f-values for the lines involved, but also the target area for collisional excitation. Eventually, as the number and quality of theoretical calculations improve, these difficulties may be overcome.

Unfortunately, there are two built-in sources of error that cannot be handled easily: (1) distribution of atoms among different stages of ionization, and (2) influence of fluctuations in density and temperature involved with stratification and filamentary structure within the nebula.

The distribution of atoms among various stages of ionization poses problems much more difficult than those in a stellar atmosphere where, if the concentration of atoms in a single ionization stage can be determined, the total number of atoms of that element usually can be computed by the Saha equation provided we know the temperature and density. These quantities may be uncertain for a given star, but once they are established, the relative chemical composition can be fairly well estimated.

The situation is otherwise for a planetary nebula. No simple ionization theory is available since we are dealing with a dilute, far-ultraviolet radiation field of unknown characteristics. Normally, we must proceed on an empirical basis to estimate the fraction of atoms present in unobservable stages, using a procedure suggested by Bowen and Wyse (1939). From the forbidden and/or recombination lines available, we observe the concentrations of ions of a few elements, for example, n(He I), n(He II), n(O I), n(O II), n(O III), n(Ne III), n(Ne IV), n(Ne V), n(S II), n(S III), and so forth, plot relative concentrations against ionization potential, and construct a mean curve relating relative numbers of ions with ionization potential. We assume that all elements follow the same curve and use this curve to extrapolate for missing ions of any element. (For an illustrative example, see Aller 1961*a*, p. 71, where a determination of the chemical composition of NGC 7009 [Plates 2, 12] is worked out in detail.)

An objection to this simple procedure is that it takes no account of the differences between the photoionization functions or statistical weights of ions of various atoms. A superior technique (if it could be applied) would be to solve for the radiation field throughout the nebula. It would be necessary to know not only the spatial structure of the nebula but also the detailed energy distribution in the spectrum of the central star.

Hence we are still dependent on some empirical approach. One might expect the permitted lines of carbon, nitrogen, oxygen, neon, etc., to supply helpful data. On the assumption that lines of O III, O IV, and O V arose from recombination (with due allowance for fluorescence effects in O III), Burgess and Seaton (1960*b*) calculated the O/H ratio in NGC 7027. Their methods were applied to eight additional planetaries (Aller 1964*a*). More recently, Seaton (1967) has concluded that the permitted lines of various oxygen and carbon ions are excited mostly by the stellar radiation field. The relative importance of stellar excitation and of recombination certainly varies spatially and from one nebula to another.

To appreciate the complexities involved, we discuss the problem of stratification and filamentary structures. In a typical ring nebula such as NGC 7662 or IC 418, the more highly ionized atoms are concentrated toward the inner region, the less ionized ones to the outer parts. The higher energy quanta become absorbed in the inner part of the nebula; the lower energy quanta persist to the outer regions (Bowen 1928). The problem can be handled quantitatively by generalizing the hydrogen ionization theory developed by Strömgren (1939) for nonuniform density distributions and for other gases such as nitrogen and oxygen (see, for example, Aller 1956, pp. 234–66; see also, Seaton 1960*a*).

Some of the complications that can be produced by stratification effects are illustrated by the [O I], [O II], and [O III] lines in IC 418. From measurements of the expansion velocities of these ions, Wilson (1953) concluded that no two of them could overlap appreciably in the nebula. Stratification effects may be less severe in other nebulae, but we must recognize that whenever information is deduced from lines of a given ion it applies properly only to some limited zone in the nebula. The Strömgren theory appears to give a good semi-quantitative interpretation of stratification effects where it can be applied, but it is not certain that it predicts correctly the extent of overlap zones between two stages of ionization.

Since theories of physical processes in gaseous nebulae all refer to the emission per unit volume, in principle we should convert intensities to emissivities. For a symmetrical nebula, the intensity distribution across the disk can be converted to the emission as a function of distance from the central star. Absolute flux measurements and a knowledge of the nebular distance are required to obtain absolute volume emissivities. On the assumption that the structure of IC 418 could be interpreted in the first approximation as a spherically symmetrical object, Wilson and Aller (1951) obtained relative volume emissivities. A similar attempt for IC 3568 (Aller 1956, p. 265) illustrates that great uncertainties are involved, partly because of observational errors (effects of bad seeing) even in this almost ideally symmetrical object. Spherical symmetry is not required; it is possible to construct a model for an object which is approximately cylindrical, such as IC 4406, which has been discussed by Evans (1950) and by Zanstra and Brandenburg (1951). It might be possible to derive realistic models for other, more complex, objects such as NGC 650–1 or NGC 6720 (Minkowski and Osterbrock 1960).

In many instances, it is not possible to allow for stratification as easily as for a symmetrical nebula; we can still measure the intensities at a fixed point in the ring or shell structure and carry out a theoretical discussion for this point.

Complex filamentary structure poses a much more difficult problem. Consider, for example, NGC 7027, (Plate 7) which possesses the richest line spectrum of any planetary and has therefore been the favorite candidate for abundance studies since the pioneer work of Bowen and Wyse. From a quantitative discussion of the Hβ surface brightness and the forbidden-line intensities, the conclusion was reached (Aller 1954*a*) that these data can be understood in terms of a nebula consisting of numerous filaments, knots, and tenuous regions such that the density may range from perhaps $<10^4$ ions/cm^3 to $\sim$200,000 ions/cm^3. Subsequently Seaton and Osterbrock (1957) strengthened this appraisal, although Osterbrock (1960) finds that NGC 7027 represents an extreme example of density fluctuations.

10.2. Filamentary Structure

Although the filamentary structures of many nebulae are obvious from direct inspection of good photographs (see e.g., Plates 8, 9, and 11), indirect evidence may be obtained by comparing electron temperatures and densities found from the following sets of data: (1) surface brightness in Hβ or the Balmer continuum; (2) Balmer discontinuity; (3) ratios of intensities of auroral and nebular lines of [N II], [O II], [O III], [Ne III], and eventually other elements, particularly [S II]; (4) the [O II] 3726 A/3729 A intensity ratio; and (5) radio-frequency data.

Each datum (1) through (4) gives a relation between N_e and T_e (we assume that all measurements have been corrected for space absorption using hydrogen—Paschen and Balmer—line data and radio-frequency data [5]). In practice we often find that the results are mutually inconsistent (see, for example, Seaton 1960*a*). Some of the trouble arises from poor observational data, some from inaccuracies in the cross-sections. Methods (1), (2), and (5) give mean values $\langle N_i \rangle$, $\langle N \rangle$, and $\langle T_e^{-3/2} \rangle$; method (4) refers only to regions where [O II] radiation originates, whereas method (3) applies to regions where various forbidden lines are produced. Lines of [O III], [O II], [N II], and [S II] tend to be produced predominantly in different strata in many nebulae.

Different filaments may possess not only different densities but also different electron temperatures. In abundance studies it is probably better to avoid objects that show effects of extreme density fluctuations and concentrate on those with relatively uniform structures. In practice, detailed studies can be made only for objects of moderately high surface brightness which are not necessarily uniform-density objects.

Early attempts to obtain chemical compositions of planetary nebulae (Bowen and Wyse 1939, Aller and Menzel 1945) were based on rough physical parameters and rough line intensities. Consequently they could yield only crude results. Improvements in transition probabilities and collisional target areas for forbidden lines, especially by Garstang, Osterbrock, Seaton, Czyzak, and their co-workers, have greatly improved the means of analysis. Observational data are also greatly improved. Refined analyses of NGC 7027 (Aller and Minkowski 1956) and of 10 other planetaries including NGC 7662 (Plate 8) (Aller 1957) emphasized the need for accurate determinations of space absorption and proper allowance for density fluctuations. The capricious behavior of forbidden nitrogen and iron lines in many planetaries (Wyse 1942, Walker and Aller 1967) indicates the importance of filamentary structure.

10.3. Abundance Ratios

The He/H ratio is probably the most accurately determined abundance in the planetaries (Mathis 1957). More recent determinations (O'Dell 1963*b*, Aller 1964*b*) give a ratio of about 0.17 (see also Sec. 5). The He/H ratio probably does not show large fluctuations from one nebula to another.

To the extent that the permitted oxygen lines are affected by a direct stellar radiation field, the O/H ratio is rendered uncertain. There may be an intrinsic variation in this ratio among the "field" planetaries.

The Ne/O ratio almost certainly varies intrinsically from one nebula to another. In NGC 2022 (Plate 1) and Anon $21^{h}31^{m}$ (VV_{I} 267) it appears to be 0.5 and 0.33, respectively, and in NGC 7662 (Plate 1) it seems to be only 0.10. All three objects are high-excitation planetaries.

The red nebular-type transitions of [N II] show striking intensity fluctuations from one nebula to another of the same excitation class. On the other hand, the pronounced variations in the [S III] line intensities must certainly be due primarily to density fluctuations.

Table 15 gives the estimated chemical composition of a planetary nebula. The value for carbon comes entirely from recombination-line data; the nitrogen value reflects an uncertain allowance for distribution of atoms among various stages of ionization; the errors involved may tend to bias the estimates toward values that are too large. Data for elements such as neon or oxygen reflect the influence of probable intrinsic abundance fluctuations. For comparison we give the estimated composition of the Orion Nebula (Liller and Aller 1959) that of γ Pegasi (Aller and Jugaku 1959), the mean

TABLE 15

COMPARISON OF CHEMICAL COMPOSITION OF PLANETARY NEBULAE WITH THOSE OF SEVERAL CELESTIAL OBJECTS

	Planetaries log N	Orion Nebula	γ Peg	Mean B Stars	Sun
H	12.00	12.0	12.00	12.00	12.00
He	11.25	11.1	11.17	11.20	....
C	8.7	8.4	8.58	8.3	8.72
N	8.5	7.7	8.02	8.2	7.98
O	9.0	8.6	8.63	8.8	8.96
F	5.2	....	6.5	6.5	....
Ne	8.2:	8.8	8.73	8.7	....
Na	6.0	....	....	....	6.30
S	8.0:	8.0	7.8	7.5	7.30
Cl	6.5:	5.9	6.2	6.8	....
Ar	6.9:	6.6	6.9	6.9	....
K	5.8	....	....	....	4.70
Ca	6.2	....	....	....	6.15

value derived for several B stars, and the solar values (Goldberg, Müller, and Aller 1960). The intrinsic uncertainties involved in each of these determinations have been discussed in detail (Aller 1961*a*).

That intrinsic abundance differences do exist between planetary nebulae is demonstrated by O'Dell, Peimbert, and Kinman's (1964) analysis of the planetary in M15. They found that although the He/H ratio, 0.18, is essentially the presently accepted cosmic one, oxygen is deficient by a factor of about 40 as compared with other planetaries. Hence, it may be meaningless to seek a mean composition for planetary nebulae, and efforts should be concentrated on improving analyses of individual objects.

The planetaries with low surface brightnesses discussed by Abell from Palomar 48-inch Schmidt plates show extremely weak hydrogen emission. The other elements observed in these nebulae and their central stars are alpha-particle nuclei, suggesting that the stars belong to a carbon sequence, are hydrogen deficient, and could be residues of giant star cores (Greenstein and Minkowski 1964). If we adopt this point of view, the nebulae would represent the envelopes of defunct giant stars, and their compositions could be mixtures of normal cosmic material and varying portions of matter that had been subjected to nuclear reactions.

11. CENTRAL STARS OF PLANETARY NEBULAE

11.1. Types of Spectra

Although the central star is visible in most planetary nebulae and is postulated to exist in all of them, it is often outshone by the overlapping nebulosity. Hence, magnitudes and colors are difficult to measure; in nebulae of high surface brightness it is frequently necessary to use special filters, as has been done, for example, by Liller (1955). Even in spectral regions relatively free of nebular lines, it is necessary to allow for the nebular continuum which can be evaluated from the hydrogen-line intensities and theory (see Sec. 5) or from slitless spectrograms taken for the purpose.

Observations of the spectra of planetary nuclei are often likewise made difficult by the superposed nebular spectrum. The nebular hydrogen and helium lines fill in the corresponding stellar absorption lines and make it difficult to establish spectral types. For the brighter nuclei, we can employ coudé dispersions and suppress part, but not all, of the nebular contamination. The spectra range from rather conventional types to exotic, high-excitation types. For illustrative examples, see, for example, Aller (1954*b*, Chap. 5, Fig. 11).

The spectra of the central stars of the planetaries fall into the following broad categories: (1) Wolf-Rayet–type characterized by broad emission lines (for example, the nuclei of IC II 1747 or NGC 6751); (2) Of-type with (often variable) emission features (for example, nucleus of NGC 2392 [Plate 1]); (3) absorption-line O-type with apparently no emission (for example, the nuclei of IC II 2149, IC II 3568, and NGC 1535); (4) Continous spectra with apparently no absorption or emission features that can be established with available dispersions (for example, the nuclei of NGC 3242, NGC 6309, NGC 6567, NGC 6807, NGC 7009, IC II 4732, and Anon 18^h15^m); and (5) high-excitation type showing such features as O VI lines in emission (for example, nucleus of NGC 246). Detailed descriptions of the spectra have been published by a number of observers including Struve and Swings (1940), Swings (1940, 1941), Swings and Struve (1941), Aller (1943, 1948, 1951), Minkowski (1943), Swings and Swensson (1952), Wilson and Aller (1954), and Andrillat and Andrillat (1959); for a summary, see Aller (1956).

The best-known examples of Wolf-Rayet nuclei are the central star of NGC 40 and Campbell's hydrogen-envelope star BD$+30°$3639, both of which have been classified as WC8. Although "classical" Wolf-Rayet stars fall in well-defined carbon-oxygen (WC) and nitrogen (WN) sequences, planetary nebulae nuclei often show features of both sequences (Struve and Swings 1940). Some Wolf-Rayet nuclei lean toward the carbon sequence; the nucleus of NGC 1501 seems to be a pure WC star (Andrillat and Andrillat 1959). Other nuclei do not closely resemble any classical type (Smith).

Nuclei of the Of-type show both absorption lines (usually of H and He) and emission features which are variable in intensity. Besides "conventional" nuclei of planetaries with high surface brightness, such as IC 418, NGC 2392, NGC 6210, and NGC 6543, the classification normally embraces Of exotic high-temperature objects, which we prefer to group with objects of the same type as nucleus of NGC 246. One of the brightest of these conventional objects, the nucleus of IC 418, shows prominent emission lines of C III, C IV, N II, N III, Si IV, and He II. The nucleus of NGC 6543, classified as W6 by Swings (1940), probably is an object of a type intermediate between an Of and a Wolf-

Rayet star. It is difficult to separate the nebular and stellar lines of IC 4997, whose central star has been classified by Swings and Struve as W7.

For absorption-line objects, it is possible to assign spectral classes on a conventional scale and estimate surface gravities from hydrogen-line profiles. The most likely absolute magnitudes for these stars then lead to masses which are comparable with that of the Sun (Aller 1956, p. 217).

The prototype of a number of very hot planetary nuclei is the central star of NGC 246, whose spectrum exhibited absorption lines of ionized helium, C III, and C IV, and conspicuous emission lines of O VI 3811 A, 3838 A (Aller 1948). Greenstein and Minkowski (1964) regarded H as weak and noted absorption by O V. They carried out a detailed study of the nuclei of a number of planetaries of low surface brightness which had been observed by Abell. He had found that the colors of some of these stars approach the limiting value corresponding to an infinite temperature. The spectrum of Abell 36 resembles that of the subdwarf BD+28°4211, except that it has a higher temperature. The hottest stars are Abell 30, 0.5fep, $M_v \sim +2$, which had broad weak H and C IV lines with a strong O VI emission, and Abell 78. The latter had broad and weak H-lines, variable C IV Of-lines, and O VI emission lines stronger than He II (Greenstein and Minkowski 1964). Chromospheric temperatures corresponding to between 60-ev and 130-ev excitations are indicated. They conclude that the radii fall in the range $0.01 < R < 0.1$. Presumably these stars have hot atmospheres at high pressure with unstable chromospheres that produce a continuous loss of mass, since most have emission lines.

11.2. Temperatures

The temperatures of the central stars of planetary nebulae may be estimated by a variety of methods:

a) The color temperature which may be derived from the energy distribution in the continuum is inconclusive because of the low sensitivity of the slope of the energy distribution to the temperature and because of effects of interstellar extinction.

b) An ionization temperature may be derived from the relation between spectral class and temperature.

c) By taking advantage of Zanstra's principle that the nebula acts as a photon counter for the far-ultraviolet radiation of the central star, we may compare the integrated intensities of certain nebular lines with the stellar radiation in the ordinary visible range to derive what amounts to a long-baseline color temperature of the star. Variations on the theme include a comparison of the stellar energy emitted in different regions of the ultraviolet and certain energy-balance arguments.

11.2.1. *Ionization temperature.*—The use of spectral classes requires that we calibrate them in terms of ionization temperatures. Petrie (1948) carried out such a program for the classical O stars of population Type I. It is possible to establish temperature-sensitive ratios of equivalent widths of lines of H, He I, and He II. If helium is much more abundant than in normal B stars, the temperature calculated from the ratios of hydrogen to helium line intensities will be too high. One can avoid this difficulty by using intensity ratios of He I and He II lines only. On the other hand, the method does not require any assumption about the optical thickness of the surrounding nebular shell,

TABLE 16

SUMMARY OF METHODS OF TEMPERATURE DETERMINATIONS

Method	Data Required	Theory	Ratios from Which T Is Determined	Remarks	Reference
Recombination Quanta					
(1) H I........	$I^n(\mathrm{H}n)$, $I^s(\mathrm{H}n)\Delta\nu$	Number of Balmer quanta = number of Lyman continuum quanta	$q(\nu_{\mathrm{H}},T)/\Delta q(\nu_{\mathrm{H}n},T)$	(1)	Zanstra (1931*a*)
(2) H I........	$I^n(\mathrm{Bac})$, $I^s(\mathrm{Bac})\Delta\nu$		$q(\nu_{\mathrm{H}},T)/\Delta q(\nu_{\mathrm{H}n},T)$	(1)	Wurm (1951)
(3) He II......	$I^n(4686)$, $I^s(4686)\Delta\nu$	(1) and (2) applied to He II	$q(4\nu_{\mathrm{H}},T)/\Delta q(\nu_{4686},T)$	(1)	Zanstra (1931*a*)
(4) He I.......	$I^n(4471)$, $I^s(4471)\Delta\nu$	Recombination theory, for helium	$q(1.81\nu_{\mathrm{H}},T)/\Delta q(\nu_{4471},T)$	(1)	Zanstra (1931*a*)
(5) He........	$I^n(4686)/I^s(4861)$	Same as (1) and (3)	$q(4\nu_{\mathrm{H}},T)/q(\nu_{\mathrm{H}},T)$	(1)	Ambartsumian (1932)
Energy Conservation Methods					
(6)............	$\Sigma I^n(FL)$, $I^s(\mathrm{H}\beta)\Delta\nu$	Energy of photoelectrons is dissipated in excitation of forbidden lines	$E(\nu_{\mathrm{H}},T)/q(\nu_{\mathrm{H}},T)$	(1)(2)	Zanstra (1931*a*)
(7)............	$\Sigma I^n(FL)$, $I^n(\mathrm{H}\beta)$		$E(\nu_{\mathrm{H}},T)/q(\nu_{\mathrm{H}},T)$	(1)(2)	Stoy (1933)
(8)............	$\Sigma I^n(FL)$, $I^n(\mathrm{H}\beta)$, T_e	Energy-balance equation (91)	$E(\nu_{\mathrm{H}},T)/q(\nu_{\mathrm{H}},T)$	(1)	Menzel and Aller (1941)
Conventional Methods (available only for cooler stars)					
(9)............	Equivalent widths of absorption lines of H, He I, He II	Conventional stellar atmosphere theory	Line-intensity ratios yield an ionization temperature	(3)	e.g., Aller (1948)
(10)...........	Bright-line intensities (Wolf-Rayet spectra only)	Boltzmann equation	Line-intensity ratios give excitation temperatures	(4)	e.g., Aller (1943)
(11)...........	Energy distribution in continuous spectra	....	Slope of energy distribution gives a color temperature	(5)	....

Remarks

General References:

Wurm (1951); Aller (1956, pp. 217–31); Seaton (1960*a*); Zanstra (1960); Harman and Seaton (1966)

Notation:

$$q(\nu_i,T) = \frac{8\pi^2R^2}{c^2}\int_{\nu_1}^{\infty}\frac{\nu^2}{e^{h\nu/kT_1}-1}\,d\nu$$

Number of quanta emitted by central star beyond limit ν_i. $\nu_i = \nu_{\mathrm{H}}(\mathrm{H}) = 4\nu_{\mathrm{H}}$ (He II) $= 1.81\nu_{\mathrm{H}}$(He I).

$$E(\nu_i,T) = \frac{8\pi^2R^2}{c^2}\int_{\nu_i}^{\infty}\frac{h\nu^3}{e^{h\nu/kT_1}-1}\,d\nu$$

Energy emitted by central star of radius R beyond limit ν_i.

$$\Delta q(\nu_j,T) = \frac{8\pi^2R^2}{c^2}\frac{\nu^2}{e^{h\nu/kT_1}-1}\,d\nu$$

Number of quanta emitted in visual region of central star at some particular frequency ν to $\nu+\Delta\nu$ near a

$I^n(\lambda)$ = total intensity of nebular line of wavelength λ.
$I^s(\lambda)\Delta\nu$ = total emission from central star in interval ν to $\nu+\Delta\nu$.

(1) Nebula must be optically thick in resonance continuum. It is usually also assumed that central star radiates as a blackbody, although method can be adapted to allow for any specified energy distribution.
(2) In the earlier formulation of theory, effect of finite electron temperature was neglected, although it was taken into account in later discussions by Zanstra (1960) and for Stoy's method by Aller (1956, p. 221). Other sources of energy dissipation, such as collisional excitation of H-lines and free-free emissions, are neglected.
(3) Accuracy is limited by uncertainties in the He/H ratio and in model atmosphere theory.
(4) Excitation temperatures differ from one ion to another and can be regarded only as crude approximations.
(5) Very inaccurate for high-temperature objects until observations from space vehicles become available, and proper correction for interstellar extinction can be made.

nor about the star's ultraviolet energy distribution. The method may be improved with the aid of appropriate model-atmosphere calculations.

11.2.2. *Indirect temperature determination.*—The temperatures of the central stars of planetary nebulae could be obtained from a comparison of the total number of quanta in the Balmer lines with those in the underlying stellar spectrum as recognized independently by Zanstra (1926) and by Menzel (1926). If the nebular shell is optically thick, the number of stellar quanta emitted beyond the Lyman limit equals the number of Balmer (line plus continuum) quanta emitted by the nebula. Therefore, by comparing the number of Balmer quanta with the number of quanta in the underlying continuum, we determine a kind of color temperature for the star. Following Zanstra (1927, 1931*a*, 1960), let $L_P(l)$ = measured integrated intensity of a monochromatic image; ν_l = frequency of line, l; and $(dL_s/d\nu)_l$ = intensity of star's spectrum per unit frequency in neighborhood of line l. Then the observed quantity

$$A_\nu(l) = \frac{L_p(l)}{\nu_l \left(\frac{dL_s}{d\nu}\right)_l} \tag{167}$$

is independent of the units in which the intensities are measured of the interstellar extinction, and of the distance of the nebula. The total number of quanta emitted by the star beyond the Lyman limit will be

$$4\pi R^2 \int_{\nu_0}^{\infty} \frac{\pi F_\nu}{h\nu}\, d\nu = \frac{8\pi^2 R^2}{c^2} \int_{\nu_0}^{\infty} \frac{\nu^2}{e^{h\nu/kT_1} - 1}\, d\nu . \tag{168}$$

The number of Balmer quanta (line plus continuum) will be

$$\Sigma \frac{L_p}{h\nu} = \Sigma A_\nu(l)\nu \frac{8\pi^2 R^2}{c^2} \frac{\nu^2}{e^{h\nu/kT_1} - 1} . \tag{169}$$

In practice, we need measure only one line of the Balmer series and use the recombination theory (Sec. 5) to calculate the ratio

$$q_n = \frac{\text{Number of Balmer quanta (line + continuum)}}{\text{Number of quanta in H}n} , \tag{170}$$

where Hn can be Hα, Hβ, etc. Then, defining $x = h\nu/kT_1$, where T_1 is the temperature of the central star, Zanstra's basic equation becomes

$$\int_{x_0}^{\infty} \frac{x^2}{e^x - 1}\, dx = \Sigma \frac{x^3}{e^x - 1} A_\nu = q_n \frac{x_n^3}{e^{x_n} - 1} A_\nu(\text{H}n) . \tag{171}$$

The great advantage of this method is that we need measure the quantity A_ν for only one line. Space absorption imposes no difficulty, and q_n varies slowly with T_e.

Measurement of $A_\nu(l)$ by photoelectric photometry is straightforward, although for planetaries with faint stars we measure, for example, the monochromatic brightness in Hα or Hβ and the broad-passband magnitude of the central star. Equation (171) is then replaced by an equation of the form

$$0.4M_i(s) + \log F(\text{H}n) = \log \int_{x_0}^{\infty} \frac{x^2 dx}{e^x - 1} - \log kT \int_0^{\infty} \frac{S_i x^3 dx}{e^x - 1} + \text{const.}, \tag{172}$$

where $F(\mathrm{H}n)$ denotes the flux received in Hα or Hβ, $M_i(s)$ in the magnitude of the star in the particular color system employed, and S_i is the sensitivity function which has to be known for each detector-filter combination employed. The constant depends on the units used, the line, the magnitude, and the color system. Applications of this method by Berman (1937), O'Dell (1962), and Abell (1966) have yielded "Zanstra" temperatures for many objects.

We must emphasize that the temperatures obtained from equations (171) and (172) involve two assumptions: (1) the absorption in the Lyman continuum is complete, and (2) the central star radiates as a blackbody beyond the Lyman limit.

If condition (1) is not fulfilled (as is often the case), the equality sign in equation (171) must be replaced by > and the temperatures derived will be minimum temperatures. Condition (2) is almost certainly not fulfilled (Aller 1956, p. 229; Aller and Zanstra 1960; Gebbie and Seaton 1963; Böhm and Deinzer 1966), and the effects have to be evaluated by theory.

In addition to the Balmer lines, we may also use the Balmer continuum (Wurm 1951, Wurm and Singer 1952). The fraction of quanta emitted in the interval ν to $\nu + \Delta\nu$ to the total number of Balmer recombination quanta and the ratio of these, in turn, to the total number of quanta in the underlying continuum (bound-free + free-free + 2-photon emissions) and to the total number of Balmer continua are known. That is, we could measure

$$\frac{F_{\kappa 2}\Delta\nu}{\Sigma F_{\kappa n}\Delta\nu + \mathcal{Q}_{2p}\Delta\nu + F_{\kappa\kappa'}\Delta\nu} = \frac{\text{Balmer continuum}}{\text{underlying nebular continuum}} \tag{173}$$

and

$$\frac{F(\text{star})\Delta\nu}{F_{\kappa 2}(\text{star})\Delta\nu} = \frac{\text{stellar continuum}}{\text{nebular continuum}} \tag{174}$$

and evaluate T_e from the first ratio and T_1 from the second. The ratios of the number of Balmer continuum quanta to the total number of Balmer quanta is known.

We can carry over the hydrogenic theory for ionized helium (noting that 4686 A corresponds to Paschen α) provided we are willing to assume that the nebula is optically thick in this region and that the star radiates in that region as a blackbody.

Thus q_{4686} = (No. of ionized helium Balmer quanta)/(No. of ionized helium Paschen α quanta),

$$\begin{aligned} q_{4686} &= \frac{\mathrm{Ba(He^+)} + \mathrm{Bac(He^+)}}{\mathrm{P\alpha(He^+)}} = \frac{\mathrm{Ba(He^+)} + \mathrm{Bac(He^+)}}{\mathrm{H\beta(He^+)}} \frac{\mathrm{H\beta(He^+)}}{\mathrm{P\alpha(He^+)}} \\ &= q_2(T_e/4)\frac{\Sigma N(4, l) A_{4,2}}{\Sigma N(4, l) A_{4\,3}} = 0.936 q_2(T_e/4) , \end{aligned} \tag{175}$$

where

$$\begin{aligned} q_2(\mathrm{He^+}, T_e) &= q_{\mathrm{H}\beta}(\mathrm{H}, T_e/4), \\ I(\mathrm{H}\beta)/h\nu &= \sum_l N(n = 4, l) A_{4,2}, \end{aligned} \tag{176}$$

and

$$I(\mathrm{P}\alpha)/h\nu = \sum_l N(n = 4, l) A_{4,3} . \tag{177}$$

Now the dependence of b_n on l should be taken into account, since more levels contribute to Pα than to Hβ. Since $A_\nu(l)$ is usually measured for some one line l (for example, Hβ), other than 4686 A, we follow Zanstra (1960) and write

$$q\,(\text{He II}, \lambda 4686)\frac{L_p(\lambda 4686)}{L_p(l)}\,\frac{\lambda(4686)}{\lambda_l}\,A\,(l) = \frac{e^{x_l}-1}{x_l^3}\int_{x_0'}^{\infty}\frac{x^2}{e^x-1}\,dx\,, \quad (178)$$

where $x_0' = x_0$ (He II) $\approx x_0(T_e/4)$. When this program is carried out, the temperatures derived from ionized helium are always larger than the hydrogen temperatures. One interpretation is that the nebula is optically thick at $\lambda < 228$ A and not optically thick at $\lambda < 912$ A; so the He II temperature corresponds more closely to the "real" one. Another interpretation is that the star has an ultraviolet energy excess.

Various combinations of H and He^+ methods are available. In his "composite" method, Zanstra assumed an excess in the far-ultraviolet. In Ambartsumian's (1933) method, we use the ratio $I(4868\text{ A})/I(4861\text{ A})$ to get a kind of a far-ultraviolet color temperature defined by

$$\int_{\nu_1}^{\infty} B_\nu(T)\,d\nu \Big/ \int_{4\nu_1}^{\infty} B_\nu(T)\,d\nu\,.$$

If the hydrogen Lyman-continuum absorption is incomplete and the corresponding ionized helium absorption is complete, the derived temperature will be too high.

The neutral helium-line spectrum may also be used. We calculate the ratio for a convenient line, for example, 5876 A, 4471 A, or 4026 A,

$$q_\lambda = \frac{\text{Number of recombinations on all levels}}{\text{Number of quanta emitted in } \lambda}\,, \quad (179)$$

using the absorption coefficients a_ν from Goldberg (1939) and Huang (1948), and the b_n's from Mathis (1957). The theory is similar in principle to that used for hydrogen. Helium temperatures tend to be systematically higher than hydrogen temperatures by about 5,000° K (cf. Aller and Zanstra 1960), a result which apparently can be understood in terms of fluxes predicted by model atmospheres.

Zanstra also recognized that the temperature of the central star could be estimated from conservation-of-energy arguments. In the simplest formulation, it is assumed that all of the energy brought into the continuum by each electron, viz.,

$$\tfrac{1}{2}mv^2 = h(\nu - \nu_0) \quad (180)$$

is radiated away in forbidden-line emissions. Then if $L_p(f)$ denotes the energy in a monochromatic image of a forbidden line f and $L_p(l)$ is the energy radiated in a line l, the temperature T_{ee} is obtained by solving the equation (cf. Zanstra 1931*b*, 1960)

$$A_\nu(l)\Sigma\frac{L_p(f)}{L_p(l)} = \frac{e^{x_l}-1}{x_l^4}\left(\int_{x_0}^{\infty}\frac{x^3}{e^x-1}\,dx - x_0\int_{x_0}^{\infty}\frac{x^2}{e^x-1}\,dx\right). \quad (181)$$

Since the summation must be taken over all forbidden lines, it is sensitive to the amount of interstellar reddening assumed and to the contributions from the infrared and ultraviolet spectral regions which are difficult to evaluate.

In Stoy's (1933) method, one compares the intensities of the nebular Balmer and forbidden lines without referring to the stellar continuum. The basic assumption is the same as the Zanstra nebulium method. The method should be modified to take into account the fact that the final mean energy of the electron is not zero, but is

$$\tfrac{1}{2}mv_0^2 = \tfrac{3}{2}kT_e\,. \qquad (182)$$

The temperatures derived by the Stoy method are then raised appreciably (Aller 1956, p. 221). Zanstra shows that in his nebulium method it is then necessary only to replace the left-hand side of equation (181) by

$$\frac{T}{T-T_e}\,A_\nu(l)\frac{L_p(f)}{L_p(l)}\,.$$

The equation is solved by iterations. Finally one may set up the detailed energy-balance equation (see Sec. 6) taking into account free-free energy losses and so forth as well as collisional effects. Given T_e and $\Sigma I(f)/I(\mathrm{H}\beta)$, one can calculate T_l, but it was emphasized (Menzel and Aller 1941*a*) that small uncertainties in T_e could lead to large fluctuations in T_l. Furthermore, at high temperature one must take into account collisional excitations of hydrogen levels (Miyamoto 1938, Chamberlain 1953, Parker 1964, Burgess 1964).

Harman and Seaton developed a systematic treatment of the Zanstra-type methods for hydrogen, helium, and ionized helium. Utilizing 47 planetaries for which measurements had been made of angular radii, Hβ fluxes, relative line intensities, and central star magnitudes, they find 42 of these objects to satisfy criteria for complete absorption in the continua of H I, He I, and He II. They define a luminosity parameter $\Lambda = L/d^2L_0$, where d is the nebula's distance in kiloparsecs, a parameter $\eta(\tau_1)$ which provides a measure of completeness of absorption in the H I continuum, and $\xi = \Omega/4\pi$, where Ω is the solid angle filled by the nebula as seen from the star. They then develop a series of equations:

$$\text{Star} \quad \Lambda = 2.59\times10^9[\nu_\beta F(\nu_\beta)]t^4[\exp(2.96/t)-1] \qquad (183)$$

$$\text{H I} \quad \Lambda = 5.89\times10^{11}\frac{F(\mathrm{H}\beta)}{\xi}\,\frac{1}{\eta(\tau_1)}\,\frac{t}{F_1(T_1)} \qquad (184)$$

$$\text{He II} \quad \Lambda = 3.64\times10^{11}\frac{F(4686\mathrm{A})}{\xi}\,\frac{t}{F_4(T_1)} \qquad (185)$$

$$\text{He I} \quad \Lambda = 1.36\times10^{12}\frac{F(4471\mathrm{A})}{\xi}\,\frac{t}{F_{1.807}(T_1)}\,, \qquad (186)$$

where $t = 10^{-4}T$.

Here $F(\nu_\beta)$ is the absolute flux of the stellar radiation as received at the top of the Earth's atmosphere. It may be derived from $A_\nu(\mathrm{H}\beta)$, equation (167), and the measured value of the nebular Hβ flux. Similarly F(4686 A) and F(4471 A) are the corresponding fluxes in λ 4686 (He I) and λ 4471 (He I). All fluxes are corrected for interstellar extinction and expressed as ergs cm^{-2} sec^{-1}. Here $\xi = 1$ for a shell nebula and about 0.3 for a ring nebula; $\xi\eta$ is the fraction of stellar quanta with $\nu > \nu_1$ which is absorbed by the nebula. Nebulae showing [O I] lines (see Khromov 1965) have complete absorption by H I. Equation (185) is valid if the nebula is optically thick in the He II continuum; similarly, equation (186) holds if the nebula is thick in the He I continuum. Harman and Seaton tabulate the functions $F_1(T_1)$, $F_{1.807}(T_1)$, and $F_4(T_1)$ in their Table 6.

For each nebula they plot Λ(star), Λ(H I), and Λ(He II) or Λ(He I) as a function of

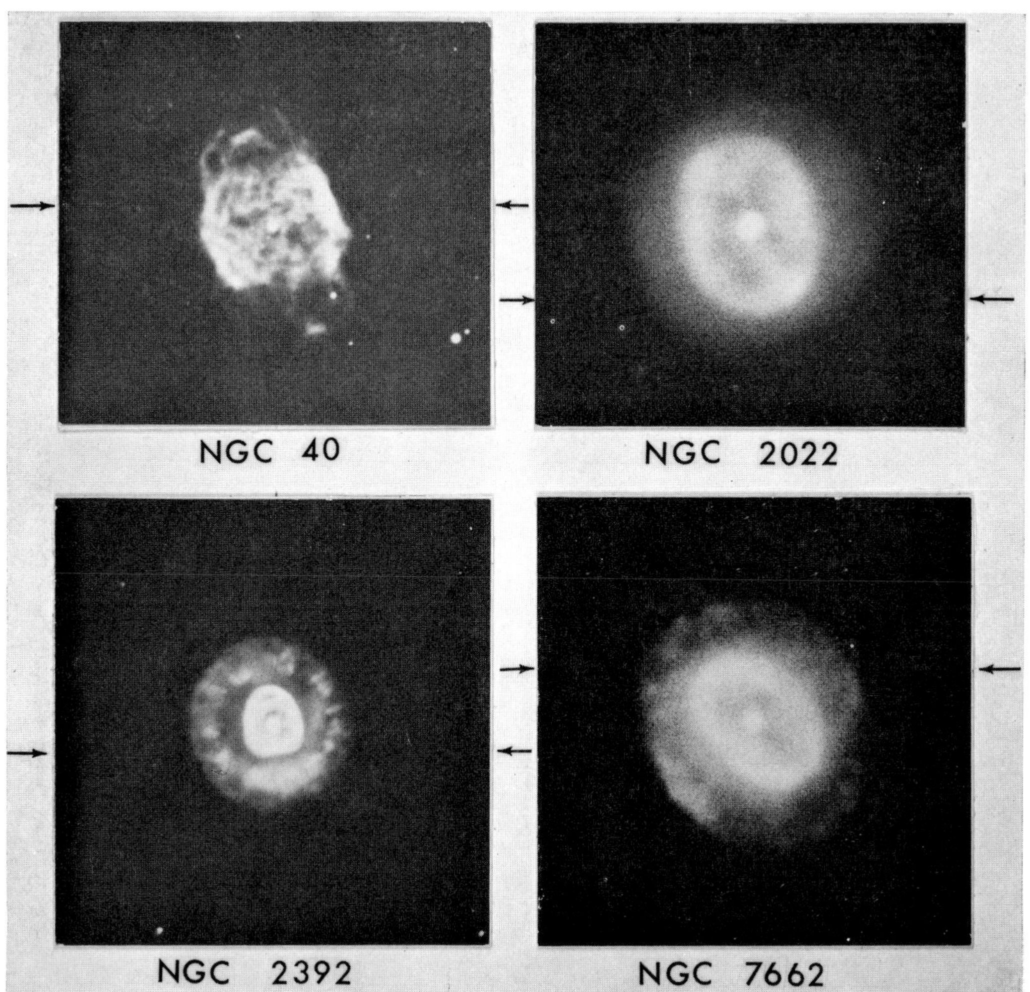

PLATE 1.—Four planetary nebulae: NGC 40, 2022, 2392, 7662. The *arrows* indicate the position of the spectrograph slit in the spectrophotometric studies of these objects by Minkowski and Aller (1956). (Courtesy Mount Wilson and Palomar Observatories.)

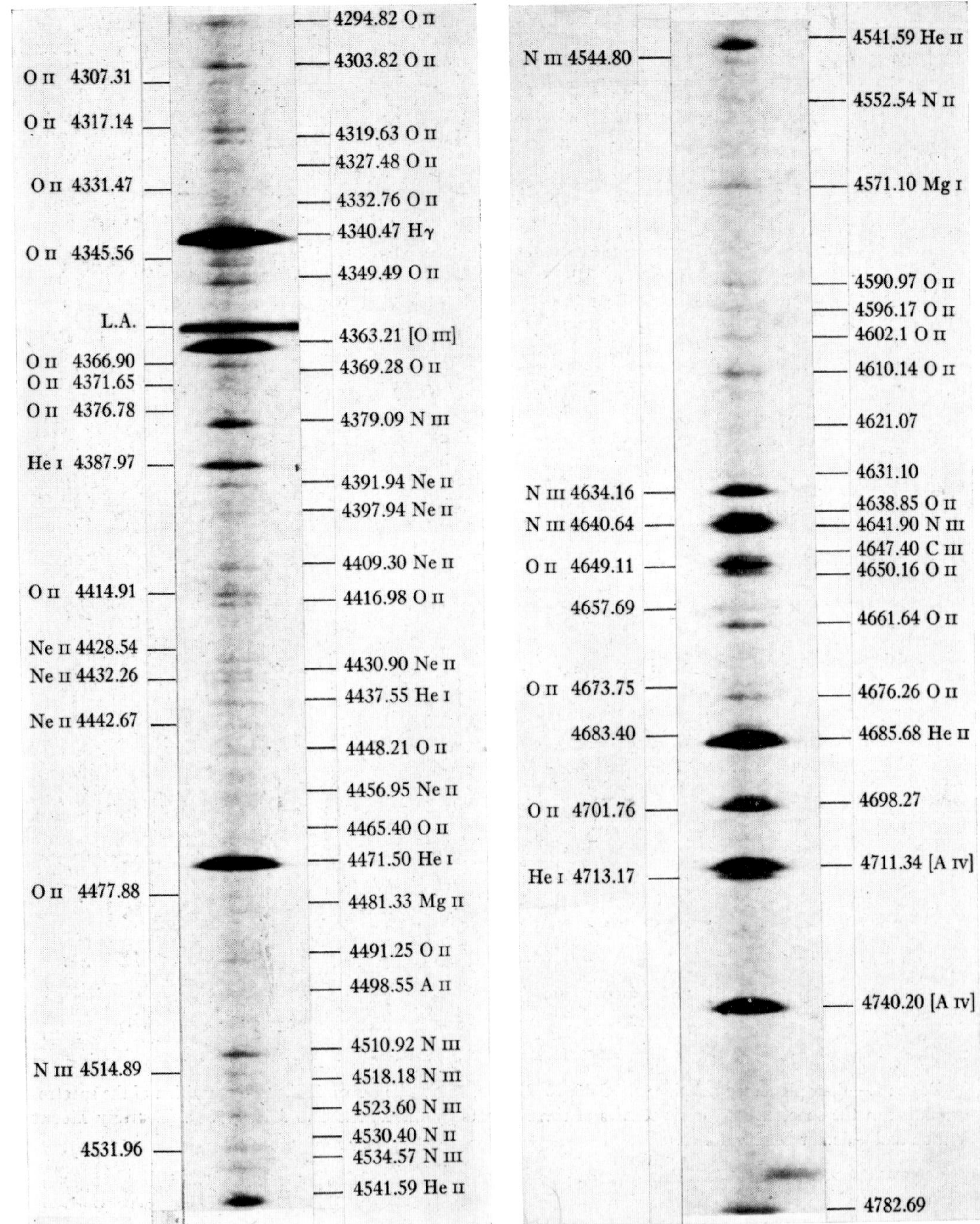

PLATE 2.—The spectrum of NGC 7009. This nebula of moderately high excitation has a rich recombination spectrum of O II and N II (cf. Wyse 1942; Aller and Kaler 1964*a*). Photographed with the coudé spectrograph at the 100-inch telescope of Mount Wilson Observatory.

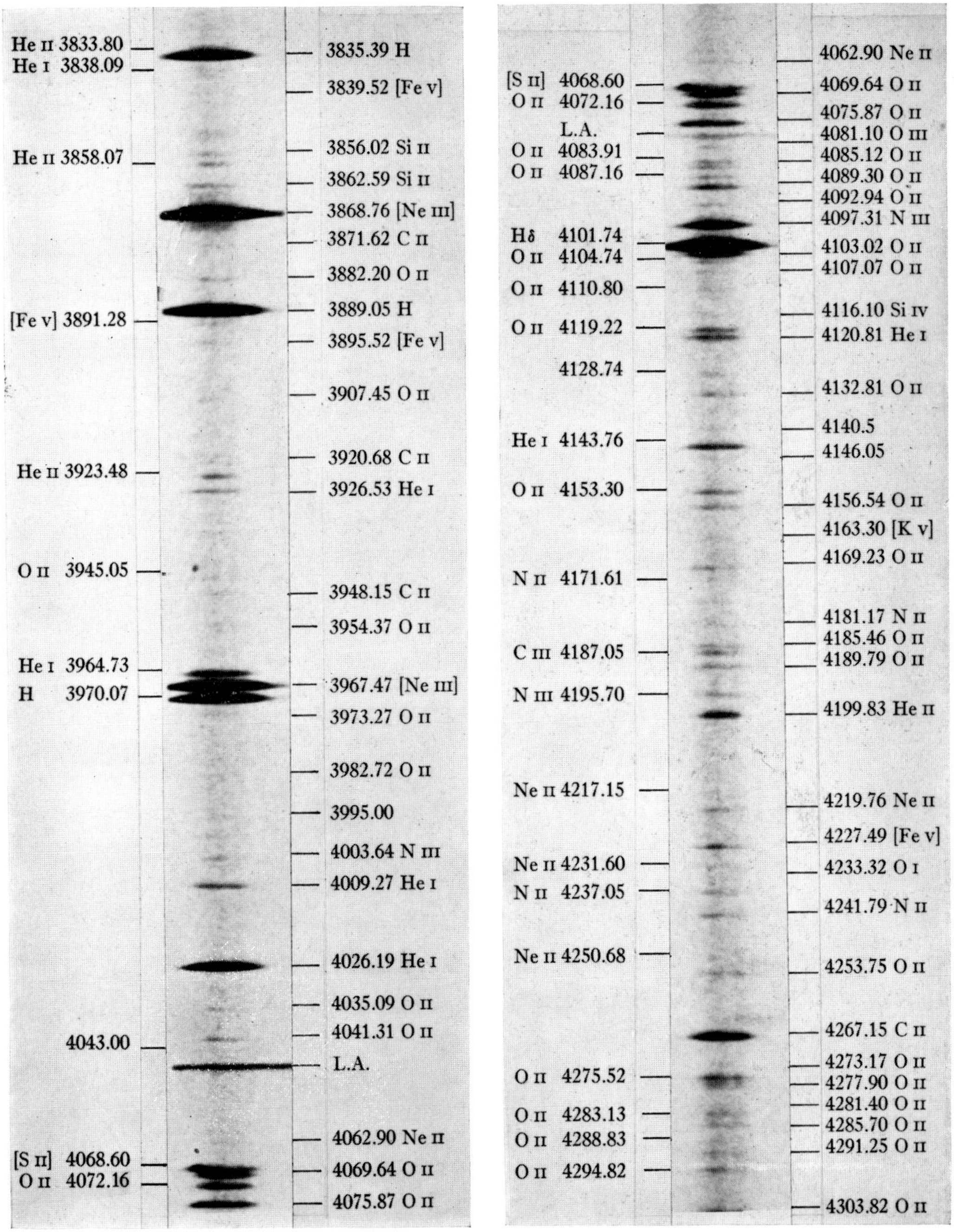

He II 3833.80
He I 3838.09
He II 3858.07
[Fe v] 3891.28
He II 3923.48
O II 3945.05
He I 3964.73
H 3970.07
4043.00
[S II] 4068.60
O II 4072.16

3835.39 H
3839.52 [Fe v]
3856.02 Si II
3862.59 Si II
3868.76 [Ne III]
3871.62 C II
3882.20 O II
3889.05 H
3895.52 [Fe v]
3907.45 O II
3920.68 C II
3926.53 He I
3948.15 C II
3954.37 O II
3967.47 [Ne III]
3973.27 O II
3982.72 O II
3995.00
4003.64 N III
4009.27 He I
4026.19 He I
4035.09 O II
4041.31 O II
L.A.
4062.90 Ne II
4069.64 O II
4075.87 O II

[S II] 4068.60
O II 4072.16
L.A.
O II 4083.91
O II 4087.16
Hδ 4101.74
O II 4104.74
O II 4110.80
O II 4119.22
4128.74
He I 4143.76
O II 4153.30
N II 4171.61
C III 4187.05
N III 4195.70
Ne II 4217.15
Ne II 4231.60
N II 4237.05
Ne II 4250.68
O II 4275.52
O II 4283.13
O II 4288.83
O II 4294.82

4062.90 Ne II
4069.64 O II
4075.87 O II
4081.10 O III
4085.12 O II
4089.30 O II
4092.94 O II
4097.31 N III
4103.02 O II
4107.07 O II
4116.10 Si IV
4120.81 He I
4132.81 O II
4140.5
4146.05
4156.54 O II
4163.30 [K v]
4169.23 O II
4181.17 N II
4185.46 O II
4189.79 O II
4199.83 He II
4219.76 Ne II
4227.49 [Fe v]
4233.32 O I
4241.79 N II
4253.75 O II
4267.15 C II
4273.17 O II
4277.90 O II
4281.40 O II
4285.70 O II
4291.25 O II
4303.82 O II

PLATE 2.—*Continued*

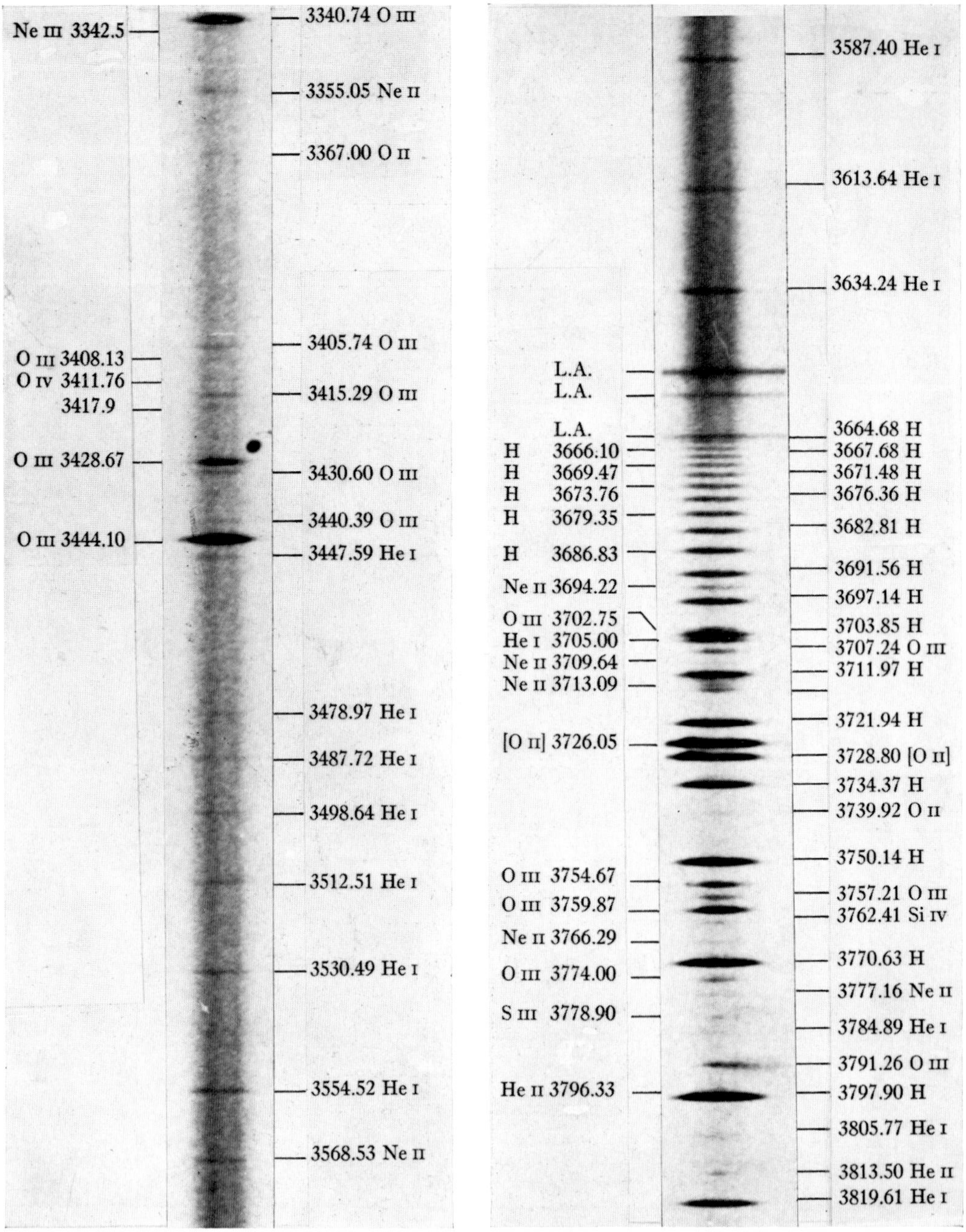

Ne III 3342.5
3340.74 O III
3355.05 Ne II
3367.00 O II
O III 3408.13
O IV 3411.76
3417.9
3405.74 O III
3415.29 O III
O III 3428.67
3430.60 O III
3440.39 O III
O III 3444.10
3447.59 He I
3478.97 He I
3487.72 He I
3498.64 He I
3512.51 He I
3530.49 He I
3554.52 He I
3568.53 Ne II

3587.40 He I
3613.64 He I
3634.24 He I
L.A.
L.A.
L.A.
H 3666.10
H 3669.47
H 3673.76
H 3679.35
H 3686.83
Ne II 3694.22
O III 3702.75
He I 3705.00
Ne II 3709.64
Ne II 3713.09
[O II] 3726.05
O III 3754.67
O III 3759.87
Ne II 3766.29
O III 3774.00
S III 3778.90
He II 3796.33
3664.68 H
3667.68 H
3671.48 H
3676.36 H
3682.81 H
3691.56 H
3697.14 H
3703.85 H
3707.24 O III
3711.97 H
3721.94 H
3728.80 [O II]
3734.37 H
3739.92 O II
3750.14 H
3757.21 O III
3762.41 Si IV
3770.63 H
3777.16 Ne II
3784.89 He I
3791.26 O III
3797.90 H
3805.77 He I
3813.50 He II
3819.61 He I

PLATE 2.—*Continued*

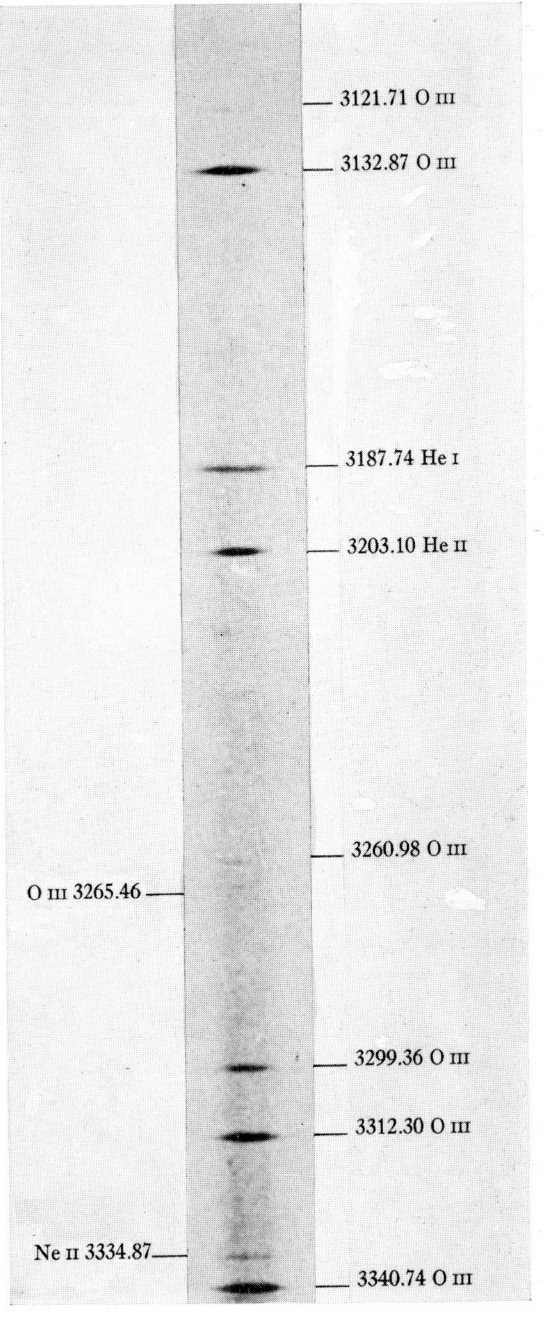

PLATE 2.—*Continued*

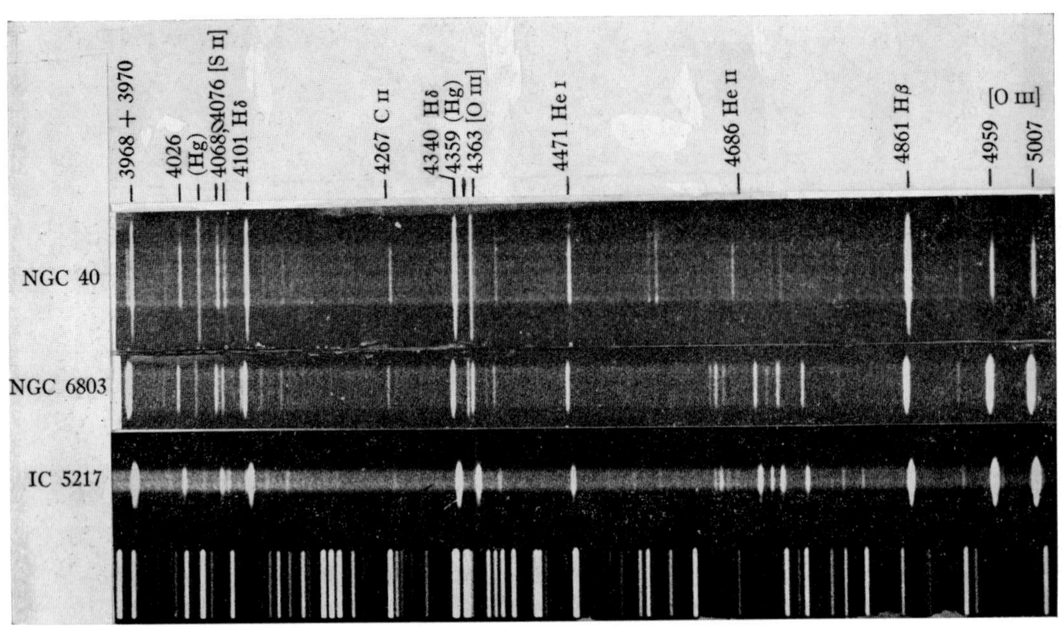

PLATE 3.—Comparison of spectra of three planetary nebulae in the blue-violet region. We compare the low-excitation nebula NGC 40 with objects of medium and moderately high excitation, NGC 6803 and IC 5217 (Lick Observatory photographs).

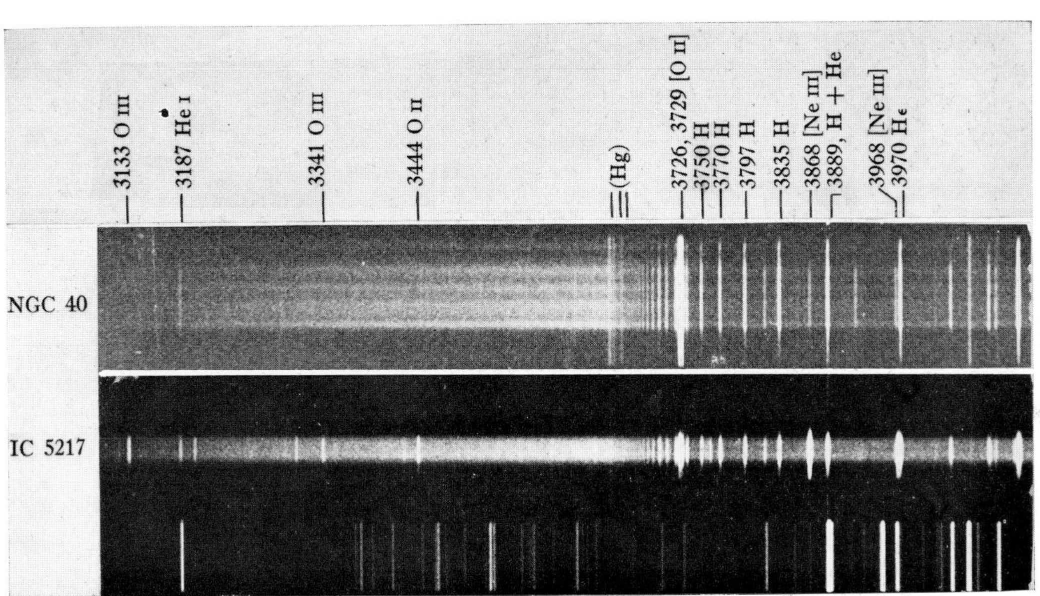

PLATE 4.—The spectra of two planetaries in the near ultraviolet. We compare the spectrum of the low-excitation planetary NGC 40 with that of the moderately high-excitation planetary IC 5217. Note the paucity and weakness of lines in the ultraviolet of NGC 40, where only λ 3187 He I is prominent and the strong Balmer continuum (Lick Observatory photographs).

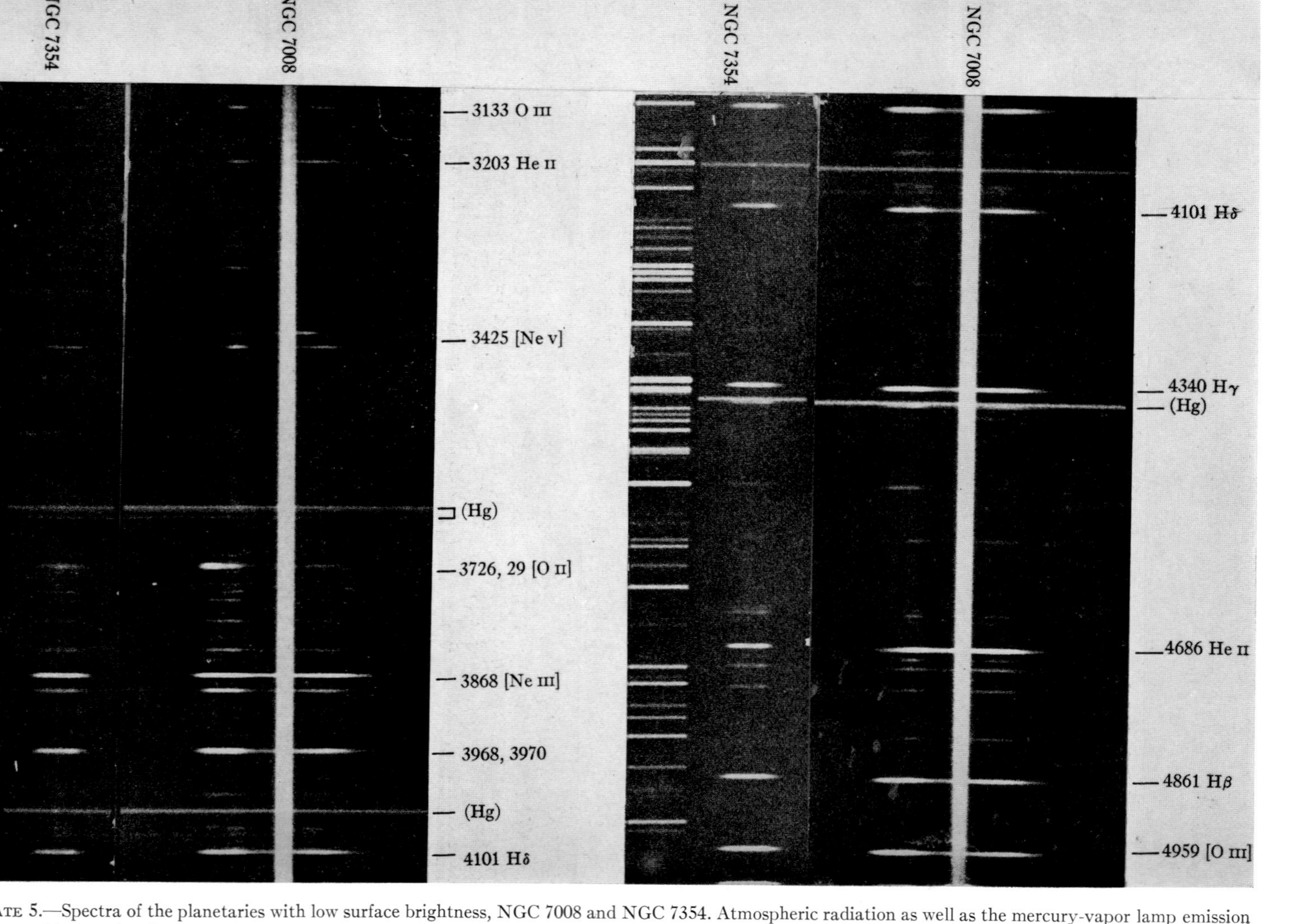

Plate 5.—Spectra of the planetaries with low surface brightness, NGC 7008 and NGC 7354. Atmospheric radiation as well as the mercury-vapor lamp emission became troublesome for these objects. Note the variations in excitation from point to point in NGC 7008 (Lick Observatory photographs).

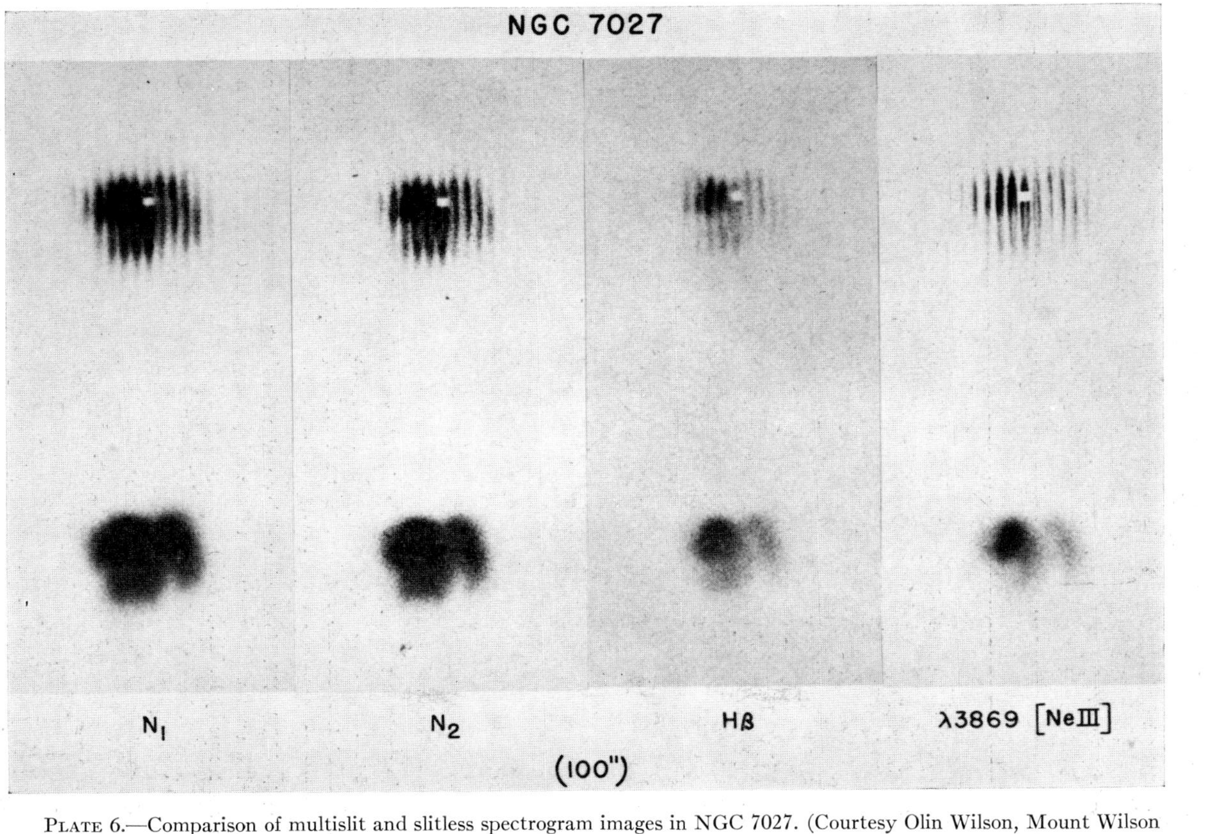

Plate 6.—Comparison of multislit and slitless spectrogram images in NGC 7027. (Courtesy Olin Wilson, Mount Wilson and Palomar Observatories.)

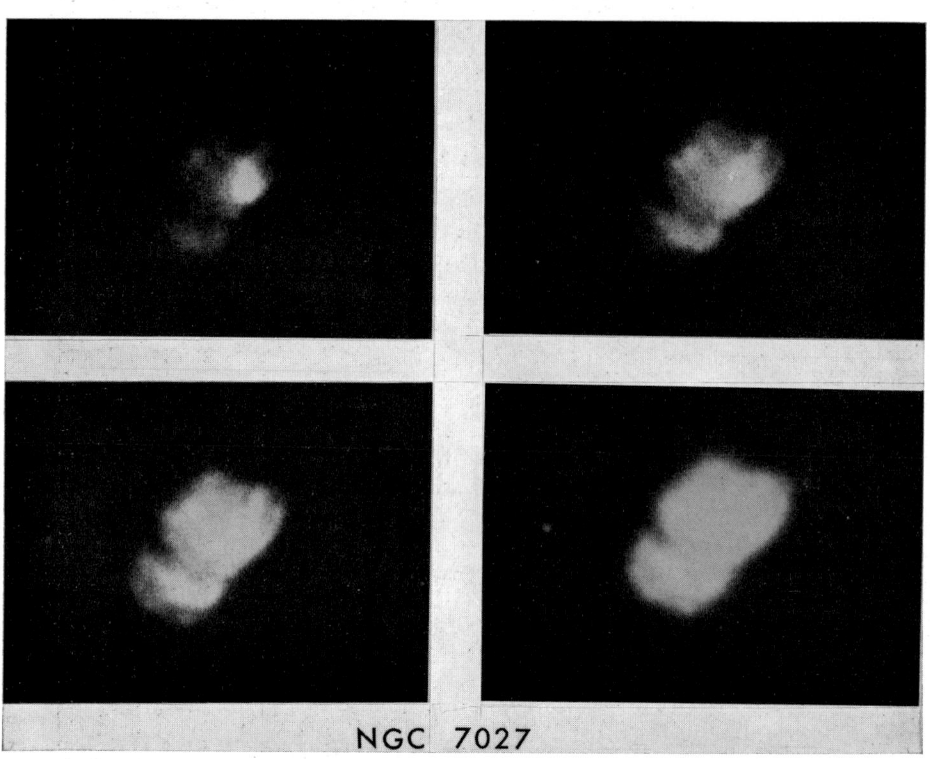

Plate 7.—NGC 7027. No star is visible on even the shortest exposures. The nebula consists of numerous small filaments. (Compare with Plate 6.) (Photographed in the light of Hα with the 200-inch Palomar telescope by R. Minkowski.)

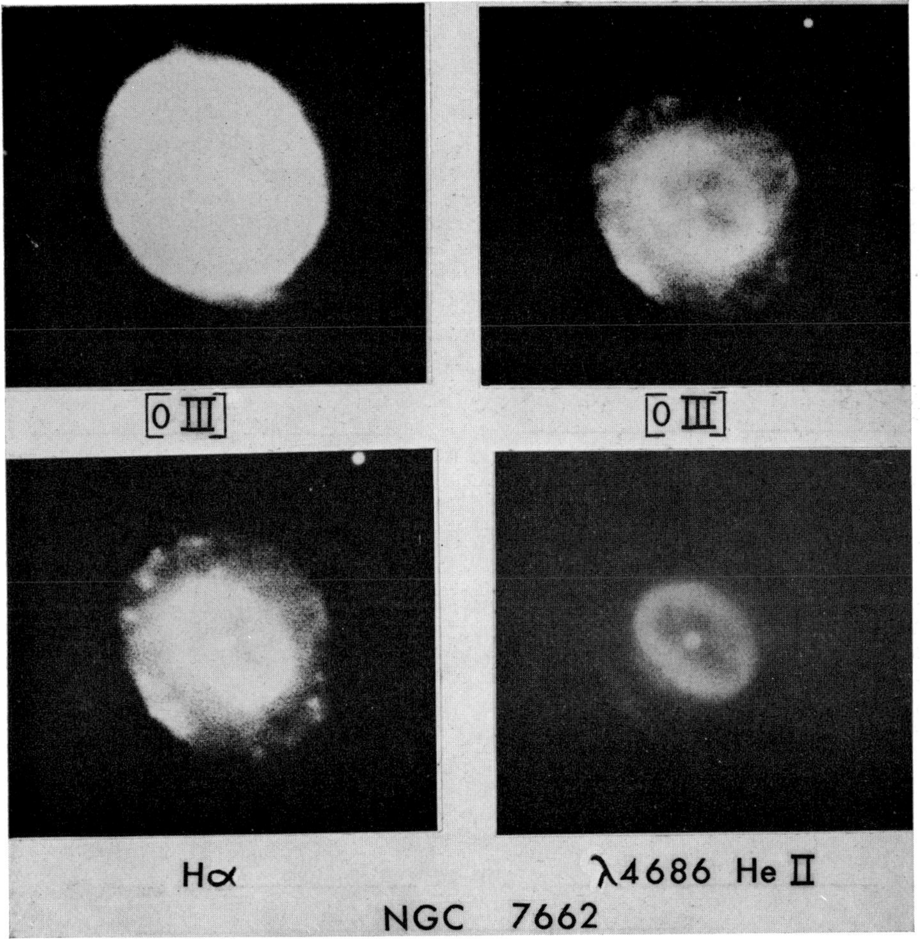

PLATE 8.—The double-ring planetary NGC 7662. Notice the tufts at the ends of the major axis in the stronger image of [O III]. Notice also the small condensations in the outer ring in Hα and [O III]. The radiation of ionized helium is rather well confined to the inner ring. Isophotic contours for this nebula are given by Aller (1956, pp. 30, 255; 1965, p. 672). (Photographed with the Palomar 200-inch telescope by R. Minkowski.)

PLATE 9.—C.D. − 29°13998. This series of exposures was secured in the light of Hα + [N II] with the 200-inch Palomar telescope by R. Minkowski. An isophotic contour has been published by Aller (1956, p. 242).

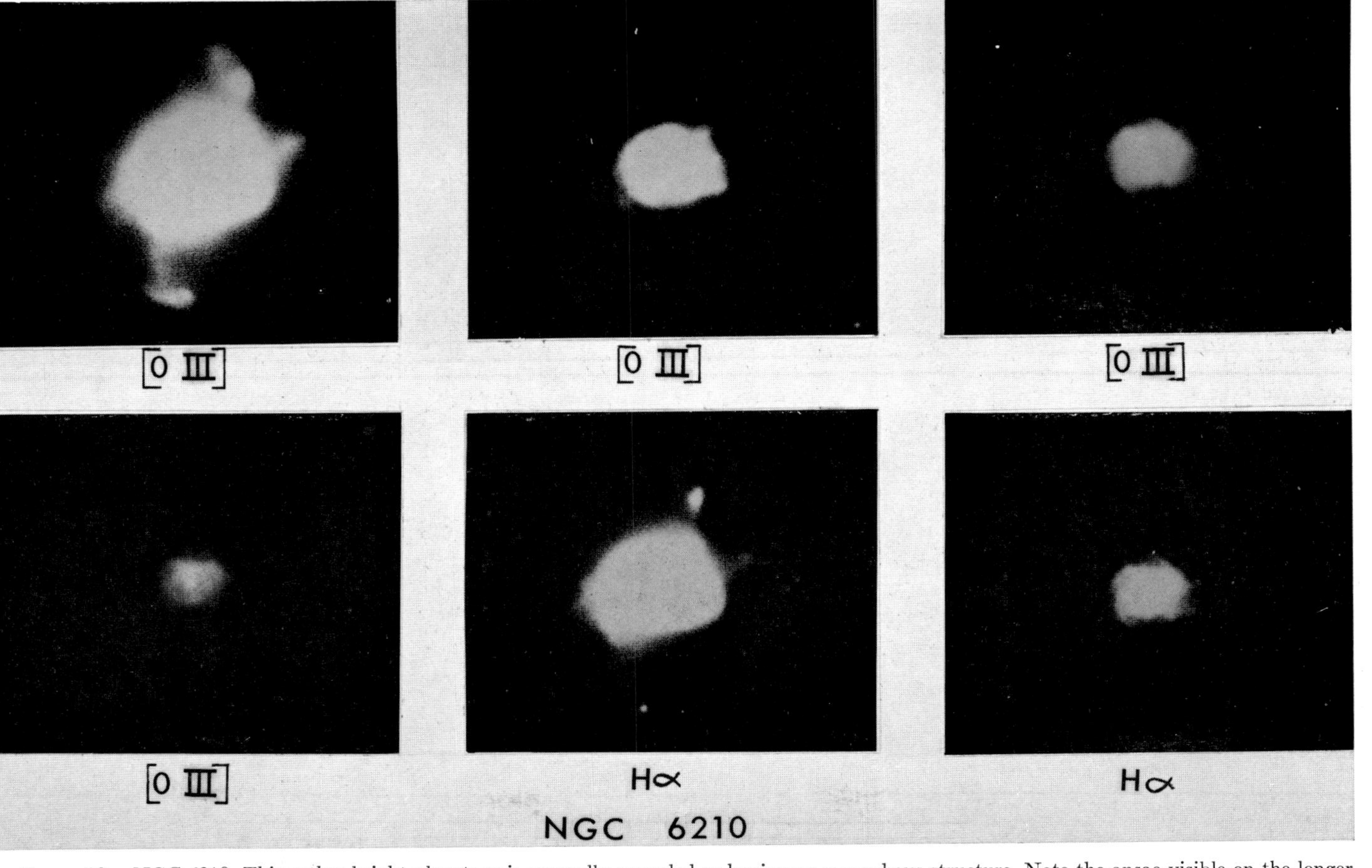

PLATE 10.—NGC 6210. This rather bright planetary is generally regarded as having an amorphous structure. Note the ansae visible on the longer exposures in Hα and [O III]. An isophotic contour is available only for the inner regions (Aller 1956, p. 253). The central star is very bright. (Photographed with the Palomar 200-inch telescope by R. Minkowski.)

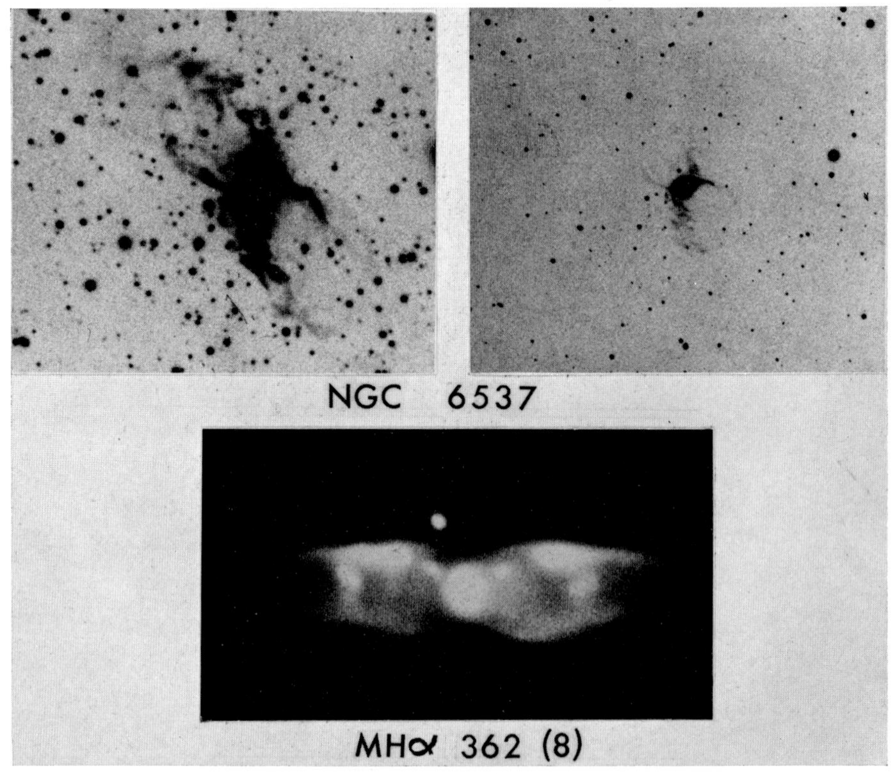

PLATE 11.—Two remarkable planetaries. NGC 6537, photographed in the light of Hα + [N II], show remarkable filamentary structure suggestive of magnetic fields. Isophotic contours have been measured (Aller 1956, p. 243). The nebula MHα 362 (designated in Table 1 as M2-9, $\alpha = 17^h02^m.9$, $\delta = -10°04$ [1950]) possesses a unique symmetry. An isophotic contour is given in Aller (1956, p. 242). (Photographed with the Palomar 200-inch telescope by R. Minkowski.)

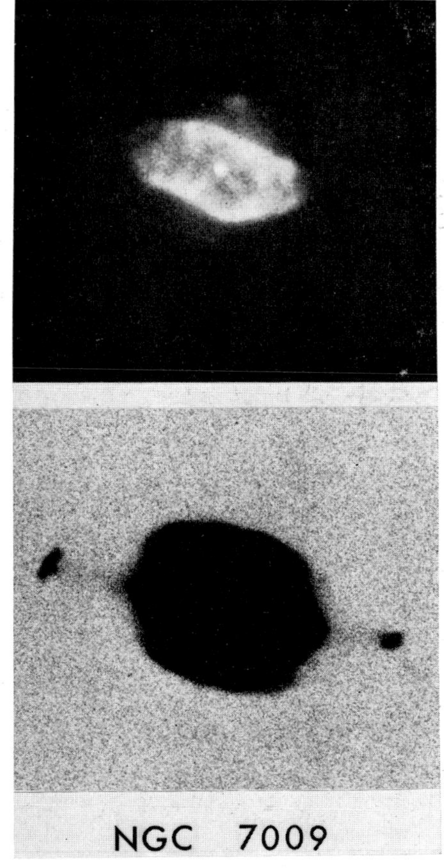

PLATE 12.—The double-ring planetary NGC 7009. This beautiful planetary has sometimes been called the Saturn Nebula because of its symmetrical ansae (see lower "negative," otherwise heavily overexposed, print). Note the elongation and complicated structure of the inner ring. The outer ring is also nonuniform. The ansae have a much lower excitation spectrum (Aller 1941, Table 4) than does the main body of the nebula (see Plate 2). Isophotic contours are given by Berman (1930) and by Aller (1956, p. 253). The relatively bright central star appears to show a perfectly continuous spectrum. (Photographed with the Palomar 200-inch telescope by R. Minkowski.)

T_1. If the absorption is complete for an ion, one may deduce the temperature from the intersection of the curve given by equation (183) with that of the ion. Likewise, a temperature determination is yielded by curves for H I and He I if the nebula is optically thick for both ions.

Table 17 gives the results of 11 nebulae discussed by Zanstra (1960). Successive columns give the reference line with respect to which Zanstra measured $A_p(l)$ on slitless spectrograms obtained at Lick Observatory (Aller 1941, 1951). The adopted $L_p(\beta)/L_p(l)$ equaled 1.98 for Hγ and 3.37 for Hδ. For 4471 A and 4026 A in NGC 6572, values of this ratio were 12.25 and 35.6, respectively. Column 4 gives the sum of the forbidden-line intensities in terms of that of Hβ after correction for interstellar extinction. Except for NGC 40, the largest contribution comes from the green nebular [O III]-lines. New photoelectric data have been used in several instances; hence we derive temperatures in columns 7 and 8 that differ slightly from those of Zanstra (except for J320).

TABLE 17

TEMPERATURES OF THE CENTRAL STARS OF ELEVEN PLANETARY NEBULAE

Nebula	Line	$A_p(l)$	$\Sigma[L_p(f)/L_p(\beta)]$	T_e	$T_{\rm H}$	T_n	T_{ee}	$T_{\rm He}$	$T_{\rm spec}$	$T_{\rm O'D}$	$T_{\rm H,S}$
NGC 40	3727A	0.211	....	10	46.4	......	......	......	......	34	34
IC 418	Hγ	.0463	6.40	11	39.9	32.7	35	43.8	33	53	43
	Hδ	.0151	....	......	35.8	29.7	32	41.3	......	......	......
NGC 6210	Hδ	.0310	16.16	11.7	42.8	39.8	42.4	48.8	32	58	50
NGC 6543	Hδ	.0319	11.67	10	43.1	37.7	39.7	46.6	(36)	53	66
NGC 6572	4471A	.019	16.4	11.3	52.2	48.6	51.0	56.6	......	64	62
	4026A	.006	....	......	54.0	46.0	48.4	54.6	......	......	......
NGC 6826	Hγ	.0210	12.05	11.6	33.4	31.2	33.7	41.8	32	42	69
NGC 1535	Hδ	.0092	18.2	15	32.0	32.5	36	40.7	......	37	74
J320	Hδ	.0076	15.7	17.7	30.8	30.8	34.9	40.0	......	36	......
NGC 3242	Hδ	.0386	21.04	14	45.4	43.9	48.2	47.9	......	50	93
NGC 6891	Hδ	.0085	12.37	15	31.5	30.2	33.4	40.6	32	36	56
IC 5217	Hδ	0.0142	18.5	11.5	35.2	35.3	37.9	42.8	......	48	74

Column 5 gives the adopted electron temperature in units of thousands of degrees. Column 6 gives $T_{\rm H}$, the temperature derived by Zanstra's original hydrogen argument. Column 7 gives the temperature T_n obtained by Zanstra's (1931*b*) nebulium argument (cf. eq. [181]), and column 8 gives the "corrected" nebulium temperature, in which the influence of a finite temperature is incorporated into equation (181). Column 9 gives the helium temperatures; column 10 gives the temperature estimated from the spectral class, and the next column gives O'Dell's estimate of the Zanstra temperatures. The last column gives the temperatures as derived by Harman and Seaton. Except for IC 418, all of these temperatures are higher than those found by Zanstra.

Zanstra noted that since the energy of photoelectrically liberated electrons rises more rapidly with T than does the number of ionizing quanta, incomplete hydrogenic absorption will cause $T_{\rm H}$ to be lower than T_{ee}. This situation holds for NGC 1535, J320, NGC 3242, NGC 6891, and IC II 5217. For IC 418, NGC 6543, and NGC 6572, $T_{\rm H}$ exceeds T_{ee}, whereas for NGC 6210 and NGC 6826 the temperatures are comparable. Error in line intensities, corrections for interstellar extinction (in the nebular method), or deviations from blackbody energy distributions may explain some of the discordances.

Harman and Seaton regarded these discordances as arising mostly from effects of incomplete absorption in hydrogen and helium shells and concluded that the higher temperatures were the valid ones. Note, however, the discrepancies between the temperature deduced from spectral classes and those given in the last column for objects for which comparisons are possible. To these we may add the nuclei of IC 2149 and NG 2392 for which spectral class temperatures of 33,000° K (Wilson and Aller 1954, Aller 1948) are to be compared with Harman and Seaton's values of 49,000 and 68,000° K, respectively! More recent observations of most of these objects secured with the Lick 120-inch coudé spectrograph have substantiated these spectral-class estimates. For example, there is no way in which the spectrum of the central star of NGC 6826 can be

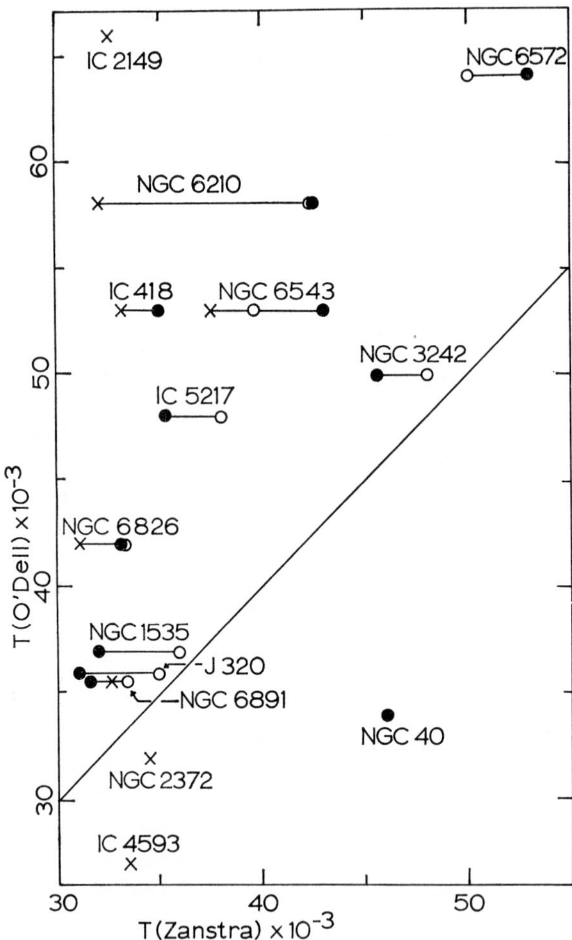

FIG. 6.—Comparison of central star temperatures for various planetary nebulae determined by different methods. We plot the temperature determined by O'Dell against those found by Zanstra's methods and from spectral classes (see Table 17). *Solid circles* = hydrogen temperatures; *open circles* = temperatures from forbidden lines (the finite electron temperature is taken into account); *crosses* = temperatures from spectral classes (when available).

reconciled with a temperature of 69,000° K. Perhaps these planetary nuclei are all binaries in which the visible star of temperature 30,000–35,000° K is paired with a much hotter companion. NGC 1514 offers a more extreme example; it is almost certainly a binary. See Kohoutek (1967). Clearly the methods must be redeveloped.

11.3. Model Atmospheres

Perhaps some of the difficulties may be resolved by detailed model-atmosphere calculations. Consider the helium temperatures given in column 9 of Table 17. The brightness temperatures predicted from model atmospheres of early-type stars at wavelengths just shorter than the helium limit are 2400–5700° K higher than at those at wavelengths slightly shorter than the hydrogen limit (see Table 18). This result is in qualitative agreement with Zanstra's results.

TABLE 18

Brightness Temperatures from Model Atmospheres

References	Visible		Hydrogen Continuum	Helium Continuum
Saito and Uesugi (1959)	37,200° K	5081 A	36,400° K	38,800° K
Underhill (1951)	39,500° K	4234 A	38,100° K	42,800° K

It should be noted that the energy distribution short of the Lyman limit depends not only on the surface gravity and effective temperature of the star, but also upon its chemical composition (Aller 1956, p. 230). In general, the ultraviolet fluxes deviate from Planckian curves by an amount that diminishes with rise in temperature. Thus Gebbie and Seaton (1963) concluded that no large errors result from assuming that hot central stars radiate as blackbodies.

Böhm and Deinzer (1965) developed a program for calculating non-gray model atmospheres in radiative and hydrostatic equilibrium for 50,000° K $< T_{\text{eff}} <$ 250,000° K. They took into account the continuous absorption due to H, He, C, N, O, and Ne in appropriate stages of ionization and allowed for radiation pressure and effects of electron scattering. A detailed model, corresponding to a star with $T_{\text{eff}} =$ 100,000° K, log $g = 6.2$, showed that although the effect of the Lyman limit in atomic hydrogen was small, there was a tenfold intensity decrease across the Lyman limit of ionized helium; an energy excess occurred just to the redward of this limit at 228 A. A curious result of the calculation was that the Ne IV and N V series limit appeared in emission.

These models may have too high a surface gravity. Such high gravities (values of g) might be checked by relativistic shifts between nebular and stellar lines in a manner envisaged and calculated by Abell for some of the stars of low surface brightness nebulae which resemble white dwarfs. A similar approach has been proposed by O'Dell. Additional calculations would be extremely worthwhile, especially if the effects of varying He/H ratios and line-blanketing phenomena are taken into account. An attempt should also be made to calculate theoretical line profiles and total intensities in order that the absorption-line spectra may be interpreted properly.

The Wolf-Rayet type and, to some extent, the Of-nuclei, represent stars with unstable envelopes which are not amenable to a quantitative discussion.

Variability of some planetary nuclei in magnitude and color has long been suspected and seems to be substantiated by Abell's observations, where fluctuations of the order of $0^{m}.5$ are reported. A striking example may be the planetary at $8^{h}55^{m}.7$, $-28°45'$ found to have $M_{pg} = 15.9$ (March 9, 1956) and 17.4 (March 10, 1956) (Kohoutek 1964). For many planetaries, observations are difficult because of the influence of the bright nebular background.

11.4. Evolution of the Central Star

The question of the evolution of the central stars of the planetary nebulae and the place of these objects in the over-all scheme of stars and stellar systems has received increased attention in recent years. That these hot stars were small objects approaching a white-dwarf state appears to have been recognized first by Menzel (1926), but after forty years the detailed evolutionary tracks are not yet established.

Assuming for the moment that we can establish the distances of the planetaries to sufficient accuracy, we can plot a conventional luminosity color diagram, for example, M_B vs. $B - V$. Such a plot would represent a distortion in two ways. First, the V or B magnitude is a poor index of the total brightness of the star, because of the large bolometic correction, and second, the color (even after correction for interstellar extinction) is insensitive to the temperature when it is very high.

A more meaningful plot would be bolometric absolute magnitude or log $L/L_{\odot}$ vs. log T_{eff}. In order to obtain the bolometric correction, we must know the temperature of the star, which can be estimated by the procedures just described. Hertzspung-Russell diagrams have been given for planetary nebulae by Shklovsky (1956), Aller (1960), O'Dell (1962), Harman and Seaton (1962), Seaton (1966), and Abell (1966).

Most of these investigations have yielded patterns in which the stars are spread in a broad scatter extending from objects of higher luminosity and lower temperature to those of lower luminosity and higher temperature. O'Dell noted a rapid decrease in luminosity and radius in the 30,000-year time-scale characteristics of the dissipation of the outer shell. He finds a dispersion corresponding to $\Delta \log T = 0.3$ and Δ (luminosity) $= 1$ mag.

Figure 7 gives a plot of log T_{eff} vs. log $L/L_{\odot}$ for nuclei of planetary nebulae. Where possible, the temperatures have been taken from spectral class calibrations or determinations by Zanstra. The distances of the classical planetaries are mostly from Minkowski (their photographic magnitudes are due to O'Dell); otherwise, we have taken O'Dell's values. For the nebulae observed by Abell, we have used this data directly. The adopted bolometric corrections correspond closely to those given by Aller (1963) for the lower end of the scale and merge with those suggested by O'Dell at the higher temperatures (see Table 19). In this way we tried to allow for the fact that the hotter the star, the more closely its energy distribution approaches that of a blackbody.

Notice that the plot shows a very large scatter. Part of this arises from the circumstance that the temperatures are often only lower limits. In such instances the points are shifted in the direction indicated by the arrow. Some scatter arises from uncertainties in the distance and improper corrections for interstellar extinction, and some distortion must certainly arise from errors in the bolometric correction. As pointed out by Abell, these data do not fit a well-defined evolutionary track.

Harman and Seaton (1964) and more recently Seaton (1966) have derived new tem-

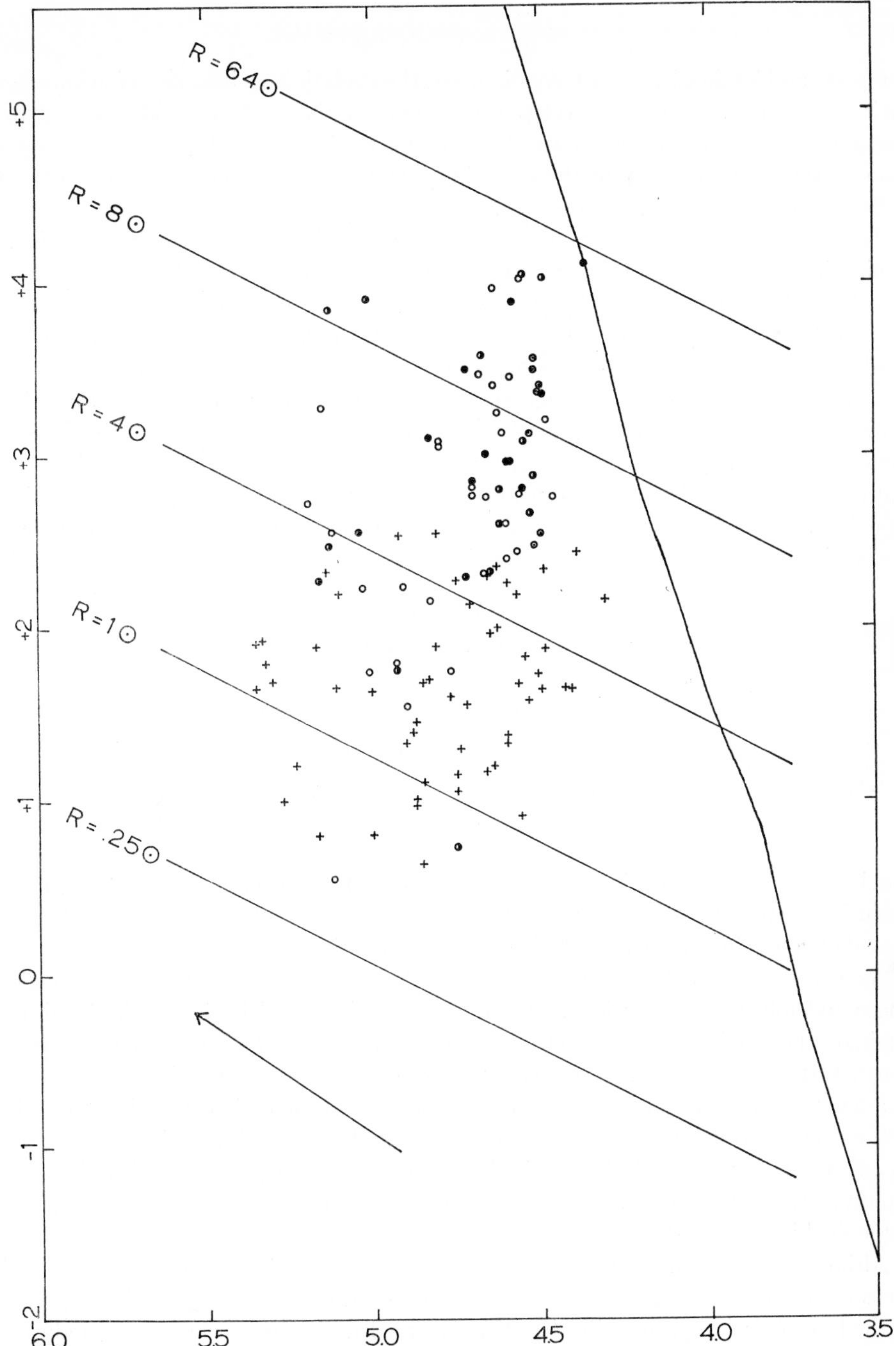

FIG. 7.—The luminosity-surface temperature relationship for planetary nebulae. We plot the bolometric brightness, log $L/L_{\odot}$, against the logarithm of the effective temperature. The solid line at the right denotes the main sequence; the diagonal lines represent the loci of points of constant radius. The arrow indicates the direction any point will be moved if the assumed temperature is too low. Data are obtained from different sources as follows:

Open circles, m_{pg}, T(O'Dell 1963*a*), distance, *d*, (Minkowski 1965*b*); *half-filled circles*, m_{pg}, (O'Dell), *d*(Minkowski), *T*(Zanstra 1960); *dot in center of circles*, m_{pg} (O'Dell 1963*a*), *d*(Minkowski 1965*b*), *T*(spectral class); *crosses*, log $(L/L_{\odot})_{\min}$ and log $T_{\min}$ from Abell (1966).

peratures and luminosities and have discussed the evolutionary behavior of nebulae and their central stars. The entire development occurs on a time scale of 50,000 years. Seaton concludes that the nebulae are optically thick for radii less than 12,000 a.u., or greater than 120,000 a.u. (because of the declining luminosity of the central star); they are optically thin for all intermediate sizes.

The stars show a steady evolution toward the white-dwarf stage from an initial luminosity $L = 60L_\odot$ and temperature $T = 32{,}000°$ K. The star evolves to $L = 25{,}000$ $L_\odot$, $T = 60{,}000°$ K; then the luminosity remains nearly constant as the temperature rises to 100,000° K. Thereafter the star fades, the temperature remaining at 100,000° K until the luminosity declines to 100 $L_\odot$. The final luminosity drop is a consequence of the onset of degeneracy. Seaton suggests that the average mass of the stars from which planetaries are formed is 1.2 $\mathfrak{M}_\odot$, with ages of the order of 5×10^9 years; the mass becomes about equally divided between the ejected shell and the central star. From the equivalent width of the system of nebulae near the Sun, that is 500 pc, and Schmidt's model of the Galaxy, O'Dell (1963*a*) also suggested ancestral stars of 1.2 $\mathfrak{M}$ but nebular shells with mean masses of about 0.2. Existing observations do not appear to establish

TABLE 19

SUGGESTED BOLOMETRIC CORRECTIONS (B.C.)

T(° K)	B.C.	T(° K)	B.C.	T(° K)	B.C.	T(° K)	B.C.
30,000	$2^m_.6$	60,000	$4^m_.1$	90,000	$5^m_.3$	120,000	$6^m_.3$
40,000	3.3	70,000	4.5	100,000	5.7	130,000	6.6
50,000	3.7	80,000	4.9	110,000	6.0	150,000	7.2

conclusively either the evolutionary track proposed by Seaton or that by O'Dell. The evolutionary picture may be more complicated. We cannot exclude that a given star may provide two or three planetary shells in the course of its life.

As a working hypothesis we may suppose that a star enters the planetary-nucleus phase and follows an evolutionary track of a type indicated by Hayashi, Hōshi, and Sugimoto (1962) for the evolution of a star of solar mass whose nuclear fuels are exhausted. It shrinks from a star of about 1 solar diameter to one a hundred times smaller in a few tens of thousands of years, and it experiences profound changes in its interior. The nuclei of old nebulae studied by Greenstein and Minkowski (1964) may represent an advanced stage in which the core is already degenerate and the atmosphere is cooling rapidly. The final transition to a white-dwarf stage is difficult to recognize because the outer shell is gone by then.

Although calculations by Upton *et al.* (1960) showed that helium-burning stars fell in the correct part of the $\log L/L_\odot - \log T_e$ diagram, the evolutionary track is not parallel to the locus of points defined by the planetary nuclei. Similar conclusions are indicated by Cox and Salpeter (1961) and by Divine (1965). Probably planetary nuclei have completed the nuclear fuel-burning phase and have entered the stage of final contractions, but we are not sure that all stars, even of those in the correct mass range, pass through the planetary-nebulae phase.

The immediate progenitors of the planetary nebulae are much harder to identify. The small range of velocities characteristic of planetary nebular shells cannot be reconciled with any ejection process from the surface of a star of small radius (see Goldreich and Abell 1966). Presumably, they must originate from giant stars, as has been suggested by a number of authors, although their immediate antecedents may be combination variables such as Z And, AX Persei, or RW Hydrae. It has also been suggested that stars with magnetic fields are involved, although fields suggested by nebular shells may have been developed *in situ*. Association of a planetary-nebula development with a helium flash (Sakashita and Tanaka 1962) or a carbon flash (if it occurs) (Hyashi, Hōshi, and Sugimoto 1962) appears unlikely (Osterbrock 1964).

Much work, both theoretical and observational, needs to be done to solve the problem of the evolution of planetaries. The very earliest stages need to be identified and studied; accurate distances are a must. Temperatures, radii, and bolometric corrections for the central stars must be established. The present situation is far from satisfactory.

12. STRUCTURE AND MAGNETIC FIELDS

Using the system of Vorontsov-Velyaminov (1934*a*), Perek and Kohoutek (1966) list the classification according to form of 505 planetaries in their catalogue. In Table 20 we summarize the Vorontsov-Velyaminov system and present the statistics of nebular forms as derived from the Perek-Kohoutek catalogue. Certainly several selection effects operate; most important are those of distance and discovery. Clearly, many of the Type I (stellar) nebulae would be classified otherwise if they were closer to the Earth. Moreover, many Type I planetaries remain undiscovered, since spectral surveys do not extend fainter than about twelfth magnitude.

TABLE 20

CLASSIFICATION OF PLANETARY NEBULAE

Type	Number	Per Cent	Description
I	102	20.2	Stellar image
II	177	35.0	Disk
II*a*	20	4.0	Disk brighter toward the center
II*b*	16	3.2	Disk of uniform brightness
III	29	55.7	Disk with irregular structure
III*a*	16	3.2	Disk with quite irregular distribution of brightness
III*b*	27	5.3	Disk showing traces of a ring
IV	105	20.8	Ring nebula
V	8	1.6	Irregular nebula intermediate between planetary and diffuse nebula
VI	5	1.0	Anomalous form
Total	505	100.0	

Omitting from consideration nebulae of Types I, V, and VI, we find that about one-third of the classified planetaries are ring nebulae. Moreover, an examination of the Curtis catalogue of photographs (1918) reveals that at least 80 per cent of those nebulae with perceptible surfaces show some symmetry about an axis lying in the plane of the sky. These nebulae are called *bipolar* by Gurzadian (1962), who suggests that magnetic fields are operating to produce this appearance. It would seem that an understanding of the processes that result in ring and bipolar forms would be a large step toward an understanding of the mechanisms which produce planetary nebulae initially.

That correlations exist between type and location in the sky becomes apparent when we examine a chart such as Figure 8, where we indicate the Vorontsov-Velyaminov classification of each nebula at different galactic coordinates. Here the breakdown is into stellar, disk, and ring nebulae only. The predominance of stellar and disk planetaries in the direction of the galactic center is striking, and the north-south asymmetry at high latitudes is amusing though hardly significant. The large number of planetaries presumably in or very near the galactic nucleus, and hence at a relatively large distance, accounts for the high concentration of stellar planetaries in the direction of the galactic center.

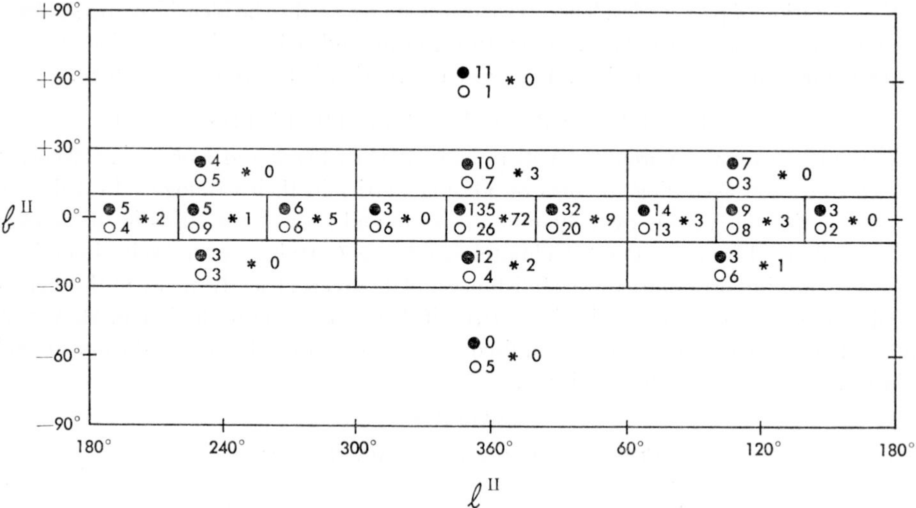

FIG. 8.—The distribution of the three main types of planetary nebulae as a function of galactic latitude and longitude. Given are the numbers of disk (*solid circles*), ring (*open circles*), and stellar (*asterisks*).

Various three-dimensional models which would result in a nebula with the two-dimensional appearance of a ring can be imagined. The more obvious possibilities are (1) a hollow shell or envelope of low opacity; (2) a torus or toroid viewed normally, or nearly so; (3) a uniform sphere or cylinder with decreased emission in the central regions; and (4) a uniform sphere or cylinder with increased opacity in the central regions. In order to examine quantitatively possibility (4), Wilson (1958) compared the intensities of the Doppler-separated emission-line components arising from behind and in front of the central stars of planetary nebulae. His study showed conclusively that the approaching or near sides of the nebulae were no brighter than the receding or distant regions, thus ruling out any absorption of light in the central parts of nebulae.

All three remaining forms may be well represented in fact. An inability to represent a planetary by a spherically or ellipsoidally symmetric model could very well mean that the true three-dimensional model is a torus or toroid. The qualitative considerations of Curtis (1918) led him to conclude that such objects as the Ring Nebula, NGC 6720, could not have a uniform shell structure, but Green (1917) arrived at models for NGC 6543 and NGC 7009 consisting of several shells of rotating gas. More precise studies of NGC 6572 and NGC 6720 by Vorontsov-Velyaminov (1936, 1937) used calibrated material and resulted in the first reasonably reliable three-dimensional nebular models, but of the *radiating* matter only.

Modern analyses have been made of IC 418 by Wilson and Aller (1951), of IC 3568 by Aller (1956, p. 264), and of NGC 650–1 and NGC 6720 by Minkowski and Osterbrock (1960). The conclusion of these studies is that two of the nebulae can be represented by spherical or ellipsoidal shells (IC 418 and IC 3568), while the other two cannot. Certainly, if toroidal structures do exist, some should be seen edge-on, that is, normal to the axis of symmetry. NGC 650–1 seems to be one excellent example, while perhaps a better case is IC 4406 (Evans 1950), which received detailed study from Zanstra and Brandenburg (1951). This latter nebula seems to be a hollow cylinder seen from the side.

Basically, the procedure used to find the emission per cm³ $E(r)$ in a spherically symmetric nebula is to measure the nebular intensity I at varying distances x from the central star, where x is measured in the plane of the sky along a line of reference. $I(x)$ is the sum of all the $E(r)$'s along a line of sight, and

$$I(x) = 2\int_x^\infty E(r)\frac{r\,dr}{\sqrt{(r^2-x^2)}}. \qquad (187)$$

Techniques for solving this equation have been given by Wallenquist (1936), Smart (1938), and Chandrasekhar and Münch (1950).

Khromov (1962) has taken a different approach to the problem of deriving the appropriate three-dimensional model. He carefully inspected numerous direct photographs of nebulae and decided that a very large percentage of all planetaries can be represented by a single model consisting of a high-density toroid at the equator of a more or less ellipsoidal shell of lower density (see Fig. 9*a*). Khromov constructed isophotal diagrams depicting the appearance of the resulting nebula as seen at different orientations, and these he compared with the actual appearances of nebulae. At a conference on gas dynamics, both Zanstra (1958) and Minkowski (1958*a*) described models which again have as a basis a high-density toroidal ring. Zanstra suggested that material escapes from this toroid, mainly from the side facing the central star (see Fig. 9*b*). He believed that the high opacity of the toroid keeps the material in the outer side neutral and relatively cold. These neutral atoms receive no acceleration from the absorption of Lyman-α photons and relatively little from the pressure of the gas itself. On the other hand, the material on the inside of the ring escapes by means of these two mechanisms. The radiation pressure of the central stars sends the escaped material out of the inner hole along the hyperbolic paths illustrated in Figure 9*b*, and a low-density H II cloud, similar to what must exist about NGC 650–1 and NGC 6720, results.

Minkowski (1958*a*) criticized the narrow ring of Zanstra's diagram and proposed a model like the one shown schematically in Figure 9*a*. However, Minkowski (1955) agreed that the low-density material surrounding the toroid has escaped from the denser nebula in much the way Zanstra described. The discrepancy in points of view would thus seem to be small.

In other papers, Minkowski (1964) and Minkowski and Osterbrock (1960) point out that NGC 650–1, after which Figure 9*c* is modeled, must look much like NGC 6720 as seen from a point close to the axis of the toroid. The very faint material which appears around not only NGC 650–1 and NGC 6720 but also around many other planetaries, as inspection of the red Palomar Sky Survey shows, might also represent earlier releases of matter from the central star, as O'Dell (1962) has suggested.

We now focus attention on the important topic of the ionization structure of the nebulae. One important unanswered question is whether the material within a shell or toroidal nebula is at a low density or whether it is ionized to a degree such that few measurable radiations occur. Certainly the radiation density is higher in the inner parts

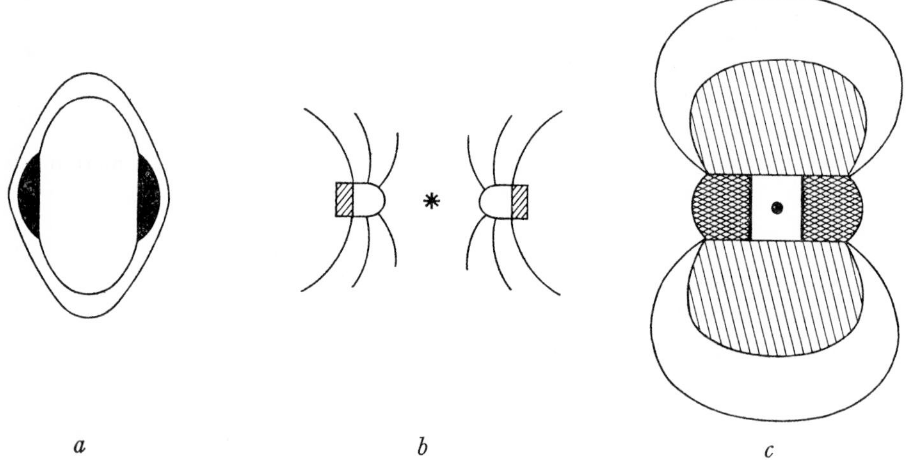

FIG. 9.—Models of planetary nebulae as derived by Khromov (*a*), Zanstra (*b*), and Minkowski (*c*)

of a nebula than farther out, but Bowen (see Wilson 1958) asserts that the temperature in the central regions is not particularly high since a whole series of ions could serve as cooling agents. According to Bowen, we should see hydrogen or helium but, of course, we often do not. Photographs taken in the light of various radiations show, in general, that the lower the ionization and excitation potential of the radiating ion, the larger the image. See Plate 8. The slitless spectra taken by Wilson (1950) and the photographs and detailed isophotal contours in Aller (1956) show this relationship very clearly. Therefore, there must be *some* increase of energetic radiation as we approach the central star. However, in some nebulae, such as NGC 7662 (see Aller, 1956, Fig. VII: 11), the ring appearance remains quite crisp, even in the light of 3426 A [Ne v]. Perhaps someday, photographs taken in the light of extreme ions whose radiations appear only in the vacuum ultraviolet will show a filled center just as the [Ne v] image of NGC 6720 occurs only inside the ring seen in the light of ions of lower excitation.

Further evidence is provided by the spectra of the central stars (see Sec. 11). When bright emission occurs on the violet side of absorption lines, indicating ejection of material at velocities of the order of 100 km/sec (Wilson 1948; Aller 1956, p. 210), we must conclude that very likely material is still being ejected from the star. The best examples are those nebulae with Wolf-Rayet type of nuclei, such as NGC 40 and NGC 6751. On the other hand, the spectrum of the central star of the Ring Nebula appears perfectly continuous, suggesting that the star is dormant and that no new material is being added to surrounding space.

If it were true that the central stars of most ring nebulae exhibited no violet-shifted emission lines in their spectra, we might conclude that the space in the inner region was, in fact, hollow. A study of the Perek-Kohoutek catalogue (1967) reveals that of the 17 ring nebulae for which the spectrum of the central star has been observed, the nuclei of 7 (41 per cent) do not show violet-shifted emission lines. Consideration of all planetaries with spectrally observed central stars leads to the result that 15 out of 26 (58 per cent) have no detectable emission lines. Hence it would seem that some ring nebulae have dormant nuclei, but this condition is hardly the general rule.

The central star of NGC 2392, which has received careful study from Wilson (1948), is one of the most active nuclei known; but even in radiations like 4686 A He II, the inner parts show a sharp ring (see Aller 1956, Fig. VII: 12).

Perhaps a definite answer to the problem of how hollow the inner parts of ring nebulae actually are will come with further detailed theoretical studies of the transfer of radiation through planetary nebulae, such as have been begun by Hummer and Seaton (1961, 1963; see also Sec. 6). Certainly another fruitful approach would be the one begun by Mathews (1966), who has considered the dynamics of nebular shells (see Sec. 13). His results give support to the concept of Zanstra that the escape of gas from a high-density toroid would tend to fill up any hollow. In fact, to maintain a ring for any appreciable time, Mathews has to postulate a continuous mass loss by the central star to provide sufficient dynamical pressure on the inner surface of the shell to prevent inward expansion. It is interesting that Mathews omits neutral hydrogen from consideration in his models.

Compared to radiations like [Ne V] and He II, the Balmer lines, [O III], and [Ne III] are of medium excitation, and [O II] and [N II] are of low excitation (see Sec. 4). In the majority of planetaries, the green nebular lines of [O III] and the Balmer lines contribute about half of the photons recorded by ordinary blue-sensitive plates and about 90 per cent of the radiation to which the dark-adapted eye responds. Hence, early photographs (and drawings) usually summarize well the appearance of the nebulae as seen in the light of medium-excitation radiation. Whether the outer edges of most planetaries as seen in the light of excited hydrogen result from a lack of matter or a lack of exciting radiation (that is, a Strömgren sphere) is still difficult to decide.

Slitless spectra and narrow-band filter photography in the light of [O II] and [N II] (when neighboring Hα is relatively weak) show images usually somewhat larger than those taken, for example, in the green (see, for example, Wilson 1950; Aller 1956, Fig. VII: 6). Often there also appears a striking increase in the visibility of the fine structure in the nebulae, particularly in the outer regions (see Aller 1956, Fig. VII: 11). Such photographs remind us that our simple shell and toroidal models must be regarded only as crude first approximations to reality. Moreover, electron densities derived from the ratio of the 3726 A, 3729 A [O II] fluxes (Osterbrock 1960) should be regarded as pertaining more to these local condensations than to the nebulae as a whole.

Why condensations should form at all remains unclear. Zanstra (1955) suggests that the relative cooling efficiencies of various ions play the important role. In particular, he suggests that where there are initially slight local increases in density, the abundance of O^{++} increases at the expense of O^{+++}. Primarily because of the green nebular lines, O^{++} radiates away much more energy than O^{+++} and cools down these more dense regions, thereby causing the pressure to drop and the density to rise even further. Daub (1963) believes that the effect is not important because of the presence of ions other than O^{+++} and O^{++}. However, since the primary cooling agents in planetary nebulae are the forbidden transitions of oxygen ions, Zanstra's general argument would seem to be correct. It may be, for example, that the often conspicuous enhancement of 3727 A [O II] in the condensations may produce the additional cooling that the Zanstra mechanism needs. Sofia's (1966) studies indicate that the origin of condensations cannot be explained by thermal instabilities at the observed stage of nebular development.

As described earlier, beyond the edges of the denser parts of many planetaries there appears on very long exposures faint wispy material (Duncan 1937). In some nebulae, this matter can be seen as a complete envelope surrounding the main body of the nebula (see, especially, Minkowski 1964). Slit spectra reveal that the outer regions are most conspicuous in the light of medium-excitation lines (Pollack 1962, Minkowski and Osterbrock 1960) and not, as we might expect, in the light of low-excitation radiation. As Minkowski (1964) suggests, the exciting radiation must, in these cases, leak out through the interspaces of the main nebula.

Gurzadian (1962) suggests that condensations are produced by the splintered remains of a magnetic field which once existed as a large dipole throughout the nebula. According to him, the bright spots arise between the magnetic fields.

Condensations of a different kind, long ropelike filaments seen, for example, in the Ring Nebula, might also be formed through the action of internal magnetic fields, although it is not difficult to imagine an irregular ejection mechanism producing the same sort of structure. Measures (to be published) by Liller and Liller indicate that the rope filaments in the Ring Nebula move slowly outward from the central star, suggesting that these elongated condensations may be a piling up of outward-streaming material crossing a magnetic constriction.

The importance of magnetic fields in planetary nebulae is almost completely unknown. Noting the predominance of "bipolar" nebulae, Gurzadian (1962) argues, often effectively, that such forces must play a role in the general shaping of a nebula. He points out that even if the central star possessed a dipole field of about 10^6 gauss, the cube law of attenuation would reduce the intensity of the field to about 10^{-10} gauss in the main body of the surrounding nebula. Because, as his study indicates, the axes of symmetry of the bipolar nebulae show no alignment with the galactic equator, the intergalactic magnetic field cannot be an important factor. To obtain an estimate of the nebular field **H**, he suggests that the magnetic energy M of a uniform spherical nebula must be of the same order as the thermal energy in the nebula. Hence,

$$M = \frac{\mathbf{H}^2}{8\pi} \cdot V = \tfrac{1}{6}\mathbf{H}^2 R^3 \sim \tfrac{1}{2}\mathfrak{M}u^2 \tag{188}$$

where $\mathfrak{M}_0$, R, and V are the mass, radius, and volume of the nebula. This expression leads to a mean value for the magnetic-field intensity $\mathbf{H} \sim 10^{-3}$ gauss. By postulating equilibrium everywhere in a nebula between the magnetic and thermal energies such that

$$\frac{\mathbf{H}^2}{8\pi} + 2N_e kT = \text{constant}, \qquad (189)$$

Gurzadian is able to construct isophotes for different models since the volume emission is proportional to N_e^2. By comparing these theoretical diagrams with actual isophotal contours (such as presented by Aller 1956, chap. VII), he again arrives at values of $\mathbf{H}$ in the neighborhood of 10^{-3} gauss for such diverse shapes as those possessed by NGC 3587 (the Owl Nebula) and NGC 6720 (the Ring Nebula). Further arguments yield explanations for elongated nebulae, rectangular shapes, helical forms, and even the existence of ring nebulae. As we stated earlier, the small knots and condensations found frequently in planetaries are explained by the disintegration of a dipole magnetic field into local, splintered fields.

Menzel (1967) has also considered the role of magnetic fields in planetary nebulae and arrives at conclusions that lend strong support to those of Gurzadian.

A conclusive argument for the existence of a magnetic field in a planetary nebula would be the detection of nonthermal emission at radio frequencies in one or more objects. To date the existence of such radiation has not been established.

13. INTERNAL MOTIONS

In their pioneering work on the spectra of planetary nebulae, Campbell and Moore (1918) observed 46 objects with prism spectrographs and found that 25 show ". . . internal motion effects. Nineteen and possibly 21 of these are interpretable as rotations about axes approximately or roughly perpendicular to the line of sight, and 4 appear to be not so interpretable. . . . The most elongated planetary nebulae show the highest rotational speeds." Campbell and Moore also found the green nebular lines and Hβ, when visible, to be doubled, at least for the inner structures of the nebulae. They conclude, "In some cases, perhaps not in all, a reversal due to an outer, cooler, more slowly rotating stratum of nebulosity appears to be the most probable cause."

It seems odd that the correct interpretation of the line doubling did not come for many years. While Perrine (1929) suggested radial expansion as the cause, it was Zanstra (1931*b*) who first gave conclusive arguments in favor of this hypothesis. He pointed out that the forbidden lines would almost never appear in absorption and that hydrogen would rarely be found in a bound, excited state.

As for the rotation hypothesis, Wilson (1958), from a study of recent data, finds that asymmetric line forms such as Campbell and Moore observed can be explained completely if different amounts of material are ejected from the nucleus in different directions with different velocities. We would expect such a condition if ellipsoidal shells were present.

Wilson's (1948, 1950) re-examination of the Doppler splitting of nebular lines has led to some most interesting as well as curious results. Most striking is the strong correlation between the measured expansion velocities and the stage of excitation of the measured lines. But image size is also well correlated with the stage of excitation of the emitting radiation, and expansion velocity is often well correlated with monochromatic radius.

Two possible interpretations exist. Either selective forces sort out the ions in such a way that the particles are accelerated inversely proportional to their level of excitation, or there is an increase in the velocity of the nebular material irrespective of its breed as we move farther away from the central star. Wilson gives convincing arguments that the latter mechanism operates. He points out that ions such as O II, O III, Ne III, and Ne V can be accurately regarded as impurities carried along by the outward-streaming hydrogen. As observation demonstrates, the relationship can be very nearly linear, and for the inner ring of NGC 7662, the radial expansion

$$\dot{r}(\text{km/sec}) = 8''.4 + 3.17r'' \tag{190}$$

(r'' is the angular radius of the monochromatic ring in arc seconds) to an accuracy of about 5 per cent. Similar expressions exist for other planetaries; in general, the better the observations, particularly of image size, the closer the correlation.

The widths of the lines themselves may be ascertained from high-resolution spectra such as those obtained by Livingston and Lynds (1964) and particularly by Osterbrock, Miller, and Weedman (1966), who observed the relatively symmetrical planetaries NGC 6210, NGC 6572, NGC 6826, NGC 7009, and NGC 7662. In each nebula they found the hydrogen lines to be the broadest, the [N II] and [O III] lines to be the sharpest, and the He II lines to be intermediate in width, a result to be expected if thermal broadening was important. The actual velocities much exceeded the thermal velocities, however. They found they could fit the observed profiles by assuming triangular-shaped velocity distribution functions, having a maximum of the emission coefficient at a definite expansion radial velocity and declining on either side of this peak velocity. Osterbrock *et al.* suggest that this distribution function could be interpreted in terms of a gradient of expansion velocity, as indicated by Wilson's (1950) data, although turbulence may still be important.

Wilson (1958) and Münch (1964) have amplified investigations of Doppler-displaced line components by using a multiple slit (see Plate 6), thus giving a detailed picture of the line-of-sight velocities in a planetary as a function of location in the nebula. From such spectrograms, Wilson was able to construct three-dimensional models on the assumption that the radial-expansion velocities must be proportional to the distance from the nucleus. His results indicate that at least some ring nebulae are ellipsoidal shells which have been and still are subjected to forces such as are produced by the gas pressure within the shells.

The determination of radial-expansion velocities allows us to make estimates of the rate of increase in angular size of planetary nebulae. Specifically, the centennial increase in angular radius

$$\dot{\theta} = \frac{100\,\dot{r}}{4.74d}\left(\frac{\text{arc seconds}}{\text{century}}\right), \tag{191}$$

where $\dot{r}$ is the spectroscopically derived expansion velocity in km/sec and d is the distance to the planetary in parsecs.

Several difficulties complicate this simple prediction. First, the spectroscopic observations refer only to the material along a line of sight passing near the center of the nebula. Asymmetric expansion velocities, as would be expected from ellipsoidal shells or toroidal rings, would invalidate any precise relation between $\dot{\theta}$ and $\dot{r}$. Second, the apparent edge of the nebular material may instead be the limit of the ionizing radiation. A prediction of the exact rate of expansion of the luminous zone can result only from an accurate knowledge of the mass distribution within the nebula. Differentiation of the formula for the radius of a Strömgren (1939) sphere,

$$S_0 = \kappa N_e^{-2/3} f(T_s,L_s) , \tag{192}$$

leads to the conclusion that the rate of expansion of the ionization sphere will be twice that of the gas. The assumptions here are that $f(T_s/L_s)$ remains constant, and that the electron density is always uniform, which requires that the outward material velocity is proportional to the distance from the central star.

If for the moment we ignore these difficulties and assume that planetaries are simple, optically thin nebulae, we find that some nebulae should have expanded by measurable amounts since early in this century when the first high-quality plates were taken. We tabulate $\dot{\theta}$'s for several of the more interesting candidates (see Table 21); here we have

TABLE 21

PREDICTED CENTENNIAL RADIAL EXPANSIONS

	d (pc)	$\dot{r}$ (km/sec)	$\dot{\theta}$ (arc secs/ 100 yr)
NGC 7293.......	145	(25)	3.6
NGC 6853.......	240	27.9	2.5
NGC 3587.......	520	44.5	1.8
NGC 2392.......	790	53.5	1.4
NGC 246........	400	(25)	1.3
NGC 6720.......	755	29.3	0.8
NGC 7662.......	1095	26.1	0.5

used Minkowski's distances (1965*b*) and, except for NGC 7293 and NGC 246, Wilson's mean Doppler expansion rates (1950, 1964). In the absence of data for these two planetaries, we have adopted the listed values.

The first attempt to measure the internal motions of a planetary nebula on direct plates was by Koslov (1935) for NGC 6543. His results, derived from refractor plates (32″/mm) with a time base of 27 years, suggest small, rather random motions of several knots and condensations, but no clear evidence of expansion.

The first well-determined increase in size came from measures described by Latypov (1957) who found the radius of NGC 6720 to be expanding at a centennial rate of 0″.9 ± 0″.1 (p.e.). His plates were taken with astrographic telescopes (60″/mm) with a time base of 50 years. Moreover, he found the greatest expansion rates to be along the major axis of the elongated ring.

More recently, Chudovicheva (1964) has compared astrograph plates of NGC 6853 (the Dumbbell Nebula) and of NGC 7662. Although the time bases of her plates were less than 30 years, she reports centennial expansions of 6″.8 ± 1″.8 (p.e.) and 1″.0 ± 0″.6 (p.e.), respectively. The sizable growth rate of NGC 6853 appears, however, to be in error, as direct comparison of suitable plates shows (see Liller 1965).

The most extensive work to date is that of Liller (1965) and of Liller, Welther, and Liller (1966), who used plates taken with the Crossley reflector (39″/mm) at Lick Observatory with time bases averaging about 50 years. They measured a total of fourteen planetaries and found that four of these showed definite expansion and six had undetectable size changes (usually $<0\rlap{.}''4$). Perhaps the most significant result of all was that some planetaries, such as NGC 6853 (the Dumbbell Nebula), showed a mean centennial radial expansion of $0\rlap{.}''64 \pm 0\rlap{.}''24$ (m.e.), considerably less than we would calculate from equation (191). On the other hand, NGC 3242 and NGC 6572 clearly showed larger expansion rates than predicted. The authors noted that nebulae that had large apparent ages and that lay near the plane of the Galaxy were without exception those with lower than expected expansion rates. It would seem that the outer edges of these planetaries were beginning to show the effects of an interaction with the interstellar medium. As for NGC 3242 and NGC 6572, we are probably seeing ionization spheres which are expanding at roughly twice the rate of the gas. This latter conclusion receives strong support from the high electron densities within these objects as reported by Liller and Aller (1954), O'Dell (1962), and others.

A very important conclusion of the Liller, Welther, and Liller paper is that the distances derived by Minkowski (1965*b*) are the most reliable, although Shklovsky's results may be somewhat better for optically thin nebulae.

The Lillers are extending this work through the use of early and recent Mount Wilson plates. Some 100-inch and 60-inch Cassegrain plates (5″/mm and 8″/mm) of the first epoch show superbly the filamentary structure in the nearer planetaries such as NGC 6720 (Ring Nebula). Initial results have given an explanation of the reason for the discrepancy between Latypov's (1957) result ($0\rlap{.}''9 \pm 0\rlap{.}''1$) and the Lillers' (no detectable expansion). The ropelike filaments exhibit less than half the expansion found by Latypov, while the edges show more than twice that measured by Latypov. Thus, it would seem that Latypov's measures refer to the mean of edges and filaments.

Expansion of stellar-appearing nebulae presumably has been detected spectroscopically by Liller and Aller (1957, 1966) for IC 4997 and possibly by Razmadze (1960) for VV8. In the spectra of both these objects, 4363 A [O III] appears abnormally strong relative to the green nebular lines, indicating either a high electron density or a high electron temperature (see Sec. 7). In the spectrum of IC 4997 Liller and Aller (1957) found 4363 A to be weakening in a secular fashion, and they have given strong arguments favoring the hypothesis that a decrease in both electron density and electron temperature is occurring. Their case has received further support from the Doppler expansion result published by Wilson and O'Dell (1962). On the other hand, Razmadze (1960) and Vorontsov-Velyaminov (1960) have supported the explanation that the central star is evolving so fast that its decrease in luminosity shows up in the nebula as a decrease in electron temperature.

Dynamical models of planetaries have only very recently been attempted (see Osterbrock 1964), although Zanstra (1931*c*, 1934, 1958), Chandrasekhar (1945), and Khromov (1964) have all tried to evaluate the sizes of the forces within the nebula. A summary of these investigations seems to favor gas pressure within the nebula as the predominant force. Next in importance is the radiation pressure in the Lyman continuum produced by nebular hydrogen. Less important is the radiation pressure from Lyman α emitted in the nebula, and this last mechanism quickly decreases in importance when motions within the nebula produce significant Doppler shifts.

Beginning with a spherical nebular shell of density 1700 cm^{-3} and inner and outer radii of 0.08 and 0.10 pc, Mathews has calculated appearances resultant from thermodynamic forces only. He finds that such a model dissipates rapidly (within a few thousand years) unless the massive ejection forming the bulk of the nebula is followed by a less continuous mass loss by the central star. This less dense matter creates just enough gas pressure on the inside of the shell to prevent inward expansion and hence disruption of the ring in this manner. Surprisingly, the velocity of the gas near the outer edge increases only slightly, and hence a relatively sharp outer edge is maintained for quite a few thousand years. Such a result would seem to be the solution to the problem, stated most clearly by Osterbrock (1964), of how a planetary maintains a sharp outer edge. No longer would a dense outer shell of neutral hydrogen nor the existence of magnetic fields (Gurzadian 1962) be required.

The importance of Mathews' work is clear, and calculations adopting other initial conditions would be immensely interesting. What would later appearances of a uniform sphere be like? How would the nebula behave in a medium of low-density neutral hydrogen instead of in a vacuum?

From an observational standpoint, it would seem that careful study should be given to the nearest planetaries where the fine structure of the nebula can be observed. Of particular interest is NGC 7293 which, according to Minkowski, is only 145 pc away. Under good conditions we should be able to resolve details less than 100 a.u. in size. Around the inner edge of this ring nebula, hundreds of comet-like structures appear on the best photographs, and Vorontsov-Velyaminov (1965) has predicted that these features should show very high rates of outward motion. However, very tentative results by the Lillers (1966) reveal small movement on 100-inch and 200-inch plates taken some 17 years apart ($0\overset{''}{.}9 \pm 0\overset{''}{.}4$/century).

On old plates of NGC 7293 it is difficult to measure the angular motions of nebular features because of the low surface brightness of this object. Photographs taken with plate-filter combinations which enhance the contrast are necessary if we are to make precise astrometric measurements, and these have been taken only relatively recently. Because of the extremely low surface brightness within the ring, high-dispersion spectrograms are even more difficult to take, and none has been attempted as yet. Nevertheless, it is hoped that angular motions and Doppler expansion rates can some day be combined to give us a clearer picture of the movements of the knots and condensations, inner and outer edges, and the ropelike filaments—all of which exist in NGC 7293.

REFERENCES

Abell, G. O. 1955, *Pub. A.S.P.*, **67**, 258.
———. 1966, *Ap. J.*, **144**, 259.
Abell, G., and Goldreich, P. 1966, *Pub. A.S.P.*, **78**, 232.
Allen, C. W. 1963, *Astrophysical Quantities* (London: Athlone Press).
Aller, L. H. 1941, *Ap. J.*, **93**, 236.
———. 1943, *ibid.*, **97**, 135.
———. 1948, *ibid.*, **108**, 462.
———. 1951, *ibid.*, **113**, 125.
———. 1953, *ibid.*, **118**, 547.
———. 1954*a*, *ibid.*, **120**, 401.
———. 1954*b*, *Nuclear Transformations, Stellar Interiors, and Nebulae* (New York: Ronald Press Co.).
———. 1956, *Gaseous Nebulae* (London: Chapman-Hall).
———. 1957, *Ap. J.*, **125**, 84.

Aller, L. H. 1960, *Mém. Soc. R. Sci. Liège*, Ser. 15, Vol. **3**, 41.
———. 1961*a*, *Abundance of the Elements* (New York: Interscience).
———. 1961*b*, *Les Spectres des Astres dan l'Ultraviole Lointain*, p. 535. Institute d'Astrophys., Liège.
———. 1963, *Atmospheres of the Sun and Stars* (New York: Ronald Press Co.).
———. 1964*a*, *Pub. A.S.P.*, **76**, 279.
———. 1964*b*, *Astrophys. Norvegica*, **9**, 293.
———. 1965, *Research Frontiers in Fluid Dynamics*, eds. R. J. Seeger and G. Temple (New York: Interscience), chap. xix.
Aller, L. H., and Faulkner, D. J. 1964, *The Galaxy and the Magellanic Clouds*, eds. A. Rodgers and F. Kerr (Canberra: Australian Academy), p. 45.
Aller, L. H., and Jugaku, J. 1959, *Ap. J. Suppl.*, **4**, 109.
Aller, L. H., and Kaler, J. B. 1964*a*, *Ap. J.*, **139**, 1074.
———. 1964*b*, *ibid.*, **140**, 621.
———. 1964*c*, *ibid.*, p. 936.
Aller, L. H., and Menzel, D. H. 1945, *Ap. J.*, **102**, 239.
Aller, L. H., and Minkowski, R. 1956, *Ap. J.*, **124**, 110.
Aller, L. H., and Walker, M. F. 1965, *Ap. J.*, **141**, 1318.
Aller, L. H., and White, M. L. 1949, *A.J.*, **54**, 181.
Aller, L. H., and Zanstra, H. 1960, *B.A.N.*, **15**, 249.
Aller, L. H., Bowen, I. S., and Minkowski, R. 1955, *Ap. J.*, **122**, 62.
Aller, L. H., Bowen, I. S., and Wilson, O. C. 1963, *Ap. J.*, **138**, 1013.
Aller, L. H., Kaler, J. B., and Bowen, I. S. 1966, *Ap. J.*, **144**, 291.
Aller, L. H., Ufford, W., and Van Vleck, J. H. 1949, *Ap. J.*, **109**, 42.
Ambartsumian, V. A. 1932, *M.N.*, **93**, 50.
———. 1933, *Pulkovo Obs. Bull.*, **13**, 3.
Andrillat, Y. 1954, *Comptes rendu*, **238**, 1781.
———. 1955, *Ann. d'Ap. Suppl.*, **1**, 1.
Andrillat, Y., and Andrillat, H. 1959, *Ann. d'Ap.*, **22**, 104.
———. 1961, *ibid.*, **24**, 139.
Baade, W. 1955, *A.J.*, **60**, 151.
Baade, W., Goos, F., Koch, R., and Minkowski, R. 1933, *Zs. f. Ap.*, **6**, 355 (see Minkowski 1934).
Bahcall, J. 1966, *Ap. J.*, **143**, 259.
Bahcall, J., and Salpeter, E. E. 1966, *Ap. J.*, **144**, 847.
Baker, J. G., Menzel, D. H., and Aller, L. H. 1938, *Ap. J.*, **88**, 422.
Barbier, D., and Andrillat, H. 1954, *Comptes rendu*, **239**, 1099.
Bates, D. R., and Damgaard, A. 1949, *Phil. Trans. Roy. Soc.*, **A242**, 101.
Baum, W., and Minkowski, R. 1960, quoted by Bowen, I. S., *Carnegie Inst. Yearbook*, **59**, 18.
Berman, L. 1930, *Lick Obs. Bull.*, **15**, 86.
———. 1936, *M.N.*, **96**, 890.
———. 1937, *Lick, Obs. Bull.*, **18**, 73.
Bethe, H., and Salpeter, H. 1957, *Hdb. d. Phys.*, **35**, 352.
Billings, A., Czyzak, S. J., Krueger, T. K., de Martins, A. P., Saraph, H. E., Seaton, M. J., and Shemming, J. 1967, International Astronomical Union Symposium on Planetary Nebulae (in press).
Böhm, K. H., and Deinzer, W. 1965, *Zs. f. Ap.*, **61**, 1.
———. 1966, *ibid.* (in press).
Bohm, D., and Aller, L. H. 1947, *Ap. J.*, **105**, 1.
Bowen, I. S. 1928, *Ap. J.*, **67**, 1.
———. 1934, *Pub. A.S.P.*, **46**, 146.
———. 1938, *Ap. J.*, **88**, 115.
Bowen, I. S., and Wyse, A. B. 1939, *Lick Obs. Bull.*, **19**, 1.
Brussard, P. J., and van de Hulst, H. C. 1962, *Rev. Mod. Phys.*, **34**, 507.
Burgess, A. 1958, *M.N.*, **118**, 477.
———. 1964, *Mem. R.A.S.*, **69**, Part 1.
Burgess, A., and Seaton, M. J. 1960*a*, *M.N.*, **120**, 121.
———. 1960*b*, *ibid.*, **121**, 76.
———. 1960*c*, *ibid.*, p. 471.
Burgess, A., Seaton, M. J., and Pengelly, C. D. 1966 (in press).
Campbell, W., and Moore, J. H. 1918, *Pub. Lick Obs. Bull.*, **13**, 75.
Capriotti, E. R. 1964*a*, *Ap. J.*, **139**, 225.
———. 1964*b*, *ibid.*, **140**, 632.
———. 1966, *ibid.*, **146**, 709.
Capriotti, E. R., and Daub, C. T. 1960, *Ap. J.*, **132**, 677.
Carroll, J. A. 1930, *M.N.*, **90**, 588.
Chamberlain, J. W. 1953, *Ap. J.*, **117**, 387.
Chandrasekhar, S. 1945, *Ap. J.*, **102**, 402.
Chandrasekhar, S., and Münch, G. 1950, *Ap. J.*, **111**, 142.
Chopinet, M. 1963, *Journal des Observateurs*, **46**, 27.

Chudovicheva, O. N. 1964, *Izv. Pulkovo Obs.*, **23**, 154.
Cillié, G. 1932, *M.N.*, **92**, 820.
———. 1936, *ibid.*, **96**, 771.
Clarke, W. 1965, thesis, U.C.L.A.
Code, A. D. 1960, *A.J.*, **65**, 278.
Collins, G. W., Daub, C. T., and O'Dell, C. R. 1961, *Ap. J.*, **133**, 471.
Courtès, G. 1954, *Comptes rendu*, **238**, 877.
———. 1955, *ibid.*, **241**, 364.
———. 1957, *A.J.*, **62**, 10.
———. 1958, *J. Phys. Rad.*, **19**, 342.
———. 1959, *Comptes rendu*, **248**, 2953.
———. 1960, *Ann. d'ap.*, **23**, 115.
Cox, R., and Salpeter, E. 1961, *Ap. J.*, **133**, 764.
Curtis, H. D. 1918, *Lick Obs. Bull.*, **13**, 55.
Czyzak, S. 1966, this volume, chapter 8.
Czyzak, S., Aller, L. H., Kaler, J. B., and Faulkner, D. J. 1966, *Ap. J.*, **143**, 327.
Daub, C. T. 1963, *Ap. J.*, **137**, 184.
Davies, J. G., Ferriday, R. J., Haslam, C. G. T., Moran, M., and Thomasson, P. 1965, *Nature*, **206**, No. 4986, 809.
Davies, L. B., Ring, J., and Selby, M. J. 1964, *M.N.*, **128**, 399.
Divine, N. 1965, *Ap. J.*, **142**, 824.
Duncan, J. C. 1937, *Ap. J.*, **86**, 496.
Elwert, G. 1948, *Zs. f. Naturforsch.*, **3a**, 477.
Evans, D. S. 1950, *M.N.R.A.S.*, **110**, 37.
Fabry, C., and Buisson, H. 1911, *Ap. J.*, **33**, 406.
Garstang, R. 1952, *Ap. J.*, **115**, 506, 569.
———. 1956, *Proc. Camb. Phil. Soc.*, **52**, Part 1, 107.
Geake, J. E., Ring, J., and Woolf, N. 1959, *M.N.*, **119**, 616.
Gebbie, K. B., and Seaton, M. J. 1963, *Nature*, **199**, 580.
Goldberg, L. 1939, *Ap. J.*, **90**, 414.
———. 1941, *ibid.*, **93**, 244.
Goldberg, L., and Aller, L. H. 1943, *Atoms, Stars, Nebulae* (Cambridge: Harvard University Press).
Goldberg, L., Müller, E. A., and Aller, L. H. 1960, *Ap. J. Suppl.*, No. 45, **5**, 1.
Gould, R. J. 1966, *Ap. J.*, **143**, 603.
Grant, I. P. 1958, *M.N.*, **118**, 241.
Green, W. K. 1917, *Lick Obs. Bull.*, **9**, 92.
Greenstein, J. L., and Minkowski, R. 1964, *Ap. J.*, **140**, 1601.
Griem, H. 1964, *Plasma Spectroscopy* (New York: McGraw-Hill).
Gurzadian, G. 1962, *Planetary Nebulae* (Moscow State Publisher of Physics—Mathe. Lit.) Eng. translation by D. H. Hummer and C. M. Varsavsky (New York: Gordon & Breach).
Gurzadian, G., and Razmadze, N. A. 1959, *Pub. Obs. Burakan*, **26**, 19.
Haro, G. 1952, *Bol. Obs. Tonantzintla y Tacubaya*, No. 1.
Harman, R. J., and Seaton, M. J. 1964, *Ap. J.*, **140**, 824.
———. 1966, *M.N.*, **132**, 15.
Hatanaka, T. 1944, *Ap. J.*, **20**, 505.
Hayashi, C., Hōshi, R., and Sugimoto, D. 1962, *Suppl. Progr. Theoret. Phys., Osaka*, **22**, 183.
Henize, K. 1951, *Publ. Univ. Mich.*, **10**, 25.
Henize, K., and Westerlund, B. 1963, *Ap. J.*, **137**, 747.
Hiltner, W. A. 1962, *Astronomical Techniques*, ed. W. A. Hiltner (Chicago: University of Chicago Press), p. 340.
Hoglund, B., and Mezger, P. G. 1965, *A.J.*, **70**, 679.
Huang, S. S. 1948, *Ap. J.*, **108**, 354.
Hummer, D. G. 1963, *M.N.*, **125**, 461.
Hummer, D. G., and Seaton, M. J. 1961, *Colloq. Intern. Astrophys., Liège*, p. 539.
———. 1963, *M.N.*, **125**, 437.
Johnson, Harold. 1966, this volume.
Kaler, J. B. 1964, *Pub. A.S.P.*, **76**, 231.
———. 1966, *Ap. J.*, **143**, 722.
Kaler, J. B., Aller, L. H., and Bowen, I. S. 1965, *Ap. J.*, **141**, 912.
Kapanay, N. S. 1957, *A.J.*, **62**, 20.
Karzas, W. L., and Latter, R. 1961, *Ap. J. Suppl.* **6**, 167.
Khromov, G. S. 1962, *Soviet Astron.*, **6**, 370.
———. 1964, *ibid.*, **7**, 609.
———. 1965, *A.J. (U.S.S.R.)*, **42**, 918.
———. 1966, *Soviet Astron.*, **9**, 431.

Khromov, G. S., Indisov, O. S., Matveyenko, L. I., Turevsky, V. M., and Sholomitsky, G. B. 1965, *A.J.* (*U.S.S.R.*), **42**, 1120.
Kohoutek, L. 1961, *Bull. Astron. Czech.*, **12**, 213.
———. 1962, *ibid.*, **13**, 71.
———. 1964, *ibid.*, **15**, 164, **15**, 161.
———. 1965, *ibid.*, **16**, 221.
———. 1967, *ibid.*, **18**, 103.
Koslov, V. 1935, *Astron. Nach.*, **254**, 137.
Lallemand, A. 1936, *Comptes rendu*, **203**, 243, 990.
Latypov, A. A. 1957, *Pub. Astron. Obs. Tashkent* (2), **5**, 31.
Liller, W. 1955, *Ap. J.*, **122**, 240.
———. 1957, *Pub. A.S.P.*, **69**, 511.
———. 1965, *ibid.*, **77**, 25.
Liller, W., and Aller, L. H. 1954, *Ap. J.*, **120**, 48.
———. 1957, *Sky and Telescope*, **16**, 222.
———. 1959, *Ap. J.*, **130**, 45.
———. 1963, *Proc. Nat. Acad. Sci.*, **49**, 675.
———. 1966, *M.N.*, **132**, 337.
Liller, M. H., and Liller, W. 1966, in preparation.
Liller, M. H., Welther, B. L., and Liller, W. 1966, *Ap. J.*, **144**, 280.
Livingston, W. C., and Lynds, C. R. 1964, *Ap. J.*, **140**, 818.
Lynds, C. R. 1961, *Pub. Nat. Rad. Obs.*, **1**, No. 5.
MacRae, D., and Stock, J. 1954, *Nature*, **173**, 589.
Marriot, R. 1955, *Proc. R. Soc. London*, **70**, 288.
Mathews, W. G. 1966, *Ap. J.*, **143**, 173.
Mathis, J. 1957, *Ap. J.*, **125**, 518; **126**, 493.
Mayer, M. G. 1931, *Ann. d'Phys.*, **9**, 273.
McLaughlin, D. B. 1960, in *Stellar Atmospheres*, ed. J. L. Greenstein (Chicago: University of Chicago Press).
Menon, T. K., and Terzian, Y. 1965, *Ap. J.*, **141**, 745.
Menzel, D. H. 1926, *Pub. A.S.P.*, **38**, 295.
———. 1937, *Ap. J.*, **85**, 330.
———. 1962, *Physical Processes in Ionized Plasmas* (New York: Dover).
Menzel, D. H., and Aller, L. H. 1941*a*, *Ap. J.*, **94**, 30.
———. 1941*b*, *ibid.*, **94**, 436.
———. 1941*c*, *ibid.*, **93**, 195.
———. 1945, *ibid.*, **102**, 239.
———. 1967, International Astronomical Union Symposium on Planetary Nebulae (in press).
Menzel, D. H., and Baker, J. G. 1937, *Ap. J.*, **86**, 70.
———. 1938, *ibid.*, **88**, 52.
Menzel, D. H., and Hebb, M. H. 1940, *Ap. J.*, **92**, 408.
Menzel, D. H., Hebb, M. H., and Aller, L. H. 1941, *Ap. J.*, **93**, 230.
Menzel, D. H., and Pekeris, C. L. 1935, *M.N.*, **96**, 77.
Meyer, W. F. 1920, *Lick Obs. Bull.*, **10**, 68.
Minkowski, R. 1934, *Zs. f. Ap.*, **9**, 202.
———. 1942, *Ap. J.*, **95**, 243.
———. 1943, *ibid.*, **97**, 162–Footnote No. 41.
———. 1946, *Pub. A. S. P.*, **58**, 305.
———. 1947, *ibid.*, **59**, 257.
———. 1948, *ibid.*, **60**, 386.
———. 1951, *Pub. Obs. of Univ. of Michigan*, **10**, 25.
———. 1955, *I.A.U. Symp.*, No. 2, 3.
———. 1958*a*, *Rev. Mod. Phys.*, **30**, 905.
———. 1958*b*, *ibid.*, p. 1031.
———. 1964, *Pub. A.S.P.*, **76**, 197.
———. 1965*a*, in *Galactic Structure*, ed. A. Blaauw and M. Schmidt (Chicago: University of Chicago Press).
———. 1965*b*, *Numerical Data and Functional Relationships in Science and Technology*, ed. H. H. Voigt (Berlin: Springer-Verlag) Group *VI*, **1**, 566.
———. 1966, private communication.
Minkowski, R., and Abell, G. O. 1963, *Pub. A.S.P.*, **75**, 488.
Minkowski, R., and Aller, L. H. 1954, *Ap. J.*, **120**, 261.
———. 1956, *ibid.*, **124**, 93.
Minkowski, R., and Osterbrock, D. 1960, *Ap. J.*, **131**, 537.
Miyamoto, S. 1938, *Contr. Astrophys. Int. Kyoto*, No. 38.
Münch, G. 1964, *Ann. Rept. Mt. Wilson and Palomar Obs.*, p. 25.
Münch, G., and Wilson, O. C. 1962, *Zs. f. Ap.*, **56**, 127.

O'Dell, C. R. 1962, *Ap. J.*, **135**, 371.
———. 1963*a*, *ibid.*, **138**, p. 67.
———. 1963*b*, *ibid.*, p. 1018.
———. 1963*c*, *ibid.*, p. 293.
———. 1964, *Pub. A.S.P.*, **76**, 308.
———. 1965, *Ap. J.*, **142**, 1093.
O'Dell, C. R., Peimbert, M., and Kinman, T. 1964, *Ap. J.*, **140**, 119.
Oster, L. 1961, *Ap. J.*, **134**, 1010.
Osterbrock, D. 1951, *Ap. J.*, **114**, 469.
———. 1955, *ibid.*, **122**, 235.
———. 1960, *ibid.*, **131**, 541.
———. 1962, *ibid.*, **135**, 195.
———. 1963, *Planetary and Space Science*, **11**, 621.
———. 1964, *Ann. Rev. Astron. Astrophys.*, **2**, 95.
———. 1965*a*, *Ap. J.*, **141**, 1285.
———. 1965*b*, *ibid.*, **142**, 1423.
Osterbrock, D. E., Capriotti, E. R., and Bautz, L. P. 1963, *Ap. J.*, **138**, 62.
Osterbrock, D. E., Miller, J. S., and Weedman, D. W. 1966, *Ap. J.*, **145**, 697.
Page, T. L. 1936, *M.N.*, **96**, 604.
———. 1942, *Ap. J.*, **96**, 78.
Parenago, P. P. 1946, *A.J.* (*U.S.S.R.*), **22**, 150; **23**, 69.
Parker, R. A. R. 1964, *Ap. J.*, **139**, 208.
Pasternack, S. 1940, *Ap. J.*, **92**, 129.
Pengelly, R. M., 1964, *M.N.*, **127**, 145.
Pengelly, R. M. and Seaton, M. J. 1964, *M.N.*, **127**, 165.
Perek, L. 1960, *Bull. Astron. Czech.*, **11**, 256.
Perek, L., and Kohoutek, L. 1967, *Catalogue of Galactic Planetary Nebulae* (Prague: Academy of Sciences).
Perrine, C. D. 1929, *Astron. Nach.*, **237**, 89.
Petrie, R. M. 1948, *Pub. Dom. Ap. Obs.*, **7**, 321.
Plaskett, H. H. 1928, *Harvard Circular*, No. 335.
Pollack, J. B. 1962, unpublished.
Pottasch, S. R. 1960*a*, *Ann. d'Ap.*, **23**, 749.
———. 1960*b*, *Ap. J.*, **131**, 202.
———. 1962, *ibid.*, **135**, 385.
Razmadze, N. A. 1960, *A.J.* (*U.S.S.R.*), **37**, 342, 1005.
Sahade, J. 1960, *Stellar Atmospheres*, ed. J. L. Greenstein (Chicago: University of Chicago Press), pp. 466–503.
Saito, S., and Uesugi, A. 1959, *Kyoto Contributions*, No. 78.
Sakashita, S., and Tanaka, Y. 1962, *Progr. Theoret. Phys., Kyoto*, **27**, 127.
Schmidt, M. 1963, *Ap. J.*, **137**, 758.
———. 1965, *ibid.* **141**, 1295.
Searle, L. 1958, *Ap. J.*, **128**, 489.
Seaton, M. J. 1951, *Proc. R. Soc. London, A*, **208**, 418.
———. 1953*a*, *ibid.*, **218**, 400.
———. 1953*b*, *Ann. d'Ap.*, **17**, 74.
———. 1954, *M.N.*, **114**, 154.
———. 1955*a*, *Proc. R. Soc. London, A*, **68**, 457.
———. 1955*b*, *ibid.*, **70**, 620.
———. 1955*c*, *ibid.*, **231**, 37.
———. 1955*d*, *M.N.*, **115**, 279.
———. 1957, *Proc. Cambridge Phil. Soc.*, **53**, 654.
———. 1958, *Rev. Mod. Phys.*, **30**, 979.
———. 1959*a*, *M.N.*, **119**, 81.
———. 1959*b*, *ibid.*, **119**, 90.
———. 1960*a*, *Rept. Prog. Phys.*, **23**, 313.
———. 1960*b*, *M.N.*, **120**, 326.
———. 1962, in *Atomic Molecular Processes*, ed. D. Bates (New York: Academic Press).
———. 1964, *M.N.*, **127**, 177.
———. 1966, *ibid.*, **132**, 113.
———. 1967, International Astronomical Union Symposium on Planetary Nebulae (in press).
Seaton, M. J., and Hummer, D. G. 1963, *M.N.*, **125**, 437.
Seaton, M. J., and Osterbrock, D. 1957, *Ap. J.*, **125**, 66.
Seaton, M. J., Czyzak, S. J., and Krueger, T. K. 1966, unpublished.
Shortley, G., Aller, L. H., Baker, J. G., and Menzel, D. H. 1941, *Ap. J.*, **93**, 178.
Shklovsky, I. S. 1956, *A.J.* (*U.S.S.R.*), **33**, 222, 315.
Slee, O. B., and Orchiston, D. W. 1965, *Australian J. Phys.*, **18**, 187.
Smart, W. M. 1938, *Stellar Dynamics* (New York: Cambridge University Press), p. 297.

Sofia, S. 1966, *Ap. J.*, **145**, 84.
Spitzer, L. 1948, *Ap. J.*, **107**, 6.
Spitzer, L., and Greenstein, J. L. 1951, *Ap. J.*, **114**, 407.
Stoy, R. 1933, *M.N.*, **93**, 588.
Strömgren, B. 1939, *Ap. J.*, **89**, 526.
Struve, O., and Swings, P. 1940, *Proc. Nat. Acad. Sci.*, **26**, 548.
Swings, P. 1940, *Ap. J.*, **92**, 289.
———. 1941, *Proc. Nat. Acad. Sci.*, **27**, 225.
Swings, P., and Struve, O. 1941, *Ap. J.*, **93**, 362.
Swings, P., and Swensson, J. W. 1952, *Ann. d'Ap.*, **15**, 290.
Swope, H. 1963, *A.J.*, **68**, 470.
Terzian, Y. 1966, *Ap. J.*, **144**, 657.
———. 1967, International Astronomical Union Symposium on Planetary Nebulae (in press).
Thackeray, A. D. 1956, *Observatory*, **76**, 154.
The, Pik-Sin. 1962, *Contr. Bosscha Obs.*, No. 14.
Thompson, A. R. 1967, *Ap. J.* (in press); see also IAU Symposium on Planetary Nebulae (in press).
Thompson, A. R., Colvin, R. S., and Stanley, G. J. 1967, *Ap. J.*, **148**, 429.
Underhill, A. B. 1951, *Pub. Dom. Astrophys. Obs. Victoria*, Vol. **8**, No. 12.
Unno, W. 1955, *Pub. Astr. Soc. Japan*, **7**, 81.
Upton, E. K. L. 1959, private communication.
Upton, E. K. L., Mutschlecner, P., Tull, R., and Kumar, S. 1960, *Mém. Soc. R. Sci. Liège*, Ser. 5, Vol. **3**, p. 41.
Visvanathan, N. 1965, unpublished thesis, Australian National University.
Vorontsov-Velyaminov, B. A. 1934*a*, *A.J.* (*U.S.S.R.*), **11**, 40.
———. 1934*b*, *Zs. f. Ap.*, **8**, 195.
———. 1936, *ibid.*, **12**, 247.
———. 1937, *A.J.* (*U.S.S.R.*), **14**, 194.
———. 1950, *Gasnebel u. neue Sterne* (Berlin: Verlag Kultur und Fortschrift).
———. 1960, *A.J.* (*U.S.S.R.*), **37**, 994.
———. 1965, private communication.
Vorontsov-Velyaminov, B. A., Kostyakova, E. B., Dokvemaeva, O. D., Arkhipova, V. P. 1966, *Soviet Astron.-A.J.*, **9**, 564.
Walker, M. F., and Aller, L. H. 1965, *Ap. J.*, **141**, 1318.
Wallenquist, A. A. E. 1936, *Uppsala Medd.*, No. 65.
Wampler, E. J. 1961, *Ap. J.*, **134**, 861.
Whitford, A. E. 1948, *Ap. J.*, **107**, 102.
———. 1958, *A.J.*, **63**, 201.
Wilson, O. C. 1939, *Pub. A.S.P.*, **9**, 274.
———. 1948, *Ap. J.*, **108**, 201.
———. 1950, *ibid.*, **111**, 279.
———. 1953, *ibid.*, **117**, 264.
———. 1958, *Rev. Mod. Phys.*, **30**, 1025.
———. 1964, unpublished monograph with R. Minkowski.
Wilson, O. C., and Aller, L. H. 1951, *Ap. J.*, **114**, 421.
———. 1954, *ibid.*, **119**, 243.
Wilson, O. C., and O'Dell, R. C. 1962, *Pub. A.S.P.*, **74**, 511.
Wurm, K. 1951, *Die planetarischen Nebel* (Berlin: Akademie-Verlag).
Wurm, D., and Singer, O. 1952, *Zs. f. Ap.*, **30**, 287.
Wyse, A. B. 1942, *Ap. J.*, **95**, 356.
Zanstra, H. 1926, *Phys. Rev.* (2), **27**, 644.
———. 1927, *Ap. J.*, **65**, 50.
———. 1931*a*, *Zs. f. Ap.*, **2**, 1.
———. 1931*b*, *ibid.*, p. 329.
———. 1931*c*, *Pub. Dom. Ap. Obs. Victoria*, **4**, 209.
———. 1931*d*, *Zs. f. Ap.*, **2**, 1239.
———. 1934, *M.N.*, **95**, 84.
———. 1955, *Vistas in Astron.*, **1**, 256.
———. 1958, *Rev. Mod. Phys.*, **30**, 1030.
———. 1960, *B.A.N.*, **15**, 237.
Zanstra, H., and Brandenburg, W. J. 1951, *B.A.N.*, **11**, 350.

CHAPTER 10

Radio-Line Emission and Absorption by the Interstellar Gas

F. J. KERR
Radiophysics Laboratory, CSIRO, Sydney, Australia

1. INTRODUCTION

THE RADIATION studied by radio astronomers is almost all produced in nonstellar sources. The strongest emission seems to originate in the synchrotron mechanism; it is spread over a broad continuum, and is always associated with large-scale processes in a very extensive plasma. One of the main problems in any investigation is to understand the mechanism of origin; little can yet be said about the physical conditions inside the source regions or about the details of the relationship between the radio emitters and the other constituents of a galaxy.

The emphasis is quite different for radio-frequency lines, such as that emitted by neutral hydrogen atoms. Here, the mechanism of origin is well understood, and the emitting material is intimately related to other interstellar and stellar constituents. By studying the hydrogen line, which occurs at a wavelength of 21 cm, we can directly explore the properties of the interstellar gas, and we can indirectly examine the structure of a whole galaxy by using the hydrogen as a tracer element. These radio studies have an added importance because neutral hydrogen is the major component of the interstellar medium, and it cannot be observed by optical means. Further, the relatively low opacity of the galactic disk to this radiation makes it possible to investigate the whole depth of the Galaxy, rather than only our immediate surroundings.

In comparing different methods of studying interstellar matter, we must remember that radio telescopes all suffer from very limited angular resolution compared with that of an optical telescope. In the Galaxy, radio telescopes give a broad and a distant view, but the results are harder to interpret in nearby regions. Optical instruments, on the other hand, reveal an immense amount of detailed structure, but they cannot reach far before the obscuring dust limits their view. The radio and optical approaches are clearly complementary because they are most effective at large and small distances respectively. Difficulties are often experienced, however, in fitting together the two sets of results in the middle-distance region where they overlap, but where each is not fully effective.

The first observations in radio astronomy were carried out in the continuum, near 15 m (Jansky 1932) and at 1.85 m (Reber 1944). While the science was still very young, van de Hulst (1945) predicted that the 21-cm hyperfine transition line from interstellar hydrogen should be detectable. He foresaw the great advantages that would follow from the availability of a sharp line at radio wavelengths that would combine the possibility of radial velocity measurements with the penetrating power of radio waves in the Galaxy. The density of the interstellar gas is very low (averaging about 1 atom/cm^3), and each atom only radiates a quantum of 21-cm emission at very rare intervals (about once every 11 million years), but the number of atoms in the line of sight is so large that a detectable amount of energy is produced. The prediction was supported by a similar analysis made later by Shklovsky (1949).

Following van de Hulst's suggestion, the Leiden group began building equipment to search for the line, but their progress was delayed by a fire which destroyed their first receiver in 1950. The first successful detection was achieved at Harvard by Ewen and Purcell (1951), who reported an antenna temperature of 25° K over a region of the Milky Way. Very shortly afterward the line was detected in the Netherlands (Muller and Oort 1951) and in Sydney (Pawsey 1951).

Since that time, the distribution of the 21-cm line emission over the sky has been widely studied by these groups and others. The line has proved to be a powerful tool for investigating the structure of our Galaxy and of other galaxies, and for studying the general properties of the interstellar gas.

Several unsuccessful attempts have been made to detect the corresponding line from deuterium at 92 cm. The only other radio lines, apart from the hydrogen line, that have so far been detected from the interstellar gas are the 18-cm group from the hydroxyl radical (OH). The manner in which the first discovery of the OH lines (by Weinreb, Barrett, Meeks, and Henry 1963) was followed very rapidly by successful observations by several other groups was reminiscent of the early days of the hydrogen line.

There are several other lines that may be detected in the next few years. These are expected to be weak, and will mainly be of interest in giving abundance ratios of various interstellar constituents. The hydrogen line is likely to remain the only one that can be used in exploring the large-scale structure of galaxies.

This chapter will cover physical processes and observational results as they relate to the properties of the interstellar gas and its small-scale structure. The application of hydrogen-line observations to galactic-structure studies was the subject of a chapter in Volume V of this compendium (Kerr and Westerhout 1965).

2. SHORT REVIEW OF OBSERVATIONS

In this section, we shall briefly discuss the broad features of the observational results of the hydrogen line before considering the emission process and the manner in which the results can be interpreted.

The 21-cm line is detectable all over the sky. A major concentration is found along the Milky Way, but radiation from galactic hydrogen is visible everywhere. The distribution across the sky shows considerable fine structure, some of which can be related to optical features connected with the interstellar medium or with young stars. In addition to the galactic emission, the line has been detected from hydrogen in about 100 external galaxies, with the likelihood of many more to follow. These usually appear at wavelengths that are well Doppler-shifted, so that a complete separation of the galactic and extragalactic contributions can be effected quite simply.

The observed brightness temperature reaches 100° to 140° K at the peak of the profile in most parts of the Milky Way, with lower values in other parts of the profile. At higher latitudes, peak values of about 20° to 30° K are common. Observed temperatures are high in the Magellanic Clouds also, but, in other external galaxies, the values are seldom above a few degrees, because the source usually occupies only a small fraction of the beam area; in many areas temperatures well below a degree must be measured. Most galactic and Magellanic Cloud work to date has been done with receivers having a sensitivity of about 1° K, which has proved adequate for large-scale structure work. However, a sensitivity of around 0.1° K is required for the much weaker radiation from external galaxies or for studying the wings of galactic profiles.

The natural frequency of the hydrogen line, as measured in the laboratory, is 1420.406 Mc/s, with a very small natural width. Observed line profiles in the Galaxy vary in width from almost 3 Mc/s near the galactic center to 50 kc/s at high latitudes. We will see in Section 4.3 that the Doppler effect is the main cause of line broadening. With this interpretation, the line widths quoted above correspond to radial velocity spreads of about 600 and 10 km/sec. Integrated profiles from external galaxies have widths of several hundred kilometers per second.

In the Milky Way, the profiles are often quite complex in shape, with six or more separate peaks. At higher latitudes, the profiles are much simpler, being usually single-peaked. The receiver bandwidth that is required for tracing out profile details is in the range of 5 to 40 kc/s, the requirements varying somewhat with the region involved and the type of investigation.

In addition to the hydrogen-line emission, absorption effects are observed in the same frequency range in the directions of discrete radio sources. These sources radiate over a broad continuum; absorption takes place over a narrow frequency range when the radiation passes through foreground hydrogen. The effects are very striking for some of the stronger sources when observed with a high-gain telescope (see Sec. 6.2).

No case has yet been found in which a hydrogen-line source has shown any variability. No polarization effects have been observed in hydrogen-line emission, but polarization phenomena have been looked for in absorption in attempts to observe an interstellar Zeeman effect (see Sec. 6.3).

The OH lines were at first seen only in absorption and only in limited regions, the direction of the source Cassiopeia A and a region surrounding the galactic center. Subsequently some very sharp emission features have been observed in a number of H II regions.

3. EXCITATION OF THE HYDROGEN LINE AND OH LINES

3.1. The Hydrogen Line

The 21-cm line arises from a hyperfine transition in the ground level ($1^2S_{1/2}$) of atomic hydrogen. This state is split into two closely spaced sublevels through interaction between the magnetic moments of the proton and the electron. In the higher-energy state the two magnetic moments are parallel, and the emission of a quantum of 21-cm radiation is associated with the change from the parallel orientation to antiparallel. This type of transition is often referred to as a *spin flip*. The lower sublevel is a singlet, but the higher one is split into a triplet in the presence of a magnetic field.

For the ground level of hydrogen, the quantum numbers are as follows, in units of $\hbar$:

$$\begin{aligned} \text{orbital angular momentum } L &= 0 \\ \text{spin of electron } S &= \tfrac{1}{2} \\ \text{total angular momentum } J = L + S &= \tfrac{1}{2} \\ \text{spin of nucleus } I &= \tfrac{1}{2}. \end{aligned}$$

The hyperfine sublevels are characterized by the total-spin angular momentum, F, which is the sum of the electronic and nuclear spins:

$$F = S \pm I\,. \tag{1}$$

Thus F can take the value 0 or 1. The smallness of the energy difference between these two sublevels arises from the small size of the nuclear magnetic moment.

The energy associated with the hyperfine structure is given by the following formula derived by Bethe (1933) from an earlier result of Fermi:

$$W = \frac{h\nu_0}{N^3}\left[\frac{F(F+1) - I(I+1) - J(J+1)}{J(J+1)(2L+1)}\right] \tag{2}$$

where

$$\nu_0 = g(i)\alpha^2\, cR\,. \tag{3}$$

N is the total quantum number, $g(i)$ is the Landé factor for the proton, α is the fine-structure constant, and R is Rydberg's constant.

Inserting the appropriate values for the quantum numbers for the ground level of hydrogen, we find that the energy separation between the two hyperfine sublevels is

$$h\nu = \tfrac{8}{3}h\nu_0\,. \tag{4}$$

The value of ν has been measured with great precision in the laboratory. The earlier results were obtained with the atomic-beam magnetic-resonance method and the later ones (Crampton, Kleppner, and Ramsey 1963; Menoud, Racine, and Kartaschoff 1964; Vessot, Peters, and Vanier 1964; Peters, Holloway, Bagley, and Cutler 1965) with atomic hydrogen masers. The following value, which is rounded off to the nearest cycle per second, is consistent with all the recent determinations:

$$\nu = 1420.405752\ \text{Mc/s}\,.$$

3.2. Transition Probability

The Einstein transition probability, A_{10}, for a spontaneous transition between an upper state $F = 1$ and a lower state $F = 0$ is given by

$$A_{10} = \frac{64\pi^4\nu_{10}^3}{3hc^3 g_1} S_{10}\,, \tag{5}$$

where

$$g_1 = 2F + 1 = 3 \tag{6}$$

is the statistical weight of the upper state, and

$$S_{10} = 3\beta^2 \tag{7}$$

is the "strength" of the line. (β is the Bohr magneton, $eh/4\pi mc$.)

The value is found to be

$$A_{10} = 2.85 \times 10^{-15} \text{ sec}^{-1} ,$$

which is about 10^{22} times smaller than typical values for permitted optical transitions. This very low value arises in part from the low frequency, since $A_{10} \propto \nu^3$, and in part from the characteristics of magnetic-dipole transitions.

The natural half-width of the line ($A_{10}/2\pi$) has the insignificant value of 5×10^{-16} c/s, which is always negligible in comparison with other causes of line broadening.

3.3. Spin Temperature

The relative populations of the two hyperfine sublevels, n_0 and n_1, can be expressed in terms of a temperature by the Boltzmann equation:

$$\frac{n_1}{n_0} = \frac{g_1}{g_0} \exp\left(-\frac{h\nu_{10}}{kT_s}\right) \tag{8}$$

where $g = 2F + 1$ is the statistical weight of each sublevel. Therefore,

$$\frac{n_1}{n_0} = 3 \exp\left(-\frac{h\nu_{10}}{kT_s}\right). \tag{9}$$

The temperature T_s is known as the *spin temperature*, and its relation to other measures of temperature, such as the kinetic temperature of the atoms, T_K, has to be considered.

Because the energy difference between the two sublevels is so small, the quantity $h\nu_{10}/k$ is also small (0.07° K), and the exponential in equation (9) is close to unity for all reasonable values of T_s. This means that the relative populations change only slightly with temperature, the proportion in the upper level being always close to $\frac{3}{4}$. For this reason, a measurement of the 21-cm emission from a body of gas can immediately tell us something about the total number of hydrogen atoms in the region concerned (provided the gas is optically thin), but it can say little about the temperature. On the other hand, as we shall see later, absorption effects are very dependent on the small differences in population from a 3:1 ratio with increasing excitation; in consequence, an absorption measurement can lead under some circumstances to an estimate of the spin temperature of the absorbing gas.

The factors that determine the spin temperature have been studied by Purcell and Field (1956) and by Field (1958) for a variety of astronomical situations. They showed the importance of atomic collisions in controlling the relative population of the two states. With such a very low probability for spontaneous transitions, a given atom will normally undergo many collisions before it spontaneously falls from the upper to the lower sublevel, with emission of a quantum. Collisions between electrons can also induce transitions, but their effect always seems to be outweighed by atomic collisions or radiation processes.

In a collision between two hydrogen atoms, an electron spin flip can be produced by the magnetic interaction between the two atoms, but this process has a low probability. Much more likely is an electron-exchange process in which the two atoms exchange their electrons through an electrostatic interaction. Sometimes an atom will receive a "new" electron with the same spin orientation as that of the "old" electron so that there is no change in the atomic state. On other occasions, the new electron will have the reverse-spin orientation, and there will be a small increase or decrease in the energy associated with the atom. This energy difference is taken from, or given to, the relative kinetic energy of the atoms. The process is a very efficient one, and the effective collision cross-section is high. It is not necessary to transmit much energy, because the hyperfine energy difference represents such a small proportion of the thermal energy of the interacting atoms.

When collisions are the dominant factor in controlling the spin states, the spin temperature will be effectively equal to the kinetic temperature. This seems to be the situation in the greater part of the Galaxy. A typical interstellar hydrogen atom will change "up" or "down" about every 400 years, due to collisions; ultimately, after 11 million years, there will be a spontaneous downward transition, and a 21-cm quantum will be emitted.

Radiation processes which affect the spin states are relatively unimportant in most of the galactic disk, but they assume a greater importance where the radiation flux is high or where the density is low and hence collisions are infrequent. The spin temperature then depends on the balance between the various processes of excitation and de-excitation.

One such process involves the absorption and re-emission of a quantum of Lyman-α radiation, which is associated with a transition between the ground level $n = 1$ and the next higher level, $n = 2$. When Lyman-α is scattered in this way, the electron of the atom concerned sometimes returns to its original hyperfine state, but, on other occasions, it goes to the alternate one. For this and other reasons, the Lyman-α line is a complex multiplet. The detailed effect on the hydrogen spin temperature depends rather critically on the intensity profile of the incident ultraviolet radiation in the region of Ly-α (Field 1958, 1959*c*).

When a beam of 21-cm radiation passes through a cloud of hydrogen atoms, some of the incident quanta are absorbed, exciting some atoms from the singlet to the triplet state. In other atoms, the interaction effect produced by a passing quantum can lead to a downward transition, with the emission of an additional quantum. This process is known as stimulated emission, which can be regarded as a negative absorption. These effects are important in observations of 21-cm absorption lines in the spectra of radio continuum sources. In these cases, the effective spin temperature may be well above the kinetic temperature.

In the galactic halo and in intergalactic space, where the density is low, collisions are not very effective in exciting the triplet state, and the spin temperature depends primarily on the background radiation. In the Galaxy, the Lyman-α flux will also play a part in the immediate vicinity of hot stars. Field (1958) illustrated the interplay of the various effects by considering several typical situations in detail. The particular case of intergalactic hydrogen will be discussed in Section 6.4.

3.4. OH Lines

At interstellar temperatures, only the lowest rotational level ($J = \frac{3}{2}$) of the $^2\pi_{3/2}$ electronic state is significant. This ground state is split by Λ-doubling into two levels, and further split by hyperfine interactions with the hydrogen nucleus (see Fig. 1). The physical mechanisms involved in the transitions have been described in detail by Barrett

(1964). The main characteristics of the four lines are given in Table 1. The frequencies quoted are laboratory values as measured by Radford (1964); the transition probabilities are values calculated by Barrett, Meeks, and Weinreb (1964) and Barrett (1964).

A low value is to be expected for the OH excitation temperature, because radiative processes will be more important than collisions in determining the population distribution. The excitation temperature T_s may in fact be close to the effective radiation tem-

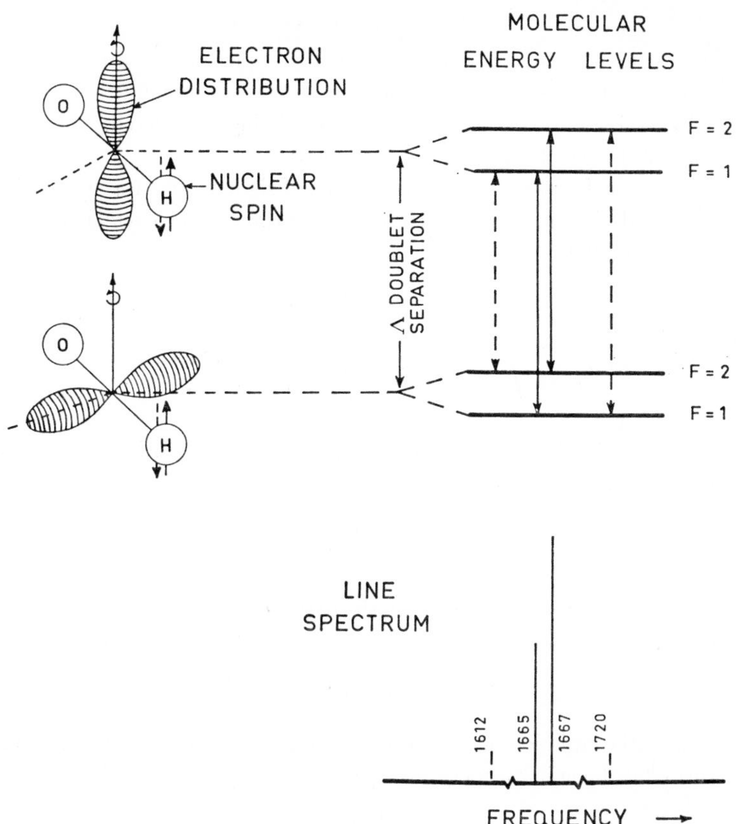

FIG. 1.—Arrangement of energy levels for the ground state of the OH radical. (After Barrett 1964.)

TABLE 1

THE MAIN GROUP OF OH LINES

Transition	Frequency (Mc/s) (±0.002)	Expected Relative Intensity	Transition Probability (sec^{-1})
$F=2\rightarrow1$........	1612.231	1	4.50×10^{-12}
$F=1\rightarrow1$........	1665.401	5	2.47×10^{-11}
$F=2\rightarrow2$........	1667.358	9	2.66×10^{-11}
$F=1\rightarrow2$........	1720.533	1	3.24×10^{-12}

perature T_r. Emission would then be difficult to detect, because the maximum emission temperature would be $T_s - T_r$ (Barrett and Lilley 1957*a*). The various processes involved have not yet been considered in detail, but Weinreb *et al.* (1963) suggest that slow-moving positive ions may be very effective in inducing transitions between the two states. As a first estimate, they adopted 10° K for the excitation temperature.

4. THE FORMATION OF EMISSION AND ABSORPTION LINES

4.1. Emission Lines

We will now consider the strength and nature of the spectral line produced by a body of gas under various circumstances, beginning with the case where the only radiation present is the line emission produced in the gas body being considered. This subject was first discussed by Wild (1952) and Shklovsky (1956).

The equation of transfer of radiation along a line of sight relates the specific intensity, I_ν, at frequency ν and distance r from the observer to the volume coefficients of emission and absorption, j_ν and K_ν:

$$\frac{dI_\nu}{ds} = j_\nu - K_\nu I_\nu . \tag{10}$$

The general solution is

$$I_\nu = \int_0^\infty j_\nu \exp\left(-\int_0^r K_\nu dr'\right) dr . \tag{11}$$

The observed intensity is usually expressed in terms of the brightness temperature T_b, which is given by the Rayleigh-Jeans formula,

$$T_b = \frac{I_\nu c^2}{2k\nu^2} . \tag{12}$$

Correspondingly, the emission coefficient can be expressed in temperature units (degrees per centimeter) as

$$J_\nu = \frac{j_\nu c^2}{2k\nu^2} . \tag{13}$$

By Kirchhoff's law, the emission and absorption coefficients are related through the temperature, in this case the spin temperature T_s:

$$J_\nu = K_\nu T_s . \tag{14}$$

Substituting from equations (12) and (13) in equation (11), we find that

$$T_{b\nu} = \int_0^\infty T_s K_\nu \exp\left(-\int_0^r K_\nu dr'\right) dr , \tag{15}$$

$$= \int_0^{\tau'_\nu} T_s e^{-\tau_\nu} d\tau_\nu , \tag{16}$$

where

$$\tau_\nu = \int_0^r K_\nu dr' \tag{17}$$

is the optical depth to a distance r at frequency ν and τ'_ν is the optical depth to infinity along the line of sight.

If the spin temperature is constant along the line of sight,

$$T_{b\nu} = T_s \left(1 - e^{-\tau'_\nu}\right) . \tag{18}$$

This is the general relation for an emission line. A more complex, but analogous, expression is required for a nonuniform T_s.

4.2. Number of Emitting Atoms

To find the number of atoms responsible for an emission line of given strength, we must consider the atomic absorption coefficient, α_ν, which can be expressed in terms of K_ν by the relation (Milne 1930):

$$K_\nu = \alpha_\nu \left(n_0 - n_1 \frac{g_0}{g_1}\right) \tag{19}$$

where the numbers of atoms in the two sublevels and their weights are related by the Boltzmann formula (eq. [8]). In the radio case, where $h\nu \ll T_s$, equation (19) reduces to

$$K_\nu = \alpha_\nu n_0 \frac{h\nu}{kT_s} . \tag{20}$$

The true absorption is largely compensated by the stimulated emission, and the effective absorption coefficient is dependent on the small residual variation of the relative populations with the excitation temperature T_s. The smallness of the population variation in this case can be seen from the following examples (for hydrogen):

T_s (°K)	10	100	1000
$\frac{n_1}{n_0}$	2.9806	2.9981	2.9998

The coefficient α_ν is related to the transition probability A_{01} by

$$\alpha_\nu = A_{01} \frac{c^2}{8\pi\nu^2} \frac{g_1}{g_0} f(\nu) \tag{21}$$

where $f(\nu)d\nu$ is the probability that a transition occurs in the frequency range ν to $\nu + d\nu$.

Substituting numerical values for hydrogen, we can now find $n(\approx 4n_0)$, the total number of ground-level atoms in unit frequency interval (1 c/s) in a cylinder of cross-section 1 cm^2 extending along the whole line of sight, that is

$$n = 3.88 \times 10^{14}\, T_s \tau_\nu . \tag{22}$$

In velocity units, the number of atoms in an interval of 1 km/sec is

$$n_v = 1.823 \times 10^{18}\, T_s \tau_v . \tag{23}$$

When the gas is optically thin ($\tau \ll 1$), the number of atoms can be derived from the observed brightness temperature without knowledge of T_s, since then

$$T_b \approx T_s \tau \tag{24}$$

and

$$\left.\begin{aligned} n &\approx 3.88 \times 10^{14}\, T_b \\ n_v &\approx 1.823 \times 10^{18}\, T_b \end{aligned}\right\} . \tag{25}$$

If we wish to know the total number of atoms in the line of sight at all velocities, we can measure the *integrated brightness*, that is, the area under the observed line profile,

$$B_{\text{int}} = \frac{2k}{\lambda^2} \int T_v d\,v\,. \tag{26}$$

Numerically

$$N = 1.823 \times 10^{18}\, B_{\text{int}} \tag{27}$$

where B_{int} is expressed in units of (1° K × 1 km/sec).

When the gas is not optically thin, the line profile must be replaced by a plot of optical depth against velocity for some assumed value of the spin temperature T_s. The corresponding relationship is then

$$N = 1.823 \times 10^{18} T_s \int_{-\infty}^{+\infty} \tau\, d\,v\,. \tag{28}$$

The quantities n and N are limited in their usefulness. Generally we can only obtain a value for the total number of atoms in a column of unit cross-section in a unit frequency or velocity range. We cannot locate these atoms along the line of sight or derive a space density unless we have other information. This is because our antenna beam explores a cone-shaped region, not a cylinder, and we cannnot distinguish, for example, between one atom at distance d and four atoms at distance $2d$.

The actual hydrogen mass can, however, be simply derived from the number of atoms in a unit column, if we have a limited concentration of gas at a known distance. Typically this is possible in isolated gas clouds in the Galaxy and nearby external galaxies.

4.3. Line Broadening

Because the natural width of the 21-cm line is so small (5 × 10^{-16} c/s), the width of the observed line is almost entirely determined by Doppler broadening. At 1420.4 Mc/s, a radial velocity of V km/sec produces a frequency shift of $-4.74\,V$ kc/s. The broadening is partly due to thermal motions of the atoms inside a single hydrogen cloud, but mainly to larger-scale motions of the clouds or cloud complexes.

The line produced by an assembly of atoms with a Maxwellian distribution of velocities, corresponding to a kinetic temperature T_K, is gaussian in shape, with a width between half-intensity points equal to

$$\delta\nu = 1.67\, \frac{\nu}{c} \sqrt{\frac{2kT_K}{m}} \tag{29}$$

where m is the mass of a hydrogen atom. Numerically

$$\left.\begin{array}{l} \delta\nu = 1.015\sqrt{T}\ \text{kc/s} \\ \text{or} \\ \delta\nu = 0.215\sqrt{T}\ \text{km/sec} \end{array}\right\}. \tag{30}$$

In general, we observe a number of thermally spread clouds, one behind the other, with different radial velocities. The line shape can be quite complex, but away from the Milky Way region the profile is usually simple in form, with a width determined mainly by the velocity dispersion of the gas clouds. Near the galactic equator, the line is considerably broadened by mass motions of cloud complexes, including galactic differential rotation and systematic departures from circular motion. Galactic rotation produces the largest effects, but the observations can provide information on all the various types of motion.

4.4. Absorption Lines

So far we have restricted the discussion to the case where emission and self-absorption take place inside a body of gas but no radiation enters from outside. The situation is more complicated when there is another emitting source in the background. When this is another line source at the same velocity, the optical depths are simply additive, and there is no formal difference from the instance where all the gas is in a single concentration. The over-all behavior is different, however, when the background emission is from a continuum source. This may be a small-diameter discrete source occupying only part of the antenna beam, or it may be a broadly distributed source, such as a portion of the Milky Way.

Radiation from the background will then add to the total received intensity, but it will be partially absorbed while passing through the gas. The observed brightness temperature at frequency ν is given by the relation

$$T_{b\nu} = T_s(1 - e^{-\tau_\nu}) + T_0\, e^{-\tau_\nu} \tag{31}$$

where the first term represents the emission from the gas concentration (from eq. [18]) and the second term the attenuated radiation from the background, whose brightness temperature is taken as T_0 considered independent of frequency over the small range occupied by the line. A formal proof of the additive character of the effects is given by Shklovsky (1956).

In observations, we are generally interested in the excess brightness temperature at the line frequency compared with that outside the line, as this is the method of recognizing a contribution from line emission. This contribution is given by

$$\Delta T = T_{b\nu} - T_0\,, \tag{32}$$

$$= (T_s - T_0)(1 - e^{-\tau_\nu})\,. \tag{33}$$

If $T_s > T_0$, the line appears in emission; if $T_s < T_0$, in absorption.

Part of the line-emitting gas may be behind the continuum source. Then, if the further and nearer sections of the gas are indicated by the subscripts 1 and 2, respectively, the observed brightness temperature is

$$T_{b\nu} = T_{s1}(1 - e^{-\tau_{\nu 1}})e^{-\tau_{\nu 2}} + T_0 e^{-\tau_{\nu 2}} + T_{s2}(1 - e^{-\tau_{\nu 2}})\,. \tag{34}$$

Absorption of line radiation in the continuum source can be neglected.

In this expression we have still imposed the condition that T_s should be constant inside each concentration of gas. This must be unreal, but we have no observational method so far for studying the variation of T_s inside a single body of gas.

Unlike the emission case, absorption measurements can give the optical depth τ of a body of gas without the spin temperature T_s. On the other hand, n (the number of atoms in unit column) and T_s cannot be separated in the absorption case, whereas they can be for an optically thin emitting gas.

When the same cloud of gas can be observed at adjacent points in both emission and absorption, it is possible to derive T_s and τ separately (and hence also n). The method is, however, limited by the extent to which the cloud can be taken as uniform from point to point.

5. OBSERVATIONAL TECHNIQUES

5.1. Receivers

The basic problem in observing line emission is to determine the intensity of the radiation as a function of frequency (that is, radial velocity) and two coordinates of position in the sky. The intensity is always low, and it is consequently necessary to use receivers that are highly sensitive and very stable.

Superheterodyne receivers of rather conventional type, with crystal mixers, have been used at many observatories, with special arrangements to achieve the utmost possible stability, as the radiation being measured is very much weaker than the noise generated inside the receiver. Lower-noise receivers using masers or parametric amplifiers

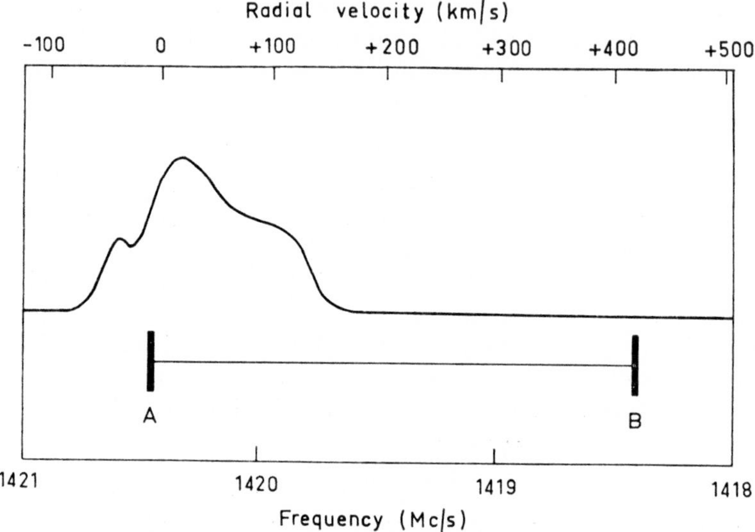

Fig. 2.—Switched-frequency method of recording a line profile. The receiver is switched rapidly between frequencies *A* and *B*.

are now being applied to hydrogen-line observations by many workers (e.g., Giordmaine *et al.* 1959, Jelley and Cooper 1961, Robinson 1963), and several such systems already in operation have not yet been described in the literature.

The most direct observing method is to measure in an absolute manner the intensity received at each frequency of interest. Such a measurement, however, includes the continuum background as well as the line emission, and there are also very serious problems of stability in a direct radiometer.

The common method for eliminating the continuum is to make rapid comparisons between two frequencies, one on the line and the other displaced from it (see Fig. 2). Receiver stability is greatly improved at the same time, because internal gain variations will be approximately the same at the two frequencies, and thus will be largely cancelled out by the comparison procedure. The usual system is to switch rapidly between two narrow frequency bands of about the same bandwidth, 1.5 to 2 Mc/s apart, but in some receivers a narrow band has been compared with a much wider one.

The simple switching arrangement loses sensitivity, because the line radiation is being received for only half the total observing time. This loss can be made up with a double-switching system (Fig. 3), which insures that line energy is then being received throughout the switching cycle (Muller 1956; Williams and Davies 1956).

To trace out the line profile at a point in the sky, the fundamental frequency of the receiver can be scanned slowly across the relevant part of the spectrum (Muller and Westerhout 1957). Alternatively, the whole profile can be recorded in a single integration period by providing a bank of filters which produce a large number (~50) of outputs at frequencies spread across the whole spectral region (Burke *et al.* 1959, McGee and Murray 1963). Intermediate systems have also been used, in which the full profile is built up from observations with a smaller number of channels, each of which is switched to a series of frequencies (Kerr, Hindman, and Robinson 1954) or scanned over a portion of the frequency range (Muller 1966).

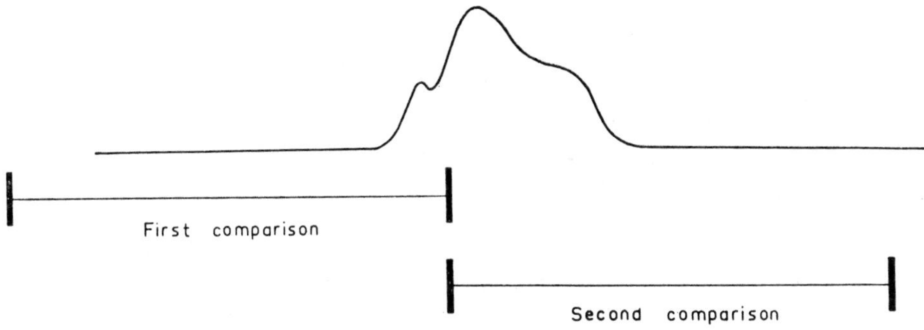

FIG. 3.—Double-switching system

The main consideration here is that a much longer time is needed to cover a given area of sky in a line survey than in the corresponding continuum study. This is a consequence of the much narrower bandwidth required for tracing out a line profile. This time disparity is enormous with a single channel, but is considerable even when a multichannel system is used.

Although the continuum is effectively suppressed in the line observation, a strong continuum signal can interfere with the line measurement, either by producing a differential effect in the two narrow channels of a switched-frequency system, or by overloading the detector. These troubles are greatest in absorption observations on a strong discrete source with a large telescope; under these conditions, the performance of the receiver is difficult to calibrate.

An alternative method of reception has recently been introduced, in which the autocorrelation function of the radiation is studied instead of its frequency spectrum. These two functions are Fourier transforms of each other, and should contain the same information. The method was first discussed by Blum (1960), and has been developed by Weinreb (1963), who handled the receiver output noise in the form of clipped noise pulses by digital techniques. This type of receiver should have higher stability against the effects of gain fluctuations and in consequence should be capable of higher sensitivity than the conventional type. However, it is noteworthy that the first detection of the OH line with a correlation receiver was followed immediately by an almost identical performance with several frequency-scanning receivers. The comparative performance of the two types of receiver has not yet been precisely established, but the correlation receiver has one clear advantage in the greater ease of changing bandwidth in multiple channels.

Receiver techniques have recently been reviewed by Drake (1960) and Robinson (1964), and their papers should be consulted if more detail is required on receiver operation and performance. They also describe several variants of the main receiving systems discussed here.

5.2. Antennas

Line work is usually done with a single parabolic dish. Diameters from 21 to 300 feet have been used, corresponding to beamwidths of 144 to 10 minutes of arc between half-intensity points. Because large collecting areas are needed for the very weak signals involved, crosses and other forms of unfilled-aperture antennas have not been used for line studies, but this is likely to happen in the future as antenna apertures become larger and receiver sensitivities improve.

The technique of two-element interferometry has been applied to line work (Radhakrishnan, Morris, and Wilson 1961; Clark, Radhakrishnan, and Wilson 1962; Clark 1965), using the Caltech 90-foot paraboloids. This system can give increased resolution of hydrogen absorption lines in the spectra of continuum sources. The interference patterns assist in distinguishing absorption and emission effects when the former are limited to the small solid angle of a discrete source and the latter are distributed over the full antenna beam. However, the system is only effective for an isolated source, and is difficult to use in a complex region such as in the direction of the galactic center.

Systems for measuring polarization are of interest in hydrogen-line work only in attempts to observe a Zeeman effect, where differences between two circularly polarized components are sought (see Sec. 6.3).

5.3. Over-All System

The frequency-switching system which is commonly used to suppress the continuum level and the effects of gain variations has an additional advantage. It gives considerable protection from broad-band interference and from radiation from the Sun which might be picked up on antenna sidelobes or "spillover" because these unwanted signals are almost identical on the signal and comparison frequencies. For this reason, line work can usually be carried out equally well by day or by night, whereas high-sensitivity continuum observations during the day are sometimes troubled by stray pickup from the Sun.

All observed line-emission sources are extended objects, and therefore the observations are concerned with brightness temperature rather than flux density. The most serious calibration problem is the measurement of antenna efficiency, which is always the most difficult part of fixing an absolute temperature scale in radio astronomy. The relative scale is usually set by comparison with an attenuated signal from a local noise source, such as a gas-discharge tube.

Muller and Westerhout (1957) made a classical calibration measurement by pointing the 24-foot Kootwijk dish at a nearby pine forest, which was assumed to be a complete absorber at ambient temperature. All subsequent Dutch measures have been expressed in terms of the scale derived from this calibration, and several other groups have fitted their own scales to it, through comparison observations of standard regions. This scale leads to a value of 125° K for the highest brightness temperature anywhere in the sky, and this value has often been quoted as the derived spin temperature of the hydrogen in the Galaxy. It is doubtful however that the beam of the Kootwijk dish was completely intercepted by the pine forest in this measurement; if it were not, the true brightness would be higher than 125° K. The Sydney group has obtained a figure of 160° K for the maximum brightness temperature anywhere in the Galaxy. Attempts are being made through the IAU to reconcile the various temperature calibrations, through more careful intercomparisons on a standard region, with due allowance for different beamwidths and bandwidths in the various observations. However, it is not easy to find a completely suitable region for this purpose.

In a frequency-scanning receiver, the zero reference level is obtained by interpolating from frequencies on either side of the observed line—assuming that the baseline is linear across the frequency range. In a multichannel system, measurements are usually referred to the coldest known part of the sky. In either method, there could be an unknown error, if there is a low level of radiation that is widely distributed in position and frequency. Such widespread weak radiation could arise in extragalactic sources.

Taking all factors into account, the accuracy of brightness temperature measurements can be expressed in the following way. For low values, the relative uncertainty is about 1° K in the older conventional receivers, or 0.1° K in the low-noise types; at higher levels, it is about ± 10 per cent for relative values, or about ± 25 per cent for absolute values.

Frequency measurements can easily be made with high precision. When we specify the radial velocity of a feature, the accuracy is limited only by the receiver bandwidth and the velocity dispersion of the hydrogen, and not at all by the frequency-measuring system.

In observations of hydrogen inside the Galaxy, the measured radial velocity is normally corrected to the local standard of rest; that is, corrected for the Earth's orbital motion and for the standard solar motion of 20.0 km/sec toward R.A. 18^h, Dec. $+30°$ (1900). Optical radial velocities are corrected only to the Sun, and an appropriate adjustment is required when comparing radio and optical velocities. Some authors have unfortunately not realized this difference and have confused the two velocity systems in their radio-optical comparisons (e.g., Dieter 1960, Davis 1962, Davies and Tovmassian 1963*a*, *b*). In extragalactic observations, velocities are usually corrected to the Sun, but this is not a universal practice, and some authors have unwisely corrected for the solar motion also, although the local standard of rest has no special relevance outside the Galaxy.

The information-gathering capacity of a multichannel hydrogen-line receiver is very high. In consequence, several observatories have made extensive use of digital techniques for recording and reducing the data. The original papers describing the main surveys should be consulted for details of reduction procedures.

6. OBSERVATIONAL RESULTS, H AND OH

6.1. Hydrogen Emission

Many surveys of hydrogen-line emission have been carried out over various parts of the sky. Since the radiation can be seen everywhere, and often over a wide frequency range, it is only possible to cover the full sky at a low resolution. Two surveys of this type have been done, but otherwise a sampling procedure is necessary.

Table 2 lists the main surveys, in chronological order of publication date, and sum-

TABLE 2

SURVEYS OF HYDROGEN EMISSION FROM THE GALAXY

Authors	Publication Date	Beamwidth (degrees)	Bandwidth (kc/s)	Region	Usual Sampling Interval	Form of Presentation
Christiansen and Hindman	1952	2.3	50	δ+40 to −66	α, δ, 5°	Sky map, peak T_b
van de Hulst, Muller, and Oort	1954	1.9 ×2.7	40	old equator, l^{II} 354–252	l, 5°	Profiles
Muller and Westerhout	1957	1.85×2.8	37	disk {l^{II} 350–252; $b^{II} \approx \pm 10$}	l, b, 2°.5	Profiles
Bolton, Stanley, and Harris	1958	1.2 ×1.4	25	disk {l^{II} 326–360; b^{II} −7 to +9}	l, b, 2°	l, v diagrams
Burke, Ecklund, Firor, Tatel, and Tuve	1959	2.2 ×3.5	12	disk {l^{II} 349–252; $b^{II} \approx 0$ to +2}	l, 3°; b, 1°	Profiles
Menon	1958	1.7	15	Orion {l^{II} 192–216; b^{II} −26 to − 11}	l, b, 3°	Profiles
Helfer and Tatel	1959	1.8 ×2.8	10	Pleiades	α, 1°; δ, 2°	α, v and α, δ diagrams
Kerr, Hindman, and Gum	1959	1.4	40	disk {l^{II} 207–37; $b^{II} \approx \pm 5$}	l, 5°; b, continuous	b, v, diagrams
Davies	1960	1.5 ×1.7	100, 25	whole sky, lat 53° N	≤5°	B_{int} map
Erickson and Helfer	1960	1.8 ×2.8	12	whole sky {lat 39° N; $\|b\| > 20°$}	l, b, 10°	Profiles
Grahl	1960	0.6	30	l^{II} 132–152, b^{II} ±2	l, 1°; b, 2°	Gaussian components
Kaftan-Kassim	1961	1.7	15	Cygnus {l^{II} 67–92; b^{II} ±10}	2°	Profiles
McGee and Murray	1961	2.2	38	whole sky, lat 34° S	l, 30°; b, 2°	Sky maps, N_H, v at max
Sorochenko	1961	1.8 ×0.75	20	Cygnus {α 20.04–20.32; δ +39 to +46}	1°	Profiles
Davis	1962	0.8	15	disk {l^{II} 200–265; b^{II} ±15}	5°	Profiles

TABLE 2—*Continued*

Authors	Publication Date	Beamwidth (degrees)	Bandwidth (kc/s)	Region	Usual Sampling Interval	Form of Presentation
van Woerden, Takakubo, and Braes	1962	0.56	10	l^{II} 352–242; $b^{II} \approx \pm 10$, 15, 20, 25	l, 10°; b, 5°	Profiles
Braes	1963	0.56	20	nucleus l^{II} 355–5; b^{II} ±5	l, 10°; b, continuous	b, v, diagrams
Höglund	1963	2.0 ×2.5	30	anticenter l^{II} 155–200; b^{II} −16 to +14	l, 5°; b, 2°.5	l, v diagrams
McGee, Murray, and Milton	1963	2.2	38	whole sky, lat 34° S	δ, 1°; α, 2^m	Maps, N_H, v at max
Burke, Turner, and Tuve	1964	0.17	10	outer parts l^{II} 11–50; b^{II} −0.8 to +2.0	l, 5°; b, 12′	Sample b, v and l, v diagrams
Burke and Tuve	1964	0.8	50	nucleus l^{II} 333–2; b^{II} ±2	l, 2°; b, 1°	Sample b, v and l, v diagrams
Dieter	1964	0.9	80, 15	NGP, $b^{II} > +80°$	1°	Maps, N_H, v at max
Girnstein and Rohlfs	1964	0.6	15, 30	I Mon l^{II} 200.5–212.5; b^{II} ±3.5	1°	Maps and l, v diagrams
Locke, Galt, and Costain	1964*a*	0.6	10	anticenter l^{II} 162.5–192.5; b^{II} −11 to +9	δ, 30′; drift scans	Maps, constant v
Makarova	1964	1.8 ×0.75	10	equator, l^{II} 0–240	l, 10°	Profiles
McGee and Milton	1964*b*	2.2	38	whole sky, lat 34° S	α, 2^m; δ, 1°	Maps, N_H, v at max, for higher velocity features
Rougoor	1964	0.56	10, 20, 40	nucleus l^{II} 352–22; b^{II} −0.6 to +0.9	0°.25 or 0°.5	Profiles
Raimond	1966	0.56	10	Monoceros l^{II} 192–212; b^{II} −9.5 to +8.5	0°.5 or 1°	Profiles
Kerr and Vallak	1966	0.23	38	nucleus b^{II} 0, l^{II} 355–5; b^{II} ±1, l^{II} 359–1	l, 0°.1; b, 0°.1 (359°–1°)	l, v and b, v diagrams

marizes for each the equipment parameters, the region covered, and the form of presentation of results. The list is restricted to investigations of the Galaxy and to papers that contain extensive observational results, either as a catalogue of line profiles, or in some other consolidated form. A large body of data is already available, but this will be greatly extended by high-resolution surveys now in progress (1966).

Figure 4 shows an array of observed profiles, selected to illustrate the range of possibilities. The figure contains samples from various types of location in the Galaxy, from each Magellanic Cloud and from the bridge between them, and from several other external galaxies. Many profiles can of course be found in the published literature. The group of profiles near $l^{\rm II} = 330°$, $b^{\rm II} = 0°$ has been included to show the rapid variation in detailed shape which occurs along the galactic equator. These profiles were taken at points 12 minutes of arc apart, observing with a 14-minute beam.

A brief review of the observational results was given in Section 2. The galactic structure aspects are discussed in detail in Volume V of this compendium (Kerr and Westerhout 1965). The physical state and fine structure of the hydrogen, and its relation to other galactic subsystems, will be considered in later sections.

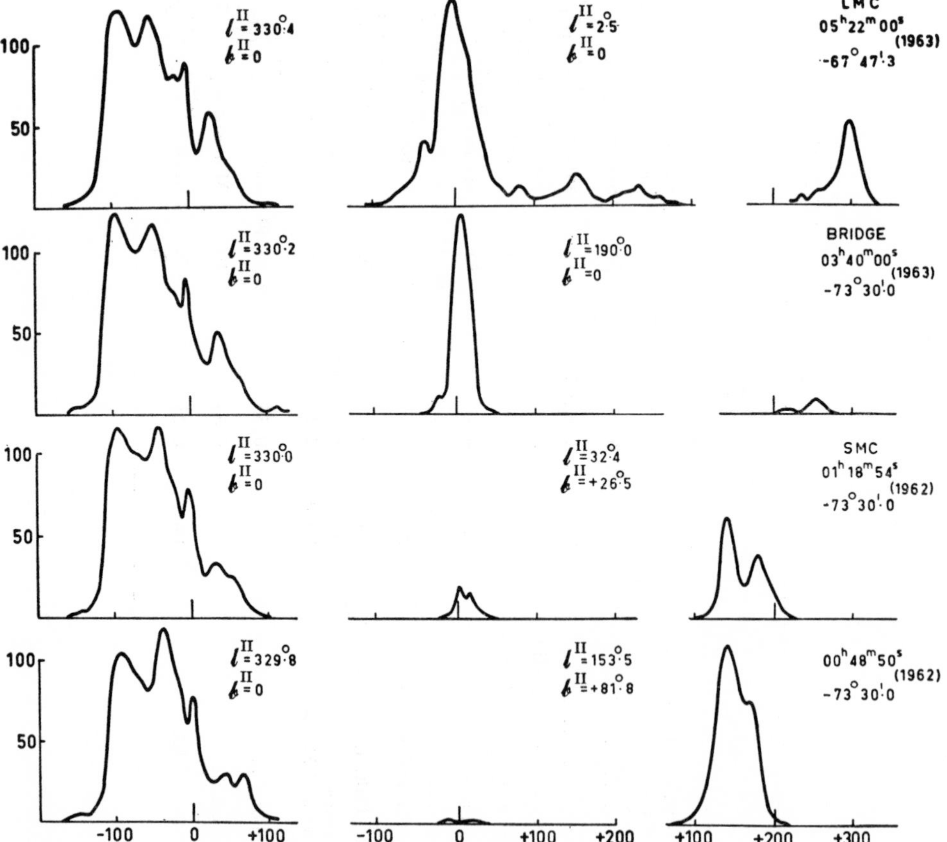

FIG. 4.—An array of hydrogen-line profiles, from various locations in the Galaxy and the Magellanic Clouds.

6.2. Hydrogen Absorption

As we have seen in Section 4.4, radiation from a source that emits over a wide continuum can be absorbed in a narrow frequency range as it passes through neutral hydrogen lying in the propagation path. The continuum background at 21 cm is only strong enough to show significant absorption effects over a portion of the Milky Way region near the galactic center and in the positions of the stronger discrete sources.

In most circumstances, absorption lines cannot be observed separately because hydrogen-line emission is present at the same time in the recorded profile; these effects are

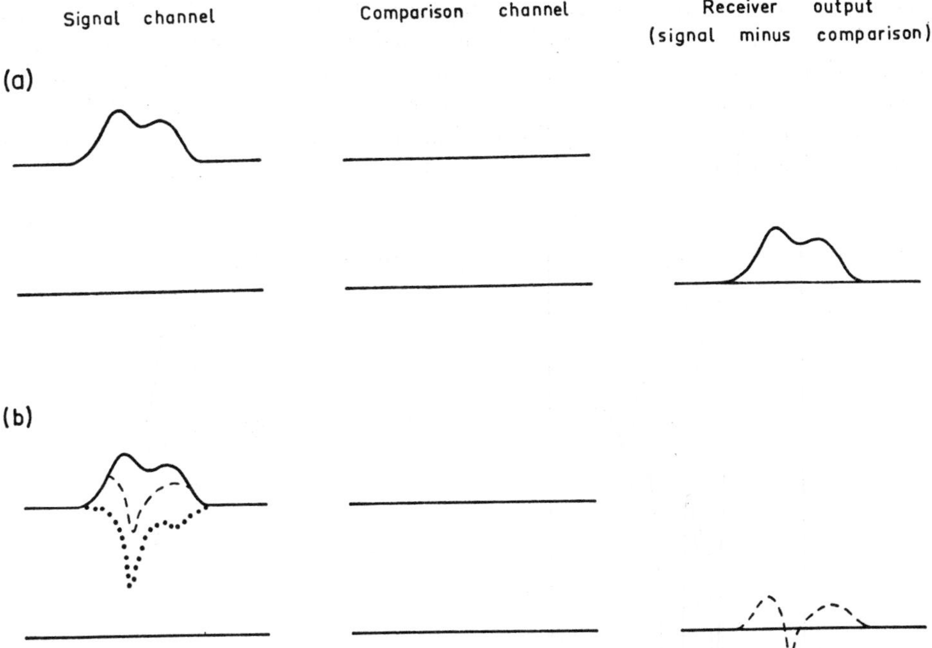

Fig. 5.—The composition of an observed absorption profile from the combination of line and continuum radiation. (*a*) Without absorption (nearby continuum source in front of all the hydrogen). (*b*) With absorption (source beyond some of the hydrogen). ——— Line emission in absence of absorption. •••••• Continuum source spectrum attenuated by hydrogen absorption at some frequencies. - - - - - Observed profile, combination of absorption and emission.

sometimes difficult to disentangle. In a switched-frequency receiver, which is the most commonly used type, the relationship between the signal and comparison channels must also be considered. In the absence of absorption, any continuum emission will give approximately equal responses in the two channels, leaving only the line emission in the output (see Fig. 5*a*). The effect of absorption is to cut down the continuum level in the signal channel, which then contains the line emission plus the attenuated continuum. The comparison channel contains the full continuum level as before. The net switched output then gives a profile which is below the normal emission profile by an amount which represents the true absorption (Fig. 5*b*). In general, the absorption varies with radial velocity, and the shape of the absorption curve is often quite different from that of the corresponding emission profile, indicating that a different distribution of hydrogen atoms is responsible. Some sample absorption profiles are shown in Figure 6.

The ever-present difficulty in interpreting absorption observations is that the degree of absorption can only be derived when we know the emission profile that would have been recorded if there had been no absorption. This profile, commonly called the *expected profile*, can only be estimated by interpolating between profiles observed at points just away from the source whose radiation is being absorbed. In some cases, the hydrogen emission varies only slowly across the sky, but there are often sizeable irregularities, which impose a corresponding uncertainty on the expected profile. In addition to the effect of general variations over the sky, there are also problems arising from angular scale differences. Emission can only be investigated on the scale of the antenna beam-width, whereas for a small-diameter source the absorption may be taking place over a much smaller solid angle. We may say that the accuracy of the expected profile is always limited by the extent to which conditions within the small solid angle of the source are representative of the average conditions over the whole antenna beam.

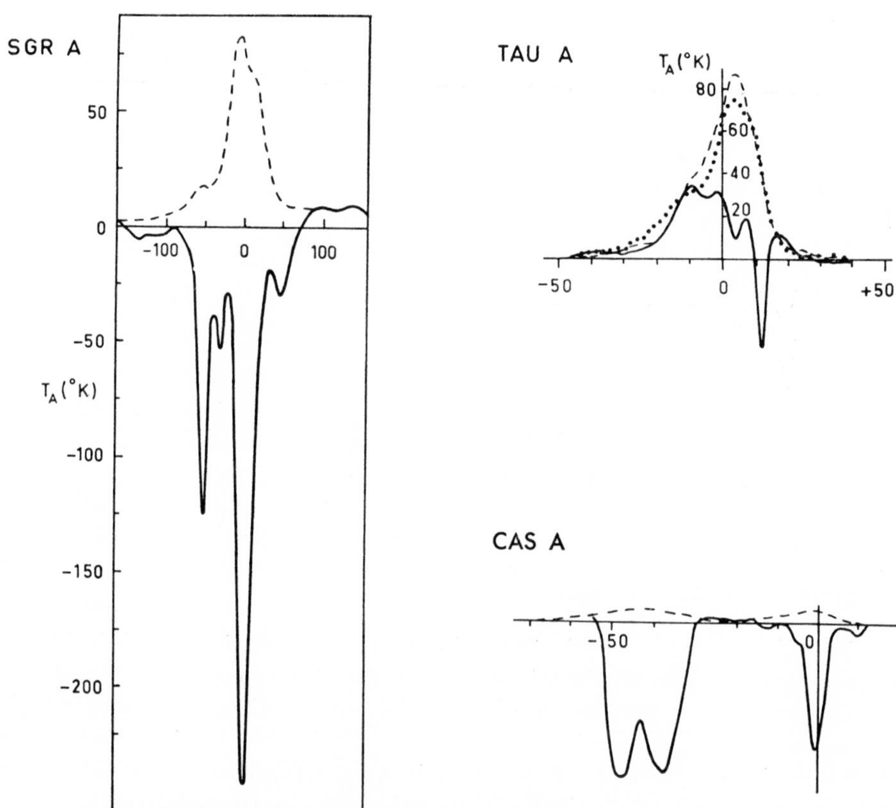

FIG. 6.—Hydrogen absorption profiles for the strong sources. (*a*) Sagittarius A (Kerr and Vallak 1966). (*b*) Taurus A (Shuter and Verschuur 1964). (*c*) Cassiopeia A (Shuter and Verschuur 1964). For Sgr A and Cas A, the dashed lines represent the "expected" emission profile in the absence of absorption. For Tau A, the dashed and dotted lines are profiles at two adjacent points.

In spite of these limitations, absorption observations have several important applications. These have become more attractive with the introduction of large telescopes, because the absorption effects on small sources are directly proportional to antenna gain. In addition, the disparity between the solid angles of beam and source is now smaller, reducing the uncertainty in the expected profile.

The main applications can be listed as follows:

a) A higher resolution can be obtained in studying the cloud structure of the interstellar hydrogen, because the effective solid angle of a small source is much less than that of any antenna beam yet available.

b) The absorbing hydrogen must always be in front of the source. Therefore, if we know the approximate location of the hydrogen, we can place limits on the distance of the source.

c) Similarly, if the source is at a known distance, this information can sometimes assist in locating the absorbing hydrogen. This application is particularly useful when a source is behind some hydrogen concentrations, and in front of others, in the same line of sight.

d) Absorption observations provide the only method for determining the spin temperature of any concentration of hydrogen, as discussed in Section 4.4. To do this, we need measurements of both absorption and emission brightness temperatures. Optical depth and spin temperature can then be separated, to the accuracy with which the absorption and emission measures apply to the same body of hydrogen.

e) Absorption studies provide the most sensitive method for detecting hydrogen at very low temperatures, for example, in the intergalactic gas. Also, as we shall see in Section 6.5, so far it has only been possible to detect the low-temperature OH radicals through absorption.

Historically, the first absorption observations were reported by Hagen and McClain (1954) and Hagen, Lilley, and McClain (1955). Other early studies were carried out by Williams and Davies (1954, 1956) and Kerr and Hindman (1954). More systematic observations were undertaken by Muller (1957, 1959) and extended by Clark, Radhakrishnan, and Wilson (1962), Clark (1965), and Shuter and Verschuur (1964).

Clark *et al.* have studied the greatest number of sources for absorption effects so far, finding absorption in 15 out of 21 sources. Their work was done partly with a single 90-foot dish, but mainly with two 90-foot dishes acting as an interferometer to gain improved resolution in separating absorption and emission effects. In some sources it was possible to detect angular structure in the absorption pattern, but, as in all absorption studies, it is difficult to draw completely unambiguous conclusions. Shuter and Verschuur studied only the three strongest sources, but their observations are noteworthy for the high resolution in frequency that was used (3 kc/s, or 0.6 km/sec). This made possible a better separation of the individual hydrogen clouds lying along each line of sight.

Some studies of particular sources by other workers are discussed below, and some extragalactic absorption studies are mentioned in Section 6.4. Results relating to cloudiness and spin temperature will be discussed later. We will here consider only the application of absorption studies to distance estimates.

The early observations confirmed that Cygnus A must be an extragalactic source, because its radiation is absorbed over the whole range of velocities covered by the hydrogen emission in the same direction. On the other hand, Cassiopeia A, Sagittarius A, and Taurus A must be inside the Galaxy, because absorption only occurs over part of the velocity range of the emission profile. Akabane and Kerr (1964) have applied the same method to a source whose identification is otherwise unknown. They found that the source W49 (in the catalogue of Westerhout 1958) appears to be a distant H II region in the Galaxy, which is large in size and optically obscured. Observations with large telescopes can be expected to give similar distance information on 100 or more sources before long.

The complex source Sagittarius A has offered particular difficulty in distance determinations because conditions are very complex in the direction of the galactic center. From small-dish absorption observations, Davies and Williams (1957) and McClain (1955) both concluded that the source is only 3 to 4 kpc from the Sun and is therefore not at the center; however it seems well established now that the main source is at, or very close to, the center. More recently, Ryzhkov, Egorova, Gosachinskii, and Bystrova (1963) have studied the absorption effects related to the various components of the source, and consider that one component must be much closer than the center. The pencil-beam observations of Kerr (1964*b*) and Kerr and Vallak (1966), however, disagree with this conclusion.

Little use has been made so far of the possibility of deriving the distances of absorbing bodies of hydrogen, as there are not enough sources at known distances in the Galaxy. Even without precise distances, however, the method is often useful in galactic structure studies for confirming whether or not one concentration of hydrogen is in front or back of another.

In addition to the absorption of radiation from continuum sources, there may be situations where line emission from one hydrogen cloud is absorbed by a cooler hydrogen cloud in front of it. This question is discussed in Section 7.1.

6.3. Zeeman Splitting and the Galactic Magnetic Field

The hydrogen line is split into three components in the presence of a weak longitudinal magnetic field, H. The frequencies of the two outlying components, which are circularly polarized in opposite senses, are $\nu_0 \pm eH/4\pi mc$. Numerically, their separation is 2.8 Mc/s per gauss.

Bolton and J. P. Wild (1957) first pointed out that this splitting might be used to measure the galactic magnetic field. The frequency displacement would be very small (of the order of cycles), but it might be detectable in one of the narrow absorption lines observed in the spectra of strong continuum sources. The proposed method, based on that used by Babcock (1953) for solar optical lines, was to set a receiver on the steep side of an absorption line and then switch an antenna to receive the two circularly polarized components alternately. For a symmetrical line, the temperature difference between the two polarizations would follow an S-shaped curve as a function of frequency.

This method has been used by Davies, Slater, Shuter, and P. A. T. Wild (1960); Galt, Slater, and Shuter (1960); Davies, Verschuur, and P. A. T. Wild (1962); Davies, Shuter, Slater, and P. A. T. Wild (1963); Weinreb (1962*b*); and Morris, Clark, and Wilson

(1963). Negative results were reported, except by Davies *et al.* (1962, 1963). These authors obtained a weak asymmetrical curve from their switched-polarization observations of a Taurus A absorption line with a very narrow (3 kc/s) filter. In interpreting this result, they suggest that the absorption line consists of two components, only partially resolved; one of these shows a Zeeman splitting corresponding to a field of 2.5×10^{-5} gauss away from the observer, and the other shows no measurable effect. They also discuss similar but weaker Zeeman features in the absorption spectrum of Cassiopeia A and consider that their results on the two sources indicate a mean galactic magnetic field in the solar vicinity of 1×10^{-5} gauss, inclined at about 20° to the neutral hydrogen spiral arms.

Verschuur (1967) has repeated these observations with equipment that was more carefully adjusted to detect circular polarization. He obtained negative results for each source, leading to an estimated field strength $<5 \times 10^{-6}$ gauss. Negative results have also been reported by Weinreb (1962*b*) and Morris, Clark, and Wilson (1963).

The observations to date imply that the galactic magnetic field is smaller than some values that have been suggested in the past, but it must be remembered that the Zeeman measurements refer to the field inside the denser clouds which are responsible for the main absorption and need not necessarily apply to the general field.

Clark (1963) has suggested an alternative method of measuring the galactic magnetic field through hydrogen-line observations. His proposal is to look for Faraday rotation in the Crab Nebula absorption, arising from the presence of anomalous dispersion in the neighborhood of the line.

6.4. Neutral Hydrogen in Other Galaxies

We will not consider here the extensive work on 21-cm line emission from other galaxies that has been done on the internal distribution and motions of the gas in these other systems. Several aspects of extragalactic studies are however of interest in a general study of the interstellar gas. Except for the Magellanic Clouds and a few spirals, the received signal is always low, and highly sensitive receivers are required to detect it.

The Magellanic Clouds were the first external systems in which the 21-cm line was detected (Kerr, Hindman, and Robinson 1954). Observations in the Clouds overcome some of the difficulties inherent in galactic work. In each Cloud, we are viewing the system from outside and can be sure that all the constituents are at approximately the same distance from us. The LMC in particular is an ideal location for studying interrelationships between neutral hydrogen, ionized hydrogen, and associations of young stars (McGee and Milton 1964*a*, McGee 1964, Bok 1964). These constituents are found together, often in isolated concentrations where it is possible to compare their spatial distribution in detail.

For example (Bok 1964), the mass of the stars in the OB association NGC 1929–37 is about 24,000 $\mathfrak{M}_\odot$, and that of the ionized hydrogen in the vicinity of the association is 50,000–60,000 $\mathfrak{M}_\odot$. The corresponding mass of neutral gas in the same complex is 5×10^6 $\mathfrak{M}_\odot$. Bok concludes that active star formation will go on here for a long time, since there is sufficient gas available for 200 generations of stars. An alternative interpretation is that the very special conditions that are required for star formation are only found in very limited regions of the whole mass of gas.

Detailed comparisons of radio and optical velocity measurements in the LMC and SMC have shown that the motion of the supergiant stars and ionized hydrogen nebulae, for which velocities are available, usually is closely related to the motion of the neutral hydrogen in the same direction (McGee 1964; Bok, Gollnow, Hindman, and Mowat 1964).

The maximum brightness temperature observed anywhere in the Clouds is 150° K; this value occurs in the main body of the SMC (Hindman 1967). We conclude that the kinetic temperature of the gas must be at least 150° K, but may be considerably higher. There is no evidence, through absorption or otherwise, of any variation in the kinetic temperature through the Magellanic System. Such differences are, however, not easy to detect unambiguously for the reasons outlined in Section 7.1 for the galactic case. The random motions of the gas clouds can also be measured as a function of positions in the Clouds. The width of the observed profile between half-density points is typically about 25 to 30 km/sec in regions where single-peaked profiles are observed.

The 21-cm line has been detected from nearly 100 galaxies so far, in observations at Leiden, Harvard, Sydney, and Green Bank. These systems are almost all spirals and irregulars; hydrogen has been observed in only one elliptical system (Robinson and Koehler 1965). The hydrogen observations for each galaxy can provide a value for the integrated mass of hydrogen in the galaxy (from the total received energy) and also for the total mass (from the gas motions). From these results the relative proportion of hydrogen can be studied as a function of morphological type (Heidmann 1961; Epstein 1964). Epstein finds mean values for the ratio of H I mass to total mass of 0.01, 0.08, and 0.16 for Sb, Sc, and Ir systems, respectively—in accord with values to be expected if the gas is a constituent of Population I.

The measured radio and optical velocities have been compared for a sample of 29 galaxies by Dieter, Epstein, Lilley, and Roberts (1962). This comparison provides a good test of the Doppler interpretation of spectral line displacements, because a very large wavelength range is involved. Dieter *et al.* obtained a slope of 0.98 ± 0.04 (standard error) for the regression of $(c\Delta\lambda/\lambda)_{\mathrm{radio}}$ on $(c\Delta\lambda/\lambda)_{\mathrm{vis}}$, indicating that $\Delta\lambda/\lambda$ is independent of wavelength over the velocity range available for the comparison, -350 to $+700$ km/sec.

Several attempts have been made to detect intergalactic neutral hydrogen, with but little success, as far as the general intergalactic region is concerned. It is, however, quite possible that intergalactic hydrogen is largely ionized. The neutral gas has been looked for in absorption by Field (1959*a*, 1962) in the spectrum of the continuum source Cygnus A, in emission by Goldstein (1963), and both in absorption and emission by Davies and Jennison (1964). Emission measurements can be interpreted more directly, and the negative results indicate that the mean density of neutral hydrogen is less than 5×10^{-29} gm/cm^3. The interpretation of absorption observations requires assumptions about the spin temperature of the hydrogen and the spectrum of the source—usually Cygnus A. Field (1962) and Davies (1964) have derived upper limits for the density in the region of 10^{-29} to 10^{-30} gm/cm^3.

Although it has been difficult to find gas in general intergalactic space, absorption lines have been reported in the spectra of the sources Virgo A and 3C273, attributable to intergalactic hydrogen inside the Virgo cluster of galaxies (Robinson, van Damme, and

Koehler 1963; Koehler and Robinson 1966). The observed absorption was at +1100 km/sec, with a linewidth ≤ 500 km/sec. Since emission was not detectable in neighboring positions, the spin temperature must be low, but the temperature and density cannot be derived directly from the single absorption observation.

Field (1959*b*) has considered the likely spin temperature under intergalactic conditions. He estimates that collisions alone cannot raise T_s above about $\frac{1}{2}°$ K. The failure so far to detect intergalactic emission anywhere implies a very low value of T_s, probably $<1°$ K.

6.5. OH Observations

The first suggestions that OH lines might be found in the radio spectrum of the interstellar gas were made by Shklovsky (1953) and Townes (1957). An early search by Barrett and Lilley (1957*b*) was unsuccessful, but at that stage the precise frequencies of the lines were not known. The detection problem was simplified when reliable laboratory frequencies were obtained for two OH lines by Ehrenstein, Townes, and Stevenson (1959).

The two stronger OH lines at 1667 and 1665 Mc/s were first detected by Weinreb, Barrett, Meeks, and Henry (1963). Successful confirmatory observations were carried out very shortly afterward by Bolton, van Damme, Gardner, and Robinson (1964); Dieter and Ewen (1964); and Weaver and Williams (1964). In each study, absorption due to OH was observed in the spectrum of a strong discrete source, Cassiopeia A or Sagittarius A. The 1667 Mc/s absorption profiles for these two sources are shown in Figures 7 and 8. Early attempts by the above authors and by Penzias (1964) to observe emission from OH were unsuccessful.

The weaker lines at 1612 and 1720 Mc/s have been detected by Gardner, Robinson, Bolton, and van Damme (1964) in the spectrum of Sagittarius A—again in absorption. On this occasion, the lines were found in the interstellar gas before their detection in the laboratory.

In the spectrum of Cassiopeia A, several distinct components of OH absorption have been observed in the velocity range between zero and -50 km/sec. The reported optical depths lie in the range 0.010 to 0.016. Taking into account the observed linewidth in each case, these values lead to figures of 10^{14} to 10^{15} for the number of OH radicals in a 1-cm^2 column along the line of sight.

The abundance ratio of OH relative to H near the Sun was thus found to be of the order of 10^{-7}, as expected from optical considerations. (The ratio of the observed absorption dips for OH and H is much larger, about 1 to 2 per cent, because of the higher transition probability for OH.) However, there was already clear evidence in the early results that the OH/H abundance ratio varies considerably from place to place. A component that is strong in the H profile may be absent or weak in the OH case, and vice versa. Some of the observed differences between the two absorption profiles must arise from the greater opacity of hydrogen, as a result of which the nearer portions of a cloud complex are relatively more important for H measurements than for OH, but the observations do imply real variations in relative abundance.

More surprising results were found for the galactic center region in the observations of Sagittarius A. Robinson, Gardner, van Damme, and Bolton (1964) reported an intense concentration of OH in this region, which gave a broad absorption line centered on a velocity of +40 km/sec. Shortly afterward, Goldstein, Gundermann, Penzias, and Lilley (1964) found another intense concentration centered on −120 km/sec. These two features can be followed in absorption over several degrees of longitude, and have only minor counterparts in the hydrogen pattern (Bolton, Gardner, McGee, and Robinson 1964). Detailed comparison is difficult, but clearly the OH and H distributions are very different in the galactic nucleus; also, the ratio of OH to H seems to be two or three orders of magnitude greater in parts of the central region than in the solar neighborhood. This might be due to some process that leads to a greater production of OH radicals near the center, or, alternatively, the oxygen in the solar neighborhood might be largely removed from the gas by attachment to dust grains.

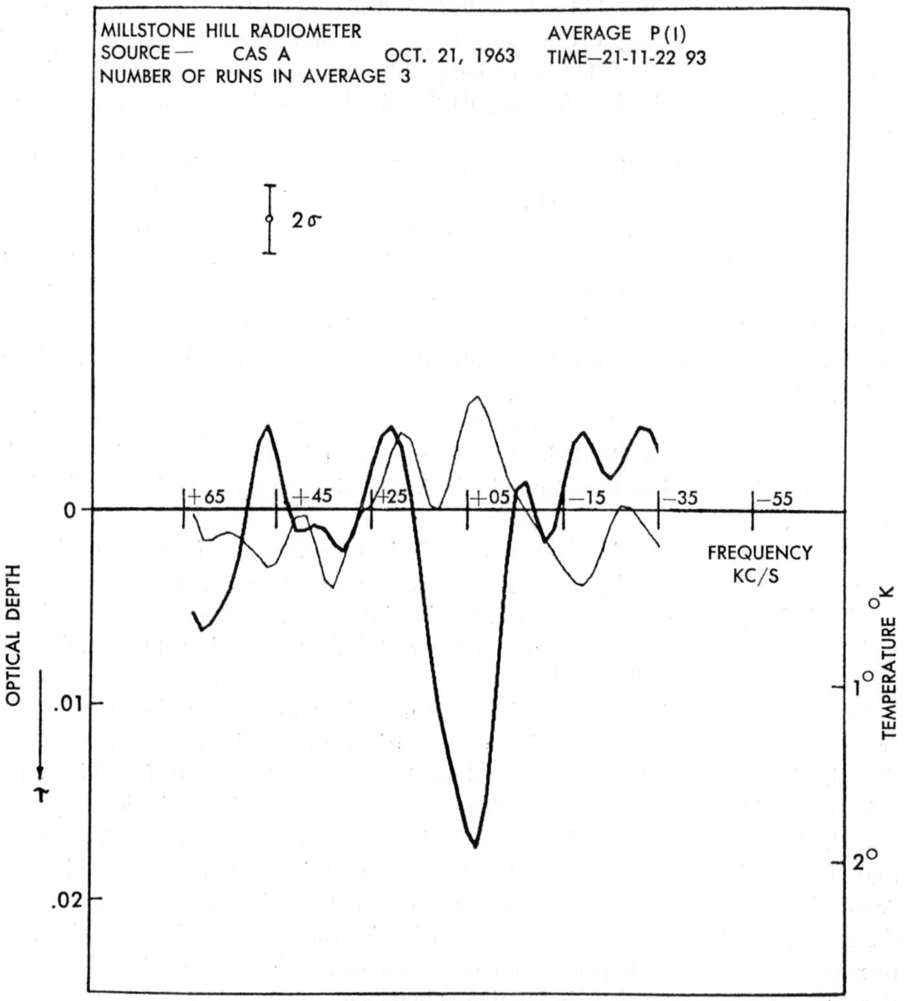

FIG. 7.—OH absorption profile for Cassiopeia A for the 1667 Mc/s line (Weinreb, Barrett, Meeks, and Henry 1963). The lighter line is a comparison profile for a point slightly displaced from the source. The frequency scale is specified with respect to the local standard of rest assuming the line rest frequency to be 1667.357 Mc/s.

The profiles of the four lines observed in the Sagittarius A direction show only minor differences in shape, but the intensity ratios are only 1:2.2:2.7:1 (Gardner *et al.* 1964), as compared with the theoretically expected values (Table 1) of 1:5:9:1. Robinson *et al.* (1964) proposed that this is an opacity effect arising from a distribution of the OH in a large number of dense blobs, of optical depth $\sim$3, arranged in such a way as to give the observed over-all apparent optical depth of 0.5 to 0.9. However, rather special velocity and spatial distributions may be necessary to account for the whole profile in this way. Barrett and Rogers (1964) pointed out that resonant coherent scattering has never been considered in the radiation theory for the radio domain, but it may be

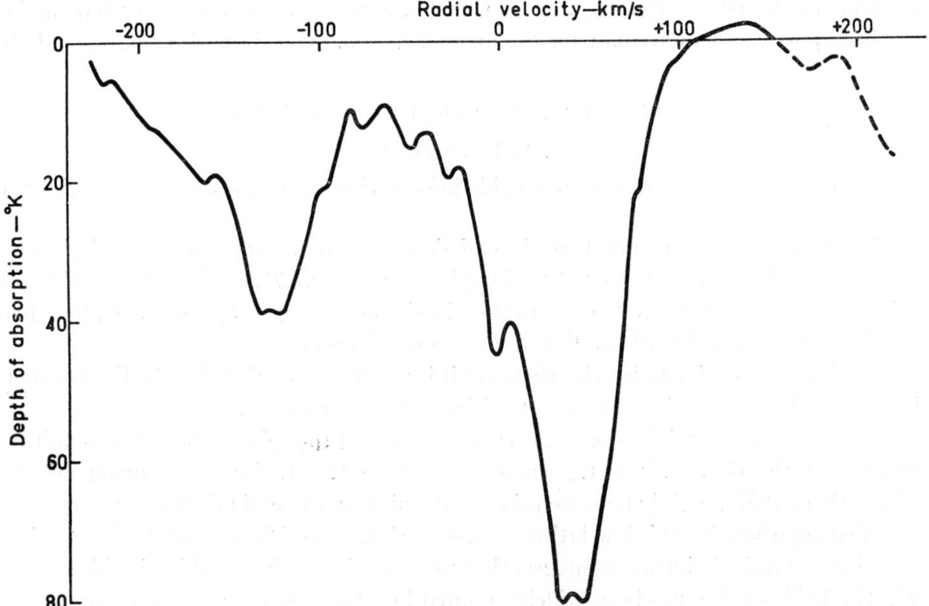

FIG. 8.—OH absorption profile (1667 Mc/s line) for Sagittarius A (Bolton, Gardner, McGee, and Robinson 1964).

important when the distance between absorbers is large compared with the wavelength and the radiation field is not isotropic. The scattering parameter would be different for each of the four lines.

The linewidths reported for OH vary from 4.2 kc/s (between half-intensity points) for a narrow component in Cassiopeia A to about 300 kc/s for one of the broad features in Sagittarius A. For a given temperature, the thermal velocities will be less for OH than for H by a factor 3.5, owing to the greater mass of the OH radical. Barrett, Meeks, and Weinreb (1964) have compared the OH and H linewidths for two narrow and closely spaced components in Cassiopeia A, and have separated out the contributions to the linewidths that arise from thermal and turbulent motions. For this source, it seems reasonable to assume that the OH and H lines are coming from the same gas clouds, and that the kinetic temperatures and the turbulent motions are the same for OH and H. The derived thermal widths correspond to kinetic temperatures of 90° K and 120° K for the two clouds; these values are close to the spin temperatures (and hence kinetic temperatures) commonly derived from neutral hydrogen observations. Barrett *et al.* point out that the smaller thermal broadening gives OH observations an intrinsically higher resolution for studies of turbulent and other motions.

OH emission was not detected in the early observations but was found in several H II regions when narrow-band (2 kc/s) measurements were undertaken by Weaver, Williams, Dieter, and Lum (1965). The emission components studied by these and other authors are very narrow in frequency and are often quite strong; they also exhibit complicated polarization effects. The intensity ratios of the four lines show large variations from one emission component to another. The populations of the various states appear to show big departures from equilibrium values, and an interstellar maser-type action has been postulated to account for the anomalous line ratios in both emission and absorption.

7. PHYSICAL PROPERTIES OF THE GAS

7.1. Temperature

Hydrogen-line observations can provide information on temperature in two different ways:

(1) The brightness temperature (in emission measurements) and the optical depth (from absorption) are both related to the spin temperature, T_s. In the galactic disk, collisions are believed to keep T_s close to the kinetic temperature, T_K (Sec. 3.3), but there are circumstances under which T_s and T_K can be different.

(2) T_K can be derived directly whenever it is possible to distinguish the line broadening due to thermal motions from that arising from other causes.

In most work so far, it has been the custom to regard the spin temperature as uniform throughout the Galaxy. The Leiden workers chose 125° K for T_s, assumed uniform (Westerhout 1957), mainly because this was the highest value of brightness temperature recorded anywhere in their low-latitude survey (Muller and Westerhout 1957) and there was circumstantial evidence to suggest that the galactic opacity was high in this range of T_b. The Leiden value has been widely adopted by other workers; all results to date on the distribution of hydrogen in the Galaxy depend on the uniform-temperature assumption because there is no easy way to handle a varying temperature when there are so many other uncertainties in the problem.

It is most unlikely however that T_s will be the same everywhere, and under all conditions to be found in the Galaxy. There could for example be a large-scale variation of T_s with distance from the galactic center; there could be a systematic difference between spiral arm situations and inter-arm regions; and there could be small-scale variations from cloud to cloud.

Several authors have reported small regions in which the brightness temperature over a small velocity range is lower than that in neighboring regions (Heeschen 1955; Davies 1956; Radhakrishnan 1960; Locke, Galt, and Costain 1964*a*). They all attribute the reduced values of T_b to self-absorption produced by a cool cloud with a T_s of about 60° K. Figure 9 shows one such case, in which a scan across a "cold" region is compared with a record of discrete-source absorption. Davies and Radhakrishnan state that their regions of low T_b coincide in position with dense dust clouds, which might be expected to cool the hydrogen in the vicinity.

In assessing these results, we may say that self-absorption is a quite possible mechanism, but in individual cases it is very difficult to distinguish between a cool cloud and a local deficiency of hydrogen. There is not yet enough statistical evidence on the frequency of occurrence of such dips and their distribution in relation to other galactic constituents. Absorption observations give the best possibility of establishing temperature variations, because absorption lines are sharper and might therefore show thermal effects directly; also T_s can in principle be derived from absorption measurements of opacity. Muller (1959) was the first to record an absorption line which was sufficiently

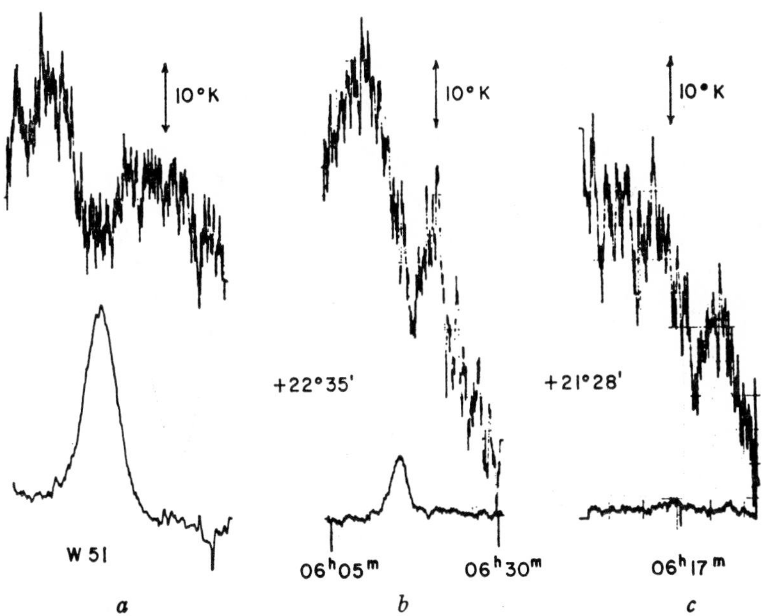

FIG. 9.—Samples of different types of 21-cm absorption: line temperature above and continuum temperature below, with gains adjusted to give roughly equal scales. (*a*) Absorption of continuum radiation from source W51. (*b*) Absorption of line radiation from background hydrogen by a cool cloud near IC443. (*c*) Same as (*b*), but at a different declination.

narrow to suggest that thermal broadening alone might be involved. On this interpretation, the width of one component in his absorption profile for Taurus A corresponded to a temperature of 125° K.

Muller also introduced the method of analyzing an absorption spectrum into a number of components, each having a gaussian distribution of optical depth with velocity. This method has been extended by Shuter and Verschuur (1964) and Clark (1965). An example of Shuter and Verschuur's resolution process is shown in Figure 10, in which the profile for Taurus A has been split into nine components. The separation process is a rather subjective one, especially in regard to the number of components to be used; also, the results depend rather critically on the derivation of the "true" absorption from the observed profile, and thus on the accuracy of the expected profile (Sec. 6.2).

From a total of 21 components in three sources, Shuter and Verschuur quote a median value of 3.5 kc/s for the velocity dispersion in a single component, corresponding to

67° K if only thermal effects are present. Clark observed with a broader bandwidth and also used a more conservative method in fitting his gaussian components. He obtained a median dispersion of 8 kc/s (Fig. 11), equal to 345° K on a thermal interpretation, but he concluded that some of the dispersion is probably due to microturbulence. Some of his components would however correspond to temperatures of 50° to 100° K. It is notable that in each case the distribution curve for component dispersions shows a pronounced maximum near the bandwidth used, 3 and 6 kc/s, respectively.

The other possible method of obtaining temperature information from absorption

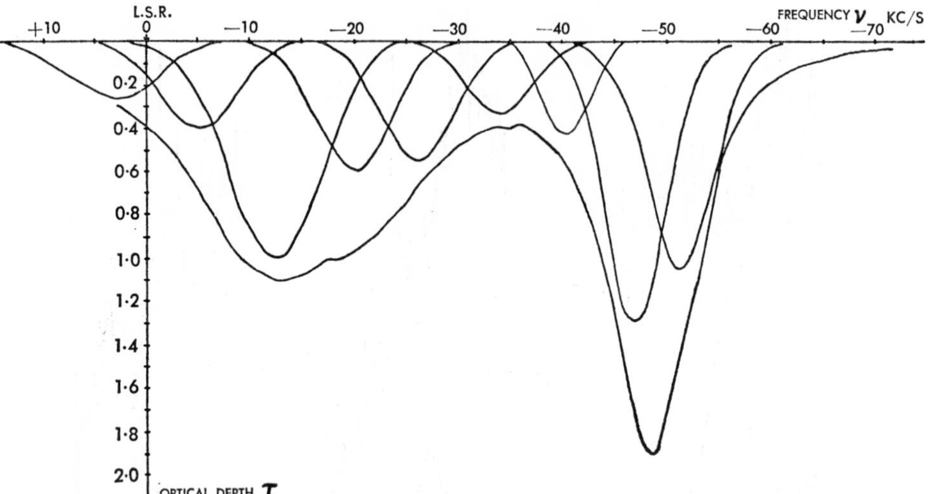

Fig. 10.—Taurus A absorption optical-depth profile with suggested resolution into gaussian components (Shuter and Verschuur 1964).

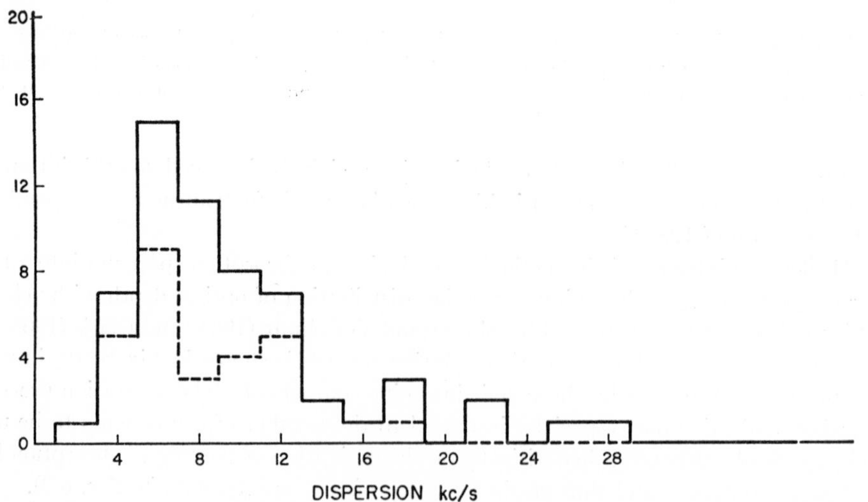

Fig. 11.—Histogram of velocity dispersion of gaussian components of absorption profiles (Clark 1965).

observations is to derive T_s from a comparison between the absorption opacity and the corresponding emission brightness temperature. This method has had only partial success. In some cases (e.g., Cassiopeia A, Fig. 7), widely different values of T_s are indicated by the data at different velocities, and it is clear that the absorption and emission results are not really comparable.

The very large differences between absorption and emission profiles are in fact the best available evidence for the existence of temperature variations from point to point, as the other evidence discussed above cannot be given high weight. Part of the difference between absorption and emission must arise from variations across the antenna beam, but this cannot explain the whole difference in most cases. If the clouds were small enough in angular size for beam effects to be important, and if the same clouds were effective in absorption and emission, we would expect sharper peaks in the distribution of emission than are in fact observed. Striking differences can however be expected if hydrogen clouds do exhibit a range of temperatures, because it is clear that absorption observations will preferentially pick out the cooler clouds; they will have the highest opacity, and emission observations will give greater weight to the hotter clouds.

Shuter and Verschuur (1964) have considered one possible cloud model in which a wide range of temperature exists. Clark (1965) has developed a "raisin pudding" model, with a population of cool, dense clouds ($T_s < 100^\circ$ K) immersed in a general medium at a high temperature ($T_s > 1000^\circ$ K). In this model the cool clouds are responsible for all absorption, but only a part of the emission; about half the gas is in the cool clouds, and half in the hot medium.

The over-all conclusion at the present time is that there probably are variations in temperature between different clouds, but a precise model cannot yet be set up. The value of 125° K should now be regarded as some suitable mean of the individual cloud temperatures, and the integrated opacity of any concentration of gas would in general be higher than that derived from a uniform-temperature model. We can conclude that existing figures for hydrogen densities and masses will all have to be revised upward.

Observations of OH have added to the picture derived from the hydrogen results. The main evidence comes from the unexpectedly low intensity ratios among the four lines of the OH group in the galactic center region. A likely interpretation is that the OH absorption takes place in small clouds, which at the galactic center distance fill only a fraction of the antenna beam (see Sec. 6.5). These may be analogous to the cool clouds of hydrogen discussed earlier; so that both H and OH observations point to the existence of an absorbing population of cool and dense clouds of gas. The OH excitation temperature seems to be much lower ($<10^\circ$ K) than that for hydrogen, but the corresponding kinetic temperature could well be in the 40° to 60° region suggested by some of the hydrogen data.

One result seems to contradict the picture of cool absorbing clouds. This comes from the separation (Barrett, Meeks, and Weinreb 1964) of thermal and turbulent velocities in Cassiopeia A absorption, through an H/OH comparison, which leads to kinetic temperatures of 90° and 120° K. These figures may however have been raised by lack of detailed agreement in the H and OH distributions, and saturation effects in the dense clouds may have produced more broadening than had been taken into account in the normal method of reduction.

7.2. Number of Atoms; Mean Density

One of the simplest quantities which can be obtained from hydrogen-line emission observations is the integral under the profile. As long as the gas is optically thin, this integrated brightness directly gives the number of atoms N in the line of sight without any need to know the temperature of the gas. However, as we have seen in the last subsection, there is evidence for temperature variations throughout the gas, which implies that there will generally be a substantial opacity in some of the contributing clouds. Under these circumstances, any value of N derived from integrated brightness must be regarded as a lower limit to the total number of atoms. With this proviso, Table 3 lists some representative values of N that indicate orders of magnitude from H and OH observations in terms of the number of atoms in a 1-cm² cylinder stretching to infinity; other information is required before the distance or space density of any part of the gas can be estimated. As the OH data are from absorption only, it was necessary to assume an excitation temperature (taken as 10° K) to derive values for N_{OH}.

TABLE 3

Representative Values of N in a 1-cm² Column, Assuming Low Opacity

	N (cm^{-2})
Hydrogen emission	
Minimum detectable value	10^{19}
Single cloud	10^{20}
Individual gaussian components	10^{20}
Whole profile, high latitudes	3×10^{20}
Whole profile, intermediate latitudes	10^{21}
Whole profile, anticenter	10^{21}
Whole profile, galactic plane, $l^{\mathrm{II}}\approx90°$	10^{22}
Whole profile, center	3×10^{21}
OH absorption	
Minimum detectable value	10^{13}
Path from Cas A	2×10^{14}
Path from Sgr A	3×10^{17}

The mean hydrogen density in the solar neighborhood can be derived from observed N-values, but only if we can introduce a distance scale. Westerhout (1957) and McGee and Murray (1961) have done this by assuming that the gas is uniformly stratified and adopting Schmidt's value (1957) of 220 pc for the thickness between half-density points. Westerhout obtained 0.6 H atom/cm³ and McGee and Murray, 0.5 atom/cm³, but again, these values must be taken as lower limits. Higher-resolution observations (Kerr 1964*a*) indicate that the layer thickness is smaller than was previously thought, but the recent increase in the conventionally adopted scale of the Galaxy approximately compensates for this.

Comparison between N values for H and OH in the direction of Cassiopeia A indicates that the mean OH density in the solar neighborhood is about 10^{-7} radical/cm³.

Small-scale variations of hydrogen density are considered in Section 7.4, and large-scale structure effects in Volume V of this compendium (Kerr and Westerhout 1965).

7.3. Motions

Neutral-hydrogen velocity measurements give information on four types of motion: galactic rotation, mass motions of large concentrations, peculiar motions of individual gas clouds, and internal motions within clouds. We are not concerned here with the first two of these, which are discussed in Volume V of this series, but only with the smaller-scale effects.

Several authors have investigated the distribution of the peculiar motions of a group of clouds, usually referred to as random motions. From analyses of Adams' (1949) measurements of calcium clouds, Blaauw (1952) and Takakubo (1958) found that an exponential function

$$\tau_0 \exp\left(-\frac{|V-V_0|}{\eta}\right)$$

gave the best fit to the observed distribution, with η between 5 and 7 km/sec. For hydrogen, the available evidence does not give a clear choice between alternative distribution functions. Pottasch (quoted by Westerhout 1957) found a slight preference for a gaussian function

$$\tau_0 \exp\frac{(V-V_0)^2}{2\sigma^2}$$

with $\sigma = 6$ km/sec in a study of the "unpermitted" wings of low-latitude profiles. However, Clark (1965) could find no significant preference for an exponential or gaussian distribution in the velocities of the individual components in his absorption profiles, and Takakubo (1963*a*) could not decide between exponential, gaussian, and turbulence functions in the distribution of a large number of gaussian components of emission profiles. His best values for the functional parameters were $\eta = 4.5$ and $\sigma = 7$ km/sec.

The difficulty in fitting a distribution function to the hydrogen velocities largely arises from the lack of a clear separation between the random motions and the larger systematic motions. Also there are indications that the distribution in fact may be a complex one. Some otherwise narrow profiles have long tails, which suggests that there may be two or more populations of clouds, with different patterns of motion.

Internal motions in clouds seem to follow a gaussian distribution. This is best shown by the shape of the narrowest and most isolated absorption lines, but isolated peaks on emission profiles also commonly have a gaussian form. This is the justification for the attempts that have been made by various authors to split profiles into gaussian components.

We have already discussed the velocity spread of some of these components to consider whether we ever reach the point where only thermal broadening effects are present. Most components that have been isolated so far seem to show some turbulent, or non-thermal, motion in addition.

One of the narrowest emission profiles yet reported is that described by Goldstein (1964*b*) and Goldstein and Welch (1966). Over an area of several square degrees, they find a single peak, approximately gaussian in shape, with a width between half-intensity points of only 25 kc/s, or 5.3 km/sec. As the receiver bandwidth was only 0.5 km/sec, instrumental broadening can be neglected. The linewidth is too large to be wholly thermal in origin but can be taken as a good indication of the motion inside one type of hydrogen cloud. Goldstein looked for a systematic pattern in the internal motion but could not draw a firm conclusion.

As we have seen, some absorption components show dispersions in the range from 0.6 to 2 km/sec. Similar values have been obtained from studies of gaussian components derived from emission profiles. Takakubo and van Woerden (1966) found dispersions ranging upward from 1 km/sec (width between half-intensity points, 2.4 km/sec) in their intermediate-latitude survey, after correction for an instrumental dispersion of 0.8 km/sec. A few measurements with a narrower bandwidth seemed to confirm that 1 km/sec was a real lower limit. In a study of the Orion region, van Woerden and Schwarz (1966) found a pronounced maximum at 2 to 3 km/sec in the distribution of dispersions, with none as low as 1 km/sec. Mrs. Dieter (1965), working near the north galactic pole, derived components with dispersions between 1.0 and 5 km/sec, averaging 2.4 km/sec. Her bandwidth ($\sigma_{\text{instr}} = 0.2$ km/sec) was well below the lowest dispersion quoted.

Within the limitations of the gaussian component technique, evidence is accumulating that the lowest dispersion for emission components is around 1.0 km/sec, implying a kinetic temperature for these clouds $\leq 120°$ K. Although the motion may be substantially thermal, the dispersions for most components probably contain both thermal and turbulent contributions.

Narrow-band hydrogen observations have now outstripped the major optical interstellar line studies in resolving power, but Livingston and Lynds (1964) have recently gained a large factor in resolution through the use of a large telescope and an image tube. They have found some components of Ca II lines with dispersions below 1 km/sec. There are of course various reasons why circumstances might be different for H and Ca lines, but more high-resolution H observations should be carried out.

7.4. Cloud Structure

There is abundant evidence for the existence of density fluctuations in the interstellar gas on several different scales. The word cloud is often used to describe a localized concentration, but the usage is generally rather loose, both as regards the order of size of a cloud and its degree of discreteness. We can ask, for example, whether there is a single preferred size (or a hierarchy of preferred sizes) of rather discrete clouds, or whether there is a continuous size spectrum of features which may be regarded as eddy cells in a turbulent medium.

We have seen in previous sections that rather discrete hydrogen clouds can be seen, especially in absorption observations. It seems that at least the denser clouds are distinct features, standing out sharply from the general medium. This conforms to the picture obtained from the optical appearance of dust patches and bright emission nebulae. Whether the word cloud should be used for the dense absorbing features, or for smaller elements still, is a matter of usage, and knowledge has not yet reached the stage where it can dictate the most logical usage.

It is more difficult to say whether the hotter and less dense units that are important for emission can be regarded as discrete clouds with rather empty space in between. One definite, isolated cloud is the sharp-line feature studied by Goldstein (1964*b*) and Goldstein and Welch (1966) and discussed above. As this feature in Serpens and Libra extends over at least 6° × 18° and its peak brightness is high (46° K), it is presumably quite close. Goldstein suggests that its distance is between 5 and 50 pc; its shorter dimension would then be in the range of 0.5 to 5 pc, and the mass of hydrogen in the cloud would equal 1 to 100 $\mathfrak{M}_{\odot}$.

Mrs. Dieter (1964), in her north galactic pole study, found several major condensations with typical dimensions of about 5° and $N_H = 2 \times 10^{20}$ atoms/cm². If such a cloud is at 100-pc distance, it would have a diameter of 7 pc, a density of 10 atoms/cm³, and a hydrogen mass of 30 $\mathfrak{M}_{\odot}$, but its distance is of course quite uncertain.

The most extensive investigation of the cloud structure so far published is that carried out by the Groningen group. In breaking up their intermediate-latitude profiles into gaussian components, they attempted to isolate the smallest discrete features. Gaussian analysis of emission profiles has been used by several workers (Matthews 1957; Davis 1957; Muller 1957, 1959; Menon 1958; Grahl 1960), usually by simple hand-computation methods. The Groningen workers developed the method further, by successive approximations on an electronic computer, with a least-squares check on the fit of each trial solution. This made the process less subjective, but there were still cases where it was found that very different gaussian solutions gave approximately equal fits to the data. The mean weakness of the method is that it is necessary to assume a constant temperature in converting to optical depth for the separation. In consequence, the detailed parameters cannot be regarded as very precise, but the technique does provide statistical data of fairly uniform style for a large-scale treatment.

An example of these results is shown in Figure 12 (Takakubo 1963*a*). The continuity of the various components indicates the reality of the various features and also provides typical angular sizes for clouds. As usual there is no direct knowledge of distance, but systems of model clouds were examined in attempts to get the best fit to the observational data.

Takakubo (1963*a*) first assumed a system of spherical clouds, all of equal size and uniform density, and found the following cloud parameters: diameters, 6.8 pc; density, 14 atoms/cm³, mass, 55 $\mathfrak{M}_{\odot}$; average angular diameter, 1°2; number of clouds along a line of sight, 11/kpc; fraction of space filled, 5 per cent. A second model, in which the radius and density of the clouds are inversely proportional, gave results which were not greatly different. Terauti (1963) examined a model in which the clouds departed from uniform density by having a central condensation, but he could not obtain good agreement with the observations.

The assumption of equal clouds is clearly an artificial one in view of the evidence on variations in size and internal velocity dispersion, but it has given useful orders of magnitude. A disturbing factor, however, is that the cloud dimensions deduced from this model are close to the beamwidth that was used; therefore, one wonders how significant the value really is. Smaller structures certainly exist in ionized hydrogen regions—some of the finest filaments and "cometary" condensations are as narrow as 200 a.u., or 0.001 pc (van de Hulst 1958). Elements of this size may not necessarily exist in neutral hydrogen, but no one has yet looked for them.

Absorption observations cannot give direct evidence on cloud diameters, because they are limited to a small number of points in the sky. They can, however, give a figure for the number of clouds along each line of sight. Adding up the number of absorption com-

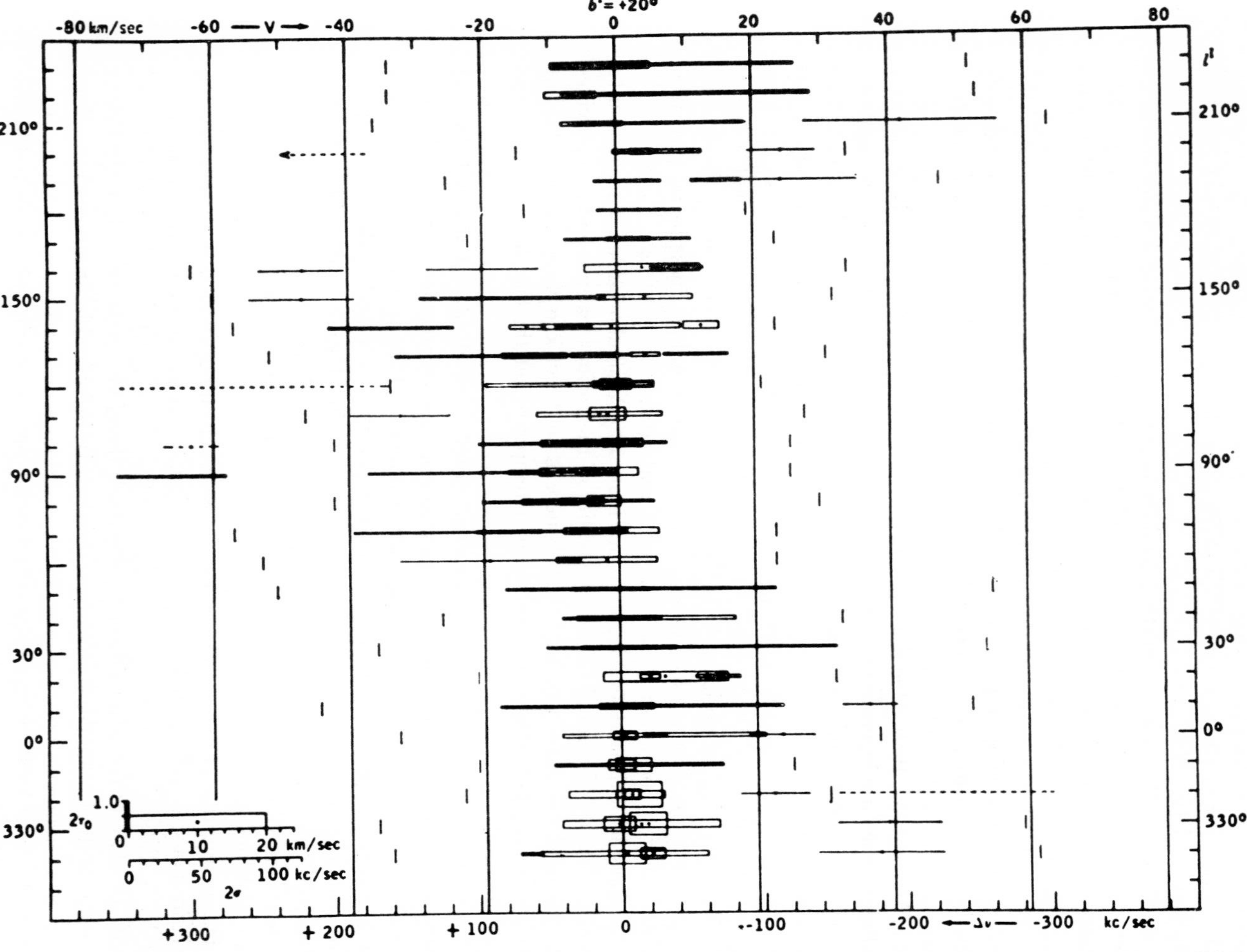

FIG. 12.—Gaussian components of emission profiles as a function of galactic longitude, l^{I}, and radial velocity, V, for $b^{\mathrm{I}} = +20°$. A component is represented by a rectangle with a dot at the center whose position shows l^{I} and V; the height of the rectangle is proportional to the central optical depth, and the half-width is equal to the velocity dispersion (Takakubo 1963*a*).

ponents in all his sources of known distance, Clark (1965) obtained a mean of 4.1 clouds of all observable densities per kiloparsec.

The absorption data also show the variations between clouds. Both Shuter and Verschuur (1964) and Clark (1965) found that the optical depth of their absorption components followed an exponential frequency distribution (see Fig. 13).

We have so far been concerned with the small, fairly discrete, clouds. Both radio and optical evidence show that these are frequently associated in larger complexes from 50

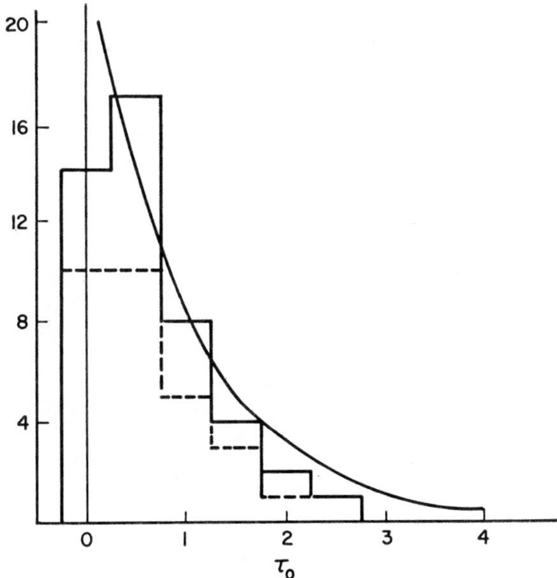

Fig. 13.—Histogram of central optical depths of gaussian components of absorption profiles (Clark 1965).

to 100 pc in diameter. There are also larger condensations of hydrogen, elongated in form, and about 1000 to 1500 pc in length (McGee 1964; Kerr 1964*b;* Burke, Turner, and Tuve 1964). These condensations are in turn the major elements of a spiral arm.

8. CORRELATION WITH OTHER POPULATION I CONSTITUENTS

8.1. Star Groupings

Many studies have been made of the neutral hydrogen content of galactic clusters and associations. Raimond (1966) discusses 22 papers on this subject in the introduction to his own extensive work on two associations in Monoceros. We will only be concerned here with general conclusions from the earlier work and with the most recent studies.

A difficulty in investigations of this type is to separate the hydrogen connected with a cluster or association from the effects of background irregularities. Some of the earlier studies can be criticized from this point of view, either for too low a telescope resolution or for insufficient exploration of the background surrounding the cluster. Davies and Tovmassian (1963*a*) attempted to minimize these uncertainties by selecting clusters of small diameters (10′–20′). They also had the advantage of using a large telescope. Raimond's two associations in Monoceros have large diameters, but he extended his observations over a large area to investigate the scale of the irregularities.

Davies and Tovmassian reported neutral hydrogen at the position of four galactic clusters out of the five observed. There was a tendency for the more massive clusters to retain a greater proportion of their gas, but the background irregularities make estimates of the hydrogen masses rather uncertain. Evidence was also reported for expanding shells surrounding the ionized hydrogen in the Trapezium-Orion cluster, NGC 2244, and NGC 6910. Expanding shells of neutral hydrogen around clusters were also described by earlier workers (Wade 1957, Menon 1958).

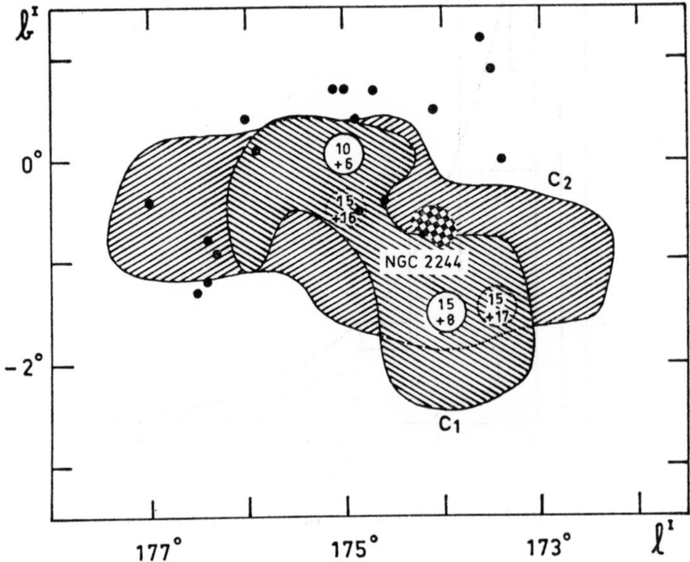

Fig. 14.—Two hydrogen clouds C_1, C_2 (*hatched areas*) in the region of the cluster NGC 2244 (Raimond 1966). For the two clouds, the upper number of the ringed pairs (*hatched circles*) indicates the maximum surface density of hydrogen atoms in units of 10^{20} cm^{-2}; the lower number indicates the radial velocity of the peak. Secondary maxima (*open circles*) are also indicated. The average surface density over the hatched area is half the peak value in each instance. The cluster (*cross checks*) and some members of the association I Mon (*filled circles*) are also shown.

Raimond (1966), in his study of a large region in Monoceros, was able to identify 11 concentrations of neutral hydrogen of a rather discrete character. Three of these appeared to be related to the association II Mon. The total mass of neutral hydrogen was about 20,000 $\mathfrak{M}_{\odot}$, an order of magnitude greater than the estimated stellar mass of the associated cluster NGC 2264.

In I Mon, Raimond found two clouds in different velocity ranges covering the areas shown in Figure 14. The total mass of the hydrogen complex in this case is 150,000 $\mathfrak{M}_{\odot}$, with an over-all diameter of 110 pc and an average density of 8.5 atoms/cm^3. Girnstein and Rohlfs, in an earlier study (1964), suggested that the two components were part of an expanding shell, but Raimond was less certain of this interpretation, mainly because the detailed arrangement of the hydrogen in the region could not be fitted to the parameters of a simple shell model. However, analogy with shell-like structures in the Magellanic Clouds (Hindman 1967, McGee and Milton 1966) suggests that too much regularity should not be expected in an expanding shell that is several million years old.

Inside the region of I Mon are the cluster NGC 2244 and the Rosette Nebula. In some observations made in cooperation with McGee, Raimond found evidence for a thin shell of neutral hydrogen around the Rosette Nebula whose age is $\lesssim 10^5$ years. Thus there seem to be two subgroups of stars and hydrogen clouds, possibly arising from two explosions at different epochs.

The Perseus region has been examined in detail by Grahl (1960), and Hack, van Woerden, and Blaauw (1966), who find a region of rather low density close to the large association h and χ Persei. The latter authors, however, discuss two isolated features at moderate velocities which they suggest may have been ejected from the association 10^7 years ago—perhaps by a thermal expansion or by the explosion of a large number of type II supernovae. If the features are at the distance of the association (2.2 kpc), they would have masses of 45,000 and 140,000 $\mathfrak{M}_\odot$.

So far we have been considering OB associations, but associations of T Tauri stars would also be expected to contain substantial amounts of gas. Sholomitskii (1962) studied the T-association Tau T2 and reported hydrogen at a velocity which agrees with results obtained from optical absorption lines. The angular size of the gas subsystem considerably exceeds the optical size, and the estimated mass of hydrogen, 200 $\mathfrak{M}_\odot$, exceeds the stellar mass by at least an order of magnitude.

The spatial relationships between stars and gas on a larger scale have been examined on numerous occasions through comparisons between the spiral structure patterns derived for each. Most of these comparisons have been inconclusive, largely because neither spiral pattern is very clearly defined. Also, the stellar and hydrogen patterns are often found to have been drawn on the basis of different distance scales. Muhleman and Walker (1964), for example, found a negative correlation between stars and gas in the Cygnus region, but their results were affected by distance problems. On the other hand, Fletcher (1963) based his comparison entirely on velocities, independent of any model, and found a strong correlation between the observed radial velocities of stars and gas.

The most direct comparison has been obtained by McGee (1964) under the somewhat simpler conditions in the Large Magellanic Cloud. He showed that there was a close relationship between the velocities of 54 supergiant stars and those of the hydrogen in the same positions. This result gives strong evidence that young stars and gas are found together in the LMC, and by analogy in the Galaxy also, although it must always be remembered that conditions are not necessarily the same in the two systems.

Globular clusters are usually believed to contain little or no gas because of their great age. However, they are very convenient objects for observation, and several attempts have been made to find neutral hydrogen in a globular cluster (Roberts 1959, Goldstein 1964*a*, Robinson 1966), all without success. Goldstein has set an upper limit of about 150 $\mathfrak{M}_\odot$ on the H I content of M17, and Robinson a limit of 50 $\mathfrak{M}_\odot$ for 47 Tuc and ω Cen.

8.2. Dust

A general correspondence is found between low-velocity (i.e., nearby) hydrogen and regions of obscuration (Christiansen and Hindman 1952; Davies 1956; McGee, Murray, and Milton 1963). Both distributions show a strong concentration toward a thin galactic disk, with local bulges toward higher latitudes in Ophiuchus–Scorpio and in Taurus–Orion–Auriga.

When more detailed relationships are studied, however, the situation becomes more complicated, with correlations obtained in some cases and not in others. The main difficulty, as usual, is that the radio and optical observations that are being compared do not refer to the same regions of space. A radio-line profile is averaged over a line of sight that traverses the whole Galaxy, whereas the optical data, especially those specifically concerned with obscuration, are derived from very nearby regions.

For example, Lilley (1955), in a low-resolution study of the anticenter region, found a good correlation between the peak brightness temperature of the hydrogen and the long-distance optical obscuration, as indicated by the galaxy counts of Hubble and Shapley, whereas the correlation was only fair with star counts, which refer to a smaller distance through the Galaxy.

A positive correlation for a smaller area was obtained by Raimond (1966). Several of the discrete hydrogen clouds that he isolated from the background in Monoceros coincided in position with dust clouds. Davies (1956) reported a correlation of a different kind, in which holes in the hydrogen distribution seemed to be related to dark clouds with very similar positions and angular sizes; one of these relationships was in Auriga, and six were in Cygnus. Davies argued that a true hole in the hydrogen would not show so prominently, because the large contribution from the background would be unaffected. Instead he concluded that a high density of hydrogen was associated with the dust cloud, but the temperature was much lower than normal, say, 60° K. Possible temperature variations were considered in Section 7.1.

On the negative side, McGee, Murray, and Milton (1963) found that the 21-cm contours from their general sky survey did not agree in detail with areas of obscuration. Also, some individual dense dust clouds do not show increased hydrogen emission compared with neighboring areas (van de Hulst, Muller, and Oort 1954; Bok, Lawrence and Menon 1955). Some of these differences may arise because only the atomic hydrogen is seen, and the hydrogen may be largely molecular in the densest clouds.

Several authors have made estimates of the density ratio of gas to dust; their results are summarized in Table 4. Most of these estimates show reasonable agreement, but it is clear from the earlier discussion that there can be considerable uncertainty in any comparison between gas and dust. In particular there may be large variations from cloud to cloud. By far the best method of comparison is to consider transmission through the whole Galaxy by using counts of external galaxies. This method has not yet been applied in a systematic way to high-resolution radio observations and the most detailed optical data.

Schmidt (1964) has estimated the gas-to-dust ratio along the major axis of M31 from the available measurements. He obtained a value of 200 for an outer region corresponding to the Sun's position in the Galaxy, with the ratio decreasing to 10 near the center.

8.3. Interstellar Calcium

Relationships between velocities of hydrogen emission lines and calcium absorption lines were first found by Lawrence (1956), and more extensively studied by Howard, Wentzel, and McGee (1963) and Takakubo (1963*b*, 1965). One of the major objects of such comparisons is to see whether the neutral hydrogen and other gaseous constituents exist in the same clouds, but complete agreement can never be expected between the two sets of results. A Ca-II line observation refers to the negligible solid angle presented by the disk of a star, whereas a hydrogen profile is always averaged over the whole antenna beam. Also, the optical line of sight is shorter, as it extends only from the star. Finally, Ca II may be found in H II as well as H I regions.

TABLE 4

Estimates of Density Ratio of Gas to Dust

Author	Region	Method		Mean ρ_{gas}/ρ_{dust}
		Gas	Dust	
Lilley (1955)	Taurus–Orion–Perseus	21-cm emission	Galaxy counts	100
Davies (1956)	Auriga and Cygnus	21-cm emission	Model cloud	300
Ampel and Iwaniszewska (1962)	Cygnus	21-cm emission	Obscuration	8–35
Metik (1963)	Aquila–Sagitta and Cassiopeia	21-cm emission	Obscuration	100
Pronik (1963)	NGC 6618	21-cm emission	Reddening of 6 O–B0 stars	70–150
Brodskaya (1963)	Crab Nebula	21-cm absorption data of Clark *et al.* (1962)	Reddening	120

Howard *et al.* surveyed the directions toward 39 stars. To keep conditions as simple as possible, he selected stars having $|b| > 15°$, and single-component Ca II lines in the list of Adams (1949). In spite of the angular disparity between a stellar disk and a 2°2 beam, the difference between hydrogen and calcium velocities measured in the same direction had a dispersion of only 3 km/sec over this sample.

A more detailed comparison was carried out by Takakubo (1966), using hydrogen data with a higher resolution in both angle and velocity, from the Groningen-Dwingeloo intermediate-latitude survey. One such comparison is shown in Figure 15, where a hydrogen profile is compared with a Ca II profile also recorded by Takakubo (1963*b*). Both profiles have been broken up into gaussian components.

The main comparison was between the list of H I gaussian components and Adams' list of Ca II *K*-components. For 55 stars, where clear comparisons could be made, the low-velocity peaks in Ca and H agreed very well in velocity. The root-mean-square difference between V_{Ca} and V_H was 1.7 km/sec, or 1.05 km/sec when four special cases were excluded. In many of these comparisons there was a displacement of several degrees between the directions of the star and the nearest available hydrogen profile. The close agreement of the velocities implies that the gas clouds concerned must have a substantial angular extent. Galactic rotation considerations show that the Ca and H producing these low-velocity components must be statistically at the same distance, the mean value being 180 pc. As the agreement is well within the dispersion of the hydrogen velocities over the antenna beam, we can also conclude that even the *K*-components must have had contributions from several individual clouds.

The higher-velocity Ca and H components do not agree so well, which probably means that the gas clouds concerned are smaller in size. They cannot be at great distances, because they must all be closer than the corresponding stars. The higher-velocity components are usually rather weak, again suggesting that the clouds are small.

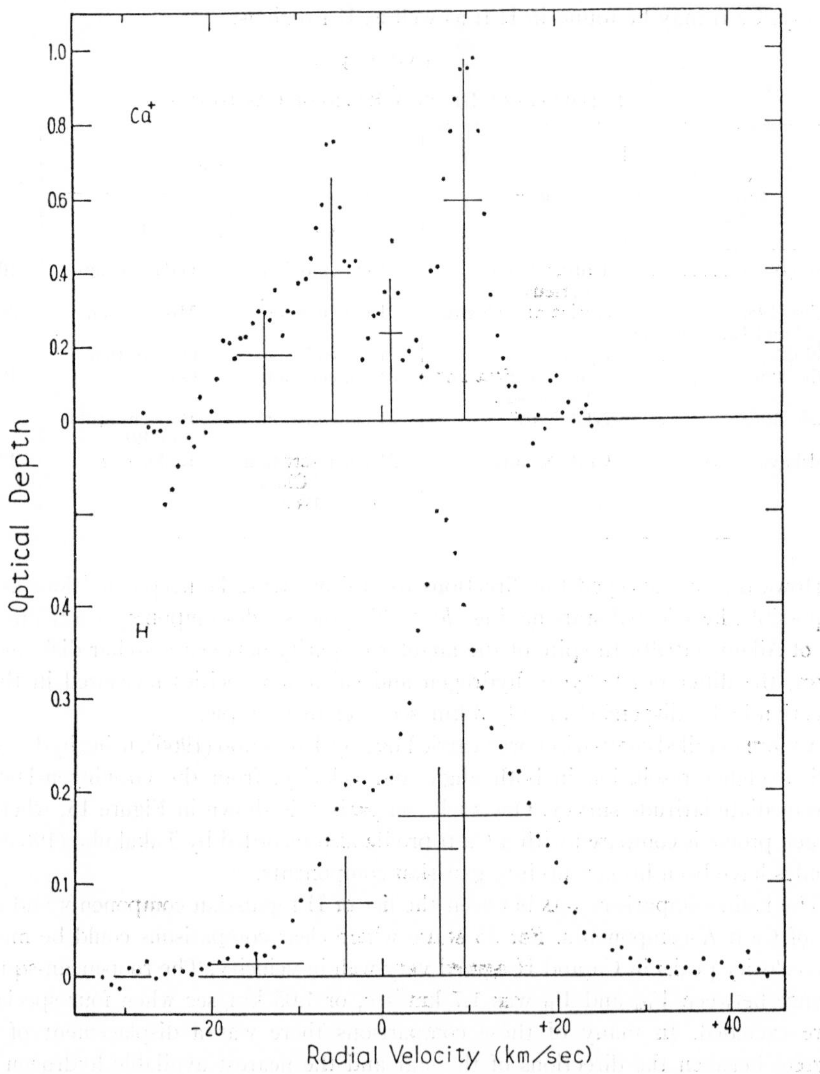

FIG. 15.—Comparison between the *K*-line in ϵ Ori (*above*) and the hydrogen line at $l^{II} = 205°$, $b^{II} = -17°$ (Takakubo 1963*b*).

A similar analysis can be made for sodium interstellar lines, but there are at present very few cases where sodium and hydrogen observations are available in the same positions.

8.4. Ionized Hydrogen

Several authors have studied the relationship between neutral and ionized hydrogen. Locke, Galt, and Costain (1964*b*) found an H I cloud centered on the IC443 H II nebulosity, with a velocity dispersion of 5 km/sec and a density of about 10 atoms/cm^3. Raimond (1966) reported a general correspondence between H I and H II in his region in Monoceros, but there was not detailed agreement between them, either in velocity or position. Davies and Tovmassian (1963*b*) have briefly examined one strip across the Cygnus X region, which is an area of complex thermal radio emission from ionized hydrogen, and have reported some H I–H II correlation, but no really detailed comparison between neutral and ionized hydrogen has yet been made for this or other, similar regions.

Again we have the situation where comparisons in the Galaxy are often made difficult by dominating background effects, and the best available evidence comes from the Magellanic Clouds. In the LMC there is good positional agreement for the major concentrations of neutral and ionized hydrogen, and the radial velocities of 42 H II regions agree with the corresponding H I velocities to an accuracy better than the velocity dispersion of the nebulae (McGee 1964). Similar results were obtained for the SMC by Bok, Gollnow, Hindman, and Mowat (1964).

We can sum up this section by saying that there is certainly a close general relationship between neutral hydrogen and all the other Population I constituents that we have considered, but that there seem to be substantial disagreements in detail, both in position and velocity. The extent of these detailed disagreements is not at present known, partly through the lack of sufficiently detailed observations, and partly through inherent difficulties in the comparisons.

9. OTHER LINE POSSIBILITIES

The only constituents of the interstellar gas that have been found by radio spectroscopy so far are the hydrogen atom and the hydroxyl radical. The only other line on which serious work has been done is the 92-cm hyperfine transition line from deuterium, but the several attempts to detect it have been unsuccessful. Several other atomic and molecular lines have been suggested as possibly detectable, and some of these may be found in the near future.

The deuterium line is directly analogous to the 1420 Mc/s line from hydrogen. The most accurate laboratory measurement of the frequency is 327.384349 Mc/s $\pm$ 0.000005 (Anderson, Pipkin, and Baird 1960). The calculated transition probability is 4.65×10^{-17} sec^{-1} (Field 1962).

Unsuccessful attempts to detect this line have been made by Getmantsev, Stankevich, and Troitskii (1955), Stankevich (1958), Stanley and Price (1956), Adgie and Hey (1957), Adgie (1959), and Weinreb (1962*a*). The greatest sensitivity was obtained by Weinreb, who integrated over a 9-week observing period with his autocorrelation receiver, in a search for deuterium absorption in Cassiopeia A. Weinreb concluded that the abundance ratio N_D/N_H is less than 1/13,000 for this sample of the interstellar gas; this value should be compared with the terrestrial abundance ratio of 1/6700. In estimating the relative abundances from measurements of optical depth, Weinreb assumed that (1) half the hydrogen-line broadening is due to thermal and half to turbulent motions, and (2) the spin temperatures for hydrogen and deuterium are equal. The latter assumption is quite uncertain.

Radio-frequency lines are associated with various other atomic and molecular species that are known or suspected to be present in the interstellar gas. The most extensive discussions of the possibility of detecting other lines have been given by Shklovsky (1956), Townes (1957), Twiss (1957), and Barrett (1958). Townes in particular gives long lists of radio-frequency transitions that may be of astronomical interest. The transition probabilities can usually be calculated rather precisely, but the frequencies of the lines are often very poorly known unless the line has actually been detected in the laboratory. In estimating the likelihood of observing a particular line astronomically, the likely abundance of the atom or molecule concerned is also very uncertain, but a determination of these abundances may be regarded as one of the primary objectives in any search for other lines.

One group of possibilities is atomic hyperfine transitions in atoms other than hydrogen and deuterium. Such transitions occur only in the relatively few nuclei that have a spin greater than zero. The most interesting of these is ionized He^3, which has an electron configuration similar to that of hydrogen. On terrestrial abundances, this line would be about as favorable as the deuterium one, and thus would be a marginal proposition (Townes 1957). However, recent observations on peculiar A stars and on solar plasma suggest that the interstellar abundance of He^3 may be higher than the terrestrial value. The expected value for the line frequency is 8665.66 ± 0.18 Mc/s (Novick and Commins 1958). Another atom with nonzero spin which might be detectable is N^{14}. Townes discusses a number of hyperfine transitions, associated with various states of ionization, but most of the frequencies are only very roughly known.

Many atomic fine-structure levels are sufficiently close to give transitions at radio frequencies, but these states are usually very short-lived. The possibility of detecting such lines from hydrogen has been considered by Wild (1952) and Kardashev (1959). Wild showed that the $2\,^2P_{3/2}$–$2\,^2S_{1/2}$ line at 9850 Mc/s might be detectable from the Sun, but the linewidth would be very large ($\sim$200 Mc/s). Kipper and Tiĭt (1958) and Field and Partridge (1961) have considered the behavior of the line in ionized hydrogen regions and found that the stimulated emission, due to Lyman-α quanta, should be detectable with a broad-band (100 Mc/s) highly sensitive receiver.

Kardashev on the other hand has discussed transitions between levels with very high total quantum number n. Transitions of the type $n \rightarrow n - 1$ would have the highest probability; they might be detectable, in spite of the small number of atoms likely to be in each highly excited state, because their oscillator strength is high. For example, the transition $n = 167 \rightarrow 166$ has a frequency of 1424.66 Mc/s. Dravskikh, Dravskikh, and Kolbasov (1964) have reported a weak line from the Omega Nebula, corresponding to the transition $n = 105 \rightarrow 104$; the line was at 5763Mc/s, with a half-power width of 1.2 Mc/s. Subsequently, Höglund and Mezger (1965) have detected the line corresponding to $n = 110 \rightarrow 109$ at 5009 Mc/s in M17, the Omega Nebula, and other H II regions.

Turning to molecular lines, we find several possibilities in addition to the OH lines already observed. Interstellar molecules will be predominantly in the lowest electronic, vibrational, and rotational states. A molecular spectrum will therefore consist of sharp lines rather than of the broad bands that are found in the optical and infrared regions.

Molecules are most likely to form in regions where the gas density is above average. Since association processes must work very slowly in the low density of interstellar space, diatomic molecules will be the commonest species, and especially hydrides, such as OH, CH, SiH, and SH. These molecules all exhibit Λ-type doubling in which the orbital angular momentum of the electron interacts with the rotation of the molecule. This type of transition involves an electric dipole moment, and should therefore be especially favorable for detection. The observation of the OH lines at unexpected strength increases the likelihood that the other molecules will be detected shortly. The expected frequencies of the main transitions are around 3400 Mc/s for CH (Douglas and Elliot 1965), 2395 Mc/s for SiH (Shklovsky 1956), and 111 Mc/s for SH (Radford and Linzer 1963). Barrett and Rogers (1964) have also suggested that lines from the isotopic species $O^{18}H$ should be detectable. A number of other possible molecular lines have been discussed by Townes (1957). Early unsuccessful attempts to detect CH have been described by Chikhachev and Sorochenko (1955). Several groups are now (1966) engaged in a search for CH and other molecules.

The molecular hydrogen ion H_2^+ may be the most interesting of the hydrides, because it may be rather abundant in some regions. Several authors have calculated the radio-frequency spectrum of this ion, but their results are rather divergent (Burke 1960; Dalgarno, Patterson, and Somerville 1960; Mizushima 1960; Stankevich 1960). The main transitions are in the 1000 to 1500 Mc/s region.

Malville (1964) has discussed two transitions that can occur in the metastable $c^3\pi$ state of molecular hydrogen at 4928 and 5898 Mc/s. However, they would be detectable only if the density of hydrogen metastables is at least 10^{-4} times that of neutral hydrogen atoms.

Although outside the normal radio-frequency range, the 28-μ and 85-μ rotational transitions of molecular hydrogen are of special interest (Twiss 1957, Takayanagi and Nishimura 1961). Molecular hydrogen may comprise a high proportion of the total interstellar mass, but it is at present unobservable. These lines could only be observed from above the atmosphere, but their successful detection would be of great astrophysical importance.

REFERENCES

Adams, W. S. 1949, *Ap. J.*, **109**, 354.
Adgie, R. L. 1959, *Paris Symposium on Radio Astronomy* (I.A.U. Symp. No. 9), ed. R. N. Bracewell (Stanford: Stanford University Press), p. 352.
Adgie, R. L., and Hey, J. S. 1957, *Nature*, **179**, 370.
Akabane, K., and Kerr, F. J. 1964, *Australian J. Phys.*, **18**, 91.
Ampel, R., and Iwaniszewska, C. 1962, *Stud. Soc. Sci. Torun.* (F), **3**, 87.
Anderson, L. W., Pipkin, F. M., and Baird, J. C., Jr. 1960, *Phys. Rev. Letters*, **4**, 69.
Babcock, H. W. 1953, *Ap. J.*, **118**, 387.
Barrett, A. H. 1958, *Proc. Inst. Radio Eng.*, **46**, 250.
———. 1964, *I.E.E.E. Trans. Ant. Prop.*, **AP-12**, 822.
Barrett, A. H., and Lilley, A. E. 1957*a*, *A.J.*, **62**, 4.
———. 1957*b*, *ibid.*, **62**, 5.
Barrett, A. H., Meeks, M. L., and Weinreb, S. 1964, *Nature*, **202**, 475.
Barrett, A. H., and Rogers, A. E. E. 1964, *Nature*, **204**, 62.
Bethe, H. 1933, *Hdb. d. Phys.*, ed. H. Geiger and K. Scheel (Berlin: Springer-Verlag), **24/1**, chap. 3, 385.
Blaauw, A. 1952, *B.A.N.*, **11**, 459.

Blum, E. J. 1960, *C. R. Paris*, **250**, 3279.
Bok, B. J. 1964, *Proc. I.A.U. Symp. No. 20* (Canberra–Sydney, 1963), p. 335.
Bok, B. J., Gollnow, H., Hindman, J. V., and Mowat, M. 1964, *Australian J. Phys.*, **17**, 404.
Bok, B. J., Lawrence, R. S., and Menon, T. K. 1955, *Pub. A.S.P.*, **67**, 108.
Bolton, J. G., Damme, K. J. van, Gardner, F. F., and Robinson, B. J. 1964, *Nature*, **201**, 279.
Bolton, J. G., Gardner, F. F., McGee, R. X., and Robinson, B. J. 1964, *Nature*, **204**, 30.
Bolton, J. G., Stanley, G. J., and Harris, D. E. 1958, *Pub. A.S.P.*, **70**, 544.
Bolton, J. G., and Wild, J. P. 1957, *Ap. J.*, **125**, 296.
Braes, L. L. E. 1963, *B.A.N.*, **17**, 132.
Brodskaya, E. S. 1963, *Izvest. Crimean Astr. Obs.*, **30**, 126.
Burke, B. F. 1960, *Ap. J.*, **132**, 514.
Burke, B. F., Ecklund, E. T., Firor, J. W., Tatel, H. E., and Tuve, M. A. 1959, *Paris Symposium on Radio Astronomy* (I.A.U. Symp. No. 9), ed. R. N. Bracewell (Stanford: Stanford University Press), p. 374.
Burke, B. F., Turner, K. C. and Tuve, M. A. 1964, *Proc. I.A.U. Symp No. 20* (Canberra–Sydney, 1963), p. 131.
Burke, B. F., and Tuve, M. A. 1964, *Proc. I.A.U. Symp. No. 20* (Canberra–Sydney, 1963), p. 183.
Chikhachev, B. M., and Sorochenko, R. L. 1955, *Trudy Soveshch. Vopr. Kosmog.*, **5**, 546.
Christiansen, W. N., and Hindman, J. V. 1952, *Australian J. Sci. Res.*, **A5**, 437.
Clark, B. G. 1963, *Nature*, **197**, 474.
———. 1965, *Ap. J.*, **142**, 1398.
Clark, B. G., Radhakrishnan, V., and Wilson, R. W. 1962, *Ap. J.*, **135**, 151.
Crampton, S. B., Kleppner, D., and Ramsey, N. F. 1963, *Phys. Rev. Letters*, **11**, 338.
Dalgarno, A., Patterson, T. N. L., and Somerville, W. B. 1960, *Proc. R. Soc. London, A*, **259**, 100.
Davies, R. D. 1956, *M.N.*, **116**, 443.
———. 1960, *ibid.*, **120**, 483.
———. 1964, *ibid.*, **128**, 133.
Davies, R. D. and Jennison, R. C. 1964, *M.N.*, **128**, 123.
Davies, R. D., Shuter, W. L. H., Slater, C. H., and Wild, P. A. T. 1963, *M.N.*, **126**, 353.
Davies, R. D., Slater, C. H., Shuter, W. L. H., and Wild, P. A. T. 1960, *Nature*, **187**, 1088.
Davies, R. D., and Tovmassian, H. M. 1963*a*, *M.N.*, **127**, 45.
———. 1963*b*, *ibid.*, **127**, 61.
Davies, R. D., Verschuur, G. L., and Wild, P. A. T. 1962, *Nature*, **196**, 563.
Davies, R. D., and Williams, D. R. W. 1957, *Radio Astronomy* (I.A.U. Symp. No. 4), ed. H. C. van de Hulst (Cambridge: Cambridge University Press), p. 71.
Davis, R. J. 1957, *Ap. J.*, **125**, 391.
———. 1962, *Smithsonian Contr. Astrophysics*, **5**, 209.
Dieter, N. H. 1960, *Ap. J.*, **132**, 49.
———. 1964, *A.J.*, **69**, 288.
———. 1965, *ibid.*, **70**, 552–58.
Dieter, N. H., Epstein, E. E., Lilley, A. E., and Roberts, M. S. 1962, *A.J.*, **67**, 270.
Dieter, N. H., and Ewen, H. I. 1964, *Nature*, **201**, 279.
Douglas, A. E., and Elliot, G. A. 1965, *Canadian J. Phys.*, **43**, 496.
Drake, F. D. 1960, *Telescopes*, ed. G. P. Kuiper and B. M. Middlehurst (Chicago: University of Chicago Press), p. 210.
Dravskikh, Z. V., Dravskikh, A. F., and Kolbasov, V. A. 1964, *Astr. Tskirk. Akad. Nauk U.S.S.R.*, No. 305, p. 2.
Egorova, T. M. 1963, *A.J. (U.S.S.R.)*, **40**, 382 (= *Sov. Astr.-A.J.*, **7**, 290).
Ehrenstein, G., Townes, C. H., and Stevenson, M. J. 1959, *Phys. Rev. Letters*, **3**, 40.
Epstein, E. E. 1964, *A.J.*, **69**, 490.
Erickson, W. C., and Helfer, H. L. 1960, *A.J.*, **65**, 1.
Ewen, H. I., and Purcell, E. M. 1951, *Nature*, **168**, 356.
Field, G. B. 1958, *Proc. Inst. Radio Eng.*, **46**, 240.
———. 1959*a*, *Ap. J.*, **129**, 525.
———. 1959*b*, *ibid.*, **129**, 536.
———. 1959*c*, *ibid.*, **129**, 551.
———. 1962, *ibid.*, **135**, 684.
Field, G. B., and Partridge, R. B. 1961, *Ap. J.*, **134**, 959.
Fletcher, E. S. 1963, *A.J.*, **68**, 407.
Galt, J. A., Slater, C. H., and Shuter, W. L. H. 1960, *M.N.*, **120**, 187.
Gardner, F. F., Robinson, B. J., Bolton, J. G., and Damme, K. J. van. 1964, *Phys. Rev. Letters*, **13**, 3.
Getmantsev, G. G., Stankevich, K. S., and Troitskii, V. S. 1955, *Dokl. Akad. Nauk U.S.S.R.*, **103**, 783.
Giordmaine, J. A., Alsop, L. E., Mayer, C. H., and Townes, C. H. 1959, *Proc. Inst. Radio Eng.*, **47**, 1062.
Girnstein, H. G., and Rohlfs, K. 1964, *Zs. f. Ap.*, **59**, 83.
Goldstein, S. J., Jr. 1963, *Ap. J.*, **138**, 978.
———. 1964*a*, *ibid.*, **140**, 802.
———. 1964*b*, *Proc. I. E. E. E.*, **52**, 1046.

Goldstein, S. J., Jr., Gundermann, E. J., Penzias, A. A., and Lilley, A. E. 1964, *Nature*, **203**, 65.
Goldstein, S. J., Jr., and Welch, B. J. 1966, *A.J.*, **71**, 297.
Grahl, B. H. 1960, *Mitt. Univ.-Sternw. Bonn*, No. 28.
Hack, M., Woerden, H. van, and Blaauw, A. 1966, unpublished.
Hagen, J. P., Lilley, A. E., and McClain, E. F. 1955, *Ap. J.*, **122**, 361.
Hagen, J. P., and McClain, E. F. 1954, *Ap. J.*, **120**, 368.
Heeschen, D. S. 1955, *Ap. J.*, **121**, 569.
Heidmann, J. 1961, *B.A.N.*, **15**, 314.
Helfer, H. L., and Tatel, H. E. 1955, *Ap. J.*, **121**, 585.
———. 1959, *ibid.*, **129**, 565.
Hindman, J. V. 1967, *Australian J. Phys.*, **20**, 147.
Höglund, B. 1963, *Ark. f. Astr.*, **3**, 217.
Höglund, B., and Mezger, P. G. 1965, *Science*, **150**, 339–40 and 347–48.
Howard, W. E. III, Wentzel, D. G., and McGee, R. X. 1963, *Ap. J.*, **138**, 988.
Hulst, H. C. van de. 1945, *Ned. T. Natuurk.*, **11**, 201.
———. 1958, *Rev. Mod. Phys.*, **30**, 913 (I.A.U. Symp. No. 8).
Hulst, H. C. van de, Muller, C. A., and Oort, J. H. 1954, *B.A.N.*, **12**, 117.
Jansky, K. G. 1932, *Proc. Inst. Radio Eng.*, **20**, 1920.
Jelley, J. V., and Cooper, B. F. C. 1961, *Rev. Sci. Inst.*, **32**, 166, 202.
Kaftan-Kassim, M. A. 1961, *Ap. J.*, **133**, 821.
Kardashev, N. S. 1959, *A.J. (U.S.S.R.)*, **36**, 838 (= *Sov. Astr.-A.J.*, **3**, 813).
Kerr, F. J. 1964*a*, *Proc. I.A.U. Symp. No. 20* (Canberra–Sydney, 1963), p. 187.
———. 1964*b*, *ibid.*, p. 187.
Kerr, F. J., and Hindman, J. V. 1954, *U.R.S.I. Gen. Assem. Proc.*, **10**, Pt. 5, 99.
Kerr, F. J., Hindman, J. V., and Gum, C. S. 1959, *Australian J. Phys.*, **12**, 270.
Kerr, F. J., Hindman, J. V., and Robinson, B. J. 1954, *Australian J. Phys.*, **7**, 297.
Kerr, F. J., and Vallak, R. 1966, in press.
Kerr, F. J., and Westerhout, G. 1965, *Galactic Structure*, ed. A. Blaauw and M. Schmidt (Chicago: University of Chicago Press), chap. 9.
Kipper, A. J., and Tiĭt, V. M. 1958, *Voprosy Kosmogonii*, **6**, 98.
Koehler, J. A., and Robinson, B. J. 1966, *Ap. J.*, **146**, 488.
Lawrence, R. S. 1956, *Ap. J.*, **123**, 30.
Lilley, A. E. 1955, *Ap. J.*, **121**, 559.
Livingston, W. C., and Lynds, C. R. 1964, *Ap. J.*, **140**, 818.
Locke, J. L., Galt, J. A., and Costain, C. H. 1964*a*, *Ap. J.*, **139**, 1066.
———. 1964*b*, *ibid.*, **139**, 1071.
McClain, E. F. 1955, *Ap. J.*, **122**, 376.
McGee, R. X. 1964, *Australian J. Phys.*, **17**, 515.
McGee, R. X., and Milton, J. A. 1964*a*, *Proc. I.A.U. Symp. No. 20* (Canberra–Sydney, 1963), p. 289.
———. 1964*b*, *Australian J. Phys.*, **17**, 128.
———. 1966, *ibid.*, **19**, 343.
McGee, R. X., and Murray, J. D. 1961, *Australian J. Phys.*, **14**, 260.
———. 1963, *Proc. Inst. Radio Eng. (Australia)*, **24**, 191.
McGee, R. X., Murray, J. D., and Milton, J. A. 1963, *Australian J. Phys.*, **16**, 136.
Makarova, S. P. 1964, *A.J. (U.S.S.R.)*, **41**, 608 (= *Sov. Astr.-A.J.*, **8**, 485).
Malville, J. M. 1964, *Ap. J.*,, **139**, 198.
Matthews, T. A. 1957, *Radio Astronomy* (I.A.U. Symp. No. 4), ed. H. C. van de Hulst (Cambridge: Cambridge University Press), p. 48.
Menon, T. K. 1958, *Ap. J.*, **127**, 28.
Menoud, C., Racine, J., and Kartaschoff, P. 1964, *Microtecnic*, **18**, 150.
Metik, L. P. 1963, *Izvest. Crimean Astr. Obs.*, **29**, 315.
Milne, E. A. 1930, *Hdb. d. Ap.*, **3**, chap. 2, 159.
Mizushima, M. 1960, *Ap. J.*, **132**, 493.
Morris, D., Clark, B. G., and Wilson, R. W. 1963, *Ap. J.*, **138**, 889.
Muhleman, D. O., and Walker, R. G. 1964, *A.J.*, **69**, 95.
Muller, C. A. 1956, *Phillips Tech. Rev.*, **17**, 351.
———. 1957, *Ap. J.*, **125**, 830.
———. 1959, *Paris Symposium on Radio Astronomy* (I.A.U. Symp. No. 9), ed. R. N. Bracewell (Stanford: Stanford University Press), p. 360.
———. 1966, in press.
Muller, C. A., and Oort, J. H. 1951, *Nature*, **168**, 357.
Muller, C. A., and Westerhout, G. 1957, *B.A.N.*, **13**, 151.
Novick, R., and Commins, E. D. 1958, *Phys. Rev.*, **111**, 822.
Pawsey, J. L. 1951, *Nature*, **168**, 358.
Penzias, A. A. 1964, *A.J.*, **69**, 146.
Peters, H. E., Holloway, J., Bagley, A. S., and Cutler, L. S. 1965, *Appl. Phys. Letters*, **7**, 34.
Pronik, I. I. 1963, *Izvest. Crimean Astr. Obs.*, **30**, 118.

Purcell, E. M. 1953, *Proc. Amer. Acad. Arts Sci.*, **82**, 347.
Purcell, E. M., and Field, G. B. 1956, *Ap. J.*, **124**, 542.
Radford, H. E. 1964, *Phys. Rev. Letters*, **13**, 534.
Radford, H. E., and Linzer, M. 1963, *Phys. Rev. Letters*, **10**, 443.
Radhakrishnan, V. 1960, *Pub. A.S.P.*, **72**, 296.
Radhakrishnan, V., Morris, D., and Wilson, R. W. 1961, *A.J.*, **66**, 51.
Raimond, E. 1966, *B.A.N.*, **18**, 191.
Reber, G. 1944, *Ap. J.*, **100**, 279.
Roberts, M. S. 1959, *Nature*, **184**, 1555.
Robinson, B. J. 1963, *Proc. Inst. Radio Eng.* (*Australia*), **24**, 119.
———. 1964. *Ann. Rev. Astr. Ap.*, **2**, 401.
———. 1966, unpublished.
Robinson, B. J., Damme, K. J. van, and Koehler, J. A. 1963, *Nature*, **199**, 1176.
Robinson, B. J., Gardner, F. F., Damme, K. J. van, and Bolton, J. G. 1964, *Nature*, **202**, 989.
Robinson, B. J., and Koehler, J. A. 1965, *Nature*, **208**, 993.
Rougoor, G. W. 1964, *B.A.N.*, **17**, 381.
Ryzhkov, N. F., Egorova, T. M., Gosachinskii, I. V., and Bystrova, N. V. 1963, *A.J.* (*U.S.S.R.*), **40**, 17 (= *Sov. Astr.-A.J.*, **7**, 12).
Schmidt, K. H. 1964, *A.N.*, **288**, 19.
Schmidt, M. 1957, *B.A.N.*, **13**, 247.
Shklovsky, I. S. 1949, *A.J.* (*U.S.S.R.*), **26**, 10.
———. 1953, *Dokl. Akad. Nauk U.S.S.R.*, **92**, 25.
———. 1956, *Cosmic Radio Waves* (Moscow: State Publishing House); English translation by R. B. Rodman and C. M. Varsavsky (Cambridge, Mass.: Harvard University Press, 1960).
Sholomitskii, G. B. 1962, *A.J.* (*U.S.S.R.*), **39**, 765 (= *Sov. Astr.-A.J.*, **6**, 595).
Shuter, W. L. H., and Verschuur, G. L. 1964, *M.N.*, **127**, 387.
Sorochenko, R. L. 1961, *A.J.* (*U.S.S.R.*), **38**, 478 (= *Sov Astr.-A.J.*, **5**, 355).
Stankevich, K. S., 1958, *A.J.* (*U.S.S.R.*), **35**, 157 (= *Sov. Astr.-A.J.*, **2**, 136).
———. 1960, *ibid.*, **37**, 983 (*ibid.*, **4**, 917).
Stanley, G. J., and Price, R. 1956, *Nature*, **177**, 1221.
Takakubo, K. 1958, *Pub. Astr. Soc. Japan*, **10**, 187.
———. 1963*a*, *Sci. Rep. Tôhoku Univ.*, ser. I, **47**, 65.
———. 1963*b*, *ibid.*, ser. I, **47**, 108.
Takakubo, K., and Woerden, H. van. 1966, *B.A.N.*, **18**, 488.
Takayanagi, K., and Nishimura, S. 1961, *Rep. Ionosph. Space Res. Japan*, **15**, 81.
Terauti, R. 1963, *Sci. Rep. Tôhoku Univ.*, ser. I, **47**, 114.
Townes, C. H. 1957, *Radio Astronomy* (I.A.U. Symp. No. 4), ed. H. C. van de Hulst (Cambridge: Cambridge University Press), p. 92.
Twiss, R. Q. 1957, *Pub. CSIRO* (*Melbourne*) (Symposium on Radio Astronomy, CSIRO Radiophysics Laboratory, Sydney), p. 66.
Verschuur, G. L. 1967, *Proc. I.A.U. Symp. No. 31* (Noordwijk, 1966), in press.
Vessot, R. F. C., Peters, H. E., and Vanier, J. 1964, *Frequency*, **2**, 33.
Wade, C. M. 1957, *A.J.*, **62**, 148.
Weaver, H. F., and Williams, D. R. W. 1964, *Nature*, **201**, 380.
Weaver, H. F., Williams, D. R. W., Dieter, N. H., and Lum, W. T. 1965, *Nature*, **208**, 29.
Weinreb, S. 1962*a*, *Nature*, **195**, 367.
———. 1962*b*, *Ap. J.*, **136**, 1149.
———. 1963, *M.I.T. Res. Lab. Electron. Tech. Rep.*, 412.
Weinreb, S., Barrett, A. H., Meeks, M. L., and Henry, J. C. 1963, *Nature*, **200**, 829.
Westerhout, G. 1957, *B.A.N.*, **13**, 201.
———. 1958, *ibid.*, **14**, 215.
Wild, J. P. 1952. *Ap. J.*, **115**, 206.
Williams, D. R. W., and Davies, R. D. 1954, *Nature*, **173**, 1182.
———. 1956, *Phil. Mag.*, ser. 8, **1**, 622.
Woerden, H. van, and Schwarz, U. 1966, unpublished.
Woerden, H. van, Takakubo, K., and Braes, L. L. E. 1962, *B.A.N.*, **16**, 321.

CHAPTER 11

Nonthermal Galactic Radio Sources

R. MINKOWSKI
Radio Astronomy Laboratory, University of California, Berkeley

1. INTRODUCTION

THE first object identified as a radio source was the Crab Nebula (Bolton and Stanley 1949), the remnant of the supernova of +1054. That supernova remnants are one class of galactic radio sources thus was established from the beginning. Initially, the flux density of the Crab Nebula seemed not to depend on frequency; this suggested thermal emission. The radio emission, however, cannot be considered an extension of the optical continuous spectrum (Greenstein and Minkowski 1953; Shklovsky 1953*a*) whose interpretation as thermal emission requires physical conditions that are not readily acceptable. Shklovsky's (1953*b*) suggestion that the radio spectrum is nonthermal and due to synchrotron emission and that the optical continuum is the continuation of the radio spectrum found a striking confirmation when Dombrovsky (1954) discovered high polarization of the optical continuum. At that epoch, extension of the radio observations to higher frequencies also confirmed that the radio spectrum is nonthermal, although with an unusually slow decrease of flux density with increasing frequency.

The remnants of Tycho's (Sec. 5.1) and Kepler's (Sec. 5.2) supernovae were later found to be nonthermal sources, and the nebulosity identified with the radio source Cassiopeia A (Sec. 6), at low frequencies the most intense nonthermal source in the sky, turned out to have the physical characteristics appropriate for the remnant of an unobserved supernova. Extended nonthermal sources have been identified with large nebulosities, such as the Cygnus Loop (Sec. 8.1). Oort's suggestion that such objects are supernova shells slowed down by interaction with the interstellar medium provides a satisfactory interpretation of these objects. Eventually the assumption was generally accepted that all galactic nonthermal sources are supernova remnants, most of them probably so old that no records of the outbursts can be expected to exist. The radio emission of the remnants must contain some contribution of thermal radiation corresponding to the observed line emission. The available data suggest, however, that the contribution of thermal radiation to the total radio emission is usually negligible.

Except for the supernova remnants, galactic emission nebulae—H II regions and planetary nebulae—are thought to be purely thermal emitters. Recent observations (Menon and Terzian 1965; Slee and Orchiston 1965) seemed at first to indicate that some planetary nebulae showed nonthermal emission; however, for the apparently best example, NGC 3242, a detailed investigation by Kaftan-Kassim (1966) has shown that, within the errors involved, the excess of radio emission at low frequencies, which suggested nonthermal emission, results from the presence of other sources in the beam. This same situation probably exists in all instances in which nonthermal emission from planetary nebulae has been suspected. Thus, supernova remnants are at this time the only objects established as galactic nonthermal sources, and therefore the only objects with which this chapter is concerned.

It is, of course, not impossible that such objects as planetary nebulae, the shells of ordinary novae, or even H II regions might have weak nonthermal emission in addition to the thermal emission, but at this time there is no good reason to suspect such situations. The radio emission of flare stars is not within the scope of this chapter.

2. THERMAL EMISSION AND ABSORPTION

All optically observed supernova remnants show emission lines and must therefore show some contribution of thermal radiation to the radio emission. In the majority of objects, the line emission is weak and the contribution of thermal emission is negligible. In some remnants, for instance the Cygnus Loop, bright emission features are present, and the similarity of optical and radio brightness distribution is so pronounced that careful attention must be given to the question of whether part of the radio emission is thermal. Connected with thermal emission is absorption which increases inversely in proportion to the square of the frequency and will therefore always become important at sufficiently low frequencies. The optical depth of an ionized gas is proportional to the surface brightness in the light of permitted emission lines. Where the surface brightness is high, as in planetary nebulae, substantial opacity may occur even at frequencies of 10^3 Mc/s and higher. Intense emission filaments in some strong radio sources may have high opacity, but their effect on the radio spectrum of the source as a whole will be small because they cover only an insignificant fraction of the source.

The radio-frequency emission $j(\nu)$ per unit volume due to free-free transitions is given by

$$j(\nu) = \frac{32Z^2e^6}{3m^2c^3}\left(\frac{2\pi m}{kT}\right)^{1/2} N_e N_i \left(\frac{3}{2}\ln\frac{kT}{h\nu_0} - \ln\frac{\nu}{4\nu_0} - 2\ln Z - 1.443\right), \quad (1)$$

where N_e and N_i are the numbers per unit volume of electrons and positive ions with charge Z, and ν_0 is the frequency of the hydrogen ionization limit (Brussaard and van de Hulst 1962; Osterbrock 1964). For a nebula of pure hydrogen, N_i equals the number of protons N_p, and $Z = 1$. Substituting the numerical values, we obtain for pure hydrogen

$$j(\nu) = 3.75 \cdot 10^{-38} T^{-1/2} N_e N_p \left(17.74 + \ln\frac{T^{3/2}}{\nu}\right) \quad (2)$$

in erg cm^{-3} sec^{-1} (c/s)$^{-1}$. For the line emission of hydrogen in the Balmer line H_n, the energy emitted per unit volume may be written as

$$j(H_n) = G_n(\text{exc.}, T) N_e N_p \text{ erg cm}^{-3}\text{ sec}^{-1}, \quad (3)$$

where G_n(exc., T) depends on the temperature and on the mode of excitation. Samples of numerical values are given in Table 1 for a nebula of great optical thickness in the

TABLE 1

RELATION OF THERMAL RADIO EMISSION AND ABSORPTION TO HYDROGEN LINE EMISSION FOR A HOMOGENEOUS NEBULA OF PURE HYDROGEN

	$T=10^4$				$T=2\times10^4$			
	Case *B*		Coll.		Case *B*		Coll.	
$j(H\alpha)/N_eN_p$ erg cm^3 sec^{-1}	30.6 [−26		1.15 [−26		16.5 [−26		100 [−26	
$j(H\beta)/N_eN_p$ erg cm^3 sec^{-1}	12.2 [−26		0.15 [−26		6.3 [−26		19.3 [−26	
	100 Mc/s	1400 Mc/s	100 Mc/s	1400 Mc/s	100 Mc/s	1400 Mc/s	100 Mc/s	1400 Mc/s
$j(\nu)/j(H\alpha)$(c/sec)$^{-1}$	1.61 [−14	1.29 [−14	4.28 [−13	3.42 [−13	2.28 [−14	1.82 [−14	3.76 [−15	3.01 [−15
$j(\nu)/j(H\beta)$(c/sec)$^{-1}$	4.03 [−14	3.22 [−14	3.33 [−12	2.66 [−12	6.01 [−14	4.81 [−14	1.95 [−14	1.56 [−14
$S(H\alpha)/\tau_\nu$ erg cm^{-2} sec^{-1} ster^{-1}	4.00 [− 3	9.77 [− 1	1.51 [− 4	3.70 [− 2	5.65 [− 3	1.39	3.42 [− 2	8.41
$S(H\beta)/\tau_\nu$ erg cm^{-2} sec^{-1} ster^{-1}	1.60 [− 3	3.92 [− 1	1.93 [− 5	4.72 [− 3	2.14 [− 3	2.57 [− 1	6.58 [− 3	1.62

Note: Case *B* is Menzel and Baker's Case *B*; [$-n$ denotes 10^{-n}.

Lyman lines with radiative excitation (Menzel and Baker's Case *B;* see chap. 9, Sec. 5, also Burgess 1958) and with collisional excitation (Parker 1964*a*). Case *B* is considered to represent satisfactorily the conditions in H II regions and planetary nebulae; it is known, however, that there are as yet unexplained deviations of the observed Balmer decrement from Case *B* (O'Dell 1963, 1964). The ratio $j(\nu)/j(H_n)$ gives the ratio of the flux density $F(\nu)$ of the radio-frequency emission to $F(H_n)$, that of the Balmer lines for a nebula whose optical depth at the frequency ν is small. The flux density of the Balmer line must be corrected, of course, for interstellar absorption. Numerical values of $j(\nu)/j(H_n)$ for Hα and Hβ, frequencies of 100 Mc/s and 1400 Mc/s, and temperatures of 10^4 ° K and 2×10^4 ° K are given in Table 1.

To take into account the He content of a nebula, the values of $j(\nu)/j(H_n)$ must be multiplied by the factor $n_i/n_p = 1.15$ if the helium is singly ionized (Osterbrock 1964). If, in nebulae of high excitation, He begins to become doubly ionized, the factor becomes larger owing to the factor Z^2 in equation (1); e.g., 1.3 if $N(He^{++}) = N(He^+)$. The abundance of all other elements is too low to play a role.

To the volume emissivity $j(\nu)$ corresponds the absorption coefficient

$$\kappa_\nu = \frac{c^2 j(\nu)}{8\pi k\nu^2 T} \tag{4}$$

or

$$\kappa_\nu = \frac{2.59 \cdot 10^{35} j(\nu)}{\nu^2 T}. \tag{5}$$

The optical depth $\tau(\nu)$ equals $\int \kappa(\nu) dl$, l being the geometrical depth. For a nebula of pure hydrogen this is proportional to $\int N_e N_p \, dl$, the "emission measure." The emission measure is often thought of as a measure of surface brightness. But this is correct only if objects are compared in which the excitation and the electron temperature are identical. It is better to introduce the surface brightness itself. The observed flux $F(H_n)$ in a hydrogen line from a homogeneous nebula with radius r at distance d is

$$F(H_n) = \frac{r^3 j(H_n)}{3d^2}. \tag{6}$$

Since r/d is the angular radius φ of the nebula, the surface brightness

$$S(H_n) = \frac{F(H_n)}{\varphi^2} \tag{7}$$

becomes

$$S(H_n) = \frac{l j(H_n)}{6}, \tag{8}$$

where $l = 2r$ is the diameter of the nebula.

The optical depth $\kappa_\nu l$ of an homogeneous nebula can be written as

$$\tau(\nu) = \frac{2.59 \cdot 10^{35}}{\nu^2 T} \frac{j(\nu)}{j(H_n)} j(H_n) l, \tag{9}$$

and this becomes

$$\tau(\nu) = \frac{1.55 \cdot 10^{36}}{\nu^2 T} \frac{j(\nu)}{j(H_n)} S(H_n), \tag{10}$$

where $S(H_n)$ is in erg cm^{-2} sec^{-1} $ster^{-1}$. For $j(\nu)$, the value adjusted for the He content is to be used. For pure hydrogen, Table 1 gives the values of $S(H_n)$ for which $\tau(\nu)$ is unity.

Characteristic for thermal emission is a blackbody spectrum at low frequencies where $\tau \gg 1$. At higher frequencies, where $\tau \ll 1$, the intensity follows equation (1) and decreases very slowly with increasing frequency.

[A similar analysis is presented in the Aller-Liller chapter with special reference to planetary nebulae.—Eds.]

3. NONTHERMAL EMISSION

The radio spectra of supernova remnants are clearly nonthermal in most sources, with an intensity distribution following a simple power law

$$I(\nu) \sim \nu^{x} \tag{11}$$

with a negative spectral index x. This spectrum is typical for synchrotron radiation by relativistic electrons with the energy distribution

$$N(E) \sim K \cdot E^{-\gamma}, \tag{12}$$

where $\gamma = 1 - 2x$. Except for the Crab Nebula, which is a source of high flux density with an unusually flat spectrum, synchrotron radiation cannot be expected to be observable in the optical region on the basis of a simple extrapolation of the radio spectrum. Where optical synchrotron radiation is not observed, there is no clue on the high-energy cutoff of the energy distribution corresponding to the critical frequency

$$\nu_c = 1.61 \cdot 10^{13}\, H_{\perp} E^2(\text{Bev})\ \text{c/s}, \tag{13}$$

where $H_{\perp}$ is the component of the magnetic field normal to the direction of vision.

Outstanding among the many unusual properties of the Crab Nebula is the presence of optical synchrotron radiation and the continuation of the observable spectrum into the X-ray region where the spectrum still suggests synchrotron radiation (Grader *et al.* 1966). The X-ray emission arises in a small part of the nebula, but there ν_c must be of the order 10^{19} c/s. The most unusual feature of the Crab Nebula, however, is the presence of a very small condensation with steep spectrum (Hewish and Okoye 1964, 1965; Andrew *et al.* 1964) that contributes a substantial part of the total emission around 20 MHz [20 Mc/s]. The emission of this source cannot be interpreted as synchrotron radiation. Ginzburg and Ozernoy (1966) suggest coherent plasma emission as the process responsible for the emission of the compact source. It seems convincingly demonstrated that synchrotron radiation is not the only nonthermal process that may play a role. There seems to be no reason to doubt, however, that the radio emission of the galactic nonthermal sources is mainly due to synchrotron radiation.

Synchrotron self-absorption (Slysh 1963; Williams 1963) becomes important only in sources of much higher brightness temperature than that shown by supernova remnants. A maximum of the flux density at low frequencies may be the result of thermal absorption (Sec. 2) or of the existence of a low-energy cutoff in the energy distribution of the relativistic electrons (Kellermann 1964*a*).

Since supernova remnants are expanding objects, their radiation flux and its spectrum must depend on time (Shklovsky 1960*a*). The time dependence has been discussed by

Kardashev (1962) for a variety of assumptions concerning the energy losses and gains by the relativistic electrons. When synchrotron losses play a role, a cutoff or break appears in the energy spectrum of the particles, and the emitted spectrum shows a break at which the spectral index changes at a frequency that decreases with time.

The model considered by Shklovsky (1960*a*) assumes adiabatic expansion and energy losses due solely to synchrotron radiation. This model is valid for some time after the outburst, when injection of relativistic electrons has stopped. With the exception of the Crab Nebula, all recent supernova remnants are probably in that phase. The assumption of adiabatic expansion becomes invalid when interaction with the interstellar medium retards the expansion. Since this process plays an important role in all remnants, with the exception of the Crab Nebula, the model cannot be expected to give a quantitative description of the time dependence. The prediction most easily accessible to observation is that of a decrease of the emitted power proportional to $t^{-2\gamma}$, where t is the time elapsed since the outburst. The decay of the flux density at low frequencies is thus

$$\frac{dF_\nu}{F_\nu} = -\frac{2\gamma dt}{t}. \qquad (14)$$

A decay of this order is indeed observed in the Cassiopeia A radio source (Sec. 6), whose age, however, is not precisely known.

4. SUPERNOVAE

The present status of our knowledge of supernovae, as far as it is relevant for the discussion of the remnants, may be summarized as follows on the basis of a recent discussion of the available data (Minkowski 1964).

The supernovae of Type I form a group with closely similar light curves and spectra. At maximum, the color index is small. Corrected for interstellar absorption in the Galaxy and in the external galaxies, the photographic and visual absolute magnitudes are nearly equal with the mean value $M_o = -19.0 \pm 0.3$, $\sigma = 0.7$, based on a Hubble constant of 100 km/sec per Mpc. After an initial drop of from 2 to 3 magnitudes in about a month, the photographic light curve shows an almost exponential decay. This pattern of decay has been observed in several supernovae for a time of about one year and in the supernova in IC 4182 for 635 days. The decay constant seems to have no unique value; a sizable dispersion seems to exist. No accurate observations of the visual decay exist, but the visual and the photographic decay constants do not seem to differ greatly (Mihalas 1963). The total light emitted is 3.6×10^{49} ergs. The bolometric correction is unknown, however, and the total radiation may be substantially larger. Supernovae of Type I occur in spheroidal galaxies and therefore belong to Population II.

Most supernovae that are not of Type I seem to form a group with light curves and spectra which show some similarity, but also very marked individual differences. At maximum, these supernovae of Type II are blue, $B - V$ being about -0.3 mag. The photographic absolute magnitude at maximum, corrected for interstellar absorption, is $M_o = -17.7 \pm 0.3$, $\sigma = 0.8$. After an initial drop of about 1.5 mag in about 30 days, the light curves tend to show a shoulder followed by a rapid decline. In one supernova, that in NGC 4725, the observations, interpolated across a gap of 125 days when the supernova was in conjunction with the Sun, suggest an exponential decay with a decay constant similar to that of supernovae of Type I until the end of the observations 265 days after maximum. The initial decay and the spectrum of this supernova, however, were typical for a supernova of Type II. The interpolation may well be misleading, and this single example cannot be accepted as proof that the light curve of a supernova of Type II can simulate closely that of a supernova of Type I. The total light emitted ranges downward within about one order of magnitude from 3.6×10^{49} ergs; the bolometric correction is not known. Supernovae of Type II do not occur in spheroidal galaxies but tend to be in spiral arms. They are, therefore, members of Population I.

Some supernovae have been reported whose light curves and spectra differ so much from those of supernovae of Types I and II that Zwicky (1964, 1965) considers them as new and additional types. The number of supernovae of these varieties is small; they seem to be very rare. It cannot be considered as established that they are indeed representatives of new types and not extreme examples of the diversity of supernovae, in particular those of Type II. The supernova of 1961 in NGC 4303 which Zwicky designates as Type III, shows properties that suggest strongly a supernova of Type II with unusually large ejected mass. One group of objects, Zwicky's Type V, best exemplified by the supernova in NGC 1058 (Bertola 1964), is best described as a peculiar irregular variable which reaches the absolute magnitude of supernovae at maximum; η Carinae might belong to this group. [See chapter by Zwicky in Volume VIII, this series.]

5. THE REMNANTS OF OBSERVED SUPERNOVAE

The remnants of four supernovae with observed outbursts are now known: +1006, +1054 (Crab Nebula), +1572 (Tycho), and +1604 (Kepler). It is probable that the remnant of the supernova of +185 is observed (Sec. 8.3). The following discussion proceeds from the well known to the uncertain.

The light curves of Tycho's nova of 1572 (Baade 1945) and of Kepler's nova of 1604 (Baade 1943) show conclusively that both these objects were supernovae of Type I. Doubts have been expressed occasionally on the validity of this interpretation because of the dissimilarity between the remnants of these two novae and the Crab Nebula, which has usually been considered as a typical remnant of a supernova of Type I.

The derivation of the light curves by Baade actually does not involve more than the reduction of visual comparisons of supernova and of nearby planets and stars to magnitudes. This procedure is basically equivalent to the reduction of photoelectric deflections to magnitudes. Whether the reduction is carried out a few days or a few centuries after the observations is not important. The visual comparisons are, of course, less precise than photoelectric data, but the small scatter of the observed points of the light curves shows the accuracy that eye estimates can reach. There can be no doubt that Tycho's and Kepler's novae were supernovae of Type I.

The assignment of the Crab Nebula to Type I goes back to a time when the existence of other types of supernovae was not yet known. A critical review of the scanty data on the supernova of +1054 (Sec. 5.4) shows that the assignment to Type I is not fully convincing. Moreover, the characteristics of the Crab Nebula are not easily reconciled with that assignment (Minkowski 1964). The true nature of the supernova of +1054 cannot be established convincingly, but an unbiased assessment of the evidence leads to the conclusion that the Crab Nebula is not a remnant of a supernova of Type I.

The available information on the supernova of +1006 is not adequate to establish its type beyond all doubt. Most likely it was Type I (Minkowski 1966*a*), and it will here be included with that type.

Even less is known about the supernova of +185; it may have been Type II.

5.1. Tycho's Nova

Tycho's nova is now the Type I remnant for which the information is most complete. It should be considered as the prototype.

The search for the remnant of Tycho's nova (Baade 1945) remained unsuccessful until its discovery as a radio source (Brown and Hazard 1952). Radio positions showed that Tycho's right ascension had an unexpectedly large error of −5 seconds and guided the search to the correct position (Minkowski 1959). The few exceedingly faint details of the remnant that could be found did not permit a unique determination of size and position. The optical position has low accuracy. The position given in Table 2 follows if the remnant is assumed to be circular, as has now been established (Ryle 1966). The optical position agrees well with the right ascension by Fomalont *et al.* (1964) and the declination

TABLE 2

POSITIONS OF TYCHO'S SUPERNOVA

ν	α(1950)	δ(1950)	Remarks
Optical..........	$00^h22^m32^s.4$	$+63°51'.4$	1
960 MHz........	$00^h22^m30^s.3 \pm 0^s.6$	$+63°51'53'' \pm 7''$	2

1. Assumed to be circular. Minkowski (1959).
2. Right ascension: Fomalont *et al.* (1964); declination: Read (1963).

by Read (1963). Radio positions in the 3C catalogue (Edge *et al.* 1959; Elsmore, Ryle, and Leslie 1959) and in the revised 3C catalogue (Bennett 1962) are not listed; they need a correction (Fomalont *et al.* 1964) which brings them into agreement.

For the size and structure of the source, various descriptions have been derived from interferometric observations by Elsmore *et al.* (1959), Lequeux (1962), and Maltby and Moffet (1962). Thompson (1965) has shown that these observations are consistent with a description of the source as a shell whose diameter at the point of maximum brightness is 6'.7 and whose thickness is 0.25 of its radius. Such a description is now definitely established by the observations of the brightness distribution with the aperture-synthesis telescope (Ryle 1966). The source is a shell with some fine-structure, having a diameter of 6'.2 at maximum intensity and apparently a rather sharp outer boundary; with an appropriate correction for the beamwidth of 80″, the outer diameter is about 7'.5. The optically visible details (Minkowski 1959) are at a distance of about 3'.7 from the center, close to the outer boundary, in agreement with the low radial velocity of the only filament bright enough to permit spectroscopic observation.

Because the age of the remnant is known, the average outward motion at the boundary can be stated to be 0″.56/yr. In the absence of observable details near the center of the nebulosity, the radial velocity of expansion cannot be observed. Thus, the distance cannot be derived from motion and radial velocity. The distance, however, has recently been established from observations of 21-cm absorption lines (Menon and Williams 1966*a*, *b*), which show a distance of 5000 pc. The linear radius thus is 5.4 pc, and the average velocity of expansion—the ratio of radius to age—is 13,400 km/sec. Distances and velocities comparable to the values now established were suggested by Minkowski (1964) and by Woltjer (1964) on the basis of the brightness of the supernova at maximum. With the distance known, the absolute magnitude at maximum can be determined from the distance modulus $m - M = 13.5$ and the interstellar absorption $A_V = 2^{\mathrm{m}}.1$ that follows from the observed colors of the supernova (Minkowski 1964). The observations of the interstellar absorption by Brodskaya and Grigoreva (1962) do not extend to sufficiently large distances, but admit without difficulty the value $2^{\mathrm{m}}.1$ at 5000 pc. The apparent magnitude $-4^{\mathrm{m}}.0$ at maximum (Baade 1945) then leads to the absolute magnitude $-19^{\mathrm{m}}.6$ at maximum, which is in good agreement with the average $-19^{\mathrm{m}}.0$ for supernovae of Type I.

The interstellar mass contained in a sphere of radius 5.4 pc is 29 n_{H} solar masses if n_{H} is the number of H atoms/cm^3 and if the interstellar medium has normal He abundance $n_{\mathrm{H}}/n_{\mathrm{He}} = 7$. With the galactic coordinates $l^{\mathrm{II}} = 120^{\circ}.08$, $b^{\mathrm{II}} = 1^{\circ}.41$, the distance above the galactic plane is $Z = 116$ pc. At the location of the remnant, the 21-cm observations (Westerhout 1957) indicate a value of about 0.1 for n_{H}. Thus, about 3 solar masses of interstellar material have been swept up by the remnant. Since supernovae of Type I seem to belong to Population II (Minkowski 1964), it may be assumed that the originally ejected mass was much smaller than the accreted interstellar mass.

If the accreted mass is substantially larger than the original mass, the expansion must have been slowed down considerably. The original velocity must have been higher than the average velocity of 13,400 km/sec, and the present velocity must be much lower. Initial velocities of 20,000 km/sec or higher are quite consistent with the observed spectra of supernovae of Type I before, during, and after maximum (Minkowski 1939). During that phase, the spectra show very broad and indistinct bands. Ejection velocities of 20,000 km/sec or higher offer the simplest interpretation. It is, however, not possible to rule out the possibility that the diffuseness of the main features arises from the presence of many features with more moderate width. Some structure of moderate width actually seems to be present even around maximum and to become more distinct later.

The present velocity can be ascertained whenever the motion of the visible details has been determined. This motion is noticeable but too small to be determined accurately from available material. The mere fact that motion is noticeable suggests that the present velocity must be larger than 1500 km/sec.

The conditions in the remnant of a supernova of Type I are clearly not basically different from those assumed to prevail in the remnants of supernovae of Type II, and the discussions of the interaction with the interstellar medium (Shklovsky 1962) should be applicable to the remnants of Type I.

A simple model for the deceleration of a fast-expanding gas mass was first suggested by Oort (1951). The deceleration is described as the consequence of the sweeping up of of interstellar material. Conservation of momentum requires that the velocity of expansion decreases as the mass of the expanding material increases. If m_1 and v_1 are the initial mass and velocity, r the radius at time t after the ejection, and ρ the interstellar density,

$$m_1 v_1 = (m_1 + m_2)\frac{dr}{dt}, \tag{15}$$

where

$$m_2 = \frac{4\pi}{3}\rho r^3 \tag{16}$$

is the accreted mass. When $m_2 \gg m_1$, integration leads to

$$r = \left(\frac{3 m_1 v_1}{\pi \rho}\right)^{1/4} t^{1/4}. \tag{17}$$

The velocity v_2 at time t is

$$v_2 = \frac{m_1 v_1}{m_2}, \tag{18}$$

and the average velocity is

$$\frac{r}{t} = 4 v_2. \tag{19}$$

The criticism by Poveda (1964) does not seem to be valid. The expanding mass has, indeed, no translational momentum, but in strictly radial expansion, the momentum in an element of solid angle must be conserved if all pressure effects are negligible as Oort's model assumes. The kinetic energy is, indeed, not conserved, but as the mass dm is accreted, its kinetic energy $\frac{1}{2}v_2^2\,dm$ relative to the border of the expanding mass is added to the internal energy. The integrated addition to the internal energy is

$$Q = \frac{m_1 m_2 v_1^2}{2(m_1 + m_2)}, \tag{20}$$

and the total energy

$$E = \frac{m_1 v_1^2}{2} + \frac{m_2 v_2^2}{2} + Q \tag{21}$$

is conserved. The simple model of deceleration by accretion is usable for the initial phase of the explosion but must be replaced for the later phases by a treatment that does not disregard the shock wave associated with the explosion.

The similarity solution for a strong explosion in a gas of constant heat capacity has been given by Taylor (1950) in numerical form and by Sedov (1959) in analytical form. The application of Sedov's results to supernova remnants was first suggested by Shklovsky (1962). It is assumed that the radiation losses of the gas behind the shock front are negligible. This adiabatic condition ceases to be satisfied when the radius of the object becomes larger than 30 pc (Heiles 1964); for all observed remnants, the adiabatic condition is probably fulfilled. If r is the radius of the shock front, E the energy released in the explosion, and ρ the interstellar density, Sedov finds

$$r = \left(\frac{2.2E}{\rho}\right)^{1/5} t^{2/5}; \tag{22}$$

or, if r_p is in parsecs, t_y in years, n_H the number of H atoms/cm³, and $n_H/n_{He} = 7$,

$$r_p = 2.0 \cdot 10^{-11}\left(\frac{E}{n_H}\right)^{1/5} t_y^{2/5}. \tag{23}$$

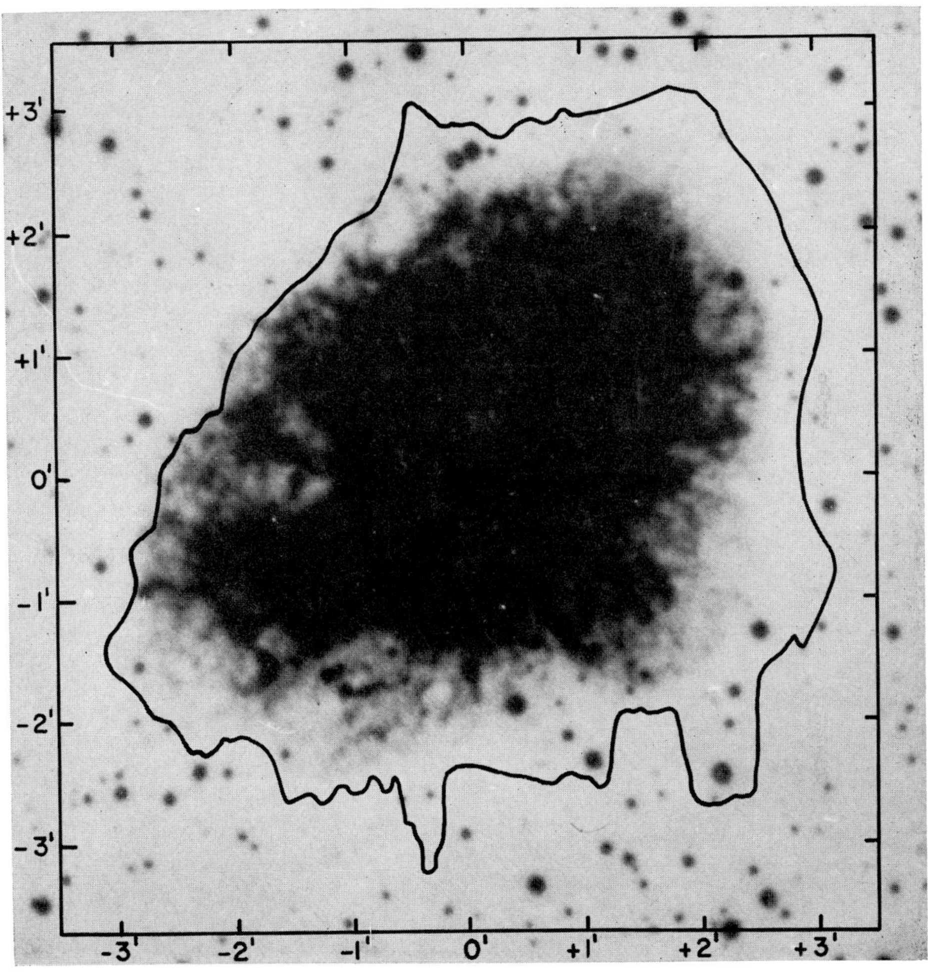

PLATE 1.—The filamentary system of the Crab Nebula photographed in the red (6300 A to 6600 A, 48-inch Schmidt telescope) and the outermost isophote by Woltjer (1957*a*).

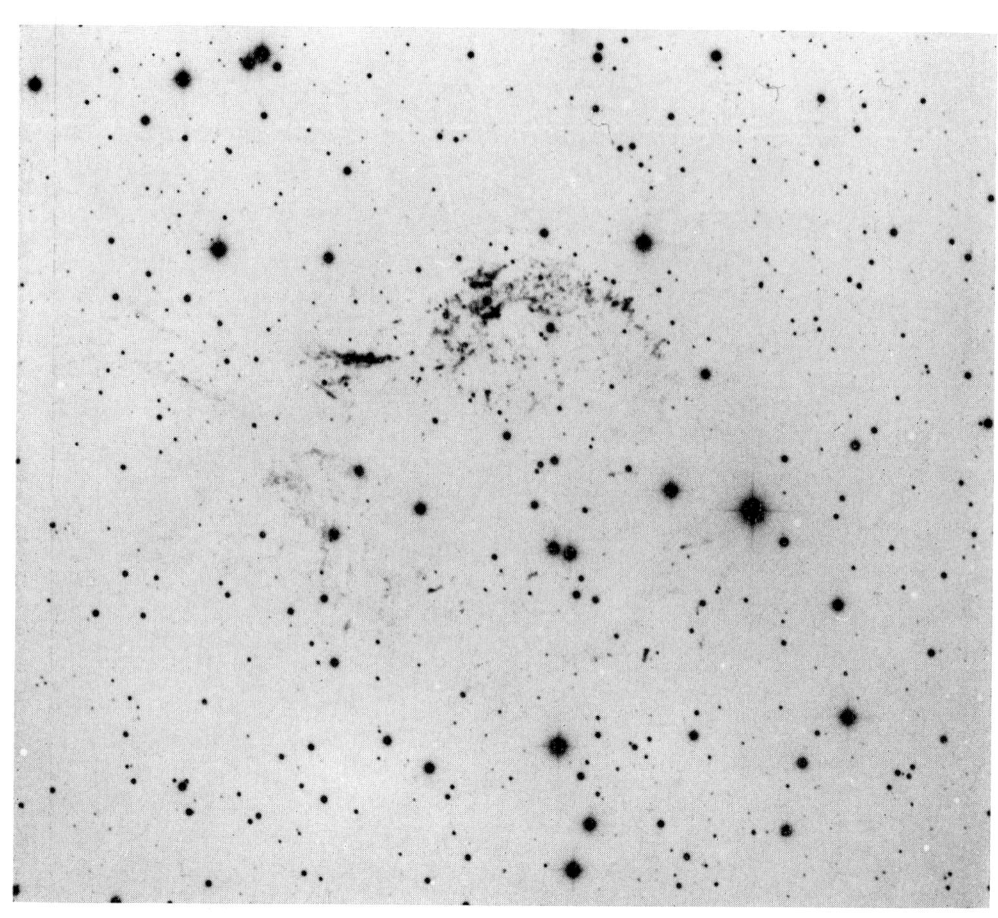

PLATE 2.—Cassiopeia A. 6300 A–6800 A; photographed with the 200-inch Palomar telescope. Scale (3″.66/mm) is the same as for Plate 3.

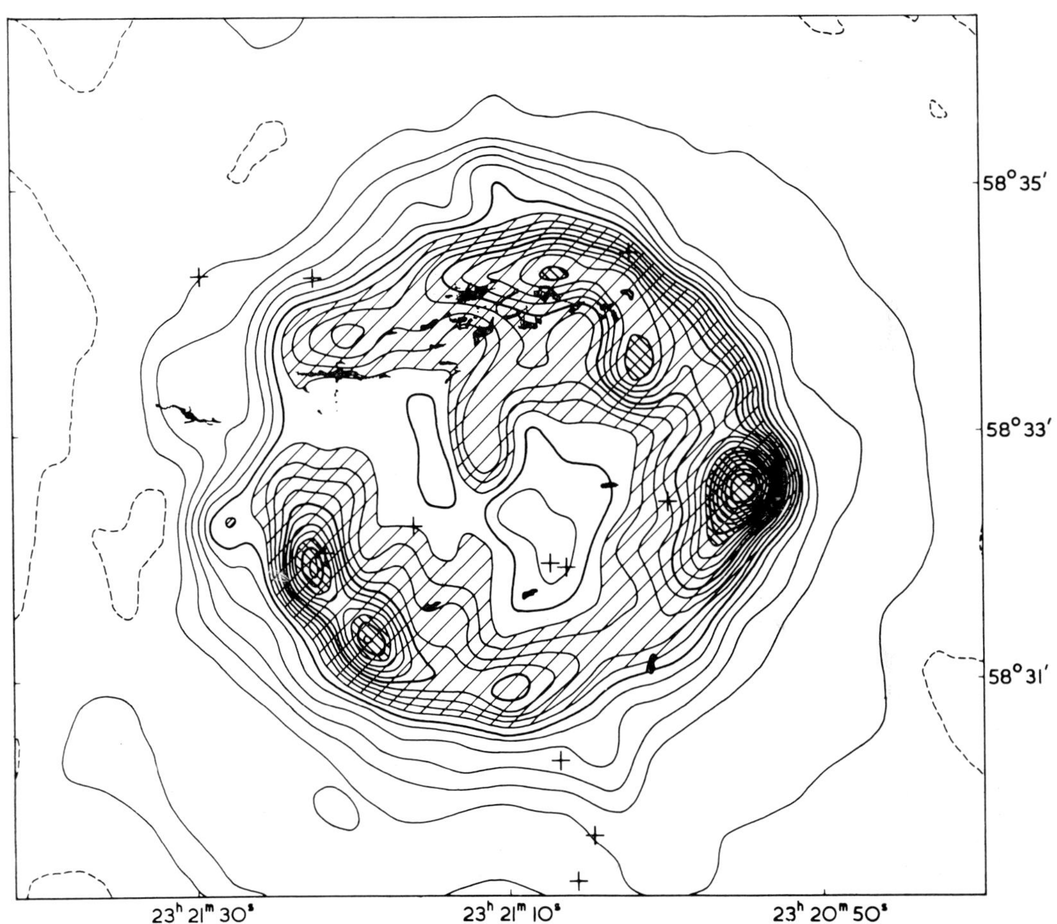

PLATE 3.—Contour map of Cassiopeia A at 1410 MHz also showing the main optical filaments. The contour interval is 3000° K. Positions of stars in the field are indicated by crosses (Ryle, Elsmore, and Neville 1965).

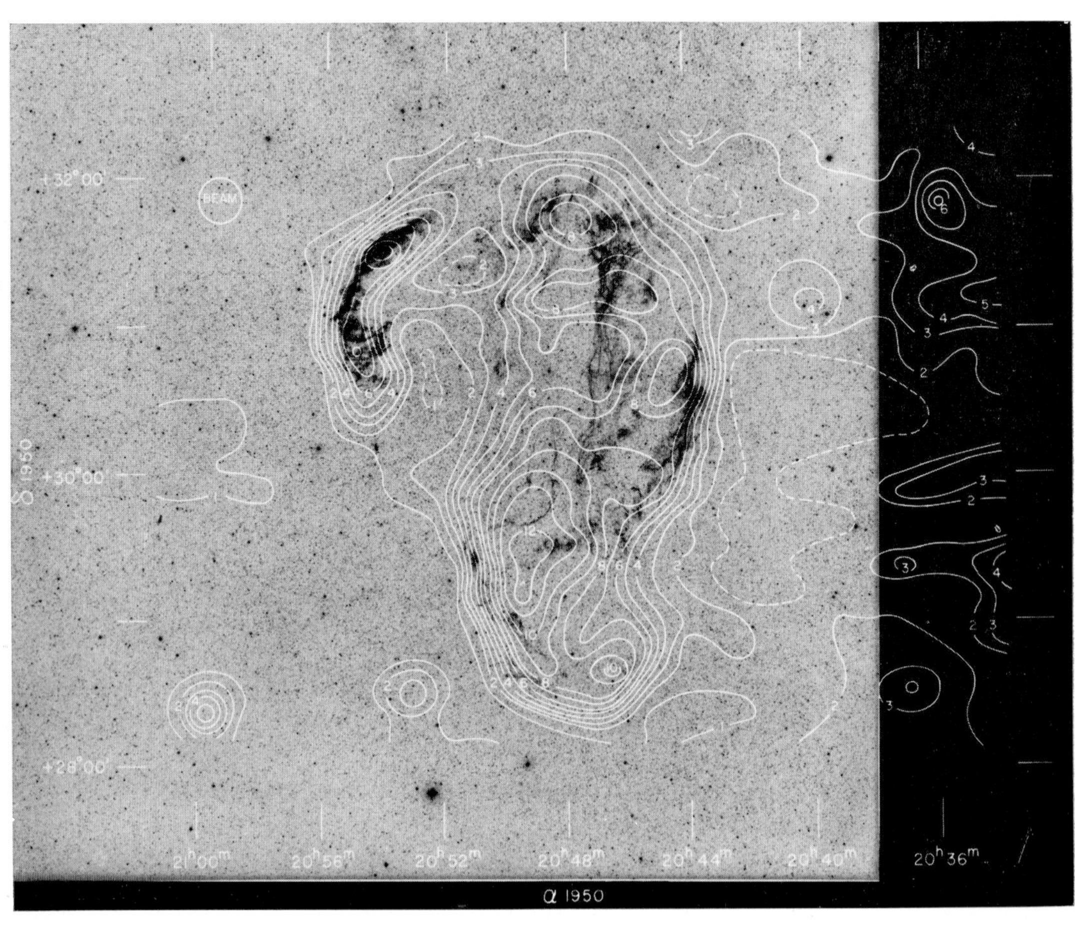

PLATE 4.—Contour map of the Cygnus Loop at 750 MHz obtained with the 300-foot telescope of the National Radio Astronomy Observatory (Hogg 1966). The contour unit is 0.75° K in antenna temperature.

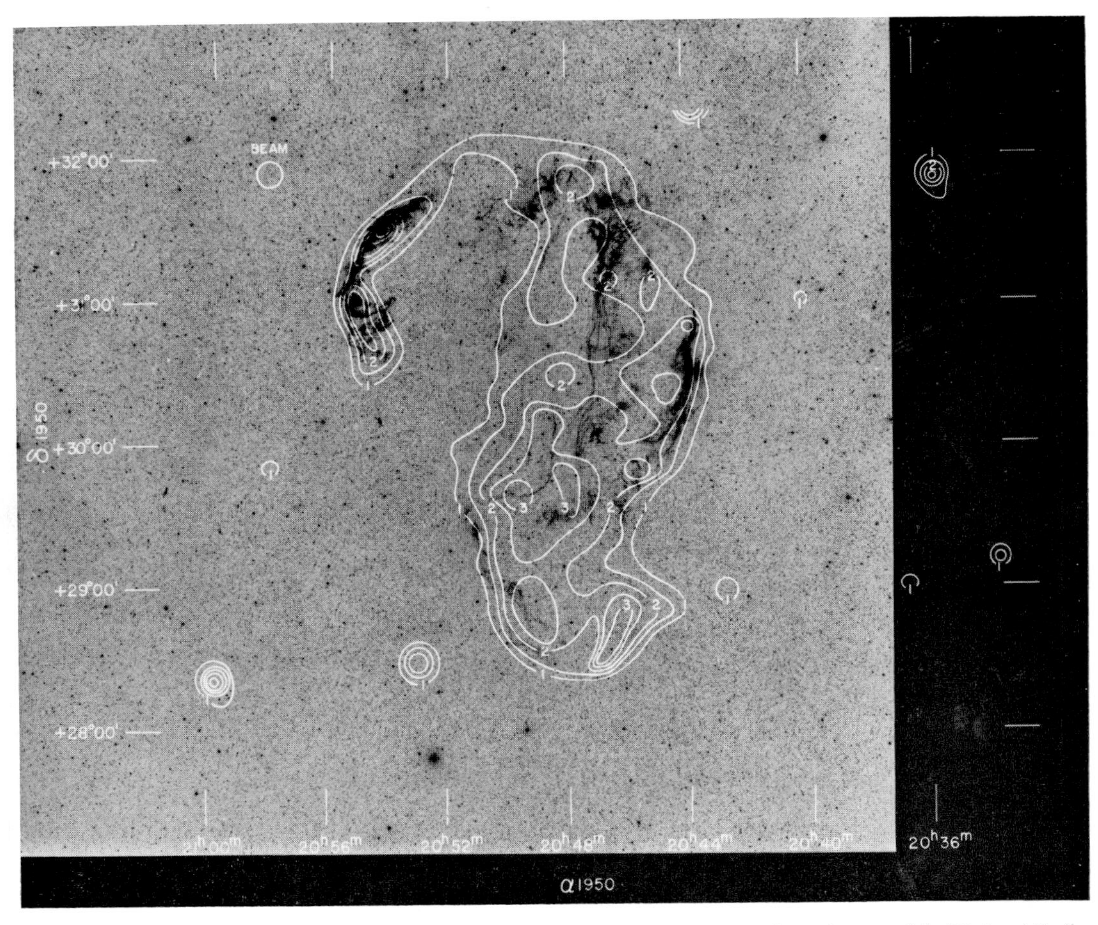

PLATE 5.—Contour map of the Cygnus Loop at 1400 MHz obtained with the 300-foot telescope of the National Radio Astronomy Observatory (Hogg 1966). The contour unit is 0.50° K in antenna temperature.

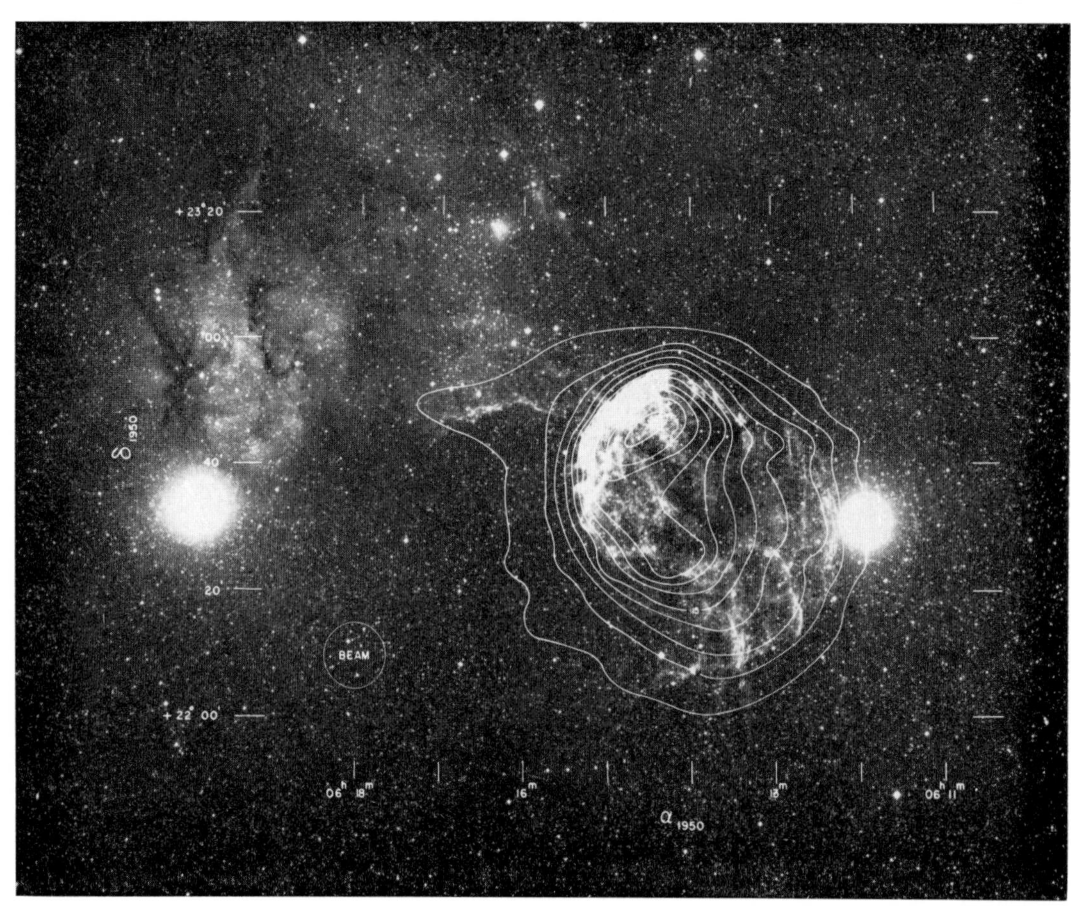

PLATE 6.—Contour map of IC 443 at 1400 MHz obtained with the 300-foot telescope of the National Radio Astronomy Observatory (Hogg 1964). The contour unit is 0.91°K in brightness temperature.

At time t_y, the velocity in km/sec is

$$v_2 = 7.6 \cdot 10^{-6} \left(\frac{E}{n_H}\right)^{1/5} t_y^{-3/5}. \tag{24}$$

The average velocity is

$$\frac{r}{t} = 2.5\, v_2. \tag{25}$$

The temperature T immediately behind the shock front is, with a correction of the numerical constant according to a comment by Spitzer,

$$T = \frac{0.015}{k}\, m_H \left(\frac{2.2E}{\rho}\right)^{2/5} t^{-6/5}, \tag{26}$$

where m_H is the proton mass, or

$$T = 0.8 \cdot 10^{-41} \left(\frac{E}{n_H}\right) r_p^{-3}. \tag{27}$$

With the radius of 5.4 pc and the age of 394 years, equation (23) gives the value 1.1×10^{52} erg cm^3 for E/n_H of Tycho's supernova. With $n_H = 0.1$, the initial energy is 0.1×10^{51} ergs. If the kinetic energy of expansion was initially $\frac{1}{3}$ of the total energy (Khare 1954), the corresponding initial kinetic energy of 4×10^{50} ergs would be consistent with an initial velocity of 20,000 km/sec, which is suggested by the spectrum around maximum, with an ejected mass of 0.1 $\mathfrak{M}_\odot$. From equation (25) the present velocity of expansion is 5400 km/sec; the corresponding motion of 0″.23/year is consistent with the observed (but not yet measured) motion. The temperature behind the shock front from equation (27) is 2.5×10^9 ° K. The optical emission at temperatures of this order is very small. The correct explanation for the optical faintness of the remnant is its high temperature. What may need to be explained is the presence of the exceedingly faint, but observable, features at the outer edge of the remnant.

The spectrum of the radio source 3C10 (Fig. 1) follows a simple power law with a flux density of 99.9 (± 2.6) $\times 10^{-26}$ watt m^{-2} Hz^{-1} at 400 MHz and a spectral index -0.67 ± 0.03 (Kellermann 1964*a*). Recent observations by Bondar *et al.* (1965) give flux densities that are about 10 per cent larger than those listed by Kellermann (1964*b*), but the claim seems to be unfounded that the spectral index between 178 and 427 is larger than Kellermann's value. The integrated flux between 10 MHz and 10^4 MHz is 29×10^{-16} watt m^{-2}. With the distance of 5000 pc, the corresponding total emitted power is 7.8×10^{33} erg sec^{-1}.

5.2. Kepler's Nova

The data for Kepler's nova of +1604 are far more sketchy than those for Tycho's nova. The light curve (Baade 1943) leaves no doubt that the supernova was of Type I. Some emission patches discovered by Baade are not adequate to give a clear picture of the size and shape of the remnant. Radial velocities of the order of -200 to -300 km/sec have been observed (Minkowski 1959). Such velocities would be appropriate if the observed features were close to the edge of a fast-expanding shell, but radio data now seem to rule out this interpretation. The radio source 3C358 is a difficult object to observe owing to its closeness to the galactic center. The situation is depicted in Figure 2. Kepler's position, with its error of $\pm 1'$, as given by Baade, is indicated by a straight cross and a broken circle. The locations of the features of the remnant are indicated by the shaded areas. The most precise radio position, right ascension by Fomalont *et al.* (1964) and declination by Wyndham and Read (1965), is indicated by the small open circle. The E-W extent of 3′.0 ± 0′.4 by Maltby and Moffet (1962) is in essential agreement with the results of observations of lunar occultations by Talen (1965*a*), who gives the size as 4′ × 7′ with the major axis about N-S. The position given by Talen is indicated by the lower cross. The solid ellipse fitted to this description seems to be the maximum extent consistent with the occultation observations. The details of the observations made available by Talen (1965*b*) show that some of the occultation curves are of

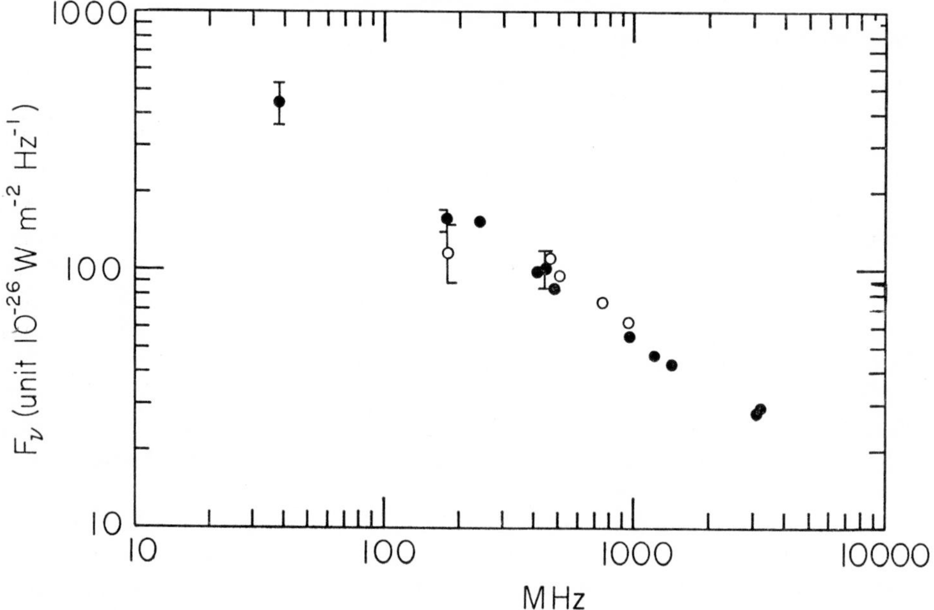

FIGURE 1.—Radio spectrum of Tycho's nova (3C10). *Solid circles:* Kellermann (1964*b*); *open circles:* Bondar *et al.* (1965).

very low quality owing to interference and the low altitude of the source at the time of the occultation. If such observations are omitted, the remaining data are consistent with the broken ellipse that has a major axis of 5′ and that would lead to a position very close to the interferometric position by Fomalont *et al.* and by Wyndham and Read.

There are no observations that permit a determination of the distance. The observed details are too diffuse to observe transverse motions with sufficient precision to obtain the distance by combining them with the radial velocities, and at the galactic longitude $l^{II} = 4^\circ.51$, the velocity of 21-cm absorption lines cannot be used as a distance indicator. It is also not possible to obtain a value for the ambient interstellar density. If the supernova were of average absolute magnitude −19.0, the observed magnitude at maximum −2.25 mag (Baade 1943) with the uncertain value −2.6 mag for the interstellar absorption (Minkowski 1964) leads to a distance of 6700 pc. At this distance and at galactic latitude b^{II} + 6°81, the supernova is 790 pc above the galactic plane. A major axis of 5′ leads to a linear semimajor axis of 4.8 pc and to an average velocity of 12,900 km/sec.

With these data and the known age, equation (24) gives the value 7×10^{51} erg cm^3 for E/n_H. Assuming the value $E = 1.1 \times 10^{51}$ ergs, obtained earlier for Tycho's nova as typical for supernovae of Type I, we find $n_H = 0.15$ cm^{-3}. This value is perhaps more than might be expected at 790 pc above the plane, but in view of the obvious uncertainty it does not argue against close similarity of Tycho's and Kepler's novae. With $n_H = 0.15$ cm^{-3}, the accreted mass becomes about 3 $\mathfrak{M}_\odot$.

The present velocity of expansion from equation (25) is 5200 km/sec. This is inconsistent with the low velocities observed in the emission patches, some of which are close to the center of the remnant. Woltjer (1964) has suggested that these features are not part of the expanding shell, but interstellar clouds in the neighborhood of the remnant

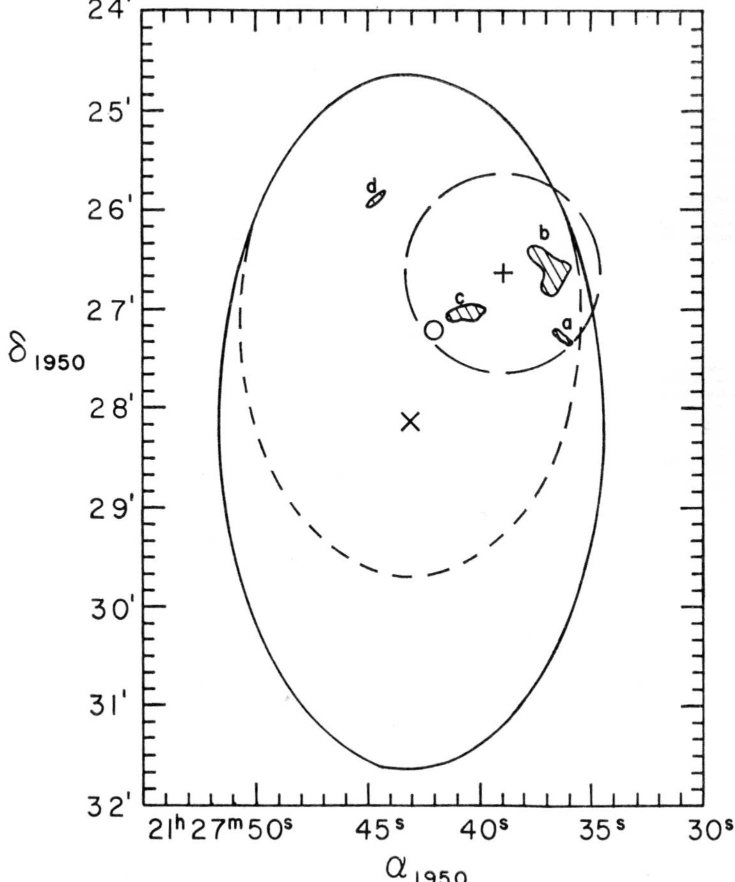

FIGURE 2.—Kepler's nova (3C358). Shaded areas show observed emission features. *Upper cross* and *broken circle:* position of the nova and its error (Baade 1943). *Circle:* position of 3C10 (Fomalont *et al.* 1964; Wyndham and Read 1965). *Lower cross* and *solid ellipse:* position and size of 3C10 (Talen 1965*a*). *Broken ellipse:* size from Talen's observations excluding results of low signal-to-noise ratios.

and possibly excited by it. The assumption that such clouds are present at 730 pc above the galactic plane is, however, not entirely satisfactory. The implication that the remnant itself is not optically observable is readily acceptable as a consequence of the high temperature behind the shock front from equation (27).

The radio spectrum of the source 3C358 follows a simple power law with a flux density of 33.6 (± 1.0) $\times 10^{-26}$ watt m^{-2} Hz^{-1} at 400 MHz and a spectral index of -0.64 ± 0.2 (Kellermann 1964*a*). The integrated flux between 10 MHz and 10 GHz is 10×10^{-16} watt m^{-2}. With a distance of 6700 pc, the corresponding total power is 5.3×10^{33} erg sec^{-1}; this is similar to the power 7.6×10^{33} erg sec^{-1} of Tycho's nova, which is 32 years older. With the time dependence given by equation (14), the power of Kepler's nova at the same age as Tycho's nova becomes 3.7×10^{33}. The use of equation (14), however, is not justified because it is based on the assumption of adiabatic expansion, and both remnants are decelerated. In any case, the emitted powers of the two remnants are reasonably similar.

5.3. The Supernova of +1006

The supernova of +1006 was probably the brightest nova ever recorded. The historical data have recently been made accessible in detail (Goldstein 1965; Goldstein and Ho 1965). They are not adequate for a definite conclusion on the nature of the nova. Salient points seem to be (Minkowski 1966*a*):

1. The supernova was visible for much longer than one year. This excludes Type II.
2. The ancient records contain some statements suggesting that the supernova may have shown sudden or irregular brightness changes and might have been visible for more than $2\frac{1}{2}$ years. If this is true, it could only have been an object with an irregular light curve, such as the supernova in NGC 1058 (Bertola 1964, 1965) designated as Type V by Zwicky (1964). Such objects, however, are rare.
3. Probably the details of the ancient records should not be taken too literally. It is quite possible that the supernova was not visible for more than $2\frac{1}{2}$ years and that there were no sudden or irregular brightness changes. The interpretation as Type I would then be permissible, and is suggested as the most probable solution. If this solution is accepted, the maximum apparent brightness may have been -8 mag. If the absolute magnitude at maximum was average, a distance of 1300 pc follows.

A search for the remnant by Baade (Minkowski 1965) was unsuccessful. Optically, the remnant thus has very low surface brightness. As a radio source, the remnant has been discovered by Gardner and Milne (1965). Its position, $\alpha = 14^{h}59^{m}.6$, $\delta = -41°42'$ (1950) agrees well with the historical evidence. The source has the structure of an incomplete, possibly toroidal, slightly elliptical shell with a major axis of about 40′, or a linear radius of 7.5 pc, if the distance is 1300 pc. The corresponding average velocity of expansion is 7600 km/sec, and the present velocity of expansion from equation (25) is 2800 km/sec. The ambient interstellar density is not known. The value $E/n_{\rm H} = 1.0 \times 10^{52}$ follows from equation (23). Assuming the value $E = 1.1 \times 10^{51}$ ergs obtained for Tycho's nova as typical for a supernova of Type I, the value $n_{\rm H} = 0.11$ cm^{-3} follows. This is reasonable for a distance of 330 pc above the galactic plane, particularly in view of the great uncertainty of the distance. The interpretation as a supernova of Type I seems satisfactory.

The flux density of the radio source BGM 1459-41 (Bolton, Gardner, and Mackey

1964) is 17.4×10^{-26} watt m^{-2} Hz^{-1} at 400 MHz with a spectral index of about 0.6 (Gardner and Milne 1965). The integrated flux between 10 MHz and 10 GHz is 5.3×10^{-16} watt m^{-2}. With a distance of 1300 pc, the corresponding total power is 1.1×10^{32} erg sec^{-1}. If equation (14) is used to represent the time dependence, the total power at the time when the age was the same as that of Tycho's nova now becomes 6.0×10^{33} erg sec^{-1}. Because of the uncertainties and of the illegitimate use of equation (14) for nonadiabatic expansion, the agreement with the present total power of Tycho's nova is quite satisfactory.

5.4. The Crab Nebula

The Crab Nebula (NGC 1952) was long known as a remarkable object. Many of its properties are unique. As a radio source (Taurus A; 3C144), it is so strong that we may exclude the possibility that another object like it can be found within somewhat more than that half of the Galaxy that is nearest to us.

The earlier discussions and observations, summarized by Mayall (1962), culminated in the discussions by Duyvendak (1942) and by Mayall and Oort (1942) and led to general acceptance of the Crab Nebula as a remnant of the nova of +1054 and to the definitive recognition of that nova as a supernova.

5.4.1. The Supernova of +1054.—Chinese and Japanese records provide rather limited information. At +23 days, the nova ceased to be visible in daytime. Under the circumstances described in detail by Mayall and Oort, the apparent visual magnitude at that time was −3.5.

At +650 days the object was reported to be invisible to the naked eye. The conclusion seems to be incorrect that the object became fainter than visual magnitude +6 at that time. The supernova was then approaching conjunction with the Sun and followed it by only 2½ hours. The zodiacal light thus contributed noticeably to the sky brightness, and at the end of twilight the altitude was below 20°. The limiting magnitude for visibility must have been much brighter than +6. Actually, the limiting magnitude +7 is suggested by the number of visible stars stated for the Chinese observations about 900 years earlier (Needham 1959), but for observations near the horizon, the limit is not certain. The brightness at +650 days cannot be established precisely. The assumption that, reduced to the zenith, it was +5 is suggested on the basis of computations by Tousey and Koomen (1953).

This discussion indicates that the brightness declined by 8.5 mag from +23 days to +650 days. This value is to be compared to the decay of supernovae of Type I. The supernova in IC 4182 is the only one observed for a comparable length of time, +635 days (Baade *et al.* 1956). On the basis of the revision of Baade's photometric data by Mihalas (1963), and with a slight extrapolation, the photographic decline from +23 to +653 days is 10.2 mag. The $(B - V)$ color at +23 days was +0.9 mag (Mihalas 1963). The color at +650 days is not known, but it is plausible that during the nearly exponential decline it did not change much from the value +0.5 at +100 days. The visual decline from +23 days to +650 days then is 10.6 mag. Extrapolation of the light curves of Tycho's and Kepler's supernovae is in accord with that value. The difference of 2 mag between the supernova of +1054 and supernovae of Type I is not conclusive evidence in view of the many uncertainties, but it tends to contradict the interpretation of the supernova of +1054 as Type I and certainly does not make it mandatory. The interpretation as Type I is strongly contradicted by the obvious dissimilarity between the Crab Nebula and the remnants of Tycho's and Kepler's supernovae, which were undoubtedly Type I. The available data are too scanty to permit the assignment of a type to the supernova of +1054.

If the type is not known, a valid determination of the magnitude at maximum from the magnitude −3.5 at +23 days is not possible. It is not more than a plausible guess to assume the apparent visual magnitude −5 at maximum. With a distance of 1800 pc (Sec. 5.4.3) and the interstellar absorption of 1.07 mag/kpc (O'Dell 1962), the absolute visual magnitude at maximum is −18.21. This would be an acceptable value for supernovae of Type I as well as of Type II (Sec. 4). In any case, the supernova was definitely not superluminous, as it seemed to have been before the revision of the extragalactic distance scale which raised the average photographic brightness of supernovae at maximum; for example, for the supernovae of Type I, the value −14.3 mag (Baade 1938) was raised to −18.9 mag.

5.4.2. General Remarks on the Crab Nebula.—It is well known that the Crab Nebula consists of two parts whose physical characteristics are quite different: a mass of filaments that surrounds the central part and a diffuse mass that fills the center of the nebula. The filaments show a spectrum of emission lines essentially similar to those of all gaseous nebulae. There is no reason to believe that the filaments contribute more to the radio emission of the nebula than their negligibly weak thermal emission. The diffuse mass, on the other hand, shows a continuous spectrum whose strongly polarized emission proves that it is due to synchrotron radiation. Observations of the continuous spectrum now cover, with large gaps in the infrared and far ultraviolet, the frequency range from 1.25×10^7 Hz to the X-ray region, with a cutoff above 10^{19} Hz.

Since the filaments show emission lines and the diffuse mass shows a continuous spectrum, they can be photographed separately with suitably chosen combinations of plates and filters. Plate 1 shows a photograph strongly exposed in the red where Hα emission from the filaments dominates, together with the outermost isophote obtained by Woltjer (1957*a*) from a photograph in the visual region which shows the continuous spectrum of the diffuse mass only. In a general way, the outlines of the filamentary system and of the diffuse mass are similar. Except for the north-following edge, where both systems seem to have a common border, the diffuse mass seems to extend slightly farther from the center.

It is obvious that "the" position and "the" size of an object as complex as the Crab Nebula are data that have no meaning unless the method used to determine them is specified accurately and in detail, particularly with regard to the smoothing of observations and corrections for limited resolving power. The position frequently quoted as "the" optical position and ascribed to Baade and Minkowski (1954) is, as stated in that reference, merely the NGC position precessed to 1950.0; this position clearly has low accuracy. The most precise positions available are those for the two 16th-magnitude stars near the center of the nebula in Table 3. The north-following star is not physically connected with the nebula, but the south-preceding star could be the central star of the nebula and would then be the most appropriate point of reference. Other suitable, but less precise, data in Table 3 are samples of positions that were derived from Woltjer's isophotes (1957*b*). It should be kept in mind that minor changes in the brightness distribution are known to occur. Variability of positions thus might occur, but it has not yet been observed.

5.4.3. The Filamentary System.—The expansion of the filamentary system was discovered by Duncan (1921) and he investigated it later in more detail (Duncan 1939). It has been discussed by Baade (1942), who has also tried to answer the question of whether one of the two stars near the center of the nebula is the central star—the stellar remnant of the supernova. The results of the discussion may be summarized as follows: The angular expansion in the direction of the major axis is 0″.235 ± 0″.008 per year, larger than the average expansion of 0″.201 ± 0″.006 obtained from the size of the major axis and the known age. The north-following star has no observable proper motion; association of this star and the nebula thus can be considered as excluded. The nebula and the south-preceding star have observable and similar proper motions. The accuracy of the proper motions is low, however, and for reasons discussed in detail by Baade, a final decision on whether the nebula and the south-preceding star are associated must be deferred.

TABLE 3

POSITIONS IN THE CRAB NEBULA

	α(1950)	δ(1950)	References
NGC 1952	$5^h31^m30^s$	+21° 59′.3	Baade and Minkowski (1954)
S-prec. star*	31.46	58′ 54″.8	G. Pels (Seeger and Westerhout 1957)
N-foll. star	31.64	58′ 58″.9	
Diffuse mass			
Maximum	30.5	58′ 55″	Derived from Woltjer's isophotes (1957*b*)
Maximum of E-W strip distribution	30.9	....	
Center of E-W strip distribution at half intensity	30.1	....	

* A position by Dewhirst, quoted by Costain, Whitfield, and Elsmore (1956), does not refer to the N-following star, as stated, but to the S-preceding star (Dewhirst 1965) and agrees with the position by Pels.

The apparent visual magnitude of the south-preceding star is 15.45, its observed $(P - V)$ +0.41 (Baade 1942). If the distance is 1800 pc and the visual interstellar absorption is 1.07 mag/kpc (O'Dell 1962), the absolute visual magnitude and the $(B - V)$ color, corrected for interstellar absorption, are +2.24 and −0.13, respectively. These values indicate a region of the color-magnitude diagram in which, for instance, nuclei of planetary nebulae are found. Minkowski (1942) and Zwicky (Bowen 1958) have found neither absorption nor emission features in the spectrum of this star. The lack of these features is not necessarily inconsistent with the absolute visual magnitude and $(B - V)$ color values. A report in a preprint, but omitted from the published article (Hoyle *et al.* 1964), that Kraft has found an F-type spectrum is premature (Kraft 1963). If the star were indeed a normal F-type star not associated with the nebula, the color would suggest negligible reddening, but the apparent magnitude would require a substantial distance. The possibility would remain that the star is double, one component being the stellar remnant of the supernova with very blue color, the other an F-type star (Shklovsky 1965), but before the observational data are definitely established, the question of whether the south-preceding star is in any way associated with the nebula or whether the stellar remnant is a fainter star, if it is observable at all, must remain open.

The angular expansion of the filamentary system in the direction of the major axis, combined with the angular size of the major axis, places the date of the outburst in 1184.

The difference of 130 years between that date and the actual date 1054, if real, would indicate an accelerated expansion. Baade rejects this conclusion because the acceleration is too large to be explained as the result of radiation pressure from the central star; today, possible effects of magnetic pressure and the pressure of highly energetic particles might be considered (Woltjer 1957*b*). It is not at all unlikely, however, that the angular expansion derived from Duncan's measures is too high and that Baade's reduction underestimates the error of the expansion, which he derived from only 8 measured points that were selected because they are closest to the major axis. In judging the reliability of the result, we must keep in mind that the transverse motions must show large irregularities similar to those of the radial velocities, and that the simple picture of an expanding elliptical shell is not more than a crude approximation. Recent work at the Pulkovo Observatory (Deutsch 1966) contradicts the presence of acceleration. The average value 0″.201 per year for the expansion in the direction of the major axis deserves most confidence. (*Added in proof:* V. Trimble, at the 123d Meeting of the American Astronomical Society, Los Angeles, December 27–30, 1966, reports that a preliminary reduction of recent observations of the motion of the filaments leads to a much smaller acceleration than that derived from Duncan's observations.)

The spectroscopic observations by Mayall (1962) show clearly that the filamentary system is expanding. The doubling of the lines at the center of the nebula shows a velocity of expansion of about 1150 km/sec. This velocity cannot be stated with great precision, owing to the large irregularities of the radial velocities. Substantial radial velocities are observed at the very edge of the nebulosity (Minkowski 1964), where a radially expanding homogeneous shell would show not more than the systemic velocity. A diagram of the filamentary system and its radial velocities by Mayall shows how much the conditions deviate from those in a homogeneous expanding shell. If it is assumed that the expansion is strictly radial, this diagram can be interpreted as a representation of the spatial distribution of the filaments. The filaments appear to form a very irregular net around the nebula; some of them may be embedded in the interior of the shell.

As Woltjer (1957*b*) remarked first, a model must be adopted for the approximately elliptical nebula so that its distance can be determined from the radial velocity of expansion observed at the center of the nebula and from the angular rate of expansion at the end of the major axis. Baade (1942) assumes tacitly that the nebula is an oblate spheroid with the equatorial plane normal to the plane of the sky. If this is true, the observed radial velocity of expansion is equal to that in the major axis, and the distance is then 1200 pc. If, however, the assumption is made that the nebula is a prolate ellipsoid with the major axis in the plane of the sky, the observed radial velocity of expansion is in the minor axis, and the radial velocity in the major axis must be larger by the axial ratio 1.5 of the nebula, that is, 1720 km/sec, and the distance then is 1800 pc. It is not necessary to assume that the nebula is actually a prolate spheroid; for a mass that is expanding under the constraint of a magnetic field, an elongated shape that can be approximated by this description, however, seems much more likely than an oblate form. The main argument in favor of the greater distance arises from difficulties in the

the intensity ratio of the [O III] and [O II] lines that led Woltjer (1957*b*) to suggest a prolate ellipsoid as a description. The [O III]/[O II] ratio also argues against much greater distances than 1800 pc.

With a distance of 1800 pc, the Crab Nebula at $l^{II} = 184^\circ\!.55$, $b^{II} = -5^\circ\!.78$, is 180 pc below the galactic plane.

As noted above, at the time of Baade's discussion, the absolute brightness of supernovae was severely underestimated owing to the use of an erroneous extragalactic distance scale. The supernova of A.D. 1054 thus seemed to be overluminous, and there was no reason to suspect that the distance derived from the oblate model might be too small.

The larger velocity of expansion in the major axis that corresponds to the prolate model is still much below the velocity of expansion of 20,000 km/sec that is typical of Type I supernovae. If the velocity of expansion has not been reduced by interaction of the material with the interstellar medium, the Crab Nebula cannot be the remnant of a supernova of Type I. That the present expansion is certainly not much smaller than the average expansion is clear evidence against the presence of strong deceleration. Unless one wants to make ad hoc assumptions, such as that the supernova of +1054 was in the center of a small dense cloud just adequate to reduce the velocity of expansion to a few per cent of its initial value in a short time, the low velocity of expansion is a strong argument against the assignment of Type I to the supernova of +1054. Supplementary evidence is given by the mass of the filaments and by their low hydrogen abundance.

The mass of the filaments is best determined from the flux density of a hydrogen line (Woltjer 1957*b*); distance, electron temperature, and electron density must be known. The most reliable value is that by O'Dell (1962). It must, however, be revised to conform to the larger distance now ascribed to the nebula. The value by O'Dell is too large by a factor of about 2, because the temperature dependence of the numerical factor in his equation (7) was overlooked. With the correct numerical factor for $T_e = 17{,}000^\circ$ K (Burgess 1958), the value of the mass for the distance of 1030 pc assumed by O'Dell, corrected for the presence of faint filaments of low density, becomes 0.32 $\mathfrak{M}_\odot$. This value is found by determining the volume of the filaments from the ratio of the total flux in Hβ, corrected for interstellar absorption, to the volume emissivity, which is computed from the observed electron density under the assumption that the excitation follows Menzel's Case *B*. The mass is then obtained from the volume and the mass density that follows from the observed electron density. The basic data are the photoelectrically determined flux, $F(\text{H}\beta) = 1.24 \times 10^{-11}$ erg cm^{-2} sec^{-1}; the electron density of 10^3 cm^{-3} for the bright filaments (Osterbrock 1957); an electron temperature of 17,000° K; and a total mass to hydrogen mass ratio of 3.3 (Woltjer 1957*b*). In a similar way, Woltjer finds a mass of 0.093 $\mathfrak{M}_\odot$ for the bright filaments; with the correction for the presence of faint filaments, the mass becomes 0.19 $\mathfrak{M}_\odot$. The value by O'Dell is larger and probably more reliable, owing to the use of a photoelectrically determined $F(\text{H}\beta)$. Previous values based on estimates of numbers of filaments and their volumes, such as that of 0.1 $\mathfrak{M}_\odot$ by Osterbrock (1957), seem to be more uncertain, relatively crude estimates.

If the distance is increased to 1800 pc, the value of the mass increases because the total flux is proportional to the square of the distance, and because the correction for interstellar absorption becomes larger. The mass given by O'Dell's data is then 1.6 $\mathfrak{M}_\odot$. Osterbrock's estimated mass, proportional to the third power of the distance, becomes 0.34 $\mathfrak{M}_\odot$. Osterbrock's value is directly proportional, but O'Dell's is inversely proportional, to the electron density. Thus, if Osterbrock's estimated mean value $n_e = 10^{-3}$ cm^{-3} were too small by a factor of 2.2, the values obtained by the two different methods would agree, and the mass then would be 0.75 $\mathfrak{M}_\odot$. About one solar mass is probably the most realistic value. It is in agreement with the mass needed to explain the frequency dependence of the polarization of the continuous emission due to the Faraday effect in the filaments (Burn 1966).

Some part of this mass must be interstellar material swept up by the expanding nebula. With a volume of 1.9×10^{56} cm^3 for the distance 1800 pc, the accreted mass is 0.26 n_H $\mathfrak{M}_\odot$ if n_H is the interstellar density of hydrogen, and if $n_\mathrm{H}/n_\mathrm{He} = 7$ (Minkowski 1964, where the volumes have been incorrectly given too small). The Crab Nebula is too close to the galactic anticenter to permit a determination of the density distribution in its neighborhood from the 21-cm line, but the trend of the density distribution (Westerhout 1957) shows that 180 pc below the galactic plane, n_H is unlikely to be as large as 0.3 cm^{-3}. The accreted mass thus is not quite negligible, but the suggestion by Hoyle *et al.* (1964) that the entire mass of the filaments is of interstellar origin does not seem to be acceptable. The high abundance of He reported by Woltjer (1957*b*), who finds $n_\mathrm{H}/n_\mathrm{He}$ to be about 2 (a value that does not depend strongly on the electron temperature), also argues against such an assumption. The overabundance of helium is actually strong enough to be recognized by comparing the relative intensities of H and He lines in the Crab Nebula with those in planetary nebulae (Minkowski 1942). If the interstellar ratio $n_\mathrm{H}/n_\mathrm{He}$ is 7, and if the ejected mass initially contained no H, about half of the present mass would be accreted mass. This is an upper limit; the accreted mass would be smaller if some H had been present initially.

If only a small fraction of the present mass is accreted, the deceleration must be small, and the present velocity of expansion must be close to the initial velocity. This assumption tends to confirm the same conclusion reached from the close similarity of the present and the average expansion—that the nebula is not the remnant of a supernova of Type I. An ejected mass of the order of one solar mass also seems high for a Population I object.

The spectrum of the filaments has been discussed in detail by Woltjer (1957*b*). The main results of that investigation, which have been already mentioned and used above, may be quoted here from Woltjer with some indicated omissions:

> The excitation of the filaments is shown to be mostly due to the ultraviolet radiation from the inner part of the nebula, with possibly a contribution from the central star. Collisional excitation plays only a minor role. The temperature of the nebula is found to be 17,000° or lower. The electron density in a few strong filaments is about 1500 cm^{-3}. The abundances of H, He, N, O, Ne, and S have been derived. The ratio of H and He to the light elements is probably not very abnormal. Helium, however, is at least a factor of two overabundant relative to hydrogen. . . . From the ionization conditions in the filaments the strength of the far-ultraviolet emission from the inner parts of the nebula has been derived. The differences between the line-intensity ratios in the various filaments may be ascribed to changes in the excitation conditions. The intensity ratio of the [O III] and [O II] lines shows a systematic dependence on the position in the nebula. This may be due either to anisotropy of the ultraviolet radiation field or to an underestimate of the distance of the nebula, which should possibly be doubled. The nebula would then resemble a prolate instead of an oblate spheroid. . . .

5.4.4. The Diffuse Mass.—The diffuse mass that emits the continuous spectrum of the Crab Nebula is by no means devoid of structure. In its full complexity, the brightness distribution can be studied only in the optical region. Time variations of the brightness distribution are well established.

As seen in the visual region, the diffuse mass has nearly the same extent as the filamentary system (Plate 1), and therefore must have expanded at very nearly the same rate as the filaments. An attempt to measure the expansion of the diffuse mass by Johnson (1963) shows that the expansion of the outer parts indeed matches the expansion of the filamentary system very satisfactorily. The difficulty of precise measures of diffuse features is obvious. In the central part, changes are more chaotic—interpreted by Johnson as the result of acceleration. If any weight can be attributed to individual values, some points actually show inward motions. It seems likely that variations of the brightness distribution that are not connected with the expansion play a role near the center, where the effects of the expansion must be small. Such variations were first noticed and described by Lampland (1921). In addition to these changes, moving features—"ripples" —were found by Baade (Oort and Walraven 1956).

The moving ripples were first seen 7″ to the west of the line joining the two stars near the center. They moved westward and seemed to disappear before or upon reaching a permanent ridge somewhat more than 8″ westward of the baseline. The observed motion of 1″.01 in 67.8 days, combined with the distance of 1800 pc, give a transverse velocity of 4.7×10^4 km/sec. Woltjer (1957*b*) has suggested that the moving ripples be interpreted as hydromagnetic waves.

The changes first noticed by Lampland, and later changes described by Oort and Walraven, seem to be mainly variations of the shape and intensity of various features of the diffuse mass. Motions, however, also seem to occur. One feature in the northwestern part of the nebula, for instance, seems to have moved northward by 15″ between 1924 and 1938. With a distance of 1800 pc, this corresponds to a transverse motion of 10^4 km/sec.

The brightness distribution has been determined by Woltjer (1957*a*) from a plate obtained by Baade in 1955 in the visual region where the light of the filaments is quite negligible. Woltjer's isophotes should be used to compute positions and sizes or brightness distributions that are fully equivalent to radio data that give strip distributions or have low resolving power. As an example, the E-W strip distribution computed from the isophotes is shown in Figure 3. Such a procedure has been used very rarely. For the size radio results give such data as, for example, the half-intensity width of the distribution, assumed to be Gaussian for the reduction of interferometric observations; and for the position that of the intensity maximum or of the center at half-intensity. Information on the procedure used to smooth observations and to correct for limited resolving power is often inadequate.

At low frequencies, an important difference between the radio and the optical features is established by the interferometric observations by Hewish and Okoye (1964) at 38 MHz and by observations of a lunar occultation by Andrew *et al.* (1964) at 26.5, 81.5, 178, and 400 MHz. A localized area, about 1′.2 south preceding the center, shows a spectrum with the very steep spectral index of −1.2. Recent observations by Slee (1966) place this source within 1′ from the center. Observations by Gower (1966) in progress at the Mullard Radio Astronomy Observatory indicate that the small diameter low-frequency source is close to the center of the larger radio source. The original observations by Andrew *et al.* were confused by interference and are obviously in error. To think of this remarkable source as connected in some way with the stellar remnant is tempting, and many discussions consider this as true. At 81.5 MHz, the localized source contributes about 10 per cent of the flux of the nebula. The contribution increases toward lower frequencies; at 38 MHz it is about 30 per cent. In view of the steep spectrum, it is not surprising that no trace of this source can be seen in the optical region. The discovery by Hewish and Okoye (1965) of interplanetary scintillation at 38 MHz, which must be attributed to this area, indicates a diameter of 0″.1 and implies a brightness temperature of 10^{14} ° K, which cannot be reconciled with synchrotron radiation. Coherent plasma oscillation has been suggested by Ginzburg and Ozernoy (1966) as the process

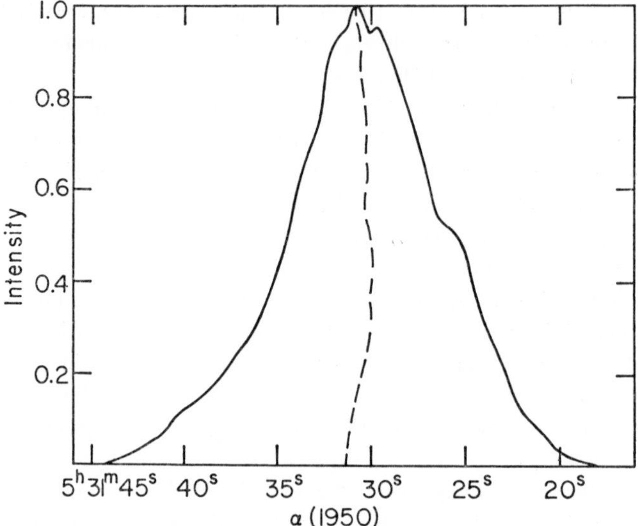

FIGURE 3.—E-W strip brightness distribution of the Crab Nebula computed from isophotes y Woltjer (1957*a*).

responsible for the radio emission. The presence of this source is not clearly shown by the observation of a lunar occultation by Costain *et al.* (1956) at 38 MHz; these observations, however, are strongly confused by interference.

Many source positions of high accuracy have been obtained from observations at frequencies above 169 MHz (listed by Howard and Maran 1965). In general, they agree well enough with positions in the expected range. Exceptions to be noted are at 1420 MHz, α(1950) $5^h31^m27^s.3 \pm 0^s.7$ (Labrum *et al.* 1964); at 3000 MHz, α(1950) $5^h31^m28^s.5 \pm 1^s$ (Matveenko and Khromov 1965); at 3300 MHz, α(1950) $5^h31^m28^s.4 \pm 0^s.7$ (Little 1963); and at 37.4 GHz, α(1950) $5^h31^m35^s \pm 0^s.5$ (Kuzmin and Salomonovich 1961). There is no reason to believe that these results have anything to do with real changes of the brightness distribution.

Determinations of the size, carried out with fan beams, interferometers, and lunar occultation, refer to strip distributions, with the majority of the data representing results from scans in the E-W direction that are directly comparable with the visual strip distribution in Figure 3. At one time it was believed that size increases with the frequency, but the data now available show clearly that over the range from 38 MHz (where the compact source must be excluded) to 94 GHz, the halfwidth of the E-W strip distribution remains constant at about $3'.5$. This is twice the halfwidth of $1'.8$ of the visual strip distribution and confirms the long-known fact that in the radio region the intensity distribution is flatter than in the optical region (Seeger and Westerhout 1957; Woltjer 1957*b*). The radio halfwidth is about $4'$ in the N-S direction, $4'.3$ in the major axis, and about $3'$ in the minor axis; appropriate optical data for a comparison have not been computed.

More detailed information on the appearance of the nebula in the radio region is now becoming available. Contour maps at 81.5 MHz, derived from lunar occultations, and at 1407 MHz, observed with the Cambridge aperture-synthesis telescope (Ryle 1962), have been presented by Branson (1965, 1966). At both frequencies, the outermost contour agrees roughly in size and shape with the outermost isophote by Woltjer (1957*a*); the bulge in the southwest seems to be less pronounced in the radio region. At 81.5 MHz, the central isophotes show some elongation toward the southeast. The conspicuous dark bays in the nebula are clearly present at 1407 MHz, but only barely indicated at 81.5 MHz; this difference, however, may result from the different shape of the antenna beam at the two frequencies. In accord with the results from strip distributions, the distribution seems to be less peaked in the radio than in the optical region. Both contour maps fail to show fine structure. Observations of a lunar occultation by Matveenko and Khromov (1965) at 920 MHz, however, seem to show the presence of fine structure. According to these observations, 83 per cent of the flux of the source comes from an elliptical area that is within, but close to, Woltjer's outermost isophote. The similarity of the outer border of the nebula in the radio and optical region thus is confirmed. A central condensation agrees well in position and extent with the brightest cloud of the nebula northwest of the two stars. In addition, five condensations are listed. Some of them seem to coincide closely with the bright clouds of the diffuse mass shown clearly by Woltjer's isophotes. The diameters of the radio condensations are given at less than $15''$; this is consistent with the size of details shown on photographs. The detailed structure of the diffuse mass as well as its extent thus seem to be very similar in the radio and in the optical regions. The relative brightness, however, decreases outward more strongly in the optical than in the radio region. Similar results have recently been reported by Davies *et al.* (1966) and by Kronberg (1966).

The observation of a lunar occultation showed that the X-ray emission of the Crab Nebula has an angular width of $1'$ (Bowyer *et al.* 1964). That these observations place the X-ray source close to the brightest cloud northwest of the two stars and comparable to it in size seems to have escaped attention.

The radio spectrum of the whole Crab Nebula is shown in Figure 4. Between 80 MHz and 20 GHz, the spectrum is well represented by a power law with the spectral index of -0.25. The spectrum seems to become gradually steeper for frequencies below 80 MHz. This increase of slope must be due to the effect of the compact source with a spectral index of -1.2 which furnishes an increasing fraction of the total radiation with decreasing frequency. The spectrum remaining after subtraction of the compact source flattens with a broad maximum near 20 MHz. The analysis of occultation observations by Andrew *et al.*

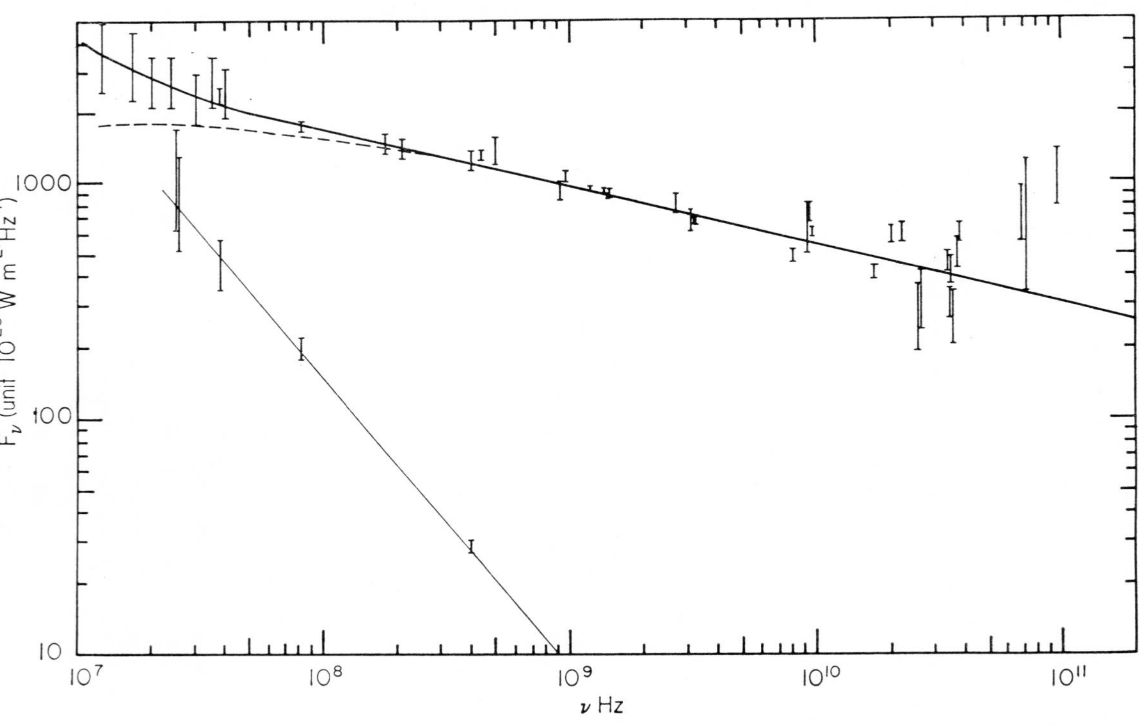

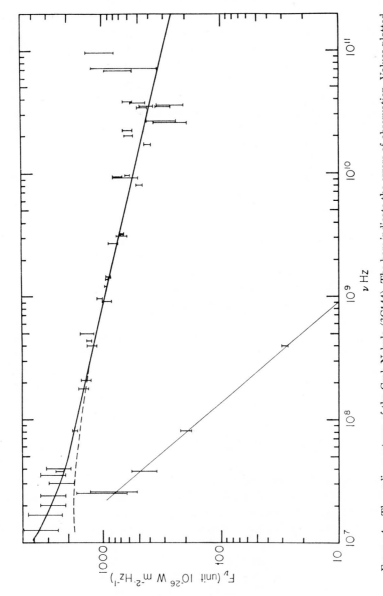

FIGURE 4.—The radio spectrum of the Crab Nebula (3C144). The bars indicate the errors of observation. Values plotted are by Kellermann (1964*b*) with addition of results by Bazelian *et al.* (1963) between 12.5 and 40 MHz, by Erickson and Cronyn (1965) at 26.3 MHz, by Troitskii, Tseitlin, and Porfir'ev (1964) between 550 and 1200 MHz, by Williams *et al.* (1965) at 19.6 GHz, and by Tolbert and Straiton (1965) and Tolbert (1965) between 35 and 94 GHz. *Thin line:* spectrum of the compact source; *broken line:* spectrum after subtraction of the compact source.

(1964) gives spectra of individual strips that show the low-frequency curvature more pronounced than that shown in Figure 4. In two strips the maximum appears near 100 MHz; in one it seems to be at a much higher frequency. Thermal absorption by the filaments may play a role in shaping the spectrum at low frequencies. The brightest filaments have a brightness in $H\beta$ of about 1.3×10^{-3} erg cm^{-2} sec^{-1} (Woltjer 1957*b*). With a surface brightness of about 10^{-4} erg cm^{-2} sec^{-1} ster^{-1}, a filament would have, according to equation (10), unit optical thickness at a frequency of about 30 MHz, if the electron temperature is 17,000° K and the excitation follows Case *B*. The brightest filaments, however, cover only a small fraction of the nebula, and thermal absorption should not play a noticeable role at frequencies above 100 MHz. A curvature such as that shown for the main part of the nebula in Figure 4 might be explained as an effect of thermal absorption, but the much stronger downward curvature reported by Andrew *et al.* needs a different interpretation, such as a low-energy cutoff of the energy spectrum of the relativistic particles.

Above 20 GHz, observations by Tolbert and Straiton (1965) and by Tolbert (1965) show a steep rise, which approaches a spectral index of +2, in the spectrum. The same series of observations shows an identical spectrum for the Orion Nebula, a well-established thermal source. An independent confirmation of this unexpected behavior is urgently needed. Shklovsky (1966) interprets the rise of the spectrum as indication of a maximum near 10^{12} Hz arising from an accumulation effect of relativistic electrons in the nebula. This interpretation could hardly be applied to a thermal source such as the Orion Nebula. (*Added in proof:* J. P. Oliver, E. E. Epstein, R. A. Schorn, and S. L. Soter, at the 123d meeting of the American Astronomical Society, Los Angeles, December 27–30, 1966, reported an integrated flux of approximately 250×10^{-26} W m^{-2} Hz^{-1} at 88 GHz. This value is one-quarter that reported at 94 GHz by Tolbert and agrees with the value extrapolated without change of spectral index from lower frequencies.)

An observation at a wavelength of 1 mm—3×10^{11} Hz—by Low (Bowen 1965*a*) gives an upper limit of 10^{-24} W m^{-2} Hz^{-1} for the central area of 60″ diameter and does not help to define the character of the spectrum in the large gap between the radio and the optical region.

The optical spectrum is shown in Figure 5. The infrared observations below 3×10^{14} Hz by Moroz (1964) and the observations by O'Dell above that frequency to 10^{15} Hz cannot be fitted without an inflection near 4.5×10^{14} Hz. The shape of the spectrum in the visual region has been confirmed by Oke (Bowen 1965*b*), who finds a rather abrupt change of the slope above 4300A or below 7×10^{14} Hz. Corrections for interstellar absorption and reddening must be applied. From photoelectric observations of 8 B-type stars close to the Crab Nebula, O'Dell (1962) finds a visual absorption $A_v = 1.07$ mag/kpc. A photographic investigation by Brodskaya (1963) of 86 stars of types B5 to F5 shows a visual absorption of $A_v = 0.70$ mag/pc, but at distances greater than 1300 pc, the absorption rises more steeply. For a distance of 1800 pc, both investigations actually agree with A_v equaling 1.9 mag. On this basis, the correction for reddening can be computed from data on interstellar absorption by Whitford (1958), used by O'Dell, or, without a significantly different result, by Johnson and Borgman (1963). The corrections used by O'Dell contain some numerical errors in the infrared region, which have been eliminated in Figure 5 where the spectrum is shown corrected for $A_v = 1.10$ and

$A_v = 1.59$, the values used by O'Dell. With the distance of 1800 pc, both values may be too small. They illustrate, however, the degree of uncertainty arising from the correction for reddening and demonstrate that the correction for interstellar reddening tends to emphasize the inflection near 4.5×10^{14} Hz. With $A_v = 1.59$, the spectrum actually shows a minimum and a maximum that become more pronounced if a larger value than $A_v = 1.59$ is correct.

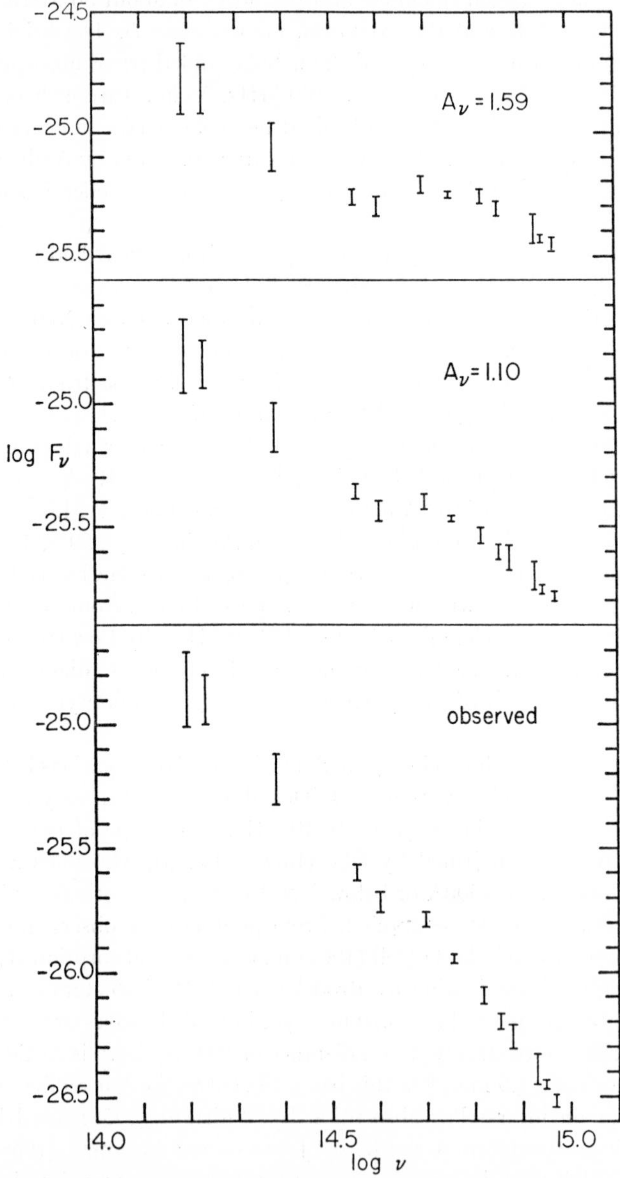

FIGURE 5.—The optical spectrum of the Crab Nebula. Values plotted are by Moroz (1962) between 10^{14} and 2.5×10^{14} Hz, by O'Dell (1962) above 3×10^{14} Hz. The bars indicate the errors of observation.

The entire spectrum is shown in Figure 6. The flux density at 1.2×10^{16} Hz follows for the distance of 1800 pc from the determination of the far-ultraviolet emission from the ionization conditions in the filaments by Woltjer (1957*b*). The X-ray data are by Bowyer *et al.* (1964), Clark (1965), Peterson, Jacobson, and Pelling (1966), Grader *et al.* (1966), and Haymes and Craddock (1966). A cutoff seems to occur near 2×10^{19} Hz. Upper limits for the flux density at still higher frequencies, determined by various authors, are at levels of flux density where they do not have much significance.

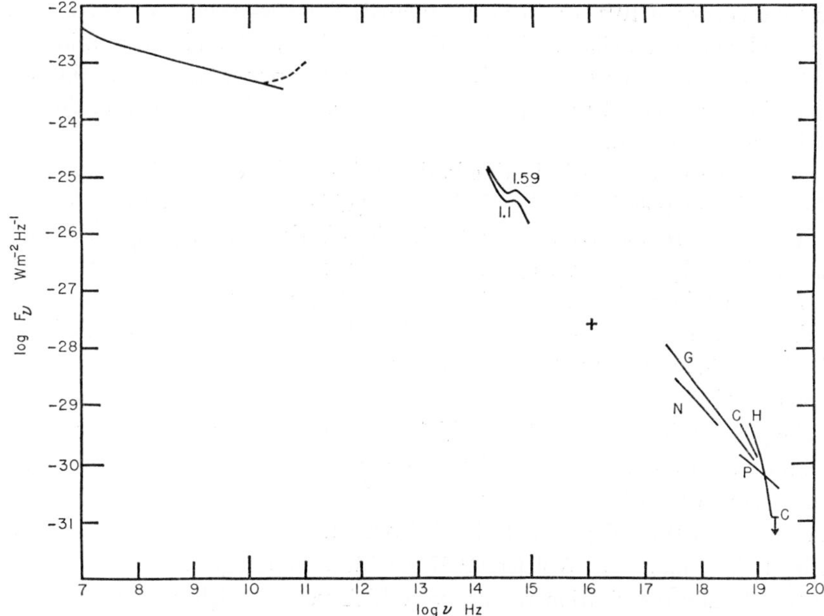

FIGURE 6.—The spectrum of the Crab Nebula. Radio region from Figure 4; optical region from Figure 5; far ultraviolet: *cross*, Woltjer (1957*b*) adjusted to distance of 1800 pc; X-ray region: N = Bowyer *et al.* (1964), C = Clark (1965), P = Peterson, Jacobson, and Pelling (1966), G = Grader *et al.* (1966), H = Haymes and Craddock (1966).

The total flux in the radio spectrum between 10 MHz and 10 GHz is 5.4×10^{-14} W m^{-2}. With a distance of 1800 pc, the corresponding total power is 2.1×10^{34} erg sec^{-1}. Integration over the whole spectrum to 10^{19} Hz gives 1×10^{38} erg sec^{-1} for the total emitted power.

It is obviously not difficult to connect the three observed parts of the spectrum, but the inflection near 4×10^{14} Hz may not be the only such feature. Another inflection between 10^{15} Hz and 10^{17} Hz seems possible. It should be pointed out that an inflection in the spectrum of the whole nebula does not necessarily indicate a similar feature in the volume emissivity of any part of the nebula. In the range from 10^{15} to 10^{17} Hz, for instance, the emitting volume depends on the frequency and shrinks to the much smaller size of the X-ray source. The frequency dependence of the volume emissivity thus depends on location in the nebula; the critical frequency of equation (13) might, for instance, increase outward. Integration over the whole nebula may lead to a spectrum with an inflection that would not indicate an inflection in the volume emissivity. If the spectrum shows maxima, they are most likely features of the volume emissivity in some part of the nebula.

A detailed interpretation of the spectrum may require detailed knowledge of the brightness distribution and its dependence on frequency. The over-all spectrum can be roughly described as smooth with a spectral index of -0.26 from 4×10^7 Hz to at least 5×10^{11} Hz, and a spectral index not far from -1 between 10^{14} Hz and 10^{19} Hz. The over-all spectrum may be interpreted as a synchrotron spectrum extending from the radio to the X-ray region (Shklovsky 1966). Permanent injection of relativistic electrons with a total energy of 10^{38} ergs/sec is postulated. The low-frequency compact source is interpreted by plasma oscillation in the injection region. Such an interpretation tends to connect the compact source with activity of the stellar remnant, a connection which is suggested in many remarks on the compact source. More information on the compact source is urgently needed.

The discovery of strong polarization in the optical region by Dombrovsky (1954) and the observation of polarization by Vashakidze (1954) were followed by a detailed investigation by Oort and Walraven (1956). The distribution of polarization over the nebula was investigated photoelectrically with three diaphragms, one with an aperture of 13.77 square minutes of arc shaped to include the total light of the nebula and to exclude the sky as much as possible; one of $1'$; and one of $0\overset{'}{.}5$ diameter. The integrated nebula showed 9.2 per cent polarization with the electric vector in position angle $159\overset{\circ}{.}6$. Local values of the polarization observed with the $0\overset{'}{.}5$ diaphragm range up to 33 per cent. The polarization was found to be highest in the central region. The most detailed observations are by Baade (1956), who photographed the nebula in the spectral region 5200 A to 6400 A (which almost completely excludes light from the filaments) with the 200-inch telescope through polaroid with different orientations of the electric vector. The analysis of these plates by Woltjer (1957*a*) with a diaphragm 5 secs of arc diameter shows that locally polarization reaches very high values—higher than 60 per cent in parts of the nebula bright enough for photometric errors to be considered small. Baade notes that the structural features tend to disappear when the electric vector is parallel to them. The features thus tend to be elongated in the direction of the magnetic field.

In the radio region, polarization is less than 1 per cent at frequencies below 1000 MHz. At higher frequencies, the polarization increases slowly, and the position angle of the electric vector increases (Gardner and Whiteoak 1963; Mayer, McCullough, and Sloanaker 1964, where complete references are given). Illustrative values of the degree of polarization and position angle are, respectively, 7 per cent and $143° \pm 2°$ at 9500 MHz, and 2.6 per cent and $125° \pm 3°$ at 2600 MHz. The slow decrease of the position angle is inversely proportional to the square of the frequency. This indicates Faraday rotation.

Oort and Walraven (1956) have shown that the shell of filaments can indeed produce a large Faraday effect that is likely to obliterate the polarization at low frequency because the magnetic field and the electron densities vary greatly over the shell. The problem has recently been discussed in detail by Burn (1966). Passage through the interstellar medium may produce a Faraday rotation of the observed order, but this is unlikely to cause much depolarization, because it is unlikely to vary much over the surface of the nebula. Extrapolation to infinite frequency gives the intrinsic polarization and position angle. At 9500 MHz, the degree of polarization of 7 per cent is slightly, although not significantly, smaller than the degree of 9.2 per cent observed by Oort and Walraven. Extrapolation of the data observed by Mayer *et al.* (1964) leads to an intrinsic position angle of 146°. This value, based on observations at high frequencies, should be more reliable than the position angle of 157° derived by Gardner and Whiteoak from observations at lower frequencies where the polarization is small. The intrinsic position angle 146° is 14° smaller than the position angle of 159°6 observed by Oort and Walraven. The difference may not have the deep physical significance suggested by Slysh (1966). Since the brightness distribution in visual light is more strongly peaked than that in the radio region, the outer parts of the nebula contribute more to the integrated radiation in the radio region than in the optical region. The observations by Oort and Walraven show less polarization and a smaller position angle for the outer parts. Thus, the integrated nebula should show a smaller degree of polarization and a smaller position angle in the radio region, even if the intrinsic polarization and its position angle in individual parts of the nebula are identical in the radio and visual regions. Whether the observations can be explained in this way has not yet been tested.

The distribution of polarization in the radio regions has been investigated by Soboleva, Prozorov, and Pariskii (1963) from observations of a lunar occultation at 9400 MHz and by Morris, Radhakrishnan, and Seielstad (1964) by variable-spacing interference polarimetry at 2840 MHz. The results agree in showing the polarization to be more strongly concentrated toward the center than is the radio brightness—rather similar to the concentration of visual brightness. Soboleva *et al.* find polarization of 17.5 per cent in a central region of about 1′ radius, practically the same as was observed by Oort and Walraven in a central region of similar size. An investigation of the distribution of polarization by Gardner (1965) with the 210-foot Parkes radio telescope at 5000 MHz shows maximum polarization of 7.5 per cent in an area 0′.5 ± 0′.2 to the west and 1′.1 ± 0′.2 to the south of the center. The half-intensity width of the beam is 4′.1 and includes so large a part of the source that it is not easy to compare the result with that to be expected with results from similarly low resolution in the optical region.

The discussions of the physical conditions in the diffuse mass and the detailed interpretation of the synchrotron radiation of the Crab Nebula by Oort and Walraven (1956) and Woltjer (1957*b*) need revision in the light of later observational evidence. Some basic information on the diffuse mass is likely to remain unchanged; for instance, the magnetic field is in the range of 10^{-3} to 10^{-4} gauss with the larger fields near the center, and the mass is small. The most important new facts are the presence of the compact source and the extension of the synchrotron spectrum into the X-ray region. In fields of the order 10^{-3} gauss, electrons with energy of 10^{13} ev are needed to give synchrotron emission at 10^{18} Hz. The half-life of such an electron is about 0.8 years. The explanation of the small size of the X-ray emitting region, suggested by Shklovsky (Bowyer *et al.* 1964), is that these high-energy electrons are formed in a small region of the nebula from which, traveling at the speed of light, they could get no farther than about 1 light year (0.3 pc). This is 0′.6 with the distance of 1800 pc and thus agrees with the observed size of the

X-ray source. The total energy injected into the nebula must be of the order 10^{38} erg sec^{-1}. The suggestion is plausible that the compact source is the region of injection. Better positions of the X-ray source and of the compact source are required to establish where the injection occurs and the role of the compact source.

6. CASSIOPEIA A

Cassiopeia A (3C461) is, at low frequencies, by far the strongest radio source in the sky. Another source like it anywhere in the Galaxy still would be among the six strongest nonthermal sources. Cassiopeia A is unique and must be the result of a rather rare event.

The nebulosity with which the source has been identified (Baade and Minkowski 1954) is shown in Plate 2. This plate is from a previously unpublished photograph by Baade in the wavelength region that includes the [S II] lines 6717/31A and shows the nebulosity best. Expansion with high velocity indicates clearly that the nebulosity is the remnant of a supernova of moderate age, but no outburst in its position has been recorded. The nebulosity is obviously different from the remnants of Type I, such as those from Tycho's and Kepler's novae, and from the Crab Nebula. It has been suggested that the nebulosity is the remnant of a supernova of Type II. The frequency of occurrence of Type II supernovae is not well known (Minkowski 1964), but it seems to be roughly of the same order as that of supernovae of Type I. The uniqueness of Cassiopeia A thus is not quite consistent with the assumption that it originates from an average supernova of Type II.

The nebulosity is a circular object of radius 2'.0 with a flare that extends to a distance of 3'.8 from the center in position angle 70°. A distance of 3400 pc follows from the observed transverse motions and from the radial velocity of expansion (Minkowski 1959), in agreement with the value found from 21-cm absorption lines (Muller 1959). The radius of the nebulosity thus is 2.0 pc; the flare extends to 3.8 pc from the center.

The nebulosity consists of small condensations of two types that differ radically in appearance, spectrum, and motion. The majority shows large transverse motions and is part of an expanding shell. These fast-moving condensations are diffuse, with apparent diameters mostly between one to two seconds of arc, or linear diameters around 0.025 pc. Analysis of the spectrum, which shows only forbidden nebular lines of [O I], [O II], [O III], and [S II], suggests an electron density of about 5×10^3 cm^{-3}, an electron temperature of 20,000° K or higher, and a visual interstellar absorption of 3 or more magnitudes in the brightest region of the nebulosity in its northern arch, which is the least obscured part of the shell. The visual absorption near the center is higher, perhaps about 5 mag (Minkowski 1957). The mass of an average fast-moving condensation is about 10^{-3} $\mathfrak{M}_\odot$, the combined mass in the northern arch about 0.2 $\mathfrak{M}_\odot$. Since the arch covers not more than 10 to 15 per cent of a complete shell, a total mass of 1.5 to 2.0 $\mathfrak{M}_\odot$ is suggested for the complete shell.

A relatively small number of condensations have small, as yet unobserved, transverse motions and small, prevalently negative, radial velocities. These semistationary condensations tend to be sharply bounded, with apparent diameters of one to two seconds of arc, or about 0.025 pc; some are elongated and slightly irregular. Their spectra show only Hα and [N II] lines. Spectrum and surface brightness are consistent with collisional excitation with an electron temperature possibly below 10,000° K and a total density of 10^5 to 10^6 H atoms cm^{-3}. The average mass of a semistationary condensation is 0.05 $\mathfrak{M}_\odot$, and the combined mass of the relatively few condensations of this type may be close to one solar mass. The origin of the semistationary condensations is still unexplained. They might result from the interaction of the expanding shell with the interstellar medium (Woltjer 1964).

It seems unlikely that the large mass of the fast-moving shell is mainly accreted interstellar material. The interstellar mass in a sphere of 2-pc radius is 1.2 n_H solar masses, with $n_H/n_{He} = 7$. At the location of the nebulosity, 3400 pc from the Sun with $l^{II} = 111.^\circ7$, $b^{II} = -2.^\circ1$, or 130 pc below the galactic plane, n_H is 0.2 cm^{-3} (Westerhout 1957). The interstellar mass thus is 0.24 $\mathfrak{M}_\odot$, small compared to the estimated mass of the shell, unless the nebulosity happens to be in a cloud of much greater than average density. An initial mass of 1 or 2 solar masses would conflict with the interpretation of the nebulosity as a remnant of a supernova of Type II if such remnants have the low mass of 0.02 $\mathfrak{M}_\odot$ suggested by Poveda (1964). A weakness of Poveda's model is the assumption of constant density in the expanding mass. An inward increase of density would lead to a larger mass. The apparent uniqueness of the Cassiopeia A source, however, suggests that it may not originate from an average supernova of Type II.

Except for the flare, the radial velocities show the fast-moving condensations to be contained in a shell of about 0.3-pc thickness. At the outer edges, the velocity of expansion is 7440 km/sec in the rear and 6000 km/sec in front. The asymmetric expansion might be interpreted as the result of inhomogeneities of the interstellar medium that might lead to stronger deceleration of the front than of the rear. The flare shows large transverse motions which could not be measured because the features are too diffuse and do not maintain their shape. Features of the flare extend radially, although the features of the shell show no strongly preferential direction. Radial velocities of features of the flare are relatively small and suggest that the flare moves in a direction at an angle of about 80° from the line of sight, almost in the plane of the sky. The optical observations give no clue for the origin of the flare.

If the fast-moving condensations have expanded without deceleration, the outburst occurred in $+1702 \pm 14$. This is the maximum possible age of the remnant.

Earlier observations of the size and structure of the radio source are now superseded by the high-resolution observations by Ryle, Elsmore, and Neville (1965) at 1420 MHz with the Cambridge aperture-synthesis telescope. The brightness distribution in Plate 3 shows the complexity of the source and, for the first time, details that suggest an interpretation of the flare (Minkowski 1966*b*). The source appears as a ring containing localized small areas of enhanced surface brightness. The source then is a shell with a ridge of 2′ diameter and a thickness of about $0.'5$, with local regions of enhanced volume emissivity. The diffuse fast-moving condensations that form the shell of the nebulosity are all inside the ridge. One filament shown in Plate 3 on the outer edge of the radio shell in the southwest is not part of the expanding shell, but one of the semistationary condensations. The elongated features shown in Plate 3 on the east side of the source, and easily identified in Plate 2, are the brightest features of the flare, which falls exactly in the conspicuous gap of the radio shell. The brightest condensation in the source seems to be diametrically opposite to the gap in the projection on the sky. The questions of whether this is true in space and whether it is of any significance must be left open.

If the radio emission is the result of the compression of the surrounding interstellar medium, as van der Laan (1962*a, b*) has proposed, both the gap and the flare may be interpreted as caused by very low interstellar density locally. If the material ejected in the direction of the gap and flare encounters only very low density, compression will be negligible, and the shell will show a gap while the ejected material will move outward with negligible deceleration.

If it is assumed that only one shell was ejected with a high velocity of expansion, which still persists almost undiminished in the flare, the rest of the shell must have been slowed down considerably to reach its present velocity of expansion. If this is true, the age of the shell estimated on the assumption of constant velocity of expansion is too high. The age may be low enough to apply Oort's model. Both this model and the similarity solution may be used to explore the possible size of the error. Since the present velocity of expansion is known, equations (19) and (25) may be used to determine the average velocity for the two models. The ages then can be obtained from the known radius. For Oort's model, the average velocity is 30,000 km/sec and the age 65 years; for the similarity solution the average velocity is 18,300 km/sec and the age 102 years. With these ages the velocity of expansion for the flare, assumed to be the original velocity of all ejected matter, is 57,000 km/sec and 35,000 km/sec, respectively. A velocity of this order has not been observed in any supernova. It seems excessive and makes the one-shell model unattractive.

The presence of multiple absorption lines in the spectra of many supernovae of Type II suggests that more than one shell has been ejected. This justifies the suggestion that two shells with different velocities were ejected in Cassiopeia A. The flare then can be interpreted as a surviving part of a shell ejected with high velocity which has in all other directions encountered interstellar material of higher density, has compressed it, and has thus formed the radio shell. A second shell of material, ejected with lower velocity, follows the fast shell which has swept up the interstellar material so that the slow shell moves without deceleration. Eventually, the slow shell must catch up with the decelerated fast shell and will then be slowed down. The similar size of the outside of the nebulosity and the ridge of the radio shell may suggest that the object is now approaching that phase. The bulk of the circular nebulosity, however, is still well inside the radio shell and, therefore, is likely to be still in the initial phase during which it moves with constant velocity and when the age derived under the assumption of uniform expansion is correct. The kinetic energy of the slow shell with a velocity of 7400 km/sec and a mass of about 2 $\mathfrak{M}_\odot$ is 2×10^{51} ergs. The average velocity at the outer edge of the radio shell is about 8000 km/sec, its present velocity from equation (25) about 3200 km/sec, and the initial energy involved in the ejection of the fast shell from equation (24) with $n_H = 0.7$ is 2×10^{50} ergs. The initial velocity of the fast shell indicated by the flare then becomes 14,500 km/sec. This velocity is comparable with the velocity of expansion of 12,000 km/sec observed in the supernova of 1961 in NGC 4303 (Zwicky 1964, 1965).

The supernova in NGC 4303 is an unusual object. It has some similarity to Type II supernovae but differs so much from typical supernovae of that type that Zwicky assigns it to a new type, Type III. The early spectrum of a supernova of this type is continuous and blue, as are the spectra of Type II supernovae, but the supernova remains in this phase for several weeks, much longer than supernovae of Type II. It differs from these also by a slower initial decline of the light and a much later leveling-off of the light curve after a much greater decline. Except for the greater width of the lines due to greater velocity of expansion, the spectrum is similar to that of supernovae of Type II. The velocity of expansion eventually decreases from 12,000 km/sec to the order of 6000 km/sec. The slower time scale of all changes suggest a much larger ejected mass than that ejected by supernovae of Type II (Greenstein; see Zwicky 1965).

The two-shell model thus seems to lead to a satisfactory interpretation of the high velocity indicated by the flare and of the large mass of the nebulosity. At the same time, the uniqueness of Cassiopeia A is explained by the connection with a supernova of a rare kind.

The radio spectrum of the source 3C461 above 30 MHz follows a power law with spectral index -0.77 ± 0.02 and a flux density of $6.42\ (\pm 0.18) \times 10^{-23}$ W m^{-2} Hz^{-1} at 400 MHz (Kellermann 1964*a*). The average surface brightness is 6×10^{-17} W m^{-2} Hz^{-1} ster^{-1}. If the spectrum would continue into the optical region without change of spectral index, the average brightness in the visual region at 5×10^{14} Hz would be 1.2×10^{-21} W m^{-2} Hz^{-1} ster^{-1}. This is approximately half the average brightness of the night sky (Liller 1964) and would be observable. Interstellar obscuration, however, must be expected to reduce the brightness by about 4 mag to an unobservably low level. No continuous radiation is observed in the optical region. X-Ray emission from Cassiopeia A has recently been observed in the range 3×10^{17} to 3×10^{18} Hz (Byram, Chubb, and Friedman 1966). The flux density at 10^{18} Hz is about 1×10^{-27} erg cm^{-2} sec^{-1} Hz^{-1}. An extrapolation of the radio spectrum with constant spectral index -0.77 leads to an expected flux density of 3.5×10^{-27} erg cm^{-2} sec^{-1} Hz^{-1} at 10^{18} Hz. It is to be expected that the spectrum steepens toward the X-ray region as the high-frequency cutoff is approached. The interpretation of the X-ray emission as a continuation of the synchrotron emission in the radio region thus does not meet any difficulty. The integrated flux in the radio spectrum between 10 MHz and 10^4 MHz is 1.7×10^{-13} watt m^{-2}. With the distance of 3400 pc, the corresponding total emitted radio power is 2.4×10^{35} erg sec^{-1}. From the observed X-ray flux of 3×10^{-9} erg cm^{-2} sec^{-1} between 3×10^{17} and 3×10^{18} Hz, the total power is 4×10^{36} erg sec^{-1}. The total power in the complete spectrum cannot be reliably estimated, but is probably about an order of magnitude smaller than that of the Crab Nebula.

Cassiopeia A is the only source in which the secular decrease in the flux density predicted by Shklovsky (1960*a*) has been observed. The decrease expected from equation (14) with the spectral index -0.77 and an age of 260 years is 2 per cent per year. Observed values of the decrease are 1.06 ± 0.14 per cent per year at 81 MHz (Högbom and Shakeshaft 1961) and 1.14 ± 0.26 per cent per year at 3200 MHz (Mayer *et al.* 1965). Less accurate evidence has been reported at 1420 MHz by Heeschen and Meredith (1961) and at 940 MHz by Lastochkin and Stankevich (1964). Since the theoretical prediction is based on the assumption of uniform expansion, which is not fulfilled, the lack of close agreement between prediction and observation is not unexpected.

7. SUPERNOVAE OF TYPE II

No remnant of moderate age is known that can be assigned to an average supernova of Type II, and no known outburst can be assigned to such a supernova. It is possible that the supernova of +185 and the associated radio source MSH 14−63 are of Type II (Sec. 8.3). It is reasonable to think that such outbursts are unrecognized among the many novae in ancient historical records because these give inadequate information, even for very bright objects such as the supernova of +1054. The frequencies of supernovae are not well known, and the fact that three remnants of supernovae of Type I have been discovered does not necessarily imply that a similar number of remnants of supernovae of Type II should be open to discovery. If Cassiopeia A is the remnant of a rare variant of a supernova of Type II, remnants of average supernovae of that type should be less rare, but it is obviously impossible to predict their frequency.

No solution can be offered for the mystery of the missing remnants. Supernovae of Type II are known to have expansion velocities of about 6000 km/sec. If their mass is even approximately as small as Poveda (1964) suggests, their kinetic energy would be small. As Population I objects they should be strongly concentrated toward the galactic plane, where the interstellar density is likely to be high. Low kinetic energy and high density would lead to a relatively fast decay. As radio sources, the remnants might be smaller and possibly weaker and thus more difficult to recognize than remnants of supernovae of Type I of comparable age. Optically, the chance of being hidden by obscuration would be larger. A search for these remnants would meet the difficulty that it is not clear what is to be found.

8. OLD SUPERNOVA REMNANTS

Some radio sources identified with faint large nebulosities are believed to be old supernova remnants. Diameters for sources and objects in this group are of the order of one degree. Many of them are sources of substantial flux density, but their surface brightness is low. They present difficult observational problems because the background radiation must be subtracted and the flux density must be determined by integration over the source. For most of these objects, the available data are not very accurate. Most of the nebulosities are very faint, and since they are very large, the optical investigation is time consuming. For most of the nebulosities, not more than their appearance is known. The object most accessible to investigation is the Cygnus Loop. The interpretation of objects in this class as old supernova remnants is mainly based on the investigation of the Cygnus Loop that showed that its properties can be satisfactorily interpreted as those belonging to a very old remnant that has been slowed down by interaction with the interstellar medium, as Oort (1951) first suggested. Some smaller objects in this class could well be more recent, however, and better information is very desirable.

It has been generally assumed that these objects are remnants of supernovae of Type II. This assumption goes back to the interpretation of the Crab Nebula as a typical remnant of a supernova of Type I. The interpretation of these objects seemed to require more energetic outbursts, such as that in Cassiopeia, which was then considered as a typical supernova of Type II. It is now clear that supernovae of Type I produced very energetic remnants. The conditions in remnants of supernovae of Type II are not known, but there is no reason to doubt that they are sufficiently energetic to produce objects such as the Cygnus Loop. The future evolution of the Crab Nebula cannot be predicted, but it might well be observable as a large faint object in the future. It seems not impossible that more than one variety of supernovae may be involved in the formation of the old remnants. The information now available, however, is inadequate for decisions about the origin of each individual object.

8.1. The Cygnus Loop

The Cygnus Loop is a large nebulosity with a diameter of about 3°. An illustration showing the whole nebula from a red photograph with the 48-inch Schmidt telescope is in *Sky and Telescope*, **17**, center, 1958. Photographs of details with the 100-inch telescope have been published by Duncan (1923, 1937). From the spectroscopically determined velocity of expansion and the transverse motion at the edge, the distance is 770 pc and the diameter 40 pc (Minkowski 1958). To a first approximation the nebula is a spherical shell of about 20-pc outer radius and 10-pc thickness. An extension beyond the sphere in the southern part is conspicuous. No central star is present. The spectrum shows H lines and the forbidden lines common to all gaseous nebulae. No continuous spectrum is observable. The gradient of the H Balmer lines suggests collisional excitation. This conforms to and supports Oort's hypothesis for the origin of the nebula. The most detailed observations and discussions of the spectrum by Parker (1964*b*) show "that the physical conditions of the filaments can be described in terms of a temperature stratification picture similar to that which might be expected behind a shock wave. Thus the spectrum of the Cygnus Loop can be explained in terms of the [O III] lines originating predominantly in a region with a temperature of the order 10^5 ° K, while the other forbidden lines and the Balmer lines arise primarily from a region with a temperature of about 1.7×10^4 ° K."

The interstellar density at the location of the Cygnus Loop, $l^{II} = 72.^{\circ}6$, $b^{II} = -8.^{\circ}2$, 110 pc below the galactic plane, has been estimated as 0.1 H atom cm^{-3} (Sholomitskii 1963). The interstellar mass in a sphere of 20 pc radius is then about 100 $\mathfrak{M}_\odot$. The mass of the bright filaments has been estimated by Osterbrock (1958) as about 0.1 $\mathfrak{M}_\odot$ on the basis of an average electron density of 200 cm^{-3}, determined from the relative intensities of the [O II] lines. From the total emission of 1.2×10^{35} erg sec^{-1} following from observations by Chamberlain (1953), and with an electron temperature of 1.7×10^4 ° K, Parker finds a mass of about 0.5 $\mathfrak{M}_\odot$ for NGC 6960 and NGC 6992, the brightest parts of the object, and he estimates the total mass as 1–2 solar masses. A lower electron temperature would lead to a much higher mass. A considerable mass could be hidden in regions of lower brightness. Sholomitskii estimates that the mass of the invisible ionized hydrogen might be as high as 400 $\mathfrak{M}_\odot$.

If the similarity solution in equation (22) is applied, the average velocity from equation (25) is 290 km/sec. The radius of 20 pc gives 67,000 years as the age of the Cygnus Loop. With this age, equation (23) gives for E_o/n_H the value 2.5×10^{50}, and if n_H is 0.1 cm^{-3}, E_o becomes 2.5×10^{49} ergs. This is so much smaller than the total energy of 1.3×10^{51} ergs for Tycho's nova that the Cygnus Loop cannot be an old remnant of a supernova of Type I. With the low mass he suggested for supernovae of Type II, Poveda

(1964) estimates their total energy as 5×10^{49} ergs. It would then be reasonable to interpret the Cygnus Loop as an old remnant of a supernova of Type II. If the mass of a supernova of Type II were larger than Poveda's estimate, the total energy of a supernova of Type II would also seem too large. An essential weakness of the argument is that the similarity solution is valid only as long as radiation losses are negligible. Results of a discussion by Heiles (1964) show that radiation losses by heavy ions in the shock front may be so large that the adiabatic condition is no longer satisfied for the Cygnus Loop. Some doubt as to which type of supernova formed the Cygnus Loop seems justified.

The Cygnus Loop is a radio source with a flux density of $260 \pm 50 \times 10^{-26}$ W m^{-2} Hz^{-1} at 408 MHz. Matthewson *et al.* (1960) find a spectral index -0.4, which is confirmed by the value -0.47 by Kenderdine (1963). The spectrum thus is essentially nonthermal. The spectral index of -0.1, given by Harris (1962), is too high, probably owing to uncertainties concerning how much of the fainter parts has been included at each frequency and concerning the background intensity in older observations with low signal-to-noise ratios. Local variations of the spectral index seem to occur. With the increasingly good quality of the radio observations it has become more and more evident that the radio brightness distribution is remarkably similar to the appearance of the nebula. Contours of the source at 750 MHz and at 1400 MHz, obtained with the 300-foot telescope of the National Radio Astronomy Observatory, are shown superimposed on a picture of the nebula in Plates 4 and 5, respectively (Hogg 1966).

The great similarity between the distribution of the radio brightness and the distribution of the thermal radiation in the optical region has led to the question of whether the radio emission could be thermal or have a large thermal component. The discussions by Matthewson *et al.* (1960), Harris (1962), Kenderdine (1963), and Parker (1964*b*) lead to the conclusion that the thermal contribution to the radio emission is of the order of a few per cent. Poveda (1965) suggests that the filaments may actually be thin sheets tangential to the surface of the shell where, seen edgewise, they would appear as filaments. Seen face on, their brightness would be too low to be observable, but their contribution might be large enough to explain the entire radio emission as thermal. Poveda's interpretation seems to be contradicted by the radio spectrum which seems to be essentially nonthermal without the flattening out at high frequencies that would be expected if a large part of the radiation were thermal.

One explanation of the resemblance of the optical features to a nonthermal radio source, first noted by Rishbeth (1956) in IC 443 (Sec. 8.2), is that both optical and radio emission might be the result of interaction with the interstellar medium—collisional heating leading to the optical emission and compression of matter and magnetic fields resulting in strong nonthermal emission (Minkowski 1959). Van der Laan's investigation (1962*a*) of the effect of the expansion of a remnant on the interstellar medium shows that the expansion forms a compressed region outside the ejected envelope and that this region emits synchrotron radiation. Results of the theory are applied to parts of the Cygnus Loop, and satisfactory agreement is found between calculated and measured flux densities. The compression process is shown to provide a direct explanation for the near-coincidence of radio and optical features. The formation of a shell source by a supernova remnant is discussed by van der Laan (1962*b*).

8.2. IC 443

The nebula IC 443 is nearly circular, with a diameter of about 45′. It is similar to the Cygnus Loop, although filaments on the southwest of the nebula are more nearly radial, with a tendency to spiral shape. An H II region on the eastern side merges with the nebula and heavy obscuring clouds are conspicuous.

Spectroscopic observations by Parker (1964*b*) reveal conditions not drastically different from those in the Cygnus Loop. If, as seems likely, abundances are the same as in the Cygnus Loop, the electron temperature in IC 443 is somewhat lower. The average electron density is about 350 cm^{-3}.

No reliable distance for the nebula is known. Velocity of expansion and transverse motion have not been measured. Various estimates of the distance—all unreliable—range from 800 pc to 2000 pc. The nebula is at $l^{\mathrm{II}} = 188\overset{\circ}{.}9$, $b^{\mathrm{II}} = +3\overset{\circ}{.}1$, where observations of the 21-cm H line cannot be used to obtain information on the distance and on the local interstellar density.

The radio source 3C157 has a flux density of 250×10^{-26} W m^{-2} Hz^{-1} at 400 MHz. Values given for the spectral index are -0.39 ± 0.3 (Howard and Dickel 1963), -0.30 ± 0.4 (Kellermann 1964*a*), -0.41 ± 0.02 (Higgs 1964), and -0.41 ± 0.04 (Hogg 1964). The observations by Hogg show that the spectral index does not vary much over the source. To the east, where the nebula merges with the H II region, the spectral index increases gradually to zero, appropriate for the thermal emission of the H II region. There is no doubt that IC 443 is a nonthermal source.

Investigations of the brightness distribution with the best resolution are by Howard and Dickel (1963) (beamwidth 6′ at 8000 MHz), and Hogg (1964) (beamwidth 10′ at 1400 MHz). Hogg also has observations with beamwidth $18\overset{'}{.}5$ at 750 MHz and $16\overset{'}{.}0$ at 3000 Hz. Hogg's brightness distribution at 1400 MHz, superimposed on a photograph of the nebula, is shown in Plate 6. The similarity of source and nebula is conspicuous.

8.3. Miscellaneous Objects

Some extended nonthermal sources have been identified with large and faint nebulosities. Lists of these objects have been given by Harris (1962) and Boischot and Lequeux (1964). There is no observational proof that these objects are the remnants of supernovae. Most of them have nearly circular symmetry, and this suggests an explosive origin. They are clearly not remnants of ordinary novae or planetary nebulae, and the most obvious interpretation is that they are remnants of supernovae. By analogy with the Cygnus Loop, many or even most of these objects may be very old, and it should not be expected that the supernovae can be found in historical records such as those compiled by Ho (1962) and Xi and Bo (1965). Usually, crude positions are the only available information. If any estimate of the distance is possible, suggested identifications can sometimes be rejected by simple arguments.

If IC 443 (Sec. 6.2) is at a distance of about 1500 pc, as seems likely, its assignment to a supernova seen in the middle of the fifth century would imply an average velocity of 6500 km/sec, which seems too high for any remnant with an age of 1500 years.

The identification of the source MSH 14−6*3* with the supernova of +185, suggested by Shklovsky (1954), may be acceptable, although available information is not adequate to prove it. The brightness distribution has been observed by Hill (1964), and the optical remnant has been located by Westerlund (1964). The supernova at $14^h38^m - 62°30'$ (1950) can have been visible in China only a few degrees above the horizon. Moreover, as Xi and Bo point out, it must have been a daytime object at maximum. It must have been an extraordinarily bright object. The original record is ambiguous concerning the duration of visibility. A key word could mean either "next year" or "the year following the next" (Yang 1966). Ho chooses the first interpretation. In that case the object ceased to be observable after about $7\frac{1}{2}$ months. This suggests a steeper decline of the light than that of supernovae of Type I. Tycho's and Kepler's novae declined by 6 mag during that time. It would thus be most probable that the supernova of +185 was of Type II. Xi and Bo prefer the second interpretation which gives 20 months as the duration of visibility. This would admit the interpretation as a supernova of Type I. An assumed apparent visual brightness at maximum of −7 mag and an interstellar absorption of 1.6 mag/kpc (Sharov 1962) leads to a distance of 725 pc, a radius of 6.3 pc from the apparent diameter of 1° of the source, and an average velocity of 3500 km/sec if the supernova was of Type II. If it was of Type I, the corresponding values are 1100 pc, 9.5 pc, and 5300 km/sec, respectively. Neither set of values seems unacceptable; the brightness at maximum might well have been greater than assumed, and distance, radius, and average velocity would then be smaller.

Some sources are shells, which supports the interpretation as remnants; others are not. The most definite characteristic is a relatively flat spectrum with a spectral index near −0.4. Larger, even positive, values have been reported, but they are probably the result of inadequate observational data. The sizes of the sources agree well with those of the nebulae. In detail, there is usually no similarity, and none would be expected if a source were nonthermal and the nebula were seen in the light of emission lines. Occasionally, a situation similar to that at the edge of the Cygnus Loop seems to exist. The nebula connected with the source HB 21, for instance, is embedded in the widespread emission nebulosity in Cygnus, and not all parts of the nebula can be clearly recognized. The most clearly visible details are sharp edges on the west and east sides. These are clearly shown by the brightness distribution of the source (Boischot 1962; Hogg 1963; Yang and Dickel 1965).

The nebulae differ greatly in appearance. Puppis A shows only small clouds and short irregular filaments (Baade and Minkowski 1954). Radial velocities are small and irregular. These filaments might be related to the semistationary filaments in Cassiopeia A. The nebula S147 is a mass of long looping filaments. The nebula HB9 shows mainly broad and irregular streaks; a few sharp filaments are in the west. The nebula HB 21 shows a few small clouds and short filaments in the northwest, and W28 (Courtès *et al.* 1964) has a double envelope of filaments. The nebula CTB1 is almost a complete circular shell, somewhat reminiscent of a planetary nebula. The physical conditions in these nebulosities thus seem to have great individual differences for which no interpretation can be offered.

No reliable method that permits the determination of distances of the old remnants is known. The determination from the transverse motion and the radial velocity of expansion is almost impossible or entirely impossible. Harris (1962) suggests the diameter of the narrowest filaments as a distance criterion. There is, however, no physical reason why these diameters should be independent of the physical conditions in the objects which differ in appearance. The filaments themselves differ in character. In some objects they are long and rather smooth, in others very short and irregular. Some objects have too few filaments to consider their diameter as statistically significant. The filaments thus are not good distance indicators. A relation that connects the surface brightness of an expanding source with its radius has been suggested by Shklovsky (1960*b*). With a correction (Minkowski 1964), the distance is given by

$$d \sim W^{(\beta+3)/[3(\beta+2)]} H_{n,o}{}^{-(18+5\beta)/[12(\beta+2)]} I_{\nu}{}^{-1/(\beta+2)} \varphi^{-1} \tag{28}$$

where $\beta = 2(1 - 2x)$, W is proportional to the energy released in the outburst, $H_{n,o}$ is the magnetic field normal to the velocity vector of the relativistic electrons at some initial time, I_ν is the radio surface brightness, and φ is the angular size. The spectral index for the old remnants is usually not well determined. If an average of -0.5 is assumed, the distance is

$$d \sim W^{0.39} H_{n,o}^{-0.53} I_{v}^{-0.17} \varphi^{-1} . \tag{29}$$

This is not very insensitive to the values of W and H_n, and the relation suggested by Shklovsky,

$$d \sim I_{v}^{-0.17} \varphi^{-1} \tag{30}$$

which follows if W and H_n are identical in all sources, cannot be expected to hold for all remnants, independent of their individual characteristics. The method thus cannot claim high accuracy. It is not valid for remnants slowed down by interaction with the interstellar medium, because the absence of such an interaction is a basic assumption.

8.4. Supernova Remnants in the Large Magellanic Cloud

Four nonthermal sources in the Large Magellanic Cloud have been identified with the emission nebulae N49, N63A, N132D, and N157 in the catalogue by Henize (1956). One of them, N157, seems to be associated with 30 Doradus, a region similar to the central region of the Galaxy in its radio properties. For the other three, interpretation as supernova remnants has been suggested (Matthewson, Healey, and Westerlund 1963; Matthewson and Healey 1964). Observations of these objects by Westerlund and Matthewson (1966) show that their optical and radio characteristics are consistent with this interpretation.

Since the distance of the Large Magellanic Cloud is known, the dimensions and the emitted powers are easily stated. The properties of these remnants are shown in Table 4. The line emission is consistent with collisional excitation. Intensity ratios of characteristic lines are most similar to those observed by Parker (1964*b*) in the filament 6′(4) (Parker's notation) in IC 443. An electron temperature of 15,000° K–20,000° K is suggested. The intensity ratio of the [S II] lines 6171/31A leads to the high electron density of about 4×10^4 cm^{-3} in all three objects.

It is obvious that N63A and N132D have some similarity to the Cassiopeia A source, although they are larger, and the emitted power is smaller. The integrated line-emission spectrum of Cassiopeia A has not been observed, but it probably differs from the spectra of N63A. The spectral index of −0.5 for both remnants is larger than that of −0.75 of Cassiopeia A and is similar to the spectral indices of the old remnants.

For a discussion of the remnants it does not seem justified to use the relations derived by Shklovsky (1960*a*, *b*) for the dependence of the flux density on the radius and for the distance determination of supernova remnants (equation 28). These relations are valid under the assumption of negligible interaction with the interstellar medium and cannot be expected to give realistic results for remnants in regions with appreciable interstellar density. For N63A and N132D, the available data are inadequate for an application of the similarity solution (equation 22), unless an assumption is made on the total energy

TABLE 4

SUPERNOVA REMNANTS IN THE LARGE MAGELLANIC CLOUD

	N49	N63A	N132D
Maximum dimension in Hα light	18 pc	7 pc	6 pc
Flux density at 400 MHz units 10^{-26} W $m^{-2}Hz^{-1}$	9	3	7
Spectral index	−1.0	−0.5	−0.5
Total power in erg sec^{-1} between 10 and 10^4 MHz	1.2×10^{35}	4.0×10^{35}	1.0×10^{35}

Source: Westerlund and Matthewson (1966).

of the explosion. Instead of this, one may compare the remnants to the Cassiopeia A source. If n, r, and E are interstellar density, radius, and total energy with index s for the remnant and c for Cassiopeia A, equation (22) leads to

$$t_s = t_c \left(\frac{n_s}{n_c}\right)^{1/2} \left(\frac{r_s}{r_c}\right)^{5/2} \left(\frac{E_s}{E_c}\right)^{-1/2}. \quad (31)$$

With $n = 1.8\ cm^{-3}$ for N63A and $n = 1.4\ cm^{-3}$ for N132D, according to the data of McGee and Milton (1964), the ages are 3200 $(E_c/E_{63A})^{1/2}$ years for N63A and 1900 $(E_c/E_{132D})^{1/2}$ years for N132D. A valid interpretation of the radio emission might be obtained following van der Laan (1962*a*, *b*), but the necessary data are missing.

The appearance of N49 is rather similar to that of the Cygnus Loop. The outstanding differences are (1) the electron density and the emitted radio power, both about 200 times larger in N49 than in the Cygnus Loop, and (2) the much steeper spectrum of N49 with a spectral index of −1.0, definitely smaller than that in the Cygnus Loop and in all other old remnants. This could be an indication that a coincidence with a source of different origin might be involved.

Radial velocities observed in N49 range from values close to the local velocity in the Large Magellanic Cloud of +308 km/sec (McGee and Milton 1964) to a value of +404 km/sec in Section 3, which seems to refer to the location that is nearest to the center of the nebula. This suggests that the nebula is expanding with a velocity similar to that in the Cygnus Loop. If this is assumed, equations (23) and (24) give for the age 31,000 years; and with $n_H = 2.5$ for N49 from McGee and Milton, the total energy of the explosion for N49 is 2.5 times that of the Cygnus Loop. Thus N49 seems indeed to be an object that has great similarity to the Cygnus Loop. The greater emitted radio power is at least partly to be ascribed to the lower age of N49. In detail, a discussion following van der Laan should be applied for the interpretation. The data available now are inadequate for this, but the apparent size of N49 is large enough to permit the necessary more detailed investigation that seems very desirable.

The assignment of the remnants in the Large Magellanic Cloud to supernovae of Type II is mainly suggested by the similarity of N49 to the Cygnus Loop, but the observational arguments in favor of that assignment are not very strong. The presence of remnants of Type II is to be expected from the richness of young stars in the Large Magellanic Cloud. There may be some indication that the initial total energy and possibly the ejected mass of the observed remnants are relatively high. This would not be unexpected in a system that contains many young stars of high luminosity and high mass. Still undiscovered remnants of less energetic outbursts should then exist.

REFERENCES

Andrew, B. R., Branson, N. J. B., and Wills, D. 1964, *Nature*, **203**, 171.
Baade, W. 1938, *Ap. J.*, **88**, 285.
———. 1942, *ibid.*, **96**, 188.
———. 1943, *ibid.*, **97**, 119.
———. 1945, *ibid.*, **102**, 309.
———. 1956, *B.A.N.*, **12**, 312.
Baade, W., Burbidge, G. R., Hoyle, F., Burbidge, E. M., Christy, R. F., and Fowler, W. A. 1956, *Pub. A.S.P.*, **68**, 296.
Baade, W., and Minkowski, R. 1954, *Ap. J.*, **119**, 206.
Bazelian, L. L., Brande, S. J., Bruk, J. M., Zhuk, I. N., Men, A. V., Raibov, B. P., Sodin, L. G., and Sharykin, N. K. 1963, *Radiofizika*, **6**, 897 (*Sov. Radiophysics*, **6**, 32).
Bennett, A. S. 1962, *Mem. R.A.S.*, **68**, 163.
Bertola, F. 1964, *Ann. d'ap.*, **27**, 319.
———. 1965, *Asiago Contr.*, No. 171.
Boischot, A. 1962, *C.R.-Paris*, **255**, 3374.
Boischot, A., and Lequeux, J. 1964, *Ann. d'ap.*, **27**, 514.
Bolton, J. G., Gardner, F. F., and Mackey, M. B. 1964, *Australian J. Phys.*, **17**, 340.
Bolton, J. G., and Stanley, G. T. 1949, *Australian J. Sci. Res.*, **2A**, 139.
Bondar, L. N., Krotikov, K. S., Stankevich, K. S., and Tseitlin, N. M. 1965, *Radiofizika*, **8**, 347 (*Sov. Radiophysics*, **8**, 310).
Bowen, I. S. 1958, *Carnegie Inst. Yearbook*, **57**, 60.
———. 1965*a*, *ibid.*, **64**, 38.
———. 1965*b*, *ibid.*, **64**, 24.
Bowyer, S., Byram, E. T., Chubb, T. A., and Friedman, H. 1964, *Science*, **147**, 394.
Branson, N. J. B. A. 1965, *M.N.*, **85**, 250.
———. 1966, *ibid.*, **86**, 40.
Brodskaya, E. S. 1963, *Izv. Crimean Ap. Obs.*, **30**, 126.
Brodskaya, E. S., and Grigoreva, N. B. 1962, *A.J.* (*U.S.S.R.*), **39**, 754 (*Sov. Astr.-A.J.*, **6**, 586).
Brown, R. H., and Hazard, C. 1952, *Nature*, **170**, 364.
Brussaard, P. J., and van de Hulst, H. C. 1962, *Rev. Mod. Phys.*, **34**, 507.
Burgess, A. 1958, *M.N.*, **118**, 477.
Burn, B. J. 1966, *M.N.*, **133**, 67.
Byram, E. T., Chubb, T. A., and Friedman, H. 1966, *Science*, **152**, 66.
Chamberlain, J. W. 1953, *Ap. J.*, **117**, 399.
Clark, G. W. 1965, *Phys. Rev. Letters*, **14**, 91.
Costain, C. H., Whitfield, G. R., and Elsmore, B. 1956, *M.N.*, **116**, 380.
Courtès, G., Véron, Ph., and Viton, M. 1964, *Ann. d'ap.*, **27**, 330.
Davies, R. D., Gardner, F. F., Hazard, C., and Mackey, M. B. 1966, *Australian J. Phys.*, **19**, 409.
Deutsch, A. N. 1966, private communication.
Dewhirst, D. W. 1965, private communication.
Dombrovsky, V. A. 1954, *Akad. Nauk U.S.S.R.*, **94**, 1021.
Duncan, J. C. 1921, *Proc. Nat. Acad. Sci.*, **7**, 179.
———. 1923, *Ap. J.*, **57**, 137.
———. 1937, *ibid.*, **86**, 496.
———. 1939, *ibid.*, **89**, 482.

Duyvendak, J. J. L. 1942, *Pub. A.S.P.*, **54**, 91.
Edge, D. O., Shakeshaft, J. R., McAdam, W. F., Baldwin, J. E., and Archer, S. 1959, *Mem. R.A.S.*, **68**, 37.
Elsmore, B., Ryle, M., and Leslie, P. P. R. 1959, *Mem. R.A.S.*, **68**, 61.
Erickson, W. C., and Cronyn, W. M. 1965, *Ap. J.*, **142**, 1156.
Fomalont, E. B., Matthews, T. A., Morris, D., and Wyndham, J. D. 1964, *A.J.*, **69**, 772.
Gardner, F. F. 1965, *Australian J. Phys.*, **18**, 385.
Gardner, F. F., and Milne, D. K. 1965, *A.J.*, **70**, 754.
Gardner, F. F., and Whiteoak, J. B. 1963, *Nature*, **197**, 1162.
Ginzburg, V. L., and Ozernoy, L. M. 1966, *Ap. J.*, **144**, 599.
Goldstein, B. R. 1965, *A.J.*, **70**, 105.
Goldstein, B. R., and Ho Peng Yoke. 1965, *A.J.*, **70**, 748.
Gower, J. F. R. 1966, private communication.
Grader, R. J., Hill, R. W., Seward, F. D., and Toor, A. 1966, *Science*, **152**, 1499.
Greenstein, J. L., and Minkowski, R. 1953, *Ap. J.*, **118**, 1.
Harris, D. E. 1962, *Ap. J.*, **135**, 661.
Haymes, R. C., and Craddock, W. L., Jr. 1966, *J. Geophys. Res.*, **71**, 3261.
Heeschen, D. S., and Meredith, B. L., 1961, *Nature*, **190**, 705.
Heiles, C. 1964, *Ap. J.*, **140**, 470.
Henize, K. G. 1956, *Ap. J. Suppl.*, **2**, 315.
Hewish, A., and Okoye, S. E. 1964, *Nature*, **203**, 171.
———. 1965, *ibid.*, **207**, 69.
Higgs, L. A. 1964, *J.R.A.S. Canada*, **59**, 56.
Hill, E. R. 1964, *The Galaxy and the Magellanic Clouds*, eds. F. J. Kerr and A. W. Rodgers (Canberra: Australian Academy of Science), p. 107.
Ho Peng Yoke. 1962, *Vistas in Astronomy*, ed. A. Beer (London, Oxford, New York, Paris: Pergamon Press), **5**, 127.
Högbom, J. A., and Shakeshaft, J. R. 1961, *Nature*, **189**, 561.
Hogg, D. E. 1963, *A.J.*, **68**, 76.
———. 1964, *Ap. J.*, **140**, 992.
———. 1966, private communication.
Howard, W. E., III, and Dickel, H. R. 1963, *Pub. A.S.P.*, **75**, 149.
Howard, W. E., III, and Maran, S. P. 1965, *Ap. J. Suppl.*, **10**, 1.
Hoyle, F., Fowler, W. A., Burbidge, G. R., and Burbidge, E. M. 1964, *Ap. J.*, **139**, 909.
Johnson, H. L., and Borgman, J. 1963, *B.A.N.*, **17**, 115.
Johnson, H. M. 1963, *Pub. Nat. Rad. Astr. Obs.*, **1**, 261.
Kaftan-Kassim, M. A. 1966, *Ap. J.*, **145**, 658.
Kardashev, N. S. 1962, *A.J.* (*U.S.S.R.*), **39**, 393 (*Sov. Astr.–A.J.*, **6**, 317).
Kellermann, K. I. 1964*a*, *Ap. J.*, **140**, 969.
———. 1964*b*, *Pub. Owens Valley Rad. Obs.*, **1**, No. 1.
Kenderdine, S. 1963, *M.N.*, **126**, 55.
Khare, R. C. 1954, *Zs. f. Ap.*, **35**, 115.
Kraft, R. P. 1963, private communication.
Kronberg, P. P. 1966, *Nature*, **212**, 1557.
Kuzmin, A. D., and Salomonovich, A. E. 1961, *Doklady Akad. Nauk U.S.S.R.*, **140**, 81.
Laan, H. van der. 1962*a*, *M.N.*, **124**, 125.
———. 1962*b*, *ibid.*, p. 179.
Labrum, N. R., Krishnan, T., Payten, W., and Harting, E. 1964, *Australian J. Phys.*, **17**, 323.
Lampland, C. O. 1921, *Pub. A.S.P.*, **33**, 79.
Lastochkin, V. P., and Stankevich, K. S. 1964, *A.J.* (*U.S.S.R.*), **41**, 769 (*Sov. Astr.–A.J.*, **8**, 612).
Lequeux, J. 1962, *Ann. d'ap.*, **25**, 221.
Liller, W. 1964, *Science*, **143**, 437.
Little, A. G. 1963, *Ap. J.*, **137**, 164.
McGee, R. X., and Milton, J. A. 1964, *The Galaxy and the Magellanic Clouds*, eds. F. J. Kerr and A. W. Rodgers (Canberra: Australian Academy of Science), p. 289.
Maltby, P., and Moffet, A. T. 1962, *Ap. J. Suppl.*, **7**, 141.
Matthewson, D. S., and Healey, J. R. 1964, *The Galaxy and the Magellanic Clouds*, eds. F. J. Kerr and A. W. Rodgers (Canberra: Australian Academy of Science), p. 283.
Matthewson, D. S., Healey, J. R., and Westerlund, B. E. 1963, *Nature*, **199**, 681.
Matthewson, D. S., Large, M. I., and Haslam. C. G. T. 1960, *M.N.*, **121**, 543.
Matveenko, L. I., and Khromov, G. S. 1965, *Astr. Tsirk.*, No. 343.
Mayall, N. U. 1937, *Pub. A.S.P.*, **49**, 101.
———. 1962, *Science*, **137**, 91.
Mayall, N. U., and Oort, J. H. 1942, *Pub. A.S.P.*, **54**, 95.
Mayer, C. H., McCullough, T. P., and Sloanaker, R. M. 1964, *Ap. J.*, **139**, 248.
Mayer, C. H., McCullough, T. P., Sloanaker, R. M., and Haddock, F. T. 1965, *Ap. J.*, **141**, 867.

Menon, T. K., and Terzian, Y. 1965, *Ap. J.*, **141**, 745.
Menon, T. K., and Williams, D. R. W. 1966*a*, *A.J.*, **71**, 392.
———. 1966*b*, private communication.
Mihalas, D. 1963, *Pub. A.S.P.*, **75**, 256.
Minkowski, R. 1939, *Ap. J.*, **89**, 156.
———. 1942, *ibid.*, **96**, 199.
———. 1957, *Radio Astronomy* (*I.A.U. Symposium No. 4*), ed. H. C. van der Hulst (Cambridge: Cambridge University Press), p. 114.
———. 1958, *Rev. Mod. Phys.*, **30**, 1048.
———. 1959, *Paris Symposium on Radio Astronomy* (*I.A.U. Symposium No. 9*), ed. R. N. Bracewell Stanford: Stanford University Press), p. 315.
———. 1964, *Ann. Rev. Astr. Ap.*, **2**, 247.
———. 1965, *A.J.*, **70**, 755.
———. 1966*a*, *ibid.*, **71**, 371.
———. 1966*b*, *Nature*, **209**, 1339.
Moroz, V. I. 1964, *A.J.* (*U.S.S.R.*), **40**, 982 (*Sov. Astr.–A.J.*, **7**, 755).
Morris, D., Radhakrishnan, V., and Seielstad, G. A. 1964, *Ap. J.*, **139**, 551.
Muller, C. A. 1959, *Paris Symposium on Radio Astronomy* (*I.A.U. Symposium No. 9*), ed. R. N. Bracewell (Stanford: Stanford University Press), p. 360.
Needham, J. 1959, *Science and Civilisation in China* (Cambridge: Cambridge University Press), **3**, 269.
O'Dell, C. R. 1962, *Ap. J.*, **136**, 809.
———. 1963, *ibid.*, **138**, 1018.
———. 1964, *Pub. A.S.P.*, **76**, 308.
Oort, J. H. 1951, *Problems of Cosmical Aerodynamics* (Dayton, Ohio: Central Air Documents Office), p. 118.
Oort, J. H., and Walraven, T. 1956, *B.A.N.*, **12**, 285.
Osterbrock, D. E. 1957, *Pub. A.S.P.*, **69**, 227.
———. 1958, *ibid.*, **70**, 180.
———. 1964, *Ann. Rev. Astr. Ap.*, **2**, 95.
Parker, R. A. R. 1964*a*, *Ap. J.*, **139**, 208.
———. 1964*b*, *ibid.*, p. 493.
Peterson, L. E., Jacobson, A. S., and Pelling, R. M. 1966, *Phys. Rev. Letters*, **16**, 142.
———. 1964, *Ann. d'ap.*, **27**, 522.
———. 1965, *Bull. Obs. Tonantzintla y Tacubaya*, **4**, 49.
Read, R. B. 1963, *Ap. J.*, **138**, 1.
Rishbeth, H. 1956, *Australian J. Phys.*, **9**, 494.
Ryle, M. 1962, *Nature*, **194**, 517.
———. 1966, private communication.
Ryle, M., Elsmore, B., and Neville, A. C. 1965, *Nature*, **205**, 1259.
Sedov, L. I. 1959, *Similarity and Dimensional Methods in Mechanics*, translated by M. Friedman (New York: Academic Press).
Seeger, C. L., and Westerhout, G. 1957, *B.A.N.*, **13**, 312.
Sharov, A. S. 1962, *A.J.* (*U.S.S.R.*), **40**, 900 (*Sov. Astr.–A.J.*, **7**, 689).
Shklovsky, I. S. 1953*a*, *A.J.* (*U.S.S.R.*), **30**, 15.
———. 1953*b*, *Doklady Akad. Nauk U.S.S.R.*, **90**, 983.
———. 1954, *ibid.*, **94**, 417.
———. 1960*a*, *A.J.* (*U.S.S.R.*), **37**, 256 (*Sov. Astr.–A.J.*, **4**, 243).
———. 1960*b*, *ibid.*, p. 369 (*ibid.*, p. 355).
———. 1962, *ibid.*, **39**, 209 (*ibid.*, **6**, 162).
———. 1965, *Doklady Akad. Nauk U.S.S.R.*, **160**, 54.
———. 1966, *A.J.* (*U.S.S.R.*), **43**, 10 (*Sov. Astr.–A.J.*, **10**, 6).
Sholomitskii, G. B. 1963, *A.J.* (*U.S.S.R.*), **40**, 223 (*Sov. Astr.–A.J.*, **7**, 172).
Slee, O. B. 1966, private communication.
Slee, O. B., and Orchiston, D. W. 1965, *Australian J. Phys.*, **18**, 187.
Slysh, V. I. 1963, *Nature*, **199**, 682.
———. 1966, *A.J.* (*U.S.S.R.*), **42**, 689 (*Sov. Astr.–A.J.*, **9**, 533).
Soboleva, N. S., Prozorov, V. A., and Pariskii, Yu. N. 1963, *A.J.* (*U.S.S.R.*), **40**, 3 (*Sov. Astr.–A.J.*, **7**, 1).
Talen, J. L. 1965*a*, *A.J.*, **70**, 332.
———. 1965*b*, private communication.
Taylor, G. I. 1950, *Proc. R. Soc. London A*, **101**, 159.
Thompson, A. R. 1965, private communication.
Tolbert, C. W. 1965, *Nature*, **206**, 1304.
Tolbert, C. W., and Straiton, A. W. 1965, *A.J.*, **70**, 177.
Tousey, R., and Koomen, M. J. 1953, *J. Opt. Soc. Amer.*, **43**, 177.
Troitskii, V. S., Tseitlin, N. M., and Porfir'ev, V. A. 1964, *A.J.* (*U.S.S.R.*), **41**, 446 (*Sov. Astr.–A.J.*, **8**, 354).

Vashakidze, M. A. 1954, *Astr. Tsirk.*, **147,** 11.
Welch, W. J. 1966, private communication.
Westerhout, G. 1957, *B.A.N.*, **13,** 201.
Westerlund, B. E. 1964, *The Galaxy and the Magellanic Clouds*, eds. F. J. Kerr and A. W. Rodgers (Canberra: Australian Academy of Science), p. 114.
Westerlund, B. E., and Matthewson, D. S. 1966, *M.N.*, **131,** 371.
Whitford, A. E. 1958, *A.J.*, **63,** 204.
Williams, D. R. W., Welch, W. J., and Thornton, D. D. 1965, *Pub. A.S.P.*, **77,** 178.
Williams, P. J. S. 1963, *Nature*, **200,** 56.
Woltjer, L. 1957*a*, *B.A.N.*, **13,** 301.
———. 1957*b*, *ibid.*, **14,** 39.
———. 1964, *Ap. J.*, **140,** 1309.
Wyndham, J. D., and Read, R. B. 1965, *A.J.*, **70,** 120.
Xi Ze-zong, and Bo Shu-ren. 1965, *Acta Astr. Sinica*, **13,** 1 (*NASA Tech. Trans.* TT F-388).
Yang, K. S. 1966, private communication.
Yang, K. S., and Dickel, J. R. 1965, *A.J.*, **70,** 300.
Zwicky, F. 1964, *Ann. d'ap.*, **27,** 300.
———. 1965, *Stellar Structure*, eds. L. H. Aller and D. B. McLaughlin (Chicago: University of Chicago Press), chap. 7.

CHAPTER 12

The Theory of Synchrotron Radiation

R. C. BLESS
Washburn Observatory, University of Wisconsin

1. INTRODUCTION

DURING the years immediately following World War II, physicists investigated the limitations set by various processes on the energies attainable in the particle accelerators then being designed. One of the most important of these processes is the radiative energy loss by relativistic electrons accelerated in the magnetic field of a betatron or synchrotron. Several authors (e.g., Artsimovich and Pomeranchuk 1946; Schwinger 1949) derived descriptions of this process, which was first brought to the attention of astronomers by Alfvén and Herlofson (1950). Alfvén and Herlofson suggested that the recently discovered "radio star emission is produced by cosmic ray electrons in the trapping field of a star." This mechanism thus provided a link between radio sources and cosmic rays, a connection proposed earlier by Ryle (1949). Soon after, Kiepenheuer (1950) put their suggestion in its modern form by arguing that the general galactic radio emission could be accounted for by the radiation of relativistic electrons moving in the interstellar magnetic field.

The importance of the synchrotron mechanism in an astronomical source was clearly demonstrated for the Crab Nebula. Several difficulties had been encountered in attempting to account for its observed optical and radio emission on the basis of free-free processes (see, e.g., Greenstein and Minkowski 1953). Perhaps the most serious of these were the spectral energy distribution (hundreds of times as much energy is radiated per frequency interval in the radio region as in the optical) and the large polarization observed in the amorphous central mass (Dombrovsky 1954; Baade 1956). No thermal process is known which can produce such polarized radiation, and free-free emission theory predicts only a few times more energy in the radio region than in the optical or, in other words, the radio radiation is much more intense than would be expected from thermal radiation at even the 10^5 to 10^6 ° K required to account for the optical radiation. Other difficulties included the large nebular mass (20–30 solar masses) required if the radiation were by free-free transitions and the absence of any line radiation from the amorphous region. Shklovsky (1953) pointed out that these features have a natural explanation if both the radio and optical emission from the amorphous mass are produced by the synchrotron mechanism. Such radiation is continuous and strongly polarized. Its spectrum can be made to fit that observed by making plausible assumptions about the energy distribution of the radiating particles. These particles emit such large amounts of energy that only about 10^{-6} to 10^{-7} solar masses are required to produce the observed radiation.

The recognition that this mechanism of radiation is important in astronomical sources has been one of the most fruitful recent developments in astrophysics. For example, it has made possible the inference that high-energy particles exist in some types of astronomical objects, it has given additional evidence for the existence of extensive magnetic fields, and it has indicated that enormous amounts of energy may indeed be stored in such modes. Furthermore, it has contributed appreciably to our understanding of the origin of cosmic rays.

Two general approaches have been considered recently in attempts to account for nonthermal radiation, that is, radiation produced by processes other than that involving electrons with a Maxwellian velocity distribution undergoing free-free transitions. The total power radiated by moving charges becomes greater with increasing velocity or with an increasing number of charges moving coherently. In the latter case, n electrons move systematically, thus producing n^2 times as much power. These electrons may have a Maxwellian velocity distribution, and each radiates by the free-free mechanism. The coherency requirement leading to large kinetic energies of the plasma blob (for example) and thence to high radiation intensities is the nonthermal aspect of this process. Plasma oscillations and a radio transmitting antenna are examples of this mechanism. In the class of processes involving relativistic particle velocities, the electrons may have neither a Maxwellian velocity distribution nor radiate by free-free transitions. Coulomb interactions are unimportant. The Cerenkov and synchrotron radiation mechanisms are examples of this class.

Only high-energy charged particles moving in a magnetic field are necessary to produce synchrotron radiation. As long as the field is static it does no work on the charges since the Lorentz force is perpendicular to the particle velocity. It is the self-force of the particle that does the work and causes the total particle energy to be slowly radiated away. Wentzel (1957) shows, however, that since the rate of energy loss by the self-force is so small compared to the total energy of the particles, only the Lorentz force need be included in the equation of motion, at least in astronomical situations.

So far, only the synchrotron mechanism is known to be important in sources outside the Solar System, and we shall now examine the theory briefly.

2. THE THEORY OF SYNCHROTRON RADIATION

Since the theory is easily available in the literature (e.g., Schwinger 1949; Jackson 1962; Landau and Lifshitz 1962) we will give only an outline of Schwinger's calculations. The theory is entirely classical, quantum effects being negligible; in fact, similar expressions for the radiation produced by high-energy electrons in a circular trajectory were derived by Schott in 1912.

2.1. The Total Power Radiated

The total power radiated by an accelerated electron moving at nonrelativistic velocities is given by the familiar Larmor formula

$$P = \frac{2}{3} \frac{e^2}{m^2 c^3} \left(\frac{dp}{dt}\right)^2 . \tag{1}$$

Since radiated energy and elapsed time transform similarly in a Lorentz transformation, the power must be a Lorentz invariant. Equation (1) may therefore be generalized to an arbitrary electron velocity by finding from the many possible Lorentz-invariant relations the one that reduces to the Larmor formula for small velocity. The correct relation is found by remembering that the general expression for the field of a moving charge contains terms in v and $\dot{v}$ only. The correct expression is

$$P = \frac{2}{3} \frac{e^2}{m^2 c^3} \gamma^2 \left[\left(\frac{dp}{dt}\right)^2 - \frac{1}{c^2}\left(\frac{d\epsilon}{dt}\right)^2\right] \tag{2}$$

where $p = \gamma m v$, $\gamma = \epsilon / mc^2$, and ϵ is the total energy. It is easy to see that the energy loss by radiation for a particle in rectilinear motion is small, since in this case $c^2 p dp/dt = \epsilon d\epsilon/dt$ and equation (2) becomes

$$P = \frac{2}{3} \frac{e^2}{m^2 c^3} \left(\frac{dp}{dt}\right)^2 = \frac{2}{3} \frac{e^2}{m^2 c^3} \left(\frac{d\epsilon}{dx}\right)^2 . \tag{3}$$

The radiated power does not depend on the electron energy but rather on the external force applied to the electron. A simple calculation (taking the ratio of radiated power to power provided by the external force) shows that radiative losses are unimportant unless $d\epsilon/dx \simeq 2 \times 10^{14}$ Mev/meter, a condition not encountered in astronomical situations (or in any other for that matter).

The situation is quite different for a particle moving in a circular path, since now, compared with its energy change per revolution, the momentum changes much more rapidly as the particle revolves; that is, $dp/dt = \omega_0 p \gg (1/c)\, d\epsilon/dt$. Thus to a good approximation,

$$P = \frac{2}{3} \frac{e^2}{m^2 c^3} \gamma^2 \left(\frac{dp}{dt}\right)^2 = \frac{2}{3} \frac{e^2 \omega_0}{R} \gamma^4 \beta^3 \tag{4}$$

where ω_0 and R are the instantaneous angular velocity and radius of curvature and where $\beta = v/c$. The energy radiated per revolution is

$$\delta\epsilon = \frac{2\pi}{\omega_0} P = \frac{4\pi}{3} \frac{e^2}{R} \gamma^4 \beta^3 \text{ ev/rev} .$$

For relativistic electrons, $\beta \simeq 1$ and

$$\delta\epsilon = 8.85 \times 10^4 \frac{[\epsilon(\text{Bev})]^4}{R(m)} \text{ ev/sec} .$$

For a 1-Bev electron or positron gyrating in a magnetic field with $H = 10^{-5}$ gauss, $R = \gamma mc^2/eH \simeq 3 \times 10^9$ meters, $\nu \simeq 0.02$ c/s, $\delta\epsilon \simeq 2.5 \times 10^{-5}$ ev/rev, or about 4×10^{-7} ev/sec. Note that a 1-Bev proton would radiate at only about 1/2000 of this rate. Thus in astronomical sources, only electrons and positrons are important for the synchrotron mechanism.

2.2. The Angular Distribution of the Radiation

To derive expressions for the angular and spectral energy distributions radiated by an electron moving along an arbitrary trajectory, Schwinger considers the rate at which the current density $\boldsymbol{j}$ does work on the retarded electromagnetic field $-\int \boldsymbol{j}\cdot \boldsymbol{E}_{\mathrm{ret}} dV$. Now the work that is done is of two kinds: in one, energy is contributed irreversibly to the electromagnetic field through radiation; in the other, energy is stored reversibly in the field. Only the dissipative part is of interest here, and Schwinger separates the two types (with one qualification) by writing

$$\boldsymbol{E}_{\mathrm{ret}} = \tfrac{1}{2}(\boldsymbol{E}_{\mathrm{ret}} + \boldsymbol{E}_{\mathrm{adv}}) + \tfrac{1}{2}(\boldsymbol{E}_{\mathrm{ret}} - \boldsymbol{E}_{\mathrm{adv}}) \tag{5}$$

and noting that the first term on the right changes sign with a change in the sense of time, while the second does not. Thus the first term determines the nondissipative, reactive power stored in the field; the second determines the power radiated away so that

$$P = -\int \boldsymbol{j}\cdot \boldsymbol{E} dV = \int \boldsymbol{j}\cdot \left(\frac{1}{c}\frac{\partial \boldsymbol{A}}{\partial t} + \nabla\varphi\right) dV \tag{6}$$

with

$$\boldsymbol{E} = \tfrac{1}{2}(\boldsymbol{E}_{\mathrm{ret}} - \boldsymbol{E}_{\mathrm{adv}}).$$

$\boldsymbol{A}$ and φ are the usual vector and scalar potentials written in terms of a δ-function of the reference and retarded times. The qualification mentioned above is that the second term in equation (5) still contains terms which represent essentially no energy dissipation. They involve the time derivative of acceleration-dependent electron energies which, for any reasonable accelerations, are small compared to the electron kinetic energies and so may be neglected. Simplifying equation (6) by means of the charge conservation equation, introducing the retarded and advanced forms of the potentials according to equation (6), and particularizing the arbitrary charge distribution to a point charge, Schwinger finds that the power radiated into unit solid angle in the direction $\boldsymbol{n}$ at time t is

$$P(\boldsymbol{n}, t) = \frac{e^2}{4\pi m^2 c^3}\left(\frac{1}{\gamma}\right)^2 \left[\frac{\dot{\boldsymbol{p}}^2 - \frac{1}{c^2}\dot{\epsilon}^2}{\left(1 - \boldsymbol{n}\cdot\boldsymbol{p}\frac{c}{\epsilon}\right)^3} - \frac{1}{\gamma^2}\frac{\left(\boldsymbol{n}\cdot\dot{\boldsymbol{p}} - \frac{1}{c}\dot{\epsilon}\right)^2}{\left(1 - \boldsymbol{n}\cdot\boldsymbol{p}\frac{c}{\epsilon}\right)^5}\right]. \tag{7}$$

Integrated over all solid angles, this equation reduces to equation (2).

So far, the electron trajectory is arbitrary. Consider now the angular distribution of radiation produced in the special case of circular motion. If the instantaneous velocity and acceleration are in the z- and x-directions, respectively, and the direction of observation is defined by the usual polar angles θ and φ, then equation (7) becomes

$$P(\boldsymbol{n}, t) = \frac{e^2}{4\pi m^2 c^3}\dot{\boldsymbol{p}}^2(1-\beta^2)\left[\frac{1}{(1-\beta\cos\theta)^3} - (1-\beta^2)\frac{\sin^2\theta\cos^2\varphi}{(1-\beta\cos\theta)^5}\right].$$

For electron velocities small compared with that of light, $\beta \cong 0$ and the angular dependence of the emission goes as $1 - \sin^2\theta\cos^2\varphi$. As the particle velocity increases, however, the radiation is increasingly emitted in the forward ($\theta \to 0$) direction in the plane of motion (see Fig. 1). A measure of the angle of emission is easily found by considering the mean value of $\sin^2\theta$ to be defined as

$$\langle \sin^2\theta \rangle = \int \sin^2\theta \, P(\boldsymbol{n}, t)\, d\Omega / P(t)$$

which for the relativistic case reduces to

$$\langle \sin^2\theta \rangle = 1 - \beta^2 = \frac{1}{\gamma^2}$$

so that the angle between the electron's direction of motion and its direction of emission is

$$\langle \theta^2 \rangle^{1/2} \cong \frac{1}{\gamma} \quad \text{or} \quad \theta = 1.76/\epsilon(\text{Bev})\,\text{min of arc}\,.$$

This type of result is typical of any relativistic radiation pattern, not just of the case of circular motion considered here. Furthermore, the radiation emitted by a relativistic particle in arbitrary motion is essentially equivalent to that emitted by such a particle

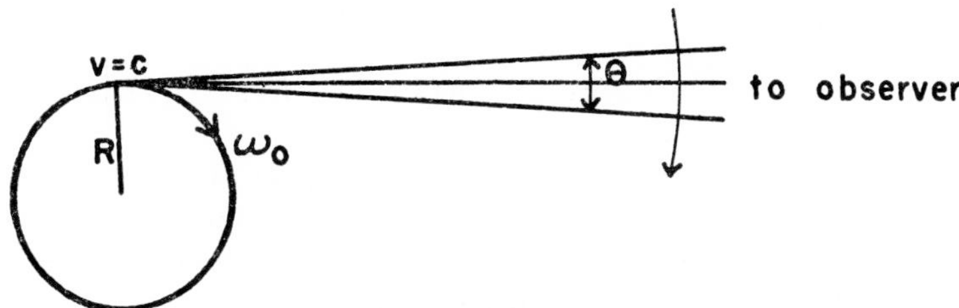

FIG. 1.—Cone of radiation emitted by a relativistic electron

in circular motion, as may be seen by comparing equations (3) and (4). For the same applied force, the power radiated when the acceleration is parallel to the velocity is only equal to $1/\gamma^2$ that radiated when the acceleration is perpendicular to the velocity; so circular motion is a good approximation to arbitrary motion in this respect. For this reason, we will henceforth consider only circular motion.

2.3. The Spectral Distribution of the Radiation

That a relativistic particle will emit high-frequency radiation is easily seen from the following consideration. Since the radiation is confined to a very narrow cone, it will sweep past an observer as a pulse of duration $\Delta t' \sim R/c\gamma$ measured in its own time. Because of the Doppler shift, however, the observer will receive the pulse for a time interval of $\Delta t \sim \Delta t'/\gamma^2$; so the pulse appears to him to last for a time $\Delta t \sim R/c\gamma^3$. From Fourier theory, $\Delta\omega\Delta t \simeq 1$; so that the spectrum of such a pulse will contain frequencies up to a critical frequency of the order of $\omega_c \sim \omega_0\gamma^3$ where $\omega_0 = c/R$, the "fundamental" or gyro-frequency. If the motion is strictly periodic, all the Fourier components are harmonics of the fundamental. Thus a relativistic particle emits energy in a broad spectral band at frequencies up to γ^3 times the gyro-frequency, with negligible radiation emitted at higher frequencies. For a 1-Bev electron moving in a field of 10^{-5} gauss, $\omega_0 = 0.1$ rad/sec whereas $\omega_c \simeq 8 \times 10^8$ rad/sec (or about 100 mc/s for ν_c compared with 0.02 c/s for

ν_0). The high-frequency radiation emitted by a relativistic electron is perhaps the most striking aspect of the synchrotron mechanism.

The spectrum radiated by a high-energy electron will now be considered in more detail. In a manner completely analogous to that used for equation (7), Schwinger derives an expression for the power radiated into angular frequency ω at time t. He uses the Fourier integral form of the delta function in the potentials to display the frequency dependence. His result is

$$P(\omega, t) = -\frac{e^2\omega}{\pi}\int_{-\infty}^{\infty}\left[1 - \frac{1}{c^2}\, v(t)\cdot v(t+\tau)\right]\cos\omega t\,\frac{\sin\frac{\omega}{c}|R(t+\tau) - R(t)|}{|R(t+\tau) - R(t)|}\,d\tau \quad (8)$$

where $\tau = t' - t$, t' is the retarded time, and R is the position of the moving electron. For this expression to be correct, $P(\omega, t)$ must change very little over a time $\simeq 1/\omega$; that is, the radiative reaction effects must be small. This condition is seen to be true from the previously calculated rate at which an electron loses energy.

For the reason given in Section 2.2, only circular motion need concern us; for this case it is easy to see that

$$\boldsymbol{v}(t)\cdot\boldsymbol{v}(t+\tau) = v^2\cos\omega_0\tau \quad\text{and}\quad |R(t+\tau) - R(t)| = 2R\left|\sin\frac{\omega_0\tau}{2}\right|.$$

Since the motion is periodic with angular frequency ω_0, we may write $\omega_0\tau = \varphi + 2\pi k$ where k is an integer, so that the integral in equation (8) is replaced by an integral over φ and a summation over k. This summation may be transformed to a summation over the harmonic number n $(=\omega/\omega_0)$, so that finally we have

$$P(\omega) = \sum_{n=1}^{\infty}\delta(\omega - n\omega_0)P_n$$

where the power radiated into the nth harmonic is

$$P_n = -n\omega_0\frac{e^2}{R}\frac{1}{2\pi}\int_{-\pi}^{\pi}(1 - \beta^2\cos\varphi)\cos n\varphi\cdot\frac{\sin(2n\beta\sin\varphi/2)}{\sin\varphi/2}\,d\varphi. \quad (9)$$

Equation (9) may be expressed in terms of Bessel functions of order $2n$,

$$P_n = n\omega_0\frac{e^2}{R}\left[2\beta^2 J'_{2n}(2n\beta) - (1-\beta^2)\int_0^{2n\beta}J_{2n}(x)\,dx\right]. \quad (10)$$

The harmonic corresponding to the critical frequency is $n_c = \omega_c/\omega_0 \simeq \gamma^3$. Given $\beta \simeq 1$ and considering only the power radiated into harmonics very much lower than n_c, we see that equation (10) becomes

$$P_n = 2n\omega_0\frac{e^2}{R}J'_{2n}(2n) \simeq \frac{3^{1/6}}{\pi}\Gamma(2/3)\,\omega_0\frac{e^2}{R}n^{1/3}. \quad (11)$$

At low frequencies, the radiated power slowly increases with n (see Fig. 2). To evaluate equation (10) for frequencies comparable to the critical frequency, that is, for both the order and argument of J and J' large, we shall express J and J' in terms of the modified Bessel function,

$$P_n = \frac{3^{3/2}}{4\pi}\,\omega_0\,\frac{e^2}{R}\,\frac{1}{(1-\beta^2)^2}\,\frac{n}{n_c^2}\int_{n/n_c}^{\infty} K_{5/3}(\eta)\,d\eta\,. \qquad (12)$$

Here $n_c = 3/2(1-\beta^2)^{3/2}$. For $n \gg n_c$, equation (12) may be approximated by integrating the asymptotic expansion of $K_{5/3}$ and keeping only the first term. Then

$$P_n \cong \frac{3^{1/2}}{2\sqrt{2\pi}}\,\omega_0\gamma\,\frac{e^2}{R}\left(\frac{n}{n_c}\right)^{1/2} e^{-n/n_c}\,,$$

and the radiation emitted into the very high-order harmonics is cut off sharply. Since for $n \ll n_c$, $P_n \sim n^{1/3}$, while for $n \gg n_c$, $P_n \sim (n/n_c)^{1/2}e^{-n/n_c}$, the spectrum has its maximum at $n \cong n_c = 3/2(1-\beta^2)^{3/2}$. For relativistic electrons, n_c is very large compared to the fundamental n_0, so that high-order harmonics separated by n_0 form a nearly continuous spectrum. (For a 1-Bev electron, n_c is the 8×10^9th harmonic.) It can also be shown (Jackson 1962) that radiation in the higher harmonics is increasingly confined to the orbital

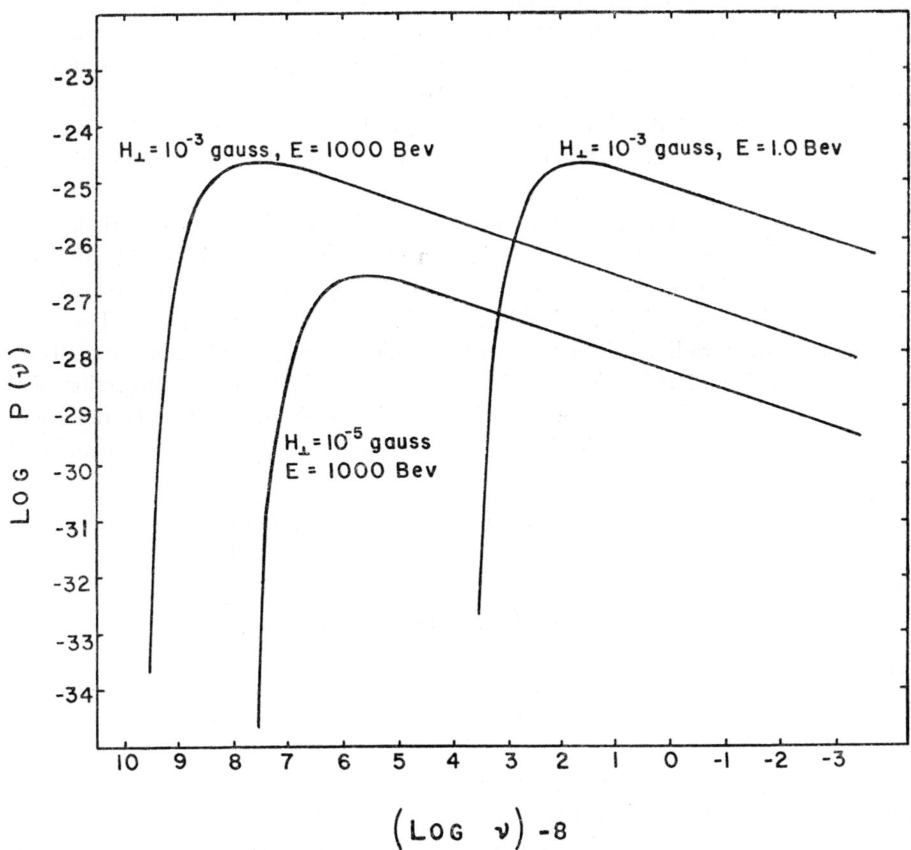

Fig. 2.—Energy radiated by a single electron in a magnetic field

plane. For $n \ll n_c$, the mean angle of emission is $\simeq \gamma^{-1}(n_c/n)^{1/3}$, a much wider angle than the average $1/\gamma$. For $n \gg n_c$, however, the radiation is confined to an angle $\simeq \gamma^{-1}(n_c/n)^{1/2}$, which is much smaller than the average.

2.4. The Polarization

One would expect that the radiation emitted by a relativistic electron traveling in a circular trajectory would be polarized primarily in the orbital plane. That it is not in general 100 per cent linearly polarized may be seen from the following argument. Consider first a nonrelativistic electron in circular motion. In the orbital plane, such a system looks like a dipole and the radiation is completely linearly polarized. Normal to the orbital plane, however, the motion is represented by two dipoles at right angles and the radiation is circularly polarized; at intermediate orientations, the polarization is, of course, elliptical. Now consider the relativistic case. The polarization depends not only on the angle of orientation of the orbit as in the nonrelativistic case but also on the confinement of the radiation to a cone; therefore the observer sees radiation from only a segment of the orbit. The component of the electric vector out of the orbital plane depends upon both of these factors. The radiation observed in the orbital plane is still completely plane-polarized. Because the angle of the emission cone is frequency-dependent, the polarization will depend upon the frequency; for a given orientation, the radiation at high frequencies is more nearly plane-polarized than that at low frequencies, because the latter radiation is emitted in a cone of larger vertex angle. Thus in general the radiation is elliptically polarized with components nearly in and perpendicular to the plane of the electron's orbit.

For his purposes, Schwinger did not need to calculate explicitly the electric-field vectors and so did not consider the polarization of the radiation. Westfold (1959) has recently made such calculations, from which we shall quote only the results. (Much of his treatment is of necessity identical in principle with that already given.) The expression for the power radiated into the nth harmonic in equation (12), has actually two components, $P_n^{(1)}$ and $P_n^{(2)}$ lying parallel to and perpendicular to the projection of the magnetic field vector in the plane normal to the line of sight. Because the radiation is confined to a cone of small angle, $P_n^{(1)}$ and $P_n^{(2)}$ lie essentially perpendicular to and in the orbital plane, respectively.

$$ {}_n = P_n^{(1)} + P_n^{(2)} \sim \left[F^{(1)}\left(\frac{n}{n_c}\right) + F^{(2)}\left(\frac{n}{n_c}\right) \right] $$

where

$$ F^{(1)}(x) = \tfrac{1}{2}x\left[\int_x^\infty K_{5/3}(\eta)\,d\eta - K_{2/3}(x) \right] $$

and

$$ F^{(2)}(x) = \tfrac{1}{2}x\left[\int_x^\infty K_{5/3}(\eta)\,d\eta + K_{2/3}(x) \right]. $$

If one sets $F(x) = F^{(1)}(x) + F^{(2)}(x)$ and $F_p(x) = F^{(2)}(x) - F^{(1)}(x)$, then $F(x)$ and $F_p(x)$ are proportional to the radiation polarized in the plane and perpendicular to the plane, respectively, and $F_p(x)/F(x)$ gives the polarization of the radiation. The functions $F(x)$ and $F_p(x)$, taken from Westfold (1959), are given in Table 1.

2.5. Quantum Effects

Quantum effects in the theory outlined above will not become important if the momentum of the electron is much greater than the momentum carried by the emitted photon, that is, if $\epsilon/c \gg h\nu_c/c$. Since $\nu_c \cong \gamma^3 c/R$ and the radius of gyration depends on both the particle energy and the external magnetic field, the above condition becomes

$$\frac{\epsilon}{mc^2} \ll \frac{mc^2}{(eh/2\pi mc)H}.$$

For no known astronomical magnetic field is this requirement violated. For example, for $H = 10^3$ gauss, quantum effects are unimportant for electron energies less than 10^7 Bev. For more detailed considerations of this point, see Vladimirsky (1948).

TABLE 1

$F(x)$ AND $F_p(x)$ TAKEN FROM WESTFOLD (1959)

x	F_p	F	x	F_p	F
0.000......	0.000	0.000	1.0.........	0.494	0.655
.001......	.107	.213	1.2.........	.439	.566
.005......	.184	.358	1.4.........	.386	.486
.010......	.231	.445	1.6.........	.336	.414
.025......	.312	.583	1.8.........	.290	.359
.050......	.388	.702	2.0.........	.250	.301
.075......	.438	.772	2.5.........	.168	.200
.10.......	.475	.818	3.0.........	.111	.130
.15.......	.527	.874	3.5.........	.0726	.0845
.20.......	.560	.904	4.0.........	.0470	.0541
.25.......	.582	.917	4.5.........	.0298	.0339
.30.......	.596	.919	5.0.........	.0192	.0214
.40.......	.607	.901	6.0.........	.0077	.0085
.50.......	.603	.872	7.0.........	.0031	.0033
.60.......	.590	.832	8.0.........	.0012	.0013
.70.......	.570	.788	9.0.........	.00047	.00050
.80.......	.547	.742	10.0.........	0.00018	0.00019
0.90.......	0.521	0.694			

3. ASTRONOMICAL APPLICATION

3.1. Radiation from a Single Electron

Before considering the synchrotron emission produced by a distribution of electrons, we shall summarize the equations so far developed for an individual electron, explicitly including the magnetic field through the Lorentz force. The total power radiated by a relativistic electron is, from equation (4),

$$P = \tfrac{2}{3}\,\frac{e^2}{m^2c^3}\,\gamma^2\left(\frac{e^2}{c^2}\,v^2H^2\sin^2\theta\right)$$

where θ is the angle between the velocity vector $\boldsymbol{v}$ and the field direction $\boldsymbol{H}$. Thus

$$P \cong \tfrac{2}{3}\,\frac{e^4}{m^2c^3}\,\gamma^2 H_\perp^2 = 1.58\times 10^{-15}\gamma^2 H_\perp^2\ \text{ergs/sec}$$

$$= 3.79\times 10^{-6}\epsilon^2 H_\perp^2\ \text{Bev/sec}\ (\epsilon\ \text{in Bev}) \qquad (13)$$

where $H_{\perp} = H \sin \theta$. Such a particle loses half of its energy in a time

$$T_{1/2} = \frac{8.35 \times 10^{-3}}{H^2 \epsilon (\text{Bev})} \text{ years},$$

found by integrating equation (13). Since $P(\nu) = (2\pi/\omega_0)P_n$, we may rewrite equation (12) to get the radiated power per unit frequency interval in the form

$$P(\nu) = \frac{3^{1/2} e^3}{m c^2} H_{\perp} F(x) = 2.34 \times 10^{-22} H_{\perp} F(x)$$

where

$$F(x) = x \int_x^{\infty} K_{5/3}(\eta)\, d\eta$$

and

$$x = n/n_c = \nu/\nu_c .$$

In the limit of $\nu \ll \nu_c$, we have from equation (11)

$$P(\nu) \cong 2^{2/3} \sqrt{3}\, \Gamma(2/3)\, x^{1/3} \frac{e^3 H_{\perp}}{m c^2} .$$

Comparing this with equation (13), we see that in this approximation

$$F(x) \cong 2^{2/3} \Gamma\ (2/3) x^{1/3} = 2.15\ x^{1/3}$$

and

$$P(\nu) \cong 5.04 \times 10^{-22} H_{\perp}\ x^{1/3} .$$

For $\nu \gg \nu_c$,

$$P(\nu) \cong \left(\frac{3\pi}{2}\right)^{1/2} \frac{e^3}{m c^2} H_{\perp} x^{1/2} e^{-x}$$

so that now

$$F(x) \cong (\pi/2)^{1/2} x^{1/2} e^{-x} = 1.26\ x^{1/2} e^{-x}$$

and

$$P(\nu) \cong 2.94 \times 10^{-22} H_{\perp}\ x^{1/2} e^{-x} .$$

We saw earlier that both the maximum of the spectral energy distribution and the frequency above which little energy is radiated may be approximated by the critical frequency $\omega_c = 3\ \omega_0 \gamma^3/2 = 3\ c/R\gamma^3/2$; so

$$\nu_c = \frac{3 e}{4\pi m c} H_{\perp} \gamma^2 = 1.61 \times 10^{13} H_{\perp} \epsilon^2 (\text{Bev})^2 \text{ c/s.} \qquad (14)$$

The corresponding harmonic is $n_c = 3/2\ \gamma^3$. Thus a relativistic electron will radiate γ^2 times as much power as a nonrelativistic electron, the radiation being largely concentrated near frequencies γ^3 times the gyro-frequency of the slow electron. Table 2 gives electron energies and lifetimes for various magnetic field strengths and critical frequencies.

3.2. Synchrotron Radiation from an Assemblage of Electrons

The power radiated at a given frequency by an assemblage of electrons with a differential energy spectrum $N(\epsilon)d\epsilon$ is of the form

$$J(\nu) \sim \int P(\nu) N(\epsilon) d\epsilon .$$

Since for a given magnetic field an electron of given energy has a specific critical frequency, the above expression may be written as

$$J(\nu) \sim \int_0^\infty P(\nu)\, N(\nu_c)\, d\nu_c \tag{15}$$

where $N(\nu_c)d\nu_c$ is the number of electrons with critical frequencies between ν_c and $\nu_c + d\nu_c$. The particle energy spectrum is generally represented by a power law for two reasons: a spectrum of this form represents the cosmic ray flux, and the observed spectrum of radiation can be reproduced by such a particle distribution. Thus we have

$$\begin{aligned} N(\nu_c)d\nu_c &= A\epsilon^{-\gamma}d\epsilon, \; \epsilon_1 \leq \epsilon \leq \epsilon_2 \,, \\ N(\nu_c)d\nu_c &= 0, \quad \epsilon < \epsilon_1, \quad \text{and } \epsilon > \epsilon_2 \,. \end{aligned} \tag{16}$$

TABLE 2

ELECTRON ENERGIES AND LIFETIMES FOR VARIOUS MAGNETIC FIELD STRENGTHS

H	$\nu_c = 10^7$ c/s		$\nu_c = 10^{10}$ c/s		$\nu_c = 6\times10^{14}$ c/s	
	E (Bev)	$t_{1/2}$ (years)	E (Bev)	$t_{1/2}$ (years)	E (Bev)	$t_{1/2}$ (years)
10^{-3}	0.025	3.4×10^5	0.794	1.1×10^4	193	43.5
10^{-4}	.079	1.1×10^7	2.50	3.4×10^5	610	1.38×10^3
10^{-5}	.25	3.4×10^8	7.94	1.1×10^7	1930	4.35×10^4
10^{-6}	0.79	1.1×10^{10}	25.0	3.4×10^8	6100	1.38×10^6

Expressing ϵ in terms of ν_c and $H_\perp$, equation (15) becomes

$$\begin{aligned} J(\nu) &= (1.17\times10^{-22})(1.61\times10^{13})^{(\gamma-1)/2} H_\perp^{(\gamma+1)/2} \nu^{-(\gamma-1)/2} \\ &\times A \int_x^\infty \xi^{(\gamma-3)/2} F(\xi)\, d\xi \,. \end{aligned} \tag{17}$$

Although the intensity of the radiation depends upon $H_\perp$, the spectral energy distribution depends only upon $\nu^{-(\gamma-1)/2}$ as long as the value of the integral is constant. If this is so and if the observed energy spectrum is given by $J(\nu) \sim \nu^{-m}$, then

$$\nu^{-m} \sim \nu^{-(\gamma-1)/2} \text{ and } \gamma = 2m + 1 \,.$$

Setting

$$G(x) = \int_x^\infty \xi^{(\gamma-3)/2} F(\xi)\, d\xi = \int_x^\infty \xi^{(\gamma-1)/2} \int_\xi^\infty K_{5/3}(\eta)\, d\eta\, d\xi$$

we may write equation (17) as

$$J(\nu) = (1.17\times10^{-22})(1.61\times10^{13})^{(\gamma-1)/2} H^{(\gamma+1)/2} \nu^{-(\gamma-1)/2} A \left[G(\nu/\nu_{c_2}) - G(\nu/\nu_{c_1}) \right]$$

where ν_{c_2} and ν_{c_1} are the critical frequencies corresponding to the highest and lowest energy electrons, respectively. For sources now known it appears that the slope of the synchrotron radiation spectrum continues unchanged to the lowest observable frequencies, which indicates that $\nu \gg \nu_{c_1}$ and $G(\nu/\nu_{c_1}) \simeq 0$. As long as $\nu \ll \nu_{c_2}$, then $G(\nu/\nu_{c_2}) \simeq G(0)$, the spectral index remains constant, and $\gamma = 2m + 1$. Any increase in the spectral index would indicate that $G(\nu/\nu_{c_2})$ is no longer constant, that is, that ν is approaching ν_{c_2}. Thus observations of the departure of the spectrum from a power-law dependence at high frequencies can in principle give the critical frequency of the most energetic source electrons that follow a $E^{-\gamma}$ dependence. Higher energy electrons probably exist in the source but should contribute relatively little to the total particle energy. Westfold shows that

$$G(x) = \frac{\gamma + 7/3}{\gamma + 1} G_p(x) - \frac{2x^{(\gamma-1)/2}}{\gamma + 1} [F(x) - F_p(x)]$$

where

$$G_p(x) = \int_x^{\infty} \xi^{(\gamma-1)/2} K_{2/3}(\xi) d\xi .$$

Dr. George Collins computed $G(x)$ and $G_p(x)$ for values of x from 0 to 10 and for nine values of γ from 5/3 to 21/3. His results are given in Table 3. $G_p(x)$ represents the com-

TABLE 3

$G(x)$ AND $G_p(x)$

γ	5/3		7/3		9/3		11/3	
x	G_p	G	G_p	G	G_p	G	G_p	G
0.0	1.69	2.53	1.31	1.83	1.21	1.61	1.30	1.67
0.001	1.61	2.41	1.31	1.83	1.21	1.62	1.26	1.63
0.005	1.60	2.38	1.30	1.82	1.21	1.62	1.26	1.63
0.010	1.58	2.34	1.30	1.81	1.21	1.62	1.26	1.62
0.025	1.54	2.24	1.28	1.78	1.21	1.61	1.26	1.62
0.050	1.46	2.11	1.25	1.73	1.20	1.59	1.26	1.62
0.075	1.40	2.00	1.23	1.68	1.19	1.57	1.26	1.61
0.10	1.35	1.90	1.20	1.64	1.18	1.55	1.25	1.60
0.15	1.25	1.73	1.15	1.55	1.15	1.51	1.24	1.58
0.20	1.16	1.59	1.10	1.47	1.12	1.46	1.22	1.55
0.25	1.08	1.46	1.06	1.40	1.09	1.42	1.20	1.53
0.30	1.01	1.36	1.01	1.33	1.06	1.37	1.19	1.50
0.40	0.890	1.17	0.924	1.20	1.00	1.28	1.14	1.43
0.50	0.787	1.02	0.845	1.08	0.943	1.19	1.10	1.36
0.60	0.697	0.893	0.772	0.978	0.883	1.11	1.05	1.29
0.70	0.620	0.784	0.705	0.884	0.825	1.02	0.996	1.22
0.80	0.552	0.692	0.643	0.800	0.769	0.948	0.946	1.15
0.90	0.492	0.613	0.587	0.725	0.716	0.876	0.895	1.09
1.00	0.439	0.538	0.536	0.652	0.665	0.806	0.845	1.02
1.20	0.351	0.426	0.444	0.536	0.571	0.686	0.748	0.893
1.40	0.282	0.339	0.368	0.441	0.489	0.581	0.658	0.779
1.60	0.226	0.271	0.305	0.363	0.416	0.493	0.576	0.677
1.80	0.182	0.215	0.253	0.297	0.354	0.414	0.501	0.584
2.00	0.147	0.172	0.209	0.244	0.300	0.384	0.434	0.503
2.50	0.0864	0.0971	0.130	0.146	0.196	0.221	0.299	0.338
3.00	0.0510	0.0560	0.0803	0.0887	0.127	0.141	0.202	0.225
3.50	0.0302	0.0318	0.0496	0.0529	0.0816	0.0880	0.135	0.146
4.00	0.0180	0.0185	0.0306	0.0320	0.0521	0.0553	0.0893	0.0955
4.50	0.0107	0.0110	0.0188	0.0196	0.0331	0.0350	0.0586	0.0623
5.00	0.00639	0.00676	0.0116	0.0123	0.0210	0.0225	0.0382	0.0410
6.00	0.00228	0.00234	0.00436	0.00452	0.00833	0.00870	0.0159	0.0168
7.00	0.000820	0.000944	0.00164	0.00185	0.00327	0.00366	0.00655	0.00727
8.00	0.000296	0.000293	0.000614	0.000619	0.00128	0.00130	0.00265	0.00273
9.00	0.000107	0.000113	0.000230	0.000243	0.000494	0.000524	0.00106	0.00113
10.0	0.0000386	0.0000417	0.0000857	0.0000922	0.000191	0.000204	0.000424	0.000453

ponent of the radiation parallel to the projection of H in the plane normal to the direction of observation, while $G(x)$ is proportional to the intensity of the radiation perpendicular to the projection of H. Thus the polarization is

$$p_\nu(\boldsymbol{n}) = \frac{G_p(\nu/\nu_{c2})}{G(\nu/\nu_{c2})}.$$

The table shows that, for low frequencies, the polarization is about 2/3 for $\gamma = 5/3$ and increases with γ; for all values of γ, the polarization asymptotically approaches 1 with increasing frequency. The observed polarization will of course decrease with increasing randomness of the field orientation within the observed volume. Inadequate instrumental resolution as well as Faraday rotation within the source or between the source and the observer also decrease the observed polarization. Both of these effects are much more marked in the radio region of the spectrum than in the optical.

TABLE 3—*Continued*

γ	13/3		15/3		17/3		19/3		21/3	
x	G_p	G	G_p	G	G_p	G	G_p	G	G_p	G
0.0	1.42	1.78	1.75	2.13	2.25	2.70	3.05	3.60	4.30	5.02
0.001	1.44	1.80	1.75	2.14	2.26	2.71	3.05	3.61	4.31	5.03
0.005	1.44	1.80	1.75	2.14	2.26	2.71	3.05	3.61	4.31	5.03
0.010	1.44	1.80	1.75	2.14	2.26	2.71	3.05	3.61	4.31	5.03
0.025	1.44	1.79	1.75	2.14	2.26	2.71	3.05	3.61	4.31	5.03
0.050	1.44	1.79	1.75	2.14	2.26	2.71	3.05	3.61	4.31	5.03
0.075	1.43	1.79	1.75	2.14	2.25	2.71	3.05	3.61	4.31	5.03
0.10	1.43	1.79	1.75	2.13	2.25	2.70	3.05	3.61	4.31	5.03
0.15	1.42	1.78	1.74	2.13	2.25	2.70	3.05	3.60	4.31	5.03
0.20	1.42	1.76	1.74	2.12	2.25	2.70	3.05	3.60	4.31	5.02
0.25	1.41	1.74	1.73	2.11	2.25	2.69	3.05	3.60	4.31	5.02
0.30	1.39	1.73	1.72	2.10	2.24	2.68	3.04	3.59	4.30	5.02
0.40	1.36	1.68	1.70	2.07	2.23	2.66	3.03	3.58	4.30	5.01
0.50	1.33	1.63	1.68	2.03	2.21	2.63	3.02	3.55	4.28	4.99
0.60	1.29	1.57	1.64	1.98	2.18	2.59	2.99	3.52	4.27	4.96
0.70	1.24	1.51	1.61	1.93	2.15	2.55	2.97	3.48	4.24	4.93
0.80	1.20	1.45	1.56	1.87	2.11	2.49	2.93	3.44	4.21	4.89
0.90	1.15	1.38	1.52	1.81	2.06	2.44	2.89	3.38	4.17	4.83
1.00	1.10	1.32	1.47	1.74	2.02	2.37	2.84	3.32	4.13	4.77
1.20	1.00	1.19	1.37	1.61	1.91	2.23	2.73	3.18	4.01	4.63
1.40	0.902	1.06	1.26	1.47	1.79	2.09	2.61	3.01	3.87	4.45
1.60	0.808	0.945	1.15	1.34	1.67	1.93	2.47	2.84	3.71	4.25
1.80	0.718	0.834	1.05	1.21	1.54	1.78	2.31	2.65	3.53	4.02
2.00	0.636	0.734	0.942	1.08	1.42	1.62	2.16	2.46	3.33	3.79
2.50	0.459	0.518	0.711	0.802	1.11	1.25	1.76	1.98	2.82	3.16
3.00	0.324	0.360	0.522	0.581	0.849	0.944	1.39	1.55	2.30	2.55
3.50	0.225	0.245	0.376	0.410	0.632	0.692	1.07	1.17	1.82	2.00
4.00	0.154	0.165	0.265	0.287	0.461	0.499	0.804	0.872	1.41	1.53
4.50	0.104	0.111	0.185	0.198	0.331	0.356	0.594	0.640	1.07	1.16
5.00	0.0696	0.0750	0.127	0.138	0.234	0.253	0.432	0.466	0.798	0.863
6.00	0.0306	0.0323	0.0588	0.0623	0.113	0.120	0.219	0.233	0.425	0.453
7.00	0.0131	0.0145	0.0264	0.0289	0.0530	0.0580	0.107	0.116	0.216	0.235
8.00	0.00553	0.00572	0.0115	0.0120	0.0241	0.0251	0.0505	0.0527	0.106	0.111
9.00	0.00230	0.00243	0.00496	0.00526	0.0107	0.0113	0.0232	0.0246	0.0504	0.0534
10.0	0.000944	0.00101	0.00210	0.00224	0.00469	0.00499	0.0105	0.0111	0.0234	0.0248

The total power radiated by an assemblage of electrons is, from equations (13) and (16),

$$J = (3.79 \times 10^{-6}) A \int_{\epsilon_1}^{\epsilon_2} H_\perp^2 \epsilon^{2-\gamma} d\epsilon . \tag{18}$$

Given J from the observed spectrum, the total number of radiating particles A may be found as a function of $H_\perp$ from equation (18). The total energy in the electrons is

$$\epsilon_e = \int_{\epsilon_1}^{\epsilon_2} N(\epsilon) \epsilon d\epsilon = A \int_{\epsilon_1}^{\epsilon_2} \epsilon^{1-\gamma} d\epsilon ,$$

or, eliminating A,

$$\epsilon_e = \frac{J}{(3.79 \times 10^{-6}) H_\perp^2} \frac{\int_{\epsilon_1}^{\epsilon_2} \epsilon^{1-\gamma} d\epsilon}{\int_{\epsilon_1}^{\epsilon_2} \epsilon^{2-\gamma} d\epsilon} . \tag{19}$$

The equation above and equation (14) show that $\epsilon_e \propto H_\perp^{-3/2}$.

3.3. Determination of the Energy in the Particles and in the Magnetic Field

Given appropriate observations of the synchrotron spectrum, we have seen how to find the particle energy spectrum and hence the total particle energy as a function of magnetic field strength. Unfortunately, independent measurements of the field strength cannot now be made; therefore the total particle energy cannot be unambiguously determined. It has been the practice, following Oort and Walraven (1956), to take that value of H for which the total magnetic energy ϵ_H equals the total particle energy. Since the particle energy varies roughly as $H^{-3/2}$ and the magnetic energy as H^2, the total energy in these two modes is given by $\epsilon_T \cong C_1 H^{-3/2} + C_2 H^2$. The field strength $H = (C_1/C_2)^{-7/2}$ for $\epsilon_H = \epsilon_e$. This is almost equal to the value for which the total energy is a minimum. Thus the assumption of equipartition of energy between the particles and magnetic field gives an order-of-magnitude estimate of the field corresponding to minimum total energy. Even apart from the questionable assumption of equipartition between the particle and magnetic energies, the physical reality of such an estimate is uncertain. Two general difficulties are involved: one, primarily observational, lies in determining the energy in the particles which produce the observed radiation; the second, of a more theoretical nature, has to do with relating this energy to the total particle energy of the system. Two simple calculations, both of which may be uncertain, must be carried out to find the total particle energy. The first is to integrate the observed spectrum over frequency to find the total synchrotron luminosity of the object; so the relevant frequency limits must be known. The spectra of most radio sources are less steep than ν^{-1}, so that it is relatively unimportant if the low-frequency cutoff is not observed. For spectra steeper than ν^{-1}, this may not be true; the situation may be further complicated by greater energy losses taking place at low frequencies by other mechanisms (see Appendix 1). For most sources, the high-frequency limit is of greater concern. In only a few sources has the high-frequency cutoff been observed, for example, in the Crab Nebula (O'Dell 1962) and M87 (Bless 1962); therefore considerable uncertainty is involved in estimates of the total synchrotron luminosity of most objects. The second step is to find the total particle energy by evaluating equation (19), and here again the limits of integration over the energies corresponding to ν_{c_2} and ν_{c_1} are uncertain. The relative importance of the maximum and minimum electron energies depends upon the spectral index as is clear from equation (19). Error in the distance to the object also introduces uncertainties in the total luminosity and, through the calculated volume of the source, in the magnetic energy.

Ignorance of the origin of the high-energy particles probably causes the greatest uncertainty in our estimates of the total particle energy, however. It is difficult to imagine a mechanism which supplies large amounts of energy to the electrons but not to the nucleons. Radioactive decay is such a mechanism, but the abundances of the heavy, unstable elements required are totally improbable. If the electrons are secondary particles produced in nucleon-nucleon collisions, then a common estimate (Burbidge 1956; Peters 1959) is that ten times more energy must exist in the primary particles than in the secondary electrons and positrons. If, on the other hand, some acceleration process provides the particle energies, then protons, which lose energy by radiation and collisions at lower rates than do electrons, should be accelerated to much greater energies than electrons. The ratio f of proton to electron energies may be as great as m_p/m_e.

The factors upon which the total particle energy depends may be written as follows: $\epsilon_t \sim JH^{-3/2}(1+f) + VH^2$, where V is the emitting volume in which the field is H and where J is the total synchrotron luminosity. We find that

$$\epsilon_e(\min) \sim J^{4/7}V^{3/7}(1+f)^{4/7}$$

and

$$H(\min) \sim J^{2/7}V^{-2/7}(1+f)^{2/7}\,.$$

The total particle energy consistent with the observations depends more strongly upon the distance and particle origin than does the field. The determination of the appropriate V not only depends upon the distance to the object but also upon its angular size, which is often poorly determined in the radio region. Furthermore, the assumed constancy of the field over V may be in error.

I would like to acknowledge the helpful criticisms and suggestions of Drs. A. D. Code, L. R. Doherty, J. S. Mathis, and D. E. Osterbrock. I am grateful to Dr. George Collins for computing the G-functions.

APPENDIX I

Electron Energy Losses by Other Processes

A few remarks concerning other modes of energy losses are appropriate here, since on occasion these modes are competitive with the synchrotron mechanism. Considerations of this matter may be found in Ginzburg (1958) and Morrison (1961), and we shall only summarize their discussions. In addition to the synchrotron mechanism, electrons may also lose energy by ionization, bremsstrahlung, and the inverse Compton effect, that is, the scattering of electrons by photons. The following approximate expressions give the relevant loss rates for relativistic electrons in a medium of number density n and radiation density ρ:

$$\text{Ionization loss: } \frac{d\epsilon}{dt} = (8 \times 10^{-9})\, n[\ln \gamma - \ln(n) + 74.6] \text{ ev/sec}$$

$$\text{Bremsstrahlung: } \frac{d\epsilon}{dt} = (8 \times 10^{-16})\, \epsilon n \quad \text{ev/sec}$$

$$\text{Photon collision: } \frac{d\epsilon}{dt} = (2 \times 10^{-14})\, \rho \gamma^2 \quad \text{ev/sec.}$$

The equation for ionization loss is for electrons moving in a medium of ionized hydrogen where the loss rate is slightly higher than that in neutral hydrogen. For energies relevant to astronomical objects, the rate of energy loss by ionization is essentially independent of energy, depending only upon n. The expression for energy loss by photon collisions is for photon energies small compared to mc^2 in the rest frame of the electron. Such losses are unimportant unless the radiant energy density is much higher than that in the galactic plane ($\simeq$1 ev/cm^3).

The loss rates by the synchrotron, ionization, and bremsstrahlung processes are shown in Figure 3. If the field of the galactic corona is 10^{-5} gauss and $n \simeq 0.1$, then electrons with $\epsilon < 10^9$ ev lose about as much energy by bremsstrahlung and ionization as by synchrotron radiation; if the field is as low as 10^{-6} gauss, then bremsstrahlung losses are greater than synchrotron losses for all ϵ up $\cong 10^{10}$ ev. Particularly at radio frequencies, then, synchrotron radiation may not be the dominant energy loss, and the total particle energy should be changed accordingly.

APPENDIX II

Application to the Crab Nebula

We shall briefly illustrate the application of some of the relations derived earlier by considering the synchrotron radiation from the Crab Nebula. The observations of the continuous spectrum compiled by O'Dell (1962) may be represented by $J(\nu) \sim \nu^{-0.31}(\gamma = 1.62)$ up to frequencies of about 10^{14} c/s. At higher frequencies, the spectrum falls off much more rapidly. This change in the spectrum is presumably caused by a deficiency in the numbers of high-energy electrons compared to the number required by a particle energy spectrum with $\gamma = 1.6$; the argument of G, ν/ν_{c_2}, is no longer zero but is becoming significant. By finding the fractional decrease in the energy emitted at a given frequency as compared to that which would be emitted if the spectrum continued to the ultraviolet with the same γ, equating this to $G(0)/G(x)$ for the appropriate γ (5/3 in this case), x and thus ν_c may be found. The value of ν_c (6.4×10^{14} c/s), so determined, varies by only about ± 5 per cent over a frequency range of a factor of 4 in the cutoff region. This result strongly suggests that the high-energy cutoff in the electron spectrum is indeed sharp and also supports the power-law representation of the particle energy spectrum.

The total energy radiated, found by integrating the observed spectrum from 10^8 c/s to the critical frequency, is about 10^{37} ergs/sec. The exact value of the lower limit is of no importance here, but increasing the upper limit to 10^{15} c/s increases the radiated energy by 10 per cent. Next the energy in the electrons is found by integrating equation (19) as a function of $H_\perp$. For each value of $H_\perp$, the integration limits, that is, the energies

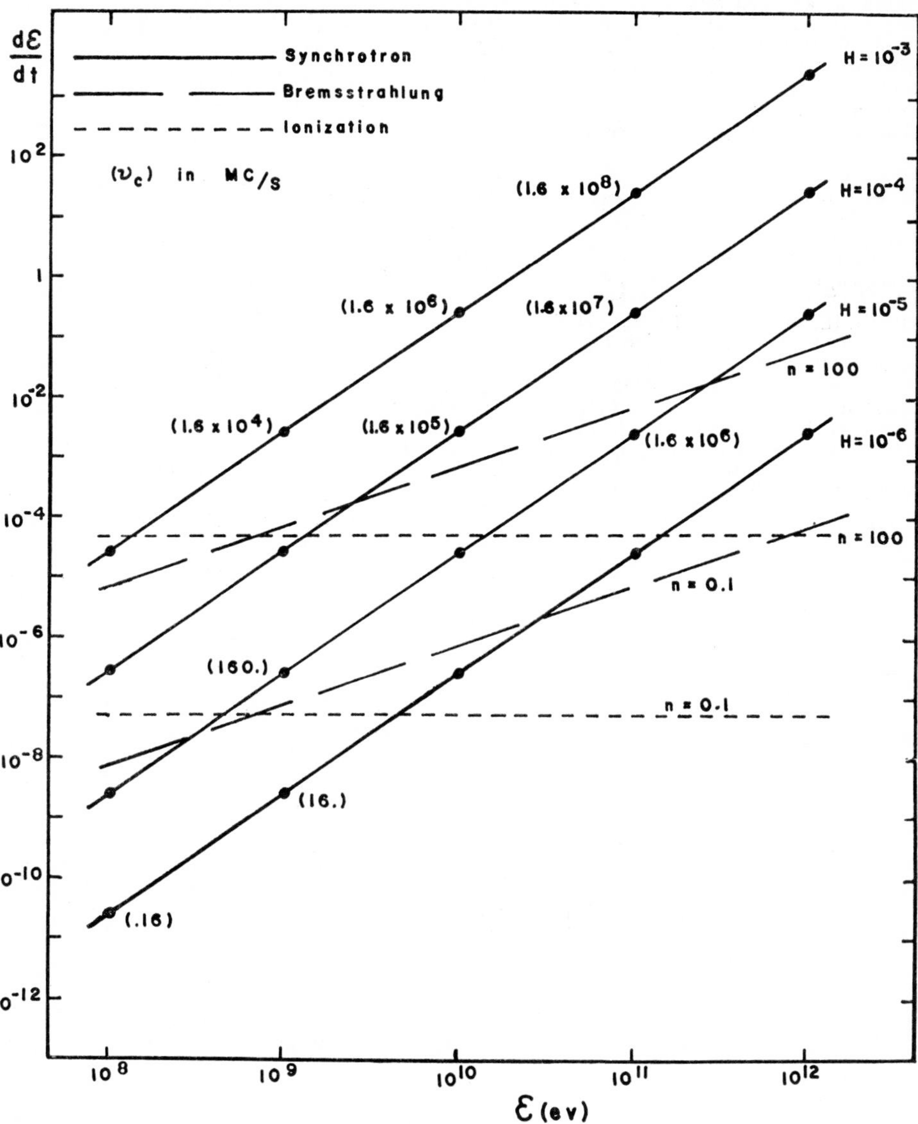

FIG. 3.—Rates of electron energy loss by various processes

corresponding to the appropriate frequency limits, are found from equation (14). With the radiating volume taken to be a sphere 0.8 pc in radius, the total magnetic energy $H_{\perp}^2/8\pi$ is found—again as a function of $H_{\perp}$. On the basis of equipartition between the magnetic and particle energies, $H_{\perp}$ is about 2×10^{-4} gauss. Thus the most energetic electrons in the Crab have energies of about 400 Bev and lose one-half of their energy in about 500 years.

REFERENCES

Alfvén, H., and Herlofson, N. 1950, *Phys. Rev.*, **78**, 616.
Artsimovich, L. A., and Pomeranchuk, I. Ya. 1946, *Zh. Eksperim. i Teor. Fiz.*, **16**, 379.
Baade, W. 1956, *B.A.N.*, **12**, 312.
Bless, R. C. 1962, *Ap. J.*, **135**, 187.
Burbidge, G. R. 1956, *Ap. J.*, **124**, 416.
Dombrovsky, V. A. 1954, *Dokl. Akad. Nauk, U.S.S.R.*, **94**, 1021.
Ginzburg, V. L. 1958, *Progress in Elementary Particle and Cosmic Ray Physics* (New York: Interscience Publishers, Inc.), Vol. **4**.
Greenstein, J., and Minkowski, R. 1953, *Ap. J.*, **118**, 1.
Jackson, J. D. 1962, *Classical Electrodynamics* (New York: John Wiley & Sons, Inc.), pp. 464–88.
Kiepenheuer, K. O. 1950, *Phys. Rev.*, **79**, 738.
Landau, L. D., and Lifshitz, E. M. 1962, *The Classical Theory of Fields* (2d rev. ed.; Oxford: Pergamon Press; Reading, Mass.: Addison-Wesley Publishing Co., Inc.), pp. 221–31, 241–45.
Morrison, P. 1961, *Hdb. d. Phys.*, **46**, Part 1, ed. S. Von Flügge (Berlin: Springer-Verlag).
O'Dell, C. R. 1962, *Ap. J.*, **136**, 809.
Oort, J. H., and Walraven, T. 1956, *B.A.N.*, **12**, 285.
Peters, B. 1959, *J. Geophys. Res.*, **64**, 155.
Ryle, M. 1949, *Proc. Phys. Soc. London A*, **62**, 491.
Schwinger, J. 1949, *Phys. Rev.*, **75**, 1912.
Shklovsky, I. S. 1953, *Dokl. Akad. Nauk, U.S.S.R.*, **90**, 983.
Schott, G. A. 1912, *Electromagnetic Radiation* (Cambridge: Cambridge University Press), pp. 103–10.
Vladimirsky, V. 1948, *Zh. Eksperim. i Teor. Fiz.*, **18**, 392.
Wentzel, D. 1957, *Ap. J.*, **126**, 559.
Westfold, K. C. 1959, *Ap. J.*, **130**, 241.

CHAPTER 13

*Discrete X-Ray Sources**

HERBERT FRIEDMAN

The E. O. Hulburt Center for Space Research, Naval Research Laboratory, Washington, D.C.

1. INTRODUCTION

X-RAY ASTRONOMY has provided the most unexpected results of the first generation of observations with instruments carried above the atmosphere in rockets and satellites. Approximately thirty discrete sources have been detected thus far. Most of the sources appear to lie within the galactic disk, and strong clustering occurs in the general directions of the galactic center and the Cygnus Region. There appear to be at least two classes of sources—those accompanied by detectable radio and optical emission and those as yet unidentified with any particular radio or optical object. Only two sources have been positively identified—Taurus XR-1 in the Crab Nebula and Scorpius XR-1—which fit a starlike object with many of the characteristics of a recurrent nova. Within the uncertainty ($\sim 1.5°$) in position measurements, sources are observed in the directions of Cas A and the Tycho Brahe Supernova of 1572. Signals have been detected from the directions of two extragalactic sources, Cygnus A and M87, and the fluxes are one to two orders of magnitude greater than their radio emissions. If the extragalactic identifications are correct, we should call these objects "X-ray galaxies" rather than radio galaxies. At least one source, Cygnus XR-1, appears to have varied greatly in brightness in the course of a year, and evidence of variability in several other sources is developing. Finally, a diffuse X-ray background exists which may be integrated radiation from extragalactic sources.

The opacity of the Earth's atmosphere permits 1 A radiation to penetrate to about 60 km, 10 A X-rays to about 100 km, and 50 A X-rays to about 140 km. To observe soft X-rays from space therefore requires rockets capable of reaching heights exceeding 150 km. The observations thus far have been made with gas-filled proportional or Geiger counters, or with crystal scintillation counters, over a wavelength range from 0.1 A to 50 A. Most of the softer X-ray measurements have been made from small rockets of the Aerobee class, and the harder X-rays have been observed with balloon-borne instruments.

* Work described in this chapter sponsored jointly by the Office of Naval Research and the National Science Foundation.

2. EARLY OBSERVATIONS

X-ray astronomy is still so young that any review may start from the earliest observations. Evidence for a diffuse background of X-ray emission from outside the atmosphere was obtained from a rocket flight instrumented by Kupperian and Friedman (1958) in 1956. The detector was a scintillation counter, and pulse amplitudes were telemetered continuously to give the spectrum from 20 to 300 kev as a function of altitude. As shown in Figure 1, a strong response was observed due to cosmic-ray secondaries in the Pfotzer maximum near 23 km. This cosmic-ray albedo still predominated at 42 to 57 km, but above 66 km a strong component of softer flux appeared and then remained unchanged with increasing altitude.

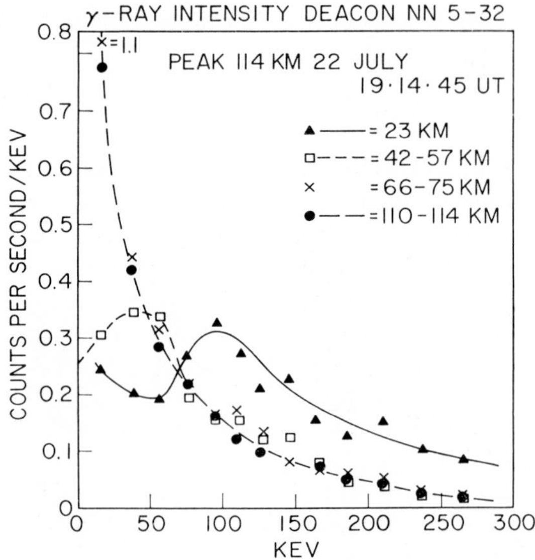

FIG. 1.—Variation with altitude of overhead flux of X-rays measured with scintillation counter, July 22, 1956.

In subsequent years, the author and his colleagues at Naval Research Laboratory made several unsuccessful attempts to obtain evidence for extra-Solar System X-ray sources and concluded that if such sources existed, the fluxes must have been less than 10^{-8} erg/cm^2 sec (Friedman 1959). In 1960, Giacconi, Gursky, Paolini, and Rossi (1962) undertook a program of observational X-ray astronomy, their first objective being an attempt to detect X-rays from the Moon, presumably excited by fluorescence under the stimulation of solar X-rays. The detectors were Geiger-Mueller counters, much larger than those previously used by the NRL group. Mica windows set the sensitivity limits between 2 and 8 A, and the effective area was about 10 cm^2. In June, 1962, they detected a source of X-rays clearly displaced from the direction to the Moon, and X-ray astronomy was born. In flight, the rocket was given a high spin which stabilized its long axis in the direction of the zenith. As the rocket spun, the detectors looked out the side and mapped the azimuthal variation of flux. A broad maximum was observed (Fig. 2) displaced from the direction to the Moon, with a peak toward 16 to 17 hours right ascension and approximately −40 degrees declination. The position of the source was described as "close to the center of the galaxy, but apparently not quite coincident with it." The maximum flux was about 5 photons cm^{-2} sec^{-1}. Even if this source were as close as the nearest stars, its X-ray emission would be 10 to 100 million times greater than that of the quiet Sun.

A repeat in October, 1962 (Gursky *et al.* 1963), when the galactic center was below the horizon, did not reveal this strong source. A third flight, in June, 1963, with the galactic center once more above the horizon, again repeated the observation of 1962. These flights, although not equipped to resolve individual sources, hinted at the existence of additional X-ray sources broadly distributed in the Cygnus region and in the general vicinity of Taurus.

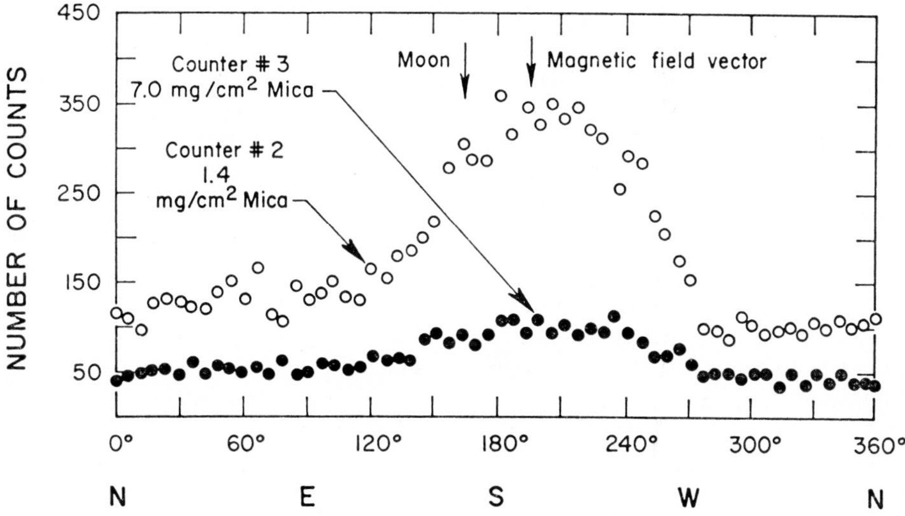

FIG. 2.—Azimuthal distribution of recorded counts from Geiger counters flown by Giacconi *et al.* (1962), June, 1962.

In April, 1963, Bowyer, Byram, Chubb, and Friedman (1964*a*) instrumented an Aerobee rocket with a proportional counter of 65 cm^2 window area. A honeycomb collimator confined the angular field of view to about 10 degrees at half-maximum intensity. The counter window of beryllium, 0.005 inch thick, transmitted X-rays up to about 8 A in wavelength. By spinning the rocket very slowly, a large precession cone was attained, and the combined spin and precession served to cover the entire sky above the horizon. Only two discrete sources were revealed—one in Scorpius and the other in the direction of the Crab Nebula. Eight scans through the Scorpius source permitted its position to be fixed to within about 0.5 degree at 15^h 15^m R.A. and $-15\overset{\circ}{.}2$ dec. The source in the direction of the Crab Nebula was only $\frac{1}{8}$ as intense as the Scorpius source. Undoubtedly, the strong source in Scorpius was included in the broad signal observed by Giacconi *et al.* in 1962 together with, as we now know, several additional sources in the general vicinity of the galactic center.

3. X-RAY EMISSION MECHANISMS

Following the detection of these sources, there was much speculation as to the possible mechanisms of X-ray production. Among the various processes under consideration were (1) bremsstrahlung from a hot plasma, (2) inverse Compton effect, (3) synchrotron radiation, (4) characteristic line emission, and (5) the blackbody radiation of a neutron star. In the direction of the Scorpius source there was no clearly related optical or radio emission. In contrast, the Crab Nebula is known to be the remnant of a supernova and is one of the most spectacular optical and radio sources in the Galaxy. It was not entirely surprising that it also produced detectable X-ray emission. Since the Crab is a supernova remnant, it was proposed that the Scorpius source might also be associated with a supernova phenomenon, and to account for the apparent absence of radio and optical emission, the neutron-star model seemed very attractive.

It is generally believed that a pre-supernova evolves from a common star somewhat more massive than the Sun which lives through successive stages of nuclear burning until it has synthesized a core made up largely of iron group elements. When the core reaches a temperature in excess of a few billion degrees, it disintegrates into neutrons and alpha particles and collapses catastrophically to a density comparable with that encountered in the atomic nucleus. The collapse results in a compacted star, possibly as small as 10 km in radius with a density of about 10^{15} gm/cm^3. This hypothetical neutron star (Chiu and Salpeter 1965; Morton 1964) would have a uniform temperature almost out to its edge. A photosphere only a few meters thick would separate the billion degree core from outer space, and in this thin skin the temperature would drop rapidly to about 10×10^6 °K in the outermost centimeter. At this high temperature, the stellar photosphere would be a powerful source of X-rays with its blackbody radiation concentrated in the 1 to 10 A region and peaking near 3 A. The tail of emission reaching toward the visible spectrum would contain so little energy that the star would be visually unobservable unless it were extremely close to the Solar System.

Further, it is believed that in the pre-supernova stage, the core is surrounded by an envelope containing appreciable amounts of unburned, lighter nuclei. During collapse of the core, this envelope would also fall inward and incidentally get heated to a temperature at which it would ignite in thermonuclear fashion. The debris would spread rapidly into space to form a nebula. A more recent model of the supernova process (Colgate and White 1966) postulates that a shock wave moves outward from the central part of the highly condensed core. As it reaches the outer portions of the star, it turns into a blast wave that carries away material at very high velocity. Some of the gas reaches relativistic velocities, and the energy spectrum of the particles is characteristic of cosmic rays.

4. OCCULTATION OF THE CRAB NEBULA

The Crab Nebula exploded in A.D. 1054. At the present time, the gaseous debris fills a roughly ellipsoidal volume about 6 light years (1.8 pc) across its major axis. It is now expanding outward at a speed of about 1000 km/sec, which explains the present size quite well if the expansion began about 900 years ago and the speed is regarded as constant.

A test of the stellar hypothesis for the X-ray source in the Crab Nebula was afforded in 1964 by the occurrence of a lunar occultation. Such occultations of the Crab Nebula are repeated every 9 years and present a unique opportunity to measure the size of the source with high angular resolution. On July 7, 1964, the NRL group launched a stabilized rocket (Bowyer *et al.* 1964*b*) during a lunar occultation which covered the nebula at the rate of about $\frac{1}{2}$ minute of arc per minute of time. Since the nebula measures 6 minutes across its longest dimension, it would require 12 minutes of X-ray measurements to observe a full eclipse. But the Aerobee rocket affords no more than about 5 minutes of time above 100 km. The interval for the experiment was chosen, therefore, so that the eclipse of the central portion of the nebula, over a range of about 2 minutes of arc, would be observed. If a neutron star X-ray source existed in the center of the Crab, the occultation would be expected to produce an abrupt disappearance of the X-rays. A gradual disappearance would mean that the source was an extended cloud. The observation of the occultation indicated that the total angular width of the X-ray source was about 2 minutes of arc, which corresponds to about 2 light years (0.6 pc) at a distance of 1100 pc.

This result clearly ruled out a neutron star as the major contributor to the X-ray flux from the Crab Nebula. The accuracy of the observation was such that it did not completely deny the possibility that a neutron star existed in the midst of the radiating cloud, but its contribution would need to be less than 10 per cent of the observed flux.

5. COSMIC X-RAY SURVEYS

Beginning in 1964, the NRL group began a series of broad sky surveys using the largest counters that could be accommodated in the instrumentation section of an Aerobee rocket. A survey on June 16, 1964, mapped the galactic plane from the southern region of Scorpius through Cygnus to the northern part of Perseus (Bowyer *et al.* 1965). A flight on November 25 extended the survey through Taurus to the southern portion of Puppis. With a honeycomb collimator in place, the total effective area of counter in the view direction was about 900 cm^2, an increase of about 14 times over the window area first used in the April, 1963, survey. Two sets of paired counters were used in each rocket and were oriented 180° apart. In the June, 1964, survey, satisfactory data were obtained from a pair of counters which swept across most of the accessible sky. As a result of this survey, the positions of 9 sources were published. The sources were labeled XR for X-ray and numbered with the various constellations according to brightness. In the survey of November, 1964, the only source observed was the Crab Nebula, despite the wide expanse of sky that was searched from Perseus to Puppis.

Clark (1965) made an important extension of the X-ray observations of the Crab Nebula with a balloon-borne apparatus on July 21, 1964. The X-ray detector was a scintillation counter employing a thin crystal of NaI (Tl) almost 100 cm^2 in area. A collimator of metal slats provided an aperture of $\pm 16°$ in the horizontal direction and $\pm 55°$ in the vertical. When the balloon was at altitude, the directional detector rotated about the vertical axis. A peak in the counting rate occurred when the Crab was in the field of view. Four channels of pulse amplitude information resolved the spectrum in the intervals from 9 to 15 kev, 15 to 18 kev, 28 to 42 kev, and 42 to 62 kev. To the observed pulse amplitude distribution, it was possible to fit various theoretical spectra based on thermal and nonthermal models. A thermal fit required a temperature in excess of 10^8 °K, which provided additional evidence against the neutron-star model for the Crab Nebula.

Whereas the NRL surveys were intended primarily to cover a large region of sky and were particularly well suited to the discovery of new sources, other observers devoted their efforts to surveys of more limited areas and the attainment of higher positional accuracies. Giacconi and his collaborators, whom we shall refer to as the ASE group (American Science and Engineering Co.), joined with a team of MIT scientists, which included M. Oda and B. Rossi, to conduct further observations with fast-spinning rockets. However, they now reduced the field of view by using cellular collimators in the form of narrow slits, some only $1\overset{\circ}{.}5$ wide at half-maximum. P. Fisher and his collaborators at the Lockheed Company employed automatically stabilized rockets to perform slow scans with fan beam collimators. On August 1, 1964, the Lockheed group (Fisher *et al.* 1966) used a $2\overset{\circ}{.}9$ by $26\overset{\circ}{.}5$ collimator to scan along the galactic equator and identified several source peaks at galactic longitudes close to the positions of the sources listed by the NRL group in Table 1. Agreements were found with the positions of Sco XR-1, Cyg XR-1, Sgr XR-1, Sco XR-2, and Sco XR-3, but Oph XR-1 was outside the position limits for the Kepler SN 1604, and two sources were resolved symmetrically about the position of Sgr XR-2. The Lockheed position for Sco XR-1 was given as $\alpha = 16^h14^m \pm 1^m$, $\delta = 15° \, 36' \pm 15'$.

Two fast-spinning rockets were flown by the ASE-MIT group in August and October, 1964 (Giacconi *et al.* 1964). Two sources, Sgr X-1 and Sco X-2, were located with an accuracy of about $0\overset{\circ}{.}5$; these were outside the position limits assigned by the first NRL survey (Table 1). No evidence was found for sources at the NRL positions of Oph XR-1 and Sco XR-3. On April 26, 1965, Byram, Chubb, and Friedman (1966) of NRL obtained a considerably improved sensitivity in a broad sky survey and were able to detect sources approximately four times weaker than previously; 0.2 photons cm^{-2} sec^{-1} produced a signal-to-background noise ratio of 3:1. For the first time, signals were also detected from the directions of extragalactic sources. The results of this survey are presented in the map (Fig. 3) and in the representative scans of Figures 4, 5, and 6. Tentative source positions are given in Table 1.

Referring first to the Cygnus region, the new survey revealed many more discrete sources and the resolution of Cyg XR-1 into two components, as shown in the traces of Figure 4. One of the components fits the original position of Cyg XR-1 as derived from the June, 1964, flight, but its intensity had decreased by a factor of 4. Here, then, was evidence of a high degree of variability in the course of 1 year. The second component lay within $1\overset{\circ}{.}5$ of the direction of the radio galaxy Cyg A. Had Cyg XR-1 remained at the high level of brightness observed in 1964, it would have masked Cyg A. Cyg XR-2 showed no change from 1964 to 1965. The Lockheed group (Fisher *et al.* 1966) had scanned Cyg XR-1 in October, 1964, and observed a flux only $\frac{1}{6}$ that found by NRL in June, 1964. Some of this reduction in flux count may be attributable to the lower long-wavelength sensitivity of the Lockheed detector, but a substantial portion of the reduction may have been real. It would appear, then, that Cyg XR-1 underwent a decrease in brightness of as much as 50 per cent in the space of only 4 months.

Figure 7 includes the results of the NRL surveys of June, 1964, and April, 1965, the ASE scans of August and October, 1964, and the Lockheed scan of October, 1964. The dashed circles identify the positions first reported by the NRL group, as given in Table 1. Broad, shaded rectangles represent the source positions found in the NRL 1965 survey.

The half-width of each box in the roll direction is the estimated 1°.5 uncertainty in position. The long dimension of the box represents the 8° half-width of the mechanical honeycomb baffle. Intersecting boxes imply that the same source was in the field of view on adjacent roll scans. Careful inspection of all the data represented in Figure 7 indicates many apparent disagreements between the results obtained by different observers as

TABLE 1

X-RAY SOURCE POSITIONS

NAME	α		δ	FLUX(6) (counts cm^{-2} sec^{-1})	NOTES
	h	m	deg		
Cas XR-1 (Cas A)....	23	21	+58.5	0.3	....
Tau XR-1 (Tau A)...	5	31	+22.1	2.7	2
Cep XR-1 (SN 1572)..	0	15	+66	0.3	....
Ser XR-1............	18	45	+ 5.3	0.7	....
Ser XR-1............	18	45	+ 5.3	0.7	1
Ser XR-2...........	18	10	−12.9	2.0	....
Sgr XR-1...........	17	55	−29.2	1.6	1
Sgr XR-1...........	17	48	−30	1.6	5
Sgr XR-2...........	18	10	−17.1	1.5	1
Sgr XR-3...........	17	56	−21.6	2.8	....
Sgr XR-4...........	18	26	−31	0.6	5
Sco XR-1...........	16	15	−15.2	18.7	1
Sco X-1.............	16	17.07	−15.53	....	3
Sco XR-1...........	16	15	−15.0	21	....
Sco XR-2...........	17	15	−38.4	2.6	1
Sco XR-2...........	17	8	−36.4	1.4	....
Sco XR-3...........	17	23	−44.3	1.1	1
Sco XR-4...........	16	25	−40	0.8	....
Sco XR-5...........	17	37	−40.4	0.7	5
Sco XR-6...........	18	7	−35.6	0.3	5
Oph XR-1...........	17	32	−20.7	1.3	1
Oph XR-2...........	17	14	−23.5	0.5	5
Ara XR-1...........	17	5	−45.9	1.9	....
Nor XR-1...........	16	24	−51	0.8	....
Nor XR-2...........	15	38	−57	0.4	5
Lup XR-1...........	15	2	−52	0.2	5
Cen XR-1...........	14	28	−63	0.17	5
Lyr XR-1...........	18	18	+36	0.17	5
Aql XR-1...........	19	12	0.0	0.13	5
Vul XR-1...........	20	38	+29	0.15	....
Lac XR-1...........	22	34	+53.8	0.17	5
Cep XR-2...........	22	42	+62	0.18	5
Cyg XR-1...........	19	53	+34.5	3.6	1
Cyg XR-1...........	19	57	+34.5	0.9	....
Cyg XR-2...........	21	43	+38.8	0.8	1
Cyg XR-2...........	21	42	+38.8	1.0	....
Cyg XR-3 (Cyg A)...	19	58	+40.4	0.4	....
Cyg XR-4...........	21	2	+42	0.3	5
Vir XR-1 (M-87).....	12	28	+12.7	0.2	5
Leo XR-1...........	9	35	+ 8.6	0.2	5
Sco X-2.............	16	50	−39.6	....	....
Sgr X-1.............	17	44	−23.2	....	....

1 NRL survey, June, 1964.
2 NRL survey, November, 1964.
3 Position of optical source identified with Sco X-1.
4 ASE-MIT survey, October 26, 1964.
5 Source observed on only one scan; pitch uncertainty ±4°.
6 1 count = 1.2×10^{-8} ergs (1 to 10A) for bremsstrahlung, $T = 5 \times 10^7$ degrees K or synchrotron, $\gamma = 1.0$.

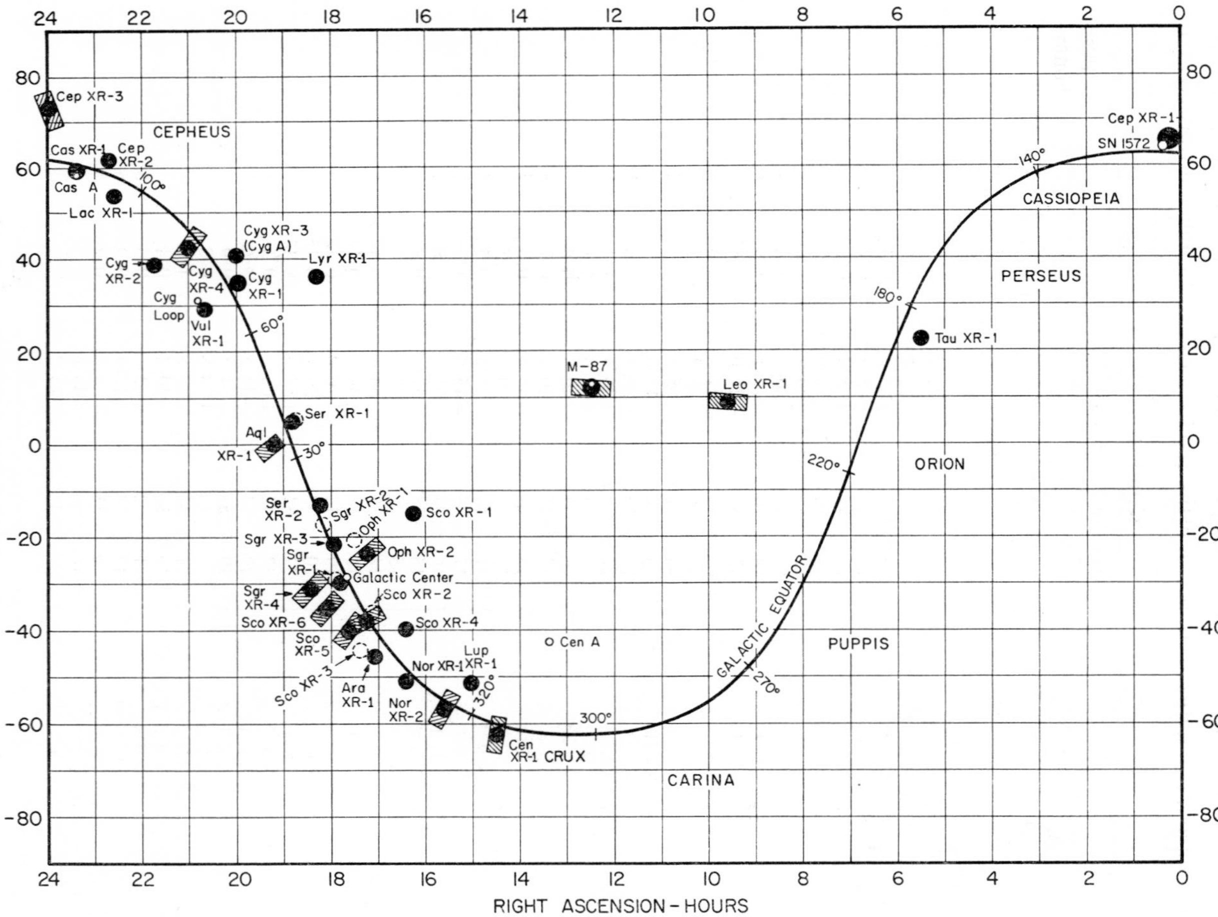

FIG. 3.—Map of sources detected in NRL surveys of 1964 and 1965. Estimated position uncertainties are generally 1°.5. Shaded rectangles indicate 4° pitch uncertainty of positions of sources detected on only one roll scan.

well as between the two NRL surveys. With regard to the latter, it should be recognized that the complex of sources in this region was much better resolved in the 1965 survey, and it is clear from the sample traces shown (Figs. 4, 5, and 6) that much subjective judgment can enter into the decomposition of an envelope of poorly resolved sources into its constituents. Even with such reservations, however, it is difficult to explain some of the discrepancies. In some cases, the original estimate of 1°5 in position uncertainty may have been somewhat optimistic. The sources, Sco XR-2, Sco XR-3, and

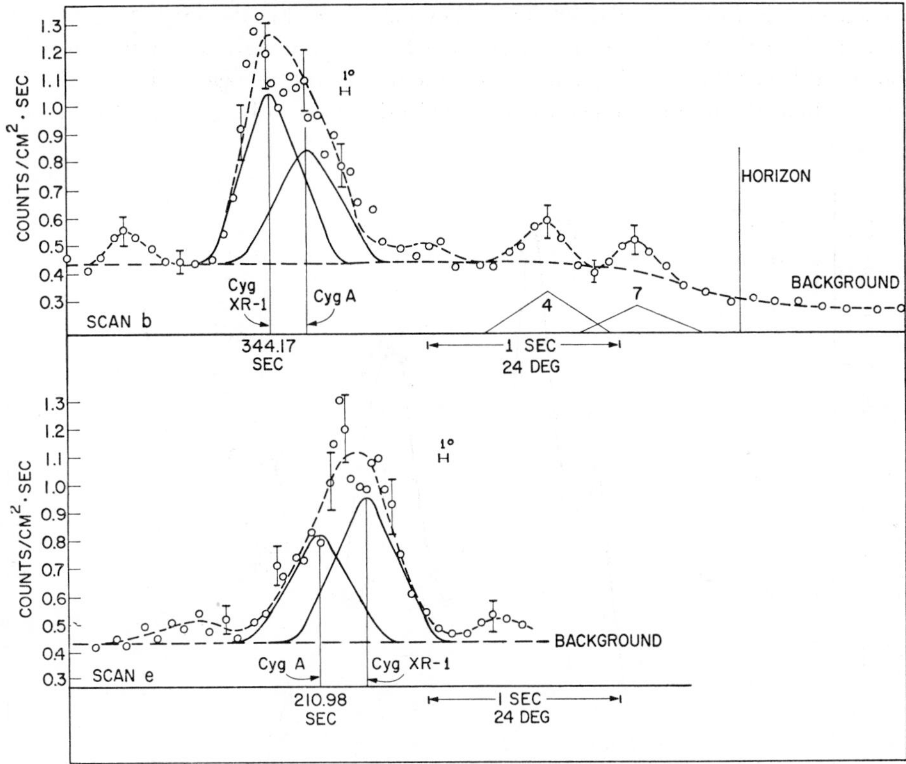

FIG. 4.—Two roll scans (NRL, April 25, 1965, survey) through Cygnus region

Sgr XR-1, appear to be present in both surveys. When one compares the NRL results with the ASE-MIT positions for Sco X-2 and Sgr X-1, there appears to be no correlation within the estimated uncertainty of the respective positions. With such conflicting evidence, one is forced to conclude that much more accurate position measurements are necessary before meaningful comparisons can be made, or that marked variations take place in the brightness and position patterns over the time intervals between sets of observations.

6. DISTRIBUTION OF X-RAY SOURCES

Of the 25 X-ray sources that lie within ±15° of the galactic plane (Fig. 3), three are observed in the directions of supernova remnants—the Crab Nebula, Cassiopeia A, and the Tycho Brahe SN 1572. Assuming the correctness of these identifications and the generally accepted distances of 1.1 kpc for the Crab and 3.4 kpc for Cas A, the computed X-ray luminosities (1 to 10 A) are roughly equal, about 4×10^{36} erg/sec. The distance to SN 1572 is more uncertain, but if it is taken to be about 3 kpc, it is comparable in X-ray luminosity to the Crab and Cas A.

Although the number of discrete X-ray sources detected thus far is still very small for a statistical analysis of the distribution, some gross features appear to be significant. Referring again to Figure 3, the sources seem to group in two broad clusters. Eight sources, which lie in the Cygnus-Cassiopeia region (l^{II} 60°–120°), are spaced within an average of $\pm 7°$ from the galactic plane. They may lie within the Orion–Cygnus spiral arm, of which the Sun is a member. If these X-ray sources are distributed in height above and below the galactic plane in the same way as the average stellar distribution, we may take the mean distance from the galactic plane to be about 400 light years

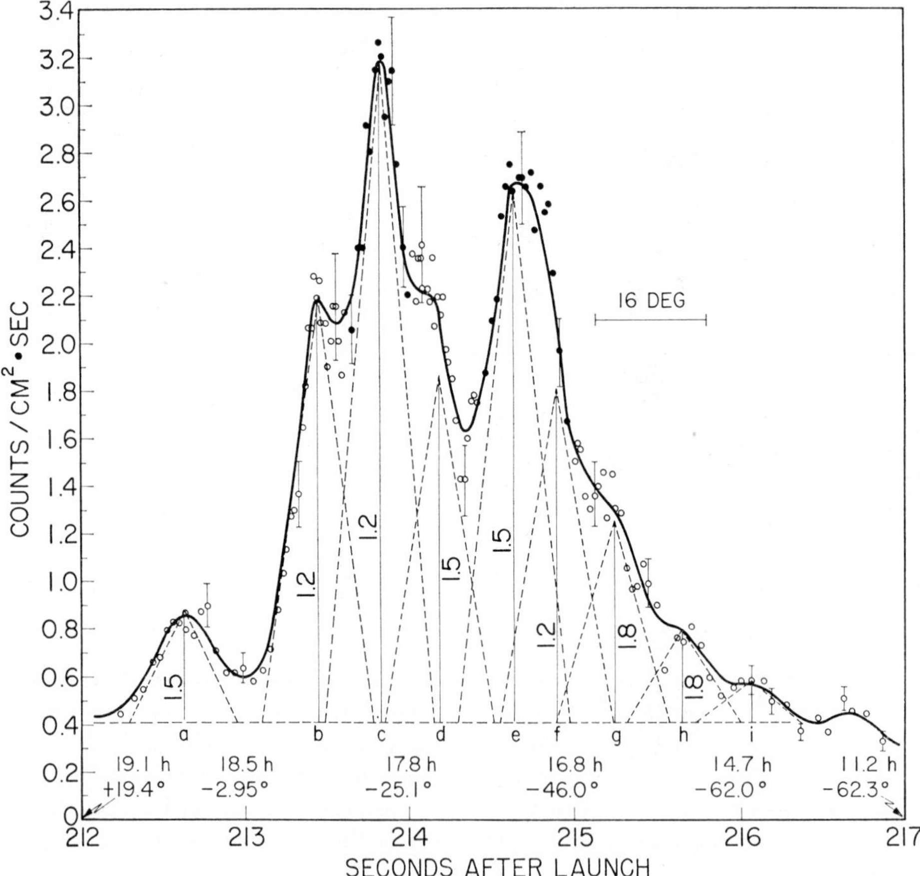

FIG. 5.—Scan (NRL, April 25, 1965) along galactic equator. Composite envelope is decomposed into individual sources, each characterized by triangular transmission pattern of honeycomb collimator. Numbers on vertical lines marking positions of discrete sources are indicative of spectral quality; larger numbers represent softer spectra.

(~120 pc). The average distance of this group of X-ray sources from the Sun is thus about 3,400 light years (1,040 pc). The second broad grouping of X-ray sources centers about the direction to the galactic center (l^{II} 315°–340°). Fifteen sources lie at an average of ±3.5° from the galactic plane, and the average distance would appear to be about 7,000 light years (2,146 pc). These sources may lie within the Sagittarius Arm.

If the observed X-ray sources lie at an average distance of only 2 kpc, they occupy a small fraction of the volume of the Galaxy. Assuming that the Galaxy is a uniform, flat, disk-shaped distribution of stars, 15 kpc in radius, the presently observable range of X-ray sources is confined to only about one-fiftieth of the Galaxy. We therefore estimate the total number of X-ray sources in the Galaxy as $25 \times 50 \approx 1{,}250$.

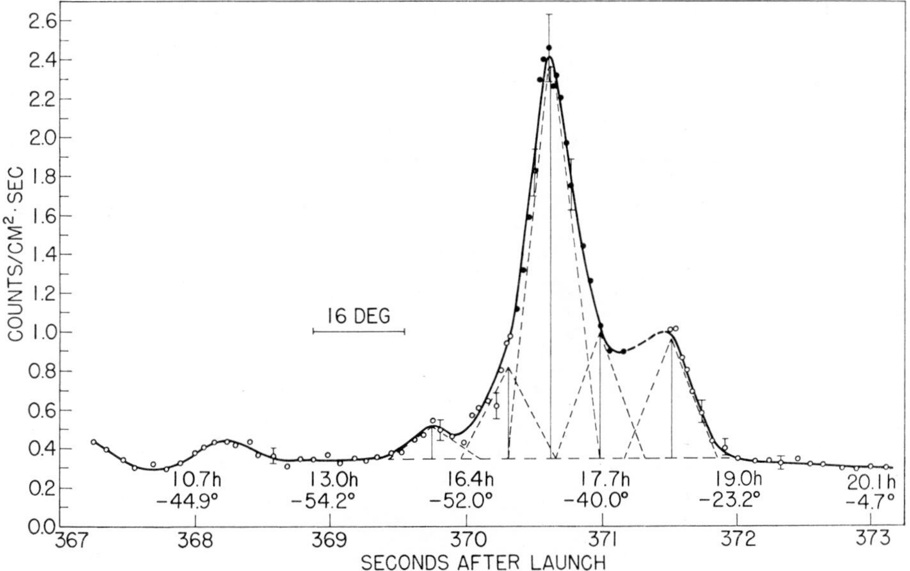

FIG. 6.—Representative scan, NRL survey, April 25, 1965

The average brightness of the X-ray sources in the Sagittarius group is about 41 per cent of that of Tau XR-1. If the mean distance is about twice as great as that of the Crab, the average luminosity must be about 1.6 times that of Tau XR-1. Similarly, we find for the Cyg-Cas group of X-ray sources an average luminosity of about 16 per cent of that of Tau XR-1, if the mean distance is 3,400 light years (1,040 pc).

The observational data, because of resolution limitations, tend to favor the detection of weak sources in the Cyg-Cas region as compared to the Sgr region. In the latter direction, the sources are more densely concentrated and many weaker sources may be hidden by the brighter sources.

The average luminosity of the combined groups of sources, as estimated above, is about 1.2 $L_{\mathrm{Tau\ XR-1}}$. If we take $L_{\mathrm{Tau\ XR-1\ (1-10A)}} = 4 \times 10^{36}$ ergs/sec, then $L_{\mathrm{Gal\ (1-10A)}} \approx 1{,}250 \times 5 \times 10^{36} \approx 6 \times 10^{39}$ ergs/sec. The X-ray luminosity of the Galaxy thus appears to be an order of magnitude greater than the radio luminosity, which is about 3×10^{38} ergs/sec.

If the distribution of X-ray sources resembles that of galactic novae, the mean distance from the galactic plane would be substantially greater than the 400 light years (120 pc) which was used above. According to Oort (1958), the mean distance of galactic novae is about 1,300 light years (~400 pc). The estimated source distances would thus be about three times as great and the luminosities about ten times higher. It does not

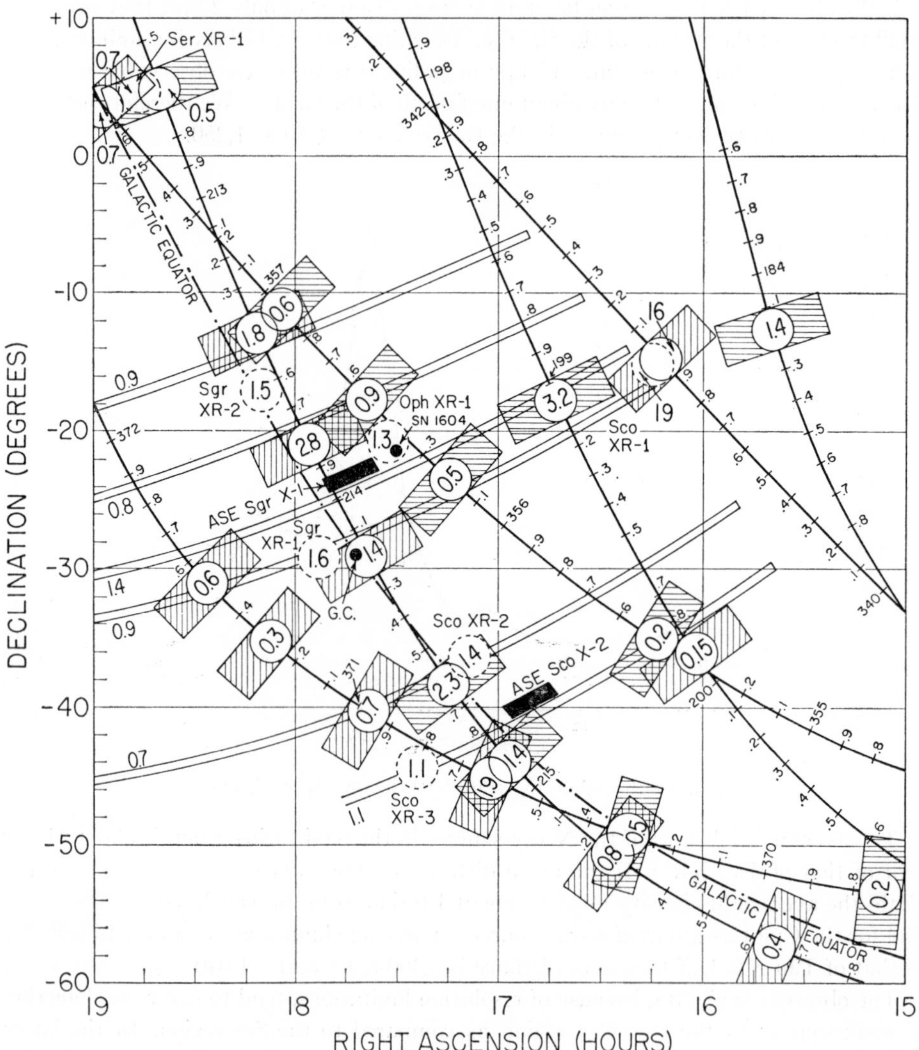

FIG. 7.—Scan data from various X-ray surveys. NRL survey, April 25, 1965, shows six roll scans at various times (seconds from start of flight) during Aerobee flight. Each circle gives position of peak obtained from analysis of signals such as shown in Figures 5 and 6. Uncertainty of position in roll direction is estimated to be about 1°.5. Pitch uncertainty of source position is indicated by 8° widths of rectangular boxes. Numbers in circled positions are observed counts/sec. Dashed circles are positions established by earlier NRL survey (June 16, 1964). Narrow bands mark positions obtained by Lockheed fan beam scan (Oct. 1, 1964; Fisher *et al.* 1966). Black bands give positions determined by ASE-MIT (Giacconi *et al.* 1964) for two sources.

seem likely, however, that the average galactic X-ray source could be 10 to 100 times as luminous as the Crab or Cas A.

If the Galaxy is typical of all spiral galaxies, an isotropic diffuse background X-ray flux, F, should exist which may be estimated by integrating over an appropriate cosmological distance, R (Gould and Burbidge 1963). Then

$$F_{(1-10\text{A})} = \frac{1}{4\pi} n_g (L_{\text{Gal}}) R , \qquad (1)$$

where $n_g \approx 2 \times 10^{-75}$ per cm^3 is the density of galaxies and $R \approx 10^{28}$ cm. Formula (1) assumes a Euclidian universe and negligible absorption of X-rays in intergalactic space. The numerical result is $F_{(1-10\text{A})\ \text{diffuse}} \approx 10^{-8}$ erg cm^{-2} sec^{-1} sterad^{-1}. The observed diffuse flux is 9×10^{-8} erg cm^{-2} sec^{-1} sterad^{-1}. The observed flux may represent primarily the contribution of peculiar galaxies, which, though relatively few in number, may make a major contribution to the background integral. As in the accounting of radio galaxies, if only one in ten thousand galaxies is 10^5 times as bright in X-ray emission as our Galaxy, the integral flux would be ten times as great as the contribution from all normal galaxies. The observational data on background flux strongly suggest a lumpiness in distribution that is indicative of a multitude of poorly resolved discrete sources. In arriving at the estimate, the assumptions all favored a higher value for the flux computed according to equation (1); intergalactic absorption was neglected but may be appreciable, and if the distribution of sources is confined to spiral arms, the sampling of observed sources would represent a greater fraction of the total Galaxy.

7. THE CRAB NEBULA

A number of observations have been made which provide information about the spectrum of the Crab Nebula X-ray source, Tau XR-1. Filter photometry by the NRL group in November, 1964, showed an important contribution to the spectrum from wavelengths differentiated between $\frac{1}{4}$-mil and $\frac{1}{2}$-mil Mylar filters (Bowyer *et al.* 1965). However, all that could be concluded from such a measurement was that the source may fit (1) a synchrotron spectrum with an index of -1.1, (2) a bremsstrahlung spectrum with a temperature of 10^7 degrees K, or (3) a blackbody spectrum with a temperature of 5×10^6 degrees K. Similarly, Clark's balloon observations of Tau XR-1 at energies from 15 to 62 kev could be fitted to a synchrotron spectrum with an index of -2, or a thermal spectrum with a temperature in excess of 6×10^7 degrees K for a blackbody and about 2×10^8 degrees K for bremsstrahlung. The blackbody temperature is far greater than that derived from neutron-star models because neutrino cooling rates would not permit the attainment of a surface temperature above 1.6×10^7 degrees K (Morton 1964).

Three more recent measurements have been made of the spectrum of Tau XR-1, two from balloons and one from an Iris rocket. Peterson and Jacobson (1966) observed that the spectrum from 20 to 100 kev was characterized by a spectral index of -0.9. Haymes and Craddock (1966) observed the spectrum from about 30 kev to 80 kev and found an index closer to -2. Grader *et al.* (1966) found that the spectrum from 1 to 20 kev was characterized by an index of -1.3. The differences in spectral shape and the extent of the spectrum to high energies may be, in part, attributable to experimental uncertainties but may equally likely be true evidence of variability.

When the available spectral data on the Crab are plotted, as in Figure 8, it appears that the X-ray fluxes lie along a simple extrapolation of the visible synchrotron spectrum. The fit suggests strongly that the X-rays are also synchrotron radiation. Any synchrotron hypothesis, however, must also provide a mechanism for the continuous regeneration of high-energy electrons. If the magnetic field strength is about 10^{-4} gauss, as is generally assumed for the Crab Nebula, the electron energy required to produce X-rays of about 3 A wavelength is 3×10^{13} ev. Such high-energy electrons are degraded rapidly by synchrotron emission, the lifetime being inversely proportional to the square of the

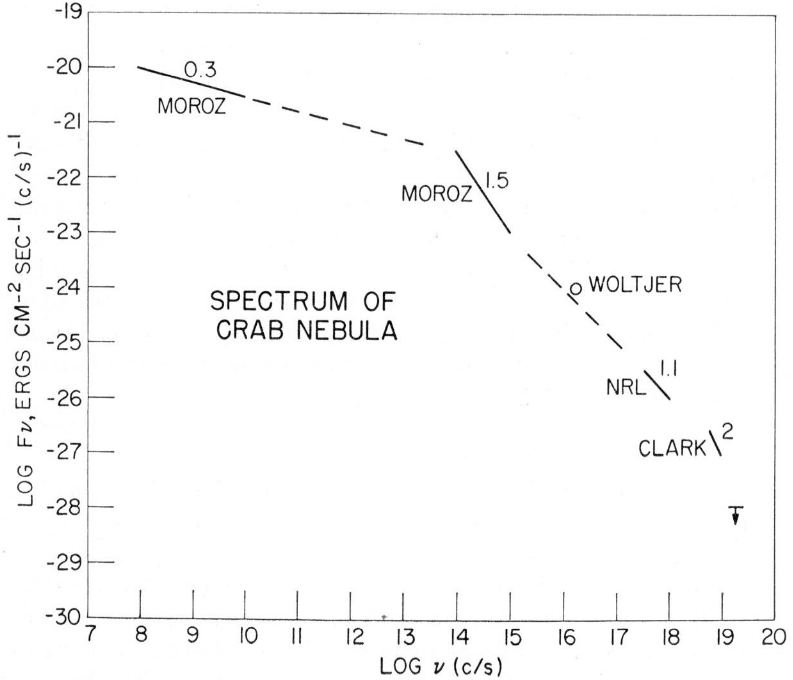

FIG. 8.—Spectrum of Crab Nebula from radio- to X-ray wavelengths. Synchrotron radiation indices in the radio region are from review by Moroz (1964). Point marked Woltjer is based on theoretical estimate (Woltjer 1958) of requirement to excite filamentary emission lines by ultraviolet synchrotron radiation.

frequency. For a magnetic field of 10^{-4} gauss and a frequency of 10^{18} c/s, the lifetime is about 30 years—far less than the age of the nebula. Shklovsky (1964) has suggested that the field in the central region of the Crab may approach 10^{-3} gauss and that the lifetime would then be of the order of 1 year. If relativistic electrons were produced at the very center of the nebula and projected outward, they could travel only one light year (0.3 pc) before losing their energy. This would account for the 1 light year diameter of the X-ray emission volume.

Over a span of half a century, major changes are known to have occurred in fine details of the amorphous mass of the Crab Nebula, giving the impression, according to Shklovsky (1960), that the Crab is "breathing"—that "ripples" are running through it.

Wisps of brightness, or several bright knots, suddenly appear and move at 20 to 25 times the expansion velocity of the nebula. The dimensions of a wisp are so great that 2 to 3 months would be required for a relativistic particle to traverse it. Yet, the feature may develop within a few weeks. The luminosity per unit volume of a wisp can reach 300 times that of the amorphous central region. Such behavior may be associated with variations in the magnetic field and in the density of relativistic electrons. If we can attribute the luminosity variations to local augmentations of the magnetic field, we have the possibility of explaining the survival of electrons since the original explosion. The

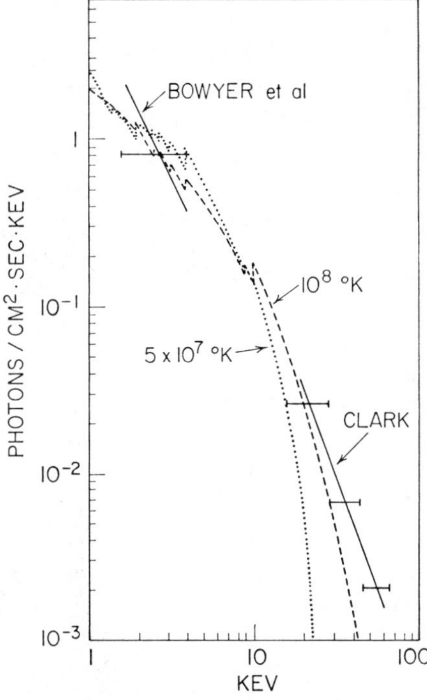

FIG. 9.—Observed spectrum of X-radiation from Crab Nebula compared to calculated free-free and free-bound emission from gaseous nebula at $T = 5 \times 10^7$ degrees K and 10^8 degrees K (Hayakawa, Matsuoka, and Sugimoto 1966).

bright optical emission, and presumably the X-rays, could be emitted when electrons from other regions of the nebula wander into regions with higher magnetic fields. Thus, the same electrons which supply the radio synchrotron radiation would be capable of supplying the optical and X-ray emission.

Hayakawa, Matsuoka, and Sugimoto (1966) prefer a thermal explanation for the X-rays from the Crab. They suggest that the X-ray region may be a single hot condensation or a collection of smaller condensations embedded in the H II region of the visible nebula. In either case, if the temperature is in the range 10^7 to 10^8 degrees K, the contribution to the optical and radio emission would be negligible. The condensations may be heated by hydromagnetic shocks repeatedly generated by some mysterious mechanism near the center of the nebula. In Figure 9, Hayakawa *et al.* attempt to match the

X-ray observations to the combined bremsstrahlung and free-bound spectrum of a hot gas having the average cosmic abundance of the elements.

The energy radiated in X-rays by the Crab, about 4×10^{36} ergs/sec, is comparable to the energy supplied by radioactive debris from the original supernova explosion. After the initial flash of a Type I supernova, the light output decays nearly exponentially. Because the half-life of the optical emission is comparable to that of Cf^{254}, about 55 days, Hoyle and Fowler (1960) proposed that Cf^{254} is formed by rapid neutron capture, together with a chain of the *r*-process nuclei, some of which have much longer lives. Burbidge, Burbidge, Fowler, and Hoyle (1957) calculated that spontaneous fission of Cf^{254} could have produced energy at the rate of about 10^{41} ergs/sec during the early luminosity of the Crab explosion. Woltjer (1958) extended the calculations to the present time and found that (1) if the neutron flux were sufficient to transport all the processed material to the range of atomic weights 110 to 260, and cycled in this range, the energy production after 1,000 years would be about 5×10^{36} ergs/sec; (2) if the steady-flow was reached in the neutron-capture process leading to Solar System abundances, the energy developed after 1,000 years would be 1.3×10^{37} ergs/sec. In the first case, the present activity would be associated with fission of Cm^{250}, and in the second case with the beta decay of Si^{32} and Ar^{39}. In comparison with the observed X-ray output of the Crab, the theoretical radioactive energy production appears to be adequate to supply the heat source for X-ray emission.

Morrison and Sartori (1965) have also examined the possibility that the Crab X-ray source is bremsstrahlung from a hot gas cloud. They estimate that about 1 per cent of the mass of the exploding star is converted to heavy elements by the *r*-process. This heavy material forms the core of a spherical cloud expelled by the explosion. Most of the mass of the original star forms a more rapidly expanding shell outside the core. Early in the lifetime of the nebula, the energy of luminosity must be supplied by radioactive decay, primarily that of Cf^{254}. As the gas cloud becomes more and more dilute, cooling by free-free transitions becomes less and less effective. Energy begins to store in the nebula, and the temperature rises until the gas is almost fully ionized and the bremsstrahlung hardens into X-rays. Morrison and Sartori find that a present temperature of 10^8 degrees K for the Crab is consistent with their model of radioactive heating.

8. SCO XR-1

A variety of evidence has been obtained for the X-ray spectrum of Sco XR-1. It is difficult to conclude from the observations whether the spectrum is thermal or synchrotron, but it is possible to make a fairly good fit to the observations with a bremsstrahlung spectrum in the range from 1 to 20 kev. Giacconi, Gurskey, and Waters (1965) found $T = (3.8 \pm 1.8) \times 10^7$ degrees K over the range from 1 to 15 kev. Chodil *et al.* (1965) obtained $T = 5.8 \times 10^7$ degrees K in the energy range from 2 to 20 kev. More recently, Grader *et al.* (1966) found $T = 4.6 \times 10^7$ degrees K for 2 to 20 kev but observed a marked fall-off at the low-energy end of the spectrum. They noted that the indication of a peak in the spectrum near 2 kev was suggestive of blackbody radiation at $T \approx 9 \times 10^6$ degrees K. The NRL group (Friedman, Byram, and Chubb 1966), using filter photometry in the range from 2 to 10 kev, derived a temperature of about 5×10^7 degrees K.

If we accept a 5×10^7 degrees K bremsstrahlung model for the spectrum, extrapolation to the visible range can be made on the assumption that the power per unit frequency interval is constant with decreasing frequency. The visible luminosity would then be only one-thousandth of the X-ray luminosity and the predicted visible brightness about 13th magnitude. To be detectable, it would be necessary that the source be smaller than 2 minutes of arc in diameter. If indeed the object were starlike, it could be any one of many faint stars in the general vicinity of the X-ray position, poorly established prior to 1966.

In order to increase the accuracy of the determination of the size of a source and its position by means of slit collimators, the field of view must be reduced. At the same time, the number of counts recorded by the detector placed behind the collimator as it scans the field of view is directly proportional to the field of view. It is difficult, therefore, with a simple slit collimator to achieve high sensitivity and high resolution at the same time. On the other hand, a modulation collimator designed by Oda (1965) permits the combination of high resolution and wide field of view. It consists essentially of two plane grids arranged like a double picket fence, one behind the other. The closer the spacing of the grids and the larger the distance between the two planes, the higher the angular resolution. In a parallel beam of X-rays, the front grid throws a shadow on the rear grid. As the collimator turns, the shadow shifts across the rear grid and the transmission changes, being at a maximum when the shadow of the front grid falls exactly on the back grid, and at a minimum when the shadow falls in the apertures of the back grid. A source whose angular dimension is small compared to the separation between adjacent maxima will produce a sequence of triangularly shaped maxima typical of a point source. If the angular dimensions of the source are larger than the separation between neighboring maxima, no modulation is observed.

An early attempt by the ASE-MIT group to employ the modulation collimator placed an upper limit of 7′ on the size of Sco XR-1. This effort was superseded on March 8, 1966, by a much more precise measurement (Gursky *et al.* 1966). The upper limit on size was reduced to 20″, and the source was located within a field of only four square minutes. In this field, there is one very peculiar, blue starlike object of magnitude 12.5. It is interesting to note that Johnson and Stephenson (1966) independently proposed the same star without the benefit of knowing the more accurate X-ray position, $16^h17^m4.3^s$ R.A. and $-15°31'33''$ dec. If the optical continuum is part of the bremsstrahlung spectrum, the X-ray source, like the optical, is probably less than 1.3 seconds of arc in diameter. The optical properties suggest a recurrent nova. Observations over 71 years show that the star has been secularly stable but that its intensity may fluctuate by several per cent in a matter of minutes and by as much as one magnitude in the course of a day. Emission lines of H_β, H_γ, H_δ, He, He II, C III, N III, and possibly O II appear in the spectrum. These emission lines are also variable.

Assuming that the optical spectrum is flat, the distance of Sco XR-1 may be estimated from its color excess (Matsuoka, Oda, and Ogawara 1966). The intrinsic color index for the flat spectrum is 0.10, and the observed color indicates an excess of 1.13. Nearby stars at distances of 300 and 500 pc show color excesses of 1.2 and 0.3, respectively. If interstellar extinction is uniform over these distances, it appears that Sco XR-1 must be at a distance of about 200 pc.

A hot plasma of the required particle density confined to a sphere of about 10^{15} cm radius would cool in a time of the order of a few years. If the volume is smaller, the cooling time becomes rapidly shorter. The mass of the cloud is insufficient to confine itself by its own gravity. It may be gravitationally trapped by a massive central star or it may be confined by intense magnetic fields. Energy would need to be continuously supplied, or the cloud may be continuously produced by ejection from an active central source. The optical emission lines require that some of the gas be cooler than 50,000 degrees K. Perhaps a planetary nebula surrounds the hot plasma cloud. It has also been suggested (Matsuoka, Oda, and Ogawara 1966) that the hot plasma is confined to the vicinity of a binary star pair and that the plasma streams from one gravitational center

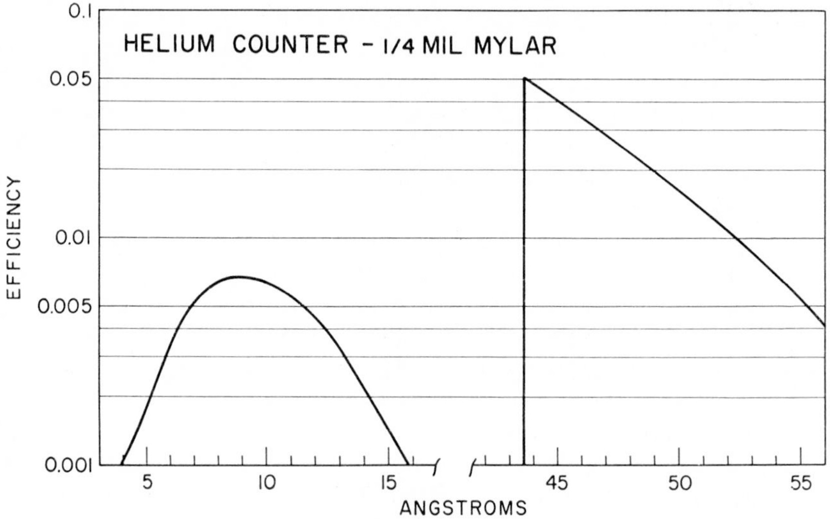

FIG. 10.—Spectral efficiency of helium-filled counter used in April, 1965, Aerobee flight

to the other. Concentrated heating may be produced by shock waves at singular points, and these would provide the X-ray emission.

The NRL group has used a helium-filled counter with a Mylar window to obtain the spectral sensitivity curve shown in Figure 10. In the survey flight of April, 1965, the only source detected with this counter was Sco XR-1. Analysis of the response relative to that obtained with counters whose sensitivities peaked in the 3 A region indicated that most of the response of the helium counter came from wavelengths longer than 44 A.

Independently of the optical estimates of the distance to Sco XR-1, the detection of a strong flux at wavelengths as long as 44 to 60 A implies that the source must be rather close. The Galaxy should be relatively transparent to X-rays of wavelength less than 10 A, but the opacity increases very rapidly toward longer wavelengths. If Aller's (1961) cosmic abundances are used and all the hydrogen is assumed to be atomic, the atomic absorption cross-section of interstellar gas at 44 A is found to be 7×10^{-21} cm^2 per hydrogen atom (Felten and Gould 1966). If the average density of hydrogen in the galactic disk is assumed to be about 1/cc, unit optical thickness would be about 140 light years (43 pc). The half-thickness of the galactic disk near the Sun, which lies close to the galactic equator, is about 4.2×10^{20} hydrogen atoms (Felten and Gould 1966), or 420 light years (130 pc) at one hydrogen atom/cc. Sco XR-1 is at galactic latitude 24° so that the pathlength through the disk in the direction of the source is about 1,260 light years (386 pc)—an optical thickness of 9. If the source lies outside the disk, the X-ray attenuation in the disk alone would exceed a factor of 10,000. This would imply a low temperature component ($\sim 2 \times 10^6$ degrees K) of the source, fantastically brighter than the 50 million degree component. It is necessary to conclude from the soft X-ray evidence alone that Sco XR-1 must lie well inside the disk and most likely within one or two optical thicknesses; otherwise the luminosity becomes implausibly high.

Gursky *et al.* (1966) ruled out the possibility that the source is a supernova remnant within the Galaxy on the assumption that a hot gas cloud at a temperature of about 50 million degrees would expand unhindered and reach a diameter in excess of 20 seconds of arc in less than 50 years. If a supernova had occurred within a distance of even a few thousand light years in the past 50 years, they argue that it should have been observed. They also rejected the neutron-star possibility because the 1- to 10-A spectrum is inconsistent with the required blackbody distribution. Their argument against the supernova, however, assumes that the event produces all its energy in one flash and that the hot nebulous material then expands steadily into interstellar space. It is equally plausible to propose an active source which continues to deliver energy from the collapsed stellar core to a magnetically confined corona or a magnetosphere in such a way that most of the X-ray emission remains highly concentrated about the central source.

The evidence for the long wavelength emission of Sco XR-1 suggests a combination of high and low temperature plasmas, or of thermal and nonthermal emission. Such complex source distributions have been suggested by Cameron (1965) as a consequence of an oscillating neutron star. According to his hypothesis, the neutron star is capable of storing gravitational potential energy in the form of radial oscillations with vibration periods in the millisecond range. The remnant of a Type I supernova may thus store up to 10^{51} or 10^{52} ergs of mechanical vibrational energy, many orders of magnitude greater than the thermal energy content. If a magnetic field is embedded in the star, the vibrations will produce hydromagnetic waves, and these will accelerate charged particles to high energies as they travel along the field lines. Woltjer (1964) has estimated that neutron stars can have surface magnetic fields up to about 10^{14} gauss as a result of the compression of the original stellar field along with the matter that forms the neutron star. A corona around the neutron star, expanding into a stellar wind, could draw the lines of force out radially. Rotation of the neutron star might then twist the lines of force in the inner region until they reconnected to form a magnetosphere. Radial oscillations of the neutron star could generate hydromagnetic waves which would travel through the magnetosphere, and electrons could be accelerated to high enough energies to radiate synchrotron X-rays. It is also possible that a neutron star may be surrounded by a corona of higher temperature, which would produce X-ray bremsstrahlung. The observed spectrum would thus be a superposition of the blackbody radiation of a neutron star and the bremsstrahlung, or synchrotron radiation, of its outer atmosphere.

Although a neutron star, theoretically, cools rapidly from $\sim 10^7$ degrees to a temperature of about 2 million degrees or less by neutrino radiation processes, it can remain in the 1- to 2-million degree range for tens of thousands of years. The estimate of the X-ray flux in the 44 to 60 A range is about 10^{-8} erg cm^{-2} sec^{-1} (about 10^{-7} erg over the full wavelength range), which is consistent with the blackbody radiation of a neutron star of 10 km radius and temperature between 1.5 and 2 million degrees K at a distance of 50 to 100 light years (15 to 30 pc). The short wavelength emission source ($\lambda < 6$ A), which is characterized by a temperature of about 50 million degrees, radiates an energy flux of the same order of magnitude.

To summarize the foregoing discussion, it appears that Sco XR-1 could be an ancient supernova remnant. At a distance of about 100 light years (30 pc), its 1 to 10 A X-ray luminosity is somewhat less than 1 per cent of that of Tau XR-1 in the Crab Nebula. The optical counterpart is starlike, but the debris of the original explosion could have expanded to very great distances and long ago have become undetectable as an X-ray nebula of the type seen in the Crab. The presently observed X-ray activity would be confined to the stellar remnant of the supernova, possibly a vibrating neutron star, and its surrounding corona or magnetosphere. Since the X-ray luminosity in the 44 to 60 A band is compatible with a blackbody temperature from only 1 to 2×10^6 degrees K, the neutron star could be as much as 50,000 years old.

9. EXTRAGALACTIC SOURCES

Although Cyg A is the brightest extragalactic radio source and was discovered in 1946, five years passed before Baade and Minkowski (1954) identified it with a weak optical galaxy about 700 million light years (200×10^6 pc) distant. The central portion of the galaxy appeared to consist of two bright condensations separated by about 2 seconds of arc, and it was proposed that these were two galaxies in collision. Subsequent studies indicated that the radio emission came from two vast regions whose centers lie about 100 seconds of arc apart. It is generally believed that Cyg A is a synchrotron source, but with the synchrotron hypothesis, it becomes necessary to explain how the billion-volt electrons, needed to produce the radio emission, are accelerated and to account for the total energy contained in electrons and magnetic fields. Because the synchrotron process is relatively inefficient, the total energy content must be in excess of 10^{60} ergs, and perhaps as great as 10^{62} ergs.

In the thermonuclear burning of stars, the efficiency of conversion of mass to energy is of the order of 1 per cent, and the burning of one solar mass produces about 10^{52} ergs. It would, therefore, take the nuclear conversion of 10^{10} solar masses, or the entire mass of a medium-sized galaxy, to produce the energy content of Cyg A. Attempts to explain the energy in terms of gravitational collapse of a superstar have thus far failed. If the identification of X-ray emission with Cyg A is confirmed, it would indicate X-ray luminosity more than an order of magnitude greater than radio plus optical, and would correspondingly increase the difficulty of explaining the total energy content of radio galaxies if the X-ray emission were also synchrotron. It seems necessary to find a thermal explanation for Cyg A.

Shklovsky (1966) has pointed out that the line spectrum of Cyg A, which includes O III, Ne III, and Ne V, indicates a very high state of ionization. The kinetic energy content of the radiating gas is about 10^{56} ergs, which could supply the ionization energy for no more than a few tens of thousands of years. It is possible that the line emission is excited by an X-ray source, but the hot plasma that would produce the X-rays could certainly not coincide with the radio emitting clouds. The region that produces the emission lines is about 5″ in diameter, but it is enveloped by a larger region measuring 30″ × 18″ that emits a continuum of white light. Shklovsky suggests that the continuum is the long-wavelength tail of the bremsstrahlung spectrum that produces the X-ray emission. This hypothesis requires that hot and cold plasma coexist in the central region of Cyg A. As pointed out by Ginzburg (1966), such a condition is possible if an enormous mass of gas, $>10^{11}\,\mathfrak{M}_{\odot}$, existed in the nuclear region of Cyg A at the time of its explosion. The relativistic particles ejected in the explosion could transfer energy to the gas and heat it strongly. The cooling-off process may take place very slowly, and as a result of the inhomogeneities that exist prior to the explosion, the heating process may occur very irregularly over the volume of the gas cloud.

M87 (Virgo A) is an elliptical galaxy and one of the brightest in the Virgo cluster at a distance of 11 Mpc. It is about 5″ in angular size, and its brightness is highly concentrated toward the center. Its total mass may be about $10^{12}\,\mathfrak{M}_{\odot}$, 10 times the mass of the Galaxy. A highly polarized luminous jet, 20″ long (about 1,000 pc) projects from its center. M87 provided the first evidence for the role of synchrotron radiation in radio galaxies. The radio power of Virgo A is about a thousand times weaker than that of Cyg A, but it ranks immediately behind Cyg A and Cen A in flux received at the Earth. The situation in Virgo A is very different from that of Cyg A, and perhaps a stronger case can be made for a synchrotron X-ray source than for a thermal source. However, the objection with regard to the limited lifetime of the relativistic electrons is even stronger than in the case of the Crab. Even the source of optical synchrotron radiation may have a lifetime of the order of a few years, which is very short compared to the several thousand years life of the jet.

10. CONCLUSION

Five years after the first detection of discrete X-ray emission regions, X-ray astronomy has advanced to the stage where identification of the stronger sources with their optical counterparts is feasible, even within the limitations of small rocket instrumentation. Evidence has accumulated which indicates major differences in spectral character and between the various sources. The observed fluxes reveal that the X-ray luminosity generally exceeds the radio and optical luminosity. Some evidence exists for detectable X-ray emission from radio galaxies. But out of all the early evidence available for approximately 30 sources, it is not possible to find a unique astrophysical model to satisfy any one of these sources.

The observations attempted thus far have been accomplished with relatively primitive means. It is technically feasible to orbit detectors 1,000 times as large as those flown on rockets with concomitant gains in sensitivity from both increased detector area and prolonged observing time. X-ray telescopes, based on grazing incidence reflection, have been successfully demonstrated on a small scale, and their construction on as large a scale as could be carried in Apollo class vehicles is technically feasible. A telescope of 3-m aperture and 80-m focal length could detect sources a million times weaker than Tau XR-1 and resolve details to a few seconds of arc. With the Moon as a base, occultation observations over the lunar horizon could provide resolution of a fraction of a second of arc for sources a thousand times weaker than the Crab.

From the primitive evidence already observed with the limited sensitivity of rocket instruments, it appears that X-ray sources may be as abundant as radio sources and perhaps even more diverse in physical characteristics. X-Ray astronomy clearly deserves a very high priority in space research.

REFERENCES

Aller, L. H. 1961, *Abundance of the Elements* (New York: Interscience Publishers), p. 179.
Baade, W., and Minkowski, R. 1954, *Ap. J.*, **119**, 206.
Bowyer, S., Byram, E. T., Chubb, T. A., and Friedman, H. 1964*a*, *Nature*, **201**, 1307.
———. 1964*b*, *Science*, **146**, 912.
———. 1965, *ibid.*, **147**, 394.
Burbidge, E. M., Burbidge, G. R., Fowler, W. A., and Hoyle, F., 1957, *Rev. Mod. Phys.*, **29**, 547.
Byram, E. T., Chubb, T. A., and Friedman, H. 1966, *Science*, **152**, 166.
Cameron, A. G. W. 1965, *Nature*, **205**, 787.
Chiu, H. Y., and Salpeter, E. E., 1965, *Phys. Rev. Letters*, **14**, 343.
Chodil, G. R., Jopson, R. C., Mark, H., Seward, F. D., and Swift, C. D. 1965, *Phys. Rev. Letters*, **15**, 605.
Clark, G. W. 1965, *Phys. Rev. Letters*, **14**, 91.
Colgate, S. A., and White, R. H. 1966, *Ap. J.*, **142**, 626.
Felten, J. E., and Gould, R. J. 1966, *Phys. Rev. Letters*, **17**, 401.
Fisher, P. C., Johnson, H. M., Jordan, W. C., Meyerott, A. J., and Acton, L. W. 1966, *Ap. J.*, **143**, 203.
Friedman, H. 1959, *Proc. Inst. Radio Engrs.*, **47**, 272.
Friedman, H., Byram, E. T., and Chubb, T. A. 1966, *Science*, **153**, 1527.
Giacconi, R., Gursky, H., Paolini, F. R., and Rossi, B. B., 1962, *Phys. Rev. Letters*, **9**, 439.
Giacconi, R., Gursky, H., and Waters, J. R. 1965, *Nature*, **207**, 572.
Giacconi, R., Gursky, H., Waters, J. R., Clark, G., and Rossi, B. B. 1964, *Nature*, **204**, 981.
Ginzberg, V. L. 1966, *Sov. Astr.-A.J.*, **9**, 877.
Gould, R. J., and Burbidge, G. R., 1963, *Ap. J.*, **138**, 969.
Grader, R. J., Hill, R. W., Seward, F. D., and Toor, A. 1966, *Science*, **152**, 1499.
Gursky, H., Giacconi, R., Gorenstein, P., Waters, J. R., Oda, M., Bradt, H., Garmire, G., and Sreekantan, B. V., 1966, *Ap. J.*, **144**, 1249.
Gursky, H., Giacconi, R., Paolini, F. R., and Rossi, B. B. 1963, *Phys. Rev. Letters*, **11**, 530.
Hayakawa, S., Matsuoka, M., and Sugimoto, D. 1966, *Space Sci. Rev.*, **5**, 109.
Haymes, R. C., and Craddock, W. L., Jr. 1966, *J. Geophys. Res.*, **71**, 3261.
Hoyle, F., and Fowler, W. A. 1960, *Ap. J.*, **132**, 565.
Johnson, H. M., and Stephenson, C. B. 1966, *Ap. J.*, **146**, 602.
Kupperian, J. E., Jr., and Friedman, H. 1958, "Gamma Ray Intensities at High Altitudes," *Proc. 5th CSAGI Assembly*, Moscow, USSR.
Matsuoka, M., Oda, M., and Ogawara, Y. 1966, *Nature*, **212**, 885.
Moroz, V. I. 1964, *Sov. Astr.-A.J.*, **7**, 755.
Morrison, P., and Sartori, L. 1965, *Phys. Rev. Letters*, **14**, 771.
Morton, D. C. 1964, *Nature*, **201**, 1308.
Oda, M. 1965, *Appl. Optics*, **4**, 143.
Oort, J. H. 1958, *Rech. Astr. Specola Vaticana*, **5**, 415.
Peterson, L. E., and Jacobson, A. S. 1966, *Ap. J.*, **145**, 964.
Sandage, A. R., Osmer, P., Giacconi, R., Gorenstein, P., Gursky, H., Waters, J. R., Bradt, H., Garmire, G., Sreekantan, B. V., Oda, M., Osawa, K., and Jugaku, J. 1966, *Ap. J.*, **146**, 316.
Shklovsky, I. S. 1960, *Cosmic Radio Waves* (Cambridge, Mass.: Harvard University Press), p. 307.
———. 1964, "On the Energy Spectrum of Relativistic Electrons in the Crab Nebula," preprint.
———. 1966, *Sov. Astr.-A.J.*, **10**, 6.
Woltjer, L. 1958, *B.A.N.*, **14**, 39.
———. 1964, *Ap. J.*, **140**, 1309.

CHAPTER 14

Dynamical Properties of Cosmic Rays

E. N. PARKER
Enrico Fermi Institute for Nuclear Studies and Department of Physics, University of Chicago

1. INTRODUCTION

Cosmic rays and X-rays were each first detected as a minor nuisance in the laboratory. X-Rays caused an annoying fogging of photographic plates stored too near electrical discharge tubes; the obvious solution to the problem would have been to store the plates elsewhere. Cosmic rays first demonstrated themselves through an annoying leak of charge from the laboratory electroscope; the obvious solution to the problem would have been to insulate the electrode more completely. In both cases a stubborn curiosity led to fascinating discovery.

It was evident that the charge leakage was through the air. Air can conduct only to the extent that it is ionized, and natural radioactivity was first thought to be responsible for the ionization. A historic balloon flight by V. F. Hess in 1912 to investigate the leakage of charge showed that the ionization increased with altitude. Hence the radiation presumably responsible for the ionization was deduced to be coming from outside the atmosphere with enough penetrating power to reach all the way down to the laboratory electroscope. The term *cosmic rays* was introduced on this basis.

It is known today that cosmic rays are fast particles passing through the Solar System on some mysterious journey through interstellar space. Their penetrating power is a consequence of their high speed and the spray of weakly interacting secondary particles produced in their passage through matter. An excellent account of the experiments which have led to the present understanding of cosmic rays is contained in a recent book by Rossi (1964).

Cosmic-ray studies have not been part of traditional astronomy, and only very recently has instrument development made real advances possible. It has suddenly become apparent that cosmic rays play the role of steam in the galactic boiler. Supernovae and other violent plasma phenomena are the fires which generate the steam. The boiler is run at near bursting capacity. But this picture is reached only after careful consideration of the observed properties of cosmic rays. Extrapolation from local observations, all in the vicinity of the Earth over the last few decades, to general conditions throughout the Galaxy is necessary. We shall consider the facts from several points of view. In this introductory section the reader will be given some perspective in preparation for a survey of the arguments which led to the present state of understanding of cosmic-ray phenomena.

The origin of cosmic rays in the Galaxy poses an interesting astrophysical question. The energy production rate seems to be enormous, about 10^{41} ergs/sec. It has become evident in recent years that fast particles are generated whenever and wherever ionized gases and magnetic fields suffer violent agitation. The mechanism (or mechanisms) actually responsible is not fully understood, and it is not possible to assess the relative cosmic-ray contributions from the many obvious candidates, for example, novae, supernovae, galactic nuclei, and so forth. Even more puzzling are those few *very* high-energy cosmic-ray particles which have so much momentum that they could not be contained in the Galaxy and which are, therefore, presumably of extragalactic origin. Altogether the origin of fast particles is of general and fundamental theoretical interest.

A second aspect of interest is that the cosmic-ray particles represent a sample of matter from other places in the Galaxy, and thus tells us something of the composition of their sources. The most striking analysis of this type so far was carried out recently by Biswas and Fichtel (1964) on energetic particles ejected from the Sun. They showed that among the energetic particles the relative abundances of He and heavier nuclei were the same as in the Sun. This indicated that the acceleration of fast particles preserves the relative abundances of all nuclei with the same charge-to-mass ratio. The abundance of elusive elements such as neon was measured directly among the energetic solar particles and gave an estimate of the abundance in the Sun. The question of abundances will be discussed again in considering the origin of cosmic rays, which show a remarkably high abundance of heavy nuclei.

A third aspect of cosmic rays is their importance in probing conditions in the Solar System (see discussion in Simpson 1960). The cosmic-ray intensity observed at the Earth is lower than the cosmic-ray intensity in interstellar space because the solar wind contains magnetic fields whose outward motion sweeps the cosmic rays out of the Solar System. The reduction of the cosmic-ray intensity varies with the strength of the wind and fields. Observation of this cosmic-ray variation first showed that magnetic fields are carried with the "solar corpuscular radiation," as the wind was then called. The interpretation of the cosmic-ray variations required magnetic fields of from 10^{-5} to 10^{-4} gauss with irregularities on a scale of from 10^5 to 10^7 km. We will not go into this interesting local aspect of cosmic rays here. An account of the development of the ideas is given elsewhere (Parker 1963, 1964*b*, 1965*a*). Local cosmic-ray effects are important on a galactic scale largely because the local variation is a correction which must be applied to cosmic-ray properties observed at the Earth in order to deduce the properties in the surrounding interstellar space. More will be said on this point later. Presumably similar winds, fields, and cosmic-ray variations occur around other stars (Parker 1960, 1963), although unfortunately such activity cannot be seen directly.

Fourth, cosmic rays produce particle interactions in the terrestrial atmosphere. The primary particles—mostly nuclei—which suffer the rare misfortune of approaching the Earth enter the terrestrial atmosphere and, after traversing about 100 gm/cm^2, collide with the nucleus of an air atom. The collision produces a number of secondary particles, including fragments of the initial cosmic-ray particle and/or the struck nucleus, mesons, and so forth. Some of these secondary particles have a considerable fraction of the primary-particle energy and themselves take part in further collisions producing more particles. The result is a complicated variety of particles appearing deep in the atmosphere; some of these—particularly the μ-mesons—penetrate to great depth because they slip through without interacting in any strong way. For this reason, cosmic rays in the atmosphere have been, and continue to be, of great interest in the field of particle physics. Many of the so-called fundamental particles were first discovered in studies of the secondary particles in the atmosphere (see, for instance, Rossi 1952).

In practice, direct observations of primary cosmic-ray particles can be made only above altitudes of about 80,000 ft, requiring balloon-borne equipment or space vehicles. Variations of the primary cosmic-ray intensity can be monitored at ground level because the number of secondary particles is directly proportional to the number of primary particles striking the top of the atmosphere. But quantitative measures of the intensity require complicated extrapolation from ground level to the top of the atmosphere at a number of geomagnetic latitudes (Simpson, Fonger, and Treiman 1953; Webber 1962).

Finally, cosmic rays appear in the instruments as individual particles but in their astrophysical context must be thought of as a hot tenuous gas. Their detection and study in space is based principally upon the track of ionization and the electromagnetic "shock" wave (Cerenkov radiation) that they produce in their passage through matter. Instrumentation for the study of cosmic rays is described in the extensive literature on the subject (see, for instance, Rossi 1952; Simpson 1958; Elliot 1958). As a consequence of their detection as individual particles, the density of cosmic rays in space is usually described by the particle flux per steradian across 1 cm^2, in which units the cosmic-ray intensity is about 0.5 particles cm^{-2} sec^{-1} steradian^{-1} with energy above 10^8 ev in the Solar System. This is equivalent to an omnidirectional intensity of about 6 particles/cm^2 sec, since there is no detectable departure from isotropy.

In contrast to the usual terrestrial appearance of cosmic rays as individual particles, on an astrophysical scale the cosmic rays must be thought of as a gas (Biermann and Davis 1958, 1960; Parker 1958*b*, 1965*c*, *d*) if their dynamical effects in the Galaxy are to be appreciated in a simple way. The cosmic-ray component of the interstellar gas is exceedingly hot and exceedingly tenuous, with a hydrostatic pressure comparable to the galactic magnetic-field pressure which is much larger than the pressure of the thermal component of interstellar gas. The dynamical importance of the cosmic-ray component is evident from this comparison of pressures. The enormous mobility of the cosmic-ray gas allows it to extend throughout the galactic arms and halo, presumably channeled along the galactic magnetic fields. The cosmic-ray gas has very little direct interaction with the rest of the interstellar medium; the collision cross-sections are of about nuclear dimensions, 10^{-26} cm^2, rather than the familiar atomic cross-sections of from 10^{-17} to 10^{-14} cm^2. The small ionization loss to individual cosmic-ray particles passing through the interstellar medium is evidently important neither to the cosmic-ray particles nor to the interstellar medium (Hayakawa 1962; Ginzburg and Syrovatskii 1964). The very few nuclear collisions which take place are only enough to leave traces of some otherwise rare nuclei which then permit the cosmic-ray age to be determined (see Sec. 5.1).

The purpose of this chapter is to present cosmic rays in their astrophysical role. The interstellar medium is to be thought of as a composite fluid, consisting of the optically detected interstellar gas and the suprathermal cosmic-ray gas that dominates many dynamical properties of the Galaxy. The two components would move quite independently of each other were it not for the presence of weak magnetic fields (from 10^{-6} to 10^{-5} gauss). The general magnetic field ties the two gases together into a composite medium with dynamical consequences that have been pointed out only very recently. After a survey of the observations, we shall discuss implications for the age, origin, and extent of the cosmic-ray gas. The final step is to point out some immediate dynamical consequences of the presence of the cosmic-ray gas in the Galaxy.

2. LOCAL PROPERTIES OF THE COSMIC-RAY GAS

The basic physical properties of the interstellar cosmic-ray gas, as it extends into the Solar System at the present time, are presented briefly in this section. More detailed discussions of the known properties, and of the observations by which the properties are

TABLE 1

COSMIC-RAY AND COSMIC NUCLEAR ABUNDANCES

Atomic Number Z	Relative Nuclear Abundance in Cosmic-Ray Gas	Fraction of Total Mass in Cosmic-Ray Gas	Cosmic Abundance
1	1	0.698	1
2	0.07	.189	0.15
3–5	0.0015	.010	10^{-9}
6–9	0.004	.042	1.5×10^{-3}
10–19	0.0014	.042	1.5×10^{-4}
≥20	0.0005	0.019	7×10^{-6}

known, can be found in several review articles (Rossi 1955; Ginzburg 1958; Parker 1958*c;* Morrison 1961; Ginzburg and Syrovatskii 1961, 1964; Webber 1965). The information given here is sufficient foundation for the discussion of the origin and dynamics of the cosmic-ray gas which follows.

Basic data (Peters 1959) are: a particle density of $10^{-9}/\text{cm}^3$; a mean energy of 10^9 ev/particle (10^{13} ° K); an energy density[1] of about 10^{-12} ergs/cm³. The hydrostatic pressure of the cosmic-ray gas is thus about 10^{-12} dynes/cm². The speed of sound in the cosmic-ray gas alone is comparable to the speed of light. The interparticle distance is about 10^3 cm. The average radius of gyration of the individual particles in an interstellar field of 10^{-5} gauss is about 10^{12} cm; the cosmic rays may be treated as a fluid on scales larger than 10^{12} cm (10^{-6} pc).

The cosmic-ray gas consists mostly of protons with a smaller number of heavier nuclei and electrons. The relative abundances of the nuclei are summarized in Table 1 from the work of Waddington (1960), Aizu *et al.* (1961), O'Dell, Shapiro, and Stiller (1962), and Daniel and Durgaprasad (1962) and compared with the estimated mean cosmic abundances (Cameron 1959). The outstanding feature in the composition of the

[1] Coincidentally equal to the energy density of starlight.

cosmic-ray gas is the enormous abundance of the light nuclei—Li, Be, B—and the heavier nuclei, in contrast to the usual cosmic abundance.

The high abundance of the heavier nuclei is probably not a result of selective acceleration of heavy ions in cosmic-ray sources, as was once conjectured (Korchak and Syrovatskii 1958). There are no known dynamical processes which would favor heavy nuclei to so large a degree. It is observed that the near-relativistic gas from solar flares strictly preserves the relative abundances of the nuclei heavier than hydrogen (Biswas and Fichtel 1964). Hence, presumably the high abundance of heavy nuclei reflects the nuclear composition of the source. The only places known where heavy nuclei are so abundant are the interiors of senile stars (Hayakawa 1956; Burbidge *et al.* 1956). This strongly suggests that at least some part of the cosmic rays may come from novae and supernovae (Ginzburg and Syrovatskii 1964).

The light nuclei, Li, Be, B, pose a somewhat different problem, since they are found naturally in significant abundance only in such rare stars as T Tauri variables and a few others. Consequently, they are believed to represent fragments of heavier cosmic-ray nuclei produced by collisions with other nuclei in interstellar space. After such a collision the fragments generally continue on with the initial velocity of the original nucleus. The abundance of the light nuclei relative to the parent heavy nuclei thus permits an estimate of the amount of material through which the heavy nuclei have passed. The more material penetrated, the fewer heavy nuclei survive and the more light nuclear fragments there are. The observed Li, Be, B abundance indicates that the heavier nuclei have penetrated from 3 to 7 gm/cm^2 of material (Hildebrand *et al.* 1963; Appa Rao *et al.* 1963). It is then usually assumed that the cosmic-ray protons have penetrated about as much matter as the heavy nuclei, since to assume otherwise leads to additional special assumptions.

From the penetration of matter it is possible to express the length of the path traversed by the heavy nuclei in terms of the average density of the matter along the path (see Sec. 4.2.1).

The cosmic-ray–gas nuclei are completely stripped of their electrons. The stripping is a direct consequence of the velocities, which are near the speed of light. Since these velocities exceed the orbital velocity of the K-electrons, there is little chance of picking up an electron and every chance of losing them in passage through matter. The preservation of the solar abundance in the near-relativistic gas from the Sun is presumably the result of early stripping; so all nuclei except hydrogen have the same 1:2 charge-to-mass ratio.[2] Presumably the same is true for the galactic cosmic rays.

Electrons are present among the cosmic-ray particles (Earl 1961; Meyer and Vogt 1961) with a relative number abundance of a few per cent in the energy range from 10^8 to 10^9 ev. It has been shown recently (DeShong, Hildebrand, and Meyer 1964) that, at most, one in three of the electrons is positively charged. Presumably all the positive electrons and an associated smaller number of the negative electrons are secondary particles produced by the collision of cosmic-ray nuclei with interstellar material (Hayakawa and Kobayashi 1953; Hayakawa and Okuda 1962; Jones 1963). The remainder, that is,

[2] In the near-relativistic gas from the Sun, the abundance of hydrogen (charge-to-mass ratio of 1:1) relative to all heavier nuclei (charge-to-mass ratio 1:2) varies markedly from one solar flare to the next (Biswas and Fichtel 1964).

most negative electrons, are accelerated from thermal energies in violent magnetic plasmas in a manner similar to the acceleration of the cosmic-ray nuclei.

It has been emphasized by Shklovsky (1953*a*, *b*, 1960*b*) that the particular interest of the electrons of the cosmic-ray gas, in spite of their small numbers, is that the large charge-to-mass ratio of the electrons causes them to emit synchrotron radiation (magnetic bremsstrahlung), thereby permitting cosmic rays to be "seen" with radio telescopes, and, in extreme cases such as the Crab Nebula, to be seen optically. Thus, so far as relativistic electrons are associated with relativistic nuclei throughout the Galaxy, the electrons may act as a detectable tracer.

It should be evident from the foregoing remarks that the cosmic-ray gas, consisting mainly of atomic nuclei, is not electrically neutral by itself. It may, however, be considered electrically neutral in space, because the electrons missing from the nuclei are present, although without necessarily having high individual energies, and effectively neutralize the charge of the nuclei.

The energy distribution of the cosmic-ray particles is interesting. As already noted, the mean particle energy is about 10^{10} ev, in terms of which it is possible to assign a temperature of 10^{14} ° K, but the energy distribution is so far from Maxwellian that the temperature defined from the average energy does not have the usual thermodynamic significance. The energy distribution of the cosmic-ray nuclei is shown in Figure 1. Below 10^{10} ev/nucleon, the spectrum is depressed by varying amounts as a consequence of the solar wind; so we know only that the distribution in interstellar space lies somewhat higher than the distribution at the Earth at sunspot minimum (Meyer and Simpson 1955) employed in Figure 1, but probably only a very little higher. Above 10^{10} ev/nucleon, the energy distribution fluctuates relatively little as a consequence of the solar wind, and the distribution observed at the Earth (Lal 1953; Greisen 1960; Cocconi 1961; Nikolsky 1962; Clark, Brandt, and LaPointe 1963; Linsley 1963) is a close approximation to the distribution in interstellar space. The energy distribution above 10^{10} ev/nucleon is approximately the same for all types of nuclei (so far as one can tell) and is closely approximated by a power law with constant exponent. Thus if $n(E)dE$ is the number of cosmic-ray nuclei per cm^3 with energy per nucleon in the interval $(E, E + dE)$, $n(E) \propto E^{\gamma}$. The exponent γ seems to lie between 2.5 and 3.0 over the entire interval from 10^{10} to 10^{20} ev! The values of $\gamma - 1$ are indicated in Figure 1.

It is evident from Figure 1 that most of the cosmic-ray particles have energies in the general vicinity of 10^9 ev/nucleon. It is principally in this energy range, up to about 10^{11} ev, that the relative abundances are determined. The energy distribution of the heavier nuclei (up to 10^{11} ev) is usually expressed in terms of energy per nucleon because in this form they closely resemble the proton-energy spectrum and a single total spectrum can be presented, as in Figure 1. The similarity of the distributions of the various nuclei, where these have been separately determined, is consistent with the idea already established for the Sun (Biswas and Fichtel 1964)—that the light and heavy nuclei are accelerated equally, presumably because they are fully stripped and have about the same charge-to-mass ratio.[3]

For energies higher than 10^{11} ev/nucleon, the relative abundances are known only to

[3] It also may be argued that Coulomb scattering, which gives losses proportional to Z^2, is not important during the initial acceleration of the particles.

an order of magnitude, at best. Such very energetic particles are detected, and their energy measured, by the number of secondary particles that each energetic particle sends showering down through the terrestrial atmosphere. The size of the shower depends only upon the total energy of the incoming particle. Speculation on the composition of the very energetic cosmic-ray particles ranges from all protons to all heavy nuclei. The question is an important one because the composition provides important clues about where the particles originated, how long their path through the interstellar medium has been, and to what extent they are confined to the Galaxy.

The confinement of cosmic-ray particles to the Galaxy is an open question at the high-energy end of the distribution. For instance, a proton with an energy of 10^{18} ev moving perpendicular to an interstellar magnetic field of 10^{-5} gauss traverses a circle of 200-pc radius, which is comparable to the thickness of the galactic disk. The highest particle energy so far observed is 10^{20} ev (Clark *et al.* 1961; Linsley 1963). For a proton, the radius of gyration would be 2×10^4 pc, but if the particle were a uranium or lead

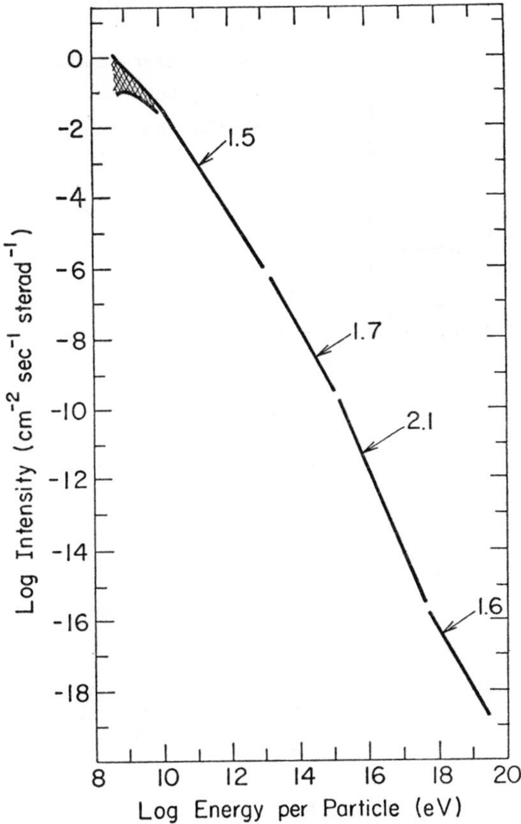

FIG. 1.—A summary plot of the observed cosmic-ray integral-energy spectrum (the total number of particles above the given energy E) from various observational studies. The empirical power law $E^{-\alpha}$ best fitting the data is indicated by the values of α given by the numbers along the curve, from Clark *et al.* (1963). The cross-hatched area below 10^{10} ev is a rough indication of the uncertainty introduced by the exclusion of the low-energy particles from the Solar System as a consequence of the solar wind.

atom this would be 200 pc. Thus the question of confinement to the Galaxy becomes an important consideration at somewhere more than 10^{16} ev/nucleon, and the answer depends very much on the charge of the nuclei at these very high energies.

The energy distribution of the electrons among the cosmic rays is known from recent work (Earl 1961; Meyer and Vogt 1961; Cline, Ludwig, and McDonald 1964; Schmoker and Earl 1965; Freier and Waddington 1965) and is summarized in Figure 2, taken from a recent paper by Meyer (1964). The solid and broken curves represent the electron distribution generated by collisions of the cosmic-ray nuclei with the interstellar medium

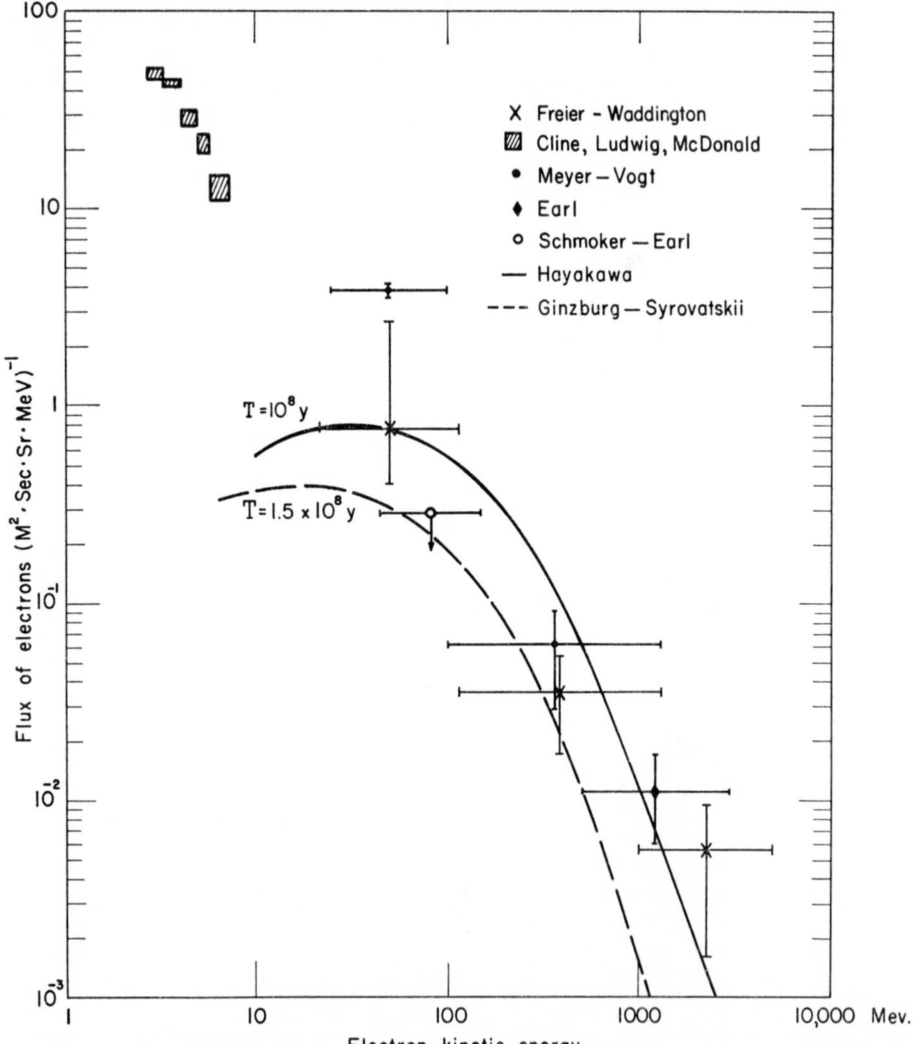

FIG. 2.—The equilibrium spectra of electrons produced by collision processes in the Galaxy calculated by Ginzburg and Syrovatskii (1964), *dashed line*, and by Hayakawa (1963), *continuous line*, with various experimental results for the differential electron-energy spectrum near the Earth.

calculated by Hayakawa (1963) and by Ginzburg and Syrovatskii (1964). Several remarks are in order concerning Figure 2. The first is the question whether the low-energy ($<10^7$ ev) electrons, observed by Cline, Ludwig, and McDonald (1964), are from interstellar space or whether they are produced locally in the Solar System (Parker 1965*l*). It is clear that the low velocity and low magnetic rigidity of *protons* below 10^7 ev would make it very unlikely that they would penetrate into the Solar System as far as the Earth. But electrons are a different problem. The electron velocity at 10^7 ev is essentially as high as for any higher energy. The magnetic rigidity of the electrons is so low as to facilitate, perhaps, rather than impede, their inward passage along the magnetic lines of force (Parker 1964*a*). So it is not unreasonable to assume, as pointed out by Cline *et al.* (1964), that the electrons observed near the Earth at the present sunspot minimum are from interstellar space. We shall for the present, then, take the observed electron spectrum in Figure 2 as an approximate representation of the spectrum in interstellar space, as we did for the proton spectrum in Figure 1.

The observed electron distribution and the theoretical secondary electron distribution are fairly close above 100 Mev. It can be concluded that secondary electrons make an important contribution to the total number of energetic electrons in interstellar space. However, the observed ratio of positive to negative electrons (DeShong, Hildebrand, and Meyer 1964) shows clearly that the secondary contribution is less than half of the total.

The final point to be made in this section is the remarkable degree of isotropy apparent for the cosmic rays. The implications of isotropy are discussed in Section 5.2. The observations (Crawshaw and Elliot 1956; Greisen 1956, 1960; Devaille, Kendziekski, and Greisen 1960; Rossi 1961; Ginzburg and Syrovatskii 1961) show that any anisotropy must be less than 1 per cent for all cosmic-ray particles up to 10^{14} ev, less than 3 per cent up to 10^{17} ev, and that there is no discernible large anisotropy even at 10^{19} ev. These observations imply[4] that the cosmic-ray intensity in the two directions perpendicular and parallel to the large-scale galactic magnetic field must be equal to within these limits, and they imply that there is a limit of less than 500 km/sec on the rate at which cosmic-ray gas is streaming by in interstellar space (see Sec. 5.2).

3. GALACTIC ENVIRONMENT OF THE COSMIC-RAY GAS

The individual stars in the galactic disk may generate cosmic rays (see Sec. 6.1), but the collective stellar cross-section is so small that the cosmic-ray loss from collisions with stars can be neglected in comparison with interaction with the interstellar gas and fields. Gravitational forces have negligible effect on the cosmic-ray gas. The scale height of the cosmic-ray gas in the galactic gravitational field is more than 10^6 times the radius of the Galaxy. The interstellar magnetic field is more important; magnetic forces cause the individual cosmic-ray particles to move in helices or circles with radii of the order of 10^{12} cm for the average particle. The magnetic field controls the fluid behavior of the cosmic rays over long distances and periods of time and ties the cosmic-ray gas and the interstellar gas into a composite fluid (Parker 1958*b*, 1965*c*, *d*). The novel properties

[4] The magnetic fields in the Solar System could not obliterate external anisotropy in cosmic-ray particles at energies higher than about 10^{12} ev/nucleon.

of this composite interstellar fluid are discussed in Section 7.2. The interstellar gas contributes to the inertia of the composite fluid and plays an important role in the calculated cosmic-ray life and production rate.

3.1. The Galactic Magnetic Field

The existence of a large-scale interstellar magnetic field was first suggested as a means of containing the cosmic-ray particles within the Galaxy (Fermi 1949, 1954). Theoretical interpretation (Davis 1951; Davis and Greenstein 1951; Spitzer and Tukey 1951) of the observed polarization of the light of distant stars (Hiltner 1949, 1951) provided a more satisfactory basis for the idea, and subsequent analyses (Chandrasekhar and Fermi 1953; Davis 1954; Pikelner 1956) seem to confirm the existence of large-scale magnetic fields of from about 10^{-6} to about 10^{-5} gauss in the galactic disk, with a statistical tendency for the lines of force to lie parallel to the disk.

Unfortunately, direct observation of Zeeman effect and Faraday rotation caused by a large-scale galactic field has not yet led to a precise picture (compare Davies *et al.* 1960; Davies, Verschuur, and Wild 1962; Davies and Shuter 1963; Morris, Clark, and Wilson 1963; Morris and Berge 1964, and the references therein). The consensus appears to be only that, with the possible exception of a few isolated regions, the general galactic field is of the order of 0.6×10^{-5} gauss in the galactic disk, but still with some serious uncertainty. The more recent observations of Faraday rotation (Morris and Berge 1964) suggest that the sign of the galactic field reverses across the galactic plane, the magnetic field being directed toward galactic longitude $l^{\text{II}} = 250°$ above ($b^{\text{II}} > 0$) and $l^{\text{II}} = 70°$ below ($b^{\text{II}} < 0$). These general directions agree rather well with that suggested by studies of the polarization of starlight (Smith 1956; Behr 1959) and with the direction of the local spiral arm (Morgan, Sharpless, and Osterbrock 1952; Weaver 1953; Westerhout 1957; Becker 1963, 1964; Blaauw 1964). The apparent reversal of the field across the plane of the Galaxy seems to rule out the simple picture of a galactic arm as a straight tube of flux (Chandrasekhar and Fermi 1953), but the observations are not incompatible with a helical field along the arms (Ireland 1961) or with some other suggested configurations.

Speculation ranges from magnetic fields with lines of force directed along each galactic arm playing an intimate part in maintaining the structure of the spiral arm (Chandrasekhar and Fermi 1953; Piddington 1964; Woltjer 1965) to ideas in which the magnetic field plays no significant role in forming the spiral arm (Lin and Shu 1964). Hoyle and Ireland (1960*a*, *b*) suggest that the magnetic fields in the galactic halo are derived from the magnetic lines of force in the spiral arms in contrast to other theories (for example, Hoyle and Ireland 1961) in which the halo field is essentially distinct from the fields in the spiral arms, magnetic lines of force from the halo crossing the galactic plane between the spiral arms. Ginzburg and Syrovatskii (1964) suggest a picture somewhere between these (see also Pikelner and Shklovsky 1957, 1958); they suggest some flux leakage, and hence some cosmic-ray leakage, from the spiral arm into the halo, but the halo field is not considered to be derived mainly from the field in the spiral arms. Most discussions are based on the idea that the halo field is somewhat irregular in form (Biermann and Davis 1960; Ginzburg and Syrovatskii 1964); if it were not, the nonthermal radio emission would be strongly polarized, contrary to observation. Hoyle and Ireland (1961) suggested that the halo field may be wound up by turbulent gas motions from a much weaker large-scale field. Sciama (1962) proposed that the magnetic lines of force of the halo are a part of a general magnetic field of the Local Group of galaxies (see also Kahn and Woltjer 1959). Review papers by Wentzel (1963*a*) and Woltjer (1963, 1965) give more detailed discussions.

It is evident that these speculations do not exhaust all the possibilities. There is simply not enough observational information to give a picture of the general topology of the galactic magnetic field. Consequently, any discussion of the origin and dynamics of the cosmic-ray gas must be restricted to those considerations which do not involve details of the structure of the galactic magnetic field.

3.2. The Interstellar Gas

The interstellar gas is believed to be largely hydrogen, in the molecular, atomic, and ionized state, depending upon the density and temperature. Perhaps 2 per cent of the observed gas is ionized (Westerhout 1957; Wilson 1963) with a temperature of about 10^4 ° K. The remainder of the observed gas is neutral atomic hydrogen with a temperature of about 10^2 ° K. The neutral atomic hydrogen is observed directly from thermal 21-cm emission, which indicates an average density throughout the disk of the Galaxy of about 1 atom/cm^3. Unfortunately, molecular hydrogen cannot be observed directly at the present time, for reasons which will become apparent later.

The atomic hydrogen distribution extends above and below the galactic plane with a characteristic scale of 100 pc[5] (Schmidt 1957; van de Hulst 1958; Rougoor 1964). Radio observations and studies of the interstellar Ca, H, and K absorption lines (see Livingston and Lynds 1964) show that interstellar gas, assumed to be chiefly atomic hydrogen, exists largely in dense clouds of about 10 atoms/cm^3, with dimensions of about 10 pc, occupying 10 per cent of interstellar space and defining the spiral arms. The root-mean-square cloud velocity is about 12 km/sec. The sizes of the individual clouds suggest, but do not define, a characteristic scale for variations in the over-all galactic magnetic field. The main concern so far as cosmic rays are concerned is the mean gas density of about 1 atom/cm^3.

A rather high density of molecular hydrogen has been suspected for many years. Recently Gould and Salpeter (1963) and Gould, Gold, and Salpeter (1963) have taken up the question, pointing out that the earlier analyses of the observed spatial distribution of K giants away from the galactic plane suggest a much stronger gravitational field than can be accounted for by stars and atomic hydrogen. They attribute the stronger field to the presence of molecular hydrogen with a mean density of about three times that of atomic hydrogen in the region within 100 pc of the plane of the Galaxy. If they are correct, the total interstellar gas density should be estimated at about 5 hydrogen atoms/cm^3 to a distance of 100 pc on either side of the galactic plane.

3.3. Theoretical Considerations

Some simple theoretical dynamical considerations may be used to supplement the meager observational information on the interstellar field and gas. These are based on

[5] The recent observation of gas clouds far above the galactic plane (Münch and Zirin 1961; Muller *et al.* 1963) is extremely interesting, but it is not evident yet that the observations imply a revision to the 100-pc scale.

the assumption that the interstellar field, gas, and cosmic rays form a system in dynamical equilibrium with the same gravitational field as the stars. The gravitational field is known from studies of the motion and distribution of the various classes of stars perpendicular to the plane of the Galaxy (Gould, Gold, and Salpeter 1963; Gold and Salpeter 1963). The interstellar magnetic field and cosmic-ray gas would expand outward and be lost immediately to the Galaxy were it not for the weight of the interstellar gas which holds the field down in the disk and nucleus of the Galaxy. This requirement places an upper limit on the interstellar magnetic field and an associated lower limit on the interstellar gas density. Since the interstellar gas density figures directly in calculating the cosmic-ray life and the rate of generation of cosmic rays in the Galaxy, the lower limit on the gas density provides a lower limit on the rate of cosmic-ray production in the Galaxy. All these effects must be kept in mind when making estimates of any one of the related quantities. The numbers, as presently estimated, define the situation rather tightly; little leeway can be allowed in the field and gas densities if we are to avoid calculated rates for cosmic-ray production far in excess of the apparent capabilities of the richest possible sources—novae, supernovae, the galactic nucleus—for production of energetic particles.

It is readily seen that the interstellar gas must be fully interwoven with the magnetic field and that the gas density cannot be too low if the gas is to perform its confining function properly. Suppose that the mean interstellar density is 1 hydrogen atom/cm^3, as indicated by 21-cm observations. The 10 km/sec motions of the gas, the energy density of a 5×10^{-6}-gauss interstellar magnetic field, and the cosmic-ray gas each contribute about 10^{-12} ergs/cm^3, giving a total of not less than 3×10^{-12} ergs/cm^3, or 2×10^{12} ergs/gm. This energy corresponds approximately to an equivalent velocity of 20 km/sec, comparable to the random velocity of the stars in the disk. Since the stars and the interstellar gas move in the same total gravitational field, it would be expected that each should extend with the same scale height out of the plane of the Galaxy. In fact, the gas has a characteristic extension of about 100 pc each way from the plane of the Galaxy (Schmidt 1957),[6] whereas only the youngest stars are so closely confined. The stars generally extend two or three times as far from the plane.

The numbers in this calculation are crude but conservative. First, if the interstellar gas is to be confined by gravity to the restricted domain observed, it must be rather denser than the 1 atom/cm^3 derived from the interstellar line intensities. A gas density of 5 atoms/cm^3, suggested by similar dynamical considerations of the K giants (Gould *et al.* 1963), would seem to be adequate. We are not free to go to much higher values because, as shown below, an average interstellar gas density of 5 atoms/cm^3 puts the calculated cosmic-ray life well below 10^6 years, requiring a cosmic-ray generation rate of 10^{41} ergs/sec in the Galaxy. A much higher interstellar density would make each of these restrictive conditions even more severe.

Second, the large-scale galactic field cannot be much stronger than about 5×10^{-6} gauss without requiring a prohibitive interstellar density, higher than 5 atoms/cm^3, to

[6] Again it is not clear to what extent the recent observations of gas clouds far above the galactic disk (Münch and Zirin 1961; Muller *et al.* 1963) may alter the older estimate of 100 pc.

hold the field down.[7] The galactic magnetic field can hardly be less than 5×10^{-6} gauss or it could not effectively confine the cosmic rays to the Galaxy. The mean value of $B^2/8\pi$ must necessarily equal or exceed the cosmic-ray pressure if the field is to confine the cosmic rays in a quasisteady manner. It does not seem possible to confine cosmic rays with a very weak field using the weight of some hypothetical interstellar gas to hold down the cosmic rays. Such configurations as a heavy fluid (the interstellar gas) supported from below by the pressure of a very light fluid (the cosmic rays) are completely unstable, being subject to the well-known Taylor instability (Biermann and Davis 1958, 1960; Chandrasekhar 1961).

Altogether, then, the observations and general dynamical considerations suggest that the average interstellar gas density in the galactic disk is perhaps 5 atoms/cm^3 and the galactic magnetic field is about 0.5×10^{-5} gauss. To assume otherwise leads immediately to complications.

3.4. Nonthermal Radio Emission

Shklovsky (1960*b*) has suggested that the background nonthermal radio emission from the Galaxy is magnetic bremsstrahlung (synchrotron emission) from the observed cosmic-ray electrons. A differential electron-energy spectrum E^{-n} predicts a nonthermal radio-emission spectrum of the form $\nu^{-\alpha}$ with $\alpha = \frac{1}{2}(n - 1)$. The observed $n \simeq 2.5$–3 gives $\nu \simeq 0.75$–1.0 in disagreement with radio observations, which indicate $\alpha \lesssim 0.5$. Recent calculations (Shklovsky 1960*b;* Hayakawa and Okuda 1962; Lund *et al.* 1963; Ginzburg and Syrovatskii 1964; Scanlon and Milford 1965; Pollack and Fazio 1965) indicate that the relativistic electron density observed (Earl 1961; Meyer and Vogt 1961) at the Solar System, or calculated by considering the effect of collision processes in the Galaxy (see Fig. 2), and extending for a distance of 10^4 pc (comparable to the radius of the galactic disk or halo) through a field of 1×10^{-5} gauss (which is the apparent upper limit consistent with the observations and the virial theorem) can give a radio brightness of several degrees Kelvin at 180 Mc/sec. Such brightness is comparable to the radio brightness usually attributed to the galactic halo, but is a factor of about ten less than the observed brightness of the disk to which the calculation was intended to apply.[8] The calculated and observed radio emission are shown in Figure 3, based on calculations by Ginzburg and Syrovatskii (1964) and the summary of the observations by Walsh, Haddock, and Schulte (1964). The calculated emission can be increased by assuming stronger fields, but the emission increases only about as fast as the field energy density; so the necessary increase seems unreasonable in view of the earlier discussion on the magnetic-field strength in the Galaxy. In the same way, it does not seem reasonable to suppose that the actual relativistic electron density in interstellar space is ten times

[7] An alternative assumption allowing for a much stronger field would be that the galactic field is approximately force-free throughout the galactic disk and is held down at the galactic nucleus. But there would then be an increase of field toward the galactic nucleus at least as fast as $1/r^2$ in order to contain the stresses in the field out in the disk (Lüst and Schlüter 1954; Chandrasekhar 1956). It is not clear how a $1/r^2$ increase toward the nucleus can be reconciled with the observed galactic structure and nonthermal radio emission.

[8] The calculated nonthermal emission from the galactic halo is considerably smaller because of the reduced cosmic-ray and magnetic-field densities assumed there.

higher than that observed in the Solar System, unless, of course, we suppose that the cosmic rays of the Solar System are simply not typical of the Galaxy. This suggests that the background nonthermal radio emission is not due to magnetic bremsstrahlung from cosmic-ray electrons. The emission may be due to a superposition of supernovae remnants. There is no way of knowing at the moment. Whether the radio telescope can, or

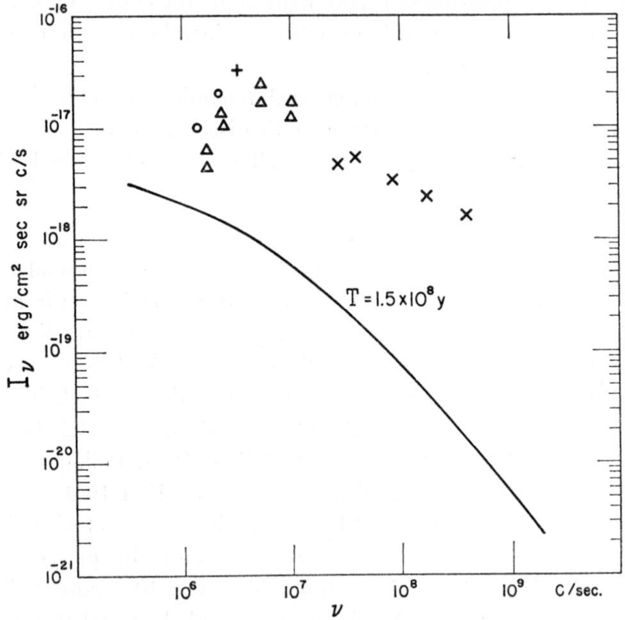

FIG. 3.—The average cosmic radio-noise frequency spectrum summarized by Walsh *et al.* (1964) compared with the calculated emission spectrum from the electrons produced by collision process (Ginzburg and Syrovatskii 1964). The crosses represent the observational points from the Mullard Radio Observatory (Baldwin 1955*b;* Pauliny-Toth and Shakeshaft 1962; Turtle *et al.* 1962; Turtle and Baldwin 1962). The triangles are from Ellis *et al.* (1962). The plus sign represents Chapman and Molozzi (1961), and the circles are from Haddock *et al.* (1963) and Walsh *et al.* (1964).

cannot, trace out the distribution of cosmic rays through the Galaxy depends on the answer to this question. It may be that the radio telescope "sees" only a superposition of unresolved localized "hot spots" (see further discussion in Sec. 4.1).

4. THE EXTENSION OF COSMIC RAYS THROUGH SPACE AND TIME

4.1. OBSERVATIONAL EVIDENCE

Observations suggest that the cosmic-ray gas extends throughout the Galaxy in a quasisteady state. Consider first the information available from studies of the unstable isotopes produced in the terrestrial atmosphere and in meteorites by cosmic-ray bombardment (Libby 1952; Singer 1953; see review by Zahringer 1964). Let the atmosphere, or meteorite, be exposed to a cosmic-ray intensity $f(t)$, producing such an isotope with decay time τ at a rate $\alpha f(t)$ where α is a constant (Bauer 1947; Huntley 1948). The abundance $F(t)$ of the isotope at a later time t is then

$$F(t) = \alpha \int_{-\infty}^{t} f(s) \exp - \left(\frac{t - s}{\tau} \right) ds .$$

Thus the present abundance, which can be measured, is a weighted average of the cosmic-ray intensity at earlier epochs. The weighting factor declines exponentially into the past with a characteristic time equal to τ. If $f(t)$ were essentially constant over the period extending from the present backward for a time τ, then

$$F(t) \cong \alpha \tau f(t)$$

regardless of earlier variations in $f(t)$. Thus, in principle, the observed abundance $F(t)$ of any isotope, with a life τ, permits one to decide whether the mean value for the cosmic-ray intensity over the last τ years is the same or different from the present cosmic-ray intensity.

Studies of short-lived isotopes such as C^{14} show that the average cosmic-ray intensity has been constant to within at least a few per cent during the last 5000 years (Libby 1952). Longer-lived isotopes in meteorites give the mean cosmic-ray intensity over longer periods of time—10^7 years (Honda and Arnold 1960; Fireman and DeFelice 1960; Arnold, Honda, and Lal 1961) from and 10^8 to 10^9 years using K^{40} (Voshage and Hintenberger 1959; Honda 1959). Although simple in principle, the meteorite determinations are complicated by a number of factors that have only recently been brought sufficiently under control for confidence in the results to develop (Geiss, Oleschger, and Schwarz 1962; Geiss 1963; Voshage 1962; Voshage and Hintenberger 1963; Zahringer 1964). It is possible to state at the present time that the average cosmic-ray intensity during the last 10^7 or even the last 10^9 years was within a factor of 1.5 of the present cosmic-ray intensity (Anders 1965; Lipschutz, Signer, and Anders 1965).

Unfortunately the assertion that the cosmic-ray intensity has averaged close to its present value for 10^4 years, 10^7 years, and 10^9 years is not equivalent to the statement that there are no Fourier components with periods of 10^4 years, 10^7 years, and 10^9 years. For instance, recent large fluctuations with a characteristic period of, say 10^5 years are not ruled out. Such fluctuations, with the restriction that the cosmic-ray intensity is now close to the averages taken over the last 10^7 and 10^9 years, could have occurred. Special possibilities of this nature are interesting, but failure to find any evidence of variation of the cosmic-ray intensity in the meteorite studies suggests they are less probable than a simple quasisteady state. This is the presently accepted view.

Let us now consider the distribution of the cosmic-ray gas throughout the Galaxy. If the flux of cosmic rays in the neighborhood of the Solar System has been approximately constant for the last 10^9 years, any assumption other than a uniform distribution of cosmic rays throughout the Galaxy seems unreasonable. The Galaxy has revolved four or five times during the period of 10^9 years, and the Sun probably has drifted a distance of 10^4 pc through the surrounding stars; there is no evidence of it ever having encountered a cosmic-ray intensity appreciably different from the present intensity. The generally accepted view is that the cosmic rays, at the present recorded intensities, are a permanent feature throughout the Galaxy.

Consider again the important idea introduced by Ginzburg (1953) and Shklovsky (1953*a*, *b*; Ginzburg and Syrovatskii 1964), based on suggestions by Alfvén and Herlofson (1950) and Kiepenheuer (1950), that the nonthermal radio emission from radio stars and from the Galaxy is produced by relativistic electrons (see discussion in Shklovsky 1960*b*; see also Ginzburg 1953, 1958; Ginzburg and Syrovatskii 1964). This idea is now generally accepted (with the reservations discussed in the previous section).

The total galactic nonthermal radio emission is about 2×10^{38} ergs/sec. The radio telescope sees a number of supernova remnants, such as the well-known Crab Nebula (age, 10^3 years), the Cygnus Loop (age, 10^4 years) (see discussion and description in Shklovsky 1960*b*), and the galactic spur that was recently tentatively identified as a nearby supernova remnant (Davies 1964) with a diameter of 40 pc at a distance of only 30 pc (age, 10^5 years). In addition to the supernova remnants, the radio telescope sees (at meter wavelengths) the general nonthermal brightness of the galactic disk; the effective brightness temperatures range from 10 to 10^3 ° K (depending upon the wavelength used; see discussions by Mills 1959; Shklovsky 1960*b*; Wilson 1963; Ginzburg and Syrovatskii 1964; Walsh, Haddock, and Schulte 1964) and a thickness of about 500 pc. Also a faint nonthermal brightness (of a few degrees Kelvin) extends over the whole sky at radio wavelengths. This brightness has been interpreted (Shklovsky 1953*a*, *b*, 1960*a*, *b*) as coming from a nearly spherical region of about 2×10^4 pc radius centered on the Galaxy and referred to as the galactic halo (Pikelner 1953). A similar galactic halo was identified tentatively around M31 (Baldwin 1954, 1955*a*, *b*, 1962), a galaxy similar to our own. The spectral distribution of the radio emission is similar to that expected as synchrotron radiation from the observed cosmic-ray energy spectrum. The nonthermal radio emission from the disk and halo is generally interpreted as produced by galactic cosmic rays (Kiepenheuer 1950; Shklovsky 1960*b*; Ginzburg and Syrovatskii 1964). Tempting as this is, however, some serious difficulties appear. As noted in the previous section, neither the cosmic-ray electron flux observed at the Solar System nor the expected secondary electron flux is adequate within a factor of ten to account for the observed general nonthermal radio emission. In addition, other galaxies do not seem to have halos of nonthermal radio emission similar to that postulated for our own (Mathewson and Rome 1963*a*, *b*; Burbidge and Hoyle 1963); the identification of a radio halo around M31 should be regarded as tentative. There is some question now about how much of the radio brightness (about 4 ° K at 178 Mc/s) attributed to the halo of our Galaxy may be extragalactic (Baldwin 1963). The degree of general polarization to be expected in the nonthermal emission, if this is due to galactic cosmic rays, has not been observed although polarization in limited regions has indeed been found (Wielebinski and Shakeshaft 1964). Altogether these considerations make it likely that the background of nonthermal radio emission is synchrotron emission from relativistic electrons, but the emission must be from something more than the cosmic-ray electrons, either in the disk or in the halo. Thus the origin of the nonthermal radio emission is not understood and unfortunately in recent years its observational study has been neglected. Recently Davies (1963) has pointed out that if supernovae remnants are as common throughout the Galaxy as his own studies suggest them to be in the neighborhood of the Sun, they might account for the entire nonthermal radio emission from the galactic disk. This would need an extraordinarily large number of supernovae in the Galaxy, about one every ten years; so the suggestion must be viewed with caution.

Similar questions concerning the origin of the nonthermal radiation from radio galaxies must be considered in any comprehensive discussion of cosmic-ray phenomena in the Universe. The cosmic-ray gas may be dense enough to account for much of the nonthermal emission and perhaps to permit radio mapping of the cosmic-ray extension throughout a radio galaxy, as suggested by Shklovsky and Ginzburg for our own Galaxy. If the radio galaxies represent a stage of evolution of a normal galaxy (Aizu *et al.* 1965, both references), their study may not be unrelated to the understanding of galactic cosmic rays. (For a description and discussion of radio galaxies and their relativistic electron distribution, see Shklovsky [1960*a*, *b*, 1962, 1963], Ginzburg and Syrovatskii [1964], Moffet and Maltby [1961], Maltby and Moffet [1962], Lequeux [1962], Hoyle and Fowler [1963], Maltby, Matthews, and Moffet [1963], Swarup, Thompson, and Bracewell [1963], Zakharenkov *et al.* [1963], Morris and Radhakrishnan [1963], Burbidge, Burbidge, and Sandage [1963], Moffet [1964], Schmidt and Matthews [1964], Sandage [1964], Ryle and Sandage [1964], Greenstein and Schmidt [1964], Matthews, Morgan, and Schmidt [1964], Burbidge, Burbidge, and Rubin [1964], Kellermann [1964], Hogg [1964], Burbidge and Burbidge [1964].)

4.2. Theoretical Considerations

From the observations, the simplest assumption about the space-time distribution of cosmic rays is that they fill the disk of the Galaxy in a more or less steady manner. This is the generally accepted view; in Section 4.2.4 possible alternatives are mentioned. There are a number of different opinions on how far cosmic rays extend beyond the Galaxy. We shall examine the various ideas and choose as a tentative hypothesis the one that involves the least drastic assumptions.

First, the remarkable isotropy of the cosmic rays suggests that they fill some region larger than the galactic disk, presumably the galactic halo (Davis 1954; Biermann and Davis 1958, 1960). This puts a lower limit on the cosmic-ray extent, and is discussed at length in the next section. The present problem is to establish an upper limit on the distance to which cosmic rays extend outward.

Second, some of the very highest energy cosmic-ray particles (10^{20} ev) may be of extragalactic origin. Even if such particles have the improbable charge of $Z = 100$, their radii of gyration are about 10^2 pc, and about 10^4 pc if they are protons. If such particles pass rather freely in and out of the Galaxy, perhaps originating elsewhere in the Universe, then their density for at least some distance outside the Galaxy is comparable to their density inside. In such an event, there is no way of deciding how far through the Universe the very high energy cosmic-ray particles extend. They are sufficiently rare, however, and possess so little (10^{-5}) of the total cosmic-ray energy that little can be inferred from them concerning the bulk of the cosmic-ray gas (from 10^8 to 10^{10} ev). We proceed, then, to other arguments.

Present ideas on the quasisteady extension of cosmic rays essentially reduce to three types: (1) cosmic rays are confined to the Galaxy and the galactic halo; (2) cosmic rays fill the Local Group of galaxies but do not extend to intergalactic space outside the Local

Group, or to others; (3) cosmic rays fill the entire Universe to the same level as they fill the Galaxy. In addition to these quasisteady suggestions, some theories suggest that cosmic rays, as now observed, are a transient phenomenon. A brief statement of each, setting out the more evident requirements and restrictions, is made here.

If cosmic rays *are* a transient phenomenon, then many current interpretations of their observed properties are open to question, and there is little basis for constructive argument. The ideas about transience are, therefore, noted but not pursued. A discussion of the quasisteady ideas centered on what now seems to be the most conservative line of thought follows.

4.2.1. *Galactic Theory.*—Perhaps the most widely held hypothesis on cosmic rays is that they are generated in the Galaxy and are confined to the Galaxy and its halo by the galactic magnetic fields. The cosmic rays presumably originate in novae, supernovae, the galactic nucleus, and so on. The necessary production rate is usually estimated from the observed energy density of 10^{-12} ergs/cm^3 and from the limit on the cosmic-ray life imposed by the 3 to 7 gm/cm^2 penetration of the helium and heavier nuclei through matter. It is usually assumed that the cosmic-ray protons have the same history as the heavier nuclei; otherwise, a number of special assumptions, such as distinct places of origin but similar energy spectra from 10^8 to 10^{11} ev/nucleon, are needed. If the average interstellar density is N hydrogen atoms/cm^3, a penetration of 5 gm/cm^2 implies a path-length of $10^6\ N^{-1}$ pc. With $N = 5$ as typical in the galactic disk (Sec. 3.3), the path-length in the disk is 2×10^5 pc, giving a life of 6×10^5 years while in the disk and requiring a production rate of 5×10^{-26} ergs/cm^3 sec. Disk dimensions of 10^4-pc radius and 200-pc thickness give 1.6×10^{66} cm^3, requiring a total cosmic-ray generation rate of 0.8×10^{41} ergs/sec in the Galaxy. Individual cosmic-ray particles eventually leak out of the Galaxy, so that the cosmic-ray gas density in intergalactic space is not identically zero, but, in view of the expansion of the Universe, the density in intergalactic space is at least three factors of ten lower than the average inside the Galaxy.

The observed cosmic-ray isotropy is presumably the result of free circulation of the individual cosmic-ray particles round and round through the galactic halo. Circulation through the halo does not affect the calculated cosmic-ray production rate because the gas density in the halo is presumably so low that little or no contribution to the 5 gm/cm^2 penetration occurs. The galactic halo gas, fields, and cosmic rays are presumably contained by the gravitational potential energy of the Galaxy. As mentioned earlier, the very highest energy particles (10^{20} ev), if they have no more total charge than a proton, probably cannot be confined to the Galaxy at all. An extensive discussion of cosmic rays confined to the Galaxy and the galactic halo is given by Biermann and Davis (1958, 1960), and Ginzburg and Syrovatskii (1964).

4.2.2. *Local Group Theory.*—Sciama (1962, 1964) has proposed that cosmic rays extend throughout the Local Group of galaxies, being confined there by intergalactic fields extending throughout the Local Group. With this view (as with any theory involving confinement) the cosmic-ray gas in intergalactic space outside the region of confinement consists of those cosmic-ray particles that have leaked from the region of confinement. Consequently, the density of the cosmic-ray gas outside the Local Group is presumed to be very low. Sciama suggests that the large-scale magnetic fields in the Local Group are perhaps 10^{-7} gauss between galaxies, rising to 10^{-5} gauss in each individual galaxy.

The cosmic-ray particles are generated in each galaxy and move freely around the Local Group along the large-scale field, and it is assumed that the magnetic moment of the individual particles is preserved during the circulation. Thus, upon leaving a galaxy, where the field is 10^{-5} gauss and the distribution is isotropic, the particles all have very small pitch angles in intergalactic space, where the field is weak. The cosmic-ray density between galaxies of the Local Group is reduced by the same factor as the solid angle occupied by the pitch angles. In this way the cosmic-ray energy density outside the local galaxies becomes less than the gravitational energy density (about 3×10^{-14} ergs/cm^3; Kahn and Woltjer 1959) of the Local Group. The assumption that there is no scattering of particles in intergalactic space implies that the particles freely enter the next galaxy they encounter without reflection from its stronger fields.

On the basis of Sciama's idea, the necessary cosmic-ray production rate is about the same as for galactic confinement, about 10^{41} ergs/sec (3×10^{-26} ergs/cm^3 sec). Intergalactic space in the Local Group represents a group cosmic-ray halo, much as a halo around our own Galaxy is postulated. The requirements of isotropy and cosmic-ray life are met as well as in galactic confinement.

4.2.3. *Universal Theory.*—Gold and Hoyle (1959) proposed that cosmic rays extend throughout the Universe with about the same intensity, energy spectrum, and so forth as presently observed at the Solar System. The idea has the appealing feature that no distinction need be made between the very high-energy particles, which are not easily contained in the Galaxy, and the average cosmic-ray particles, and it also satisfies in one sweep the conditions of isotropy and uniformity in time and space throughout the Galaxy.

In this theory, cosmic rays are an intrinsic property of space (Burbidge, Burbidge, and Hoyle 1963) and/or generated in such violent phenomena as radio galaxies, and quasistellar radio sources (Hoyle 1964). The necessary rate of cosmic-ray generation is easily estimated from the rate of expansion of the Universe, which has a characteristic time of about 10^{10} years. The input necessary to maintain the cosmic-ray energy density of 10^{-12} ergs/cm^3 is then 3×10^{-30} ergs/cm^3 sec. There are approximately 10^{74} cm^3 per large galaxy, requiring 3×10^{44} ergs/sec per large galaxy. This is to be compared with the total luminosity of a large galaxy, typically 3×10^{44} ergs/sec, most of which is electromagnetic radiation in the 1-volt region.

Felten and Morrison (1963) have pointed out that there is a restriction placed on the universal theory of cosmic rays by γ-ray observations, viz., that cosmic rays in intergalactic space must be free of the electron component observed in the Galaxy, for otherwise the scattering of starlight from the electrons would give too many inverse Compton X-rays and γ-rays (Feenberg and Primakoff 1948). Present observational limits on the γ-ray flux suggest that the intergalactic electron flux cannot be more than a few per cent of the electron flux observed in the Galaxy. Thus, the electron component would be of galactic origin, and the extragalactic component would be essentially free of electrons. Other aspects of universal cosmic rays are discussed in the literature (see, for example, Ginzburg and Syrovatskii 1963; Sciama 1964).

The composition of cosmic rays tells something about conditions at their place of origin. The large cosmic-ray abundance of heavy nuclei implies that if cosmic rays in the universal theory are from radio galaxies or quasistellar sources, an outburst resulting in a radio galaxy or quasistellar source comes from highly evolved material; this argument has important implications for the theory of nuclear synthesis in massive objects. The necessary lack (Felten and Morrison 1963) of relativistic electrons puts some restrictions on the nature of the outburst. On the other hand, if universal cosmic rays are simply a fundamental property of space, it would imply that matter is created in an evolved state, with an abundance of heavy nuclei far exceeding the cosmic abundance as presently determined.

4.2.4. *Transience Theories.*—At the other extreme from the universal theories are the ideas in which cosmic rays are of a transient nature within the Galaxy. For instance, Burbidge (1956) and Biermann and Davis (1958) have considered the possibility that the present cosmic rays are a remnant from an early active phase of the Galaxy, ranging from 0.5 to 1×10^{10} years ago. The short life for the heavy cosmic-ray nuclei in the disk of the Galaxy would now seem to rule out this possibility.

The recent suggestion by Burbidge and Hoyle (1963) that the nucleus of the Galaxy exploded at some time in the last 10^7 years, perhaps producing vast quantities of cosmic rays, does not suffer the same objection. But the possibility that cosmic rays were produced in the explosion is considered by Burbidge and Hoyle as a secondary effect, the primary effect of the explosion being to create the extended gases and fields making up the galactic halo. Burbidge and Hoyle point out that, in their picture, the cosmic rays produced in the explosion would be enormously decelerated in the expanding and newly created halo. They seem to favor the idea of universal cosmic rays. In Section 5.2 we present a different view of the halo, in which the possibility of a large cosmic-ray contribution from a hypothetical explosion of the galactic nucleus cannot easily be ruled out.

Finally, in the most transient view of all—to be blamed entirely on me—suppose that a hypothetical supernova appearing at a distance of 30 pc about 1×10^5 years ago (as suggested by Davies [1964] as the origin of the galactic radio spur) contributed the major portion of today's cosmic rays. In what position would the inquiry into the origin of cosmic rays be? The 10^{52} ergs of the supernova might easily fill a volume 300 pc on a side to the presently observed cosmic-ray energy density. It is not obvious that the transient cosmic rays would violate the meteorite evidence on the *long-term average* cosmic-ray intensity. There might be some minor difficulty in understanding the present isotropy since the average drift of the present cosmic rays would be about 30 pc in 10^5 years. But this gives an anisotropy $\delta = 0.6 \times 10^{-2}$ and this value is not so large that it could not be explained by some special assumptions. The cosmic-ray age deduced from Li, Be, B abundance is merely a measure of the time since the supernova explosion. The supernova is now too faint and too old for its past to be explored with the aid of present radio telescopes. None of us will live to see the next such nearby event. The conclusion is that, with present knowledge, the question of the origin of cosmic rays can be given no final answer.

This chapter offers no further comment on ideas that cosmic rays are a transient phenomenon in the Galaxy.

In order to make some appraisal of the three ideas on the quasisteady extension of cosmic rays through the Universe, we believe that, lacking observational evidence bearing directly on the question, the only satisfactory course is to pursue that idea which seems at the present time to involve the fewest and least drastic assumptions. Thus, the simplest idea must be considered a working hypothesis, not a final conclusion.

Consider, then, the energy requirements per large galaxy. The universal theory requires that cosmic rays be generated at an average rate of 10^{44} ergs galaxy^{-1} sec^{-1} throughout the Universe. The theories involving confinement require some 10^{41} ergs galaxy^{-1} sec^{-1}. The more conservative choice is that cosmic rays are largely confined. There is at present no observational or theoretical consideration which requires more.

The next question is whether cosmic rays are confined and circulate through the Galaxy or through the local cluster of galaxies. If cosmic rays are confined and circulate freely through the Local Group, it is necessary that they should not be scattered by irregularities in the fields between the galaxies. In the absence of scattering, the cosmic-ray pressure, or energy density, may be made smaller than the confining gravitational potential, and a quasiequilibrium imagined. Certain technical difficulties, however, must then be circumvented by special assumptions. For instance, the cosmic-ray pressure may be made arbitrarily small in the absence of scattering if the cosmic rays pass into a sufficiently weak field of strength B, as pointed out by Sciama. But the cosmic-ray pressure declines only in proportion to B (for small B), whereas the confining pressure of the field declines as B^2. The cosmic-ray and field pressures are comparable in the Galaxy; so the magnetic pressure must be much smaller than the cosmic-ray pressure outside the Galaxy. But the magnetic field is then not strong enough to contain the cosmic rays. We may imagine, of course, that the weight of the gas confines the field and the cosmic rays in spite of the relative weakness of the field; this is a highly unstable situation, as pointed out by Biermann and Davis (1958). Another difficulty is the instability which would result in the weak fields outside the Galaxy where the cosmic-ray particles are streaming with small pitch angle back and forth along the field. The well-known hose instability results, scattering the particles and thereby trapping them in the weak-field regions. Quantitative treatment of the streaming of cosmic rays along varying fields is given in the next section, where it is worked out for other purposes. The point made here is that confinement of the cosmic rays to the Galaxy and the galactic halo seems to be the simplest choice. Galactic confinement avoids the high energy requirements of the universal theory and the confinement difficulties in the Local Group. Galactic confinement represents the minimum cosmic-ray volume consistent with the observed isotropy and quasisteady state. The choice made is to overlook the problem of the very highest energy cosmic-ray particles. The remainder of this chapter contains a critical review of the dynamical consequences of galactic confinement of cosmic rays.

5. COSMIC-RAY LIFE, PRODUCTION RATE, AND ISOTROPY

5.1. Production Rate

The life of a cosmic-ray particle in the galactic disk can be deduced from the He^3 and Li, Be, and B abundances with the assumption that cosmic rays are in a quasisteady state. The value deduced, 10^6 years, is rather surprisingly short in view of the high degree of isotropy of cosmic rays observed at the Solar System at the present time.

The observed relative abundances of light and heavy nuclei indicate that both helium and heavier cosmic-ray nuclei have passed through an average of from 3 to 7 gm/cm^2 of material before observation at Earth (Waddington 1960; Aizu *et al.* 1959, 1961; O'Dell, Shapiro, and Stiller 1962; Daniel and Durgaprasad 1962; Hildebrand *et al.* 1963; Appa

Rao *et al.* 1963). If the average number density of the material is represented by N, then the pathlength associated with a mean value of 5 gm/cm^2 is $10^6/N$ pc. This is the mean pathlength of a heavy cosmic-ray nucleus in the disk of the Galaxy. Cosmic-ray protons are assumed to have the same history as the heavier nuclei, otherwise, a number of special assumptions are needed. If then cosmic-ray protons are produced in the same sources as the heavy nuclei, $1 \times 10^6/N$ pc is the mean pathlength traversed by all cosmic rays. There is no basis for supposing that much of the matter is traversed at the cosmic-ray source; so N is generally taken to be equal to the mean interstellar density. Hence in the disk of the Galaxy, where $N \cong 5/\text{cm}^3$, the average pathlength for cosmic rays is 2×10^5 pc. The time for a relativistic particle to traverse this path is 6×10^5 years (2×10^{13} sec).

The volume of the galactic disk, as defined by the gas, is not less than about 2×10^{66} cm^3, corresponding to a thickness of 200 pc and a radius of 10^4 pc. Therefore, to maintain the observed cosmic-ray energy density of 10^{-12} ergs/cm^3 assumed to extend throughout the disk, cosmic rays must be generated at the rate of 10^{41} ergs/sec.

Cosmic rays may, of course, spend much of their time in a galactic halo, where the gas density is much less than in the disk. This does not affect the mean life in the disk or the cosmic-ray production rate deduced from it, unless the gas density in the halo exceeds about 10^{-2} times the density in the disk, that is, about 0.05 atoms/cm^3. If the halo density is less than 0.05/cm^3, it means merely that cosmic rays freely circulate around the halo without much chance of colliding with matter and fragmenting the heavy nuclei. The halo volume is estimated as 10^2 larger than the disk, so the cosmic rays would presumably spend only 10^{-2} of their time in the disk (see discussion in Ginzburg and Syrovatskii 1964), where their life is limited to 6×10^5 years. Their total life in the Galaxy (including the halo) is then 6×10^7 years.

5.2. Isotropy and the Galactic Halo

The high degree of cosmic-ray isotropy observed at the Solar System puts some serious restrictions on the structure of the galactic magnetic field, particularly when combined with the short cosmic-ray life in the disk of the Galaxy. The observed isotropy was one of the original reasons for suggesting that cosmic rays circulate through a galactic halo (Biermann and Davis 1958). The argument is even stronger now through the reduction of the calculated cosmic-ray life consistent with the upward revision of the estimated interstellar gas density.

The escape of cosmic rays from the disk of the Galaxy after only 10^6 years in the disk is usually ascribed to the open nature of the galactic field. Either the galactic arms are open at their ends or the cosmic rays pass freely back and forth along the magnetic lines of force from the disk of the Galaxy to a supposed galactic halo.

The radius of gyration of most cosmic-ray particles in a typical galactic field of 10^{-5} gauss is only about 10^{12} cm. This radius is so small compared to the scale of the expected gradients in the galactic magnetic field that the particle drift velocity u across the field is very small compared to the particle velocity ($\sim c$) along the field (Parker 1964*b*). For instance, a steep gradient in the interstellar field with a scale of 3 pc gives $u = 10^3$ cm/sec and $u/c \cong 3 \times 10^{-8}$. Thus the cosmic-ray particle is essentially confined to motion along a single line of force. But a line of force never ends, because $\nabla \cdot \boldsymbol{B} \cong 0$; so a cosmic-ray particle can escape from the Galaxy only if the lines of force escape. A galactic arm can be open at its end only if the lines of force continue directly into intergalactic space, that is, only if the galactic magnetic field is a constriction in a large-scale intergalactic field. But since the Galaxy has rotated about 50 times in the last 10^{10} years, it is difficult to understand how so complete and direct a connection between the galactic and intergalactic field could be maintained.

A more immediate problem arises if cosmic rays escape at the ends of the galactic arms. Cosmic-ray angular distribution has been studied to energies of at least 10^{16} ev/nucleon. If the anisotropy δ is defined as the difference between the maximum and minimum cosmic-ray intensities observed in their respective directions divided by the average intensity, then

$$\delta = \frac{I_{\max} - I_{\min}}{\frac{1}{2}(I_{\max} + I_{\min})}.$$

No detectable anisotropy has yet been established.[9] Observations show (Greisen 1956) that δ is from rather less than 10^{-2} to about 10^{16} ev. Now suppose that cosmic rays are produced at various points scattered over a distance L along a galactic arm and that the cosmic-ray particle lifetime in the arm is t. It follows that to travel a distance L in a time t the cosmic rays must have a streaming velocity $v \simeq L/t$. But it is readily shown (Compton and Getting 1935; Parker 1964*b*) that streaming with a velocity v yields an anisotropy $\delta \simeq 6v/c$. It follows that if $\delta < 10^{-2}$, then $L < ct/600$. The pathlength ct of the individual particle in the galactic arm is estimated at 2×10^5 pc, yielding $L < 300$ pc. This distance is very short. The distance $L < 300$ pc is comparable to the thickness of the Galaxy, rather than the length of a spiral arm, suggesting that cosmic rays are generated in the disk of the Galaxy and escape directly out of the edges of the disk. For, if they did not escape at the sides, the cosmic-ray gas would generally be streaming past the Solar System from distances very much in excess of 300 pc.

But now, if the cosmic-ray particles escape directly out of the edge of the galactic disk, a new difficulty appears. If it is supposed that the cosmic rays are generated in supernovae, then the cosmic rays observed at the Solar System must have come from supernovae within a few hundred parsecs of the Solar System. It is difficult to see how cosmic rays occasionally injected on to the lines of force of the galactic magnetic field at a few separate points in this small region could lead to the steady uniform cosmic-ray distribution customarily assumed.

Let us now consider a galactic halo. It is evident that free circulation of cosmic rays through the halo (halo volume is $10^2 \times$ that of the disk) can satisfy the requirement on anisotropy. The pathlength ct before escape is 10^2 times longer than that if the disk alone is considered; therefore, L can be 10^2 times larger, or $L \leq 3 \times 10^4$ pc. Thus cosmic rays produced anywhere in the Galaxy may contribute to the flux of cosmic rays observed at the Solar System. With the possibility of such broad mixing, it is easy to understand how the cosmic-ray intensity is more or less steady in time, as seems to be required by the meteorite studies and as is assumed in deducing the cosmic-ray life from the Li, Be, B abundance. Free circulation of the cosmic rays through a galactic halo (Biermann and Davis 1958, 1960) seems necessary to account for the isotropy and uniformity.

[9] Except for the diurnal effect, which is obviously of local origin in the Solar System.

Various dynamical pictures of the galactic halo have been based on gas temperatures of 10^6 ° K (Spitzer 1956), on 10^2 km/sec turbulent gas motions (Shklovsky 1952; Pikelner 1953, 1957; Pikelner and Shklovsky 1957, 1958), and on explosions in the galactic nucleus (Burbidge and Hoyle 1963). We shall have more to say on this in Section 7. In Section 6, the little that is known about the actual process of generation of cosmic rays in various stars, in interstellar space, and in other locations is reviewed before we proceed further with the arguments on the dynamical structure of the galactic halo.

6. COSMIC-RAY ORIGIN

6.1. General Considerations

Consider where and under what circumstances cosmic rays are generated in the Galaxy. It has become evident over the last 15 years that energetic charged particles (of both signs) are produced wherever and whenever there is sufficiently vigorous agitation of a tenuous ionized gas. On a laboratory scale, fast particles are produced in the plasma pinch (Spitzer 1958; Burkhardt and Lovberg 1958; Honsaker *et al.* 1958; Smullin and Getty 1962; Alexeff *et al.* 1963). On a planetary scale, energetic particles are produced in the region of impact of the solar wind against the outer boundary of the geomagnetic field and in the geomagnetic field itself (Freeman, Van Allen, and Cahill 1963; Frank, Van Allen, and Macagno 1963; Fan, Gloeckler, and Simpson 1964). The aurora is a visible manifestation of some of these particles. There is evidence that energetic particles are produced in the violent turbulence in interplanetary space (Parker 1965*b*). On a stellar scale, energetic particles are produced in association with solar flares (Forbush, Stinchcomb, and Schein 1950; Meyer, Parker, and Simpson 1956; Reid and Leinbach 1959; Obayashi and Hakura 1960; Simpson 1960; Biswas *et al.* 1963) and in supernova ejecta (Shklovsky 1953*a*, *b*, 1960*b*; Ginzburg 1953; Baade 1956; Mayer, McCullough, and Slonaker 1957). On a galactic scale, fast particles are produced in extraordinarily high quantities in quasistellar radio sources and radio galaxies (Hazard, Mackey, and Shimmins 1963; Schmidt 1963; Moffet 1964; Ryle and Sandage 1964; Shklovsky 1960*b*). It is likely, therefore, that all cosmic rays have their origin in regions of violent agitation.

The mechanism, or mechanisms, by which the fast particles are accelerated in agitated plasmas in not known, although a number of ideas have been proposed and are discussed in the next subsection. Present knowledge indicates only that some mechanism successfully converts a significant portion of the plasma energy into the kinetic energy of a few fast particles.

The solar flare is the most prominent local plasma violence, with an average output of protons ($>10^2$ Mev) of the order of 10^{22} to 10^{23} ergs/sec. If, in this respect, the Sun is typical of the 10^{11} stars in the Galaxy, a total output of the order of 10^{33} to 10^{34} ergs/sec in the Galaxy would result. The production rate required to maintain the present cosmic-ray intensity is estimated to be about 10^{41} ergs/sec (Sec. 5); so there is no reason to believe that flare phenomena are of much importance unless the Sun is not typical.

More active stars than the Sun are much less common. Flare stars, T Tauri stars, Wolf-Rayet stars, emission-line stars, magnetic variables, giants and supergiants, stars with high rotation, very close binaries, and so on, are all possibilities for copious generation of fast particles, for much the same reasons that they are suspected of having stellar winds (Parker 1960, 1963). But estimates of their fast-particle production give no suggestion that they can contribute more than a small fraction of the necessary 10^{41} ergs/sec (see discussion in Ginzburg and Syrovatskii 1964).

Fermi suggested that the interstellar medium is in sufficiently violent motion to accelerate cosmic rays. The presently estimated short cosmic-ray life of only 6×10^5 years in the galactic disk makes this idea now appear of questionable value by reason of energy considerations. The necessary cosmic-ray production rate to maintain the 10^{-12} ergs/cm^3 is 5×10^{-26} ergs/cm^3 sec throughout the galactic disk, whereas estimates of the energy supply to the interstellar medium (Oort 1954; Kahn 1954; Oort and Spitzer 1955; Savedoff 1956; Parker 1958*a*) are mostly about 1×10^{-26} ergs/cm^3 sec or less. Other difficulties are mentioned in Section 6.2.

Novae and supernovae seem to be more effective generators of cosmic rays. The energy output of a single nova can be as high as 10^{45} or 10^{46} ergs; a nova frequency of about 10^2 per year in the Galaxy (Payne-Gaposchkin 1958; Shklovsky 1960*a*) gives a mean rate of energy release of from 3×10^{39} to 3×10^{40} ergs/sec. Supernovae are estimated to release total energies of the order of 10^{50} ergs (Minkowski 1959, 1964; Shklovsky 1960*a*, *b*). The frequency of occurrence of supernovae averages about one every 350 years over some 3000 galaxies that have been surveyed (Zwicky 1958, 1965; Minkowski 1964), but it is known that some galaxies are exceptions—NGC 3184, 4303, 4321, 5236, and 6946 having produced three each since the year 1903. The ancient records indicate supernovae in our own Galaxy visible to the naked eye in the years A.D. 185, 369, 1006, and 1054 in addition to the two well-known supernovae seen in the years 1572 and 1604. The remnants of all but the supernova of 1006 have now been identified with radio telescopes. On this basis Shklovsky (1960*b*) points out that the rate in our Galaxy must be something like one every 50 to 100 years, since only those supernovae properly placed can be seen. Presumably, then, our own Galaxy has rather more supernovae than most, and the energy output by supernovae averages from about 10^{40} to 10^{41} ergs/sec in our Galaxy (Ginzburg and Syrovatskii, 1964). Finally, the hypothetical explosion of the core of the galactic nucleus may involve 3×10^{56} ergs (Burbidge 1963; Burbidge and Hoyle 1963). One such explosion every 5×10^7 years leads to a mean output of 2×10^{41} ergs/sec.[10] Altogether, then, it would seem that there may be enough *total* available energy in stellar and galactic explosions to supply the necessary 10^{41} ergs/sec for cosmic-ray production.

The most important question is whether a large portion of the total available energy goes into fast particles. The solar flare of February 23, 1956, produced relativistic particles. Approximately 3×10^{30} ergs escaped into interplanetary space with such particles. The total flare energy was estimated at 3×10^{32} ergs; so the efficiency was approximately 10^{-2} with some considerable uncertainty (Parker 1957*a*). This does not exclude the possibility that novae, supernovae, and galactic nuclei explosions are more efficient, but it lends no certain support to that view. Presumably the question must be answered by observation.

[10] Cosmic-ray fluctuations with a period of 5×10^7 years might escape detection in studies of isotopes produced by cosmic rays in meteorites, because, so far, isotopes with half-lives of 10^7 years or less and the isotope K^{40} with a half-life of 10^9 years have been used; 5×10^7 years lies in between.

The anomalously high abundance of heavy nuclei in cosmic rays must not be overlooked. As noted earlier, this high abundance is a strong argument for a cosmic-ray origin in old stellar objects in which the matter has been well cooked (Hayakawa 1956; Burbidge *et al.* 1956).

At the present time, radio, optical, X-ray, γ-ray, and neutrino observations all show some promise for resolving the question of the origin of cosmic rays. One of the outstanding radio questions concerns the nonthermal radio emission from the disk and halo of the Galaxy. All nonthermal emission cannot be magnetic bremsstrahlung from the cosmic-ray electrons. It may include radiation from many discrete sources; such emission may come from many old nova and supernova remnants if these continue to radiate much longer than is presently estimated.[11] Is it possible to disentangle the sources of nonthermal emission, and thus perhaps directly detect some cosmic-ray electrons as well as obtain a better picture of discrete radio sources, some of which may be cosmic-ray sources?

Optical astronomy may assist in a number of ways, as with the question of the efficiency of cosmic-ray generation in supernovae. Ginzburg and Syrovatskii (1964) point out that if the magnetic bremsstrahlung contribution to the total optical emission from a supernova could be observed, perhaps through polarization studies, we would have a measure of the fast-particle production efficiency in terms of the total supernova energy.

X-ray astronomy is in a vigorous infancy at the present time (Giacconi *et al.* 1962; Gursky *et al.* 1963; Bowyer *et al.* 1964; Friedman 1964; see also Friedman's chapter in this volume). Many X-ray sources in the sky have already been observed. One X-ray source is in the supernova remnant represented by the Crab Nebula. It is too soon to say what contribution X-ray astronomy will make to the cosmic-ray origin problem, but the general association of X-rays with violently hot plasmas suggests that X-ray and cosmic-ray sources are not unrelated; X-ray emission may be magnetic bremsstrahlung and/or inverse Compton effect (Felten and Morrison 1963; Hayakawa and Matsuoka 1963).

The more difficult field of γ-ray astronomy also promises insight when a really successful observing technique can be worked out (Danielson 1960; Cline 1961; Kraushaar and Clark 1962*a*, *b*). It is hoped that secondary γ-rays produced by collisions of individual cosmic-ray particles with the nuclei of the ambient gas (Morrison 1958) and perhaps also those produced by inverse Compton effect (Kraushaar and Clark 1962; Arnold *et al.* 1962) can be directly observed.

Finally, high-energy neutrinos are produced in collisions of cosmic-ray particles with other nuclei; so, in principle, neutrino observations might disclose regions of intense fluxes of fast nucleons (Chiu 1963), as radio observations show regions of intense fluxes of relativistic electrons. Unfortunately, successful detection of the expected neutrino fluxes depends upon the existence of a resonance involving the still hypothetical intermediate boson (Bahcall and Frautschi 1964); so contributions from neutrino astronomy are at some distance in the future.

[11] Recall again Davies' (1964) conjecture, mentioned in Section 4.1, that the radio telescope sees a very large number of "supernova remnants."

More generally, the present studies of solar flares, and of the impact of the solar wind on the geomagnetic field, are of great interest because these phenomena are known to be sources of fast particles. Their proximity permits detailed study, which is impossible in the remote supernovae and other cosmic-ray sources. In the same way, fast-particle production in laboratory plasma may shed some light on the general mechanism, although we must in all cases be careful in extrapolating even the qualitative features of fast-particle production from one scale to another. Nature may have more than one acceleration mechanism for fast particles.

6.2. Theoretical Cosmic-Ray Acceleration Mechanisms

Three basic processes that have been considered as possible acceleration mechanisms in plasmas have been developed sufficiently to be worth mentioning. First, a few of the faster particles in a plasma are repeatedly reflected from the moving hydromagnetic waves present in the plasma (Fermi 1949, 1954). Second, the front of a shock wave running outward from an explosion through an atmosphere of decreasing density will increase its velocity of propagation without limit (Colgate and Johnson 1960; Colgate and White 1963). All material near the wave front achieves relativistic velocities. Third, some plasma waves may accelerate the particles that happen to pass by with the correct velocity to "see" the waves in resonance with the cyclotron frequency (Stix 1964). There may be other processes, but none has been demonstrated so far as we are aware.

6.2.1. *The Fermi Mechanism.*—This is basically a hydromagnetic process, occurring in violently agitated plasmas as a consequence of gas and field motions with characteristic frequencies well below the ion cyclotron frequency. Hence the electric field is the usual $\boldsymbol{E} = -\boldsymbol{v} \times \boldsymbol{B}/c$ where $\boldsymbol{v}$ is the plasma velocity and $\boldsymbol{B}$ the magnetic field. The non-relativitistic equation of motion of a particle with charge q, mass M, velocity $\boldsymbol{w}$ is

$$M \frac{d\boldsymbol{w}}{dt} = q \left(\boldsymbol{E} + \frac{\boldsymbol{w} \times \boldsymbol{B}}{c} \right).$$

Eliminating $\boldsymbol{E}$ and taking the scalar product with $\boldsymbol{w}$, we get the energy equation, which, upon permuting the triple product, can be written

$$\frac{d}{dt} \tfrac{1}{2} M w^2 = \boldsymbol{v} \cdot \boldsymbol{F} \tag{1}$$

where $\boldsymbol{F}$ is the Lorentz force $q\boldsymbol{w} \times \boldsymbol{B}/c$ exerted on the particle by the magnetic field. This equation shows that in hydromagnetic fields the energy of a fast particle changes *only* when the motion of the gas does work against the Lorentz force. Thus a particle may be accelerated as it "bumps" against moving fields or as it is adiabatically compressed or expanded in a changing field. Adiabatic compression has been termed "betatron acceleration" from the analogy with that laboratory device, although this appellation evidently sometimes obscures the simple nature of the effect and has led to serious computational errors (Riddiford and Butler 1952). Adiabatic compression is reversible when it occurs in its pure state; so by itself it is of little or no interest in cosmic-ray acceleration. Adiabatic compression and expansion certainly occurs in association with the bumping that a particle may do in a field of hydromagnetic waves; therefore it is closely associated with the actual operation of the Fermi mechanism. For this reason we prefer to call any acceleration of the type shown in equation (1) the Fermi mechanism.

In its simplest and original form, the Fermi mechanism is analogous to the trend toward equipartition of energy of a light particle of mass m and speed w, placed among a number of randomly moving massive elastic objects of very large mass $\mathfrak{M}$ and velocity v. It is well known that energy exchange between such different masses is very slow, and equipartition requires an enormous number of collisions. The trend toward equipartition exists because head-on collisions between the particle and the massive objects are slightly more probable and the energy transfer is slightly larger than in overtaking collisions, as a consequence of the slightly higher relative velocity in the head-on collision. The average fractional energy gain per collision is about $(v/w)^2$. For cosmic rays, w is of the order of c and the large "objects" with velocity v are irregularities in the magnetic field that move with the Alfvén velocity. Hence v/w is generally small compared to one, and $(v/w)^2$ much smaller.

Davis (1956) showed that if we do not restrict our attention to the average energy gain but instead consider the particle distribution, there are some particles which gain energy much faster than the average. Fan (1956) considered special wave forms in which the field grows exponentially with time between two approaching hydromagnetic waves. He found that the average energy gain is then first order, rather than second order, in v/w. Later it was shown that in the presence of sharp hydromagnetic waves, head-on collisions are much more probable than overtaking collisions, so that the average fractional energy gain is much larger, $O(v/w)$ (Parker 1958*b*). Even so, the acceleration rate in interstellar space seems to be unimportant. Observations of the interstellar gas do not show sufficient irregularity to accomplish the necessary high acceleration rate. Approximately, if the fractional energy gain $\Delta W/W$ per collision is α, then after n collisions the energy is

$$W(n) = W(0) \exp \alpha n ,$$

$W(0)$ being the initial energy. If the mean free path between collisions is λ, then in a total path s (measured along the cosmic-ray trajectory) there are $n = s/\lambda$ collisions and

$$\lambda = \frac{s\alpha}{\log_e[W(n)/W(0)]} . \tag{2}$$

Now if t is the mean cosmic-ray life in the acceleration region (which is usually identified with the galactic disk), it follows that $s = ct$. Optimistically speaking, $\alpha = v/c$ and $W(n) = 10\ W(0)$. Then $\lambda = 0.435\ vt$. With $v = 10$ km/sec and $t = 0.5 \times 10^6$ years, it follows that $\lambda \cong 2$ pc. There is no evidence for irregularity of sufficiently large amplitude on so small a scale. Even the latest high-resolution observations indicate scales of much more than 10 pc (Livingston and Lynds 1964). A more detailed discussion of this problem may be found in Morrison, Olbert, and Rossi (1954) and in Ginzburg and Syrovatskii (1964), although the reader may wish to revise some of the numerical values used there.

The Fermi mechanism under a variety of circumstances has been worked out in the literature (Parker 1955, 1957*a*, 1958*d*; Parker and Tidman 1958*a*, *b*; Davis 1958; Schatzman

1963; Wentzel 1963*b*, 1964; Davis and Jokipii 1964) in connection with its many possible applications, to the solar flare, aurorae, and so forth, in addition to cosmic rays.

6.2.2. *Hydrodynamic Mechanism.*—The hydrodynamic cosmic-ray acceleration mechanism proposed by Colgate is distinct from the other ideas in that the cosmic rays are accelerated to relativistic velocity as a fluid, rather than as individual particles. The explosion of the core of a supernova sends a shock wave running out through the star. The velocity of the shock increases as it propagates outward because of the declining density of the gas into which it is moving. The velocity of the material immediately behind the shock increases to c in the limit as the density goes to zero. A large amount of material may in this way achieve an energy of 10^9 ev/nucleon or more. In order to estimate how much material achieves relativistic velocities, it is necessary to make a quantitative calculation of the dynamics of the imploding-exploding star based on the known properties of nuclear matter, neutrino energy transfer, and so forth. Consequently, the study of this problem (Colgate and Johnson 1960; Colgate, Grasberger, and White 1962; Colgate and White 1963) is interesting from the point of view of stellar interiors, stellar evolution, supernova explosions, and supernova remnants, in addition to its importance for cosmic-ray production. The estimates of the cosmic-ray production from Type II supernovae are from 10^{48} ergs above 10^9 ev/nucleon for an explosion in which the total mass ejected from the star is 1 $\mathfrak{M}_\odot$ to 2×10^{51} ergs for an explosion in which the total mass ejection from the star is 8 $\mathfrak{M}_\odot$. An explosion in the Galaxy giving 10^{51} ergs of cosmic rays once each 100 years is sufficient to account for the observed cosmic-ray intensity.

The energy spectrum of the material reaching cosmic-ray energies is calculated to be not unlike the observed cosmic-ray energy spectrum. The composition of the material shows the necessary high abundance of heavy nuclei. Altogether, then, the idea of hydrodynamic acceleration looks promising in its present stage of theoretical and observational development. This mechanism may supply a major portion of the cosmic rays in the Galaxy.[12]

6.2.3. *Wave Acceleration.*—Finally, consider the idea that particles are accelerated when their velocity is such that they fall into resonance with plasma waves of one kind or another. This was proposed recently by Stix (1964) in connection with the production of fast electrons in laboratory plasmas. Stix suggested a resonance between the electron plasma wave, as seen by a fast electron, and the cyclotron frequency of the fast electron. Bernstein, Fredricks, and Scarf (1964; Scarf, Bernstein, and Fredricks 1965; Fredricks, Scarf, and Bernstein 1965) have taken this general idea and applied it to generation of the fast particles observed in the region of impact of the solar wind against the geomagnetic field (Freeman *et al.* 1963; Frank *et al.* 1963; Fan *et al.* 1964). They suggest there a resonance of fast electrons with ion acoustic waves.

The principle is simply demonstrated for acceleration of electrons in the geomagnetic field by resonance with electromagnetic waves in the whistler mode (Parker 1961).

[12] Even if hydrodynamic acceleration is the source of most cosmic rays, there may be other fast-particle production mechanisms. Hydrodynamic acceleration cannot be applied, so far as we know, to solar flares and the present Crab Nebula.

Consider an electron with charge -e, mass m, and speed w along a uniform magnetic field B in the z-direction. Suppose that transverse to the magnetic field there is an electric field $E(z, t)$ in the x-direction. It follows at once that the equations of motion are

$$\frac{d^2x}{dt^2} = -\left[\frac{e}{m}E(z, t) + \Omega\frac{dy}{dt}\right] \tag{3}$$

$$\frac{d^2y}{dt^2} = +\Omega\frac{dx}{dt} \tag{4}$$

$$\frac{d^2z}{dt^2} = 0 \tag{5}$$

where $\Omega = eB/mc$. The terms involving the magnetic field of the wave are small and of the second order; so they have been dropped. It is evident that the particle motion in the z-direction is unaffected to this order; so we may transform to the frame of reference moving in the z-direction with the speed w. In this frame, $dz/dt = 0$, $z = 0$. Integrate equation (4) and choose the origin of the coordinates so that dy/dt vanishes where $x = 0$. Then eliminate dy/dt from equation (3), obtaining

$$\frac{d^2x}{dt^2} + \Omega^2 x = -\frac{e}{m}E(0, t). \tag{6}$$

The particle motion produced by $E(0, t)$ is

$$\frac{dx}{dt} = -\frac{e}{m}\int_{-\infty}^{t} d\tau \cos\Omega(t-\tau)E(0, \tau). \tag{7}$$

Now suppose that $E(0, t)$ is of the form

$$E(0, t) = f(t)\cos\int_{-\infty}^{t} du\,\omega(u). \tag{8}$$

It is readily shown that as long as $f(t)$, $\omega(t)$ and $[\omega(t) - \Omega]/\Omega$ change by only a small fraction in each cyclotron period $2\pi/\Omega$, the velocity follows the amplitude of the wave and subsides to zero with the final passage of the exciting wave packet. If, however, the wave should pass across resonance, $\omega = \Omega$, a residual velocity remains after the passage of the wave. Suppose that $\omega(t) = \Omega(1 \pm 2\nu^2\Omega t)$ during the passage across resonance, where $\nu < 1$ and is a constant. After passage of the exciting wave there is left the residual motion

$$\frac{dx}{dt} = -\frac{ef(0)}{m}\int_{-\infty}^{+\infty} d\tau \cos\Omega(t-\tau)\cos\Omega(\tau \pm \nu^2\Omega^2\tau^2). \tag{9}$$

Then

$$\frac{dx}{dt} \cong c\,\frac{\pi^{1/2}f(0)}{2\nu B}\cos(\Omega t \pm \pi/4). \tag{10}$$

In this expression, $f(0)$ is the amplitude of the wave at the time of resonance. The slower the passage through resonance, the smaller is ν and the larger is the residual velocity.

In the calculations of Fredricks *et al.* (1965) it is estimated that ion waves with a potential variation of a few hundred volts in the transition region outside the magnetosphere are sufficient to produce 50-kev electrons. The amplitude and frequency of the ion wave are subject to direct measurement from space vehicles. Thus, of the three present ideas for particle acceleration, wave acceleration may be the only one subject to direct experimental check. Unfortunately, it is not evident how to extrapolate information on electron acceleration around the magnetosphere to ion acceleration in supernovae, and so on, which is presumably of greater relevance to cosmic rays.

Various other kinds of particle acceleration in plasma fields have been considered in the collisionless shock transition, but so little is known of the phenomenon that nothing definite can be said on their relative importance at the present time. In this connection the collisionless shock transition upstream in the solar wind from the magnetosphere, and at the head of interplanetary blast waves from the Sun, may afford the one situation directly accessible to probes. But we can only speculate on the relative importance of the various known acceleration mechanisms in supernovae and in active supernovae remnants, such as the Crab Nebula. It should be remembered that whatever the acceleration process, it must put a significant fraction of the supernova energy into cosmic rays. Otherwise we must look elsewhere for the origin of cosmic rays.

7. THE DYNAMICAL PROPERTIES OF COSMIC-RAY GAS

7.1. General Remarks

Up to this point the discussion has dealt with the usual astrophysical problems of cosmic rays, principally their observational properties, origin, and fate. The present section takes up the question of the dynamical effects of cosmic rays.

On a large scale, the cosmic rays constitute a very high-temperature gas with a pressure as large as any pressure in interstellar space (Parker 1958*l*). Thus it has been pointed out that the cosmic rays make a large contribution to the speed of sound and to the hydrostatic pressure supporting the interstellar medium against the gravitational field of the disk (see Sec. 3.3). More novel effects, however, are the propagation of interstellar disturbances and the formation of a galactic halo; these have been pointed out elsewhere (Parker 1965*c*, *d*). Approximating the cosmic-ray gas by a classical compressible fluid gives the essential features correctly.

There is some theoretical reason to believe that the cosmic-ray gas can be treated as a fluid to a useful degree of approximation for sufficiently low frequency disturbances with scales sufficiently large compared to the radius of gyration of the average cosmic-ray particle—about 10^{12} cm. This has been discussed elsewhere (Parker 1962, 1963) in connection with the approximate treatment of the large-scale motion of a collisionless plasma. The basis of the argument is simply that a collisionless gas, such as the cosmic-ray gas, behaves approximately as a fluid if its thermal motions remain nearly isotropic ($p_{\|} = p_{\perp}$) throughout the motion. Collisions accomplish the isotropy in a dense gas. In a collisionless gas, plasma instabilities and magnetic irregularities tend to maintain isotropy.

Consider first the instabilities. It has been shown (Harris 1961; Sagdeyev and Shafranov 1961; Noerdlinger 1963; Lerche 1965) that if a collisionless gas becomes anisotropic with the pressure perpendicular to the large-scale magnetic field exceeding the pressure parallel to the field by any finite amount, then instabilities (growing waves) occur which consume the isotropy. It has been shown that instability also arises when the parallel pressure exceeds the perpendicular pressure by $B^2/4\pi$ (Parker 1958*e;* Chandrasekhar, Kaufman, and Watson 1958). It has not been shown yet, so far as we are aware, that instability arises when the parallel pressure exceeds the perpendicular pressure by less than $B^2/4\pi$. But remember that small irregularities in the magnetic field tend to scatter the particles and give isotropy under all circumstances. If l is the distance between irregularities, the characteristic relaxation time for an anisotropy is l/c. Altogether, then, some approximation to isotropy is expected. It is interesting, in this connection, that the cosmic-ray gas is observed to be very nearly isotropic at the Solar System at the present time.

If it can be assumed that a collisionless gas remains essentially isotropic throughout a disturbance, it can be shown that the dynamical equations for the large-scale motion of the gas reduce to the familiar hydromagnetic equations. This is discussed in Appendix I.

With this simplified approach it is evident that the interstellar medium is a composite fluid composed of the ordinary interstellar gas, the cosmic-ray gas, and the interstellar magnetic field. The interstellar field contributes the usual magnetic stresses to the composite fluid and in addition ties the two gases together (since otherwise the two gases would move entirely independently of each other).

7.2. Hydromagnetic Waves

If we imagine that the interstellar magnetic field is disordered on a smaller scale than the over-all dimensions of a given disturbance, then for that disturbance the two gases move together as a single fluid and the speed of sound is isotropic with the value $(\gamma p/\rho)^{1/2}$. Here p is the total pressure of the composite fluid, including the mean value of $B^2/8\pi$, the gas pressure, and the cosmic-ray pressure. The density ρ is contributed by the interstellar gas. The relation between $dp/d\rho$ and p/ρ is determined by the proportions of $B^2/8\pi$ and the relativistic cosmic-ray gas (Synge 1957). It is sufficient for the present to approximate $dp/d\rho$ by p/ρ. Then, for instance, in a region where the interstellar gas density is 5 hydrogen atoms/cm^3 at a temperature of 10^2 ° K, the gas pressure is 0.07×10^{-12} dynes/cm^2. The speed of sound is 1 km/sec in the gas alone. The cosmic-ray pressure is 10^{-12} ergs/cm^3; so the actual speed of sound in the composite medium, if we neglect $B^2/8\pi$, is about 4 km/sec. Including $B^2/8\pi$ would raise the speed of sound still further. It is evident, therefore, that the cosmic-ray gas makes an important contribution to the speed of sound in the interstellar medium.

Suppose, on the other hand, that the interstellar magnetic field is of a larger scale than the scale λ of the disburbance; that is, small-scale irregularities are absent. Then, obviously, the two gases are tied together only over the two dimensions perpendicular to the large-scale magnetic field. Their motions parallel to the field are independent of each other.[13] It is a simple matter to implement this idealized picture by writing down the hydromagnetic equations for two fluids with a common uniform magnetic field. For instance, if $\boldsymbol{v}$, ρ, p, and $\boldsymbol{a}$ represent the fluid velocity, density, pressure, and speed of

[13] There is no evident way in which a two-stream instability might arise to tie the parallel motions together.

sound, with subscripts 1 and 2 to designate the two fluids, then the equations of motion perpendicular to the field are

$$(\boldsymbol{v}_1 - \boldsymbol{v}_2) \times \boldsymbol{B} = 0 , \tag{11}$$

$$\left[\rho_1 \frac{d\boldsymbol{v}_1}{dt} + \rho_2 \frac{d\boldsymbol{v}_2}{dt} + \nabla(p_1 + p_2) - \frac{(\nabla \times \boldsymbol{B}) \times \boldsymbol{B}}{4\pi} \right] \times \boldsymbol{B} = 0 . \tag{12}$$

Parallel to the field

$$\left(\rho_1 \frac{d\boldsymbol{v}_1}{dt} + \nabla p_1 \right) \cdot \boldsymbol{B} = \left(\rho_2 \frac{d\boldsymbol{v}_2}{dt} + \nabla p_1 \right) \cdot \boldsymbol{B} = 0 . \tag{13}$$

Conservation of fluid gives

$$\frac{\partial \rho_1}{\partial t} + \nabla \cdot (\rho_1 \boldsymbol{v}_1) = \frac{\partial \rho_2}{\partial t} + \nabla \cdot (\rho_2 \boldsymbol{v}_2) = 0 . \tag{14}$$

The hydromagnetic equation is

$$\frac{\partial \boldsymbol{B}}{\partial t} = \nabla \times (\boldsymbol{v}_1 \times \boldsymbol{B}) \tag{15}$$

where $\boldsymbol{B}$ is the magnetic field. Considering the simple case of a small perturbation $\delta \boldsymbol{v}$, $\delta \rho$, etc., on a uniform equilibrium state, the equations are easily linearized and the energy equation can be written simply as

$$\delta p_1 = a_1^2 \delta \rho_1 , \qquad \delta p_2 = a_2^2 \delta \rho_2 . \tag{16}$$

The calculations, which are straightforward, demonstrate a third mode of hydromagnetic wave propagation, the *suprathermal mode* (Parker 1965*d*), in addition to the familiar fast and slow hydromagnetic modes. The suprathermal mode proves to be essentially a sound wave in the cosmic-ray gas propagation along the magnetic field, except as the wave vector comes perpendicular to the magnetic field. When the wave vector is within an angle of the order of the speed of sound a in the ordinary gas divided by c, the usual fast mode suddenly collapses to zero velocity and the fast mode is taken over by the suprathermal mode. This is illustrated in Figure 4 for the special case where a is equal to the Alfvén speed; both are very small compared to the speed of sound in the cosmic-ray gas (of the order of c). The figure is a plot of the wave velocity ω/k in units of the speed of sound a as a function of the angle θ between the wave vector and the magnetic field. The suprathermal mode lies outside the figure for the most part except where it appears at $\theta = \pi/2$ or $3\pi/2$ to take over from the fast mode, which collapses to zero there.

It is evident that the first signal to arrive from a distant disturbance is the suprathermal mode. Much of the disburbance from a supernova explosion of the kind proposed by Colgate and Johnson (1960) would be in the suprathermal mode. On the other hand, an explosion of, say, 10^4 km/sec would strongly excite the suprathermal mode only over a width of about 4° centered on the direction perpendicular to the field. Parallel to the field the suprathermal mode would be but weakly excited because of its high velocity of propagation. Presumably there is heavy Landau damping of the suprathermal mode.

Observations do not yet show the degree of small-scale disordering in the galactic field. It was noted in Section 6 that there is no observational evidence for irregularities on scales up to 10 pc; so presumably the two-fluid approach, introducing the third or suprathermal mode is usually more appropriate.

7.3. Flow out of Strong Fields

It is of some interest to consider the streaming of cosmic rays as individual particles (rather than as a fluid) from a region of strong magnetic field B_o into a region of weak magnetic field B. The problem has relevance to the escape of cosmic rays from the

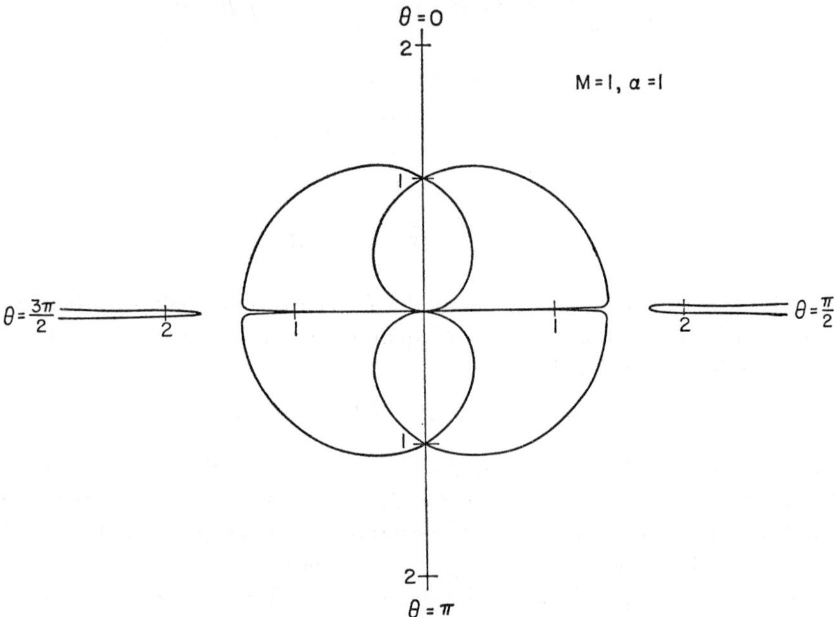

Fig. 4.—A plot of the wave velocity ω/k of hydromagnetic waves in a uniform magnetic field $\boldsymbol{B}$ in which there is a thermal gas of density ρ with speed of sound a and a suprathermal gas with speed of sound b ($\gg a$). The wave velocity is given in units of a as a function of the angle θ between the wave vector k and the magnetic field for the special case that a is equal to the Alfvén speed $B/(4\pi\rho)^{1/2}$. (See text.)

Galaxy (Secs. 4 and 7.4). It is evident that if the magnetic field is chaotic, the flow of cosmic rays will be essentially a random diffusion. The effective diffusion coefficient is $\lambda w/3$ where λ is the effective mean free path and w is the cosmic-ray particle velocity, $w \simeq c$. The cosmic-ray density satisfies a diffusion equation (see Morrison, Olbert, and Rossi 1954). If, on the other hand, the field is sufficiently smoothly varying that scattering is negligible, the cosmic-ray particles stream freely along the magnetic field, with the magnetic moment of each particle constant. For simplicity, suppose that the magnetic field is constant in time. Then the pitch angle θ of a particle varies such that $\sin^2\theta/B =$ constant. The largest pitch angle $\theta_{\max}$ to be found in the weak field B is given by

$$\sin^2\theta_{\max} = B/B_o \tag{17}$$

corresponding to a pitch angle $\pi/2$ in the strong field B_o. If we assume an isotropic distribution of cosmic rays streaming out of the strong field B_o, where there are N_0 particles per unit volume, it follows from Liouville's theorem that the number density N of the particles where the field has the weaker value B is proportional to the solid angle in $\theta_{\max}$; that is,

$$N(B) = \frac{N_o}{2}\int_0^{\theta_{\max}} d\theta \sin\theta = \frac{N_o}{2}\left[1-\left(1-\frac{B}{B_o}\right)^{1/2}\right] \cong \frac{N_o}{4}\frac{B}{B_o}$$

for $B \ll B_o$. The outward kinetic energy density, in terms of the individual particle energy $\frac{1}{2}Mw^2$, is

$$E(B) = \tfrac{1}{2}N(B)Mw^2 = \tfrac{1}{4}NMw^2\left[1-\left(1-\frac{B}{B_o}\right)^{1/2}\right] \cong \tfrac{1}{8}N_oMw^2\frac{B}{B}.$$

The pressure parallel to the field is

$$p_{||} = \tfrac{1}{6}Mw^2N_o\left[1-\left(1-\frac{B}{B_o}\right)^{3/2}\right] \cong \tfrac{1}{4}N_oMw^2\frac{B}{B_o}$$

and the pressure perpendicular to the field is

$$p_{\perp} = \tfrac{1}{4}Mw^2N_o\left[1-\left(1-\frac{B}{B_o}\right)^{1/2}\right] - \tfrac{1}{2}p_{||} \cong \tfrac{1}{16}N_oMw^2\left(\frac{B}{B_o}\right)^2.$$

Compare these quantities with the energy density and pressure $B^2/8\pi$ of the magnetic field. It is evident at once that the kinetic energy of the particles decreases only as the first power of B; so the kinetic energy completely overpowers the field where the field is weak. Going further, $p_{||}$ dominates both $p_{\perp}$ and $B^2/8\pi$. The hose instability results (Parker 1958*b*). Disorder grows in the field at the expense of $p_{||} - p_{\perp}$ with a characteristic time of the order of a few cyclotron periods and characteristic scales of a few radii of gyration.

It is evident, then, that cosmic rays cannot pass adiabatically out of a strong field B_o into an empty weak field B without (*a*) having a kinetic energy density in excess of $B^2/8\pi$ as B becomes small, and (*b*) without strong disorder and scattering in the weak field B. The result is that a particle, having once left the strong field B_o, has little chance of returning to B_o. The particles leaving the strong field are trapped in the weak field where they accumulate up to the density in the strong field. This is one example of how dynamical instabilities in the collisionless cosmic-ray gas have essentially the same effect as collisions in an ordinary dense gas—tending toward isotropy and a uniform density distribution along the magnetic field.

7.4. Inflation of a Galactic Halo

7.4.1. *Non-Existence of Static Equilibrium.*—The cosmic-ray gas prevents the interstellar gas and magnetic field from possessing a static equilibrium in the gravitational field of the Galaxy. As noted earlier, in Section 3.3, the galactic field and the cosmic rays are confined to the Galaxy by the weight of the interstellar gas. Let us assume that the system is in equilibrium in the absence of cosmic rays.

Consider a gas cloud, confined within a finite region of space by the gravitational field of the mass (including stars) distributed throughout the interior of the cloud. Suppose that the cloud is surrounded by vacuum. Let the cloud be filled with a magnetic field $\boldsymbol{B}$. The electrical conductivity of the gas is presumed to be so high that the magnetic lines of force are "frozen" into the gas.

As a general rule there will be some weak field protruding through most of the "surface" of the cloud exactly as weak fields protrude everywhere through the surface of the Sun. The protruding lines of force are largely re-entrant and the gas tends to drain down out of the re-entrant lines of force back into the cloud. Now suppose that some very hot gas, such as cosmic-ray gas, is generated throughout the cloud up to some limiting pressure p_o which is very small[14] compared to the characteristic magnetic pressure $B^2/8\pi$ in the cloud. The very hot gas communicates quickly along the field throughout the cloud, including those re-entrant loops which penetrate through the "surface" of the cloud. The theorem follows that the very hot gas, generated up to a pressure p_o in the cloud, progressively inflates the protruding re-entrant loops of field, extending them outward without limit from the cloud as time progresses. This occurs for any nonvanishing p_o, no matter how small (Parker 1965*c*).

Inflation of the protruding loops of field is irresistible no matter how small the very hot gas pressure p_o because the protruding fields must fall off to zero with increasing distance from the "surface" of the cloud. Where the field is sufficiently weak, the lines of force cannot contain the nonvanishing pressure p_o. All those lines of force passing far enough out that $B^2/8\pi$ falls below p_o must be steadily inflated there by the very hot gas, and consequently extended farther and farther out from the cloud with the passage of time. The more feeble is the pressure p_o, the fewer will be the lines of force extending sufficiently far out to suffer the unlimited inflation, of course, but no matter how small p_o is, the inflating lines of force occupy all space beyond where $B^2/8\pi$ falls to p_o.

The inflation involves exactly that streaming illustrated in the previous subsection, where it was shown that the cosmic rays streaming from strong to weak field overwhelm the weak field. The hose instability is expected and should contribute to the isotropy of the cosmic-ray particles accumulating in the weaker portions of the field. The inflation may be illustrated by a simple formal example of a periodic two-dimensional field extending upward from a plane cloud surface $y = 0$. The x and y components of the field are $\sin kx \exp(-ky)$ and $\cos kx \exp(-ky)$ before inflation and are illustrated in Figure 5. The magnetic energy density declines upward as $\exp(-2ky)$ and falls below p_o at some height y_1. Above y_1 the very hot gas distorts the field outward with the lines of force shown in Figure 6 for progressively increasing pressure $\epsilon p_o (\epsilon < 1)$. The details of the calculation are given in Appendix II.

The characteristic rate of extension of the protruding magnetic fields depends upon the rate of generation of the very hot gas up to the limiting pressure p_o. The extension may proceed rapidly if a copious supply of very hot gas is present.

In the simple quasiequilibrium treatment given here, the extension of the magnetic fields proceeds without limit. In the actual case, however, an instability must eventually develop which releases the very hot gas from the outer ends of the loops, thereby limiting the extension. A general treatment of the limiting instability is beyond the scope of the present chapter. The onset of the hose instability, which may not be the most effective one, has already been mentioned.

7.4.2. *Application to Galaxy.*—Now apply this general nonequilibrium theorem to the

[14] The smallness of the limiting pressure p_o is not necessary; it is for simplicity and generality in the exposition and will be dropped in application to the Galaxy.

interstellar gas and fields in the Galaxy. It follows that the cosmic-ray gas must inflate the magnetic fields outward from the Galaxy, at a rate depending upon the rate of generation of cosmic rays. Present observations suggest that the galactic field is from about 0.5 to 1×10^{-5} gauss; so p_o is not much less than $B^2/8\pi$, suggesting that inflation begins near the "surface" of the galactic disk.[15] The rate of inflation of the magnetic fields outward from the surface of the disk[16] can be computed from the estimated 10^6-year life of the individual cosmic-ray particles in the Galaxy. From the fact that the net outward streaming of cosmic rays must take them through the 100-pc half-thickness of the disk in

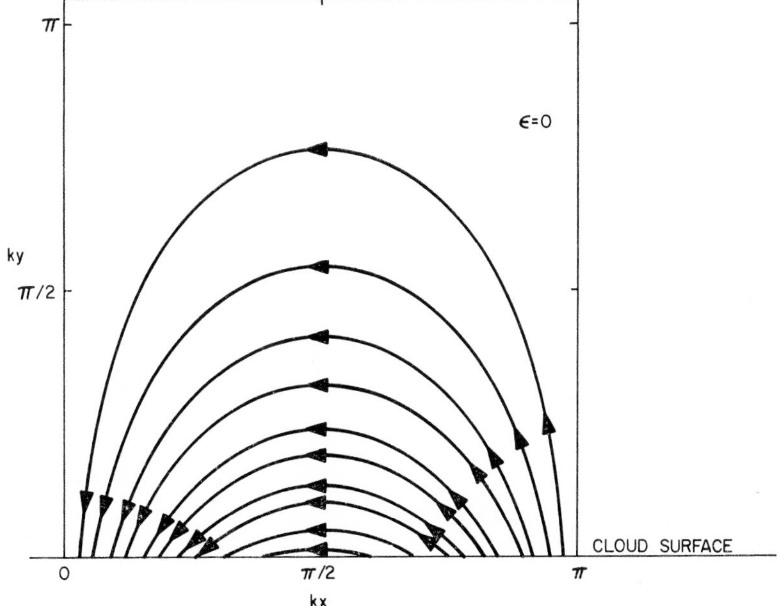

FIG. 5.—Magnetic lines of force of the undistorted two-dimensional periodic field $B_x = \sin kx \exp(-ky)$, $B_y = \cos kx \exp(-ky)$ extending above the surface $y = 0$.

10^6 years, the rate of outward inflation of the galactic fields must be of the general order of magnitude of 10^2 km/sec. This esimate is subject to revision when improved estimates of the cosmic-ray life and other facts become available.[17]

[15] Subject to revision if improved observations should show B to be greater.

[16] Cosmic rays undoubtedly inflate fields outward from the nucleus of the Galaxy, but no estimate of conditions there is available.

[17] It follows at once from energy considerations that the inflating fields do not carry a gas density in excess of 10^{-2} hydrogen atoms/cm^3 upward with them. Otherwise the 10^{-12} dynes/cm^2 cosmic-ray pressure would not be sufficient to impart the 100 km/sec outward streaming velocity inferred from observations; the cosmic rays would then not escape so quickly and the cosmic-ray energy density would increase above the observed level. There is no reason to expect that gas could be carried upward by inflating fields, of which Figure 2 is an example, because the lines of force are stretched vertically. So there is no evident objection to 10^{-2}/cm^3 or less. Estimates of the halo density usually tend to be rather lower than this (see, for instance, Woltjer 1965), although of course they are based on concepts of the halo rather different from those presented here.

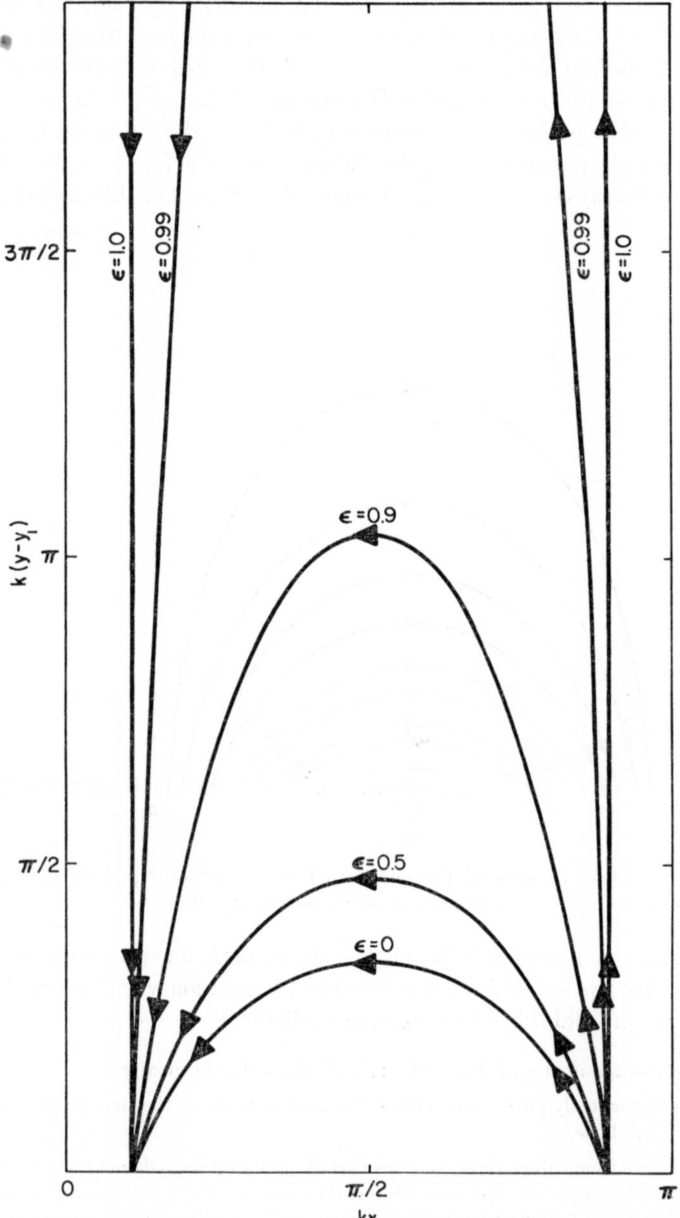

FIG. 6.—A magnetic line of force of the two-dimensional periodic field of Figure 5 above the level $y = y_1$, where the gas pressure, as measured by ϵ, inflates the field.

Thus, present estimates of the interstellar magnetic-field strength, the thickness of the gas in the galactic disk, and the rate of generation of cosmic rays lead to the conclusion, when combined with the dynamical theorem, that the galactic magnetic fields are extended outward by cosmic rays at a rate of about 10^2 km/sec to form a halo of magnetic field and cosmic-ray particles around the Galaxy. This effect does not deny the possibility that other processes may contribute to halo formation, too. It asserts that cosmic rays are sufficient by themselves to create an extensive galactic halo.

It is not possible to state how far the cosmic-ray halo may extend. It is suggested that the extension of the fields is limited only by some ultimate instability of the fields and cosmic rays.

It is evident that the theorem on inflation of galactic fields is applicable to galaxies other than our own, such as radio galaxies and other active structures. Only a lack of observational values for the relevant physical quantities prevents application here.

7.4.3. *Discussion.*—Consider some of the implications of the cosmic-ray inflation of galactic fields to form a halo. The formation of a halo of fields and cosmic rays around the Galaxy is a question of broader interest than the simple dynamical theorem on which the present development is based. It is evident that more observational information is needed. A galactic halo has been postulated and discussed for 10 years from considerations entirely different from those given here (see Sec. 5.2 and Spitzer 1956; Pikelner and Shklovsky 1957, 1959; Pikelner 1957; Field 1963; Woltjer 1965); so there is the problem to establish the degree to which the other mechanisms contribute to the halo. The cosmic rays give continual disturbance and expansion of the halo at something presently estimated as 100 km/sec. This question is closely associated with the problem raised by present observations indicating that cosmic rays have a short life (10^6 years) (and yet are remarkably isotropic) in the disk of the Galaxy. This important question has been a strong argument for believing that cosmic rays circulate freely through some kind of large galactic halo region (Biermann and Davis 1958, 1960; Ginzburg and Syrovatskii 1964). If the hypothesis of the present chapter is correct, the origin of the galactic halo is specified: the cosmic rays affect the galactic magnetic field in such a way as to manufacture a halo and automatically produce cosmic-ray isotropy in the galactic disk. Inflation of the surface fields of the Galaxy acts as a crude pressure regulator on the cosmic rays generated in the Galaxy itself. A copious supply of cosmic rays increases the cosmic-ray pressure up to $B^2/8\pi$ and no more. Increased cosmic-ray generation in the Galaxy would not increase the cosmic-ray pressure much. It would instead increase the rate of inflation of the surface fields and lead to a shorter life in the disk of the Galaxy (appearing as a decreased Li, Be, B abundance). On the other hand, greatly reducing the rate of generation of cosmic rays would lower their pressure somewhat below $B^2/8\pi$, so that only a fraction of the magnetic lines of force crossing the surface would be subject to inflation. A cosmic-ray particle would then have to circle several times before getting out into an inflating field. The cosmic-ray life would be considerably increased. Cosmic-ray anisotropy would appear in the Galaxy as a consequence of some of the longer escape paths.

Turning to other aspects of the dynamical role of cosmic rays in the Galaxy, we find the fundamental problem of quasisteady gravitational equilibrium of the disk of the Galaxy. Consider the distribution of stars and interstellar gas and fields over the distance from the plane of the Galaxy. The stars and the gas form two systems which move independently except that they are contained in the disk by the same gravitational field. We noted in the text that studies have been made elsewhere to deduce the gravitational field from the observed scale height and root-mean-square velocity of various classes of stars. A similar study of the interstellar gas should give the same gravitational field, although unfortunately the gas is sufficiently difficult to observe that at present the argument is approached from the opposite direction. Thus at present one starts with the gravitational field deduced from the stars and combines it with the observed scale height of the gas to deduce the weight of the gas. We pointed out that when such analysis is applied to the present observational picture, it leads to the conclusion that the average interstellar gas density is not less than about 5 atoms/cm^3. Otherwise, the calculated scale height of the interstellar gas, fields, and cosmic-ray system would exceed the observed 100 pc. On the other hand, a much higher gas density requires an extraordinarily short life and high rate of production of cosmic rays, since cosmic rays seem to have penetrated only about 5 gm/cm^2. It seems that 5 hydrogen atoms/cm^3 is the best compromise estimate at the moment, but improved observations need to be made. The role of the gas clouds apparently falling in toward the galactic plane with high speeds from large heights above the plane is not understood. Such clouds may be an indication of something not included in the usual picture of gravitational equilibrium in the disk of the Galaxy.

There is the question of the cosmic-ray phenomenon in galaxies other than our own. The cosmic-ray gas has important dynamical effects in our Galaxy under conditions which are quiet, if anything, when compared to many other galaxies (see, for instance, Lynds and Sandage 1963). The presence of the cosmic-ray gas means that there can be no static equilibrium of the galactic magnetic fields and interstellar gas. The absence of equilibrium appears as a 100 km/sec extension of the galactic fields outward from the Galaxy. We can imagine the enormously more rapid inflation of fields around a radio galaxy, or Seyfert galaxy, as a consequence of the more rapid generation of cosmic rays there. It is generally believed that the extended radio sources associated with such galaxies are an ejection of fields and particles from the galaxies themselves. The condition is believed to be transient, rather than quasisteady, as discussed here, but we suggest that it may be similar to the cosmic-ray inflation of the fields of our own Galaxy. In this view the extended sources represent the unavoidable inflation of the galactic fields by the rapid generation of fast particles, with perhaps little or no associated explosion and ejection of ordinary thermal matter. The extended radio sources may be the big brother of our own galactic halo and worth observing on this account alone—apart from their intrinsic interest. The interesting possibility that violent explosions may recur in a galaxy has been considered at some length by Burbidge (1963).

Finally, it must be remembered that the present cosmic-ray picture in the Galaxy stems from a number of inferences based on rather meager observational support. The outstanding questions deal mainly with the origin, history, and extension throughout space, which are dealt with at the present time with the principle that the simplest explanation is the least improbable. Presumably continued progress in cosmic-ray studies near Earth, and radio and optical astronomy, will shed further light on the extension of cosmic rays through space and time. It is to be hoped that the newer fields of X-ray, γ-ray, and neutrino astronomy will assist too.

This chapter was written to point out the dynamical role of the cosmic-ray gas and to summarize the general astrophysical knowledge of the origin and behavior of the cosmic-ray gas in the Galaxy. The cosmic-ray phenomenon is so broad a problem and the present writing so brief and selective in its views that the serious reader is urged to consult the excellent and more comprehensive works of Shklovsky (1960*b*), Morrison (1961), Burbidge (1963), and Ginzburg and Syrovatskii (1964). Special topics are also treated in such standard references as *Handbuch der Physik* and the series of volumes *Progress in Cosmic Ray Physics*.

Financial support by the Air Force Office of Scientific Research under Grant AF-AFOSR-62-23 Res. is gratefully acknowledged.

APPENDIX I

EQUATIONS OF MOTION OF COLLISIONLESS GAS

The equations of bulk motion for a collisionless plasma in a large-scale slowly varying magnetic field have been worked out in a number of ways (see, for example, Chew, Goldberger, and Low 1956; Watson 1956; Brueckner and Watson 1956; Burgers 1960) in the guiding center approximation. Our own formulation (Parker 1957*b*) was to note that the motion of an individual particle in a collisionless plasma can be computed (using the guiding center approximation) in terms of the large-scale magnetic and electric fields $\boldsymbol{B}$ and $\boldsymbol{E}$. Summing over the individual particle motions then gives the mean plasma velocity $\boldsymbol{v}$, the current density $\boldsymbol{j}$ and so on. The current density should be written $\boldsymbol{j}(\boldsymbol{r}, t, \boldsymbol{B}, \boldsymbol{E})$ to emphasize that it depends upon the motion of the individual particles, which in turn are expressed in terms of the fields $\boldsymbol{B}$ and $\boldsymbol{E}$. Substituting the expression for the current density into Maxwell's equation,

$$4\pi \boldsymbol{j} + \frac{\partial \boldsymbol{E}}{\partial t} = c\nabla \times \boldsymbol{B}, \qquad \text{(I 1)}$$

and, neglecting the displacement current because the discussion is limited to low-frequency disturbances, we obtain the equation of motion

$$NM\frac{d\boldsymbol{v}_\perp}{dt} = -\nabla_\perp\left(p_\perp + \frac{B^2}{8\pi}\right) + \frac{[(\boldsymbol{B}\cdot\nabla)\boldsymbol{B}]_\perp}{4\pi}\left(1 + \frac{p_\perp - p_\parallel}{B^2/4\pi}\right) \qquad \text{(I 2)}$$

for the component of the velocity perpendicular to $\boldsymbol{B}$. The number of particles per unit volume is N with each particle of mass M. The subscripts $\perp$ and $\parallel$ denote the components perpendicular and parallel to $\boldsymbol{B}$, respectively. The pressures $p_\parallel$ and $p_\perp$ are defined in the usual way as the sum over all particles of twice the kinetic energy in the direction parallel and in one direction perpendicular to the field, respectively. The mass velocity of the plasma is equal to $\boldsymbol{v}_\perp \equiv c\boldsymbol{E} \times \boldsymbol{B}/B^2$ if we neglect terms of the order of the radius of gyration divided by the scale of the field.

The other Maxwell equation leads to the familiar hydromagnetic equation

$$\frac{\partial \boldsymbol{B}}{\partial t} = \nabla \times (\boldsymbol{v} \times \boldsymbol{B}). \qquad \text{(I 3)}$$

The motion parallel to the field can be computed from the Liouville equation,

$$0 = \frac{\partial \Psi}{\partial t} + w \cos\theta \frac{\partial \Psi}{\partial s} + \frac{w \sin\theta}{2B} \frac{dB}{ds} \frac{\partial \Psi}{\partial \theta} \tag{I 4}$$

for particles spiraling along a magnetic line of force, where $\Psi(w, \theta, s, t) \sin\theta \, dw d\theta$ is the number of particles per unit volume with velocities in w, $w + dw$ and pitch angles in θ, $\theta + d\theta$ at a distance s along the line of force where the field strength is approximately time-independent with a value $B(s)$. This equation states that particles are conserved as they move with constant speed $w \cos\theta$ along $B(s)$ with $\sin^2\theta/B \simeq$ *constant* for each particle. The number of particles per unit volume is

$$N = 2\pi \int_0^\pi d\theta \sin\theta \Psi . \tag{I 5}$$

The bulk motion $u_\parallel$ is

$$N u_\parallel = 2\pi w \int_0^\pi d\theta \sin\theta \cos\theta \Psi , \tag{I 6}$$

and the pressure $p_\parallel$ is defined in the usual way as

$$\begin{aligned} p_\parallel &= 2\pi M \int_0^\pi d\theta \sin\theta (w \cos\theta - u_\parallel)^2 \Psi \\ &= 2\pi M w^2 \int_0^\pi d\theta \sin\theta \cos^2\theta \Psi - N M u_\parallel^2 . \end{aligned} \tag{I 7}$$

Operation on equation (I4) with $2\pi \int d\theta \sin\theta$ and integration by parts gives the equation for conservation of mass along a flux tube with cross-section inversely proportional to B,

$$\frac{\partial N}{\partial t} + B \frac{\partial}{\partial s} \left(\frac{N u_\parallel}{B} \right) = 0 . \tag{I 8}$$

Operation on equation (I4) with $2\pi w \int d\theta \sin\theta \cos\theta$ and integration by parts gives the momentum equation

$$NM \left(\frac{\partial u_\parallel}{\partial t} + u_\parallel \frac{\partial u_\parallel}{\partial s} \right) + \frac{\partial p_\parallel}{\partial s} + \frac{1}{2B} \frac{dB}{ds} [NM (w^2 - u_\parallel^2) - 3 p_\parallel] = 0 \tag{I 9}$$

after equation (I8) is used to eliminate $\partial N/\partial t$.

Now equations (I2) and (I9) are the momentum equations for a collisionless gas, giving the velocity $\boldsymbol{u}$ in terms of the pressure gradient. The system is not complete, of course, because it does not specify how the pressure is to be computed. The purpose so far is merely to show that the motion of a collisionless gas with isotropic thermal motions obeys essentially the same momentum equations as does a classical fluid. For upon putting $p_\perp = p_\parallel \equiv \frac{1}{3} NM(w^2 - u^2_\parallel) = p$ for isotropy, we can reduce (I2) and (I9) to the familiar hydromagnetic equation

$$NM \frac{d\boldsymbol{u}}{dt} = - \nabla p + \frac{(\nabla \times \boldsymbol{B}) \times \boldsymbol{B}}{4\pi} . \tag{I 10}$$

It was pointed out in the text that there are many reasons, including instabilities and small-scale magnetic irregularities, for believing that the pressure of a collisionless gas is approximately isotropic in the limit of low-frequency large-scale disturbances. So we expect the slow, large-scale motions of the collisionless gas to be described approximately by the classical hydromagnetic momentum equation (I10). What is more, we suspect that the pressure varies adiabatically with the density in the limit of large-scale disturbances, because the characteristic time for diffusion of energy (heat conduction) increases with the square of the scale of the disturbance, whereas the characteristic time of the disturbance increases only as the first power of the scale of the disturbance. This contention is not easy to prove, since it is based on thermal conditions determined by small-scale instabilities, and so forth. The usual calculations of pressure from the collisionless linearized Boltzmann equation do not and cannot properly take the randomizing effects of instabilities and small irregularities into account because the randomizing effects are essentially nonlinear. The effects could be inserted as an effective collision integral in the Boltzmann equation, but the point we are making in this discussion is that a collisionless gas behaves approximately like a classical fluid for slow large-scale variations.

APPENDIX II

INFLATION OF A SIMPLE TWO-DIMENSIONAL FIELD

The condition for quasistatic equilibrium of a very hot tenuous gas with pressure p in a two-dimensional (x, y) magnetic field $\boldsymbol{B}$ is

$$0 \cong -4\pi\nabla p + (\nabla \times \boldsymbol{B}) \times \boldsymbol{B} . \tag{II 1}$$

Writing $\boldsymbol{B} = \nabla \times [\boldsymbol{e}_z A(x,y)]$, it is readily shown that the equation becomes

$$0 = +4\pi\nabla p + \nabla^2 A \, \nabla A \tag{II 2}$$

where $\boldsymbol{e}_z$ is a unit vector perpendicular to the xy-plane. It is necessary and sufficient to put $p = F(A)$, which states merely that the gas pressure is constant along each line of force, $A = constant$. Then $\nabla p = F'(A)\nabla A$ and if $\nabla A \neq 0$, it follows that

$$\nabla^2 A + 4\pi F'(A) = 0 . \tag{II 3}$$

If the pressure is uniform, or negligible, of course $\nabla^2 A = 0$ and the field is undistorted.

Consider the case that the pressure is proportional to the minimum $B^2/8\pi$ along each line of force. Then

$$p = \frac{\epsilon k^2}{8\pi} A^2(x, y) \tag{II 4}$$

and

$$\nabla^2 A + \epsilon k^2 A = 0 \tag{II 5}$$

where ϵ is a constant. A simple example of a solution of this wave equation is

$$A(x, y) = C \sin kx \exp[-(1 - \epsilon)^{1/2} ky] \tag{II 6}$$

where C is an arbitrary constant, with $0 \leq \epsilon \leq 1$. Increasing ϵ represents progressive slow inflation of the undistorted field

$$A = C \sin kx \exp(-ky) \tag{II 7}$$

(illustrated in Fig. 5) with the condition that the y-component of the field remained fixed in the form $kC \sin kx$ at $y = 0$. In the limit as $\epsilon \rightarrow 1$, the field extends to infinity, with $B_x = 0$, $B_y = kC \cos kx$.

Now, suppose that the inflation of the field (II7) extending beyond the cloud surface $y = 0$ is brought about by the slow generation of a very hot gas up to a small pressure p_o at the cloud surface. Inertial effects are negligible for sufficiently slow inflation. The gas pressure has little effect on the field where $B^2/8\pi \gg p_o$, and hence has little effect for $y < y_1$, where

$$k y_1 = \log_e \frac{k^2 C^2}{8\pi p_o}. \tag{II 8}$$

Beyond y_1, however, the field cannot resist the gas pressure, and inflation occurs. The gas pressure does not rise to p_0 on the lines of force extending beyond y_1 because the field cannot contain this gas pressure. Instead, the field expands farther and farther with the steady generation of hot gas. To take a simple example, suppose that p rises to p_o along lines which do not reach to y_1 with little effect, whereas on the lines passing beyond y_1, p is of the form (II4) and ϵ increases slowly with time. The field extends farther and farther with the continuing generation of hot gas. Far beyond y_1, the field is given by (II6). The lines of force of this inflated portion of the field are shown in Figure 6 for progressively larger values of ϵ up to the maximum value $\epsilon = 1$, at which time the lines of force are extended parallel to the y-axis all the way to $y = \infty$.

REFERENCES

Aizu, K. Fujimoto, Y., Hasegawa, H., Kawabata, K., and Taketani, M. 1965, *Prog. Theoret. Phys. Japan*, **32,** 973.
Aizu, H., Fujimoto, Y., Hasegawa, H., Koshiba, M., Mito, I., Nishimura, J., Yokoi, K., and Schein, M. 1959, *Phys. Rev.*, **116,** 436.
———. 1961, *ibid.*, **121,** 1206.
Aizu, K., Fujimoto, Y., Hayakawa, S., Hasegawa, H., and Taketani, M. 1965, *Prog. Theoret. Phys. Japan*, **32,** 970.
Alfvén, H., and Herlofson, N. 1950, *Phys. Rev.*, **78,** 616.
Alexeff, I., Neidigh, R. V., Peed, W. F., Shipley, E. D., and Harris, E. G. 1963, *Phys. Rev. Letters*, **10,** 273.
Anders, E. 1965, *Space Sci. Rev.*, **3,** 583.
Appa Rao, M. V. K., Dahanayake, C., Kaplon, M. F., and Lavakare, P. J. 1963, *Proc. IUPAP Cosmic Ray Conference, Jaipur*, **3,** 95.
Arnold, J. R., Honda, M., and Lal, D. 1961, *J. Geophys. Res.*, **66,** 3519.
Arnold, J. R., Metzger, A. E., Anderson, E. C., and Dilla, M. A. van. 1962, *J. Geophys. Res.*, **67,** 4878.
Baade, W. 1956, *B.A.N.*, **12,** 312.
Bahcall, J. N., and Frautschi, S. C. 1964, *Phys. Rev.*, **135,** B788.
Baldwin, J. 1954, *Nature*, **174,** 320.
———. 1955*a*, *Observatory*, **75,** 229.
———. 1955*b*, *M.N.* **115,** 684.
———. 1962, *J. Phys. Soc. Japan*, **17,** Suppl. A–III, 173.
———. 1963, *Observatory*, **83,** 150.
Bauer, C. A. 1947, *Phys. Rev.*, **72,** 354.
Becker, W. 1963, *Zs. f. Ap.*, **57,** 117.
———. 1964, *ibid.*, **58,** 202.
Behr, A. 1959, *Nachr. Akad. Wiss. Göttingen, Math.-Phys. Kl.*, **185.**
Bernstein, W., Fredricks, R. W., and Scarf, F. L. 1964, *J. Geophys. Res.*, **69,** 1201.
Biermann, L., and Davis, L. 1958, *Zs. f. Naturforsch.*, **13*a*,** 909.
———. 1960, *Zs. f. Ap.*, **51,** 19.
Biswas, S., and Fichtel, C. E. 1964, *Ap. J.*, **139,** 941.
Biswas, S., Fichtel, C. E., Guss, D. E., and Waddington, C. J. 1963, *J. Geophys. Res.*, **68,** 3109.
Blaauw, A. 1964, *Ann. Rev. Astr. Ap.*, **2,** 213.
Bowyer, S., Byram, E. T., Chubb, T. A., and Friedman, H. 1964, *Nature*, **201,** 1307.
Brueckner, K. A., and Watson, K. M. 1956, *Phys. Rev.*, **102,** 19.
Burbidge, E. M., and Burbidge, G. R. 1964, *Ap. J.*, **140,** 1307.
Burbidge, E. M., Burbidge, G. R., and Hoyle, F. 1963, *Ap. J.*, **138,** 873.
Burbidge, E. M., Burbidge, G. R., and Rubin, V. C. 1964, *Ap. J.*, **140,** 942.
Burbidge, G. R. 1956, *Phys. Rev.*, **101,** 906.
———. 1963, *Proc. IUPAP Cosmic Ray Conference, Jaipur*, **3,** 229.

Burbidge, G. R., Burbidge, E. M., and Sandage, A. R. 1963, *Rev. Mod. Phys.*, **35**, 947.
Burbidge, G. R., and Hoyle, F. 1963, *Ap. J.*, **138**, 57.
Burbidge, G. R., Hoyle, F., Burbidge, E. M., Christy, R. F., and Fowler, W. A. 1956, *Phys. Rev.*, **103**, 1145.
Burgers, J. M. 1960, *Symposium on Plasma Dynamics*, ed. F. H. Clauser (Reading, Mass.: Addison-Wesley Publishing Co.).
Burkhardt, L. C., and Lovberg, R. H. 1958, *Nature*, **181**, 228.
Cameron, A. G. W. 1959, *Ap. J.*, **129**, 676.
Chandrasekhar, S. 1956, *Proc. Nat. Acad. Sci.*, **42**, 1.
———. 1961, *Hydrodynamic and Hydromagnetic Stability* (Oxford: Clarendon Press).
Chandrasekhar, S., and Fermi, E. 1953, *Ap. J.*, **118**, 113, 116.
Chandrasekhar, S., Kaufman, A. N., and Watson, K. M. 1958, *Proc. R. Soc. London, A*, **245**, 435.
Chapman, J. H. and Molozzi, A. R. 1961, *Nature*, **191**, 480.
Chew, G. F., Goldberger, M. L., and Low, F. E. 1956, *Proc. R. Soc. London, A*, **236**, 112.
Chiu, H. Y. 1963, *Proc. IUPAP Cosmic Ray Conference, Jaipur*, **6**, 131.
Clark, G., Brandt, H., and LaPointe, M. 1963, *Proc. IUPAP Cosmic Ray Conference, Jaipur*, **4**, 65.
Clark, G. W., Earl, J., Kraushaar, W. L., Linsley, J., Rossi, B., Scherb, F., and Scott, D. W. 1961, *Phys. Rev.*, **122**, 637.
Cline, T. L. 1961, *Phys. Rev. Letters*, **7**, 109.
Cline, T. L., Ludwig, G. H., and McDonald, F. B. 1964, *Phys. Rev. Letters*, **13**, 786.
Cocconi, G. 1961, *Hdb. d. Phys.*, ed. S. Flügge (Berlin: Springer-Verlag), **46**, 215.
Colgate, S. A., Grasberger, W. H., and White, R. H. 1962, *J. Phys. Soc. Japan*, **17**, Suppl. A-III, 157.
Colgate, S. A., and Johnson, M. H. 1960, *Phys. Rev. Letters*, **5**, 235.
Colgate, S. A., and White, R. H. 1963, *Proc. IUPAP Cosmic Ray Conference, Jaipur*, **3**, 335.
Compton, A. H., and Getting, I. A. 1935, *Phys. Rev.*, **47**, 817.
Crawshaw, J. K., and Elliot, H. 1956, *Proc. R. Soc. London, A*, **69**, 102.
Daniel, R. R., and Durgaprasad, N. 1962, *J. Phys. Soc. Japan*, **17**, Suppl. A-III, 15.
Danielson, R. E. 1960, *J. Geophys. Res.*, **65**, 2055.
Davies, R. D. 1963, *Observatory*, **83**, 154.
———. 1964, *M.N.*, **128**, 173.
Davies, R. D., and Shuter, W. L. H. 1963, *M.N.*, **126**, 369.
Davies, R. D., Slater, C. H., Shuter, W. L. H., and Wild, P. A. T. 1960, *Nature*, **187**, 1088.
Davies, R. D., Verschuur, G. L., and Wild, P. A. T. 1962, *Nature*, **196**, 563.
Davis, L. 1951, *Phys. Rev.*, **81**, 890.
———. 1954, *ibid.*, **96**, 743.
———. 1956, *ibid.*, **101**, 351.
———. 1958, *Nuovo Cimento Suppl.* **8**, 126.
Davis, L., and Greenstein, J. L. 1951, *Ap. J.*, **114**, 206.
Davis, L., and Jokipii, J. 1964, *Phys. Rev. Letters*, **13**, 739.
DeShong, J. A., Hildebrand, R. H., and Meyer, P. 1964, *Phys. Rev. Letters*, **12**, 3.
Devaille, J., Kendziekski, F., and Greisen, K. 1960, *Proc. IUPAP Moscow Cosmic Ray Conference*, **3**, 143.
Earl, J. A. 1961, *Phys. Rev. Letters*, **6**, 125.
Elliot, H. 1958, *Annals of the IGY* (London and New York: Pergamon Press), **4**, 374.
Ellis, G. R. A., Waterworth, M. D., and Bessell, M. 1962, *Nature*, **196**, 1079.
Fan, C. Y. 1956, *Phys. Rev.*, **101**, 314.
Fan, C. Y., Gloeckler, G., and Simpson, J. A. 1964, *Phys. Rev. Letters*, **13**, 149.
Feenberg, E., and Primakoff, H. 1948, *Phys. Rev.*, **73**, 449.
Felten, J. E., and Morrison, P. 1963, *Phys. Rev. Letters*, **10**, 453.
Fermi, E. 1949, *Phys. Rev.*, **75**, 1169.
———. 1954, *Ap. J.* **119**, 1.
Field, G. B. 1963, *Interstellar Matter in Galaxies*, ed. L. Woltjer (New York: W. A. Benjamin Co.), p. 193.
Fireman, E. L., and De Felice, J. 1960, *Geochim. et Cosmochim. Acta*, **18**, 183.
Forbush, S. E., Stinchcomb, T. B., and Schein, M. 1950, *Phys. Rev.*, **79**, 501.
Frank, L. A., Allen, J. A. van, and Macagno, E. 1963, *J. Geophys. Res.*, **68**, 3543.
Fredricks, R. W., Scarf, F. L., and Bernstein, W. 1965, *J. Geophys. Res.*, **70**, 21.
Freeman, J. W., Allen, J. A. van, and Cahill, L. J. 1963, *J. Geophys. Res.*, **68**, 2121.
Freier, P., and Waddington, J. 1965, *J. Geophys. Res.*, **70**, 5753.
Friedman, H. 1964, *Sci. Amer.*, **210**, No. 6, 36.
Geiss, J. 1963, *Proc. IUPAP Cosmic Ray Conference, Jaipur*, **3**, 434.
Geiss, J., Oeschger, H., and Schwarz, U. 1962, *Space Sci. Rev.*, **1**, 197.
Giacconi, R., Gursky, H., Paolini, F. R., and Rossi, B. 1962, *Phys. Rev. Letters*, **9**, 439.
Ginzburg, V. L. 1953, *Uspekhi Fiz. Nauk*, **51**, 343; *Doklady Akad. Nauk U.S.S.R.*, **92**, 1133.
———. 1958, *Progress in Elementary Particle and Cosmic Ray Physics* (Amsterdam: North-Holland Publishing Co.), **4**, chap. 5.

Ginzburg, V. L., and Syrovatskii, S. I. 1961, *Prog. Theoret. Phys. Japan*, Suppl. 20, 1.
———. 1963, *Sov. Astr.-A.J.*, **7**, 357.
———. 1964, *Origin of Cosmic Ray* (New York: Pergamon Press).
Gold, T., and Hoyle, F. 1959, *Paris Symposium on Radio Astronomy*, ed. R. N. Bracewell (Stanford: Stanford University Press), p. 583.
Gould, R. J., Gold, T., and Salpeter, E. E. 1963, *Ap. J.*, **138**, 408.
Gould, R. J., and Salpeter, E. E. 1963, *Ap. J.*, **138**, 393.
Greenstein, J. L., and Schmidt, M. 1964, *Ap. J.*, **140**, 1.
Greisen, K. 1956, *Progress in Cosmic Ray Physics* (Amsterdam: North-Holland Publishing Co.), **3**, chap. 1.
———. 1960, *Ann. Rev. Nucl. Sci.*, **10**, 63.
Gursky, H., Giacconi, R., Paolini, F. R., and Rossi, B. 1963, *Phys. Rev. Letters*, **11**, 530.
Haddock, F. T., Schulte, H. F., and Walsh, D. 1963, *A.J.*, **68**, 75.
Harris, E. G. 1961, *J. Nucl. Energy, Part C*, **2**, 138.
Hayakawa, S. 1956, *Prog. Theoret. Phys. Japan*, **15**, 111.
———. 1962, *J. Phys. Soc. Japan*, **17**, Suppl. A-III, 181.
———. 1963, *Proc. IUPAP Cosmic Ray Conference, Jaipur*, **3**, 125.
Hayakawa, S., and Kobayashi, S. 1953, *J. Geomag. Geoelect.*, **5**, 83.
Hayakawa, S., and Matsuoka, M. 1963, *Proc. IUPAP Cosmic Ray Conference, Jaipur*, **3**, 213.
Hayakawa, S., and Okuda, H. 1962, *Prog. Theoret. Phys. Japan*, **28**, 517.
Hazard, C., Mackey, M. B., and Shimmins, A. J. 1963, *Nature*, **197**, 1037.
Hildebrand, B., O'Dell, F. W., Shapiro, M. M., Silberberg, R., and Stiller, B. 1963, *Proc. IUPAP Cosmic Ray Conference, Jaipur*, **3**, 101.
Hiltner, W. A. 1949, *Ap. J.*, **109**, 471.
———. 1951, *ibid.*, **114**, 241.
Hogg, D. E. 1964, *Ap. J.*, **140**, 992.
Honda, M. 1959, *Geochim. et Cosmochim. Acta*, **17**, 148.
Honda, M., and Arnold, J. 1960, *Geochim. et Cosmochim. Acta*, **23**, 219.
Honsaker, J., Karr, H., Osher, J., Phillips, J. A., and Tuck, J. L. 1958, *Nature*, **181**, 231.
Hoyle, F. 1964, *Second Texas Symposium on Relativistic Astrophysics, Austin.*
Hoyle, F., and Fowler, W. A. 1963, *M.N.*, **125**, 169.
Hoyle, F., and Ireland, J. G. 1960*a*, *M.N.*, **120**, 173.
———. 1960*b*, *ibid.*, **121**, 253.
———. 1961, *ibid.*, **122**, 35.
Hulst, H. C. van de. 1958, *Rev. Mod. Phys.*, **30**, 913.
Huntley, H. E. 1948, *Nature*, **161**, 356.
Ireland, J. G. 1961, *M.N.*, **122**, 461.
Jones, F. C. 1963, *J. Geophys. Res.*, **68**, 4399.
Kahn, F. D. 1954, *B.A.N.*, **12**, 187.
Kahn, F. D., and Woltjer, L. 1959, *Ap. J.*, **130**, 705.
Kellermann, K. I. 1964, *Ap. J.*, **140**, 969.
Kiepenheuer, K. O. 1950, *Phys. Rev.*, **79**, 738.
Korchak, A. A., and Syrovatskii, S. I. 1958, *Doklady Akad. Nauk, U.S.S.R.*, **122** 792.
Kraushaar, W. L., and Clark, G. W. 1962*a*, *Phys. Rev. Letters*, **8**, 106.
———. 1962*b*, *J. Phys. Soc. Japan*, **17**, Suppl. A-III, 1.
Lal, D. 1953, *Proc. Indian Acad. Sci.*, *A*, **38**, 93.
Lequeux, J. 1962, *Ann. d'ap.*, **25**, 221.
Lerche, I. 1965, *Proc. R. Soc. London*, *A*, **283**, 203.
Libby, W. F. 1952, *Radiocarbon Dating*, 2d ed. (Chicago: University of Chicago Press, 1955).
Lin, C. C., and Shu, F. H. 1964, *Ap. J.*, **140**, 646.
Linsley, J. 1963*a*, *Phys. Rev. Letters*, **10**, 146.
———. 1963*b*, *Proc. IUPAP Cosmic Ray Conference, Jaipur*, **4**, 77.
Lipschutz, M. E., Signer, P., and Anders, E. 1965, *J. Geophys. Res.*, **70**, 1473.
Livingston, W. C., and Lynds, C. R. 1964, *Ap. J.*, **140**, 818.
Lüst, R., and Schlüter, A. 1954, *Zs. f. Ap.*, **34**, 263.
Lund, N., Swaneberg, B., Tanaka, Y., and Wapstra, A. H. 1963, *Proc. IUPAP Cosmic Ray Conference, Jaipur*, **3**, 163.
Lynds, C. R., and Sandage, A. R. 1963, *Ap. J.*, **137**, 1005.
Maltby, P., Matthews, T. A., and Moffet, A. T. 1963, *Ap. J.*, **137**, 153.
Maltby, P., and Moffet, A. T. 1962, *Ap. J. Suppl.*, **7**, 141.
Mathewson, D. S., and Rome, J. M. 1963, *Observatory*, **83**, 20; *Australian J. Phys.*, **16**, 360.
Matthews, T. A., Morgan, W. W., and Schmidt, M. 1964, *Ap. J.*, **140**, 35.
Mayer, C. H., McCullough, T. P., and Slonaker, R. M. 1957, *Ap. J.*, **126**, 468.
Meyer, P. 1964, *Second Texas Symposium on Relativistic Astrophysics, Austin.*
Meyer, P., Parker, E. N., and Simpson, J. A. 1956, *Phys. Rev.*, **104**, 768.
Meyer, P., and Simpson, J. A. 1955, *Phys. Rev.*, **99**, 1517.
Meyer, P., and Vogt, R. 1961, *Phys. Rev. Letters*, **6**, 193.
Mills, B. Y. 1959, *Paris Symposium on Radio Astronomy*, ed. R. N. Bracewell (Stanford: Stanford University Press), p. 465.

Minkowski, R. 1959, *Paris Symposium on Radio Astronomy*, ed. R. N. Bracewell (Stanford: Stanford University Press), p. 315.
———. 1964, *Ann. Rev. Astr. Ap.*, **2**, 247.
Moffet, A. T. 1964, *Science*, **146**, 764.
Moffet, A. T., and Maltby, P. 1961, *Nature*, **191**, 453.
Morgan, W. W., Sharpless, S., and Osterbrock, D. E. 1952, *A.J.*, **57**, 3.
Morris, D., and Berge, G. L. 1964, *Ap. J.*, **139**, 1388.
Morris, D., Clark, B. G., and Wilson, R. W. 1963, *Ap. J.*, **138**, 889.
Morris, D., and Radhakrishnan, V. 1963, *Ap. J.*, **137**, 147.
Morrison, P. 1958, *Nuovo Cimento*, **7**, 858.
———. 1961, *Hdb. d. Phys.*, ed. S. Flügge (Berlin: Springer-Verlag), **46**, 1.
Morrison, P., Olbert, S., and Rossi, B. 1954, *Phys. Rev.*, **94**, 440.
Münch, G., and Zirin, H. 1961, *Ap. J.*, **133**, 11.
Muller, C. A., Berkhuijsen, E. M., Brouw, W. N., and Tinbergen, J. 1963, *Nature*, **200**, 155.
Nikolsky, S. J. 1962, *Proc. Fifth Interamerican Seminar on Cosmic Rays, LaPaz*, Vol. **2**.
Noerdlinger, P. D. 1963, *Ann. Phys.*, **22**, 12.
Obayashi, R., and Hakura, Y. 1960, *J. Geophys. Res.*, **65**, 3131, 3143.
O'Dell, F. W., Shapiro, M. M., and Stiller, B. 1962, *J. Phys. Soc. Japan*, **17**, Suppl. A-III, 23.
Oort, J. H. 1954, *B.A.N.*, **12**, 177.
Oort, J. H., and Spitzer, L. 1955, *Ap. J.*, **121**, 6.
Parker, E. N. 1955, *Phys. Rev.*, **99**, 241.
———. 1957*a*, *ibid.*, **107**, 830.
———. 1957*b*, *ibid.*, **107**, 924.
———. 1958*a*, *Rev. Mod. Phys.*, **30**, 955.
———. 1958*b*, *Phys. Rev.*, **109**, 1328.
———. 1958*c*, "Cosmic Rays" in *McGraw-Hill Encyclopedia of Technology* (New York: McGraw-Hill), p. 498.
———. 1958*d*, *Phys. Fluids*, **1**, 171.
———. 1958*e*, *Phys. Rev.*, **109**, 1874.
———. 1960, *Ap. J.*, **132**, 821.
———. 1961, *J. Geophys. Res.*, **66**, 2673.
———. 1962, *Astrophysics and the Many-Body Problem* (Brandeis Lectures) (New York: W. A. Benjamin Co.), **2**, 1.
———. 1963, *Interplanetary Dynamical Processes* (New York: Interscience Publishers).
———. 1964*a*, *J. Geophys. Res.*, **69**, 1755.
———. 1964*b*, *Planet. Space Sci.*, **12**, 735.
———. 1965*a*, *ibid.*, **13**, 9.
———. 1965*b*, *Phys. Rev. Letters*, **14**, 55.
———. 1965*c*, *Ap. J.*, **142**, 584.
———. 1965*d*, *ibid.*, **142**, 1086.
Parker, E. N., and Tidman, D. A. 1958, *Phys. Rev.*, **111**, 1206; **112**, 1048.
Pauliny-Toth, I. I. K., and Shakeshaft, J. R. 1962, *M.N.*, **124**, 61.
Payne-Gaposchkin, C. 1958, *Hdb. d. Phys.*, **51**, 725.
Peters, B. 1959, *Nuovo Cimento Suppl.*, **14**, 436.
Piddington, J. H. 1964, *M.N.*, **128**, 345.
Pikelner, S. B. 1953, *Doklady Akad. Nauk, U.S.S.R.*, **88**, 229.
———. 1956, *Uspekhi Fiz. Nauk*, **58**, 285.
———. 1957, *A.J. (U.S.S.R.)*, **34**, 314.
Pikelner, S. B., and Shklovsky, I. S. 1957, *A.J. (U.S.S.R.)*, **34**, 145.
———. 1958, *Rev. Mod. Phys.*, **30**, 935.
———. 1959, *Ann. d'ap.*, **22**, 913.
Pollack, J. B., and Fazio, G. G. 1965, *Ap. J.*, **141**, 730.
Reid, G. C., and Leinbach, H. 1959, *J. Geophys. Res.*, **64**, 1801.
Riddiford, L., and Butler, S. T. 1952, *Phil. Mag.*, **43**, 447.
Rossi, B. 1952, *High-Energy Particles* (Englewood Cliffs, N.J.: Prentice-Hall).
———. 1955, *Nuovo Cimento Suppl.*, **2**, 275.
———. 1961, *Phys. Rev.*, **122**, 637.
———. 1964, *Cosmic Rays* (New York: McGraw-Hill Publishing Co.).
Rougoor, G. W. 1964, *B.A.N.*, **17**, No. 6, 381.
Ryle, M., and Sandage, A. 1964, *Ap. J.*, **139**, 419.
Sagdeyev, R. S., and Shafranov, V. D. 1961, *Sov. Phys.-JETP*, **12**, 130.
Sandage, A. 1964, *Ap. J.*, **139**, 416.
Savedoff, M. P. 1956, *Ap. J.*, **124**, 533.
Scanlon, J. H., and Milford, S. N. 1965, *Ap. J.*, **141**, 718.
Scarf, F. L., Bernstein, W., and Fredricks, R. W. 1965, *J. Geophys. Res.*, **70**, 9.
Schatzman, E. 1963, *Ann. d'ap.*, **26**, 234.

Schmidt, M. 1957, *B.A.N.*, **13**, No. 475, 247.
———. 1963, *Nature*, **197**, 1040.
Schmidt, M., and Matthews, T. A. 1964, *Ap. J.*, **139**, 781.
Schmoker, J. W., and Earl, J. A. 1965, *Phys. Rev.*, **138**, B300.
Sciama, D. W. 1962, *M.N.*, **123**, 317.
———. 1964, *Quart. J.R.A.S.*, **5**, 196.
Shklovsky, I. S. 1952, *A.J.* (*U.S.S.R.*), **29**, 418.
———. 1953*a*, *Doklady Akad. Nauk, U.S.S.R.*, **90**, 983; **91**, 475.
———. 1953*b*, *A.J.* (*U.S.S.R.*), **30**, 15.
———. 1960*a*, *ibid.*, **37**, 369 (*Sov. Astr.-A.J.*, **4**, 355).
———. 1960*b*, *Cosmic Radio Waves* (Cambridge: Harvard University Press).
———. 1962, *Uspekhi Fiz. Nauk*, **77**, 3; *Sov. Phys.–Uspekhi*, **5**, 365; *A.J.* (*U.S.S.R.*), **39**, 591.
———. 1963, *Sov. Astr–A.J.*, **6**, 465.
Simpson, J. A. 1958, *Annals of the IGY* (New York: Pergamon Press), **4**, 351.
———. 1960, *Ap. J. Suppl.*, **4**, 378.
Simpson, J. A., Fonger, W., and Treiman, S. B. 1953, *Phys. Rev.*, **90**, 934.
Singer, S. F. 1953, *Phys. Rev.*, **90**, 168.
Smith, E. von P. 1956, *Ap. J.*, **124**, 43.
Smullin, L. D., and Getty, W. D. 1962, *Phys. Rev. Letters*, **9**, 3.
Spitzer, L. 1956, *Ap. J.*, **124**, 20.
———. 1958, *Nature*, **181**, 221.
Spitzer, L., and Tukey, J. W. 1951, *Ap. J.* **114**, 187.
Stix, T. H. 1964, *Phys. Fluids*, **7**, 1960.
Swarup, G., Thompson, A. R., and Bracewell, R. N. 1963, *Ap. J.*, **138**, 305.
Synge, J. L. 1957, *The Relativistic Gas* (Amsterdam: North-Holland Publishing Co.)
Turtle, A. J., and Baldwin, J. E. 1962, *M.N.*, **124**, 459.
Turtle, A. J., Pugh, J. F., Kenderline, S., and Pauliny-Toth, I. I. K. 1962, *M.N.*, **124**, 297.
Voshage, H. 1962, *Zs. f. Naturforsch.*, **17a**, 422.
Voshage, H., and Hintenberger, H. 1959, *Zs. f. Naturforsch.*, **14a**, 828.
———. 1963, *Radioactive Dating* (Vienna: International Atomic Energy Agency).
Waddington, C. J. 1960, *Progress in Nuclear Physics* (New York: Pergamon Press), **8**, 3.
Walsh, D., Haddock, F. T., and Schulte, H. F. 1964, *Space Research*, ed. P. Muller (New York: John Wiley & Sons), **4**, 935.
Watson, K. M. 1956, *Phys. Rev.*, **102**, 12.
Weaver, H. F. 1953, *A.J.*, **58**, 177.
Webber, W. R. 1962, *Progress in Elementary Particle and Cosmic Ray Physics* (Amsterdam: North-Holland Publishing Co.), Vol. **6**.
———. 1965, *Hdb. d. Phys.*, ed. S. Flügge (Berlin: Springer-Verlag), **46/2**.
Wentzel, D. G. 1963*a*, *Ann. Rev. Astr. Ap.*, **1**, 195.
———. 1963*b*, *Ap. J.*, **137**, 135.
———. 1964, *ibid.*, **140**, 1013.
Westerhout, G. 1957, *B.A.N.*, **13**, No. 475, 201.
Wielebinski, R., and Shakeshaft, J. R. 1964, *M.N.*, **128**, 19.
Wilson, R. W. 1963, *Ap. J.*, **137**, 1038.
Woltjer, L. 1963, *Interstellar Matter in Galaxies*, ed. L. Woltjer (New York: W. A. Benjamin Co.), p. 88.
———. 1965, *Stars and Stellar Systems*, ed. A. Blaauw and M. Schmidt (Chicago: University of Chicago Press), Vol. **5**.
Zahringer, J. 1964, *Ann. Rev. Astr. Ap.*, **2**, 121.
Zakharenkov, V. F., Kaidanovskii, N. L., Pariiskii, Yu. N., and Prozorov, V. A. 1963, *A.J.* (*U.S.S.R.*), **40**, 216 (*Sov. Astr.-A.J.*, **7**, 167).
Zwicky, F. 1958, *Hdb. d. Phys.*, ed. S. Flügge (Berlin: Springer-Verlag), **51**, 766.
———. 1965. Unpublished.

CHAPTER 15

Evidence for Galactic Magnetic Fields

LEVERETT DAVIS, JR., AND G. L. BERGE
California Institute of Technology, Pasadena, California

1. INTRODUCTION

EVIDENCE bearing on the presence and structure of a galactic magnetic field has been accumulating for at least two decades. Early discussions of the implications of the properties of cosmic rays and arguments based on galactic structure were quite indirect. Later, the explanation of the polarization of starlight seemed to require a galactic magnetic field and to give evidence for its average orientation. Within the last five years, radioastronomical results relating to synchrotron radiation, Faraday rotation, and the Zeeman effect have given more direct evidence on the presence, structure, and limits of magnitude of the field. As yet, only a beginning has been made in gathering all the evidence that will be required to produce a consistent model, and different phenomena still lead to substantially different models. It seems certain, however, that a galactic magnetic field plays an important role in determining some features of galactic structure. In our vicinity this field seems to be preferentially aligned along the spiral arm, but its structure amounts to more than a smooth variation over the arm. It is difficult to obtain any reliable estimate of the field strength, but it appears plausible that it lies within a factor of 3 of $1\gamma = 10^{-5}$ gauss $= 10\ \mu$gauss in the Sun's neighborhood. In the following discussion we direct our main attention to the basic principles involved, and any conclusions drawn are to be regarded as tentative until more data are available.

2. DIRECT CONNECTIONS BETWEEN MAGNETIC FIELDS AND RADIATION

In the radio-frequency range several methods can provide information about the magnetic field within the Galaxy. All depend upon the production or modification of the radiation by the magnetic field.

The magnetic field strength in the galactic halo and disk and perhaps even an indication of the field orientation can be estimated from the observed properties of the galactic synchrotron emission at decimeter and longer wavelengths. In principle, Zeeman splitting of the 21-cm neutral hydrogen line can be used to determine the mean line-of-sight component of the field within H I regions, although to date only negative evidence has been obtained. All other methods possible at present rely on the measurement of Faraday rotation of the plane of polarization of linearly polarized radiation. Much of the Faraday rotation observed for extragalactic radio sources displaying a measurable degree of linear polarization seems to occur in the Galaxy and thus provides information about the galactic field. Galactic synchrotron radiation is also observed to exhibit Faraday rotation. In this case the emitting region is also the region in which the rotation occurs; so there is Faraday depolarization as well. The rotation and depolarization properties both tell something about the field direction immediately. However, their use to estimate the field strength requires knowledge of the electron density and distribution along the line of sight. The Faraday rotation within the 21-cm line in absorption, often called the *Clark effect*, should also, in principle, yield the mean line-of-sight component of the field within H I regions, the necessary information on the number of atoms per cm^2 along the line of sight being given by the absorption.

2.1. Synchrotron Radiation within the Galaxy

The characteristics of the spectrum and polarization of the galactic background radiation indicate that, below 1000 Mc/s at least, it is produced mainly by synchrotron emission, described by Ginzburg and Syrovatskii (1965), and by Bless in this volume. Synchrotron emission, of course, provides definite proof of the existence of a galactic magnetic field. However, we would like to go further and estimate the field strength and configuration, if possible.

Pawsey (1965) has discussed in detail the various observations of the galactic emission and their analysis. Clearly interpretation of the data presents problems. There are instrumental difficulties; it is hard to separate the nonthermal synchrotron emission from the thermal emission and absorption due to free-free transitions in H II regions; and the galactic radiation must be somehow distinguished from the extragalactic background. It has sometimes been difficult to distinguish between discrete sources and continuous distributions. There appears to be a distinct division of the galactic radio radiation into a flat disk component and a spheroidal halo component, though Baldwin (1963) and others do not agree that the halo is definitely established. Thus the discussion below of the halo field is provisional until this point has been settled. Mills (1959) found that at 85 Mc/s the halo component has a volume emissivity, $J(\nu)$, of about 1.5×10^{-39} erg sec^{-1} $cm^{-3}(c/s)^{-1}$ and that near the Sun the disk component is ten times larger. The spectral index, α, defined by $J(\nu) \propto \nu^{-\alpha}$, is about 0.5 at 85 Mc/s, but α may be a function of frequency.

Consider now the properties of the synchrotron radiation. This is produced by highly relativistic electrons moving in a magnetic field. Suppose the field to be roughly uniform in the source region with strength B μgauss and suppose a particular electron to have energy E Bev. The power radiated by the relativistic electron in the frequency interval $d\nu$ at ν is (Oort and Walraven 1956)

$$P(\nu)\,d\nu = CB \sin \phi F(3\nu/10\nu_m)\,d\nu\,, \tag{1}$$

where $C = 2.34 \times 10^{-28}$ erg $sec^{-1}(\mu gauss)^{-1}$ $(c/s)^{-1}$ and ϕ is the angle between the field and the velocity of the electron. The quantity $F(3\nu/10\nu_m)$ is a tabulated function of its argument with a maximum very near to $\nu = \nu_n$, which is

$$\nu_m = 0.3\,LB \sin \phi\, E^2\,, \tag{2}$$

where $L = 1.61 \times 10^7$ (c/s) (μgauss)$^{-1}$ Bev^{-2} and ν_m is three-tenths of the usual critical frequency. The power emitted per c/s falls to one-third the peak value at $\nu = 7\nu_m$ and at $\nu = 0.01\nu_m$. Thus highly relativistic electrons with energies many times the rest-mass energy, 5×10^{-4} Bev, in reasonable fields of 1 to 100 μgauss will radiate in the observed frequency range, and it is consistent with equation (2) to ascribe the radiation to the synchrotron mechanism.

Let the number of electrons per unit volume with energies in the range dE and helix angle ϕ in $d\phi$ be (1/2) sin ϕ $n(E,\phi)$ dE $d\phi$. Thus for an isotropic distribution, ndE is merely a particle density. If the energy distribution is given by the power law

$$n(E,\phi) = k(\phi)\, E^{-\gamma}\,, \tag{3}$$

then the flux received by an observer at a distance R from an element of volume dV in the frequency interval $d\nu$ is

$$(4\pi R^2)^{-1}\, J(\nu,\phi)\, d\nu dV = (4\pi R^2)^{-1}\, k(\phi)\, CB \sin\phi\, (LB \sin\phi/\nu)^{\alpha} g(\alpha)\, d\nu dV\,, \tag{4}$$

where $\alpha = (\gamma - 1)/2$; $g(\alpha)$ is a slowly varying function of α that is 1.04 for $\alpha = 0.5$; and ϕ is now the angle between the magnetic field and the line of sight. If the source were isotropic, J would be merely the volume emissivity. Equation (4) follows if it is assumed that each electron radiates only in the direction of its instantaneous velocity. Actually, it radiates over a very small cone about this direction, but the integrated flux is the same unless $k(\phi)$ varies extremely rapidly with ϕ. The radiation from a small volume of uniform magnetic field is strongly plane-polarized, approximately 65 to 75 per cent, depending on the energy distribution, with the electric vector normal to the plane through **B** and the line of sight.

If the angle between the field and the line of sight is not known, it seems reasonable to replace sin ϕ by $\pi/4 \approx 0.8$, its average value in an isotropic distribution. However, it must be realized that there is a small chance that sin ϕ lies near zero, in which case J will be overestimated or B underestimated. If the radiation is observed to be strongly polarized, it is reasonable to presume that sin ϕ is not excessively small, since then even small variations in the direction of **B** would lead to large variations in the plane of polarization. It is useful to consider a model in which **B** is isotropically distributed over a source of large volume. In this model the radiation is emitted isotropically and is unpolarized. For $0 \leq \alpha \leq 0.5$, a good approximation to J is then obtained by setting sin ϕ equal to 0.8 in equation (4). The converse argument does not follow. If radiation is observed to be unpolarized but to have the spectral dependence of synchrotron radiation, **B** could still be fairly uniform in direction. Faraday rotation in a large source easily produces the same effect. If the magnitude of **B** varies over the volume, equation (4) still holds, with a suitable average value of B. Although observations of the energy distribution of electrons often are fitted by a power law, one cannot expect equation (3) to hold over too extended an energy range, hence there will be a variation of α with ν in equation (4). Useful examples of this are given by Turtle (1963).

We see from the above that, given observed values of $J(\nu)$, α, and k, we can find B. L'Heureux and Meyer (1965) find that, near the Earth at a period near solar minimum and for energies of 0.5 to 3 Bev, $k = 5 \times 10^{-13}$ cm^{-3} (Bev)$^{\gamma-1}$, with an uncertainty that seems to be less than 50 per cent and with $\gamma = 1.6 \pm 0.5$. This result is essentially the same as, but more precise than, the earlier determination of Earl (1961) and the upper limit of Critchfield, Ney, and Oleksa (1952), which, however, were made at a time when the Sun was much more active. Thus, there is less reason to multiply this latest result by the conventional factor of 2 or 3 to allow for the screening out from the Solar System of particles of energy around 1 Bev that obviously occurs when the Sun is active and may occur to a lesser extent when it has minimal activity. Using the value of $J(\nu)$ from Mills, that of k from L'Heureux and Meyer, $\alpha = 0.5$, and $\sin \phi = 0.8$, we get as an upper limit for the galactic field strength $B = 50$ μgauss in the disk and 10 μgauss in the halo. If we use $\alpha = 0.3$ to correspond to $\gamma = 1.6$, we get for the upper limits 75 μgauss in the disk and 12 μgauss in the halo, showing that the results are not particularly sensitive to α in this case. The same procedure was used earlier by Woltjer (1965) using Earl's data, leading to estimates of 35 μgauss and 7 μgauss for the two regions and by Biermann and Davis (1960), using the upper limit of Critchfield, Ney, and Oleksa to get lower limits of 20 μgauss and 5 μgauss, respectively. These lower values of B stem from the use of larger values of k for the Galaxy as a whole rather than the use of that observed in the Solar System at 1 a.u. from the Sun.

Our results above are upper limits because there is no assurance that the total galactic electron flux penetrates to the Earth's orbit even at solar minimum and because there is still some question (see, e.g., Baldwin 1963; Davies and Shuter 1963) as to how much of the observed radiation arises from a continuous distribution as assumed above and how much arises from small, high-intensity regions. In any case, the simple model with constants α and B used above will doubtless have to be revised (Turtle 1963).

Mills (1959, 1964) attempted to deduce the structure of the galactic spiral arms from the variations of radio brightness with galactic longitude. However, he assumed isotropic radiation and hence an isotropic distribution of magnetic-field directions. As discussed below, a more realistic model would be a field directed generally along the spiral arms, but with substantial fluctuations. If a substantial part of the galactic radio emission arises from synchrotron radiation in such a field, it should be possible to learn something of its structure from reasonably precise observations.

2.2. Faraday Rotation

Faraday rotation is the rotation of the plane of polarization of an electromagnetic wave as it passes through a medium in which a magnetic field changes the velocities of the two circularly polarized components. Three cases are considered: (1) the simplest is the case in which rotation is due to free electrons and occurs outside the source; (2) the case in which rotation is distributed in depth through the source, one cause of depolarization; and (3) the case in which the rotation is produced in the absorption line of neutral hydrogen (the *Clark effect*).

In a region containing N_e free electrons per unit volume, each of charge $-e$ and mass m, a right-hand circularly polarized wave has a velocity (Jackson 1962; Ginzburg 1964) which is greater by

$$v_R - v_L = N_e e^3 \lambda^3 \frac{B \cos \phi}{2\pi^2 c^3 m^2}$$

than that of the corresponding left-hand circularly polarized wave, where λ is the wavelength, B the magnetic field strength, and ϕ the angle from **B** to **n**, the direction of propagation. As a result, the plane of the electric vector of plane-polarized radiation and the major axis of elliptically polarized radiation rotate to the right through the angle

$$d\psi = (v_R - v_L)\frac{\pi ds}{\lambda c} = \frac{e^3\lambda^2}{2\pi m^2 c^4} N_e B \cos\phi ds = 2.62 \times 10^{-17}\lambda^2 N_e B \cos\phi ds \tag{5}$$

as the wave propagates a distance ds, all units being in the cgs system and ψ being in radians. Using astronomically convenient units, we may integrate this to give for the total rotation over the distance L

$$\psi = (R.M.)\lambda^2 \equiv \left(0.81\int_0^L N_e B \cos\phi ds\right)\lambda^2 , \tag{6}$$

where now λ is in meters, N_e in cm^{-3}, B in μgauss, s and L in parsecs. The quantity designated by $R.M.$ is called *rotation measure.* Units are radians per meter squared, and it is independent of the wavelength. The sign convention is that ψ is positive for right-hand rotation along **n**, that is, for counterclockwise rotation when viewed against the plane of the sky, which corresponds to the usual astronomical convention of measuring position angle from north through east in the sky. Note that if one prefers to measure ϕ from **B** to the line of sight, that is, to $-$**n** so that $B \cos \phi$ is positive rather than negative when **B** is directed away from the observer, a minus sign must be inserted in all the above expressions.

The rotation measure cannot be determined from a single observation of the position angle of the plane of polarization because there is almost never any way to estimate the position angle of the plane of polarization at the source and because there is no way to distinguish between values of ψ that differ by 180°. Thus, it is necessary to observe the source at several frequencies, plotting the observed position angles as a function of λ^2 and fitting the points by a straight line whose slope gives $R.M.$ and whose extrapolation to $\lambda^2 = 0$ gives the position angle at the source. In principle, for each observation, several points should be plotted, differing by multiples of 180°, and the set which best fits a single straight line selected to determine $R.M.$ If observations are available at only two frequencies, or if the values of λ^2 are essentially equally spaced, the choice will not be unique. Some choices can be excluded as requiring implausibly large values of $\langle N_e BL\rangle$, but one must be careful not to force the data into a preconceived pattern when using this criterion.

As may be seen from equation (6), a determination of the rotation measure does not give direct information on the galactic magnetic field. Possible corrections for Faraday rotation in the source, outside the Galaxy if the source is extragalactic, and in the Earth's ionosphere must be allowed for. Although the total electron content of the ionosphere fluctuates considerably, it may be of the general order of 2×10^{13} cm^{-2} at night and 5×10^{13} cm^{-2} by day. Thus, depending on the direction of the line of sight, the magnetic latitude, and the ionospheric conditions, the ionospheric contribution to the rotation measure can be as large as 5 or 10 or more. After these corrections have been made, determination of B requires some assumption about averages over the line of sight and an estimate of N_e, L, and $\cos \phi$. Any conclusions about B are only as accurate as these estimates.

It is known that a large fraction of the extragalactic radio sources exhibit an observable degree of linear polarization at decimeter and shorter wavelengths. Gardner and Whiteoak (1963) suggested that there is a strong correlation between observed rotation measures and galactic latitude. This was later confirmed by Seielstad, Morris, and Radhakrishnan (1964). For high-latitude sources the magnitudes of the rotation measures are, on the average, much smaller than for low-latitude sources. Thus, it seems that

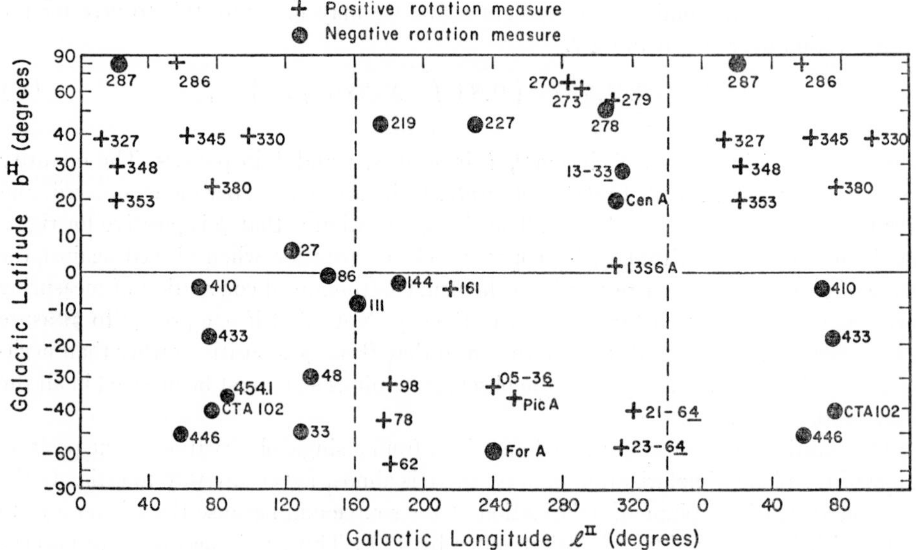

FIG. 1.—The dependence of sense of Faraday rotation on galactic coordinates. (Morris and Berge 1964.)

the major part of the rotation occurs within the Galaxy rather than in the sources themselves or in the intergalactic medium. Therefore, the observed Faraday rotation can be used to investigate the galactic medium. The rotation must also be produced in a flattened structure in order to exhibit the strong latitude dependence.

Morris and Berge (1964) reported two additional relations between the observed rotation measures and galactic coordinates based on a sample of 37 sources. In general, as shown in Figure 1, the rotation measures changed sign from one side of the galactic plane to the other and also when crossing the longitudes $l^{II} = 160°$ and $340°$. The simplest model consistent with the observations is that, in the neighborhood of the Sun, the magnetic field is generally parallel to the galactic plane and directed toward $l^{II} = 250°$ above the plane and toward $l^{II} = 70° = 250° - 180°$ below the plane. Sizable local irregularities are also needed to provide the observed scatter of the rotation measures. There are some possible variations of this model, and future observations may change it somewhat.

The magnitudes of the rotation measures indicated by these observations are +25 to 30 rad/m^2 for the data near $l^{\text{II}} = 70°$, $b^{\text{II}} = +45°$ and a negative value perhaps 20 to 50 per cent greater in magnitude for the data near $l^{\text{II}} = 70°$, $b^{\text{II}} = -45°$. Magnetic field strengths calculated on this basis can run from 1 μgauss to several tens of μgauss, depending on the assumed values of N_e and L, the distribution of electron density over the field structures, and whether the field is predominantly uniform in direction, or is composed of a number of anti-parallel filaments, or is rather irregular so that we are merely detecting a slight residual in the average value of $B \cos \phi$.

Part of the observed radio noise has a continuous power-law frequency spectrum and an angular distribution that suggest that it is galactic synchrotron emission. However, theory shows that such radiation emitted in a uniform field should be approximately 70 per cent plane-polarized with the plane of the electric vector normal to the plane containing the magnetic field and the line of sight. The observed degree of polarization is always much less than this. The most reasonable explanation is that the radiation has been depolarized, as happens when, at the Earth, it is a superposition of beams with different planes of polarization. This can occur because the planes of polarization in the different source elements are different or because there are differences in the Faraday rotation between the sources and the observer. These differences can arise either because the finite antenna beam width allows the beam paths to differ slightly in direction or because they vary in the length to the source elements even when they start in exactly the same direction.

As a convenient model, assume that the radiation observed is a superposition of plane-polarized elements with planes of polarization uniformly distributed between ϕ_0 and $\phi_0 + \Delta\phi$, where $\Delta\phi$ may be much larger than π. Then a simple analysis using Stokes' parameters shows that the polarization is $p\,|(\sin \Delta\phi)/\Delta\phi|$, where p is the percentage of polarization of the individual elements. If the main cause of depolarization is nonuniformity in the sources, the depolarizing factor should be essentially independent of frequency, but if the main cause is Faraday rotation within the source region, $\Delta\phi$ will be the maximum variation in ψ over the source and will be proportional to λ^2.

Large-scale surveys of the galactic background polarization have been conducted at Leiden (Brouw, Muller, and Tinbergen 1962; Westerhout, Seeger, Brouw, and Tinbergen 1962; Berkhuijsen and Brouw 1963; Berkhuijsen, Brouw, Muller, and Tinbergen 1964), at Cambridge University (Wielebinski and Shakeshaft 1964), and at the CSIRO Radiophysics Laboratory (Mathewson and Milne 1964). Until recently, almost all the observations were made at a wavelength of 75 cm. The measurements are difficult because there are many instrumental problems, and it is necessary to correct for the ionospheric Faraday rotation which is quite large at such long wavelengths. Over most of the Galaxy there is little polarized radiation, and the plane of polarization is almost random for what there is. However, there are a few spots where this is not true. The Dutch survey, in particular, shows a region near the galactic plane and at longitude 140° where there is radiation with a polarization of the order of 6 per cent and with its electric vector perpendicular to the galactic plane. Mathewson and Milne (1964), whose southern hemisphere survey includes the parts of the Galaxy not accessible to the northern observatories, have combined all the available data and find that 90 per cent of the polarized radiation comes from within 25° of the great circle that passes through the galactic poles and cuts the plane at longitudes 160° and 340°. The small depolarization in directions passing through this circle is ascribed to a small Faraday rotation which is a consequence of small values of cos ϕ. They conclude that there is a large-scale magnetic field perpendicular to this great circle, running in the $l^{\text{II}} = 70° - 250°$ direction. Thus, the agreement with the conclusions of the extragalactic source measurements is very good.

Recently, an extensive survey has been conducted at a wavelength of 50 cm (Berkhuijsen *et al.* 1964), so that it is now possible to get rotation-measure information in a few directions in addition to depolarization information for the galactic background radiation. Near the direction $l^{\text{II}} = 140°$, $b^{\text{II}} = 6°$ the polarization is a maximum, about 6 per cent; the direction of the electric vector is normal to the plane of the Galaxy; and the rotation measure is zero for one particular direction and shifts smoothly from positive to negative values in the immediate neighborhood of this direction. All this is compatible with a model in which the field along this line of sight is in the plane of the Galaxy and runs in the general direction of $l^{\text{II}} = 50°$. However, the polarization is much lower

TABLE 1

INTENSITIES OF THE ZEEMAN COMPONENTS

Line Component	σ	π	σ
Frequency	$\nu_0 - \Delta\nu_0$	ν_0	$\nu_0 + \Delta\nu_0$
Intensity (*R.H.*)	$I(1-\cos\phi)^2/8$	$I(\sin^2\phi)/4$	$I(1+\cos\phi)^2/8$
Intensity (*L.H.*)	$I(1+\cos\phi)^2/8$	$I(\sin^2\phi)/4$	$I(1-\cos\phi)^2/8$

than the 70 per cent expected for synchrotron radiation under these circumstances, and the rotation measure varies much more slowly with galactic longitude than would be expected from the rotation measure found for extragalactic sources. It is not clear whether these difficulties are to be ascribed to instrumental problems, to dilution by unpolarized thermal radiation, to local irregularity in the field structure that introduces randomness in the planes of polarization and irregularity in the Faraday rotation, or to an interlacing of filaments in which the field has opposite senses.

2.3. Zeeman Effect

In the presence of a magnetic field, the 21-cm line of neutral hydrogen is split into the three components of a normal Zeeman triplet. Bolton and Wild (1957) suggested that it might be practical to use this effect to measure the magnetic field within hydrogen clouds where the line is produced. The frequencies, intensities, and polarizations of the various components of the line are shown in Table 1, which is needed for a careful derivation of equation (7) below, although a plausible brief argument gives the same result. Here ϕ is the angle between $\mathbf{B}$, the magnetic field, and $\mathbf{n}$, the direction of propagation ($-\mathbf{n}$ is the line of sight). I is the intensity of the unpolarized emission line (or the intensity removed in an absorption line) that would be observed if there were no magnetic field, ν_0 equals 1420.406 Mc/s, and $\Delta\nu_0$ equals $eB/4\pi$ mc or $\Delta\nu_0/B$ equals 1.400 (c/s)(μgauss)$^{-1}$. The polarization is described in terms of a superposition of right- and left-handed circular polarization ([*R.H.*] and [*L.H.*]), the phases being such that the unshifted, or π, component is linearly polarized with its electric vector in the plane of **B** and **n**, and the two shifted, or σ, components are elliptically polarized with the major axis normal to this plane. Thus, for radiation emitted in the direction of **B**, where $\phi = 0$, right-hand circular polarization is found with non-zero intensity only for the high-frequency component and the left-hand polarization only for the low-frequency component, both intensities being $I/2$. For radiation in the direction opposite to **B**, where $\phi = \pi$, the right-hand polarization is concentrated at the low frequency and the left-hand polarization at the high. For $\phi = \pi/2$, the normal triplet is found for both senses of circular polarization. According to the convention used in radio astronomy, the direction of polarization, *R.H.* or *L.H.*, is the direction in which the electric vector at a fixed point rotates as time progresses when looking in the direction in which the beam propagates; that is, right-hand polarization corresponds to positive helicity in the terminology of modern physics. In classical optics the opposite convention is used, the direction of rotation being that observed when looking backward along the beam to the source, and the helix considered is the locus in space of the electric vector of the wave at a fixed time.

If the typical galactic field were of the order of 10 μgauss, the shifted components would be separated by only 28 c/s. This is very small compared to the width of the narrowest absorption features known (about 10,000 c/s). We may note that a velocity of 6×10^{-3} km/sec produces a Doppler shift of 28 c/s. Although it is impossible to detect such a small shift in the line profile directly, it is possible to detect the resulting difference in power in the left- and right-hand polarized components where the intensity varies rapidly with frequency. A narrow-band receiver is scanned over the line, switching alternately between left- and right-handed polarization, and the sum and difference of the two powers received are plotted as a function of frequency. We now derive the formula by which the component of **B** along the line of sight, averaged over the region producing the line, can be determined from these curves.

Let $f(\nu)d\nu$ be the intensity that would be produced in the interval $d\nu$ if there were no magnetic field. Radiation can be produced at a frequency in the observer's frame of reference that is different from ν_0 because of the Doppler shift due to motion of the source. With a field present, the radiation that would otherwise be emitted at a frequency $\nu + \Delta\nu_0$ is split into components, and right- and left-hand beams of intensity $f(\nu + \Delta\nu_0)(1 \pm \cos\phi)^2/8$ are shifted back to ν. The total intensity in the beam with right-handed polarization is

$$I_R(\nu)\,d\nu = [\tfrac{1}{8}f(\nu+\Delta\nu_0)(1-\cos\phi)^2 + \tfrac{1}{4}f(\nu)\sin^2\phi + \tfrac{1}{8}f(\nu-\Delta\nu_0)(1+\cos\phi)^2]\,d\nu\,,$$

and the intensity in the beam with left-hand polarization, I_L, is obtained by changing the signs of the $\cos\phi$ terms. Now expand $f(\nu \pm \Delta\nu_0)$ in a Taylor's series to get

$$I(\nu) = \tfrac{1}{2}(I_L + I_R) \cong \tfrac{1}{2}f(\nu)$$

and

$$D(\nu) = I_L - I_R \cong f'(\nu)\Delta\nu_0 \cos\phi = 2I'(\nu)\frac{eB\cos\phi}{4\pi m c} = 2.80I'(\nu)B\cos\phi\,,$$

where $f' = df/d\nu$, $-e$ is the charge on the electron, and B is in μgauss in the last term. In the usual terminology of radio astronomy, in which the powers received in the right-

and left-hand beams are described in terms of the antenna temperatures, T_R and T_L, respectively, and where $T = \frac{1}{2}(T_L + T_R)$ and $\Delta T = T_L - T_R$, $T' = dT/d\nu$, we have

$$\Delta T(\nu) = 2.80\, B \cos \phi\, T'(\nu) \ . \tag{7}$$

For emission features the signs are automatically accounted for in this equation if $\cos \phi$ is taken to be negative when **B** is directed away from the observer and ΔT and T' are given the correct signs. For absorption features, which are more commonly observed, the sign must be changed. It is very desirable that both T and ΔT be determined from the same observations, since any effects of the finite band width and of the shape of the passband then cancel out. If an attempt is made to determine $B \sin \phi$ in a similar way from the difference in intensities of plane-polarized beams, the effect is found to be proportional to $(\Delta\nu_0)^2 I''$, and hence to be unobservably small.

The hydrogen atoms that contribute to each frequency in equation (7) are those having the proper velocity, and the value of $B \cos \phi$ determined from equation (7) for each ν is the average over the volume occupied by these atoms. If the width of the feature is due only to random thermal motion of the atoms, this volume and the resulting average, $\langle B \cos \phi \rangle$, will be independent of ν. But if the width is mainly due to bulk motion of the gas, and if $B \cos \phi$ varies from one region to another, perhaps changing signs (the regions could even lie in different clouds), then the curve for ΔT will not be simply proportional to that for T', and it may be difficult to tell whether this lack of consistency is due to inhomogeneities or to other effects, such as noise, in the system. A common procedure is to determine an average value of $B \cos \phi$ for the entire region by finding the value that gives the best least-squares fit between the observables, $\Delta T(\nu)$ and $T'(\nu)$, in equation (7). A standard deviation is also obtained and, based on some assumed value of $\cos \phi$, a value of B is deduced.

Bolton and Wild (1957) concluded that with a 150-ft parabolic antenna using the best receivers available at the time, a field strength of $| B \cos \phi | = 3$ μgauss could be detected in a feature 10 kc/s wide. This argument was based only on signal-to-noise considerations and did not take account of instrumental effects such as instrumental polarization. Better receivers are now available, but because of the other problems, the detection limit remains at about 3 μgauss.

Three independent groups have attempted the determination to date. The Manchester University observations at Jodrell Bank have been reported by Galt, Slater, and Shuter (1960), Davies, Slater, Shuter, and Wild (1960), Davies, Verschuur, and Wild (1962), and Davies, *et al.* (1963, papers I, II, III). Weinreb (1962) reports work done at M.I.T. and Morris, Clark, and Wilson (1963) report work done at C.I.T. Weinreb's and Morris' groups have confined their measurements to absorption features in front of the strong radio sources Cassiopeia A and Taurus A. Neither group claims to have detected a magnetic field within the clouds studied, and the results indicate an upper limit of about 5 μgauss. The first group has studied these features as well as a few others and has reported the detection of somewhat larger magnetic fields, but it has recently revised its results to lower values. We must conclude that there has been no positive detection of a magnetic field by the method of Zeeman splitting and that the upper limit of $\langle B \cos \phi \rangle$ for the regions which have been studied is only a few microgauss.

It is difficult to interpret the results so far obtained, and we will mention only the pitfalls involved. First of all, the regions under study are very localized in space and have a much higher than average hydrogen density. There is no justification for supposing that the magnetic field within these regions is representative of the galactic medium. In addition, only a few regions have been investigated; so no general conclusions can be reached by statistical arguments. Finally, the magnetic fields within these regions may not be uniform. For example, the transition region between oppositely directed fields will tend to have a high gas density, a cos ϕ that ranges over both positive and negative values, and a low B.

Since the field strength measured by this method is merely the average of one component, it represents only a lower limit to the field strength within the hydrogen cloud forming the absorption line.

Because of the well-known connection between the index of refraction (that is, the phase velocity) and the absorption coefficient, the various components of the Zeeman triplet, with their slightly differing frequencies, have different velocities in a narrow absorption feature. Consequently there is a rotation of the plane of polarization that varies rapidly with frequency. It has been pointed out by Clark (1963) that this allows a determination of the magnetic field in the feature without requiring a knowledge of N_e or the dimension of the clouds producing the feature. It is necessary that the radiation falling on the cloud be partially linearly or elliptically polarized. Basically, the procedure is the same as that described for the determination of the field from the Zeeman effect, except that the variation with frequency in the position angle of the plane of polarization is used instead of the difference in intensity of the circularly polarized components. It is not clear, a priori, which of the two methods has greater precision and sensitivity. Presumably, it will be desirable to try both to see which is more effective in practice.

The Clark effect formula gives the rotation of the plane of polarization to be

$$\psi(\nu) = 17.4\, B \cos\, \phi\, T_s \frac{d}{d\nu} \int_{-\infty}^{\infty} \frac{\tau(\nu_0)\, d\nu_0}{\nu_0 - \nu}, \tag{8}$$

where the Cauchy Principal Value is used for the integral which can be converted to a variety of equivalent forms, and where $\tau(\nu_0)$ is the optical depth of the absorption feature at the frequency ν_0, where T_s is the spin temperature, assumed constant, and where B is the field in μgauss. It will be noted that determination of B from equation (8) requires a computational procedure that is more complicated and more sensitive to observational noise than that used in equation (7). Also, equation (7) is simpler when the optical depth is large, all the terms needed to convert from the observed antenna temperature to optical depth cancelling out in equation (7) but not in equation (8). Both the Zeeman effect and the Clark effect procedures suffer from the major disadvantage that they give the average of the field along the line of sight for rather special features that may not be typical of the Galaxy as a whole.

3. POLARIZATION OF STARLIGHT

An important souce of information on the structure of the galactic magnetic field is provided by the interstellar polarization discovered by Hiltner and by Hall (for an extensive analysis, see Serkowski [1962], and Hall and Serkowski [1963]). This polarization is found only for those stars whose light has been substantially reddened by interstellar absorption, and the plane of polarization (defined here, contrary to the older usage, as the plane of the electric vector) shows a marked preference for the galactic plane. Both factors indicate that the polarization is not a stellar phenomenon and that it originates in interstellar space. It is now generally accepted that the polarization is due to a partial alignment of nonspherical dust grains and that a galactic magnetic field whose average direction tends to be in the plane of the Galaxy is the large-scale feature that produces this alignment with some uniformity over large volumes.

The method of alignment now seems to be that proposed by Davis and Greenstein (1951). Some details of this analysis were refined by Davis (1958) and Miller (1962) without significantly changing the conclusions. The grains are kept spinning by the bombardment of the interstellar gas, while dissipative interactions with the magnetic field damp motions in which the magnetization of a grain varies with time. This tends to align the angular velocity with the field vector, since the magnetization is then constant in the grain's frame of reference. In free rotation the angular velocity can remain stationary in a rigid body only if it coincides with a principal axis of inertia. The only stable mode of rotation is about the shortest principal axis, the damping of any precessional motion tending to reduce the kinetic energy more rapidly than the square of the angular momentum and thus to align the short axis of the grain along the spin axis and the magnetic field. Although the spinning motion does not yield complete alignment, it does yield an anisotropic distribution of grain orientations that can polarize light. The degree of alignment produced depends on the magnetic field strength, the density and temperature of the surrounding gas, the size and temperature of the grains, and the dissipative processes, which depend on the imaginary part of the magnetic susceptibility, and hence on the chemical composition of the grain. The grain temperature must be less than that of the gas or the system will be in thermodynamic equilibrium and there will be no alignment. The ratio of polarization to absorption depends on the degree of alignment, the eccentricity of the grains, their sizes, and indexes of refraction.

When the theory was first developed, it seemed that the maximum degree of alignment that could be produced in this way was, at best, barely adequate to explain the observed polarization if the grains had the expected optical properties. Fields of 10 μgauss to 100 μgauss seemed to be required to produce a substantial degree of alignment. Subsequent investigations of the properties of grains containing free radicals (Platt 1956), of ferromagnetic grains (Henry 1958), and of graphite flakes (Cayrel and Schatzman 1954; Wickramasinghe 1962) indicate that with such grains either less perfect alignment is required to produce the polarization or a weaker field is required to produce the alignment. The work of Greenberg, Lind, Wang, and Libelo (1963) showed that ordinary dielectric particles could produce adequate polarization, provided the alignment was not too incomplete. (See chapter by Greenberg, this volume, for further discussion.)

Thus, it now seems that although comparison of the observations with the theory may give some information on the properties of possible types of grains, several different models may be able to explain the observations, and it seems unlikely that a useful estimate of the strength of the field can be obtained when there are so many other undetermined parameters in the theory.

The observations do give considerable information on the structure of the field. If the light is resolved into two orthogonal plane-polarized components, a grain scatters or absorbs more of the component whose electric vector is parallel to the long axis of the grain, that is, normal to the magnetic field. Thus the light passing through a homogeneous dust cloud has its plane of polarization in the plane of the magnetic field and line of sight. If all clouds had the same field strength and similar distributions of grain sizes, shapes, and compositions, it would be possible, in principle, to determine approximately the angle between the magnetic field and the line of sight from the ratio of polarization to absorption in different directions. In practice, uncertainty on these points, uncertainty in the absorption as deduced from the color excess, and uncertainty in possible depolarization due to variation in the alignment in successive clouds through which the light may have passed mean that such an analysis has, at most, only statistical significance. When the light passes through several clouds in which the fields lie in different planes, the absorption is additive but the polarizations partially cancel out. Such cases are best analyzed by describing the radiation in terms of Stokes' parameters.

In spite of all these difficulties in the way of straightforward, quantitative analyses, some valuable results can be obtained. All surveys show some regions where the plane of the electric vector is nearly parallel to the galactic plane for almost all stars and other regions where there is great irregularity. If the galactic field is assumed to be nearly uniform averaged over a spiral arm, but on a smaller scale to have substantial fluctuations, then the greatest regularity in the plane of polarization would be observed in directions normal to the average field. In directions parallel to the average field, the fluctuations produce randomly oriented field components normal to the line of sight, and hence great irregularity in the planes of polarization. On this basis, Hiltner (1956) concluded that the average field in the part of the Galaxy near the Sun is essentially parallel to the galactic plane and runs in the direction of $l^{\mathrm{II}} \approx 45°$ or 225°. The 25° difference between this result and the directions deduced from the Faraday rotation should not be regarded as significant because fluctuations in the amplitude of the irregularities from region to region would modify the analysis, and variations in the density of dust and of O and B stars results in great variations in the number of observations of the plane of polarization that can be made in different regions.

From an analysis of the polarization of the light from 216 stars whose distance from the Sun is between 50 and 250 pc, Behr (1959) arrived at a field direction whose l^{II} was 62° (or 242°). The statistical error was $\pm 5°$, but the total uncertainty of the analysis was larger. There was also some evidence that the magnetic-field lines were wound in a right-handed helix about the spiral arm.

It may well be that although observations of the polarization of starlight will not provide simple, direct evidence for the structure of the galactic magnetic field, they will be very useful in fixing the parameters of models of the Galaxy and its field and in excluding many otherwise possible models. Thus, the original suggestion of Fermi that the turbulent gas motions in the Galaxy would produce a randomly oriented magnetic field whose strength was determined by an equipartition of the magnetic and kinetic energies is inconsistent with the observed polarizations. The alternative model, in which there is an average field that is uniform or varies smoothly over distances of the order of the radius of a spiral arm but has waves or oscillations over distances of the order of those between gas clouds, fits the polarization observations much better. Based on these observations, estimates ranging from 7 to 100 μgauss for the field strength in the spiral arm have been made by Davis (1951), Chandrasekhar and Fermi (1953), and Stranahan (1954). However, some caution must be observed in accepting a model in which there are such substantial fluctuations that there will be substantial depolarization for most stars whose light passes through several clouds unless in such models one has rather complete alignment of the dust grain in each cloud and a size distribution that is efficient in producing a large ratio of polarization to absorption. Otherwise, if one allows too much inefficiency in the model, it is likely to be difficult to explain the observed ratio of polarization to absorption.

4. INDIRECT ARGUMENTS

Discussions of the distribution of neutral and ionized gas in the Galaxy, of the dynamics of galactic features involving this gas, of star formation, and of galactic cosmic rays all involve estimates of the strength and structure of the galactic magnetic field. An analysis of any of these topics may provide indirect evidence on the galactic magnetic field, if enough is known about the other parameters of the problem. To date it has been difficult to get definite information on the magnetic field in this way, because the other parameters are not known with adequate accuracy or because it has not been possible to distinguish between alternative models. For a more detailed discussion, see reviews by Wentzel (1963) and Woltjer (1965), and for an attractive recent model see Parker (1966).

A brief summary of forces and pressures is useful in deciding what are the important features of any particular model. For magnetic fields of 3 μgauss (or 30 μgauss), the energy density is 4×10^{-13} erg cm^{-3} (or 4×10^{-11} erg cm^{-3}). The pressures normal to and tensions along the lines of force in dyne cm^{-2} are numerically equal to the energy densities. The cosmic-ray energy density in the Solar System at 1 a.u. from the Sun is about 1 ev cm^{-3} = 1.6×10^{-12} erg cm^{-3}, and the pressure is essentially one-third as great numerically. Outside the Solar System, these values are at least this large and probably somewhat larger. The gas pressure in a 100° K H I region where the particle density is 1 cm^{-3} is 1.4×10^{-14} dyne cm^{-2}. Thus, as pointed out by Chandrasekhar and Fermi (1953), magnetic forces and cosmic-ray pressure (to the extent that it is not balanced by an equal pressure in the halo) seem to be more important than kinetic-theory gas pressure in balancing the gravitational forces that flatten the galactic disk. This is possible only if the magnetic-field direction is predominantly in the plane of the Galaxy.

The virial theorem (see, for example, Biermann and Davis 1960) gives useful necessary conditions that must be satisfied by any static model, but it must always be remembered that the complete system of equations of motion imposes further conditions.

Information derived from cosmic rays has significant bearing on the structure of the galactic magnetic field. The relative abundance of primary heavy particles and fragments produced by collisions with the interstellar gas shows that most cosmic-ray particles are lost by escape from the disk rather than by nuclear interactions. Diffusion, or drift, of the cosmic-ray particles normal to the field lines to regions of weaker field is very slow. Hence, the cosmic rays must escape from the disk along the field lines. This implies that a reasonable fraction of the field lines lead out of the disk and allow the cosmic rays to escape to intergalactic space. Not all field lines can make closed loops in the disk. It is usual to assume, implicitly or explicitly, that small-scale irregularities, such as shocks, lead to a diffusion of the cosmic rays along the field lines. For a 10-μgauss field, the scale of the irregularity must be less than 10^{12} cm. This will produce the observed isotropy of cosmic rays, reduce the rate at which they escape from the Galaxy, and allow the observed density to be maintained with a smaller, more nearly reasonable, production rate.

It is also usual to assume that the density of cosmic rays in the 10^9 to 10^{13} ev range is substantially greater in the galactic disk than outside the Galaxy. This has the important implication that the magnetic field must be reasonably strong, with an energy density of the order of that of the cosmic rays, so that they may be contained. It can be argued that a much weaker field would couple the cosmic rays to the thermal gas and that gravitational forces would provide the containment. However, it seems that this model would be unstable, any fluctuation in density leading to rapid expansion of the field in low-density fluctuations and to collapse downward in high-density fluctuations. The cosmic-ray pressure must be included in discussions of the dynamics of the galactic gas and magnetic fields.

If the cosmic rays were to fill all the space between the galaxies, they would produce a uniform pressure everywhere, and there would almost be no effects on the structure of the galactic field, except that it must be connected to the intergalactic field. Such models remove many difficulties in finding force systems that will contain the galactic cosmic-ray pressure. They also avoid the necessity for finding field configurations that produce a high degree of cosmic-ray isotropy and the proper mean storage time. However, the enormous total cosmic-ray energy of the Universe that is implied makes such models most unattractive, and they are usually rejected in treating the galactic magnetic field.

REFERENCES

Baldwin, J. E. 1963, *Observatory*, **83**, 153.
Behr, A. 1959, *Nach. Akad. Wissensch. Göttingen Math. Phys. Kl.*, p. 185.
Berkhuijsen, E. M., and Brouw, W. N. 1963, *B.A.N.*, **17**, 185.
Berkhuijsen, E. M., Brouw, W. N., Muller, C. A., and Tinbergen, J. 1964, *B.A.N.*, **17**, 465.
Biermann, L., and Davis, L., Jr. 1960, *Zs. f. Ap.*, **51**, 19.
Bolton, J. G., and Wild, J. P. 1957, *Ap. J.*, **125**, 296.
Brouw, W. N., Muller, C. A., and Tinbergen, J. 1962, *B.A.N.*, **16**, 213.
Cayrel, R., and Schatzman, E. 1954, *Ann. d'Ap.*, **17**, 555.
Chandrasekhar, S., and Fermi, E. 1953, *Ap. J.*, **118**, 113.
Clark, B. G. 1963, *Nature*, **197**, 474.
Critchfield, C. L., Ney, E. P., and Oleksa, S. 1952, *Phys. Rev.*, **85**, 461.
Davies, R. D., and Shuter, W. L. H. 1963, *M.N.*, **126**, 369 (paper III).
Davies, R. D., Shuter, W. L. H., Slater, C. H., Verschuur, G. L., and Wild, P. A. T. 1963, *M.N.* **126**, 343 (paper I).
Davies, R. D., Shuter, W. L. H., Slater, C. H., and Wild, P. A. T. 1963, *M.N.*, **126**, 353 (paper II).
Davies, R. D., Slater, C. H., Shuter, W. L. H., and Wild, P. A. T. 1960, *Nature*, **187**, 1088.
Davies, R. D., Verschuur, G. L., and Wild, P. A. T. 1962, *Nature*, **196**, 563.
Davis, L., Jr. 1951, *Phys. Rev.*, **81**, 890.
———. 1958, *Ap. J.*, **128**, 508.
Davis, L., Jr., and Greenstein, J. L. 1951, *Ap. J.*, **114**, 206.
Earl, J. A. 1961, *Phys. Rev. Letters*, **6**, 125.
Galt, J. A., Slater, C. H., and Shuter, W. L. H. 1960, *M.N.*, **120**, 187.
Gardner, F. F., and Whiteoak, J. B. 1963, *Nature*, **197**, 1162.
Ginzburg, V. L. 1964, *Electromagnetic Waves in Plasmas* (London: Pergamon Press), p. 94.
Ginzburg, V. L., and Syrovatskii, S. I. 1965, *Ann. Rev. Astr. and Ap.*, **3**, 297.

Greenberg, J. M., Lind, A. C., Wang, R. T., and Libelo, L. F. 1963, *Proceedings of ICES I* (New York: Pergamon Press), p. 123.
Hall, J. S., and Serkowski, K. 1963, *Basic Astronomical Data*, ed. K. Strand (Chicago: University of Chicago Press), p. 293.
Henry, J. 1958, *Ap. J.*, **128**, 497.
Hiltner, W. A. 1956, *Ap. J. Suppl.*, **2**, 389.
Jackson, J. D. 1962, *Classical Electrodynamics* (New York: Wiley), p. 228.
L'Heureux, J., and Meyer, P. 1965, *Phys. Rev. Letters*, **15**, 93.
Mathewson, D. S., and Milne, D. K. 1964, *Nature*, **203**, 1273.
Miller, C. R. 1962, *On the Orientation of Dust Grains in Interstellar Space*, Ph.D. thesis, California Institute of Technology.
Mills, B. Y. 1959, *Paris Symposium on Radio Astronomy* (*IAU Symposium*, No. 9), p. 431.
———. 1964, *Ann. Rev. of Astr. and Ap.*, **2**, 185.
Morris, D., and Berge, G. L. 1964, *Ap. J.*, **139**, 1388.
Morris, D., Clark, B. G., and Wilson, R. W. 1963, *Ap. J.*, **138**, 889.
Oort, J. H., and Walraven, Th. 1956, *B.A.N.*, **12**, 285.
Parker, E. N. 1966, *Ap. J.* **145**, 811.
Pawsey, J. L. 1965, *Galactic Structure*, eds. A. Blaauw and M. Schmidt (Chicago: University of Chicago Press), p. 219.
Platt, J. R. 1956, *Ap. J.*, **123**, 486.
Seielstad, G. A., Morris, D., and Radhakrishnan, V. 1964, *Ap. J.*, **140**, 53.
Serkowski, K. 1962, *Adv. in Astr. and Ap.*, **1**, 289.
Stranahan, G. 1954, *Ap. J.*, **119**, 465.
Turtle, A. J. 1963, *M.N.*, **126**, 405.
Weinreb, S. 1962, *Ap. J.*, **136**, 1149.
Wentzel, D. G. 1963, *Ann. Rev. of Astr. and Ap.*, **1**, 195.
Westerhout, G., Seeger, C. L., Brouw, W. N., and Tinbergen, J. 1962, *B.A.N.*, **16**, 187.
Wickramasinghe, N. C. 1962, *M.N.*, **125**, 87.
Wielebinski, R., and Shakeshaft, J. R. 1964, *M.N.*, **128**, 19.
Woltjer, L. 1965, *Galactic Structure*, eds. A. Blaauw and M. Schmidt (Chicago: University of Chicago Press), p. 531.

CHAPTER 16

Primordial Stellar Evolution

E. K. L. UPTON
University of California, Los Angeles

1. GENERAL CONSIDERATIONS

THE FORMATION of a star from the interstellar medium involves a change in the mean density from about 10^{-24} to 10^{0} gm/cm^{3}. At the same time, there must be striking increases in the temperature, pressure, and optical thickness of the body, as well as great changes in the ionization, angular momentum, and magnetic field strength. It has long been clear that no one or two factors can dominate the whole of the contraction through such a wide range of the physical conditions. Indeed, it is by no means certain which factors are important at each step of the way. Eventually, however, a stage is reached at which the conditions are basically stellar and we can reasonably expect to identify all dominant factors governing the contraction. This final stage of star formation has been given most detailed consideration, and we will confine our attention to it here. The broader problem of the complete formation of a star has been reviewed by G. R. Burbidge (1960), F. D. Kahn (1960), and Ebert, von Hoerner, and Temesváry (1960).

Theoretically, a star in the final stages of formation from the interstellar medium may be considered as an object that is like a main-sequence star in all essential respects, except that it lacks nuclear energy sources. Lord Kelvin (1887) and others pointed out long ago that such a body will contract following leakage of energy that is not replaced. Details of this process can be calculated by an application of the usual principles of stellar structure if the following two conditions are satisfied: (1) the optical thickness of the star must be large, so that radiation escapes only from a thin surface layer whose detailed structure may be crudely approximated and (2) the hydrodynamic time scale must be substantially shorter than the thermal time scale, so that hydrostatic equilibrium can be assumed in calculating the heat flow.

The first of these conditions is satisfied when

$$\kappa \gg 10^{-11} \frac{R^2}{\mathfrak{M}}, \tag{1}$$

where κ is the suitably averaged absorption coefficient per gram, and $\mathfrak{M}$ and R are the mass and radius in solar units.

An order-of-magnitude examination of the second condition may be performed as follows. The hydrodynamic time scale is approximately (Schwarzschild 1958, p. 32)

$$\tau_{\text{hyd}} = \left(\frac{2G\mathfrak{M}}{R^3}\right)^{-1/2} = 10^3 \frac{R^{3/2}}{\mathfrak{M}^{1/2}} \text{ seconds}. \qquad (2)$$

The thermal or Kelvin time scale is approximately

$$\tau_{\text{Kel}} = \frac{\text{energy}}{\text{luminosity}} = \frac{G\mathfrak{M}^2}{R} \cdot \frac{1}{L}. \qquad (3)$$

If energy transport is radiative, then L can be approximated to an order of magnitude by

$$L = 4\pi R^2 \cdot \frac{4acT^4}{3\kappa\rho R}, \qquad (4)$$

where T and ρ are average values for the star. These may also be estimated to an order of magnitude (e.g., Schwarzschild 1958, p. 32) by the approximate relations

$$\rho \cong \frac{3}{4\pi} \frac{\mathfrak{M}}{R^3} \qquad (5)$$

$$T \cong G\mu \frac{m_{\text{H}}}{k} \frac{\mathfrak{M}}{R}. \qquad (6)$$

Upon combining equations (3) and (6), with $\mathfrak{M}$ and R in solar units, we obtain

$$\tau_{\text{Kel}} = 7 \times 10^{10} \frac{\kappa}{\mu^4 \mathfrak{M} R} \text{ seconds}. \qquad (7)$$

For main-sequence stars, the hydrodynamic time scale is much less than the thermal, but with increasing radius, the former increases while the latter decreases. The desired inequality is maintained only as long as

$$\kappa \gg 1.4 \times 10^{-8} \mu^4 \mathfrak{M}^{1/2} R^{5/2}. \qquad (8)$$

Inequality (1) states that the opacity must be high enough to block the outflow of radiation below the surface layers; inequality (8) provides the more stringent condition that it must, in addition, be high enough to cause heat to flow at a slower pace than pressure waves do.

For ordinary stellar masses at any radius larger than the main-sequence value, condition (8) is substantially more stringent than (1) and accordingly limits the range of the ordinary treatment. Hayashi and Hōshi (1961) have shown (see Sec. 6) that stellar radii cannot increase beyond about 10 times their main-sequence values without a change from radiative to convective energy transport. When this occurs, the thermal time scale cannot be calculated by (7), and inequality (8) loses some of its significance. A full discussion of the thermal time scale for a convective star will not be attempted here, but the time scale can hardly be longer than that based on radiative transport alone. Thus condition (8) must still be satisfied.

An extensive examination of opacity mechanisms in very early stages of star formation has been made by Gaustad (1963), who found that condition (1) would be satisfied in the stage where $T < 1000°$ K and where the opacity is due to ice or mineral grains. Condition (8) is not satisfied in this domain, however. In the temperature range of $1000° < T < 2000°$ K, the ices evaporate and the residual opacity, due mainly to H_2, is too small for either condition (1) or (8) to be satisfied. Above 2000° K the dissociation of H_2 leads to a thermodynamic instability (discussed in Section 9) that rules out hydrostatic models until both dissociation and ionization of hydrogen are complete, with mean temperatures of the order of 10^5 ° K and with radii of the order of 100 times the main-sequence values. Here we are in the well-studied domain of stellar opacities, where both conditions (1) and (8) are easily satisfied.

Thus it appears that the combination of opacity deficiency and thermodynamic instability rules out treatment of the contracting star by the usual methods of stellar structure—at least for a star of approximately 1 $\mathfrak{M}_\odot$—until the radius is down to about 100 times the main-sequence value. Gaustad finds, however, that the opacity criteria may be satisfied earlier for stars with masses $< 0.1\ \mathfrak{M}_\odot$.

Additional factors might introduce special complexities into the pre-main-sequence problem. For example, rotational and magnetic effects certainly play prominent roles at some stage in the formation of a star. The reasoning on these points has been reviewed many times; see, for example, Hoyle (1960); Spitzer (1963, and this volume); Burbidge (1960); Kahn (1960); and Ebert, von Hoerner, and Temesváry (1960). See also Volume 8 of this compendium, chapters 5, 8, and 9. A brief summary follows.

An interstellar cloud from which a star condenses must have an angular velocity about its center of gravity of at least 10^{-15} radians/sec owing to galactic rotation. If part of such a cloud contracted uniformly from a density 10^{-24} to 1 gm/cm^3 without loss of angular momentum, the angular velocity must increase by a factor of about 10^{16}, leading to a final value of about 10 radians/sec. It is clear that rotational instability will affect the course of events long before such a stage is reached.

In a spherically symmetrical contraction where the magnetic field is dragged along with the gas, the magnetic energy of the contracting object increases in proportion to $1/R$. This is the same as the rate of increase of the gravitational and thermal energies; so the ratios of magnetic to these other forms of energy remain constant in such a contraction.

The thermal energy in an H I region with $\rho = 100$ atoms/cm^3 and $T = 100°$ K is about $U_{\text{thermal}} = 2 \times 10^{-12}$ ergs/cm^3. The same value applies to an H II region with $\rho = 1$ atom/cm^3 and $T = 10^4$ ° K. On the other hand, if the magnetic field in these regions is 10^{-5} gauss, the magnetic energy is $U_{\text{magnetic}} = 3 \times 10^{-12}$ ergs/cm^3. Thus the magnetic energy is at least comparable with the thermal energy, and this ratio will be maintained unless the gas can slip across the lines of force as suggested by Mestel and Spitzer (1956).

According to the Jeans criterion the self-gravitating energy of a stellar condensation must exceed its thermal energy. For 1 $\mathfrak{M}_\odot$ at $\rho = 10^{-22}$ gm/cm^3, the gravitational energy density amounts to $U_{\text{grav}} = G\mathfrak{M}\rho/R \sim 10^{-14}$ ergs/cm^3. Thus in the interstellar materials which at first sight seem likely sources of star formation, thermal and magnetic energies are both far in excess of the gravitational energy and will prevent any condensation unless they can somehow be removed from the gas. These problems are evidently present at the very beginning of the contraction process, but the various theories on their solution lie outside the scope of this chapter.

Schatzman (1962) has considered the problem of contraction onto the main sequence with mass loss due to rotational instability. Other workers have preferred to assume that the magnetic and rotational problems are solved at a relatively early stage, so that these factors may be neglected when a star is in the ordinary regions of the Hertzsprung-Russell diagram. This is essentially a hopeful assumption made for the sake of convenience, but it is supported to a certain extent by the lack of high rotational velocities in the T Tauri stars studied by Herbig (1962). These stars are certainly in the last stages of contraction before reaching the F to M regions of the main sequence. Of four stars whose rotation was studied by Herbig, the largest value of $V \sin i$, that of RY Tauri, does not exceed 50 km/sec. The radius of this star is about 3 $R_\odot$, according to Herbig, and on the basis of its luminosity, its mass may be 1 to 2 $\mathfrak{M}_\odot$. (A mass estimate of this kind requires only the assumption of radiative transport and a temperature structure roughly comparable to the main-sequence stars. The question of whether masses can be found more precisely by a detailed comparison with theoretical contraction paths will be taken up in Sec. 5.) Assuming $\mathfrak{M} = 1$ and $R = 3$, we find an escape velocity of 360 km/sec. Thus we may conclude that the stars studied by Herbig are far from rotational instability. It must be borne in mind, however, that these stars cover only about the last factor 3 in radial contraction.

Let us then neglect rotational and magnetic complications and limit ourselves to situations where the hydrodynamic time scale is short compared to the thermal time scale. We can then regard the contraction process in the following way. At any instant the star has some total energy E, a sum of gravitational and thermal contributions that is negative if the zero of gravitational energy is taken to correspond to matter at infinity. For the given negative energy, there is some radius of order

$$R \sim \frac{G\mathfrak{M}^2}{-E}, \tag{9}$$

where the star will be in hydrostatic equilibrium. The exact value of this radius and also the detailed structure of the star, are determined by the distribution of energy throughout the interior. Given this distribution and the total energy, the star will find its equilibrium radius in a few pulsation times, during which E may be considered constant. Contraction to a smaller radius then comes about as a result of the subsequent slow decrease in energy (that is, increase in $|E|$), and the rate of contraction is governed by the rate at which energy is lost. This in turn is governed by the opacity in those regions of the star where radiative transport prevails.

Radiative energy transport is simpler to study theoretically than convective transport. Partly for that reason, and partly because radiative transport was the first of the two mechanisms shown to be important in modern stellar astrophysics, most studies of stellar contraction have been based on the assumption that radiative transport prevails throughout the star. In the present review we shall consider the results of those studies first and turn to questions of convection in Section 6.

2. CONTRACTION WITH RADIATIVE TRANSPORT

Jeans (1919) formulated the equations governing the behavior of a contracting body with radiative transport in a way that included hydrodynamic effects. His equations apply to the behavior of a star throughout its life, starting with any arbitrary pressure and temperature distribution. The hydrodynamic terms in these equations are usually omitted for the reasons mentioned above, and when this is done, the equations may be expressed as follows:

$$\frac{\partial P}{\partial r} = -\rho \frac{Gm}{r^2}, \tag{10}$$

$$\frac{\partial m}{\partial r} = 4\pi r^2 \rho, \tag{11}$$

$$T \frac{\partial S}{\partial t} = -\frac{1}{\rho} \operatorname{div} \mathbf{H}, \tag{12}$$

$$\mathbf{H} = \frac{-c}{K\rho} \operatorname{grad} P_r, \tag{13}$$

where r = distance from center of star; m = mass interior to r; P = total pressure; P_r = radiation pressure; ρ = density; T = temperature; S = entropy per unit mass; $\mathbf{H}$ = heat flow (energy per unit time per unit area); G = gravitational constant; and t = time.

These equations correspond to the four basic equilibrium equations as listed by Schwarzschild (1958, p. 96). Here we use partial derivatives because there are two independent variables, r and t. The equation of radiative transfer, equation (13), is given here in the form used by Eddington (1926, p. 101) which has the advantage of simplicity. The vector $\mathbf{H}$ is introduced here only for the purpose of exhibiting the basic divergence property of (12). In terms of the running luminosity L_r used by Schwarzschild,

$$H = \frac{L_r}{4\pi r^2}. \tag{14}$$

The time derivative in equation (12) distinguishes the present set from the equations for a steady-state star. Instead of equating the divergence of the flux to an energy source, we equate it to the rate at which energy is given up in the form of heat, and thus to a rate at which the entropy changes:

$$T \frac{\partial S}{\partial t} = \frac{dQ}{dt}. \tag{15}$$

The fundamental role of entropy in this equation, and in the physics of stellar contraction generally, was pointed out by Thomas (1931*a*), and before him, by Emden (1907). The advantage of the entropy concept is apparent when this problem is described in the following terms.

We are dealing basically with a problem of time-dependent heat flow. The run of temperature and opacity at any instant governs the pattern of heat flow, in accordance with equation (13). Since there is a net outward flow of energy, the heat-flow pattern has a divergence; and as a result of that divergence, there is a decrease in entropy at a rate given by equation (12). The entropy values are thus determined for a later time, and in a numerical calculation, the new values of S can be found by (12) without any complications due to changing pressure or density. Thus it is often useful in these calculations to adopt the entropy as one of the independent-state variables of the gas.

There are eight dependent variables in equations (10)–(13); hence four auxiliary equations must be introduced for a solution to be possible. The first equation is a formula for the opacity in terms of density and temperature:

$$\kappa = \kappa(\rho, T) . \tag{16}$$

The other three will be provided by three relations among the five state variables S, T, P, P_r, ρ. We will always have, for example,

$$P_r = \tfrac{1}{3}aT^4 , \tag{17}$$

and assuming full ionization with no electron degeneracy, we may write

$$P = \frac{k}{\mu m_{\mathrm{H}}} \rho T + \tfrac{1}{3} a T^4 , \tag{18}$$

$$S = \frac{k}{\mu m_{\mathrm{H}}} \left(\tfrac{5}{2} + \log \frac{T^{3/2}}{\rho} + C \right) + \tfrac{4}{3} \frac{aT^3}{\rho} , \tag{19}$$

where a = Stefan's constant, k = Boltzmann's constant, m_{H} = mass of hydrogen atom, and μ = molecular weight. The constant C in equation (19) is related to the zero point of entropy. Its value depends on the composition of the star, but in nondegenerate cases, it can be arbitrarily taken as zero since only entropy differences matter. Where it is necessary to deal with electron degeneracy or with a variable degree of ionization, equations (18) and (19) must be replaced by more general equations. The forms given here include the effects of radiation as the second term in each equation.

If the usual boundary conditions are added to the equations listed above, we can obtain numerical solutions. It is important, however, to recognize that the numerical solution will not, in general, be unique in the sense that a unique model can be found for main-sequence stars or other steady-state cases. We are dealing here with a non–steady-state situation, and the Vogt-Russell theorem on the existence of unique solutions does not apply to such cases. The solutions to the time-dependent equations (10)–(13) will, in general, be unique only in the following sense: given some initial model, it is possible to determine its development uniquely. Any initial model can be assumed, as far as the thermal structure is concerned, and the equations then give an evolutionary sequence from that point on. That is, the contracting-star problem as described by these equations is necessarily one of sequences rather than of an individual model taken in isolation.

In his original discussion of the contraction problem, Jeans pointed out that calculation of sequences for arbitrary initial models would not be a simple matter, and in fact, it was not accomplished until fast electronic computers came into use. But Jeans also pointed out that under certain conditions a homologously contracting sequence, in which the star changes only by scale factors while the model remains constant in dimensionless form, is possible. Jeans furthermore guessed, but was unable to prove, that stars which differ initially from the homologous model should approach it as their contraction progresses. Partly because of the intuitive appeal of this idea, but also largely because of the great saving of labor in the calculations, the idea of homologous sequences gained and has maintained a prominent position in studies of stellar contraction.

3. CONDITIONS FOR HOMOLOGOUS CONTRACTION

The conditions for homologous contraction can most readily be demonstrated by a plot of the star's structure in a log ρ–log T diagram (Fig. 1).

The two curves in the figure show schematically the run of temperature and density from center (C) to surface (S) at two stages in a homologous contraction. Such a contraction may be defined as one in which curve C_2S_2 is generated by a constant displacement of curve C_1S_1; that is, the motion of the center of the star along the line C_1C_2 is the

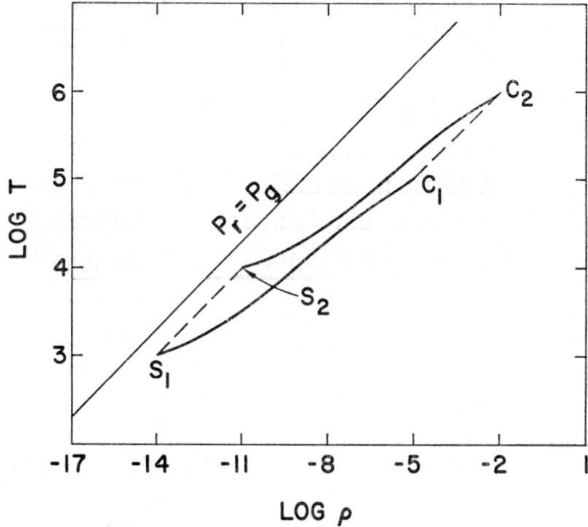

FIG. 1.—General schematic illustration of homologous contraction. Curve S_1C_1, representing the initial structure of the star, undergoes uniform displacement to later position S_2C_2, each point following a path parallel to the line $P_r = P_g$.

same as the motion of the surface along the line S_1S_2, and likewise for any other point in the star. The slope of these contraction lines in the log ρ–log T diagram is

$$\frac{d \log T}{d \log \rho} = \tfrac{1}{3}, \tag{20}$$

a result known as Lane's law, which can be derived from elementary dimensional analysis. (See, for example, Jeans [1929], p. 64.)

In Figure 1, it can be seen that the slope for homologous contraction is the same as the slope for constant ratio of radiation pressure to gas pressure, since

$$\frac{P_{\text{rad}}}{P_{\text{gas}}} = \tfrac{1}{3}\,\frac{a\mu m_{\text{H}} T^3}{k\rho}. \tag{21}$$

That is, the lines S_1S_2, C_1C_2, and so forth are parallel to the line on which $P_{\text{rad}} = P_{\text{gas}}$. Expressed otherwise, Eddington's β, defined as $P_{\text{gas}}/P_{\text{total}}$, remains constant at any point in the star. Furthermore, the radiative entropy per unit mass,

$$S_{\text{rad}} = \tfrac{4}{3} a \,\frac{T^3}{\rho}, \tag{22}$$

remains constant as well. Thus there arises the useful circumstance that radiation terms do not affect the conditions for homologous contraction. That is, in equations (10)–(13) we can make the substitutions

$$P = \frac{P_{\text{gas}}}{\beta}, \tag{23}$$

$$P_r = P_{\text{gas}} \frac{(1-\beta)}{\beta}, \tag{24}$$

and

$$\frac{\partial S}{\partial t} = \frac{\partial S_{\text{gas}}}{\partial t} = \frac{k}{\mu m_{\text{H}}} \frac{\partial}{\partial t} \log\left(\frac{T^{3/2}}{\rho}\right), \tag{25}$$

thereby eliminating all radiation terms except β, which remains constant and has no effect on the type of solution found. This result was noted by Jeans (1919) and also by Thomas (1931*a*).

We next consider what type of opacity law may be compatible with homologous contraction. Equations (10)–(13) show that, just as ρ and T must increase by constant scale factors in all parts of the star, so too must the luminosity and, hence, the opacity; otherwise the flux distribution in the star would change and the homology conditions would not be met. Therefore the opacity must be a function of T and ρ such that equal displacements along the lines C_1C_2, S_1S_2, and so forth lead to equal changes in log κ. Functions which meet this condition are of the form (Thomas 1931*a*)

$$\kappa = T^{-n} \lambda \left(\frac{\rho}{T^3}\right), \tag{26}$$

where n is any constant and λ is any function of ρ/T^3. Electron scattering, Kramers formula, and the modified Kramers formula

$$\kappa = \kappa_0 \rho^a T^{-b}, \tag{27}$$

are of the right form to permit homologous contraction, whereas an opacity formula containing both scattering and Kramers-type terms is not.

Consider next the entropy, S. We have already seen that radiation entropy remains constant in homologous contraction. The gas entropy for an ideal monotonic gas is of the form given in (19):

$$S_{\text{gas}} = \frac{\kappa}{\mu m_{\text{H}}} \left(\tfrac{5}{2} + \log \frac{T^{3/2}}{\rho} + C\right). \tag{28}$$

It is evident that an entropy of this form increases uniformly throughout the star during homologous contraction. The same would be true if the temperature exponent in (28) were replaced by the more general expression $1/(\gamma - 1)$ for a gas of any constant γ (ratio of specific heats).

Under these conditions, equation (12) may be written

$$\frac{1}{T} \frac{\partial}{\partial m} (4\pi r^2 H) = \epsilon_0, \tag{29}$$

where ϵ_0 is a constant whose value depends on the total luminosity of the star.[1] Equation (29) then expresses the condition of homologous contraction as far as the entropy is concerned.

[1] ϵ_0 may be regarded as an energy-generation constant for gravitational energy.

Equation (28) does not represent the most general entropy relation for which homologous contraction can occur, but an exhaustive discussion of other possibilities will not be given here. There is probably only one homologous case of any practical interest involving a more complicated expression for the entropy, and that is a partially degenerate star with an adiabatic structure and convective heat transport. Such a star should contract homologously as long as convection maintains an adiabatic structure against the tendency of electron heat conduction to develop an isothermal one; but in such a case, the relations among the scale factors for P, ρ, and T are continually changing as degeneracy increases. The red-dwarf stars very likely are in this condition for part of their contraction. Other common deviations from the entropy formula (28), such as those associated with a variable degree of ionization, can easily be seen to preclude anything approaching a homologous contraction.

Let us look finally at the outer boundary conditions in relation to homologous contraction. If it is sufficiently accurate, in relation to the main body of the star, to use the zero boundary conditions, viz.,

$$P = 0 \,, \qquad T = 0 \tag{30}$$

at the surface, then homology is possible all the way out to this boundary, provided the other requirements are met. If, however, it is necessary to use more realistic photospheric boundary conditions such as

$$P_{\text{phot}} = \tfrac{2}{3} \frac{G\mathfrak{M}}{R^2 \kappa} \tag{31a}$$

$$4\pi R^2 \sigma T^4_{\text{phot}} = L \,, \tag{31b}$$

then homology will be violated in the atmospheric layers. For homology requires that P and T increase in proportion to $1/R^4$ and $1/R$, respectively, and the photospheric values calculated by equations (31) will never conform to this rule in any realistic case. For example, in contraction of the Sun from 100 $R_\odot$ to its present size, the ratio of surface temperature to central temperature should remain constant in a truly homologous series. However, since homologous radiative contraction tends to proceed at nearly constant luminosity, the surface temperature increases by only a factor of 10 in this process, while the central temperature increases by a factor of 100. The ratio T_{phot}/T_c therefore goes from about 1/300 to the present value, 1/3000. When the effect of convective transport is taken into account, the change in T_{phot} is still smaller, and the departure from homology is correspondingly greater. Common sense, of course, suggests that this effect cannot be important for the star as a whole as long as the ratio is in the region of these small values. Calculations by the author and by Brownlee and Cox (1961), although directed primarily to other questions, support the intuitive idea that lack of homology in the photospheric layers is not inconsistent with homology in the star as a whole.

To summarize, we can say that homologous contraction is possible when the star is composed of a perfect gas in a constant state of ionization, plus radiation in any desired proportion, provided the opacity has the right sort of dependence on density and temperature. That necessary condition on the opacity, found by Thomas and reproduced as equation (26), is met in the interiors of the stars to the extent that they follow Kramers formula or a modified Kramers formula (eq. [27]). There will inevitably be serious deviations from homology near the stellar photosphere, but as long as T_{phot}/T_c is small, these deviations have little or no effect on the homology of the star as a whole.

4. RESULTS FOR HOMOLOGOUS CONTRACTION

The first detailed study of a star in homologous contraction was conducted by Levée (1953) for the case of Kramers opacity and no radiation pressure. The equations governing the structure of the star are obtained from (10), (11), (13), and (29) by elimination of the density with the help of (18), and the radiation pressure with the help of (17). We also replace H by the total flux through the spherical surface of radius r, $L_r = 4\pi r^2 H$. The following set of equations is obtained:

$$\frac{dP}{dm} = -\frac{G}{4\pi}\frac{m}{r^4}, \tag{32}$$

$$\frac{dr}{dm} = \frac{k}{4\pi\mu m_{\text{H}}}\frac{T}{\beta P r^2}, \tag{33}$$

$$\frac{dLr}{dm} = \epsilon_0 T, \tag{34}$$

$$\frac{dT}{dm} = -\frac{3}{64\pi^2}\frac{\kappa}{ac}\frac{L_r}{r^4T^3}, \tag{35}$$

$$\beta = 1 - \frac{a}{3}\frac{T^4}{P}, \tag{36}$$

$$\kappa = \frac{\kappa_0 \mu m_{\text{H}} \beta}{k} P T^{-4.5}. \tag{37}$$

These equations are the same as those of Levée, except that here we have kept the Lagrangian form with the mass as independent variable. We have also kept the contribution of the radiation pressure, expressed by β, which was taken equal to one in Levée's model.

The appearance of total instead of partial derivatives in these equations reflects the fact that by means of the homology assumption a separation has been effected between the time and space aspects of the problem. To solve the spatial aspect, we must find the march of the physical variables which satisfies the four differential equations (32)–(35), together with the radiation pressure relation (36) and the opacity formula (37), and which at the same time satisfies the four appropriate boundary conditions. The two boundary conditions at the center are

$$r = 0\,, \quad L_r = 0 \quad \text{at} \quad m = 0\,, \tag{38}$$

while for the surface either (30) or (31) can be used with little difference for radiative models.

A number of different methods can be used to work out this spatial solution numerically. One of the simplest in concept is to integrate the equations simultaneously from the surface toward the center. To start the integrations, we must specify a mass $\mathfrak{M}$ and a radius R corresponding to the type of star and the stage of contraction desired. The total luminosity L and the energy-generation constant ϵ_0 must be guessed. We then have the following starting values for the surface: $m = \mathfrak{M}$, $r = R$, $L_r = L$, $T = 0$, and $P = 0$. We then integrate inward to $m = 0$ and check whether the two boundary conditions (38) are satisfied. If they are not, new guesses at ϵ_0 and L are made, and the process is repeated until the correct model is found.

The preceding description passes over certain questions of practical technique, such as the need for a series development when the zero boundary conditions are used at the surface, and the fact that divergences in the trial solutions which occur as the center is approached make it preferable to run the integrations both from the surface and from the center, fitting at some midway point. For discussion on these points of technique, the reader is referred to the original paper of Levée or to Schwarzschild's discussion of these questions for stellar models in general (1958, chap. 3).

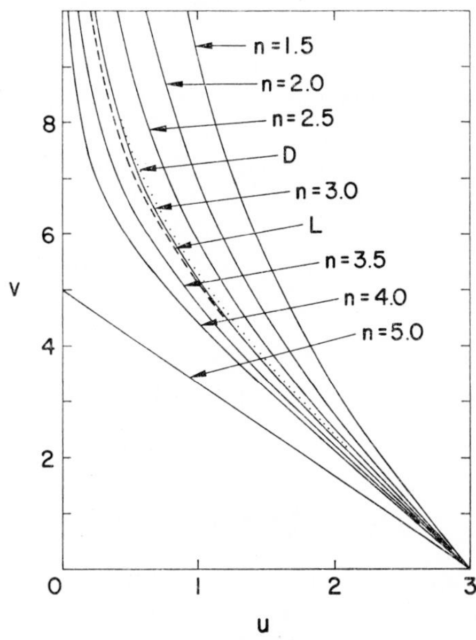

FIG. 2.—Structure of the Lane-Emden polytropic models in the UV diagram of Milne. Values of the polytropic index, n, are given. Curve L (*dashed line*) is Levée's model for radiative homologous contraction. Curve D (*dotted line*) is a model discussed in Section 5.

Levée's solution of the foregoing equations for the case of no radiation pressure turned out to be very nearly the Emden polytropic model with index $n = 3$.[2] This is illustrated in Figure 2, in which a number of stellar models are shown in the UV diagram of Milne. The polytropic models are designated by their n values, and curve L is Levée's model. Curve D is an approximately homologous model by the author which will be discussed at the end of Section 5.

The reason for the close resemblance of Levée's model to Eddington's standard model can be understood when it is recalled that Eddington's model was derived from the assumption that

$$\eta\kappa = \text{constant}, \tag{39}$$

where

$$\eta = \frac{L_r}{m}\frac{\mathfrak{M}}{L} = \frac{\bar{\epsilon}_{\text{inside}}}{\bar{\epsilon}_{\text{star}}}. \tag{40}$$

[2] Called the *standard model* by Eddington.

The quantities $\bar{\epsilon}_{\text{star}}$ and $\bar{\epsilon}_{\text{inside}}$ are mass averages of the energy source per unit mass for the whole star and for the region interior to the point in question. When the energy generation rate ϵ is proportional to temperature (eq. [34]) and when the opacity follows Kramers formula (37), the run of temperature and pressure (or density) in the standard model is such that $\eta\kappa$ is nearly constant. For, since P is proportional to $T^{n+1} = T^4$ in the standard model, we have

$$\kappa = \text{const}\, T^{-1/2}\,, \tag{41}$$

which may be combined with (34) and (40) to obtain

$$\eta\,\kappa = \text{const}\,(T^{-1/2})\left(\frac{\bar{T}_{\text{inside}}}{\bar{T}_{\text{star}}}\right). \tag{42}$$

The factor $\bar{T}_{\text{inside}}/\bar{T}_{\text{star}}$ slowly decreases from 1.708 at the center to 1.000 at the surface in the standard model, approximately balancing the increase in $T^{-1/2}$, except in the outermost layers.

The time aspect of the homologous contraction problem is solved much more readily than the space aspect, requiring little more than the Lane-Emden type of homology relation based on dimensional analysis. In terms of the changes in the stellar radius R, other quantities change as follows (the subscript m denotes a fixed mass element):

$$d\,\log\,r_m = d\log R \tag{43}$$

$$d\,\log\,T_m = -d\,\log R \tag{44}$$

$$d\,\log\,\rho_m = -3\,d\,\log R \tag{45}$$

$$d\log P_m = -4\,d\,\log R \tag{46}$$

$$d\,\log\,\kappa_m = (b - 3a)\,d\log R \tag{47}$$

$$d\log L_m = -(b - 3a)\,d\,\log R\,, \tag{48}$$

where a and b are the exponents in the opacity formula (27).

The actual time rate of these changes is found from the relation

$$\frac{dE}{dt} = -L\,, \tag{49}$$

where E is the star's total energy, gravitational plus thermal. The zero point of E is generally set at infinite radius, so that E is a negative number which represents the binding energy of the star. Let Ω be the gravitational potential energy, also a negative quantity. Then from the virial theorem we may derive Eddington's (1926) equation 103.5, relating E to Ω:

$$E = \frac{\bar{\beta}\Omega(\gamma - 4/3)}{\gamma - 1}\,, \tag{50}$$

where $\beta = P_g/P$ and

$$\frac{1}{\gamma - 1} = \frac{\text{gas energy density exclusive of radiation}}{\text{gas pressure}}\,. \tag{51}$$

Both β and γ as used here are averaged throughout the star. The potential energy Ω can be expressed as

$$\Omega = -C_1\frac{G\mathfrak{M}^2}{R}\,, \tag{52}$$

where C_1 is a constant depending on the model.

Combining (50) and (52), with $\gamma = 5/3$ for a monatomic gas and $C_1 = 3/2$ for the standard model, and giving the mass and radius in solar units, we can express the total energy as

$$E = -2.85 \times 10^{48} \beta \frac{\mathfrak{M}^2}{R} \text{ ergs}. \tag{53}$$

In the standard model, β is a constant, and its value may be found from Eddington's quartic equation (84.6):

$$\frac{1-\beta}{\beta^4} = 0.00309 \mu^4 \mathfrak{M}^2. \tag{54}$$

Table 1 is an extension of Eddington's Table 9, giving values of $1 - \beta$ or β as a function of $\mathfrak{M}$ for $\mu = 0.693$, corresponding to the composition whose properties have been extensively tabulated by Unsöld (1955). The large masses have been included here because of recent speculations concerning the contraction of objects in this mass range.

TABLE 1

β AS FUNCTION OF MASS FOR $\mu = 0.693$

$\mathfrak{M}/\mathfrak{M}_\odot$	$1-\beta$	$\mathfrak{M}/\mathfrak{M}_\odot$	β	$\mathfrak{M}/\mathfrak{M}_\odot$	β
0.0625	2.78×10^{-6}	4	0.989	10^3	0.185
0.125	1.11×10^{-5}	8	.962	10^4	.0612
0.250	4.45×10^{-5}	16	.885	10^5	.0192
0 500	1.78×10^{-4}	32	.760	10^6	.00612
1.000	7.11×10^{-4}	64	.606	10^7	.00194
2.000	2.82×10^{-3}	128	0.465	10^8	0.000612

Now, combining (49) with (53), and expressing luminosity in solar units (denoted by L) and time in millions of years (denoted by τ), we obtain a convenient expression for the time rate of contraction,

$$d\tau = 23.4 \frac{\beta \mathfrak{M}^2}{L} d\left(\frac{1}{R}\right). \tag{55}$$

The standard model has been adopted in these energy calculations because of its apparent general applicability to cases of homologous contraction, as remarked above. We can see the smallness of the difference between it and the Levée model by comparing the constants in (53) and (55) with the corresponding values for the Levée model, 2.94×10^{48} and 24.1, respectively.

Equations (43)–(48), together with (53)–(55), describe the time variations of all the important quantities in the star relative to their initial values. It remains to fix the magnitude of a few key quantities at some stage in the contraction. The physical variables at the center are related to the mass, radius, and molecular weight as follows:

$$\rho_c = 76.4 \frac{\mathfrak{M}}{R^3} \text{ gm/cm}^3, \tag{56}$$

$$P_c = 0.125 \times 10^{18} \frac{\mathfrak{M}^2}{R^4} \text{ dynes/cm}^2, \tag{57}$$

$$T_c = 19.8 \times 10^6 \beta \mu \frac{\mathfrak{M}}{R} \,^\circ\text{K}. \tag{58}$$

These are simply the standard Lane-Emden homology relations with constants evaluated for the standard model, and for $\mathfrak{M}$ and R in solar units.

The luminosity is given as follows by Levée for the case of Kramers opacity:

$$L = \frac{9.92 \times 10^{24}}{\kappa_0} \frac{\mu^{7.5} \mathfrak{M}^{5.5}}{R^{0.5}}, \tag{59}$$

where the large constant is of the same order of magnitude as κ_0. A typical value of κ_0 for stars near the main sequence, and of Unsöld's composition referred to previously, is 1.8×10^{24}. The origin and applicability of this value will be explained presently. Putting this value of κ_0 into (59) with $\mu = 0.693$, and allowing for $\beta \neq 1$, we obtain

$$L = 0.35 \frac{\mathfrak{M}^{5.5}}{R^{0.5}}. \tag{60}$$

In Figure 3 the dashed lines are tracks for homologous contraction in the theoretical Hertzsprung-Russell diagram, calculated by equation (60). The zero-age main sequence and the position of the Sun are also shown. It is perhaps worth repeating here the three

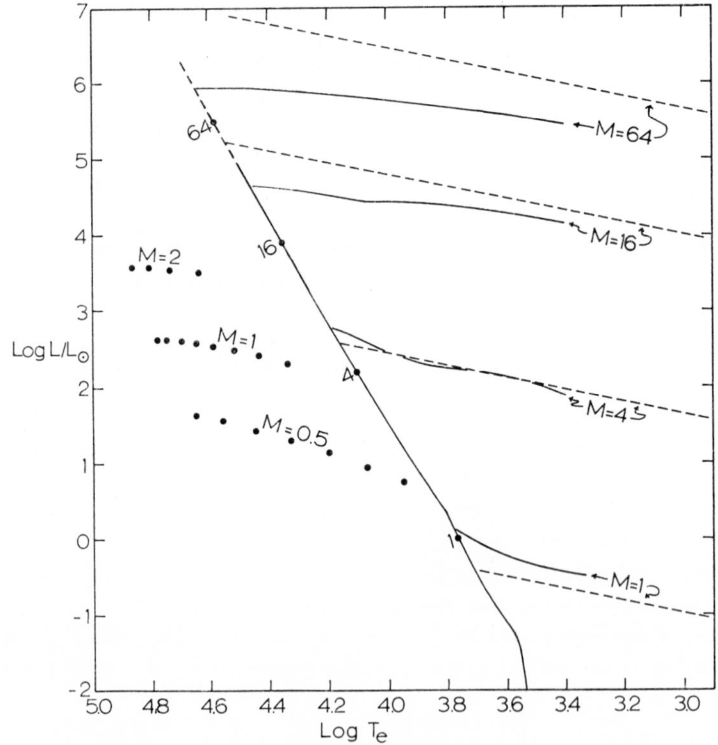

FIG. 3.—Pseudo-homologous pre-main-sequence evolutionary tracks based on Levée's model for purely radiative transport. Tracks and main-sequence positions are shown for 1, 4, 16, and 64 $\mathfrak{M}_\odot$ and Unsöld's composition ($X = 0.561$, $Y = 0.406$, $Z = 0.033$). Kramers opacity with constant coefficient, - - - - -; opacity for mean point in star taken from Keller-Meyerott tables, ———; helium star contraction tracks, ...

main assumptions on which these tracks are based. These are (*a*) Kramers opacity, (*b*) radiative transport as the method of energy transfer throughout, and (*c*) initial conditions such that contraction is homologous. Any or all of these assumptions may be invalid in a particular star, so that the tracks in Figure 3 should be regarded only as a basis for comparison with more exact theories. We will explore here the effect of dropping assumption (*a*), and in later sections, the effects of dropping assumptions (*b*) and (*c*).

Let us consider the actual opacity to be expected in a contracting star built on the standard model. Figures 4 and 5 show the situation in the log P_g–log T diagram for stars of 1, 4, 16, and 64 solar masses, each at approximately ten times its main-sequence

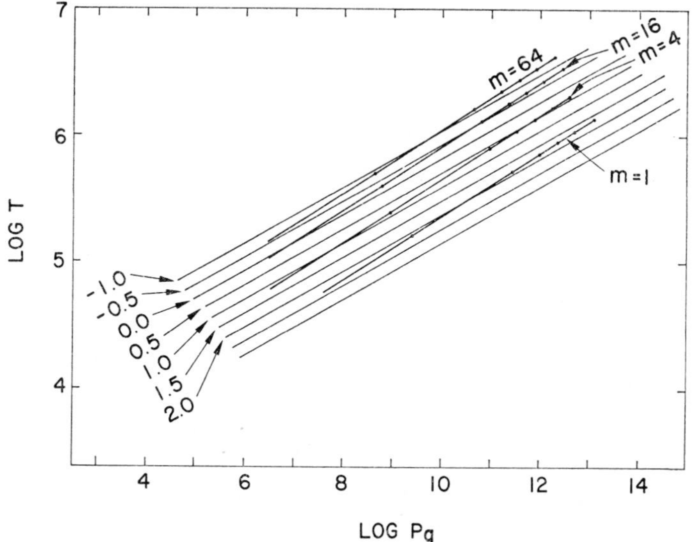

FIG. 4.—Temperatures, pressures, and opacities in the standard model for four stellar masses (unit: solar mass) with radii 10 × the main-sequence values. Dots correspond to mass fractions q = 0.0, 0.2, 0.4, 0.6, 0.8, and 0.99 starting at top right. The opacity contours are based on the Kramers formula, and are labeled with the values of log κ.

radius. The precise radii are 10, 26, 57, and 126 in solar units. The molecular weight $\mu = 0.693$ has again been adopted. The heavy straight lines show the variation of gas pressure with temperature in these stars. Starting at top right, the first five marked points on these lines divide the stars into five shells of equal mass, and the lowest point marks off the shell containing 99 per cent of the mass. The curves labeled with their log κ values (at left) are opacity contours. In Figure 4 the opacity contours are calculated by Kramers formula $\kappa_0 = 1.8 \times 10^{24}$, while in Figure 5 actual Rosseland mean opacities for the Unsöld composition are shown. The latter are taken from Unsöld's diagram (1955, p. 198) for low temperatures and are obtained by interpolation in the Keller-Meyerott tables (1955) for high temperatures. The value of κ_0 for the Kramers case was chosen to agree as well as possible with the Keller-Meyerott opacities in the model shown for 4 $\mathfrak{M}_\odot$.

An examination of Figures 4 and 5 shows several features of interest. First, the Kramers opacities assumed in the Levée model do not vary by more than a factor of 3 from the center of the star to the 99 per cent mass shell. The variations of the more exact opacities are greater than this for masses 1 and 4, but are less for masses 16 and 64 because of the lower limit on the opacity set by electron scattering. Relative to mass 4, the Kramers formula gives too high a value for the opacity at mass 1 and too low a value at masses 16 and 64, with the consequence that the luminosities calculated by equation (60) and shown by the dashed lines in Figure 3 are too low for the Sun and too high for the massive stars.

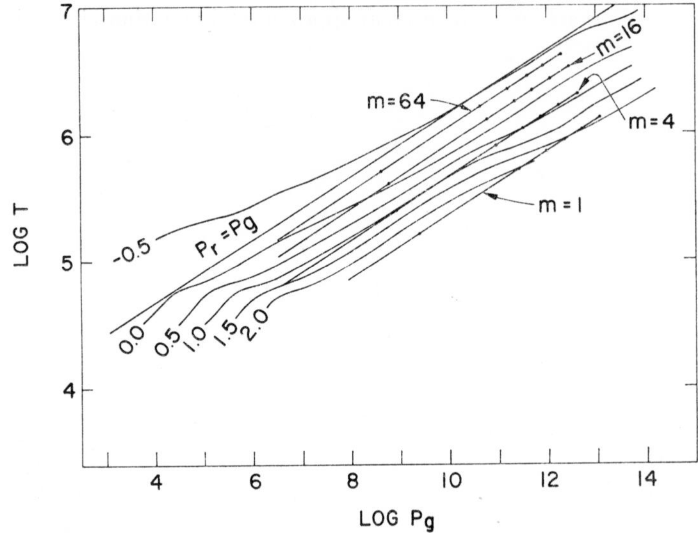

FIG. 5.—Same as Figure 4, except that the opacity contours are based on an interpolation for Unsöld's composition in the Keller-Meyerott tables.

One way to calculate better luminosities would be to find κ_0 separately for each star in the same way as was done here for mass 4, and put the resulting values into equation (59). But the small variations of opacity in the Levée model suggest a somewhat more direct procedure. It is evident from the radiative-transfer equation that the luminosity can be related to some suitable mean value of the opacity throughout the star by an equation of the form

$$L = C_2 R^2 \frac{T_c^4}{\bar{\kappa} \rho_c R} \tag{61}$$

or, by virtue of equations (56) and (58),

$$L = C_3 \frac{\mathfrak{M}^3 \beta^4 \mu^4}{\bar{\kappa}}, \tag{62}$$

where the constant C_2 or C_3 depends on the model and on the method of averaging the opacity. A detailed discussion of relations of this kind can be found in Eddington (1926, chap. 6).

Now, in view of the small range of κ actually encountered in the stars in Figures 4 and 5, it is not a matter of critical importance how κ is averaged over the star. $\bar{\kappa}$ may be replaced by the value at some typical point in the model and the value of C_3 calculated accordingly. For example, we can replace $\bar{\kappa}$ by the central value as Eddington did, in which case the evaluation of the constants in his equation (90.1) with $\alpha = 1.74$ leads to

$$L = 23.0 \frac{\mathfrak{M}^3 \beta^4 \mu^4}{\kappa_c} \tag{63}$$

as the mass-luminosity relation for the standard model. In Levée's model the corresponding value of the constant is 21.9.

When the variations of opacity throughout the star follow some law other than that of Kramers, it is better to calculate the luminosity in terms of opacity at some midway point rather than at the center. The shell containing 60 per cent of the mass and 32 per cent of the radius seems well suited to this purpose. Let the opacity at this point be called κ_1. In the standard model with Kramers opacity,

$$\kappa_1 = 1.38 \, \kappa_c \, . \tag{64}$$

This is the case for which Eddington's and Levée's luminosities were calculated. Therefore, in place of (63), we can write

$$L = 31.7 \frac{\mathfrak{M}^3 \beta^4 \mu^4}{\kappa_1} \, . \tag{65}$$

The significance of equation (65) is that it gives a good approximation to the luminosity even when the opacity does not follow the Kramers formula, provided the standard model is approximately valid.

We can use equation (65) to calculate improved tracks for homologously contracting stars in the H-R diagram. We adopt the standard model ($n = 3$) in all radiative cases. For any mass and composition we first find β by equation (54) or Table 1. Then for any given radius the values of ρ_c, P_c, and T_c are given by equations (56)–(58), and the values at the 60 per cent mass shell are given by the tabulated solutions of the $n = 3$ model (*Brit. Assoc. Adv. Sci.* 1932):

$$\rho_1 = 0.149 \, \rho_c \, , \qquad P_1 = 0.0788 \, P_c \, , \qquad T_1 = 0.529 \, T_c \, . \tag{66}$$

We find from detailed tables the opacity κ_1 for our composition at temperature T_1 and density ρ_1 and use this value to find the luminosity by equation (65). This gives us one point in the H-R diagram. The process may be repeated for other values of the radius to obtain an evolutionary track. The time scale is calculated by equation (55).

The results of such calculations are shown by the solid lines in Figure 3 for several masses using the opacities of Figure 5. The tracks for helium stars of $\frac{1}{2}$, 1, and 2 $\mathfrak{M}_\odot$ contracting onto the helium main sequence are also shown. These tracks, previously described by Aller (1959), were calculated by the method described here with opacities taken from the Keller-Meyerott tables for composition mixture 3 ($Y = 0.99$, $Z = 0.01$). It will be evident that in the procedure described here we have abandoned the assumption of strict homologous contraction as far as the opacity is concerned. We are, in effect, assuming that the opacity variations, although not satisfying equation (26) as required for strict homologous contraction, are still close enough to that form to allow the star to approximate to the standard model. Under that assumption all that is needed in regard to the opacity is the instantaneous midway value κ_1, which fixes the luminosity and hence the rate of contraction. We shall call this the assumption of pseudo-homologous contraction. Although such an assumption obviously cannot be valid in all conceivable circumstances, some results to be presented in the next section (Fig. 6) lend support to it in the present application. The luminosities calculated for pseudo-homologous contraction with real opacities (solid lines in Fig. 3) are certainly more realistic than those calculated by applying the Kramers formula to stars of all masses and radii. It will be noted that the pseudo-homologous tracks approach a constant luminosity as the stars move into the electron-scattering region of the opacity diagram, and that this occurs earlier for the more massive stars.

5. NONHOMOLOGOUS CONTRACTION AND THERMAL RELAXATION

We now turn to calculations of contraction sequences in which the homology assumption is dropped and the full set of equations in Section 2 is used to generate time sequences. In such a calculation we assume some temperature distribution for the initial model (see Sec. 2) and calculate subsequent configurations by integrating equations (10)–(13) in both space and time. At least three different approaches to the numerical solution of these equations have been devised.

These techniques are discussed in full in chapter 11 of Volume 8, this compendium, by Brownlee and Sears. Here, we shall consider only one broad aspect of the treatment of the equations in which two fundamentally different approaches lead to significant differences in the size of the time steps between successive models and in the nature of the mathematical problem for an individual model in the sequence. In numerical calculations of time-dependent heat flow, of which the contracting star is an example, there is a choice between what are generally known as the explicit and the implicit methods. The distinction between them is in the way in which equations (12) and (13) are brought into the calculation.

In the explicit method, we start each model with a known or assumed distribution of specific entropy in the star. (It will be seen that entropy distribution plays a more fundamental role here than temperature distribution.) The hydrostatic equations (10) and (11) are then soluble by themselves, the entropy distribution providing the necessary relation between P and ρ. By one of the various techniques described in chapter 11, Volume 8, we find the distribution of P, ρ, and T that satisfies equations (10) and (11) together with the equation of state (18), the specified entropy distribution related to T and ρ by equation (19), and the two boundary conditions:

$$r = 0 \quad \text{at} \quad m = 0\,, \tag{67}$$

$$P = 0 \quad \text{at} \quad m = \mathfrak{M}\,. \tag{68}$$

For example, we could integrate the two differential equations from the center out to the surface, trying various values of the starting value P_c until a solution is found which satisfies condition (68). Whatever method is actually used, there are only two differential equations involved at this stage. Then, with the distribution of P, ρ, and T determined, we calculate the opacities, the heat flow by equation (13), and the instantaneous rate of entropy change by equation (12). Thus equations (12) and (13) are treated here not as differential equations which determine the model, but as formulae for calculating the heat flow and the rate of entropy change. We extrapolate this rate of change ahead in time to obtain the entropy distribution in the next model of the sequence.

This extrapolation forward in time is the distinguishing feature of the explicit method. It leads to an important limitation on the length of time step that can be taken between successive models of a sequence. If the time step exceeds a certain critical value Δt_{crit}, then any small space fluctuations in the temperature distribution are overstable in the numerical treatment and develop into oscillations of increasing amplitude. This phenomenon, common to all time-dependent diffusion problems, is discussed in most modern treatises on finite difference methods. A particularly thorough discussion is given by Richtmyer (1957). In a star the critical time may be regarded as the thermal relaxation time of one of the individual shells into which the star is divided for purposes of the heat-flow calculation. Its value is found by an extension of Richtmyer's equation (1.16) to the stellar case:

$$\Delta t_{\text{crit}} = \frac{1}{2\rho C_p}\left(\frac{\kappa\rho}{4acT^3}\right)\Delta r^2 , \tag{69}$$

where C_p is the specific heat at constant pressure, the quantity in parentheses is the reciprocal of the radiative conductivity, and Δr is the thickness of the shell. If numerical instabilities are to be avoided, the time step used must not exceed the shortest of these shell relaxation times encountered in the star. Because of the dependence on Δr^2, this limiting time for a star divided into n shells is about $1/n^2$ multiplied by the relaxation time for the star as a whole, which may be identified with the Kelvin time τ_{Kel} discussed in Section 1. For example, if the star is divided into 50 shells, the calculation will have to proceed in time steps of such size that it takes 2500 of them to cover one Kelvin time of the star. Evidently this method is extravagant in its use of the computing machines.

In the implicit method, instead of extrapolating entropy derivatives calculated in the *old* model forward in time to find entropies in the new model, we require that the entropy changes between the two models be consistent with the derivatives calculated in the *new* model. That is, we approach the calculation of the new model without knowing its entropy distribution in advance; it must be found from the condition that equations (12) and (13) are satisfied, with

$$\frac{\partial S}{\partial t} = \frac{S_{\text{new}} - S_{\text{old}}}{\Delta t} . \tag{70}$$

Thus the two thermal differential equations are introduced at the outset into the calculation of the model, and we have a fourth order set of differential equations to solve simultaneously by one of the methods of Sears and Brownlee (Vol. 8, chap. 11, this compendium). The full set of four boundary conditions is now used, rather than only the hydrodynamic boundary conditions as in the explicit method. This approach is formally the same as that followed in the calculation of main-sequence or other steady-state models, but despite the formal similarity an important difference in substance should not be overlooked. There is no calculation of an isolated model here—only the calculation of a model which follows another model whose entropy distribution was specified. The implicit method is free of the numerical instability problem discussed in the preceding paragraph.

Calculations of the kind discussed here were first carried out by Henyey, LeLevier, and Levée (1955) on a UNIVAC computer at the Livermore Radiation Laboratory. They used the implicit method, coupled with a numerical relaxation technique for solving the four simultaneous differential equations, to calculate pre–main-sequence contraction sequences for eight stars of selected masses and compositions. The zero boundary conditions were used and radiative transport was assumed, as in other cases discussed thus far. For the opacity an interpolation formula was used, based on Morse's (1940) tables, with electron scattering and other contributions added.

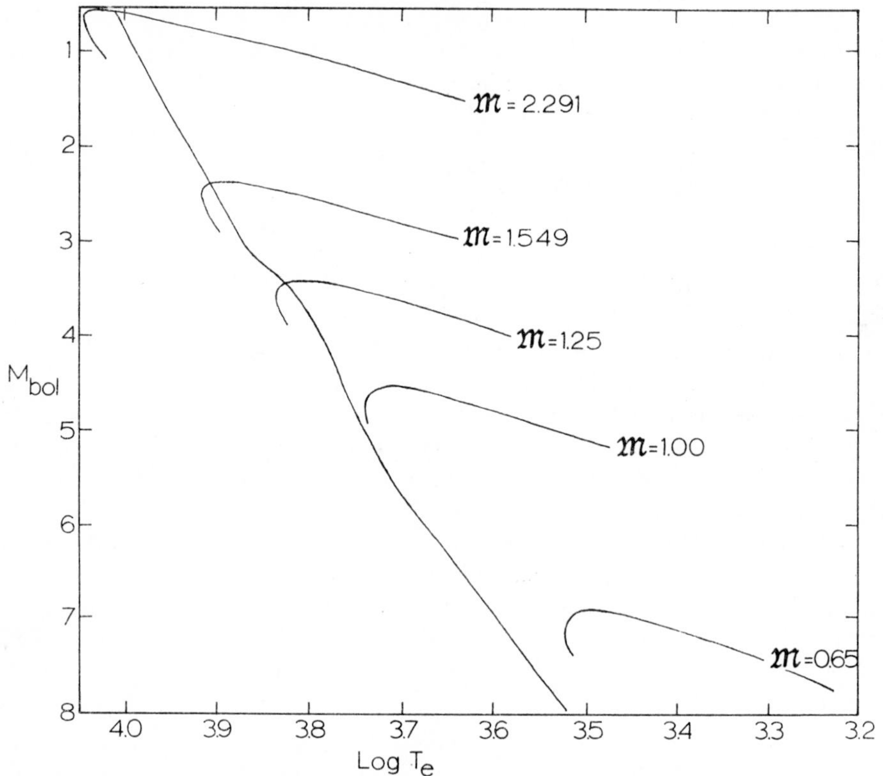

FIG. 6.—Pre–main-sequence evolutionary tracks for purely radiative transport, according to Henyey, LeLevier, and Levée (1955). Composition $X = 0.74$, $Y = 0.25$, $Z = 0.01$. The masses in solar units are given. The diagonal curve is a version of the Keenan-Morgan main sequence used by them for comparison.

The Levée model was adopted as the initial state for each star, and it was then allowed to develop according to the dictates of the time-dependent equations. The results of Henyey *et al.* for stars of composition $X = 0.74$, $Y = 0.25$, $Z = 0.01$ are reproduced in Figure 6.

It will be seen that the tracks in the Hertzsprung-Russell diagram are very similar to the pseudo-homologous ones in Figure 3. This is a consequence of having started with the Levée model, which is able to contract in a nearly homologous manner, as these calculations show, despite the failure of the opacity to satisfy the exact condition (26). The demonstration of this possibility can be regarded as the chief value of these calculations, since it provides a sound basis for constructing sequences by the method described in Section 4.

A calculation of a somewhat different type was performed for the Sun in its pre–main-sequence contraction by Brownlee and Cox (1961) using the IBM 704 computers of the Los Alamos Scientific Laboratory. A form of the explicit method was employed in which time dependence was introduced into the pressure equation (10) as well as into the thermal equation, and the initial model was allowed to be out of hydrostatic equilibrium. An artifically high viscosity was introduced to make the hydrodynamical time scale comparable with the Kelvin time scale. As a result, the early part of the sequence consists of nonhydrostatic models, whereas the later parts satisfy the hydrostatic condition quite closely.

In physical aspects the Brownlee and Cox calculations are quite close to those of Henyey *et al.* (1955), except for one additional phenomenon which provides the chief interest in this work apart from method. Allowance was made for the nuclear energy released by the conversion of H^2 to He^3, with two abundances of H^2 assumed. In one sequence, the H^2 content was assumed to be zero, and in the other, the abundance was taken to be two parts in 10^4 by mass. Figure 7 shows the two sequences in the H-R diagram. The numbers along the two tracks are times before the main sequence is reached in millions of years. The track for no deuterium burning is the smoother of the two curves. It does not differ greatly from the tracks of Henyey *et al.*, of which one is shown on the same diagram. The starting configuration was a scaled main-sequence model that did not differ greatly from the Levée model.

The contraction of the star containing deuterium shows features of greater interest, of which the most striking is the brief halt on a deuterium main sequence (just below point A) when the central temperature is 800,000° K. After the exhaustion of deuterium, there occur readjustments of the model which take the star for 40 million years to higher luminosities than its counterpart which contained no deuterium. Here is an indication of a relaxation phenomenon which will be described presently in greater detail.

An attempt was made in this work to allow for convective heat transport in the outer layers. A small and relatively insignificant convective envelope was found throughout most of the sequence, which should therefore be regarded as a radiative one. The later work of Hayashi, summarized in the next section, shows, however, that stars must be primarily convective during most of the stages covered by the calculations of both the Los Alamos and the Livermore groups. It will be shown that this result depends very much on close attention to the photospheric boundary condition, which was apparently not sufficiently precise in the Brownlee and Cox work to show the true extent of the convection zone.

In both the Livermore and Los Alamos calculations, the transition from a contracting to a hydrogen-burning model was investigated by including the energy production by the p-p and CNO chains in equation (12). In both cases a drop in luminosity by about $\frac{1}{2}$ mag was found as shown in Figures 6 and 7. The reason for this drop can be understood in terms of the dimensionless parameters of the models, particularly Schwarzschild's parameter C. An explanation in these terms is given by Schwarzschild (1958, p. 162). The drop can also be understood in terms of the greater central concentration of the energy source in the main-sequence model and the associated behavior in the log P–log T diagram.

I calculated (Upton 1962) a third set of radiative contraction sequences for stars of 10 $\mathfrak{M}_\odot$ in the pre–main-sequence state. These calculations were done by the explicit method with the aid of the IBM 7090 computers of the Goddard Space Flight Center. Owing to the limitation of the explicit method explained above, as many as 7000 time steps per sequence were required. Again pure radiative transport was assumed, with an opacity formula comprising about equal parts of Kramers opacity and electron scattering; that is,

$$\kappa = \kappa_0 \rho T^{-3.5} + 0.19\,(1 + X)\,. \tag{71}$$

Nuclear energy generation was not considered.

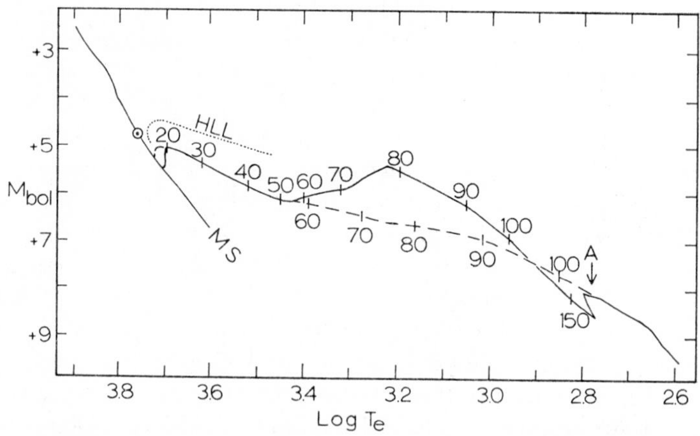

FIG. 7.—Pre-main-sequence evolutionary tracks for a star of one solar mass, according to Brownlee and Cox (1961). Case of no deuterium, - - - - -; initial deuterium abundance 2 × 10⁻⁴ by mass, ———; track of Henyey, LeLevier, and Levée's model (*HLL*) is shown also. The numbers refer to times before the main sequence in millions of years. The point marked *A* is discussed in the text. The circle indicates the position of the Sun; main sequence is labeled *MS*.

The primary purpose of these calculations was to determine what happens when the initial configuration is far from the Levée or the standard model. Specifically, does a kind of thermal relaxation occur whereby any initial configuration progressively changes toward the standard model, with a resulting convergence of tracks in the H-R diagram? If so, what is the time scale for the convergence? Closely related is the question whether there is such a thing as a unique pre–main-sequence track for a star of given mass and composition. Jeans (1919, p. 198) supposed that such a convergence would take place but was unable to prove it. Thomas (1931*a*, *b*) investigated the same question in more detail but was still not able to reach any conclusions.

For this convergence question, four initial models were made, each having about one-tenth the negative binding energy of Kushwaha's (1957) main-sequence model for the same mass and composition and therefore having a mean temperature $\bar{T}$ (mass average) of about one-tenth the main-sequence value. Four different entropy distributions were chosen, all subadiabatic and therefore stable against convection, which led to the four temperature distributions shown in Figure 8. Figures 9 and 10 show the same models in the log ρ, log T diagram and the UV diagram. (For definitions and discussion of the UV diagram, see Schwarzschild, 1958, pp. 108 ff.) A comparison of these figures with similar diagrams for the Lane-Emden polytropic models will show that model A of this set is close to the standard $n = 3$ model, while the other three have characteristics not met in any polytropic model. Model B has approximately an isothermal center, Model C

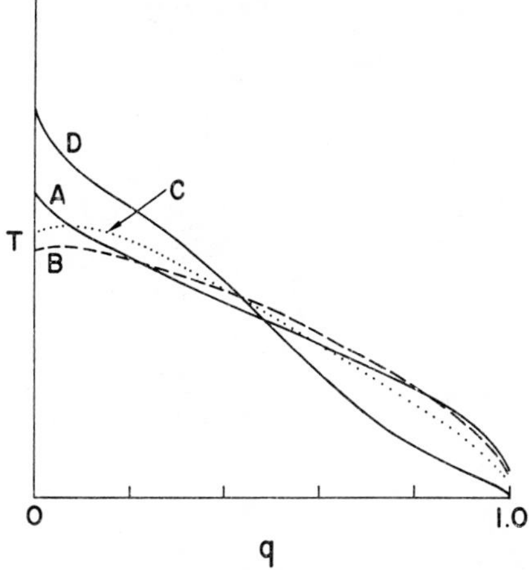

FIG. 8.—Temperature distribution as a function of the mass fraction q in four initial models, A, B, C, D (see text), studied for relaxation during radiative contraction. Temperatures are shown scaled to same mean value (mass average).

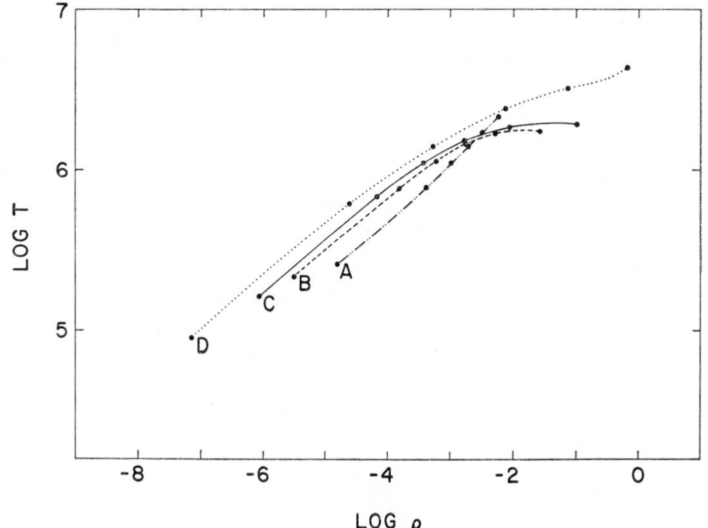

FIG. 9.—Temperature-density distribution in four initial models studied for thermal relaxation. Dots correspond to mass fractions $q = 0.0, 0.2, 0.4, 0.6, 0.8$, and 0.99 from right to left.

has a cold spot in the center, and model D illustrates the possibility of achieving a hotter center than in the $n = 3/2$ model without introducing any superadiabatic regions. The radii and luminosities of these initial models can be seen in or deduced from Figure 11. Both radius and luminosity increase in the order A, B, C, D. Total energy is considered here to be the most fundamental measure of a star's progress toward the main sequence, and hence all four stars are considered to be at the same stage of contraction despite the disparities in radii.

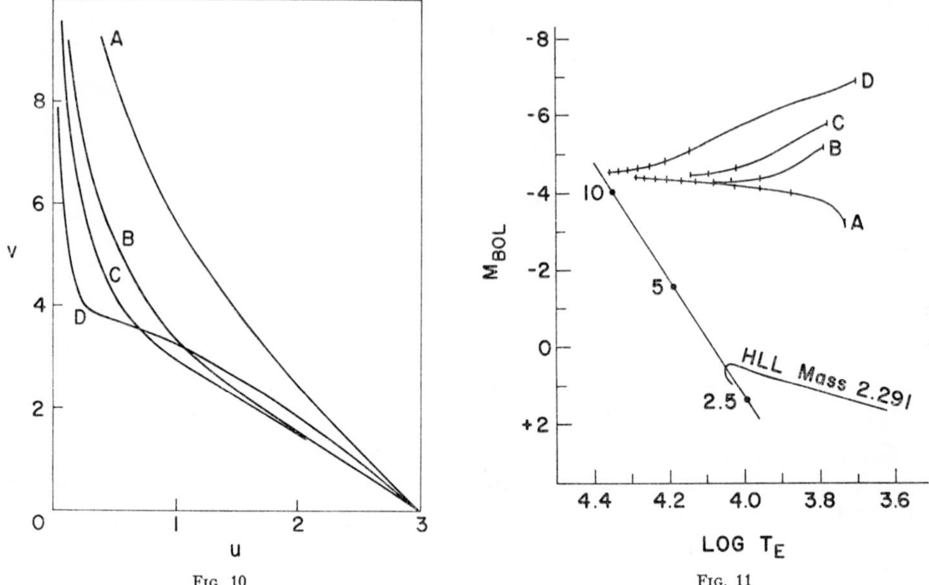

FIG. 10 FIG. 11

FIG. 10.—Structure in the UV diagram of four initial models studied for thermal relaxation. Compare with Figure 2 for polytropic models.

FIG. 11.—Convergence of evolutionary tracks in the Hertzsprung-Russell diagram for radiative models of 10 solar masses with diverse initial conditions. Tracks are divided into equal time intervals of 10^4 years. Dots are main-sequence positions calculated by Kushwaha for 2.5, 5, and 10 solar masses. One track of Henyey, LeLevier, and Levée is shown for comparison.

The contractive evolution of the four stars is shown in Figures 11, 12, and 13. Figure 11 shows tracks in the H-R diagram, divided into 10,000-year segments. Here track A resembles the pseudo-homologous tracks of Figure 3, as expected. The other three tracks are covered more quickly than A in the early stages, since the luminosities are higher. A convergence of the type anticipated by Jeans is plainly evident. Three theoretical main-sequence models of Kushwaha have been included in the diagram as reference points. That the models A–D converge near the main sequence rather than somewhere else is a fortuitous result of the energy with which they were arbitrarily started off. The significant result to be derived from this diagram is not the place where convergence occurs, but the rate at which it proceeds.

Figure 12 shows this convergence rate from a point of view closer to the fundamental thermodynamics of the situation. Here coordinates are the logarithm of the star's total energy (absolute value) and the total entropy per particle in natural or Boltzmann units. As the stars contract they lose both energy and entropy by radiation, and consequently they move from upper left to lower right in this diagram. The brief rise in entropy at the beginning of some sequences is associated with transient readjustments of the outer 1 or

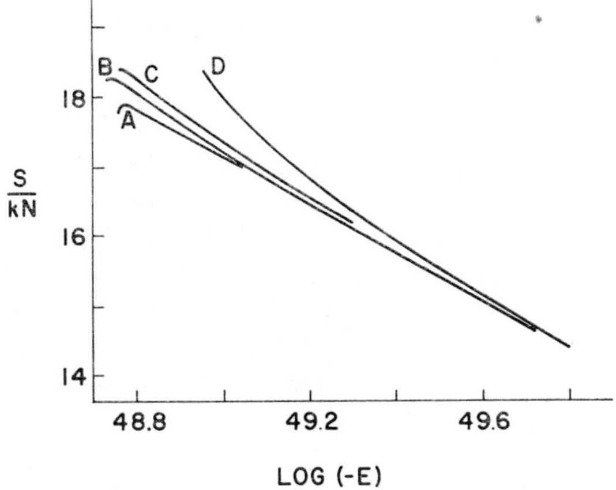

FIG. 12.—Convergence of evolutionary tracks in the energy-entropy diagram for radiative models of 10 solar masses with diverse initial conditions. Energy ($-E$) is binding energy in ergs. Ordinate (S/kN) is entropy per particle in natural units. Zero point is arbitrary. Direction of evolution is downward and to the right.

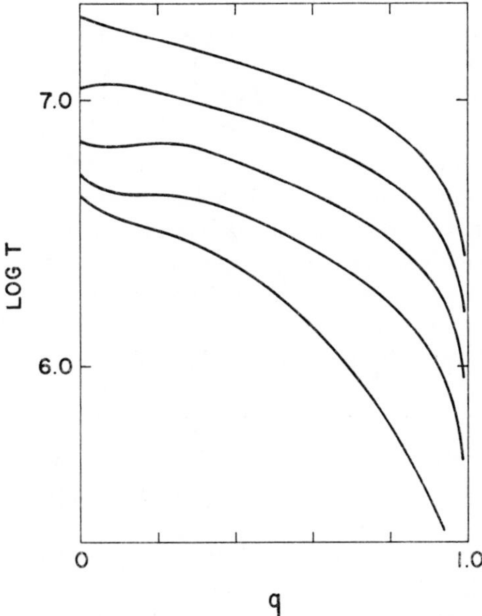

FIG. 13.—Successive temperature distributions as function of mass fraction q during radiative contraction of a particular model. The fifth stage (*top curve*) is very close to the distribution in the standard model ($n = 3$).

2 per cent of the mass to a more stable thermal condition than the one originally specified. We see here briefly the law of increase of entropy at work within the star as if it were a closed system, whereas during most of the contraction the transfer of heat from the star into the cold surrounding space dominates in the second law of thermodynamics. The evolutionary tracks converge in a regular manner in the energy-entropy diagram, and from the behavior exhibited there we can tentatively formulate the general law of convergence. Entropy differences between models decay in an approximately exponential fashion according to the law

$$d \log (\Delta S) = -1.5\, d \log (-E) \,, \tag{72}$$

where the constant 1.5 has been determined empirically from the results shown in Figure 12. This corresponds to a decrease of entropy differences by a factor e while the total energy changes by a factor of 2, so that for radiative stars, the time scale for thermal relaxation is the Kelvin time. This result might well have been anticipated on intuitive grounds, since the process by which heat is lost and the process by which it is redistributed within the star are the same. A different result would be expected when the primary transport mechanism inside the star is convection or conduction.

Figure 13 shows the temperature distribution in star D at five successive stages in the contraction. Of the four cases studied, this star undergoes the greatest changes in its dimensionless properties but by the last stage shown here it is settled into a nearly constant dimensionless structure with a nearly uniform polytropic index around 2.9. In the UV diagram (Fig. 3) model D can be seen to deviate from the standard model to about the same extent as Levée's model does, but in the opposite direction. In both cases the deviations are minute.

The contraction of a star is an interesting process thermodynamically speaking. Here is a system that is isolated from outside influences, except for the 0° heat sink with which it is in contact at the surface; yet it does not go to equilibrium, nor does it even go to a steady state (as long as nuclear reactions or degeneracy do not come into play). Nonetheless, diverse initial temperature configurations converge. Loosely speaking, it is possible to say that they converge toward the homologous sequence or the standard model. But this description is not entirely satisfactory, because there are no strictly homologous sequences with real opacities, and the standard model is only an approximation to the real star. It is hard to resist the impression that the convergence results from an attempt on the part of the star to maximize or minimize something that is thermodynamically significant, but no such quantity has yet been identified. Entropy production does not seem to fit this role.

This concludes the discussion of contraction with purely radiative transport. The remainder of this chapter will show that the results derived for this process apply to the Sun only during the very end of its contraction and do not apply at all to stars much less than the Sun in mass.

Radiative transport does seem to be applicable, however, to stars more massive than the Sun during a substantial part of their contraction, and hence these studies are relevant for such stars. We may summarize by saying that radiative contraction may take a variety of courses for a few Kelvin times, depending on the initial conditions; but after that, the standard model is always achieved and is thereafter maintained. Luminosities and time scales during the standard-model stage can be calculated with tolerable accuracy by the method of the typical opacity discussed in Section 4. These calculations show luminosity slowly increasing in stars between about 1 and 10 $\mathfrak{M}_\odot$ in which absorption by atoms dominates and nearly constant luminosity for the higher masses in which scattering by free electrons is the main opacity process.

6. HAYASHI'S THEOREM AND CONVECTIVE TRANSPORT

A major revolution in the concept of pre–main-sequence contraction was brought about by Hayashi (1961), Hayashi and Hōshi (1961), and Hayashi, Hōshi, and Sugimoto (1962) when they put forward their arguments in support of convective transport during the early stages of stellar contraction. This conclusion was reached, surprisingly, not through a detailed study of transport mechanisms, but through a more careful consideration of the outer boundary conditions and their implications for the internal structure than had previously been done. These implications are not limited to contracting stars, but constitute in fact an important limit theorem applicable to all hydrostatic stars. That the theorem was so long overlooked was due to the nearly universal use of the zero boundary conditions in stellar-structure calculations. Otherwise it might have been discovered by anyone since the days of Milne, and in fact nearly was on at least one occasion. Hoyle and Schwarzschild (1955) found that the evolutionary tracks of stars in the H-R diagram are sometimes quite different from those calculated on the basis of the zero boundary conditions. They found that the correct evolutionary track for red giants could be calculated only if the finite temperature and pressure conditions for an actual photosphere were taken into account. When this was done, deep convective zones were found.

It remained, however, for Hayashi and Hōshi (1961) to point out that some long-recognized properties of complete and incomplete convective star models, when coupled with the photospheric boundary conditions, have the effect of dividing the H-R diagram into a permitted and a forbidden region for stars of a given mass and composition, and that the closer a star is to the dividing line the more convective it must be. This division applies to both the early and late stages of evolution. Hayashi's theorem does not say where in the permitted region a star should go, but the red giants seem to hug the dividing line, and Hayashi has suggested that stars in the pre–main-sequence stage should do the same.

Since Hayashi's theorem has not yet been accorded a full exposition in the literature, I shall give one in this section. (The line of proof given here was suggested to me by Schwarzschild in 1962.) I shall then show the separate effects of photospheric conditions and of superadiabatic convection on the location of the permitted and forbidden domains.

Hayashi's theorem may be stated as follows: For stars of given mass and chemical composition, certain combinations of the surface parameters L and T_e can be ruled out on the ground that they require one of the following inadmissible conditions: (*a*) violation of photospheric boundary conditions, (*b*) superadiabatic temperature gradients, or (*c*) departure from hydrostatic equilibrium. These impossible combinations of L and T_e make up a forbidden region in the H-R diagram, the limits of which depend somewhat on the mass and composition of the star. The existence of the Hayashi limits does not depend on the peculiar properties of partially ionized hydrogen to which the onset of convection is often attributed, but these properties have an important effect on where the limits fall in the H-R diagram.

The argument, in brief, is that for completely convective stars of given mass and composition there are certain relationships which are limiting for stars of any arbitrary structure which violate none of the three conditions listed above. If the mass and composition of a completely convective star are specified, there remains only one free parameter which may conveniently be taken to be the radius. All other properties of the star can be given, in principle, as functions of its radius. For example, there exist a radius-luminosity (*R-L*) relation and a radius-entropy (*R-S*) relation.[3] After reviewing these relations we shall show that they are limiting ones.

In our present examination of these relations, we assume that convective stars have a strictly adiabatic structure and that, as far as the *R-S* relation is concerned, this adiabatic structure is well represented by the Emden polytropic model of index 1.5. It will be evident, however, that the existence of the Hayashi limits does not depend on these simplifications. By introducing them, we neglect the effect of a superadiabatic convection zone at the surface and the effect of the departures from full ionization on the radius. The first effect will be introduced at the end of the discussion as a correction to the values of the limits. The latter effect is known to be small in ordinary stars, and we can well afford to omit it from the discussion.

6.1. The Radius-Entropy Relation

A convenient expression of the *R-S* relation for the Emden model makes use of the adiabatic constant K for a monatomic gas:

$$K = \frac{P}{T^{2.5}} \tag{73}$$

and of its dimensionless equivalent E, introduced by Osterbrock (1953):

$$E = 4\pi \left(\frac{m_{\rm H}}{k}\right)^{2.5} G^{1.5} \mu^{2.5} \mathfrak{M}^{0.5} R^{1.5} K \,. \tag{74}$$

The relation between $\mathfrak{M}$, R, and K is fixed by (74) together with the condition that E has the value

$$E_1 = 45.48 \tag{75}$$

in the complete Emden model. These two equations constitute the *R-S* relation.

This expression of the relation is appropriate when the regions of interest have a constant degree of ionization so that constancy of K is equivalent to constancy of entropy. When the surface layers must be considered, as in the present discussion, it is preferable to go over to the true specific entropy denoted here by S^* when expressed as a logarithm of thermodynamic probability per atomic nucleus. This is the quantity called S/kN by Unsöld (1955, pp. 228 ff.). In the region of constant ionization, the relation between S^* and K is

$$S^* = -\frac{2.3026}{a} \log K + C \,, \tag{76}$$

[3] These relations are essentially the ones worked out by Emden (1907) for the models with constant polytropic index $n = 1.5$ ($P \propto T^{n+1}$, $\rho \propto T^{n}$).

where α is the ratio of the number of nuclei to the number of all free particles and C is a constant depending on the composition and on the entropy zero point adopted. We can express the $\mathfrak{M}$-R-S relation in terms of S^* directly by combining equations (74)–(76) to obtain

$$\tfrac{1}{2}\log \mathfrak{M}+\tfrac{3}{2}\log R-\frac{\alpha}{2.3026}\,S^*+C'=\log E_1=1.658\,, \tag{77}$$

where C' is a combination of physical constants, solar dimensions, and composition parameters.[4] The $\mathfrak{M}$-R-S relation as expressed in equation (77) has a direct meaning for all parts of the star, whether the ionization is constant or not. Strictly speaking, the constant on the right of (77) should not be precisely E_1, because of the departures from the Emden model in the zone of variable ionization. This effect will generally be small, and we shall ignore it. It does not alter the general argument.

6.2. The Radius-Luminosity Relation

The luminosity of a convective star depends on the opacity of the atmospheric layers. In the approximation of no superadiabatic convection, we assume that all the radiation escapes from a "photospheric" surface at optical depth 2/3, where $T = T_{\rm eff}$. The layers above this surface may be represented for present purposes by a gray atmosphere for the corresponding g and T_e, while the layers below follow an adiabatic curve. The fitting of the atmosphere to the adiabatic interior can be done by a graphical technique. Figure 14 shows a pressure-temperature diagram for the Unsöld composition. One set of curves (solid lines), taken from the gray-atmosphere calculations of de Jager and Neven (1957), shows the loci of possible photospheres with log g as a parameter. An important feature of the photospheric curves is that, for a given g, the effective temperature always (or nearly always) increases monotonically with entropy. There is no obvious fundamental reason why a monotonic relation of this kind should exist, but the Hayashi limits will be seen to depend upon its existence in actual stellar atmospheres. The other set (dotted lines) are adiabatic curves for the interior from the calculations of Unsöld (1955, p. 229). For a given mass and radius, a single curve is determined in each family—the adiabatic curve by equation (77) and the photospheric curve by the value of g—and their intersection gives the effective temperature and photospheric pressure. The effective temperature then determines the luminosity and the locus in the H-R diagram. These loci form a series of curves, one for each mass, the different points along the curve corresponding to different radii, and are shown in Figure 15 for the Unsöld composition. These curves are not yet the $\mathfrak{M}$-R-L relation for real convective stars, however, because the effect of superadiabatic convection has not been considered. That modification will be discussed presently.

6.3. The Completely Convective Model as a Limit

We now consider stars of any arbitrary structure, subject only to the three restrictions: (*a*) that they are in hydrostatic equilibrium; (*b*) that $dP/d\rho$ does not exceed the adiabatic value at any point below the photosphere; and (*c*) that there are no composition inhomogeneities such that the molecular weight decreases inward.

The first point to be proved is that a star satisfying these conditions cannot have a

[4] For Unsöld's composition and entropy zero point, $\alpha/2.3026 = 0.2020$ and $C' = 6.552$.

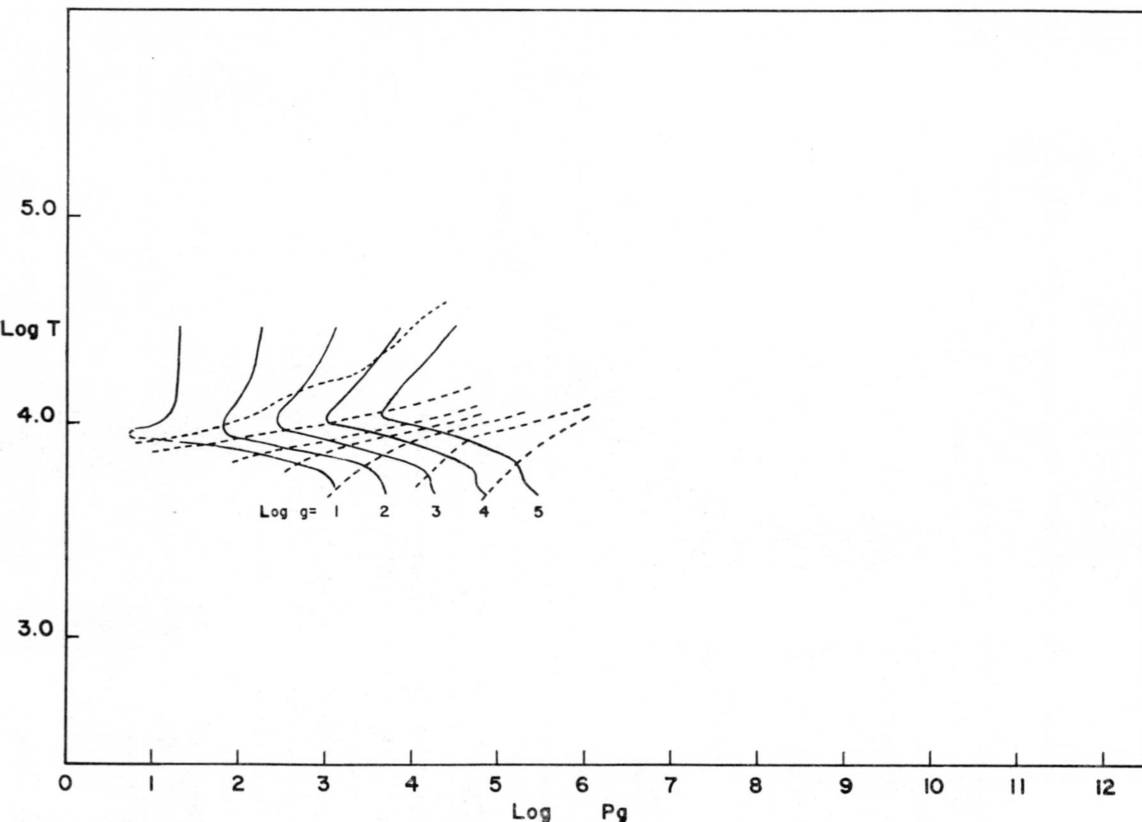

FIG. 14.—Photospheric conditions(———) and adiabatic curves (............) for Unsöld's composition. Photosphere curves correspond to level where $T = T_e$ in models of de Jager and Neven for various effective temperatures and surface gravities (log g).

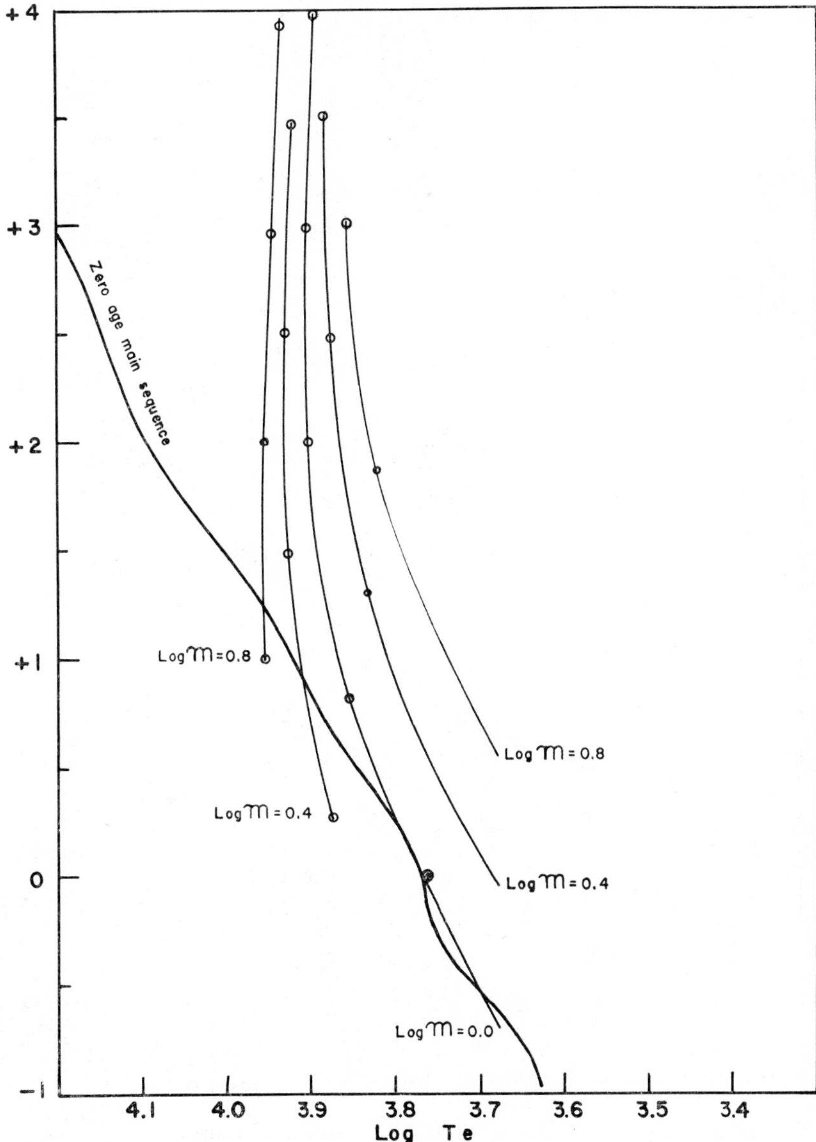

FIG. 15.—Hayashi limits in the Hertzsprung-Russell diagram when no superadiabatic zone of any kind is permitted. Curves refer to different masses (in solar units) with Unsöld's composition. [NOTE.—The two values of log $\mathfrak{M}$ on the right should be negative, viz., -0.8 and -0.4, *not* 0.8 and 0.4 as shown.]

lower photospheric entropy than a completely convective star of the same mass and radius. The proof is most easily given in terms of the Milne homology variables U and V defined as

$$U = \frac{d \log m}{d \log r}, \tag{78}$$

$$V = -\frac{d \log P}{d \log r}, \tag{79}$$

In a UV diagram, all possible convective envelopes comprise a one-parameter family, for which Osterbrock's E is a convenient parameter. The completely convective stars

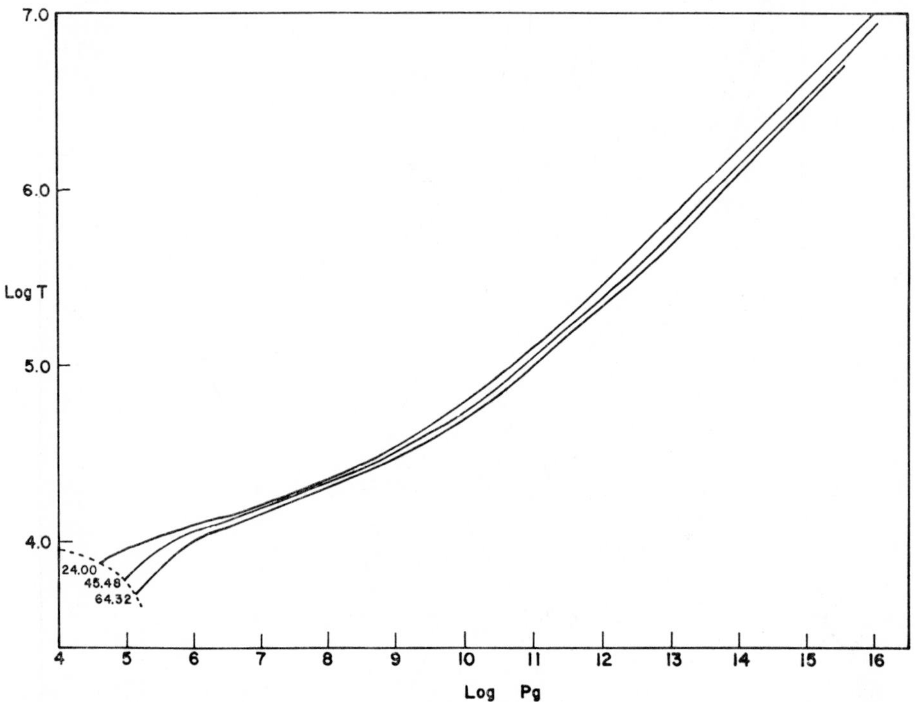

FIG. 16.—Structure of three adiabatic convection zones for the outer regions of a star of solar mass and radius. The dashed curve is the locus of possible photospheric conditions with $\log g = 4.44$. Values of the Osterbrock parameter E are shown. The middle curve corresponds to the completely adiabatic model. Significance of other curves is discussed in the text.

have the value E_1 mentioned above. For a given mass and radius, E is a measure of the entropy in the convective region, in accordance with equation (77). It will be useful to regard E as a measure of the entropy at the photosphere in particular and even to extend this definition of E to stars that have no outer convection zone.

Thus the completely convective star of given $\mathfrak{M}$ and R has a certain photospheric entropy S^* corresponding to the value E_1. This star is represented by the middle curves in Figures 16 and 17. In Figure 16, the pressures and temperatures are scaled for $\mathfrak{M} = \mathfrak{M}_\odot$ and $R = R_\odot$.

Stars with photospheric entropies $S^* > S_1^*$ have $E < E_1$. If such stars have convective envelopes at all, they belong to the class called "centrally condensed" configurations by Milne (1930). An envelope of this class is represented by the upper curve in Figure 16 and the left-hand curve in Figure 17. In these envelopes, as we proceed inward, the center of the star is reached before all the mass has been accounted for. Thus they may be described as stars with a point mass in the center.

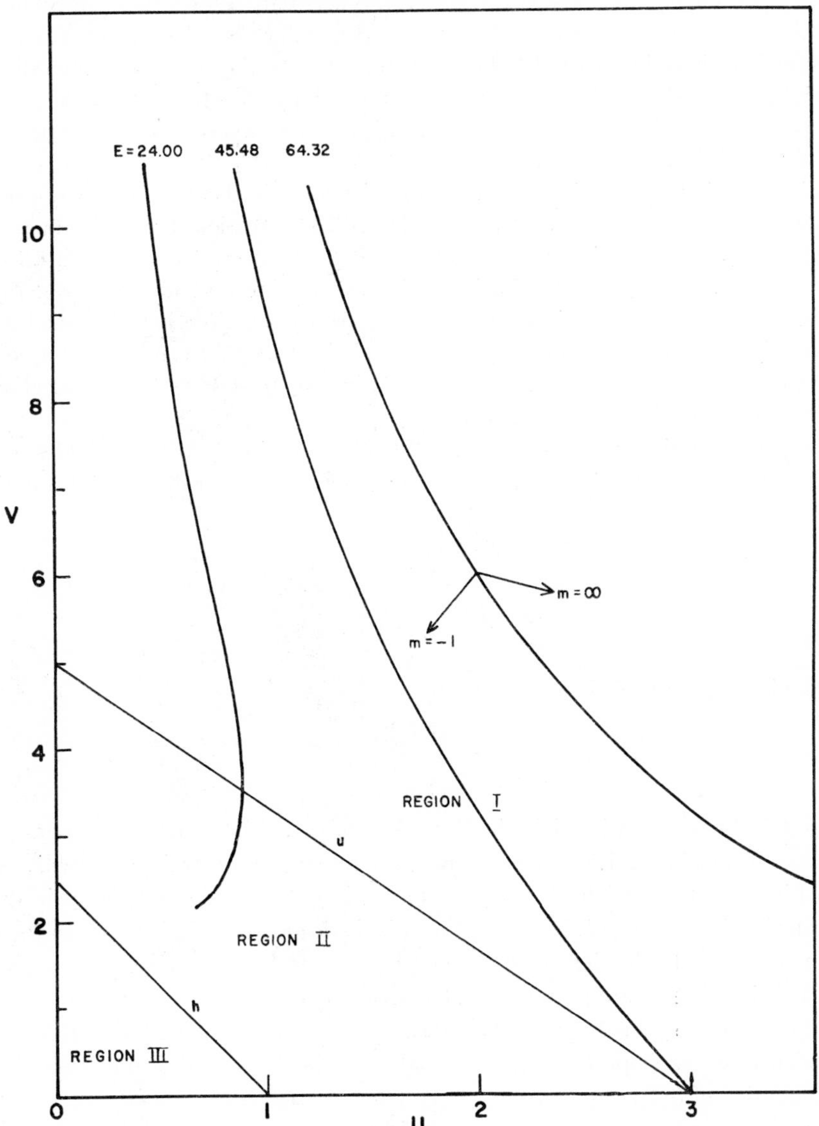

FIG. 17.—Structure in the UV diagram of the three adiabatic regions shown in Figure 16. $E = 45.48$ corresponds to the complete polytropic model of index 1.5. $E = 24.00$ and 64.32 correspond to the centrally condensed and collapsed types, respectively. Significance of other features is discussed in the text.

The alternative case (lower curve in Fig. 16 and right-hand curve in Fig. 17) consists of stars with photospheric entropy $S^* < S_1^*$ and $E > E_1$. If such a star has a convective envelope, it belongs to the class called "collapsed" configurations by Milne. Proceeding inward in these envelopes, we find that all the mass is used up before the center of the star is reached. These may be described as stars with a hole in the center.

If we start with $E \neq E_1$ at the photosphere and set out to construct a star that has neither a point mass nor a hole at the center, it is necessary to depart at some point from the adiabatic curve. Of course, it is not necessary to follow an adiabatic curve at all, but it is useful to think in terms of following it part way to the center and then diverging from it. It can be shown that for envelopes of the collapsed type ($E > E_1$, $S^* < S_1^*$) the departure required is a superadiabatic one, whereas for envelopes of the centrally condensed type, it is subadiabatic.

Before turning to the rigorous demonstration of this theorem, let us consider its qualitative plausibility. In a centrally condensed configuration, the excess mass at the center can be spread out in an acceptable way if the temperature of the central regions is lowered so that a higher density is required to provide the same pressure. In other words, a lower than adiabatic temperature gradient is required in the central regions. Conversely, in a collapsed configuration, the hole in the center can be filled up only by raising the central temperature, that is, by raising the temperature gradient above the adiabatic value in the central regions.

For a mathematical proof we make use of the fact that in the Milne variables the combined hydrostatic and Poisson equations reduce to a first-order differential equation:

$$\frac{U}{V}\frac{dV}{dU}=\frac{U-1+\dfrac{1}{n+1}V}{3-U-\dfrac{n}{n+1}V}=\frac{U}{V}\lambda(U,V\ n), \qquad (80)$$

where n is the polytropic index defined by

$$\frac{n+1}{n}=\frac{d\log P}{d\log\rho}.$$

If $n = 1.5$, the structure is adiabatic. For $-1 \leq n < 1.5$, the structure is superadiabatic and unstable, while for $1.5 \leq n \leq \infty$, the structure is subadiabatic and stable.

In Figure 17 the straight line h is the locus of horizontal slopes ($\lambda = 0$) in the UV diagram for the case $n = 1.5$, and the line u is the locus of vertical slopes. In the entire region above and to the right of u (Region I) the slopes are negative for this adiabatic case. This is the region of present interest, since the completely convective model with $E = E_1$ lies entirely in this region, as do all convective envelopes with $E > E_1$. The dependence of the slope λ on n, at a fixed point in the UV diagram, is obtained by differentiation of (80):

$$-\frac{1}{\lambda}\frac{\partial\lambda}{\partial n}=\frac{V}{(n+1)^2}\frac{2U+V-4}{\left(U-1+\dfrac{1}{n+1}V\right)\left(-3+U+\dfrac{n}{n+1}V\right)}. \qquad (81)$$

Since all factors on the right of (81) are positive in the region concerned, increasing n (a subadiabatic change) makes the slope less strongly negative, or flatter. Conversely, a superadiabatic change makes the slope steeper. The limits of subadiabatic and superadiabatic slopes for a typical point in Region I are indicated in Figure 17 by the arrows corresponding to $n = \infty$ and $n = -1$, respectively. The situation shown there is qualitatively the same everywhere in Region I.

Since a solution that is well behaved at the center must run to the point $U = 3$, $V = 0$, and since a convective envelope with $E > E_1$ lies to the right of the E_1 curve in Figure 17, it follows that only a superadiabatic modification can make such an envelope well behaved at the center. Therefore, if superadiabatic gradients are ruled out, stars with $E > E_1$ at the photosphere cannot exist at all. By virtue of the monotonic relation between S^*_{phot} and T_{phot} noted above, it follows that photospheric temperatures, and hence luminosities lower than those of the completely adiabatic model, are impossible. Therefore the curves of Figure 15 are limiting curves, each for a particular mass, such that stars of that mass cannot lie to the right of the curve without violating one of the conditions. (Again we note, however, that the curves must be somewhat modified to allow for superadiabatic convection near the surface.)

A summary of the physical basis of this theorem may be useful at this point. If we set out to make a star of a given mass and radius, working from the outside inward, we must begin with a physically realistic photosphere, that is, one that lies somewhere on the appropriate log g curve in Figure 14. There is one special point on this curve where the dimensionless entropy E has the correct value E_1 to enable us to follow an adiabatic curve all the way to the center and come out with a good hydrostatic model. The photospheric temperature that satisfies this condition determines the luminosity of the completely convective model. If we take the photosphere at some other point on the appropriate log g curve, we cannot make a good hydrostatic model by following an adiabat because there will be difficulties at the center. If we start with T_{phot} higher than the value corresponding to E_1 we have Milne's centrally condensed case with extra mass left over. We can take care of this by following a subadiabatic curve instead of an adiabatic one, and hence build a variety of acceptable models of this kind. But if we start with T_{phot} lower than the value corresponding to E_1, we have Milne's collapsed case with a hole in the center, and there is no way to fill up the hole with any combination of adiabatic and subadiabatic gradients. It takes a superadiabatic gradient to eliminate the hole, and this is not allowed because it is convectively unstable.

6.4. The Effect of Superadiabatic Convection

In many stars a superadiabatic region of limited extent actually does occur near the surface, the density being too low for quasiadiabatic convection to transport an appreciable fraction of the energy. The occurrence of this zone relaxes somewhat the limits set in the discussion thus far and allows stars to lie to the right of the curves shown in Figure 15 by an amount depending on the entropy increase ΔS^* permitted in the superadiabatic region. This quantity is not known with any certainty, since a complete theory of superadiabatic convection has not been developed. An approximation to ΔS^* can be found, however, from calculations using the mixing-length theory as developed by Böhm-Vitense (Vitense 1953, Böhm-Vitense 1958). Calculations with mixing length equal to pressure-scale height have been done for the Unsöld composition by Böhm-Vitense (1958) and by Kippenhahn and Baker (1962). The values of ΔS^* taken from these calculations are indicated by the family of curves shown in Figure 18. The largest values occur at low temperatures and low gravities, that is, for red giant conditions.

The effect of these permitted entropy rises on the location of the Hayashi limits in the H-R diagram may be found most readily if we think of them as mass limits in a (g, T_e)

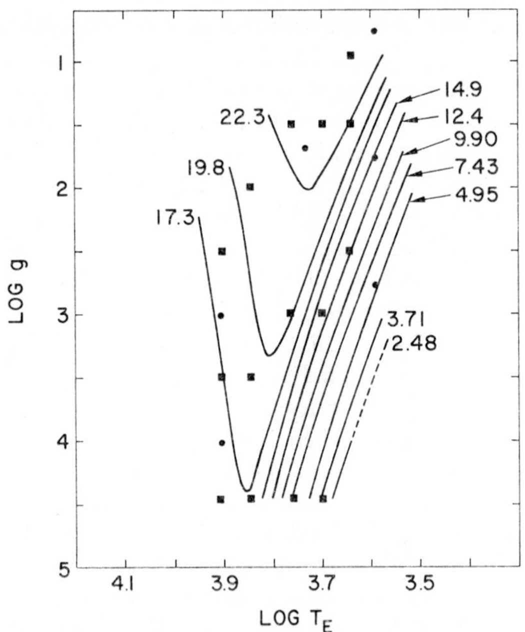

FIG. 18.—The effect of superadiabatic convection zones as a function of effective temperature and gravity. The numbered contours refer to the entropy change ΔS^* in the superadiabatic regions (units given in text). Symbols indicate the points for which data were available to construct the contours. Data are from calculations of Böhm-Vitense (■) or Baker and Kippenhahn (●). All data are based on the Unsöld composition and with the mixing length, l, equal to the pressure scale height, H_p.

diagram. These mass limits, $\mathfrak{M}_1(g, T_e)$, are found in the case of no superadiabatic zone by expressing equation (77) in terms of $\mathfrak{M}$ and g instead of $\mathfrak{M}$ and R:

$$\tfrac{5}{4}\log \mathfrak{M}_1(g, T_e) = \tfrac{3}{4}\log g + 0.2020\, S^*_{\text{phot}}(g, T_e) + \log E_1 - 9.884\,, \qquad (82)$$

where $S^*_{\text{phot}}(g, T_e)$ is the entropy at the photospheric point read from Figure 14. To include the effect of the superadiabatic region we simply add the value of $\Delta S^*(g, T_e)$ read from Figure 18:

$$\log \mathfrak{M}_2(g, T_e) = \log \mathfrak{M}_1(g, T_e) + 0.1616\, \Delta S^*(g, T_e)\,, \qquad (83)$$

where $\mathfrak{M}_2(g, T_e)$ is the mass limit for this superadiabatic case. The resulting Hayashi limits, based on the ΔS^* curves in Figure 18, are shown in Figure 19. When these are compared with the corresponding curves for the strictly adiabatic case (Fig. 15), we see that the superadiabatic convection zone brings about a very large displacement of the curves to the right in the red giant region. It is evident that the exact location of the limiting curves in this region is going to be sensitive to the value of the mixing length. On the other hand, at the lower luminosities where photospheric densities are higher the superadiabatic effect is small, as has been pointed out previously, for example, by Osterbrock (1953) or Temesváry (1959).

The dashed curve in Figure 19 shows one of the limiting curves as originally calculated by Hayashi and Hōshi (1961). This calculation was based not on the mixing-length theory but on the Hoyle-Schwarzschild (1955) treatment of the extent of the superadiabatic zone in terms of the velocity of sound, in which an undetermined efficiency factor β plays a role analogous to that of the unknown ratio of mixing length to scale height in the Böhm-Vitense method. These curves of Hayashi and Hōshi were based on the value $\beta = 0.15$ and a composition $X = 0.61$, $Y = 0.37$, $Z = 0.02$. To what extent the differences between the two sets of curves are due to the different methods of treating the

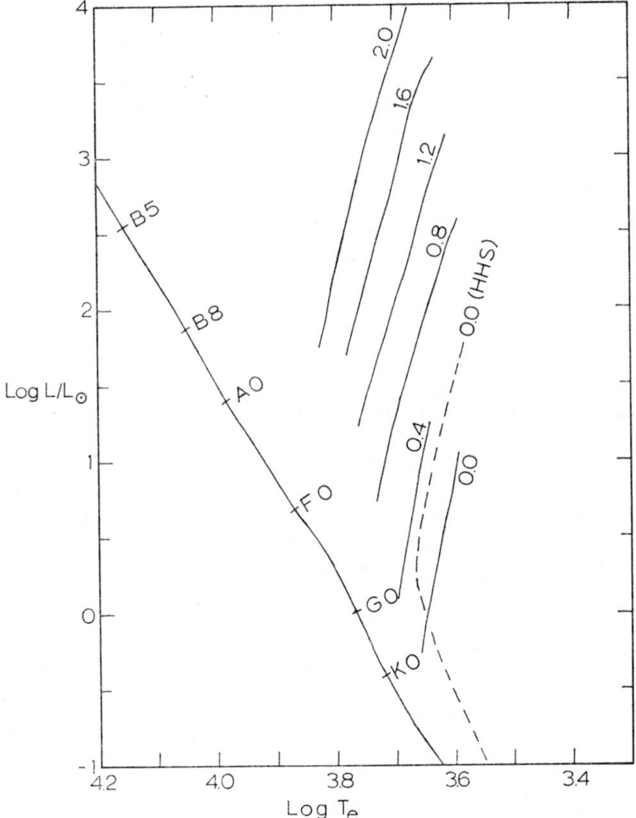

FIG. 19.—Hayashi limits, including the effect of superadiabatic convection, based on mixing-length calculations for Unsöld's composition with $l = H_p$. Numbers on the curves refer to the logarithm of the mass in solar units. The dashed line is the curve found by Hayashi *et al.* for one solar mass with a lower metal abundance, using the Hoyle-Schwarzschild method for the superadiabatic region. The diagonal line on the left shows my interpretation of the observed main sequence.

superadiabatic region, and to what extent they are due to differences of composition and opacity, is not easy to determine.

The position of the Hayashi limit curves is not only sensitive to the extent of the superadiabatic region, as indicated by the shifts between Figure 15 and Figure 19. They are also sensitive to the metal abundance that determines the atmospheric opacity. Indications of the sensitivity to both factors were provided in a study of red giant evolution by Kippenhahn, Temesváry, and Biermann (1958). The evolutionary tracks obtained by these authors in the red-giant region may be taken as a close approximation to the Hayashi limits for the various conditions assumed.

In working out the Hayashi limits shown by the solid curves in Figure 19, I have limited the range of effective temperatures to $T_e > 4000°$ K. For lower temperatures the opacities are not known with sufficient precision for a reliable calculation, mainly because of the formation of molecules. The importance of further studies in this low-temperature region is obvious. A considerable amount of work has been done already on the molecular opacities, for example, by Yamashita (1962), Gaustad (1963), Somerville (1964), and Vardya (1964). It appears that our knowledge of opacities in the temperature range from 4000° K to about 2000° K may soon be reasonably complete.

Unfortunately, however, the uncertainty which still pervades the theory of superadiabatic convection bars the way to a definitive calculation of the Hayashi limits, regardless of the accuracy of the opacity calculations. The effect of the superadiabatic zone is so great in red giant models that there is no immediate prospect of obtaining accuracy in this region by purely theoretical calculations. The color-magnitude diagrams of star clusters give the best guide at present to the position of the Hayashi limits in this region.

7. RESULTS FOR CONVECTIVE TRANSPORT

Hayashi (1961) suggested that contracting stars hug the edge of the forbidden region discussed in Section 6 during the large-radius stages, as do the giants in old clusters. Evolutionary tracks would then be identical with the Hayashi limit curves (Fig. 19) down to a point where the limiting curve crosses the pseudo-homologous track for radiative contraction (Fig. 3). Such an evolution is homologous in the sense that the rate of entropy change is the same in all parts of the star. This condition is in accord with the usual definition of homology if the degree of ionization is constant (cf. Sec. 3). If the ionization is not constant, the energy available per unit mass still has the same temperature dependence as in the Levée model, but the resulting changes of density and temperature are not homologous.

In contrast to the radiative case, the opacity now need satisfy no rigid conditions for such a distribution of the flux divergence to be possible. Rather, convection is counted upon to keep the entropy constant throughout the star and to carry heat in the required pattern without departing sensibly from the constant entropy structure. All that is required of the opacity is that it be high enough to inhibit radiative transport, i.e., to make the radiative temperature gradient steeper than the adiabatic. This condition will be easily satisfied in the high-luminosity models of the early part of the sequence, but as the star moves down the Hayashi curve, the diminishing flux will eventually fall below what is carried by radiation alone. This must occur at approximately the intersection of the Hayashi curve with the track for homologous radiative contraction but slightly above it because the radiative flux in the adiabatic $n = 1.5$ model is somewhat greater than in the standard $n = 3$ model of the same radius.

These qualitative arguments show the possibility of a completely convective sequence until the star nears the line for pseudo-homologous radiative contraction. At this point the star can be expected to switch to radiative transport and thereafter to evolve more or less along the tracks of Figure 3. A comparison of the Hayashi lines with the main-sequence positions of stars of the same masses shows that the stars of large mass have the greatest distance to travel on the radiative portion of their contraction paths, and that stars bound for the lowest parts of the main sequence have no reason to turn radiative at all. These same trends can also be seen in terms of the increasing extent of the convective envelope for main-sequence stars of successively lower masses that has been established by the work of Osterbrock (1953) and Limber (1958). Of particular interest is Limber's conclusion that main-sequence stars of spectral class M4 and later are probably convective throughout.

Details of the switchover from convective to radiative structures have been calculated by Hayashi *et al.* (1962) for a variety of stellar masses, and by Weymann and Moore (1963) and by Ezer and Cameron (1963) for the Sun. All three studies are approximate because the time-dependent thermal equation (12) has been replaced by equation (29), corresponding to an assumption of homologous contraction. Since the contraction is not homologous in this stage, a degree of error has thus been introduced whose magnitude is not easy to estimate a priori. The effect of this approximation is to exaggerate the sharpness of the turn onto the radiative line.

The calculation of these models by Hayashi and his associates differs from the other two calculations (1) in the treatment of the superadiabatic convection zone and (2) in the opacity approximation for the radiative core. The first point is discussed at the end of Section 6. The velocity-of-sound method leads to the limiting surface conditions shown by the dashed curve in Figure 19, whereas the mixing-length method leads to the limiting surface conditions shown by the solid curves. The slight differences between these two sets of curves are not likely to give any great differences in the results. On the second point, Hayashi has approximated the opacity by a Kramers formula $\kappa = \kappa_0 \rho T^{-3.5}$, while Weymann and Moore used a modified Kramers formula, originally devised by Schwarzschild, Howard, and Härm (1957), of the form $\kappa = \kappa_0 \rho^{0.75} T^{-3.5}$. Ezer and Cameron used tables computed by the program developed by Cox at Los Alamos, which may be assumed to yield the most accurate opacities of the three sets. (See Vol. 8, chap. 3, this compendium, by Cox.)

The results of Hayashi *et al.* are probably the least accurate for the Sun, but they are more readily applicable to other stars by suitable scaling of the models. In addition, their method gives the greatest insight into the relative roles of radiative transport and the surface boundary condition. The loss of precision in the opacity is not a serious drawback, provided the values are reasonable on the average, since precision was already given up when the time-dependent thermal equation was replaced by the assumption of homologous contraction. Furthermore, we shall show how the Japanese models lend themselves readily to a luminosity correction when opacities more precise than those given by the Kramers formula are available.

The calculations of Hayashi *et al.* are based on the consideration that the conditions of homologous flux divergence (eq. [29]) and Kramers opacity can lead not only to the purely radiative Levée model but also to a whole series of alternative models which have convective envelopes of greater or lesser extent. Out of the whole series of possible models for the star at any given radius, only one satisfies the photospheric boundary conditions (31a) and (31b), although they all satisfy the zero boundary conditions. The situation is exactly analogous to the one for the lower main-sequence stars discussed by Limber (1958), and most of his discussion is applicable to the present problem. We calculate two separate curves in the Hertzsprung-Russell diagram for each type of stellar model and each mass. One curve is based on the internal structure and one on the photospheric boundary condition. The actual star of that mass and model must lie at the intersection of the two curves.

TABLE 2

PROPERTIES OF RADIATIVE-CONVECTIVE MODELS

E......	0.00	4.35	9.58	13.62	20.77	32.22	39.66	42.62	45.11	45.44	45.48
$C_1/10^{24}$.	9.92	9.86	9.47	9.27	9.31	10.55	13.69	17.54	30.30	43.06	....
$\rho_c/\bar{\rho}$....	55.7	44.9	33.7	28.0	20.7	13.0	9.35	7.92	6.49	6.12	5.99
T_c/T_a..	....	11.46	6.69	5.15	3.66	2.37	1.77	1.54	1.25	1.119	1.000
q_a......	1.000	0.998	0.982	0.962	0.906	0.751	0.564	0.430	0.191	0.073	0.000
x_a......	1.000	0.851	0.785	0.748	0.695	0.608	0.526	0.468	0.331	0.239	0.000
n_c	2.972	2.972	2.972	2.972	2.972	2.970	2.950	2.900	2.600	2.200	1.500
$a_T/10^6$.	20.0	18.46	16.79	15.79	14.40	12.74	11.97	11.83	12.02	12.28	12.46
a_ρ......	78.5	63.3	47.5	39.5	29.2	18.39	13.18	11.17	9.15	8.63	8.45
$a_p/10^{16}$.	12.9	9.64	6.58	5.14	3.47	1.932	1.301	1.090	0.907	0.874	0.868
$a_E/10^{48}$.	2.94	2.658	2.419	2.278	2.082	1.843	1.720	1.673	1.635	1.624	1.628
C_3......	21.9	23.09	23.19	23.41	23.98	26.29	30.41	34.40	46.05	57.26	....

The curve based on internal structure is found as follows: for the inner radiative region the equations governing the structure are (32)–(37), the same as in the Levée model. An extra degree of freedom is now introduced into the solutions, however, because P and T are not required to go to zero simultaneously when the radiative temperature gradient is carried out to the surface. The solutions in which T goes to zero first are now admissible, because they can be joined smoothly onto convective envelopes at the point where $n = 1.5$. These are the solutions illustrated in the lower section of Schwarzschild's (1958) Figure 11.1. Each of the possible solutions has a certain value of E in the convective region, the permissible range of E values running from 0 for the completely radiative model to 45.48 for the completely convective model. The model with $E = 0$ is the Levée model.

For each E in this range a model is found which, like the Levée model, can be scaled homologously in both radius and mass, provided we take Eddington's $\beta = 1$ (no radiation pressure). If the mass is fixed, then the models of various radii lie along a line whose slope in the H-R diagram is that shown in Figure 3 for the Levée model; that is, it is governed by the equation

$$L = \frac{C_1}{\kappa_0} \frac{\mu^{7.5} \mathfrak{M}^{5.5}}{R^{0.5}}, \tag{84}$$

where the constant C_1 depends on the model. The relation between C_1 and E, taken from Hayashi *et al.* (1962), is given in the first two lines of Table 2. Lines 3–7 give other important properties of the models, defined as follows:

$$\frac{\rho_c}{\bar{\rho}} = \frac{\text{central density}}{\text{mean density}}$$

$$\frac{T_c}{T_a} = \frac{\text{central temperature}}{\text{temperature at inner edge of convection zone}}$$

q_a = fraction of mass in radiative core

x_a = fractional radius of radiative core

n_c = polytropic index at center .

Lines 8–11 give coefficients defined as follows for the central temperature, central density, central pressure, and total energy in terms of the mass and radius expressed in solar units:

$$T_c = a_T \beta \mu \frac{\mathfrak{M}}{R} \tag{85}$$

$$\rho_c = a_\rho \frac{\mathfrak{M}}{R^3} \tag{86}$$

$$P_c = a_P \frac{\mathfrak{M}^2}{R^4} \tag{87}$$

$$E = -a_E \frac{\mathfrak{M}^2}{R} . \tag{88}$$

Line 12 gives an alternative coefficient C_3 for the luminosity in terms of the central opacity according to the relation

$$L = \frac{C_3}{\kappa_c} \mathfrak{M}^3 \beta^4 \mu^4 . \tag{89}$$

C_3 was derived from C_1 in the same manner as was done earlier for the Levée model (cf. eqs. [59], [63], [65]), but here we take the central opacity rather than an intermediate value as standard. The significance of (89) is the same as that of (65) for the purely radiative case: a better approximation to the luminosity than that given by Kramers opacities can be obtained by inserting an accurate opacity for the central point into (89). Of course, as we proceed from the purely radiative (Levée) model at the left of the table to the purely convective model at the right, the luminosity becomes less and less dependent on any opacity except the atmospheric value, and this is not taken into account in these calculations. Nevertheless, the luminosity calculated by the above formulae has a significance across the entire table, because the core can contract homologously with the envelope only when the luminosity has the prescribed value. Only when the radiative core completely vanishes does this condition lose its meaning, and hence only the model at the extreme right of the table is unrestricted by a luminosity condition.

We now apply these models of Hayashi *et al.* to the star of 1 $\mathfrak{M}_\odot$ with the Unsöld composition discussed earlier. For each model in Table 2 we assume a variety of radii. For each radius we calculate the central temperature and density by equations (85) and (86), taking the appropriate constants from the table. Then we find the opacity corresponding to T_c and ρ_c by interpolation in the Keller-Meyerott tables (cf. Fig. 5). Finally we calculate the luminosity by (89). The result is a set of curves in the H-R diagram, one curve for each of the Hayashi models. These curves are shown in Figure 20, which is modeled after Hayashi's (Hayashi *et al.* 1962) Figure 10–1. The difference between Hayashi's curves and the ones shown here lies solely in the opacity values assumed.

We now take into account the photospheric boundary condition, which was ignored in constructing the first set of curves. This condition is the one discussed in Section 6 and illustrated in Figure 19, except that we now consider values of E other than 45.48; that is, we consider convective envelopes that do not reach to the center of the star. Thus we obtain a family of surface boundary-condition curves for each mass corresponding to convective envelopes of different extent. The curves for one solar mass are

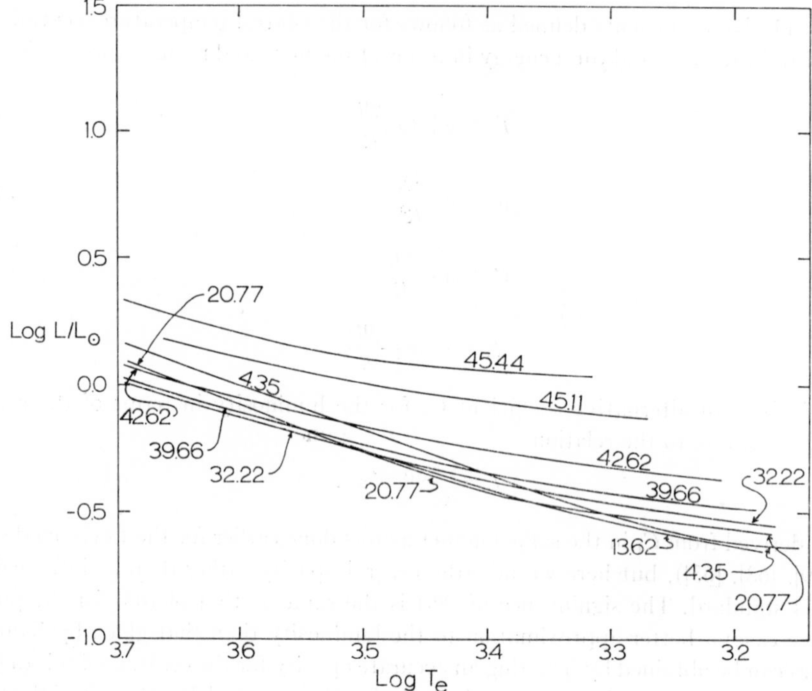

FIG. 20.—First, or interior, radius-luminosity relation for partially convective contracting stars based on Hayashi's models. The numbers on the curves refer to the value of E in the convection zone. All curves are for one solar mass and Unsöld composition. The crossings of the curves are due to the complexities of the Keller-Meyerott opacities used.

shown in Figure 21. Here, as before, we take the superadiabatic convection zone into account by means of the mixing-length theory, using the published calculations for $l/H_p = 1$. The E values are a measure of the specific entropy in the adiabatic part of the convection zone according to equation (77).

Now, following Hayashi, we can construct the sequence of solar models that are in homologous contraction (in the sense discussed at the beginning of this section). The permitted models are those which lie at the intersection of an internal-structure curve and a boundary-condition curve having the same E value. There is only one such model for any given radius. The intersection points are circled in Figure 21, and the evolutionary track is a curve which joins the circles. Figure 22 shows the paths in the log ρ–log T

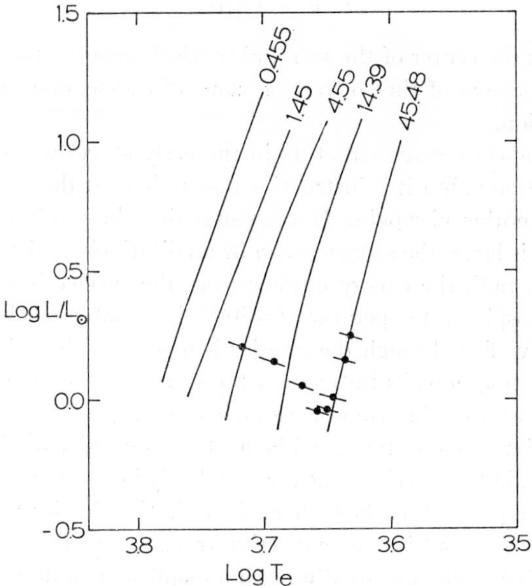

FIG. 21.—Second, or exterior, radius-luminosity relation for partially convective contracting stars based on convective-envelope calculations with $l/H_p = 1$. Mass and composition are the same as in Figure 20. Numbers on curves are E values as in Figure 20. Dots show the positions of complete models, determined by the intersection of interior and exterior curves with the same E. The totality of dots comprises the evolutionary sequence. A short section of the appropriate interior curve is shown at each intersection.

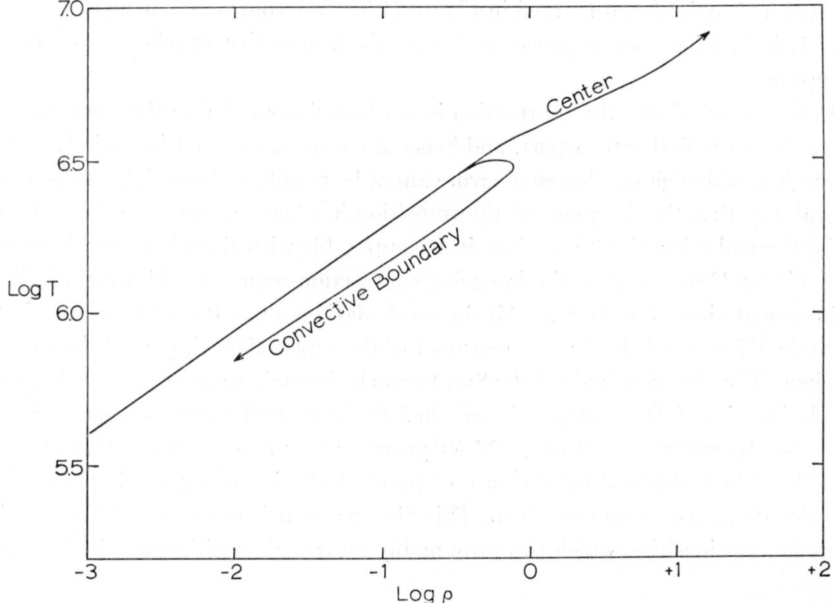

FIG. 22.—Progression of temperature and density in a contracting star of one solar mass at the center and at the lower boundary of the convective envelope. The evolutionary sequence is the same as in Figure 21.

diagram followed by the center of the star and by the bottom of the convection zone in this sequence. The bottom of the convection zone, of course, does not correspond to a constant mass fraction.

The reason for the convective structure in the early stages was described from one point of view in Section 6, but it is instructive now to look at the transport processes in this sequence from another viewpoint. If we assume that the star has an adiabatic structure when its radius is large, the sequence can be qualitatively understood in the following terms. For large radii the energy outflow from the surface is high because of the relatively high photospheric temperature (~3000° K) established by the atmospheric opacity. The radiative flux through the interior is much less than the surface flux, and consequently, convection prevails throughout the star, maintaining the adiabatic structure. As the radius decreases, the atmospheric opacity varies so as to hold the photospheric temperature approximately constant, and hence the luminosity diminishes inversely as R^2. The radiative flux in the interior is meanwhile slowly increasing because of decreasing opacity. Eventually a point is reached where the radiative flux in the interior is comparable to the surface flux, and hence convective transport is no longer required. More specifically, the radiative flux eventually becomes sufficient to diminish the entropy in the interior faster than that at the surface, thus destroying the adiabatic structure and choking off convection. The change occurs first in the center of the star, because there the opacity is lowest compared to the Levée model. Thus a radiative core develops, beginning at the first intersection point in Figure 21, and works its way gradually outward in the star. If the contraction proceeds far enough, a new homologous condition is eventually reached, based on radiative transport with $n \simeq 3$ instead of the original $n = 1.5$. This point is not marked in Figure 21 because contraction in a star of one solar mass is halted by nuclear processes before the convective envelope has completely disappeared.

As was noted above, the contraction is not homologous during the changeover from convective to radiative transport, and hence the sequence cannot be entirely correct in this region. Although the degree of error cannot be readily estimated, it can be seen in a general way that the sharpness of the transition has been exaggerated through neglect of the thermal relaxation time, since it is comparable with the whole transition period.

In Figure 23 we compare the foregoing contraction sequence with those of other authors quoted above for the Sun. All these calculations show the same general features, the main differences being in the treatment of the superadiabatic part of the convective envelope. The time required for the Sun to reach the main sequence, according to these calculations, is as follows: Hayashi *et al.*, 26.5×10^6 years; Ezer and Cameron, 5.7×10^6 years; and Weymann and Moore, 2×10^6 years. These times are all substantially shorter than the values obtained for the case of purely radiative transport, but there is considerable disagreement among them. This disagreement is due to the differences in the computed luminosities, which can presumably be traced to differences in the opacities assumed.

Figure 24, reproduced from Hayashi *et al.* (1962), shows the contraction paths found by them for stars of several masses. This diagram shows the trend toward shorter radiative sections for stars of progressively smaller masses. The times required for these stars to reach the main sequence were found to be as follows:

Mass (solar units):	2	1	0.6
Time (10^6 years):	1.22	26.5	143

8. PRE-MAIN-SEQUENCE NUCLEAR REACTIONS

Except for the isotopes of helium, the atomic nuclei with masses between 2 and 11 will combine with protons more readily than protons will combine with one another.

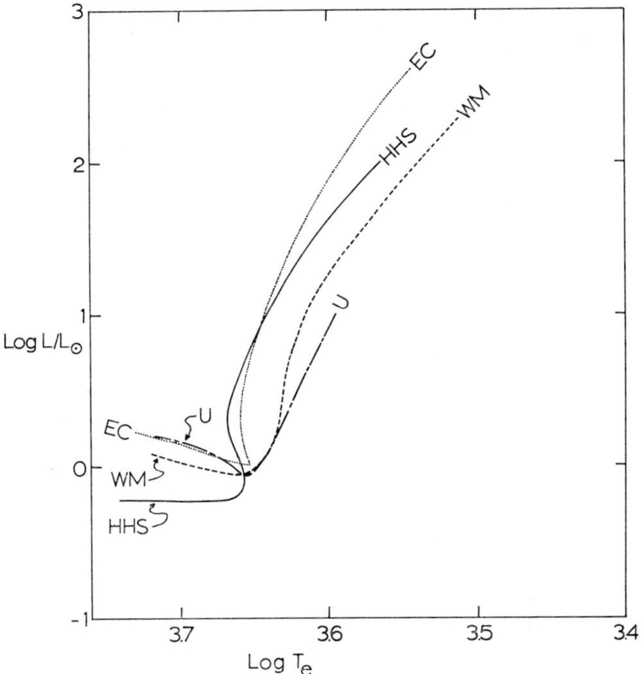

FIG. 23.—Four pre-main-sequence contraction paths for the Sun. *EC* = Ezer and Cameron (1963); *WM* = Weymann and Moore (1963); *HHS* = Hayashi, Hōshi, and Sugimoto (1962); *U* = Upton, sequence derived here from Hayashi's models. The differences between the sequences are due mainly to differences in interior opacities on the horizontal sections and to differences in convective boundary conditions on the vertical sections.

This circumstance has often been called upon to explain the low abundance of these constituents in the Sun and other stars. The possibility of reactions involving these isotopes in the pre-main-sequence stage of evolution is an obvious one, and several theoretical studies on this question have been carried out.

The stable isotopes which may be presumed initially present are H^2, Li^6, Li^7, Be^9, B^{10}, and B^{11}. The details of the reactions between these particles and protons, with their alternative branches, have been summarized by Reeves (Vol. 8, this compendium). Omitting intermediate stages, reactions are as follows:

$$H^2 + H^1 \rightarrow He^3 + 5.49 \text{ Mev}$$

$$Li^6 + 2H^1 \rightarrow 2He^4 + 23.80 \text{ Mev}$$

$$Li^7 + H^1 \rightarrow 2He^4 + 17.33 \text{ Mev}$$

$$Be^9 + 3H^1 \rightarrow 3He^4 + 25.93 \text{ Mev}$$

or

$$Be^9 + 3H^1 \rightarrow C^{12} + 33.20 \text{ Mev}$$

$$B^{10} + 2H^1 \rightarrow 3He^4 + 19.34 \text{ Mev}$$

or

$$B^{10} + 2H^1 \rightarrow C^{12} + 26.61 \text{ Mev}$$

$$B^{11} + H^1 \rightarrow 3He^4 + 8.66 \text{ Mev}$$

or

$$B^{11} + H^1 \rightarrow C^{12} + 15.93 \text{ Mev}.$$

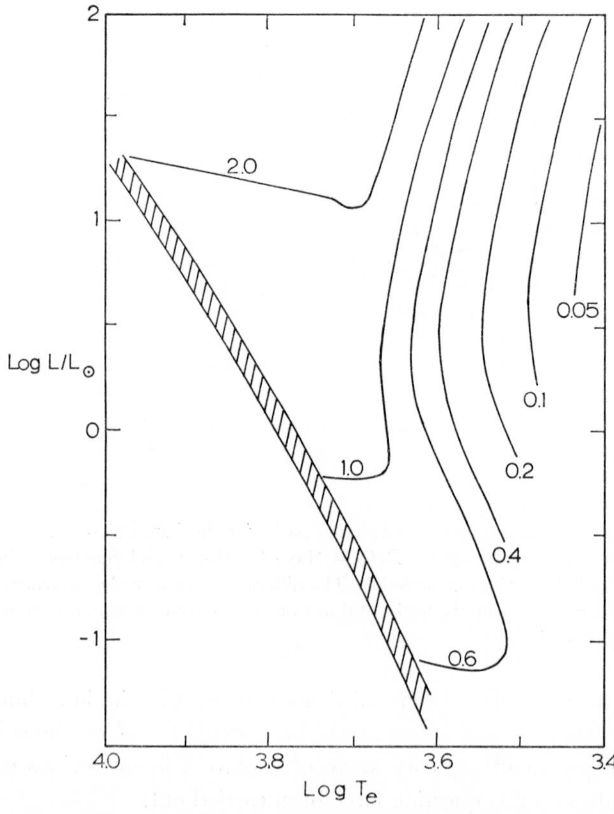

FIG. 24.—Pre-main-sequence evolutionary tracks for several masses, according to Hayashi, Hōshi, and Sugimoto (1962). The shaded band is the main sequence used by them for comparison. Masses are shown in solar units. Composition $X = 0.61$, $Y = 0.37$, $Z = 0.02$.

The energies listed here do not take into account neutrino losses, which are generally small. However, in the reactions involving Li^6, Be^9, and B^{10} there are possible branches leading through B^8, in which case an average neutrino loss of 7.2 Mev is expected.

Salpeter (1954) pointed out that if these light isotopes were present in sufficient quantities in the pre-stellar material, the energy derived from their consumption should halt the contraction of the star at a certain central temperature. He gave the estimates reproduced here in Table 3. The upper and lower estimates for each element correspond to mean reaction times of 10^7 and 10^3 years, respectively. Such a halt, if it occurs, should give rise to a deuterium main sequence, for example, in the color-magnitude diagram, of which some part would be always populated in a cluster containing young contracting stars. So far the observations of young clusters, e.g. by Walker (1956, 1957, 1959, 1961) show no clear indications of such sequences; so we can tentatively conclude that the

TABLE 3

CENTRAL TEMPERATURES AT WHICH CONTRACTION OF STAR IS HALTED*

	H^2	Li^6	Li^7	Be^9	B^{10}	B^{11}
$T_7/10^6\,°K$	0.8	2.7	3.2	4.2	6.4	6.6
$T_3/10^6\,°K$	1.4	4.1	5.0	7.7	9.5	9.9

* From Salpeter (1954). The upper and lower estimates correspond to mean reaction times of 10^7 and 10^3 years, respectively.

amounts of the light elements required to produce such sequences are not present in pre-stellar material. However, there are still a number of unsolved problems concerning the membership of many of the non-main-sequence objects in the young clusters and the determination of their radii from the color and magnitude data. (See, for example, The 1960, Underhill 1960, Varsavsky 1960.)

What abundances of these light isotopes would be required to halt the contraction? Since the effect of a new energy source cannot make itself felt throughout the star in less than the thermal relaxation time, which may be taken equal to the Kelvin time, the energy available from the nuclear reactions must at least equal the energy lost by the star in that length of time. This energy is identical with the thermal energy content of the star at the time the reactions are taking place. Thus the energy source must be sufficient to heat the star to its mean temperature, which is 0.53 to 0.60 times the central temperature in the relevant models. That is, per gram of stellar material there must be an energy

$$E_{\rm nuc} = \frac{3}{2}\,\frac{k\bar{T}}{\mu m_{\rm H}} \qquad (90)$$

available. Taking the energies per reaction listed above (with the larger value for alternative branches), and taking the mean temperatures equal to 0.55 the average values given by Salpeter, we obtain the required abundances of the various isotopes given in Table 4. In deriving the numbers required relative to hydrogen and silicon we use the composition of Unsöld in which $N_{\rm Si}/N_{\rm H} = 56 \times 10^{-6}$. It is of interest to compare the figures in this table with experimentally determined abundances of the same isotopes in terrestrial and astronomical sources, as summarized, for example, by Aller (1961). The required abundances of Li, Be, and B are not known to occur anywhere in nature except in cosmic rays. The required abundance of deuterium is below the present limit of detection in stars, cosmic rays, and the interstellar medium, but it is exceeded by about a factor of 3 on Earth. Thus a main sequence based on one or more of these isotopes would be possible if terrestrial or cosmic-ray abundances were representative of the pre-stellar material.

In the calculation of the required abundances by this method, only total energies are involved; in particular, the rate of contraction with respect to time does not enter the calculations. The reason lies in the identity of the contraction time scale and the thermal relaxation time scale, which would enter in mutual cancellation into the calculation if it were done in time-dependent terms. Likewise the rate of the nuclear reactions need not be considered except to fix the approximate temperature at which they can proceed at a rate comparable with the contraction rate. Of course this type of calculation has only an order-of-magnitude validity.

TABLE 4

ABUNDANCES OF LIGHT ISOTOPES REQUIRED TO HALT CONTRACTION

	H^2	Li^6	Li^7	Be^9	B^{10}	B^{11}
N/N_H	4	3	4	3	6	9 $\times 10^{-5}$
N/N_{Si}	0.7	0.5	0.8	0.6	1.0	1.7

A more detailed investigation of the effects of deuterium burning was included by Brownlee and Cox (1961) in the study referred to in Section 5 (see Fig. 7). With an assumed deuterium abundance five times larger than the critical value given in Table 4, their calculations showed a brief halt on the deuterium main sequence at a central temperature of 8×10^5 ° K, in good agreement with Salpeter's estimate and with the critical abundance calculated here. This investigation needs to be repeated, however, with stellar models that include the outer convection zone.

Calculations of lithium burning have been done for the convective envelope models. Although the possibility of a lithium main sequence seems remote in view of the low abundance of this element, there is considerable interest in whether during the contraction the surface material in stars of the middle main sequence passes through a region of sufficiently high temperature to destroy the primordial lithium. This interest has arisen largely because of the high lithium abundance in T Tauri stars as compared with that of the Sun. Bonsack and Greenstein (1960) and Bonsack (1961) found Li/Ca ratios of 4×10^{-4} to 3×10^{-3} in T Tauri stars, whereas the same ratio in the solar atmosphere is 6×10^{-6} (Goldberg, Müller, and Aller 1960).

The studies of solar contraction reviewed in Section 7, where the convective envelope was included, have all been concerned with determining whether lithium would be significantly depleted in the envelope. The answer depends basically on whether the bottom of the envelope ever reaches the temperatures estimated by Salpeter, when account is taken of the fact that the envelope retreats toward the surface as the star contracts. The maximum envelope temperatures found in the studies referred to are reproduced in

Table 5. Hayashi *et al.* concluded that Li^7 would be significantly depleted. Ezer and Cameron found a significant depletion of Li^6 but not of Li^7, whereas Weymann and Moore found that neither isotope would be significantly depleted. The depletion is obviously highly sensitive to the maximum temperature, which in turn depends quite critically on how the superadiabatic part of the convection zone is treated—specifically, on what is assumed about the ratio of mixing length l to pressure scale height H_p. The second and third results above were based on an assumed $l/H_p = 1$. With a greater l/H_p ratio, the convection zones would reach to higher temperatures and lithium would be more readily depleted, as has been pointed out by all these authors. Thus it appears that a reliable calculation of this process must await a more definitive theory of superadiabatic convection.

9. THERMODYNAMIC COLLAPSE

When the pre–main-sequence history of a star is extended back to larger and larger radii, what is the practical limit to the hydrostatic models? According to what has been said thus far, the opacity considerations discussed in Section 1 should set the limit; but

TABLE 5

MAXIMUM TEMPERATURE IN CONVECTION ZONE OF THE SUN

Source	T_{max}(°K)
Hayashi *et al.* (1962)	3.5×10^6
Ezer and Cameron (1963)	3.4×10^6
Weymann and Moore (1963)	2.5×10^6

Cameron (1962) has pointed out that another type of limitation is to be expected when the radius is only $\backsim$100 times the main-sequence value. This limitation is a thermodynamic instability associated with the onset of ionization or of molecular dissociation in the star.

The thermodynamic stability of a star is governed by the way in which the total energy, E_{hyd}, required for the hydrostatic model, changes with the radius R. The star is stable when a decrease in R is accompanied by a decrease in E_{hyd}. In this instance the radius shrinks slowly at a rate governed by the leaking out of energy. A faster rate would lead to a model in which the energy exceeds the value E_{hyd} required for hydrostatic equilibrium, so that the star would expand back to its equilibrium radius. If, on the other hand, E_{hyd} increases with decreasing radius, then a small diminution of R at constant energy would lead to a model in which the energy is less than the value required for equilibrium. Then a runaway collapse can be expected to occur on the hydrodynamic or free-fall time scale, that is, on a time scale of a few days to a year for convective models with radii of order 100 $R_\odot$. The collapse can halt only when a domain of thermodynamic conditions is reached such that E_{hyd} again decreases with decreasing R. These considerations were described by Eddington (1926, pp. 142 ff.).

Eddington showed that the equilibrium energy E_{hyd} can be related to the potential energy Ω by the equation

$$E_{hyd} = \beta\Omega \frac{(\gamma - 4/3)}{\gamma - 1} . \tag{91}$$

In this equation, Ω is defined as the negative quantity

$$\Omega = -\int \frac{Gm\,dm}{r}, \tag{92}$$

and β is the ratio of gas pressure to total pressure, the value depending primarily on the mass of the star and the molecular weight (cf. Sec. 4). The quantity γ is called the ratio of specific heats by Eddington, but it can be seen from his discussion that the basic definition of the γ in (91) is

$$\gamma = \frac{1+a}{a}, \tag{93}$$

where a is the ratio of gas energy density to gas pressure. This γ would be the same as the ratio of specific heats if it were independent of temperature, but actually it is not. An alternative expression for the equilibrium energy in terms of a is

$$E_{\text{hyd}} = \beta\Omega\left(1 - \frac{a}{3}\right). \tag{94}$$

Equations (91) and (94) are not actually suitable for the calculation of the energy in the interesting cases, since the quantities a and β must be averaged over the star. These equations do illustrate, however, the kind of condition required for instability. Since Ω always decreases in a contraction, E_{hyd} can have the unstable increasing behavior only when a is close to 3 or greater. If $a > 3$, then the equilibrium energy is positive according to (94). Equilibrium is clearly impossible in this situation, since the zero point of energy is chosen so that the star has zero energy when its radius is infinite.

In terms of equipartition ideas, a can be considered as half the total number of degrees of freedom of the gas molecules, the normal value being $2a = 3$, corresponding to the three translational degrees of freedom. The energy in these three degrees of freedom holds the star in equilibrium when supplied with $\beta/2$ times the gravitational energy released by contraction $(-\Omega)$, a result derivable from the virial theorem. From this point of view, instability arises when there are three or more additional degrees of freedom which make an equal or greater demand on the energy supply. Such a demand cannot be met, since the radiation in the star holds an amount $(\beta - 1)\Omega$, leaving a surplus of $-(\beta/2)\Omega$—exactly enough to supply three degrees of freedom if there were no leakage. The result of the impossible demand on the energy supply is the collapse of the star.

Eddington (1926) pointed out that during the onset of ionization a gas can have $\gamma < \frac{4}{3}$ or $a > 3$; that is, the ionization energy is often equivalent to more than three degrees of freedom at the temperature at which ionization occurs. He suggested that a thermodynamic collapse might occur at an early stage when some abundant element undergoes ionization, but this suggestion could not be developed until the composition of the stars was better determined. Cameron (1962) revived the idea in connection with the formation of the Sun, suggesting that not only the ionization of H but also the dissociation of H_2 would give rise to Eddington's thermodynamic collapse.

In a rough calculation, the stage at which this phenomenon occurs can be deduced from the consideration that the total energy of the star at that stage must be about equal to the ionization and dissociation energy of the entire hydrogen content of the star.

Applying this calculation to the Sun, we find that the critical energy is about 100 times smaller (in absolute value) than the present solar energy, so that the critical radius is about 100 $R_\odot$.

S. J. Little and I have carried out a more exact calculation for pure hydrogen composition. At the large radii involved, the stars can be assumed to be completely convective, so that the structure is governed by an adiabatic relation. The adiabatic curves are shown in Figure 25, where log T is plotted as a function of log P for constant entropy. This diagram does not show γ directly, but the unstable domain ($\gamma < \frac{4}{3}$) is located on

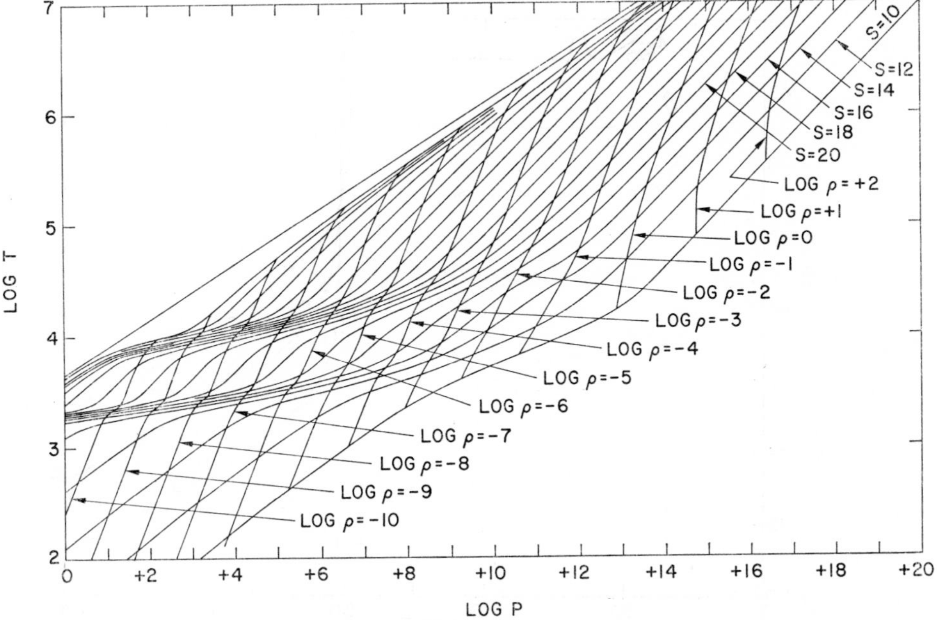

FIG. 25.—Pressure-temperature diagram for hydrogen with adiabatic curves and density contours calculated by the author. Convective models investigated for stability were based on these adiabatic curves. S (S^* in text) denotes entropy per hydrogen atom in natural units discussed in text.

the flat portions of the curves associated with ionization and dissociation. Notice that for many of the curves, radiation pressure dominates at both very high and very low temperatures; so γ approaches $\frac{4}{3}$ at both limits.

Figure 26 shows the relation between the E_{hyd} and central temperature for stellar models constructed on these adiabatic relations. Each curve in this diagram corresponds to a fixed stellar mass. On the basis of the considerations discussed above, the parts of the curves that slope downward and to the right correspond to stable models, and those parts which slope upward and to the right correspond to unstable models. The dashed lines show the expected collapse of the stars through the unstable domain.

These results show that the instability occurs at a very low temperature and, by implication, a very large radius for all bodies within the normal range of stellar masses. Actually the models corresponding to the onset of instability are not realistic, since the opacities in those models have been shown by Gaustad (1963) to be too low to meet the criteria discussed in Section 1. In particular, some of the large-radius models would be optically thin and would therefore not be able to hold the radiation that was included in the adiabatic relation and in the energy. In addition, Gaustad's calculations show that the free-fall velocities in these early models are too small to enable them to contract at a rate commensurate with the leakage of energy. Thus Gaustad's criteria, rather than the thermodynamic criterion illustrated in Figure 26, are the relevant ones for the models with mean temperatures below about 2000° K. The conclusion is the same, however: The stars collapse on a free-fall time scale.

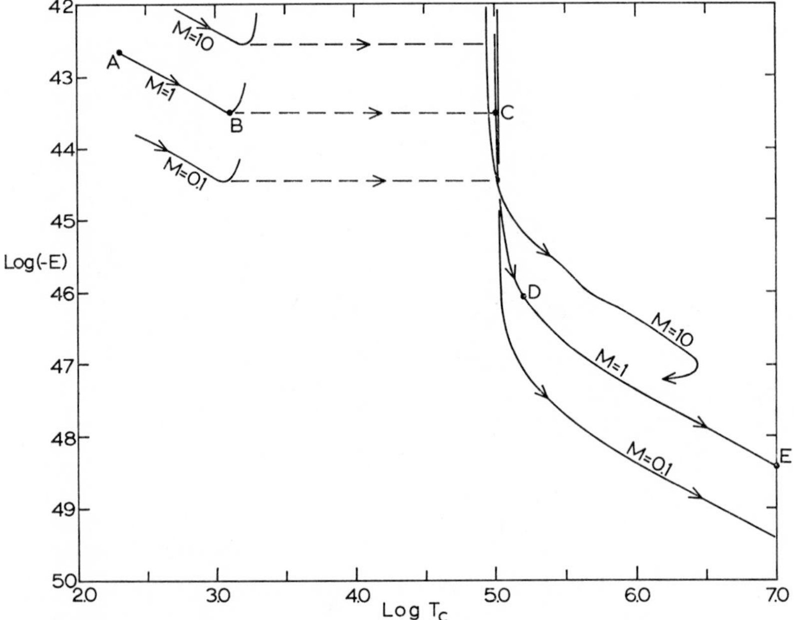

FIG. 26.—Log of total negative energy vs. log of central temperature for contracting stars in convective equilibrium (adiabatic models). [NOTE.—M = mass of star in units of solar mass.] (*Erratum.* Masses on right should read, top to bottom, 0.1, 1, 10.)

In summary then, the contraction of a star in the absence of rotation or magnetic forces is expected to be essentially a free-fall process throughout the early stages corresponding to mean temperatures in the range from 10^3 to 10^5 ° K. The free fall begins as a result of an opacity deficit, but it does not end when Gaustad's opacity criteria are met, because the star is then in the temperature range of thermal instability. The details of the collapse through this domain can be found only by a calculation based on the hydrodynamical equations, the results of which may depend to a considerable extent on the initial conditions assumed. The approximate time required for the collapse can be found from Kepler's third law, which yields 2×10^6 years for the Sun if the initial density is 10^{-22} gm/cm^3.

The collapse is expected to end at approximately the first stable radius, shown in Figure 26, where a hydrostatic structure becomes possible. A few oscillations about this equilibrium radius may be expected, perhaps giving rise to some interesting phenomena as suggested by Cameron (1962). At this stage the Hayashi limits apply, and a convective structure is indicated for stars with masses a few times the solar mass down to the smallest known values. There may be a significant time spent in relaxing to the convective model, however, depending on the details of the preceding collapse stage. This relaxation process could conceivably last from about 100 $R_\odot$ to about 10 $R_\odot$ for a star of 1 $\mathfrak{M}_\odot$. The permitted nonconvective models would have to have luminosities higher than the completely convective model of the same radius, so that contraction down the Hayashi line provides an upper limit to the time scale at this stage.

Except in stars with the smallest masses, a sufficiently low luminosity is eventually reached that convection begins to die out, a radiative core being formed and spreading gradually outward from the center. The star then evolves more or less horizontally to the left in the Hertzsprung-Russell diagram. The extent of this radiative portion of the contraction path diminishes with decreasing stellar mass. At 30 $\mathfrak{M}_\odot$ it should comprise nearly the whole post-collapse stage of contraction, whereas at 0.2 $\mathfrak{M}_\odot$ it is not expected to occur at all. At 1 $\mathfrak{M}_\odot$ the radiative portion occurs only from about 2 $R_\odot$ to 1 $R_\odot$, but it accounts for the bulk of the time required to reach the main sequence.

REFERENCES

Aller, L. H. 1959, *Mém. Soc. R. Sci. Liège*, **3**, Series 5, 41.

———. 1961, *The Abundance of the Elements* (New York and London: Interscience Publishers, Inc.).

Böhm-Vitense, E. 1958, *Zs. f. Ap.*, **46**, 108.

Bonsack, W. K. 1961, *Ap. J.*, **133**, 340.

Bonsack, W. K., and Greenstein, J. L. 1960, *Ap. J.*, **131**, 83.

Brownlee, R. R., and Cox, A. N. 1961, *Sky and Telescope*, **21**, 252.

Burbidge, G. R. 1960, *Die Entstehung von Sternen durch Kondensation diffuser Materie*, eds. G. R. Burbidge, R. Ebert, S. von Hoerner, and S. Temesváry (Berlin-Göttingen-Heidelberg: Springer-Verlag), chap. 1.

Cameron, A. G. W. 1962, *Icarus*, **1**, 13.

Ebert, R., Hoerner, S. von, and Temesváry, S. 1960, *Die Entstehung von Sternen durch Kondensation diffuser Materie*, eds. G. R. Burbidge, R. Ebert, S. von Hoerner, and S. Temesváry (Berlin-Göttingen-Heidelberg: Springer-Verlag), chap. 3.

Eddington, A. S. 1926, *The Internal Constitution of the Stars* (Cambridge: Cambridge University Press).

Emden, R. 1907, *Gaskugeln* (Leipzig and Berlin: B. G. Teubner).

Ezer, D., and Cameron, A. G. W. 1963, *Icarus*, **1**, 422.

Gaustad, J. E. 1963, *Ap. J.*, **138**, 1050.

Goldberg, L., Müller, E. A., and Aller, L. H. 1960, *Ap. J. Suppl.*, **5**, 1.

Hayashi, C. 1961, *Pub. Astr. Soc. Japan*, **13**, 450.

Hayashi, C., and Hōshi, R. 1961, *Pub. Astr. Soc. Japan*, **13**, 442.

Hayashi, C., Hōshi, R., and Sugimoto, D. 1962, *Prog. Theoret. Phys. Suppl.*, No. 22.

Henyey, L. G., LeLevier, R., and Levée, R. D. 1955, *Pub. A.S.P.*, **67**, 154.

Herbig, G. H. 1962, *Advances in Astronomy and Astrophysics*, ed. Z. Kopal (London and New York: Academic Press), **1**, 47.

Hoyle, F. 1960, *Quart. J.R.A.S.*, **1**, 28.

Hoyle, F., and Schwarzschild, M. 1955, *Ap. J. Suppl.*, **2**, 1.

Jager, C. de, and Neven, L. 1957, *Rech. Astr. Obs. Utrecht*, **13**, No. 4.

Jeans, J. H. 1919, *Problems of Cosmogony and Stellar Dynamics* (Cambridge: Cambridge University Press).

———. 1929, *Astronomy and Cosmogony*, 2d ed. (Cambridge: Cambridge University Press).

Kahn, F. D. 1960, *Die Entstehung von Sternen durch Kondensation diffuser Materie*, eds. G. R. Burbidge, R. Ebert, S. von Hoerner, and S. Temesváry (Berlin-Göttingen-Heidelberg: Springer-Verlag), chap. 2.

Keller, G., and Meyerott, R. E. 1955, *Ap. J.*, **122**, 32.

Kelvin, Alfred Lord. 1887, *Phil. Mag.*, **23**, Series 5, 287.

Kippenhahn, R., and Baker, N. 1962, private communication.

Kippenhahn, R., Temesváry, S., and Biermann, L. 1958, *Zs. f. Ap.*, **46**, 257.

Kushwaha, R. S. 1957, *Ap. J.*, **125**, 242.

Levée, R. D. 1953, *Ap. J.*, **117**, 200.
Limber, D. N. 1958, *Ap. J.*, **127**, 387.
Mestel, L., and Spitzer, L., Jr. 1956, *M.N.*, **116**, 503.
Milne, E. A. 1930, *M.N.*, **91**, 4.
Morse, P. M. 1940, *Ap. J.*, **92**, 27.
Osterbrock, D. E. 1953, *Ap. J.*, **118**, 529.
Reeves, H. 1965, *Stellar Structure*, eds. L. H. Aller and D. B. McLaughlin (Chicago: University of Chicago Press), chap. 2.
Richtmyer, R. D. 1957, *Difference Methods for Initial-Value Problems* (New York: Interscience Publishers).
Salpeter, E. E. 1954, *Mém. Soc. R. Sci. Liège*, **14**, Series 4, 116.
Schatzman, E. 1962, *Ann. d'ap.*, **25**, 18.
Schwarzschild, M. 1958, *Structure and Evolution of the Stars* (Princeton: Princeton University Press).
Schwarzschild, M., Howard, R., and Härm, R. 1957, *Ap. J.*, **125**, 233.
Sears, R. L., and Brownlee, R. R. 1965, *Stellar Structure*, eds. L. H. Aller and D. B. McLaughlin (Chicago: University of Chicago Press), chap. 11.
Somerville, W. B. 1964, *Ap. J.*, **139**, 192.
Spitzer, L., Jr. 1963, *Origin of the Solar System*, eds. R. Jastrow and A. G. W. Cameron (London and New York: Academic Press), p. 39.
Temesváry, S. 1959, *Mém. Soc. R. Sci. Liège*, **3**, Series 5, 403.
The, Pik-Sin. 1960, *Ap. J.*, **132**, 40.
Thomas, L. H. 1931*a*, *M.N.*, **91**, 122.
———. 1931*b*, *ibid.*, **91**, 619.
Underhill, A. B. 1960, *Ap. J.*, **131**, 524.
Unsöld, A. 1955, *Physik der Sternatmosphären* (Berlin-Göttingen-Heidelberg: Springer-Verlag).
Upton, E. K. L. 1962, thesis, University of Michigan.
Vardya, M. S. 1964, *Ap. J. Suppl.*, **8**, No. 80, 277.
Varsavsky, C. M. 1960, *Ap. J.*, **132**, 354.
Vitense, E. 1953, *Zs. f. Ap.*, **32**, 135.
Walker. M. F. 1956, *Ap. J. Suppl.*, **2**, 365.
———. 1957, *Ap. J.*, **125**, 636.
———. 1959, *ibid.*, **130**, 57.
———. 1961, *ibid.*, **133**, 438.
Weymann, R., and Moore, E. 1963, *Ap. J.*, **137**, 552.
Yamashita, Y. 1962, *Pub. Astr. Soc. Japan*, **14**, 390.

Index of Subjects and Definitions

[*Page numbers for definitions are in italics*]

Absolute-magnitude calibrations, 182
Absorption, 109; *see also* Extinction, interstellar
  in dark nebulae, 125
  interstellar line, 7, 36; *see also* Interstellar lines
Absorption effects
  in discrete radio sources, 577
  21-cm; *see* Hydrogen
Abundances
  beryllium, 8
  calcium, H and K lines, 13
  calcium/sodium, 8
  cosmic, 398, 700, 702, 710, 711, 726
  cosmic ray, 710 f., 818
  deuterium; *see* Deuterium
  of elements, 69 f., 238, 239, 247, 681, 708, 727, 729, 735, 745
  He, 642
  He^3, 618
  hydrogen, 71; *see also* Hydrogen
  interstellar, 397, 576
  light isotopes, 815–18
  lithium, 818
  nebular, 69
  terrestrial, 817, 818
  Unsöld composition, 783 f., 799 f., 806, 811, 812, 817
Activation coefficient, *524*
Alfvén velocity (speed), 734, 739, 740
Alfvén waves, 18
Allowed lines, 403 f.
Allowed transitions, selection rules, 439
Associations, 56, 58
  extinction in, 174 f., 215
  H I in, 611 f.
  new stars in, 3
  O-, 2, 57, 104
  OB-, 241, 597, 613
  origin of, 56 f.
  Orion, 15
  T-, 149, 150, 160, 164, 613
A stars (peculiar), 618
Atlas of Galactic Dark Nebulae, 121
Atlas of the Milky Way, 121
Atom-electron collisions, 454 f.
Aurorae, 730, 735

Balloon-borne instruments, 685
Balmer continuum, 550
Balmer decrement, 91, 96, 97, 99, 101 f. 112
  in planetary nebulae, 497, 505 f., 513, 514, 531 f., 626
Born approximation, 454, 458, 460, 462
Born-Oppenheimer approximation, 524
Bremsstrahlung, 682, 688, 700, 703, 719
  magnetic, 732
  spectrum, 697, 701
Bright nebulae, 7, 67, 68, 119
  catalogue, 74
  emission; *see* Emission nebulae
  filaments; *see* Filaments
  reflection; *see* Reflection nebulae
Bright rims, 7, 16, 27 f.; *see also* Nebular forms
Brightness temperatures, 388, 582, 589, 602
B stars, 3, 7, 26 f., 647.
Bulbs (in nebulae); *see* Nebular forms

Cameras, electronic, 492, 499
Carbon flash, 559
Cassiopeia A (3C461); *see* Radio sources; Supernovae
Catalogues
  of bright nebulae, 74
  of dark nebulae, 74, 120
    Lundmark, 121
    Lynds, 123
    Schoenberg, 123
  Herschel's (of nebulae), 119
  of planetary nebulae, 484 f., 563
    Perek-Kohoutek, 484, 563
    Vorontsov-Velyaminov, 484
  of reflection nebulae, 74
  3C catalogue of radio sources, revised, 630
Central stars; *see* Planetary nebulae
Cerenkov radiation, 668, 709
Chemical composition, 71; *see also* Abundances of elements; separate entries
  of stars and nebulae, 405
Clark effect, *756*, *758*, 765
Clouds; *see* Interstellar clouds
Cluster diameters, 170, 218
  angular, 184
  apparent, 167, 168
Cluster distances, 167, 168, 171, 173, 174, 215 f.
Clusters, 172 f., 181 f.
  absolute magnitudes, 181

Clusters—*Continued*
  age sequence, 77
  distance determination, 167
    photometric, 170
  extinction in; *see* Extinction, interstellar
  globular, 613
  H I in, 611, 612
  main sequence, 174
  new stars in, 3
  radial velocity, 181, 182
  reddening, 171
Coalsack (Southern), 129 f., 137
Collimators
  honeycomb, 687, 689, 691, 694
  modulation, 701
  slit, 701
Collision cross-sections, 405, 479
  allowed, 405
  calculation, 404, 405, 454 f., 458
  Fe (XVI), 406
  forbidden lines, 405
Collision strength, *524*
  calculations, 524
    distorted wave method, 455, 459, 470, 477, 525
    exact resonance method, 455, 459, 477, 525
Collisional excitation, 542, 552, 657
Collisions
  atom-electron, 454 f.
  grains, 241
Color excess, 221, *223*, 369
Color-excess ratios, 185, 193
  in Aquila, 200
  in ι Aurigae, 203
  in α^1 Capricorni, 198
  in Cepheus, 201
  in Cygnus, 199
  in early-type stars, 185
  in 10 Lacertae, 212, 213
  in α Leonis, 209, 210
  in M stars, 185
  in NGC 2244, 202
  in NGC 6530, 197
  in NGC 6611, 196
  in Ophiuchus, 185, 195
  in Orion, 206 f.
  in Perseus, 185, 194, 210
  in Scorpius, 204, 208
  in Taurus-Orion region, 205, 208
Configuration interaction; *see* Wave functions
Contraction; *see* Stellar contraction
Convection, superadiabatic, 805 f.
Convective energy transport; *see* Energy transport
Convective heat transport; *see* Energy transport
Convective star models; *see* Star models
Core-mantle particles; *see* Interstellar grains
Coriolis force, 42
Corona; *see* Galactic halo
Cosmic abundance; *see* Abundances
Cosmic-ray flux, 677
Cosmic-ray gas, 723, 738, 741 ff.
  components, 710
  dynamics of, 717, 737 f.
  origin, 717, 747
  physical properties, 710
Cosmic rays, 71, 238, 243, 667, 686, 688, *707* f., 727, 742, 768, 769
  acceleration mechanism, 733 f.
  age, 709, 726
  density, 740
  energies, 19, 729, 735, 769
  energy density, 724 ff., 741, 743
  energy distribution, 712
  energy production rate, 708
  high-energy, 723
  intensity, 52, 708, 721, 735
  interstellar, 709
  isotropy, 715, 723 f., 769
  lifetime, 718, 724–34, 745
  origin, 668, 708, 726, 730 f., 732, 737
  primary, 709, 710
  production rate, 725, 727 f., 735, 745
  properties, 737, 755
  scattering in magnetic fields, 727
  secondary, 709, 713
  sources, 711, 715
    Crab Nebula, 712
    galactic nucleus, 718
    novae, 718, 731
    supernovae, 711, 731
Coulomb scattering, 712
Coulomb wave functions, 405, 469
Crab Nebula (NGC 1952), 15, 16, 33, 34, 69, 101, 597, 627 ff., 637 f., 641, 642, 644 f., 652, 657, 681, 684, 685, 687 f., 693, 697 f., 704, 722, 735, 737; *see also* Radio sources
  abundances in, 111
  cosmic rays from, 712
  density, 69
  diffuse mass, 642, 667, 698
  filaments, 638 f., 647 f.
  isophotes, 645
  magnetic field, 35, 698, 699
  models, 689, 697
  occultation, 644, 645, 688
  polarization, 667
    optical, 650, 651
    radio, 650, 651
  radial velocities, 640
  radio spectrum, 623, 645, 646
  spectral energy distribution, 667
  supernova remnant, 732
  synchrotron radiation, 651, 682
  temperature, 69, 642
  X-ray emission, 645 f., 732
Cygnus Loop, 15, 16, 33, 34, 623, 624, 656, 657 f., 722
  age, 657
  magnetic fields, 658
  mass of filaments, 657
  spectrum, 657
  total energy, 657
Cygnus X radio spectrum, 87

Dark nebulae (clouds), 67, 76, 77, 119 f., 221; *see also* Globules
  absorption by, 126
  catalogues, 74, 120, 121, 123

densities, 135
distances, 124
distribution of, 124
dust in, 120
masses, 135
proper motions, 133
reddening by, 128
shapes, 133
Deactivation coefficient, *524*
Degeneracy in stars, 558
Density function, *126*
Deuterium
abundance, 818
burning, 818
main sequence, 791, 817, 818
92-cm, 617
Dielectric grains; *see* Interstellar grains
Diffuse nebulae, 65 f., 110, 395, 403
exciting star, properties, 83
spectra, 98 f.
Doppler broadening, 584
Doppler effect, 577
Doppler shift, 763
Doppler splitting of nebular lines, 565
Double galaxies, 111
Dust grains; *see* Interstellar grains
Dust/neutral hydrogen, 7

Early-type stars, 7, 11, 21, 22, 185, 191
B, 3, 7, 26 f., 647, 767
color-excess ratios, 185
H and K lines in spectra, 11
intrinsic colors, 191
model atmospheres, 555
O, 3, 8, 22 f., 26 f., 767
unknown companions, 208, 209
Einstein probability coefficients, 385, 502, 510; *see also* Transition probabilities
Electric-dipole, 404; *see also* Transition probabilities
Electric-quadrupole, 405, 432 f.; *see also* Transition probabilities; Forbidden transitions
calculations, 405
contributions, 450
radiation, 432, 438
transitions, 447, 496
Electron degeneracy, 776
Electron densities, 650
Electron energies, 670 f.
Electronic cameras, 499; *see also* Cameras
Elephant-trunk structures; *see* Nebular forms
Emission nebulae, 68, 73, 78, 82 f., 136, 396, 608; *see also* H II regions
cosmic abundances, 111
filamentary structure, 78
radial velocities, 106
radio spectrum, 86
Energy transport
convective, 772, 774, 779, 791, 797 f., 808 f., 814
radiative, 772, 774 f., 784, 785, 790, 796, 808, 809, 814
Entropy, 775 f., 788, 789, 792, 795, 796, 799, 802, 804 f., 808, 812, 821
Exciting stars; *see* Planetary nebulae; Diffuse nebulae
Expansion of the Universe, 724
Extinction, interstellar, 83, 86, 99, 167–219, 221 f., 229, 235, 237, 238, 240, 243, 250, 308 f., 339, 340, 349, 490, 491, 547, 556, 647
in associations and clusters, 174 f.
cluster-diameter method, 167, 170 f., 184
cluster-distance method, 212
color-difference method, 168, 170, 185 f., 212, 218; *see also* Color-excess ratios
determination of, 167
infrared 192 f., 358
"λ^{-1} law," 221, 224 f.
law, 169, 191 f., 218
"abnormality," 208
for Orion, 209
optical, 83
in other galaxies, 235
photographic, 67
total-to-selective, 168
variable-extinction method, 167, 168, 170, 174 f., 177, 181, 184, 212, 218
variation, 213, 227
wavelength dependence, 221, 223, 225, 227, 228, 339, 355 f., 361
in Aquila, 183, 184, 192, 200
in Ara, 178, 181
in Auriga, 178, 181, 183, 192, 203
in Capricornus, 191, 198
in Cassiopeia, 175, 181, 183, 184
in Cepheus, 176, 177, 181, 182, 192, 201, 226
in Cygnus, 175, 181, 183, 184, 192, 199, 226
in Gemini, 181, 183, 184
for 10 Lacertae, 213
in I Mon, 180, 183, 192, 226
in NGC 2244, 202
in NGC 6530, 197
in NGC 6611, 196
in Ophiuchus, 191, 195
in Orion, 180, 209, 226
in Orion Belt, 206
in Orion Sword, 207
in Perseus, 175, 181, 183 f., 194, 226
ϕ Persei, 211
in Scorpius, 204
in Taurus-Orion region, 205
in Vulpecula, 183, 184
Extinction, other galaxies, 235

f-values, 424, 436 f., 443, 542
Thomas-Kuhn sum rule, 437
Fabry-Perot interferometer (etalon), 106, 107, 370, 492, 493
Faraday effect, 642, 650
Faraday rotation, 233, 597, 650, 679, 716, 755–62, *758*, 767
apparent absence of, 10
Fermi mechanism, 733, 734
Fiber optics, 493

Filaments, 7, 11, 34 f., 78, 106, 134, 232, 490, 543, 544, 564, 568, 569, 624, 639, 641, 642, 658, 659, 660, 661
distance of, 641
electron density, 641
electron temperature, 641
mass, 641, 657
shells, 650
Flare stars, *141* f., 624, 730
in clusters and aggregates, 141
energies, 160
and interstellar material, 155, 160
light curves, 148, 161
in Orion, 143 f., 160, 161
in Pleiades, 150 f., 160
proper motions, 151, 154
radio emission, 160
repeated flares, 143, 149
in solar neighborhood, 157
spatial distribution, 164
spectra, 148, 158 f., 161, 163
spectral types, 148, 155, 158, 162
Fluorescence, 496
lunar (presumed), 686
Forbidden lines, 387, *496*, 523 f., 530, 531, 657
collisional excitation, 520 f.
collision cross-section calculations, 479
emissions, 551
transition probability, 432–33
Forbidden transitions, 403 f., 439, 454, 564
electric quadrupole, 405, 432 f., 447, 496
magnetic dipole, 404–5, 432, 435 f., 447, 450, 496
theory, 405
magnetic quadrupole, 432
selection rules, 439
Free radicals, 259, 265, 340, 766; *see also* Platt particles
optical properties, 303

Galactic center (nucleus), 487, 599, 726
distance 214 f.
Galactic gravitational fields, 65
Galactic halo, 7, 36, 709, 723, 726, 727, 729, 737, 745, 746, 758
inflation of, 741 f.
Galactic magnetic fields; *see* Magnetic fields
Galactic radio emission, 667
Galactic radio spur, 722, 726
Galactic rotation; *see* Interstellar lines
Galactic structure, theories of, 221, 222
Galaxies
double, 111
elliptical, 109 f.
energy requirements, 727
extinction in, 235
Local Group, 717, 723 f.
magnetic fields, 724
peculiar, 111
radio; *see* Radio galaxies
spheroidal, 628
spiral, 109
Seyfert, 110, 746
Galaxy, age of, 244
Gas, interstellar; *see* Interstellar gas
Gaseous nebulae, 403, 493; *see also* Bright nebulae, etc.
classification, 68, 69
photoionization in, 67
spectra, 523
thermal balance, 518 f.
Gaunt correction factor, 536
Geomagnetic field, 730, 733, 735
Globular clusters, 67, 613
Globules, 5, 30, 36, 39, 47, 57, *104*, 124, *134*, 136, 138
gravitational contraction of, 57
Gould's Belt, 77, 79, 130
Grains: *see* Interstellar grains
Grains/hydrogen, 47
Graphite flakes (particles); *see* Interstellar grains
Gravitational contraction
of globules, 57
in stars, 40, 48 f.
Great Loop (in Orion), 68
Great Rift (in Cygnus), 130, 131

Hayashi limits, 797 f., 801, 806 f., 823
Hayashi (star) models, 811, 815; *see also* Star models
Hayashi's theorem, 797 f.
Helium, Lyman limit, 555
Helium flash, 559
Helium stars, 787
Helmholtz contraction, 48
Herbig-Haro prestellar objects, 98
Hertzsprung-Russell (H-R) diagram, 774, 787, 790, 794, 797, 799, 801, 810, 823
deuterium main sequence, 791, 817, 818
lithium main sequence, 818
for planetary nebulae, 556
theoretical, 784
Homologous contraction; *see* Stellar contraction
Horsehead Nebula, 120
Hose instability, 727, 742
Hubble constant, 628
Hydrogen, 576, 577; *see also* Deuterium
abundances, 71
clouds, distances, 596
collisional excitations, 552
density in interstellar space, 4, 606
density in solar neighborhood, 4, 58
distribution, 600
expected profile, *594*
interstellar, 3
Lyman limit, 555
molecular, 7, 717
negative ion, 497
neutral; *see* H I
ratio to dust, 5
to grains, 47
to helium, 69
to lighter elements, 7
to metals, 238
wave functions, 406 f.

Hα emission regions, 68

Hα surveys, 69, 82

H I, 80
  in clusters and associations, 611, 612
  density, 4, 6
  and dust, 614 f.
  expanding shells, 612, 613
  H II relationship, 617
  and interstellar calcium, 615 f.
  polarization, 231, 336
  regions, 8, 9, 36, 68, 71, 104, 134, 244, 251, 396, 398, 575, 755, 756, 773
  temperatures, 8, 9, 38, 57
  21-cm, 387, 575 f., 642, 659, 717
    absorption, 9, 14, 593 f., 652, 756
    emission, 4, 7, 11, 13 f., 71, 539, 755
      surveys of, 589 f.
    observations, 83, 112, 113, 394, 718
    polarization of, 10
  velocity, 607

H II, 80
  H I relationship, 617
  regions, 36–37, 68, 71, 80, 82, 83, 86 f., 91, 98, 99, 104, 105, 107, 109, 111, 112, 134, 384, 385, 396, 398, 596, 623, 626, 659, 699, 756, 773; *see also* Strömgren spheres
    dynamics of, 112
    electron density in, 87
    interstellar reddening in, 112
    isophotes, 565
    masses in, 87, 88, 110
    polarization, 231, 336
    radial velocities, 108, 112
    temperatures, 9, 38, 258

H_2, 38, 58, 71

H^+ abundances, 71

Hydromagnetic shocks, 699

Hydromagnetic waves, 703
  propagation, modes, 739

Hydroxyl (OH) radical; *see* Interstellar gas, radio spectrum

Image converter, 492
  Lallemand tube, 492

Image rotator, 492

Image slicer, 492, 493, 499

Image-tube techniques, 68

Infrared anomalous extinction, 192, 199 f., 358

Infrared photometry, 169 f.

Infrared radiation, excess, 208

Intermediate coupling, 439, 446 f.

Interstellar absorption lines; *see* Interstellar lines, absorption

Interstellar abundances; *see* Abundances

Interstellar clouds, 210, 259, 349, 372, 773
  accretion of, 39
  collision rate, 39, 67
  dark; *see* Dark nebulae
  D-critical ionization front, 25 ff.
  dust, 109, 119, 602, 608
  energy sources, 22
  evolution, 39
  formation of, 35 f.
  gas, 238, 240, 584, 601, 717, 741, 742, 746
    abundance variations, 401
    diameters, 609
    motions, 584
    structure, 110, 608
  high-velocity, 372, 399 f.
  kinematics, 16, 400–401
  low-velocity, 401
  mean density, 17
  models, 44 f., 605
    discrete-cloud (Ambartsumian), 5, 6, 372
    standard, 5, 7
  motions (velocities), 11 f.
    turbulence, 48, 53 f.
  temperature, 242

Interstellar dust, 4, 6, 7, 38, 65, 71, 575, 767

Interstellar extinction; *see* Extinction, interstellar

Interstellar gas, 4, 65, 222, 240, 243, 575 f., 597, 599, 710, 716 f., 728, 734, 738, 741, 743, 745, 746
  chemical composition, 397
  clouds; *see* Interstellar clouds, gas
  composition, 7
  cosmic-ray component, 709
  density, 68, 576, 609, 611, 746
  electron density, 83
  electron temperature, 83
  ionic abundances, 83
  ionization, 7
  possible molecular lines, 618, 619
  radial velocity, 81, 83, 613
  radio spectrum, 599
    density (solar neighborhood), 606
    distributions, 600
    He^3, 618
    H, 21-cm; *see* Hydrogen
    OH (hydroxyl radical) absorption lines, 576 f., 599 f.
  temperature, 7, 717
  thermal instability, 27 f.

Interstellar gas clouds; *see* Interstellar clouds, gas

Interstellar grains, 38, 65, 79, 80, 221 f., 401
  albedo, 235, 343
  alignment, 222, 318
    Davis-Greenstein, 10, 268, 271, 319 f., 327 f., 346, 766
    by magnetic fields, 232
    perfect, 319, 320, 337
  birefringence, 302, 303
  chemical composition, 247, 397
  coated, 260 f.
  core-mantle, 261, 304 f., 337 f., 360
  cylinders, 329 f.
    perfect alignment, 334 f.
    wavelength dependence of extinction, 329
  density, 4
  destruction of, 240, 243
  dielectric, 222, 224, 274, 276 f., 300, 308, 323, 328, 332, 337, 343, 347, 351, 354, 360, 361, 766; *see also* dirty ice, graphite, etc., below
    magnetic properties, 258
  dirty ice, 243, 312, 313, 356, 360, 362
  elongated, 250
  evaporation, 241 f., 259

Interstellar grains—*Continued*
extinction cross-section, *235*
extinction theory, 235
ferrite, 274
ferromagnetic 260, 267, 269, 274, 766
graphite, 222, 224, 240, 261 f., 269, 274, 337 f., 341, 360, 361, 766
core, 360
temperature of, 260
growth and formation, 222, 229, 238 f., 250, 260, 309
condensation nuclei, 239, 240
theory, 222, 244
ices, 222
metallic, 79, 222, 250, 293, 300, 332, 337, 347, 351, 354, 360, 361
models, 243, 358
nonspherical, 249, 275, 290, 360
scattering by, 308
optics, analog methods, 275 f., 360
orientation 232–33, 238, 267 f.
physical properties (characteristics), 225, 238 f.
platelets, 240, 261
Platt-type; *see* Platt particles; Free radicals
polarization by, 80, 222, 276
polarization theory, 235
refractive index, 247
scattering, 79, 82, 236–63, 266, 275, 308, 323
by very small particles, 300
size distribution, 308, 312 f., 323, 332, 349, 355
Oort–van de Hulst, 309, 340
spherical, 237, 239, 258, 263, 275, 313, 332, 340, 349
spherical dielectric, 337, 353
sputtering of, 242, 262, 267
temperature, 241, 250, 254 f., 260, 263, 266, 271, 273, 299, 766
Interstellar lines
absorption features, 36, 81, 229, 365 f., 631, 639, 717
structure, 370 f., 399
calcium (H and K), 5, 13, 81, 107, 210, 365, 366, 370, 375, 376, 393 f., 615, 616.
catalogues, 369
curves of growth, 373, 377 f., 389
theoretical, 375, 376
diffuse, 355 f.
doublet-ratio method, 366, 374, 388 f.
galactic rotation effects, 11, 365, 366, 371, 373, 379, 388 f., 485
hydrogen, 21-cm; *see* Hydrogen
hydroxyl radical, 576 f., 599 f.
intensities (strengths), 365, 395
intensity-distance relations, 391 f.
laboratory identification, 369
molecular, 366, 367
nebular absorption, *367*
radial velocity, 107, 365, 370
at radio wavelengths, 576; *see also* Hydrogen, 21-cm; Interstellar gas, radio spectrum
sodium, 5, 365, 366, 370, 375, 376, 393 f., 617
stationary, 365
unidentified, 366
Interstellar magnetic fields; *see* Magnetic fields
Interstellar medium, 568, 631, 642, 650, 653, 656, 661, 662; *see also* Plasma
density, 396, 397
ionization equilibrium, 380 f., 394
solar neighborhood, 4
Interstellar polarization; *see* Polarization
Interstellar reddening, 167, 168, 177, 212, 221, 229, 235, 395, 493, 507, 513, 551, 629, 647
in clusters, 171
color-difference method; *see* Extinction, interstellar
in H II regions, 112
in Orion, 168, 169
Interstellar reddening law, 225, 244, 310, 336
infrared, 169
variations, 169, 226
Interstellar spectra; *see also* Interstellar lines
ionic, 369
laboratory data, 369
molecular, 366
Inverse Compton effect, 682, 688, 725, 732
Ionized hydrogen; *see* Hydrogen; H II

Keller-Meyerott opacities; *see* Opacity
Kelvin time-scale; *see* Star formation, time scales
Kepler's supernova (3C358); *see* Radio sources; Supernovae
Kepler's third law, 822
Kirchhoff's law, 582
Kramer's formula, 778, 779, 782, 785 ff., 809
modified, 809
Kramer's opacity; *see* Opacity

Lallemand tube; *see* Image converter
Lane's law, 777
Levée model; *see* Star models
Line blanketing, 555
Line broadening, 388, 577, 584
Lithium burning, 818
Lithium main sequence, 818
Local Group of galaxies; *see* Galaxies
Lorentz force, 668, 675
LS-coupling, 405, 439, 446
deviations from, 437, 446, 449
Lunar occultation (of radio sources), 634, 644, 645
Lyman α scattering, 510, 511
Lyman continuum, 550, 551
Lyman limit, 550, 555

Magellanic clouds, 488, 577, 592, 597, 598, 617
Large, 110, 111, 613, 617, 661, 663
Small, 110, 111, 617
Magnetic bremsstrahlung, 732
Magnetic fields, 233, 559 f., 564, 565, 569, 699, 702 f., 708, 733, 734, 745
Clark effect, *756*, *758*, 765
in clouds, 764
energy, 680
galactic, 10, 35, 48, 65, 134, 222, 232, 233, 238, 267 f., 596, 597, 650, 709, 710, 715 f., 724, 727 f., 739–40, 743, 746, 755 f.
models, 755, 758, 762, 768, 769

intergalactic, 769
interstellar, 16, 18, 27 f., 31, 42, 47, 53, 55, 71, 80, 667, 715, 738, 741, 745
in Local Group of galaxies, 724
model, 760
scattering by, 727
in solar neighborhood, 755, 758, 760, 767
in solar wind, 708

Magnetosonic (compressional) waves, 18
Magnetosphere (terrestrial), 737
Masers, 586
Maxwellian distribution of velocities, 584, 668
Meteorites, unstable isotopes in, 720
Microwave analog method, 274, 275, 360
Mie theory, 251, 258, 277, 312
Milky Way, 36, 68, 120, 138, 171, 173, 218, 221, 576, 577, 585
Mixing length, *54*
Model atmospheres, nongray, 555
Model stars; *see* Star models
Molecular hydrogen; *see* Hydrogen
Monoceros Nebula, 6
M supergiants; *see* Stars

Nebulae; *see also* individual nebulae *and* Table, p. xi
bright; *see* Bright nebulae
catalogues; *see* Catalogues
colors, 80, 81
dark; *see* Dark nebulae
diffuse; *see* Diffuse nebulae
distances, 83 f.
emission; *see* Emission nebulae
filamentary structure, 80, 82
gaseous; *see* Gaseous nebulae
Hubble's classification, 65, 66
magnetic fields, 35, 564
planetary; *see* Planetary nebulae
polarization, 79 f.
radial velocity, 106 f.
reflection; *see* Reflection nebulae
spectral type, dependence on involved stars, 66
spectrum, 65, 66, 67, 76, 91, 98, 404, 523
emission, 92–95
optical, 98
theory, 91
structure; *see* Nebular forms

Nebular condensations, 564, 565, 569; *see also* Interstellar grain formation
Nebular continuum
negative hydrogen ion, 497
radiative equilibrium, 520
Nebular forms, 102
bulbs, 76, 77, 79
bright rims, 7, 27, 29, 31, 104, 105
comet shapes, 105, 106
elephant trunks, 7, 27, 31, 104, 105, *134*
filaments; *see* Filaments
globules, *104*
symmetric, 106
Nebular lines; see other entries
Doppler splitting, 565
Nebular spectrograph, 68
Negative hydrogen ion, 497; *see also* Hydrogen
Neutral hydrogen, 575; *see also* Hydrogen
Neutrinos, 697, 703, 732, 747
losses, 817
Neutron star, 688, 689, 703, 704
Nonthermal emission, 627, 658, 668, 703
radio, 541, 716, 722, 723, 732; *see also* Radio sources
galactic disk, 722
origin, 722
spectrum, 719 f.
suspected sources, 497, 541
Novae, 71, 624, 659, 708, 724, 731
distances, 487
mass ejection, 20
old, 732
spectra, 404, 523
Nuclear energy distributions, 712
Nuclear synthesis, 726

O-associations; *see* Associations
OB-associations; *see* Associations
Objective prism, 484, 494
OH lines; *see* Radio spectrum of interstellar gas
Oort's constant, 214 f.
Opacity, 775, 778 f., 785, 790, 822
Cox, 809
Keller-Meyerott, 785, 811, 812
Kramers, 780, 784, 785, 787, 792, 810, 811
molecular, 808
Rosseland mean, 785
stellar, 773, 774, 778, 787 f., 796 f., 808–22
Orion Nebula (NGC 1976), 3, 6 f., 14, 15, 26 f., 57, 68, 73, 77, 98, 102, 168, 169, 499, 647
anomalous absorption law, 97
color-excess ratios, 207
flare stars in, 143 f.
interstellar extinction, 207
polarization, 101
proper motions, 112
spectrum
optical, 92–95
radio, 83
Trapezium
reddening, 168, 169
red stars in, 209
21-cm observations, 26, 83
O stars; *see* Early-type stars

Palomar Observatory Sky Survey, 76, 79, 82, 123, 484, 562
Particles; *see* Interstellar grains
Pauli exclusion principle, 407, 410, 455, 459, 469
Perfect alignment; *see* Interstellar grains
Photographic Atlas of the Southern Milky Way, 121
Photoionization in gaseous nebulae, 67
Photometry
B stars, 168
infrared, 169
two-color, 168

Photometry—*Continued*
seven-color, 169, 182
six-color, 168
UBV, 169, 174

Pinch effect, 34

Planck's law, 250, 432

Planetary nebulae, 14, 68, 69, 97, 107, 110, 403, 483 f., 623 f., 639, 642, 659, 660, 702
catalogues, 484 f., 563
Perek-Kohoutek, 484, 563
Vorontsov-Velyaminov, 484
central (exciting) stars, 71, 483, 486 f., 515, 546 f., 561
age, 485
chemical composition, 555
evolution, 556 f.
ionization temperature, 547
Lyman limit, 550
magnetic fields, 559 f., 564, 565
magnitudes, meaning of, 487, 494
mass, 485
mass loss, 547, 563, 569
model atmospheres, 555
Of-nuclei, 555
spectra, 546, 554, 563
temperatures, 500, 547, 551, 555, 556
Wolf-Rayet type, 555, 563
chemical composition, 541 f.
intrinsic differences, 545
classification, 559
condensations, 564, 565, 569
densities, 523 f.
distances, 483, 486 f., 556
distribution, 560
electron temperatures, 523 f., 541
expansion, 566 f.
in external galaxies, 487
filamentary structure, 568
fluorescence mechanism, 496
and galactic rotation, 485
and galactic structure, 483
Hertzsprung-Russell (H-R) diagrams, 556
identification, 483
isophotes, 494, 541, 565
models, 560 f., 569
number known, 484
parallaxes, 486, 487; *see also* Planetary nebulae, distances
photoelectric photometry, 493
polarization in, 493
proper motions, 486
radial velocities, 483 f., 566
radio-frequency emission, 536 f.
spectra, 483, 492, 496 f.
Balmer decrement, 497, 505 f., 513, 514, 531 f.
continuous, 100, 531 f.
helium, 514 f.
total number, 484

Plasma, 575, 618, 644, 702 f., 712, 730 f.; *see also* Interstellar medium
cloud, 702
equations of motion, 747 f.
in laboratory, 733, 735
particle acceleration in, 737
waves, 735

Platt particles, 4, 265, 266, 274, 340, 343; *see also* Free radicals
scattering, 343

Pleiades, 174
distance modulus, 174
filamentary structure, 80, 81
nebulae, 77 f.

Polarization, 230 f., 269, 335, 343, 588, 650, 679, 722, 756, 761 f.
interstellar, 10, 80, 222, 229–38, 261, 308 f., 493, 766
linear, 760
nebular, 80
optical, 232, 233, 623
other galaxies, 222
phenomena, 577
radio, 233, 651
ratio to extinction, 328
by relativistic electrons, 674 f.
in solar neighborhood, 10
of starlight, 716, 755, 765 f.
of stars, 716
wavelength (frequency) dependence, 230, 231, 323, 325, 328, 329, 336 f., 360, 361, 642

Population I, 2 f., 8, 109, 403, 485, 598, 610, 617, 642

Population II, 2 f., 109, 403, 628, 631

Pre–main-sequence
contraction, 790, 791
evolutionary track, 792
nuclear reactions, 815 f.
stars; *see* Stars

Protostars, 48, 53, 55, 56
formation of, 35 f.

Quadrupole transition probability, 435

Quadrupole transitions, 452

Quasi-stellar objects (sources), 499, 725, 726, 730

Radial velocities, 662; *see also* Interstellar lines
of emission nebulae, 106 f.
galactic rotation, 397
of H II regions, 112
of interstellar gas, 613
local standard of rest, 589
of planetary nebulae, 483 f., 566
radio and optical, 589, 598
stars, 613

Radiation fields
electric-dipole, 435–36
electric-quadrupole, 435–36
magnetic-dipole, 435–36

Radiative energy transport; *see* Energy transport

Radiative equilibrium, 520

Radioactive decay, 681

Radio galaxies, 697, 704, 705, 723 f., 730, 745, 746

Radio receivers, 586 f.

Radio sources, 667, 706
Cassiopeia A (3C461), 577, 596, 597, 600, 601, 605, 606, 623, 628, 652 f., 661, 662, 685, 764; *see also* Supernovae
distance, 652
radial velocities, 265, 653

radio spectrum, 654, 655, 662
X-ray emission, 655
Crab Nebula (NGC 1952), 597, 637; *see also* Crab Nebula
Cygnus X, radio spectrum, 87
discrete, absorption effects, 577
extended, 746
extragalactic, 596, 685, 704
galactic, nonthermal, 623 f.
Kepler's supernova (3C358), 635; *see also* Supernovae
radio spectrum, 636
lunar occultation of, 634, 644, 645
NGC 1976, radio spectrum, 83
nonthermal; *see* Nonthermal radio emission
Sagittarius A, 596, 599, 601; *see also* Sagittarius A
Taurus A (3C144), 596, 597, 637, 764; *see also* Crab Nebula
Virgo A, 598
W 49, 596
3C273, 598
Radio spectrum of emission nebulae, 86
Radio telescopes, angular resolution, 575
Radius-entropy relation (for stars), 798, 799
Radius-luminosity relation (for stars), 798, 799
Rayleigh-Jeans formula, 582
Rayleigh-Taylor instability, 27
Recurrent novae, 701
Reddened stars, 130
observed colors, 185
Reddening; *see* Interstellar reddening; individual regions
Reflection nebulae, 68, 73 f., 78, 222, 235, 346 f., 396
catalogue, 74
color, 79, 235
extinction, 235
model, 348, 349
scattering in, 79, 235
Relative abundances; *see* Abundances
Relativistic effects in atoms, 404
Relativistic electrons, polarization of, 674 f.
Relativistic shifts, 555
Resolving power
optical, 71
radio, 71
Rocket
Aerobee, 685 f.,
Iris, 697
Rocket and satellite research, 71, 225, 395, 690
Rosette Nebula, 26, 613
Rotation measure, *759* f.
Rotational instability, 773, 774
RW Aurigae stars; *see* Variable stars

Sagittarius A (radio source); *see* Radio sources
Saha equation, 380, 542; *see also* Interstellar grains
Satellite research; *see* Rocket and satellite research
Scattering, 82, 222
by dielectric spheres, 308
elastic, 460
by electrons, 360, 497, 725, 778, 788, 790, 792, 797
by grains, 99, 100, 101, 236
inelastic, 460
Lyman α, 510, 511
by magnetic fields, 727
in other galaxies, 222
Rayleigh approximation, 320, 326 f., 339
by small particles, 343
Self-absorption, 603
Self-reversal effects, 513
Shells, expanding, 613
Shock front, 632, 633
Shock waves, 107, 518, 657, 688, 702, 733, 735
Solar flares, 730, 731, 733, 735
Solar models, 812; *see also* Star models
Solar motion, standard, 589
Solar neighborhood, 4, 600, 755, 758, 760, 767
Solar wind, 708, 712, 730, 733, 735, 737
Spectrographs
coudé, 366, 369, 370, 499
image rotator, 492
image slicers, 492, 493, 499
multislit, 107, 492, 493, 494
nebular, 492, 499
slit, 493
slitless, 492, 499
Spectrum scanner, 493, 494
infrared, 494
Speed of sound, 738, 809
isothermal, *24*, 28
Spherical grains; *see* Interstellar grains
Spillover, *588*
Spin-flip, *578*
Spin temperature, *579*, 583, 599, 618
relation with other temperatures, 602
Spiral arm, 232
structure, 232, 758
Standard cloud; *see* Interstellar clouds, models
Star counts, 124, 130
reduction in, 119
Wolf diagram, 124, 125, 128, 131
Star formation, 22, 35 f., 56, 771; *see also* Stellar contraction
time scales, 771 f., 789, 791, 796, 817 f., 823
Star models, 781, 789, 792, 793, 810, 817, 818, 821
adiabatic, 805
boundary conditions, 781, 789, 790, 797, 809, 810, 812, 815
convective, 797 f., 810, 811, 819, 823
convective envelope, 818
Hayashi, 809, 811, 812, 815
Henyey, LeLevier, and Levée, 792
high-luminosity, 808
homologous, 781
Kushwaha, 792, 794
Levée, 781–88, 790 f., 796, 808, 810, 811, 814
polytropic, 781, 792, 798, 799, 803
pre–main-sequence, 792
red giant, 808
standard, 781, 783, 785, 787, 792 f.

Stars
ages, 558
A peculiar, 618
B; *see* Early-type stars
binary, 56
boundary conditions; *see* Boundary conditions
central; *see* Planetary nebulae
contraction; *see* Stellar contraction
degeneracy, 558, 779
early-type; *see* Early-type stars
evolutionary track, 558, 787
flare; *see* Flare stars
formation of, 1 f., 35 f., 48, 57
gravitational contraction in, 40, 48
intrinsic colors, 174
M, 809
color-excess ratios, 185
supergiants, 192, 215, 218
water-vapor bands, 192
pre–main-sequence, 773
radial velocities, 613
red, 209
red dwarf, 779
red giant, evolution, 808
reddened, 130
standard model; *see* Star models
variable; *see* Variable stars
white dwarfs, 483, 555, 556, 558
Wolf-Rayet, 546, 730
spectrum, 546
young, 576
Stellar associations; *see* Associations
Stellar atmospheres
in early-type stars, 555
formation of grains in, 240
nongray, 555
Stellar contraction, 773 f.
homologous, 777–84, 808, 809
nonhomologous, 788 f.
pre–main-sequence, 790, 791, 797
theoretical paths, 774
time scale, 818, 823
Stellar energy transport; *see* Energy transport
Stellar motions, 2
Stellar opacity; *see* Opacity
Stellar winds, 703, 730–31
Stokes' parameters, 767
Strömgren radius, *384*, 567
Strömgren spheres, *8*, 241, 384, 385, 397, 484, 486, 488, 563; *see also* H II regions
Sun, pre–main-sequence contraction, 791
Supernova explosions, 104, 739
Supernova remnants, 14, 15, 65, 69, 82, 109, 623, 624, 627–33, 638, 652, 656 f., 662, 688, 703, 704, 720, 722, 732, 735, 737
Crab Nebula; *see* Crab Nebula; Radio sources
interaction with interstellar medium, 631, 653, 656, 661, 662
in Large Magellanic Cloud, 661 f.
nonthermal emission, 627
spectrum, 627
Supernova shells, 21, 22, 33 f.
Supernovae, 57, 703, 707, 708, 718, 723 f., 729, 731 f.
absolute brightness, 641
acceleration mechanism, 737
ancient records, 636, 659, 731
Cassiopeia A; *see* Radio sources
Chinese observations, 637
as cosmic-ray sources, 711, 718, 731
frequency 20, 731
ion acceleration, 737
Kepler's (3C358), 623, 629, 635, 636, 652
light curve, 633
light curves, 628
of 185 A.D., 659 f.
radial velocities, 633
spectra, 628
Tycho's of 1572 (3C10), 623, 629, 630, 633 f., 652, 685, 693
age, 630
distance, 631
position, 630
remnants, 638
total energy, 657
Type I, 20, 628 f., 637, 642, 652, 656, 700
absolute magnitude at maximum, 631
Type II, 16, 19, 20, 628, 630, 652, 655 f., 663
cosmic-ray production by, 735
Zwicky classes, 629, 655
Symmetric nebulae, 106
Synchrotron radiation (emission), 10, 71, 232, 575, 623, 627, 628, 638, 651, 658, 667 f., 703 f., 712, 719, 722, 755 f., 761, 762
optical, 705
radio, 699
from single electrons, 675, 680
spectrum, 670 f., 678, 680, 683, 697, 698
theories of, 232

T-associations; *see* Associations
Taurus A (3C144); *see* Radio sources
Temperature; *see* Brightness temperatures, etc.
Terrestrial atmosphere, 713
particle interactions, 708
unstable isotopes, 720
Thermal instability, 819 f.
Transition probabilities, 424, 443, 450 ff.; *see also* *f*-values
allowed, 405
calculations, 404, 405, 432 f.
Einstein coefficients, 385, 431, 502, 510, 578
electric-dipole, 432 f., 436 f.
electric-quadrupole radiation, 433, 435
Fe (XVI), 405 f.
forbidden lines, 403 f., 432 f.
magnetic-dipole radiation, 433 f.
theoretical, 426 f., 454
Turbulence
in interstellar clouds, 48, 53 f.
in interstellar gas, 767
in nebulae, 566
T Tauri stars; *see* Variable stars
Tycho's star; *see* Supernovae

UBV photometry; *see* Photometry
Universe, expansion of, 724

Variable stars, 484; *see also* Flare stars
  Orion-type, 141 f., 148, 163
  RW Aurigae-type, 141, 142, 163
  T Tauri-type, 141 f., 148 f., 155, 160 f., 613, 711, 730, 774, 818
    lithium abundance, 818
  U Geminorum-type, 142, 162
  UV Ceti-type, 141, 157
  W Ursa Majoris-type, 142, 162, 163
Virial theorem, 782, 820
Viscosity, 42
Vogt-Russell theorem, 776

Water-vapor bands, 192
Wave functions (calculation of), 405–79
  atomic, 404, 478
  Born approximation, 454, 458, 460, 462
  Born-Oppenheimer approximation, 524
  central-field approximations, 407
  configuration interaction, 424, 479
  Coulomb, 405, 469
  Coulomb-Born-Oppenheimer approximation, 468, 469
  distorted wave method, 455, 459, 468 f., 525
  exact resonance method, 455, 458, 459, 468 f., 525
  Fe (XVI), 405 f., 413 f.
  Hartree methods, 404 f., 411–26, 440, 453, 458
  for hydrogen, 406 f.
  orthonormal, 412 f.
White dwarfs; *see* Stars
Wolf diagram; *see* Star counts
Wolf-Rayet stars; *see* Stars
Wolf-Rayet-type spectrum, 546

X-rays, 685 f., 707, 725, 747
  association with plasma, 732
  diffuse background, 685, 686, 697
  solar, 686
  sources, 685 f., 732
    discrete, 685 f.
    distribution, 692 f.
    extragalactic, 704
    table of, 691
  spectrum, 627

Young stars, 576

Zanstra principle, 547
"Zanstra" temperature, 550 f.
Zeeman effect, 238, 577, 588, 716, 755, 762 f., 765
Zeeman splitting, 596, 597
Zero-age main sequence, 174, 784
Zodiacal light, 637